Lecture Notes in Electrical Engineering

Volume 730

The book series *Lecture Notes in Electrical Engineering* (LNEE) publishes the latest developments in Electrical Engineering - quickly, informally and in high quality. While original research reported in proceedings and monographs has traditionally formed the core of LNEE, we also encourage authors to submit books devoted to supporting student education and professional training in the various fields and applications areas of electrical engineering. The series cover classical and emerging topics concerning:

- Communication Engineering, Information Theory and Networks
- Electronics Engineering and Microelectronics
- Signal, Image and Speech Processing
- Wireless and Mobile Communication
- Circuits and Systems
- Energy Systems, Power Electronics and Electrical Machines
- Electro-optical Engineering
- Instrumentation Engineering
- Avionics Engineering
- Control Systems
- Internet-of-Things and Cybersecurity
- Biomedical Devices, MEMS and NEMS

For general information about this book series, comments or suggestions, please contact leontina.dicecco@springer.com.

To submit a proposal or request further information, please contact the Publishing Editor in your country:

China

Jasmine Dou, Editor (jasmine.dou@springer.com)

India, Japan, Rest of Asia

Swati Meherishi, Editorial Director (Swati.Meherishi@springer.com)

Southeast Asia, Australia, New Zealand

Ramesh Nath Premnath, Editor (ramesh.premnath@springernature.com)

USA, Canada:

Michael Luby, Senior Editor (michael.luby@springer.com)

All other Countries:

Leontina Di Cecco, Senior Editor (leontina.dicecco@springer.com)

**** This series is indexed by EI Compendex and Scopus databases. ****

More information about this series at http://www.springer.com/series/7818

Ahmad Fakhri Ab. Nasir ·
Ahmad Najmuddin Ibrahim · Ismayuzri Ishak ·
Nafrizuan Mat Yahya · Muhammad Aizzat Zakaria ·
Anwar P. P. Abdul Majeed
Editors

Recent Trends in Mechatronics Towards Industry 4.0

Selected Articles from iM3F 2020, Malaysia

Springer

Editors
Ahmad Fakhri Ab. Nasir
Faculty of Manufacturing and Mechatronic Engineering Technology
Universiti Malaysia Pahang
Pekan, Malaysia

Ahmad Najmuddin Ibrahim
Faculty of Manufacturing and Mechatronic Engineering Technology
Universiti Malaysia Pahang
Pekan, Malaysia

Ismayuzri Ishak
Faculty of Manufacturing and Mechatronic Engineering Technology
Universiti Malaysia Pahang
Pekan, Malaysia

Nafrizuan Mat Yahya
Faculty of Manufacturing and Mechatronic Engineering Technology
Universiti Malaysia Pahang
Pekan, Malaysia

Muhammad Aizzat Zakaria
Faculty of Manufacturing and Mechatronic Engineering Technology
Universiti Malaysia Pahang
Pekan, Malaysia

Anwar P. P. Abdul Majeed
Faculty of Manufacturing and Mechatronic Engineering Technology
Universiti Malaysia Pahang
Pekan, Malaysia

ISSN 1876-1100 ISSN 1876-1119 (electronic)
Lecture Notes in Electrical Engineering
ISBN 978-981-33-4596-6 ISBN 978-981-33-4597-3 (eBook)
https://doi.org/10.1007/978-981-33-4597-3

This Springer imprint is published by the registered company Springer Nature Singapore Pte Ltd.
The registered company address is: 152 Beach Road, #21-01/04 Gateway East, Singapore 189721, Singapore

Preface

The Innovative Manufacturing, Mechatronics and Materials Forum 2020 (iM3F 2020) is the first edition of the forum organised by the Faculty of Manufacturing and Mechatronic Engineering Technology, Universiti Malaysia Pahang (UMP). This forum is aimed at building a platform that allows academics as well as other relevant stakeholders within the region to share, discuss and deliberate their latest research findings in the domain of manufacturing, mechatronics and materials, respectively.

With the latest trend in mechatronics engineering that is gearing towards Industry 4.0, iM3F provides an excellent avenue for the community to keep abreast with the current technological advancements. This volume hosts 92 papers from the mechatronics tracks of the forum. The papers published in this proceeding have been thoroughly reviewed by the appointed technical review committee that consists of various experts in the field of mechatronics.

A sincere thanks to all members of the organising committee for making the conference a success. Not forgetting our sponsors, CREST and Cisco Webex, as well as our partners, Smart Manufacturing Research Institute, innovationlabs.my, SEAIC and BioMeC for their kind gesture and continuous support. We also would like to extend our appreciation to authors for contributing valuable papers to the proceedings. We hope this book will intensify the knowledge sharing among colleagues in the field of mechatronics engineering.

Pekan, Pahang, Malaysia
August 2020

Ahmad Fakhri Ab. Nasir
Ahmad Najmuddin Ibrahim
Ismayuzri Ishak
Nafrizuan Mat Yahya
Muhammad Aizzat Zakaria
Anwar P. P. Abdul Majeed

Contents

Hybrid Manta Ray Foraging—Particle Swarm Algorithm for PD Control Optimization of an Inverted Pendulum 1
Mohd Falfazli Mat Jusof, Shuhairie Mohammad, Ahmad Azwan Abd Razak, Nurul Amira Mhd Rizal, Ahmad Nor Kasruddin Nasir, and Mohd Ashraf Ahmad

Multi-objective Particle Swarm Optimization with Alternate Learning Strategies 15
Wee Sheng Koh, Wei Hong Lim, Koon Meng Ang, Nor Ashidi Mat Isa, Sew Sun Tiang, Chun Kit Ang, and Mahmud Iwan Solihin

Self-directed Mobile Robot Path Finding in Static Indoor Environment by Explicit Group Modified AOR Iteration 27
A. A. Dahalan and A. Saudi

Position and Swing Angle Control of Nonlinear Gantry Crane System 37
Abdulbasid Ismail Isa, Mukhtar Fatihu Hamza, Yusuf Abdullahi Adamu, and Jamilu Kamilu Adamu

The Classification of Heartbeat PCG Signals via Transfer Learning 49
Omair Rashed Abdulwareth Almanifi, Mohd Azraai Mohd Razman, Rabiu Muazu Musa, Ahmad Fakhri Ab. Nasir, Muhammad Yusri Ismail, and Anwar P. P. Abdul Majeed

The Classification of Wink-Based EEG Signals: An Evaluation of Different Transfer Learning Models for Feature Extraction 61
Jothi Letchumy Mahendra Kumar, Mamunur Rashid, Rabiu Muazu Musa, Mohd Azraai Mohd Razman, Norizam Sulaiman, Rozita Jailani, and Anwar P. P. Abdul Majeed

Development of Polymer-Based Y-Branch Symmetric Waveguide Coupler Using Soft Lithography Technique 71
M. S. M. Ghazali, F. R. M. Romlay, and A. A. Ehsan

Hybrid Flow Shop Scheduling with Energy Consumption in Machine Shop Using Moth Flame Optimization 77
Mohd Fadzil Faisae Ab. Rashid, Ahmad Nasser Mohd Rose, and Nik Mohd Zuki Nik Mohamed

Sustainability of Fertigation in Agricultural Crop Production by IoT System: A Review 87
Nur Syahirah Mohd Sabli, Mohd Faizal Jamlos, and Fatimah Dzaharudin

DSRC Technology in Vehicle-to-Vehicle (V2V) and Vehicle-to-Infrastructure (V2I) IoT System for Intelligent Transportation System (ITS): A Review 97
Aidil Redza Khan, Mohd Faizal Jamlos, Nurmadiha Osman, Muhammad Izhar Ishak, Fatimah Dzaharudin, You Kok Yeow, and Khairil Anuar Khairi

Real-Time and Predictive Analytics of Air Quality with IoT System: A Review 107
Nurmadiha Osman, Mohd Faizal Jamlos, Fatimah Dzaharudin, Aidil Redza Khan, You Kok Yeow, and Khairil Anuar Khairi

Near-Infrared Spectroscopy for Ganoderma Boninense Detection: An Outlook 117
Mas Ira Syafila Mohd Hilmi Tan, Mohd Faizal Jamlos, Ahmad Fairuz Omar, Fatimah Dzaharudin, Mohd Azraie Mohd Azmi, Mohd Noor Ahmad, Nur Akmal Abd. Rahman, and Khairil Anuar Khairi

Investigation of Features for Classification RFID Reading Between Two RFID Reader in Various Support Vector Machine Kernel Function 127
Chun Sern Choong, Ahmad Fakhri Ab. Nasir, Anwar P. P. Abdul Majeed, Muhammad Aizzat Zakaria, and Mohd Azraai Mohd Razman

Modeling and Analyzing of Traveling Wave Gait of Modular Snake Robot 141
Areej G. Abdulshaheed, Mohamed Bin Hussein, Mohd Azuwan Mat Dzahir, and Shaharil Mad Saad

Use of Artificial Neural Network for Analyzing the Contributions of Some Kinematic Parameters in the Polishing Process of Porcelain Tiles 153
Ricardo Halla II, Adeilson de Oliveira Souza, and Fábio José Pinheiro Sousa

Magnetorheological Damper Control for Semi-active Suspension System Using Skyhook-Differential Evolution 161
Mat Hussin Ab Talib, Mohd Ariff Durranie Muhammad Afandi, Intan Zaurah Mat Darus, Hanim Mohd Yatim, Zainab Asus, Ahmad Hafizal Mohd Yamin, and Muhamad Sukri Hadi

PID Controller Based on Flower Pollination Algorithm of Flexible Beam System 173
Ali Akhbar Mohd Fadzli, Muhamad Sukri Hadi, Rickey Ting Pek Eek, Mat Hussin Ab. Talib, Hanim Mohd Yatim, and Intan Zaurah Mat Darus

An Ultraviolet C Light-Emitting Robot Design for Disinfection in the Operating Room 185
Nguyen Duy Minh Phan, Ngo Quoc Huy Tran, Le Anh Doan, Duy Chung Tran, Van Sanh Huynh, Quang Truong Vo, Duc Long Nguyen, and Qui Tra Phan

Experimental of Multi-holes Drilling Toolpath Using Particle Swarm Optimization and CAD-CAM Software on PCB 197
N. W. Z. Abidin, N. Salim, Mohd Fadzil Faisae Ab. Rashid, N. M. Z. N. Mohamed, A. N. M. Rose, and A. Mokhtar

Parameter Identification of Horizontal Flexible Plate System Using Cuckoo Search Algorithm 205
Nurul Nisha Mawarnie Mohd Rawi and Muhamad Sukri Hadi

Bearing Fault Detection Using Discrete Wavelet Transform and Partitioning Around Medoids Methods 219
Gigih Priyandoko, Diky Siswanto, Istiadi, Dedy U. Effendi, and Eska R. Naufal

The Implementation of a Novel Augmented Reality (AR) Based Mobile Application on Open Platform Controller for CNC Machining Operation 227
Anbia Adam, Yusri Yusof, Kamran Latif, Aini Zuhra Abdul Kadir, Toong-Hai Sam, Shamy Nazrein, and Danish Ali Memon

Firefly Algorithm for Modeling of Flexible Manipulator System 235
Hazeem Hakeemi Baseri, Hanim Mohd Yatim, Muhamad Sukri Hadi, Mat Hussin Ab. Talib, and Intan Zaurah Mat Darus

Cubic Spline Interpolations in CNC Machining 253
W. R. W. Yusoff, I. Ishak, and F. R. M. Romlay

Modified Particle Swarm Optimization with Unique Self-cognitive Learning for Global Optimization Problems 263
Koon Meng Ang, Wei Hong Lim, Nor Ashidi Mat Isa, Sew Sun Tiang, Chun Kit Ang, Cher En Chow, and Zhe Sheng Yeap

The Role of 3D-Technologies in Humanoid Robotics: A Systematic Review for 3D-Printing in Modern Social Robots 275
Jayesh Saini and Esyin Chew

A Survey on the Contributions of 3D Printing to Robotics Education—A Decade Review 289
Adamu Yusuf Abdullahi, Mukhtar Fatihu Hamza, and Abdulbasid Ismail Isa

H_∞ Filter with Fuzzy Logic Estimation to Refrain Finite Escape Time 303
Bakiss Hiyana Abu Bakar and Hamzah Ahmad

The Identification of Significant Mechanomyography Time-Domain Features for the Classification of Knee Motion 313
Tarek Mohamed Mahmoud Said Mohamed, Muhammad Amirul Abdullah, Hasan Alqaraghuli, Rabiu Muazu Musa, Ahmad Fakhri Ab. Nasir, Mohd Azraai Mohd Razman, Mohd Yazid Bajuri, and Anwar P. P. Abdul Majeed

Parameter Estimation of Lorenz Attractor: A Combined Deep Neural Network and K-Means Clustering Approach 321
Nurnajmin Qasrina Ann, Dwi Pebrianti, Mohamad Fadhil Abas, and Luhur Bayuaji

Design and Performance Analysis of Body Worn Textile Antenna Using 100% Polyester at 2.4 GHz for Wireless Applications 333
Shehab Khan Noor, Nurulazlina Ramli, Najah Najibah Zaini, and N. H. Abd Rahman

Simulation on Circularly Polarization Cotton Textile Antenna for Wireless Communication System 345
Taher Khalifa, Nurulazlina Ramli, Anis Fariza Md. Pazil, N. H. Abd Rahman, and Ahmad Jais Alias

Development of a 6-DOF 3D Printed Industrial Robot for Teaching and Learning 355
Adamu Yusuf Abdullahi, Hwa Jen Yap, Mukhtar Fatihu Hamza, and Musa Mohammed Bello

Mathematical Model for Planar Reflectarray Antenna Design 369
M. Inam and M. Y. Ismial

Broadband Reflectarray Antenna Based on Highly Conductive Graphene 381
M. Y. Ismial and M. Inam

Stability Derivative Identification Using Adaptive Robust Extended Kalman Filter for Multirotor Unmanned Aerial Vehicle (M-UAV) 391
Danial Rosli, Erwin Sulaeman, Ari Legowo, and Alia Farhana Abdul Ghaffar

A Novel BiGRUBiLSTM Model for Multilevel Sentiment Analysis Using Deep Neural Network with BiGRU-BiLSTM 403
Md. Shofiqul Islam and Ngahzaifa Ab Ghani

A Comparative Study on Nonlinear Control of Induced Sit-to-Stand in Paraplegia with Human Mass Variation 415
Mohammed Ahmed, M. S. Huq, B. S. K. K. Ibrahim, Nura Musa Tahir, Zainab Ahmed, and Garba Elhassan

The Study of Time Domain Features of EMG Signals for Detecting Driver's Drowsiness 427
Faradila Naim, Mahfuzah Mustafa, Norizam Sulaiman, and Noor Aisyah Ab Rahman

The Classification of Skateboarding Tricks by Means of the Integration of Transfer Learning Models and *K*-Nearest Neighbors 439
Muhammad Nur Aiman Shapiee, Muhammad Ar Rahim Ibrahim, Mohd Azraai Mohd Razman, Muhammad Amirul Abdullah, Rabiu Muazu Musa, Noor Azuan Abu Osman, and Anwar P. P. Abdul Majeed

An Improved Grey Wolf Optimizer with Hyperbolic Tangent Updating Mechanism for Solving Optimization Problems 451
Mohd Zaidi Mohd Tumari, Mohd Ashraf Ahmad, and Mohd Helmi Suid

Non-dominated Sorting Manta Ray Foraging Algorithm with an Application to Optimize PD Control 463
Ahmad Azwan Abdul Razak, Ahmad Nor Kasruddin Nasir, Nor Maniha Abdul Ghani, Shuhairie Mohammad, Mohd Falfazli Mat Jusof, and Nurul Amira Mhd Rizal

Review on Effects of Adverse Sonic Environment in Hospital and Control Solutions 475
Khin Fai Chen

Evaluation of the Convolutional Neural Network's Performance in Classifying Steel Strip's Surface Defects 485
Tan Kai Wen, Nur Safwati Mohd Nor, Teoh Wen Kang, Nor Akmal Fadil, Intan Zaurah Mat Darus, Ahmad Hafizal Mohd Yamin, and Fazila Mohd Zawawi

Investigation of Time-Domain and Frequency-Domain Based Features to Classify the EEG Auditory Evoked Potentials (AEPs) Responses 497
Md. Nahidul Islam, Norizam Sulaiman, Mamunur Rashid, Mahfuzah Mustafa, and MohdShawal Jadin

Suppression of Tremors in Parkinson's Patients Using a Dynamic Vibration Absorber 509
Naheed Rihan, Asan G. A. Muthalif, Thaer M. I. Syam, and N. H. Diyana Nordin

Design of Condition Monitoring System on Stairlift Based on Internet of Things (IoT) for Physical Data Acquisition Using Multi Sensors 517
Rohmat Setiawan, Danardono A. Sumarsono, and Wahyu Sulistiyo

Non-vibrate Palm Oil Tree Harvesting Cutter Using DC Motor 529
Zamzuri Hamedon, Ammar Zakwan Abdullah, Ismayuzri Ishak, and Hasnulhadi Jaafar

Feasibility Study of CO, CO_2, NO_2, and O_2 Sensors for Hazardous Gas Detection System in Vehicle Cabin 537
Cheow Shek Choon and Ismayuzri Bin Ishak

Impact Analysis of Harassment Against Women in Bangladesh Using Machine Learning Approaches 549
Busrat Jahan, Bahar Uddin Mahmud, Abdullah Al Mamun, Md. Mujibur Rahman Majumder, and Mahbubul Alam

Fuzzy Logic Controller Optimized by MABSA for DC Servo Motor on Physical Experiment 561
Nurainaa Elias and Nafrizuan Mat Yahya

Effect of Different Signal Weighting Function of Magnetic Field Using KNN for Indoor Localization 571
Caceja Elyca Anak Bundak, Mohd Amiruddin Abd Rahman, Muhammad Khalis Abd Karim, and Nurul Huda Osman

The Classification of Electrooculogram (EOG) Through the Application of Linear Discriminant Analysis (LDA) of Selected Time-Domain Signals 583
Farhan Anis Azhar, Mahfuzah Mustafa, Norizam Sulaiman, Mamunur Rashid, Bifta Sama Bari, Md Nahidul Islam, Md Jahid Hasan, and Nur Fahriza Mohd Ali

Experimental Study of the Effect of Vehicle Velocity on the Ride Comfort of a Car on a Road with Different Types of Roughness 593
Kazem Reza Kashyzadeh and Nima Amiri

Performance Evaluation of BPSO & PCA as Feature Reduction Techniques for Bearing Fault Diagnosis 605
Atik Faysal, Ngui Wai Keng, and M. H. Lim

A Study on Different Techniques in ALPR System: The Systems Performance Analysis 617
Gan Vi Vi and Ahmad Afif bin Mohd Faudzi

Disturbance Rejection Performance Evaluation of GA Optimized PI Controller for Brushed DC Motor for Cart Follower Application 629
A. A. M. Zahir, S. S. N. Alhady, W. A. F. W. Othman, A. A. A. Wahab, and M. F. Ahmad

Localised Muscle Contraction Predictor for Steering Wheel Operation in Simulated Condition 647
Nor Kamaliana Khamis, Dieter Schramm, Mohd Anas Mohd Sabri, and Muhamad Syukri Abdul Khalid

Experimental of CVT Ratio Control Using Single Actuator Double Acting Electro-mechanical Continuously Variable Transmission 659
Nur Rashid Mat Nuri, Khisbullah Hudha, and Muhammad Luqman Hakim Abd Rahman

A Summarization of Image and Video Databases for Emotion Recognition 669
Arselan Ashraf, Teddy Surya Gunawan, Farah Diyana Abdul Rahman, and Mira Kartiwi

Speech Emotion Recognition Using Feature Fusion of TEO and MFCC on Multilingual Databases 681
Syed Asif Ahmad Qadri, Teddy Surya Gunawan, Mira Kartiwi, Hasmah Mansor, and Taiba Majid Wani

Normal Forces Effects of a Two In-Wheel Electric Vehicle Towards the Human Body 693
Nurul Afiqah Zainal, Muhammad Aizzat Zakaria, K. Baarath, Anwar P. P. Abdul Majeed, Ahmad Fakhri Ab. Nasir, and Georgios Papaioannou

On the Effect of Feature Compression on Speech Emotion Recognition Across Multiple Languages 703
Muhammad Fahreza Alghifari, Teddy Surya Gunawan, Nik Nur Wahidah Nik Hashim, Mimi Aminah binti Wan Nordin, and Mira Kartiwi

Real-Time Power Quality Disturbance Classification Using Convolutional Neural Networks 715
Budi Yanto Husodo, Kalamullah Ramli, Eko Ihsanto, and Teddy Surya Gunawan

Power Quality Disturbance Classification Using Deep BiLSTM Architectures with Exponentially Decayed Number of Nodes in the Hidden Layers 725
Teddy Surya Gunawan, Budi Yanto Husodo, Eko Ihsanto, and Kalamullah Ramli

Machine Vision and Convolutional Neural Networks for Tool Wear Identification and Classification 737
Tiyamike Banda, Bryan Yeoh Wei Jie, Ali Akhavan Farid, and Chin Seong Lim

TOM: The Assistant Robotic Tutor of Musicianship with Sound Peak Beat Detection 749
Gareth Hawkins and Esyin Chew

Investigation on Accuracy of Sensors in Sensor Fusion for Object Detection of Autonomous Vehicle Based on 2D Lidar and Ultrasonic Sensor 761
Mohammad Hazrul Ashraf Bin Rosdi and Ahmad Shahrizan Abdul Ghani

Decision Support System on Determination of Contraception Tools as an Effort to Suppress the Number of Growth Ratios in Indonesia 771
Agus Perdana Windarto and Tutut Herawan

***K*-Means Algorithm with Rapidminer in Clustering School Participation Rate in Indonesia** 779
Agus Perdana Windarto and Tutut Herawan

Investigation on Integration of Sensors and Vision-Based Vehicle Detection System for Autonomous Vehicle 795
Muhammad Aizzat Iqbal Bin Abd Rashid and Ahmad Shahrizan Abdul Ghani

Study of Linear-Correlation Based Solar Irradiance Measurement Device Photovoltaic Application 805
Amirah Nurhafizah Abu Bakar, Nuratiqah Mohd Isa, Mohamad Shaiful Abdul Karim, Ruhaizad Ishak, Suazlan Mt Aznam, and Ahmad Syahiman Mohd Shah

3D Traffic Sign Detection Using Camera-LiDAR Projection 821
Wonho Song and Hyun Myung

Development of Physiotherapy-Treadmill *(PhyMill)* as Rehabilitation Technology Tools for Kid with Cerebral Palsy 829
Mohd Azrul Hisham Mohd Adib, Rabiatul Aisyah Arifin, Mohd Hanafi Abdul Rahim, Muhammad Rais Rahim, Muhammad Shazzuan Sharudin, Afif Awaluddin Othman, Ahmad Hijran Nasaruddin, Afiq Ikmal Zahir, Idris Mat Sahat, Nurul Shahida Mohd Shalahim, Narimah Daud, and Nur Hazreen Mohd Hasni

Pediatrics Technology Applications: Enhance the Bilirubin Jaundice (*BiliDice*) Device for Neonates Using Color Sensor 839
Mohd Azrul Hisham Mohd Adib, Mohd Hanafi Abdul Rahim, Idris Mat Sahat, and Nur Hazreen Mohd Hasni

A Supervised Learning Neural Network Approach for the Prediction of Supercapacitive Energy Storage Materials 849
Varun Geetha Mohan, Mohamed Ariff Ameedeen, and Saiful Azad

Two-Steps Approach of Localization in Humanoid Robot Soccer Competition 859
Anhar Risnumawan, Miftahul Anwar, Rokhmat Febrianto, Cipta Priambodo, Mochamad Ayuf Basthomi, Puguh Budi Wasono, Hendhi Hermawan, and Tutut Herawan

Evaluation of the Transfer Learning Models in Wafer Defects Classification 873
Jessnor Arif Mat Jizat, Anwar P. P. Abdul Majeed, Ahmad Fakhri Ab. Nasir, Zahari Taha, Edmund Yuen, and Shi Xuen Lim

Transitioning into a Deregulated Energy Market for Sabah: Strategies and Challenges for Generators 883
Tze Wei Lim, Andrew Huey Ping Tan, Eng Hwa Yap, Kim-Yeow Tshai, and Wei Kong

Rain Classification for Autonomous Vehicle Navigation Using Machine Learning 895
Abdul Haleem Habeeb Mohamed, Muhammad Aizzat Zakaria, Mohd Azraai Mohd Razman, Anwar P. P. Abdul Majeed, Mohamed Heerwan Bin Peeie, Choong Chun Sern, and Baarath Kunjunni

Development of an Innovative Inferior Alveolar Nerve Block (IANB) Simulator Kit with Data Visualization and Internet of Things (IoT) 905
S. Z. Zainudin, M. H. M. Ramli, T. I. B. T. Jamaluddin, and S. A. Abdullah

Optimization of CNG Direct Injector Parameters Using Model-Based Calibration Framework ... 917
Mohamad Hafidzul Rahman Alias, Mohd Fadzil Abdul Rahim, and Rosli Abu Bakar

The Application of Modified Equipment in Retention of Motor Task Performance Amongst Children of Low and High Working Memory Capacity ... 931
Rabiu Muazu Musa, Mohsen Afrouzeh, Pathmanathan K. Suppiah, Anwar P. P. Abdul Majeed, Mohammad Sadegh Afroozeh, and Mohamad Razali Abdullah

Firefly Algorithm for Functional Link Neural Network Learning ... 941
Yana Mazwin Mohmad Hassim, Rozaida Ghazali, Norlida Hassan, Nureize Arbaiy, and Aida Mustapha

Kinematic Variables Defining Performance of Basketball Free-Throw in Novice Children: An Information Gain and Logistic Regression Analysis ... 949
Mohsen Afrouzeh, Ferman Konukman, Rabiu Muazu Musa, Pathmanathan K. Suppiah, Anwar P. P. Abdul Majeed, and Mohd Azraai Mohd Razman

The Identification of Significant Time-Domain Features for Wink-Based EEG Signals ... 957
Tang Jin Cheng, Jothi Letchumy Mahendra Kumar, Mamunur Rashid, Rabiu Muazu Musa, Mohd Azraai Mohd Razman, Norizam Sulaiman, Rozita Jailani, and Anwar P. P. Abdul Majeed

Hyper-Heuristic Strategy for Input-Output-Based Interaction Testing ... 967
Fakhrud Din and Kamal Z. Zamli

Forecasting Daily Travel Mode Choice of Kuantan Travellers by Means of Machine Learning Models ... 979
Nur Fahriza Mohd Ali, Ahmad Farhan Mohd Sadullah, Anwar P. P. Abdul Majeed, Mohd Azraai Mohd Razman, Chun Sern Choong, and Rabiu Muazu Musa

The Classification of Hallucination: The Identification of Significant Time-Domain EEG Signals ... 989
Chin Hau Lim, Jothi Letchumy Mahendra Kumar, Mamunur Rashid, Rabiu Muazu Musa, Mohd Azraai Mohd Razman, Norizam Sulaiman, Rozita Jailani, and Anwar P. P. Abdul Majeed

The Classification of Blinking: An Evaluation of Significant Time-Domain Features 999
Gavin Lim Jiann Kai, Jothi Letchumy Mahendra Kumar, Mamunur Rashid, Rabiu Muazu Musa, Mohd Azraai Mohd Razman, Norizam Sulaiman, Rozita Jailani, and Anwar P. P. Abdul Majeed

The Classification of Electrooculography Signals: A Significant Feature Identification via Mutual Information 1005
Phua Jia Hwa, Jothi Letchumy Mahendra Kumar, Mamunur Rashid, Rabiu Muazu Musa, Mohd Azraai Mohd Razman, Norizam Sulaiman, Rozita Jailani, and Anwar P. P. Abdul Majeed

The Classification of Skateboarding Tricks: A Support Vector Machine Hyperparameter Evaluation Optimisation 1013
Muhammad Ar Rahim Ibrahim, Muhammad Nur Aiman Shapiee, Muhammad Amirul Abdullah, Mohd Azraai Mohd Razman, Rabiu Muazu Musa, and Anwar P. P. Abdul Majeed

Hybrid Manta Ray Foraging—Particle Swarm Algorithm for PD Control Optimization of an Inverted Pendulum

Mohd Falfazli Mat Jusof, Shuhairie Mohammad, Ahmad Azwan Abd Razak, Nurul Amira Mhd Rizal, Ahmad Nor Kasruddin Nasir, and Mohd Ashraf Ahmad

Abstract This paper presents a hybrid Manta ray foraging—particle swarm optimization algorithm. Manta Ray Foraging Optimization (MRFO) algorithm is a recent algorithm that has a promising performance as compared to other popular algorithms. On the other hand, Particle Swarm Optimization (PSO) algorithm is a well-known and a good performance algorithm. The proposed hybrid algorithm in this work incorporates social interaction and elitism mechanisms from PSO into MRFO strategy. The mechanisms help search agents to determine their new search direction. The proposed algorithm is tested on various dimensions and fitness landscapes of CEC2014 benchmark functions. In solving a real world engineering problem, it is applied to optimize a PD controller for an inverted pendulum system. Result of the benchmark function test is statistically analyzed. The proposed algorithm has successfully improved the accuracy performance for most of the test functions. For optimization of the PD control, result shows that the proposed algorithm has attained a better control performance compared to MRFO.

Keywords Manta ray foraging optimization · Particle swarm optimization · PD control · Inverted pendulum system

1 Introduction

Manta Ray Foraging Optimization (MRFO) is a recently introduced optimization algorithm [1]. It is developed inspired from the foraging strategy of a manta ray population. Manta ray is known to find its food location by an individual or in a group form. However, they are frequently observed searching their food location in a group. This helps them to harvest their food optimally. From literature, it is known

M. F. M. Jusof (✉) · S. Mohammad · A. A. A. Razak · N. A. M. Rizal · A. N. K. Nasir · M. A. Ahmad
Faculty of Electrical and Electronics Engineering, Universiti Malaysia Pahang, 26600 Pekan, Pahang, Malaysia
e-mail: mfalfazli@ump.edu.my

A. N. K. Nasir
e-mail: kasruddin@ump.edu.my

A. F. Ab. Nasir et al. (eds.), *Recent Trends in Mechatronics Towards Industry 4.0*, Lecture Notes in Electrical Engineering 730, https://doi.org/10.1007/978-981-33-4597-3_1

that plankton is the name of manta ray food. Two unique features of the manta ray foraging strategy are the linear and spiral movements. They are the trajectories of manta ray toward the targeted plankton location. For a linear movement, several manta rays form a line where a younger manta ray through its head is connected to the tail of an older manta ray. The first manta ray in the line leads the others and moves toward the location of the plankton. For the spiral movement, manta rays in a group based, approach the targeted plankton in a spiral motion. The strategy is used when they found a high concentration of plankton. The good part of this trajectory is that the motion of a manta ray is guided by the position of the plankton at the center of spiral. It is also guided by another manta ray that is located in front of the updating manta ray.

At this stage, the application of MRFO as a tool for solving real world problems can be found in a limited number of publications. As part of proving the MRFO as an algorithm that has a good performance, Weiguo et al. [1] have tested the algorithm to solve eight different mechanical engineering problems. These include problems in spring, pressure vessel, welded beam, speed reducer, bearing and disc clutch brake. Another application of MRFO is parameter estimations of proton exchange membrane fuel cells (PEMFCs) [2]. The authors concluded that the MRFO is a promising algorithm due to its capability to give a good result. Due to its recent introduction, modification of the MRFO strategy as part of improving the algorithm performance is hardly found in literature.

On the other hand, Particle Swarm Optimization (PSO) is a well-known optimization algorithm developed inspired from the strategy of bird flocking behaviour. PSO is known as an algorithm with a competitive performance. One of the unique strategy offered by the PSO is the social interaction of three birds. The interaction gives information about a fitness of a local and the global best birds to the updating bird. It enables the updating bird to consider the fitness prior to its movement to a new possible location. There are a lot of modifications have been proposed by other researchers with the aim to improve the accuracy performance of PSO. These include incorporation of an adaptive equation into position update equation of a search agent. Adaptive type PSO includes an incorporation of a sigmoid function into PSO [3]. The cognitive and social components of PSO are made adaptable with respect to the algorithm iteration. Another adaptive scheme found in literature was position update equation formulated based on an agent fitness [4]. There are also numerous applications of PSO in solving a real world problem can be found in literature. These include optimization of a convolutional neural network for classifying image [5], parameter estimation and structure optimization of deep learning model [6] and optimization of a convolution neural network for various image dataset [7].

An optimization algorithm also can be applied to acquire an optimum set of value of controller parameter. In the work of controlling an inverted pendulum system, Sasa Vrkalovic et al. [8] optimized a state feedback type Takagi–Sugeno fuzzy control using PSO, Simulated Annealing and Gravitational Search algorithms. Result of the work showed that all the algorithms satisfactorily controlled the system. Wang and Liu [9] applied heterogeneous comprehensive learning particle swarm optimization algorithm to optimize hierarchical sliding-mode control for trajectory tracking of cart position. Application of the algorithm resulted in a high controller performance. Several other algorithms that were applied to optimize control scheme for an inverted pendulum includes Genetic Algorithm, Artificial Bee Colony and Ant Colony optimization algorithms [10]. In the work, the authors optimized weighting matrix Q of a Linear Quadratic control. It was found that the pendulum response was significantly improved using Ant Colony algorithm. Spiral dynamic algorithm was found as a tool to optimize type-2 fuzzy control for a triple-link inverted pendulum. The algorithm was compared with PSO. Result showed that PSO outperformed Spiral dynamic algorithm and had a better controller performance [11].

Motivated by all the aforementioned works, this paper presents a hybrid Manta Ray Foraging—Particle Swarm Optimization (MRFPSO) algorithm in optimizing parameters of a Proportional-Derivative (PD) controller for an inverted pendulum system. The paper structure is organized as follows. Sections 2 and 3 present the MRFO and the proposed MRFO respectively. Performance test on CEC2014 benchmark functions is presented in Sect. 4. Section 5 gives a detail description and the control mechanism for the inverted pendulum. Result and discussion of the benchmark functions performance test and controller performance are presented in Sect. 6. Finally, Sect. 7 concludes the paper and the work.

2 Manta Ray Foraging Algorithm

Manta Ray Foraging Optimization (MRFO) algorithm consists of 3 main phases. The first phase is called the Chain foraging. In this phase, the manta rays form a line towards a position that consists of plankton. The plankton is considered as a food source for the manta rays. Mathematically, the Chain foraging is represented as (1a) and (1b).

$$x_i^d(k+1) = x_i^d(k) + r.(x_{best}^d(k) - x_i^d(k))x_i^d(k+1) + \alpha.(x_{best}^d(k) - x_i^d(k)) \quad i = 1 \tag{1a}$$

$$x_i^d(k+1) = x_i^d(k) + r.(x_{i-1}^d(k) - x_i^d(k)) + \alpha.(x_{best}^d(k) - x_i^d(k)) \quad i = 2, ..., N \tag{1b}$$

where is the position of the ith manta ray in the kth iteration for d^{th} dimension problem. $x_{best}^d(k)$ is the position of the best manta ray in the kth iteration for d^{th} dimension problem. r is a random vector that is defined in the range [0,1] and N is the maximum number of searching agent. α is defined as (1c).

$$\alpha = 2.r.\sqrt{|\log(r)|} \tag{1c}$$

The second phase of the MRFO proses is called a Cyclonic foraging. In this phase, the manta rays move towards the plankton in a spiral form. The spiral trajectory is generated based on plankton position and a relative position of an agent to its front agent. Mathematically, it can be represented as Eqs. (2a) and (2b).

$$\begin{aligned} x_i^d(k+1) = {} & x_{best}^d + r.(x_{best}^d(k) - x_i^d(k)) \\ & + \beta.(x_{best}^d(k) - x_i^d(k)) \quad i = 1 \end{aligned} \tag{2a}$$

$$\begin{aligned} x_i^d(k+1) = {} & x_{best}^d(k) + r.(x_{i-1}^d(k) - x_i^d(k)) \\ & + \beta.(x_{best}^d(k) - x_i^d(k)) \quad i = 2, \ldots, N \end{aligned} \tag{2b}$$

where α is defined similar to (1c), β is an adaptive equation defined with respect to current and maximum iterations. β is defined as Eq. (2c).

$$\beta = 2e^{r_1 \frac{T-t+1}{T}} \sin(2\pi r_1) \tag{2c}$$

where t and T are the current and maximum iterations respectively, r_I is a random number between [0,1]. Alternatively, the Eqs. (2a) and (2b) can be replaced by Eqs. (2d) and (2e). Here, the spiral trajectory is generated based on a random position rather than the best agent position.

$$\begin{aligned} x_i^d(k+1) = {} & x_{rand}^d + r.(x_{rand}^d(k) - x_i^d(k)) \\ & + \beta.(x_{rand}^d(k) - x_i^d(k)) \quad i = 1 \end{aligned} \tag{2d}$$

$$\begin{aligned} x_i^d(k+1) = {} & x_{rand}^d(k) + r.(x_{i-1}^d(k) - x_i^d(k)) \\ & + \beta.(x_{rand}^d(k) - x_i^d(k)) \quad i = 2, \ldots, N \end{aligned} \tag{2e}$$

where x_{rand}^d is a random position defined within the feasible searching area. The x_{rand}^d is defined as (2f).

$$x_{rand}^d = x_{\min} + r \times (x_{\max} - x_{\min}) \tag{2f}$$

where $x_{\min}$ and $x_{\max}$ are the minimum and maximum boundaries of the searching area while the and r is a random number between [0,1].

The third phase of the MRFO operation is the Somersault foraging. It is a phase where a searching agent rolls around the plankton repeatedly. The operation can be mathematically represented as (3).

$$x_i^d(k+1) = x_i^d(k) + S.(r_2.x_{best}^d - r_3.x_i^d(k)) \quad i = 1 \ldots N \tag{3}$$

where S is the somersault constant, r_1 and r_2 are random number between [0,1]. All these three operations are continuously repeated until the algorithm reaches a stopping condition.

3 Hybrid Manta Ray Foraging—Particle Swarm Algorithm

Hybrid Manta Ray Foraging - Particle Swarm Optimization (MRFPSO) is a synergy between MRFO and PSO algorithms. The MRFO is considered as the main body of algorithm while an improved version of acceleration update equation of PSO algorithm is incorporated into the Somersault phase of the MRFO algorithm. This is to complement the drawback of the MRFO. The improved acceleration update equation for the Somersault phase is defined as (4).

$$x_i^d(k+1) = x_i^d(k) + \lambda \times \left|x_{gbest}^d - x_i^d(k)\right| + \lambda \times \left|x_{lbest}^d - x_i^d(k)\right| \tag{4}$$

where λ is an adaptive equation with respect to iteration and it is defined as (5). x_{gbest}^d and x_{lbest}^d are the global best and the local best of the searching agents.

$$\lambda = \frac{T-t}{T} \times \sin(2 \times \pi \times rand) \tag{5}$$

where $rand$ is a random number defined in the range of [0, 1].

In general the MRFPSO algorithm consists of six steps. The step-by-step description of the proposed MRFPSO is shown as follows.

Step 1: Initialize populations, x_N^d
Step 2: Compute fitness, f_i and determine x_{best}^d.
Step 3:
    For i = 1 to N
        If $rand < 0.5$
            Adopt Cyclone foraging. Apply equations (2a) – (2f).
        Else
            Adopt chain foraging. Apply equations (1a) – (1c).
        End
    End
Step 4: Compute fitness, f_i and determine x_{best}^d.
Step 5: Adopt Somersault foraging.
    For i = 1 to N
        If $rand < 0.5$
            Apply equations (4) and (5).
        Else
            Apply equation (3).
        End
    End
Step 6: Compute fitness, f_i and determine x_{best}^d.

4 Benchmark Functions Test

Benchmark functions are used to test the performance of a newly developed algorithm. The proposed algorithm was tested on CEC2014 benchmark functions. Table 1 shows the mathematical representation of the functions [12].

In order to analyze the performance of the algorithm, it was statistically evaluated. A total of 51 independent runs were conducted for each function. The best accuracy

Table 1 Benchmark functions

Function No.	Equation of the benchmark functions
1	$f_1(x) = \sum_{i-1}^{D} (10^6)^{\frac{i-1}{D-1}} x_i^2$
2	$f_2(x) = x_1^2 + 10^6 \sum_{i=2}^{D} x_i^2$
3	$f_3(x) = 10^6 x_1^2 + \sum_{i=2}^{D} x_i^2$
4	$f_4(x) = \sum_{i=1}^{D-1} (100(x_i^2 - x_{i+1})^2 + (x_i - 1)^2)$

achievement of each run was recorded and their corresponding mean value was calculated. Graphical representation of the algorithm in locating theoretical optima point of the functions was also plotted. It shows the convergence trend of the algorithm from the beginning until the end of the searching operation.

5 Inverted Pendulum System

An inverted pendulum system consists of a cart and a pendulum. The cart moves linearly back and forth in a horizontal direction while the pendulum rotates freely in 360° direction. The pendulum is directly attached on the cart's body. Therefore, the pendulum motion is directly affected by the cart's motion. Schematic diagram of the inverted pendulum system is shown in Fig. 1. It shows that the pendulum is pointing vertically upward and it has an angle 0°. The centre point of the horizontal axis represents the initial position at 0 cm. If 10 cm desired position is defined for the cart, a dc motor will drive the cart to move horizontally, and thus causes the pendulum to rotate. During control mode, the pendulum is required to maintain its angle at 0° regardless the position of the cart. The physical parameters of the cart and pendulum are shown in Table 2.

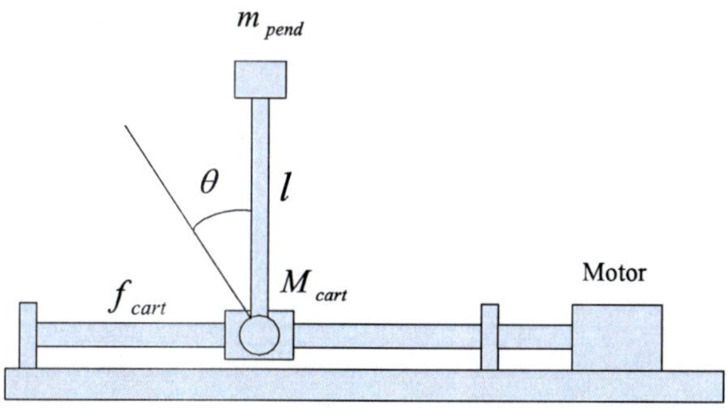

Fig. 1 Schematic of the inverted pendulum system

Table 2 Physical parameter of inverted pendulum system

Parameter	Values
Mass of cart, M_{cart}	0.1 kg
Mass of pendulum, m_{pen}	0.05 kg
Friction or cart, f_{cart}	0.1 $Nm^{-1}s^{-1}$
Length of pendulum, l	0.3 m
Inertia of the pendulum, I	0.006 kg m^2
Motor torque constant, K_m	4.9 N cm A^{-1}
Motor back *emf* constant, K_b	0.0507 V $rad^{-1}s^{-1}$
Motor armature resistant, R	0.3 Ω

5.1 PD Control for Inverted Pendulum System

Proportional-Derivative (PD) control scheme for the inverted pendulum system is shown in Fig. 2. In the work, the desired position of the cart and pendulum angle are defined as 10 cm and 0° respectively. While moving the cart from initial position at 0 cm to the desired position, the pendulum is needed to maintain its angle at 0°. In order to satisfy the objective, two Proportional-Derivative (PD) control schemes are considered for controlling the cart's position and the pendulum's angle. PD_1 is adopted to attenuate the error of the cart's position while the PD_2 is applied to reduce the error of the pendulum's angle. One output gain is introduced at each pendulum's angle and cart's position responses and they are represented as K_1 and K_2 respectively. Those two errors can be reduced if the parameters for both controllers and output gains have a suitable set of value. It can be achieved if both errors are inputted into the proposed algorithm and is considered as the cost function of the algorithm. Error of the cart position is defined as (6).

$$e(t) = (x_{act}(t) - x_{des}(t)) + \theta(t) \tag{6}$$

where $e(t)$ is the error, $x_{act}(t)$ is the actual cart's position, $x_{des}(t)$ is the desired cart's position and $\theta(t)$ is the pendulum's angle at t sampled data. The general PD control is presented as (7).

$$c(t) = e(t)K_p + \frac{de(t)}{dt}K_d \tag{7}$$

where $c(t)$ is the control output, K_p is the proportional gain and K_d is the derivative gain.

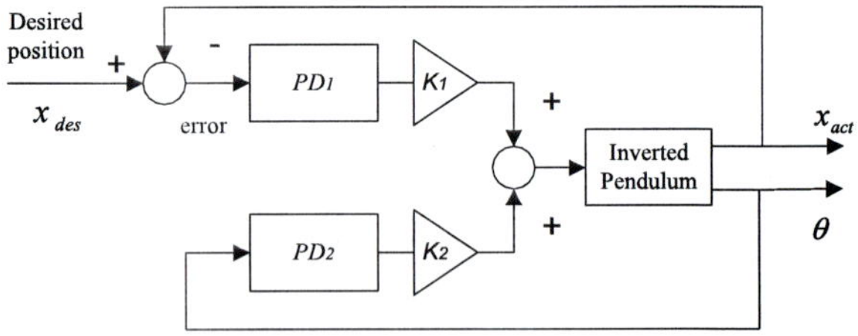

Fig. 2 Block diagram of the control system

Table 3 Fitness cost represents accuracy achievement

Function No.	Algorithm	
	MRFO	MRFPSO
1	3.86×10^4	$\mathbf{1.03 \times 10^4}$
2	1.11×10^3	$\mathbf{5.95 \times 10^2}$
3	5.43×10^2	$\mathbf{3.76 \times 10^2}$
4	4.19×10^2	4.19×10^2

6 Result and Discussion

6.1 Result of the Benchmark Functions Test

Result of the proposed algorithm performance in comparison with the original algorithm tested on four different benchmark functions is presented in both numerical and graphical forms. The numerical result shown is the average value of the accuracy achievement of the algorithm to find the theoretical optima point of each function. The average value is calculated based on 51 independent runs. In the test, the number of searching agents was defined as 10 and the maximum iteration was defined as 100. Result of the performance test is shown in Table 3. Smaller value indicates that the algorithm has achieved a higher accuracy and this is highlighted in bold font. Based on the result shown in the table, it shows that the proposed algorithm has achieved a better accuracy for functions 1–3. Both algorithms show the same performance for function 4. Graphical result of the performance test is shown in Fig. 3. The graphs were generated based on the average value of the 51 independent runs. It represents convergence trend of the algorithm for the whole searching process. The red dotted line represents convergence plot of the MRFO while the dark dashed line represents convergence plot of the MRFPSO. From the graphs, it can be noted that the convergence speed of the algorithms for all functions are almost the same. However, towards the end of iteration, MRFPSO has shown a steeper convergence trend and consequently achieved a better accuracy for functions 1–3.

6.2 Performance of the MRFPSO-PD Controller

The result of the proposed MRFPSO and MRFO convergence plot is shown in Fig. 4. The dark dashed-line is the graph for the MRFPSO while the red dotted-graph is the graph for the MRFO. It is noted from the graphs that the MRFO has shown a faster convergence speed at the early operation. However, the MRFO graph has failed to converge further starting at the iteration 11. On the contrary, the MRFPSO graph has shown a good trend of convergence speed until iteration 45. After the iteration, it has shown a slower speed until the end of iteration. It also has shown that the

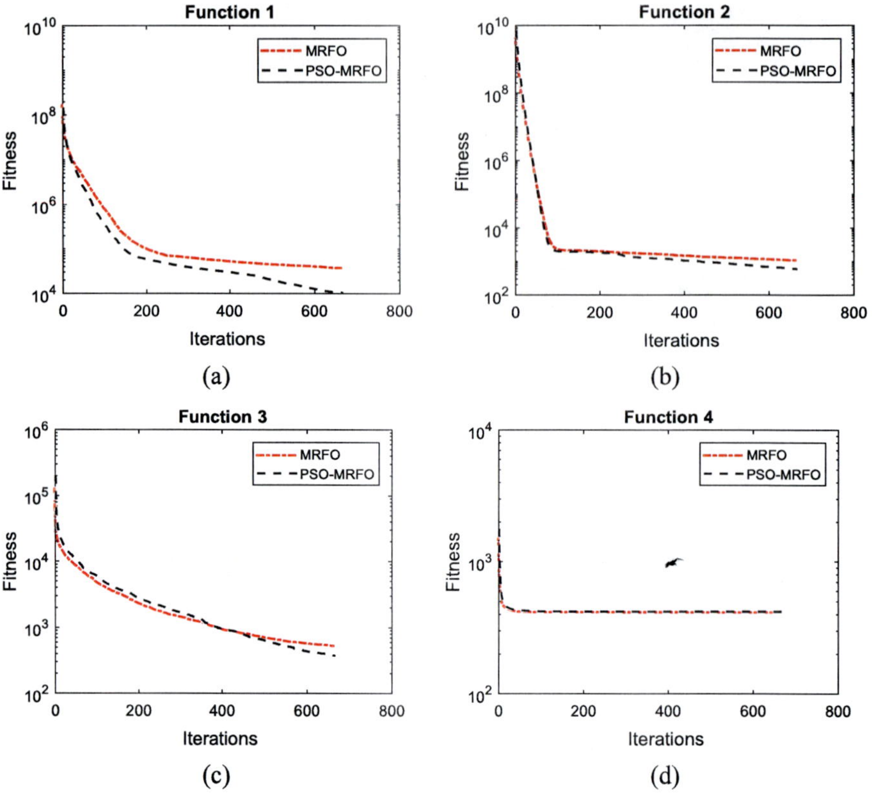

Fig. 3 Convergence plots, **a** function 1, **b** function 2, **c** function 3 and **d** function 4

Fig. 4 Convergence plot for PID optimization

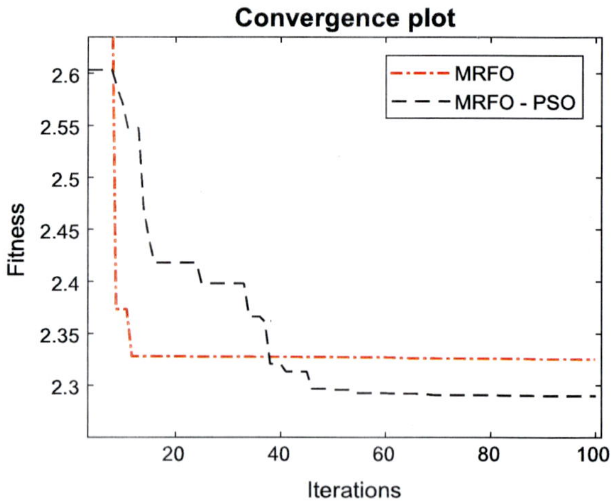

MRFPSO successfully converged at fitness cost 22.9 and it is better than the MRFO result which has converged at fitness cost 23.3.

The responses of the PD control for the pendulum's angle and cart's position are shown in Fig. 5a–d. It is noted that the response of the cart's angle controlled by the MRFPSO-PD controller is more stable than the MRFO-PD controller. From the zoomed-in figure as shown in Fig. 5c, it is noted that the pendulum's angle controlled by MRFO-PD has more ripple between [−0.1 and 0.1] radians. It causes the pendulum to oscillate for about 10 s during the motion. On the other hand, the response of the cart's position optimized by the MRFPSO has faster rise time about 0.13 s compared to the response of the cart's position optimized by the MRFO. The comparison is clearly shown in Fig. 5d. It is noted the graphs show a straight horizontal line after 1.5 s and it indicates that both controller controls the cart's position very well.

The time-domain analysis for the pendulum's angle and cart's position is presented in Tables 4 and 5 respectively. It is noted from the result that the pendulum's angle based on MRFO-PD has a lower maximum overshoot, *max os* compared to the

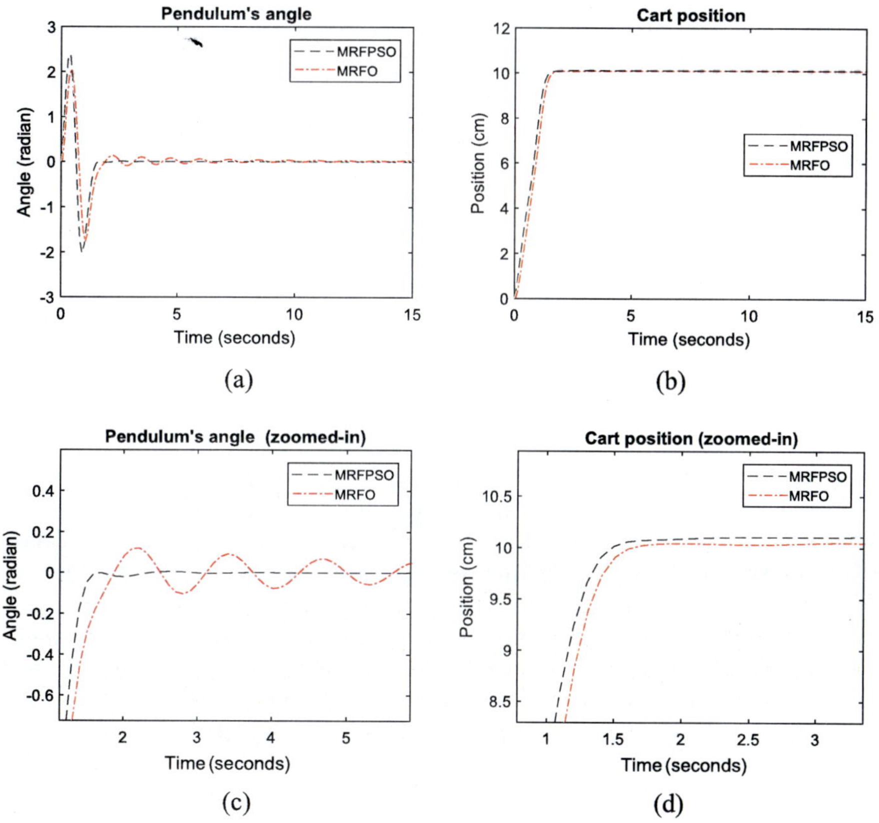

Fig. 5 Output response of the inverted pendulum system, **a** pendulum's angle, **b** cart's position, **c** zoomed-in pendulum's angle and **d** zoomed-in cart's position

Table 4 Pendulum's angle

Algorithm	Time-domain analysis			
	Max. os	e_{ss}	t_r	t_s
MRFO	2.04	0	0.51	9.70
MRFPSO	2.36	0	0.38	1.99

Table 5 Cart's position

Algorithm	Time-domain analysis			
	Max. os	e_{ss}	r_t	s_t
MRFO	10.05	0	1.03	1.44
MRFPSO	10.11	0	1.00	1.35

MRFPSO-PD. The difference of maximum overshoot between the two responses is 0.32 cm. As a consequence, it affects the rise time and settling time performances of the pendulum responses. Notice that, the response based on MRFPSO-PD has shown a faster rise time, t_r and settling time, t_s compared to the response of the MRFO-PD. The MRFO-PD could settle the pendulum's angle 7.71 s after the MRFPSO-PD. For the steady state error e_{ss}, both output responses have shown zero error. The response of the cart's position has shown the similar trend as portrays by the response of the pendulum's angle. The cart's position based on MRFPSO-PD has 0.03 s faster rise time. It is also noted that due to the faster rise time, the cart's position based on MRFPSO-PD has 0.06 cm overshoot relatively larger and 0.09 s settling time relatively faster than the cart's position controlled by the MRFO-PD.

7 Conclusion

A hybrid Manta Ray Foraging—Particle Swarm Optimization (MRFPSO) has been proposed in this paper. A social interaction equation of PSO has been integrated into Somersault foraging phase of MRFO. This is to enhance the exploration capability of the original MRFO. To prove the strategy, the proposed hybrid algorithm has been tested on various CEC2014 benchmark functions in comparison to the original MRFO algorithm. The functions consist of various fitness landscapes and complex features. Statistical analysis has been conducted to evaluate and compare the result produced by the proposed and the original MRFO algorithms. Result of the accuracy performance test has shown that the proposed hybrid strategy has significantly outperformed the predecessor algorithm for most of the benchmark functions. Application to optimize parameters of PD control has shown that both algorithms have produced a satisfactory controller performance. However, the proposed technique has shown a significant improvement for controlling both pendulum's angle and cart's position.

In conclusion, the result of the analysis has proven that the proposed hybrid strategy has improved the performance of MRFO.

Acknowledgements This research is financially supported by the Fundamental Research Grant Scheme (FRGS/1/2019/ICT05/UMP/03/1) with the RDU number RDU1901217. It is awarded by the Ministry of Higher Education Malaysia (MOHE) through Research and Innovation Department, Universiti Malaysia Pahang (UMP) Malaysia.

References

1. Zhao W, Zhang Z, Wang L (2020) Manta ray foraging optimization: an effective bio-inspired optimizer for engineering applications, engineering applications of artificial intelligence 87:103300. ISSN 0952-1976. https://doi.org/10.1016/j.engappai.2019.103300
2. Selem SI, Hasanien HM, El-Fergany AA (2020) Parameters extraction of PEMFC's model using manta rays foraging optimizer. Int J Energy Res 44(6):4629–4640
3. Junior FEF, Yen GG (2019) Particle swarm optimization of deep neural networks architectures for image classification. Swarm Evol Comput 49(2019):62–74
4. Tian D, Zhao X, Shi Z (2019) Chaotic particle swarm optimization with sigmoid-based acceleration coefficients for numerical function optimization. Swarm Evol Comput 51(2019):100573
5. Liu H, Zhang X-W, Liang-Ping Tu (2020) A modified particle swarm optimization using adaptive strategy. Expert Syst Appl 152(2020):113353
6. Li Y, Xiao J, Chen Y, Jiao L (2019) Evolving deep convolutional neural networks by quantum behaved particle swarm optimization with binary encoding for image classification. Neurocomputing 362(2019):156–165
7. Kang L, Chen R-S, Cao W, Chen Y-C (2020) Non-inertial opposition-based particle swarm optimization and its theoretical analysis for deep learning applications. Appl Soft Comput J 88(2020):106038
8. Vrkalovic S, Teban T-A, Borlea I-D (2017) Stable Takagi-Sugeno fuzzy control designed by optimization. Int J Artif Intell 15(2):17–29
9. Wang J-J, Liu G-Y (2019) Hierarchical sliding-mode control of spatial inverted pendulum with heterogeneous comprehensive learning particle swarm optimization. Inf Sci 495:14–36
10. Singhal NK, Swarup A (2019) Performance improvement of inverted pendulum using optimization algorithms. In: 2019 3rd international conference on electronics, communication and aerospace technology (ICECA), 12–14 June 2019 Coimbatore, India, pp 1–6
11. Masrom MF, Ghani NMA, Tokhi MO (2019) Particle swarm optimization and spiral dynamic algorithm-based interval type-2 fuzzy logic control of triple-link inverted pendulum system: a comparative assessment. J Low Freq Noise Vibr Active Control:1–16
12. Liang JJ, Qu BY, Suganthan PN (2014) Problem definitions and evaluation criteria for the CEC 2014 special session and competition on single objective real-parameter numerical optimization. Techn Rep 201311:1–32

Multi-objective Particle Swarm Optimization with Alternate Learning Strategies

Wee Sheng Koh, Wei Hong Lim, Koon Meng Ang, Nor Ashidi Mat Isa, Sew Sun Tiang, Chun Kit Ang, and Mahmud Iwan Solihin

Abstract An improved multi-objective particle swarm optimization (MOPSO) variant known as the MOPSO with alternate learning strategies (MOPSOALS) is proposed to overcome the drawbacks of most existing MOPSO variants that can only solve the selected categories of optimization problems with good performance due to the limited directional information brought by search operators. Particularly, both of the current and memory swarm evolution are incorporated into MOPSOALS as the more robust mechanisms in handling different types of problems. Two search operators are introduced in current swarm evolution to determine the particle's new velocity, while three operators are proposed to fine tune the particle's personal best position. These five proposed search operators are anticipated to guide all MOPSOALS particles to perform thorough searching in the solution search spaces with various exploration and exploitation strengths by fully utilizing all useful information contained in the non-dominated solution set. The proposed MOPSOALS is reported to have better performance in solving all selected test functions than the five peer algorithms.

Keywords Alternative learning strategies · Multi-objective optimization · Particle swarm optimization

1 Introduction

Multi-objective optimization problems (MOPs) are commonly encountered in the real-world engineering applications because these problems consist of multiple conflicting objectives that need to be satisfied simultaneously [1]. For instance, the process parameters optimization of machining processes involve the maximization of

W. S. Koh · W. H. Lim (✉) · K. M. Ang · S. S. Tiang · C. K. Ang · M. I. Solihin
Faculty of Engineering, Technology and Built Environment, UCSI University, 56000 Kuala Lumpur, Malaysia
e-mail: limwh@ucsiuniversity.edu.my

N. A. M. Isa
School of Electrical and Electronic Engineering, Universiti Sains Malaysia, 14300 Nibong Tebal, Malaysia

A. F. Ab. Nasir et al. (eds.), *Recent Trends in Mechatronics Towards Industry 4.0*, Lecture Notes in Electrical Engineering 730, https://doi.org/10.1007/978-981-33-4597-3_2

material's removal rate and minimization of material's surface roughness simultaneously [2, 3]. Particle swarm optimization (PSO) [4] is a popular nature-inspired metaheuristic search algorithm (MSA) commonly applied for handling various types of optimization problems such as those in [5–9]. Numerous PSO variants with improved optimization performances have been proposed via the modifications of parameters, neighborhood structure, search operator, hybridization and etc. [10–12].

Given that PSO is primarily designed to solve the single objective problems, additional mechanism such as Pareto dominance was incorporated into PSO to handle the conflicting objectives of MOPs. A MOPSO variant was proposed in [13], where an external archive used to store the non-dominated solutions was utilized to guide particles searching towards the true Pareto front of a MOP. In [14], the concepts of Pareto dominance and crowding factor were used by OMOPSO to obtain the list of leader's solutions used to guide the search process, while two mutation operators were designed to improve the population diversity. In order to overcome the swarm explosion issue, a velocity construction procedure was introduced in SMPSO to obtain new effective particles when their velocities become too high [15]. Both of the sharing-learning and dynamic crowding distance (SDCD) mechanism were incorporated into MOPSO-SDCD [16] to improve the accuracy and diversity of Pareto fronts. In [17], SAMOPSO was proposed to produce the well-distributed Pareto fronts via circular sorting and elitist-preserving methods on external archive. Decomposition approach is another popular method used to tackle MOPs by transforming it into a set of sub-problems and then optimize them in a collaborative manner [18]. MOPSO/D was first the MOPSO variant employed decomposition framework to tackle MOPs [19], where the personal and global best particles are updated via the aggregation values of all objectives and an epsilon-dominance method was used to manage the external archive members. In [20], the decomposition method was used by AgMOPSO to handle MOPs and an archive-guided search operator is used to guide each particle to solve the sub-problem effectively. For dMOPSO [21], a set of global best particles obtained via scalar aggregated values were used to update the particle's position and a memory re-initialization scheme was introduced to preserve the population's diversity. The MMOSPO particles in [22] were assigned to optimize each sub-problems with multiple search strategies, while an evolutionary search strategy was designed to further exploit useful information stored in external archive. Although numerous MOPSO variants were developed in past decades, most of these works are not able to deliver convincing optimization results in solving different categories of MOPs because of the limited information brought by a single search operator in guiding the search process [11].

A multi-objective particle swarm with alternate learning strategies (MOPSOALS) is designed in this paper to tackle the aforementioned challenges. The main contributions of current work are summarized as: (a) Both of the current and memory swarm evolutions are incorporated into MOPSOALS for enabling more robust mechanisms to handle different types of MOPs. (b) Two search operators are introduced for current swarm evolution to update the velocity of each particle, while three search operators are used to further evolve the personal best positions of all MOPSOALS particles. (c) The five proposed search operators are able to guide all particles to search around the solution search spaces with different levels of exploration and exploitation strengths by fully utilizing all useful information contained in the non-dominated solution set. (d) The optimization performances of MOPSOALS are evaluated rigorously by using 12 MOP benchmark functions.

For the remaining parts of current research, the mechanisms of MOPSOALS are provided in Sect. 2, followed by the performance evaluation in Sect. 3. The conclusion and future study of this works are presented in Sect. 4.

2 Mopsoals

2.1 *Construction of External Archive*

An initial population of MOPSOALS denoted as $\mathbf{P} = [X_1, \ldots, X_n, \ldots, X_N]$ with N particles is randomly generated to solve a MOP with M objectives. Let $f_m(X_n)$ be the m-th objective function value of the n-th particle, where $m = 1, \ldots, M$ and $n = 1, \ldots, N$. Referring to the objective function values of all particles, the non-dominated solutions obtained so far are identified using Pareto dominance relation [1] and stored into a finite size external archive denoted as $\mathbf{A}$. Essentially, the external archive $\mathbf{A}$ is an M dimensional objective space that has been explored so far and formed by multiple equally-spaced hypercubes using adaptive grid approach to construct the Pareto fronts as explained in [13]. Each of the non-dominated solution is inserted into an appropriate hypercube based on its objective function values. The useful information of these non-dominated solutions are used to adjust the trajectory of each particle in search space.

2.2 *Current Swarm Evolution of MOPSOALS*

For current swarm evolution, two search operators are used to determine the particle's new velocity. Let $r_x, r_y \in [0, 1]$ be two random numbers generated from uniform distribution. If $r_x \leq r_y$, conventional scheme is utilized for updating the new velocity of each n-th particle in the d-th dimension as follow:

$$V_{n,d} = \omega V_{n,d} + c_1 r_1 \left(P_{n,d}^{best} - X_{n,d}\right) + c_2 r_2 \left(G_{n,d}^{best} - X_{n,d}\right) \tag{1}$$

where ω is an inertia weight;c_1 and c_2 are the acceleration coefficients; $r_1, r_2 \in [0, 1]$ are two uniformly distributed random numbers; $P_{n,d}^{best}$ and $G_{n,d}^{best}$ are the d-th dimension of personal and global best positions assigned to each n-th particle.

For MOPs, all members in external archive $\mathbf{A}$ are non-dominated to each other and it is not trivial to identify the global best position for Eq. (1). A selection scheme is designed in MOPSOALS to assign a unique global best position $G_{n,d}^{best}$ for each

n-th particle from the existing members of $\mathbf{A}$ based on the density of each occupied hypercube. Let H be the total number of occupied hypercube in $\mathbf{A}$ and κ_h be the number of non-dominated solutions contain in each h-th occupied hypercube. Define $\Lambda_h = \upsilon/\kappa_h$ as the probability of each h-th occupied hypercube to be selected by roulette-wheel method, where $\upsilon > 1$ is a constant. The likelihood for each occupied hypercube in $\mathbf{A}$ being chosen to offer the $G_{n,d}^{best}$ of n-th particle increases with the lower κ_h and vice versa in order to produce an evenly distributed Pareto front. Let $G_{n,d}^{best}$ be a randomly chosen non-dominated solution stored in the selected h-th occupied hypercube for determining the new velocity of each n-th particle using Eq. (1).

For $r_x > r_y$, a search operator inspired from [23] is utilized for updating the particle's velocity by leveraging the useful information of stored in $\mathbf{A}$. Let $r_a \in [0, 1]$ be a random number assigned to each a-th non-dominated solution of A_a in the h-th occupied hypercube, where $a = 1, \ldots, \kappa_h, h = 1, \ldots, H$ and $A_a \in \mathbf{A}$. The weighted mean position of $\tilde{X}_n^{mean}$ assigned to update the velocity of selected n-th particle is then obtained as:

$$\tilde{X}_n^{mean} = \frac{\sum_{h=1}^{H}\sum_{a=1}^{\kappa_h} r_a A_a}{\sum_{h=1}^{H}\sum_{a=1}^{\kappa_h} r_a} \tag{2}$$

From Eq. (2), a weighted mean position is formulated by considering the unique contribution or weightage r_a assigned to each a-th non-dominated solution A_a. For each n-th particle, different $\tilde{X}_n^{mean}$ is generated to determine its new velocity without resulting the rapid population diversity loss as follow:

$$V_{n,d} = \omega V_{n,d} + c_1 r_1 \left(\tilde{X}_{n,d}^{mean} - X_{n,d}\right) \tag{3}$$

Referring to the updated $V_{n,d}$, the corresponding new position for every n-th particle is obtained as:

$$X_{n,d} = X_{n,d} + V_{n,d} \tag{4}$$

For each n-th particle, the objective function values of X_n corresponds to the m-th objective is evaluated as $f_m(X_n)$ and compared with that of personal best position, i.e., $f_m\left(P_n^{best}\right)$ for $m = 1, \ldots, M$, using Pareto dominance relation. The updated X_n is used to replace P_n^{best} if $X_n \succ P_n^{best}$. Otherwise, the current value of P_n^{best} is maintained. Random selection is used to determine the latest personal best position of each n-th particle when the both compared solutions of X_n and P_n^{best} are non-dominated to each other.

2.3 Memory Swarm Evolution of MOPSOALS

For memory swarm evolution, a mutation operator with the probability of $P_{mut} = 1/D$ is incorporated as diversity maintenance scheme. Let $\psi \in [-1, 1]$ be a random number obtained from uniform distribution; $P_{n,d_r}^{best,new}$ is the d_r-th component of the perturbed personal best position; $X_{d_r}^U$ and $X_{d_r}^U$ are the upper and lower boundaries of solutions in the d_r-th dimension, respectively, where $d_r \in [1, D]$. For each n-th particle selected for mutation, the d_r-th dimension in the personal best position of n-th particle is perturbed:

$$P_{n,d_r}^{best,new} = P_{n,d_r}^{best} + \psi\left(X_{d_r}^U - X_{d_r}^L\right) \tag{5}$$

For the particles that are not selected for mutation, two new search operators are used by the memory swarm evolution of MOPSOALS to produce new personal best position by exploiting the promising information of non-dominated solution set stored in the external archive **A**. Let $r_x, r_y \in [0, 1]$ be two uniformly distributed random numbers. If $r_x \leq r_y$, two non-dominated solutions of A_i and A_j, where $i \neq j$, are randomly selected from **A** to produce the new personal best position of n-th particle, i.e.,:

$$P_n^{best,new} = P_n^{best} + r_1\left(A_i - P_n^{best}\right) + r_2\left(A_j - P_n^{best}\right) \tag{6}$$

For $r_x > r_y$, three non-dominated solutions of A_i, A_j and A_k, where $i \neq j \neq k$ are randomly selected from **A** to produce the new personal best position of n-th particle. Specifically, each d-th component of $P_n^{best,new}$ is computed as:

$$P_{n,d}^{best,new} = \begin{cases} A_i + \psi\left(A_j - A_k\right), & \text{if } r_3 > r_4 \\ P_{n,d}^{best}, & \text{otherwise} \end{cases} \tag{7}$$

where $r_3, r_4 \in [0, 1]$ are two uniformly distributed random numbers. Based on the new $P_n^{best,new}$ obtained, all objective function values $f_m\left(P_n^{best,new}\right)$ for $m = 1, \ldots, M$ are evaluated and compared with those of $f_m\left(P_n^{best}\right)$ for $m = 1, \ldots, M$ with Pareto dominance relation. Similar mechanisms as explained in Sect. 3.2 are then utilized to determine the personal best position of each n-th particle.

2.4 Archive Controller of MOPSOALS

For each iteration of MOPSOALS, a new set of personal best positions, i.e., P_n^{best} for $n = 1, \ldots, N$, are obtained from the current and memory swarm evolutions. Pareto dominance concept is then applied to update the archive **A** by comparing each P_n^{best} with respect to each a-th non-dominated solutions stored in **A**, where $a = 1, \ldots, |\mathbf{A}|$.

An archive controller is integrated into MOPSOALS to manage the new incoming solutions or discard those extra non-dominated solution members when **A** is fully occupied.

For the archive controller, the rule of thumbs applied in managing the archive members stored in **A** are explained as follows: (a) A new solution is rejected by **A** if it is dominated by at least one archive member. (b) A new solution is added into **A** if at least one of its archive members is dominated and removed. (c) A new solution is added into **A** if it is non-dominated with all archive members. (d) An adaptive grid approach is used to rearrange the segmentation of objective space if the new solution obtained exceeds the hypercube regions covered by existing **A**. (v) The redundant archive members need to be discarded if **A** is fully occupied.

The density of occupied hypercube in **A** is also considered in eliminating the redundant archive member, aiming to generate a uniformly distributed Pareto front. Suppose that $B_h = e^{\gamma \kappa_h}$ refers to the probability of each h-th occupied hypercube being selected by roulette-wheel method to discard its archive members, where $\gamma > 1$ and $h = 1, \ldots, H$. The probability of an occupied hypercube being selected to randomly discard one of its archive member increases with the larger values of κ_h and vice versa.

2.5 The Complete MOPSOALS Algorithm

The pseudocode of MOPSOALS is presented in Fig. 1. A population **P** consists of N particles and an external archive **A** are initialized. After evaluating the M objective function values of these N particles, Pareto dominance concept is utilized to obtain all non-dominated solutions and stored them into **A**. Iterative processes of current swarm evolution, memory swarm evolution and external archive updates of MOPSOALS are executed until all termination conditions are satisfied, i.e., $\gamma \leq \Gamma^{\max}$, where γ and $\Gamma^{\max}$ refer to the current and maximum fitness evaluations (FEs), respectively.

3 Performance Evaluation

3.1 Simulation Settings

The performance of MOPSOALS is evaluated with 12 MOP test functions, where the first five functions are high-dimensional bi-objective problems (i.e., ZDT1-ZDT4 and ZDT6) [24], whereas the remaining six functions are tri-objective problems (i.e., DTLZ1-DTLZ5) [25]. Inverted generational distance (IGD) is defined to measure the diversity and accuracy of the approximated Pareto front in order to solve a given MOP [26]. Smaller IGD implies that the approximated Pareto front obtained is more

```
Main Algorithm: MOPSOALS
Input: N, M, D, X^U, X^L, Γ^max
Initialize γ = 0 and A = { };
for n = 1 to N do
    Randomly initialize the V_n and P_n of each n-th particle;
    Evaluate all M objective functions;
    γ ← γ + M ;
    Update personal best position of each n-th particle (Refer Section 2.2);
end for
Archive controller to update the members of A (Refer Section 2.4);
while γ ≤ Γ^max do
    for n = 1 to N do
        Current Swarm Evolution (Refer Section 2.2);
        Evaluate all M objective functions;
        γ ← γ + M ;
        Update personal best position of each n-th particle (Refer Section 2.2);
        Memory Swarm Evolution (Refer Section 2.3);
        Evaluate all M objective function;
        γ ← γ + M ;
        Update personal best position of each n-th particle (Refer Section 2.2);
    end for
    Archive controller to update the members of A (Refer Section 2.4);
end while
Output: Pareto Front generated from A ;
```

Fig. 1 Pseudo-code for complete framework of MOPSOALS

desirable because it is not only uniformly distributed but also closer to the true Pareto front.

The performances of MOPSOALS in tackling all selected MOPs were compared to five multi-objective algorithms known as: MOPSO [13], DDMOPSO [22], OMOPSO [14], multi-objective teaching learning based optimization (MOTLBO) [27] and multi-objective grey wolf optimizer (MOGWO) [28]. Referring to the recommendations of respective authors, the parameter settings of all involved algorithms are shown in Table 1. The same population size and archive size are set for all compared algorithms, where $N = |\mathbf{A} = 200|$. The maximum FEs used by an algorithm to solve each test function is set as $\Gamma^{\max} = 200{,}000$. Each simulation is performed using Matlab with 30 independent runs on a personal workstation equipped with Intel®Core i7-7500 CPU @ 2.0 GHz.

Table 1 Parameter settings of all compared algorithms

Algorithm	Parameter settings
MOPSO	$\omega \in [0.1, 0.5]$, $c_1, c_2 \in [1.5, 2.0]$, $P_{mut} = 1/D$
DDMOPSO	$\omega \in [0.1, 0.5]$, $c_1, c_2 \in [1.5, 2.0]$
OMPSO	$\omega \in [0.1, 0.5]$, $c_1, c_2 \in [1.5, 2.0]$
MOTLBO	$T_f \in [1.5, 2.0]$
MOGWO	$\vec{c} \in 2 \to 0$, $\vec{C} \in [0, 2]$, $\vec{A} \in [-1, 1]$
MOPSOALS	$\omega \in [0.1, 0.5]$, $c_1, c_2 \in [1.5, 2.0]$, $P_{mut} = 1/D$

3.2 Performance Comparisons

The mean IGD (IGD_{mean}) and standard deviation (*SD*) produced by MOPSOALS and five peers in tackling all MOP functions are reported in Table 2. The best and second best results are presented as the boldface and underline texts, respectively. The comparison between MOPSOALS and its five peers are also summarized as *w/t/l* and *#BR*. *w/t/l* implies that MOPSOALS is better than a peer in *w* functions, ties in *t* functions and worse in *l* functions. *#BR* is the number of best IGD_{mean} produced by each method.

From Table 2, it is observed that MOPSOALS has demonstrated the most promising search performance among all compared algorithms due to its ability to produce 8 best IGD_{mean} values in 12 MOP functions, i.e., ZDT1, ZDT4, ZDT6, DTLZ1 to DTLZ4 and DTLZ7. The Pareto fronts produced by MOPSOALS in most tested functions are more uniformly distributed and have closer proximity to the true Pareto front as compared to its peers. Despite of being outperformed by MOPSOALS in eight test functions, OMPOPSO has solved the ZDT2, ZDT3, DTLZ5 and DTLZ6 functions with the best IGD_{mean} results. The proposed MOPSOALS is also observed to dominate both MOPSO and MOGWO completely in tackling all test functions. The single search operator scheme of MOPSO and MOGWO has limited the exploration and exploitation strengths of these algorithms to search the true Pareto fronts of MOPs with different characteristics. Although both of the MOTLBO and DDMOPSO employ more than one search operator to solve MOPs, the proposed MOPSOALS still outperforms MOTLBO and DDMOPSO in 11 and 10 test functions, respectively.

4 Conclusions

In this paper, a multi-objective particle swarm optimization with alternate learning strategies (MOPSOALS) is proposed to solve different types MOPs more effectively. Multiple search operator scheme is introduced into the current and memory swarm evolution of MOPSOALS to enhance its robustness in tackling various challenging MOPs by enabling each particle to search with different intensity levels of exploration and exploitation. For each search operator, different exemplars are selected from the

Table 2 Performance comparison between MOPSOALS with five peers in 12 test functions

Function	Metrics	MOPSO	DDMOPSO	OMOPSO	MOTLBO	MOGWO	MOPSOALS
ZDT1	IGD_{mean}	4.58E−03	3.42E−03	*1.86E−03*	3.89E−03	2.40E−02	**1.66E−03**
	SD	1.98E−03	7.26E−04	2.11E−05	1.56E−03	1.02E−02	7.71E−04
ZDT2	IGD_{mean}	9.20E−03	2.26E−01	**1.92E−03**	9.28E−03	2.90E−02	*7.08E−03*
	SD	1.21E−02	2.98E−01	1.76E−05	1.21E−02	5.62E−03	1.23E−02
ZDT3	IGD_{mean}	6.03E−03	6.06E−03	**2.24E−03**	2.93E−02	3.50E−02	*5.44E−03*
	SD	1.65E−03	1.78E−03	8.61E−05	9.01E−03	9.17E−03	3.06E−03
ZDT4	IGD_{mean}	5.04E−01	1.66E+00	4.44E+00	*2.13E−01*	3.15E−01	**1.02E−01**
	SD	3.35E−01	1.33E+00	2.02E+00	7.80E−02	1.96E−01	1.73E−01
ZDT6	IGD_{mean}	3.41E−03	3.75E−03	*1.56E−03*	3.25E−03	1.41E−02	**1.46E−03**
	SD	1.76E−03	2.16E−03	8.72E−05	1.00E−03	2.44E−03	5.02E−04
DTLZ1	IGD_{mean}	5.54E−01	3.22E+00	2.47E+01	*5.32E−01*	5.66E−01	**5.23E−01**
	SD	3.40E−02	2.95E+00	8.01E+00	8.14E−02	3.59E−01	3.91E−02
DTLZ2	IGD_{mean}	*1.18E−02*	2.63E−02	2.81E−02	1.59E−02	1.31E−02	**8.61E−03**
	SD	1.79E−03	3.91E−04	4.88E−04	2.35E−03	2.03E−03	1.67E−03
DTLZ3	IGD_{mean}	6.89E+00	1.26E+01	8.64E+01	*5.85E+00*	9.59E+00	**5.63E+00**
	SD	7.74E−01	1.19E+01	2.15E+01	1.97E−01	4.56E+00	1.01E+00
DTLZ4	IGD_{mean}	8.16E−03	3.01E−02	2.78E−02	8.36E−03	*7.98E−03*	**7.94E−03**
	SD	1.88E−03	2.56E−03	4.93E−03	1.96E−03	3.41E−03	1.86E−03
DTLZ5	IGD_{mean}	9.84E−03	*1.07E−03*	**6.81E−04**	1.09E−02	8.00E−03	5.20E−03
	SD	3.70E−03	1.53E−04	2.18E−05	4.43E−03	2.53E−03	2.89E−03
DTLZ6	IGD_{mean}	9.28E−03	4.08E−03	**6.62E−04**	*2.52E−03*	5.54E−02	4.47E−03
	SD	1.06E−02	8.75E−04	3.47E−05	1.21E−03	5.00E−02	1.16E−02
DTLZ7	IGD_{mean}	*8.47E−03*	3.64E−02	2.93E−02	1.52E−02	1.68E−02	**7.76E−03**
	SD	4.43E−03	2.08E−03	8.02E−04	1.39E−03	1.33E−03	4.24E−03
w/t/l		12/0/0	10/0/2	8/0/4	11/0/1	12/0/0	–
#BR		0	0	4	0	0	8

non-dominated solution sets to ensure all of the valuable information contained are fully utilized to guiding the search process. Extensive simulations were performed to measure the performance of MOPSOALS. The proposed algorithm is proven to have promising optimization performance for being able to produce the uniformly distributed Pareto optimal solution sets that are close to the true Pareto fronts of the tested MOPs.

References

1. Deb K, Pratap A, Agarwal S, Meyarivan T (2002) A fast and elitist multiobjective genetic algorithm: NSGA-II. IEEE Trans Evol Comput 6(2):182–197
2. Natarajan E, Kaviarasan V, Lim WH, Tiang SS, Tan TH (2018) Enhanced multi-objective teaching-learning-based optimization for machining of Delrin. IEEE Access 6:51528–51546
3. Natarajan E, Kaviarasan V, Lim WH, Tiang SS, Parasuraman S, Elango S (2019) Non-dominated sorting modified teaching–learning-based optimization for multi-objective machining of polytetrafluoroethylene (PTFE). J Intell Manuf:1–25
4. Kennedy J, Eberhart R (1995) Particle swarm optimization. In: Proceedings of ICNN'95 international conference on neural networks, vol 4, pp 1942–1948
5. Yu LJ, Sahrim AH, Kong I, Mouad AT (2012) Microwave absorbing properties of nickel-zinc ferrite/multiwalled nanotube thermoplastic natural rubber composites. Adv Mater Res 501:24–28
6. Tarawneh MA, Yu LJ, Tarawni MA, Ahmad SH, Al-Banawi O, Bathiha MA (2015) High performance thermoplastics elastomer (TPE) nanocomposite based on graphene nanoplates (GNPs). World J Eng 12(5):437–442
7. Yao L, Lim WH (2018) Optimal purchase strategy for demand bidding. IEEE Trans Power Syst 33(3):2754–2762
8. Yao L, Yao L, Lim WH (2018) A soft curtailment of wide-area central air conditioning load. Energies 11(3):492
9. Ang KM, Lim WH, Isa NAM, Tiang SS, Wong CH (2020) A constrained multi-swarm particle swarm optimization without velocity for constrained optimization problems. Expert Syst Appl 140:112882
10. Lim WH, Isa NAM (2015) Particle swarm optimization with dual-level task allocation. Eng Appl Artif Intell 38:88–110
11. Lim WH et al (2018) A self-adaptive topologically connected-based particle swarm optimization. IEEE Access 6:65347–65366
12. Bonyadi MR, Michalewicz Z (2017) Particle swarm optimization for single objective continuous space problems: a review. MIT Press
13. Coello CAC, Pulido GT, Lechuga MS (2004) Handling multiple objectives with particle swarm optimization. IEEE Trans Evol Comput 8(3):256–279
14. Sierra MR, Coello CAC (2005) Improving PSO-based multi-objective optimization using crowding, mutation and ∈-dominance. In: Coello CAC, Aguirre AH, Zitzler E (eds) Lecture notes in computer science, vol 3410. Springer, Berlin, Heidelberg
15. Nebro AJ, Durillo JJ, Garcia-Nieto J, Coello CC, Luna F, Alba E (2009) SMPSO: a new PSO-based metaheuristic for multi-objective optimization. In: 2009 IEEE symposium on computational intelligence in multi-criteria decision-making (MCDM). IEEE, pp 66–73
16. Peng G, Fang Y-W, Peng W-S, Chai D, Xu Y (2016) Multi-objective particle optimization algorithm based on sharing–learning and dynamic crowding distance. Optik 127(12):5013–5020
17. Tang B, Zhu Z, Shin H-S, Tsourdos A, Luo J (2017) A framework for multi-objective optimisation based on a new self-adaptive particle swarm optimisation algorithm. Inf Sci 420:364–385
18. Zhang Q, Li H (2007) MOEA/D: a multiobjective evolutionary algorithm based on decomposition. IEEE Trans Evol Comput 11(6):712–731
19. Peng W, Zhang Q (2008) A decomposition-based multi-objective particle swarm optimization algorithm for continuous optimization problems. In: 2008 IEEE international conference on granular computing, Hangzhou, China
20. Zhu Q et al (2017) An external archive-guided multiobjective particle swarm optimization algorithm. IEEE Trans Cybern 47(9):2794–2808
21. Zapotecas Martínez S, Coello Coello CA (2011) A multi-objective particle swarm optimizer based on decomposition. In: Proceedings of the 13th annual conference on Genetic and evolutionary computation, pp 69–76

22. Moubayed NA, Petrovski A, McCall J (2014) D2MOPSO: MOPSO based on decomposition and dominance with archiving using crowding distance in objective and solution spaces. Evol Comput 22(1):47–77
23. Mendes R, Kennedy J, Neves J (2004) The fully informed particle swarm: simpler, maybe better. IEEE Trans Evol Comput 8(3):204–210
24. Zitzler E, Deb K, Thiele L (2000) Comparison of multiobjective evolutionary algorithms: empirical results. Evol Comput 8(2):173–195
25. Deb K, Thiele L, Laumanns M, Zitzler E (2005) Scalable test problems for evolutionary multi-objective optimization. In: Abraham A, Jain L, Goldberg R (eds) Advanced information and knowledge processing. Springer, London
26. Li H, Zhang Q (2009) Multiobjective optimization problems with complicated pareto sets, MOEA/D and NSGA-II. IEEE Trans Evol Comput 13(2):284–302
27. Lin W et al (2015) Multi-objective teaching–learning-based optimization algorithm for reducing carbon emissions and operation time in turning operations. Eng Optim 47(7):994–1007
28. Mirjalili S, Saremi S, Mirjalili SM, Coelho LDS (2016) Multi-objective grey wolf optimizer: a novel algorithm for multi-criterion optimization. Expert Syst Appl 47:106–119

Self-directed Mobile Robot Path Finding in Static Indoor Environment by Explicit Group Modified AOR Iteration

A. A. Dahalan and A. Saudi

Abstract The main concerns in handling a self-directed path finding is that we must address the obstacle avoidance issue in which the mobile robot has to create a bump-free route in order to surpass the efficiency of its movement from any departure position to the destination position in the areas concerned. This study seeks to solve the problem by elucidating it iteratively through a numerical approach. The solution builds on the potential field approach, that uses the equation of Laplace to restrict the formation of potential functions across regions in which the mobile robot operates. This article proposes an iterative method for the resolution of mobile robot path finding problem, namely Explicit Group Modified Accelerated Over-Relaxation (EGMAOR). The experiment demonstrates that, by applying a finite difference method, the mobile robot is competent to produce a collision-free trail from any start to specific target point. In addition, the findings of the model verified that numerical techniques could provide an accelerated solution and have produced smoother path than earlier work on the same issue.

Keywords Robot path finding · Harmonic potential · Explicit group · Optimal path · Collision free

1 Introduction

Path finding applications like the moving machines and autonomous agents has turned into a popular research field in recent years. Mobile robot navigation typically involves spotting for collision-free movement in the environment defined with obstacles to end up at a specific position. In this article, founded on the principle of heat transfer, the path finding of a mobile robot is executed via numerical potential

A. A. Dahalan (✉)
Centre for Defence Foundation Studies, National Defence University of Malaysia, Kuala Lumpur, Malaysia
e-mail: a.qilah@upnm.edu.my

A. Saudi
Faculty of Computing and Informatics, Universiti Malaysia Sabah, Kota Kinabalu, Sabah, Malaysia

A. F. Ab. Nasir et al. (eds.), *Recent Trends in Mechatronics Towards Industry 4.0*, Lecture Notes in Electrical Engineering 730, https://doi.org/10.1007/978-981-33-4597-3_3

function in a known environment. The heat transfer problem is designed by means of Laplace's equation, and the solutions to the formula are the harmonic functions. The ambient temperature values in the designated area are used for the simulation of routes resulting after harmonic functions. Numerical techniques were used to generate harmonic functions because of the accessibility of fast processing devices and the competence to solve the problem. This article performed a number of tests to investigate the efficiency of the iterative scheme of Explicit Group Modified Accelerated Over-Relaxation (EGMAOR) for generating mobile robot paths with varying numbers of obstacles in several sizes of environment.

2 Related Studies

In the past, Connolly et al. [1] and Akishita et al. [2] initiated a global method independently by means of Laplace's path planning equations to construct a smooth path with no collision. These two studies prove the harmonic functions give a rapid way of generating routes in a robot configuration area and prevent local minima from being spontaneously generated. The practice of numerical approach in solving the problem of path navigation was then shown by Sasaki [3]. It says the new computational method to motion planning operated very well by simulating complicated problems with the maze. Apart from that, Barraquand et al. [4] and Connolly and Grupen [5] have addressed the path finding problem in global manner by integrating the iterative approaches with path searching processes. Later, Khatib [6] implemented the application of potential functions for robot navigation. His interpretations see every obstacle exercising a repulsive force on the final effector, whereas the target applies an appealing force. In the meantime, Karonava et al. [7] applied the Dijkstra's algorithm with the image processing in a labyrinth for mobile robot track design. The algorithm discovers the shortest route to an end goal, and shows that it can pass an object for a minimum number of time in a labyrinth with a large scale. Whereas Hachour [8] suggested a policy of a self-directed mobile navigation built in a grid-map setup of an undesignated area using hybrid intelligent with static unidentified obstacles. The key aspect here is the usage of the finest approach of biological genetic theory in conjunction with networks in the role of fuzzy reasoning and inference capturing by human expert knowledge in order to adopt on the optimum escaping course in obtaining safer route from dangerous hazards. The simulation of the study is verified by the use of two programming lingos: the robot reaches the goal in visual basic language by avoiding all obstacles, while the robot is taking the shortest trail in Delphi language to meet the goal.

The ordinary numerical approach for instance Successive Over-Relaxation (SOR) and Gauss–Seidel (GS) iterative schemes was examined in the measurement of the harmonic potentials in the previous studies [1, 3, 5, 9]. The SOR block variants technique has recently been used for rapid computing [10, 11]. Along with autonomous robot movement planning, the harmonic potentials for a number of other applications has also been extended like marine navigation [12], UAV motion planning [13, 14], vessel direction-finding [15], space exploration [16], flight control [17], etc.

3 Laplacian Potentials

Self-directed robot navigation can be modelled in the form of steady-state heat transfer problem. In the sense of thermal conductivity, the Laplace equation is often referred to as a steady-state heat equation [18], and the solutions to this equation are always signified as harmonic functions. By consuming the analogy of heat transfer, the thermal sources derive from the borders and the heat dissipation hauls the heat in. The target position is seen as a sink heat-pulling in. Meanwhile, the obstacles, internal barriers and external borders are identified as thermal sources fixed with constant temperature values. During a heat transmission cycle, temperature dispersals reflecting the values of Laplacian potential fill the configuration space, evolving and flowing from the heat flux lines into the sink. It is then easier to see the path through the heat stream. This cycle assures a path toward the target minus facing local minima and positively escaping any barriers as discovered by Connolly et al. [1].

With the intention of measuring the temperature dispersal, the harmonic function is exerted to design the configuration space paradigm. In mathematical terms, harmonic function in domain $\Omega \subset R^n$ fulfils the equation of Laplace's, with x_i = i-th Cartesian coordinate, n = dimension. For this setup, the domain Ω composed of the external boundaries, obstacles, starting points and target point, for the construction of robot path.

$$\nabla^2\phi = \sum_{i=1}^{n} \frac{\partial^2\phi}{\partial x_i^2} = 0 \tag{1}$$

In this design, a point or spot in the configuration area depicts a robot. The area is constructed in mesh pattern, and iteratively computed the coordinates and the function values of every node using numerical approach to comply with Eq. (1). The potential value allocated for starting spot is set to high, the target spot to be set as lowest, while different initial temperature values for the external borders and obstacles. For this article, the solution of Laplace's equation has been subject to Dirichlet constraint, $\Phi|\partial\Omega = c$, c = constant. As soon as the harmonic function is established within the boundary conditions, the appropriate route can be certainly identified by tailing the heat stream carried out by the gradient descent scheme on calculated potential values. The descending search leads to the spot with the least potential value representing the target spot. This descends process is a sequence of spots with lower potential values. The coordinates and nodal temperature gradients derived from the analysis of finite difference gives the line of the pathway. In summary, the harmonic potentials are measured in the space of configuration carrying obstacles across the region and employing the solutions to sense a trail lines for a mobile robot as of any initial location to a specific target position.

4 Explicit Group Modified Accelerated Over-Relaxation (EGMAOR) Iterative Scheme

The Laplacian as Eq. (1), can be solved by numerical technique efficiently. The standard GS [1] and SOR [9–11] were used for solving Eq. (1) in robotics literature. Other than that, several techniques for block iteration have been implemented using Laplace's equation and produce impressive performance [19, 20]. To figure the solutions of Laplace's Eq. (1), this article offered more rapidly numerical solver by engaging EGMAOR iterative scheme. The Modified Acceleration Over-Relaxation (MAOR) technique essentially is the generalization of the Acceleration Over-Relaxation (AOR) technique. It can be seen that MAOR technique reduces the extrapolation of the Modified SOR and Jacobi scheme with different parameters parallel to the blocks of matrix row for particular acceleration and relaxation matrices choices. Earlier work on block iterative methods [19–23] uses multiple points of Explicit Group (EG) approaches, showing that block iterative methods are superior than the conventional point methods.

Taking into account the two-dimensional Laplace's equation in Eq. (1) stated as

$$\nabla^2 U = \frac{\partial^2 U}{\partial x^2} + \frac{\partial^2 U}{\partial y^2} = 0 \tag{2}$$

Equation (2) can be generalized into 5-point second-order central difference approximation as usually described in the subsequent equation, also called Gauss–Seidel iterative scheme

$$u_{i+1,j+1} + u_{i-1,j-1} + u_{i+1,j-1} + u_{i-1,j+1} - 4u_{i,j} = 0 \tag{3}$$

An approach which contains the execution of red–black ordering scheme strategy, called MAOR method is appended to Eq. (3) to enhance convergence speed, and can generally be expressed as

$$\begin{aligned} U_{i,j}^{(k+1)} &= \frac{\omega}{15}[4S_1 + S_2] + (1-\omega)U_{i,j}^{(k)}, \\ U_{i+1,j+1}^{(k+1)} &= \frac{\omega}{15}[S_1 + 4S_2] + (1-\omega)U_{i+1,j+1}^{(k)}, \end{aligned} \tag{4}$$

where

$$S_1 = U_{i-1,j-1}^{(k+1)} + U_{i-1,j+1}^{(k)} + U_{i+1,j-1}^{(k)},$$

$$S_2 = U_{i+2,j}^{(k)} + U_{i,j+2}^{(k)} + U_{i+2,j+2}^{(k)},$$

for red nodes, while for black nodes are given as

$$\begin{aligned} U_{i,j}^{(k+1)} &= \frac{1}{15}\left[4\omega' S_1 + 4r S_2 + \omega' S_2 + r S_3\right] + \left(1 - \omega'\right) U_{i,j}^{(k)}, \\ U_{i+1,j+1}^{(k+1)} &= \frac{1}{15}\left[\omega' S_1 + r S_2 + 4\omega' S_2 + 4r S_3\right] + \left(1 - \omega'\right) U_{i+1,j+1}^{(k)}, \end{aligned} \tag{5}$$

where

$$\begin{aligned} S_1 &= U_{i-1,j-1}^{(k)} + U_{i-1,j+1}^{(k)} + U_{i+1,j-1}^{(k)}, \\ S_2 &= U_{i-1,j-1}^{(k+1)} + U_{i-1,j+1}^{(k+1)} + U_{i+1,j-1}^{(k+1)} - S_1, \\ S_3 &= U_{i+2,j}^{(k)} + U_{i,j+2}^{(k)} + U_{i+2,j+2}^{(k)}, \\ S_4 &= U_{i+2,j}^{(k+1)} + U_{i,j+2}^{(k+1)} + U_{i+2,j+2}^{(k+1)} - S_3. \end{aligned}$$

The optimum relaxation parameters, r, ω and ω', are specified within [1,2). In deciding the optimal values of r, ω and ω', there is no exact formula in gaining the minimum iteration counts. The r and ω' values are commonly chosen, agreeing by Hadjidimos [24], to get to the nearest value ω of the analogous SOR.

5 Results and Discussion

This study conducted experiments on three different sizes, i.e. 300 × 300, 600 × 600, 900 × 900, of the motionless environment, consisting of various obstacles in four different sceneries. In the initial setting of the configuration space, high temperature values are placed onto the obstacles and external boundaries. The target spot was specified to lowest temperature, whereas there is no initial value at the starting spot. The temperature value for the open/empty space inside the environments was fixed to zero. The experiment was conducted on a 2.50 GHz speed PC with 8 GB of RAM. The measurement of the temperature values proceeded numerically at each spots before the stoppage circumstances were encountered. The loop ends when the temperature values no longer shows any changes, with iterations k and k + 1 has a very insignificant distinction between harmonic potentials, i.e. 1.0^{-16}. To avert the incidence of failure in generating route and to avoid flat areas a.k.a. saddle points in the configuration area, this elevated accuracy was required. Tables 1 and 2 display the iteration counts and the time taken (in seconds) required to compute entire temperature values in the space for each method associated with the experiments. As stated above, in the range of [1,2), the optimal values of weighted parameters, r, ω and ω', are selected. Clearly, EGMAOR iterative method provided high performance in

Table 1 Performance of the proposed methods in terms of iteration counts

	Methods	N × N		
		300 × 300	600 × 600	900 × 900
Case 1	EGSOR	1258	5899	12,844
	EGAOR	1042	4994	10,928
	EGMSOR	1182	5552	12,871
	EGMAOR	1079	5305	11,289
Case 2	EGSOR	1729	6782	14,874
	EGAOR	1610	6368	13,953
	EGMSOR	1657	6559	15,276
	EGMAOR	1557	6154	13,543
Case 3	EGSOR	2666	11,076	24,519
	EGAOR	2480	10,389	22,995
	EGMSOR	2552	10,709	23,608
	EGMAOR	2370	10,002	21,996
Case 4	EGSOR	1629	6487	14,194
	EGAOR	1392	5648	12,367
	EGMSOR	1568	6255	13,701
	EGMAOR	1317	5433	12,863

Table 2 Performance of the proposed methods in terms of execution time (in seconds)

	Methods	N × N		
		300 × 300	600 × 600	900 × 900
Case 1	EGSOR	6.88	163.72	871.66
	EGAOR	6.05	137.87	751.78
	EGMSOR	6.74	165.65	983.56
	EGMAOR	6.66	169.96	880.50
Case 2	EGSOR	7.67	199.59	1009.48
	EGAOR	8.25	185.36	926.49
	EGMSOR	7.99	205.17	1150.19
	EGMAOR	8.25	205.08	1022.43
Case 3	EGSOR	13.24	315.87	1602.81
	EGAOR	13.83	301.27	1633.35
	EGMSOR	12.64	334.98	1828.16
	EGMAOR	12.28	327.72	1801.35
Case 4	EGSOR	7.80	187.33	990.20
	EGAOR	7.56	167.65	891.51
	EGMSOR	6.94	196.36	1059.32
	EGMAOR	6.41	170.40	1041.04

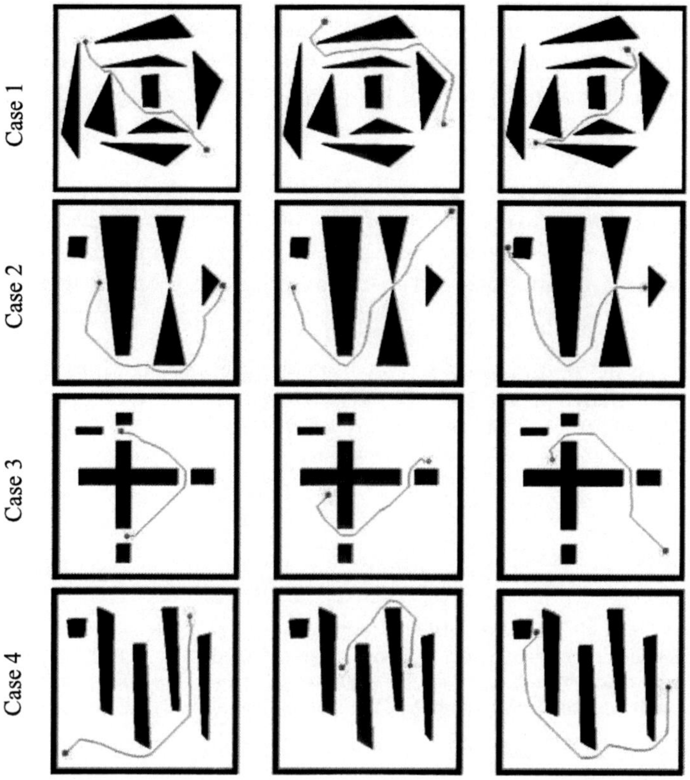

Fig. 1 The developed trails for four environments from several different initial and goal positions

terms of iteration number compared with other proposed methods. While, time taken for the modified families were slightly quicker than those for standard approaches.

When the temperature values are acquired, the appropriate trail was produced by executing the steepest descent search from starting spots to target spot. Figure 1 illustrates the trails successfully created in a known stationary environment by numerical computation based on the gained Laplacian potential. At the specified target spot (red dot/circle dot), all the starting spots (green dot/square dot) were effectively completed, overcoming various types of obstacles in different environment. Because no interpolation is done, some paths have a jagged appearance in some situations. The idea that gradient interpolation will offer the smoother paths.

6 Conclusion and Future Works

The studies in this analysis demonstrate that solving Laplace's Eq. (1) numerically to answer the problem of finding mobile robot paths was certainly very appealing and realistic owing to recent innovative and newly discovered numerical approaches, along with the convenience of faster machines today. As demonstrated in Tables 1 and 2, the iterative scheme of EGMAOR has proven to be very effective compared with previous existing methods. The growth in the amount and various types of obstacles does not adversely influence the effectiveness, in reality the computation is getting quicker because the regions engaged by obstacles are overlooked throughout the calculation. In addition to the concept of full-sweep iteration, an advance exploration into half [9, 25, 26] and quarter-sweep [27–30] iterations is likewise possible to measure, with the intention of speed up the convergence rate of the suggested iterative scheme.

Acknowledgements This research was supported by **Ministry of Education (MOE) through Fundamental Research Grant Scheme (FRGS/1/2018/ICT02/UPNM/03/1)**. The researchers declare that the publication of this study has no conflict of interest.

References

1. Connolly CI, Burns JB, Weiss R (1990) Path planning using Laplace's equation. In: Proceedings of IEEE international conference of robotics automation, pp 2102–2106
2. Akishita S, Hisanobu T, Kawamura S (1993) Fast path planning available for moving obstacle avoidance by use of Laplace potential. In: Proceedings of IEEE international conference of intelligent robots system, pp 673–678
3. Sasaki S (1998) A practical computational technique for mobile robot navigation. In: Proceedings of IEEE international conference of control applications, pp 1323–1327
4. Barraquand J, Langlois B, Latombe JC (1992) Numerical potential field techniques for robot path planning. IEEE Trans Syst Man Cybern 22(2):224–241
5. Connolly CI, Gruppen R (1993) On the applications of harmonic functions to robotics. J Robot Syst 10(7):931–946
6. Khatib O (1985) Real-time obstacle avoidance for manipulators and mobile robots. IEEE Trans Robot Autom 1:500–505
7. Karonava M, Zhelyazkov D, Todorova M, Penev I, Nikolov V, Petkov V (2015) Path planning algorithm for mobile robot. Recent Res Appl Comput Sci:26–29
8. Hachour O (2008) Path planning of autonomous mobile robot. Int J Syst Appl Eng Develop 4(2):178–190
9. Saudi A, Sulaiman J (2013) Robot path planning using Laplacian behaviour-based control (LBBC) via half-sweep SOR. In: Proceedings of the international conference on technological advances in electrical, electronics and computer engineering. Konya, Turkey, pp 424–429
10. Saudi A, Sulaiman J, Hijazi MHA (2014) Fast robot path planning with laplacian behaviour-based control via four-point explicit decoupled group SOR. Recent J Appl Sci 9(6):354–360
11. Saudi A, Sulaiman J (2012) Robot path planning using four point-explicit group via nine-point Laplacian (4EG9L) iterative method. Int Symp Robot Intell Sens 41:182–188
12. Pedersen MD, Fossen TI (2012) Marine vessel path planning and guidance using potential flow. In: Proceedings of 9th IFAC conference of manoeuvring AMD control of marine craft, pp 188–193
13. Liang X, Wang H, Li D, Liu C (2014) Three-dimensional path planning for unmanned aerial vehicles based on fluid flow. In: Proceedings of IEEE aerospace conference, pp 1–13
14. Motonaka K, Watanabe K, Maeyama S (2014) 3-dimensional kinodynamic motion planning for an X4-Flyer using 2-dimensional harmonic potential fields. In: 14th international conference of control automation system, pp 1181–1184

15. Shi C, Zhang M, Peng J (2007) Harmonic potential field method for autonomous ship navigation. In: 7th international conference of telecommunication (ITST'07), pp 1–6
16. Vallvé J, Andrade-Cetto J (2015) Potential information fields for mobile robot exploration. Robot Autonom Syst 69:68–79
17. Masoud AA, Al-Shaikhi A (2015) Time-sensitive, sensor-based, joint planning and control of mobile robots in cluttered spaces: a harmonic potential approach. In: 54th IEEE conference of decision control (CDC), pp 2761–2766
18. Evans LC (1998) Partial differential equations. American Mathematical Society, Providence
19. Ibrahim A (1993) The study of the iterative solution of boundary value problem by the finite difference method. PhD thesis, Universiti Kebangsaan Malaysia
20. Sulaiman J, Hasan MK, Othman M (2007) Red-Black EDGSOR iterative method using triangle element approximation for 2D Poisson equations. In: Gervasi O, Gavrilova M. (eds) Computer science applied 2007. Lecture notes in computer science (LNCS 4707). Springer-Verlag, Berlin, pp 298–308
21. Evans DJ (1985) Group explicit iterative methods for solving large linear systems. Int J Comput Math 17:81–108
22. Evans DJ, Yousif WS (1986) Explicit group iterative methods for solving elliptic partial differential equations in 3-space dimensions. Int J Comput Math 18:323–340
23. Martins MM, Yousif WS, Evans DJ (2002) Explicit group AOR method for solving elliptic partial differential equations. Neural Parall Sci Comput 10(4):411–422
24. Hadjidimos A (1978) Accelerated overrelaxation method. Math Comput 32(141):149–157
25. Dahalan AA, Sulaiman J (2016) Half-sweep two parameter alternating group explicit iterative method applied to fuzzy poisson equation. Appl Math Sci 10(2):45–57
26. Dahalan AA, Shattar NA, Sulaiman J (2016) Implementation of half-sweep age method using seikkala derivatives approach for 2D fuzzy diffusion equation. J Eng Appl Sci 11(9):1891–1897
27. Dahalan AA, Sulaiman J (2015) Approximate solution for 2 dimensional fuzzy parabolic equations in QSAGE iterative method. Int J Math Anal 9(35):1733–1746
28. Dahalan AA, Aziz NSA, Sulaiman J (2016) Performance of quarter-sweep successive over relaxation iterative method for two-point fuzzy boundary value problems. J Eng Appl Sci 11(7):1456–1463
29. Sulaiman J, Othman M, Hassan MK (2009) A new quarter-sweep arithmetics mean (QSAM) method to solve diffusion equations. Cham J Math 2(1):93–103
30. Muthuvalu MS, Sulaiman J (2010) Quarter-sweep arithmetic mean (QSAM) iterative method for second kind linear Fredholm integral equations. Appl Math Sci 4(59):2943–2953

Position and Swing Angle Control of Nonlinear Gantry Crane System

Abdulbasid Ismail Isa, Mukhtar Fatihu Hamza, Yusuf Abdullahi Adamu, and Jamilu Kamilu Adamu

Abstract Crane systems are widely used in logistics due to their efficiency of transportation. The major control problem of gantry crane system is some oscillations while crying load to the desired location. This work developed Radial Basis Function Neural Network (RBFNN) supervised PID controller for position and swing (swing angle suppression) angle of crane system. The supervising RBFNN for position control has two inputs namely instantaneous values of position controller and crane position; while the supervising RBFNN for swing angle control is based on angular acceleration of the swing and anti-sway controller output. The simulation result showed that the proposed control is more robust to the testing conditions in terms of tracking position, swing angle suppression as compared to conventional PID and LQR controllers. Although LQR controller takes less time to settle to its final value for swing angle control under all testing conditions.

Keywords PID · Radial basis function · Gantry crane · Gantry crane system

1 Introduction

Carrying heavy loads from one point to another using human strength and other duty animals is very challenging and inefficient. Gantry crane systems can effectively replace use of animals and humans' power in transporting heavy loads efficiently. This makes it very useful carrying hazardous materials in shipyards, factories, nuclear installations, and high building constructions sites [1, 3]. Crane system's control

A. I. Isa
Department of Electrical and Electronics Engineering, Usmanu Danfodiyo University, Sokoto, Nigeria

M. F. Hamza (✉) · Y. A. Adamu
Department of Mechatronic Engineering, Bayero University, Kano, Nigeria
e-mail: Mfhamza.mct@buk.edu.ng

J. K. Adamu
Department of Engineering Services, Federal Ministry of Works and Housing Headquarters, Abuja, Nigeria

A. F. Ab. Nasir et al. (eds.), *Recent Trends in Mechatronics Towards Industry 4.0*, Lecture Notes in Electrical Engineering 730, https://doi.org/10.1007/978-981-33-4597-3_4

design requires effective position tracking, disturbance rejection, fast response, high robustness, stability and effective damping swing oscillations of the payload throughout the process by minimizing swing angle [2, 4].

Gantry crane system (GCS) has a complex nonlinear dynamics, highly unstable with under actuated configuration. The under-actuated nature of the system leads to huge oscillations due to suspended under-actuated payload which may affect the overall safety operators and operation.

Researches were done to address control problems associated with GCS using both conventional and intelligent control methods. Kwangseok et al. in [5] developed adaptive LQR based steering control for of a multi-axle crane which solves the optimal steering angles to improve the driver's steering efficiency, the simulation result showed efficiency of steering control by improving the steering efficiency and decreasing the driver's steering effort, and dynamics stability by reducing the yaw rate. Abhya and Agrawl in [6] compared fractional order and conventional MPC controllers for position tracking control at minimum swing angle, in which the proposed strategy was found to be very effective in achieving the control objectives. Ning et al. cited in [7] designed SMC nonlinear control strategy based on some practical constraints, namely; swing amplitudes and maximum velocity. The controller object is to follow planned trajectories with continuous control efforts, ensuring asymptotic stable tracking in the presence of perturbations/uncertainties. Also, Wang et al. in [8] developed an SMC controller using a sigmoidal function based switching meant for load position control with a small load swing at minimum chattering. Muhammad et al. in [9] used LMI based state feedback control for double pendulum crane system, in which the proposed controller was able to track the trolley position relatively fast with minimum hook and payload swing angles hence, reduce the main problem of a double pendulum crane.

Moreover, intelligent controllers ware extensively used for control of GCS system. Notably, He et al. in [10] developed an emergency braking control and swing suppression technique using FL based gain scheduling approach, the result showed that FL is fast in trolley braking with minimum swing angle. Also Mahmud et al. [3] proposed anti-sway fuzzy tuned PID for gantry crane system. Performance of the proposed controller proves more robustness as compared to conventional PID controller Hussien et al. in [2] developed priority fitness scheme based PSO scheme (PFPSO) used to optimize PID and PD controller (KP, KI, KD, KPS and KDS), in which the simulation result proves the effectiveness of controller in moving the trolley as fast as possible with minimum payload oscillation. Mahmud et al. in [11] uses GA tuned PID controller with inlet derivative for double inverted pendulum crane, the algorithm was found to be effective in minimizing pay load swing and hook angles respectively. Naif and Muhammad in [12] uses two PD-fuzzy controllers to control the motion of a gantry crane system while suppressing the swings of the payloads. The proposed controllers were implemented on a test-bed apparatus; the simulation results as well as the experimental results confirm the effectiveness of the proposed controllers. Mansour et al. in [13] uses H∞ based adaptive fuzzy coupled with a VS to develop a robust control scheme system of nonlinear crane system, the controller proves to efficient in the presence of uncertainties, disturbance and time delays.

Chen et al. in [14] developed NN based on Lyapunov stability theory for anti-sway control of double inverted pendulum crane. The simulations result showed that the proposed NN controller has an excellent performance of trolley position tracking and payload anti-sway controlling. In this work, radial basis function based neural network supervised PID controller was applied to a nonlinear gantry crane system cited in [15] including the system parameters. The proposed control will be assed due to disturbance rejection and robustness to parameter variation, also conventional PID and LQR controllers will be used to validate the proposed control strategy.

2 RBF Supervised PID Controller

Generally, RBFNN has 3 layers: namely input layer, hidden layer, and output layer. The hidden layer activation functions are radial basis functions and each neuron in this layer include a central neuron [17, 18]. Here we proposed two RBFNN-PID supervised controllers; dedicated for position control and swing angle suppression. The control law of RBF NN supervised PID controller for position tracking was designed to have two inputs, namely position of the crane and PID controller output. The equation governing the control law is presented as follows:

The nonlinear activation function used for RBF supervised controller is function of crane position as given in [15].

$$h_j = \exp\left(-\frac{y_p(k) - \mathbf{c}_j^2}{2b_j^2}\right) \tag{1}$$

The output of RBF is given by:

$$u_n(k) = h_1 w_1 + \cdots h_j w_j \ldots h_m w_m \tag{2}$$

where

$w_1 \ldots w_j \ldots w_m, y_p$ are the weights and position of the crane respectively.

The chosen criterion of the update is given as:

$$E(k) = \frac{1}{2}\big(u_n(k) - u_p(k)\big)^2 \tag{3}$$

Applying descent (gradient) method, the learning algorithm is as

$$\left.\begin{aligned} \Delta w_j(k) &= \eta(u_n(k) - u(k))h_j(k) \\ w(k) &= w(k-1) + \Delta w_j(k) + \alpha(w(k-1) - w(k-2)) \end{aligned}\right\} \tag{4}$$

The overall control law for position tracking is given by:

Table 1 NN parameters

Constants	Values
η	0.005
α	0.001
b	$\begin{bmatrix} 0.05\ 0.05\ 0.05\ 0.050.050.050.050.05 \end{bmatrix}^T$
w	$\begin{bmatrix} 1.0\ 1.0\ 1.0\ 1.01.01.01.01.0 \end{bmatrix}^T$

$$u_t(t) = K_p e(t) + K_i \int_0^t e(\tau)d\tau + K_d \frac{de(t)}{dt} + u_n(k) \tag{5}$$

The control law of RBF NN supervised PID controller for payload swing suppression was designed to have two inputs, namely swing acceleration of the payload and anti-sway PID controller output. The same activation function and is used, although input of the activation function is swing acceleration, based on the same activation function (Table 1).

3 Simulation Results

The performance of the proposed controller was investigated by simulations in Simulink. The simulation results are compared with that of LQR and PID-PID controllers. The parameters of PID position controller and that swing angle were obtain through trial and error, which were given as: $K_P = 115.78$, $K_I = 0.00012$, $K_D = 74.31$, while the parameters of PID swing controller are: $K_P = 115.78$, $K_I = 0.00012$, $K_D = 74.31$. The detailed information of LQR design was given [16]; LQR parameters used in this work are given as follows:

$$K = \begin{bmatrix} 8.7636\ 12.5629 & -32.4760 & -5.0768 \end{bmatrix}$$

$$N = \begin{bmatrix} 8.7636\ 23.9087 & -56.7680 & -5.0768 \end{bmatrix}$$

3.1 Position Control

The simulation was carried out with the desired position of 1 m and the desired swing angle of 0 rad. Figures 1 through 3 showed the simulation results of crane position at various testing conditions. Figure 1 and Table 2 showed the system response and summary of controller's performances.

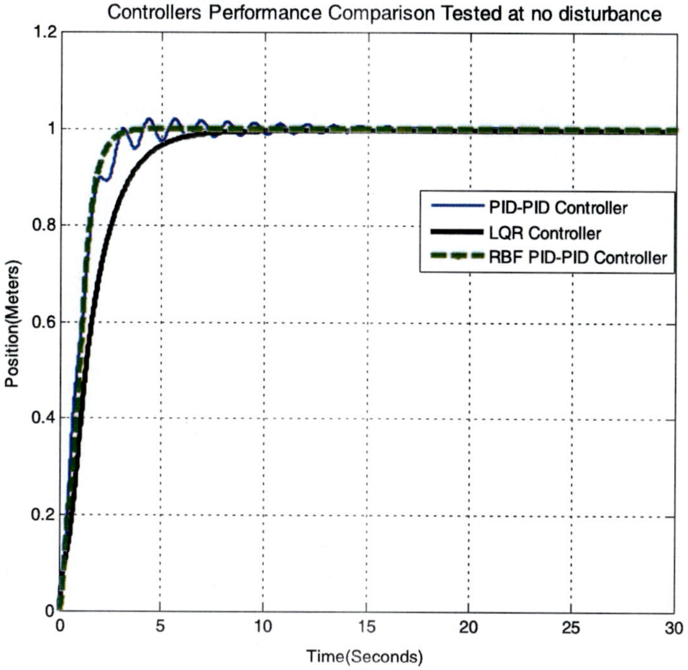

Fig. 1 Crane position response at no disturbance

Table 2 Performance indices of the controllers at no disturbance

Controllers	Settling time (s)	Overshoot	Rise time (s)	IAE
RBF PID-PID	2.637	0	1.5048	1.007
LQR	5.593	0	3.1381	1.742
PID-PID	6.273	1.9386	2.1441	1.143

The robustness of the proposed controller due to application of disturbance was studied; the system response was shown in Fig. 2 and Table 3 present the performance of controllers.

Also, the robustness of the proposed controller was investigated due to parameter variation, the system response and performance indices of the controllers were showed in Fig. 3 and Table 4 respectively.

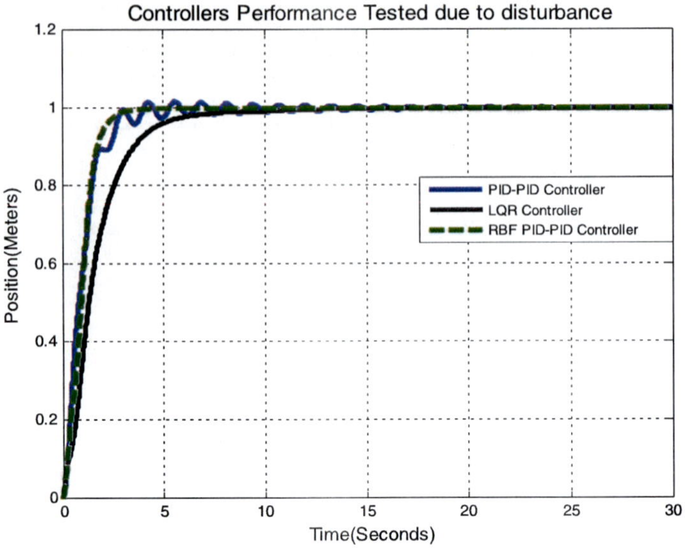

Fig. 2 Crane position response due to disturbance

Table 3 Performance indices of the controllers due to disturbance

Controllers	Settling time (S)	Overshoot	Rise time (s)	IAE
RBF PID-PID	2.6592	0	1.5074	1.021
LQR	6.0313	0	3.2221	1.804
PID-PID	6.2815	1.8282	2.1448	1.155

3.2 Swing Angle Control

This section presents swing angle suppression capacity of the developed controllers. Figure 4 showed the swing angle response of the developed controllers and Table 5 summarized the performance indices of swing angle controllers.

The robustness of the proposed controller due to disturbance rejection was also investigated and the result was compared with that of conventional controllers. The effect of disturbance injected into the system ware clearly shown in Fig. 5 and Table 6 respectively.

The robustness of the proposed controller is investigated due to parameter variation and the simulation result ware shown in Fig. 6 and Table 7 respectively.

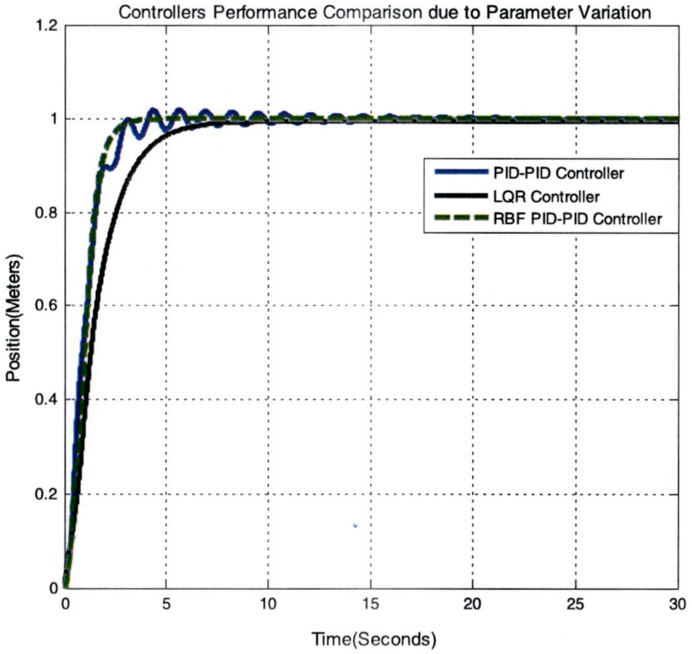

Fig. 3 Crane position response due to parameter variation

Table 4 Performance indices of the controllers due to parameter variation

Controllers	Settling time (s)	Overshoot	Rise time (s)	IAE
RBF PID-PID	2.6327	0	1.5048	1.007
LQR	5.5930	0	3.1381	1.742
PID-PID	6.2730	1.9386	2.1441	1.143

4 Conclusion

RBF supervised PID controller was designed for position tracking and swing angle control of a gantry crane system, PID-PID and LQR were developed to achieve the same control objectives for validation of results. Controllers' performance was assessed in terms of fast response, robustness against disturbance and parameter changes.

The proposed supervising controller proves to be very effective in disturbance rejection, it was observed the proposed controller takes 41% of LQR and 42% of PID-PID values of settling time to track the desired position, with 56% of LQR and 88.4% of PID-PID values of IAE (Integral of absolute error), while for swing angle control under the same testing condition, the proposed controller is slower than LQR

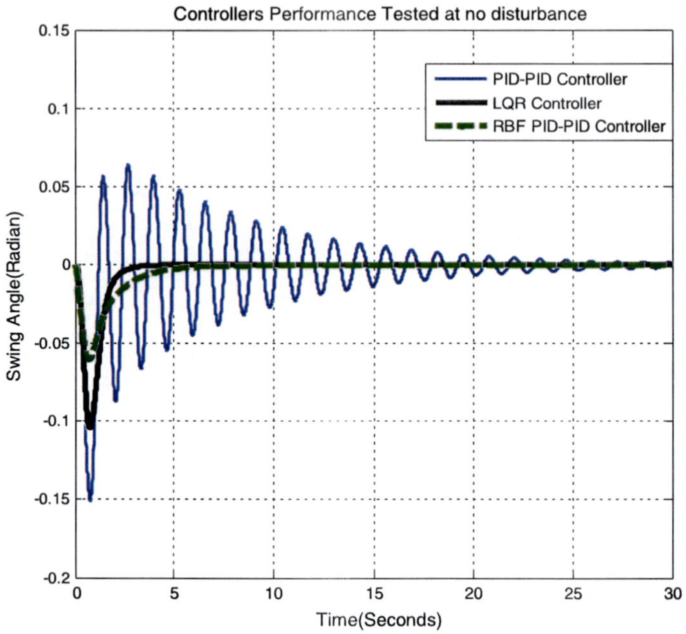

Fig. 4 Swing angle response at no disturbance

Table 5 Performance indices of the controllers at no disturbance

Controllers	Settling time (s)	Overshoot	Rise time (s)
RBF PID-PID	5.8577	6.0309	2.5147 1e−4
LQR	5.5930	10.4624	0.0028
PID-PID	28.0410	15.1744	0.0024

by 107%, at 21.3% of PID-PID values of settling time to suppress the swing angle to the minimum value, with 57.5% of LQR and 40.2% of PID-PID values of overshoot.

Similarly, it was showed that RBF PID-PID is very robust to parameter variation. Although the variation between the controllers' performance under no disturbance condition.

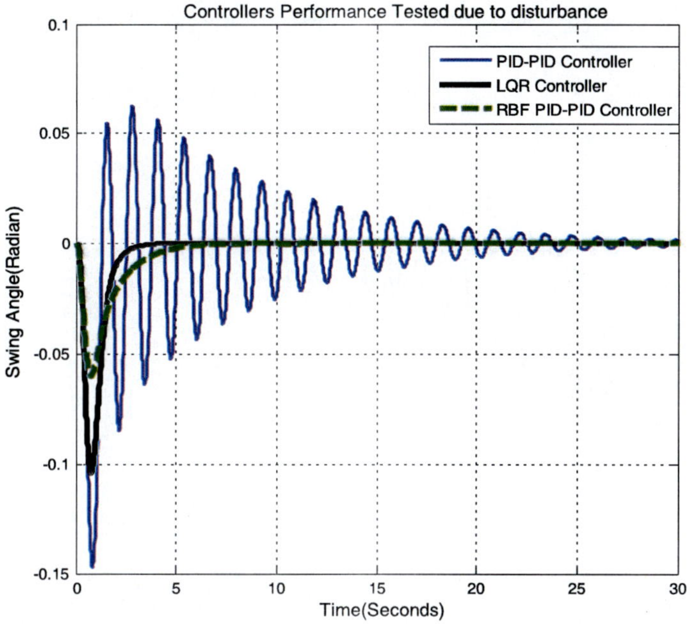

Fig. 5 Swing angle response due to disturbance

Table 6 Performance indices of the controllers due to disturbance

Controllers	Settling time (s)	Overshoot	Rise time (s)
RBF PID-PID	5.9742	5.9926	1.7281e−4
LQR	2.8841	10.4280	0.0031
PID-PID	28.0375	14.9047	0.0024

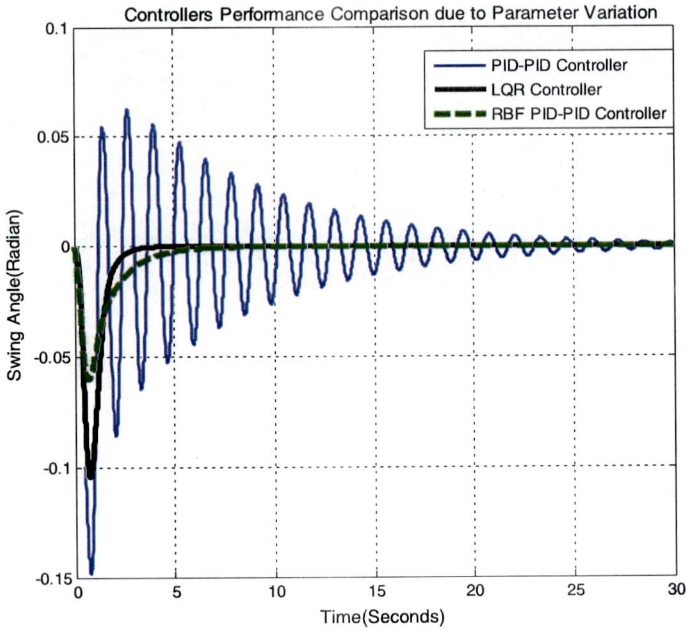

Fig. 6 Swing angle response to parameter variation

Table 7 Performance indices of the controllers due to parameter variation

Controllers	Settling time (s)	Overshoot	Rise time (s)
RBF PID-PID	5.8525	6.0179	2.2102e−4
LQR	2.8859	10.4162	0.0023
PID-PID	28.0495	14.9502	0.0025

References

1. Qian D (2018) Anti-sway control for crane design and implementation using MATLAB. Germany Walter de Gruyter, Berlin
2. Hussien SYS, Ghazali R, Jaafar HI, Soon CC (2015) Robustness analysis for PID controller optimized using PFPSO For underactuated gantry crane system. In: 2015 IEEE international conference on control system, computing and engineering, Penang, Malaysia
3. Solihin MI, Wahyudi, Legowo A (2010) Fuzzy-tuned PID anti-swing control of automatic gantry crane. J Vibr Control 16:127–145
4. Xiao R, Wang Z, Guo N, Wu Y, Shen J, Chen Z (2018) Multi-objective motion control optimization for the bridge crane system. Appl Sci 8:1–19
5. Oh K, Seo J, Han J-W (2016) LQR-based adaptive steering control algorithm of multi-axle crane for improving driver's steering efficiency and dynamic stability. In: 16th international conference on control, automation and systems (ICCAS 2016), Gyeongju, Korea, pp 792–796
6. Singh AP, Agrawl H (2018) A fractional model predictive control design for 2-D gantry crane system. J Eng Sci Technol 13:2224–2235
7. Sun N, Fang Y, Chen H (2016) A continuous robust anti-swing tracking control scheme for underactuated crane systems with experimental verification. J Dyn Syst Measure Control

8. Wanga T, Tana N, Zhoc C, Zhanga C, Zhid Y (2018) A novel anti-swing positioning controller for two dimensional bridge crane via dynamic slidingmode variable structure. In: 8th international congress of information and communication technology (ICICT-2018), pp 626–632
9. Muhammad M, Abdullahi AM, Bature AA, Buyamin S, Bello MM (2018) LMI-based control of a double pendulum crane. Appl Modell Simul 2:41–50
10. Chen H, Fang Y, Sun N (2017) A payload Swing suppression guaranteed emergency braking method for overhead crane systems. J Vibr Control 1
11. Abdel-razak MH, Aata AA, Mohamed KT, Haraz EH (2018) Proportional–integral-derivative controller with inlet derivative filter fine-tuning of a double-pendulum gantry crane system by a multi-objective genetic algorithm. Eng Optim
12. Almutairi NB, Zribi M (2016) Fuzzy controllers for a gantry crane system with experimental verifications. Math Prob Eng:1–17
13. Karkoub M, Wu TS, Chen CT (2012) H∞ based adaptive fuzzy control of a tower crane system. In: Proceedings of the ASME 2012 international mechanical engineering congress and exposition (IMECE2012), Houston, Texas, USA, pp 1–6
14. Chen Q, Cheng W, Gao L, Fottner J (2019) A pure neural network controller for double pendulum crane anti sway control: based on Lyapunov stability theory. Asian J Control:1–12
15. Jaafar HI, Mohamed Z, Ahmad MA, Ghazali R, Kassim AM (2016) Linear and nonlinear dynamic model of a gantry crane system. In: Mechanical engineering research day, pp 41–42
16. Isa AI, Hamza MF, Muhammad M (2019) Hybrid fuzzy control of nonlinear inverted pendulum system. Bayero Univ J Eng Technol (BJET) 14:200–208
17. Mahjoub S, Mnif F, Derbel N, Hamerlain M (2014) Radial-basis-functions neural network sliding mode control for underactuated mechanical systems. Int J Dyn Control
18. Miao Y, Xu F, Hu Y, An J, Zhang M (2019) Anti-swing control of the overhead crane system based on the harmony search radial basis function neural network algorithm. Adv Mech Eng 11:1–10

The Classification of Heartbeat PCG Signals via Transfer Learning

Omair Rashed Abdulwareth Almanifi, Mohd Azraai Mohd Razman, Rabiu Muazu Musa, Ahmad Fakhri Ab. Nasir, Muhammad Yusri Ismail, and Anwar P. P. Abdul Majeed

Abstract Cardiovascular auscultation is a process of listening to the sound of a heartbeat to pick up on any abnormalities. One of these abnormalities is heart murmurs, which are the result of blood turbulence, in or near the heart. Heart murmurs can be innocent, or they can indicate the existence of very serious diseases. Normally the process is performed with a stethoscope, by a medical professional, where murmurs are identified by the subtle difference in timing and pitch from a normal heartbeat. These professionals, however, are not always available; hence, the need for the automation of this process rises. This paper aims at testing the performance of pre-trained CNN models at the classification of heartbeats. A database of phonocardiogram (PCG) heartbeat recordings, under the name of the PASCAL CHSC database was used to train four pre-trained models: VGG16, VGG19, MobileNet, and inceptionV3. The data was processed, and the features were extracted using Spectrogram signal representation. They were then split into training and testing data, and the results were compared using the metrics of accuracy and loss. The classification accuracies of the VGG16, VGG19, MobileNet, and inceptionV3 models are 80.25%, 85.19%, 72.84% and 54.32%, respectively. The findings of the paper indicate that the use of different transfer learning models can, to a certain extent, enhance the overall accuracy at detecting the murmurs of the heart.

O. R. A. Almanifi · M. A. Mohd Razman · A. F. Ab. Nasir · A. P. P. Abdul Majeed (✉)
Innovative Manufacturing, Mechatronics and Sports Laboratory, Faculty of Manufacturing and Mechatronics Engineering Technology, Universiti Malaysia Pahang (UMP), 26600 Pekan, Pahang Darul Makmur, Malaysia
e-mail: amajeed@ump.edu.my

R. M. Musa
Centre for Fundamental and Liberal Education, Universiti Malaysia Terengganu (UMT), 21030, Kuala Nerus, Terengganu Darul Iman, Malaysia

M. Y. Ismail
Faculty of Mechanical and Automotive Engineering Technology, Universiti Malaysia Pahang (UMP), 26600 Pekan, Pahang Darul Makmur, Malaysia

A. P. P. Abdul Majeed
Centre for Software Development & Integrated Computing, Universiti Malaysia Pahang (UMP), 26600 Pekan, Pahang Darul Makmur, Malaysia

A. F. Ab. Nasir et al. (eds.), *Recent Trends in Mechatronics Towards Industry 4.0*, Lecture Notes in Electrical Engineering 730, https://doi.org/10.1007/978-981-33-4597-3_5

Keywords PCG · Heart murmur · Machine learning · Classification · Feature selection · Transfer learning

1 Introduction

A stethoscope is a medical device used to facilitate listening to the sounds generated by animals or humans, and this process is often known as auscultation. A type of auscultation widely performed with a stethoscope is Cardiovascular auscultation [1] where a medical professional listens to the acoustic patterns made by the heart. This process aids in detecting abnormalities in heartbeats, such as murmurs [2], which can be early indicators of cardiovascular diseases (CVDs). CVDs were on the top of the list as the first cause of death globally [3]. In fact, in 2016, it was responsible for the deaths of 31% of the entire counts of death globally, noting that almost 70% of these deaths affected low-to-middle income countries. Therefore, it is evident that the early detection of such diseases can help reduce their fatality, which illustrates the importance of such a process.

Although it might sound easy, auscultation of the heart requires a great deal of training, as it involves having the ability to differentiate between the subtle differences and changes in both timing and pitch. This can be performed by trained medical professionals; however, these professionals are not always available, especially in rural areas, where a shortage of trained medical doctors is present. Giving rise to the need for alternative ways to perform such a procedure.

The use of artificial intelligence, specifically machine learning (ML), is deemed to be a solution to the problem. This can be done by recording the heartbeat signals in a digital format, customarily referred to as phonocardiogram (PCG) [4]. The signals are then processed and prepared for training, and this normally involves cleaning the data, processing the wave signals, and extracting its features. The features extracted are then used to train the model, in which the model would eventually be able to classify different signals. The results are often measured with two metrics of analysis, namely accuracy and loss. It is worth noting that different ML architectures, as well as different feature extraction methods, does directly affect the classification accuracy (CA) of the model.

The classification or the identification of murmur has been reported in the literature; for instance, Ahmad et al. [5] employed the Mel-frequency cepstral coefficients (MFCC) as a means for feature extraction. The PCG data were recorded at the cardiology department of Ayub Teaching Hospital, Abbottabad, Pakistan, with a total number of 283 samples. The Support Vector Machine (SVM) and *k*-Nearest Neighbour (*k*NN) models were used to classify the murmur based on the extracted features. The authors utilised the five-fold cross-validation technique to train the models. It was shown from the study that the SVM model is able to provide a CA of 92.6%.

Lubis et al. [6] used data acquired from the PASCAL CHSC 2011 database, specifically database B, to train the ML model, which is based on backpropagation neural network. The study compared the effect of three methods of feature extraction, namely, Discrete Cosine Transform (DCT), MFCC and Basilar-membrane Frequency-band Cepstral Coefficient (BFCC). The authors used the k-fold cross-validation as the main cross-validation technique, and it was shown that the MFCC achieved the highest results of average accuracy at 63.54%. Conversely, the Modified MFCC managed to record an accuracy of 95.83%, but the average accuracy was merely 61.45%.

All the aforementioned applications utilised machine learning algorithms and trained its models on the datasets available to them. However, in a recently suggested idea, the weights of models can be transferred to be trained on different datasets, with the aim of increasing the accuracy of the model. This technique is called Transfer Learning, and it has gained considerable attention over the past few years [7, 8]. While the performance of the pre-trained model on PCG data has not been studied, it has been used widely in other applications, especially in the biomedical field.

One of these transfer learning applications was carried out by Fukae et al. [9], in a recent study using pre-trained CNN for the diagnosis of rheumatoid arthritis (RA). The model chosen was AlexNet, tested on 1037 images, with 252 having RA. The clinical information of patients was represented in two-dimensional images and was then used to train the model with different methods of fine-tuning. The results obtained were very promising, achieving 98% testing accuracy by the 3rd fine-tuned AlexNet model.

In a recent study, Gherardini et al. [10] developed a lightweight version of U-net CNN using transfer learning for the segmentation of catheters and guidewires in 2d X-ray images. The model was trained on three data sets of fluoroscopic images, 9000 synthetic images, 2000 images of experiments done on a silicon aorta phantom, and 1207 frames of In-vivo procedures. The first dataset was used in experiment 1, and the second in the second experiment, as for the third data set, was used for the fine-tuning of both experiments. The highest dice coefficient recorded was 0.78 achieved by the model in the first experiment on the second test.

As noted above, the performance of ML models used for PCG detection depends heavily on many factors, leaving significant room for improvements. The promising performance of TL techniques in several areas and fields is the primary motivation behind the present investigation. This study aims at testing the effect of different transferred pre-trained ML architectures to perform PCG murmur detection. Four pre-trained models were compared in this study, namely VGG16, VGG19, MobileNet, and inceptionV3, all of which are Convolutional neural networks (CNN) based algorithm. Conversely, the features were extracted through the use of spectrogram signal representation technique.

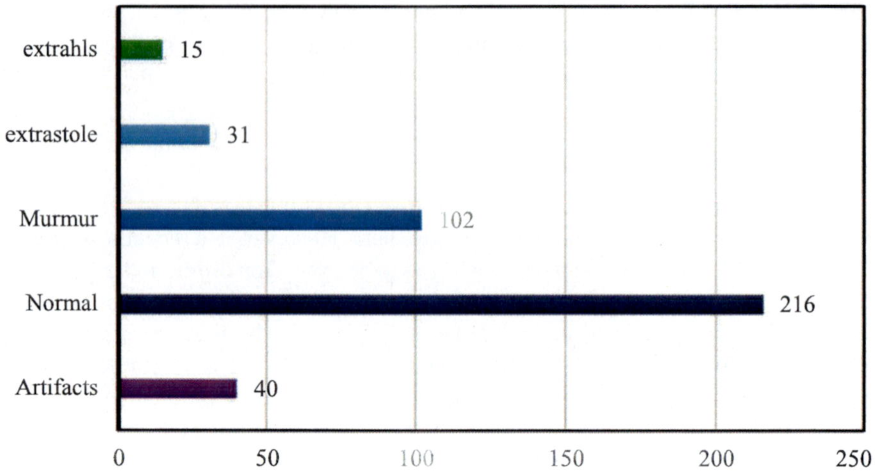

Fig. 1 Dataset distribution

2 Methodology

2.1 Data Acquisition

There are many open-sourced datasets available online and the PASCAL CHSC 2011 database [11] is one of such datasets. It is a PCG dataset that consists of heartbeat recordings in WAV format. The dataset is divided into two sections, A and B. The former contains data that were acquired using iStethoscope Pro iPhone app, and the latter recorded at hospitals using DigiScope. It consists of 5 classes: Artifacts, refereeing to non-heartbeat sounds, normal heartbeats, murmur, extrasystole, and extrahls, which is normal heartbeats with additional noise. The multiple sources the data was acquired from gives it great diversity. However, it is important to note that the total number of tracks is not very large and is not evenly distributed, as it has 832 tracks, only 404 of which were useable, that is to say, are larger than 4 s and are playable. This distribution of the data can be shown in Fig. 1.

2.2 Signal Representation

In this study, the time-domain signals are transformed into spectrograms as CNN works best on image-based data. A spectrogram is a two-dimensional visual representation of the frequencies spectrum of a certain signal, as the name implies [12]. Spectrogram representation is a flexible method and has be employed in many different areas related to signal processing. Essentially, the time-domain signal is transformed into frequency-domain via Fast Fourier Transform prior to the generation of the

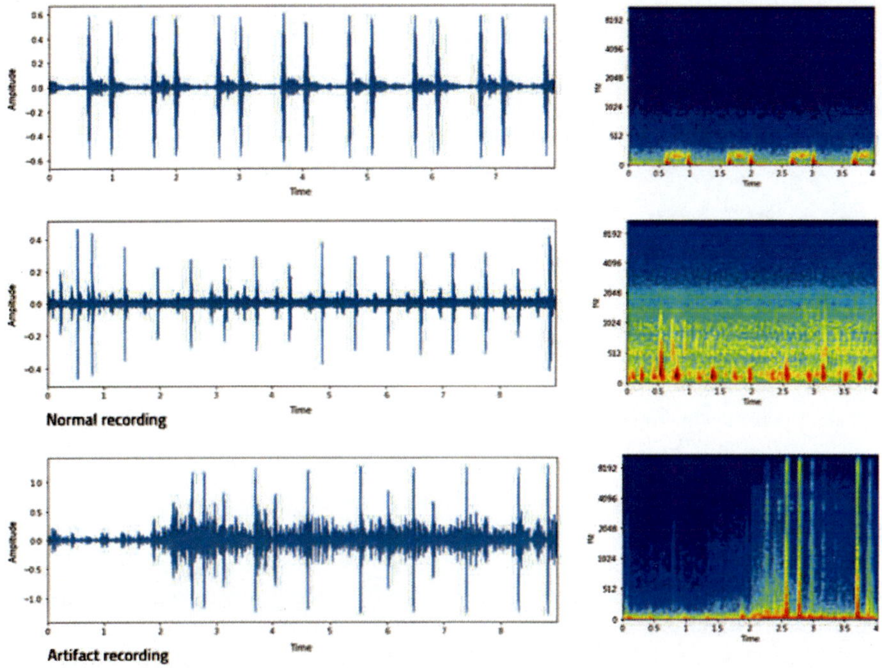

Fig. 2 Audio representation in spectrograms

spectrogram. A Python library called Librosa [12] is employed to produce the spectrograms. The sampling rate is fixed at 22,050 Hz with each spectrogram has 97 Mel-scale bins, and a length of windowed signal padded with zeros of 2048 samples. Figure 2, some examples of signals represented in the form of spectrograms are shown.

2.3 Model Development via Transfer Learning

Conventional machine learning models will need to go through a pipeline that consists of feature extraction prior to the implementation of a given classifier. More often than not, extensive feature extraction topology is required to ascertain the ideal method. Moreover, the hyperparameter tuning is also required to attain the best model for a given set of features is inadvertently an arduous task. Conversely, a full-fledge Convolution Neural Network (CNN) model, requires extensive tuning on both the convolution part for feature extraction as well as the dense part for classification as depicted in Fig. 3. However, it is worth noting that a relatively recent technique of transferring weights from a pre-trained model, particularly for feature extraction, eliminating the need to ascertain the convolution part of a CNN model has been

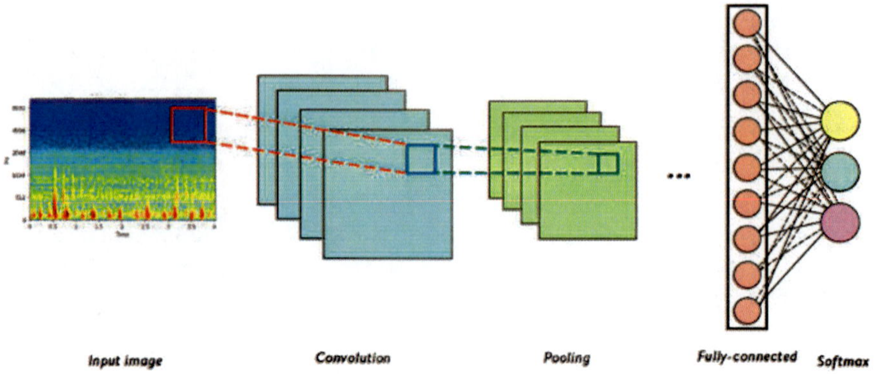

Fig. 3 An illustration of a conventional CNN model

reported in the literature. This method, in turn, required only the dense part of the CNN be trained and tuned. This approach is known as Transfer Learning [13]. The basis of this notion is the transference of knowledge from a model to another to expand on the previously acquired knowledge from the previous data, i.e., the rich information of the features initially trained. Figure 4 further illustrates the process.

Four pre-trained models were selected for this study, the first of which is VGG16 [14], developed at the University of Oxford. It consists of 16 layers overall of convolution and pooling, with a kernel size of 3 * 3. The second on the list is VGG19, which is an enhanced version of VGG16, with 19 layers instead. In addition, the

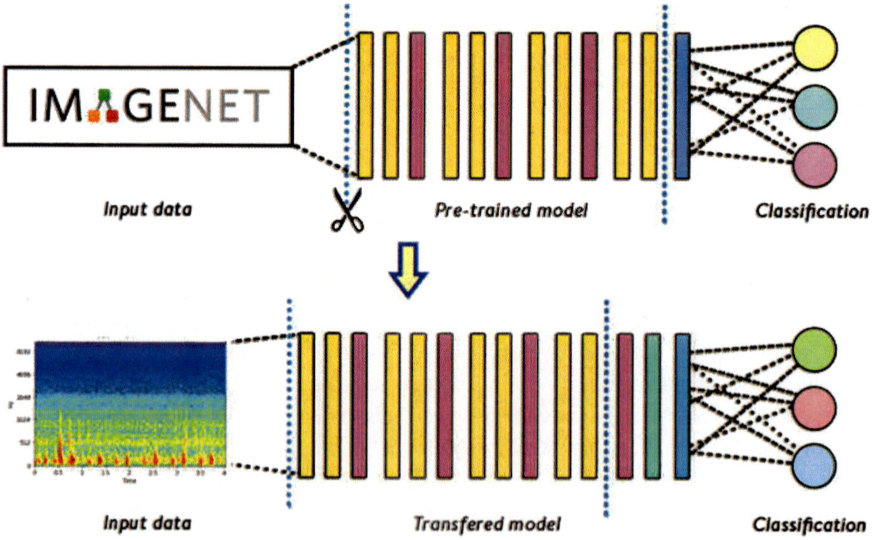

Fig. 4 An illustration of transfer learning models

MobileNet [15] which is another popular lightweight pre-trained CNN, due to its streamlined structure that uses depth-wise separable convolutions, is also evaluated. Finally, the efficacy of InceptionV3, which was developed by Google as a part of the series GoogleNet [16] is also appraised. It has 42 layers, with kernels of two sizes 3 * 3, and 5 * 5. Its unique architecture as it reduces the parameters trained while maintaining good accuracy, using efficient grid size reduction with is a process where both convolution and pooling are performed at once and then concatenated to avoid pooling greediness [17].

All the models are pre-trained on a dataset called ImageNet [18], an open-source dataset made to facilitate research on object detection, and it contains 14 million images. The pre-trained models were acquired from the Keras [19] with a library with Tensorflow [20] as its backend. The present study utilises several activation functions, one of which is the Softmax activation function [21]. Which gives a value between 0 and 1, and was used for the last layer of classification. Moreover, a rectifier linear unit (ReLU) was also used in the model prior to the Softmax layer, that is connected to a dropout layer that randomly gets rid of half of the produced parameters [22]. The results of each model were then compared together based on accuracy, loss, as well as the confusion matrix [23]. The cross-entropy loss function is evoked during the training phase and the ratio of the train to test set at 80:20 [24].

3 Result and Discussion

The performance measures of the TL models evaluated in the study are tabulated in Table 1, whereby Figs. 5 and 6 illustrate its classification accuracy and loss. It is evident from the results presented in the present investigation that, in general, the VGGs models performed well in both the training and testing phase in comparison to the rest of the evaluated models. The highest training accuracy was achieved by VGG16 at 99.38%, whilst the lowest to be observed by InceptionV3 at 54.32% for the particular dataset, suggesting that the features identified via VGG16 are rather significant.

However, it could be seen that the VGG19 model provides a better classification accuracy on the test dataset. The average accuracy for the VGG19 model is slightly better than that of the VGG16 model. Figure 7 depicts the confusion matrix of the

Table 1 The resulted training and test accuracy (%) and loss for each transfer learning model

Dataset	Metric	Evaluated models			
		VGG16	VGG19	MobileNet	Inception V3
Training	Accuracy (%)	99.38	96.59	78.02	53.25
	Loss	0.0863	2.3372	0.7493	7.5351
Testing	Accuracy (%)	80.25	85.19	72.84	54.32
	Loss	13.5604	8.6889	1.0699	7.3626

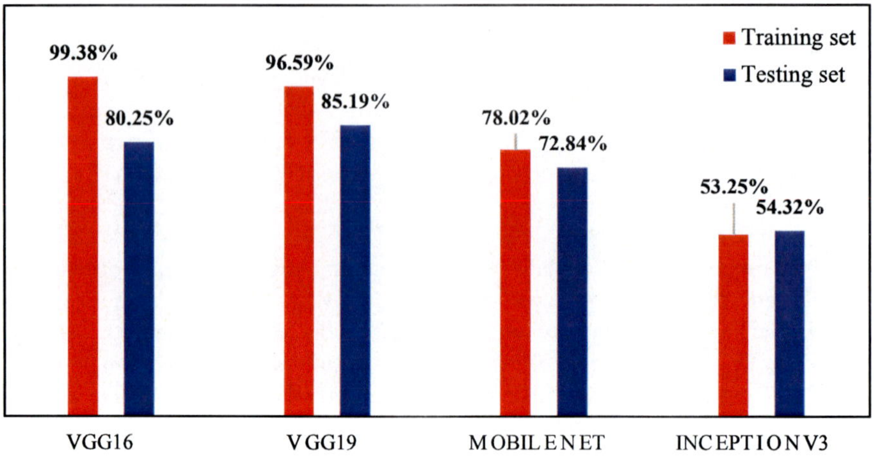

Fig. 5 The accuracy achieved by each model

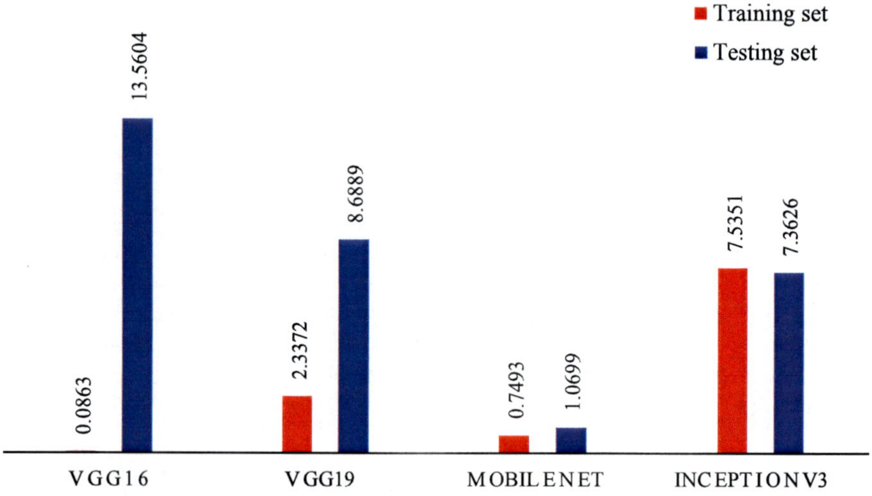

Fig. 6 The loss achieved by each model

evaluated models on the test dataset. It is apparent that the misclassification often transpires for the 'normal' and 'murmur' class across the developed models; this is non-surprising owing to the skewed (imbalanced) nature of the dataset. Nonetheless, it is worth noting that the model performed quite well in comparison to models that do not adopt this approach as reported in [6].

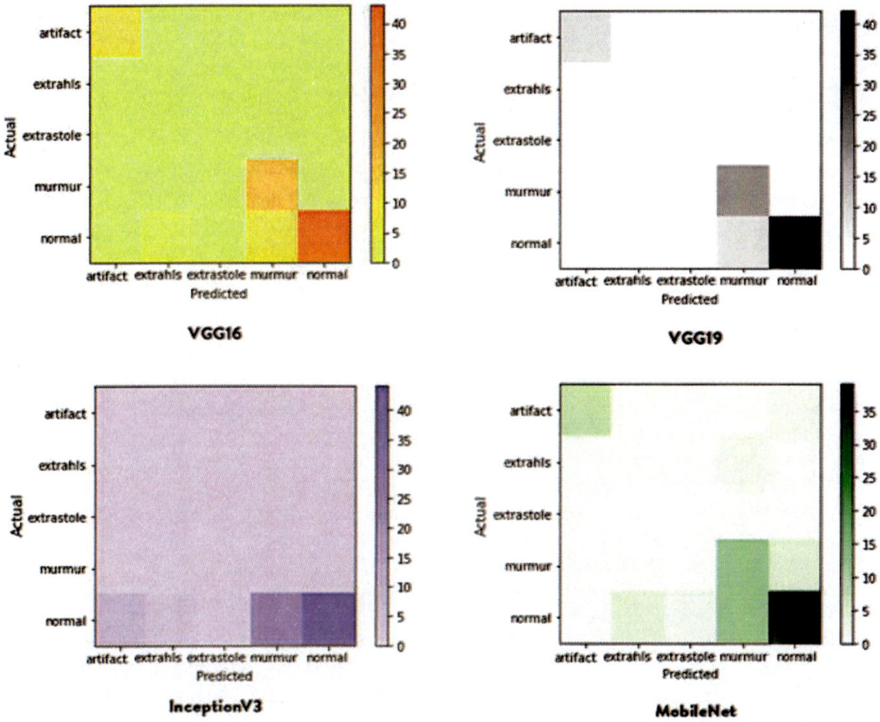

Fig. 7 Confusion matrices for each model on the test dataset

4 Conclusion

This study evaluated a number of TL models in classifying PCG heartbeat signals acquired from the PASCAL CHSC 2011 database. The signals were initially converted into spectrogram prior allowing the dense part to be trained. It was demonstrated from the present investigation that the VGG models are well capable in classifying the heartbeat classes in comparison to other models evaluated. The findings from the study are rather in its preliminary stage; therefore, future works shall investigate the efficacy of other TL models as well as evaluating the classifiers with a more balanced nature of data.

Acknowledgements The authors would like to acknowledge Universiti Malaysia Pahang for funding this study via RDU180321.

References

1. Chizner MA (2008) Cardiac auscultation: rediscovering the lost art. Curr Probl Cardiol. https://doi.org/10.1016/j.cpcardiol.2008.03.003
2. The gale encyclopedia of children's health: infancy through adolescence. Choice Rev Online (2012). https://doi.org/https://doi.org/10.5860/choice.49-3617
3. American Hearth Association (2014) What is cardiovascular disease
4. Goodman G (2019) Cardiovascular techniques and technology. Clinical Engineering Handbook
5. Ahmad MS, Mir J, Ullah MO, Shahid MLUR, Syed MA (2019) An efficient heart murmur recognition and cardiovascular disorders classification system. Australas Phys Eng Sci Med. https://doi.org/10.1007/s13246-019-00778-x
6. Lubis C, Gondawijaya F (2019) Heart sound diagnose system with BFCC, MFCC, and backpropagation neural network. In: IOP conference series: materials science and engineering
7. Rashid M, Sulaiman N, Majeed APP, Musa RM, Nasir AFA, Bari BS, Khatun S (2020) Current status, challenges and possible solutions of EEG based brain-computer interface: a comprehensive review. Front Neurorobot 14:25
8. Shapiee MNA, Ibrahim MAR, Mohd Razman MA, Abdullah MA, Musa RM, Hassan MHA, Majeed APPA (2020) The classification of skateboarding trick manoeuvres through the integration of image processing techniques and machine learning, 1st edn. In: Nasir ANK, Ahmad MA, Najib MS, Wahab YA, Othman NA, Ghani NA, Irawan A, Khatun S, Ismail RMTR, Saari MM, Daud MR, Faudzi AAM (eds) InECCE2019 proceedings of the 5th international conference on electrical, control and computer engineering, Kuantan, Pahang, Malaysia, 29th July 2019. Springer Singapore
9. Fukae J, Isobe M, Hattori T, Fujieda Y, Kono M, Abe N, Kitano A, Narita A, Henmi M, Sakamoto F, Aoki Y, Ito T, Mitsuzaki A, Matsuhashi M, Shimizu M, Tanimura K, Sutherland K, Kamishima T, Atsumi T, Koike T (2020) Convolutional neural network for classification of two-dimensional array images generated from clinical information may support diagnosis of rheumatoid arthritis. Sci Rep. https://doi.org/10.1038/s41598-020-62634-3
10. Gherardini M, Mazomenos E, Menciassi A, Stoyanov D (2020) Catheter segmentation in X-ray fluoroscopy using synthetic data and transfer learning with light U-nets. Comput Methods Programs Biomed. https://doi.org/10.1016/j.cmpb.2020.105420
11. Gomes EF, Bentley PJ, Coimbra M, Pereira E, Deng Y (2013) Classifying heart sounds: approaches to the PASCAL challenge. In: HEALTHINF 2013—proceedings of the international conference on health informatics
12. McFee B, Raffel C, Liang D, Ellis D, McVicar M, Battenberg E, Nieto O (2015) Librosa: audio and music signal analysis in python. In: Proceedings of the 14th python in science conference
13. Transfer W, Now L, Scenarios TL, Methods TL (2017) Transfer learning—machine learning's next frontier. PPT
14. Hassan MU (2018) VGG16—convolutional network for classification and detection. Neurohive
15. Howard AG, Zhu M (2017) MobileNets: open-source models for efficient on-device vision. Google AI Blog
16. Szegedy C, Liu W, Jia Y, Sermanet P, Reed S, Anguelov D, Erhan D, Vanhoucke V, Rabinovich A (2014) GoogLeNet. Proc IEEE Comput Soc Conf Comput Vis Pattern Recognit. https://doi.org/10.1109/CVPR.2015.7298594
17. Zeng G, He Y, Yu Z, Yang X, Yang R, Zhang L (2016) InceptionNet/GoogLeNet—going deeper with convolutions. CVPR. https://doi.org/10.1002/jctb.4820
18. Krizhevsky A, Sutskever I, Geoffrey EH (2012) Imagenet. Adv Neural Inf Process Syst 25.https://doi.org/10.1109/5.726791
19. Chollet F (2015) Keras documentation. Keras.Io
20. Abadi M, Barham P, Chen J, Chen Z, Davis A, Dean J, Devin M, Ghemawat S, Irving G, Isard M, Kudlur M, Levenberg J, Monga R, Moore S, Murray DG, Steiner B, Tucker P, Vasudevan V, Warden P, Wicke M, Yu Y, Zheng X (2016) TensorFlow: a system for large-scale machine learning. In: Proceedings of the 12th USENIX symposium on operating systems design and implementation, OSDI 2016

21. Gupta DS (2017) Fundamentals of deep learning—activation functions and their use. Anal Vidhya
22. He K, Zhang X, Ren S, Sun J (2015) Delving deep into rectifiers: surpassing human-level performance on imagenet classification. In: Proceedings of the IEEE international conference on computer vision
23. Ting KM (2017) Confusion matrix. In: Encyclopedia of machine learning and data mining
24. Neapolitan RE, Neapolitan RE (2018) Neural networks and deep learning. Artif Intell

The Classification of Wink-Based EEG Signals: An Evaluation of Different Transfer Learning Models for Feature Extraction

Jothi Letchumy Mahendra Kumar, Mamunur Rashid, Rabiu Muazu Musa, Mohd Azraai Mohd Razman, Norizam Sulaiman, Rozita Jailani, and Anwar P. P. Abdul Majeed

Abstract Electroencephalogram (EEG) is non-trivial in the diagnosis and treatment of neurogenerative diseases. Brain-Computer Interface (BCI) that utilises EEG is often used to improve the activities of daily living of patients with the aforesaid disorder. In this study, the efficacy of different Transfer Learning (TL) models, i.e., ResNet50, ResNet101 and ResNet152 in extracting features to classify wink-based EEG signals is evaluated. The time–frequency spectrum transformation of the Right-Wink, Left-Wink, and No-Wink based on EEG signals was achieved via Discrete Wavelet Transform (DWT). The extracted features were then fed into different variation of Support Vector Machine (SVM) classifiers to evaluate the performance of the different feature extraction method in classifying the wink class. The data are divided into training, validation, ad test, with a stratified ratio of 60:20:20. It was shown from the study, that the features extracted via ResNet152 were better than that of ResNet50 and ResNet101. The overall validation and test accuracy attained through the ResNet152 model is approximately 92%. Henceforth, it could be concluded that

J. L. Mahendra Kumar · M. A. Mohd Razman · A. P. P. Abdul Majeed (✉)
Innovative Manufacturing, Mechatronics and Sports Laboratory, Faculty of Manufacturing and Mechatronics Engineering Technology, Universiti Malaysia Pahang (UMP), 26600 Pekan, Pahang Darul Makmur, Malaysia
e-mail: amajeed@ump.edu.my

M. Rashid · N. Sulaiman
Faculty of Electrical and Electronics Engineering Technology, Universiti Malaysia Pahang (UMP), 26600 Pekan, Pahang Darul Makmur, Malaysia

R. M. Musa
Centre for Fundamental and Liberal Education, Universiti Malaysia Terengganu (UMT), 21030, Kuala Nerus, Terengganu Darul Iman, Malaysia

R. Jailani
Faculty of Electrical Engineering, Universiti Teknologi MARA (UiTM), 40450 Shah Alam, Selangor Darul Ehsan, Malaysia

A. P. P. Abdul Majeed
Centre for Software Development & Integrated Computing, Universiti Malaysia Pahang (UMP), 26600 Pekan, Pahang Darul Makmur, Malaysia

A. F. Ab. Nasir et al. (eds.), *Recent Trends in Mechatronics Towards Industry 4.0*, Lecture Notes in Electrical Engineering 730, https://doi.org/10.1007/978-981-33-4597-3_6

the proposed pipeline suitable to be adopted to classify wink-based EEG signals for different BCI applications.

Keywords EEG · BCI · DWT · Transfer learning · SVM · Classification

1 Introduction

Stroke is a neurological disorder that affects the arteries that lead to and within the brain. In a recent report by the Global Burden of Disease Study 2016, suggests that stroke will be the second leading cause of mortality in 2040 [1]. Patients that are suffered from stroke are often left with limited motor functions, which limits their activities of daily living (ADL) [2]. The advancement of technology has allowed for the exploitation of brain signals, particularly Electroencephalogram (EEG) through Brain-Computer Interfaces (BCI) that could grant patients to regain their ADL [3, 4]. Different BCI applications have been reported in the literature, namely, virtual keyboards, mouse, neuro-prosthesis biofeedback therapy, and rehabilitation devices, amongst others [5]. It is imperative to mention at this juncture that the ability to classify accurately different actions is crucial for the successful implementation of BCI applications [6].

Various research has applied motor imagery signals to implement into BCI applications. It is worth noting that there is limited research on the classification of facial expression-based EEG signals. Domrös et al. [7] evaluated eye blink detection signals from EEG signals. The authors extracted time-domain features to classify the EEG signals. The samples were divided into two sets of datasets, which are 70% for training and 30% for testing, respectively. Different classifiers were compared, namely multilayer perceptron (MLP) with Feed Forward Back Propagation (FFBP), MLP-Cascade Forward Back Propagation (CFBP) and Radial Basis Function (RBF) Binary Classifier. The classification accuracy (CA) attained were 96.69%, 99.83%, and 100% for FFBP, CFBP, and RBF, respectively.

A multimodal emotion recognition framework by combining facial expression and EEG signals based on a valence-arousal emotional model was investigated by Huang et al. [8]. They have implemented Transfer Learning (TL) approach for facial expression through multi-task Convolutional Neural Network (CNN) architectures to detect the state of valence and arousal. The targets were detected by the Support Vector Machine (SVM) classifier. They have demonstrated that their proposed method could yield a CA of 69.75% and 70.00% for the valence space and arousal space, respectively.

It is evident from existing literature that the employment of Transfer Learning models in the classification of wink-based EEG signals is limited. Therefore, the present investigation aims at evaluating the effectiveness of different pre-trained CNN architectures (TL models) in extracting features from wink-based EEG signals. Subsequently, a number of SVM classifiers were developed in identifying the best

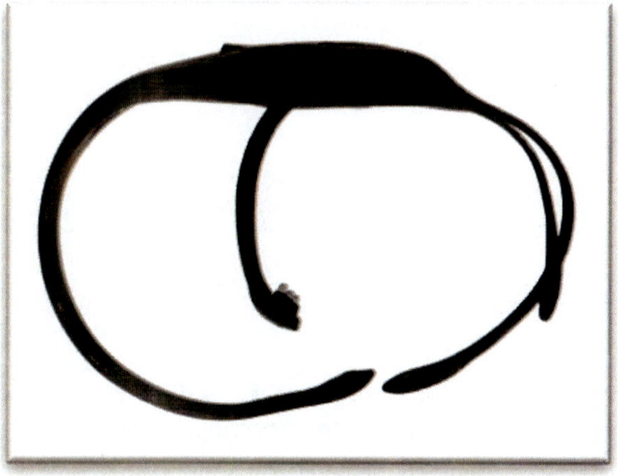

Fig. 1 Emotiv insight signal acquisition device

hyperparameters that are attained via the grid search algorithm that allows for a better classification of the winks.

2 Methods

2.1 Signal Acquisition Device

A Wireless Emotiv Insight (EI) mobile headset with five (5) channels were utilised to collect the EEG-based eye winking signals in the present investigation. Figure 1 shows the device used to collect the signals which has a bandwidth between 0.5 and 43 Hz. The resolution of each channel is 0.51 μV. The signals retrieved are from the Left-Wink, Right-Wink, and Relaxed actions. All these signals were collected at the sampling rate of 128 samples per second of each channel. The electrodes are positioned at AF3, AF4, T7, T8, and Pz, according to the International 10–20 EEG system. Whereas, the placement of the reference electrode is positioned at the left mastoid.

2.2 Subjects

Four males and two females aged between the age of 23 to 27 years old participated in the experiment. The chosen subjects do not have any known medical history, and all of them have normal vision. The background of the subjects clearly shows that they do not have any history of neurological disorders. The subjects do not have any

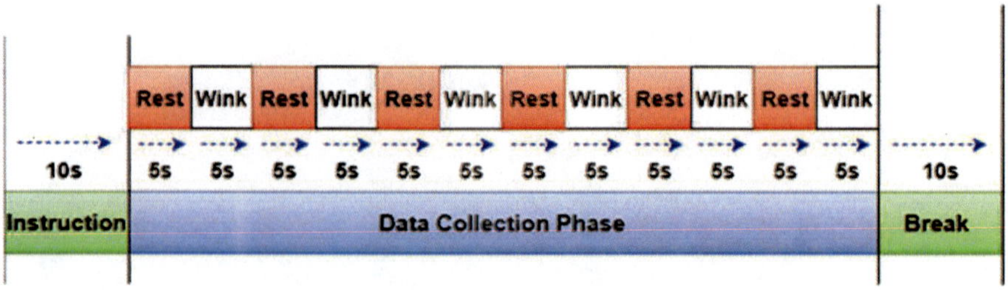

Fig. 2 The experimental paradigm for data collection

knowledge of the experiment that will be carried out. The ethical approval for this study was obtained via an institutional research ethics committee (FF-2013–327).

2.3 Experimental Setup

The data collection took place at the Applied Electronics and Computer Engineering (AppaECE), University Malaysia Pahang. The E1 was placed on the subjects' head as per the Standard Operating Procedure reported on their official website. Penetrating gel was applied to the sensor nodes for the sensor to penetrate properly through the hair of the subjects. The subjects were instructed to sit on an ergonomic chair and in a relaxed position. They were instructed to sit in such a manner in order to avoid any extra physical movements. The instruction of the actions was given through a cue that was displayed on the LCD. The display was placed one meter away from the subject. The experiment paradigm consists of five (5) trials of the actions that were utilised throughout the experiment. The paradigm consists of No-Wink and Right-Wink/Left-Wink for every five seconds. This was repeated for one minute. The experimental paradigm is shown in Fig. 2. The subjects were instructed to adhere to the paradigm and act in accordance to it to collect the signals required. The rest (No-Wink) slide is displayed a big black dot in the middle. The subjects were asked to concentrate on the dot in order to avoid the eyeball movements (EEG artifacts).

2.4 Signal Processing

The raw signal that was collected from EI through Emotiv BCI needs to undergo a number of pre-processing steps. The signals collected were filtered via a digital notch filter at 50 Hz in order to reduce the power line noise [9–11]. The sample of signals collected from Left-Wink, Right-Wink, and No-Wink actions are depicted in Fig. 3. The collected signals were segmented into five segments. The selected

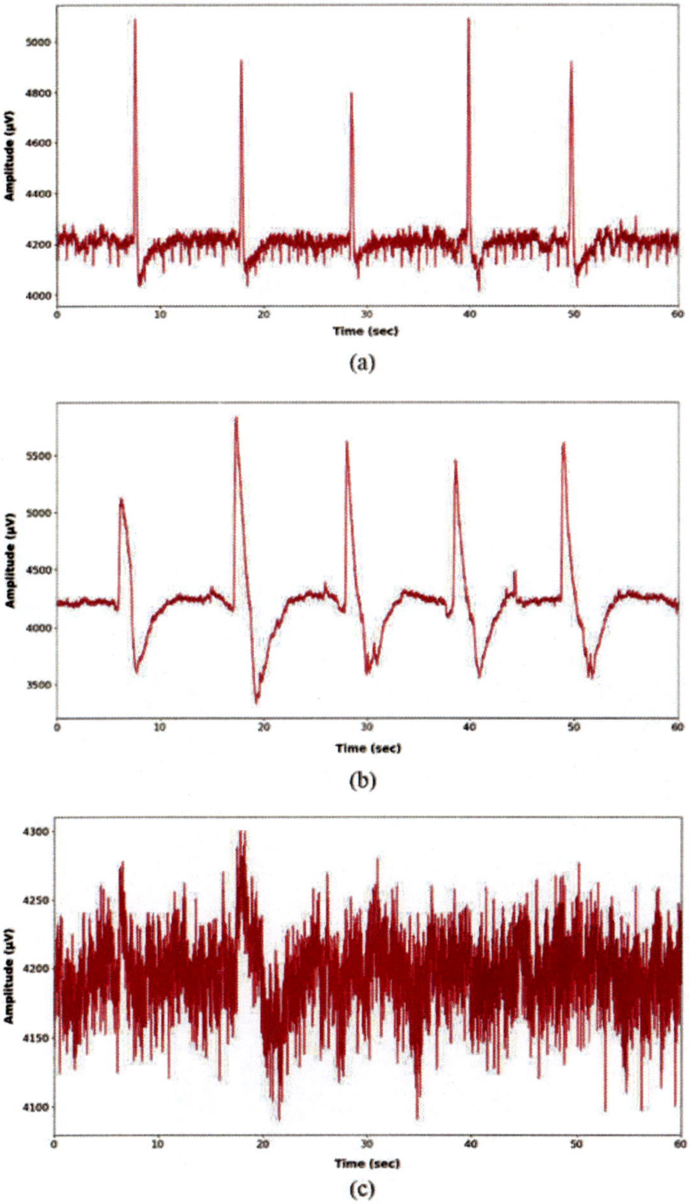

Fig. 3 EEG signals of **a** left-wink, **b** right-wink and **c** no-wink

segments consist of winking actions. Therefore, the total number of ninety samples were segmented.

The segmented signals were then fed into the Discrete Wavelet Transform (DWT) algorithm to transform the signals into a scalogram. This method has been employed in [12]. DWT signal is represented as the linear amalgamation of a specific set of functions attained by shifting and extending the mother wavelet. The wavelet that was used in this experiment is the Daubechies wavelet. Figure 4 depicts the scalogram obtained through the DWT algorithm for Left_Wink, Right_Wink, and No-Wink.

The features of the scalogram images were extracted through TL models. TL is a popular method in computer vision as it allows for the development of accurate models in a time-effective way. TL models are pre-trained CNN models and hence reducing the time required to train the models. In this study, the following TL models, namely ResNet50, ResNet102, and ResNet152, are utilised to extract the features of the images. The flatten size of all the three TL models are 7 * 7 * 2048. The input image size for the models is 224 × 224.

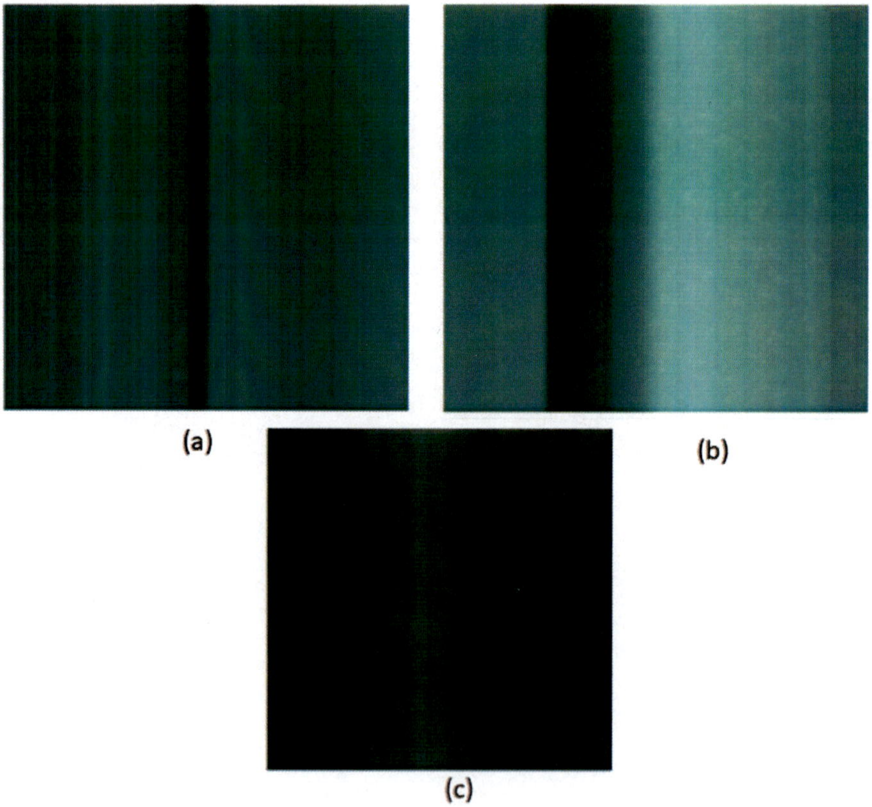

Fig. 4 Scalogram of **a** left-wink, **b** right-wink, and **c** no-wink

Table 1 Hyperparameters of SVM algorithm

Hyperparameters	Values
Kernel	Linear, radial basis function (RBF), polynomial, and sigmoid
Regularization, c	0.001, 0.01,0.1, 1, 10, 100, 1000
Gamma, γ	0.001, 0.01,0.1, 1, 10, 100, 1000
Degree of polynomial	2 and 3

2.5 *Classification*

The Support Vector Machine (SVM) model was employed to evaluate the effect of the features extracted through the TL models. SVM exploits the use of the kernel trick to transform input signal into a higher dimensional space, which in turn, isolates the data through a hyperplane with a maximum margin [13]. This type of supervised ML algorithm is widely used owing to its minimal number of hyperparameters that are required to be tuned [14–16]. The SVM algorithm was fine-tuned through the grid search method. The hyperparameters that were tuned are the kernel, gamma, regularisation, and the polynomial degree. Table 1 illustrates the hyperparameters that were tuned in this study.

The scikit-learn Python library v0.22.2 was utilised to build 245: models. The datasets were divided into training, validation, and test datasets. The ratio for training and testing datasets was set to a 60:40 ratio from a total of 90 samples. The test datasets were further divided into a stratified ratio of 20:20, which are test datasets and validation datasets. The performance of the models was evaluated through Classification Accuracy (CA) and via the confusion matrix.

3 Result and Discussion

All the converted signals were implemented into ResNet50, ResNet101, and ResNet152 and classified through a hyperparameter tuned SVM algorithm. The bar chart in Fig. 5 illustrates the classification accuracy of training, validation, and test datasets. It could be seen from Fig. 5, ResNet101 and ResNet152 models achieved 100% CA via training datasets, while ResNet50 has only achieved 98% CA. On the validation dataset, the ResNet152 demonstrated the highest accuracy amongst other TL model with a CA of 100%. Hence it can be inferred that ResNet152 would be the best TL model for this set of pre-processed datasets. Finally, the models were evaluated with the test dataset, and it could be observed that the features extracted through the ResNet152 model, and trained on the optimised SVM model, yield a CA of 83%. The average CA on the validation and test dataset of the ResNet152 pipeline is approximately 92%, suggesting that this pipeline is the best amongst the evaluated models. The best parameters for SVM classifier are listed in Table 2. Figure 6

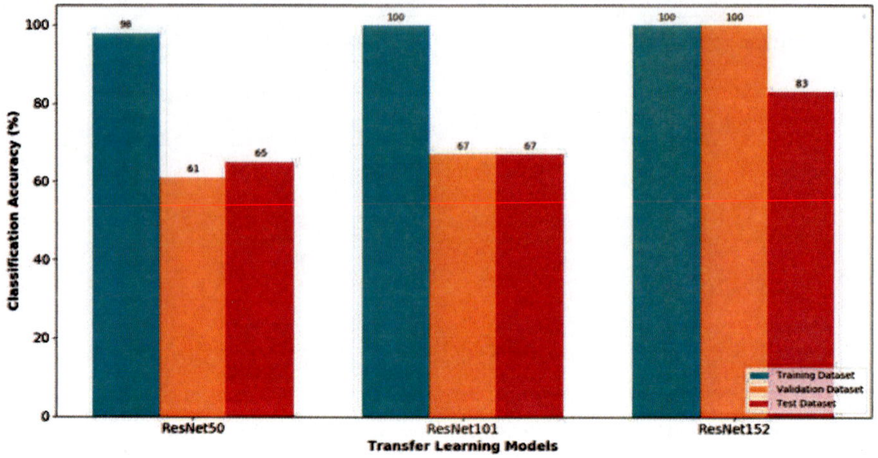

Fig. 5 Bar chart of the TL models CA

Table 2 SVM hyperparameters that achieved the highest CA

Hyperparameters	Values
Kernel	Linear
Regularisation, c	0.01
Gamma	0.01

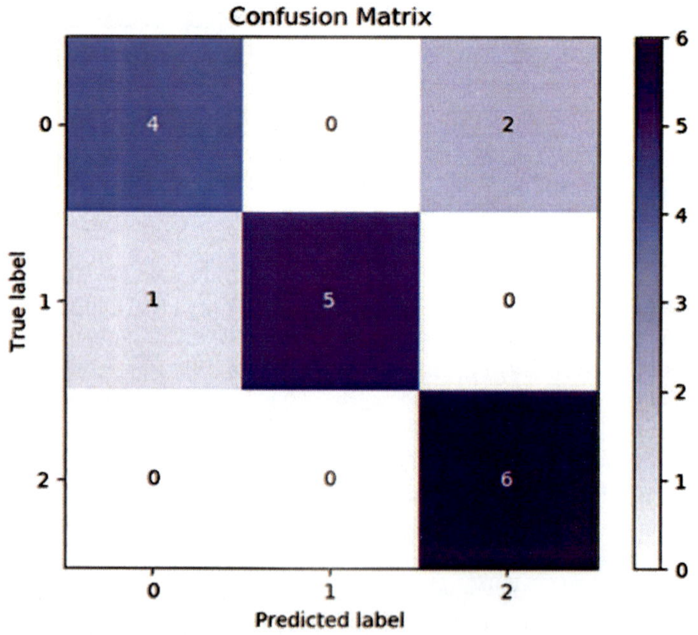

Fig. 6 Confusion matrix of the test dataset of the ResNet152-SVM pipeline

depicts the confusion matrix of the test datasets of the best model, where 0, 1 and 2 are referred to Left-Wink, Right-Wink and No-Wink, respectively.

4 Conclusion

In this study, a set of pre-trained CNN models, also known as TL models were utilised to extract features from wink-based EEG signals that were converted into a scalogram through the DWT algorithm. The classification of the models was done through fine-tuned or optimised SVM models. It was demonstrated from the study that the ResNet152 TL model is better in extracting features as compared to ResNet50 and ResNet101. The findings are non-trivial, mainly towards real-time BCI real-time implementation as the processing expense could be reduced by considering TL models to extract significant features. Future studies shall attempt on evaluating other TL models coupled with different traditional classifiers.

Acknowledgements The present study is funded by Universiti Malaysia Pahang via RDU180321.

References

1. Ganasegeran K, Fadzly M, Jamil A, Sivasampu S (2019) Discover! Malaysia's stroke care revolution—special edition. ResearchGate 2:1–32
2. Ab Patar MNA, Said AF, Mahmud J, Majeed APPA, Razman MA (2014) System integration and control of dynamic ankle foot orthosis for lower limb rehabilitation. In: ISTMET 2014—1st international symposium technology management emerging technology Proceedings, vol 2, pp 82–85. https://doi.org/10.1109/ISTMET.2014.6936482
3. Shih JJ, Krusienski DJ, Wolpaw JR (2012) Brain-computer interfaces in medicine. Mayo Clin Proc 87:268–279. https://doi.org/10.1016/j.mayocp.2011.12.008
4. Vaughan TM (2003) Brain-computer interface technology: a review of the second international meeting. IEEE Trans Neural Syst Rehabil Eng 11:94–109. https://doi.org/10.1109/TNSRE.2003.814799
5. Lin JS, Hsieh CH (2016) A wireless BCI-controlled integration system in smart living space for patients. Wirel Pers Commun 88:395–412. https://doi.org/10.1007/s11277-015-3129-0
6. Rashid M, Sulaiman N, Majeed APPA, Musa RM, Ahmad AF, Bari BS, Khatun S (2020) Current status, challenges, and possible solutions of EEG-based brain-computer interface: a comprehensive review. Front Neurorobot 14:1–35. https://doi.org/10.3389/fnbot.2020.00025
7. Domrös F, Störkle D, Ilmberger J, Kuhlenkötter B (2013) Converging clinical and engineering research on neurorehabilitation. Converg Clin Eng Res Neurorehab 1:409–413. https://doi.org/10.1007/978-3-642-34546-3
8. Huang Y, Yang J, Liu S, Pan J (2019) Combining facial expressions and electroencephalography to enhance emotion recognition. Futur Internet 11:1–17. https://doi.org/10.3390/fi11050105
9. Choy TTC, Leung PM (1988) Real time microprocessor-based 50 Hz notch filter for ECG. J Biomed Eng 10:285–288. https://doi.org/10.1016/0141-5425(88)90013-1
10. Jayant HK, Rana KPS, Kumar V, Nair SS, Mishra P (2006) Efficient IIR notch filter design using minimax optimisation for 50 Hz noise suppression in ECG. In: Proceedings of 2015

international conference on signal processing computing control. ISPCC 2015, pp 290–295. https://doi.org/10.1109/ISPCC.2015.7375043
11. Leske S, Dalal SS (2019) Reducing power line noise in EEG and MEG data via spectrum interpolation. Neuroimage 189:763–776. https://doi.org/10.1016/j.neuroimage.2019.01.026
12. Bekbalanova M, Zhunis A, Duisebekov Z (2019) Epileptic seizure prediction in EEG signals using EMD and DWT. In: 2019 15th international conference on electronics comput. Comput. 1–4 (2019)
13. Gholami R, Fakhari N (2017) Support vector machine: principles, parameters, and applications. Elsevier Inc. https://doi.org/10.1016/B978-0-12-811318-9.00027-2
14. Yang J, Singh H, Hines EL, Schlaghecken F, Iliescu DD, Leeson MS, Stocks NG (2012) Channel selection and classification of electroencephalogram signals: an artificial neural network and genetic algorithm-based approach. Artif Intell Med 55:117–126. https://doi.org/10.1016/j.artmed.2012.02.001
15. World Health Organization (2008) Neurological disorders. Public Health Challenges. J Nerv Ment Dis 196:176. https://doi.org/10.1097/nmd.0b013e31816372ab
16. Musa RM, Majeed APPA, Taha Z, Chang SW, Nasir AF, Abdullah MR (2019) A machine learning approach of predicting high potential archers by means of physical fitness indicators. PLoS One 14:1–12. https://doi.org/10.1371/journal.pone.0209638

Development of Polymer-Based Y-Branch Symmetric Waveguide Coupler Using Soft Lithography Technique

M. S. M. Ghazali, F. R. M. Romlay, and A. A. Ehsan

Abstract The development of a Y-branch symmetric waveguide coupler based on soft lithography approach was presented. This paper focused on the design, fabrication and testing of the symmetric waveguide coupler that produces an optic with power output. The fabrication was done by engraving acrylic to produce a master mould using milling machining tools for optical devices. The device was constructed via soft lithography which duplicated the pattern from the master mould onto a second mould to produce an actual device. Afterward, optical polyme epoxy OG142 was injected into the second mould, of which the product was then put on top of acrylic. The device was completed after curing the optical polymer glue, epoxy OG142 by exposing the assembly on the second mould under UV light until both parts bonded. The results of the tap off ratio testing for asymmetric waveguide coupler were ranged from 21.1 to 49.2% within +2% error.

Keywords Waveguide · Optical · Soft lithography

1 Introduction

Plastic optical fibres (POFs) and waveguides have been of interest since the late 1960s to early 1970s [1]. POFs are often used as a medium for optical data communication. Couplers and splitters as passive devices are often necessary to split or combine optical signals for an optical device to complete its designated function [2–4].

Coupler and splitter constructions had been reported by [5–7] through polishing, embossing and micro- injection process. However, the processes require advance machinery and high cost. Therefore, this paper reports a low cost fabrication of a polymer-based Y-branch symmetric waveguide POFs coupler.

M. S. M. Ghazali · F. R. M. Romlay (✉)
Machine Manufacturing Union in Mechatronics Laboratory, Manufacturing Focus Group, Universiti Malaysia Pahang, 26600 Pekan, Pahang, Malaysia
e-mail: fadhlur@ump.edu.my

A. A. Ehsan
Institute of Microengineering and Nanoelectronics (IMEN), Universiti Kebangsaan Malaysia, 43600 Bangi, Selangor, Malaysia

A. F. Ab. Nasir et al. (eds.), *Recent Trends in Mechatronics Towards Industry 4.0*, Lecture Notes in Electrical Engineering 730, https://doi.org/10.1007/978-981-33-4597-3_7

The soft lithography is a manufacturing process that utilized patterns with a polydimethylsiloxane (PDMS) master stamp. This technique adopts a pattern transfer using a master mould that deforms the substrate material. The details process is presented in Sect. 3.

2 Device Design

The design of the asymmetrical waveguide coupler is based on a simple TOFR (Tap Off Ratio) technique to tap off power from the main bus line [5]. The concept of asymmetric Y-junction splitter in the form of a simple power tap can be achieved in multimode device [6]. Figure 1 shows the illustrated of 1 × 2 symmetric Y-junction splitter with the bus and tap line design by computer-aided design (CAD). The splitting angle is set at 18°. Both the spatial and s-bend splitters provide a good splitting ratio of 3 dB [7]. From the design, several types of asymmetric waveguide couplers with any different size of tap width are accessible.

Design of 50% TOFR tap line sizes are created as shown in Fig. 2 The widths of the bus line is x mm and tap size is y mm.

The design of the 1 × 2 asymmetric waveguide coupler was based on simple tap of ratio technique to tap off power from the main bus line. The ratio of the power exiting through the tap line to the total power incident in the bus line is given by Eq. (1), where y is the tap width size and x is the bus width size [8].

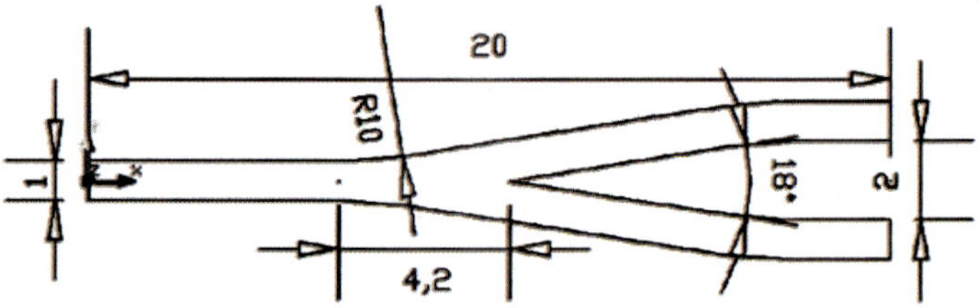

Fig. 1 Y-junction symmetric waveguide coupler parameter design by CAD (unit in mm)

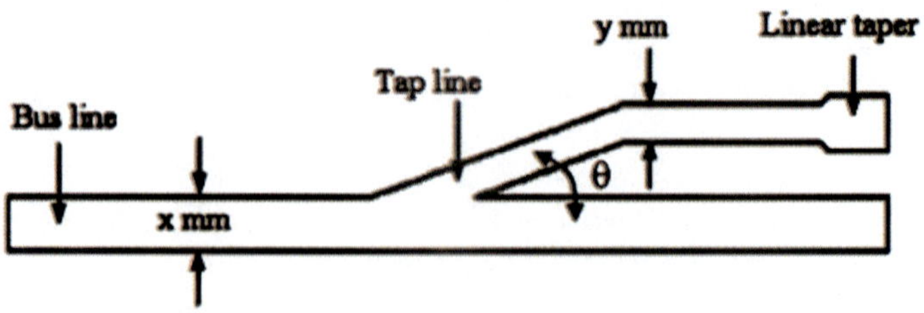

Fig. 2 1 × 2 asymmetric waveguide coupler; at 50% TOFR

$$TOFR = \frac{y}{y + x} \tag{1}$$

Mirroring of the asymmetric waveguide coupler design indicates symmetric design of the device.

3 Materials and Methodology

Several chemicals and materials were applied throughout the soft lithography fabrication of the Y-branch splitter, which included optical polymer OG142 as the core for the splitter, elastomer material of Polydimethylsiloxane (PDMS) and curing agent for the PDMS, Epo-Kwick.

The optical polymer OG142 has a refractive index of 1.57, a haze of 0.2% and light transmissions of 92%. The material used as a substrate for the splitter was acrylics. It has a refractive index of 1.49, a haze of 0.5% and light transmissions of 92% [9].

The master mould is fabricated from acrylic plastic, through milling process. The acrylic plastic-based device has an optically clear polymer as the waveguide core by Epo-Tek OG142. The epoxy mixture was pour into the mould and allowed to harden before the sample was released from the holder. The sample was obtained by separating the acrylic master mould and the Epo-Kwick material. The acrylic was placed on top of the PDMS pattern to acquire the Y-branch symmetric waveguide coupler, which was then cured under UV light with an intensity of 60 W. Upon the exposure to UV-light, the optical polymer (OG142) would bond to the acrylic and create a permanent sealing.

The design of asymmetric coupler provided 1 × 2 coupler splitter as shown in Fig. 3. The Y-branch coupler was cascaded with another asymmetric waveguide coupler to indicate more splitters.

Upon completion of the soft lithography procedures, the Y-branch symmetric waveguide devices were tested and analysed accordingly.

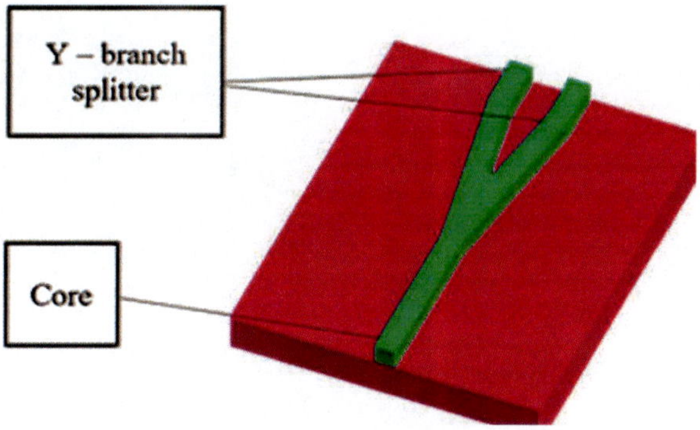

Fig. 3 3D dimensional of final device

4 Testing and Discussion

A test on the Y-branch symmetric waveguide couplers was carried on. The fabricated waveguide coupler was proceed for ray tracing using Zemax equipment. The optical source for this simulation was from a rectangular source with a wavelength of 650 nm and input power of 1.0 mW. For ray tracing, the core of the waveguide was the optical polymer Epo-Tek OG142 with an index of refraction of 1.58. The two-dimensional (2D) ray tracing diagram of the Y-branch splitter is shown in Fig. 4.

The cladding for the waveguides was PMMA or acrylic with an index of refraction of 1.49. Figure 5 shows the actual final device of the Y-branch symmetric waveguide coupler with the proposed materials for the core and cladding. The ray tracing tested was conducted on the assembled coupler for performance measurement.

The output signals for both devices obtained from output Ports 1 and 2 were 0.402 and 0.390 mW as shown in Fig. 6.

Based on device outputs from Port 1 and Port 2, the insertion loss for the device 1 was about 4.1 dB The value of the symmetric output ratio for section Y- splitter is 50%. If the input power was 100%, then the output power is 50% of the input.

The fabricated Y-branch asymmetric waveguide coupler in this research may serve as an alternative to those devices developed by Takezawa et al. [5], Mizuno et al.

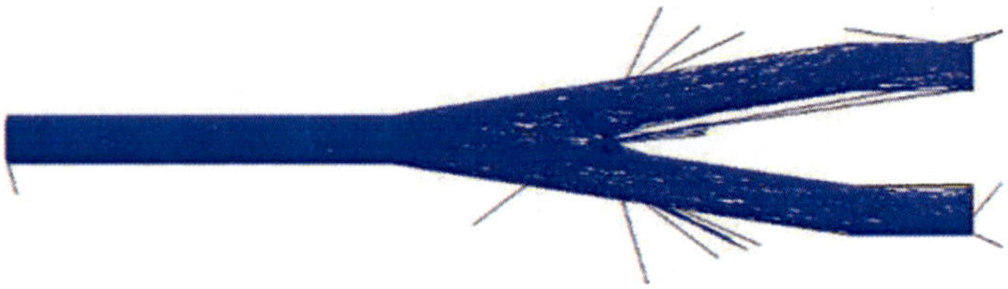

Fig. 4 2D ray tracing diagram for Y-branch symmetric waveguide coupler device

Fig. 5 The fabricated coupler devices were assembled and tested

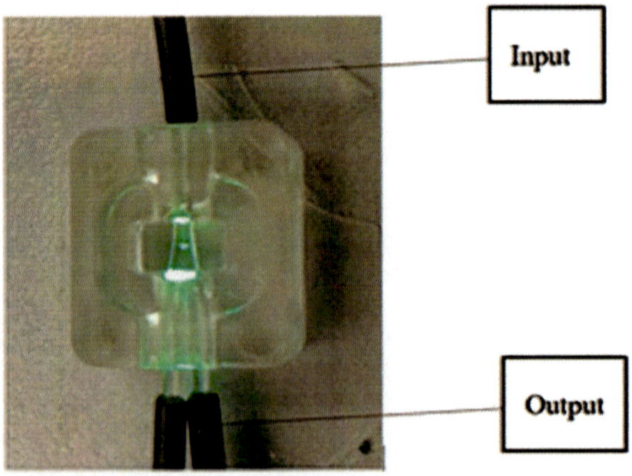

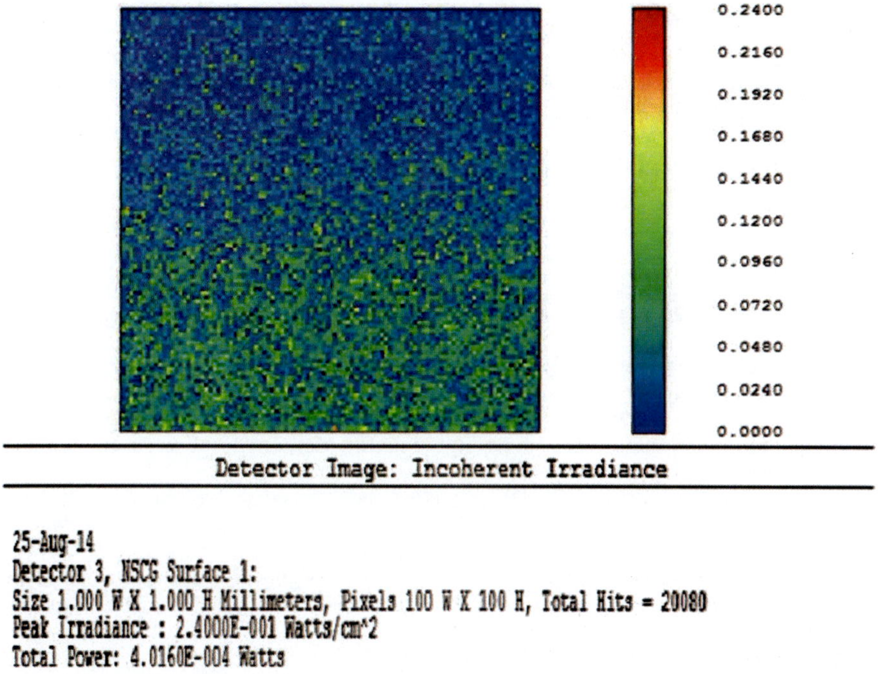

Fig. 6 Detector image for the Y-branch asymmetric waveguide coupler performance

[6], and Kotzbuecher et al. [7]. These devices, as aforementioned, require expensive production equipment and have time-consuming procedural steps.

The proposed fabrication technique that incorporates the soft lithography technique can produce the master mould, PDMS, optical polymer and UV-curable polymer for the waveguide. These results showed that the technique was capable of providing reliable Y branch optical splitter at acceptable optical performance.

5 Conclusions

The fabrication and simulation of polymer-based Y-branch symmetric and asymmetric waveguide couplers as splitters were presented. The Y-branch splitters were designed and fabricated using soft lithographic. The device gave a good performance as an optic splitter.

Moreover, the proposed device has simple fabrication procedures, low-cost production techniques and high device throughput. A single master mould would even allow for the fabrication of the device multiple times.

Acknowledgements This research was partially supported by research, development and commercialization grant RDU151006, RDU172206 and UIC180302 of University Malaysia Pahang. Fundamental Research Grant Scheme; FRGS/2/2013/TK01/UMP/02/6 from Ministry of Higher Education Malaysia is also acknowledged.

References

1. Tricker R (2002) Optoelectronic and fiber optic technology, 1st edn. Elsevier Science; Oxford, U.K.
2. Ayesta I, Azkune M, Arrospide E, Arrue J, Illarramendi MA, Durana G, Zubia J (2019) Fabrication of active polymer optical fibers by solution doping and their characterization, polymers 11
3. Hernaez M, Zamarreno CR, Melendi-Espina S, Bird LR, Mayes AG, Arregui FJ (2017) Optical fibre sensors using graphene-based materials: a review. Sensors 17
4. Zieman O, Krauser J, Zamzow PE, Daum W (2008) POF handbook, 2nd edn. Springer, Berlin
5. Burtscher C, Seyringer D, Kuzma A (2017) Modeling and optimization of 1 × 32 Y-branch splitter for optical transmission systems. Opt Quant Electron 49
6. Mizuno H, Sugihara O, Okamoto KT, Ohama M (2005) Compact Y-branch-type polymeric optical waveguide devices with large-core connectable to plastic optical fibers. Jpn J Appl Phys 44
7. Klotzbuecher T, Braune T, Dadic D, Sprzagala M, Koch A (2003) Fabrication of optical 1x2 POF couplers using the laser-LIGA technique. Proc SPIE 4941
8. Loch M (2004) Plastic optical fibers: properties and practical applications, optical transmission systems and equipment for WDM networking III, proceedings of SPIE 5596
9. Ehsan AA, Shaari S, Abd Rahman MK, Kee Zainal Abidin KMR (2009) Hollow optical waveguide coupler for portable access card system application. J Opt Commun 30

Hybrid Flow Shop Scheduling with Energy Consumption in Machine Shop Using Moth Flame Optimization

Mohd Fadzil Faisae Ab. Rashid, Ahmad Nasser Mohd Rose, and Nik Mohd Zuki Nik Mohamed

Abstract Hybrid flow shop with energy consumption (HFS-EC) combine the flow shop scheduling and parallel machine scheduling problem with the aim to optimize energy utilization, besides regular makespan in the production scheduling. This paper optimizes an HFS-EC case study using Moth Flame Optimization (MFO). The case study has been conducted in a machine shop concentrating on three machining types; lathe, milling and deburring. The objectives were to optimize makespan and total energy consumption in the machine schedule. Optimization using MFO has been conducted and the results was compared with well-established algorithm like Genetic Algorithm, Ant Colony Optimization and Particle Swarm Optimization. The results were also compared with relatively recent algorithm such as Whale Optimization Algorithm and Harris Haws Optimization. Based on the optimization results, the MFO outperformed other comparison algorithms for the mean fitness and also the best fitness. Although there were other solutions with better individual optimization objectives, but results obtained by MFO compromised between minimum makespan and energy consumption. The proposed HFS-EC model and MFO algorithm has a great potential to be implemented in other scheduling case study due to benefit of reducing carbon emission and at the same time maintain the production output.

Keywords Hybrid flow shop · Energy consumption · Scheduling optimization · Moth flame optimization

1 Introduction

Hybrid flow shop scheduling (HFS) problem has been well studies because of its complexity and representing real application in industry. In simple word, HFS is a combination of flow shop and parallel machine scheduling. Flow shop scheduling refers to scheduling of a set of machine or process in serial order. Each process must undergo every machine stage in predetermined sequence order. There is only a single

M. F. F. Ab. Rashid (✉) · A. N. Mohd Rose · N. M. Z. Nik Mohamed
Department of Industrial Engineering, College of Engineering, Universiti Malaysia Pahang, Lebuhraya Tun Razak, 26300 Kuantan, Pahang, Malaysia
e-mail: ffaisae@ump.edu.my

A. F. Ab. Nasir et al. (eds.), *Recent Trends in Mechatronics Towards Industry 4.0*, Lecture Notes in Electrical Engineering 730, https://doi.org/10.1007/978-981-33-4597-3_8

machine for each stage. Meanwhile, parallel machine scheduling is the problem to assign job with single process to a set of similar machines. Although the machine can conduct similar process, but the capacity of machines might be different because of model variety.

Hybrid flow shop scheduling is a combination of these two scheduling problems variant. HFS involve scheduling for multiple processes in different stages, whereas the number of machines for each stage can be more than one (parallel). For HFS, there are n jobs to be processed at S stages in the similar route. In each stage, there are M machines that can performed similar operation with different capabilities.

In previous research, the most prominent objective in HFS optimization is to minimize makespan. The makespan is the total time required to accomplish all the jobs. A smaller makespan means that the faster jobs will be processed. Besides that, researchers also consider other objectives such as lateness, penalty and machine utilization in their research.

Recently, researchers also embedded sustainable factors objective in the HFS optimization. Lu et al. for example considered the noise pollution in the HFS, besides the makespan in their work [1]. On the other hand, Schulz considered the energy consumption in the HFS to identify the optimum configuration for energy cost, for the different power rate period [2, 3]. Besides finding the optimum job-machine assignment to minimize energy consumption, researcher also minimize non-processing energy with the aim to reduce the machine idle between processing [4]. Other than that, the total carbon emission was another sustainable factor embedded in the HFS. The carbon emission was optimize together with makespan and energy consumption [5].

Various optimization algorithms have been implemented to optimize HFS with energy consumption (HFS-EC). Yan et al. optimize HFS with energy consumption using Genetic Algorithm (GA) [6]. Since their work involve multiple objectives, the weighted approach was used. Besides that, there were also a few recent publications on HFS-EC using GA-based algorithm such as [2, 7]. Researchers also implemented Particle Swarm Optimization (PSO) to optimize HFS-EC problem. One of the work was dated back in 2009 [8]. Instead of considering regular HFS-EC, this work modelled the dynamic HFS-EC, whereby the input for the scheduling was in the dynamic mode. Another well-established algorithm used to optimize HFS-EC is Ant Colony Optimization (ACO) [9]. The ACO also was combined with Tabu Search to enhance its performance during HFS-EC optimization [10].

Besides the well-established metaheuristics above, researchers also implemented other algorithm such as Artificial Bee Colony (ABC) algorithm [11], Decomposition Evolutionary Algorithm [12], Ant Lion Optimization [13], Imperialist Competitive Algorithm [14] and Multi-Verse Optimization [15]. The main factor contributed to this variety was because of metaheuristic algorithm performance is dependent on the problem. Performance of algorithm might be dissimilar even within similar HFS variety.

This paper optimizes an HFS-EC case study problem. A case study has been conducted in a machine shop that offers machining services for metal-based product for the manufacturer. The problem has been optimized using Moth Flame Optimization (MFO). To the best of authors' knowledge, none of existing publication implemented this algorithm to optimize HFS problem. MFO is a metaheuristic inspired by moth flying and navigation in the present of artificial light [16]. The MFO has been successfully implemented in different field such as engineering, medical and financial.

2 Case Study

A case study has been conducted at a machine shop located in Batu Pahat, Malaysia. The shop processed multiple customized metal base product to be used in different products. There were several types of machines in the premise. However, the most critical machining processes were lathe and milling. For the study purpose, the scope was limited by considering only three machining processes; lathe, milling and deburring. In addition, the jobs that considered in the study must be processed in these three-machining processes starting with lathe, milling and finally deburring.

In the first and second stage of the process, three lathe and milling machines were available respectively. While only two machines available for deburring process in the third stage. The scheduling process will only consider these resources. The jobs to be processed come in different part and volume. Therefore, the processing time for each job will be different. In addition, the machines considered in this study were from different model. Even though similar process can be conducted in any machine at each stage, the capacity and power rate will be different.

There are 14 jobs need to be scheduled in the available machines. The detail of processing time and power rate was presented in Table 1.

In this study, two optimization objectives are considered; minimize makespan, $C_{\max}$ and minimize total energy consumption, *TEC*.

$$C_{\max} = \max_{j \in n}\{c_{S \cdot j}\} \tag{1}$$

$$TEC = \sum_{s=1}^{S}\sum_{m=1}^{M} t_{s.m} \cdot P_{s.m} \tag{2}$$

In Eq. (1), $c_{s.j}$ refers to completion time for the jth job at the last stage, S. In this equation, j was the index of job from 1 to n. On the other hand, $t_{s.m}$ in Eq. (2) refers to the duration (in hour) of utilization for machine m in stage s. Meanwhile, $P_{s.m}$ was the power rate (in kW) for machine m in stage s. For optimization purpose, a weighted sum approach was used, since the problem involved more than one optimization objective. To avoid bias in the fitness function, both of the optimization objectives were normalized into [0,1] range. The normalized $C_{\max}$ and TEC noted as $\hat{C}_{\max}$ and $\hat{T}EC$.

Table 1 Processing time for each machine (in minute)

	Stage 1 (Lathe)			Stage 2 (Milling)			Stage 3 (Deburring)	
Job	M1_1	M1_2	M1_3	M2_1	M2_2	M2_3	M3_1	M3_2
1	480	300	240	360	300	180	240	150
2	400	250	200	300	250	150	200	125
3	160	100	80	120	100	60	80	50
4	960	600	480	720	600	360	480	300
5	384	240	192	288	240	144	192	120
6	192	120	96	144	120	72	96	60
7	320	200	160	240	200	120	160	100
8	800	500	400	600	500	300	400	250
9	640	400	320	480	400	240	320	200
10	576	360	288	432	360	216	288	180
11	1280	800	640	960	800	480	640	400
12	768	480	384	576	480	288	384	240
13	960	600	480	720	600	360	480	300
14	400	250	200	300	250	150	200	125
Power (Watt)	1200	1600	2000	1200	1500	1800	700	1250

$$\min f = 0.5\hat{C}_{\max} + 0.5\hat{T}EC \quad (3)$$

3 Moth Flame Optimization

Moth flame optimization (MFO) is an optimization algorithm inspired by behavior of moth in the presence of flame or artificial light. During night, moth will fly in straight line guided by moon position. The traverse orientation in the flying navigation allow the moth to maintain the flying direction, since it utilizes constant 90° angle to the moon. However, in the presence of flame or other light source, moth will fly in spiral direction to maintain 90° angle to the light source. This algorithm has been introduced by Mirjalili in 2015. The flowchart of MFO algorithm is shown in Fig. 1.

The MFO algorithm begin with the initialization step, where an initial population is randomly created. In this work, size of population is 30, while the dimension is equivalent to $n \times M$. Each of the created dimension represents the weightage for particular machine and job. The initial population is then decoded into machine schedule using smallest element heuristic. There are two decisions that will be decoded from initial population; job processing sequence, and job-machine assignment for each stage. For the job processing sequence, the smallest element for each job representation is extracted. Then the sequence is determined by prioritizing the job with the

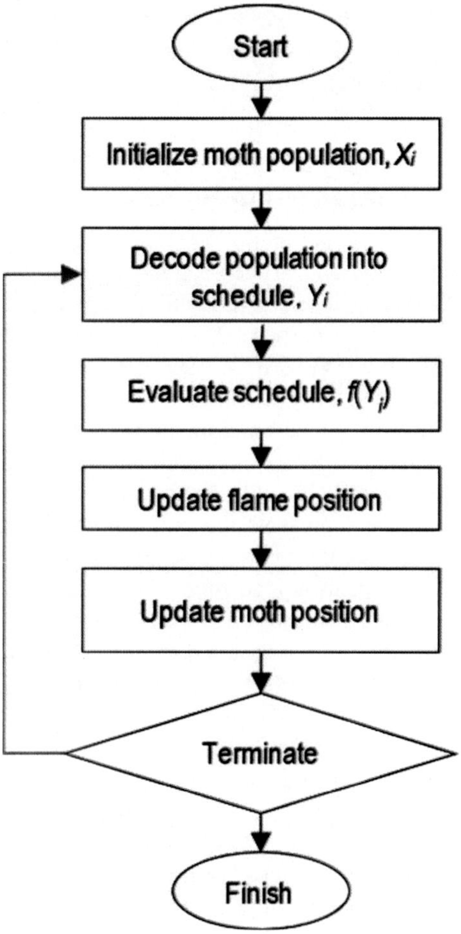

Fig. 1 Flowchart of moth flame optimization for HFS-EE

smallest element. The job-machine assignment is also made using similar heuristic. A specific job in a particular stage will be assigned based on the smallest element in the representation.

Next, the decoded population will be evaluated using fitness function as stated in Eq. (3). Then, the flame position is updated. In MFO, flame position represents the individual best solution. In each iteration, the obtained fitness value will be compared with the respected flame and being updated. The MFO algorithm then undergone moth position update using the following formula.

$$S(X_i, F_j) = D_i \cdot e^{bt} \cdot \cos(2t) + F_j \tag{4}$$

In this formula, X_i is the ith moth position, while F_j is the jth flame position. D_i is the absolute distance between X_i and F_j. Meanwhile b and t are constants that determine the logarithm spiral shape.

4 Optimization Results

Optimization has been conducted using MFO algorithm for the case study problem. The MFO has been run with 30 repetitions to obtained the best quality solution. For comparison purpose, MFO performance has been compared with well-established algorithms; GA, ACO, PSO, ABC and Firefly Algorithm (FA). In addition, relatively new algorithms also used to assess the MFO performance against recent algorithm. They are Whale Optimization Algorithm (WOA) and Harris Hawks Optimization (HHO). All the comparison algorithms utilize similar parameter as MFO; such as population size = 30, maximum iteration = 500 and number of reputations = 30.

Table 2 presents the optimization results for case study problem. The second and third columns show mean fitness and standard deviation from 30 runs for each of optimization algorithm. The fourth column meanwhile presents the best fitness obtained from 30 run. Finally, the fifth and sixth columns indicated the $C_{\max}$ and *TEC* for the best fitness obtained by each algorithm.

Based on Table 2, MFO algorithm obtained the best mean fitness compared with comparison algorithms. It shows that MFO able to search for better solution consistently. The MFO performance was followed by FA, PSO, GA, ABC, WOA, HHO and ACO. This result also showed that the well-established algorithm like PSO and GA also capable to produce relatively good results. This was the reason behind the popularity of GA and PSO although they were introduced more than 25 years ago.

In term of the best fitness, MFO also came out with the minimum fitness value compared with other algorithms. The obtained results confirm the performance of MFO in optimizing the studied problem. Again, FA, GA and PSO were just behind MFO in term of the best fitness. For MFO, the C_{max} for the best fitness was 2704 min,

Table 2 Optimization results for machine shop problem

Algorithm	Mean fitness	Standard deviation	Best fitness	$C_{\max}$ (minute)	*TEC* (kWh)
GA	0.3500	0.0243	0.3163	2896	294.92
ACO	0.3790	0.0149	0.3594	2831	299.81
PSO	0.3485	0.0164	0.3275	2640	305.05
ABC	0.3614	0.0081	0.3455	2794	299.5
FA	0.3384	0.0128	0.3157	2722	298.34
WOA	0.3667	0.0168	0.3387	2760	300.42
HHO	0.3731	0.0298	0.3276	2682	303.39
MFO	0.3313	0.0174	0.3028	2704	296.96

while the *TEC* was 296.96 kWh. Figure 2 shows the schedule obtained from this solution. The number inside the schedule represents the job number.

However, based on individual C_{max} optimization objective, the schedule in Fig. 2 is not the schedule with the smallest makespan obtained in the optimization. The smallest C_{max} is belong to solution by PSO with 2640minutes. However, for this solution, the *TEC* was 305.05 kWh which was the highest based on Table 2. On the other hand, the schedule with the smallest *TEC* was obtained by GA with 294.92 kWh, but the C_{max} is 2896 kWh. The machine schedule with the best C_{max} and best *TEC* individually presented in Figs. 3 and 4 respectively.

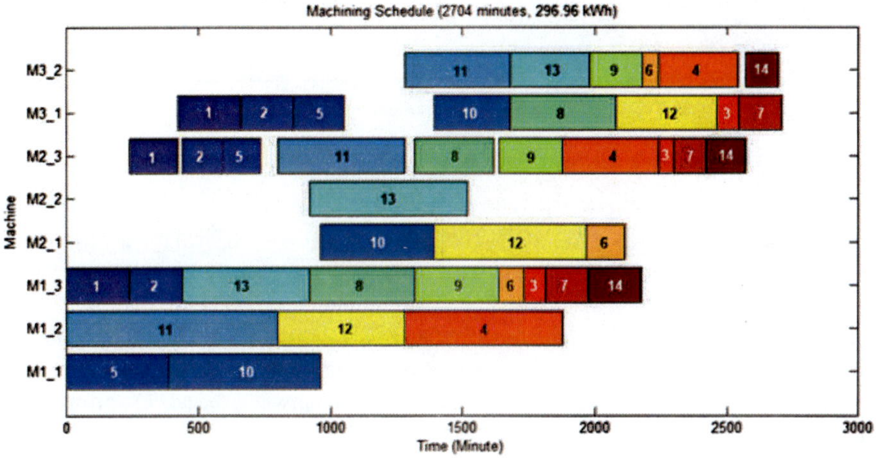

Fig. 2 Machining schedule with minimum fitness value by MFO

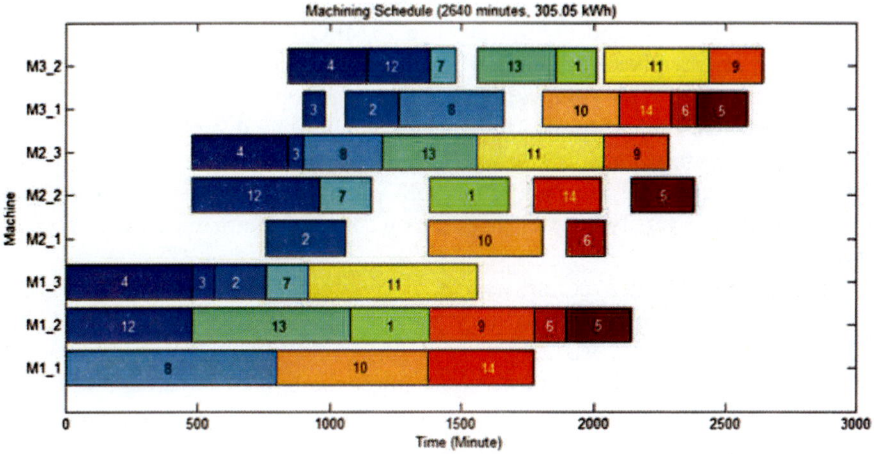

Fig. 3 Machining schedule with minimum C_{max} by PSO

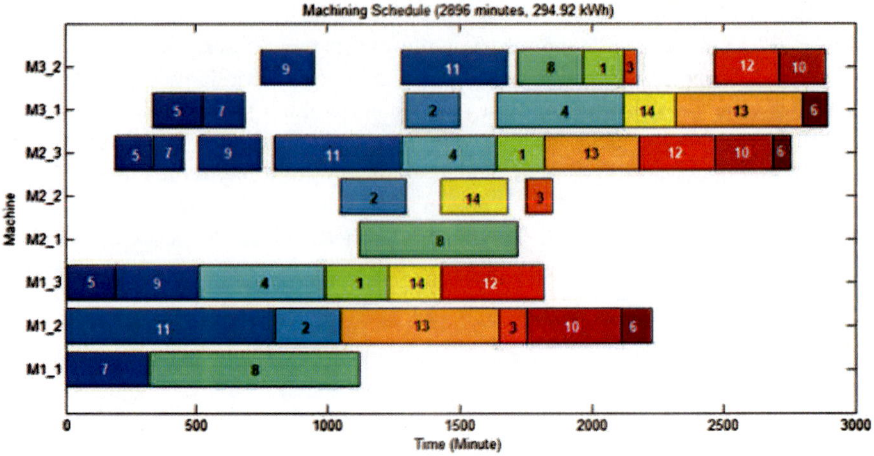

Fig. 4 Machining schedule with minimum *TEC* by GA

Although the $C_{\max}$ or *TEC* obtained by MFO was not the smallest, this result compromised between the $C_{\max}$ and *TEC* value. The MFO solution provided relatively small optimization objectives in both $C_{\max}$ and *TEC*. In comparison with PSO solution, although the $C_{\max}$ was the smallest, but the *TEC* was the highest among the solutions. In contrast, GA solution that provided the smallest *TEC* required the longest maskespan to complete all jobs. The different between the MFO optimization objectives and the smallest value were 2.4% and 0.7% respectively.

Figure 5 presents the mean convergence plot for all algorithms to optimize the case study problem. For this plot, the mean of convergence for 30 repetition were recorded. Based on the plot, it was found that the FA has the fastest convergence, whereby the slope was flattened roughly at 35 iterations. This is followed by WOA at 170 iterations. On the other hand, the convergence for MFO occurred until 270th iteration. For other algorithms like ACO and HHO, the convergence still occurred until end of iteration. This plot not only shows how fast each algorithm able to find the optimum solution, but also capability of algorithm to maintain the diversity of population. An algorithm with better diversity will have slower convergence and theoretically able to avoid trapped in local optimum. MFO able to compromise between faster convergence and diversity maintenance, since it converged until middle of maximum iteration.

Based on the results, superiority of MFO to optimize machine shop scheduling problem was proven by obtaining minimum fitness and also good $C_{\max}$ and *TEC* values. This performance was because of unique sorting flame mechanism in MFO. In MFO, the flame number will be sorted form the best to the worst, while the moth number is maintained throughout iteration. It means that for each iteration, the moth will be assigned to different flame. This mechanism made the search direction become diverse and exploration become more effective.

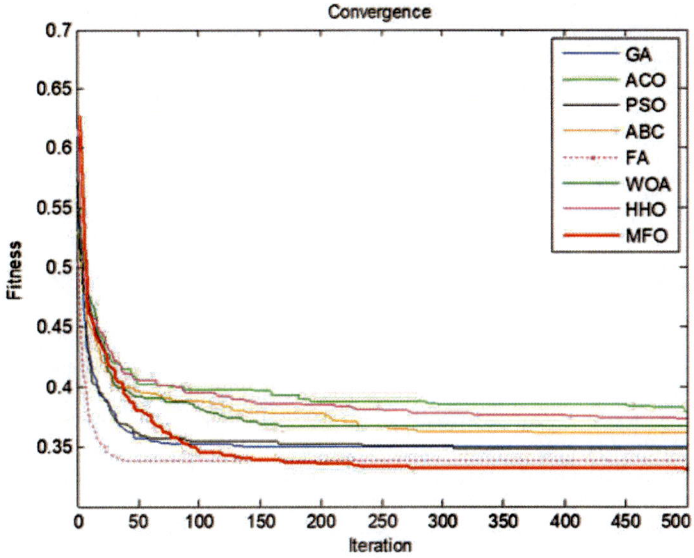

Fig. 5 Convergence plot for case study optimization

The production schedule obtained by MFO able to be implemented by the studied machine shop in order to minimize $C_{\max}$ and *TEC* at the same time. However, in the case that manufacturer feel individual $C_{\max}$ is far more important than *TEC*, schedule in Fig. 3 can be used. The HFS-EE optimization not only important to reduce the carbon emission, but also maintain the completion time for the jobs in the studied shop.

5 Conclusion

This paper optimizes hybrid flow shop scheduling with energy consumption (HFS-EC). A case study has been conducted in a machine shop to schedule the lathe, milling and deburring machines utilization. Since the machine model is dissimilar, the energy consumption for similar job processing will also be different. Moth flame optimization (MFO) algorithm has been implemented for optimization purpose.

Based on the optimization results, the MFO obtained minimum mean fitness and also minimum best fitness, compared with comparison algorithms including GA, PSO, ACO, FA, ABC, WOA and HHO. The result obtained by MFO compromise the total makespan and total energy consumption in the solution. Although there were different solutions with better individual optimization objective, these solutions sacrifice the importance of another objective. As a conclusion, the schedule proposed by MFO is the best because minimized both optimization objectives in a single schedule.

Acknowledgements The authors would like to acknowledge Universiti Malaysia Pahang for funding this research under research grant RDU190317.

References

1. Lu C, Gao L, Pan Q, Li X, Zheng J (2019) A multi-objective cellular grey wolf optimizer for hybrid flowshop scheduling problem considering noise pollution. Appl Soft Comput J 75:728–749
2. Schulz S (2019) A genetic algorithm to solve the hybrid flow shop scheduling problem with subcontracting options and energy cost consideration. Adv Intell Syst Comput 854:263–273
3. Schulz S, Neufeld JS, Buscher U (2019) A multi-objective iterated local search algorithm for comprehensive energy-aware hybrid flow shop scheduling. J Clean Prod 224:421–434
4. Jiang S-L, Zhang L (2019) Energy-oriented scheduling for hybrid flow shop with limited buffers through efficient multi-objective optimization. IEEE Access 7:34477–34487
5. Shi L, Guo G, Song X (2019) Multi-agent based dynamic scheduling optimisation of the sustainable hybrid flow shop in a ubiquitous environment. Int J Prod Res (in Press)
6. Yan J, Wen J, Li L (2014) Genetic algorithm based optimization for energy-aware hybrid flow shop scheduling. In: Proceedings of the 2014 international conference on artificial intelligence, ICAI 2014—WORLDCOMP, pp 358–364
7. Chen T-L, Cheng C-Y, Chou Y-H (2020) Multi-objective genetic algorithm for energy-efficient hybrid flow shop scheduling with lot streaming. Ann Oper Res 290:813–836
8. Zeng L-L, Zou F-X, Xu X-H, Gao Z (2009) Dynamic scheduling of multi-task for hybrid flow-shop based on energy consumption. In: 2009 IEEE international conference on information and automation, ICIA 2009, pp 478–482
9. Luo H, Du B, Huang GQ, Chen H, Li X (2013) Hybrid flow shop scheduling considering machine electricity consumption cost. Int J Prod Econ 146(2):423–439
10. Wang S, Wang X, Chu F, Yu J (2020) An energy-efficient two-stage hybrid flow shop scheduling problem in a glass production. Int J Prod Res 58(8):2283–2314
11. Zhang B, Pan Q-K, Gao L, Li X-Y, Meng L-L, Peng K-K (2019) A multiobjective evolutionary algorithm based on decomposition for hybrid flowshop green scheduling problem. Comput Ind Eng 136:325–344
12. Li J-Q et al (2020) Efficient multi-objective algorithm for the lot-streaming hybrid flowshop with variable sub-lots. Swarm Evol Comput 52 (in Press)
13. Geng K, Ye C, Dai ZH, Liu L (2020) Bi-objective re-entrant hybrid flow shop scheduling considering energy consumption cost under time-of-use electricity tariffs. Complexity 2020:8565921
14. Li M, Lei D, Cai J (2019) Two-level imperialist competitive algorithm for energy-efficient hybrid flow shop scheduling problem with relative importance of objectives. Swarm Evol Comput 49:34–43
15. Geng K, Ye C, Cao L, Liu L (2019) Multi-objective reentrant hybrid flowshop scheduling with machines turning on and off control strategy using improved multi-verse optimizer algorithm. Math Probl Eng 2019:2573873
16. Mirjalili S (2015) Moth-flame optimization algorithm: a novel nature-inspired heuristic paradigm. Knowl-Based Syst 89:228–249

Sustainability of Fertigation in Agricultural Crop Production by IoT System: A Review

Nur Syahirah Mohd Sabli, Mohd Faizal Jamlos, and Fatimah Dzaharudin

Abstract The increasing of world population over the year lead to an exponential in food demanding which is the main cause towards precision agriculture. Sustainability of agriculture production is an emerging and fascinating field of IoT, predictive analytics and classifier research. It helps farmers to enrich their fertigation business and support growth of local economy. This review describes a review of an integrated research field on current fertigation technology, IoT roles with various communication protocols in agriculture. Interestingly, (i) Agricultural crop with fertigation technology required different variables area e.g. electrical conductor (EC), Total dissolve solid (TDS) for different specification in particular crops and (ii) Particular type of wireless communication protocol of real-time data transfer of end-node sensors to gateway is depending on size of farm. It reveals that crops with fertigation technology advancement like IoT capable to sustain the productions even tough in adverse environmental conditions.

Keywords Fertigation · IoT · Lora

1 Introduction

Agriculture plays an importance role in sustaining food shortage and global economy. Agriculture crops has been cultivated for many years back as it holds the key element in various cuisine around the world. High cash value input of agricultural crops in agribusiness among farmers also contributed to extensive of agricultural crop planting

N. S. M. Sabli · M. F. Jamlos (✉) · F. Dzaharudin
College of Engineering, Universiti Malaysia Pahang, 26300 Kuantan, Pahang, Malaysia
e-mail: mohdfaizaljamlos@gmail.com

N. S. M. Sabli
e-mail: ninasyahirah2504@gmail.com

F. Dzaharudin
e-mail: fatimahd@ump.edu.my

A. F. Ab. Nasir et al. (eds.), *Recent Trends in Mechatronics Towards Industry 4.0*, Lecture Notes in Electrical Engineering 730, https://doi.org/10.1007/978-981-33-4597-3_9

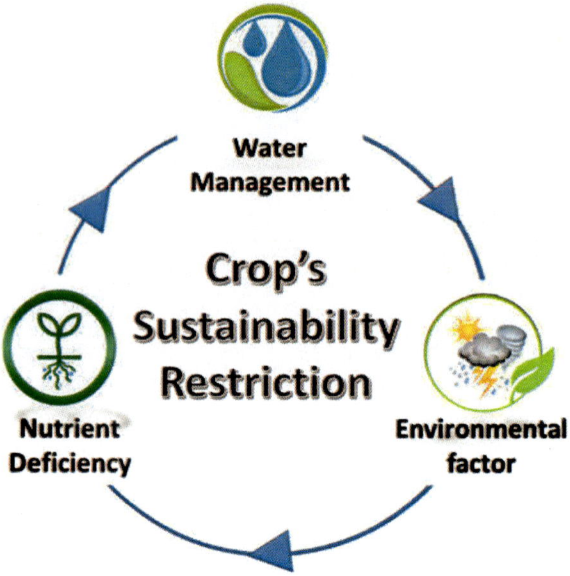

Fig. 1 The main factors restricts crop's sustainability [1]

[1]. Sustainability in agricultural production are very challenging task as main restriction input are deprived from variety factor (deficiency of nutrient, environmental changes and pest management) (see Fig. 1).

Among these three factors, deficiency of nutrient element (potassium, nitrogen, phosphorus) and problem in water management become major problem in ensuring the stability of crop's production in each of reap seasons. As an alternative, fertigation technique does emerge as a solution to tackle the nutritional and irrigation problem in plant. Fertigation technique considered as one of effective method to distributes evenly amount of water and fertilizer depending on different type of crops [2]. Nevertheless, the most common way to detect crop's nutritional status are by visualize the colour changes in plant but this is time consuming process and the result are not always accurate [3]. Thus, a variety of methods and technology for data monitoring and decision support system has been develop in effort to optimize crop yields in agriculture by emphasizing on factors that affect the growth of plants (irrigation, nutritional status etc.).

Precision agriculture is concept on supplying exact amount of resources on the exact duration of time [4]. The integration of Internet of Things (IoT) technology together with fertigation agriculture lead to precision agriculture in term of data collection and monitoring approaches to increase crop's yield while minimizing impact of excessive fertilizers and irrigation on environmental [5]. Data retrieval using via sensors gives better understanding and specific insight on physical environment (bio diversity, soil analysis and weather indicator) leading to the precise data measurement. The development of IoT data driven in agriculture sector together with variety of wireless technology had been explored for example, Bluetooth [6], Wi-Fi [7], ZigBee [8] and LoRa [9] which interestingly gives an options to farmers for farm management according to the size of farm. In many dimensions of the agricultural landscape, IoT-based solutions prove to be very helpful and these smart solutions could also be lush in smart irrigation with optimal water utilization [10]. Therefore, fertigation technology and implementation of IoT in crops are briefly discuss in Sects. 2 and 3.

2 Application of Fertigation Technique in Crops

This section exploring the details on fertigation applied in various crops with consideration of concept and system design together with fertilizer management in subsections 2.1 and 2.2.

2.1 Concept and System Design on Fertigation Technique

Fertigation system is on of popular agriculture technique of based on integration irrigation and fertilizer in one system for time controlling water management. In fertigation, fertilizers were dissolved in certain amount of water at the right time and sufficient quantity of nutrients provided and to the root zone to ensure maximum absorption of more crops per drop of water [2]. The advantages of fertigation system include (i) sufficient water and fertilizers to minimize nutrient loss and leaching (ii) costly chemical are applied efficiently (iii) reduce major cost of labour and machineries and (iv) Soil erosion minimization [11]. Drip water fertigation have more advantages compared to conventional fertigation system as the droplet water from drip nozzle ensure the wetness of soil and increasing the probability of nutrient absorption in soil [12].

Drip fertigation technique were specifically designed for specific crop, climatic and soil condition [13]. Drip fertigation architecture consist of several equipment such as fertilizer, water tank, dripper, timer, water pump and connected pipe line. Feeding of nutrient soluble to plants are by drippers. The dedicated drippers were connected to the main pipe line later connected to the main nutrient solution (combination of water and fertilizer) tank. As the dripper drop the nutrient solution according to the need of specific crops, the timer are used to control the supply amount of fertilizer a day [14]. Figure 2 shown the basic architecture of drip fertigation.

2.2 Fertilizer in Various Crops

In fertigation, specific amount and type of fertilizers are supply accordingly to the type of plant. This include of (i) type of plant and stage of growth (ii) irrigation system and soil type and (iii) quality of available water. The sensitivity of a plant

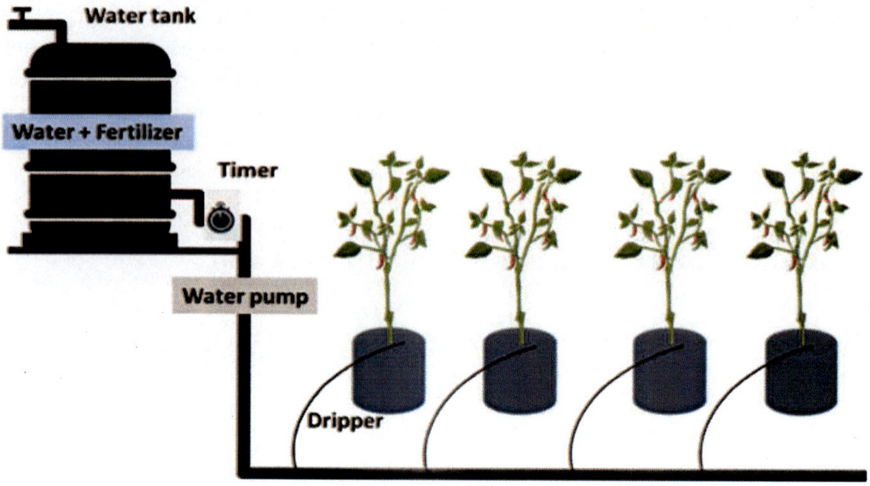

Fig. 2 The drip fertigation system [26]

to the forms of N should also be considered in varying physiological stages as most type of plants (roots area) are sensitive to the present of ammonia. Ammonia can cause damage to the root due to high root soluble sugar content. Phosphorus on the other hand is depending on pH level not to be diminished [12].

In Malaysia, AB fertilizer are extensively being used in fertigation system. AB fertilizer are composed into two types, Micro and macro nutrient. Most of the plants need macro nutrient more compared to micro nutrient. These two types of nutrients are originated from basic element of plant fertilizers known as NPK where N is Nitrogen, P is Phosphorus and K is Kalium/Potassium.

Fertilizer in fertigation system is validated trough three main indicators, ensuring the sufficient nutrient being supplied to the plants. The parameters are pH level, EC, and TDS measurement. These three indicators are greatly affected to the growth of plant fruit. EC indicates concentration of fertilizer in water and measured in μS (Siemens) [26]. Different crops required different concentration of fertilizer varying from planting seedlings to fruit maturity as shown in Table 1.

Table 1 Average EC readings for different stages of agricultural crops [15, 16]

Type of crops	Average EC measurement in each stage (μS)			Average pH readings	Plant cycle (days)
	Seedling	Middle	Fruit Maturity		
Chilli	1.8	2.0–2.8	3.0–3.2	5.5–6.5	75
Cucumber	1.6–1.8	2.0–2.4	2.4–3.0	5.5–6.5	60
Eggplant	1.8	1.8	2.0–2.4	5.5–6.5	80
Tomato	1.8	1.8	2.0–2.4	5.5–6.5	60–70
Rock melon	1.7	1.9–2.0	2.6–3.0	5.5–6.5	60–70

Though the EC are precisely measured, the efficiency of nutrient absorbance depend on pH value of crop's medium. Average pH requirement for agricultural crops varies from 5.5 to 6.5. TDS in other ways is to measure either fertilizer is completely dissolved or not and important factor in providing crops quality in term of fruit taste and firmness [17]. Mixing of more than two types of fertilizer can affect the solubility of fertilizer causing impractical of nutrient management [13]. Thus, agricultural crops using fertigation system require three crucial remote monitoring of variable, pH, EC, and TDS as it is substantially affect to the grown and maturity of agricultural crops.

Fertigation technique were proved as one of good solution in ensuring both water and fertilizer enough to cover up the growth factor depending on type of crops, however excessive water and fertilizer used in soil lead to bleaching and drainage problem that lead to increase of soil salinity and greatly affected the plant growth cycle. Thus, IoT based fertigation technique are used as initiative to control the amount of water and fertilizer supply in agricultural crops while wirelessly monitor condition of soil (soil moisture and soil temperature) and weather (temperature and humidity). Section 3 focusing on application of IoT in agriculture by deep review of architecture, sensor and wireless communication module used.

3 Overview of Internet of Things (IoT) in Agriculture

IoT based fertigation precision farming are enabling the wireless monitoring on desired variables such as soil moisture, soil temperature, water level and weather condition in crop's production while facilitate new era of modern digital technology with integrating sensors and communication standard. Radio Frequency (RF) transceiver, embedded device (microprocessor/microcontroller), sensors and power source (battery, solar panel) to operate as a complete monitoring system. Advantage of IoT are (i) simplification and major cost reduction in wiring and harness (ii) allowing impossible sensors deployment for extremely hard event monitoring (iii) speedy deployment and various sensors installation and (iv) mobility [18]. In agricultural crops fertigation system, IoT aiming to collect and monitor the growth of various crops variables for example; soil moisture, temperature, humidity, pH level of EC, pH, TDS and water level.

In this section, detail on architecture of IoT including with type of agriculture sensor and communication standard used as well as application of IoT in agriculture are explained in Sect. 3.1, subsections 3.1.1, 3.1.2 and Sect. 3.2.

3.1 Architecture of IoT

IoT hardware system has features of energy efficient, low cost processor, robust and reliable radio technology, long lifetime power source, various flexible I/O for various sensors and flexible open source development platform. IoT architecture comprise four major layers; sensing, network, analyzing and application [19]. Type of sensors, communication modules and related works of IoT implementation that have been done previously are discussed in next subsection.

3.1.1 Type of Agriculture Sensors

IoT in general view, consist of few end-nodes and base station [20], depend on networking layer for data transferring to platform layer and application platform. Sensing is a method used to collect data of real object and phase, including activities such as drop in temperature or pressure. Sensing activities activated by things called sensors which integrated with embedded device [21]. Different sensors for different applications were used according to the parameter or properties needed by the system and users. For fertigation in agricultural crops, several parameters such as soil moisture, nutrient status and water level in main tank are monitor trough sensors. Sensors selection play the important role in IoT based agriculture development in order to ensure the robustness, reliability, accuracy and common type of sensors used in agriculture. There are six factors to be consider when choosing sensors which are (1) precision and accuracy (2) identified environmental changes (3) signal conditioning (4) Response time (5) cost (6) size.

3.1.2 Revolutionary Communication Standard

RFID have been used as communication modules to collect data from sensors. RFID at the first place being developed for identification purpose that later growth more interest in possible application which led to the development of new range RFID outfit with sensor for wider range of applications [22]. Several work on application of RFID (semi passive/active tag) in agriculture such as soil temperature and salinity [23] and humidity [24] were proved to increase the productivity of plant and help farmers to manage to their plantation. Commonly, application of RFID in sensing environment used semi-passive tag or active tag rather than passive tag.

Integrating RFID technology in sensing world led to the development IoT standardize protocol for easy implementation of sensing application. This is due to installation complexity. Robustness guaranteed, cost and easiest installation has led Bluetooth as communication module in WSN with operating frequency 2.4 GHz [25]. Development of Bluetooth in IoT environment especially in agriculture productivity had been noticed such temperature [6]. Even though Bluetooth module have advantage in term of transmission reliability, however it is difficult to emerge as an alternative choice for further deployment of sensor network application due to its expansive maintenance cost and energy inefficient and relying on dual radio approaches.

The IEEE 802.15.4 standard is developed to provide a framework and lower power networks and cost of MAC and Physical layers, leaving the upper layers to be developed according to the market needs. IEEE 802.15.4 standard like ZigBee [26] are the most common to deploy and its outlined the criteria of the physical and MAC layers for Low-Rate Wireless Personal Area Networks (LR-WPAN) and basically used peer to peer and star topologies. ZigBee based 802.15.4 offer a flexible protocol, simple, short range operation, cheaper, reliable data transfer, easy installation and reasonable battery life. However, ZigBee technology is lacking in terms of security, prone to attack from unauthorized people and limited coverage [21].

Development of IEEE 802.15.4 standard (ZigBee) somehow show great interest in agriculture deployment such greenhouse [27] and storage house [28]. Since different application needs specific requirement in terms of capacity and range coverage, long power wide area (LPWA) are being introduced. Long range (LoRa) is a spread spectrum modulation technique which is based on 802.15.4 standard are the leading wireless communication module as efficient solution of connecting sensors which has features of better capacity, longer battery lifetime and lower cost [29].

3.2 Application of IoT as Monitoring System in Agriculture

IoT have great impact in agriculture sector by monitoring the most important parameters [30] as stated below:

- Environment changes (temperature, and humidity)
- Fertility condition (soil moisture, and soil temperature)
- Irrigation management (water level).

The wireless technology application does emerge in various sector such as, home appliances, vehicles and health including agriculture sector. In the earlier development of IoT technology, various wireless communication module such as, Bluetooth and ZigBee has been introduced. A number of researchers have reported developing IoT based platform including [30] who developed wireless irrigation system (WIS) for irrigating Romaine Lettuce (Lactuca Sativa) and Red Lettuce (Loollo Rosa) in Greenhouse using Bluetooth. The Romance Lettuce were plant in 64 bags. 32 of them were irrigated by WIS system and timer were set to pump water in every 15 minutes started at 7 am. Five sensing nodes were setup with Bluetooth as wireless communication at different location. As the result, the author reveals that the plant in test group grown higher and heavier compared to plant in normal irrigation system. The electricity and water consumption also were low than normal irrigation system with 60% efficiency.

While, [8] in other work develop a system that low cost and energy efficient frame work for precision agriculture in automated water supply for sweet melon using ZigBee approach. The system was placed at a pot of sweet melon plant and the senor node and BS station were setup in outdoor and indoor. The author only shown the graph on variation of temperature and humidity only but the details discussion on result were not provided. Meanwhile, using the same wireless communication standard, Ref. [31] present development of IoT system for real time temperature and humidity monitoring for Chongqing mountain citrus using ZigBee technology in China. The IoT system comprises of HA2002 (soil temperature), HA2001 (Soil moisture) and Fk-WS (air temperature and humidity) sensor that integrate on JN5139 module for data collection and transmission from node to base station. After a year of the implemented design in citrus orchard, author reveals that soluble slid content of fruit increase with 1–2%, orchard per mu yield increase 500 kg or more and 20% save in water and fertilizer management.

The trend of development IoT based platform keep increasing with new wireless technologies; LoRa, which focusing in increasing area of coverage and considerable amount of literature has been published on development of IoT with Lora technologies. These studies include [32] that develop IoT system for monitoring water level of trough in cow barn. Reference [33] discuss on work of development of Smart Mushroom House (SMH) using LoraWAN-IoT technology controlling environmental factor (temperature, humidity and Co2) that affect the growth of Shitake Mushroom. The SMH using controller are setup with humidifier and exhaust fan for controlling the wetness and c02 level in mushroom house. The transmission between SMH controller, humidifier and exhaust fan were established by LoraWAN wireless communication module with distance of 100 m of EN to Base Station (BS). Lora works better with open environment rather than in buildings (concrete walls). IoT Monitoring system indeed gives farmers deep insight on changes in factors (weather condition and soil condition) that affected crop's growth even in small changes.

4 Limitation and Challenge

As previously mentioned, the monitoring IoT system comprised of sensors and wireless communication modules located at the end-nodes as transmitting components while the gateway receiving all data from the nodes. Data transfer reliability is important as data loss from sensor nodes can cause the recidivism of crops growth. Position of sensor node and gateway play an important role in determine the reliability of conveying data to gateway. Furthermore, number of sensor nodes influenced the transmission of data, which greater nodes leads to higher possibilities of data collision between nodes and center station [32]. The communications modules have different standards and specifications, thus the user needs to identify the technical requirements at first e.g. for this review of various agricultural crop, requirements are based on the area of the farm and distance between node sensors and gateway.

Acknowledgement This project is supported in part by CREST P12C2-17 (UIC180804), RDU190349, FRGS/1/2019/STG02/UMP/02/4, UIC191205, UIC200814 and RDU202803.

References

1. Mariyono J, Sumarno (2015) Chilli production and adoption of chilli-based agribusiness in Indonesia. J Agribus Develop Emerg Econ Article Inf 5(2): 57–75
2. Sureshkumar P et al (2017) Fertigation-the key component of precision farming. J Trop Agric 54(2):103
3. Graeff S et al (2008) Evaluation of image analysis to determine the N-fertilizer demand of broccoli plants (Brassica oleracea convar. botrytis var. Italica). Adv Opt Technol
4. Dholu M, Ghodinde K (2018) Internet of things (IoT) for precision agriculture application. In: 2018 2nd international conference on trends in electronics and informatics (ICOEI). IEEE
5. Karaşahin M, Dündar Ö, Samancı A (2018) The way of yield increasing and cost reducing in agriculture: smart irrigation and fertigation. Turkish J Agric Food Sci Technol 6(10):1370–1370
6. Yue S et al (2010) The application of Bluetooth module on the agriculture expert system. In: 2010 2nd international conference on industrial and information systems, IIS, vol 1, pp 109–112
7. Nayyar A, Puri V (2017) Smart farming: Iot based smart sensors agriculture stick for live temperature and moisture monitoring using arduino, cloud computing and solar technology. In: Communication and computing systems—proceedings of the international conference on communication and computing systems, ICCCS 2016, 2017(Oct 2017), pp 673–680
8. Math RK, Dharwadkar NV (2017) A wireless sensor network based low cost and energy efficient frame work for precision agriculture. In: 2017 international conference on nascent technologies in engineering, ICNTE 2017—proceedings
9. Tapashetti S, Shobha KR (2018) Precision agriculture using LoRa. Int J Sci Eng Res 9(5):2023–2028
10. Sharma DK et al (2016) A priority based message forwarding scheme for opportunistic networks. In: IEEE CITS 2016—2016 international conference on computer, information and telecommunication systems, pp 1–5
11. Srivastava AK (2012) Advances in citrus nutrition. In: Advances in citrus nutrition, 9789400741, pp 1–477
12. Kafkafi U, Kant S (2005) Advantages of Fertigation introduction
13. Kafkafi U, Kant S (2004) Fertigation. In: Encyclopedia of soils in the environment, June 2004, pp 1–9
14. Fertigasi Cili: Penanaman Cili Menggunakan Sistem Fertigasi Terbuka—WNR Agro PLT [cited 01 June 2019]. Available https://wnragro.com/fertigasi-cili/
15. Mohd YS, Abd M (2013) Penanaman terung secara fertigasi (Planting eggplant using fertigation system), vol 3, pp 19–24
16. Yusoff K (2012) Penanaman cili menggunakan sistem fertigasi terbuka, vol 1, pp 1–8
17. Liu H et al (2013) Drip irrigation scheduling for tomato grown in solar greenhouse based on pan evaporation in North China plain. J Integr Agric 12(3):520–531
18. Wang N, Zhang N, Wang M (2006) Wireless sensors in agriculture and food industry—recent development and future perspective. Comput Electron Agric 50(1):1–14
19. Rad C-R et al (2015) Smart monitoring of potato crop: a cyber-physical system architecture model in the field of precision agriculture. Agric Agric Sci Proc 6:73–79
20. Jangra R, Kait R (2017) Principles and concepts of wireless sensor network and ant colony optimization: a review. Int J Adv Res Comput Sci 8(5)
21. Dargie W, Poellabauer C (2011) Fundamentals of wireless sensor networks: theory and practice, pp 1–311
22. Want R (2004) Enabling Ubiquit Sens RFID Title 37(4):84–86
23. Dey S et al (2016) Electromagnetic characterization of soil moisture and salinity for UHF RFID applications in precision agriculture. In: European microwave week 2016: "microwaves everywhere", EuMW 2016—conference proceedings; 46th European microwave conference, EuMC 2016, pp 616–619

24. Palazzi V et al (2019) Leaf-compatible autonomous RFID-based wireless temperature sensors for precision agriculture. In: 2019 IEEE topical conference on wireless sensors and sensor networks (WiSNet), vol 2, pp 1–4
25. Leopold M, Dydensborg MB, Bonnet P (2003) Bluetooth and sensor networks: a reality check. In: Proceedings of the first international conference on embedded networked sensor systems—SenSys'03, p 103
26. Standard I IEEE Standard for Part 15.4: wireless medium access control (MAC) and physical layer (PHY) specifications for low-rate wireless personal area networks (WPANs). In: Local and metropolitan area networks, pp 1–26 (2003)
27. Xue-fen W et al (2018) Smartphone accessible agriculture IoT node based on NFC and BLE, pp 78–79
28. Kumar S, Hiremath V, Rakhee K (2012) Smart sensor network system based on ZigBee technology to monitor grain depot. Int J Comput Appl 50(21):32–36
29. Zhang X et al (2017) Monitoring citrus soil moisture and nutrients using an IoT based system. Sensors 17(3):447
30. Hong GZ, Hsieh CL (2016) Application of integrated control strategy and bluetooth for irrigating romaine lettuce in greenhouse. IFAC-PapersOnLine 49(16):381–386
31. Zhang X et al (2017) Monitoring citrus soil moisture and nutrients using an IoT based system. Sensors (Switzerland) 17(3):1–10
32. Lukas WA, Tanumihardja, Gunawan E (2015) On the application of IoT: monitoring of troughs water level using WSN. 2015 IEEE conference on wireless sensors, ICWiSE 2015, 2016, pp 58–62
33. Nik Ibrahim NH et al (2018) LoRaWAN in climate monitoring in advance precision agriculture system. Int Conf Intell Adv Syst ICIAS 2018:1–6

DSRC Technology in Vehicle-to-Vehicle (V2V) and Vehicle-to-Infrastructure (V2I) IoT System for Intelligent Transportation System (ITS): A Review

Aidil Redza Khan, Mohd Faizal Jamlos, Nurmadiha Osman, Muhammad Izhar Ishak, Fatimah Dzaharudin, You Kok Yeow, and Khairil Anuar Khairi

Abstract Intelligent Transportation System (ITS) consisting of Vehicle Ad-hoc Networks (VANET) offers a major role in ensuring a safer environment in cities for drivers and pedestrians. VANET has been classified into two main parts which are Vehicle to Infrastructure (V2I) along with Vehicle to Vehicle (V2V) Communication System. This technology is still in development and has not been fully implemented worldwide. Currently, Dedicated Short Range Communication (DSRC) is a commonly used module for this system. This paper focuses on both V2V and V2I latest findings done by previous researcher and describes the operation of DSRC along with its architecture including SAE J2735, Basic Safety Message (BSM) and different type of Wireless Access in Vehicular Environment (WAVE) which is being labeled as IEEE 802.11p. Interestingly, (i) DSRC technology has been significantly evolved from electronic toll collector application to other V2V and V2I applications such as Emergency Electronics Brake Lights (EEBL), Forward Collision Warning (FCW), Intersection Moving Assist (IMA), Left Turn Assist (LTA) and Do Not Pass Warning (DNPW) (ii) DSRC operates at different standards and frequencies subject to the country regulations (e.g. ITS-G5A for Europe (5.875–5.905 GHz), US (5.850–5.925 GHz), Japan (755.5–764.5 MHz) and most other countries (5.855–5.925 GHz)) where the frequencies affected most on the radius of coverage.

A. R. Khan · M. F. Jamlos (✉) · N. Osman · M. I. Ishak
College of Engineering, Universiti Malaysia Pahang, 26300 Kuantan, Malaysia

M. I. Ishak
e-mail: mizhar@ump.edu.my

F. Dzaharudin
Department of Mechanical, Kuliyyah of Engineering, International Islamic University Malaysia, 53100 Jalan Gombak, Malaysia
e-mail: fatimahdz@iium.edu.my

Y. K. Yeow
School of Electrical Engineering, Faculty of Engineering, Universiti Teknologi Malaysia, 81310 Skudai, Malaysia
e-mail: kyyou@fke.utm.my

K. A. Khairi
VAT Manufacturing, Industrial Zone, 14100 Batu Kawan, Penang, Malaysia

A. F. Ab. Nasir et al. (eds.), *Recent Trends in Mechatronics Towards Industry 4.0*, Lecture Notes in Electrical Engineering 730, https://doi.org/10.1007/978-981-33-4597-3_10

Keywords DSRC · V2V · V2I

1 Introduction

Vehicular accidents increased rapidly over the years in worldwide. According to the Malaysia Ministry of Transportation, a total of 533,473 accidents involving 802,523 vehicles have been recorded in 2019 which caused 3310 severe injuries and 6740 death [1]. Accidents mostly occurs at intersection and adjacent lanes which are due to several factors such as poor observation and conditions of the surrounding area [2]. About 40% of possibility that crashes occurred at intersection [3]. Even though the safety improvements have been upgraded such as airbags, seatbelts and other vehicle passive safety technologies, yet vehicular accidents still stay on its critical endangering path. Therefore, Vehicular ad-hoc network (VANET) has been introduced to overcome this issue. This system plays a major role as part of Intelligent Transportation Systems (ITS) worldwide [4]. Nowadays, ITS is attracting remarkable investment from governments, academy, and industries in applications development on vehicles and road safety infrastructure [2]. VANET is basically a distributed self-organizing communication formed by vehicles [5, 6]. The main purpose of VANET is to ensure the safety of the driver, passenger and pedestrians [7, 8]. However, there is still room for improvement in this network.

VANET is usually classified into two main components which is Vehicle-to-Vehicle (V2V) Communication and Vehicle-to-Infrastructure (V2I) Communication [9]. In V2V communication technology, information such as heading, speed, position and brake status is exchanged among vehicles in order to help and prevent crashes between the surrounding vehicles by alerting the drivers of dangerous situation [10]. Meanwhile, V2I communication is a system where the vehicle is sending and receiving information from the infrastructure modes such as streetlights, building, and traffic light [10]. V2I acts as a wireless medium to provide real-time environmental conditions such as rain, humidity and haze from transportation management system to the vehicles. The communication is achieved through the information exchange among the On Board Unit (OBU), which is integrated inside each vehicle for V2V and Road Side Unit (RSU) which is a device installed on infrastructures for V2I Communication System.

Nowadays, GPS technology is integrated inside V2V/V2I devices to provide neighboring vehicles position. This includes information by vehicles at the blind spots or vehicles that was hidden from view [11, 12]. Hence, if there is a possibilities of accidents in the hidden sight area, the driver will be alerted by other vehicles and able to react and change the current driving mode in order to avoid dangerous situation and accidents. There are multiple ways for vehicles and infrastructures to communicate with each which will be discussed in the next section.

2 Dedicated Short Range Communication (DSRC)

Dedicated Short Range Communication (DSRC) technology is a wireless communication module specially developed for V2V and V2I communication [13]. DSRC was previously used in public safety system and also in Electronic Toll Collection (ETC). Table 1 shows evolution of DSRC which is recently used in V2V and V2I applications. DSRC provides a high speed data transfer rate with low communication delay in wireless communication service [14]. The performance of DSRC is similar with WiMax, but it provides lower cost and lower realization complexity compared to WiMax, as depicted by Table 2. DSRC is a low latency medium which is adapted for high mobile vehicle environment [15, 16]. This technology is one of the wireless technologies used to exchange information among other devices and lies in 5.9 GHz of operating frequency with IEEE 802.11p protocol [17]. In the 5.9 GHz frequency band, DSRC application has allocated 75 MHz spectrum for US [18] and 30 MHz spectrum for Europe [19]. The spectrum allocation for other countries is shown in Table 3. DSRC consists of two types of channel which are service channel (SCH) and control channel (CCH) that consist of six and one spectrum respectively such in [20, 21].

Table 1 Evolution of DSRC [22]

	Past DSRC	New DSRC
Frequency	915 MHz	5.9 GHz
Distance	< 30 m	1000 m
Data rate	0.5 Mbps	6–27 Mbps
Application	Electronic toll collector (ETC) and other applications	General internet access and electronic toll collector (ETC)
License	Single unlicensed channel	7 licensed channels
Software	Requires special chip set and software	Using open off-the shelf chip set and software
Protocol	Command-response	Command-response peer to peer

Table 2 Comparison between DSRC and other wireless module [23]

	DSRC	Wi-Fi	GSM	WiMax
Delay	< 50 ms	1 s	1 s	–
Mobility (mph)	>60	<5	>60	>60
Transfer rate (Mbps)	3–27	6–54	<2	1–32
Distance	< 1000 m	< 100 m	<10 km	15 km
Communication channel bandwidth (MHz)	10	20	<3	<10
Communication band	5.86–5.925 GHz	2.4, 5.2 GHz	800, 1.9 MHz	2.5 GHz
IEEE Standards	802.11p (WAVE)	802.11a	–	802.16e

Table 3 Spectrum allocation for its based on countries [13, 24]

Country	Spectrum (GHz)	Allocated bandwidth (MHz)
Australia	5.855–5.925	70
China	5.905–5.925 (trials)	20
Europe	5.855–5.925/5.875–5.905 (safety applications)	70/30
Japan	0.7555–0.7645 (safety applications) 5.770–5.850	9/80
Korea	5.855–5.925	70
Singapore	5.875–5.925	50
USA	5.850–5.925	75

Wireless Access in Vehicular Environments (WAVE) is a mode of operations based on IEEE 802.11 protocol in DSRC band [14]. IEEE 802.11 is used to provide reliable communications protocol includes Wi-Fi services [25]. This IEEE 802.11p, which is dedicated for V2V/V2I communication, is difference compared to other 802.11 protocols by integrating specific sub-protocols of IEEE 1609.1, IEEE 1609.2, IEEE 1609.3 and IEEE 1609.4 [26, 27].

IEEE 1609.1 is a Resource Manager to control resources and data flows [28]. It describes the main components of WAVE system architecture and how to deal with command message and data storage format to ensure that it can specify the types of components supported by the OBU. IEEE 1609.2 on the other hand is a Security Services dedicated to application and Management Messages [28]. It provides the process of message format and deals with circumstances of using secure message exchanges [27].

3 Implementation of DSRC in Research

Author in [29] conduct an experiment to investigate the performance of Vehicular Communication using Bluetooth sensor networks in terms of speed, response time, distance and RSSI. The experiment is conducted in the circuit of Cataluya where Bluetooth device is placed at 8 m of infrastructure and mobile devices is placed inside F30 Ferrari. The mobile device is controlled using laptop integrated with GPS to obtain speed and location of the host vehicle. 3 dBi omnidirectional is used in this system and line of sight distance is fixed to 1 km approximately. The vehicle is moving with speed ranging from 160 up to 250 km/h. The average connection time obtained in this work is about 230 ms. The experiment is then carried on a highway where there could be an interference to the transmission process with the occurrence of buildings and larger vehicles surrounding the target vehicles. The result shows that the respond time is almost the same as the circuit experiments which is 200 ms and the coverage area improves up to 2 km due to the higher placement of the Bluetooth devices. In this work, the latency obtained which is 200 ms is high compared to

DSRC which is about 100 ms. Therefore, Bluetooth is not suitable for V2V/V2I development as DSRC provides better performance.

Meanwhile studies on data broadcasting in V2V using visible light communication (VLC) was presented in Ref. [30]. In this technology, communication is retrieved through low cost LED laser and photodiode receiver. The author developed a custom Vehicle-to-Vehicle Visible Light Communication (V3LC) research platform using Arduino microcontroller. It is used to evaluate the V3LC system performance by measuring the noise and interference of the visible light. The evaluation of the author's prototype provide evidence that the system can successfully detect potential risks in advance and generate warnings to the driver in real-world situations. Result shows that this system is capable of achieving bit rate up to 111.607 kbps at maximum distance of 1.5 m. This development only focuses on the fundamental development and it is not suitable to be tested inside vehicles as the range does not meet DSRC requirements.

Reference [31] presents a V2V communication using Visible Light Communication (VLC) in fog. During fog conditions such in Fig. 1, drivers usually switch on the headlights, fog lamps or both. Communication is achieved by sending data through the red LED of car A's taillight to car B while fog lamp or headlight by car C is used to transmit data to car B.

Author in [32] developed a system using cameras combined with V2I communications system for vehicle safety applications at the junction or road intersections. This system aimed to generate information to alert drivers at intersections in order to avoid collisions. The system is developed using a low-cost camera unit with image processing technology for V2I communication. When an obstacle or an object such as pedestrian or vehicle is detected by the system at the collision path, alert will be generated to nearby vehicles.

This system uses image processing technique and algorithm to calculate and verify the possibilities of collisions. The warning is generated in the form of audio and even visual which is integrated inside the vehicle system. This system is deployed at unmonitored intersections or junction where all activities from all directions can

Fig. 1 Fog condition pictured in [31]

be monitored as shown in Fig. 2. In the first scenario, car 1 is turning left with no vision of crossing pedestrian. For the second scenario, car 3 and bike 4 are both going through the intersection without realizing the presence of vehicle 2. Warning messages will be transmitted to all the vehicles according to the captured images by the camera.

Author in [33] uses an IoT technology to demonstrate a V2V communication. ZigBee with 2.4 GHz and IEEE 802.15.4 protocol has been used in this development. 2dBm X-Bee nodes were tested inside floor of and seat of Sedan as shown in Fig. 3. Two people were presented inside the car during the test. Another X-Bee node is placed inside another car within the range of 2.74 m from the static car. Path loss and packet drop rate are measured. The test is then repeated using 18 dBm with 5 m of the distance between the vehicles.

From these experiments, it can be observed that the connectivity is lost at 8.22 m using 2 dBm X-Bee in the car floor. The signal blockage from inside the vehicle is too high for low power nodes. However, the signal blockage is minimal when the nodes were placed on the seat of the car. The path loss is also lower with the usage of high power node in the measurement. From this work, it can be observed that X-Bee can support limited range for vehicle communications compared to DSRC that can

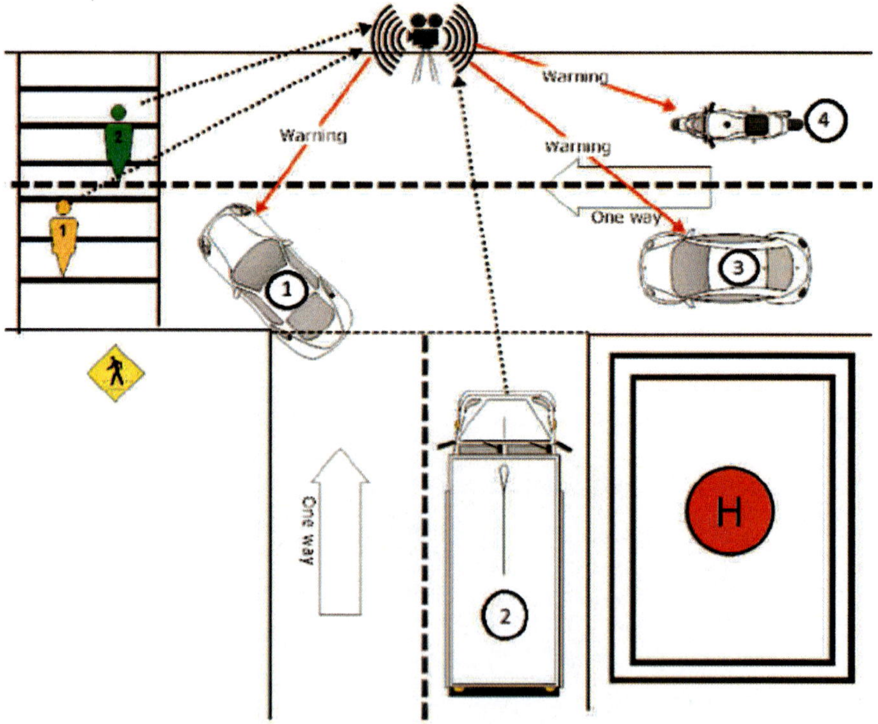

Fig. 2 Typical unmonitored intersection [32]

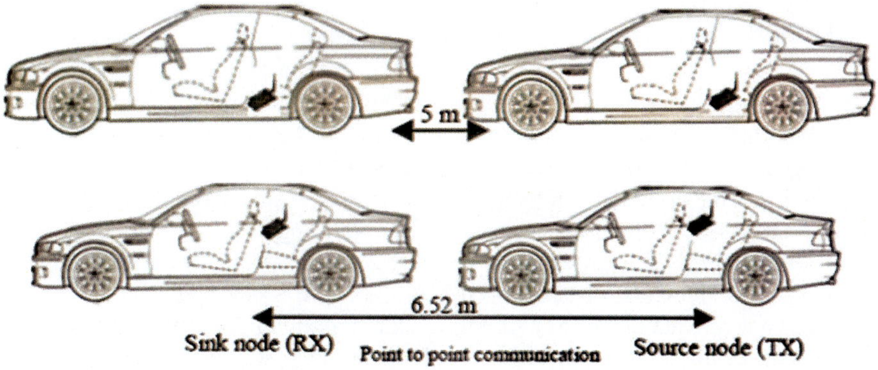

Fig. 3 Measurement setup in [33]

support at least 100 m. Therefore, it can be concluded that X-Bee is not suitable for real implementations of V2V/V2I communication system.

A system to enhance driving safety which can predict collisions is called Visibility-based Collision Warning System (ViCows). It can be characterized into two parts which is physical part and cyber part as shown in Fig. 4. Physical parts consist of DSRC module, camera, GPS and Electronic Control Unit (ECU) while cyber parts consist of four models which is headway distance prediction, velocity prediction, prediction horizon estimation and rear-end collision warning. Velocity prediction model is used to predict the future velocity of subject vehicle (SV) and leading vehicle (LV).

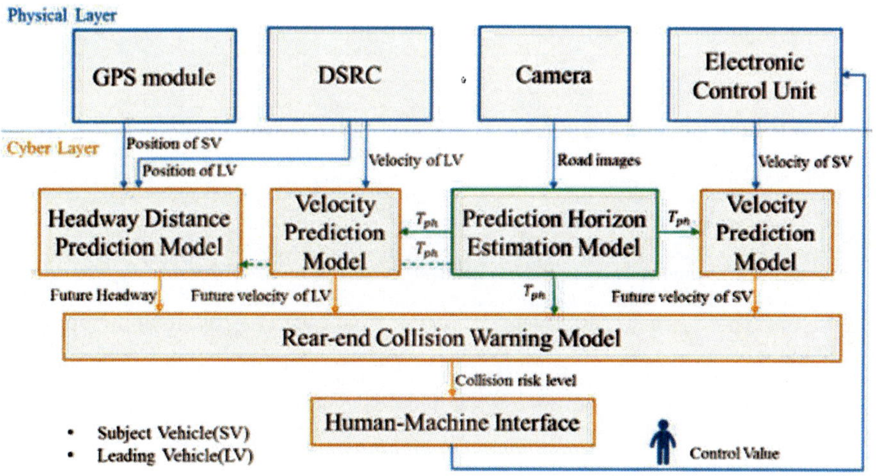

Fig. 4 Visibility-based collision warning system architecture developed in [34]

4 Conclusion

From this study, it can be concluded that V2V and V2I communication is one of the most important technology in the current era of IoT. V2V/V2I has been implemented by using several types of techniques using variety of wireless technologies that exist to ensure that the collision can be avoided. However, results from previous studies shows that DSRC technology contributes a huge impact on the vehicular safety applications in the ITS. This technology exchanges encrypted information at 70 MHz along 5.855–5.925 GHz (different according to countries) with interval of 10 MHz of each channel. Data rate of 512 Kbps is usually required in this radio but it may be different according to the application developed in the system. Previous test proved that DSRC can support up to 200 m NLOS detection and the performance is optimized up to 100 m. The latency of most applications achieved is about 100 ms. To solve hidden nodes issue in DSRC, engineers are now trying to combine DSRC with cellular for V2V/V2I applications in the next generations. This future development may also improve the NLOS detection and latency in this system for more accidents reduction involving vehicles and safe more lives. The review finding shows that implementation of V2V/V2I communication using DSRC in ITS may reduce the possibilities of accidents involving vehicles [27, 35, 36]. However, there are still flaws in this technology such as packet collision problems and intolerable latency. Advancement of time delay in DSRC module will make the V2V/V2I technology ready for vast implementations. Moreover, the enhanced V2V/V2I will support the autonomous vehicles technology to become much safer.

Acknowledgements This project is supported in part by CREST P12C2-17 (UIC180804), RDU190349, FRGS/1/2019/STG02/UMP/02/4, UIC200814 and RDU202803.

References

1. Statistik pengangkutan Malaysia 2019. http://www.mot.gov.my/my/sumber-maklumat/statistik-tahunan-pengangkutan (2019)
2. Ansari K, Feng Y, Tang M (2015) A runtime integrity monitoring framework for real-time relative positioning systems based on GPS and DSRC 16(2):980–992
3. Wu X et al (2013) Vehicular communications using DSRC: challenges, enhancements, and evolution. IEEE J Sel Areas Commun 31(9):399–408. https://doi.org/10.1109/JSAC.2013.SUP.0513036
4. Ma T (2016) Improved scheme for the problem of anti-fading of DSRC systems in IEEE 802.11p environments, vol 10, pp 632–640. https://doi.org/10.1049/iet-com.2015.0613
5. Sataraddi MJ, Kakkasageri MS (2017) BDI agent based dynamic routing scheme for vehicle-to-vehicle communication in VANETs, pp 735–740
6. Wang L, Iida RF, Wyglinski AM (2017) Performance analysis of EDCA for IEEE 802.11p/DSRC based V2V communication in discrete event system
7. Memon A, Shaikh FK, Felemban E (2015) Experimental evaluation of vehicle-to-vehicle based data transfer, July 2015. https://doi.org/10.1109/ICTRC.2015.7156475

8. Yasser A, Zorkany M, Abdel Kader N (2017) VANET routing protocol for V2V implementation: a suitable solution for developing countries. Cogent Eng 4(1):1–26. https://doi.org/10.1080/23311916.2017.1362802
9. Noori H (2013) Impact of VANET-based V2X communication using IEEE 802.11p on reducing vehicles traveling time in realistic large scale urban Area, pp 654–661. https://doi.org/10.1109/ICCVE.2013.161
10. Jawhar I, Mohamed N, Usmani H (2013) An overview of inter-vehicular communication systems, protocols and middleware, Dec 2013. https://doi.org/10.4304/jnw.8.12.2749-2761
11. Biddlestone S, Member S, Redmill K, Member S, Miucic R, Özgüner Ü (2012) An integrated 802.11p WAVE DSRC and vehicle urban (LOS and NLOS) propagation models, vol 13, no 4, pp 1792–1802
12. Hafeez KA, Anpalagan A, Member S, Zhao L (2016) Optimizing the control channel interval of the DSRC for vehicular safety applications, vol 65, no 5, pp 3377–3388
13. The V2X (vehicle-to-everything) communications ecosystem: 2019–2030—opportunities, challenges, strategies and forecasts
14. Kenney JB (2011) Dedicated short-range communications (DSRC) standards in the United States. Proc IEEE 99(7). https://doi.org/10.1109/JPROC.2011.2132790
15. Leonard K Ds-rc (dsrc) sp
16. Syahroni N, Suparno HW, Budiman H, Umam RS, Puspitorini O (2015) Performance evaluation of VANET docking guidance for AUV using DSRC, vol 2015. ICAMIMIA, pp 147–150
17. Bansal G, Kenney JB, Rohrs CE (2013) LIMERIC : a linear adaptive message rate algorithm for DSRC congestion control, vol 62, no 9, pp 4182–4197
18. Li YJ (2015) An overview of the DSRC/WAVE technology
19. Pressas A, Sheng Z, Ali F, Tian D, Nekovee M (2017) Contention-based learning MAC protocol for broadcast vehicle-to-vehicle communication, pp 263–270
20. Vivek N (2014) WAVE protocol stack for V2V and V2I communication, no 978
21. Ho K, Kang P, Hsu C, Lin C (2010) Implementation of WAVE/DSRC devices for vehicular communications, pp 1–4
22. Weigle M (2008) Standards
23. Teng XH (2014) The application of image processing technology in the intelligent transportation system. Appl Mech Mater 543–547:2678–2680. https://doi.org/10.1049/cp.2009.1941
24. Abboud K, Omar HA, Zhuang W (2016) Interworking of DSRC and cellular network technologies for V2X communications: a survey, vol 65, no 12, pp 9457–9470
25. Nwizege KS, Bottero M, Mmeah S, Emmanuel D (2014) Vehicles-to-infrastructure communication safety messaging in DSRC. Proc Proc Comput Sci 34(DPNoC):559–564. https://doi.org/10.1016/j.procs.2014.07.070
26. Peden M, Sleet D, Mohan D (2020) Vehicular communications standards 2.1, pp 13–34. https://doi.org/10.1007/978-3-319-47244-7
27. Ahmed SAM, Ariffin SHS, Fisal N Overview of wireless access in vehicular environment (WAVE) protocols and standards
28. Ryu M, Cha S, Cho K (2011) DSRC-based channel allocation algorithm for emergency message dissemination in VANETs, pp 105–112
29. De Cerio DP, Valenzuela JL (2015) Provisioning vehicular services and communications based on a Bluetooth sensor network deployment, pp 12765–12781. https://doi.org/10.3390/s150612765
30. Gupta R, Sharma A (2014) Vehicle-to-vehicle data broadcasting through visible light communication. IOSR J Electron Commun Eng 9(2):90–95
31. Kim YH (2015) Experimental demonstration of VLC-based vehicle-to-vehicle communications under fog conditions experimental demonstration of VLC-based, vol 7, no 6. https://doi.org/10.1109/JPHOT.2015.2499542
32. Vishnu S, Ramanadhan U, Vasudevan N, Ramachandran A (2015) Vehicular collision avoidance using video processing and vehicle-to-infrastructure communication, pp 387–388. https://doi.org/10.1109/ICCVE.2015.36

33. Oraibi I, Otero CE, Olasupo TO (2017) Empirical path loss model for vehicle-to-vehicle IoT device communication in fleet management, pp 8–11
34. Visibility conditions (2018). https://doi.org/10.3390/s18093026
35. Tubbene H (2015) Performance evaluation of V2V and V2I messages in C-ITS, June 2015
36. Tian D, Luo H, Zhou J, Wang Y, Yu G (2013) A self-adaptive V2V communication system with DSRC. https://doi.org/10.1109/GreenCom-iThings-CPSCom.2013.271

Real-Time and Predictive Analytics of Air Quality with IoT System: A Review

Nurmadiha Osman, Mohd Faizal Jamlos, Fatimah Dzaharudin, Aidil Redza Khan, You Kok Yeow, and Khairil Anuar Khairi

Abstract Environmental pollution particularly due to the emission of combustible gas from industry, haze, and vehicles, that has always been a major concern. Continuous monitoring of the air quality is hence essential to ensure early precaution or preventive measure can be taken in eliminating potential health risk which may be done via Smart Environmental Monitoring system with the Internet of Things (IoT), which is cost-effective and efficient way to control air pollution and curb climate change, IoT applications along with Machine Learning(ML) can make the data prediction in real-time. ML can be used to predict the previous and current data obtained by sensors. This review describes the existence of an integrated research field in the development of the environmental monitoring system and ML method. The findings of this review interestingly show that (i) various communication module is used for environmental monitoring system. (ii) Very less integration of IoT together with predictive analytics, it is separately to study for air pollution monitoring system. (iv) Data analytics for Air Pollution Index (API) prediction along with IoT, with various communication protocols can assist in the development of real-time, and continuous high precision environmental monitoring systems. (v) Machine

N. Osman · M. F. Jamlos (✉) · F. Dzaharudin · A. R. Khan
College of Engineering, Universiti Malaysia Pahang, 26300 Kuantan, Malaysia
e-mail: mohdfaizaljamlos@gmail.com

N. Osman
e-mail: nurmadihabintiosman@gmail.com

F. Dzaharudin
e-mail: fatimahd@ump.edu.my

A. R. Khan
e-mail: aidilredza0412@gmail.com

Y. K. Yeow
Faculty of Engineering, School of Electrical Engineering, Universiti Teknologi Malaysia, 81310 Skudai, Malaysia
e-mail: kyyou@fke.utm.my

K. A. Khairi
VAT Manufacturing, Industrial Zone, Batu Kawan, 14100 Penang, Malaysia
e-mail: anuar144@yahoo.com

A. F. Ab. Nasir et al. (eds.), *Recent Trends in Mechatronics Towards Industry 4.0*, Lecture Notes in Electrical Engineering 730, https://doi.org/10.1007/978-981-33-4597-3_11

Learning (ML) Regression algorithm is suitable for prediction and classification of concentration gas pollutant, while ANN and SVM algorithm is used for forecasting.

Keywords Environmental pollution · Monitoring system · Machine learning

1 Impact of Air Pollution on the Health

The air pollution impact is a serious health concern which can suffer people from heart or lung disease effect, and each pollutant gases also have their level of potential health effect. There are two types of pollutant gases [1], which is primary pollutant such as Sulphur Dioxide (SO_2), Carbon Monoxide (CO), Carbon Dioxide (CO_2), Particulate Matter (PM), and secondary pollutant like Ozone (O_3) and Nitrogen Dioxide (NO_2), each gases health effect has been discussed as shown in Table 1. From each pollutant gases will determine the Air Pollutant Index (API) which gives the status of quality of the air in the range of 1–500, the lower the API reading, the air quality is better, and the range of 0–50 ppm indicate good air quality in blue color, and above 300 ppm shows very unhealthy air quality. [2]. The API description as Table 2.

Table 1 Gas pollutants and health effects

Pollutant	Health Effects
CO_2 [3, 4]	Heart rate increases and headaches. It shows that air condition is unusual with 2000–5000 ppm gases high level. Toxicity or oxygen deprivation could occur. This is the permissible exposure limit for daily workplace exposure
CO [5–7]	Level of CO more than 10 ppm can increase in heart disease deaths and heart failure
SO_2	Bronchoconstriction high risk factors can cause asthma symptoms is increase
PM	Premature death can happen with heart and lung diseases, coughing, and difficult to breathe because of asthma effect
O_3 [8]	At level, more than 99 ppb can trigger chest pain, coughing, asthma, emphysema, and throat irritation
NO_2 [4, 9]	Could produce some negative effect in health like asthma and other respiratory problems, and at 5 ppm level can cause serious health problem

Table 2 API level indicator [10]

API level	Status	Color
0–50	Good	Blue
51–100	Moderate	Green
101–200	Unhealthy	Yellow
201–300	Very unhealthy	Orange
300–500	Hazardous	Red

2 Applications of Internet of Things (IoT) in Air Pollution Monitoring System

Based on Fig. 1 Architecture of air quality system, the environmental monitoring system is mainly composed of 3 layers, (i) perception layer which involved the field of sensor network, (ii) network layer, which data transmission or communication technologies involved and (iii) application layer, involved the predicting or forecasting of air pollution [11]. The sensor network is the basic function in the environment monitoring system to sense and measure air quality parameters. Sensor network also can be categorized to three types, Static Sensor Network (SSN), Community Sensor Network (CSN) and Vehicle Sensor Network(VSN). Usually, sensor network is comprised of sensor modules such as gas sensor, and the common types of gas sensors used such as electrochemical, infrared, and semiconductor gas. The manufacturer that supply semiconductor gas such Figaro, Synkera, and Sencera for sensors such CO, CO_2, and NO_2. The gas sensors is embedded with a microcontroller such as Arduino or microprocessor such as RaspberryPi as a platform that used suitable languages such as python, C, C++ and Java. As the speedy improvement of communication technology in data transmission, remote sensing, and network technology, many research utilized IoT in the air pollution monitoring system includes GSM and GPRS. Be that as it may, these modes are surprising expenses in both installation and maintenance and complexity, with existing of new communication technology in IoT it makes the environment monitoring system more liability, simplicity, and low cost [12].

IoT is a deployment of multiple devices equipped with sensors to perform a measurement process. Each sensor may use various protocols such as Wi-Fi use IEEE 802.11 (WLAN) protocol [13], IEEE 802.15 WPAN, ZigBee, 6LoWPAN, and Wireless HART use IEEE 802.15.4 LR-WPANS, Low Power Wide Area Network

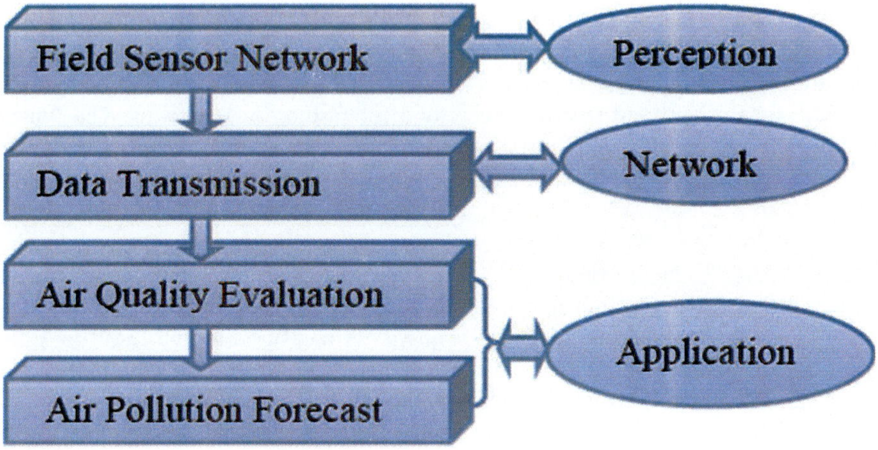

Fig. 1 Architecture of air quality system

(LPWAN) technology, like LoRa, Sigfox, NB-IoT, and LTE-M to connect with the gateway, to any system or cloud storage such as Thingspeak, ubidots-cloud service, cayenne, any mobile app, and web page as end-user platform to monitor the air quality easily in real time. Then the air quality data such as the concentration of gases can be used for predictive analytics, and air pollution forecast [14].

Smart air monitoring system in real-time with emergency alert has been developed by [15] which is being implemented in Dhaka, Bangladesh. This system is developed using gas sensors which consist of MQ-2, MQ-3, and MQ-7, Intel Edison board as microcontroller, buzzer and LED. The gas sensors are connected to the microcontroller. Microcontroller is then linked to the internet through Wi-Fi to identify its IP address. the developed device is placed in the kitchen near to the gas jar that is used for cooking. The result shows that the CH_4 level in the kitchen is harmful and precaution procedures must be taken. The buzzer alarm is triggered when the system detects a danger level of data.

In 2016, a real-time cognitive Wireless Sensor Network (WSN) system for indoor environment CO_2 monitoring has been developed in [16]. This system is implemented in researcher offices, laboratory, and meeting room This work focuses on four requirements which is the types of gas sensors, number of sensors, real-time data aggregation, and energy consumptions. In this development, CO_2 sensors combined with radio modules transmit the data towards the control room. A smoothing algorithm is used to find outliers and eliminate it from the measurement. The node will not transmit the data if a random spike occurs within the window time, which is 6 s. The result shows that the concentration of CO_2 increases when the number of people increases. The concentration of CO_2 reported is around 580 ppm up to 640 ppm when there is a presence of people in the office.

In 2017, ref. [17] developed a portable platform of monitoring system based on Unmanned Aerial Vehicle (UAV), with the integration of sensor board and communication module to monitor air quality in indoor environment, along with a mobile. The developed monitoring system has acquisition of data sensor of PM and other components of air chemical mounted to the UAV with a communication link on the ground base station. In this work, The UAV platform is integrated and mounted by a monitoring sensor by wireless transmitter to reach a flexible communication system, then the data is sent and stored in the cloud for further data analysis. The sensor board contains six calibrated sensors which are CO, $PM_{2.5}$, SO_2, O_3, PM_{10}, and NO_2 was attached to one box. The monitoring box is then attached to the lowest of the UAV. There are more or less some differences among the data of UAV monitoring compared with the ground base station. This happened when there is a larger wind occurs during the monitoring flight. The developed UAV environment monitoring system is then being used for IoT Smart Home application to measure indoor $PM_{2.5}$ and CO_2 concentration levels. If the air quality is measured by the UAV does not meet the air quality standard, air cleaner will be switch on at corresponding power level to eliminate $PM_{2.5}$ and if the UAV measured a high concentration level of CO_2, the ventilator will be opened for corresponding work time. The result shows that through the mutual-coupling working mode, the quality of air can be controlled with the ventilators and air cleaners.

In 2018, the author in [18] studies on air pollution in Bangkok, Thailand using smart sensors on low power wide area network (LPWAN) via NB-IoT. The data was sent via NB-IoT network-based LPWAN. Linkit Smart 7688 microcontroller is used to control the system. NB-IoT system has low power consumption, cost, and complexity in transceiver design, the coverage is also enhanced. IEEE 802.15.4 K is a wireless LAN standard for LPWAN technology, and the advantages of LPWAN are high coverage area, long-distance, consume less power, and less throughput things. Five type sensors have been deployed, (i) MQ131ozone (O_3) gas sensor, it operates at 5 V DC voltage. (ii) noise sensor to detect noise in dB. The function of the sensor is to observe the level of sound parameters and detects the intensity of ambient sound in the city. The sensor can be used to detect 30 up to 100 dB of noise. (iii) carbon dioxide (CO_2) sensor which operates at 6 V DC voltage with LM393 as the main chip. From 0 ppm to 10,000 ppm is the maximum level of CO_2 sensitivity, (iv) dust sensor or PM10 is used to identify in the air a very well particle dust, (v) carbon monoxide (CO) sensor is used to identify indoor air quality monitoring which a wide range of VOC. It uses ADC and MCU that contains and I2C interface. The experiment is done from Oct 7 to Oct 13 in 2017. From the experimental result, it shows that the AQI level is not above 100. This means that the location is secure for the people. AQI scale range for unhealthy state claimed from 0 to 200, while air quality in good condition is 0–50, and 51–100 is medium range, above 100 shows unhealthy to people.

As a summary in this section, Table 3 presents a comparison of the research studied. The table discuss about the various type of sensors and wireless communication technologies used, with its advantages correspondingly. Most of the research produced a real-time monitoring system which gives the reading of the concentration of gas like CO, CO_2, PM10, and O_3 in real-time. From the table also shows that most of the researcher used Wi-Fi, LoRa, Bluetooth, and NB-IoT as a communication module to send the data from sensors to the cloud.

3 Machine Learning in Air Pollution Monitoring System

S. Y. Muhammad et al. points out that machine learning algorithms are best suited for air quality prediction. Machine learning (ML) is the branch of computer science that makes computers capable of performing a task without being explicitly programmed. ML involves computational methods that improve the performance of mechanizing the acquisition of knowledge from experience. There are three types of machine learning which are supervised learning, unsupervised learning, and reinforcement learning. Supervised machine learning is the machine learns from the training data that is labeled, while unsupervised is non-labeled training data, and reinforcement learning is the machine learns by itself. Different ML approaches have been used in recent years to predict a set of air pollutants using different combinations of predictor parameters. The most used algorithm are Ensemble Learning Methods, Artificial Neural Networks, Support Vector Machines, and Linear Regressions [23].

Table 3 Comparison application of IoT

Refs.	Communication module	Collected parameter	Advantages	Disadvantages
[15]	Wi-Fi	CH4, CO_2	Real-time display, and remote monitoring	The range of Wi-Fi is limited, to add more range, additional access point is needed
[16]	CSMA/CA	CO_2	Less delay, and real-time display	
[17]	Bluetooth	PM. SO_2, NO_2, O_3,	Reduce weight of monitoring box and shorter hover time	Has low bandwidth and only allows short range communication between devices
[18]	NB-IoT	CO, CO_2, PM10 and O_3	High coverage area, long-distance, low power and low throughput things	NB-IoT has Low data rate
[19]	GPRS	Temperature, humidity, CO_2, combustible gas, hazardous gas	Low-cost, versatile, scalable, reliable system, real-time display, and remote monitoring	Data rate supported is low
[20]	GPRS	Temperature, humidity, CO_2	Simple structure remote, real-time display, automatic storage, and easy to operate	Data rate supported is low
[21]	Wi-Fi	Temperature, humidity, dust concentration, combustible gas	Good reliability, high monitoring, precision, real-time display, and remote monitoring	The range of Wi-Fi is limited, to add more range, additional access point is needed
[22]	LoRa	Temperature, humidity, CO, CO_2, NO_2	Low-cost and low-power consumption	Data transmission of LoRa is very limited, and easy to get interference

Dixon et al. [24] applied ANN to the gas monitoring data and forecast the concentration of methane gas inside the mine environment. Similarly, Park et al. [25] simulated ANN for the prediction of the PM_{10} concentration in metropolitan subway stations in Seoul. However, the main disadvantages of ANN-based methods are: (1) ANN-based methods fell very easily under the local minimum and are poorly generalized; (2) the deficiency of an analytical model selection approach; (3) it takes time to determine the efficient model of architecture and its indicated by trial and error. Zhang and Ding [26] has done research in Hong Kong on the prediction of air pollutants concentration based on an extreme learning machine (ELM). The parameters that had been studied were NO_2, NO_x, O_3, PM2.5, SO_2, and meteorological data, that obtained from 2010 to 2015 in two monitoring stations, Sham Shui Po and Tap Mun in four seasons. The authors compared the ELM with a feedforward neural network based on back propagation (FFANN-BP) and multiple linear regression (MLR), to show that ELM performs well than other algorithms. The performance of the algorithm is tested based on RMSE, R^2, IA, and MAE. R^2 have almost significant values which is 0.70 in all seasons. The RMSE and MAE show better in summer than in all three seasons, and IA shows the same reading in all seasons. This research is very helpful in machine learning study, because the authors have provided complete analysis to show that ELM is better than FFANN-BP and MLR algorithm. From this study, it might good to apply these three algorithms to predict the concentration of pollutant gases in different parameters and situations. The uniqueness of this research is there are four seasons to examine the results, based on RMSE, R^2, IA, and MAE, and it shows that the meteorological condition has significantly affected the levels of air pollution.

The author in [27] has conducted research on air pollution prediction in smart cities through machine learning methods in Murcia, Spain with several machine learning methods that had been done to predict the concentration of ozone (O_3) gas, in four different cities at the Region of Murcia. which (i) Bagging (ii) Random Committee (iii) Random Forest (iv) M5P Tree and (v) KNN algorithm. Based on 5 techniques, KNN shows the worst performance because it has the highest RMSE and MAE values. The authors stated that the Wilcoxon test confirms that the Random Forest method gives good results and fit better than other machine learning algorithm with a 90% confidence level in prediction of O_3. In predicting O_3, the author also used the decision tree method to examine which parameters are the most influenced the O_3, and the parameters are: NO_x, temperature, DD, VV, HR, SO_2, NO and $PM_{10.}$ This paper only focuses on O_3 prediction, and the advantages of this study are, the researcher can find out the influenced parameter which temperature, humidity, and wind to O_3 prediction by using the decision tree method.

Masih Adven in [28] study on modeling the atmospheric concentration of CO_2 prediction by using Ensemble Learning Algorithms in Marylebone road located in London, United Kingdom. The studied parameter was NO_2, SO_2, NO, NO_x, CO, and O_3, the data source is from the Department of Environment Food and Rural Affairs from first January 2013 to 18th June 2013. The author made a comparison between Ensemble classifiers such as Random Forest and Bagging, with ANN and SVM, and conclude that it was performed better than ANN and SVM in CO concentration prediction. Prediction model of air pollutant levels in research [29] by using Linear Model with component analysis Linear model (i) Independent component analysis (ICA), (ii) principal component analysis (PCA) located in Surabaya City, Indonesia. NO_2, PM10, and O_3, and Meteorological data. The data obtained was from March to April 2002. From the result, it shows that the regression with variables extracted from ICA was the worst model for all pollutants such as NO2, PM10, and O3 as their residual errors were highest compared with other models, and the prediction of one-step-ahead 30 min interval of each pollutant was best obtained by employing original variables combination of air pollutants and meteorological factors.

Table 4 Comparison of machine learning model

Refs.	ML model	Collected parameter
[26]	Extreme machine learning	NO_2, NO_x, O_3, PM2.5, and SO_2 and meteorological data
[27]	Random Forest, Decision Tree, Random Committee, Bagging and KNN	O_3
[28]	Ensemble learning algorithms	NO_2, SO_2, NO, NO_x, CO, O_3
[29]	Linear model Independent component analysis (ICA) principal component analysis (PCA)	NO_2, PM10, and O_3, and meteorological data
[30]	Regression support vector machine, artificial neural network and auto-regressive nonlinear neural network	PM10 and PM2.5 Meteorological data
[31]	Multiple linear regression	NO_2, O_3, PM2.5, SO_2, PM10, and CO

The authors of [30] aim to determine the prediction models to determine air pollutions based on PM10 and PM2.5 pollution concentrations in Tehran regression using support vector machine, artificial neural network, and auto-regressive nonlinear neural network. The reading of PM10, PM2.5, and meteorological data were provided by Iran Meteorological Organization in a 10 years, from 2006 to 2016 in 6 stations, which are Shemiran, Dowshan-Tappeh, Chitgar, Mehrabad Airport and Geophysics stations. The most reliable algorithm for the prediction of air pollution was autoregressive nonlinear neural network with external input using the proposed prediction model, where its one-day prediction error reached 1.79 g/m3.

Table 4 Comparison of Machine Learning model summarizes several studies on the comparison of ML applications based air quality prediction. Some researchers used more than one ML model to study the best ML model to get the accuracy of the air pollution level. In practical application, a meteorological parameter such as current weather conditions such as temperature, wind speed, humidity, wind direction have a significant effect on the air pollution level. In the process of system implementation, therefore, consideration should be taken to the influence of environmental factors on monitoring and prediction effect level of air pollution.

4 Research Challenges

There are three elements challenges to be considered when deploying IoT based on smart environmental monitoring systems such as gas sensors detection, data collection retrieved from sensors and predictive analysis to estimate the concentration of gases and to be classified by Air Pollutant Index (API) indicator. Firstly, for gas detection, the main thing is the sensitivity of the sensor itself towards the exposed gases. The calibration is needed to get accurate data from the sensor. However, if the sensor is already calibrated, it consumes high cost and maintenance. Then, for data collection, it plays an important role to indicate the quality of data and it has validation issues. Lastly, to conduct predictive analysis, it needs suitable Machine Learning (ML) algorithm to estimate the concentration of pollutant gases based on the ML performance such as RMSE, R^2, MAE, and IA.

Air pollution problems can be reduced by using proper monitoring system and machine learning involved, as for monitoring system, developers have successfully developed an efficient system which includes the application of real-time monitoring, distance remote monitoring and secured system with low cost and low power consumption. The system will continue to be developed with improvements in the future for the implementation of the next generation. The result of gas concentration and API reading can get from the sensor involved the monitoring system. Then the reading of the concentration of gases and API can be predicted by using some machine learning or statistical analysis. Therefore, the usage of IoT real-system is important in measuring air quality because it gives many benefit to society and environment, it can help people from serious health hazards and medical costs. alert and warning the people based on the concentration of gas and Air Quality Index (AQI) reading. As for predictive analytics, it helps for gas concentration prediction and for forecasting, and to get the accuracy of prediction, suitable algorithm is needed to make sure the prediction of gas concentration is accurate.

Acknowledgement This project is supported in part by FRGS/1/2019/STG02/UMP/02/4, RDU190349, UIC191205, UIC200814 and RDU202803.

References

1. Smil V (2001) Air pollution. Int Encycl Soc Behav Sci, pp 355–359
2. Yi WY, Lo KM, Mak T, Leung KS, Leung Y, Meng ML (2015) A survey of wireless sensor network based air pollution monitoring systems. Sensors 15(12):31392–31427
3. Carbon Dioxide Detection and Indoor Air Quality Control—Occupational Health and Safety. [Online]. Available: https://ohsonline.com/articles/2016/04/01/carbon-dioxide-detection-and-indoor-air-quality-control.aspx. Accessed 2 July 2019
4. Rai A, Saxena V, Kusrey S (2017) Zigbee based air pollution monitoring and control system using WSN. SSRG Int J Electron Commun Eng 4(6):7–12
5. Ayele TW (2018) Air pollution monitoring and prediction using IoT. In: 2018 Second international conference on inventive communication and computational technologies, Icicct, pp 1741–1745 (2018)
6. CDC—carbon monoxide poisoning—frequently asked questions. [Online]. Available: https://www.cdc.gov/co/faqs.htm. Accessed 27 July 2019
7. Huang WH, Lin JH, Weng CH, Hsu CW, Yen TH (2016) Environmental NO_2 and CO exposure: ignored factors associated with uremic pruritus in patients undergoing hemodialysis. Sci Rep 6(August):2–11
8. Tang Q, Hess PG, Brown-Steiner B, Kinnison DE (2013) Tropospheric ozone decrease due to the Mount Pinatubo eruption: reduced stratospheric influx. Geophys Res Lett 40(20):5553–5558
9. Scientific Facts on Air Pollution Nitrogen Dioxide. [Online]. Available: https://www.greenfacts.org/en/nitrogen-dioxide-no2/. Accessed 27 July 2019

10. Mansour S, Nasser N, Karim L, Ali A (2014) Wireless sensor network-based air quality monitoring system. In: 2014 International conference on computing, networking and communications ICNC 2014, pp 545–550
11. Xiaojun C, Xianpeng L, Peng X (2015) IOT-based air pollution monitoring and forecasting system. In: 2015 International conference on computer and computational sciences ICCCS 2015, pp 257–260
12. Kasar AR, Khemnar DS, Tembhurnikar NP (2013) WSN based air pollution monitoring system. Int J Sci Eng Appl 2(4):55–59
13. IEEE (1997) Telecommunications and information exchange Part 11 : Wireless LAN Medium Access Control (MAC) and Physical Layer (PHY) specifications
14. Chaouchi H, Bonnin J (2013) Wireless sensor networks : a survey on recent developments and potential synergies Wireless sensor networks : recent developments and potential synergies
15. Ahmed MM, Banu S, Paul B (2017) Real-time air quality monitoring system for Bangladesh's perspective based on Internet of Things. In: 2017 3rd International conference on electrical information and communication technology (EICT), pp 1-5
16. Spachos P, Hatzinakos D (2016) Real-time indoor carbon dioxide monitoring through cognitive wireless sensor networks. IEEE Sens J 16(2):506–514
17. Zhi S, Wei Y, Cao Z, Hou C (2017) Intelligent controlling of indoor air quality based on remote monitoring platform by considering building environment. In: 2017 4th International conference on systems and informatics (ICSAI), pp 627–631
18. Duangsuwan S, Takarn A, Nujankaew R, Jamjareegulgarn P (2018) A study of air pollution smart sensors LPWAN via NB-IoT for Thailand smart cities 4.0. In: 2018 10th International conference on knowledge and smart technology (KST), pp 206–209
19. Wu L, Hu J (2010) Design and implementation of production environment monitoring system based on GPRS-Internet. In: 2010 Fourth international conference on genetic and evolutionary computing, pp 818–821
20. Yang H, Qin Y, Feng G, Ci H (2013) Online monitoring of geological storage and leakage based on wireless sensor networks. IEEE Sens J 13(2):556–562
21. Li D, Bao Z, Yang Y (2016) Design of workshop environment monitoring system based on Internet of Things, pp 165–170
22. Raju V, Varma AS, Raju YS (2017) An environmental pollution monitoring system using LORA. In: 2017 International conference on energy, communication, data analytics and soft computing (ICECDS), pp 3521–3526
23. Rybarczyk Y, Zalakeviciute R (2018) Machine learning approaches for outdoor air quality modelling: a systematic review. Appl Sci 8(12):2570
24. Couillet J, Prince M, Couillet J, Prince M (2014) Application of an artificial neural network to the prediction of firedamp emissions in coal mines
25. Park S et al (2018) Predicting PM10concentration in Seoul metropolitan subway stations using artificial neural network (ANN). J Hazard Mater 341:75–82
26. Zhang J, Ding W (2017) Prediction of air pollutants concentration based on an extreme learning machine: The case of Hong Kong. Int J Environ Res Public Health 14(2):1–19
27. Martínez-España R, Bueno-Crespo A, Timón I, Soto J, Muñoz A, Cecilia JM (2018) Air-pollution prediction in smart cities through machine learning methods: a case of study in Murcia, Spain. J Univ Comput Sci 24(3):261–276
28. Masih A (2018) Modelling the atmospheric concentration of carbon monoxide by using ensemble learning algorithms. In: CEUR Workshop Proceedings, vol 2298
29. Syafei AD, Fujiwara A, Zhang J (2015) Prediction model of air pollutant levels using linear model with component analysis. Int J Environ Sci Dev 6(7):519–525
30. Delavar M et al (2019) A novel method for improving air pollution prediction based on machine learning approaches: a case study applied to the capital City of Tehran. ISPRS Int J Geo-Inf 8(2):99
31. Li LN, Gong XP, Dai LC, Zhan XH (2013) The regression models of PM2. 5 and other air pollutants in Wuhan. In: Advanced materials research, vol 864, pp 1356–1359

Near-Infrared Spectroscopy for Ganoderma Boninense Detection: An Outlook

Mas Ira Syafila Mohd Hilmi Tan, Mohd Faizal Jamlos, Ahmad Fairuz Omar, Fatimah Dzaharudin, Mohd Azraie Mohd Azmi, Mohd Noor Ahmad, Nur Akmal Abd. Rahman, and Khairil Anuar Khairi

Abstract Ganoderma boninense (G. boninense) infection reduces the productivity of oil palms and causing a serious threat to the palm oil industry. This catastrophic disease ultimately destructs the basal tissues of oil palm that causing the eventual death of the palm. Early detection of G. boninense is vital since there is no effective treatment to stop the continuing spread of the disease. This mini-review describes past and future prospects of integrated research of near-infrared spectroscopy (NIRS) towards early G. boninense detection system. This effort could reduce the cost of plantation management and avoid production losses. Remarkably, (i) spectroscopy techniques are more reliable than other detection techniques such as serological, molecular, biomarker-based sensor and hyperspectral in reacting

M. I. S. M. H. Tan · M. F. Jamlos (✉) · F. Dzaharudin
College of Engineering, Universiti Malaysia Pahang, 26300 Kuantan, Malaysia
e-mail: mohdfaizaljamlos@gmail.com

M. I. S. M. H. Tan
e-mail: masira.hilmitan@gmail.com

F. Dzaharudin
e-mail: fatimahd@ump.edu.my

A. F. Omar · M. N. Ahmad
Universiti Sains Malaysia, 11800 Gelugor, Penang, Malaysia
e-mail: fairuz_omar@usm.my

M. N. Ahmad
e-mail: amnoor6355@gmail.com

M. A. M. Azmi
Universiti Kuala Lumpur, 53100 Kuala Lumpur, Gombak, Malaysia
e-mail: mazraie@unikl.edu.my

N. A. Abd. Rahman
Advanced Communication Engineering Centre (ACE), Universiti Malaysia Perlis, 01000 Kangar, Malaysia
e-mail: akay_rahman@yahoo.com.my

K. A. Khairi
VAT Manufacturing, Industrial Zone, 14100 Batu Kawan, Penang, Malaysia
e-mail: anuar144@yahoo.com

A. F. Ab. Nasir et al. (eds.), *Recent Trends in Mechatronics Towards Industry 4.0*, Lecture Notes in Electrical Engineering 730, https://doi.org/10.1007/978-981-33-4597-3_12

with organic tissues, (ii) NIR spectrum is more precise and sensitive to particular diseases include G. boninense compared to visible light, (iii) hand-held NIRS for in situ measurement is to explore the efficacy for early detection system in real-time using machine learning (ML) classifier algorithms and predictive analytics model. This non-destructive, environmentally friendly (no chemical involved), mobile and sensitive leads the integrated hand-held NIRS with ML, and predictive analytics has significant potential as a platform towards early detection of G. boninense in the future.

Keywords Ganoderma boninense · Near-infrared spectroscopy · Machine learning

1 Introduction

In 2018, oil palm contributed 37.9% to the Gross Domestic Product (GDP) of the agriculture sector [1]. As reported by the Malaysian Palm Oil Board (MPOB) in 2019, Malaysia produces 17.18 tonnes per hectare of oil palm fresh fruit bunch for 5.9 million hectares [2]. Unfortunately, infection of G. boninense, a type of fungus has caused great losses in the production of oil palm by 80% [3]. G. boninense can infect oil palm trees at all stages, from seedlings to mature plants [4].

Various studies and approaches have been taken to control this disease in oil palm trees but as yet no effective detection of G. boninense. When the disease symptoms begin to appear, more than half of the internal tissues are already rotten [5]. The earliest visible external symptoms occur in the leafage which is almost similar to the physical condition of water stress, malnutrition drought, hyperacid soil, or high soil water salinity, as shown in Fig. 1 [6, 7]. It is, therefore, visual diagnosis is not effective.

2 Methods for Plant Diseases Detection

There are several techniques for early detection of plant disease, including G. boninense infection. Visual inspection is used to evaluate the physical signs and symptoms of the plants. This approach can detect a wide range of disease distribution [9]. However, visual inspection of infected trees is not effective and requires a great deal of labour and time [10].

The earliest attempt to detect disease on the plant is using enzyme-linked immunosorbent assay (ELISA) where Antibodies were employed to detect G. boninense in culture media [11]. Other lab-based techniques to detect BSR disease are Ganoderma Selective Medium (GSM) [12], Multiplex PCR-DNA Kit [13], GanoSken Tomography [14] and Electrochemical DNA biosensor [15]. On the other hand, direct molecular techniques which required sample preparation and DNA extraction are tedious, complex costly procedures and time-consuming.

Fig. 1 **a–d** Oil palm infected with G. boninense with **a** foliar symptom, **b** later foliar symptoms, **c** internal symptoms in oil palm stem and **d** Basidiomata on the base of oil palm [8]

Recently, the spectroscopy technique which is based on the optical approach to detect G. boninense is explored. Majority of spectroscopic application for the detection of plant diseases complies with these criteria: non-invasive, rapid, sensitive and precise to a particular disease, which has been taking into consideration for the development and design of early-stage infection detector [16]. Spectroscopy techniques assess the condition of the plant by emitting visible and non-visible light at specific wavelengths to penetrate tissues and the backscattered light with certain intensities becomes an indicator of different conditions. These light wavelengths are essential for studying various plant fungal diseases [17].

3 Spectroscopy Techniques for Ganoderma Boninense Detection with Machine Learning Algorithms

Several spectroscopy techniques have been conducted to detect the infection of G. boninense in oil palms such as Fourier Transform Infrared (FTIR) spectroscopy [18, 19], MIR spectroscopy [20], hyperspectral imaging spectroscopy [21], and visible to near-infrared (VIS-NIR) spectroscopy [22, 23] as summarised in Table 1.

Dayou et al. [18] investigate the possibility of using FTIR spectroscopy technique to detect G. boninense infection and to distinguish between healthy and infected oil palm trunk tissue. The significant resemblance of infected oil palm tissue and pure

Table 1 Previous studies on G. boninense detection using spectroscopy techniques

Spectroscopy method	Sample grouping	Models/algorithms	Significant result	References
FTIR spectroscopy	• Healthy • Infected	–	Resemblance pattern of infected oil palm with pure G. boninense is observed at a particular wavelength which can be used as biomarkers	[18]
MIR Spectroscopy	• Healthy • Mild • Moderate • Severe	LDA, QDA, kNN and NB	The highest overall classification accuracy is 92% using LDA	[20]
Imaging Spectroscopy/HSR	• Healthy • Mild • Severe	Maximum likelihood	The net accuracy is found to be 82%	[21]
VIS-NIR Spectroscopy	• Healthy • Mild • Moderate • Severe	ANN	Accuracy up to 100% is achieved without any pre-processing methods	[22]
VIS-NIR Spectroscopy	• Healthy • Mild • Moderate • Severe	LDA, QDA, kNN and NB	kNN has the highest classification accuracy of 97.3% Significant differences between each severity levels were observed in the NIR region compared to VIS region	[23]

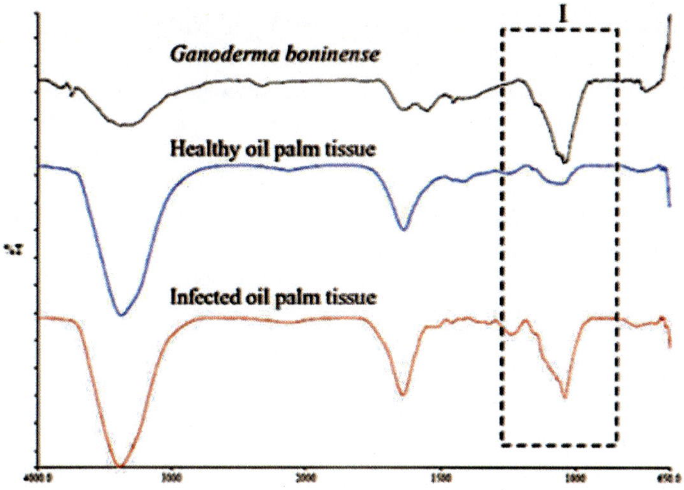

Fig. 2 FTIR spectra of G. boninense, healthy oil palm tissue and infected oil palm tissue [18]

G. boninense compared to the healthy sample can be observed in region I of the FTIR spectra as illustrated in Fig. 2. This finding corroborates the study in [24], which reported that a unique IR pattern due to the presence of fungi to discriminate between the infected tissue and uninfected tissues.

The feasibility of MIR spectroscopy for G. boninense detection by using FTIR spectrometer is assessed by Liaghat et al. [20]. Oil palm leaf samples were grinded into powder and processed into pellets for the spectroscopy measurement. In this study, linear discriminant analysis (LDA), quadratic discriminant analysis (QDA), k-Nearest Neighbour (kNN) and Naïve Bayes (NB) classifiers were used to classify the different level of disease severities. LDA classifier indicated the highest overall classification accuracy of 92% [20].

Hyperspectral remote sensing or imaging spectroscopy technique is utilised for G. boninense detection using a portable spectroradiometer [21]. Shafri et al. [21] applied this approach to identify the spectral differences between healthy and infected leaves of 6 months old oil palm seedlings. The spectral reflectance managed to identify three levels of disease severity which are healthy, mild and severe. Classification of the severity level is performed using maximum likelihood classifier based on the most significant spectral wavelength. The net accuracy is found to be 82%.

Ahmadi et al. [22] utilised the imaging spectroscopy in VIS-NIR spectral region (273–1100 nm) by a portable spectroradiometer to discriminate and classify G. boninense infection level in oil palm tree at an early stage. The samples were classified into four categories based on the level of severity: healthy, mild, moderate and severe. Accuracy up to 100% is achieved by applying artificial neural network (ANN) classifier on the raw spectral data without any pre-processing approaches.

Another similar study by Liaghat et al. [23] assessed the in-field VIS-NIR spectroscopy (325–1075 nm) to detect G. boninense infection in oil palm. It is noted significant differences between each severity level in NIR region compared to VIS region as shown in Fig. 3, which depicted the efficacy of NIRS to detect G. boninense. The reflection in the NIR region decreases drastically with increasing disease severity while the healthy leaves show the highest reflectance. Such results are possible due to the degradation of cell walls or wilting in plants [25]. Similar classification techniques as [20] were utilised to discriminate four levels of disease severity. kNN revealed to be the best classifier with the highest classification accuracy of 97.3% compared to other models. The classifier could differentiate the level of severities of Ganoderma-infected oil palm from the healthy ones [23].

Most of the past studies were carried out under laboratory condition, so it is impractical for real-time in-field measurement. Meanwhile, FTIR spectrometers used to perform FTIR and MIR spectroscopic analysis are bulky in size, forcing the measurement to be performed in the laboratory. Besides, despite a good spectral response of FTIR spectroscopy, FTIR requires the sample to be processed into pellets for the lab measurements. In addition, based on the VIS-NIR spectroscopy study in [23], spectral data in NIR region portrays significant differences between classes of samples compared to VIS region. Liang et al. [26] stated that VIS region is only useful for visual analysis; thus, it is not useful for asymptomatic detection. Therefore, further research to detect G. boninense based on NIRS alone should be considered for asymptomatic early-stage infection.

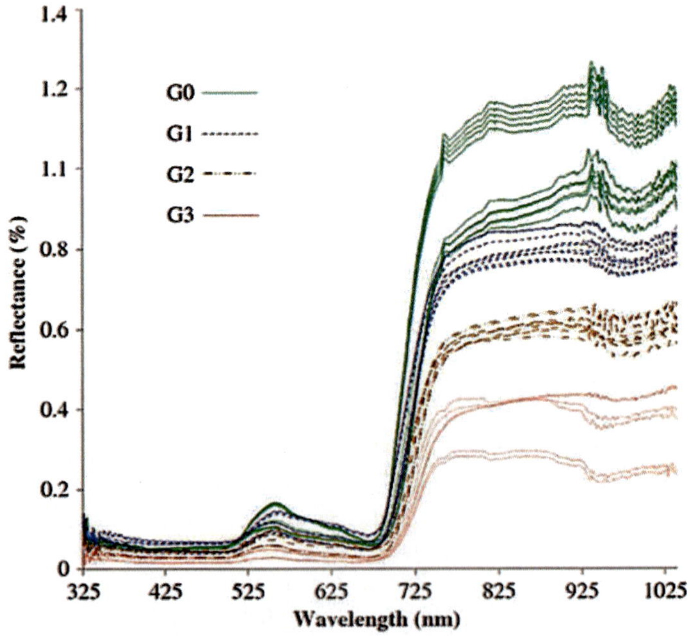

Fig. 3 Reflectance spectra of healthy (G0) and non-healthy (G1, G2 and G3) leaf samples [23]

4 NIRS for Plant Diseases Detection with Machine Learning Algorithms

NIRS is a non-destructive analytical technology that provides rapid and accurate analysis. NIRS is a reliable and non-invasive technique with potential for plant disease detection. NIRS has the highest accuracy for disease detection on different types of plants compared to MIR and VIS-NIR. NIR has a higher penetration depth into the sample compared to MIR radiation [27]. NIRS also suitable for early detection of plant disease as it associates with the interior information of the sample [28]. Utilisation of NIRS to detect various plant diseases, including fungi infections is summarised in Table 2.

NIRS technique is utilised to detect fungal infection in maize kernels [29, 30]. kNN classification used in [29] at two NIR wavelengths (i.e. 715 and 965 nm) which provided a correct classification of healthy and infected kernels with an accuracy of 98.1% and 96.6%, respectively. A similar study by Draganova et al. [30] classified healthy and Fusarium fungi infected maize grains by using Soft Independent Modelling by Class Analogy (SIMCA), Probabilistic neural network (PNN) and K-Means classifier. PNN produced the best performance with an accuracy of 99.3% and 98.7% for healthy and diseased grains, respectively. In [31], the authors applied NIRS for detection of fungal infection in chestnuts. LDA, QDA and kNN classifiers were applied to classify the healthy chestnut, medium and severely infected chestnut. The highest classification accuracy of 97% was achieved using QDA. The application of NIRS to identify Fusarium fungi in barley is investigated in [32]. Partial least squares discriminant analysis (PLS-DA) was used for discriminant prediction of normal hulled barley and Fusarium-infected hulled barley. Classification accuracy of up to 100% was achieved. A study by Xu et al. [33] was conducted to assess the NIRS for detection of leaf miner infection on tomato leaves. Significant differences in reflectance were observed at wavelengths of 1450 and 1900 nm. Regression analysis for predictive modelling was performed on both wavelengths. Single wavelength reflectance at 1450 nm showed a good prediction performance of $R^2 = 0.982$.

As discussed in Sects. 3 and 4, several machine learning and statistical approaches were utilised and proven to be a reliable tool for disease monitoring and early detection of plant diseases. Machine learning discerns data patterns by extracting information from a data set and transforming it into useful data to assist decision making. It has gained interest in the latest agriculture technologies as a promising approach for faster and efficient data analytics [34].

5 Conclusion

This mini-review paper demonstrates the utilisation of NIRS techniques along with machine learning classifier as a feasible method for early detection of the G. boninense in oil palm. However, most of the studies only focus on detection without

Table 2 Previous studies on plant disease detection using NIRS

Plant	Disease	Wavelength (nm)	Models/algorithms	Significant result	References
Maize	Fungal infection	500–1700	kNN	Classification accuracy obtained for healthy and infected kernel were 98.1% and 96.6%, respectively	[29]
Maize	Fusarium infection	400–2500	SIMCA, PNN, K-Means	PNN has the best performance with accuracy of 99.3% and 98.7% for healthy and infected grain, respectively	[30]
Chestnuts	Fungal infection	1100–2300	LDA, QDA and kNN	The highest overall classification accuracy is 97% using QDA	[31]
Barley	Fusarium fungi infection	1175–2170	PLS—DA	Classification accuracy up to 100% was achieved	[32]
Tomato	Leaf miner infection	800–2500	Regression analysis	Single wavelength reflectance at 1450 nm showed a good prediction performance of $R^2 = 0.982$	[33]

further development on the whole complete system. It is found out that NIRS is applicable instead of VIS spectroscopy and MIR for G. Boninense detection. A suitable machine learning classifier model based on kNN, ANN, NB must be executed with the NIR spectra as input data to predict and classify the healthy and infected oil palm. Nonetheless, it is identified that kNN as a potential classifier for this research prototype since kNN is easy to be implemented and exhibit decent performance for prediction. With the advancement of this study, portable NIRS devices could be used commercially in the near future for diagnosis techniques for other applications.

Acknowledgements This project is supported in part by CREST T14C2-16/006, FRGS/1/2019/STG02/UMP/01/1, UIC200814 and RDU202803

References

1. M. Department of Statistics, Department of Statistics Malaysia, 29 Nov 2019.[Online].Available:https://www.dosm.gov.my. Accessed 1 July 2020
2. Malaysian Palm Oil Board, January 2020. [Online]. Available: http://bepi.mpob.gov.my/images/overview/Overview_of_Industry_2019.pdf
3. Flood J, Hasan Y, Turner PD, O'Grady EB (2000) The spread of Ganoderma from infective sources in the field and its implications for management of the disease in oil palm. In: Ganoderma diseases of perennial crops, pp 101–112
4. Naher L, Yusuf UK, Ismail A, Tan SG, Mondal MMA (2013) Ecological status of 'Ganoderma' and basal stem rot disease of oil palms ('Elaeis guineensis' Jacq.). Aust J Crop Sci 7(11):1723
5. Kandan A, Bhaskaran R, Samiyappan R (2010) Ganoderma–a basal stem rot disease of coconut palm in south Asia and Asia pacific regions. Arch Phytopathol Plant Prot 43(15):1445–1449
6. P. D. Turner, Oil palm diseases and disorders, Oxford University Press, 1981
7. Turner PD, Gillbanks RA (1974) Oil palm cultivation and management, Kuala Lumpur. Incorporated Society of Planters, Malaysia
8. Miller R, BioScience C, Min CK (2018) Ganoderma boninense (basal stem rot of oil palm). CABI—invasive species compendium, 14 July 2018. [Online]. Available: https://www.cabi.org/isc/datasheet/24924. Accessed 11 Jan 2019
9. Parker SR, Shaw MW, Royle DJ (1995) The reliability of visual estimates of disease severity on cereal leaves. Plant Pathol 44(5):856–864
10. Wong LC, Bong CFJ, Idris AS (2012) Ganoderma species associated with basal stem rot disease of oil palm. Am J Appl Sci 9(6):879–885
11. Ananthanarayanan TV, Reddy MK (1984) Serological test for the diagnosis of Ganoderma lucidum. Curr Sci 53(2):97–98
12. Ariffin D, Idris S, Khairudin H (1995) Conformation of ganoderma infected palm by drilling technique. In: PORIM international palm oil congress (No. L-0314). PORIM
13. Idris AS, Rajinder S, Madihah AZ, Wahid MB (2010) Multiplex PCR-DNA kit for early detection and identification of Ganoderma species in oil palm. MPOB information series TS, vol 531, no 73
14. Idris AS, Rajinder S, Madihah AZ, Wahid MB (2010) GanoSken for early detection of Ganoderma infection in oil palm. MPOB information series TT, vol 442, no 4
15. Dutse SW, Yusof NA, Ahmad H, Hussein MZ, Zainal Z (2012) An electrochemical DNA biosensor for ganoderma boninense pathogen of the Oil palm utilizing a New ruthenium complex, [Ru (dppz) 2 (qtpy)]Cl2. Int J Electrochem Sci 7:8105–8115
16. López M, Bertolini E, Olmos A, Caruso PGM, Llop P, Penyalver R, Cambra M (2003) Innovative tools for detection of plant pathogenic viruses and bacteria. Int Microbiol 6(4):233–243
17. Ray M, Ray A, Dash S, Mishra A, Achary KG, Nayak S, Singh S (2017) Fungal disease detection in plants: traditional assays, novel diagnostic techniques and biosensors. Biosens Bioelectron 87:708–723
18. Dayou J, Alexander A, Sipaut CS, Chong KP, Lee PC (2014) On the possibility of using FTIR for detection of Ganoderma boninense in infected oil palm tree. Int J Adv Agric Environ Eng 1(1):161–163
19. Alexander A, Sipaut CS, Chong KP, Lee PC, Dayou J (2014) Sensitivity analysis of the detection of Ganoderma boninense infection in oil palm using FTIR. Trans Sci Technol 1(1):1–5

20. Liaghat S, Mansor S, Ehsani R, Shafri HZM, Meon S, Sankaran S (2013) Mid-infrared spectroscopy for early detection of basal stem rot disease in oil palm. Comput Electron Agric 101:48–54
21. Shafri HZ, Anuar MI, Seman IA, Noor NM (2011) Spectral discrimination of healthy and Ganoderma-infected oil palms from hyperspectral data. Int J Remote Sens 32(22):7111–7129
22. Ahmadi P, Muharam FM, Ahmad K, Mansor S, Abu Seman I (2017) Early detection of Ganoderma basal stem rot of oil palms using artificial neural network spectral analysis. Plant Dis 101(6):1009–1016
23. Liaghat S, Ehsani R, Mansor S, Shafri HZ, Meon S, Sankaran S, Azam SH (2014) Early detection of basal stem rot disease (Ganoderma) in oil palms based on hyperspectral reflectance data using pattern recognition algorithms. Int J Remote Sens 35(10):3427–3439
24. Brandl H (2012) Detection of fungal infection in Lolium perenne by Fourier transform infrared spectroscopy. J Plant Ecol 6(4):265–269
25. Knipling EB (1970) Physical and physiological basis for the reflectance of visible and nearinfrared radiation from vegetation. Remote Sens Environ 1:155–159
26. Liang PS, Slaughter DC, Ortega-Beltran A, Michailides TJ (2015) Detection of fungal infection in almond kernels using near-infrared reflectance spectroscopy. Biosys Eng 137:64–72
27. Blanco M, Villaroya I (2002) NIR spectroscopy: a rapid-response analytical tool. TrAC Trends Anal Chem 21(4):240–250
28. Wu D, Feng L, Zhang C, He Y (2008) Early detection of Botrytis cinerea on eggplant leaves based on visible and near-infrared spectroscopy. Trans ASABE 51(3):1133–1139
29. Pearson TC, Wicklow DT (2006) Detection of corn kernels infected by fungi. Trans ASABE 49(4):1235–1245
30. Draganova T, Daskalov P, Tsonev R (2010) An approach for identifying of Fusarium infected maize grains by spectral analysis in the visible and near infrared region, SIMCA models, parametric and neural classifiers. Int. J. Bioautomation 14:119–128
31. Moscetti R, Monarca D, Cecchini M, Haff RP, Contini M, Massantini R (2014) Detection of mold-damaged chestnuts by near-infrared spectroscopy. Postharvest Biol Technol 93:83–90
32. Lim J, Kim G, Mo C, Oh K, Yoo H, Ham H, Kim MS (2017) Classification of fusarium-infected Korean Hulled Barley using near-infrared reflectance spectroscopy and partial least squares discriminant analysis. Sensors 17(10):2258
33. Xu H, Ying Y, Ye Z (2007) Near-infrared spectroscopy in detecting leaf miner damage on tomato leaf. Biosys Eng 96(4):447–454
34. Singh A, Ganapathysubramanian B, Singh AK, Sarkar S (2016) Machine learning for high-throughput stress phenotyping in plants. Trends Plant Sci 21(2):110–124

Investigation of Features for Classification RFID Reading Between Two RFID Reader in Various Support Vector Machine Kernel Function

Chun Sern Choong, Ahmad Fakhri Ab. Nasir, Anwar P. P. Abdul Majeed, Muhammad Aizzat Zakaria, and Mohd Azraai Mohd Razman

Abstract Radio Frequency Identification (RFID) is the primary technology for tripartite logistics information and automation. The RFID-based logistics system able to increase logistic operating capacity and improve the efficiency of worker to minimize the logistic operation failure. However, the precise location of the RFID device is still a problem in a specific area due to the interference of the radiofrequency. An indoor positioning using RFID technology based on various kernel function of the support vector machine (SVM), and feature extraction are proposed to determine the location of the goods. SVM classifier is utilized the acquire received signal strength indicator (RSSI) data for trained the model from the indoor moving objects as well as relationship between RSSI and distance is constructed to boost RSSI accuracy. Instead, the distance verses RSSI algorithm is used to determine the objects to be located based on the distance of the tag to be located to each reader. The feature of RSSI is extracted to nine single statistical features and three combinations of different statistical features for evaluated the classification performance in different kernel functions of the SVM classifier. The Polynomial-SVM model is capable of delivering a classification accuracy of 84.81 and 20.00% of the error rate in test data by using the function MIN extracted. The experimental results show that the algorithm improves the positioning accuracy of indoor localization with select the suitable feature combination.

Keywords Radio frequency identification · Support vector machine · Feature extraction · Kernel function

---

C. S. Choong (✉) · A. F. Ab. Nasir · A. P. P. Abdul Majeed · M. A. Zakaria · M. A. M. Razman
Innovative Manufacturing, Mechatronics, and Sports Lab (IMAMS), Faculty of Manufacturing and Mechatronic Engineering Technology, University Malaysia Pahang, Pekan, Pahang, Malaysia
e-mail: cs_choong@hotmail.com
URL: http://imamslab.ump.edu.my/

A. F. Ab. Nasir et al. (eds.), *Recent Trends in Mechatronics Towards Industry 4.0*, Lecture Notes in Electrical Engineering 730, https://doi.org/10.1007/978-981-33-4597-3_13

1 Introduction

A key aspect of Radio Frequency Identification (RFID) is electromagnetic waves to perform contactless data transmission between the RFID reader and the RFID tag to accomplish the objective of detecting the tagged items [1]. This RFID is a kind of automatic identification technology with considered to be one of the most successful information systems and widely used in animal tracking, automotive antitheft alarm and anti-counterfeiting systems, access control, keyless car system, electronic toll collection, parking access control system, automation of a production line, and supply chain management [2]. In the identification of RFID system, reading and writing as well as communicating with RFID tags is carried out by electromagnetic waves. Based on the communication distance, the near field and the far-field can be divided, and the data exchange manner between the read/write device and the electronic tag is correspondingly divided into load modulation and backscatter modulation.

Previous studies have reported [3–5] Support Vector Machine (SVM) has a high classification to resolve discrimination and regression problems in an indoor environment using RFID technology. Bertoncini et al. applied RFID fingerprinting, which is linear discriminant classifiers (LDC), quadratic discriminant classifiers (QDC), k nearest neighbour (k-NN), and support vector machine (SVM) for identifying unique RFID tags [4]. SVM classification results are the best with a value of area under the curve (AUC) near 1. In the work of [3], object motion detection in an indoor environment isolates the features of the received signal strength and categorizes them using machine-learning classifiers. SVM classifier in the development of cascaded classification resulted in better scores similar to the decision tree (DT). Zoning localization [5] approach is used to locate a tag with only using received signal strength indicator (RSSI) measurements. A comparison with artificial neural networks (ANN) indicates the efficacy of proposal for location based on SVM that the proposal method achieves a classification accuracy about 90% in indoor environment.

The received signal strength (RSS) is the relative value of the transmitted signal for a system correlated with the distance between the tags and the RFID antenna. Some studies mentioned RSS approaches to detect false positives where static and moved tags were classified respectively as false positive and true positive readings [6, 7], and the studies applied both arithmetic mean, standard deviation, max-min, and difference of max-min scaling of RSS. Machine learning classifier is used to detect false positives effectively using features based on RSS as inputs. Hauser et al. researched identification systems in a variety shop to detect false-positive RFID readings to help prevent theft happened [8]. Classification models such as DT, Logistic Regression (LR), random forest (RF), ANN, and SVM were used in his study to distinguish between tag movement across the gate and near the gate using extracted RSS as features. Another study found that the integration of machine learning models with RFID could allow for automated checkout systems using RSS [9].

In summary, this paper is focused on investigating the different kernel functions in SVM model performance using various features for identifying the location of electronic product code (EPC) tag RSS in overlapping the RFID reader antenna. RFID reading from the EPC tags is measured and extracted from the reader antenna for further analysis. The challenges and strategies for increasing the classification accuracy are looking for a suitable feature and classifier to minimize the error (rate) that deducted the fraction of misclassified cases. Comparison of classification accuracy between a single feature and combine all feature is shown and determine the most suitable feature combination selected as a feature extraction method for the final model. Various of the kernel function in SVM classifier is utilized for evaluating data mining systems by separation of data into training and testing sets. The model by making predictions against the test set after a sample has been trained using the training set, and it is easy to determine if the assumptions of the model are accurate.

This paper first gives a brief overview of the recent history of RFID indoor localization technology. Section Two begins by laying out the RFID principle of the research and looks at how RFID reading influences indoor localization via feature extraction and machine learning. The third chapter is concerned with the result and discussion for this study which applies different feature selection approach and different classification algorithm of the kernel function in SVM. Finally, the conclusion gives a summary and critique of the findings.

2 RFID Principle

RFID technology's fundamental working concept relies on the tagged items accessing the reading area of the RFID reader. The radio frequency (RF) signal is received from the reader antenna and the EPC is transmitted from the tagged items stored in the chip by the energy produced by the induced current, or the active tag actively transmits a signal of a certain frequency. The reader reads and decodes the data and transfers it to the developed management system [10]. The theory of operating is that the radio wave energy reader generates a certain RF from the reader circuit, transmits internal data and sequentially transfers the reader to the reading data and sends them for the subsequent processing application [10].

The contact and energy transfer between the RFID reader and the EPC tag can be loosely divided into two types: inductive coupling and backscatter coupling. The passive tags typically connect the transmitter to the receiver by means of load modulation or backscatter, based on the tagged items are near or far away. The low RF RFID implements the inductive coupling while the higher RF RFID mostly implements the backscatter coupling [10]. Coupling is the transfer of energy from one circuit segment to another medium, like a passive tag's internal antenna draws the energy for transmitting data. The type of coupling used, inductance, or backscatter (also known as radiative) depends on the frequency and distance between the tag and the reader antenna. For instance, inductive coupling operates near-field effects, while backscatter coupling operates far-field effects. Near-field effects and far-field effects are different mechanisms of RFID for the transmission of energy in free space.

The RFID reader has the function of reading or reads/write in the device depending on the structure and technology used, and it is an RFID system information control

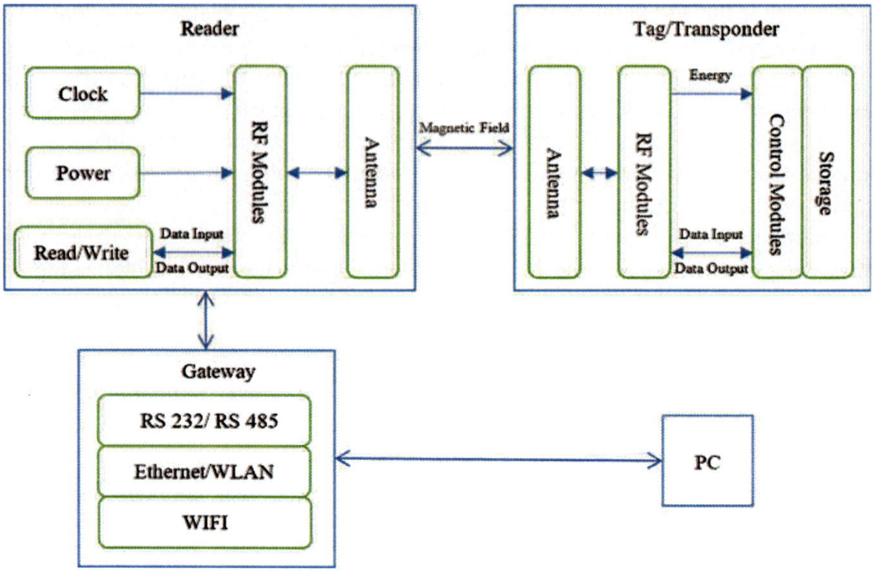

Fig. 1 RFID system structure and working principle

and processing center. The reader usually consists of a coupling module, an RF module, a control module, and an interface unit. The half-duplex communication method is commonly used for information exchange between the reader and the tag along with the reader, supplies energy, and timing by coupling to the passive tag. In practical applications, management functions such as collection, processing, and remote transmission of object identification information can be further implemented through Ethernet or WLAN. The configuration of the RFID system can be seen in Fig. 1.

2.1 *Data Extraction*

Feature extraction is a primary operation in signal processing, and it is the first arithmetic processing performed on an RFID reading [11]. It checks each received signal strength (RSS) to determine if the RSS value represents a feature. As a premise operation of feature extraction, the input RFID reading is generally smoothed from the continues-time signal to the discrete-time signal by the sampling. One or more features of the RFID reading are thereafter extracting from a continuous signal.

RSS values are intently associated with the distance between the tagged items and the reader antenna. Thus, the vary of the tag's orientation and the area will affect the RSS values of the indicator mirrored by the tagged items. These indicator variety will be leveraged to observe the fluctuations triggered by using the tagged items movement. Nine statistical features of RFID reading are extracted from the RSS values to indicate the different reading areas [7, 12]. The justifications for choosing these nine features are outlined below (Table 1).

Despite this, statistical features obtained from the RSS histograms are also used to characterize the RFID readings. The skewness and kurtosis are statistical attributes of sampling in higher-order that describes in the shape of the histograms. These parameters could be computed as being,

$$\text{Mean} = \frac{\sum_{i=1}^{n} x_i}{n} \tag{1}$$

Table 1 Extracts sample from RFID reading

Features	Description
MAX	The RFID is read to the greatest extent during the sampling process during data collection. The RSS value will higher if the tag is near to the reader antenna
MIN	The least RFID reads during sampling during data collection. The RSS value will be lower if far away from the reader antenna
Mean	The Arithmetic Mean is the average of the numbers that is the calculated "central" value of a set of RFID readings during data collection. The RSS value will change depending on the distance to the reader antenna
Mode	The mode of the RSS reading in sampling during the data collection. The most stable RSS value from the sampling in the static stage
Median	The "median" is the "middle"value in the list of RSS reading during sampling during data collection. The data separates the higher half from the lower half of the RSS value sample
Variance	The variance of the RSS reading in sampling based on the distance between the tag and the reader antenna if there is an adjustment on tag's orientation so the higher variance will occur
Range	During data collection, there is a difference between the maximum and the minimum RFID readings during sampling
Skewness	The data can be considered "skewed" meaning that it can have a long tail on one side in sampling. Skewed distributions tend to leptokurtic if the tags have movement
Kurtosis	Symmetric distributions of the RFID reading in sampling. The interest-only when dealing with distribution if heavy-tailed phase distribution present means the tags have movement

Table 2 Parameter of different kernel functions in SVM

Kernel function	C	Degree	Coef0	Gamma
Linear	1	–	–	–
Polynomial	0	3	–	1e−2
RBF	–	–	–	1e−2
Sigmoid	0	–	1/3	1e−2

$$\text{Median} = \left(\frac{n+1}{2}\right) \tag{2}$$

$$\text{Variance} = \frac{1}{n-1}\sum_{i=1}^{n}(x_i - \overline{x})^2 \tag{3}$$

$$\text{Range} = \text{MAX} - \text{MIN} \tag{4}$$

$$\text{Skewness} = \frac{\frac{1}{n}\sum_{i=1}^{n}(x_i - \overline{x})^3}{\left(\sqrt{\frac{1}{n}\sum_{i=1}^{n}(x_i - \overline{x})^2}\right)^3} \tag{5}$$

$$\text{Kurtosis} = \frac{\frac{1}{n}\sum_{i=1}^{n}(x_i - \overline{x})^4}{\left(\frac{1}{n}\sum_{i=1}^{n}(x_i - \overline{x})^2\right)^2} \tag{6}$$

where $\bar{x}$ is an average of the discrete numbers set and n is the discrete set of $\bar{x}_i$ alternatively.

The features that mentioned at above contribute essential insights into tag movement when the system is runned. From the data in the figure above, it is apparent that the RSS reading and each extracted feature capture form RFID reader. The peak of the RSS value is the local maximum that the tag is going nearest to the reader antenna. Then, it repeats it move away and going near to the reader antenna, which the tag attached to the goods is transferred by the person. A total of 550 raw data were taken from two RFID readers within 60 s, and the raw data undergoes feature engineering to extract the respected feature in each second (approximate 9 data collected in one second). Therefore, 116 data extracted from a respected feature of each second as the feature illustrated in Fig. 2. These views show above in respect of RSS's maximum, minimum, mode, mean, median, variance, range, kurtosis and skewness of RSS values have the different characters on the distance between the tag and reader antenna. Besides, the human clustering approach performs on the distinguish the tag belongs to which side of the RFID reading zone based on the algorithm of distance in [13].

2.2 *Support Vector Machine*

Support Vector Machine (SVM) is a comprehensive linear classifier that categorizes data in a supervised way of learning. The best boundary shall be the overall distance for the evaluation of the learning samples. The SVM uses the hinge loss function to quantify the analytical risk and applies to the solution method a regularization concept to maximize the systemic risk. It is a sparsely and sturdily classifier. SVM algorithms use a set of mathematical functions known as the kernel [14]. The kernel function is to take input data and convert it to the appropriate format. There are many

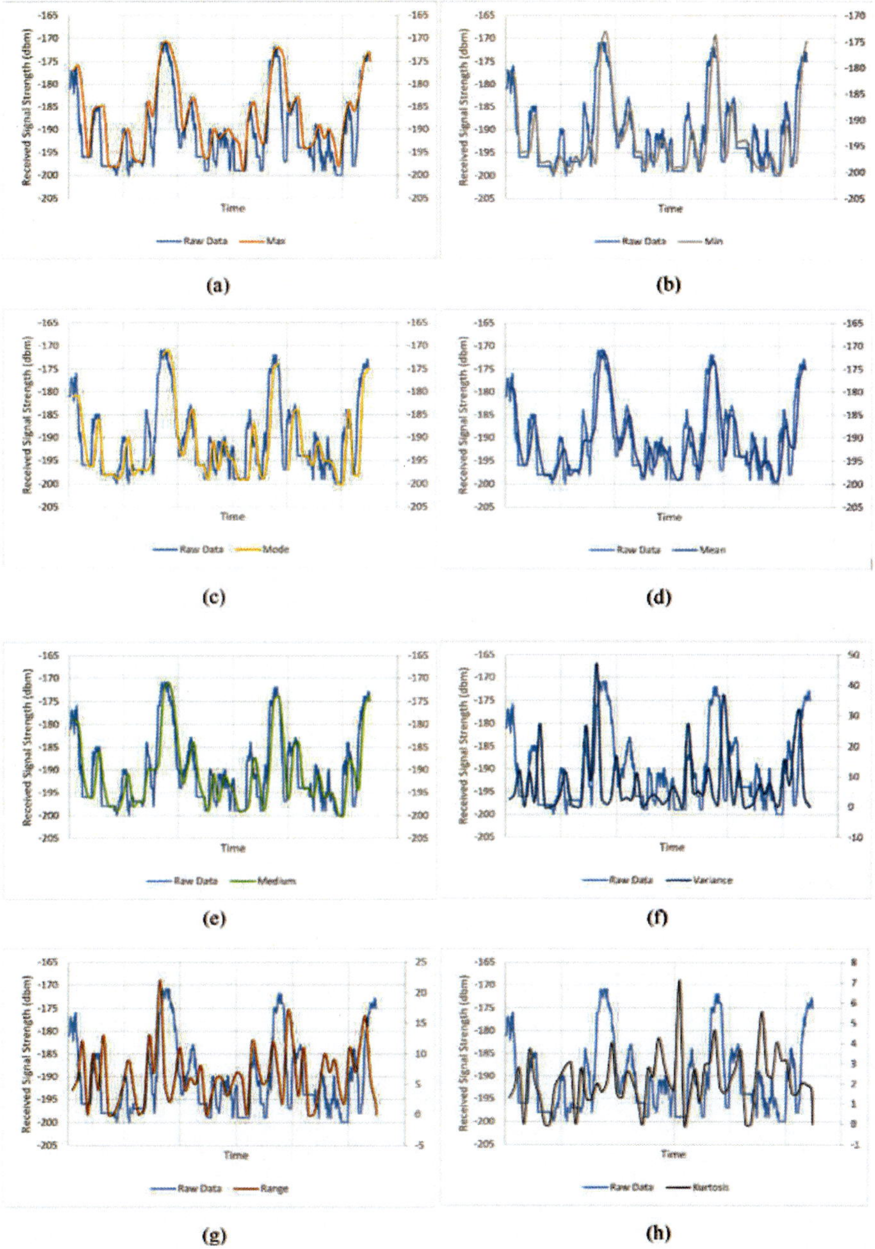

Fig. 2 RSS distribution for dataset in 60 s. **a** Maximum, **b** Minimum, **c** Mode, **d** Mean, **e** Median, **f** Variance, **g** Range, **h** Kurtosis, **i** Skewness

Fig. 2 (continued)

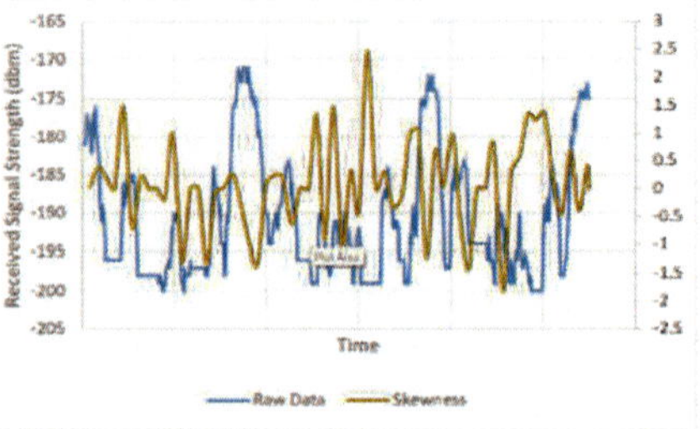

(i)

different functions of the kernel in various SVM algorithms, such as linear, nonlinear, polynomial, radial basis function (RBF), and sigmoid are simply separate in case of making the hyperplane decision boundary between the classes. The functions of the kernel are used as an inner product to a higher dimensional space to make it a linear dataset. Linear and polynomial kernels are generally less time-consuming and less reliable than the RBF or Gaussian kernels [15]. Therefore, by establishing different kernel function of SVM in those features is used to split the training set into 70–30. Then the 70 will be used for training undergoes 5-fold cross-validation, the mean value of classification accuracy, and the standard deviation is display in Table 3. The standard deviation of accuracy is verification throughout the entire training process. The slightly different sets of parameters in SVM and have different mathematical formulations (see Table 2).

It could be observed from the results of the correlational analysis that the RBF and Polynomial kernel function based SVM model is capable of producing outstanding

Table 3 Performance of classification accuracy in SVM classifier with a different kernel function

Features	Linear	Polynomial	RBF	Sigmoid
MAX	74.69 ± 2.19	78.48 ± 1.24	68.36 ± 0.85	73.42 ± 2.33
MIN	78.48 ± 2.19	**83.54 ± 1.96**	**83.54 ± 1.88**	78.48 ± 2.19
Mean	**81.02 ± 1.35**	**83.54 ± 1.64**	**82.28 ± 1.13**	74.69 ± 2.61
Mode	74.67 ± 2.35	51.90 ± 4.00	**83.55 ± 1.88**	73.41 ± 2.21
Median	70.89 ± 1.57	**81.02 ± 3.27**	78.48 ± 2.08	58.22 ± 1.69
Variance	64.55 ± 2.41	62.02 ± 2.02	70.87 ± 3.36	64.53 ± 4.47
Range	67.08 ± 3.17	54.42 ± 3.65	59.48 ± 2.70	59.50 ± 3.34
Skewness	59.50 ± 1.04	34.17 ± 2.68	60.75 ± 2.48	62.03 ± 1.30
Kurtosis	49.36 ± 2.84	48.10 ± 2.44	53.17 ± 2.66	32.91 ± 2.66
Combination of highest four single feature in Polynomial	74.69 ± 1.95	77.22 ± 2.35	73.42 ± 1.77	73.43 ± 2.80
Combination of highest four single feature in RBF	74.67 ± 3.11	74.68 ± 1.02	84.81 ± 1.60	58.22 ± 1.69
Combination of all feature	69.62 ± 3.65	75.95 ± 1.80	84.82 ± 3.50	67.09 ± 1.75

classification accuracy by testing the single feature in relation to the other kernel function [16, 17]. Thus, the four highest of single extracted feature from each RBF and Polynomial kernel function based SVM model is selected to form the combination to test classification accuracy. It is obvious that the model provides a low error rate with a reasonably accurate classification. In contrast, the SVM model based on linear and sigmoid kernel functions excluded from the analysis due to the overall classification accuracy is less than 80%.

Besides, there are few models that are rather suitable for predicting the correct classification of about 83%, as it yields the highest misclassification rate at 32% on top of a low negative class accuracy has been seen. Through the analysis, it is evident that the RBF or Polynomial-based kernel functions SVM model may be used for the classification zone based on RFID reading.

3 Result and Discussion

The cumulative data obtained during the experiment is 550. Class identification shall be accomplished by determining the rate of classification on the basis of RFID signal pattern. The selection of these four best features of the ten is based on earlier research [7]. This is carried out to differentiate the distance of the reader antenna between the designated area or non-designated area, which are not influenced by the prediction of human behavior that may be spent time looking tagged goods. The procedure begins by captured the data from two reader antenna within the 60 s, and here lies the assumption that the state of being is to be identified for 'Designated Area' as '0' or 'Non-designated Area' as '1' by detecting the working process.

In Tables 4, 5, 6, 7, 8, and 9 evaluate the percentage error rate for train and test data, respectively. The error rate is measured as the number of all incorrect predictions divided by the total number of the dataset. The comparison in the trained set determines that all the variations provide the high accuracy of classification above

Table 4 Confusion matrix of MIN in SVM based polynomial for training and testing

Error	Train set 16.46%	Test set 25.71%
Confusion Matrix	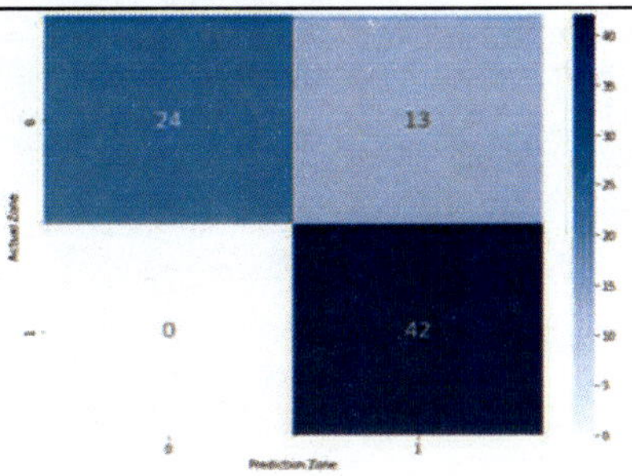	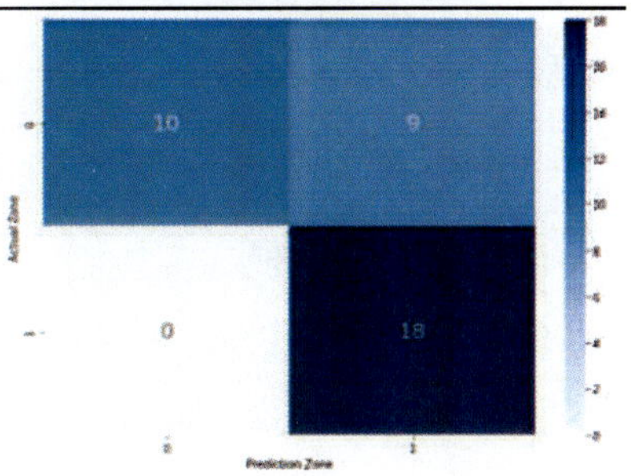

Table 5 Confusion matrix of MIN in SVM based RBF for training and testing

	Train set	Test set
Error	16.46%	28.57%
Confusion Matrix	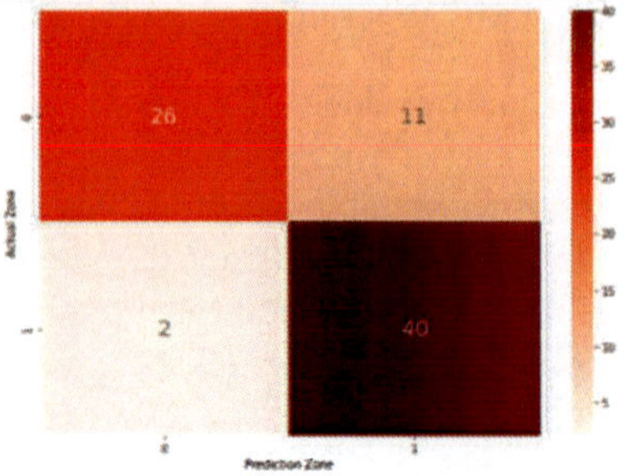	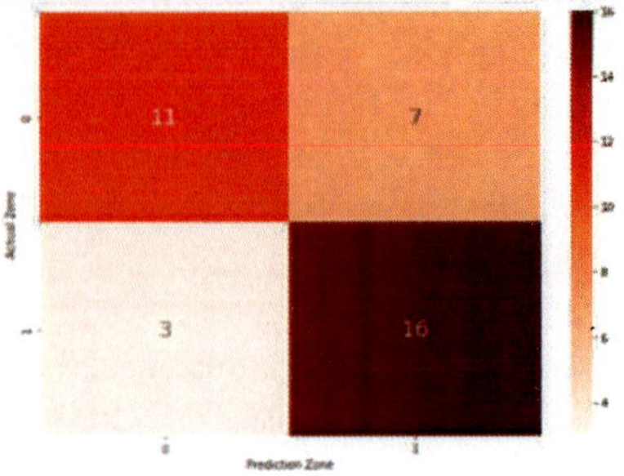

Table 6 Confusion matrix of Mean in SVM based Polynomial for training and testing

	Train set	Test set
Error	16.46%	34.29%
Confusion Matrix	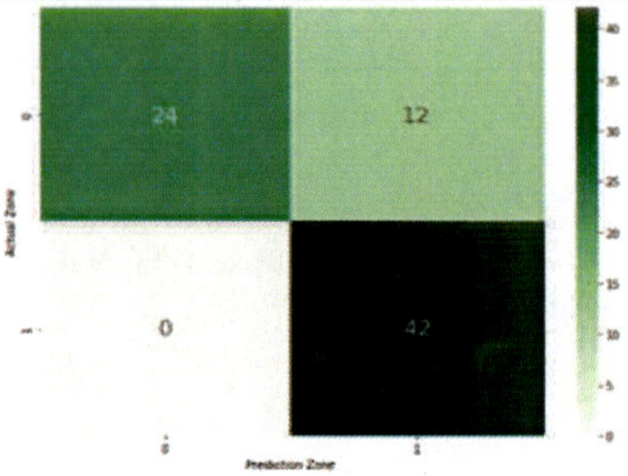	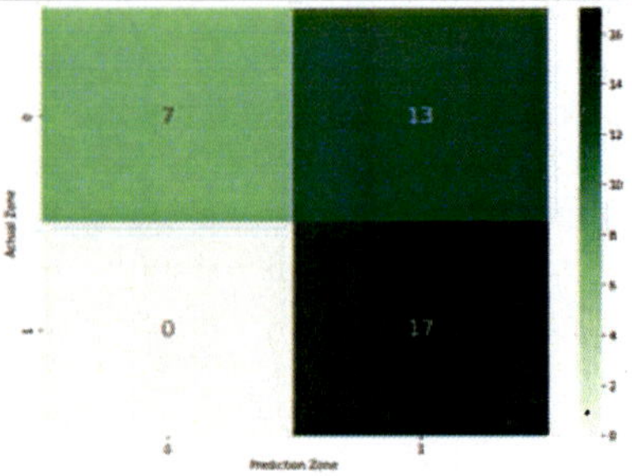

Table 7 Confusion matrix of Mode in SVM based RBF for training and testing

	Train set	Test set
Error	16.45%	25.71%
Confusion Matrix	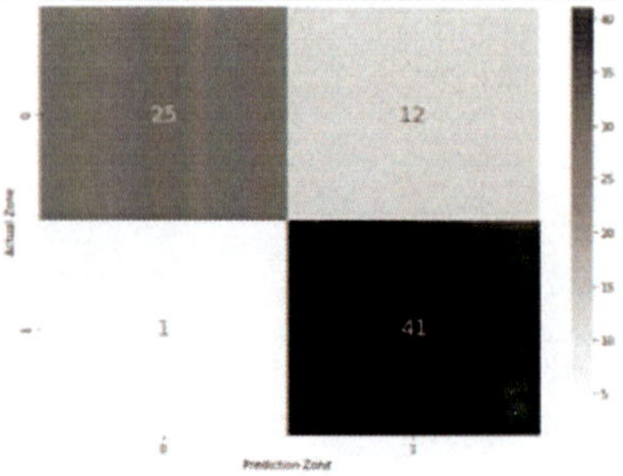	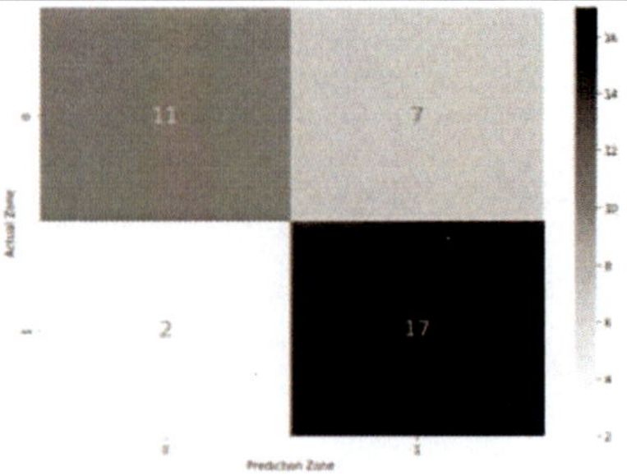

Table 8 Confusion matrix of combination of the highest single feature in SVM based RBF for training and testing

	Train set	Test set
Error	15.19%	20.00%
Confusion Matrix	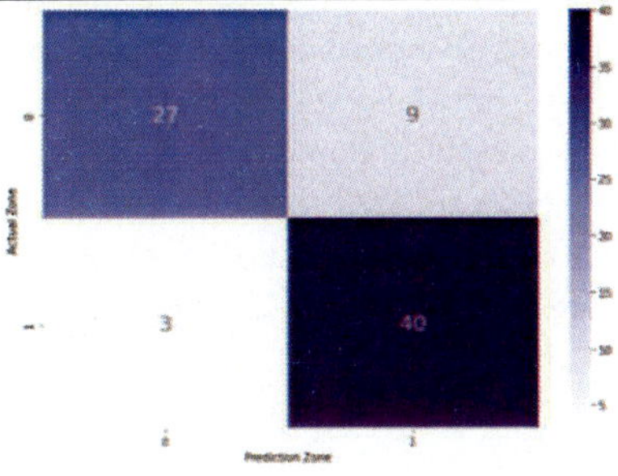	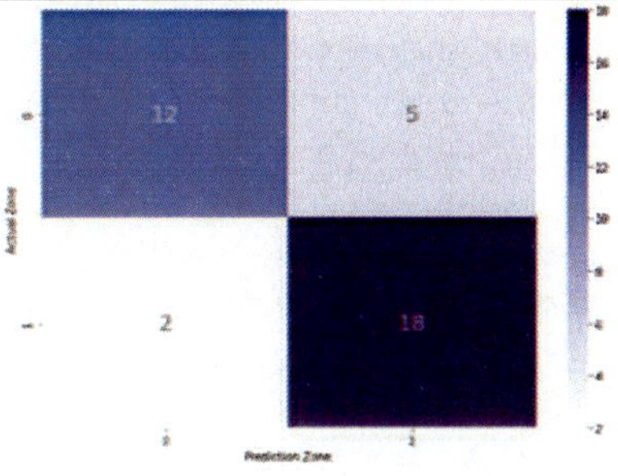

Table 9 Confusion matrix of Combination of all features in SVM based RBF for training and testing

	Train set	Test set
Error	15.18%	31.43%
Confusion Matrix	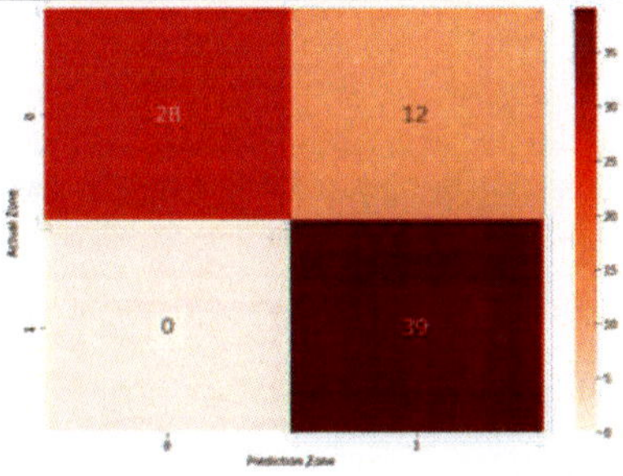	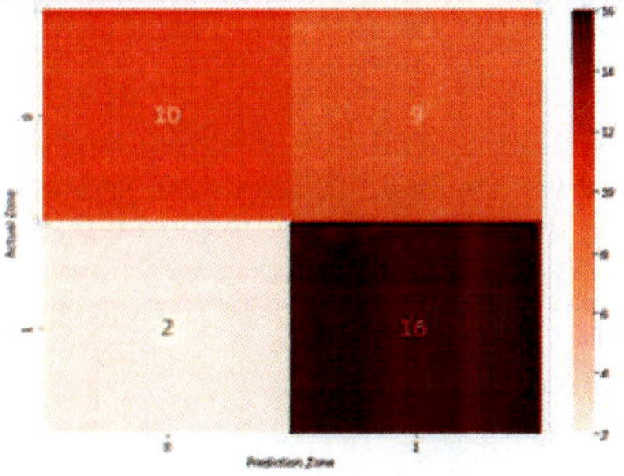

83%. From the result in Table 3 suggest that SVM based Polynomial on MIN, SVM based RBF on MIN, SVM based Polynomial on Mean, SVM based RBF on Mode, SVM based RBF on the combination the highest four features, and SVM based RBF on the combination of all feature from the trainset to test with misclassification rate. The 30% of the splitting data initially is going to test those models, and the SVM based RBF on the combination of the highest four features have the lowest misclassification rate at 20% for selected as the final model. In comparison, the other misclassification rate is slightly higher than the final model with 5.75%, 8.57%, 14.29%, 5.71%, and 11.43%, respectively [18]. As a result, the final model is discovered by the most suitable feature in terms of save more processing times in a calculated lot of features and exclude the single feature with lower classification accuracy.

4 Conclusion

In this overview, a contrast of the SVM model with different kernel functions by extracting features from RFID reading to differentiate tags for the designated area and the non-designated area has been shown. The classification procedure here evaluates a single feature and the combination of nine selected extracted features. It has been shown that the MIN derived by the Polynomial-SVM model will provide a classification accuracy of 84.81% and 20.00% of the error rate in the test results. Nevertheless, these findings may help us to understand the concept of the classification of the different zone in its preliminary study. Besides, an enhancement may be made by using a more suitable form of feature selection, such as Principal Component Analysis, Feature Importance, or univariate selection techniques, etc., to determine its effectiveness in classifying the tagged item.

Acknowledgements The authors are grateful to KPT Malaysia for providing the funding under the FRGS grant [FRGS/1/2017/ICT05/UMP/02/2] and [RDU170107].

References

1. Nayak R, Nayak R (2019) Introduction to radio frequency identification. In: Radio frequency identification (RFID) technology and application in fashion and textile supply chain, pp 1–18
2. Hunt VD, Puglia A, Puglia M (2006) History and evolution of RFID technology. In: RFID-A guide to radio frequency identification, pp 25–31
3. Parlak S, Marsic I (2013) Detecting object motion using passive RFID: A Trauma Resuscitation Case Study. IEEE Trans Instrum Meas 62:2430–2437
4. Bertoncini C, Rudd K, Nousain B, Hinders M (2012) Wavelet fingerprinting of radio-frequency identification (RFID) tags. IEEE Trans Ind Electron 59:4843–4850
5. Chriki A, Touati H, Snoussi H (2017) SVM-based indoor localization in wireless sensor networks. In: 2017 13th International wireless communications and mobile computing conference (IWCMC), pp 1144–1149. IEEE
6. Keller T, Thiesse F, Kungl J, Fleisch E (2010) Using low-level reader data to detect false-positive RFID tag reads. In: 2010 Internet of Things (IOT), pp 1–8. IEEE
7. Ma H, Wang Y, Wang K (2018) Automatic detection of false positive RFID readings using machine learning algorithms. Expert Syst Appl 91:442–451
8. Hauser M, Zügner D, Flath C, Thiesse F (2015) Pushing the limits of RFID: Empowering RFID-based electronic article surveillance with data analytics techniques. In: 2015 International conference on information systems: exploring the information frontier, ICIS 2015
9. Hauser M, Günther SA, Flath CM, Thiesse F (2019) Towards digital transformation in fashion retailing: a design-oriented IS research study of automated checkout systems. Bus Inf Syst Eng 61:51–66
10. Karmakar NC (2010) Handbook of smart antennas for RFID systems
11. Liang Z, Wei J, Zhao J, Liu H, Li B, Shen J, Zheng C (2008) The statistical meaning of kurtosis and its new application to identification of persons based on seismic signals. Sensors (Basel) 8:5106–5119
12. Zhang J, Tian GY, Zhao AB (2017) Passive RFID sensor systems for crack detection & characterization. NDT E Int 86:89–99

13. Choong CS, Fakhri A, Nasir A, Majeed APPA, Zakaria MA (2020) Automatic identification and categorize zone of RFID reading in warehouse management system. In: Advances in mechatronics, manufacturing, and mechanical engineering. Lecture notes in mechanical engineering. Springer
14. Taha Z, Razman MA, Adnan FA, Abdul Ghani AS, Abdul Majeed AP, Musa RM, Sallehudin MF, Mukai Y (2018) The identification of hunger behaviour of Lates calcarifer through the integration of image processing technique and support vector machine. In: IOP conference series: materials science and engineering
15. Homoliak I, Increasing classification accuracy in LIBSVM Using String Kernel functions, pp 2–4
16. Taha Z, Musa RM, Majeed AA, Abdullah MR, Zakaria MA, Alim MM, Jizat JA, Ibrahim MF (2018) The identification of high potential archers based on relative psychological coping skills variables: a support vector machine approach. In: IOP conference series: materials science and engineering
17. Taha Z, Musa RM, Majeed AP, Alim MM, Abdullah MR (2018) The identification of high potential archers based on fitness and motor ability variables: a support vector machine approach. Hum Mov Sci 1(57):184–193
18. Razman MA, Susto GA, Cenedese A, Majeed AP, Musa RM, Ghani AS, Adnan FA, Ismail KM, Taha Z, Mukai Y (2019) Hunger classification of Lates calcarifer by means of an automated feeder and image processing. Comput Electron Agric 1(163):104883

Modeling and Analyzing of Traveling Wave Gait of Modular Snake Robot

Areej G. Abdulshaheed, Mohamed Bin Hussein, Mohd Azuwan Mat Dzahir, and Shaharil Mad Saad

Abstract Traveling wave locomotion of snake robot considers one of the efficient locomotion that can be used in narrow spaces and unstructured environment. Few works have been done on studying and analyzing this gait. In this paper, we present the kinematic and dynamic analysis of traveling wave locomotion using the Lagrange Euler equation. The effects of a dynamic model and an elaborate model of the friction between the robot and the supporting plane are presented, were, the supporting and friction forces can be calculated by solving the equilibrium equation. The dynamic model, forward and turning motion of the traveling wave gait are verified using simulation. It was shown from the simulation investigation that joint torque was observed to be high near the robot's center of mass. It is also noticed that the joint torque increases with the increase of the friction coefficient and reduces with the increase of the initial winding angle.

Keywords Snake robot · Serpenoid curve · Dynamic · Stability · Traveling wave

1 Introduction

Snake-like robots, motivated by nature, are anticipated to aid with tasks such as rescue and search, surveillance, and investigation [1, 2]. By integrating biomimetic actuation and design approaches, snake-like robotic systems embrace beneficial locomotive approaches, which expedite mission performance in challenging environments, as demonstrated by their biological analogues. Mobility in these situations includes skilled locomotion through a range of terrain distinguished by various body-ground encounters, such as pavement, concrete, or carpet, which illustrates the potential surface types that may be identified. Previous experiments with robotic locomotion on flat surfaces utilized passive wheels to enforce cinematic restrictions to allow

A. G. Abdulshaheed (✉)
Al-Dewaniya Technical Institute, Al-Furat Al-Awsat Technical University, Al-Dewaniya, Iraq
e-mail: ghareej80@gmail.com

M. B. Hussein · M. A. M. Dzahir · S. M. Saad
Universiti Teknologi Malaysia, Johor, Malaysia

A. F. Ab. Nasir et al. (eds.), *Recent Trends in Mechatronics Towards Industry 4.0*, Lecture Notes in Electrical Engineering 730, https://doi.org/10.1007/978-981-33-4597-3_14

efficient mobility via lateral undulation [3, 4] as well as traveling wave locomotion [5–8]. Traveling wave locomotion which develops in the vertical plane orthogonal to the supporting plane is more suited to confined spaces and tight corridors. For a snake robot with wheels, a purely kinematic model can be used by assuming there is no slippage or it could be modeled with both kinematic and dynamic [9, 10]. While for a snake robot without wheels, the motion of the snake robot is affected directly by friction between the snake body and the ground because of that the locomotion patterns can be modeled by using a dynamical model [11, 12]. Euler-Lagrange equation of motion was used in [13, 14] to develop a dynamic feedback controller law. This controller which is used to guide the orientation motion of the planar snake robot to the desired path depended on the time evaluation of the dynamic compensator's state variables. Another dynamical model of sidewinding and sinus-lifting motion are presented in [10]. In [15] a dynamic model of inchworm locomotion based on the Lagrange equation is introduced. The model is simulated on bond graphics and a comparison had been made between the simulation the dynamic model to ensure the stability of the system.

This paper sets out to expound on the kinematic and dynamic of the travel wave gait, as well as to present the parameters of this gait, since changing the parameters, in turn, alter the style and form of the motion which results. It also describes the effect of the environment on joint torque.

2 Serpenoid Curve

The snake body generates continuous sine waves during its propulsion. This sine wave is called serpenoid curve, which is used to control the body motion of a snake robot to mimic the motion of a real snake. Hirose was the first one who describes this curve and used it to formulate the serpentine locomotion which is one of the most common locomotions of the biological snake [3]. For convenience, the same serpeniod curve in [12] is used to describe the curvature function.

$$\rho(s) = \frac{2K_n \pi \alpha}{L} \sin\left(\frac{2K_n \pi}{L} s\right) \tag{1}$$

Where K_n is the number of wave shapes, α is the initial winding angle, L is the whole length of the snake body, and s is the body length along the body curve. See Fig. 1. The snake robot body shape is controlled by the relative joint $\phi_i(s)$, which can be obtained from the integration of the serpentine curvature.

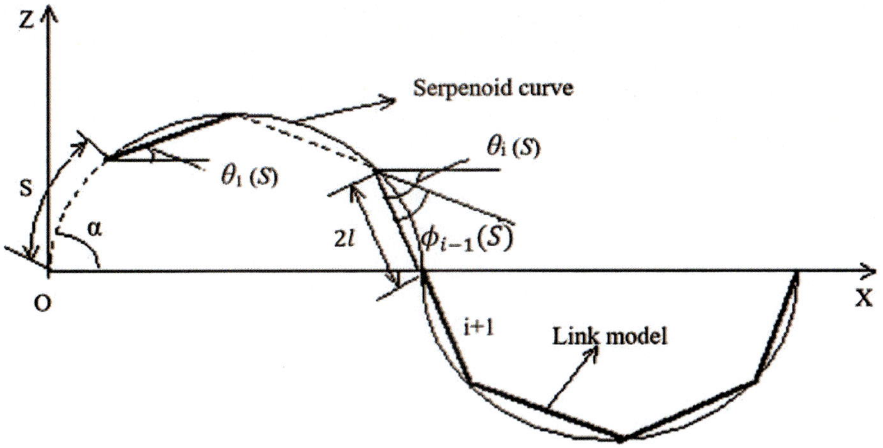

Fig. 1 Sine wave of serpentine motion

$$\phi_i(s) = \int_{s+(i-1)l}^{s+il} \rho(u)du$$
$$= -2\alpha \sin\left(\frac{K_n\pi}{n}\right). \sin\left(\frac{2K_n\pi s}{L} + \frac{2K_n\pi i}{n} + \frac{K_n\pi}{n}\right)$$
$$\mathrm{i} = 1, 2, \ldots\ldots, \mathrm{n} - 1 \tag{2}$$

where l, n is the unit length of the link, and the number of links, respectively. The relative angle can be calculated when s is given. θ_{i} (S) is the absolute joint angle. The velocity and acceleration of the relative joints can be obtained from derivative Eq. (2)

3 Kinematic of Snake Robot

The kinematic map represents the relationship between the relative joint angles ϕ_i and the absolute link angles θ_i can be introduced as.

$$\theta_i = \sum_{i=1}^{n-1} \phi_i + \theta_n \tag{3}$$

The modules of the hybrid redundant discrete mobile robot are represented by an equal length of n rigid links connected by n-1 active joints (yaw and pitch order) as shown in Fig. 2. All the links assumed to have the same moment of inertia $I = \frac{1}{3}ml^2$ and mass m. The center of the mass (CM) is located at the center point of each link.

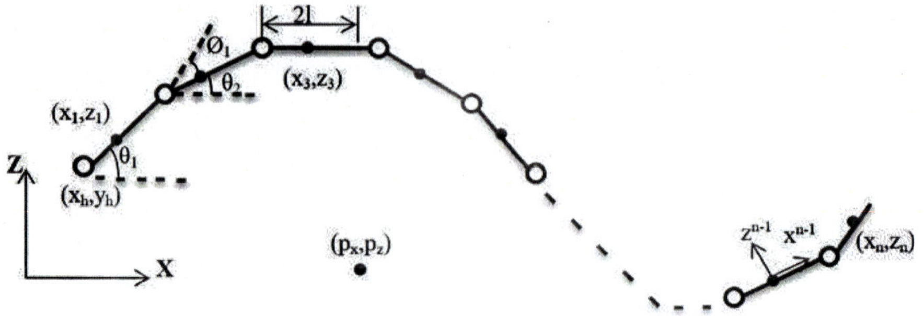

Fig. 2 Coordinate located on the robot

The mathematical model in this situation has n + 2 degree of freedom, which 1 for orientation, 2 for position, and n − 1 for the shape.

Vector-valued are defined as:

$$\begin{aligned} \cos\theta &= [\cos\theta_1, \cos\theta_2, \ldots, \cos\theta_n]^T \\ \sin\theta &= [\sin\theta_1, \sin\theta_2, \ldots, \sin\theta_n]^T \end{aligned} \tag{4}$$

The diagonal matrices functions are defined as:

$$\begin{aligned} C_\theta &= diag(\cos\theta) \in R^{n\times n} \\ S_\theta &= diag(\sin\theta) \in R^{n\times n} \end{aligned} \tag{5}$$

The CM of the link can be calculated as

$$x_i = x_h + \sum_{j=1}^{i-1} 2lcos\theta_j + lcos\theta_i \tag{6}$$

$$z_i = z_h + \sum_{j=1}^{i-1} 2lsin\theta_j + lsin\theta_i \tag{7}$$

The position, velocity, and acceleration of the CM of each link can be summarized as

$$\begin{cases} X = ex_h + lK\cos\theta \\ Z = ez_h + lK\sin\theta \end{cases} \tag{8}$$

$$\begin{cases} \dot{X} = e\dot{x}_h - lKS_\theta\dot{\theta} = [e0 - lKS_\theta]\dot{\bar{q}} = J_{qx}\dot{\bar{q}} \\ \dot{Z} = E\dot{z}_h + lKC_\emptyset\dot{\theta} = [e0lKC_\theta]\dot{\bar{q}} = J_{qz}\dot{\bar{q}} \\ \dot{\theta} = \left[0_{n\times 2}\ I_n\right]\dot{\bar{q}} \end{cases} \tag{9}$$

$$\text{where, } J_{qx} = [e0 - lKS_\theta] \in R^{n\times(n+2)},\ J_{qz} = [e0lKC_\theta] \in R^{n\times(n+2)} \tag{10}$$

$$\dot{\bar{q}} = \left[\dot{x}_h, \dot{z}_h, \dot{\theta}_1, \dot{\theta}_2, \ldots\ldots, \dot{\theta}_n\right] \in R^{n+2} \tag{11}$$

$$K = \begin{bmatrix} 1\,0 \cdots 0 \\ 2\,1 \cdots 0 \\ \vdots\,\vdots \ddots 0 \\ 2\,2\,2\,1 \end{bmatrix} \in R^{n\times n} \tag{12}$$

Where $\bar{q}$ represents the general coordinate for snake robot with n links, for more convenience, the generalized coordinate system can be defined as

$$q = \left[x_h, z_h, \theta_1, \phi^T\right] \in R^{n+2} \tag{13}$$

The relationship between the two generalize coordinate system, as follows

$$\dot{\bar{q}} = J_{q\bar{q}}\dot{q} \tag{14}$$

$$\text{Where, } J_{q\bar{q}} = \begin{bmatrix} I_2 & 0_{2\times n} \\ 0_{n\times 2} & J_{\theta\phi} \end{bmatrix} \tag{15}$$

$$J_{\theta\phi} = \begin{bmatrix} 1\,0 & \cdots & 0 \\ 1\,1 & & 0 \\ \vdots\,\vdots & \ddots & \vdots \\ 1\,1 & \cdots & 1 \end{bmatrix} \in R^{n\times n} \tag{16}$$

4 Dynamic Model of the Robot

As it noticed, there were only a few works done on describing the dynamic of the traveling wave locomotion of the snake robot. Lagrange Euler Equation is used to analyze the forces acting on the snake robot body during motion, which can be written as:

$$\frac{d}{dt}\left(\frac{\partial T}{\partial \dot{q}}\right) - \frac{\partial T}{\partial \mathrm{q}} + \frac{\partial V}{\partial \mathrm{q}} = \mathcal{F}_i^{nc} \tag{17}$$

Where T represents the kinematic energy and V represents the potential energy. While $\mathcal{F}_i^{nc}$ represents the Non-conservative forces. The Non-conservative forces represent the generating forces that give virtual displacement. These generalized forces are the actuator torque, friction, and supporting forces. Therefore, the Lagrange equation for this system can be written as:

$$M(\bar{\phi})\ddot{q} + W\left(\bar{\phi}, \dot{\bar{\phi}}\right) + G(\bar{\phi}) + F(\bar{\phi}) + N(\bar{\phi}) = B\tau \tag{18}$$

where $\bar{\phi} = [\theta_1, \phi_1, \phi_2, \ldots, \phi_{n-1}]$ represents the angular acceleration of snake robot links, $M(\bar{\phi}) \in R^{n+2}$ is the inertia matrix, $W\left(\bar{\phi}, \dot{\bar{\phi}}\right) \in R^{n+2}$ is the Coriolis and centrifugal forces. $G(\bar{\phi})$, $F(\bar{\phi})$, $N(\bar{\phi}) \in R^{n+2}$, represents the potential energy matrix, friction forces matrix, and supporting forces matrix, respectively. B is the constant matrix. Finally, $\tau = \tau_1, \ldots, \tau_{n-1} \in R^{n-1}$ is the input torque. Equation (18) can be divided into two parts.

$$M_{11}\ddot{q}_a + M_{12}\ddot{q}_u + W_1 + G_1 + F_1 + N_1 = 0_{3\times 1} \tag{19}$$

$$M_{21}\ddot{q}_a + M_{22}\ddot{q}_u + W_2 + G_2 + F_2 + N_2 = \tau \tag{20}$$

The first equation represents the three unactuated degrees of freedom, where $q_u = [x_h, z_h, \theta_1]^T \in R^3$. The second equation refers to the actuated system, where $q_a = [\phi_1, \phi_2, \ldots, \phi_{n-1}]^T \in R^{n-1}$. By solving Eq. (19), the acceleration of the head link $(\ddot{x}_h, \ddot{z}_h)$, and the angular acceleration of the snake robot's head $\ddot{\theta}_1$ can be found. Subsequently, the linear and angular position and velocity of the snake robot head can be obtained through integration. In addition, the joint torque can be calculated by solving Eq. (20). Finally, the friction forces and the supporting forces can be calculated from the next section.

5 Ground Friction Model

In linear progressing locomotion, most of the robot's body is lifted from the ground keeping only a few points in contact with the ground. This causes instant friction to last for a short time during the landing and lifting of the robot's joints. The snake robot is assumed to have 16 joints with a body shape number K_n= 2. As shown in Fig. 3, the robot contacts the ground with only two points, which is the minimum no. of contact points that keep the body of the robot balanced. These contacting joints of the robot with the ground are set as $\bar{r}, andr$. To simplify the model, the state where the links of the robot become in a horizontal state with the ground is neglected and only Coulomb friction is considered to acts between the environment and the robot.

$$F_i = -\mu sgn(v)N_i \tag{21}$$

where, F_i, μ, N_i are the friction force in each link, coefficient of friction, and x. Sgn (v) is used to denote to the signum function, where:

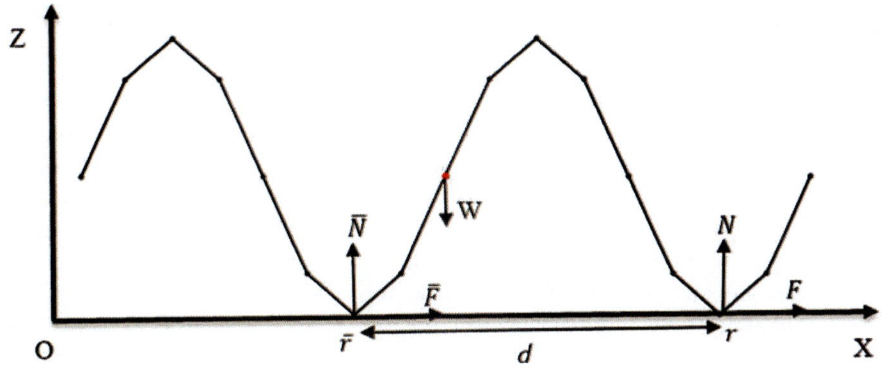

Fig. 3 Free body diagram of the snake robot during linear progressing

$$sgn(v) = \begin{cases} 1 & v > 0 \\ 0 & v = 0 \\ -1 & v < 0 \end{cases}$$

The supporting forces can be obtained from the equilibrium equation of the free body diagram

$$\bar{N} + N - W - M\ddot{Z} = 0 \tag{22}$$

$$NL - Wp_x - M\ddot{Z}p_x + M\ddot{X}p_z = 0 \tag{23}$$

$$N = \frac{1}{d}(W - M\ddot{Z})p_x - M\ddot{X}p_z \tag{24}$$

$$\bar{N} = W - N \tag{25}$$

where, N, $\bar{N}$, p_x, p_z, d are the supporting forces, the center of the mass in the x-direction, the center of the mass in the z-direction, and the displacement between two contacting points, respectively. W is the total weight of the snake robot. The driving force is a result only of the friction force because the supporting force N is balanced with the gravity force W. As can be seen from Eqs. (21), (22), and (24), the acceleration and position of the snake robot are associated with the friction force. According to the kinematic and dynamic of the snake robot, it can be controlled by its body shape and torque of the robot's joint.

6 Simulation Result

In this section traveling wave locomotion of hybrid redundant mobile robot is simulated. The effect of the environmental parameters such as the coefficient of friction and initial winding angle on the performance torque are discussed. The snake robot is assumed to have 16 links with 15 joints of 1 DoF. The model is assumed to contact the supporting plane with two points only. The simulation parameters of the snake robot and the serpenoid curve are shown in Table 1.

The change of the body shape of the snake robot is identified with respect to the modular modeled robot tail displacement along serpenoid curve. The acceleration $\ddot{s}$ is given as follows:

$$\ddot{s} = \begin{cases} a0 \leq t < T/10 \\ 0T/10 \leq t < 9T/10 \\ -a9T/10 \leq t < T/10 \end{cases} \tag{26}$$

where a = 0.0625 m/s^2, and T represents the simulation time ($T = 20$ s). The initial position and velocity of the head of the robot are set as $x_h = 0, z_h = 0, \dot{x}_h = 0, and \dot{z}_h = 0$. The joint angle and angular velocity of the first link is set as $\theta = \alpha$, and $\dot{\theta}_1 = 0$. The input torque applied to the joint of the modeled robot of traveling wave locomotion is simulated using Eq. (20) by sitting the winding angle $\alpha = \pi/6$, and the friction coefficient $\mu = 0.3$. Figure 4 shows the torque of joints 2, 3, 7, 9 respectively. It is noticed from this figure that joints 9, and 7 have the largest torque while joints 3, and 2 have the smallest torque. This is due to the different distance from these joints to the center of the gravity (CM) of the modeled robot and the position of the contacting points of the robot with the supporting plane. The joint torque increases if there is a small distance between the joint and the center of the gravity of the robot. Because joint 9 has the nearest distance to the center of gravity, it has the largest torque. It is also noticed that the torque changes periodically, where it forms three different peaks during the period 14 s. After 14 s the same torque cycle repeated again. This due to the different 9 phases generated by the robot during this period. In each phase, the robot contacts the supporting plane with different two joints, and the position of the center of gravity changes. The trace of the robot is shown in Fig. 5.

Table 1 Parameters of simulation

Robot parameters		Serpenoid curve parameters	
Link length l (m)	0.1	Initial winding angle α	π/6
Number of links n	16	Total length of snake L (m)	1.6
Mass of each link m (kg)	0.1	Number of s shape kn	2
Friction coefficient μ	0.3		

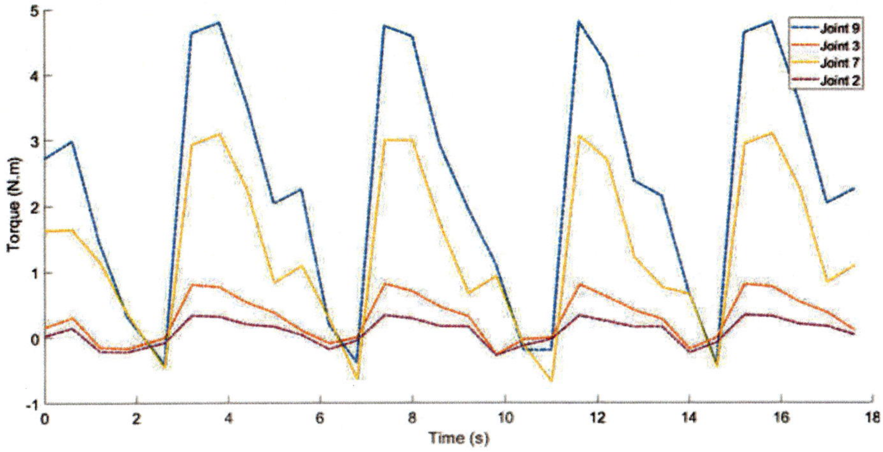

Fig. 4 Variation of torque joint ($\mu = 0.3$, $\alpha = \pi/6$)

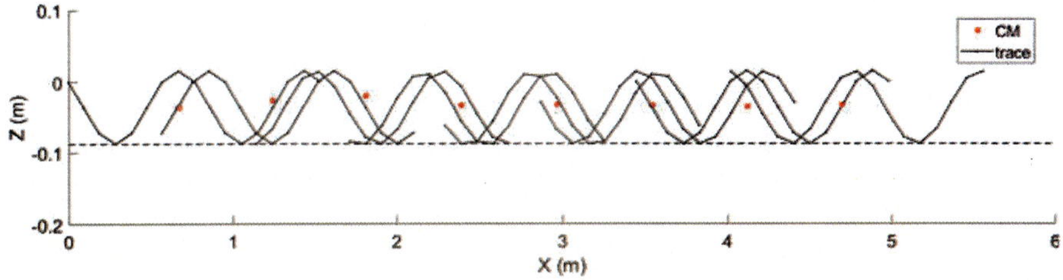

Fig. 5 Trace of the snake robot during traveling wave locomotion

In general, the input joint torque is related to the value of pressure applied on each point of the contact points, and the distance of these points to the center of gravity. Higher pressure is applied to the contact point which has the closest distance to the CM. So the joints which are nearest to this contact point have higher torque than the other joints. Controlling the input torque to the robot joint angles is important to enhance the performance of the locomotion. Higher required torque could increase power consumption and reduce the efficiency of locomotion.

Joint 9 is selected as a reference joint to study the effect of environmental parameters such as the coefficient of friction and initial winding angle on the torque joints. To study the effect of friction force on the joint torque of the modeled robot during traveling wave locomotion, a different coefficient of friction is selected. $\mu = 0.1$, 0.3, 0.5, and 0.7 respectively, and the winding angle α is set to $\pi/6$. Figure 6 shows increasing of the joint input torque with the increase of the coefficient of friction. This increase of torque belongs to the increasing of the power consumption of each joint in traveling wave locomotion with increasing of the friction coefficient. However, the Coefficient of friction may play a major role in some locomotion such as climbing, where, climbing locomotion requires a higher coefficient of friction between the

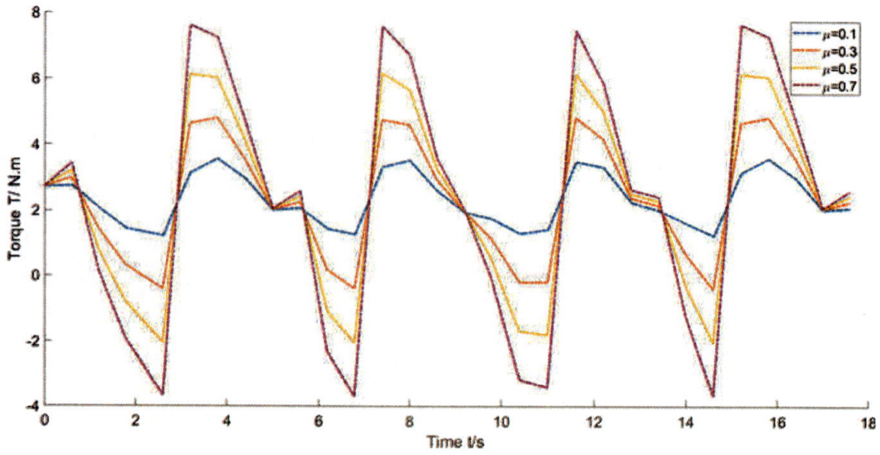

Fig. 6 Effect of the varying friction coefficient on joint torque

robot and the environment than other type of locomotions to prevent the robot body from slipping down.

Set of winding angle $\alpha = \pi/12$, $\pi/9$, and $\pi/6$ are selected to investigate the influence of the winding angle of the joint torque. As it is noticed from Fig. 7 that there is increasing in the joint torque with the reduction of the value of the initial winding angle. From this, it can be concluded that the winding angle can be used to decrease the required input joint torque for the environment with high friction coefficient by increasing the initial winding angle.

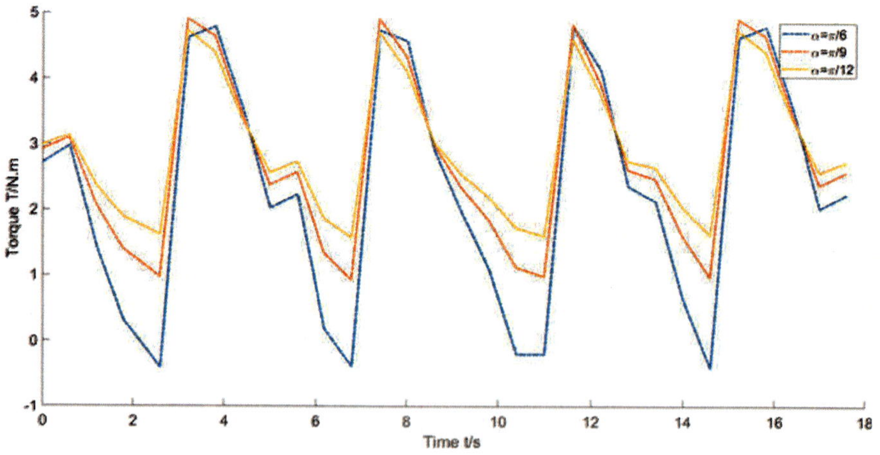

Fig. 7 Torque joint influence by the varying of the initial winding angle

7 Conclusion

In this paper, the kinematic and dynamic model of traveling wave locomotion is proposed. A description of the internal and external forces acting on the body of the robot is proposed, the relationship between these forces with the torque joint and body shape of the robot is clarified. The mathematical model is verified by simulation. According to the simulation, the joint torque value depends on the position of the robot's joint from the robot's center of gravity and the static contact point with the supporting plane. The simulation also shows the effects of the varying of the coefficient of friction and initial winding angle on the joint torque, where the joint torque increases with the increase of friction coefficient while reduces with the rising of initial winding angle.

References

1. Ansari A et al (2017) Modular platforms for advanced inspection, locomotion, and manipulation. In: Waste management conference
2. Whitman J et al (2018) Snake robot urban search after the 2017 mexico city earthquake. In: 2018 IEEE international symposium on safety, security, and rescue robotics (SSRR). IEEE
3. Hirose S (1993) Biologically inspired robots: serpentile locomotors and manipulators. Oxford University Press
4. Mori M, Hirose S (2001) Development of active cord mechanism ACM-R3 with agile 3D mobility. In: Proceedings 2001 IEEE/RSJ international conference on intelligent robots and systems. Expanding the Societal Role of Robotics in the Next Millennium (Cat. No. 01CH37180). IEEE
5. Chirikjian G, Burdick J (1995) The kinematics of hyper-redundant robot locomotion. IEEE Trans Robot Autom 11(6):781–793
6. Chang AH et al (2019) Autonomous, monocular, vision-based snake robot navigation and traversal of cluttered environments using rectilinear gait motion
7. Varlashin V, Shmakov O (2019) Comparative analysis of the virtual and physical experiments with a snake robot's rectilinear movement. J Phys Conf Ser. IOP Publishing
8. Manzoor S et al (2019) Serpentine and rectilinear motion generation in snake robot using central pattern generator with gait transition, pp 1–11
9. Liljeback P et al (2011) Controllability and stability analysis of planar snake robot locomotion. IEEE Trans Autom Control 56(6):1365–1380
10. Ariizumi R, Matsuno F (2017) Dynamic analysis of three snake robot gaits. IEEE Trans Rob 33(5):1–13
11. Chen L et al (2007) Design and modelling of a snake robot in traveling wave locomotion. Mech Mach Theor 42(12):1632–1642
12. Chen L et al (2003) Analysis of traveling wave locomotion of snake robot. In: IEEE international conference on robotics, intelligent systems and signal processing, 2003. Proceedings. 2003. IEEE
13. Rezapour E, Pettersen KY, Gravdahl JT, Liljebäck P (2014) Body shape and orientation control for locomotion of biologically-inspired snake robots. In: 2014 5th IEEE RAS & EMBS international conference on biomedical robotics and biomechatronics (BioRob). São Paulo, Brazil
14. Matsuno F, Sato H (2005) Trajectory tracking control of snake robots based on dynamic model. In: Proceedings of the 2005 IEEE international conference on robotics and automation. Barcelona, Spain

15. Patra A et al (2020) A bio-inspired climbing robot: dynamic modelling and prototype development. In: Recent advances in mechanical engineering. Springer, pp 191–209

Use of Artificial Neural Network for Analyzing the Contributions of Some Kinematic Parameters in the Polishing Process of Porcelain Tiles

Ricardo Halla II, Adeilson de Oliveira Souza, and Fábio José Pinheiro Sousa

Abstract The glossiness of porcelain tiles is a highly appreciated feature by its consumers and so it became a crucial aspect on the quality control of this product. The polishing process, which enhance the porcelain tile with the glossiness effect, is a complex process regarding the individual gloss values of each region of the tile and the glossiness final pattern. Due to this complexity, the use of artificial neural networks is a promising tool to build a regression model of glossiness prediction. This paper presents the performance of eight neural networks designed to predict glossiness in a specific point of the tile's surface. Each model has different inputs and so the aim of this work is to quantify the influence of the inputs on the performance metric of the neural network. The identified input parameters which had the highest influence on the gloss value were the number of abrasive contacts, average scratching speed and standard deviation of scratching direction. This study also intends to provide scientific basis to future works on employing neural networks to predict glossiness of porcelain tile after the polishing process.

Keywords Artificial neural network · Polishing · Porcelain tile

1 Introduction

Porcelain tiles have been widely accepted by the ceramic tiles' consumers mainly due to their appreciation for its surface glossiness. Industries have been trying to increase the competitiveness of their products by improving the glossiness feature of porcelain tiles. The glossiness enhancement of porcelain tiles occurs in a process colloquially known as polishing. Due to the complex nature of this process, the employment of neural networks represents a considerable opportunity to overcome the performance of analytical models of porcelain tiles' gloss after the polishing process.

Artificial neural networks (ANNs) are widely spread and accepted models in many fields of science and engineering due to its high capacity to treat complex problems

R. Halla II (✉) · A. de O. Souza · F. J. P. Sousa
Federal University of Rio Grande do Norte, Natal, RN, Brazil
e-mail: ricardo.halla@gmail.com

A. F. Ab. Nasir et al. (eds.), *Recent Trends in Mechatronics Towards Industry 4.0*, Lecture Notes in Electrical Engineering 730, https://doi.org/10.1007/978-981-33-4597-3_15

and not fully understood phenomena. These models are commonly applied in cases where analytical and mechanistic models do not work properly. Instead of using physical assumptions and consider the true nature of the problem, like an analytical model does, the driving force of many ANN models comes from the convenience of gathering plenty of experimental data. This leads to a model which is frequently called data-driven model, in which according to Haykin [2], the data are expected to speak for themselves. The behavior of a dependent variable is learned by means of a learning algorithm and by all the previously collected data from the process.

This paper brings the results of eight distinct ANNs which predict the gloss value on a specific point of the porcelain tile's surface based on analytical kinematic parameters which comes from already known models found on the literature [1, 5]. The difference among each ANN is only the inputs. The purpose of this work is to show what group of inputs has more influence on the gloss output variable. This study contributes with theoretical basis to further researches on modeling the polishing process of porcelain tiles by using ANN.

2 The Polishing Process of Porcelain Tiles

The industrial polishing process of porcelain tiles is a result of an abrasive process where a set of abrasive blocks, connected into a polishing head, scratches the surface of the polishing tile which is transported by a conveyor belt. The polishing head has different sources of motion, as shown in Fig. 1. The polishing head is free to perform two motions, a rotation motion regarding its center and a transverse oscillation motion. The combination of these motions in addition to the conveyor belt speed results in a complex pattern of gloss in the tile's surface.

During the polishing process, every region of the tile undergoes different kinematics conditions regarding the scratching speed V and its direction θ, as can be seen

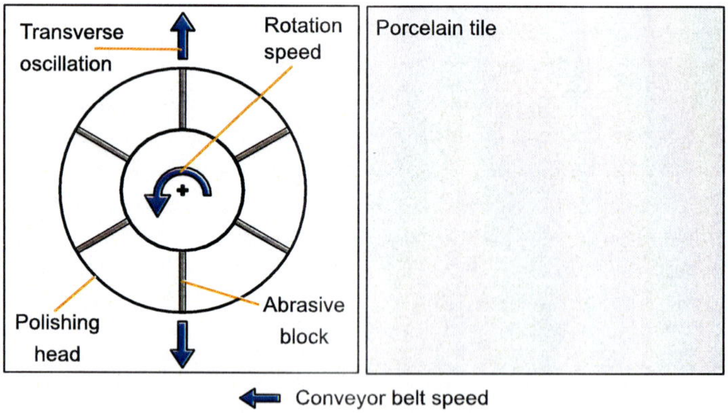

Fig. 1 Representation of the polishing process of porcelain tiles

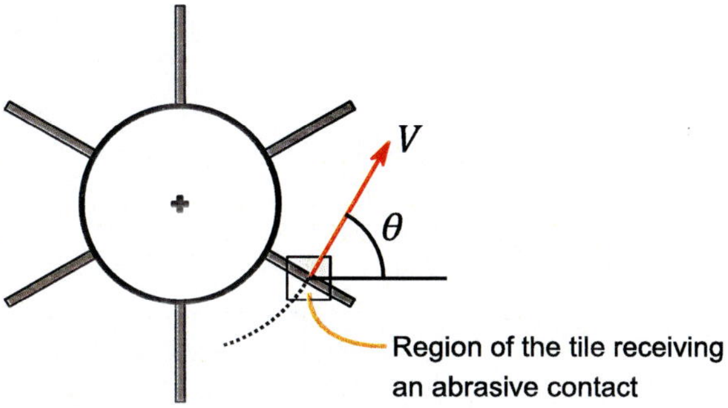

Fig. 2 Representation of the scratching speed, scratching direction and abrasive contact

in Fig. 2. The number of abrasive contacts N_A, refers to the total amount of times a specific region of the tile receives a contact from the abrasive block. With these concepts, a computational model can be built to calculate the history of scratching speed V, scratching direction θ and number of abrasive contacts N_A that every region of the tile experienced during the process. Once collected the history of scratching speed and scratching direction, some statistical parameters could be calculated such as average μ, standard deviation σ and coefficient of variation CV. Such statistical parameters in addition to the number of abrasive contacts will be used as inputs of the ANN which in turn will predict the gloss as output.

3 Method

All the results were achieved by using a neural network coded in Python with Keras library [3]. The developed model is a multilayer perceptron trained by supervised learning with Adam optimizer [4] and mean squared error as loss function. To select the proper number of hidden layers, number of neurons and activation function, a single neural network was tested varying these hyperparameters to fine tuning them in order to obtain the model which have the higher performance metric. Thus, the chosen model has two hidden layers both with 32 neurons, only one output neuron which stands for the gloss value in gloss units (G.U.) and all connected with the activation function ReLU.

Regarding the data used to train all the models, the output data consists of 578 values of gloss measured with a glossmeter over different points on the surface of two tiles. The available input data includes the cumulated number of abrasive contacts N_A, average scratching speed μ_V, standard deviation of scratching direction σ_θ, average scratching direction μ_θ and the coefficient of variation of scratching direction CV_θ. The different combinations of inputs used in each neural net are the following: $\{N_A\}$, $\{N_A, \mu_V\}$, $\{N_A, \mu_V, \sigma_\theta\}$, $\{N_A, \sigma_\theta\}$, $\{N_A, \mu_V, \mu_\theta\}$, $\{N_A, \mu_V, CV_\theta\}$, $\{N_A, \mu_\theta\}$ and $\{N_A, CV_\theta\}$. The basic idea was therefore to identify those most affecting inputs, and consequently, the more important variables regarding the prediction of gloss.

The values of these inputs came from a computational algorithm based on already known equations from the literature [1, 5] and they were calculated matching the same 578 points where the gloss was measured and considering the same specific polishing conditions used to produce the measured values of glossiness. More details on this polishing condition can be seen elsewhere [6].

The whole data was divided into two groups: a train set and a test set. The model was trained using the training set while the test set was used only for evaluating its performance in generalizing, i.e., in predicting the gloss values for unseen data. A total of 87 samples was randomly chosen to compose the test set, approximately 15% of the total 578 samples.

The mean squared error *MSE* and the coefficient of determination R^2 were chosen to be the performance metrics of the ANN. The mean squared error of a set of n samples can be calculated by applying Eq. 1, where y_i is the true output value of the i-th sample and $\hat{y}_i$ is the corresponding predicted output value. Once this metric depends on the difference between the true output and predicted output, it is a measure of the disparity between the model's predictions and the actual data.

$$MSE = \frac{1}{n}\sum_{i=1}^{n}(y_i - \hat{y}_i)^2 \tag{1}$$

The coefficient of determination can be expressed by Eq. 2, where $\overline{y}$ is the average of the true output value y_i. The coefficient of determination ranges from 0 to 1 and indicates, in percentage, how much the variability of the output is explained by the variability of the inputs. This metric measures the fitting adequacy of the regression model to the data provided.

$$R^2 = 1 - \frac{\sum_{i=1}^{n}(y_i - \hat{y}_i)^2}{\sum_{i=1}^{n}(y_i - \overline{y})^2} \tag{2}$$

4 Results and Discussion

Due to the random selection of the test set, train set and initial weights, every execution of the neural network, i.e., training and evaluation of performance metrics, gives slightly different results. Thus, a statistical treatment of the performance metrics was carried out.

The results presented in Table 1 represent the average coefficient of determination $\overline{R^2}$ and the average mean squared error $\overline{MSE}$ of 10 consecutives executions of the neural network for each group of inputs. These metrics were evaluated individually for the train set, test set and the whole dataset.

Table 1 Average of performance metrics for each ANN

Inputs	Train set		Test set		All data	
	$\overline{R^2}$	$\overline{MSE}$	$\overline{R^2}$	$\overline{MSE}$	$\overline{R^2}$	$\overline{MSE}$
$\{N_A\}$	0.104	0.022	0.098	0.020	0.105	0.022
$\{N_A, \mu_V\}$	0.388	0.015	0.323	0.017	0.381	0.015
$\{N_A, \mu_V, \sigma_\theta\}$	0.614	0.009	0.449	0.014	0.590	0.010
$\{N_A, \sigma_\theta\}$	0.476	0.013	0.367	0.015	0.462	0.013
$\{N_A, \mu_V, \mu_\theta\}$	0.461	0.013	0.280	0.018	0.434	0.014
$\{N_A, \mu_V, CV_\theta\}$	0.409	0.014	0.279	0.017	0.390	0.015
$\{N_A, \mu_\theta\}$	0.262	0.018	0.210	0.021	0.254	0.018
$\{N_A, CV_\theta\}$	0.129	0.021	0.106	0.023	0.126	0.021

The group of inputs found to have the highest influence on the prediction of gloss was $\{N_A, \mu_V, \sigma_\theta\}$ due to its highest $\overline{R^2}$ and lowest $\overline{MSE}$, as shown in Table 1. To expand the results of Table 1, all the values of R^2 and *MSE* assessed on all data is shown on a boxplot in Figs. 3 and 4, respectively. The boxplots of Figs. 3 and 4 aid a direct comparison among the performance of each designed model and quantifies the influence of each group of inputs on the output variable.

The boxplots from Fig. 3 show that the inputs $\{N_A, \mu_V, \sigma_\theta\}$ produced the highest average coefficient of determination, indicating that this group of inputs best explain the variation of the output in comparison to the others. The poorest prediction power was obtained admitting exclusively the number of abrasive contacts, in which only

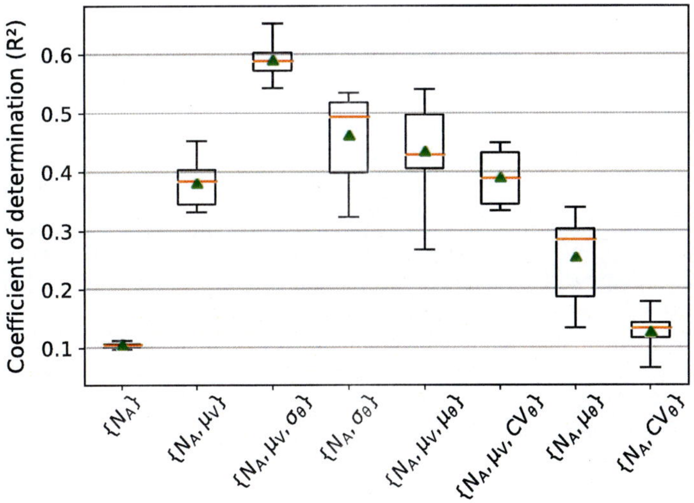

Fig. 3 Comparison of boxplots of R^2 evaluated on all data

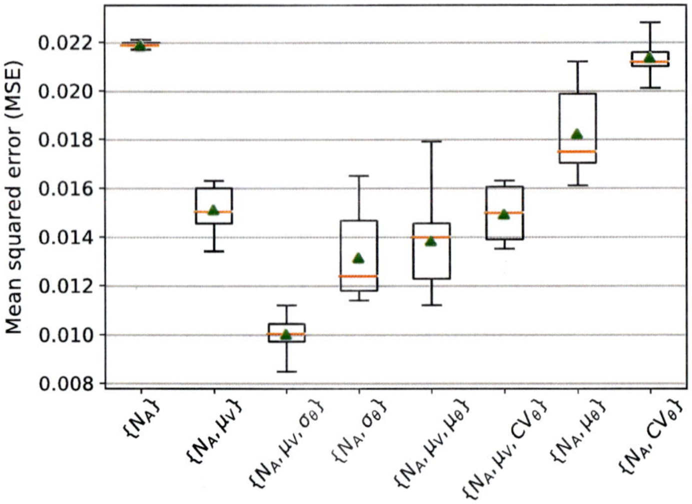

Fig. 4 Comparison of boxplots of *MSE* evaluated on all data

about 10% of the changes on the output could be explained. Including the coefficient of variation of the scratching direction as an input showed no major improvements.

The second worst input set in terms of determination coefficient was $\{N_A, CV_\theta\}$. In this case, it must be recalled that the variation coefficient CV_θ is in turn composed by the ratio between the standard deviation σ_θ and the average value μ_θ. The small R^2 when compared to the set $\{N_A, \mu_V, \sigma_\theta\}$ indicates that the main information present in those variables were cancelled during that division.

To provide a more quantitative comparison among the different input sets, a sequence of t-tests was carried out based on the values of R^2 obtained for each input set. Admitting a confidence level of 0.95 (bicaudal), all differences that reach a statistically significance are highlighted in Fig. 5.

The results regarding the *MSE* metric, shown in Fig. 4, reveal a similar behavior to that shown in the boxplots from Fig. 3. While the learning algorithm minimizes the mean squared error loss function towards 0, at the same time it implies a trend to maximize the R^2 metric towards 1. As expected, the same inputs which produced the highest average R^2 generated the lowest average *MSE*, indicating that the outputs from this model are the closest to the experimental data than the other models.

5 Conclusions

The performance results from eight multilayer perceptron neural networks with different combinations of inputs were presented. The designed ANN intends to predict the glossiness value in a specific region of a porcelain tile, based on kinematic features previously obtained from a computational polishing process model.

Set of Inputs	$\{N_A, \mu_V\}$	$\{N_A, \mu_V, \sigma_\theta\}$	$\{N_A, \sigma_\theta\}$	$\{N_A, \mu_V, \mu_\theta\}$	$\{N_A, \mu_V, CV_\theta\}$	$\{N_A, \mu_\theta\}$	$\{N_A, CV_\theta\}$
$\{N_A\}$	0.0000	0.0000	0.0000	0.0000	0.0000	0.0000	0.0691
$\{N_A, \mu_V\}$	-	0.0000	0.0095	0.0833	0.6371	0.0002	0.0000
$\{N_A, \mu_V, \sigma_\theta\}$	-	-	0.0001	0.0000	0.0000	0.0000	0.0000
$\{N_A, \sigma_\theta\}$	-	-	-	0.4631	0.0230	0.0000	0.0000
$\{N_A, \mu_V, \mu_\theta\}$	-	-	-	-	0.1582	0.0001	0.0000
$\{N_A, \mu_V, CV_\theta\}$	-	-	-	-	-	0.0001	0.0000
$\{N_A, \mu_\theta\}$	-	-	-	-	-	-	0.0001

Fig. 5 Matrix of t-tests among different input sets

The contributions of each set of inputs were implemented to quantify the influence of the inputs on the output variable.

The prediction power of the ANNs quantified in terms of coefficient of determination was found to vary greatly according to the input variables. The best case was observed for the inputs $\{N_A, \mu_V, \sigma_\theta\}$, with a $R^2 \approx 0.59$, whereas the worst case came from the input $\{N_A\}$, with $R^2 \approx 0.1$. The set of inputs which had the higher significance and provided the lowest error on the output variable could be determined by the analysis of the statistical data from the boxplots of Figs. 3 and 4. Regarding the model with the highest average R^2, nearly 59% of the variability of the output could be explained by the variability of its inputs $\{N_A, \mu_V, \sigma_\theta\}$. A matrix of statistical t-test was carried out to identify all the significant differences among the determination coefficients calculated for each input set.

The contribution of this work is to offer support to subsequent studies on building a regression model of porcelain tiles' glossiness by applying artificial neural network.

Acknowledgements The authors gratefully acknowledge the support of the Brazilian Coordination for the Improvement of High Education Personnel (CAPES) and the German Research Foundation (DFG) by funding the Project "Manufacturing System Models for Industry 4.0 based on highly heterogeneous and unstructured data sets", in the scope of the Collaborative Research Initiative - PIPC 8881.473092/2019-1 (DFG Grant Number AU 185/72). This study was financed in part by the Coordenação de Aperfeiçoamento de Pessoal de Nível Superior - Brasil (CAPES) - Finance Code 001.

References

1. Campione I, Fragassa C, Martini A (2019) Kinematics optimization of the polishing process of large-sized ceramic slabs. Int J Adv Manuf Technol 103(1–4):1325–1336. https://doi.org/10.1007/s00170-019-03623-3
2. Haykin S (1999) Neural networks: a comprehensive foundation, 2nd edn. Prentice Hall Inc.
3. Keras Homepage, https://keras.io/. Last accessed 22 May 2020
4. Kingma DP, Ba JL (2015) Adam: a method for stochastic optimization. In: 3rd International conference on learning representations, pp 1–15. ICLR, San Diego
5. Sousa FJP, Aurich JC, Weingaertner WL, Alarcon OE (2007) Kinematics of a single abrasive particle during the industrial polishing process of porcelain stoneware tiles. J Eur Ceram Soc 27(10):3183–3190. https://doi.org/10.1016/j.jeurceramsoc.2006.12.007
6. Sousa FJP, Olenburg A, Aurich JC (2011) Spatial distribution of glossiness over the surface of polished stoneware tiles. InterCeram Int Ceram Rev 60(3):214–216

Magnetorheological Damper Control for Semi-active Suspension System Using Skyhook-Differential Evolution

Mat Hussin Ab Talib, Mohd Ariff Durranie Muhammad Afandi, Intan Zaurah Mat Darus, Hanim Mohd Yatim, Zainab Asus, Ahmad Hafizal Mohd Yamin, and Muhamad Sukri Hadi

Abstract Magnetorheological (MR) damper is an effective semi-active vibration actuator which can produce controllable damping force by supplying proper voltage or electrical current as an input. By altering the input, the behaviour of MR fluid will change accordingly from free flow to semi-solid state. A semi-active suspension system consists of MR damper combines the advantages of both passive and active suspensions which is more economical, safe, mechanical simplicity and does not required a large power supply. Therefore, a lot of continuous efforts have been done to develop a vehicle suspension system incorporating with MR damper to reduce the vibration and road disturbance in order to provide a good ride comfort to driver and passenger. This study intends to investigate the performance of quarter car model in terms of body acceleration and body displacement adopted by the Skyhook-Differential Equation controller. Simulation and experimental validation data of MR damper model tests show that the modified Bouc-wen model can be used to represent the MR damper as the model is more accurately predicts the behaviour of the MR damper compared to other parametric models. Skyhook controller utilizing by Differential Evolution (DE) is used in order to optimize the system and its performance is improved with up to 63.04% as compared to the other controller responses.

Keywords Differential equation · Groundhook controller · Magnetorheological damper · Semi-active suspension · Skyhook controller

M. H. Ab Talib (✉) · M. A. D. Muhammad Afandi · I. Z. Mat Darus · H. Mohd Yatim · Z. Asus · A. H. Mohd Yamin
Faculty of Engineering, School of Mechanical Engineering, Universiti Teknologi Malaysia, 81310 Johor Bahru, Johor, Malaysia
e-mail: mathussin@utm.my

M. S. Hadi
Faculty of Mechanical Engineering, Universiti Teknologi MARA, 40450 Shah Alam, Selangor, Malaysia

A. F. Ab. Nasir et al. (eds.), *Recent Trends in Mechatronics Towards Industry 4.0*, Lecture Notes in Electrical Engineering 730, https://doi.org/10.1007/978-981-33-4597-3_16

1 Introduction

Vibration control of vehicle suspension system have been widely researched in order to provide a good comfort and performance for drivers and passengers [1]. A lot of investigation has been done in order to ensure the suspension system to work in the best and optimal conditions by altering and optimizing the parameters of the suspension system. Suspension system can be classified into three categories namely as passive, semi-active and active suspension system. The passive system basically consists of a spring and a damper that attached between the wheel and body of the vehicle. Compared with the semi-active and active suspension, passive system provides poor performance on the vehicle stability due to the components of the passive system (the spring and the damper) unable to perfectly sustain the vibration or road disturbance acted to it. Both semi-active and active suspension system performed very well in term of vehicle stability and improving ride quality but active system required a large power source and more expensive compared to semi-active system [2].

Semi active suspension system is currently the main area of this research since this system combines the advantages of both active and passive system which is economical, safe and does not required a large power supply. MR dampers is used to represent the semi-active suspension system since this type of damper is commonly used in semi-active system and has some advantages such as fast time response, mechanical simplicity, high dynamic range and low power requirement [3]. An MR damper also known as magnetorheological damper is a semi-active vibration actuator that able to change its rheological behaviour from free flow—linear viscous to semi-solid by altering the input voltage or current of the MR damper.

In order to maximize the application of MR damper, a controller system should be properly adopted to the system. Various number of researches on the implementation of intelligent controller for MR damper system have attempt to be investigated recently. The H ∞ controller [4], fuzzy logic [5] and linear quadratic controller [6] are among of the controllers that have been proposed in the last decade. Similarly, the skyhook controller is also one of them, that could possibly use to improve the vehicle ride comfort since it is presented as the basis of practical implementation of semi-active vehicle suspensions. [7]. However, improper tuning parameters has led the system is unable to obtain the high-performance level since most of the researchers have used a conventional method to tune the controller parameters [8, 9]

To improve the said problem, controller development integrated with an intelligent evolutionary algorithm such as the particle swarm optimization (PSO), firefly algorithm and artificial bee colony are normally being used in suspension system and other applications as well [10, 11]. Similarly, Differential Evolution (DE) is one of evolutionary algorithm that also highly being used as an optimizer in various applications [12, 13]. Thus, this study is attempt to use the DE algorithm integrated with the Skyhook controller as a novel controller configuration for the semi-active suspension system and also focusing on the MR damper modelling using parametric approach.

The paper will be organized as follows. Section 2 describe an experimental and simulation works for MR damper modelling. In Sect. 3, configuration of controller system using Skyhook-Differential Equation is developed. Section 4 shows the analysis and discussion and in Sect. 5, the conclusion is shown.

2 MR Damper System

2.1 Experimental Work

An MR damper system is emerging as one of the favorable types that generally being used as an actuator for semi-active suspension system. Using damper RD-1005-3 type manufactured by Lord Cooperation, it is important investigate the behaviour of the MR damper by conducting a preliminary experimental work in this study. The collected data from experimental study is compared to the simulation data using parametric modelling in the form of force-velocity and force-displacement characteristics. It is important to ensure that, the parametric MR model used in this study is in-line with the force generated from the experimental data. The model of NI instrument used is PCI-6259 which can utilize the signal input and output from the sensors. The configuration of test rig structure for validation work is depicted in Fig. 1.

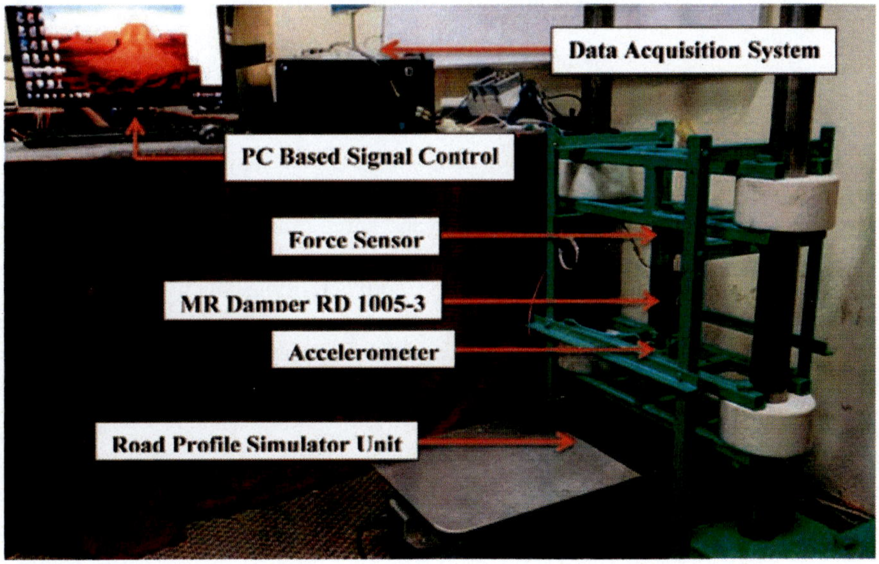

Fig. 1 Test rig structure for MR damper validation

2.2 MR Damper Parametric Modelling

The characteristics of the MR damper must be investigated in order to validate the performance of the damper to the system. Selection of the best model is needed to represent the real characteristic of MR damper in mathematical form to be used in the suspension system. In other words, the performance of the suspension system will be improved significantly by choosing the optimal model of MR damper. Few types of model were investigated by previous researcher [14] to represent the characteristic of MR damper as shown in Table 1.

The selection of the best model can be done based on the estimation error of the parameter value, *E*. The lower the error value means that the closer the characteristic of the model with the real MR damper. Referring to Table 1, Modified Bouc-Wen model is proved to be the most suitable model to represent the MR damper since the estimation error is the lowest compared to other models.

3 Controller Design and Optimization

The Skyhook controller has been proposed by Karnoop and his team in 1974 [15]. It is normally used as the fundamental of practical implementation of semi-active suspension system and promising good ride comfort for the vehicle system. The equation of Skyhook control is given by:

$$\begin{cases} v_1 v_{12} < 0 & F_d = c_{sky} v_1 \\ v_1 v_{12} > 0 & F_d = 0 F_d = 0 \end{cases} \tag{1}$$

where F_d is damping force generated form the configuration of skyhook controller. According to an Eq. (1), the damping force is directly proportional to the sprung mass velocity, v_1 when the relative velocity, v_{12} and sprung mass velocity have the same sign. The performance of the controller is depending on the parameter of coefficient, c_{sky} and the selection of method or technique will determine how well the parameter is being tuned. It is important to ensure that, the parameter will give an optimum value in order to achieve good ride comfort. To make it happen, an intelligent optimizer called as DE algorithm is used in this study.

Table 1 MR Damper comparison model [14]

MR damper model	Parameter estimation error (*E*)
Bingham	3.8096
Simple Bouc Wen	1.9245
Modified Bouc Wen	1.3770
Hyperbolic tangent function	3.3688
Nonlinear biviscous	3.8052

3.1 Differential Evolution Algorithm

Differential Evolution (DE) algorithm is a soft computing optimization method introduced by Storn and Price, back on 1995 [16]. The DE algorithm uses real coding of floating-point numbers and add the weighted difference between two population vectors to a third vector. Several rules need to be considered as it is the main element in selecting the control parameters. The listed rules are shown below:

1. The initialization of population is spread that give high possible place over
2. the objective function surfaces
3. Crossover probability, *CR* is normally within the range of zero to one
4. The mutation constant, *F* is normally in between [0.5, 1]
5. The population size, *NP* is inversely proportional to the mutation constant, *F*.

The summarization of DE work flow is depicted in Fig. 2. Referring to this figure,

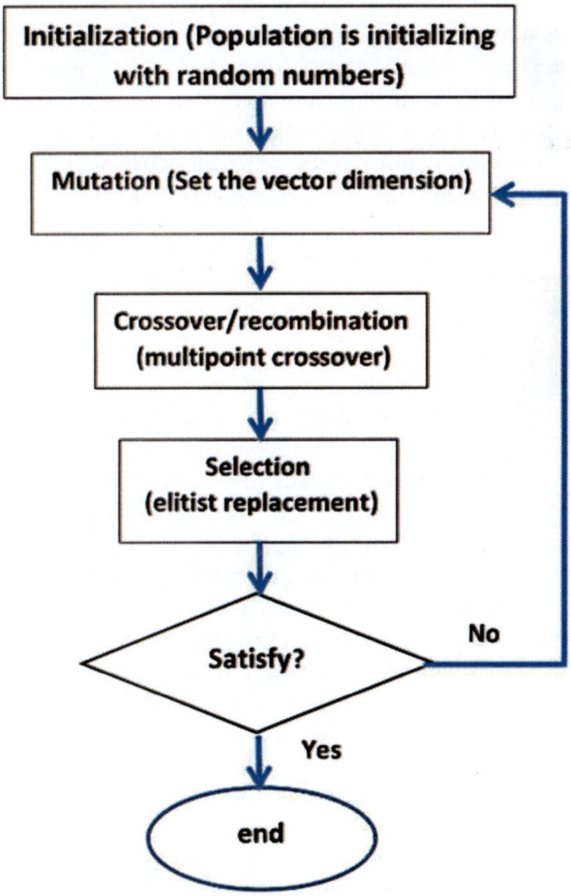

Fig. 2 Work flow of DE algorithm

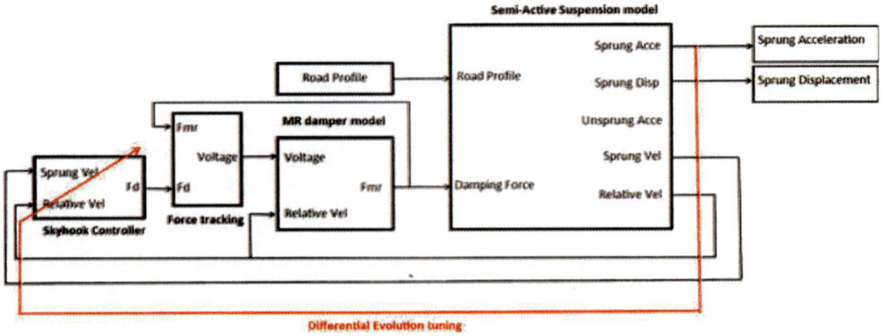

Fig. 3 Full configuration of block diagram

it is important to ensure that the selection of parameters and mutation is appropriate in order to satisfy the criteria's need.

The full configuration of controller design with Skyhook-Differential Equation including the MR damper actuator is shown as in Fig. 3. Referring to the figure, the input of the Skyhook controller is set to be sprung mass and relative velocities while the objective function of the DE algorithm is set as an acceleration of the body response. The selection of the body acceleration as the objective function is due to the vehicle ride comfort since the performance of the vehicle is depends on the reduction of body amplitude of the acceleration.

4 Result and Discussion

Experimental works in investigating the effectiveness of the Modified Bouc Wen model to be validated with the experimental data are performed in MATLAB simulation. It is worth to be investigated to make it close with the real characteristic of MR damper system. The responses of Modified Bouc Wen model are compared with the responses of the experimental data of force-velocity and force-displacement characteristics shown in Figs. 4 and 5, respectively.

It can be seen from the graphs that the model from the simulation can simulate the experimental data very well. In other words, the data collected from the simulation mostly cover the data from the experimental. Therefore, it is confirmed that the MR damper is modelled with the suitable parametric approach and the Modified Bouc-Wen model can represent the actual behaviour of the MR damper. From the graphs as well, it is clearly observed that the damping force is increase as the applied volt is also increase.

The performance of the quarter car model suspension system to provide a good ride comfort for driver and passenger can be evaluated in terms of body acceleration and body displacement by comparing the proposed Skyhook-DE controller with the Groundhook-DE controller and passive suspension system. The selection of

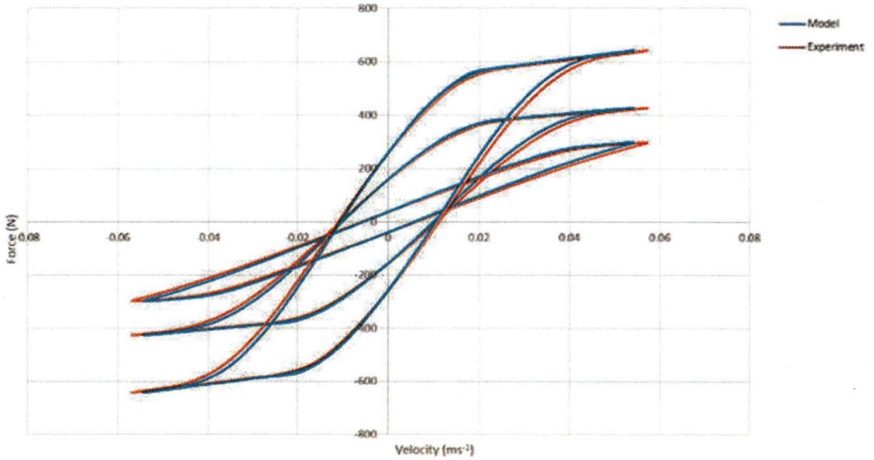

Fig. 4 Comparative experimental and simulation models for force-velocity charatersitic

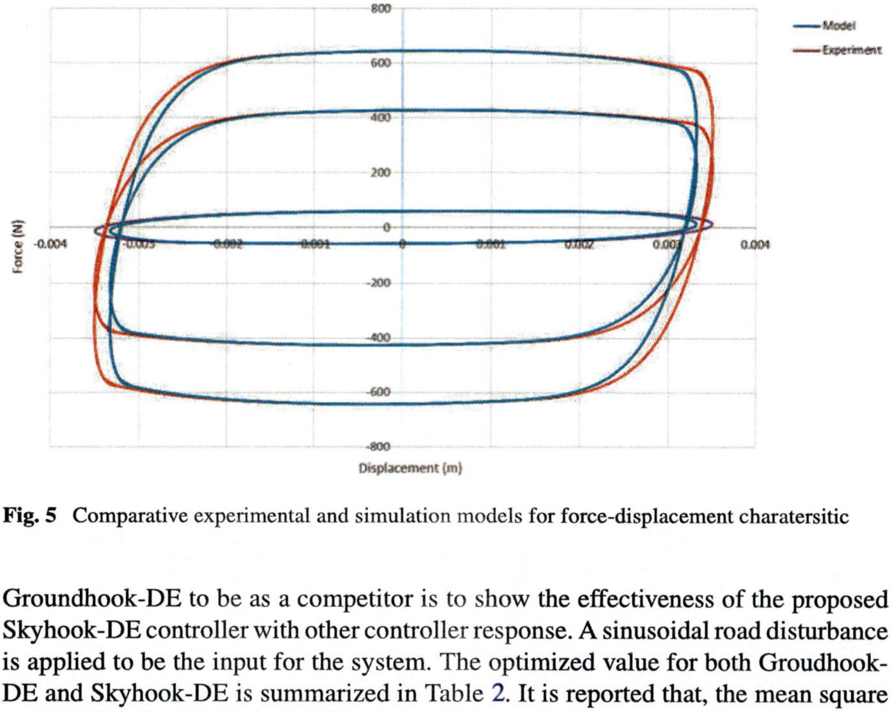

Fig. 5 Comparative experimental and simulation models for force-displacement charatersitic

Groundhook-DE to be as a competitor is to show the effectiveness of the proposed Skyhook-DE controller with other controller response. A sinusoidal road disturbance is applied to be the input for the system. The optimized value for both Groudhook-DE and Skyhook-DE is summarized in Table 2. It is reported that, the mean square error (MSE) for Skyhook-DE is better and it is worth to have a good performance as compared to the Groundhook-DE controller response. Figures 6, 7, and 8 represent the convergence results, the time series of body acceleration and body displacement, respectively. Body acceleration is said to be the primary indicator for evaluating the ride comfort of a vehicle. Based these figures, it is clearly observed that the Skyhook

Table 2 The optimized result and MSE

Controller	No. of iteration	Optimized value	MSE
Groundhook-DE	100	4317	11.45
Skyhook-DE	100	3109	5.10

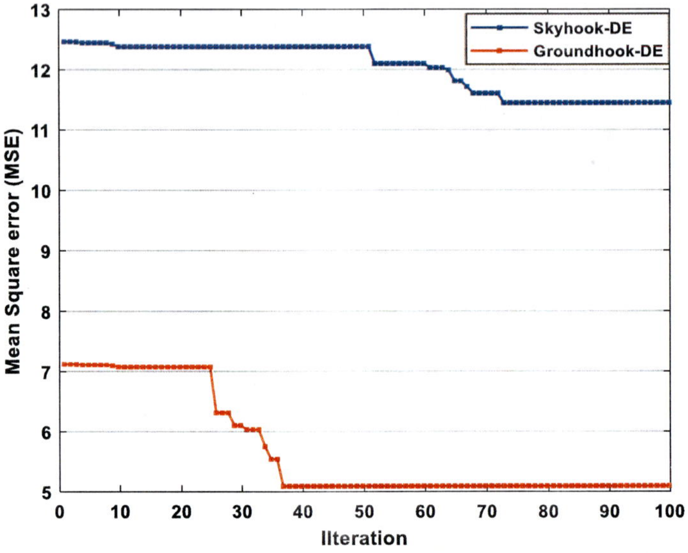

Fig. 6 Convergence curve for both Groundhook-DE and Skyhook-DE

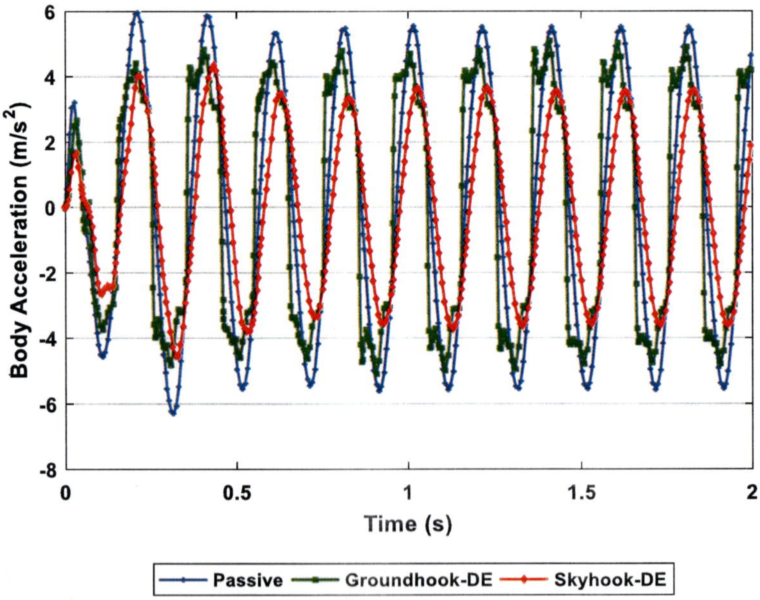

Fig. 7 Body acceleration response of semi-active suspension system

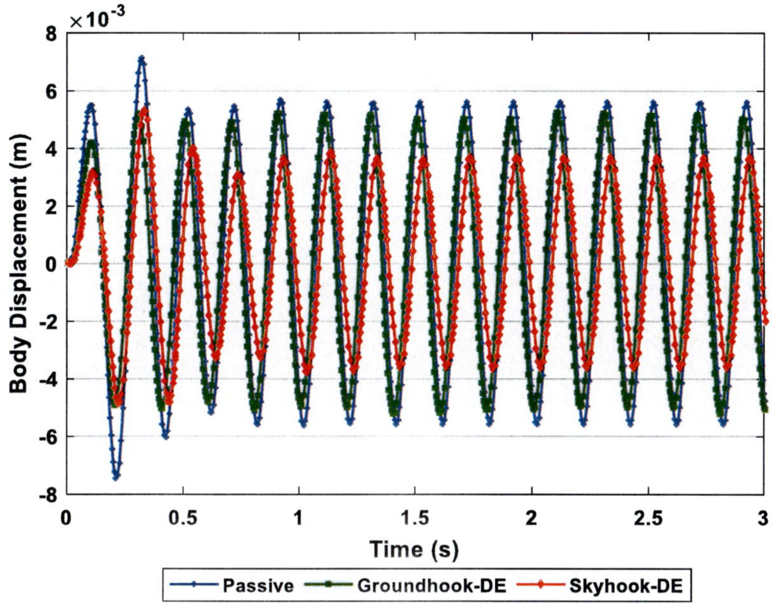

Fig. 8 Body displacement response of semi-active suspension system

Table 3 MSE and percentage improvement of all parameters of interest

Parameters of interest	Passive	Groundhook-DE	Skyhook-DE
Body acceleration (m/s^2)	13.75	11.45 (16.72%)	5.10 (63.04%)
Body displacement (m) (10^{-6})	1.28	1.17 (8.59%)	0.57 (55.4%)

tuned by DE manages to reduces the amplitude oscillation better compared to others for both body acceleration and body displacement response. For the body displacement response, it can be mentioned that the force transmitted from the unsprung to sprung masses has been reduced in order improve the ride comfort of the body vehicle. Details of the percentage improvement of all controller responses is shown in Table 3.

5 Conclusion

The model of the MR damper including the controllers tuned by DE algorithm have developed using MATLAB Simulink. Based on the results, the Skyhook controller tuned by DE showed a better performance and improvement compared to Groundhook controller tuned by DE and the passive system with 63.04% for body acceleration and 55.4% for body displacement. It can be concluded that the semi-active system using the Skyhook-DE and Groundhook-DE controllers are able to improve the vehicle performance in term of ride comfort compared to the passive suspension system and among of them, Skyhook-DE is more effective in controlling the body vehicle of suspension system.

The method used in this study can be extended to the optimal design of MR damper by using more complex approach for example the non-parametric approach. Based on the studies, non-parametric approach such as neural networks, polynomial and neuro-fuzzy modelling have a potential to be more effective in controlling the vibration of the vehicle suspension system compared to the parametric approach.

Acknowledgements The authors would like to express their gratitude to Minister of Education Malaysia (MOE) and Universiti Teknologi Malaysia (UTM) for funding and providing facilities to conduct this research. This research is supported by UTM Research University grant, Vote No. 4B447.

References

1. Kolekar S, Venkatesh K (2019) Experimental investigation of damping effect in semi-active magnetorheological fluid sandwich beam under non-homogeneous magnetic field. J Vibr Eng Technol 7(2):107–116
2. Cruze D, Hemalatha G, Jebadurai SVS, Sarala L, Tensing D, Christy SJE (2018) A review on the magnetorheological fluid, damper and its applications for seismic mitigation. Civ Eng J 4(12):3058–3074
3. Moradi Narbin M, Mirzaei M, Aghasizade S (2018) Identification of a new non-linear model for MR damper and its application in control of vehicle semi-active suspension system. J Nonlinear Syst Electr Eng 4(1):4–25
4. Yu S, Zhang J, Xu F, Chen H (2019) $\mathscr{H}\infty$ control of semi-active MR damper suspensions. In: 2019 12th Asian control conference (ASCC), pp 337–342. IEEE
5. Mohammadi RK, Haghighipour F (2017) Implementation of uniform deformation theory in semi-active control of structures using fuzzy controller. Smart Struct Syst 19(4):351–360
6. Kumar G, Kumar A, Jakka RS (2018) An adaptive LQR controller based on PSO and maximum predominant frequency approach for semi-active control scheme using MR damper. Mech Ind 19(1):109
7. Liu C, Chen L, Yang X, Zhang X, Yang Y (2019) General theory of skyhook control and its application to semi-active suspension control strategy design. IEEE Access 7:101552–101560
8. Jurkiewicz A, Kowal J, Zając K (2017) Sky-hook control and Kalman filtering in nonlinear model of tracked vehicle suspension system. Acta Mech et Automatica 11(3):222–228
9. Zhang H, Liu J, Wang E, Rakheja S, Su CY (2018) Nonlinear dynamic analysis of a skyhook-based semi-active suspension system with magneto-rheological damper. IEEE Trans Veh Technol 67(11):10446–10456
10. Ab Talib MH, Darus IZM, Samin PM (2019) Fuzzy logic with a novel advanced firefly algorithm and sensitivity analysis for semi-active suspension system using magneto-rheological damper. J Ambient Intell Humaniz Comput 10(8):3263–3278
11. Priyadarshi N, Azam F, Sharma A, Bhoi AK, Kumar MA (2018) Particle swarm optimization based fuzzy logic control for photovoltaic system. Int J Eng Technol 7(3.24):491–496
12. Moharam A, El-Hosseini MA, Ali HA (2016) Design of optimal PID controller using hybrid differential evolution and particle swarm optimization with an aging leader and challengers. Appl Soft Comput 38:727–737

13. Ajithapriyadarsini S, Mary PM, Iruthayarajan MW (2019) Automatic generation control of a multi-area power system with renewable energy source under deregulated environment: adaptive fuzzy logic-based differential evolution (DE) algorithm. Soft Comput 23(22):12087–12101
14. Kwok NM, Ha QP, Nguyen MT, Li J, Samali B (2007) Bouc-Wen model parameter identification for a MR fluid damper using computationally efficient GA. ISA Trans 46(2):167–179
15. Karnopp D, Crosby MJ, Harwood RA (1974) Vibration control using semi-active force generators, pp 619–626
16. Storn R, Price K (1997) Differential evolution—a simple and efficient heuristic for global optimization over continuous spaces. J Glob Optim 11:341–359

PID Controller Based on Flower Pollination Algorithm of Flexible Beam System

Ali Akhbar Mohd Fadzli, Muhamad Sukri Hadi, Rickey Ting Pek Eek, Mat Hussin Ab. Talib, Hanim Mohd Yatim, and Intan Zaurah Mat Darus

Abstract Flexible beam is commonly used in a wide variety of engineering applications such as mechanical, aircraft and architecture. One of the advantages of these flexible structures is its lighter as compared to rigid structures. However, flexible structures are more sensitive to vibration compared to rigid structures. Excessive long term vibration can damage parts and reduce the performance of such structures. Therefore, a suitable system should be identified to overcome the problem. In the new era of technology, there are many methods developed to suppress unwanted vibration. One well-known system to suppress unwanted vibration is active vibration control (AVC). This system is suitable for structures experiencing low frequency of vibration. A proper modelling must be developed using system identification techniques in order to achieve high vibration cancellation in the system. For system identification, evolutionary swarm algorithm is the latest technique used compared to other methods. In this study, flower pollination algorithms (FPA) was used to develop a mathematical model of a system. The main objective of this study was to develop a model for flexible beam system using FPA in order to achieve an approximate model that represents the real characteristic of the flexible beam system. The developed model will then be used as a platform for PID controller development. The model was validated using three robustness methods, which are the mean squared error (MSE), correlation test, and pole zero diagram stability. Based on the validation, it was observed that the FPA was able to exhibit the lowest MSE value, very good correlation test and high stability. The model achieved in this study was used in controller development for vibration cancellation of a flexible beam system. It was noticed that the PID controller achieved 17.7 dB of attenuation level at the first mode of vibration. The attenuation of vibration was reduced from 56.72 to 39.03 dB, which is equivalent to 31.2% of reduction when the vibration control is active.

A. A. M. Fadzli · M. S. Hadi (✉)
Faculty of Mechanical Engineering, Universiti Teknologi MARA, 40450 Shah Alam, Selangor, Malaysia
e-mail: msukrihadi@uitm.edu.my

R. T. P. Eek · M. H. Ab. Talib · H. M. Yatim · I. Z. M. Darus
School of Mechanical Engineering, Universiti Teknologi Malaysia, 81310 Johor, Malaysia

A. F. Ab. Nasir et al. (eds.), *Recent Trends in Mechatronics Towards Industry 4.0*, Lecture Notes in Electrical Engineering 730, https://doi.org/10.1007/978-981-33-4597-3_17

Keywords Flexible beam · System identification · Flower pollination algorithm · Active vibration control · PID controller

1 Introduction

Flexible beam is one of the parts in a flexible structure. A flexible beam is more preferable compared to a rigid structure. Since the flexible beam is lighter, it can reduce the use of material, cost, energy and many more. The characteristic of a flexible beam can induce unwanted vibration in the system. Vibration can be described as a repeating motion at a certain period. In some mechanical systems, it is an unwanted element that exists in the system. Vibration becomes a problem because it can damage the structure and inhibits the structures' performance. This problem has garnered plenty of attention among researchers in means to find a solution to overcome it. Over the years, there are two methods that are usually used to suppress vibration. The methods are passive vibration control and AVC [1].

Passive control method is a good technique for simple structures and incendence of high frequency vibration. However, vibration suppression on flexible beam is more effective when AVC technique is employed. AVC is more reliable and efficient for systems which experience low frequency vibration and the parameters can be tuned until a desired output is attained [2]. Closed-loop feedback system used in AVC consists of sensors, controllers and actuators while passive vibration control refers to shock mitigation by passive techniques such as mechanical springs or rubber pads that are mounted to the object that vibrates. In terms of technology, AVC is more advanced because it uses electronics in the system which provides accurate results. Based on the superposition theory, AVC generates a secondary vibration which has a similar magnitude but in the opposite direction of the unwanted vibration to destructively interfere it at the desired location. In AVC, it is important to obtain an accurate model in order to get maximum vibration suppression in the system [3, 4].

System identification is one of the approaches that has been used by researchers to develop a mathematical model of the system. System identification is a sophisticated method which uses input and output data in order to develop a mathematical model that represents a system [5]. Recently, it is a growing trend among researchers to use swarm intelligence algorithm for identification and implementation of control system. It consists of a variety of algorithms that were developed by researchers, including genetic algorithm (GA), particle swarm optimization (PSO), ant colony system (ACS) and many more [6]. Evolutionary swarm algorithm is a subset of evolutionary computation based on the collective intelligence behavior of self-organized and decentralized systems or artificial groups of simple agents [7, 8]. The purpose of this study is to develop a model for flexible beam system using swarm intelligence algorithm via flower pollination and to control the undesired vibration of the dynamic system.

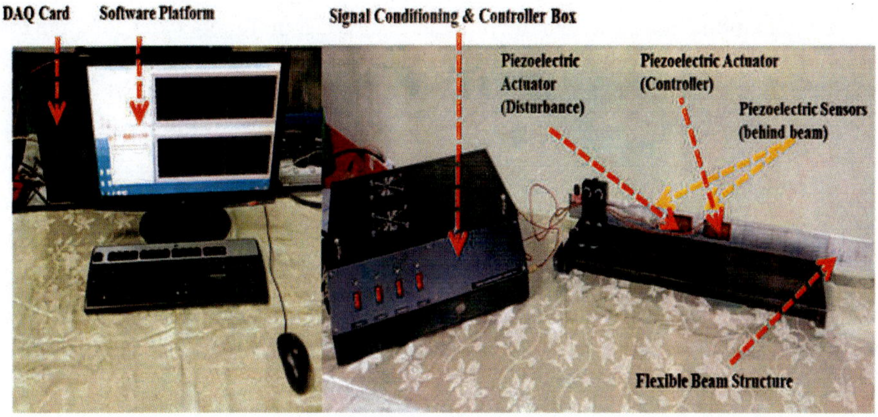

Fig. 1 Experimental setup for flexible beam structure [9]

2 Experimental Setup and Data Collection

Input-output dataset to be used for system identification as acquired via experiments. In the experiment, a cantilever beam was used as a flexible beam as shown in Fig. 1. The flexible beam is divided in 20 segments with 21 nodal points. The nodal point between 9th and 10th is considered as the detection point while nodal point between 19th and 20th is considered as an observation point. Nodal point between 8th and 9th is set as disturbing point in order to get the first dominant mode when exciting. The input of this system was data at detection point and the output was data at the observation point. The detection point must be closed to the excitation source while the observation point at maximum deflection is placed at the end of the beam. From the experiment, a total of 3334 input and output data were collected [9].

3 Flower Pollination Algorithm

Flower pollination algorithm (FPA) is one of the algorithms that can solve a problem which was inspired from the pollination process of flowering plants. FPA is a new metaheuristic intelligent algorithm introduced by British scholar Yang in 2012 [10]. Naturally, flowers reproduce through pollination, while the purpose of flower pollination is to survive. Successful reproduction is dependent on the pollinators. The advantages of this algorithm is that it has a small number of parameters, very easy to implement, and has high efficiency [11, 12]. Flower pollination is a process associated with transferring flower pollens. The transfer agent are mainly birds, bats, insect, and others animal. There is a connection between flower and flower pollinator which is known as flower-pollinator partnership. Glover in 2014 conducted a research on the behavior of flower during pollination [12]. FPA has been employed in some applications like parameter identification of DC motor model and compute the optimal parameters of an unknown infinite impulse response (IIR) system [13, 14].

The pollination behavior can be applied as a mathematical equation which shows the pattern of global pollination. Pollen and pollinators can travel a long distance because birds can fly over a long distance. This ensures optimal pollination and reproduction. The role of global pollination is shown in (1) [14].

$$x_I^{t+1} = x_i^t + L(x_i^t - g_*) \tag{1}$$

where x_i^t the pollen i at iteration t, g_* is the best solution found in the current population, and L is the Lévy flight step size corresponding to the strength of the pollination. Lévy distribution can be given by (2).

$$L \sim \frac{\lambda \Gamma(\lambda) \sin\left(\frac{\pi\lambda}{2}\right)}{\pi} \frac{1}{s^{1+\lambda}}, \quad (s \gg s_0 > 0) \tag{2}$$

here $\Gamma(\lambda)$ is the standard gamma function, and s is the step size. The local pollination takes place when the random number is lesser than the switch probability p which is represented by (3).

$$x_i^{t+1} = x_i^t + \in \left(x_j^t - x_k^t\right) \tag{3}$$

In the above equation, x_j^t and x_k^t are the pollen from different flowers of the same type. It equivalently became a local random walk if $\in$ is drawn from a uniform distribution [0, 1]. Figure 2 shows the flowchart of FPA.

4 Active Vibration Control

Based on superposition theory, AVC technique uses a secondary vibration response which has a similar magnitude but in the opposite direction with the flexible structure system in order to suppress the vibration. The simulation of AVC is conducted using MATLAB/Simulink R2018a software. Proportional-Integral-Derivative (PID) controller is developed to simulate vibration suppression of flexible beam. The PID controller parameter is tuned by using heuristic method until it effectively suppresses the vibration at first mode of vibration of the system. Figure 3 shows the block diagram of AVC based controller developed in MATLAB/Simulink software.

5 Result and Discussion

FPA was applied to obtain an approximate model of the flexible beam system. There are 3334 data from the experiment was used to develop a mathematical of flexible

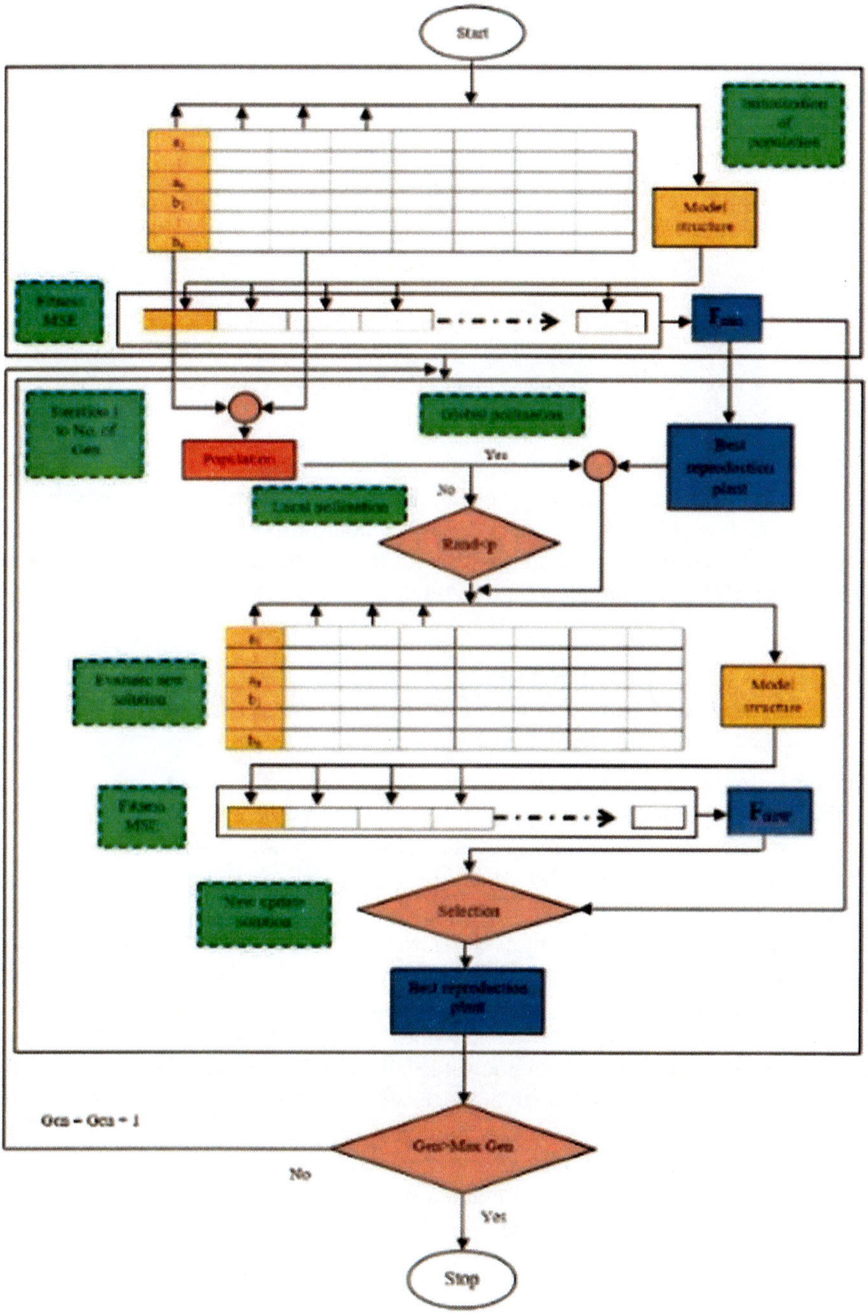

Fig. 2 Flower pollination algorithm flowchart

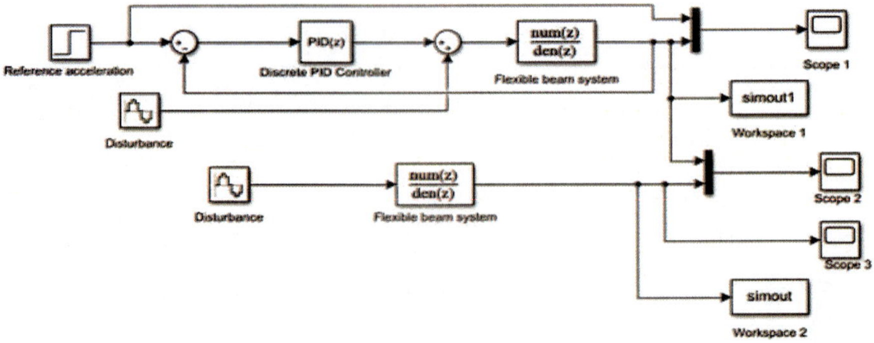

Fig. 3 Block diagram of PID controller in MATLAB/Simulink environment

beam model. 3334 experimental data were divided into two parts which are 1666 data for training and 1668 data for testing. The developed model was then validated using three criteria in order to select the best model of the system. In this study, the best model was chosen based on the lowest MSE value, achieving 95% confidence level of correlation test and has high stability.

5.1 *Modelling Using Flower Pollination Algorithm (FPA)*

The best model employing FPA) was obtained by properly tuning five parameters in the algorithm which are model order, switch probability, maximum generation, population size, lower and upper boundary using heuristic method. The process of tuning parameter was made by adjusting one parameter at a time and the other parameters were fixed. Initially, the population size was tuned between the range of 10 to 25. Then, the switch probability was tuned from 0.5 until 1.0, where 1 is the maximum limit of switch probability. According to the previous study, 0.8 is the best switch probability value for most applications [15]. Next, the value of lower boundary and upper boundary was set from 1 till 10. Basically, lower and upper boundaries have the same magnitude value but different integer. This means that the lower boundary value is in negative integer while the upper boundary is in positive integer. After that, the maximum generation value was tuned from 500 until 2500 with an increment of 500 for each tuning. The number of maximum generations were tuned to 2500 only till the result obtained no longer changes. Lastly, the parameter of developed model was tuned from 1 until 10. Table 1 presents the set of parameters used to achieve the best model for the system.

Based on the set of parameters used in Table 1, the lowest MSE achieved in FPA algorithm are 1.2549×10^{-06} and 5.3544×10^{-09} in training and testing data, respectively. The best model order was achieved in order 3 of the system. The convergence curve of FPA modelling as shown in Fig. 4. Figure 5 presents the actual and predicted

Table 1 The set of parameters used for FPA algorithm

No.	Parameter	Value
1.	Maximum generation	2500
2.	Switch probability	0.8
3.	Population size	25
4.	Upper boundary	−10
5.	Lower boundary	10
6.	Model order	3
7.	Number of parameters	6

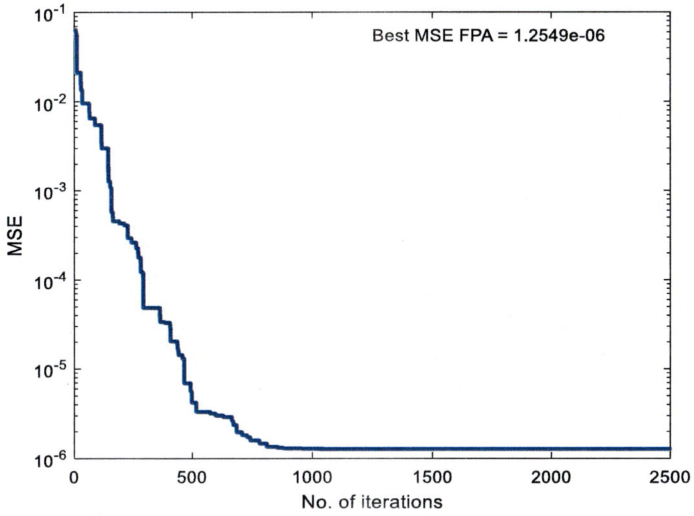

Fig. 4 The convergence curve of FPA modelling

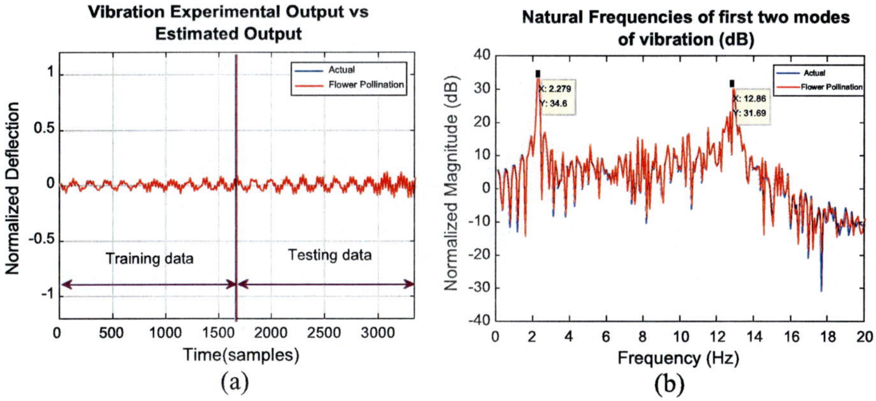

Fig. 5 Actual and prediction outputs of FPA modelling: **a** Time domain, **b** frequency domain

outputs of the system in time and frequency domains, respectively. Here, it can be observed that the developed system imitated the actual characteristic of flexible beam structure very well. Figure 6 shows the error between actual and estimated outputs of the system in FPA modeling. The robustness of the developed model was validated using pole zero diagram stability and correlation test as shown in Figs. 7 and 8, respectively. Figure 7 shows the pole zero diagram stability. It was indicated that the system has high stability since all the poles of the transfer function are within the unit circle. Figure 8 presents the correlation tests of FPA modeling. It can be seen that the system has a good correlation between input and error where the graph shows the system was unbiased and yields within 95% confidence interval. The discrete transfer function obtained using FPA modelling is described as (4):

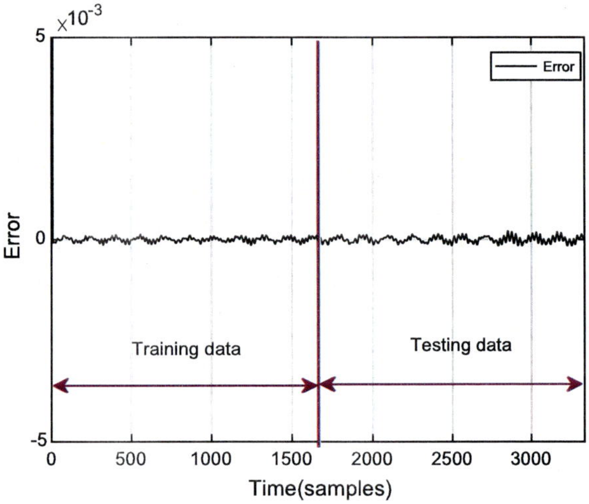

Fig. 6 Error between actual and prediction outputs of FPA modelling

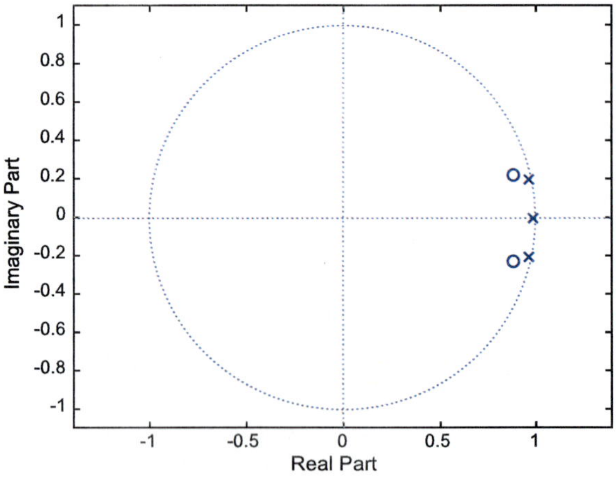

Fig. 7 Pole zero diagram stability of FPA modelling

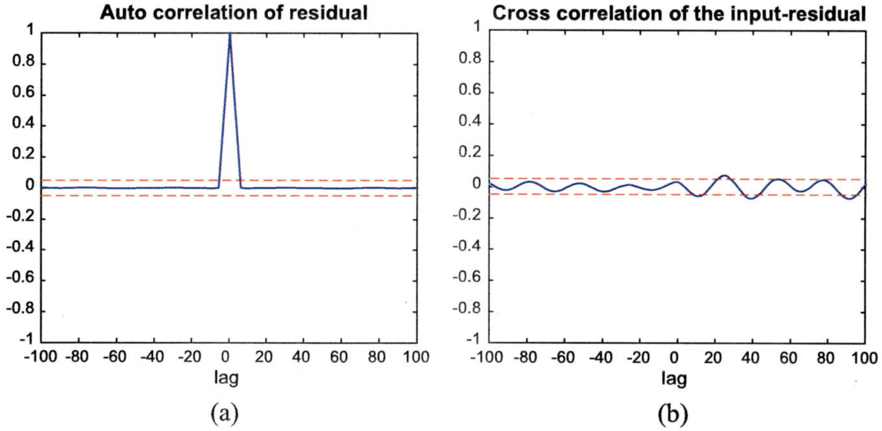

Fig. 8 Correlation tests of FPA modelling: **a** Auto correlation, **b** cross correlation

$$\frac{0.01298\ z^{-1} - 0.02299z^{-2} + 0.01083z^{-3}}{1 - 2.921z^{-1} + 2.885z^{-2} - 0.963z^{-3}} \tag{4}$$

5.2 *PID Controller Tuned by Heuristic Method*

The parameters of PID controller are obtained using heuristic method where it was based on a trial and error estimation. The proportional gain, K_p, integral gain, K_i and derivative gain, K_d are tuned till a high attenuation at first mode of vibration on flexible beam system was achieved. The simulation of PID controller is conducted by introducing sinusoidal disturbances in the system. The PID parameters used in this study are 1.38, 1.9785 and 0.494625 for K_p, K_i and K_d, respectively. Table 2 elucidates the attenuation level achieved by PID controller using heuristic method with sinusoidal disturbance in the system. Figure 9 presents the vibration suppression of the flexible beam system in time and frequency domains, respectively. It was noticed that the controller successfully achieved 17.7 dB of attenuation level in the

Table 2 The attenuation level achieved by PID controller based on FPA algorithm

Controller	Decibel magnitude (dB)	Attenuation level (dB)	Percentage of reduction
	1st mode	1st mode	1st mode
Without PID controller	56.72	Reference	Reference
With PID controller	39.03	17.7	31.2%

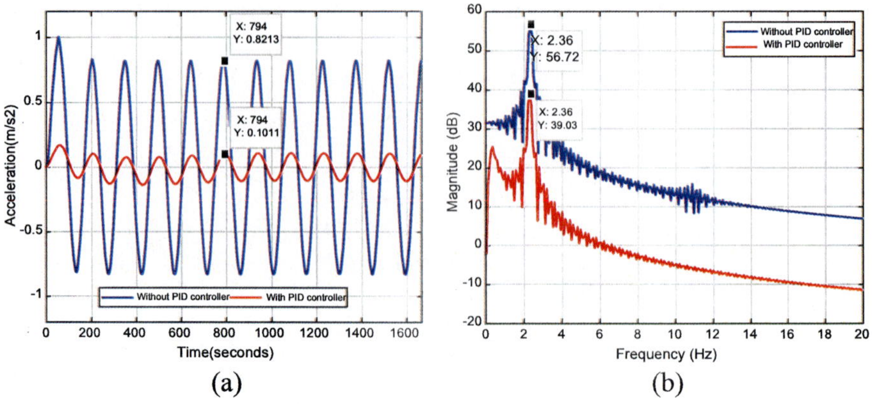

Fig. 9 PID controller based on FPA: **a** Time domain, **b** frequency domain

system. The attenuation of vibration is reduced from 56.72 to 39.03 dB, which is equivalent to 31.2% of reduction. Based on the result, the developed PID controller was successful in suppressing undesired vibration experienced by the system in a simulation environment.

6 Conclusion

The modelling of flexible beam using FPA was conducted in this study. The model was validated to determine its effectiveness in a system using MSE, correlation test and pole zero stability. The developed model successfully achieved the lowest MSE value, good correlation between input and error, and to be found within 95% confidence level with high stability. Then, the development of an AVC was successfully carried out for vibration suppression of a flexible beam system. The PID controller was selected because of its simplicity and robustness. It was indicated that the controller developed achieved good attenuation level at the first mode of vibration in the system. The controller achieved 31.2% reduction of unwanted vibration in the dominant mode of the flexible beam system.

Acknowledgements The authors would like to express their gratitude to Universiti Teknologi MARA (UiTM), Universiti Teknologi Malaysia (UTM) and Ministry of Higher Education (MoHE) for funding the research and providing facilities to conduct this research. Sponsor file number (RACER/1/2019/TK03/UITM//1).

References

1. Zhang D, Jia J, Guo Y, Duan Y (2006) Research on active and passive compound vibration control based on photoelectric displacement detection. In: 6th World congress on intelligent control and automation. IEEE, Dalian, China
2. Rahman TAZ, Darus IZM (2011) Active vibration control of a flexible plate via active force control strategy. In: 4th International conference on mechatronics (ICOM). IEEE, Kuala Lumpur
3. Hadi MS, Darus IZM, Tokhi MO, Jamid MF (2019) Active vibration control of a horizontal flexible plate structure using intelligent proportional–integral–derivative controller tuned by fuzzy logic and artificial bee colony. J Low Freq Noise Vibr Active Control 0(0):1–13
4. Hadi MS, Darus IZM, Yatim HM, Talib MHA, Jamali A (2019) Fuzzy-PID based controller for active vibration control of nonlinear dynamic systems. In: 22nd International conference on climbing and walking robots and the support technologies for mobile machines. CLAWAR Association Ltd, Kuala Lumpur, Malaysia
5. Fu L, Li P (2013) The research survey of system identification method. In: 5th International conference on intelligent human-machine systems and cybernetics. IEEE, Hangzhou, China
6. Quan H, Shi X (2009) Evolutionary swarm algorithm based on evolutionary swarm model. In: WRI global congress on intelligent systems. IEEE, Xiamen, China
7. Bonabeau E, Dorigo M, Theraulaz G (2001) Swarm intelligence: from natural to artificial systems. Q Rev Biol 76(2):268–269
8. Wahab MNA, Nefti-Meziani S, Atyabi A (2015) A comprehensive review of swarm optimization algorithms. Plos One 10(5)
9. Eek RTP, Darus IZM, Sahlan S, Samin PM, Shaharuddin NMR (2016) Implementation of swarm algorithm in modeling a flexible beam structure. J VibroEng 18(8):4914–4934
10. Nabil E (2016) A modified flower pollination algorithm for global optimization. Expert Syst Appl 57:192–203
11. Yang XS (2012) Flower pollination algorithm for global optimization. In: Unconventional computation and natural computation lecture notes in computer science, pp 240–249
12. Glover BJ (2014) Understanding flowers and flowering. Oxford University Press, Oxford
13. Singh S, Ashok A, Rawat TK, Kumar M (2016) Optimal IIR system identification using flower pollination algorithm. In: 1st International conference on power electronics, intelligent control and energy systems (ICPEICES), IEEE, Delhi, India
14. Chiroma H, Shuib NLM, Muaz SA, Abubakar AI, Ila LB, Maitama JZ (2015) A review of the applications of bio-inspired flower pollination algorithm. Procedia Comput Sci 62:435–441
15. He X, Yang XS, Karamanoglu M, Zhao Y (2017) Global convergence analysis of the flower pollination algorithm: a discrete-time Markov chain approach. Procedia Comput Sci 108:1354–1363

An Ultraviolet C Light-Emitting Robot Design for Disinfection in the Operating Room

Nguyen Duy Minh Phan, Ngo Quoc Huy Tran, Le Anh Doan, Duy Chung Tran, Van Sanh Huynh, Quang Truong Vo, Duc Long Nguyen, and Qui Tra Phan

Abstract The service robots are being ramped up rapidly in healthcare establishment, especially in the fight against the Covid-19 pandemic. This paper presents a new design and implementation of an Ultraviolet Disinfection Robot (UV Robot) which can be used in many medical institutions to disinfect patient rooms, the operating rooms, reduce cost of traditional disinfection and most importantly risks for patients and medical personnel. The robot is equipped with a high power UV-C lamps: five 100 W UV lamps covering 360° directions on the top and two 30 W lamps on two sides. The high intensity ultraviolet light emitted by the UV-C lamps makes it possible to kill 99% bacteria, germs molds, virus in around 30 s with a radius from 1 to 2.5 m depending on the type of microorganisms. A high disinfection efficiency in less time of the robot is confirmed by the experiments conducted in the hospital in Danang city, Vietnam. Compared to the available similar robots, this domestic product provides an economical and effective solution for Vietnamese medical institution in reducing the spread of harmful microorganisms.

Keywords UVGI · Service robot · Disinfection

1 Introduction

Studies have shown that there are approximately 14–17% of infections in operation, 38% of hospital infections occurs in patients who have surgery [1]. The disinfection of the operating room is a critical method in reducing and preventing the infection as well as the transmission of pathogens. Disinfection is defined as the destruction of microorganisms that can cause infection. The UV-C disinfection for air treatment and surfaces is well known and has been used as an alternative to conventional disinfection procedures for killing harmful microorganisms. This method offers several advantages compared with standard room disinfection using chemical disinfectants.

N. D. M. Phan (✉) · N. Q. H. Tran · L. A. Doan · D. C. Tran · V. S. Huynh · Q. T. Vo · D. L. Nguyen · Q. T. Phan
The University of Danang—University of Technology and Education, 48 Cao Thang, Danang 550000, Vietnam
e-mail: pndminh@ute.udn.vn

A. F. Ab. Nasir et al. (eds.), *Recent Trends in Mechatronics Towards Industry 4.0*, Lecture Notes in Electrical Engineering 730, https://doi.org/10.1007/978-981-33-4597-3_18

These include shorter time requirement for killing of bacteria, viruses, safety and eco-friendly without hazardous residual, low operating costs, and relatively simple way to set up and operate in healthcare facilities.

In recent years, there have been great increases in a number of UV-C sterilizing robots such as UVD Robot developed by the Danish company Blue Ocean Robotics [2], Xenex Germ-Zapping of Xenex [3], the IRIS 3200 m [4] robot of PowerBoost which have shown a 99% bactericidal capability. In the study [5], the authors presented the design of a disinfection robot using UV-C rays, the bactericidal effect of this robot was tested with some bacteria and molds. However, these types of robots are very expensive and difficult to access in Vietnamese health facilities. A new UV Robot design, which will be described in this paper is an attempt to reduce the risk of hospital infection. The robot equipped the high power UV-C disinfection lamps provides an economical and effective solution for Vietnamese medical institution in reducing the spread of harmful microorganisms. The robot is steered around the room and controlled by the medical personnel via wireless remote control and the time for which disinfect a room is 10–15 min depending on the size of the room.

2 Background

2.1 *Ultraviolet Disinfection*

Ultraviolet (UV) light is a form of radiation that is invisible to the human eye. It is an invisible part of the electromagnetic spectrum between X-rays and visible light. The Ultraviolet germicidal irradiation (UVGI) is a disinfection method that uses short-wavelength ultraviolet (UV-C) light from 200 to 280 nm to kill or inactivate microorganisms (i.e., bacteria, viruses, molds) by damaging nucleic acids and disrupting their DNA- thereby destroying their ability to reproduce. The main mode of inactivation occurs when the absorption of photons of light creates pyrimidine dimers between adjacent thymine bases and renders the microorganism incapable of replicating. UVGI has been a proven technology used in a variety of applications, such as food, air, water and instruments purification. Recently, the crisis of rising multidrug resistant organisms (MDROs), UV treatment has become the top list in the war against superbugs and has found a new application in hospitals providing surface disinfection of patient rooms, bathrooms, operating rooms (ORs), equipment rooms [6].

2.2 *Modeling of UV Dose*

Each pathogen is deactivated by a UV-C dosage. UV-C disinfection dosage is a function of total intensity of UV-C light and the length of exposure. Microorganisms

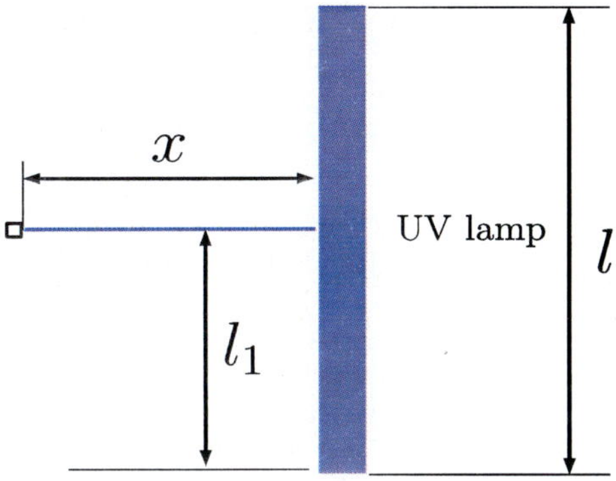

Fig. 1 The illustration of geometry for the view factor model of a UVGI lamp

have an exponential decrease in population when they are exposed to UVGI lamp. The surviving fraction of initial microbial population after exponential decay for microbes exposed to UV irradiation [7] is as follows:

$$S = e^{-kIt}, \tag{1}$$

where k is standard rate constant [cm^2/μJ], I *is* UV dose (or intensity) [μW/cm^2], and t is time of exposure [s] (seconds).

From Eq. (1), we can realize that the surviving of microorganisms depends on two factors, either the time they are exposed to UVGI or the UV dose. Therefore, UV dose is considered a key factor to evaluate the bactericidal ability of UVGI (in case exposurtime is fixed). In this manuscript, we use a model based on thermal radiation view factors[1] [8] to compute the UV dose along the axis of the UVGI lamp at any radial distances from the axis, the lamp must be divided into two segments of lengths l_1 and $l - l_1$. The sum of these sections will be the total lamp length l as illustrated in Fig. 1.

Hence, the total view factor at any point is computed from the lamp will be defined by:

$$F_{total}(x, l_1, l - l_1, r) = F(x, l_1, r) + F(x, l - l_1, r) \tag{2}$$

With $F(x, l, r)$ can be described in a general form as follows:

$$F(x,l,r) = \frac{L}{\pi H}\left[\frac{1}{L} a\tan\left(\frac{L}{\sqrt{H^2 - 1}}\right) - a\tan(M) + \frac{X - 2H}{\sqrt{XY}} a\tan\left(M\sqrt{\frac{X}{Y}}\right)\right] \tag{3}$$

[1]Which define the amount of diffuse radiation transmitted in the air or from one surface to another.

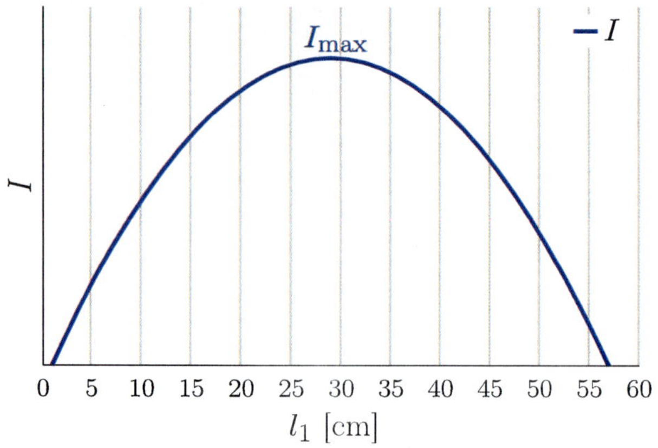

Fig. 2 The variation of the UV dose when $l_1 \in (0, l)$ in case l is taken by 58 cm

where $H = \frac{x}{r}$; $L = \frac{l}{r}$; $X = (1 + H^2) + L^2$; $Y = (1 - H^2) + L^2$ and $M = \sqrt{\frac{H-1}{H+1}}$, in which l is length of the UVGI lamp segment [cm], x is distance from the UVGI lamp [cm] and r is the radius of the UVGI lamp [cm]. Finally, the UV dose, $[\mu W/cm^2]$, will be presented as in [9]:

$$I = \frac{E_{uv}}{2\pi r l} F_{total} \tag{4}$$

where E_{uv} is considered as the UV power output of lamp, $[\mu W]$.

From Eq. (4), the UV dose obtains a maximum value at the lamp midpoint, this is also illustrated in Fig. 2.

3 Methods

3.1 Design of UV Robot

The 3D CAD model of UV disinfection robot is designed by using PTC Creo software (https://www.ptc.com/en). The UV disinfection system includes 5 UV-C lamps (100 W for each lamp) mounted in a circular pattern with 72° apart and 2 UV-C lamps (30 W lamp) fixed on the two side of the robot platform (Fig. 3). The upper UV-C lamps provide the 360° disinfection coverage and the lower UV lamps ensure the UV-C radiation can reach the lower zones. The UV-C lamps systems are protected by the protective frames.

The control circuits and the power source are stored in the robot platform with the size of 45 × 40 × 50 cm. The electrical system ensures the moving and illuminating functions of UV robot. This robot can be controlled remotely by using the NRF24L01

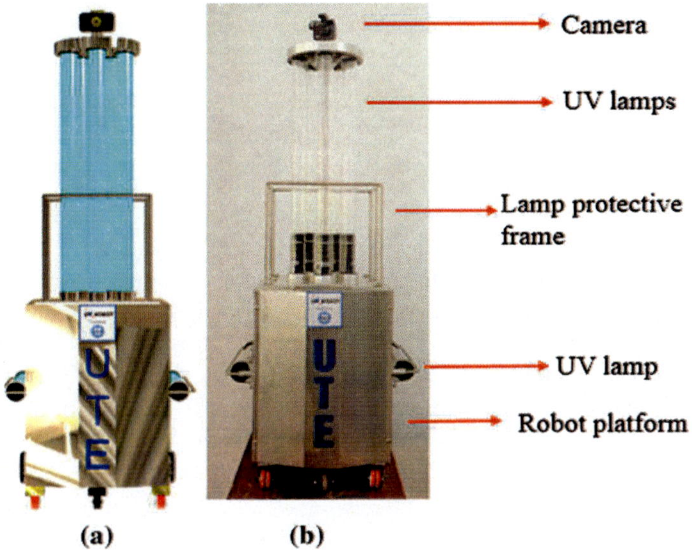

Fig. 3 **a** The design of UV Robot in PTC Creo. **b** The fabricated UV Robot

transceiver module to send signals to control the movement of the robot and the UV lamps status as depicted in Fig. 4. This wireless communication uses the 2.4 GHz

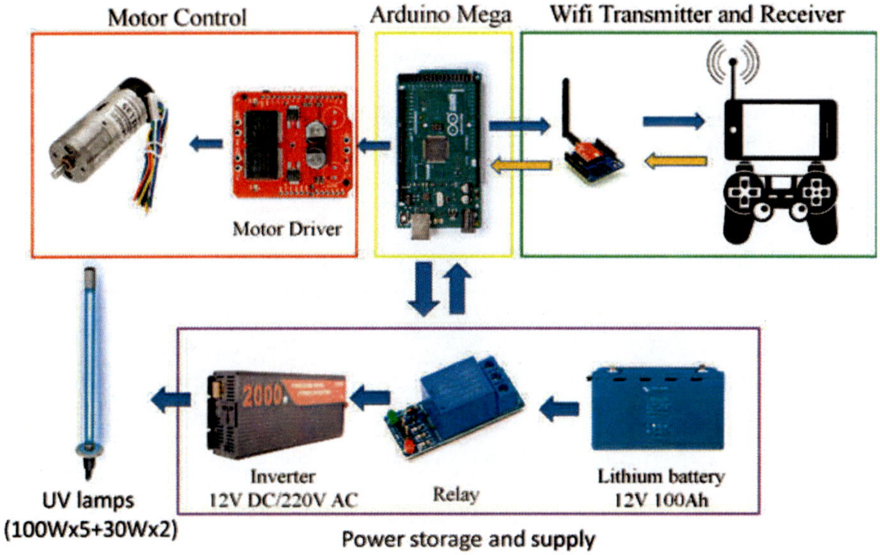

Fig. 4 Diagram of the electrical system of UV robot

band and can operate with baud rates from 250 kbps up to 2 Mbps. In case of working in an open space and with lower baud rate, its range can reach up to 100 m.

To avoid looking directly at the robot while the UV-C light is being illuminated, a smartphone mounted on the remote control to observe the robot motion through a camera fixed on top of the robot. Driving signals are processed at the Arduino Mega and sent to the motor driver (SparkFun Monster Moto Shield). This motor shield allows the current drawn by the motors up to 30 A and the PWM frequency to 20 kHz. The electrical wiring diagrams are shown in Appendix 1. For the concern of the illuminating control, with the power consumption approximates to 600 W, to ensure that the robot can operate continuously in 2.5 h, it is decided to choose the lithium battery of 12 V-100 Ah. This battery is high in capacity and light in weight. Moreover, a 2000 W inverter is used to convert direct current (DC) to alternating current (AC) which is necessary for by the UV-C lamps. It is worth noting that a disinfection process of UV-C robot in a room of 20 m^2 just takes around 15 min, therefore the 2.5 h of operating will allow the robot to handle up to 10 rooms.

3.2 Exposure-Distance Calculation For UV Disinfection

UV Robot uses 5 main UVGI lamps, each lamp owns the specifications such as l = 58 cm, r = 2.5 cm, $E_{uv} = 100\,W$, the exposure time, t is fixed by 30 s, and the coefficient k of some microorganisms are referenced in [10] as can be seen in Table 1.

In this study, the exposure-distance with respect to some viruses in Table 1, is only computed by one main lamp (even though the UV Robot owns totally 7 lamps including 5 main and 2 auxiliary lamps).

By observing the Fig. 5, the Corona-Sars is killed up to 99% within radius (is considered from UV Robot) 250 cm, whereas HIV-1 and Adenovirus (Type 2*) need a closer distance, around 70 cm.

Table 1 Standard rate constant, *k*, with respect to some microorganisms

Virus	k (m^2/J)
Corona-Sars	0.1106
HIV-1	0.00822
Avian Sarcoma	0.01047
Herpes (type 2)	0.06569
Adenovirus (type 2*)	0.00691
Rubella	0.0152
Ebola (Zaire)	0.0991
Mousepox	0.1247

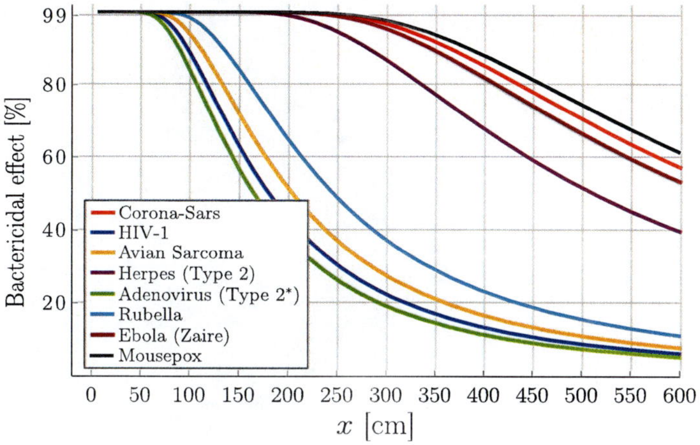

Fig. 5 Bactericidal effect with one lamp is taken into account

4 Results and Discussion

4.1 UV Disinfection Robot

The main features of robot are presented in Table 2. The robot motion and turn on–off lamp is controlled by the remote wireless control. User can observe working space thanks to the camera mounted on the top of the robot. The robot can operate in 2.5 h via a battery or connect directly to an electrical outlet. The disinfection time proposed for each room could take 5 min per spot with a radius of 1–2.5 m. Battery charging time may take up to 10 h.

Table 2 Main features of robot

Features	Parameters
Weight	55 kg
External dimensions	L:45 × W:40 × H: 150 (cm)
Operating time	2–2.5 h
Disinfection coverage	360°
Disinfection time	10–15 min per room (15–20 m^2)
Type of UV radiation	UV-C
Charging requirements	220–240 VAC, 50 Hz, 10 A

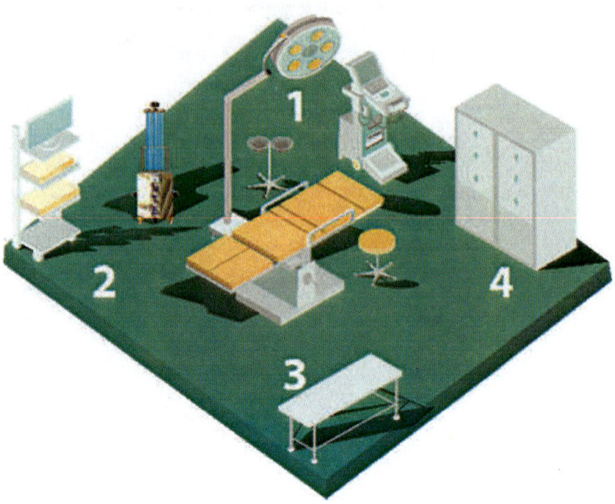

Fig. 6 Robot displacements in the room

4.2 *Disinfection Test*

The effectiveness of UV-C light on organisms is dependent on the dose and intensity, the distance from the object being disinfected, the type of surfaces, and the type of microorganisms. In this research, the efficacy of the UV Robot was assessed in two uncleaned operating rooms at Danang General Hospital. The floor area of each operating room is about 25 m^2 (5 m × 5 m). Since the UV lamp system of the robot is a 360° disinfection coverage with a radius from 1 to 2.5 m depending on the type of microorganisms, the test protocol proposed is to drive the robot to different locations in the operating rooms. The robot is steered to 4 locations in the room and stayed for 5 min at each location as can be seen in Fig. 6. The rooms are vacated and doors are closed during the disinfection process (see Fig. 7).

The quantity of microorganisms in the room is measured in CFU (Colony Forming units)/m^3 of air. For each room, the air sampling is performed using the settle plates, 4 Petri dishes containing solid nutrient medium are left open to air for 5 min. The microbes in the air will fall onto the surface of nutrient. The plates are closed and incubated at 37 °C for 48 h, the microbes grow into colonies in a number proportional to the level of microbial contamination. The visible colonies were counted. The mean number of CFUs of all plates at 4 different areas was taken. The CFU values in CFU/m^3 is determined based on the settling rate for each type of plate which is expressed as CFU/m^2/min. The 4 plates were used as the reference samples without UV exposure in comparison with the other 4 plates which were exposed to the UV light of the robot.

The total disinfection time using UV Robot for each room is about 20 min. The results of disinfection test are presented in the Table 3.

From these results, it is worth noting that the mean of CFUs counts obtained by the sampling plates after UV exposure is equal to zero. The harmful microorganisms

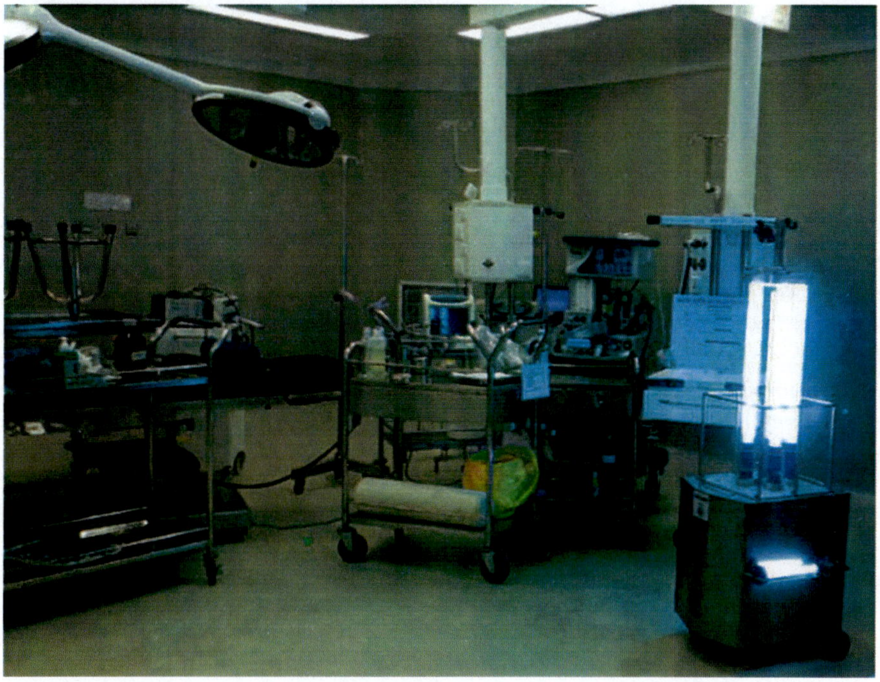

Fig. 7 UV-C light radiation of Robot in the room

Table 3 Results of disinfection test

	Operating room 1	Operating room 2
Before UV disinfection (CFUs/m^3)	264	132
After UV disinfection	0	0
Types of microorganisms	– Staphylococcus epidermidis – Gram-positive bacteria – Dermatophytes	– Staphylococcus epidermidis – Gram-positive bacteria
Disinfection time (min)	20	20

(Staphylococcus epidermidis, Gram-positive bacteria, Dermatophytes) in the operation are totally killed by UV exposure. The rooms after being sterilized by UV Robot meet the microbiological criteria of the operating room proposed by the hospital.

5 Conclusions and Perspectives

In this paper, we presented a new design and implementation of the disinfection robot which achieves a 99% disinfection rate like other similar robots but with a lower manufacturing cost. The high power UV-C lamps and the flexible mobility of the robot make it possible to improve environmental cleaning and disinfection. The test results confirmed the disinfection efficiency of the UV Robot when the most common and potentially harmful microorganisms in the operation room are killed in less time under the UV exposure. These obtained results met the operating room standard for microbiological indicators of the hospital. Although the efficiency of UV Robot needs to be tested with more harmful microorganisms such as Corona-Sars, Rubella..., the UV Robot has shown a great potential for the disinfection with the aims at preventing and reducing the spread of infectious diseases in the medical environment, especially in the fight against COVID-19 pandemic. The robot is currently driven manually by the wireless remote control and the camera. For future works, the autonomous navigation will be underway relying on simultaneous localization and mapping. Furthermore, the control, monitoring and data collection of the robot will be deployed through IoT systems.

Acknowledgements This research is funded by Funds for Science and Technology Development of University of Technology and Education- University of Danang. We would like express our gratitude to the medical staffs and doctors of Danang General hospital for their support in testing the UV disinfection.

Appendix 1. Electrical Wiring Diagram

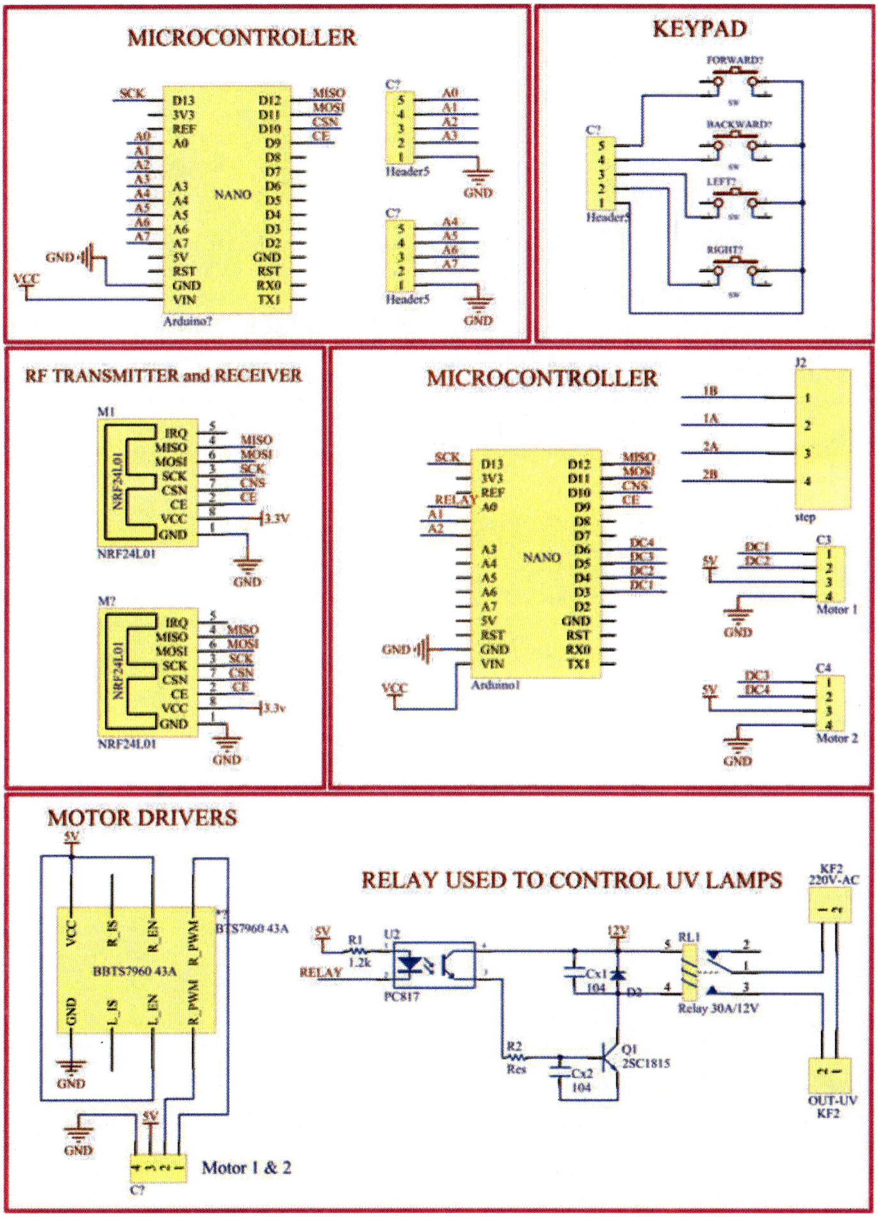

References

1. Spagnolo AM, Ottria G, Amicizia D, Perdelli F, Cristina ML (2013) Operating theatre quality and prevention of surgical site infections. J Prev Med Hyg 54(3):131
2. https://www.uvd-robots.com/
3. https://www.xenex.com/our-solution/lightstrike/
4. https://medicaldealer.com/iris-3200m-uv-c-germ-killing-robot-reaches-milestone/
5. Chanprakon P et al (2019) An ultra-violet sterilization robot for disinfection. In: 2019 5th International conference on engineering, applied sciences and Technology (ICEAST). IEEE
6. Begic A (2017) Application of service robots for disinfection in medical institutions. In: International symposium on innovative and interdisciplinary applications of advanced technologies. Springer, Cham, pp 1056–1065
7. Mitscherlich E, Marth EH (2012) Microbial survival in the environment: bacteria and rickettsiae important in human and animal health. Springer Science & Business Media
8. Modest MF (2013) Radiative heat transfer. Academic press
9. Kowalski W (2010) Ultraviolet germicidal irradiation handbook: UVGI for air and surface disinfection. Springer science & business media
10. Kowalski W, Bahnfleth JW, Hernandez MT (2009) A genomic model for predicting the ultraviolet susceptibility of viruses. IUVA News 11(2):15–28

Experimental of Multi-holes Drilling Toolpath Using Particle Swarm Optimization and CAD-CAM Software on PCB

N. W. Z. Abidin, N. Salim, Mohd Fadzil Faisae Ab. Rashid, N. M. Z. N. Mohamed, A. N. M. Rose, and A. Mokhtar

Abstract A multi-holes drilling process is widely used in electronics industry to produce printed circuit board (PCB). Nowadays, millions of PCB need to be produced in a single day to support the technological growth in all aspects of life. In this industry, the most time-consuming process is to drill the holes on the board. According to a survey, the tool movement in multi-holes drilling process spent up to 70% of the machining time. Various approaches have been proposed to optimize the toolpath in multi-holes drilling process. Previously, a computational experiment has been conducted to identify the best meta-heuristic algorithm to optimize this problem. The finding shows that Particle Swarm Optimization (PSO) has outperformed other comparison algorithm to generate the best toolpath. This paper aim to validate the PSO performance through an experiment. For this purpose, the experiment consist of nine drilling problems has been conducted to compare the toolpath that generated by PSO and commercial CAD-CAM software. The results indicated that the PSO generated toolpath is consistently faster than CAD-CAM generated toolpath, with 5% average difference. This finding confirmed that PSO has a great potential to be used in this process.

Keywords Multi-holes drilling · Particle swarm optimization (PSO) · CAD-CAM software

1 Introduction

Computer numerically controlled (CNC) machine has been introduced since 1940s. The system has evolved with the growth of computation technology. Furthermore, the customer demand with high expectation on quality and variability also push the

N. W. Z. Abidin · M. F. F. Ab. Rashid (✉) · N. M. Z. N. Mohamed · A. N. M. Rose
Department of Industrial Engineering, College of Engineering, Universiti Malaysia Pahang, Lebuhraya Tun Razak, 26300 Kuantan, Pahang, Malaysia
e-mail: ffaisae@ump.edu.my

N. Salim · A. Mokhtar
Faculty of Mechanical and Automotive Engineering Technology, Universiti Malaysia Pahang, 26600 Pekan, Pahang, Malaysia

A. F. Ab. Nasir et al. (eds.), *Recent Trends in Mechatronics Towards Industry 4.0*, Lecture Notes in Electrical Engineering 730, https://doi.org/10.1007/978-981-33-4597-3_19

CNC technology to become more robust. CNC drilling is one of the most frequently being used in industry, especially in the electronics to produce printed circuit board. According to a survey, almost 38% of the published research on CNC drilling were focusing on PCB drilling.

In the multi-holes drilling as in PCB making, the important criteria besides the quality of the product is how fast the drilling process can be completed. In this industry, a single second different in the drilling process give huge effect due to the high volume of production per month. Considers a PCB manufacturer with demand 10,000 pieces per day. In this example, if an engineer able to reduce machining time for only 1 s, the company will save for about 2.7 h per day. For multi-holes drilling operation process, on average, 70% of the machining time is because of tool movement and tool switching time. Based on this fact, the optimization of drilling toolpath could improve the productivity of the machining process by reducing the total machining time [1].

Various optimization approaches have been used to optimize the drilling toolpath. Based on a survey, the most frequently algorithm used to optimize this problem is Genetic Algorithm (GA) [2]. Kumar and Pachauri [3] for example used a binary representation in GA to optimize drilling tool path. On the other hand, Nabeel et al. [4] found that the GA able to reduce machining time up to 50% compared with manual path planning. A part from that, Al-Janan and Liu [5] combined Taguchi approach with GA to optimize the toolpath for drilling problem.

Besides GA, Ant Colony Optimization (ACO) also is another favorite algorithm for drilling toolpath optimization. Montiel-Ross et al. [6] claimed that the ACO algorithm with some improvement able to improve the solution up to 62% compared with comparison algorithms. Iberahim et al. [7] however mention that whenever the number of operations in drilling increased and the evaporation rate increased, the solution accuracy by ACO will reduced.

Recently, a lot of different algorithms have been implemented for drilling path optimization. Zhang and Zhu [8] proposed a Foraging Algorithm for this problem. Srivastava [9] meanwhile suggested an Intelligent Water Drops algorithm. A part from these works, researchers also implemented Particle Swarm Optimization (PSO) to optimize drilling toolpath [10, 11].

Since there are too many optimization algorithms were proposed to optimize drilling toolpath, and all published papers claim that their algorithm is better, the authors were conducted a computational experiment to study the algorithms performance for multi-holes drilling toolpath [12]. In that study, eight optimization algorithms were compared to optimize 15 drilling toolpath problem ranging from 10 until 150 holes. The computational results showed that the PSO algorithm had the best performance compared with 7 other algorithms.

Our existing study however limited to the computational experiment in order to identify the best optimization algorithm without physical drilling experiment. This paper therefore intends to validate the PSO performance in the actual multi-holes drilling process. Since the PSO performance has been proven in previous study, the drilling toolpath that generated by PSO will be compared with the drilling toolpath that generated by a commercial computer-aided design and computer-aided manufacturing (CAD-CAM) software.

2 Methodology

The problem of multi-holes drilling toolpath is modelled as a Travelling Salesman problem (TSP). TSP is a routing problem to visit each city exactly once and return to the starting city. In order to minimize the toolpath, the objective function is formulated as follow:

$$\min f(x) = \sum_{i}^{n} \sum_{j}^{n} X_{ij} D_{ij} \tag{1}$$

subjected to:

$$\sum_{i=1}^{n} X_{ij} = 1, j = 1, \ldots, n \tag{2}$$

$$D_{ij} = D_{ji} = \sqrt{(x_i - x_j)^2 + (y_i - y_j)^2} \tag{3}$$

In Eq. (1), n is the number of holes, D_{ij}, is the distance from hole *i* to *j*. Meanwhile, constraint (2) ensure that each hole is visited only once. In this case X_{ij}= 1, if the tool travel from point *i* to *j*, otherwise X_{ij}= 0. Meanwhile in (3), *x* and *y* are the Cartesian coordinates.

2.1 Particle Swarm Optimization

Particle Swarm Optimization (PSO) is a population-based algorithm that inspired by the behaviour of bird flocking or fish schooling. This algorithm was introduced back in 1995 by Kennedy and Eberhart [13]. The pseudo-code for PSO is as follow:

```
Begin
pop = Initial population
for t = 1 to max_iter
 for each particle x in pop, do
    f(x): evaluate fitness for particle x
    if f(x) is better than f(pBest)
       Update Pbest = x
    end
 end
 Update Gbest = best x in pop
 for each x in pop, do
    v^{t+1} = v^t + c_1*rand (Pbest – x) + c_2*rand (Gbest – x)
    p^{t+1} = x^t + v^{t+1}
 end
end
```

Based on the algorithm, PSO consist of four steps; Initialization, Evaluation, Update *Pbest* and *Gbest* and Update velocity and position. Initialization step establish an initial population of solution, known as particle or position, *x*. Besides that, the maximum iteration, *max_iter*, coefficient c_1 and c_2 are defined.

Next, particle *x* will be evaluated according to objective function as in Eq. (1). Since the drilling toolpath is a discrete combinatorial problem, a topological sort procedure is used to convert continuous particle into permutation format. The detail of this steps has been explained in earlier study [2].

The next step in PSO is to update *Pbest* and *Gbest*. *Pbest* represent the best solution for a particular particle, while *Gbest* is the global best solution. The *Pbest* is simply updated by comparing with fitness for particle *x*. Meanwhile *Gbest* is updated by comparing the fitness with the whole population. The final step in PSO is to update the velocity, *v* and position, *x*.

2.2 Experimental Design

An experiment has been conducted to compare the machining time for the toolpath that generated from PSO, and the toolpath that automatically generated using CAD-CAM software. In total, there are nine drilling problems consist of 10, 30, 40, 70, 80, 90, 130, 140 and 150 holes. The sample of the toolpath are shown in Fig. 1.

For the experimental purpose, the material use is aluminum plate 300 × 300 mm and the thickness are 3 mm. For the drilling process, a CNC milling machine, Haas VF-6 model is used with the speed rate 100, spindle speed 1500 rpm as in Figs. 2 and 3. In this experiment, the only output variable that collected is machining time.

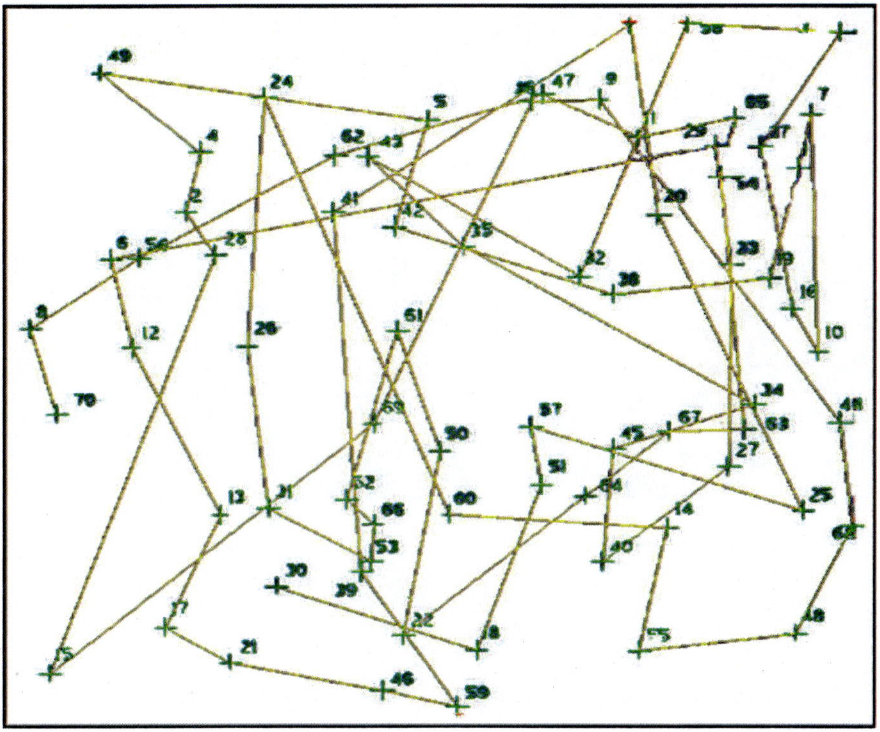

Fig. 1 Sample of 70 holes test problem

Fig. 2 CNC milling machine (Haas VF6)

Fig. 3 Drilling process using CNC milling

3 Results and Discussions

Table 1 shows the machining time for PSO and CAD-CAM toolpath. The machining time is divided into three categories that is small, medium and large problems. Based on the table, the PSO generated path is consistently faster than CAD-CAM generated path.

In small size problem, the difference between PSO and CAD-CAM generated paths were only below than 2.55%. When the problem size increased to medium, the difference is more obvious. In the medium group, the PSO paths were faster than

Table 1 Machining time for PSO and CAD-CAM generated toolpath

Problem	Number of holes	Machining time (s)		Percentage of difference (%)
		PSO	CAD-CAM	
Small	10	56.6	57.0	0.70
	30	164.1	168.4	2.55
	40	219.1	222.7	1.61
Medium	70	342.5	384.9	11.02
	80	435.6	496.3	12.23
	90	489.2	559.8	12.61
Large	130	699.2	739.8	5.48
	140	751.8	792.0	5.07
	150	806.2	842.1	4.26

CAD-CAM between 11 and 12.6%. Meanwhile in large problem size, the difference is about 5%.

This trend is related with the search space that excessively increased with the number of holes. In small size problem, the search space is still small, which make the near optimum solution is achievable by most of the algorithms. However, when the size of problem increased, the search space is excessively increased. This phenomenon makes only a good algorithm able to converge to near optimum solution. In the large size problem, the search space is too large to be handled by PSO. It makes the solution provided by PSO is still unable to reach the near optimal solution. In this case, some of the PSO parameters (i.e. number of population and maximum iteration) and coefficients (c_1 and c_2) need to be re-tuned to handle too large search space.

Even though the toolpath for large size problem were belief not at the near optimal due to the computational constraint, there were some improvement showed by the PSO solution. Considering a PCB maker with capacity only 50,000 units per day. By using a PSO generated toolpath, the capacity can be increased to 52,000 units, taken into account 5% improvement of the machining time.

4 Conclusions

This paper presents a validation experiments to prove that the PSO generated toolpath for multi-holes drilling is able to generate less machining time compared with commercial CAD-CAM software. In previous study, PSO was proven computationally to came out with better solution compared with other algorithms including Genetic Algorithm and Ant Colony Optimization.

In this experiment, nine multi-holes drilling problem is used ranging from 10 to 150 holes. Based on the results, PSO able to generate faster tool path with the average of 5% better than tool path that generated by commercial CAD-CAM software. The most obvious difference is observed in medium size problem with 70–90 holes. As a conclusion, this research confirmed that the PSO is able to generate better toolpath for multi-holes drilling compared with commercial CAD-CAM software.

Acknowledgements This research is funded by Universiti Malaysia Pahang under grant number RDU160356 and RDU180331.

References

1. Lim WCE, Kanagaraj G, Ponnambalam SG (2016) A hybrid cuckoo search-genetic algorithm for hole-making sequence optimization. J Intell Manuf 27(2):417–429
2. Zainal Abidin NW, Ab Rashid MFFF, Nik Mohamed NMZ (2017) A review of multi-holes drilling path optimization using soft computing approaches. Arch Comput Methods Eng

3. Kumar A, Pachauri PP (2012) Optimization drilling sequence by genetic algorithm. Int J Sci Res Publ 2(9):1–7
4. Nabeel P, Abid K, Abdulrazzaq HF (2014) Tool path optimization of drilling sequence in CNC machine using genetic algorithm. Innov. Syst Des Eng 5(1):15–26
5. Al-Janan DH, Liu T-K (2014) Path optimization of CNC PCB drilling using hybrid Taguchi genetic algorithm. Kybernetes 43(6):107–125
6. Montiel-Ross O, Medina-Rodríguez N, Sepúlveda R, Melin P (2012) Methodology to optimize manufacturing time for a CNC using a high performance implementation of ACO. Int J Adv Robot Syst 9:1–10
7. Iberahim F, Ramli R, Narooei KD, Qudeiri JA (2014) Tool path optimization for drilling process by CNC milling machine using ant colony optimization (ACO). Aust J Basic Appl Sci 8(April):385–389
8. Zhang WB, Zhu GY (2018) Drilling path optimization by optimal foraging algorithm. IEEE Trans Ind Inf 14(7):2847–2856
9. Srivastava PR (2015) A cooperative approach to optimize the printed circuit boards drill routing process using intelligent water drops. Comput Electr Eng 43:270–277
10. Dalavi AM, Pawar PJ, Singh TP (2015) Optimization of hole-making operations for injection mould using particle swarm optimization algorithm. Int J Ind Eng Comput 6(4):433–444
11. Othman MH et al (2011) A binary particle swarm optimization approach for routing in PCB holes drilling process. In: 1st International conference on robotic automation system, vol 2, pp 201–206
12. Zainal Abidin N, Ab. Rashid M, Nik Mohamed N (2018) Optimization of multi-holes drilling path using particle swarm optimization. In: Lecture notes in mechanical engineering: intelligent manufacturing & mechatronics, pp 101–107. Springer Singapore
13. Kennedy J, Eberhart R (1995) Particle swarm optimization. In: Proceedings of international conference on neural networks, vol 4, pp 1942–1948

Parameter Identification of Horizontal Flexible Plate System Using Cuckoo Search Algorithm

Nurul Nisha Mawarnie Mohd Rawi and Muhamad Sukri Hadi

Abstract The flexible structure offers various advantages such as lightweight, efficient, fast system response and low energy consumption. However, this structure produces too much vibration which lead to system failure. This thesis presents a modelling of the horizontal flexible plate system based on cuckoo search algorithm. The aim of this research is to eliminate vibrations produced by the flexible plate structure in order to preserve the performance and lengthen the life of the system. Initially, the system models are developed based on the vibration data collected experimentally. The dynamic models were developed based on auto-regressive with exogenous (ARX) using system identification technique. The system identification technique applied in this study to develop the mathematical models of the system are the conventional technique known as recursive least squares (RLS) and swarm intelligence algorithm known as cuckoo search algorithm (CS). Then, the developed system models were evaluated and validated by mean squared error (MSE), correlation test and also pole-zero diagram. The performance of developed model was compared between the conventional and swarm intelligence algorithms. The best system model was verified based on lowest MSE, within 95% confidence level of the correlation test and high stability of the pole-zero diagram.

Keywords Flexible plate · System identification · Recursive least algorithm · Cuckoo search algorithm · Swarm intelligence algorithm

1 Introduction

Nowadays, flexible structure such as plates, shells, beams and frames have become one of the main interests to the researchers. The advantage of flexible structure such as lightweight, lower energy consumption, safer operation, faster system response and lower maintenance have attracted the industries to use it structures instead of rigid structure [1].

N. N. M. M. Rawi · M. S. Hadi (✉)
Faculty of Mechanical Engineering, Universiti Teknologi MARA, 40450 Shah Alam, Selangor, Malaysia
e-mail: msukrihadi@uitm.edu.my

A. F. Ab. Nasir et al. (eds.), *Recent Trends in Mechatronics Towards Industry 4.0*, Lecture Notes in Electrical Engineering 730, https://doi.org/10.1007/978-981-33-4597-3_20

Recently, the flexible plate structure is mostly employed in the engineering applications, for examples solar panels, bridge decks, satellite panels, electronic circuit board design, winglet part in the airplane and automotive parts in industry [2]. However, the flexible flat, thin plates are easily affected by vibration when subjected to the disturbance's forces. This undesirable vibration may affect the stability and performance of the flexible plate system as the generating vibrations in the system will lead to noise, fatigue, wear, destruction, human discomfort and indirectly cause the failure of structure [3]. Thus, the unwanted vibrations produced on the flexible plate structure need to be removed to ensure the performance of the system is preserved.

Previously most of the researchers have been proposed passive vibration control (PVC) technique to reduce the excessive vibration in the structure. PVC is a traditional method that used viscoelastic materials such as dampers and absorbers [4]. However, due to its limitation by increase the weight structures, it cannot withstand the low frequency vibration of the flexible plate system. Therefore, the active vibration control (AVC) technique is introduced in this research to overcome the limitation of the PVC technique.

Even so, the accurate model system needs to be modeled before the effective controller can be developed for vibration suppression of the flexible plate system. Recently, system identification technique has become the researchers' choice to select the best model structure. The system identification is dealing with the building of the mathematical model of the dynamic system evolved from the vibration data collected [5]. The system identification technique has two types which are conventional and swarm intelligence algorithms. Several numbers of researchers have applied the swarm intelligence algorithm in modelling the flexible structure system such as the artificial bee colony algorithm, particle swarm optimization, artificial immune system algorithm, cuckoo search algorithm, flower pollination algorithm, ant colony algorithm and other metaheuristic optimization algorithms [6–10]. Thus, this study presents the modeling of horizontal flexible plate structure using swarm intelligence algorithm via cuckoo search approach. The performance of developed model was then validated using conventional algorithm known as recursive least square.

2 Experimental Setup for Vibration Data Collection

A complete data acquisition to compute the physical conditions of a system consists of sensors and actuators, signal conditioning hardware and a computer running DAQ system. The horizontal flexible plate experimental rig was developed with attached actuators and sensors on the rig and from the developed rig, the experimental vibration data were collected. Square, thin and flat aluminium plate with dimension of 0.7 m for each side and 0.001 m thickness is used in this experiment to represent the flexible structure. Moreover, in order to let the flexible plate to vertically vibrate, the structure was placed in the horizontal position. The characteristics of the flexible plate used in this research as referred in [1].

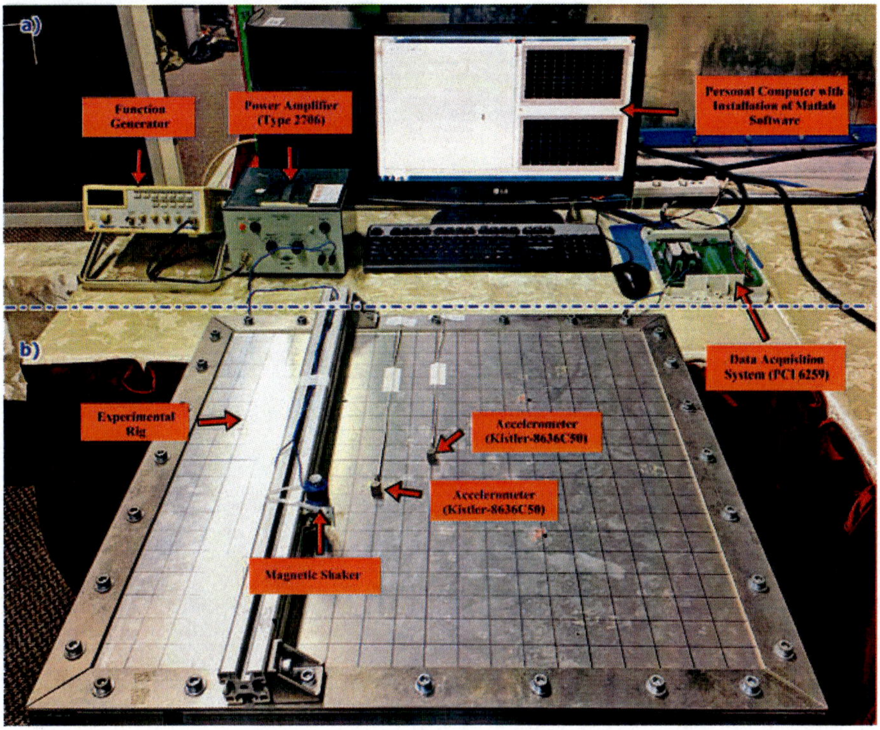

Fig. 1 Experimental setup for vibration data collection, **a** The integration of DAQ system into experimental rig, **b** view of the sensors from the bottom of the plate

On the other hand, the experimental rig was designed with all clamped edges. To generate actuation force, a magnetic shaker was placed at the excitation point on the experimental rig with distance of 1 cm parallel to the permanent magnet. The magnetic shaker then was driven by a function generator through a power amplifier to create sinusoidal force to excite the system. Two pieces of piezo-beam type accelerometers were mounted at two different positions on the experimental rig to act as sensors in order to collect the vibration data of the system. The different positions where accelerometers attached are particularly the observation and detection points. Moreover, the data acquisition system which mounted in computer was directly connected with the piezo-beam type accelerometers and the experimental data were analysed in this experiment (Fig. 1).

3 Cuckoo Search Algorithm

Cuckoo search algorithm was initially proposed by Xin-She Yang and Suash Deb in 2009 based on reproduction of cuckoo birds. In cuckoo search, the algorithm starts searching by initializing a population of host nests. The population initialization is performed as follows [11, 12]:

$$x_i = lb_i + Rand_i(ub_i - lb_i) \tag{1}$$

where x_i is a solution represents cuckoo egg in nest of the host, lb_i and ub_i are lower and upper boundary of the problem domain and $Rand_i$ is a random number from the uniform distribution within the interval. Cuckoo search performs Lévy flight as below:

$$x_{i+1} = x_i + a \oplus Lé\ vy(\lambda) \tag{2}$$

where a is a step size which a > 0, x_i is the existing search location and λ is the step length. The Lévy distribution can be described as:

$$\text{Lé vy }(\lambda) = \frac{\lambda \Gamma(\lambda)\sin\left(\frac{\pi\lambda}{2}\right)}{\pi} \frac{1}{s^{1+\lambda}} \tag{3}$$

where Γ is a Gamma function which represents random step length meanwhile s is step size. The step size, s is to determine the total number of random walk in a fixed number of iterations [13]. The step size is performed as below:

$$s = 0.01\left(\frac{Lé\ vy(\lambda).Rand_1}{Rand_2}\right)^{\frac{1}{\lambda}}.\left(x_i - x_{gbest}\right) \tag{4}$$

$$x_{new} = x_i + s.Rand_3 \tag{5}$$

where x_i is the ith solution and x_{gbest} is current global best solution. x_{new} is the new solution which will evaluated with the present solution. Then, a portion of solutions is abandoned to replace with the randomly generated new solutions or in simple terms, switching nests. The new solutions which will replace a portion of solutions is generated using equation below.

$$x_{new} = x_i + Rand_1 \oplus H(P_a - Rand_2) \oplus \left(x_j - x_k\right) \tag{6}$$

where x_i is current solution, x_j and x_k are two different solutions which randomly chosen, H(u) is the function of Heaviside meanwhile P_a is the probability of nest to be abandoned. Then, the solutions are all ranked to discover the global best solution. This global best solution is then applied in Eq. (4).

4 Result and Discussion

In this research, the system models have been developed using recursive least square (RLS) and cuckoo search (CS) algorithms. The 5000 input-output vibration data obtained from the experiment were divided equally into two parts in order to train and test the quality of developed model. Next, the developed models of the system were validated using mean squared error (MSE), pole-zero diagram and correlation test. The selection of the best model is focus on the evaluation results of the robustness tests which are the lowest MSE, high stability of the pole-zero diagram and unbiased for correlation test. Validation of the model is one of the important process to verify the superlative compatible model to personify the structure of the system. The most suitable model was chosen using heuristic approach. [14].

For modelling using conventional algorithm via RLS, two parameters were tuned properly to achieve the best model of the flexible plate system which are forgetting factor and model order. The forgetting factor value were tuned from 0.2 to 0.8 meanwhile the model order was tuned from 2 until 20. Based on the tuning results, the best model was obtained at order 14 with forgetting factor 0.8 as it has lowest value of MSE which are 1.5353×10^{-5} and 3.7392×10^{-6} for training and testing data, respectively. Figures 2 and 3 show the outputs of the flexible plate in time and frequency domains. Based on the figures, the developed model is able to follow the actual flexible plate system as the graphs of time and frequency domain are successfully overlapped the actual measured output. Furthermore, the error between the actual and predicted outputs using RLS modelling are plotted in Fig. 4.

The correlation tests and pole-zero diagram stability are carried out to determine the effectiveness of the attained model system. Figure 5 shows the correlation tests for RLS modelling meanwhile Fig. 6 shows the pole-zero diagram stability of the system model. According to Fig. 5 the auto correlation of the RLS modelling was

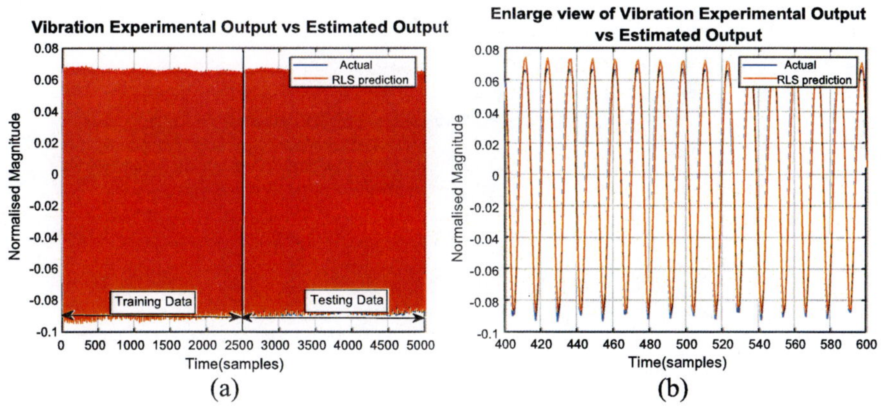

Fig. 2 The actual and prediction outputs in time domain using RLS modelling, **a** prediction output for 5000 vibration data; **b** enlarge view of data from 400 until 600

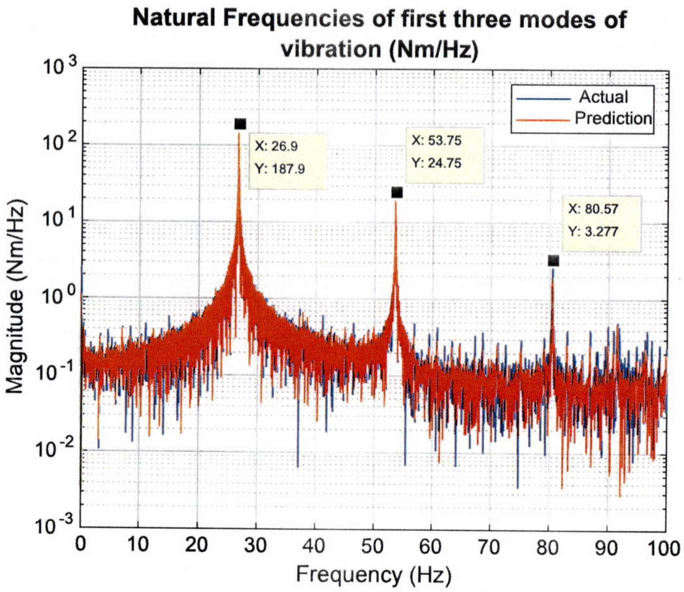

Fig. 3 Actual and estimated outputs of the flexible plate system in frequency domain using RLS modelling

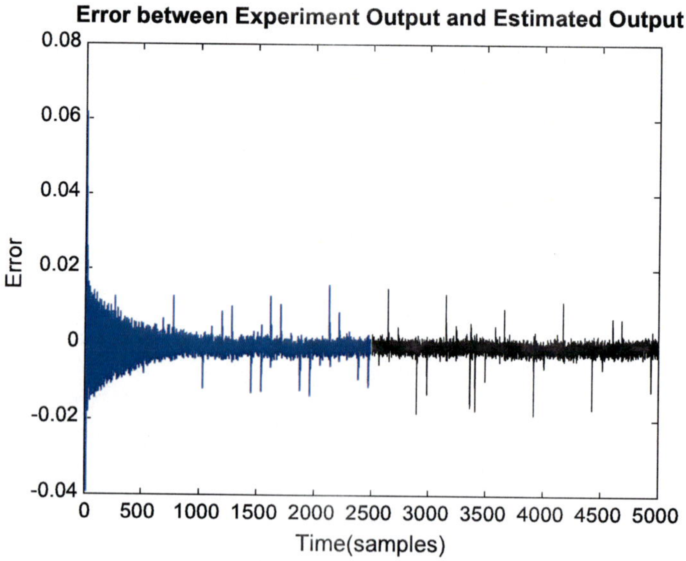

Fig. 4 Error between experimental and estimated outputs of the system using RLS modelling

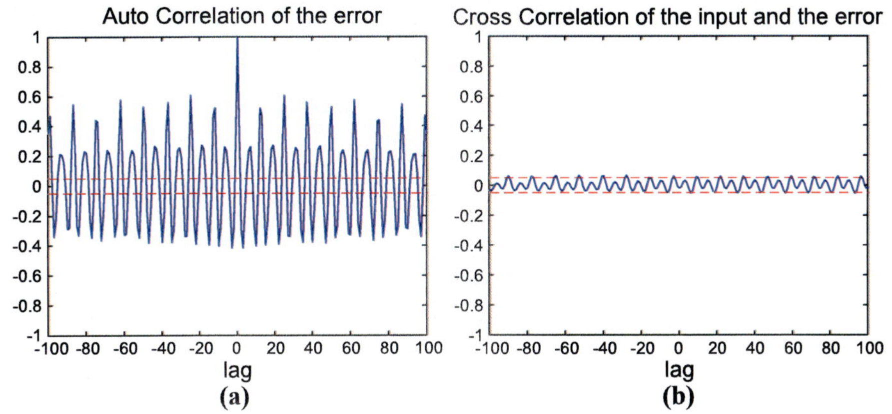

Fig. 5 The correlation tests for flexible plate system using RLS modelling, **a** auto correlation; **b** cross correlation

Fig. 6 The pole-zero diagram stability of the system using RLS modelling

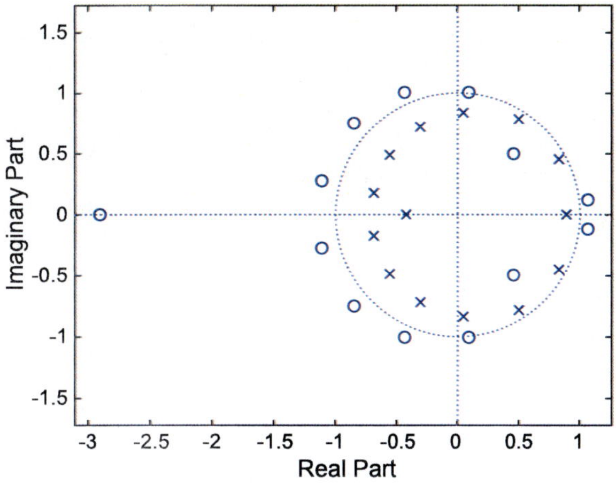

biased as the result is exceeding 95% confidence level however, the graph of the cross correlation shows that the RLS modelling was unbiased as the developed model is within the 95% confidence level. For pole-zero diagram stability, the good model must have high stability of the system. The best model of the RLS modelling was found to be stable as all the poles are located in the circle of diagram. The discrete transfer function obtained using RLS modelling from the best model is defined as Eq. (7).

$$\frac{y(t)}{u(t)} = \frac{\begin{array}{l} 0.01527z^{-1} + 0.06862z^{-2} + 0.08078z^{-3} + 0.01027z^{-4} \\ - 0.09732z^{-5} - 0.148z^{-6} - 0.107z^{-7} - 0.01457z^{-8} \\ + 0.06594z^{-9} 0.08618z^{-10} + 0.04845z^{-11} \\ + 0.002782z^{-12} + 0.002323z^{-13} + 0.04746z^{-14} \end{array}}{\begin{array}{l} 1 - 0.09507z^{-1} + 0.008669z^{-2} + 0.0723z^{-3} + 0.07151z^{-4} \\ + 0.03964z^{-5} + 0.01104z^{-6} + 0.01637z^{-7} + 0.05073z^{-8} \\ + 0.07053z^{-9} + 0.02882z^{-10} - 0.07151z^{-11} - 0.1519z^{-12} \\ - 0.1319z^{-13} - 0.03428z^{-14} \end{array}} \quad (7)$$

For swarm intelligence algorithm using CS algorithm, the best model order was achieved by tuning five parameters which are model order, number of nests, probability, lower and upper boundary and maximum generation. The parameters were properly tuned one by one using trial and error method. The number of nests were tuned from range 10 until 40. According to previous researchers, the number of nests in range of 10 until 40 provided good results for most optimization problems [15, 16, 17]. The boundary limit and probability are varied from 0.5 to 3 and 0.01 to 0.40, respectively. Other than that, the maximum generation of the CS algorithm was set from 500 until 3500 with increment of 200. The boundary limit, probability and maximum generations are stopped at certain value as the result of CS is already converged at those maximum values.

Among all the attempted order, the best model of the flexible plate system was found to be model order 4 as presented in Table 1. The MSE for the best CS algorithm model are 1.9184×10^{-5} and 5.8092×10^{-6} respectively for training and testing data. The convergence of CS algorithm modelling during simulation has been plotted in Fig. 7. Figures 8 and 9 present the actual and estimated outputs of the system in time and frequency domains, respectively. The error between actual and estimated outputs were displayed as in Fig. 10.

The accuracy of the CS modelling was shown in Fig. 11, which the result is unbiased as it is correlated within 95% of confidence level for both auto and cross correlations. On the other hand, Fig. 12 shows that all the poles in the unit circle are plotted thus, it is verified that the developed model using CS algorithm is stable. The discrete transfer function achieved using CS modelling are described as follows (8):

Table 1 The best parameters used to obtain the best model using CS modelling

Parameters	Values
Model order	4
Number of nests	14
Probability	0.01
Lower and upper boundary	[−0.5,0.5]
Maximum generations	2000

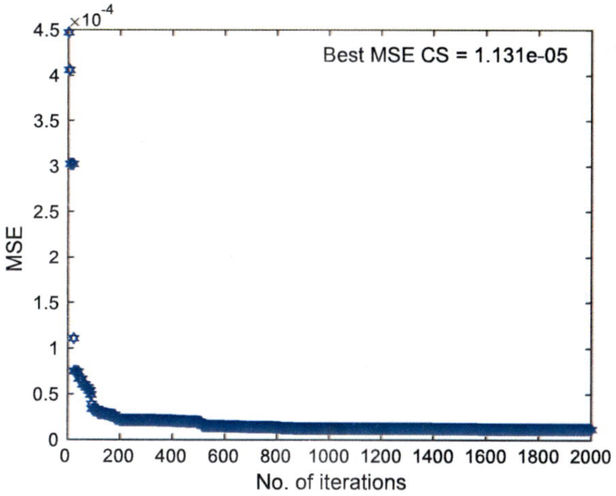

Fig. 7 Convergence of cuckoo search modelling during simulation

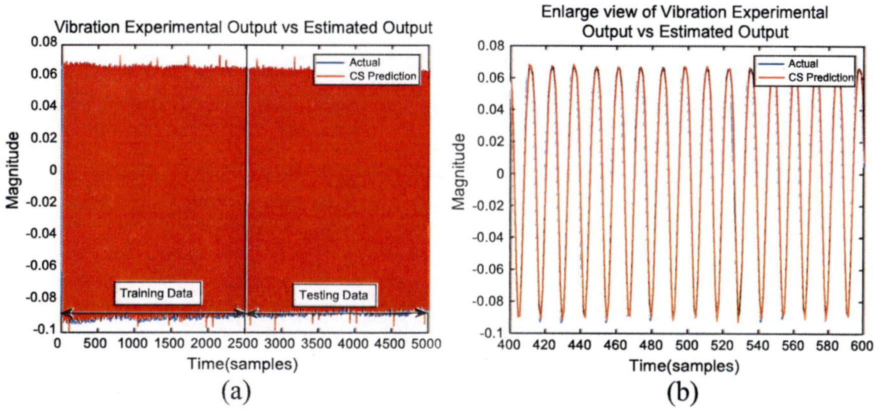

Fig. 8 The vibration actual and CS prediction outputs of system in time domain, **a** outputs of predicted model for 5000 samples; **b** enlarge view of 400 to 600 data samples

$$\frac{y(t)}{u(t)} = \frac{-0.1524z^{-1} + 0.1646z^{-2} + 0.4487z^{-3} - 0.6344z^{-4}}{1 - 1.124z^{-1} + 0.5913z^{-2} - 0.1715z^{-3} + 0.1782z^{-4}} \tag{8}$$

Based on the result presented, the comparative assessment between RLS and CS modelling can be made. The lowest mean squared error (MSE) for recursive least square (RLS) and cuckoo search algorithm (CS) were 3.7392×10^{-6} and 5.8092×10^{-6}, respectively. As recorded, the value of MSE for RLS is slightly lower than CS algorithm however, there are two more validations to be performed before choose the best one. For pole-zero diagram stability, if all the poles located in the circle of

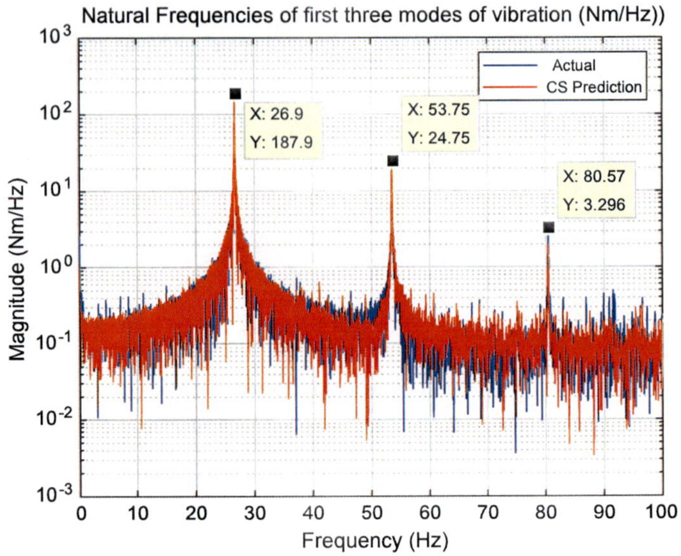

Fig. 9 The actual and estimated outputs of the system in frequency domain

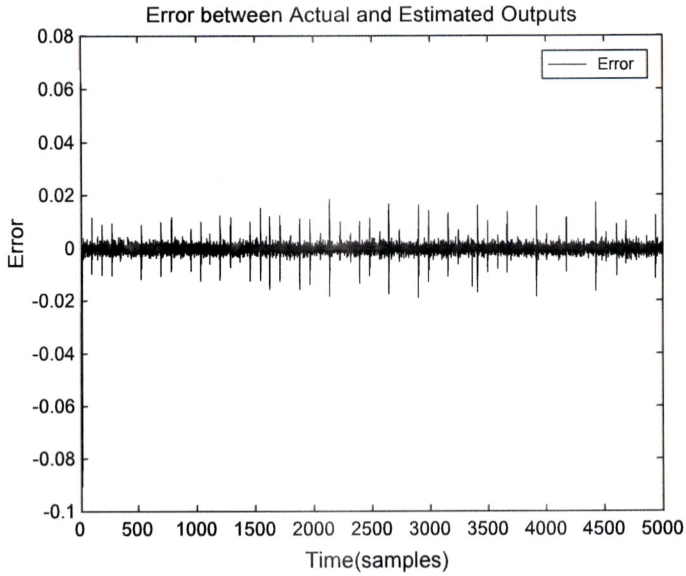

Fig. 10 Error of the output using CS modelling

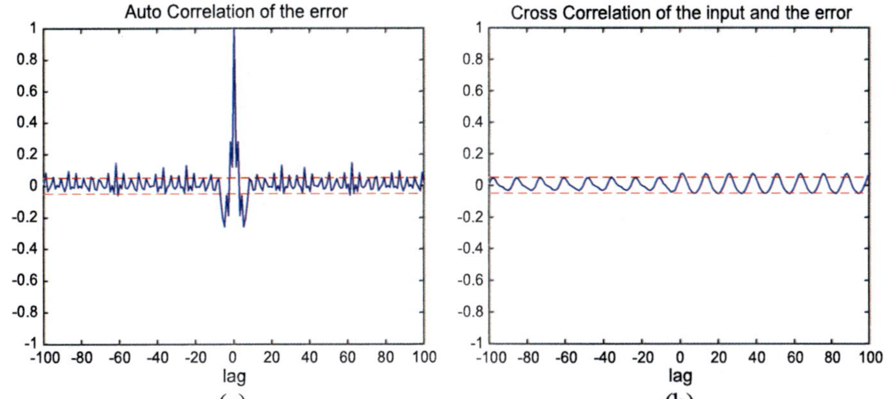

(a) (b)

Fig. 11 Correlation tests of CS modelling, **a** auto correlation; **b** cross correlation

Fig. 12 The stability diagram of flexible plate via CS modelling

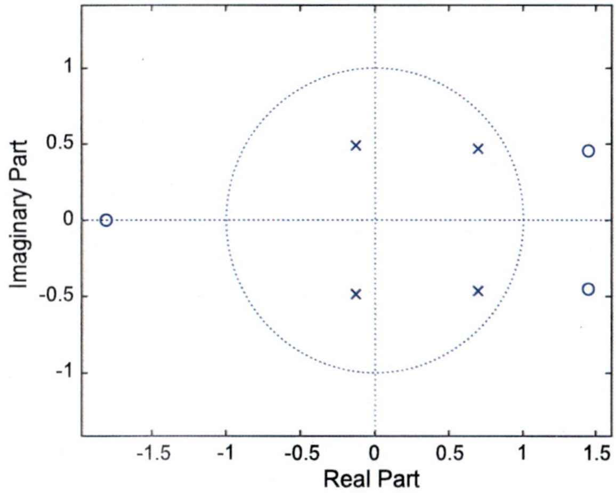

the diagram, the system is considered stable. Both model validation of RLS and CS are considered stable as all poles within the circle area manage to be plotted. As for correlation test, cuckoo search algorithm performs better as compared to the recursive least square. The correlation test for RLS was found to be biased for auto correlation as it is exceeding 95% of confidence level, meanwhile, for CS the system is found to be unbiased for both auto and cross correlation as the results were correlated within the confidence level of 95%. From all the validation methods that have been done, it is noticed that the cuckoo search model is more approximate in representing the horizontal flexible plate system as compared to the recursive least square.

5 Conclusion

In this paper, a flexible plate system was properly modelled using RLS algorithm and CS algorithm. The results of modelling and validation were presented in this paper. The best model represents the dynamic system of the flexible plate structure was obtained based on lowest mean squared error, high stability of pole zero diagram and good correlation tests. Therefore, it is noticed that the performance of the CS algorithm was outperformed the RLS algorithm performance in modelling the horizontal flexible plate system.

Acknowledgements The authors would like to express their gratitude to Universiti Teknologi MARA (UiTM), Universiti Teknologi Malaysia (UTM) and Ministry of Higher Education (MoHE) for funding the research and providing facilities to conduct this research. Sponsor file number (RACER/1/2019/TK03/UITM//1).

References

1. Hadi MS, Darus IZM, Jamid MF, Tokhi MO (2019) Active vibration control of a horizontal flexible plate structure using intelligent proportional–integral–derivative controller tuned by fuzzy logic and artificial bee colony algorithm. J Low Freq Noise Vibr Active Control, pp 1–13
2. Darus IZM, Al-Khafaji AAM (2011) Non-parametric modelling of a rectangular flexible plate structure. Eng Appl Artif Intell 25(1):94–106
3. Aizuddin M (2013) Effect of fatigue life to the natural frequency of metallic. B.S. Thesis, Faculty of Mechanical Engineering, Universiti Malaysia Pahang, Pahang
4. Salleh AM (2011) Active intelligent control of vibration of flexible plate structures. Ph. D. Thesis Department of Automatic Control and Systems Engineering, The University of Sheffield
5. Tavakolpour AR, Mailah M, Darus IZM, Tokhi O (2010) Genetic algorithm-based identification of transfer function parameters for a rectangular flexible plate system. Eng Appl Artif Intell 23(8):1388–1397
6. Hadi MS, Darus IZM (2018) Modelling of horizontal flexible plate structure using artificial bee colony algorithm. Int J Eng Technol 7:415–419
7. Hadi MS, Yatim HM, Darus IZM (2018) Modelling and control of horizontal flexible plate using particle swarm optimization. Int J Eng Technol, pp 13–19
8. Salleh AM, Tokhi O (2009) Parametric modelling of a flexible plate structure using artificial immune system algorithm. In: Andrews et al (eds) ICARIS 2009, vol 5666 LNCS, pp 301–314. Springer, Heidelberg (2009)
9. Al-Khafaji AAM, Darus IZM (2013) Cuckoo search for modelling of a flexible single-link manipulator. In: Manufacturing Engineering, Automatic Control and Robotics, pp 163–172
10. Maseri SZH (2019) Active vibration control of flexible plate structure optimized by bio-inspired flower pollination algorithm. Bachelor Thesis, Universiti Teknologi MARA
11. Yang X (2010) Nature-inspired metaheuristic algorithms, 2nd edn. Luniver Press, UK
12. Hussain K, Mohd Salleh MN, Prasetyo YA, Cheng S (2018) Personal best Cuckoo search algorithm for global optimization. Int J Adv Sci Eng Inf Technol, pp 1209–1217
13. Mohamad A, Zain AM, Bazin NEN, Udin A (2013) Cuckoo search algorithm for optimization problems - a literature review. Appl Mech Mater, pp 502–506
14. Hadi MS, Darus IZM (2013) Intelligence swarm model optimization of flexible plate structure system. Int Rev Autom Control, pp 322–331

15. Santillan JH, Tapucar S, Manliguez C, Calag V (2018) Cuckoo search via Lévy flights for the capacitated vehicle routing problem. J Ind Eng Int, pp 293–304
16. Mohanty PK, Parhi DR (2016) Optimal path planning for a mobile robot using cuckoo search algorithm. J Exp Theor Artif Intell, pp 35–52
17. Fadzli AAM (2019) Vibration suppression on flexible beam based on flower pollination algorithm. Bachelor Thesis, Universiti Teknologi MARA

Bearing Fault Detection Using Discrete Wavelet Transform and Partitioning Around Medoids Methods

Gigih Priyandoko, Diky Siswanto, Istiadi, Dedy U. Effendi, and Eska R. Naufal

Abstract Induction motor is widely used in industrial applications. The research paper presents the diagnosis of the induction motor bearing faults using Discrete Wavelet Transforms and Partitioning Around Medoids algorithm methods. The experimental test rig was developed to obtain data of the bearings on healthy or damaged conditions. Several mother-level wavelets are tried in order to get the best performance to find bearing faults. The wavelet transform results are used as an input of the Partitioning Around Medoids algorithm to cluster the bearing condition. The results showed that the methods proposed could provide an accurate diagnosis of the bearing condition.

Keywords Fault diagnosis · Bearing · Discrete wavelet transform · Partitioning around medoids

1 Introduction

Induction motors are widely used in industry to convert electric power into mechanical energy. Application of induction motors is used in various fields such as power plants, the food industry and others. Its use is mainly for the pumps drive, conveyors, press machines, elevators and much more. Due to they are strong, sturdy, cheap, reliable, easy to maintain, and power efficiency is quite high [1–7]. If damage to the induction motor is not detected early, it can cause very severe damage. So that it can result in a shutdown of the production process which causes a loss of productive time, the number of components that must be replaced, and others. The main problems in induction motors are an odd air gap, damage to the rotor shaft, bearing damage, and imbalance of the stator winding [8].

G. Priyandoko (✉) · D. Siswanto · D. U. Effendi · E. R. Naufal
Department of Electrical Engineering, Faculty of Engineering, University of Widyagama, Malang, Indonesia
e-mail: gigih@widyagama.ac.id

Istiadi
Department of Informatics Engineering, Faculty of Engineering, University of Widyagama, Malang, Indonesia

A. F. Ab. Nasir et al. (eds.), *Recent Trends in Mechatronics Towards Industry 4.0*, Lecture Notes in Electrical Engineering 730, https://doi.org/10.1007/978-981-33-4597-3_21

The Motor Current Signature Analysis (MCSA) method is used to measure the stator current in order to identify the presence of a single or a combination of many errors in an electric engine steady-state condition. This method has been introduced as an effective way to monitor electrical machines for years [2, 13–15]. In order to consider the effects of non-stationary behaviour on an induction motor, stator current signal analysis can use a variety of analysis methods in the time domain, frequency and time-frequency. Time-domain methods such as root mean square (rms) have achieved limited success in detecting local defects. Frequency analysis methods using signals have been developed to detect errors. FFT is a frequency domain analysis method and shows successful applications in the industry. Because bearing faults have relatively low energy, they are often covered by noise. Therefore, frequency-time domain analysis methods such as Wavelet Transforms [1, 7, 8], Hilbert Huang's transformation [9], empirical mode decomposition (EMD) and empirical mode decomposition (EEMD) [10] has been developed.

The goal of this research to classify the characteristics of the induction motor current amplitude when bearing is healthy and has fault conditions using a Discrete Wavelet Transform (DWT) method to find more accurate results. The stator current signal is used to detect inner and outer race defects independently and in conjunction with each other. In order to improve the accuracy of error detection results, various machine learning was used. Among many clustering methods, the PAM algorithm is proposed for this research, due to its strong robustness and its ability to handle abnormal values. Also, the PAM algorithm has good convergence and time complexity, and the effect obtained in global searches is excellent [10–12, 16].

2 Bearing

The bearing has inner race, outer race, and balls elements. The sectional view of the bearing is shown in Fig. 1. Recent studies have shown that more than 40% of induction motor failures are bearing faults. Therefore, this error type must be detected early to avoid more severe fatal damage to the motor. The most common type of bearing fault is bearing cracks, mainly due to several events such as inadequate lubrication, overload, excessive speed, misalignment, and high temperature [1]. Two types of

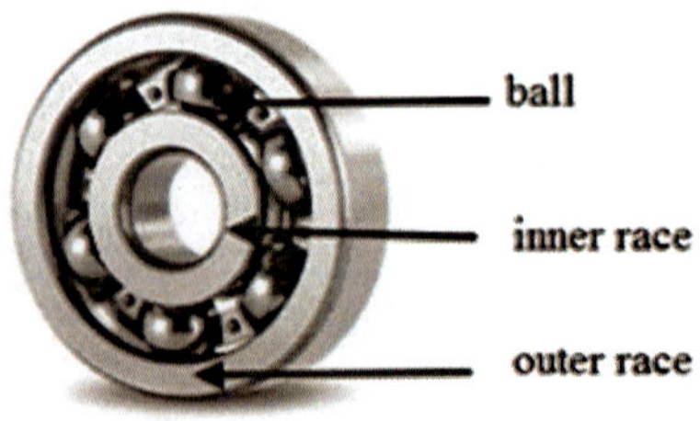

Fig. 1 The sectional view of bearing

bearing fault will be investigated in the research; namely inner race and outer race bearing defects.

3 Wavelet Transform

Wavelet transform has been successfully applied in many applications such as transient signal analysis, image analysis, communications systems, and other signal processing applications. Wavelet transforms an appropriate method for analysing non-stationary nature signals. Two types of wavelet transform methods, they are Continuous Wavelet Transform (CWT) and Discrete Wavelet Transform (DWT). The CWT is a signal processing time-scale method, and it can be written as follows,

$$\mathrm{CWT(a, b)} = \frac{1}{\sqrt{|a|}} \int_{-\infty}^{\infty} f(t)\Psi^*\left(\frac{t-b}{a}\right)dt \quad (1)$$

where a and b are scale and translation parameters, respectively and $\Psi^*(t)$ is the complex conjugate of $\Psi(t)$ represented as a mother wavelet. The CWT is the total number of signal times multiplied by the version of the wavelet function that was scaled and shifted $\Psi(t)$. The DWT which is derived from the discretisation of CWT, which can be written as follows,

$$\mathrm{CWT(a, b)} = \frac{1}{\sqrt{2^j}} \int_{-\infty}^{\infty} f(t)\Psi^*\left(\frac{t-2^j k}{2^j}\right) \quad (2)$$

where 2^j and $2^j k$ replace a and b, respectively. The CWT implementation is using filters was developed by Mallat in 1989 [11]. CWT requires much computational effort to find the coefficient in each parameter value of the scale. In DWT, to reduce the number of calculations and to have better results, calculation efficiency is carried out. The signal is passed through a series of high-pass filters to analyse high frequencies, and through a series of low-pass filters to analyse low frequencies.

4 Partitioning Around Medoids

Among all the method of clustering, PAM method proposed by Kaufman and Rousseeuw [16] is a group of partitional clustering methods that minimises the distance between the point labelled in the cluster and the point designated as the centre of the cluster. Compared to K-Mens, the PAM method selects data points as centres (medoids) of the cluster for each cluster. The K-Means uses the mean as the centre of the cluster. The PAM method is more reliable for dealing with noise and outliers because it minimises the number of paired dissimilarities, not the sum of the squares of Euclidean distances. The PAM method algorithm steps:

a. For initialisation, select *k* random objects to be used as the medoid.
b. Associate each data point with the closest medoid using distance measurements and calculating costs.
c. Randomly select a new *k* object which will serve as a medoid and keep a copy of the original set.
d. Use the new set of medoids to recalculate costs.
e. If the new costs are higher than the old ones, then stop the calculation on the algorithm.
f. Repeat steps two to five until there is no change in medoid.

5 Results and Discussion

Figure 2 shows the experimental test rig is used in this research to collect a three-phase induction motor bearing defect data. The three-phase induction motor connected to the DC generator as a load. The DAQ was connected to the laptop and the current sensor in order to send the stator current signal to the laptop for the next processing. Following are the three-phase induction motor specification.

Name	Motology
Rated output power (HP)	1.5
Rated voltage (V)	220–230
Rated frequency (Hz)	50

(continued)

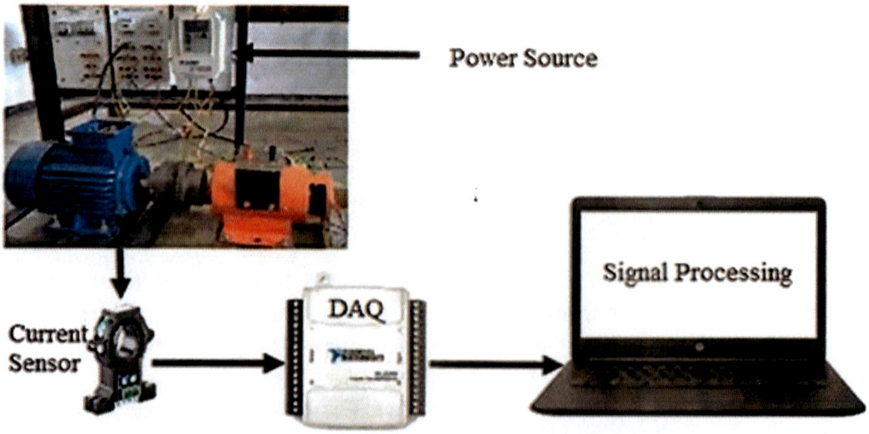

Fig. 2 The experimental set-up

(continued)

Name	Motology
Pole number	4
Rated speed (RPM)	1400
Connection	Y

The bearing type is 6203Z, with the following specification.

Inner diameter (mm)	17
Outer diameter (mm)	40
Ball diameter (mm)	6
Cage diameter (mm)	30
Number of balls	8

The procedure used in this research is as follows: apply three types of bearing to the system; they are the condition of the healthy, inner race and outer race bearings defects, respectively. The stator current was recorded with at 10 kHz sampled data from a three-phase induction motor for each case. Data is collected at a steady-state motor condition. The stator current signal for three bearing conditions shown in Fig. 3.

The main process in the research divided into two parts, namely digital signal processing and clustering of the stator current signal caused by bearing failures. There are healthy, inner race and outer race bearing defects conditions. In digital

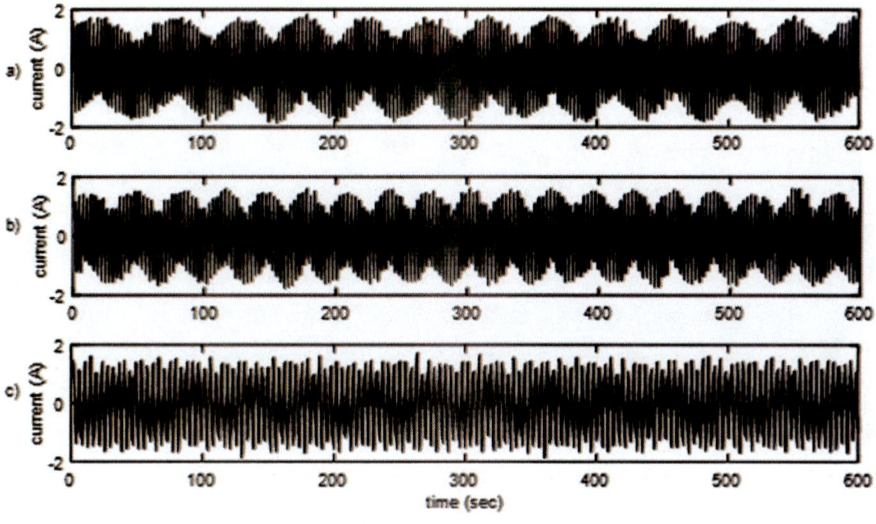

Fig. 3 Stator current **a** normal bearing **b** bearing with inner race defect **c** bearing with outer race defect

signal processing, the stator current signal needs to be processed to match the system. Firstly, dividing the signal into one cycle. Secondly, equalizing the separated signal size, and lastly, feature extraction process. The equalised signal process is using a zero-bearing technique. For the next step, the signal feature extraction process is carried out for easy identification uses the DWT method. The stator current signal is described by the DWT method in detail which is low frequency and is called high-frequency component. The data obtained from this decomposition process is smaller than the previous level. Decomposition rates from one to ten were investigated using Daubechies mother wavelets for this study. The resulting decomposition signal is used as input to the identification system. The PAM algorithm is used in the identification system for cluster signals based on bearing conditions.

A cluster is represented by its centroid using the PAM method, which is a mean (average) and energy entropy of the points within a cluster. For building the assessment model, 650 feature vectors for each case are used to train the PAM algorithm clustering method. The first step, select three random objects to be used as three medoids. After finding a set of three medoids, three clusters are created by assigning each object to the nearest medoid. The goal is to find three representative objects which minimise the sum of the dissimilarities of the objects to their closest representative object. The whole operation is performed in the offline phase. The accuracy of the PAM algorithm based on the Daubechies mother wavelet is 99.17%. The DWT-PAM algorithm results using Daubechies mother wavelet shown in Fig. 4.

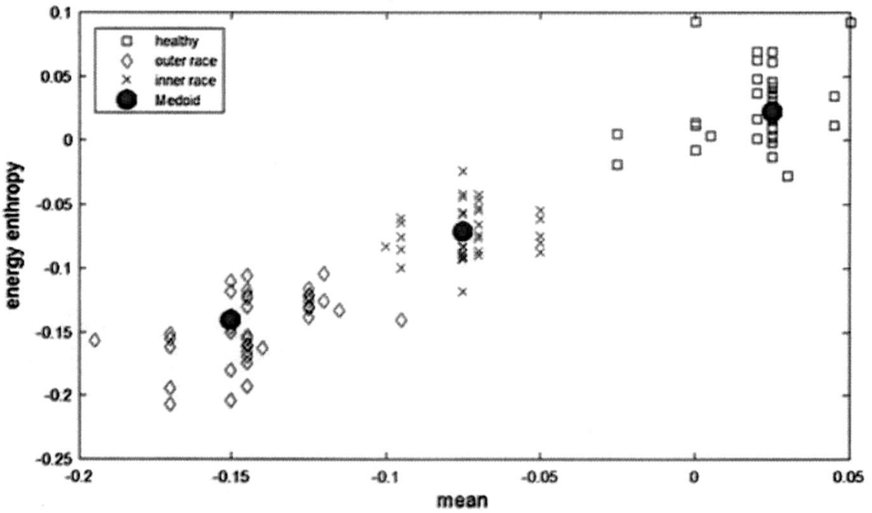

Fig. 4 The DWT-PAM algorithm clustering results

6 Conclusion

The research proposes a DWT and PAM algorithm approach for induction motor bearing defect classification, validating its effectiveness diverse fault conditions such as inner race and outer race bearing defects. The DWT is utilised to extract the features by using Daubechies mother wavelet that derives rich information from stator current signals. Based on these results is used as the PAM algorithm input to classify the bearing faults in an induction motor with 99.17% accuracy results.

Acknowledgements The authors would like to thank the Ministry of Research and Technology of the Republic of Indonesia, for their fully supported by a DRPM research grant under PDUPT Widyagama University.

References

1. Djaballah S, Meftah K, Khelil K, Tedjini M, Sedira L (2019) Detection and diagnosis of fault bearing using wavelet packet transform and neural network. Frat ed Integrità Strutt 13(49):291–301
2. Singh S, Kumar A, Kumar N (2014) Motor current signature analysis for bearing fault detection in mechanical systems. Procedia Mater Sci 6:171–177
3. Zheng H, Zhou L, Yang H (2015) Rolling bearing fault diagnosis based on wavelet packet analysis and multi-kernel learning. Hangkong Dongli Xuebao/J Aerosp Power 30(12):3035–3042
4. Al-Raheem KF, Abdul-Karem W (2011) Rolling bearing fault diagnostics using artificial neural networks based on laplace wavelet analysis. Int J Eng Sci Technol 2(6):278–290
5. Konar P, Chattopadhyay P (2011) Bearing fault detection of induction motor using wavelet and support vector machines (SVMs). Appl Soft Comput J. 11(6):4203–4211
6. Benbouzid M et al (2014) What stator current processing based technique to use for induction motor rotor faults diagnosis?
7. Bessous N, Zouzou SE, Bentrah W, Baa S, Sahraoui M (2018) Diagnosis of bearing defects in induction motors using discrete wavelet transform. Int J Syst Assur Eng Manag 9(2):335–343
8. Gaeid KS, Ping HW (2011) Wavelet fault diagnosis and tolerant of induction motor: a review. Int J Phys Sci 6(3):358–376
9. Osman S, Wang W (2016) A normalized Hilbert-Huang transform technique for bearing fault detection. J Vib Control 22(11):2771–2787
10. Mohanty S, Gupta KK, Raju KS (2016) Vibro-acoustic fault analysis of bearing using FFT, EMD, EEMD and CEEMDAN and their implications. In Advances in machine learning and signal processing, Springer, pp 281–292
11. Rai A, Upadhyay SH (2017) Bearing performance degradation assessment based on a combination of empirical mode decomposition and k-medoids clustering. Mech Syst Signal Process 93:16–29
12. Nistane VM, Harsha SP (2018) Assessment of bearing degradation by using intrinsic mode functions and k-medoids clustering. Proc Inst Mech Eng Part K J Multi-body Dyn 0(0):1–14
13. Ye Z, Wu B, Sadeghian A (2003) Current signature analysis of induction motor mechanical faults by wavelet packet decomposition. IEEE Trans Ind Electron 50(6):1217–1228
14. Immovilli F, Lippi M, Cocconcelli M (2019) Experimental evidence of MCSA for the diagnosis of ball-bearings

15. Deekshit Kompella KC, Gopala Rao MV, Rao RS, Sreenivasu RS (2015) Estimation of bearing faults in induction motor by MCSA using daubechies wavelet analysis. In: 2014 International conference on smart electric grid (ISEG), pp 1–6
16. Batra A (2011) Analysis and approach: k-means and k-medoids data mining algorithms. In: ICACCT, 5th IEEE international conference on advanced computing and communication technologies, pp 274–279

The Implementation of a Novel Augmented Reality (AR) Based Mobile Application on Open Platform Controller for CNC Machining Operation

Anbia Adam, Yusri Yusof, Kamran Latif, Aini Zuhra Abdul Kadir, Toong-Hai Sam, Shamy Nazrein, and Danish Ali Memon

Abstract Manufacturing technology has reached to a point where there is a need for online platform for users' easy access. Essentially, the trend for Industry 4.0 towards automation and Cyber Physical System (CPS) has become a compulsory addition to any system, manufacturing is not an exception to the need of that technological improvement. With the number of mobile users growing by the day, open CNC system has risen to merge the idea of controlling, monitoring, and utilizing the mobile platform to its advantage. This paper proposes the implementation of an Augmented Reality (AR) based mobile application for guiding users to improve CNC machining operation as well as remotely control the system from mobile phone. The CNC system is developed from conventional PROLIGHT 1000 Milling CNC machine from Light Intelitek based on novel controller established as UTHM Open CNC Controller (UOCC) under the ISO 14649 of STEP-NC environment. The UOCC have efficient ability to operate and monitor the CNC machine from a distance which would reduce the risk of accidents and hazards from occurring while eliminating the use of paper and increasing machining efficiency.

Keywords CNC · Industry 4.0 · STEP-NC · Augmented reality

A. Adam (✉) · Y. Yusof · S. Nazrein · D. A. Memon
Faculty of Mechanical and Manufacturing Engineering, Universiti Tun Hussein Onn Malaysia, UTHM, Parit Raja, Batu Pahat 86400, Johor, Malaysia
e-mail: anbiaadam@gmail.com

K. Latif
Faculty of Mechanical and Manufacturing Engineering Technology, Universiti Teknikal Malaysia Melaka, Jalan Hang Tuah Jaya, Durian Tunggal, Melaka 76100, Malaysia

A. Z. A. Kadir
School of Mechanical Engineering, Faculty of Mechanical Engineering, Universiti Teknologi Malaysia, UTM Skudai, 81310 Skudai, Johor Bahru, Malaysia

T.-H. Sam
INTI International University, Persiaran Perdana BBN, Putra Nilai, 71800 Nilai, Negeri Sembilan, Malaysia

A. F. Ab. Nasir et al. (eds.), *Recent Trends in Mechatronics Towards Industry 4.0*, Lecture Notes in Electrical Engineering 730, https://doi.org/10.1007/978-981-33-4597-3_22

1 Introduction

Computer Numerical Control (CNC) system functions as the decipher for machine tools thus controlling the movement of machine tools according to NC programs. However, conventional CNC machine tools are uncapable of providing data that are significant influence on product quality, productivity and cost [1]. As the competition in manufacturing industry increases, companies ought to have products that are good in quality, fast produced and reduced in terms of development cost [2]. Thus, with the great improvement of CNC systems in recent years, its functional performance, reliability, and diversity has been enhanced from the conventional ISO 6983 to IS014649 [3] to enable improvement on compatibility, language difference, and control of the CNC machine. However, many manufacturing execution still needs human involvement in its operation to complete [4]. As the Industry 4.0 moves further developed, the manufacturing industry becomes more complex thus leading to increment in complexity of daily task for operators [5]. Operators are then needed to be highly flexible and to be able to adapt to a very dynamic working environment [6]. There is a need in Industry 4.0 for self-improvement in manufacturing sector [7] thus, the use of AR can be the key to enable operator to improve the transfer of information from digital to physical world faster and more efficient [8]. AR is considered as one of the enabling technologies towards Industry 4.0 [9] through allowing users to see the real world with virtual objects superimposed or overlay which coexist at the same time [10] which could be allowed through integration of CNC manufacturing technology in ISO 14649 with mobile application.

AR technology provides a reality which the real time physical real-world environments are augmented with merged computer-generated images over a user's view of the real-world, enhancing the perception of user with useful information provided in the AR system [11]. AR frameworks require the view of the physical world and its component elements where estimating the position and orientation overlay image is paramount to properly aligned between the user and the machine [12]. The AR technology allows operator to interact with virtual animation digital contents in highly interactive way, gaining useful information and experience along the way [13]. In other words, operator's ability to perceive and act within the physical world is further enhanced by the possibility to be immersed in a virtual reality environment [14] where different contents levels superimpose between real world and virtual animation occurs and benefit the users [15].

The AR application helps reduce failure rate and faster cycle times through the real time image assist in the system [16]. It can support either experienced or non-experienced operators in real time during operations by becoming digital assistance while reducing human errors, concurrently decreases the usage of printed work instructions, and negligence [17]. Moreover, by combining the operators' intelligence and flexibility along with the assist of AR, the machining process would be able to reduce defect, rework and reduce redundancy from human error. Thus, AR is the most viable solution for manufacturing technology application to incorporate human-machine interface [18].

In manufacturing industry, the AR operators have the advantage of not only gaining experience through daily operational tasks, AR also provide value-added contents that are suited to improve skills and abilities in manufacturing working environment [6]. It allow operator to accomplish several tasks, making it possible the shift from mass production to mass customization [19] which inherently saves time and cost. Therefore, to cater to the needs of Industry 4.0, UOCC AR is developed as the solution to deal with the limitation of operating complex machining and novel systems effectively. Thus, this paper covers the development of the mobile application for UOCC which utilizes AR technology to assist users operating the novel system.

2 Development of UOCC Mobile Application

The development of mobile application based on AR is not foreign to the world of technology. However, the application of the AR itself varies and significant based on the use of the its function. For UOCC mobile application, it uses AR to reduce time taken for machining setup for UOCC system. The UOCC AR application provides users with visual Standard Operating Procedure (SOP), machine and material details to assist especially new users to operate the machine. Not only it reduces time of machining setup process, it also improves the quality of the process ergonomically for the user and mitigate accidents from occurring.

UOCC AR application is mainly developed by three separate software—Vuforia, Unity and Android Studio. The development also involves the C# language coding in several key function such as animation triggering, button function, scene switching and scene manager adopting the marker-based augmented reality technique. UOCC AR application is created in Unity3D engine using some auxiliary libraries and some custom C# scripts. In addition, the maker-based (also known as the 'Image Target') for augmented reality program can be generated by Qualcomm's Vuforia library which is integrated in Unity3D. Vuforia library deliver quick, multiple marker recognition and display information based on the marker recognized. In the other aspect, the application can display digital objects accurately when the marker is not available in the camera's field of view. The idea of this application is to let user have easy access to machining SOP and details by using the AR technology through UOCC AR application accessing the mobile phone's camera to detect the 'Image Target' and projecting the 'Model Target' to mobile screen as shown in Fig. 1.

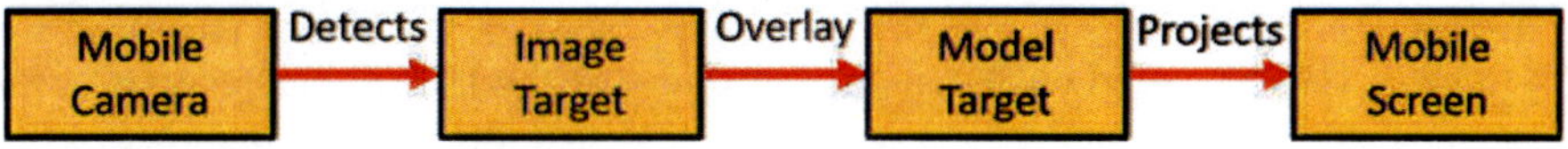

Fig. 1 AR working flow principle for UOCC application

2.1 Algorithm Design of UOCC AR Application

The development of the application requires many settings and coding scripts. The research uses Unity 2017 software with installed components like 'Vuforia AR' and 'Android-iOS Build support' to ensure that the project can utilizes the AR technology to be adapted to mobile phones through transferring the coding path to the Android Studio software where the coding file is converted to.APK format to be installed in phones. Nonetheless, to start using UOCC's assets (image of SOP and such) to be viewed on the screen, 'Download Database' on Vuforia Developer Portal is filled with the uploaded 'Image Target' and 'Model Target' that we need. The databases were activated through 'App License Key' we acquire from the website and 'Load Database' control in the AR settings.

The major part of the interface of Unity involved in the process are three named part which are 'Animation Control', 'Image Target' and 'Model Target'. The 'Animation Control' is utilized to control the coordination of the part, there are three segments with three unique headings which are X, Y and Z. Other than that, it additionally can change the size of the virtual articles. Meanwhile, 'Image Target' is the specific image that the UOCC application wanted to recognize using the camera once the AR animation begins. Thus 'Model Target' is the projection of the front layer of the AR that UOCC application shows once it detects the 'Image Target'. The dynamic between the images are the main build for this application as it only shows normal live scene as normal camera, however the users will see the AR layered image once the application detects its marker.

The UOCC AR application activates the camera once it is open, thus proceed to identify and recognize markers that have been set up as the 'Image Target'. Once users approved, the camera lens is activated, the real live scene is display as output on the screen. To add the camera deployment function in the application, the AR camera coding package was integrated into the application. After the camera is activated, the marker identification is started. The camera scans the scene for one frame per second, frame is being process by the processor. The data is acquired and processed simultaneously which is then displaying the materials that users are supposed to view- 'Model Target' which is set to assist them with the machining setup process through SOP and details on the screen. The user has the choice to end the UOCC AR application once the setup process is completed as shown in Fig. 2.

2.2 GUI of AR in UOCC Mobile Application

The markers are installed on different places namely the UOCC machine's control case, the CNC machine and various other places depending on the suitability of the location and its functions. The SOP for the machine is layered as 'Model Target' for the image of UOCC's logo on the control case and the desktop's front wallpaper

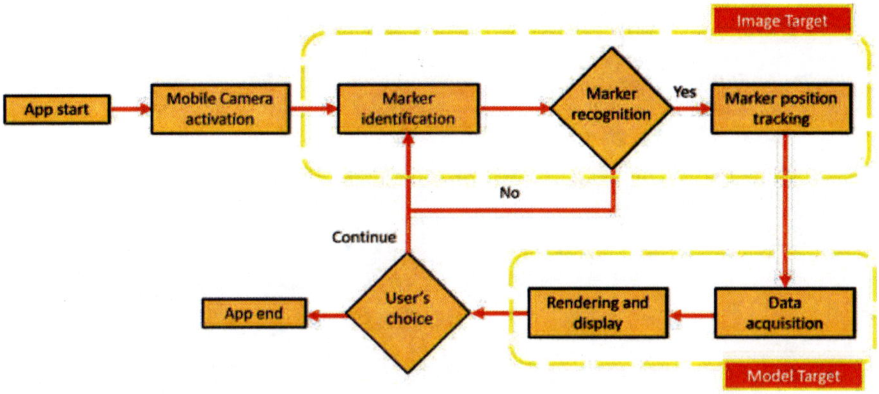

Fig. 2 Flow design of the UOCC AR application

display to ease the users to detect and setting up the machining process at the same time based on the 'Model Target' as shown in Fig. 3.

The UOCC AR application proceeds to scan with infinity loop until another set 'Image Target' is detected continuously if the application is not ended. Once the marker has recognized another target, the application proceeds to marker position tracking. The different makers have different image characteristic to acquire different data. When a maker is recognized, the image sensor in the camera generates the algorithm as signal and send it to the device processor. The processor decides to load the data from the data storage based on the signal acquired showing the 'Model Target' that UOCC provides as shown in Fig. 4. The UOCC AR operators are able to view the SOP through preferred devices such as mobile phone or even smart glasses to assist them with the operation of the CNC machining system.

Fig. 3 The images used for AR application in UOCC

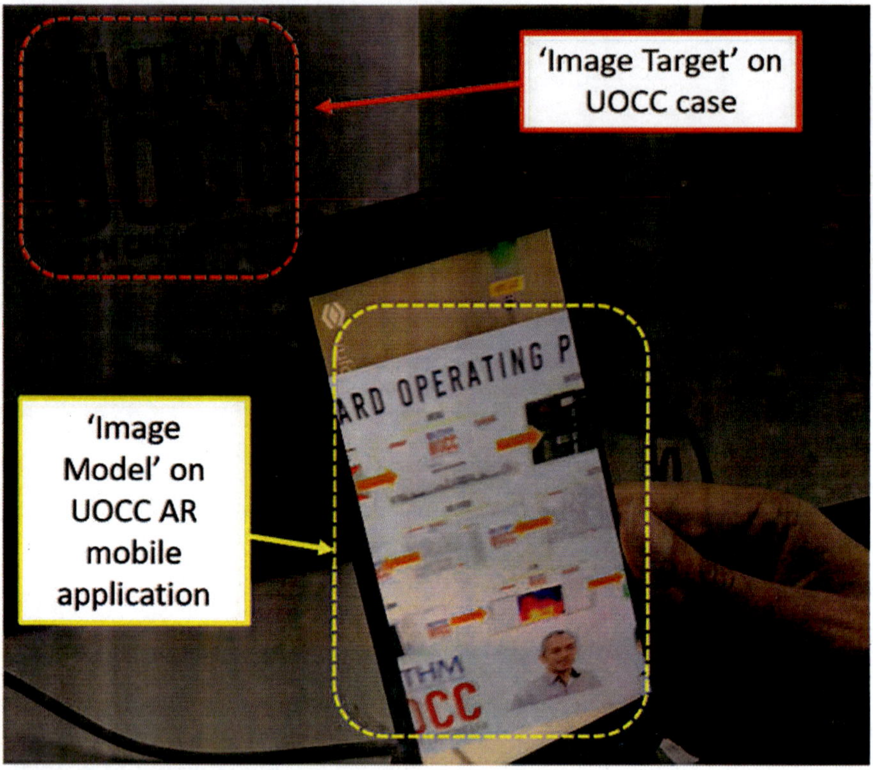

Fig. 4 GUI of UOCC AR mobile application showing the SOP

Therefore, the AR operators have the advantage of not only gaining experience through daily operational tasks, AR also provide value-added contents that are suited to improve skills and abilities in manufacturing working environment [6]. It allow operator to accomplish several tasks, making it possible the shift from mass production to mass customization which inherently saves time and cost as done by Uva [19]. AR has been widely applied to develop applications to provide guidance and user manuals to assist users in manufacturing processes [20]. AR could reduce dependence on paperwork in the changing of batches, setting up of machines or tools, quality checking, and introduction of new processes [21]. With the advantages that it has, AR is dubbed as one of the main enabling digital industrial technologies that will support the wide scope of challenges concerning Industry 4.0 and facilitate the digitization of the manufacturing sector [22]. Thus, the UOCC AR application is targeting to increase reduce time taken for machining operation of the novel system, ergo increasing the output production from the CNC machining system.

3 Conclusion

This paper has established a new approach to operate the novel CNC machining motion system through the development for of UOCC mobile application which utilizes the combination of Vuforia, Unity and Android Studio for the development of the AR system in the UOCC mobile application to improve the conventional CNC machining operation to enable adaption of Industry 4.0. The development of UOCC mobile application under Android operating system for AR have allow users the ability to operate the CNC machine efficiently which would reduce the risk of accidents and hazards from occurring while increasing machining efficiency and reducing time taken for its completion.

References

1. Liu C, Cao S, Tse W, Xu X (2017) Augmented Reality-assisted Intelligent Window for Cyber-Physical Machine Tools. J Manuf Syst 44:280–286
2. Wang ZB, Ng LX, Ong SK, Nee AYC (2013) Assembly planning and evaluation in an augmented reality environment. Int J Prod Res 51(23–24):7388–7404
3. A. Adam, "Review On Advanced Numerical Control In Manufacturing System," *2019 4th Int. Conf. Electromechanical Control Technol. Transp. Rev.*, pp. 2005–2008, 2019
4. X. Wang *et al.*, "Enhancing smart shop floor management with ubiquitous augmented reality," *Int. J. Prod. Res.*, vol. 0, no. 0, pp. 1–16, 2019
5. V. Paelke, "Augmented reality in the smart factory: Supporting workers in an industry 4.0. environment," *19th IEEE Int. Conf. Emerg. Technol. Fact. Autom. ETFA 2014*, 2014
6. Longo F, Nicoletti L, Padovano A (2017) Smart operators in industry 4.0: A human-centered approach to enhance operators' capabilities and competencies within the new smart factory context. Comput Ind Eng 113:144–159
7. Mittal S, Khan MA, Romero D, Wuest T (2018) A critical review of smart manufacturing & Industry 4.0 maturity models: Implications for small and medium-sized enterprises (SMEs). J. Manuf. Syst. 49(June):194–214
8. J. Cabero Almenara and J. Barroso Osuna, "The educational possibilities of Augmented Reality," *J. New Approaches Educ. Res.*, vol. 6, no. 1, pp. 44–50, 2016
9. F. Ferraguti *et al.*, "Augmented reality based approach for on-line quality assessment of polished surfaces," *Robot. Comput. Integr. Manuf.*, vol. 59, no. October 2018, pp. 158–167, 2019
10. Azuma RT (1997) A Survey of Augmented Reality. Chaos, Solitons Fractals 42(3):1451–1462
11. J. Carmigniani and B. Furht, *Handbook of Augmented Reality.* 2011
12. Zubizarreta J, Aguinaga I, Amundarain A (2019) A framework for augmented reality guidance in industry. Int J Adv Manuf Technol 102(9–12):4095–4108
13. Yuan ML, Ong SK, Nee AYC (2008) Augmented reality for assembly guidance using a virtual interactive tool. Int J Prod Res 46(7):1745–1767
14. Zhang J, Ong SK, Nee AYC (2011) RFID-assisted assembly guidance system in an augmented reality environment. Int J Prod Res 49(13):3919–3938
15. C. Bohil, C. B. Owen, and E. J. Jeong, "Virtual reality and Presence," *21st Century Commun. A Ref. Handb. 21st century Commun. A Ref. Handb.*, no. January, 2012
16. Z. H. Lai, W. Tao, M. C. Leu, and Z. Yin, "Smart augmented reality instructional system for mechanical assembly towards worker-centered intelligent manufacturing," *J. Manuf. Syst.*, vol. 55, no. July 2019, pp. 69–81, 2020

17. M. Funk, T. Kosch, R. Kettner, O. Korn, and A. Schmidt, "motionEAP: An Overview of 4 Years of Combining Industrial Assembly with Augmented Reality for Industry 4.0," *Proc. 16th Int. Conf. Knowl. Technol. Data-driven Bus.*, pp. 2–5, 2016
18. Weyer S, Schmitt M, Ohmer M, Gorecky D (2015) Towards Industry 4.0-Standardization as the crucial challenge for highly modular, multi-vendor production systems. Ifac-Papersonline 48(3):579–584
19. Uva AE, Gattullo M, Manghisi VM, Spagnulo D, Cascella GL, Fiorentino M (2018) Evaluating the effectiveness of spatial augmented reality in smart manufacturing: a solution for manual working stations. Int J Adv Manuf Technol 94(1–4):509–521
20. S. K. Ong, A. W. W. Yew, N. K. Thanigaivel, and A. Y. C. Nee, "Augmented reality-assisted robot programming system for industrial applications," *Robot. Comput. Integr. Manuf.*, vol. 61, no. June 2019, p. 101820, 2020
21. S. Fox, A. Kotelba, I. Marstio, and J. Montonen, "Aligning human psychomotor characteristics with robots, exoskeletons and augmented reality," *Robot. Comput. Integr. Manuf.*, vol. 63, no. June 2019, p. 101922, 2020
22. F. Bruno, L. Barbieri, and E. Marino, "An augmented reality tool to detect and annotate design variations in an Industry 4. 0 approach," 2019

Firefly Algorithm for Modeling of Flexible Manipulator System

Hazeem Hakeemi Baseri, Hanim Mohd Yatim, Muhamad Sukri Hadi, Mat Hussin Ab. Talib, and Intan Zaurah Mat Darus

Abstract Flexible manipulator is a general-purpose machine used for industrial automation in order to increase productivity, flexibility and product quality that has been widely applied to replace their rigid manipulator counterparts. Flexible manipulator is a distributed parameter system and has infinitely many degrees of freedom. However, flexible manipulator will develop unwanted vibration during manoeuvre that may reduce the efficiency of the flexible manipulator system for precise positioning requirements. Thus, the dynamics of this system are highly non-linear and complex. Therefore, an accurate model and efficient control system must be developed in order to sustain the advantages of the flexible manipulator system. This paper highlights the flexible manipulator modelling using system identification (SI) method employing Firefly Algorithm (FA). Initially, flexible manipulator test rig is developed for input output data collection. Behaviour of system response including hub angle and end-point acceleration are acquired and analyse. Later, data collected is fed into system identification method optimized by Firefly Algorithm via linear auto regressive with exogenous (ARX) model structure. Validations of the algorithm is assessed on basis of minimizing the mean-squared error (MSE) and correlation tests. It is demonstrated that FA modeling is superior that conventional algorithm known as Least Square (LS) Algorithm with lowest MSE obtained and achieved 95% confidence interval in correlation tests for both hub angle and end-point acceleration identification.

Keywords Flexible manipulator · System identification · Firefly algorithm

H. H. Baseri · M. S. Hadi
Faculty of Mechanical Engineering, Universiti Teknologi MARA, 40450 Shah Alam, Selangor, Malaysia

H. Mohd Yatim (✉) · M. H. Ab. Talib · I. Z. Mat Darus
School of Mechanical Engineering, Faculty of Engineering, Universiti Teknologi Malaysia, 81310 Skudai, Johor, Malaysia
e-mail: hanim.my@utm.my

A. F. Ab. Nasir et al. (eds.), *Recent Trends in Mechatronics Towards Industry 4.0*, Lecture Notes in Electrical Engineering 730, https://doi.org/10.1007/978-981-33-4597-3_23

1 Introduction

Previously, rigid structure with bulky design is widely employed in industrial application. However, several industries require the system to maintain the weight of the structure as low as possible, which difficult to attain by rigid structure. Higher power consumption, more bulky and reduced speed with respect to the operating payload for the existing heavy rigid structures lead to less efficient system [1]. Its dynamic deflection which continue for a period after a move has severely limits rigid structure for the operation of high precision application [2]. Due to this drawbacks, flexible manipulator system which known to be better, improvised and lighter has surpass this rigid structure in many field.

Demand for the application of flexible manipulator has become increasingly prevalent because of its lighter weight, longer service life and lower power consumption which has received attention from the industries [3]. However, flexible manipulator may exhibit vibration that may exist due to low stiffness of the structure and can severely increase after a manoeuvre [4]. This unwanted vibration can cause noise, disturbance and may impacts the effectiveness and accuracy of machine which later lead to fatigue [5].

The main purpose of modelling the flexible manipulator structure is to design a good controller including both rigid and flexible motion control of the system. Hence, a system identification (SI) technique is utilized to model this complex system. System identification required a parameter estimation and employment of intelligent optimization algorithm has extensively able to solve identification problem. Presently, various evolutionary algorithms (EA) have become a trend as a good optimization alternatives to solve the problems of global optimization [6]. Among them, firefly algorithm (FA) has been developed to be one of the important tools in engineering practice to solve design optimization problems.

Firefly Algorithm that has gain attention from many researchers is inspired by flashing light patterns like firefly. The main variables that are crucial to the efficiency of the FA includes attractiveness formulation between neighbouring fireflies and light intensity variation [7]. FA has been known to interpret a variety of problems and different variants has been developed to better suit specific types of applications [8]. Literally, FA has shows the potential to be used effectively as an optimizer in modeling and control problems for flexible manipulator. Hence, this paper will be utilized FA in optimizing an accurate parametric model of flexible manipulator system.

2 Experimental Setup and Data Acquisition

A mechatronic of flexible manipulator system consists of mechanical system, instrumentation system includes data acquisition (DAQ), sensors and actuators, signal conditioning and also computer aspects. Mechanical components were developed

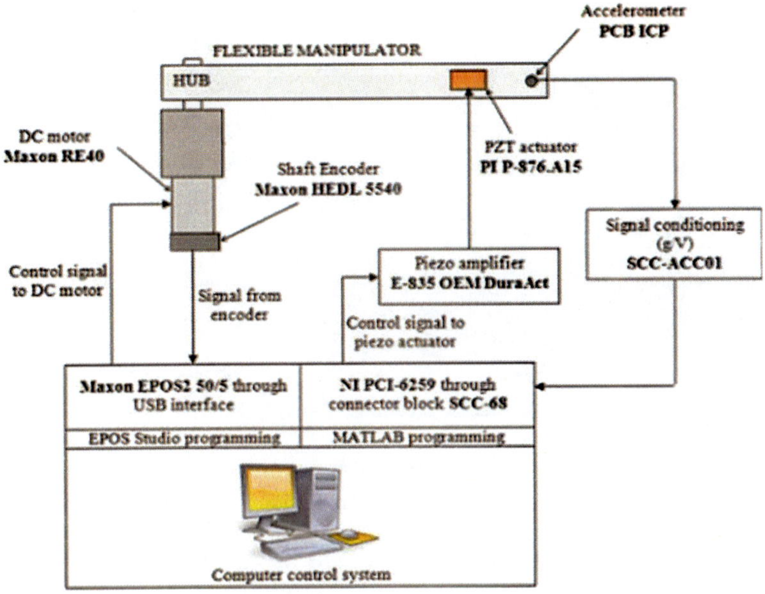

Fig. 1 Schematic diagram of flexible manipulator system [9]

which integrate with instrumentation components to conduct the experimental procedure of flexible manipulator rig in order to collect the input-output data of the system.

The flexible structure in the experimental rig was presented by a single link thin aluminum alloy with dimensions of 600 mm in length, 40 mm in width and 1.5 mm in thickness. Motor is mounted at the hub while free at the endpoint. Figure 1 shows the schematic diagram of flexible manipulator system [9].

The instrumentation components includes sensors, actuators and a computer with a programmable software was interfaced. Piezoelectric and DC motor acts as actuators while accelerometer and encoder is used as sensors. DC motor embed with encoder attached directly at the hub drives the flexible link. Thus, the angle displaced and speed could be controlled directly via PC connection. Encoder will control the position of flexible manipulator link by adjusting the angle with precision of 500 counts per turn. On the other end, vibration of flexible manipulator was sensed by an accelerometer and placed at endpoint of the flexible link where maximum vibration will occur. As for piezoelectric actuator, it will be utilize for further analysis of active vibration control strategy which responsible to reduced the endpoint vibration. Signals from accelerometer and to the piezo actuator was interfaced by data acquisition system (DAQ) and connected to the PC for data analysis. System integration of the experimental setup of flexible manipulator are shown as in Fig. 2.

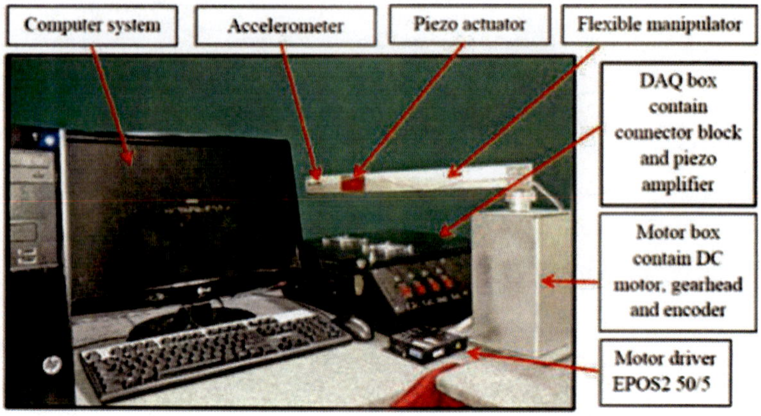

Fig. 2 Experimental setup for flexible manipulator structure [9]

3 Firefly Algorithm (FA)

Firefly Algorithm (FA) introduced by Yang inspired from flashing patterns and behavior of fireflies [10]. Major objective of fireflies is to flash in order to attract mates and each firefly species will generates its own flash pattern. They can attract mates or prey by generating natural light that is discovered in a distinctive pattern.

As the distance increases, the light brightness reduces, so does the attractiveness between fireflies that may have limited distance range. To implement FA, the brightness of a firefly was associated with the landscape of the objective function to be optimized. Basically, FA uses the following rules [11]:

- The attractiveness of the fireflies is proportional to its brightness and as the distance increase, both parameters decrease. Thus, for any two flashing fireflies, the less brighter will move toward the more brighter]. If there is no brighter one than a particular firefly, then it will move randomly. More brightness implies less space between two fireflies, as its attract to one another.
- The firefly's brightness for visibility is determined by the landscape of objective function to be optimized

Firefly Algorithm (FA) is determined by the range of light strength and the value of attractiveness, β. β is an attractiveness variable that distinct fireflies should attracted to another and going to influence the firefly to shift to a brighter firefly. Moreover, the attractiveness differs with the degree of absorption, γ where the strength of heat reduced with growing distance from its origin and relies on the mode of propagation. Therefore, β is dependent on the range between firefly i drawn by the firefly j light. The movement of a firefly i attracted to another more attractive (brighter) firefly j in fireflies' position, x in time step, t is determined by:

$$x_i^{t=1} = x_i^t + \beta_o e^{-\gamma r_{ij}^2}(x_i^t - x_j^t) + (rand - \frac{1}{2}) \tag{1}$$

Given β_0 is the fireflies attractiveness, r_{ij} is the distance between any two fireflies i and j defined as $|x_i - x_j|$. *rand* is the random number allowing the algorithm to escape from local optima.

4 System Identification

System identification is the building of mathematical models of dynamic systems from observed input-output data [12]. Relationship between input (u) and output (y) of the system will develop mathematical expression. Prior to that, an appropriate order and parameters is essential to be determine that best fits that relation. In this study, flexible manipulator was modelled using ARX model structure given by:

$$y(t) = \frac{B(z^{-1})}{A(z^{-1})}u(t) + \frac{\xi(t)}{A(z^{-1})} \tag{2}$$

where $A(z^{-1})$ and $B(z^{-1})$ expressed as

$$A(z^{-1}) = 1 + a_1 z^{-1} + \cdots + a_n z^{-n}$$
$$B(z^{-1}) = b_0 + b_1 z^{-1} + \cdots + b_n z^{-(n-1)}$$

white noise, $\xi(t) = 0$, z^{-1} is defined by a backshift operator, $[a_1,\ldots, a_n, b_1,\ldots, b_n]$ are model parameters that need to be optimized and n is the orders of the model. $y(t)$ and $u(t)$ is the output and input vector, respectively. Thus, the identified model of the system can be represented in terms of transfer function form $H(z^{-1})$ as follows:

$$H(z^{-1}) = \frac{B(z^{-1})}{A(z^{-1})} = \frac{b_0 + b_1 z^{-1} + \cdots + b_n z^{-(n-1)}}{1 + a_1 z^{-1} + \cdots + a_n z^{-n}} \tag{3}$$

The objective function for optimization formula is set to be the minimization of mean-squared error (MSE). Least squares (LS) and Firefly algorithm (FA) are employed in determining the parameter of ARX model. For both optimization, total samples observed were 20,000 and 10,000 of the samples were chosen for training and the rest was used for testing.

5 Results and Discussions

FA and LS was applied to obtain an approximate model of the flexible manipulator system. For this purpose, input output data is fed to the ARX model structure.

Input-output data were acquired from experimental study consists of two sets input-output includes hub angle and endpoint acceleration. FA algorithms will be utilized for system identification and the performance was compared to the LS algorithm. The performance was observed in input/output mapping, lowest MSE value and correlation test.

5.1 Modeling Using Firefly Algorithm (FA)

Investigations were carried out by modeling with FA algorithm using experimental input-output data obtained. FA was tuned heuristically by varying the number of fireflies, number of iterations, degree of absorption, γ and orders for each set of hub angle and endpoint acceleration respectively. Best results which obtained smallest MSE were recorded. Table 1 shown the FA parameters used to achieve best result.

The best model for both hub angle and endpoint acceleration modeling were obtained by FA with an order of 2. Therefore, FA modeling for hub angle achieved the smallest MSE of 0.0023 in 200th iteration as shown in convergence profile as in Fig. 3. Meanwhile, Fig. 4 shows the convergence profile of FA modeling for endpoint acceleration which achieved the best MSE of 0.0032 in the 100th iteration.

The FA modeling output of hub angle in time domain is shown in Fig. 5. The robustness of the developed model was validated using pole zero diagram stability and correlation test as shown in Figs. 6 and 7, respectively. Figures 8 and 9 represents the FA modeling output of endpoint acceleration in time and frequency domains, respectively. The poles and zeroes mapping are shown in Fig. 10 and its corresponding correlation test as in Fig. 11 for endpoint acceleration modeling.

It is noted that, from Figs. 5 and 8, the simulated output using FA closely match the actual output for both hub angle and endpoint acceleration, respectively and the model has successfully characterized the dynamics of the system as shown in frequency domain from Fig. 9. Figures 6 and 10 are the pole zero stability diagram which shows all the poles lied inside the circle unit which indicates the stable model for both modeling. The corresponding correlation tests as in Figs. 7 and 11 for hub angle and endpoint acceleration, respectively, were found to be unbiased and within 95% confidence interval confirming an adequate model fit.

Table 1 The set of parameters used for FA algorithm

No.	Parameters	Hub angle	Endpoint acceleration
1	Number of fireflies	60	20
2	Number of iterations	200	100
3	Degree of absorption, γ	1	1
4	Model order	2	2

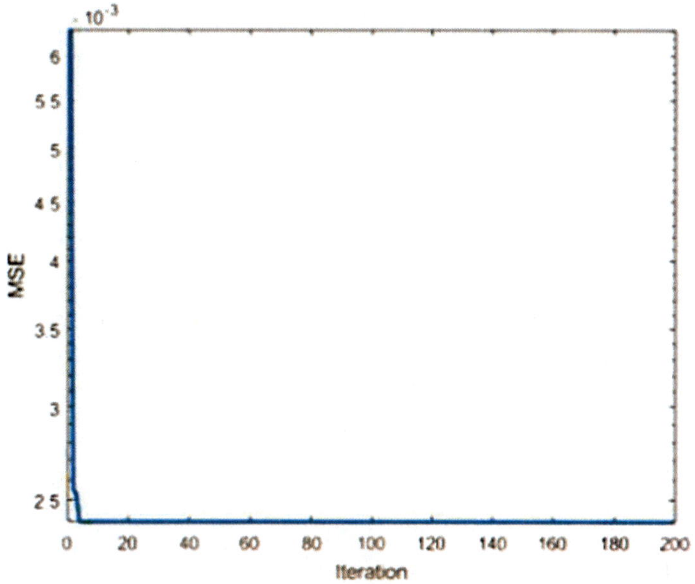

Fig. 3 Convergence profile of FA modeling for hub angle

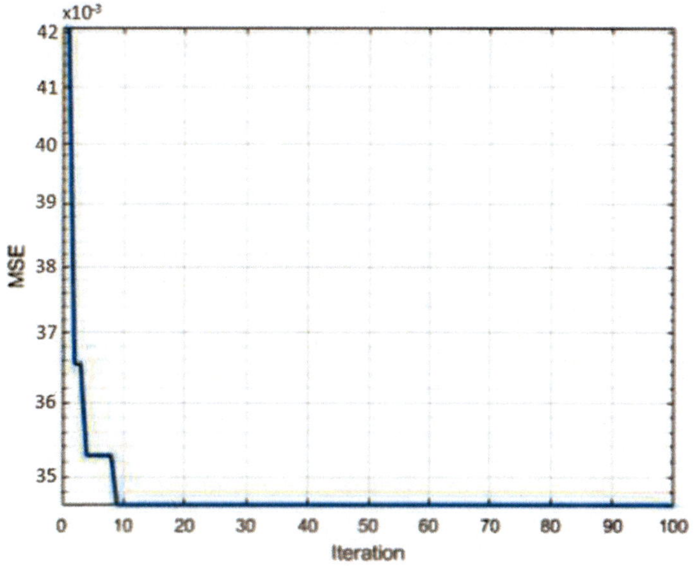

Fig. 4 Convergence profile of FA modeling for endpoint acceleration

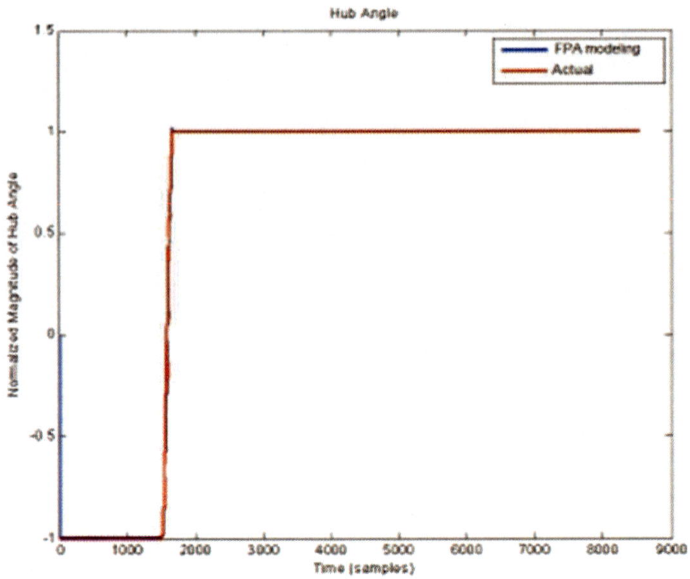

Fig. 5 Actual and FA modeling in time domain for hub angle

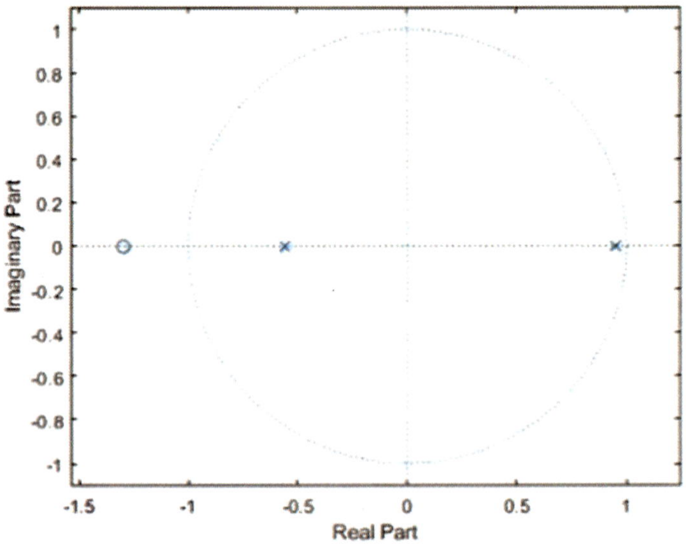

Fig. 6 Pole zero diagram of FA modeling for hub angle

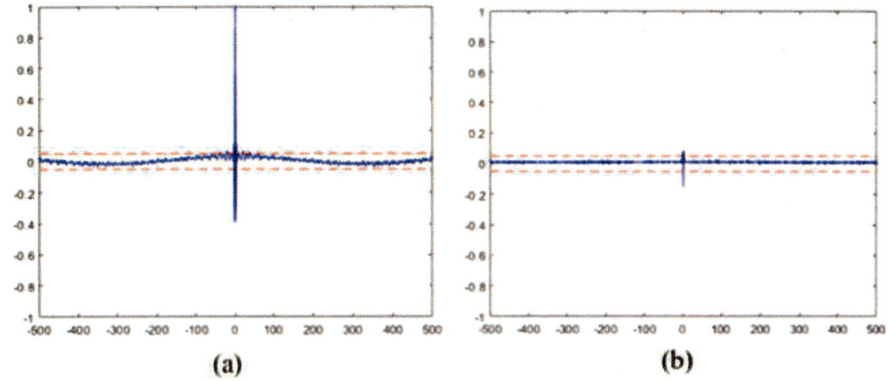

Fig. 7 Correlation tests of FA modeling for hub angle: **a** auto correlation, **b** cross correlation

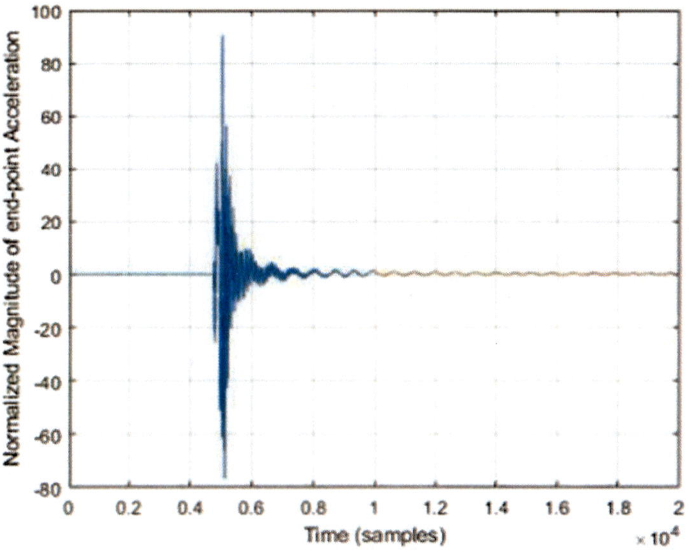

Fig. 8 Actual and FA modeling in time domain for endpoint acceleration

5.2 *Modeling Using Least Square (LS) Algorithm*

Then, flexible manipulator system was modeled using LS algorithm for both hub angle and endpoint acceleration behavior. The hub angle model was observed with an order of 8 and endpoint acceleration was successfully modelled at order of 2.

The LS modeling output of hub angle in time domain obtained is shown in Fig. 12. Figure 13 shows the poles and zeroes mapping and Fig. 14 illustrated the corresponding correlation test for hub angle modeling. Figures 15 and 16 represents the

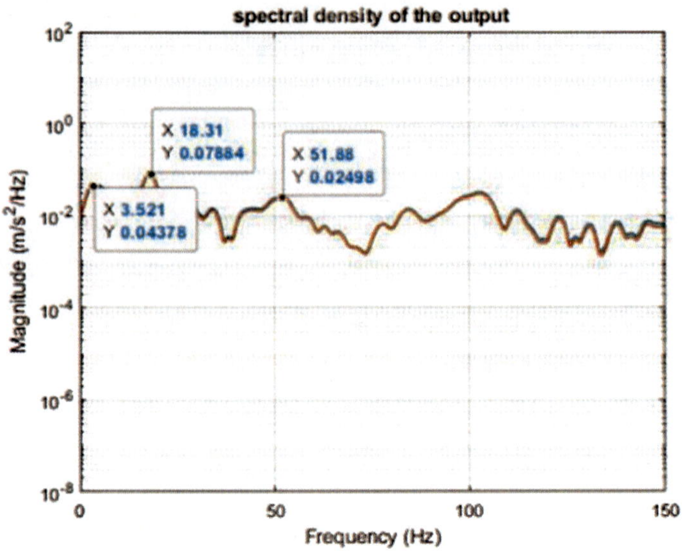

Fig. 9 Actual and FA modeling in frequency domain for endpoint acceleration

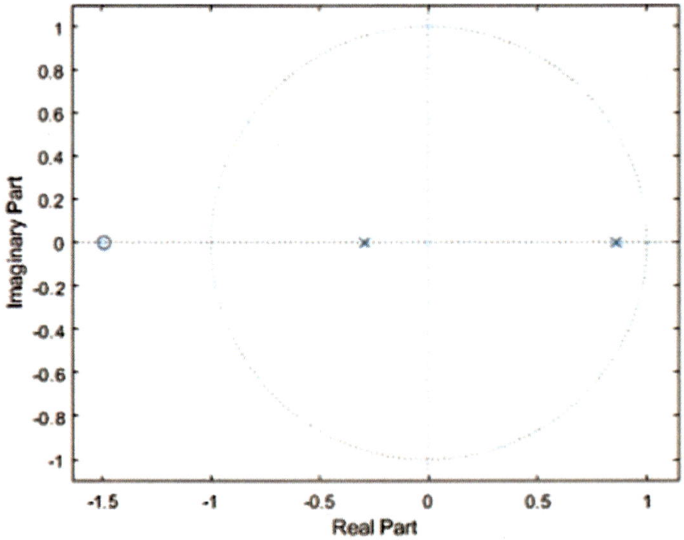

Fig. 10 Pole zero diagram of FA modeling for endpoint acceleration

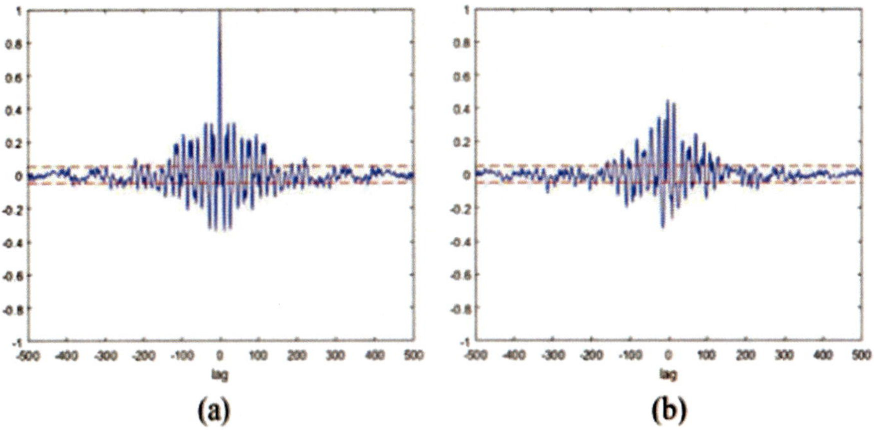

Fig. 11 Correlation tests of FA modeling for endpoint acceleration: **a** auto correlation, **b** cross correlation

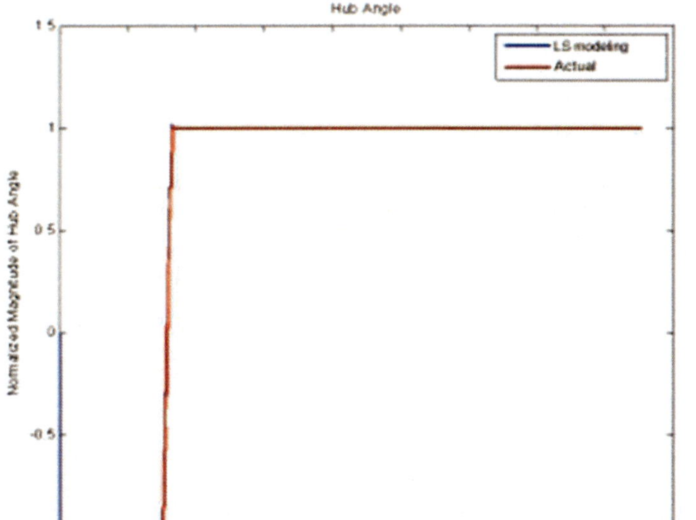

Fig. 12 Actual and LS modeling in time domain for hub angle

FA modeling output of endpoint acceleration in time and frequency domains, respectively. The poles and zeroes mapping with its corresponding correlation test are shown in Figs. 17 and 18 for endpoint acceleration modeling, respectively.

From Figs. 12 and 15, it is noted that both LS modeling output for hub angle and endpoint acceleration was matched the actual output, respectively. The model has successfully characterized the dynamics of the system as shown in frequency domain

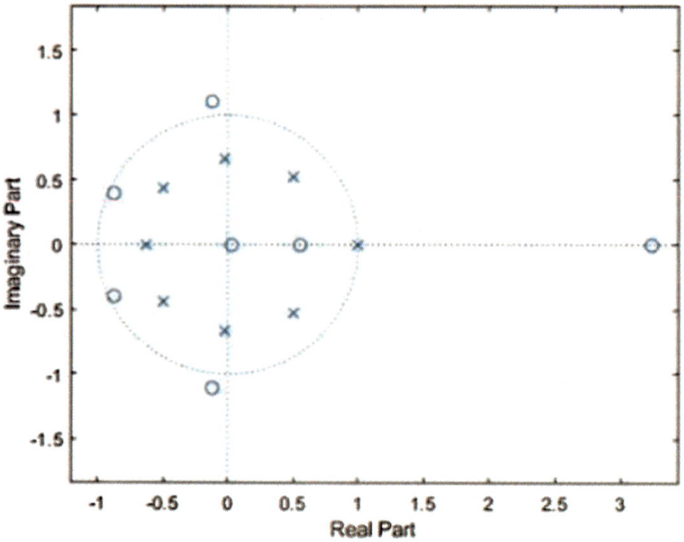

Fig. 13 Pole zero diagram of LS modeling for hub angle

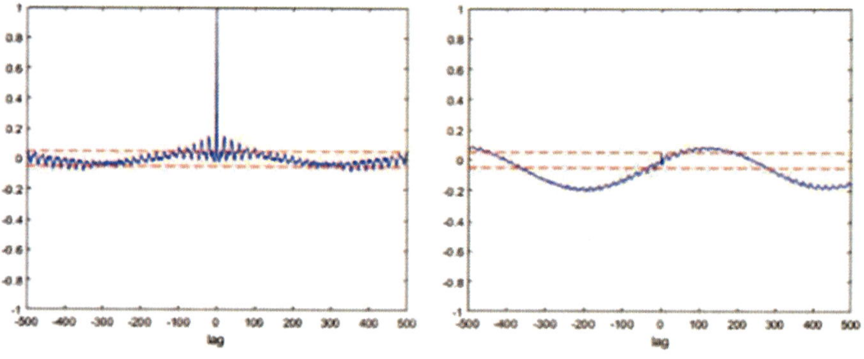

Fig. 14 Correlation tests of LS modeling for hub angle: **a** auto correlation, **b** cross correlation

from Fig. 16. For poles and zeroes mapping as in Figs. 13 and 17 for hub angle and endpoint acceleration respectively, all poles lied inside the circle unit indicating the stable model. However, the corresponding correlation test were found to be out of the range of 95% confidence interval which not satisfied the correlation test requirement as presented in Figs. 14 and 18.

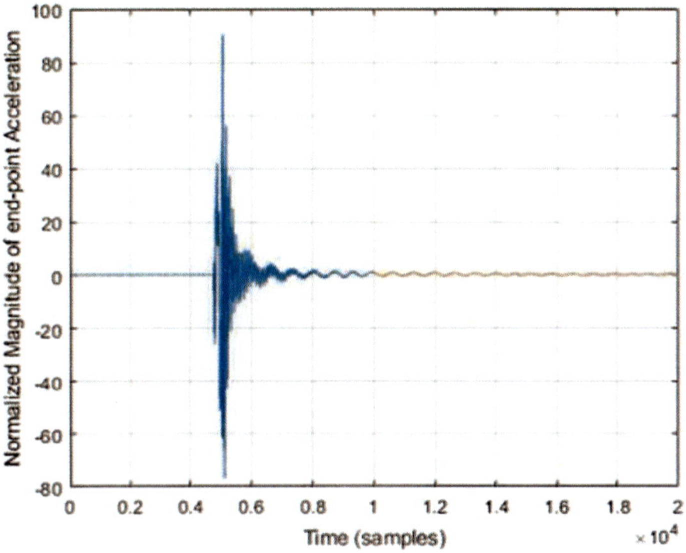

Fig. 15 Actual and LS modeling in time domain for endpoint acceleration

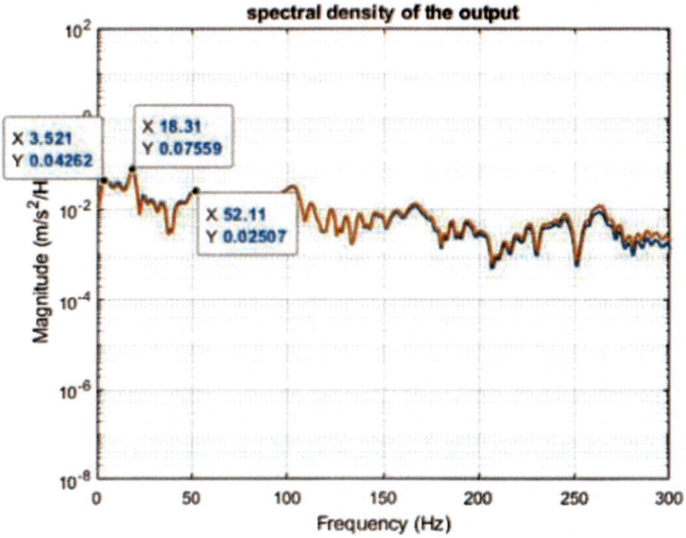

Fig. 16 Actual and LS modeling in frequency domain for endpoint acceleration

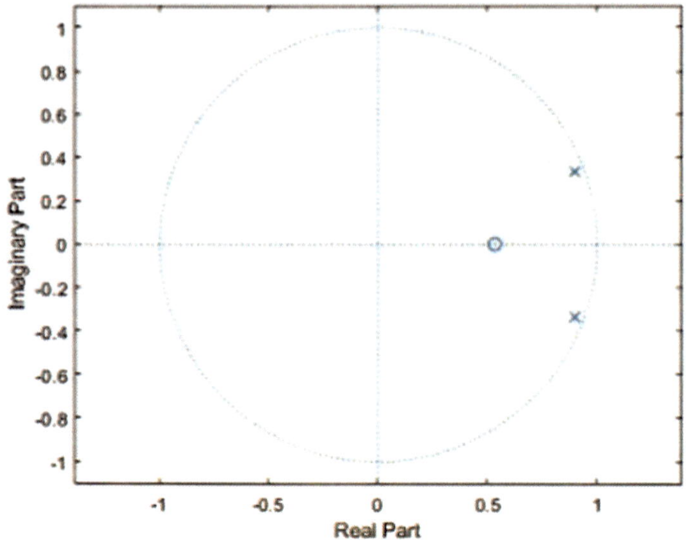

Fig. 17 Pole zero diagram of LS modeling for endpoint acceleration

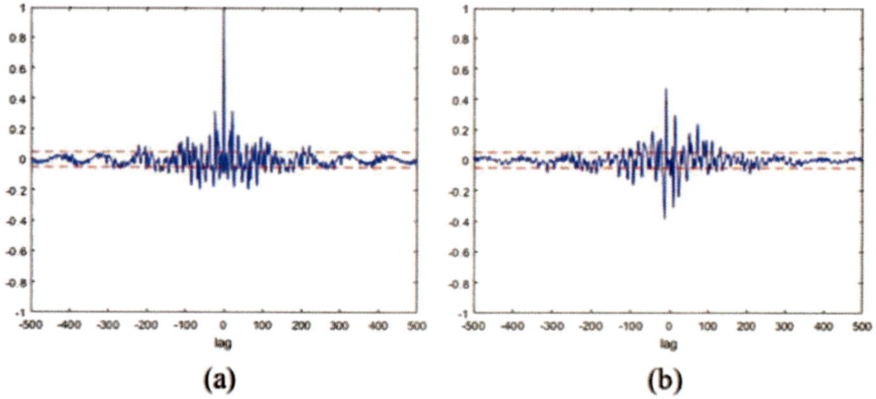

Fig. 18 Correlation tests of LS modeling for endpoint acceleration: **a** auto correlation, **b** cross correlation

5.3 *Comparative Assessment*

Table 2 shows the comparative assessment achieved by FA and LS algorithm for hub angle and endpoint acceleration modeling. It depicts the lowest MSE obtained and model parameters optimized in term of transfer function.

From Table 2, LS model for hub angle obtained lower MSE value than FA modeling. However, it is noticed that from Fig. 14, the correlation test was shown to be biased and not correlate within this 95% confidence interval which this interval

Table 2 Comparative assessment

	Modeling domain	MSE	Transfer function
Hub angle	FA	0.0023	$\frac{y(t)}{u(t)} = \frac{-0.0299z^{-1}+0.0425z^{-2}}{1-0.5341z^{-1}-0.4569z^{-2}}$
	LS	0.0019	$\frac{y(t)}{u(t)} = \frac{-0.009672z^{-1}+0.01765z^{-2}+0.0299z^{-3}+0.03624z^{-4}+0.0299+0.03072z^{-5}-0.0003352z^{-6}-0.01975z^{-7}+0.0005665z^{-8}}{1-0.3291z^{-1}-0.2153z^{-2}-0.100z^{-3}-0.0722z^{-4}-0.08532z^{-5}-0.05342z^{-6}-0.06752z^{-7}+0.06517z^{-8}}$
Endpoint acceleration	FA	0.0032	$\frac{y(t)}{u(t)} = \frac{0.185z^{-1}+0.1988z^{-2}}{1-0.7028z^{-1}-0.2731z^{-2}}$
	LS	0.0115	$\frac{y(t)}{u(t)} = \frac{-0.06323z^{-1}+0.03366z^{-2}}{1-1.793z^{-1}-0.9138z^{-2}}$

considered as a good result in real application. Additionally, FA model shows its reliability as it achieves lower MSE at model order of 2 while LS model required model order of 8. Higher model order indicates a complex system and can contribute to a difficulty in adapting with control strategy for further analysis.

While, for endpoint acceleration modeling, FA successfully model the dynamic of the system with smallest MSE value, good stability and unbiased correlation tests. Thus, it concludes that FA modeling is superior than LS in both hub angle and endpoint acceleration due to the simpler model with acceptable MSE value, good stability and unbiased correlation tests. FA algorithm has proven to performed well in modeling flexible manipulator system compared to conventional approach which provides a good benchmark to be used in control methodologies. From the results, it is observed that FA was easy to be implemented though the learning parameters need to be set appropriately. Suitable learning parameter is important in order to achieve the best fitness value and good overall performance.

6 Conclusion

Firefly algorithm (FA) has been utilize to model the flexible manipulator system in comparison with conventional LS technique. The performance of the models were assessed through minimum MSE, pole-zero diagram and correlation tests. The experimental input output data acquired for endpoint acceleration and hub angle was then fed to the system identification. The estimated models using FA and LS have been obtained, verified and acceptable to be used for further analysis of active vibration controller development. From the results, it was found that the performance of FA modelling is better in approximating the system response compared to LS in term of the lowest MSE, stable model and within confidence interval.

Acknowledgements The authors wish to thank the Ministry of Higher Education and Universiti Teknologi Malaysia (UTM) for providing the research grant and facilities. This research is supported using UTM Research Grant GUP Tier 2 Vote No. V17J07.

References

1. Dwivedy SK, Eberhard P (2006) Dynamic analysis of flexible manipulators, a literature review. Mech Mach Theory 41(7):749–777
2. Zhao ZL, Qiu ZC, Zhang XM, Da Han J (2016) Vibration control of a pneumatic driven piezoelectric flexible manipulator using self-organizing map based multiple models. Mech Syst Signal Process 70–71:345–372
3. Scaglioni B, Bascetta L, Baur M, Ferretti G (2017) Closed-form control oriented model of highly flexible manipulators. 52:174–185
4. Hu Q, Zhang J (2015) Acta Astronautica maneuver and vibration control of flexible manipulators using variable-speed control moment gyros. 113:105–119

5. Kumar P, Pratiher B (2019) Nonlinear modeling and vibration analysis of a two-link flexible manipulator coupled with harmonically driven flexible joints. Mech Mach Theory 131:278–299
6. Konstantinov SV, Baryshnikov AA (2017) Comparative analysis of evolutionary algorithms for the problem of parametric optimization of PID controllers. Procedia Comput Sci 103:100–107
7. Yelghi A, Köse C (2018) A modified firefly algorithm for global minimum optimization 62:29–44
8. Patle BK, Parhi DR, Jagadeesh A, Kashyap SK (2017) On firefly algorithm: optimization and application in mobile robot navigation. World J Eng
9. Mohd Yatim H, Mat Darus IZ (2017) Development of an experimental single-link flexible manipulator system. Int J Eng Technol 7:7–12
10. Yang XS, He X (2013) Firefly algorithm: recent advances and applications. Int J Swarm Intell 1(1):36–50
11. Wang H, Wang W, Zhou X, Sun H, Zhao J, Yu X (2017) Firefly algorithm with neighborhood attraction. Inf Sci 383:374–387
12. Ljung (1999) System identification: theory for the user. University of Linkoping Sweden. Prentice Hall Upper Saddle River NJ

Cubic Spline Interpolations in CNC Machining

W. R. W. Yusoff, I. Ishak, and F. R. M. Romlay

Abstract A cubic spline polynomial is applied to control the machine tool movements defined by the spline. This paper is an attempt to implement cubic spline interpolation in computer numerical method (CNC) machining. Three different C++ interpolation libraries were studied: Boost, Alglib and TK spline. The goals are to compare the accuracy of interpolation and the ease of implementation of the libraries. Twenty cubic spline interpolant functions were calculated using a selected test function One thousand interpolated points were calculated using the three different cubic spline interpolation libraries. Based on findings in this work, the Boost library is best on accuracy based on RMSE (root mean square error), while the TK spline library is simplest to implement in software code. The Alglib library is the most complicated in setup, and its accuracy is similar to the TK spline library. Included at the end of this report are the C++ cubic spline source codes, results of code executions and visual plots confirming the correctness of this work.

Keywords Spline · Cubic · Polynomial · Interpolation · Computer numerical method

1 Introduction on Interpolation

CNC controller works on interpolation and extrapolation principles. The prediction of a points between two known points coordinate known as interpolation. While extrapolation is an estimation of a point subjected on a known sequence of points coordinate. In CNC programming applications, most of interpolation algroithms is linear which is used for straight-line machining between two points. Circular interpolation is used for circles and arcs while helical interpolation, used for threads and helical forms.

W. R. W. Yusoff · I. Ishak · F. R. M. Romlay (✉)
Machine Manufacturing Union in Mechatronics Laboratory, Manufacturing Focus Group, Universiti Malaysia Pahang, 26600 Pekan, Pahang, Malaysia
e-mail: fadhlur@ump.edu.my

A. F. Ab. Nasir et al. (eds.), *Recent Trends in Mechatronics Towards Industry 4.0*, Lecture Notes in Electrical Engineering 730, https://doi.org/10.1007/978-981-33-4597-3_24

An interpolant is a function that passes through a set of known points and the function is used to compute values at intermediate points. In polynomial interpolation, the polynomial interpolant is the unique polynomial function of degree (n − 1) or less which passes through all of the given n knot points. Any single function which would effectively fit the data would be diffcult to obtain and highly unwieldy.

Instead of having a single polynomial function covering all the knot points, the idea of splines is introduced. A spline function covers an interval between two successive data points. By joining a series of piecewise, continuous and unique spline polynomials, the program achieves interpolation that covers all the data points. The condition for spline interpolation is that the curve obtained be continuous and smooth throughout.

Typically, the interpolation error is small even though for low degree of polynomials. Thus, spline interpolation is often favoured over polynomial interpolation. For spline interpolation, the data is fitted exactly with continuous piecewise functions at the knot points. Spline interpolants are assumed to be piecewise polynomial and globally smooth.

The cubic spline is the unique piecewise cubic polynomial such that its point values and its first two derivatives (but not the third) are continuous at the given n knot points. Normally for spline interpolation, cubic spline interpolation is selected and successfully solved the Runge's phenomenon issue. The method gives an interpolating polynomial that is smoother and has smaller error than some other interpolating polynomials such as Lagrange polynomial and Newton polynomial.

The Newton basis polynomials are defined in terms of finite differences. The Newton polynomial divided differences interpolation polynomial because the coefficients of the polynomial are based on Newton's divided differences approach. Similarly, the Newton polynomial is not piecewise like a spline since the function is formed by linear combinations of Newton basis polynomials.

Runge's phenomenon is the situation where oscillations occur between points when interpolating using high degree polynomials. It turns out that high order interpolation using a single global polynomial often exhibit serious oscillations. To reduce these oscillations, piecewise interpolation is used.

Lagrange polynomials are used for polynomial interpolation. Given distinct points (xi, yi), for the case which no two values xi is equal, the Lagrange polynomial is the lowest degree polynomial that assumes at each value of xj that resultant to the value yj, the functions correspond at each point. For large oscillation, Lagrange interpolation is inclined to Runge's phenomenon. As changing the points xj, it is often easier to use Newton polynomials to avoid necessitates looping the entire interpolant. A Lagrange polynomial is not piecewise like a spline. It is a global function formed by weighted coefficients of linear combinations of Lagrange basis functions in degree order sequences.

2 Rationale for Spline Interpolation in CNC

A cubic spline polynomial is a third degree (n = 3) polynomial order. In reference [1], a method of direct cubic spline interpolation on CNC-controller based on a single board microcomputer and the corresponding format of NC signal has been implemented. The technique typically controls the machine tool movements that defined by the spline, according to specific algorithm parameter setup. This results in a simplification of continuous path NC programming and a drastic contraction of the NC-program which provides a more efficient performance of machine controller.

A quadratic spline polynomial is a fourth degree (n = 4) polynomial order. Spline interpolation is often used to estimate the toolpath generated by G01 code (linear line segments). In reference [2], 3D points curve fitting that generated by G01 code interpolation based on quadratic B-splines are investigated. Feature points of G01 code is chosen through adaptive approach. Next, quadratic B-splines are gained through feature points fitting curve interpolation. Computational calculation is required in implementing the velocity planning algorithm and are very complicated because of the appearance of high-order curves. A better time-optimal method for the quadratic B-spline curves is presented to resolve the issue. The algorithms are verified with simulations or real-time CNC controller.

In reference [3], the interpolation of higher order profile curve in CNC controller with high speed and precision accuracy are required. Based on B spline curve theory, a revised algorithm on cubic B spline curve interpolation is presented. The proposed algorithm gives a new method which is based on pre-judgment of machining feed-rate to calculate the tangent vectors of spline curve junction. The tangential vectors are used to generate the spline curve equations as a function of time as a parameter. The trajectory with feed-rate profiling is established simultaneously with these equations. The simulation show that the algorithm has high calculation efficiency and meet the demands of trajectory accuracy with feed-rate smoothing.

In reference [4], it was said that as machine controller progressed, the linear interpolation paradigm become constraint the contour machining process. The factors are:

(a) Accuracy: Linear interpolation is not accurate. It has tendency to the prediction of the true work-piece surface geometry rather than the real thing.
(b) Smoothness: Linear interpolation is not smooth. A continuous contour is instead machined as a series of facets.
(c) Speed: Linear interpolation is not effective since the desire to keep the facets small will constraint the obtained feed rate.
(d) Efficiency: Many lines required to approximate a contoured surface will swell part programs to enormous sizes, bogging down the process of transferring files from external computers into the CNC. Thus, linear interpolation is not efficient.

Non-uniform rational B-splines (NURBS) curves and surface are very general mathematical surfaces widely used for representing complex three dimensional shapes in computer graphics [4].

As far as considering NURBS having a higher mathematical order capability, the weight of the control points can be specified. A non-uniform means that the knot vector, that indicates which portion of a curve is affected by an individual control point which is not essentially uniform. The upshot of it all is that more control factors can be applied to a formula so that considerably more complex forms can be expressed with a single curve.

As an example, the openNURBS [5] initiative provides CAD, CAM, CAE, and computer graphics software developers the tools to accurately transfer 3-D geometry between applications. openNURBS is an open-source, cross-platform library.

In reference [6], a method was developed for implementing an online non-uniform rational B-spline (NURBS) curve fitting process on CNC machines for improving the quality and efficiency of machining. Online NURBS interpolation are only available on sophisticated and high end CNC machines.

3 Properties and Implementation of Cubic Spline Interpolation

The four properties of cubic splines are as follows: (a). The piecewise functions S(x) will interpolate all data points, (b). the function S(x), will be continuous on the interval f(x1, xn), (c). the first derivative $S'(x)$, will be continuous on the interval f(x1, xn) and (d). the second derivative $S''(x)$, will be continuous on the interval f(x1, xn).

The characteristics of the spline are as follows:

1. Each cubic polynomial (degree 3) is defined by 4 coefficients (a, b, c, d) and so have a total of 4 N parameters, for N knot points.
2. The formula for spline interpolant S(x) is given by a different cubic polynomial in each interval. It is as follows:

$$S(x) = a(x - xi) * *3 + b(x - xi) * *2 + c(x - xi) + d \tag{1}$$

3. The type of spline S(x), is defined by a choice of the following 2 conditions at the 2 interval end points (xi) and (xj).
4. Natural cubic spline:

$$S'(xi) = 0 = S'(xj) \tag{2}$$

   The natural cubic spline is when the first derivative S'(x) at the end points are set to zero.
5. Clamped cubic spline:

$$S'(xi) = f'(a) \text{ and } S'(xj) = f'(b) \tag{3}$$

The clamped cubic spline is when the first derivative $S'(x)$ at the end points are set to be equal to the first derivative of the actual function at the end points. The clamped cubic spline gives more accurate approximation to the actual function f(x), but requires knowledge of the derivative at the end points.

This study report three different methods of using published open-source cubic spline libraries. (a) Boost interpolation library (Boost version 1.71 on 64-bit Linux); (b) Alglib interpolation library (Alglib version 3.15 Free Edition on 64-bit Linux) and (c). TK open-source spline.h library (Tino Kluge 2014 (ttk448@gmail.com) for 64-bit Linux).

In this report, the test function selected is a complex one-dimensional curve:

$$y = f(x) = (5.0/(x)) * \sin(5 * x) + (x - 6) \tag{4}$$

Based on the test function, 20 known points for (x, y) in the xrange [0.0: 10.0] spaced at 0.5 intervals are calculated. These 20 points are made to be (x, y) knot points.

The purpose of the interpolation is to estimate unknown equal, spaced at 0.01 points, between two sequential knot points. Each interval provides 50 interpolated points, so for the 20 intervals, there will be a total of 1000 x-valued interpolated points.

Since the test function y = f(x) is known, the exact values of (x,y) for the 1000 x-valued points can be calculated. With these exact values. the interpolation errors (differences) against the interpolated (x, y) values can be calculated. The errors at the twenty (x, y) knot points are zeros.

4 Results of Implementation

In next section, the results, the important issues and challenges encountered are discussed. Figure 1 shows the 20 (x, y) knot or control points in red circles derived from the test function y = f(x) = (5.0/(x))*sin(5*x) + (x − 6). The knot points are spaced at 0.5 apart over the x-range 0.0 to 10.0.

The blue line shows the interpolated curve using 1000 calculated points based on the Boost cubic spline interpolation library. The calculated points are spaced 0.01 apart. There are 1000 points that make up the blue curve from x = 0.0 to x = 10.0.

The interpolation involves 20 piecewise and continuous spline interpolant calculations. Each calculation results in a specific set of values for [a, b, c, d] for each spline section. The formula for each spline section is S(x) = a(x − xi)**3 + b(x − xi)**2 + c(x − xi) + d. The 1000 plotted points (blue curve) are interpolated based on these sequential spline sections.

Note that the interpolation criteria requires the blue curve pass exactly on all knot points. This will not be the case, for example, if we are calculating an approximation function that minimizes the errors for all the points. The approximation function may pass in-between the knot points.

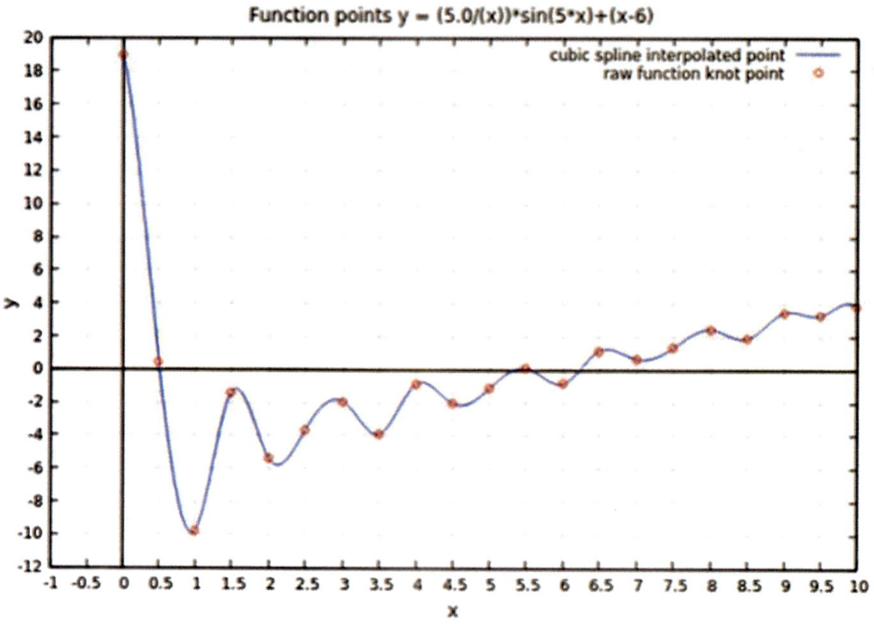

Fig. 1 Knot points for the function y(x)

The interpolated curve versus exact function curve are present in Fig. 2.

Figure 2 shows that the match between the actual exact function (blue curve) and the cubic spline interpolated function (red curve) is in good agreement. The RMSE (Root Mean Square Errors) calculation is a good measure of the fit (accuracy) of the interpolated to the exact function. In Fig. 3, the accuracy of three different interpolation cubic spline libraries were compared.

The extract output data of SSE, RSSE and RMSE calculation are displayed in Fig. 4. The data are depicted from x and y coordinate with error compensation consideration that take into account. The calculated RMSE for the following cubic spline interpolation libraries are as listed in Table 1.

From Table 1, Boost interpolation library provided the RMSE at 1.574222 E−02 which is the minimum value. RMSE for Alglib interpolation library and TK open-source spline.h library marked the same value at 2.275334 E−02 which is 44.58% higher compared to the Boost interpolation.

5 Conclusion

As a conclusion, the accuracy of interpolation for the Boost library is the best approach. From the study, the implementation code for Alglib library is most complex. However, the implementation code for TK spline.h library is the simplest

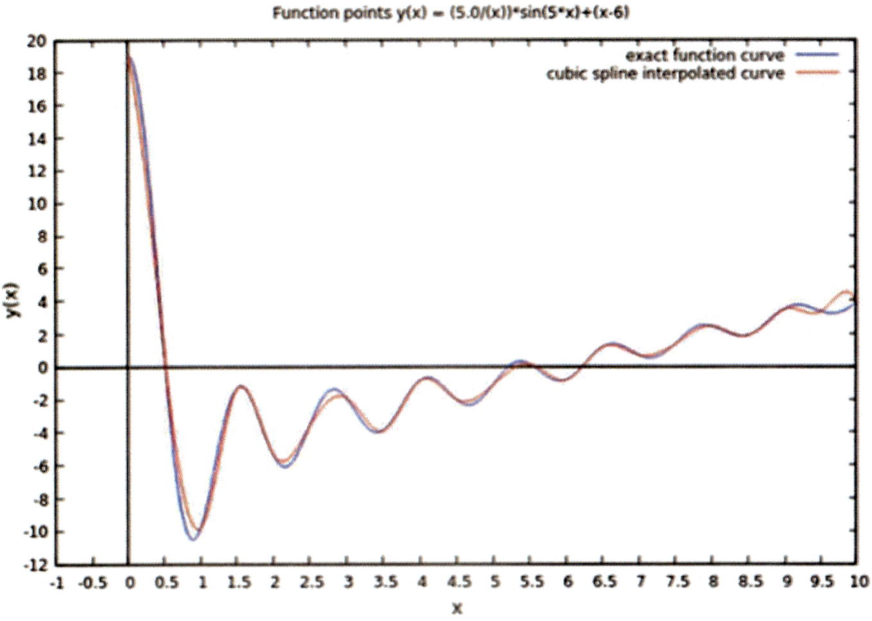

Fig. 2 Plot of cubic spline interpolated curve against the exact function curve

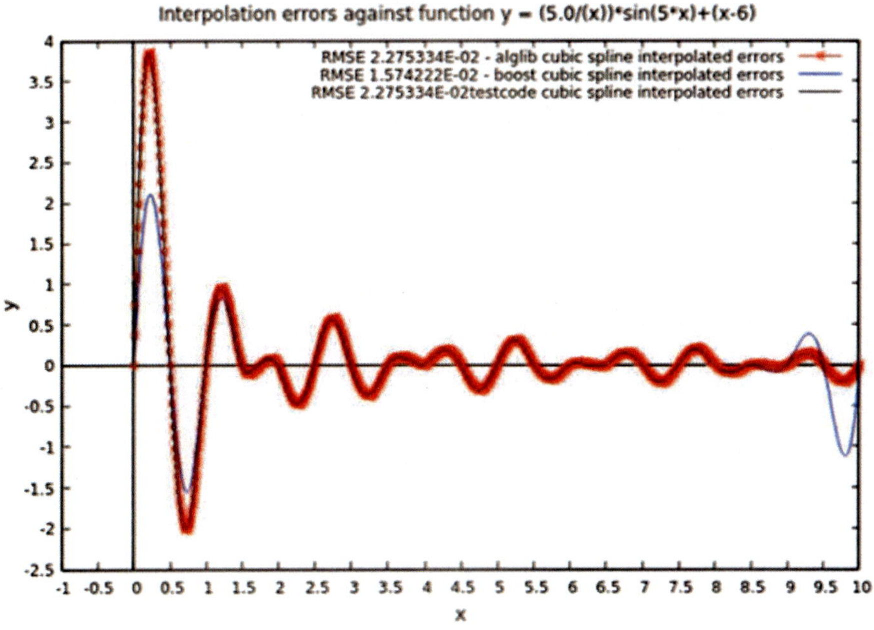

Fig. 3 Comparison of cubic spline interpolation errors

```
//
X       Yexact  Yinter  YError      PcntErr    SquareYerror
1E-18   19      19      0           0                     0
0.01    18.9996 18.8179 0.181732    0.956503   0.0330265198
0.02    18.9784 18.6217 0.356613    1.87905    0.1271728318
0.03    18.9364 18.4121 0.524297    2.76873    0.2748873442
0.04    18.8737 18.1892 0.68447     3.62659    0.4684991809
0.05    18.7904 17.9535 0.836849    4.4536     0.7003162488
....
....
9.93    3.63929 4.32175 -0.682463   -18.7526   0.4657557464
9.94    3.67047 4.27675 -0.606278   -16.5177   0.3675730133
9.95    3.70228 4.22552 -0.52324    -14.1329   0.2737800976
9.96    3.73467 4.16787 -0.433197   -11.5993   0.1876596408
9.97    3.76758 4.10358 -0.336      -8.91821   0.112896
9.98    3.80094 4.03244 -0.2315     -6.0906    0.05359225
9.99    3.83471 3.95426 -0.119549   -3.11755   0.0142919634
10      3.86881 3.86881 0           0                     0

                        248.3134708519  SSE   (Sum Square Errors)
                        15.7579653145   RSSE (Root Sum Square Errors)
        1.574222E-02    0.0157422231    RMSE (Root Mean Square Errors)
//
```

Fig. 4 Extract of RMSE calculation

Table 1 Root mean square errors (RSME) calculation

Approach	RMSE
Boost interpolation library	1.574222 E−02
Alglib interpolation library	2.275334 E−02
TK opensource spline.h library	2.275334 E−02

since the initial calculation of the predefined parameter is complete. Notice that the interpolated errors at the 20 knot points are zero, meaning the spline curves pass exactly on the knot points spaced at 0.5 intervals, satisfying the interpolation criterion.

Acknowledgements This research was partially supported by research, development and commercialization grant RDU172206 & RDU192805 of University Malaysia Pahang. Fundamental Research Grant Scheme; FRGS/2/2013/TK01/UMP/02/6 of Ministry of Higher Education Malaysia is also acknowledged.

References

1. Huan Ji (1985) Direct spline interpolation of CNC-machine tool. IFAC Proc 18(9):263–267
2. Mei Z, Wei Y, Chun-Ming Y, Ding Kang W, Xiao-Shan G (2010) Curve fitting and optimal interpolation on CNC machines based on quadratic B-splines. Sci China 53(1):1–18
3. Zhang W, Gao S, Cheng X, Zhang F (2017) An innovation on high-grade CNC machines tools for B-spline curve method of high-speed interpolation arithmetic. AIP Conf Proc 1834:040013
4. Beard T (1997) Interpolating curves, modern machine shop. https://www.mmsonline.com/articles/interpolating-curves. Gardner Business Media, Inc

5. Robert McNeel & Associates (2018) OpenNURBS the open source toolkit https://www.rhino3d.com/opennurbs
6. Yeh S, Su H (2009) Implementation of online NURBS curve fitting process on CNC machines. Int J Adv Manuf Technol 40:531–540

Modified Particle Swarm Optimization with Unique Self-cognitive Learning for Global Optimization Problems

Koon Meng Ang, Wei Hong Lim, Nor Ashidi Mat Isa, Sew Sun Tiang, Chun Kit Ang, Cher En Chow, and Zhe Sheng Yeap

Abstract Although different modified versions of particle swarm optimization (PSO) were proposed in past decades to solve global optimization problems, the appropriate mechanism used to attain proper balancing of algorithm's exploration and exploitation searches remains as an open-ended challenges. A modified PSO with unique self-cognitive learning (MPSO-USCL) is proposed in this paper to address this issue. For each particle, a unique exemplar can be generated by the proposed USCL module to replace the self-cognitive component of each particle and guide its search process towards the promising regions of search space with different levels of exploration and exploitation strengths. Extensive simulation studies are performed to compare the optimization performances of MPSO-USCL with six existing PSO variants using 12 benchmark functions. The proposed MPSO-USCL is reported to outperform its peer algorithms for all benchmark functions.

Keywords Metaheuristic search algorithm · Particle swarm optimization · Unique Self-cognitive learning

1 Introduction

Technological advancement in this Industry Revolution 4.0 era has led to the deployment of various engineering systems that require the complex mathematical modelling. Most often, these modern engineering optimization problems have challenging characteristics such as non-linearity, discontinuous, lacking of gradient information, and etc. that make them difficult for conventional mathematical programming methods to solve. Recently, metaheuristic search algorithms (MSAs) emerge as the promising optimizers to handle these complex engineering problems by leveraging

K. M. Ang · W. H. Lim (✉) · S. S. Tiang · C. K. Ang · C. E. Chow · Z. S. Yeap
Faculty of Engineering, Technology and Built Environment, UCSI University, 56000 Kuala Lumpur, Malaysia
e-mail: limwh@ucsiuniversity.edu.my

N. A. M. Isa
School of Electrical and Electronic Engineering, Universiti Sains Malaysia, 14300 Nibong Tebal, Malaysia

A. F. Ab. Nasir et al. (eds.), *Recent Trends in Mechatronics Towards Industry 4.0*, Lecture Notes in Electrical Engineering 730, https://doi.org/10.1007/978-981-33-4597-3_25

the effectiveness of their search operators inspired from various natural phenomena such as Darwinian's theory of evolution and social behavior of animals. MSAs have the advantages of having better global search ability, fast convergence rate and not require good guess of initial solution and derivative information to solve the an optimization problems competitively. Hence, these MSAs are widely used by practitioners to tackle various types of optimization problems [1–16].

Particle swarm optimization (PSO) [17] inspired by the collective behavior bird in searching for foods is a popular MSA, where the particles and food source are considered as the potential solutions and global optimum of given optimization problem, respectively. During the search process, useful information are shared among all particles to adjust their search trajectories and this collaborative behavior enables all swarm members searching towards the promising solution regions of search space. Despite of its simplicity in implementation and fast convergence speed, conventional PSO tends to suffers with premature convergence issue due to the poor regulation of exploration and exploitation strengths. Extensive researches were conducted in past decades to improve the performance of PSO via the modifications of parameter settings, population topology, learning strategy or the combinations of these approaches [18–27].

Although substantial amounts of new variants were proposed to address the demerits of conventional PSO, premature convergence is still the main deficiency [28] because the social learning aspect of PSO is only restricted to the global best particle. The poor interaction between each particle and other non-fitness solutions in the population is deemed as the main reason for particles to be trapped in local optima regions, especially if the search environment is complex. The learning strategy of global best particle is also not effective despite of its important role in leading the swarm because both of the self-cognitive and social components in its velocity update equation are essentially the same. In [22], it was advocated that both of the self-cognitive and social components of PSO should be discarded can replaced with an exemplar with each of its dimensional component is contributed by other non-fittest particles in population. These non-fittest particles contain useful information in certain dimension to guide the search process of population towards the promising solution region. Different combinations of exemplars can also be generated from these non-fittest particles and the swarm diversity is expected to increase because each particle can be guided by a unique exemplar.

Motivated by the findings of [22] and its extended works [23, 24], a modified particle swarm optimization with unique self-cognitive learning (MPSO-USCL) is proposed. The technical contributions of MPSO-UCSL are summarized as follows:

- MPSO-UCSL leverages the useful information from other non-fittest particles in population to perform searching towards the promising solution regions without experiencing the rapid loss of swarm diversity.
- An USCL module is designed to generate a unique exemplar to replace the self-cognitive component (i.e., personal best position) of each MPSO-UCSL particle in guiding its search process.

- The exemplars generated for different particles can have different search behaviors, i.e., explorative, exploitative or the combination of both, depending on its fitness value and the contribution from other non-fittest particles in population.
- Rigorous performance evaluations and comparisons are performed on the proposed MPSO-USCL with 12 global optimization benchmark functions.

The rest of this paper are presented herein. Section 2 explains the search mechanisms of MPOS-UCSL, followed by the performance evaluations in Sect. 3. The conclusion and future works of current study are presented in Sect. 4.

2 MPSO-USCL

2.1 *Construction of Exemplar with USCL Module*

The procedures of USCL module to generate the unique exemplar of each i-th particle to guide its search process are explained as follows. Initially, two trial positions are computed as the exemplar candidates for the i-th particle. Denote $X_i^{trial,1}$ as the first trial position of the i-th particle, then:

$$X_i^{trial,1} = \begin{cases} r_1 P_{a1} + r_2(P_{a2} - P_{a3}), \ f(P_{a2}) < f(P_{a3}) \\ r_1 P_{a1} + r_2(P_{a3} - P_{a2}), \text{ otherwise} \end{cases} \tag{1}$$

where P_{a1}, P_{a2}, and P_{a3} represents the personal best positions of three randomly selected particles with population indices of a_1, a_2 and a_3, respectively, with $a_1 \neq a_2 \neq a_3$; r_1 and r_2 are two uniformly distributed random numbers with the values between 0 to 1 with $r_1 + r_2 = 1; f(P_{a2})$ and $f(P_{a3})$ are the personal best fitness of a_2-th and a_3-th particles, respectively. From Eq. (1), $X_i^{trial,1}$ is generated by performing neighborhood search around a randomly selected a_1-th particle with P_{a1} that is prone to locate far away from the global best position P_g. Therefore, $X_i^{trial,1}$ tends to exhibits stronger explorative behavior to prevent the stagnation of i-th particle at local optima by guiding it to visit the unexplored solution regions.

For the second trial position of i-th particle defined as $X_i^{trial,2}$, it is obtained as:

$$X_i^{trial,2} = \begin{cases} r_3 P_g + r_4(P_{a4} - P_{a5}), \ f(P_{a4}) < f(P_{a5}) \\ r_3 P_g + r_4(P_{a5} - P_{a4}), \text{ otherwise} \end{cases} \tag{2}$$

where P_{a4} and P_{a5} refers to the personal best positions of two randomly selected non-fittest particles with population indices of a_4 and a_5, respectively, with $a_4 \neq a_5 \neq i$; r_3 and r_4 are two uniformly distributed random numbers with the values between 0 to 1 with $r_3 + r_4 = 1; f(P_{a4})$ and $f(P_{a5})$ are the personal best fitness of a_4-th and a_5-th particles, respectively. In contrary to $X_i^{trial,1}$, $X_i^{trial,2}$ is obtained through the neighborhood search around P_g, hence the latter candidate is more exploitative and tends to fine tune its search around the promising solution regions.

Since both of the $X_i^{trial,1}$ and $X_i^{trial,2}$ are generated via stochastic procedures, two scenarios are anticipated when comparing the personal best fitness of i-th particle, $X_i^{trial,1}$ and $X_i^{trial,2}$ denoted as $f(P_i)$, $f\left(X_i^{trial,1}\right)$ and $f\left(X_i^{trial,2}\right)$, respectively. For Scenario 1, the best solution found between $X_i^{trial,1}$ and $X_i^{trial,2}$ are better than P_i, therefore the fitter trial solution is assigned as the exemplar of i-th particle, i.e., E_i. For Scenario 2, the best solution between $X_i^{trial,1}$ and $X_i^{trial,2}$ are still inferior than P_i. Then, the third trial position of $X_i^{trial,3}$ is created by leveraging the useful information of $X_i^{trial,1}$ and $X_i^{trial,2}$ through a crossover mechanism. Let $W_{i,k}$ be a weightage value used to indicate the tendency of each k-th trial solution to contribute its directional information in creating each dimensional component of $X_i^{trial,3}$, where $k = 1$ and 2, i.e.:

$$W_{i,k} = \begin{cases} 1/\left[1 + f\left(X_i^{trial,k}\right)\right], & f\left(X_i^{trial,k}\right) \geq 0 \\ 1 + \left|f\left(X_i^{trial,k}\right)\right|, & \text{otherwise} \end{cases} \quad (3)$$

Based on the $W_{i,k}$ values of $X_i^{trial,1}$ and $X_i^{trial,2}$, roulette wheel selection is used to select the trial solution that contributes to each d-th component of $X_i^{trial,3}$. Evidently, the trial exemplar with better (i.e., smaller) fitness value has higher tendency to contribute its information in deriving $X_i^{trial,3}$ and vice versa according to Eq. (3). In order to prevent the formulation of $X_i^{trial,3}$ is dominated by the fitter candidate solution among $X_i^{trial,1}$ or $X_i^{trial,2}$, a random dimension index of d_r is selected and the d_r-th component of $X_i^{trial,3}$ is contributed by the inferior candidate solution among $X_i^{trial,1}$ or $X_i^{trial,2}$. The objective function value of $X_i^{trial,3}$ is denoted as $f\left(X_i^{trial,3}\right)$ and then compared with $f\left(X_i^{trial,1}\right)$ and $f\left(X_i^{trial,2}\right)$. The best solution among the trial solutions of $X_i^{trial,1}$, $X_i^{trial,2}$ and $X_i^{trial,3}$ is then assigned as the exemplar E_i of i-th particle for Scenario 2.

The implementations of the crossover operation and USCL module are presented in Figs. 1 and 2, respectively. The variable *fes* in Fig. 2 is the current fitness evaluations (FEs) numbers incurred by MPSO-USCL, whereas a vector $\mathbf{P} = [P_1, \ldots, P_i, \ldots, P_S]$ is defined to store the personal best positions of all MPSO-USCL particles.

2.2 *Complete Framework of MPSO-USCL*

In contrary to conventional PSO, the proposed MPSO-USCL updates the velocity of each particle i based on the unique exemplar E_i generated from the USCL module

$X_i^{trial,3}$ = **Crossover** ($X_i^{trial,1}$, $X_i^{trial,2}$, $f(X_i^{trial,1})$, $f(X_i^{trial,2})$)

Calculate the value of $W_{i,k}$ of each trial particle with Eq. (3);
Select random dimension d_r;
for each dimension d **do**
  if $d \neq d_r$ **then**
    Perform roulette wheel selection based on the $W_{i,k}$ of each trial position;
    $X_i^{trial,3}$:= d-th component of the selected trial position;
  else if $d = d_r$ **then**
    $X_i^{trial,3}$:= d_r-th component of $X_i^{trial,1}$ or $X_i^{trial,2}$ with inferior fitness;
  end if
end for

Fig. 1 Crossover operation in USCL module

E_i = **USCL**(particle i, **P**, P_g)

Compute $X_i^{trial,1}$ and $X_i^{trial,2}$ with Eqs. (1) and (2), respectively;
Calculate $f(X_i^{trial,1})$ and $f(X_i^{trial,2})$;
$fes := fes + 2$;
if $\min\left[f(X_i^{trial,1}), f(X_i^{trial,1})\right] < f(P_i)$ **then**
  E_i := trial positions $X_i^{trial,1}$ or $X_i^{trial,2}$ with the best fitness;
else if $\min\left[f(X_i^{trial,1}), f(X_i^{trial,1})\right] \geq f(P_i)$ **then**
  $X_i^{trial,3}$:= **Crossover** ($X_i^{trial,1}$, $X_i^{trial,2}$, $f(X_i^{trial,1})$, $f(X_i^{trial,2})$);
  Calculate $f(X_i^{trial,3})$;
  $fes := fes + 1$;
  E_i := trial positions of $X_i^{trial,1}$, $X_i^{trial,2}$ or $X_i^{trial,2}$ with the best fitness;
end if

Fig. 2 USCL module of the proposed MPSO-USCL

and the global best particle P_g. For each particle i, the d-th component of its velocity and position at the $(t + 1)$-th iteration are updated in Eqs. (4) and (5) respectively as:

$$V_{i,d}(t+1) = \omega V_{i,d}(t) + c_1 r_5 \big(E_{i,d} - X_{i,d}(t)\big) + c_2 r_6 \big(P_{g,d} - X_{i,d}(t)\big) \quad (4)$$

$$X_{i,d}(t+1) = X_{i,d}(t) + V_{i,d}(t+1) \quad (5)$$

where $i = 1, \ldots, S$ is the particle index; c_1 and c_2 refers to the acceleration coefficients used to govern the influences of E_i and P_g, respectively; r_5 and r_6 are two uniformly distributed random numbers with the values between 0 to 1;ω is an inertia weight used to identify the influence of particle's velocity in previous iteration. The P_g component is retained in Eq. (4) because it can yield efficient convergence.

The overall framework of MPSO-USCL is shown in Fig. 3. First, the velocity and position of all particles are initialized. During the optimization process, the USCL module is triggered to generate a unique exemplar E_i to produce the new velocity and position components of each i-th particle. The personal best position of i-th particle and global best particle are updated if a fitter new position is obtained. The reconstruction of E_i is only triggered if the i-th particle fails to update its personal best position for M successive times as recorded by its counter variable η_i. The threshold value set for M should not be too small or too large because the former setting tends to reconstruct E_i too frequent and prevent the convergence of particle, whereas the latter setting tends to consume excessive computational cost on the local optima with E_i that is no longer effective.

MPSO-USCL

```
Initialize the MPSO-USCL population;
while fes < FE_max do
    for each particle i do
        if η_i > M then /*Trigger USCL module to produce new E_i*/
            E_i := USCL(particle i, P, P_g);
            η_i := 0;
        end if
        Calculate V_i and X_i by using Eqs. (4) and (5), respectively;
        Calculate the fitness values of new X_i;
        fes := fes + 1;
        Update P_i, f(P_i), P_g, and f(P_g);
        if f(X_i) < f(P_i) then
            η_i := 0;
        else
            η_i := η_i + 1;
        end if
    end for
end while
```

Fig. 3 Complete framework of PSO-MSCL

Table 1 12 benchmark functions used in this study

No.	Function name	*RG*	F_{min}
F1	Sphere	$[-100,100]^D$	0
F2	Schewefel 1.2	$[-100,100]^D$	0
F3	Rastrigin	$[-5.12,5.12]^D$	0
F4	Noncontinuous Rastrigin	$[-5.12,5.12]^D$	0
F5	Griewank	$[-600,600]^D$	0
F6	Ackley	$[-32,32]^D$	0
F7	Weierstrass	$[-0.5,0.5]^D$	0
F8	Rotated schewefel 1.2	$[-100,100]^D$	0
F9	Rotated rastrigin	$[-5.12,5.12]^D$	0
F10	Rotated noncontinuous rastrigin	$[-5.12,5.12]^D$	0
F11	Rotated griewank	$[-600,600]^D$	0
F12	Rotated weierstrass	$[-0.5,0.5]^D$	0

3 Performance Analysis of Proposed Algorithm

3.1 Simulation Settings

The proposed MPSO-USCL is compared with seven conventional problems and five rotated problems summarized in Table 1 that consists of the feasible search range *RG* and objective function value in global optimum F_{min}. The performance of MPSO-USCL in solving all 12 test functions are compared with six existing PSO variants known as: APSO [26], CLPSO [22], FLPSO-QIW [23], FlexiPSO [27], FPSO [25] and OLPSO [24]. Notably, CLPSO, FLPSO-QIW and OLPSO share similarities with the proposed MPSO-USCL because the former three PSO variants also generate exemplars from non-fittest solutions using different strategies. APSO is a PSO variant improved using parameter adaptation strategy, whereas the modifications of population topology are applied to both of FlexiPSO and FPSO.

For MPSO-USCL, ω is linearly decreased from 0.9 to 0.4, whereas both c_1 and c_2 are set as 2.0. Empirical study also reveals that a threshold value of $M = 8$ can offer satisfactory performance for MPSO-USCL. The population sizes of all algorithms and the dimensional sizes of tested functions are set as $S = 30$ and $D = 50$, respectively. The maximum FEs used by an algorithm to solve each test function is set as $FE_{max} = 3.00\text{E} + 05$.

3.2 Comparison of Mean Fitness Results

The optimization performances of all involved PSO variants are evaluated in terms of search accuracy via the mean fitness (F_{mean}) and standard deviation (*SD*) values.

Wilcoxon test [29] is also utilized for pairwise comparison between MPSO-USCL and each of its peer at the significant level of $\alpha = 0.05$ and their comparison results are summarized as h values. The h values are indicated as "+" when the search accuracy of MPSO-USCL is significantly better, "=" when insignificant and "−" when significantly worse than its competitors at significance level of 0.05. The F_{mean} and SD values produced by MPSO-USCL and its six peers in solving all test functions are reported in Table 2, where the best results are indicated in boldface.

From Table 2, the proposed MOPSO-USCL has exhibited the best search accuracy because it is able to outperform all of its peer algorithms with large margins for all selected benchmark functions. The h values obtained from Wilcoxon tests are consistent with those of F_{mean} values, implying that the excellent search performances MOPSO-USCL are verified from statistical point of view and they are not achieved by any random chances. For the conventional benchmark problems of F1 to F7, the proposed MPSO-USCL is the only algorithm that have located the global optimum solution of functions F2, F3 and F4. Other PSO variants such as CLPSO, FLPSO-QIW and OLPSO also demonstrate competitive search performances in handing these conventional benchmark functions because they are able to solve the functions F1 and F5 to F7 with considerably good F_{mean} values. It is notable that the proposed MPSO-USCL, CLPSO, FLPSO-QIW and OLPSO share similarities in terms of their algorithmic framework design. Particularly, a unique exemplar is derived to guide the search process of each particle in these PSO variants towards the promising solution regions by using their respective exemplar creation mechanisms. The competitive simulation results exhibited by these PSO variants implies that the directional information from non-fittest solutions in population are indeed useful to guide the search process effectively.

Nevertheless, Table 2 shows that most of the compared algorithms suffer with different levels of performance deterioration when they are applied to address the more challenging rotated problems. For instance, the peer algorithms such as CLPSO, FLPSO-QIW and OLPSO are unable to solve all rotated problems despite of their promising search accuracies shown in simpler conventional functions. In contrary, the proposed MPSO-USCL has demonstrated impeccable robustness towards the challenging landscapes of rotated problems and successfully solve these functions. These findings imply that the exemplars produced by the USCL module have better effectiveness than those in CLPSO, FLPSO-QIW and OLPSO in guiding the MPSO-USCL particles searching towards the global optima of benchmark functions with more complex fitness landscapes.

4 Conclusions

In this paper, a new PSO variant called MPSO-USCL is designed. An USCL module is incorporated into MPSO-USCL, aiming to generate the unique exemplars for more effective guidance. Unlike conventional PSO, these unique exemplars used to replace the self-cognitive component of each MPSO-USCL particle to perform

Table 2 F_{mean}, SD and h-values for 50-D problems

		F1			F2			F3	
	F_{mean}	SD	h	F_{mean}	SD	h	F_{mean}	SD	h
APSO	2.50E−01	1.84E−01	+	1.49E+03	4.79E+02	+	5.75E−01	6.26E−01	+
CLPSO	3.30E−48	1.27E−47	+	5.14E+03	1.01E+03	+	9.09E+01	1.07E+01	+
FLPSO-QIW	2.89E−81	5.96E−81	+	2.61E+02	8.89E+01	+	2.59E+00	1.51E+00	+
FlexiPSO	1.79E−04	5.24E−05	+	1.41E+00	6.68E−01	+	2.13E−04	6.25E−05	+
FPSO	7.03E+01	6.99E+01	+	3.45E+03	1.34E+03	+	1.83E+01	1.00E+01	+
OLPSO	4.89E−34	5.18E−34	+	5.70E+2	1.84E+02	+	3.31E−01	6.02E−01	+
MPSO-USCL	**0.00E+00**	0.00E+00		**0.00E+00**	0.00E+00		**0.00E+00**	0.00E+00	
		F4			F5			F6	
	F_{mean}	SD	h	F_{mean}	SD	h	F_{mean}	SD	h
APSO	3.62E−02	3.24E−02	+	1.67E−01	8.20E−02	+	6.63E−02	2.60E−02	+
CLPSO	8.12E+01	9.77E+00	+	3.38E−11	1.71E−10	+	1.16E−14	2.57E−15	+
FLPSO-QIW	5.57E+00	2.35E+00	+	5.74E−04	2.20E−03	+	3.42E−14	1.06E−14	+
FlexiPSO	2.05E−04	7.49E−05	+	8.33E−03	9.47E−03	+	3.57E−03	5.38E−04	+
FPSO	1.61E+01	9.57E+00	+	1.88E+00	9.30E−01	+	1.81E+00	1.11E+00	+
OLPSO	1.16E+00	1.14E+00	+	**0.00E+00**	0.00E+00	=	5.08E−15	1.78E−15	+
MPSO-USCL	**0.00E+00**	0.00E+00		**0.00E+00**	0.00E+00		**0.00E+00**	0.00E+00	
		F7			F8			F9	
	F_{mean}	SD	h	F_{mean}	SD	h	F_{mean}	SD	h
APSO	5.42E−01	1.86E−01	+	1.27E+03	3.21E+02	+	1.80E+02	5.67E+01	+
CLPSO	**0.00E+00**	0.00E+00	=	5.78E+03	9.92E+02	+	3.31E+02	2.35E+01	+

(continued)

Table 2 (continued)

		F1			F2			F3	
	F_{mean}	*SD*	*h*	F_{mean}	*SD*	*h*	F_{mean}	*SD*	*h*
FLPSO-QIW	1.87E−05	8.28E−05	+	2.61E+02	7.61E+01	+	1.25E+02	1.75E+01	+
FlexiPSO	1.11E−01	1.15E−02	+	4.91E+00	3.66E+00	+	1.50E+02	3.43E+01	+
FPSO	3.33E+00	2.33E+00	+	3.21E+03	1.77E+03	+	1.81E+02	5.02E+01	+
OLPSO	**0.00E+00**	0.00E+00	=	1.91E+03	4.16E+02	+	9.82E+01	5.18E+01	+
MPSO-USCL	**0.00E+00**	0.00E+00		**0.00E+00**	0.00E+00		**0.00E+00**	0.00E+00	
		F10			F11			F12	
	F_{mean}	*SD*	*h*	F_{mean}	*SD*	*h*	F_{mean}	*SD*	*h*
APSO	2.58E+02	6.13E+01	+	2.11E+02	1.04E+02	+	6.29E+01	4.25E+00	+
CLPSO	3.22E+02	2.51E+01	+	1.44E+00	4.51E−01	+	5.66E+01	2.41E+00	+
FLPSO-QIW	1.27E+02	2.12E+01	+	1.51E+00	5.38E−01	+	4.85E+01	3.39E+00	+
FlexiPSO	2.18E+02	8.28E+01	+	2.66E+02	9.16E+01	+	6.59E+01	4.58E+00	+
FPSO	1.54E+02	3.75E+01	+	7.27E+00	5.61E+00	+	5.17E+01	3.92E+00	+
OLPSO	1.76E+02	4.92E+01	+	7.57E−01	2.67E−01	+	4.59E+01	4.78E+00	+
MPSO-USCL	**0.00E+00**	0.00E+ 00		**0.00E+00**	0.00E+00		**0.00E+00**	0.00E+00	

Bold represents the best simulation results obtained for each benchmark function searching with better diversity. Extensive performance studies indicate that MPSO-USCL has more competitive search accuracy than its competitors. It is also reported that the exemplars produced by USCL module can provide more effective guidance to particles than those with similar algorithmic structures such as CLPSO, FLPSO-QIW and OLPSO.

References

1. Ang CK, Tang SH, Mashohor S, Arrifin MKAM (2014) Solving continuous trajectory and forward kinematics simultaneously based on ANN. Int J Comput Commun Control
2. Abdullah Al-Sanabani DG, Solihin MI, Liew PP, Astuti W, Ang CK, Lim WH (2019) Development of non-destructive mango assessment using handheld spectroscopy and machine learning regression. J Phys Conf Ser 1367:012030
3. Alrifaey M, Sai Hong T, Supeni EE, As'arry A, Ang CK (2019) Identification and prioritization of risk factors in an electrical generator based on the hybrid FMEA framework. Energies 12, 649
4. Yao L, Shen J, Lim WH (2016) Real-Time energy management optimization for smart household. In: 2016 IEEE international conference on internet of things (iThings) and IEEE green computing and communications (GreenCom) and IEEE cyber, physical and social computing (CPSCom) and IEEE smart data (SmartData), Chengdu, China, pp 20–26
5. Yao L, Damiran Z, Lim WH (2017) Energy management optimization scheme for smart home considering different types of appliances. In: 2017 IEEE International conference on environment and electrical engineering and 2017 IEEE industrial and commercial power systems Europe (EEEIC/ I&CPS Europe), Milan, Italy, pp 1–6
6. Yao L, Lim WH (2018) Optimal purchase strategy for demand bidding. IEEE Trans Power Syst 33:2754–2762
7. Yao L, Lim WH, Tiang SS, Tan TH, Wong CH, Pang JY (2018) Demand bidding optimization for an aggregator with a genetic algorithm. Energies 11:2498
8. Yao L, Yao L, Lim WH (2018) A soft curtailment of wide-area central air conditioning load. Energies 11:492
9. Yao L, Chen Y, Lim WH (2015) Internet of things for electric vehicle: an improved decentralized charging scheme. In: 2015 IEEE international conference on data science and data intensive systems, Sydney, NSW, Australia, pp. 651–658
10. Natarajan E, Kaviarasan V, Lim WH, Tiang SS, Tan TH (2018) Enhanced multi-objective teaching-learning-based optimization for machining of delrin. IEEE Access 6:51528–51546
11. Natarajan E, Ang CT, Lim WH, Kosalishkwaran G, Ang CK, Parasuraman S (2019) Design topology optimization and kinematics of a multi-modal quadcopter and quadruped. In: 2019 IEEE student conference on research and development (SCOReD), pp. 214–218. Bandar Ser Iskandar, Malaysia
12. Natarajan E, Kaviarasan V, Lim WH, Tiang SS, Parasuraman S, Elango S (2019) Non-dominated sorting modified teaching–learning-based optimization for multi-objective machining of polytetrafluoroethylene (PTFE). J Intell Manuf
13. Tarawneh MA, Yu, LJ, Tarawni MA, Ahmad SHJ, Al-Banawi O, Bathiha MA (2015) High performance thermoplastic elastomer (TPE) nanocomposite based on graphene nanoplates (GNPs). World J Eng 12:437–442
14. Yu LJ, Sahrim AH, Kong I, Mouad AT (2012) Microwave absorbing properties of nickel-zinc ferrite/multiwalled nanotube thermoplastic natural rubber composites. Adv Mater Res 501:24–28
15. Yu LJ, Ahmad SH, Kong I, Tarawneh MA, Flaifel MH (2013) Preparation and characterisation of NiZn ferrite/multiwalled nanotubes thermoplastic natural rubber composite. Int J Mater Eng Innov 4:214–224

16. Yu LJ, Ahmad SH, Kong I, Appadu S, Flaifel MH (2012) Magnetic properties, microsturcture and mophology of thermoplastics natural rubber composite reinforced with NiZn ferrite/Mwnt. Sains Malaysiana 41:453–458
17. Kennedy J, Eberhart R (1995) Particle swarm optimization. In: Proceedings of ICNN'95—international conference on neural networks, Perth, WA, Australia, pp 1942–1948, vol 1944
18. Lim WH, Isa NAM (2015) Particle swarm optimization with dual-level task allocation. Eng Appl Artif Intell 38:88–110
19. Lim WH, Isa NAM, Tiang SS, Tan TH, Natarajan E, Wong CH, Tang JR (2018) A self-adaptive topologically connected-based particle swarm optimization. IEEE Access 6:65347–65366
20. Ang KM, Lim WH, Isa NAM, Tiang SS, Wong CH (2020) A constrained multi-swarm particle swarm optimization without velocity for constrained optimization problems. Expert Syst Appl 140:112882
21. Bonyadi MR, Michalewicz Z (2017) Particle swarm optimization for single objective continuous space problems: a review. Evol Comput 25:1–54
22. Liang JJ, Qin AK, Suganthan PN, Baskar S (2006) Comprehensive learning particle swarm optimizer for global optimization of multimodal functions. IEEE Trans Evol Comput 10:281–295
23. Tang Y, Wang Z, Fang J-A (2011) Feedback learning particle swarm optimization. Appl Soft Comput 11:4713–4725
24. Zhan Z, Zhang J, Li Y, Shi Y (2011) Orthogonal learning particle Swarm optimization. IEEE Trans Evol Comput 15:832–847
25. De Oca, MAM, Stutzle T, Birattari M, Dorigo M (2009) Frankenstein's PSO: a composite particle swarm optimization algorithm. IEEE Trans Evol Comput 13:1120–1132
26. Zhan Z, Zhang J, Li Y, Chung HS (2009) Adaptive particle swarm optimization. IEEE Trans Syst Man Cybern Part B (Cybern) 39:1362–1381
27. Kathrada M (2009) The flexi-PSO: towards a more flexible particle swarm optimizer. OPSEARCH 46:52–68
28. Van den Bergh F, Engelbrecht AP (2004) A cooperative approach to particle swarm optimization. IEEE Trans Evol Comput 8:225–239
29. Derrac J, García S, Molina D, Herrera F (2011) A practical tutorial on the use of nonparametric statistical tests as a methodology for comparing evolutionary and swarm intelligence algorithms. Swarm Evol Comput 1:3–18

The Role of 3D-Technologies in Humanoid Robotics: A Systematic Review for 3D-Printing in Modern Social Robots

Jayesh Saini and Esyin Chew

Abstract The novelty of Three Dimensional (3D) technologies for modeling and printing robotics parts are increasingly popular. In order to use 3D technologies in designing modern social humanoid robots and to draw convincing concluding remarks for other researchers' references, we need to critically incorporate various latest works available in the field of 3D printing and humanoid robotics. The basic principle behind 3D-printing is that it permits the creation of complex artefacts in the simplest possible way. Thus, the aim of this study is to critically evaluate the role of 3D-technologies for developing modern social robots in humanoid robotics. The research methodology is systematic review the relevant literatures on Web of Science and Scopus. Two large literature databases were screened to get the latest 3D print research and applications in modern social robots. As a result, 61 articles were analyzed, discussed and reviewed. We investigated and compared the use of 3D printing for various purposes such as humanoid robots, social robot models, hybrid robot projects and social robot auxiliaries. The finding leads to the design principles that contribute to the home-built 3D printed humanoid robot by adding all the dimensions together, following the principles of social robotics. Future research will revolve around these definitions for the applied role of 3D technology in modern social robotics and to develop a low-cost prototype of 3D printed humanoid robot, which will be reported in another paper.

Keywords 3D printing low cost · 3D printing low cost robot · 3D printing robot · 3D printing efficient robot · Robot hospitality · Social robotics

---

J. Saini (✉) · E. Chew
EUREKA Robotics Lab, Cardiff School of Technologies, Cardiff Metropolitan University, Western Avenue, Cardiff CF5 2YB, UK
e-mail: J.Saini@outlook.cardiffmet.ac.uk

E. Chew
e-mail: echew@cardiffmet.ac.uk

A. F. Ab. Nasir et al. (eds.), *Recent Trends in Mechatronics Towards Industry 4.0*, Lecture Notes in Electrical Engineering 730, https://doi.org/10.1007/978-981-33-4597-3_26

1 Introduction

1.1 The Definition and Characteristics of 3D-Printing in Modern Social Robots

The basic concept behind 3D-printing is that it helps in developing complex things in simplest ways as possible. For using 3D-printing in modern social robots, we have to combine different researches that are currently available in the field of humanoid robotics to reach at a proper conclusion. We adopted various research strategies in terms of search, study selection and quality assessment. We discussed the history of the use of 3D-printing in modern social robots and the taxonomy for humanoid robotics through the lens of 3D-printing i.e., literature review summary for 3D-printing and humanoid robots from 3 studies, 3D-printing and social robot models from 16 studies, 3D-printing and hybrid robot projects from 6 studies and 3D-printing social robot auxiliaries from 36 studies. In total, we have critically analyzed 61 articles to reach proper convincing conclusions and recommendations that can provide helpful references to other robotics researchers and practitioners in building modern social robots with 3D technologies.

1.2 Research Methodology

Two major literature databases, Web of Science and Scopus were selected for screening literatures of 3D printing in modern social robots. Each index term was searched over both databases using the flow of research methodology represented in Fig. 1. First, we have searched the respective index term aligned with the objective, i.e. to evaluate the role of available 3D-technologies for developing modern social robots in humanoid robotics. After that we have excluded total results with less relevance databases/disciplines, less relevance document types and finally results are sorted out by relevance for further analysis and inclusion.

- Search Technique

We searched the literature using Scopus and Web of Science databases/disciplines. A combination of relevant index terms was used to collect articles. Our search involves relevant printing till Jan., 2020. Figure 1, represents our total strategy of search and for reference organization, we used Mendeley developed by Elsevier.

- Research Criteria

Less relevance databases/disciplines were removed then, the name and extract of all publication were taken into consideration to get the desired criteria for incorporation. The criteria for incorporation are:

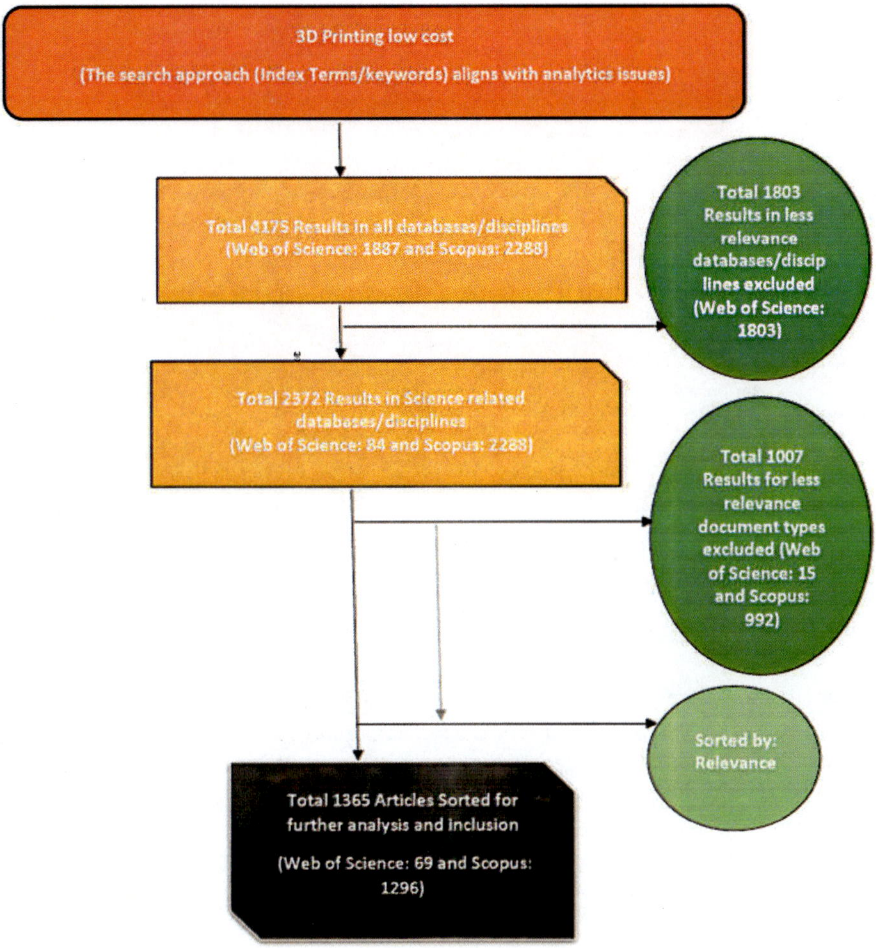

Fig. 1 Flow diagram of the research methodology

- *Articles: less relevance document types were excluded from the review, i.e., only articles are considered.*
- *Materiality: papers need to show primarily on the basis of search of index term that evaluates the importance accordingly.*
- *Communication: papers that are written or converted in English medium are considered for review.*

Where the results of the reviewed articles conflicted, rational for exclusion/inclusion was considered and for the justification of the materiality of article's, research methods and opening statements of paper's was evaluated. For all the included articles agreement on acceptability was attained. Using the corresponding inclusion criteria, included articles undergo a full text evaluation and for relevance cited reference in the review was also assessed.

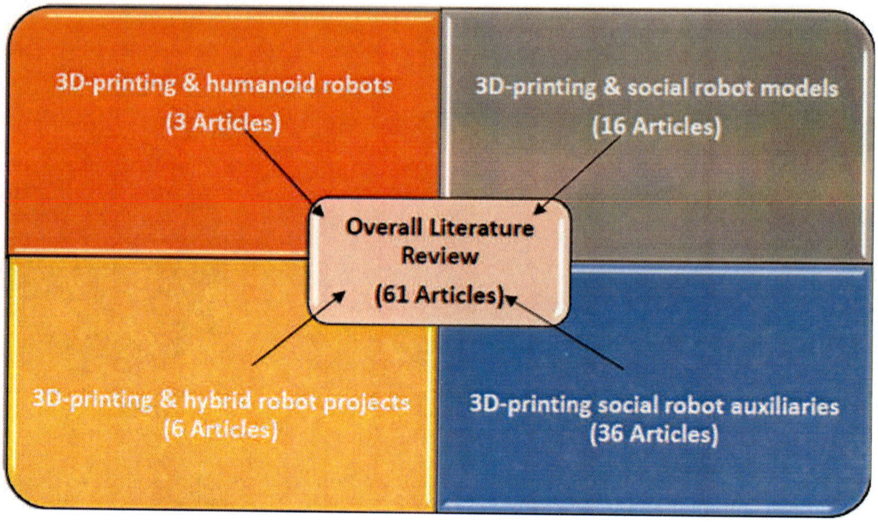

Fig. 2 The taxonomy of the systematic literature review

- Standard evaluation

Figure 1 shows the flow of research methodology that we have used to reach at relevant conclusions which also acts as a basis for quality evaluation. First, we started by searching the relevant index terms that aligns with the research problem. After that we have excluded the less relevance databases/disciplines and document types from the total results to get proper study base. Finally, we consolidated the papers by relevance for further analysis and inclusion based on the study criteria defined. Figure 2 depicts the overview of taxonomy of the systematic review.

2 The History of the Use of 3D-Printing in Modern Social Robots

2.1 3D-Printing and Humanoid Robots

In building a person i.e., human like robot, personhood criteria or thresholds that robotic agents must possess should be considered [4]. There is also a robot known as Reachy which assist in the study, design and evaluation of novel control mechanisms and interfaces on a human robot [33]. All these foster a critical theoretical ethical approach to social robotics “synthetic ethics” which aimed at motivating people to use social robots for two main purposes i.e., self-knowledge and moral growth [10] (Fig. 3).

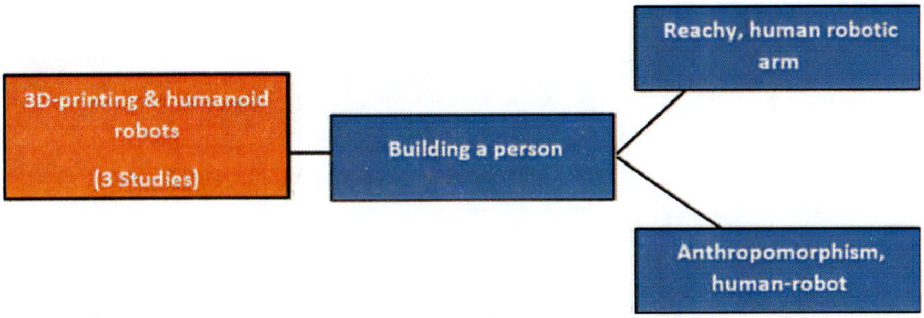

Fig. 3 Synopsis of the literature review for 3D-printing and humanoid robots from 3 studies

2.2 3D-Printing and Social Robot Models

There are several researches that enhances the existing knowledge of social robot models in terms of 3D-printing technologies, first the pedagogy behind the MIT's Beaver Works Summer Institute Robotics Program—a recent STEM robotics high school programme [21], second an approach that combines evolutionary robotics with 3D printing as a way to quickly and cheaply build autonomous mobile robots [48], third efforts at do-it-yourself to create a low-cost build framework for small robots [51] and lastly a low-cost, lightweight robotic instruction interface [36]. Wider availability of 3D printing has allowed small printable robots i.e., print bots to be directly integrated into the engineering courses. Print bots can be used in many ways to build lifelong learning skills, improve communication and promote cooperation and teamwork [2]. There are technologies that combines motion controller integration with image recognition in an opto-mechatronic system [29] and various frameworks that offer fundamental bases for future studies on what impacts market acceptance on AI robots as an emerging technology that can be applied to empirical experiments and analysis to include long-term approaches and practical tips for implementing and managing a variety of tasks [14]. In 3D printing, there is a multi-step correction algorithm based on model optimization and image correction by investigating the cause for color distortion [8] and there are various points to the relevance of spatial organization and coordination between the robot and the humans who interact with it [1]. There is an environment that centered on using robotics as a social dimension and as a tourism intermediary i.e., social robot Karotz [35]. Some authors sketched briefly the basics of a role-based approach to socio-technical innovation, and including examples of why a role-based approach could be useful for observation and interaction in social robotics [32] while some authors on the other hand examined how robots co-construct Japan's history [41]. In a real-world social robot model offers the social metaphors of personality, character, desires, and responsibilities that are interpreted and applied to promote the achievement of specific social goals [12]. There are helpful ways to develop an interdisciplinary and multi-theoretical approach to promote robotic architecture [53] and a type of robot-mediated communication that continues in the absence of potentially biased signs of identity-and explain how this

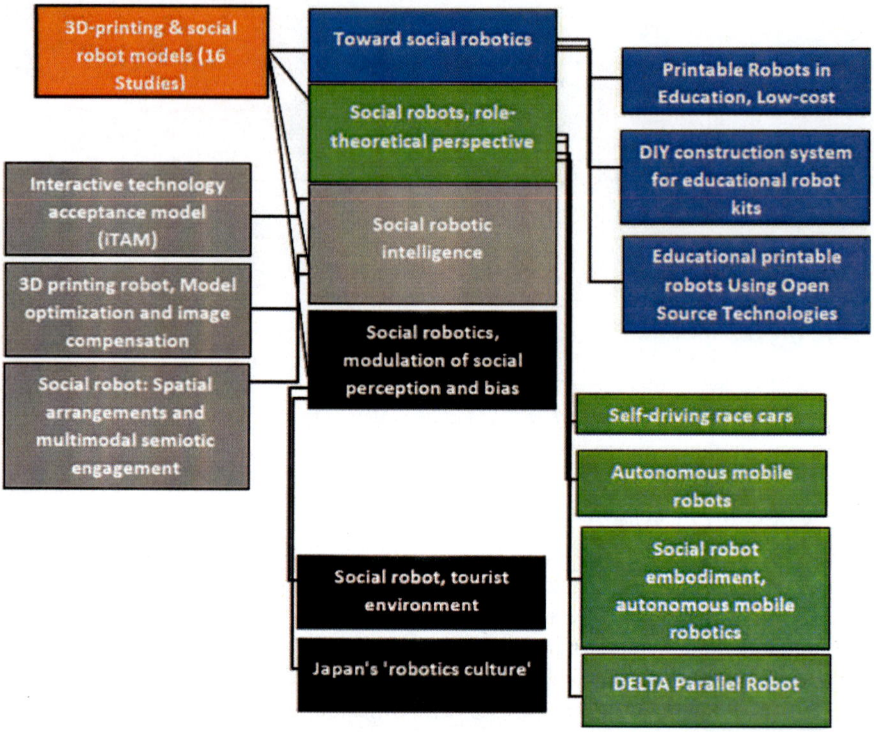

Fig. 4 Synopsis of the literature review for 3D-printing and social robot models from 16 studies

social robotics technology can be used to illuminate implicit bias in social cognition and inform new strategies for bias reduction [45]. There is a self-reference pairing of self-organizing processes as architectural basis for the autonomous collaborative systems as well [42] (Fig. 4).

2.3 3D-Printing and Hybrid Robot Projects

3D manufacturing using stereolithographic projection assisted by magnetic field technique [19] and effective optimization-based approach, producing stable, randomly defined motions for legged robots (Fig. 5).

The relation between morphological features and the resulting motions [31] i.e., the 3D printer, industrial robot, and computer vision combined to create a mobile prototype production line [17]. A scale walking robot [7], quadrupled robot [23] and flexible or partially flexible robot bodies which are capable of being more stable and adaptable to human activity and safer than traditional or rigid robots [5].

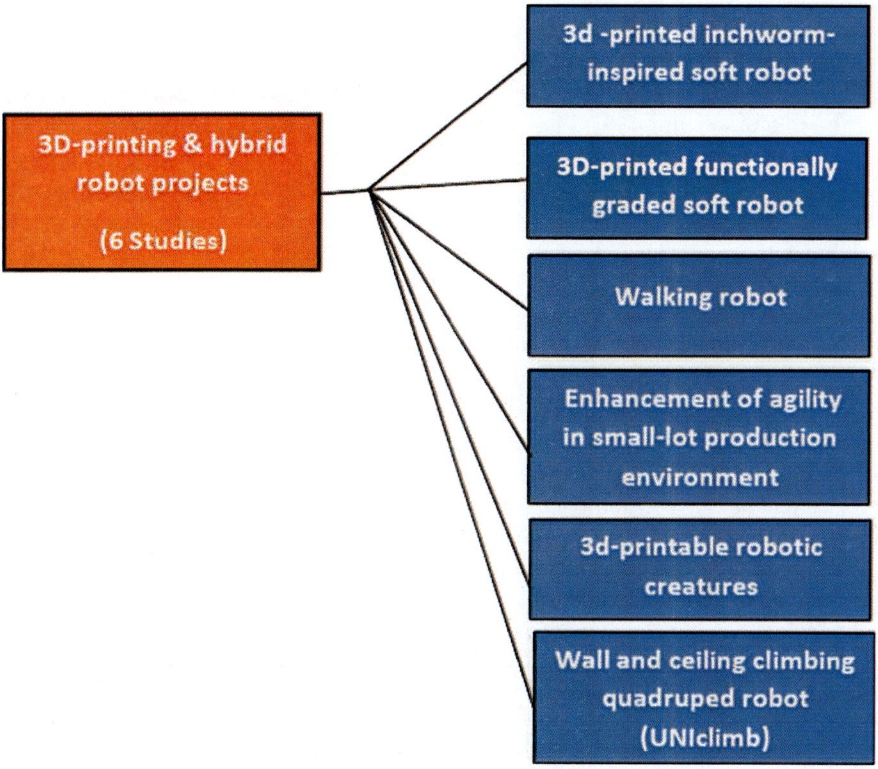

Fig. 5 Synopsis of the literature review for 3D-printing and hybrid robot projects from 6 studies

2.4 *3D-Printing Social Robot Auxiliaries*

A computationally efficient heuristic search algorithm for finding fast routes and overheads in the printing process [13] combined with cloud systems and cloud-enabled databases which would allow the production of large quantities of virtual products [18]. For the systematic and efficient customization of material formulation and different processing parameters using digital light processing-based 3D printing [60], a new concept of micromotors carried by small millimeter-sized engines for long distance crossing, and a wide variety of applications [24] was developed. Topology optimization along with four-dimensional printing is an important digital method that can be used to achieve optimum internal architecture for efficient porous soft actuators output [61]. Balloons provide anchorage into the colonic wall for a bio-inspired inchworm locomotion [30] and there is an innovative way for robotic oil recycling [28] which will act as an important part in the 3D printing process based on carbomer rheology limiter for quick ink writing of different useful hydrogels. In addition, this process unlocks new paths with its unparalleled versatility for the manufacturing of bioprinting and engineered hydrogel products [9]. The combination of 3D printing technology with shape memory apps enables smart devices with extremely complex 3D structures to be created, enabling an arbitrary transition between permanent and temporary 3D forms [59]. Not only does one-step manufacturing shorten the production cycle but it also narrows tool differences and enables large-scale production [15]. The form organizing plan that give a coherent method to completely exploit the ability i.e., to evolve a broad scale of actuators that are soft and useful for bionics, soft robotics and curative supervision [37]. The metamaterial absorber combined with 3D printing technology i.e., assisted by a swastika symbol to minimize footprint size [22] is also helpful and the same goes to truncated spherical cone structure as well [57].

The growing use of 3D printing has opened the door to low cost goods produced by individual consumers [54] which helps in creating the flexibility and low cost that allow for a fast turnaround [40]. It facilitates the development of a low cost, 3D patch antenna printing equipment [20] in that calculation of construction costs in terms of both the production property and goods on-site and offsite was also helpful [56], even the 'carbomorph' can be used to print electronic sensors capable of sensing mechanical flexing and ability changes [27]. The properties of food printing and food ingredient materials which can be used to design the 3D food matrix for developing a food production system [55] with the combination of a single, multi-scale, 3D printing technique based on multi-material nozzle, combining electrohydrodynamic jet printing with active multi-material mixing nozzles [26] would be very helpful. Even the manufacturing theory [47] and topics like gadgets that help you track missing objects, clever downside cities and soldiers mix together with their robots [16] will increase the demand for 3d printing as well. There are various examples like a robot machining tool to machine 3D printed objects like five degree-free serial robot arms were attached to the spindle frame [44].

In a large-scale 3D printing system, that consists of many collaborative robots [43], a rapid prototyping method, which can generate scaled prototypes for experimental validations from the early stages of robot creation [6] was very helpful. Optimization steps in the printing cycle [38] that uses lots of mobile robots [58] which utilizes a 6° liberty wire draped robot for arranging, accompanied by polyurethane foam as the item matter and shaving foam as the hold-up matter [3] was also helpful. Somehow it contributes to the management of hospitality and tourism [49] and introducing ever broader front-of-house restaurant service automation systems that requires a cross-cultural study of employee positions in the context of robotic operation [50]. The effect of service robot features on client hospitality experience has been examined widely in literature from a relationship building perspective [39] but there are various other factors that drive the development of service robots and applied a strategic perspective to the hotel industry regarding service innovation [25]. By using social robots in hospitality, the facilities are conceptualized and empirically evaluated through structural equation modelling and semi-structured manager interviews [11]. In the last part, it helps in developing the tourism or hospitality consequences of robonomics i.e., the positive and negative impacts of robots on tourism and vice versa [52] which helps in fostering robot capabilities that influence anthropomorphism [34]. Finally, the importance of pioneering inventions i.e., 3D printers and their impact on society and culture to the field of robotics science [46] are indeed novel (Fig. 6).

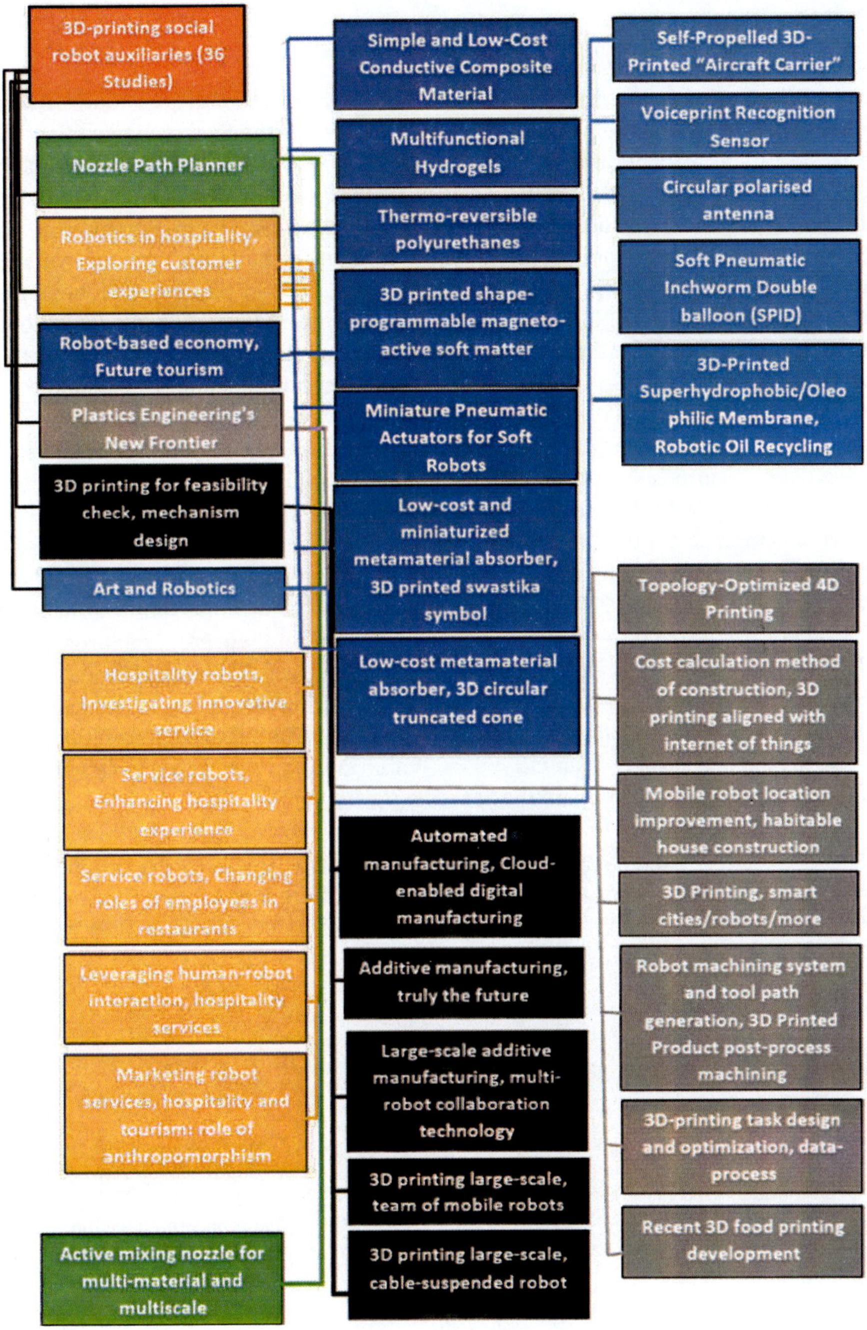

Fig. 6 Synopsis of the literature review for 3D-printing social robot auxiliaries from 36 studies

3 Conclusion

We have reached at the relevant insights by analyzing 61 studies in four different dimensions of taxonomy for humanoid robotics through the lens of 3D-printing. We analyzed all these four dimensions and divided them into 4 different spheres of studies aligned to the research problem i.e., to evaluate the role of available 3D-technologies for developing modern social robots in humanoid robotics. The first dimension, i.e., 3D-printing and humanoid robots, we find out that for building a person we need to follow specific measures which includes other researches like human robotic arm and Anthropomorphism. In the second dimension, i.e., 3D-printing and social robot models, we find out that there are four approaches to deal with social robot models with 3D-printing, first is toward social robotics, second is social robots—role theoretical perspective, third is social robotic intelligence and fourth is social robotics—modulation of social perception and bias. Within these four broad approaches we have defined various social robot models (Fig. 4). In the third dimension, i.e., 3D-printing and hybrid robot projects, we have discussed six projects - 3d printed inchworm inspired soft robot, 3d printed functionally graded soft robot, walking robot, enhancement of agility in small lot production environment, 3D printable robotic creatures and wall and ceiling climbing quadruped robot. In the fourth dimension, i.e., 3D-printing social robot auxiliaries, we find out that there are six approaches to deal with social robot auxiliaries with 3D-printing, first is nozzle path planner, second is robotics in hospitality—exploring customer experiences, robot - based economy: future tourism, Plastics Engineering's New Frontier, 3D printing for feasibility check - mechanism design and Art and Robotics. Within these six broad approaches we have defined various social robot auxiliaries (Fig. 6). By combining all these dimensions together, we will be able to make our own complete 3D printed humanoid robot that follows the principles of social robotics. Future research will revolve around these definitions for the applied role of 3D technology in modern social robotics and to develop a low-cost prototype of 3D printed humanoid robot, which will be reported in another paper.

References

1. Alač M, Movellan J, Tanaka F (2011) When a robot is social: spatial arrangements and multimodal semiotic engagement in the practice of social robotics. Soc Stud Sci 41(6):893–926. https://doi.org/10.1177/0306312711420565
2. Armesto L, Fuentes-Durá P, Perry D (2016) Low-cost printable robots in education. J Intell Robot Syst 81(1):5–24. https://doi.org/10.1007/s10846-015-0199-x (Springer Science and Business Media LLC)
3. Barnett E, Gosselin C (2015) Large-scale 3D printing with a cable-suspended robot. Add Manuf 7:27–44. https://doi.org/10.1016/j.addma.2015.05.001. (Elsevier B.V.)
4. Barresi J (2019) On building a person: benchmarks for robotic personhood. J Exp Theoret Artif Intell. https://doi.org/10.1080/0952813x.2019.1653386. (Taylor and Francis Ltd.)

5. Bartlett NW et al (2015) A 3D-printed, functionally graded soft robot powered by combustion. Science 349(6244):161–165. https://doi.org/10.1126/science.aab0129 (American Association for the Advancement of Science)
6. Cafolla D et al (2016) 3D printing for feasibility check of mechanism design. Int J Mech Control 17(1):3–12 (Levrotto and Bella)
7. Chavdarov IN (2016) Walking robot realized through 3D printing. C R L Acad Sci 69(8):1057–1064 (Academic Publishing House)
8. Chen X et al (2012) 3D printing robot: model optimization and image compensation. J Control Theory Appl 10(3):273–279. https://doi.org/10.1007/s11768-012-1122-7
9. Chen Z et al (2019) 3D Printing of multifunctional hydrogels. Adv Funct Mater 29(20). https://doi.org/10.1002/adfm.201900971. (Wiley-VCH Verlag)
10. Damiano L, Dumouchel P (2018) Anthropomorphism in human-robot co-evolution. Front Psychol 9(MAR). https://doi.org/10.3389/fpsyg.2018.00468. (Frontiers Media S.A.)
11. de Kervenoael R et al (2020) Leveraging human-robot interaction in hospitality services: incorporating the role of perceived value, empathy, and information sharing into visitors' intentions to use social robots. Tourism Manag 78. https://doi.org/10.1016/j.tourman.2019.104042. (Elsevier Ltd)
12. Duffy B (2004) Robots social embodiment in autonomous mobile robotics. Int J Adv Rob Syst 1(1):155–170. https://doi.org/10.5772/5632
13. Fok KY et al (2019) A Nozzle path planner for 3D printing applications. IEEE Trans Ind Inf. https://doi.org/10.1109/TII.2019.2962241. (IEEE Computer Society)
14. Go H, Kang M, Suh SBC (2020) Machine learning of robots in tourism and hospitality: interactive technology acceptance model (iTAM)—cutting edge. Tourism Rev. Emerald Group Publishing Ltd. https://doi.org/10.1108/tr-02-2019-0062. (Emerald Group Publishing Ltd)
15. Guo R. et al (2020) A voiceprint recognition sensor based on a fully 3D-Printed triboelectric nanogenerator via a one-step molding route. Adv Eng Mater. https://doi.org/10.1002/adem.201901560. (Wiley-VCH Verlag)
16. Hong J, Baker M (2014) 3D printing, smart cities, robots, and more. IEEE Pervasive Comput Inst 13(1):6–9. https://doi.org/10.1109/mprv.2014.1. (Institute of Electrical and Electronics Engineers Inc.)
17. Hu F et al (2016) Enhancement of agility in small-lot production environment using 3D printer, industrial robot and machine vision. Int J Simul Syst Sci Technol 17(43):32.1–32.6. https://doi.org/10.5013/ijssst.a.17.43.32. (UK Simulation Society)
18. Huang CH (2015) Continued evolution of automated manufacturing—Cloud-enabled digital manufacturing. Int J Autom Smart Technol 5(1):2–5. https://doi.org/10.5875/ausmt.v5i1.861 (Chinese Institute of Automation Engineers)
19. Joyee EB, Pan Y (2019) A fully three-dimensional printed inchworm-inspired soft robot with magnetic actuation. Soft Roboti Mary Ann Liebert Inc 6(3):333–345. https://doi.org/10.1089/soro.2018.0082. (Mary Ann Liebert Inc.)
20. Jun S et al (2017) Circular polarised antenna fabricated with low-cost 3D and inkjet printing equipment. Electron Lett 53(6):370–371. https://doi.org/10.1049/el.2016.4605 (Institution of Engineering and Technology)
21. Karaman S et al (2017) Project-based, collaborative, algorithmic robotics for high school students: programming self-driving race cars at MIT. In: ISEC 2017—proceedings of the 7th IEEE integrated STEM education conference, pp 195–203. Institute of Electrical and Electronics Engineers Inc. https://doi.org/10.1109/isecon.2017.7910242
22. Kim M et al. (2019) Low-cost and miniaturized metamaterial absorber using 3D printed swastika symbol. Microwave Opt Technol Lett. https://doi.org/10.1002/mop.32221. (Wiley)
23. Ko H, Yi H, Jeong HE (2017) Wall and ceiling climbing quadruped robot with superior water repellency manufactured using 3D printing (UNIclimb). Int J Precis Eng Manuf Green Technol 4(3):273–280. https://doi.org/10.1007/s40684-017-0033-y (Korean Society for Precision Engineering)
24. Kong L et al (2019) Self-Propelled 3D-printed "Aircraft Carrier" of light-powered smart micromachines for large-volume nitroaromatic explosives removal. Adv Funct Mater 29(39). https://doi.org/10.1002/adfm.201903872. (Wiley-VCH)

25. Kuo CM, Chen LC, Tseng CY (2017) Investigating an innovative service with hospitality robots. Int J Contemp Hospitality Manag 29(5):1305–1321. https://doi.org/10.1108/IJCHM-08-2015-0414 (Emerald Group Publishing Ltd.)
26. Lan H (2017) Active mixing nozzle for multimaterial and multiscale three-dimensional printing. J Micro Nano-Manuf 5(4). https://doi.org/10.1115/1.4037831. (American Society of Mechanical Engineers (ASME))
27. Leigh SJ et al (2012) A simple, low-cost conductive composite material for 3D printing of electronic sensors. PLoS ONE 7(11). https://doi.org/10.1371/journal.pone.0049365
28. Li G et al (2019) All 3D-Printed superhydrophobic/oleophilic membrane for robotic oil recycling. Adv Mater Interfaces 6(18). https://doi.org/10.1002/admi.201900874. (Wiley-VCH)
29. Lin J, Luo CH, Lin KH (2015) Design and implementation of a new delta parallel robot in robotics competitions. Int J Adv Robot Syst 12(10). https://doi.org/10.5772/61744. (SAGE Publications Inc.)
30. Manfredi L et al. (2019) A soft pneumatic inchworm double balloon (SPID) for colonoscopy. Sci Reports 9(1). https://doi.org/10.1038/s41598-019-47320-3. (Nature Publishing Group)
31. Megaro V et al (2015) Interactive design of 3d-printable robotic creatures. In: ACM transactions on graphics. Association for computing machinery. https://doi.org/10.1145/2816795.2818137
32. Meister M, and Schulz-Schaeffer I (2016) Investigating and designing social robots from a role-theoretical perspective: response to "social interaction with robots—three questions". AI Soc 31(4):581–585. https://doi.org/10.1007/s00146-015-0635-2. (Gesa Lindemann (this volume), Springer, London Ltd)
33. Mick S et al (2019) Reachy, a 3D-printed human-like robotic arm as a testbed for human-robot control strategies. Front Neurorobotics 13. https://doi.org/10.3389/fnbot.2019.00065. (Frontiers Media S.A.)
34. Murphy J, Gretzel U, Pesonen J (2019) Marketing robot services in hospitality and tourism: the role of anthropomorphism. J Travel Tourism Mark 36(7):784–795. https://doi.org/10.1080/10548408.2019.1571983 (Routledge)
35. Nieto D et al (2014) A social robot in a tourist environment. Lecture Notes in Computer Science (including subseries Lecture Notes in Artificial Intelligence and Lecture Notes in Bioinformatics). Springer, vol 8867, pp 21–24. https://doi.org/10.1007/978-3-319-13102-3_5
36. Pérula-Martínez R et al (2016) 'Developing Educational Printable Robots to Motivate University Students Using Open Source Technologies. J Intell Robot Syst 81(1):25–39. https://doi.org/10.1007/s10846-015-0205-3 (Springer Science and Business Media LLC)
37. Qi S et al (2020) 3D printed shape-programmable magneto-active soft matter for biomimetic applications. Compos Sci Technol 188. https://doi.org/10.1016/j.compscitech.2019.107973. (Elsevier Ltd)
38. Qian H et al (2013) Design and optimization of 3D-printing task based on data-process for printing robot. Jiqiren Robot 35(6):678–685. https://doi.org/10.3724/SP.J.1218.2013.00678
39. Qiu H et al (2019) Enhancing hospitality experience with service robots: the mediating role of rapport building. J Hospitality Mark Manag. https://doi.org/10.1080/19368623.2019.1645073. (Routledge)
40. Romeo J (2019) Plastics engineering's new frontier: embracing the brave new world of 3D printing. Plastics Eng 75(1):22–27. https://doi.org/10.1002/peng.20055. (Wiley-Blackwell)
41. Šabanović S (2014) Inventing Japan's "robotics culture": the repeated assembly of science, technology, and culture in social robotics. Soc Stud Sci 44(3):342–367. https://doi.org/10.1177/0306312713509704. (Social Studies of Science. SAGE Publications Ltd)
42. Sekiyama K, Fukuda T (1999) Toward social robotics. Appl Artif Intell 13(3):213–238. https://doi.org/10.1080/088395199117405
43. Shen H, Pan L, Qian J (2019) Research on large-scale additive manufacturing based on multi-robot collaboration technology. Add Manuf 30. https://doi.org/10.1016/j.addma.2019.100906. (Elsevier B.V.)
44. Shim J et al (2019) Design of a robot machining system and tool path generation for post-process machining of 3D printed product. Trans Korean Soc Mech Eng A 43(5):359–364. https://doi.org/10.3795/KSME-A.2019.43.5.359

45. Skewes J, Amodio DM, Seibt J (2019) Social robotics and the modulation of social perception and bias. Philosophical Trans Royal Soc B: Biological Sci 374(1771). https://doi.org/10.1098/rstb.2018.0037. (Royal Society Publishing)
46. St-Onge D (2019) Robotic art comes to the engineering community (Art and Robotics). IEEE Robot Autom Mag 26(3):103–104. https://doi.org/10.1109/mra.2019.2927198. (Institute of Electrical and Electronics Engineers Inc.)
47. Subrin K et al (2018) Improvement of the mobile robot location dedicated for habitable house construction by 3D printing. IFAC-Papers OnLine 51(11):716–721. https://doi.org/10.1016/j.ifacol.2018.08.403. (Elsevier B.V.)
48. Teo J et al (2015) 'Evolutionary robotics +3D printing=rapid and low-cost deployment of autonomous mobile robots. ARPN J Eng Appl Sci 10(18):8372–8378 (Asian Research Publishing Network)
49. Tung VWS, Au N (2018) Exploring customer experiences with robotics in hospitality. Int J Contemp Hospitality Manag 30(7):2680–2697. https://doi.org/10.1108/IJCHM-06-2017-0322 (Emerald Group Publishing Ltd.)
50. Tuomi A, Tussyadiah I, Stienmetz J (2020) Service robots and the changing roles of employees in restaurants: a cross cultural study. e-Review Tourism Res 17(5):662–673 (Texas A and M University)
51. Vandevelde C et al (2016) 'Design and evaluation of a DIY construction system for educational robot kits. Int J Technol Des Educ 26(4):521–540. https://doi.org/10.1007/s10798-015-9324-1 (Springer Netherlands)
52. Webster C, Ivanov S (2019) Future tourism in a robot-based economy: a perspective article. Tourism Rev. https://doi.org/10.1108/tr-05-2019-0172. (Emerald Group Publishing Ltd)
53. Wiltshire TJ et al (2017) Enabling robotic social intelligence by engineering human social-cognitive mechanisms. Cogn Syst Res 43:190–207. https://doi.org/10.1016/j.cogsys.2016.09.005. (Elsevier B.V.)
54. Winkless L (2015) Is additive manufacturing truly the future? Metal Powder Report 70(5):229–232. https://doi.org/10.1016/j.mprp.2015.05.003 (Elsevier Ltd)
55. Yang F, Zhang M, Bhandari B (2017) Recent development in 3D food printing. Crit Rev Food Sci Nutr 57(14):3145–3153. https://doi.org/10.1080/10408398.2015.1094732. (Taylor and Francis Inc.)
56. Yang H et al (2018) The cost calculation method of construction 3D printing aligned with internet of things. Eurasip J Wirel Commun Netw 2018(1). https://doi.org/10.1186/s13638-018-1163-9. (Springer International Publishing)
57. Yoon Y et al (2018) Low-cost metamaterial absorber using three-dimensional circular truncated cone. Microwave Opt Technol Lett 60(7), pp. 1622–1630. https://doi.org/10.1002/mop.31211. (Wiley)
58. Zhang X et al (2018) Large-scale 3D printing by a team of mobile robots. Autom Constr 95:98–106. https://doi.org/10.1016/j.autcon.2018.08.004. (Elsevier B.V.)
59. Zhang Y et al (2019) 3D printing of thermoreversible polyurethanes with targeted shape memory and precise in situ self-healing properties. J Mater Chem A 7(12):6972–6984. https://doi.org/10.1039/c8ta12428k (Royal Society of Chemistry)
60. Zhang YF et al (2019) Miniature pneumatic actuators for soft robots by high-resolution multi-material 3D printing. Adv Mater Technol. Wiley-Blackwell 4(10). https://doi.org/10.1002/admt.201900427. (Wiley-Blackwell)
61. Zolfagharian A. et al (2019) Topology-optimized 4D printing of a soft actuator. Acta Mechanica Solida Sinica. https://doi.org/10.1007/s10338-019-00137-z. (Springer International Publishing)

A Survey on the Contributions of 3D Printing to Robotics Education—A Decade Review

Adamu Yusuf Abdullahi, Mukhtar Fatihu Hamza, and Abdulbasid Ismail Isa

Abstract There has been a constant rise in interest on the subjects of robotics and additive manufacturing. Although robotics being older than additive manufacturing, both subjects have gained increased publicity in terms of research over the years. The objective of this paper is to review 3D printing, robotics education, and finally how 3D printing has contributed to robotics education over a period of ten years between 2011 and 2020. the last ten (10) years. By restricting the search to academic articles only, papers comprising of journal papers and conference proceedings were reviewed. For each paper, three information were sought which includes; robotics education, 3D printing, and finally how these two have been used together. We expect that the result of this review will serve as a pivot for schools at all levels to combine both subject areas and include the union into their STEM curricula which is believed to improve robotics education. The review shows that both tertiary and lower educational levels of robotics education benefit immensely from 3D printing. We also find out that most of the contributions of 3D printing to robotics education comes more in the form of using already printed parts rather than using the 3D printer itself. The challenges and trends are highlighted and future research directions have been suggested in this paper.

Keywords Additive manufacturing · Robotics education · 3D printing · STEM

A. Y. Abdullahi · M. F. Hamza
Department of Mechanical Engineering, University of Malaya, 50603 Kuala Lumpur, Malaysia

Department of Mechatronics Engineering, Bayero University, P.M.B 3011, Kano, Nigeria

M. F. Hamza (✉) · A. I. Isa
Department of Electrical and Electronics Engineering, Usman Danfodiyo University, Sokoto, Nigeria
e-mail: mfhamza.mct@buk.edu.ng

A. F. Ab. Nasir et al. (eds.), *Recent Trends in Mechatronics Towards Industry 4.0*, Lecture Notes in Electrical Engineering 730, https://doi.org/10.1007/978-981-33-4597-3_27

1 Introduction

As the world drive towards a new age of technology particularly the industry 4.0 and IoT, the demand for competent engineers in terms of technical and communication skills engineers is rapidly on the increase [1–3]. However, the supply for such qualified engineers is falling short of the demand [4]. New devices, machines, software, and various technologies are being born on a daily basis, which places a huge responsibility on schools and vocational centres to train and produce engineers and technician to use these new technologies and also try to catch up the pace [5, 6].

One of the areas where the above mentioned training and development need urgent attention is in the field of robotics [7]. Robots have been touching human lives for almost six decades now since the first robot was put to service in 1961 at General Motors plant. These robots impacted our lives behind the scene as we only enjoyed the already made parts without knowing how they were made [8]. This was due to the fact that the first of the robots to be put to use were the industrial robots. In recent times, robots have become parts and parcel of our lives as they can be seen in almost works of lives ranging from industries, to domestic, to entertainment, and many other aspects of lives.

Additive manufacturing (AM) is building process where parts are generated given their computer-aided design (CAD) model or any form of AM file by successively adding material layer by layer [9]. AM has many advantages over the conventional traditional methods especially in the area of producing parts with complex geometry, material management and time-to-market reduction [9]. Even though there are some limitations that hinder the growth of AM such as low strength, accuracy and surface finish, it is still becoming widely accepted and used in the modern manufacturing industries. In recent times, AM is becoming widely known as 3D printing although there are other methods of AM.

Being a technique which was discovered for over two decades, AM is gaining greater attention in the field of manufacturing in the sense that it is no longer restricted to making functional prototypes nowadays but also building tools, jigs and fixtures, replacement parts, concept models, molds and castings, and other parts, which can serve as much as the previously made types. With the increasing popularity of 3D printers, so much attention is now focused on competences of 3D modelling techniques and other related techniques [2].

In the area of education, robots have been playing important roles to develop young competent engineers to fit in today's world of rapidly growing technology. With the increased interest and recent widespread of the Science, Technology, Engineering, and Mathematics (STEM) programs, robotics education being an important component of STEM is a promising tool to get students engage in the STEM programmes [10]. The inflow of researchers and students into robotics education over the past few years cannot be overemphasized. Quite a number of researches have shown that educating students of nowadays using robots will not only boost their interests of programming, computational thinking and robotics but also improve their science literacy. Robotics education can serve to enhance students in three major ways which are; being an object of study, being a tool for cognition, and be a means of teaching, development and upbringing of students [11]. In many countries, robot education is in the process of being incorporated into the school curricula so that students in preschool, primary school, high school, and college can get involved in robotics from an early stage in life.

What this review is trying to highlight is the importance of combining 3D printing (additive manufacturing) and robotics education to see how both knowledge can improve teaching and learning in STEM education, particularly robotics education. It is believed that this will help tremendously if incorporated into the curricula of robotics education at all levels of learning. The contributions this review will avail the readers include; (1) The pedagogical level that makes most use of 3D printers in Education, (2) Some countries of active robotics education activities, (3) The nature of the 3D printed parts used in the delivering robotics education and (4) Challenges and future trend.

The paper starts in section two by discussing robotics education as a whole, then as applied in pedagogical setting, and then regions of active robotics education involvement. 3D printing is explained in section three focusing on its usage in education and highlighting some printed parts. The contributions of 3D printing to robotic education is presented in section four including challenges and future trends. The conclusion comes in section five.

2 Robot Education

Robotics being an exceptional engineering science that covers the design, modelling, and controlling of robots, is swiftly taking charge of the routine of peoples' daily lives today [12]. Robots these days go from industrial to education, to medical sciences, to military, to space exploration, and other numerous fields. Advancements in the field of robotics place high demand of well trained and skilful graduates who will handle such technologies [1].

The introduction of robotics into the educational system has really gone a long way to show how students can learn faster when they use an actual tool to support the theories they learn in class. However, it is at its initial stage of development and the application it yet to be fully efficient [11]. Arguing that interdisciplinary projects should be given great attention in robotics education, they presented three objectives that needs to be met to prepare modern graduates for the future. [7] believe learning robotics through competitions for students, and training and workshops for teachers, school children will develop critical thinking skills which is vital for the future generation of scientists and engineers. This will encourage students to take up robotics courses they believe. In order to fully engage students of science, technology, engineering, and mathematics (STEM), robotics education has shown great prospects in attracting the interest of researchers, teachers, and students alike from kindergarten up to university level over the past few years [10]. The review which was aimed at K-12 subjects checks how rich in robotics content the curricula was for teaching and learning. Some studies have suggested that using robots in schools could improve the science literacy in students [13]. Robotics is also believed to created opportunities for primary school pupils to develop language skills in addition to the technical skills [14]. Opinions of [15, 16] show particularly that the interest and self-efficacy of girl students will be boosted through robotic education. The studies of [17, 18] stresses the potentials of using robotics as an educational tool. This will help students get a better understanding of STEM concepts as seen in their results. At higher levels of learning, [19] suggests that teaching activities be organized to include basic knowledge of robotics and advanced skills of automation and control software. In another work [1], a robotic framework meant for training, vocational, and academic purposes was presented. The framework aids the proper understanding of robot manipulators. Moreover, FASTBot robot trainer was used as a tool to create awareness for school children and to also train them components of robotics like sensors, Mechatronics, embedded systems etc. [20]. This they believe will guide the students in choosing a career path. A description of how the field of wheel mobile robots have evolved is given in [21], targeted on the masters programme offered in their university. They argued that inductive learning (which involves giving students a particular task and allowing them freedom to seek for answers) aids better retention of what is taught as against the traditional learning (which involves teaching the basics before giving students tasks to perform). Through project-based learning, students have acquired working skills, robot design knowledge, and application of theoretical knowledge [22].

There are so many ways robots can benefit the educational system. Yet, it still remains a mystery to many researchers and educators how these robots can actually be of benefit to the system [6]. A study carried out by [6] highlights the essential areas where robots can be of benefit to the educational system. Five essentials areas where robots should be deployed and used were discovered by [6]. In this paper, the advantage of using 3D printing to enhance robotics education is going to be discussed.

2.1 Robotics Education in Pedagogical Settings

There are three major areas where robots can serve in the education of students: as a tool for cognition, as an object of study, and as a means of developing and tutoring [11]. The technical creativity of students is being developed by involving robots in the educational system. There is more to robotics than just robotics as pointed out by [23], they believe robotics is more about creativity and challenging one's creative ability. The technical environment of the modern world is being transformed through innovation, thus these technical changes needs to be visible in student educational content [11]. The review [10] also supports the fact that robotics education can improve students' STEM knowledge, skills, and attitudes. Furthermore, [14] have registered improvement of teachers and primary school pupils after implementing the robotics course in the curricula. Curiosity by pupils coupled with enthusiasm of teachers has

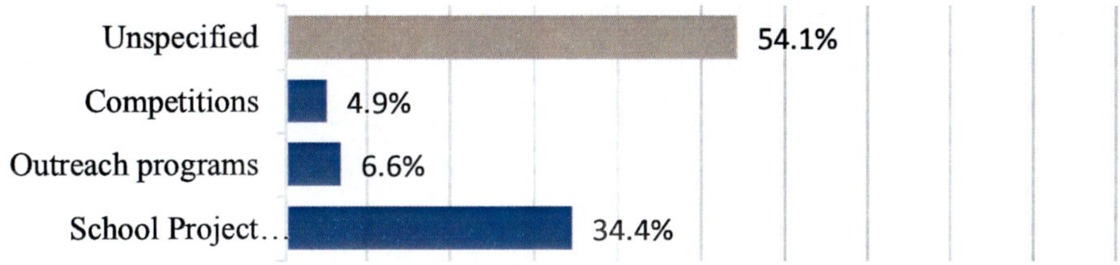

Fig. 1 Nature of activities of robotics education

made them work together, build and program robots in groups, and develop problem solving skills. This is mostly achieved through project-based learning and competitions. Increasing the number of experiment and the duration for these experiment was the proposition made by [10] in order to help students retain and easily remember what they have learnt from the robotics class. Based on experience, [22] stated that robot competitions organized for students places more demands on the teachers than on the students. However, [21] argued that when students are taking several courses together with the project-based learning, the project might become boring due to time constraints. The reasons for teachers' apathy towards robotics education have been registered in [3]. They proffer suggestions to boost teachers' enthusiasm. Also, [22] suggested that students should not be allowed too much freedom as this lowers motivation and decreases learning experience due to disorientation. Figure 1 gives the nature of the types of activities employed in robotics education. It can be seen that school project based works have the highest patronage with 34.4%. The unspecified bar are articles that have not mentioned the types of activities involved in the robotics education.

2.2 *Some Regions of Active Robotic Education*

Looking at Russia, [11] stated that the interest and creativity of students learning robotics has been on the increase since 2009 even though robotics education has been on for two decades now. Several published works of teachers, research specialists, engineers, and pedagogues of education all show how robotics education has been engulfed in the school education. In Finland, [22] through a field robot competitions have shown how robots education can be enhanced through competitions. With detailed specifications to work on agricultural fields, students' teams from different schools have built robots from scratch to carry out field operations. Another research which supports the view that competition plays an important role in robotics education is [24]. At the university of Bologna, Masters students taking automation were introduced to learning how to write efficient code and also gain knowledge of mobile robots using LEGO mindstorms Kits [19]. Collaborations between universities and lower schools in Russia have been reported [25, 26]. Summer programmes such are Robocamp have been reported to supplement the robotics knowledge taught in

Table 1 Some countries with robotics education activities based on reviewed papers

Countries	Source (s)
Greece	[27]
Russia	[11, 25, 26, 28]
Finland	[22]
Bologna	[19]
Malaysia	[7]
Japan	[24]
Italy	[14]
USA	[29–31]
China	[32]
Brazil	[33]

schools [25]. In Italy, robotics is being introduced into the primary curricula [14] which give the teachers the opportunity to be trained. Malaysia is not left out in the interest burst of robotics education as a government sponsored program developed a curriculum to include robotics into the Malaysian schools [7]. Table 1 give a brief summary of countries that have been active in robotics education activities. This is based on the reviewed papers.

3 3D Printing

Additive manufacturing as described earlier is a technology that is gaining wide acceptance in the manufacturing industry. Other areas that are recently making heavy use of additive manufacturing are the medical [34] and education [5] industries. New applications are seriously taking advantage of this technology. Advantages of additive manufacturing over the traditional manufacturing can be found in [35]. Meanwhile, the pace at which these applications are growing creates a rather wide gap between the education and skills development, as shown by several published work on additive manufacturing and 3D printing. This gap could hinder adoption if the 3D printing technology. Although there are few published literatures to affirm the foregoing claim, [5] produced a rich review to elicit where and how 3D Printing is being used and can be used for teaching and learning. 3D printing comes with numerous advantages such as complex parts creation, among others which makes it suitable for the production of robots parts. However, it comes with some disadvantages of which the major ones are poor surface finish due to stepping and low dimensional accuracy. [9] proposed a hybrid additive-subtractive manufacturing process which mitigates the disadvantages. A schematic diagram of the applications of 3D printing in education is provided in Fig. 2.

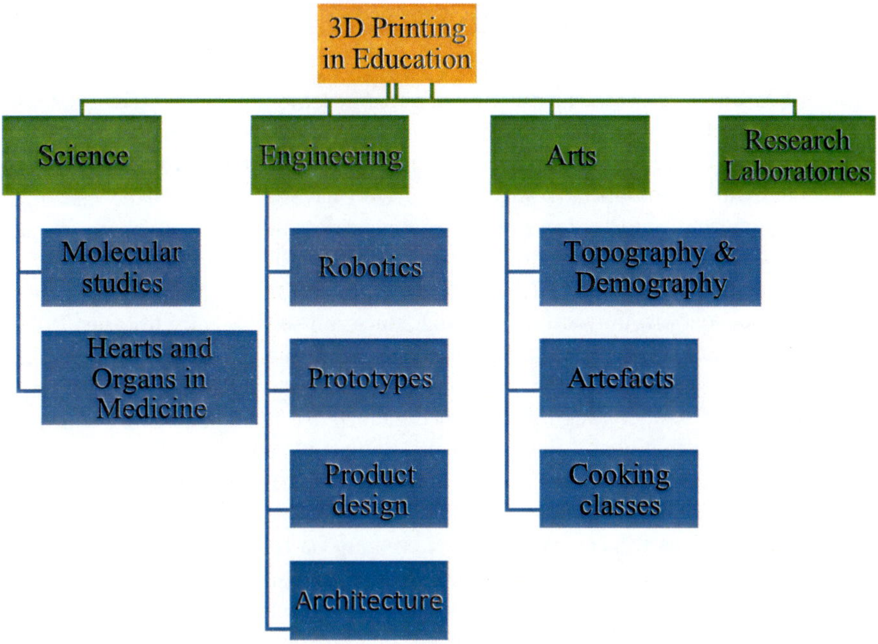

Fig. 2 Schematic diagram of the applications of 3D printing in education

3.1 3D Printed Robot Parts

Varieties of objects have been 3D printed ever since the technology was born and among those is the Poppy project [36] being the first humanoid robot to be 3D printed. Other projects are in the form of Industrial robots, artefacts, skeletons, domestic objects to mention a few. The robot arm and the gripper designs in [37] also took advantage and made heavy use of additive manufacturing technique (3D printing). Another form of 3D printing using inkjet technology was implored by [38] to design a soft gripper which will almost serve as a universal gripper. By limiting to a small budget, [39] designed and manufactured a lightweight robotic arm using a 3D printing technology. A full but small scale industrial robotic/robotic arm was built using the this new manufacturing technique [40]. Design education have also enjoyed the benefits of using 3D printers as presented in [41, 42].

Some of the disadvantages of 3D printing include health and safety of the operator due to the fumes produced by the filaments (with that of ABS being more than PLA), high cost of consumables for those with limited budget [5, 43]. Nowadays, 3D printers with superior qualities are available at affordable prices, likewise the consumables.

3.2 Some Contributions of 3D Printing to Education

Quite a number of researches have demonstrated how the 3D printing technology has benefited the educational system. Students' interest and engagement have increased, creativities are being inspired, skills are being developed, and learning is becoming more interesting. Feedbacks from students who have experienced the use of 3D printers supports the foregoing claims. Enthusiastic students take photos of made parts to show friends and parents which gets other kids motivated and join the 3D printing class. The four pedagogical environments where 3D printing is being used include; schools, universities, special education settings, and libraries [5]. 3D printing has chiefly transformed the STEM and technical education in different ways. Teaching materials produced using this technique has proven to facilitate learning in various fields of science ranging from medical science to engineering. 3D printing has improved the spatial abilities of college students [2] through a CDIO (conceive, design, implement, and operate) approach. Even though it comes with teaching and learning frustrations, 3D printing technology has found its way into the world special education as it has served well in creating parts used for cognitive, visual and motor experiments [27]. As suggested by [29], inputting the 3D printing into school curriculum will encourage innovation, experimentations, entrepreneurship, facilitation of multi-disciplinary approaches, and integration of technical knowledge.

Schools in general have gotten their fair share of the 3D printing in the form different programs including camps, outreach lessons, and workshops. Design and 3D printing of engineering projects involving have added value to students' understanding of STEM subjects and also creativity [32, 44, 45]. Outreach activities ranging from days to weeks which brings together students from all educational levels have been reported in many publications. 3D printed modular robots used during a student camp for middle school students have been found to boost their confidence in the use of robots and computers [29, 44]. To reduce the overdependence on outreach programmes, [4] have proposed training workshops for educators in order to prepare school teachers for the 3D printing technology.

The universities have also adopted 3D printing which is visible in quite a number of courses. [33, 46] have both shown how 3D printing included in courses have encouraged entrepreneurship, product innovation, and creative experimentation. Mechatronics projects involving students building a RepRap 3D printer have been carried out [47], while it has been added to mechatronics programme in [48]. In some cases, 3D printing has been used by both undergraduate and graduate students of science and engineering to carryout projects [30, 31, 49, 50]. Nonscience and business students have also benefitted from this technology. Artefacts, bones, and others printed specimens are cheaper than the real ones and be studied just like the original ones. [5] concluded that the universities are the highest users of the 3D printing technologies and promising fields are spanning from it. Figure 3 gives a summary of levels at which all stages of education benefit form 3D printing in their robotics education. Schools here includes all levels of learning below the tertiary institutions. It is seen

Fig. 3 3D printing in pedagogical level

that the robotics education in tertiary institutions benefit more from 3D printing as compared to all other levels of learning.

On a final note, additive manufacturing is not expected to fully replace the existing traditional manufacturing methods. However, it will help revolutionize the manufacturing industry when it is coupled with the traditional method [35].

4 Contributions of 3D Printing to Robotics Education

Due to the novelty of this technology, the knowledge gap between the teachers and students is lean, which leaves the duo at the mercy of internet sources which are more up-to-date and accurate than the very few publications [46]. In teaching robotics, 3D printers have served well as they have been wonderfully used to produce low cost educational robots [51, 52]. Students have had the opportunity to modify robotic designs and have printed them using the 3D printers. It is stated [5] that the areas that make heavy use of the technology are Mechatronics engineering and STEM subjects. In projects, Mechatronics students make use of 3D printing to build part or the entire chassis of their robots [53]. Students also showcase their ideas by modifying and sharing their robots design among friends and course mates. Researches have reported where students use 3D printer in robotics projects [53–55]. While others have reported where 3D printed robots such as Printbots [50, 56] and other robot parts have been used in robotics projects [51, 52, 57–59]. Other researches [42, 60] have presented the robots that have been designed and/or fabricated using 3D printers. However, in other works [61, 62], students have developed 3D printers as projects. Figures 4 provides a summary of the direct usage of 3D printers or 3D printed parts in robotics education. It is seen that 42.6% of the projects make use of parts printed from 3D printer rather than using the printer to print the robots. Others did not specify what method was used.

Figure 5 gives a summary of the kinds of printed parts used in the delivering robotics education. Based on the reviewed papers, it is seen that most of the projects did not specify the kinds of parts used or did not mention the part at all. While only

Fig. 4 The nature of application of 3D printing in robotics education

Fig. 5 Types of 3D printed parts used in robotics education

4.9% of the articles show a manipulator, 42.6% have used different parts ranging from printbots, and other complementary parts.

As 3D printing creates an avenue to harmonize students from different technical background, educators find it challenging to handle the students. One of the methods that has proven successful in tackling this issue is grouping students of complementary backgrounds together. And this has generated projects with novel results.

4.1 *Challenges and Future Trend*

There has been an increase in workshops and summer camps being organized all over the globe to promote robotics educations across all levels of formal learning. However, very little of these activities is documented. As a result, there are few literatures that have been published in this regard [5]. On the part of contribution of 3D printing to robotics education, there have been a number of published articles as highlighted. However, researches have shown that teachers are apathetic to the idea of using 3D printers or 3D printed parts in robotics education. This could be as a result of the teachers having limited knowledge themselves on the technologies. There has been a steady rise in the subject of 3D printing over the past 10 years even though there was little work done in the early part of the chosen timeline for this review which is a decade. On the other hand, works on robotics and robotics education have been ongoing with most of the work done also in 2015 as seen from timelines in the reviewed articles. However only [29] talked about including the duo into the curriculum of schools for teaching robotics education. This review has provided an insight to how 3D printing has been contributing to robotics education. It is believed that if 3D printing is incorporated in the robotics education curricula, it will go a long way in improving teaching and learning of robotics education.

5 Conclusion

3D printing is a modern manufacturing technique which allows complex geometries to be produced from plastics by melting them to the required shape in a layer-wise manner. Due to its uniqueness, it can be a very promising tool for the field of robotics which will improve robotics education. With availability of affordable 3D printers, robot models can be printed and used as tools for learning in all levels of robotics education. Considering the fact that there are few articles that discusses the relationship between 3D printing and robotics education [5], this review has been able to point out the importance 3D printing has brought to robotics in order to enhance a better STEM education targeted at robotics teaching and learning. This paper presents a review on three subject areas which are; robotics education, 3D printing, and the contributions of the former to the latter. We find that robotics education in tertiary institutions benefit the highest from 3D printing. We also find that most of the projects made use of 3D printed parts more frequently than using the 3D printer to print their projects. With the advancements made in technology so far, there is an imminent need to have a combine curricula for 3D printing and robotics education which will not only improve the understanding of the both subjects, but will also breach the gap between the needed industrial skills and what is actually obtainable in the schools.

This we believe will tremendously improve robotics education if it is incorporated in the curriculum of all institutions of learning. This review will also provide a direction to guide future researches on other methods of additive manufacturing and robotics education.

Acknowledgements This work was supported by the Tertiary Education Trust Fund (TETFund), through Bayero University Kano, Nigeria.

References

1. Manzoor S et al (2014) An open-source multi-DOF articulated robotic educational platform for autonomous object manipulation. Robot Comput Integr Manuf 30(3):351–362
2. Huang T-C, Lin C-Y (2017) From 3D modeling to 3D printing: development of a differentiated spatial ability teaching model. Telematics Inform 34(2):604–613
3. Kim C et al (2015) Robotics to promote elementary education pre-service teachers' STEM engagement, learning, and teaching. Comput Educ 91:14–31

4. Timothy WS, Christopher BW, Michael H (2017) Preparing industry for additive manufacturing and its application: summary and recommendations from a national science foundation workshop. Add Manuf 13:166–178
5. Ford S, Minshall T (2019) Invited review article: Where and how 3D printing is used in teaching and education. Add Manuf 25:131–150
6. Cheng Y-W, Sun P-C, Chen N-S (2018) The essential applications of educational robot: requirement analysis from the perspectives of experts, researchers and instructors. Comput Educ 126:399–416
7. Ramli R, Yunus MM, Ishak NM (2011) Robotic teaching for Malaysian gifted enrichment program. Procedia Soc Behav Sci 15:2528–2532
8. Hagele M, Nilsson K, Pires JN (2008) Industrial robotics, in Springer handbook of robotics. Springer, Berlin, Heidelberg
9. Li L, Haghighi A, Yang Y (2018) A novel 6-axis hybrid additive-subtractive manufacturing process: design and case studies. J Manuf Process 33:150–160
10. Xia L, Zhong B (2018) A systematic review on teaching and learning robotics content knowledge in K-12. Comput Educ 127:267–282
11. Ospennikova E, Ershov M, Iljin I (2015) Educational robotics as an inovative educational technology. Procedia Soc Behav Sci 214:18–26
12. Raza K, Khan TA, Abbas N (2018) Kinematic analysis and geometrical improvement of an industrial robotic arm. J King Saud Univ Eng Sci 30(3):218–223
13. Shih BY, Chen TH, Wang SM, Chen CY (2013) The exploration of applying LEGO NXT in the situated science and technology learning. J Baltic Sci Educ 12(1):73–91 (2013)
14. Scaradozzi D et al (2015) Teaching robotics at the primary school: an innovative approach. Procedia Soc Behav Sci 174:3838–3846
15. Gomoll A et al (2016) Dragons, ladybugs, and softballs: Girls' STEM engagement with human-centered robotics. J Sci Educ Technol 25(6):899–914
16. Master A et al (2017) Programming experience promotes higher STEM motivation among first-grade girls. J Exp Child Psychol 160:92–106
17. Benitti FBV (2012) Exploring the educational potential of robotics in schools: a systematic review. Comput Educ 58(3):978–988
18. da Neto MBS, de Mendonça JMC, de Sena APC (2015) Development and control of a prototype manipulator SCARA type as teaching tool. IFAC-Papers OnLine 48(19):209–213 (2015)
19. Raffaele G, Riccardo F, Claudio M (2014) Robotic competitions—teaching robotics and real-time programming with LEGO mindstorms. In: International federation of automatic control, Cape Town, South Africa
20. Balaji M et al (2015) Robotic training to bridge school students with engineering. Procedia Comput Sci 76:27–33
21. Zdesar A, Saso B, Gregor K (2017) Engineering education in wheeled mobile robotics. In: International federation of automatic control. Elsevier, pp 12173–12178
22. Timo O et al. (2011) Robot competition as a teaching and learning platform, in international federation of automatic control. Milano, Italy
23. Linert J, Kopacek P (2016) Robots for education (edutainment). In: International federation of automatic control
24. Akagi T et al (2015) Systematic educational program for robotics and mechatronics engineering in OUS using robot competition. Procedia Comput Sci 76:2–8
25. Sergey F et al (2017) Teaching robotics in secondary school. In: International federation of automatic control, pp 12155–12160
26. Sergey F et al (2017) Teaching of robotics and control jointly in the University and in the high school based on LEGO mindstorm NXT. In: International federation of automatic control
27. Vasilis K, Vasilis N, Christos G (2014) Open source 3D printing as a means of learning: an educational experiment in two high school in greece. Telematics Inform 32(1):118–128
28. Sergey F et al. (2017) Teaching robotics in secondary school examples and outcomes. In: International federation of automatic control

29. Montironi MA, Eliahu DS, Cheng HH (2015) A robotics-based 3D modelling curriculum for K-12 education. In: ASEE annual conference and exposition, pp 1–14
30. Williams WB, Schaus EJ (2015) Additive manufacturing of robot components for a capstone senior design experience. In: ASEE Annual conference and exposition, American social for engineering education, Seatle, WA, pp 26.157.1–26.157.15
31. Willia FC et al (2016) MAKER: applications of 3D printing and laser cutting in the development of automous robotics. In: ASEE annual conferences and exposition. ASEE: New Orleans, LA
32. Xiwei L et al (2017) A new framework of science and technology innovation education for K-12 in Qingdao, China. In: ASEE international forum, ASEE: Columbus, Ohio
33. de Sampaio CP et al (2013) 3D printing in graphic design education: educational experiences using fused deposition modelling (FDM) in a Brazilian University. In: 6th International conferences advance research virtual rapid prototyping
34. Ismianti H (2020) Adoption of 3D printing in indonesia and prediction of its application. In: 2025 IOP Conference series: materials science and engineering IoP Publishing. In Press
35. Attaran M (2017) The rise of 3-D printing: the advantages of additive manufacturing over traditional manufacturing. Bus Horiz 60(5):677–688
36. Matthieu L et al (2014) Poppy project: open-source fabrication of 3D printed humanoid robot for science. Education and Art, Digital Intell
37. Barbieri L et al (2018) Design, prototyping and testing of a modular small-sized underwater robotic arm controlled through a Master-Slave approach. Ocean Eng 158:253–262
38. Schreiber F, Manns M, Morales J (2019) Design of an additively manufactured soft ring-gripper. In: International conference on changeable, agile, reconfigurable and virtual production, pp 142–147
39. Gutierrez SC, Zotovic R, Navarro MD, Meseguer MD (2017) Design and manufacturing of a prototype of a lightweight robot arm. In: Manufacturing engineering society international conference, MESIC 2017, pp 283–290
40. Aburaia M, Markl E, Stuja K (2015) New concept for design and control of 4 Axis robot using the additive manufacturing technology. Procedia Eng 100:1364–1369
41. Stefan J, Rebecca M (2015) New approach to introduction of 3D digital technologies in design education. In: Design conferences onnovative product creation, pp 35–40
42. Krupke D et al (2015) Printable modular robot: an application of rapid prototyping for flexible robot design. Ind Robot An Int J 42:149–155
43. Andy P (2014) Building a culture of creation. Teach Libarian 41(5):12–16
44. Glen B et al (2014) Advancing Childern's engineering thorugh desktop manufacturing. In: Spector JM et al (ed) Handbook research education communication technology, Springer Science, New York, pp 675–688
45. Glen B et al (2015) An educational framework for digital manufacturing in schools. Add Manuf 2:42–49
46. Loy J (2014) e-Learning and e-making: 3D printing blurring the digital and the physical. Educ Sci 4:108–121
47. Kayfi R, Ragab D, Tutunji TA (2015) Mechatronic system design project: a 3D printer case study. In: 2015 IEEE Jordan conference on applied electrical engineering and computing technologies (AEECT), pp 1–6 (2015)
48. Jaksic NI (2014) New inexpensive 3D printersopen doors to novel experimental learning practices in engineering education. In: ASEE annual conference and exposition, pp 1–23
49. Warren R, Yalcin E, Eric MC (2014) An autonomous arduino-based racecar for first-year engineering technology students In: ASEE annual conference and exposition, ASEE: Indianapolis, IN (2014)
50. Armesto L, Fuentes-Dura P, Perry D (2015) Low-cost printable robots in education. J Intell Robot Syst Theory 81:5–24
51. Wong N, Cheng, HH (2016) CPSbot: a low cost reconfigurable and 3D printed robotics kit for education and research on cyber-physical systems. In 12th IEEE/ASME international conference on mechatronic and embedded systems and applications (MESA), pp 1–6

52. Ziaeefard S, Ribeiro GA, Mahmoudian N (2015) GUPPIE underwater printed robot; a game changer in control design education. In: American control conference (ACC), pp 2789–2794
53. Gonzalez-Gomez J et al (2012) A new open source 3D printable mobile robotic platform for education. In: 6th International symposium autonomous minirobots research edutainment
54. Kuat T, Yedige T, Almas S (2015) A low-cost open-source 3D printed three-finger gripper platform for research and educational purposes. J Rapid Open Access Publishing 3:638–647
55. Spendlove T (2016) Maker: 3D printing and designing with robot chassis. In: ASEE annual conferences and exposition, ASEE: New Orleans, LA
56. Valero-Gomez A et al (2012) Printable creativity in plastic valley UC3M. In: EDUCON 2012, IEEE, Editor, pp 1–9
57. Martinez MO et al (2016) 3D printed haptic devices for educational applications. In: IEEE haptic symposium, pp 126–133
58. Cesar V et al (2015) Design and evaluation of a DIY construction system for educational robot kits. Int J Technol Des Educ
59. Ryan WK, Chad TV (2017) Maker: a 3D printed balancing robot for teaching dynamic systems and control. In ASEE. ASEE, pp 1–8
60. Mitsuhashi K et al (2015) Production and education of the modular robot made by 3D printier. In: 2015 10th Asian control conference (ASCC), pp 1–5
61. Range K, Dana R., Tarek AT (2015) Mechatronic system design project: A 3D printer case study. In: IEEE Jordan Conference on applied electrical engineering and computing technologies (AEECT), pp 1–6
62. Chealsea S et al (2015) Open-source 3D printing technologies for education: bringing additive manufacturing to the classroom. J Visual Lang Comput 28:226–237

H_∞ Filter with Fuzzy Logic Estimation to Refrain Finite Escape Time

Bakiss Hiyana Abu Bakar and Hamzah Ahmad

Abstract In this paper, a H_∞ Filter with Fuzzy Logic (FHF) based on mobile robot localization and mapping is proposed to effectively prevent the Finite Escape Time (FET) issue in H∞ Filter (HF). The capabilities of HF to offer better solution than Kalman Filter (KF) is limited by the shortcomings of Finite Escape Time. FET define as a state that can go infinite in normal condition that lead to in accurate estimation result, with the present of FET, the estimation cannot be totally guarantee. Therefore, it is essential to guarantee no FET is observed in the state covariance on its update. A new FHF is developed for this matter by using fuzzy logic approach. The analysis is done by implementing the Fuzzy Logic technique with triangular membership. The simulation result show that Fuzzy Logic effectively avoid the FET from appear and simultaneously improve the estimation between mobile robot and the landmark.

Keywords Mobile robot · H∞ filter · Fuzzy logic · Finite escape time

1 Introduction

During the last four decade, the mobile robot in navigation system has growth rapidly and attain lots of attention among researcher. The knowledge about the position and orientation of the mobile robot is very useful in different tasks and surrounding such as obstacle avoidance and hazardous condition [1, 2]. A variety of approaches for mobile robot localization has been developed and variety of techniques used to represent about the robot current position. One of the available technique is called Simultaneous Localization and Mapping (SLAM). SLAM was originally been proposed by Smith and Cheeseman in 1987 [3]. SLAM is design based on the relationship between the robot and the landmarks of the environment around it.

B. H. Abu Bakar
Jabatan Kejuruteraan Elektrik, Politeknik Sultan Haji Ahmad Shah (POLISAS), Kuantan, Pahang, Malaysia

H. Ahmad (✉)
Department of Electrical Engineering, College of Engineering, University Malaysia Pahang, Kuantan, Pahang, Malaysia
e-mail: hamzah@ump.edu.my

A. F. Ab. Nasir et al. (eds.), *Recent Trends in Mechatronics Towards Industry 4.0*, Lecture Notes in Electrical Engineering 730, https://doi.org/10.1007/978-981-33-4597-3_28

SLAM involve the observation process which done by mobile robot at certain area and in meantime the map of the unknown area will be constructed while moving through it and at the same time simultaneously keeping track to determine its own location in the map [4, 5]. During the mobile robot autonomous navigation, there is a lot of uncertainties issue need to be considered. Due to this unsolved issue, a lot of research done that aim to produce highly accurate maps of the areas that being observe with variety techniques.

To tolerate this unsolved issue, Kalman Filter (KF) and Extended Kalman Filter (EKF) have been study immensely in SLAM using various approaches [6]. Unfortunately, EKF efficiency only capable in gaussian noise environment, SLAM demands further consideration about the environment conditions. Due to this limitation of incapability, based on previous work another filter known as H∞ Filter (HF) that surpassed the KF performance may offer better solution [7, 8]. HF offer better estimation for non-Gaussian noise environment [9] where the noise characteristics is unknown.

Several paper have being investigate the performance and capabilities of HF [8, 10–12], even previous research done by West et al. [13] that compare the performance of EKF, particle filter (PF) and HF claim that HF give more reliable estimation compare to EKF and PF but HF exhibit a FET issue during mobile robot navigation. The present of FET will lead to in accurate estimations results. FET defines as a state that can go infinite in normal condition, the state covariance suddenly becomes positive or negative that may cause lose of confidence by the mobile robot during the observation.

A lot of study have done before successfully combine the EKF and Fuzzy Logic [14–17] for better performance. In 1995, K. Kobayashi et al. founded that fuzzy logic able to tune the covariance and reset the initialization of the filter [15]. In 2015, Hamzah Ahmad and Nur Aqilah Othman proposed a combination of HF and Fuzzy Logic to create H∞ Filter- Fuzzy Logic (FHF) to avoid FET from appear during mobile robot navigation, the analysis are executed only in Gaussian membership function while others membership still not being examine yet [18].

Inspired by this reason, this paper proposed a combination between HF and Fuzzy Logic technique (FHF) to avoid the FET from occurring during mobile robot observations. The recommended technique focus on the HF innovation stage. This is done by adding few fuzzy logic in triangular membership function. The first stage will analysis the HF before the combination with Fuzzy Logic Technique and the second stage will analysis the HF with Fuzzy Logic implementation in triangular membership to prove that the combination with HF based on mobile robot localization may refrain FET.

The remained of this paper will be construct as follows. Section 2 will present the proposed technique. Next, Sect. 3 will demonstrate the comparison of the simulations result between novelty HF and FHF technique, and lastly, followed with conclusion.

2 Proposed Technique

The phenomenon of FET is unavoidable in normal HF making this filter less favorite compare to Kalman Filter (KF) or EKF although HF competent in non-Gaussian noise. The present of FET will cause the state covariance become positive or negative during the mobile robot observations and cause loses of confident to the robot to determine the location even after multiple measurement on the surrounding being recorded.

In order to overcome the FET issue and gain highly accurate maps, our focus approach is to control the measurement input of mobile robot. In this paper, the estimation measurement by HF will be compare to the measurement information through the measurement innovation. During the mobile robot observation, if the robot observe large error, it will affect the estimation and becomes erroneous. In order to reduce the error, the measurement innovation must be control to always be small in order to improve the estimation. The further detail will be discuss in next section.

2.1 SLAM General Model

For mobile robot localization and mapping, two process involve which are process model and measurement model. Process model calculate the mobile robot kinematic while the measurement model will measured the distance between robot and landmarks during observations.

The process model of mobile robot describe as follow:

$$X_{k+1} = f(X_k, \omega_k, \upsilon_k, \delta\omega, \delta\upsilon) \tag{1}$$

where X_k is state vector that consist the x, y positions and heading angle θ_k of a mobile robot. ω_k, υ_k is the angular acceleration and the velocity of the mobile robot with its associated noise δ_ω, δ_υ.

The state vector of SLAM robot is a combination between robot position X_r and landmark position X_m as follow structure:

$$X_k = [X_r X_m]^T \tag{2}$$

where the position of mobile robot is $X_r = [X_{Rk} Y_{Rk} \theta_{Rk}]^T$ is represent the center of mobile robot coordinate $(X_{Rk} Y_{Rk})$ and the heading angle θ_{Rk}. The state of the landmarks is represented by $X_m = [l_1 l_2 \ldots\ldots l_m]^T$. The landmarks is describe as the coordinate $(x_i\, y_i)$ with $i = 1, 2, 3, \ldots, m$, where m refer to the number landmarks.

The measurement model of mobile robot defined as follow:

$$Z_k = \begin{bmatrix} r_k \\ \varphi_k \end{bmatrix} = \left[H_k X_k + V_{r,k} \right] \quad (3)$$

$$Z_k = \begin{bmatrix} \sqrt{(x_i - x_{Rk})^2 + (y_i - y_{Rk})^2} + \upsilon_{r,k} \\ \tan^{-1}\left(\frac{(y_i - y_{Rk})}{(y_i - y_{Rk})}\right) - \theta_{Rk} + \upsilon_{\varphi} \end{bmatrix} \quad (4)$$

where Z_k is the sensor reading of the robot that represent the relative distance and angle measurement, r_k and φ_k among mobile robot and landmark while $\upsilon_{r,k}$ is the measurement noise that will affect the measurement reading.

2.2 H∞ Filter Algorithm

H∞ Filter (HF) algorithm almost similar to the structure of KF, only the presence of γ in the update stage give the different to KF algorithm. HF involve two stage which are prediction and update. The prediction stage is defined as bellow:

$$\widehat{X}_{k+1}^{-} = f_k\left(\widehat{X}_k, w_k, u_k, 0, 0\right) \quad (5)$$

$$P_{k+1}^{-} = f_k P_k \left(I - \gamma^{-2} P_k + H_k^T R_k^{-1} H_k P_k\right)^{-1} f_k^T + Q_k \quad (6)$$

The predicted states is represent by Eq. (5) while the state covariance represent by Eq. (6). Q_k is the control noise covariance. The update stage is defined as follow:

$$\widehat{X}_{k+1} = f_k \widehat{X}_k + f_k K_k \left(z_k - H_k \widehat{X}_k\right) \quad (7)$$

where the gain of the system is $K_k = P_k\left(I - \gamma^{-2} P_k + H_k^T R_k^{-1} H_k P_k\right)^{-1} H_k^T R_k^{-1}$.

The Eq. (7) show the measurement innovation that indicate the error of the system. The error prefer to be small or else the whole estimation will be error. Due to this aspect, the possibility of controlling the size of covariance may contribute to control the size of error as at the same time controlling the presence of FET. The research is focus to decrease the error generated by the second term of Eq. (7) from getting bigger.

2.3 Fuzzy Logic Design in HF

The main focus on designing the Fuzzy logic is to decrease the error of the input at the measurement innovation. The Fuzzy logic is design to receive two input which are angle and distance errors. By choosing appropriate input may minimize the measurement error cause by sensor inaccuracies.

Mention earlier that HF is competence in non-Gaussian noise characteristics, this will affect the sensor reading exhibits bigger error and cause bigger covariance noise, R. Due to this, if the gain K is control to be small during the observations, then the value of measurement error might be smaller. Previous research has done to reduce the size of the measurement noise by using the Fuzzy logic [16]. Same method also previously implement to KF [11] and give smaller measurement noise.

Due to this fact, fuzzy logic is proposed in order to discover most suitable value of measurement innovation. The proposed method of HF with Fuzzy Logic combination technique (FHF) is illustrated in Fig. 1.

The general design of Fuzzy logic is represent in Fig. 2. The proposed design used the Mamdani technique for analysis purpose. The fuzzy sets consists two input and two correction output which are angle and distance measurement. The Fuzzy logic rules used to define the output of measurement innovation are describe as follow.

- IF the angle error is negative and distance error is negative, THEN angle is negative
- IF the angle error is negative and distance error is normal, THEN angle is normal

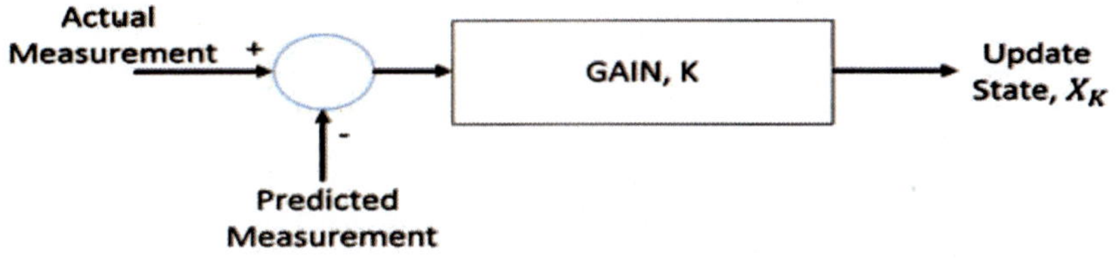

(a) Original HF measurement innovation

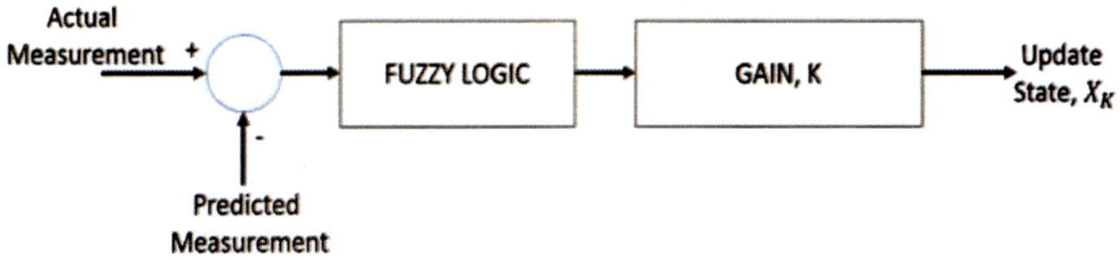

(b) FHF measurement innovation

Fig. 1 Original HF measurement innovation and proposed method for HF with fuzzy logic combination technique (FHF)

Fig. 2 Fuzzy logic general design

- IF the angle error is negative and distance error is positive, THEN angle is negative, distance is normal
- IF the angle error is positive and distance error is normal, THEN angle is negative
- IF the angle error is positive and distance error is negative, THEN distance is normal
- IF the angle error is positive and distance error is positive, THEN angle is negative, distance is normal.

In this paper, for the evaluation purpose, the triangular membership is considered. Three fuzzy sets being choose for three categories which are negative, positive and normal regions. The scale being set [−100, 100] under normal condition with high error. Each fuzzy sets had different value that already being tune for several times for the best estimation results.

The angle and the distance measurement are prior observe based on the tuning process and all uncertainties are considered. All positive and negative value of both input are being considered making it's the main different from the previous research done [11].

3 Simulation Result and Discussion

Section 3 will demonstrate the simulation result and discussion. Few assumptions on the system also being made as follow.

Firstly, the mobile robot involve in the simulation is a two wheeled mobile robot. The robot is assumed to start from global coordinate system of (0, 0). The mobile robot move arbitrary in the environment and will recognize each landmark that near to it. The measurement data will be collected and send to HF to be analyze and update the process. All landmarks are define as point landmarks and locate in specific location.

Another important setting need to be clarify is the value of γ where the value will lead to the performance of the estimation result. The γ value are justify based on [7]. The simulation is considered to be done in non-Gaussian noise conditions in order to fulfill the purpose of why HF is proposed in this research.

In order to gain good consistency, the simulations carried by MATLAB Simulink done in 1000[s] with selected initial condition using the parameters from Table 1

Table 1 Simulations parameters

Variable	Parameter value
Process noise, Q_{min}, Q_{max}	−0.002, 0.001
Measurement noise, $R_{\theta min}$, $R_{\theta max}$ R_{dist_min}, R_{dist_max}	−0.04, 0.01 −0.15, 0.3
Initial Covariance; P_{robot}, $P_{landmark}$	0.0001, 100
Simulation time (s)	1000 s

to describe the real mobile robot. The mobile robot assume at least to have one measurement sensor.

Figures 3, 4 and 5 show the comparison on mobile robot position in both HF and FHF in non-Gaussian noise environment. Figure 5 show detail the error occurred for both HF and FHF in mobile robot and landmark. The error of HF higher that FHF in mobile robot. While for state covariance, Fig. 6 show the detail. Clearly that FHF show less uncertainties and FET only occurred in initial state better that HF that FET occurred multiple times during the estimations. The consequences of applying the fuzzy logic enhance the performance of HF to refrain the presence of FET.

The result show above clearly prove that FHF perform better that original HF even though the performance of mobile robot in Fig. 3 show that HF and FHF almost equal, but in normal HF, the value of γ create the difficulty to be design and become HF disadvantage. From the analysis, bigger γ value may cause increasing of FET while smaller γ value may cause bigger error in mobile robot movement. Some

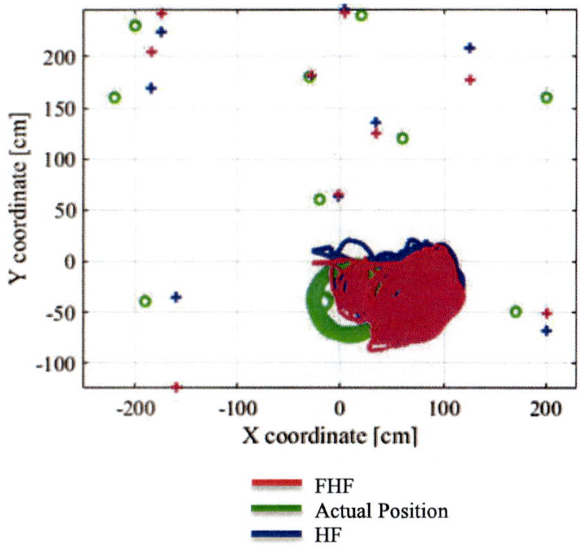

Fig. 3 The mobile robot movement through the environment

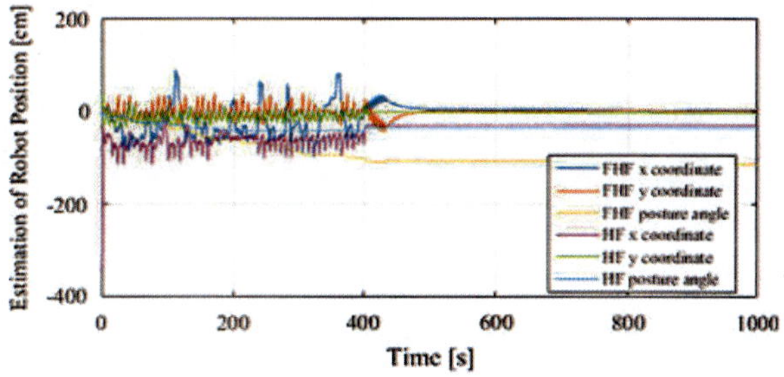

Fig. 4 Estimation of mobile robot position for both HF & FHF

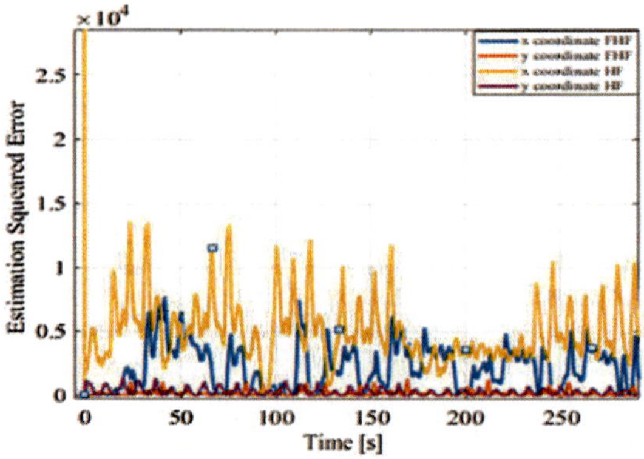

(a) Mobile robot square error

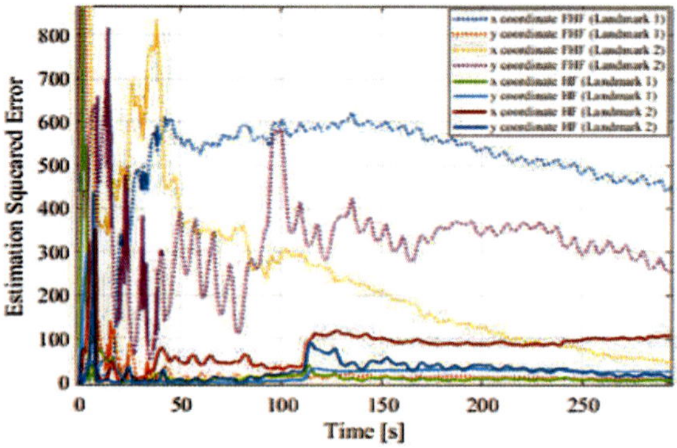

(b) Landmark square error

Fig. 5 Mobile robot and landmark square error for both HF and FHF

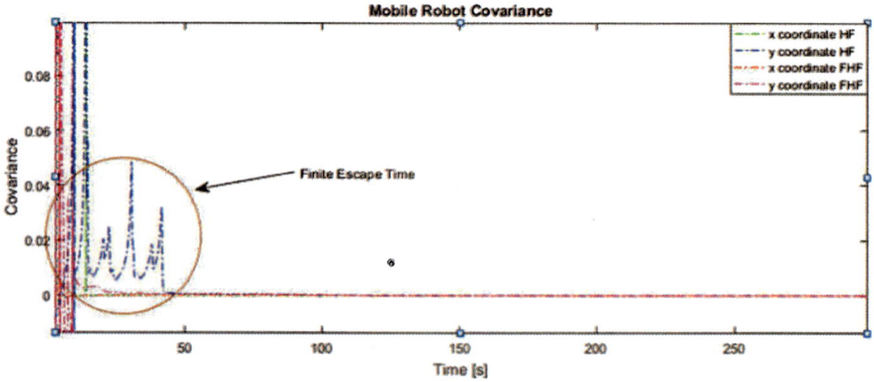

Fig. 6 The state covariance condition between original HF and FHF. FET in FHF exhibits only during initial condition compare to normal HF

design rule such as being suggested in [7] need to be follow in order to gain good measurement estimation.

Due to his limitation, the proposed technique is relevant and prove able to overcome the issue. Fuzzy logic with triangular membership prove capable to refrain FET same as Gaussian membership in [18]. Although an additional fuzzy logic just give few effect on computations time, but lot of research previously prove that fuzzy logic suitable in mobile robot navigation system. As the main objective is to refrain the FET, only triangular membership being examined while others fuzzy membership function will be examine next in order to find the best membership types.

4 Conclusions

For a non-Gaussian noise environment, HF being prove able to perform better that KF. HF become most reliable solution when the robot motion is uncertain and the environment condition is partially understood. Despite the fact that the computation cost and processing time in FHF might become slightly higher compare to original HF, but the capabilities of FHF prove able to give better estimation result and refrain the presence of FET issue during the mobile robot observations must be celebrated.

Acknowledgements The author would like to thanks Politeknik Sultan Haji Ahmad Shah (POLISAS) for continuous support in realizing this research.

References

1. Yang F, Wang Z, Lauria S, Liu X (2009) Mobile robot localization using robust extended h infinity filtering. J Syst Control Eng 223
2. Zhang S, Wang Z, Fellow (2016) Nonfragile H∞ fuzzy filtering with randomly occuring gain variations ANS channel fading. IEEE Trans Fuzzy Syst 24(3):505–518
3. Smith RC, Cheeseman P (1987) On representation and estimation of the spatial uncertainty. Int J Rob Res 5:56–68
4. Thrun S, Burgard W, Fox D (2000) A real time algorithm for mobile robot mapping with applications to multi robot and 3d Mapping. In: IEEE international conference on robotics and automation, pp 321–328
5. Thrun S, Burgard W, Fox D (2005) Probabilistic robotics. MIT Press
6. Paz LM, Neira J (2006) Optial local map size for EKF-based SLAM. In: IEEE/RSJ International conference intelligent robots and systems
7. Ahmad H, Namerikawa T (2011) H∞ filter slam: a sufficient condition for estimation. In: 18th World congress of the international federation of automatic control (IFAC), pp 3159–3164
8. Ahmad H, Namerikawa T (2011) Robotic mapping and localization considering unknown noise statistics. J Syat Des Dyn 5(1):70–82
9. Ahmad H, NA Othman (2015) a solution to finite escape time for H∞ filter based SLAM. In: IEEE international conference Asian control conference
10. Ahmad H, Othman N (2015) The impact of cross-correlation on mobile robot loclization. Int J Control Autom Syst 13–5. (In press)

11. Wang JH, Song CL, Chen JB (2010) Sigma point h infinity filter for initial alighment in marine strapwon inertial navigation system. In: 2010 2nd International conference on signal processing systems, pp 580–584
12. Gualda D, Urena J, Gracia JC, Gracia E, Ruz D, Lindo A (2014) Fusion of data from ultrasonic LPS and isolated beacons for improving MR nvigation. In: IEEE International conference on instrumentation and measurement technology conference, pp 1552–1555
13. West ME, Symos VL (2006) Navigation of an autonomous underwater vehicle (AUV) using robust SLAM. In: Proceedings 2006 IEEE CCA, pp 1801–1806
14. Choomang R, Afzulpukar N (2005) Hybrid Kalman filter/fuzzy logic based position control of autonomous mobile robot. Int J Adv Rob Syst 2(3):197–208
15. Kobayashi K, Ceok KC, Watanabe K (1995) Estimation of absolute vehicle speed using fuzzy logic rule-based Kalman filter. Am Control Conf
16. Kobayashi K, Ceok KC, Watanabe K, Munekata F (1998) Accurate differential global positioning system via fussy logic Kalman filter sensor fusion technique. IEEE Trans Ind Electron 45(3):510–518
17. Raimondi FM, Melusso M (2006) Fuzzy EKF control for wheeled nonholonomic vehicle. In: 32th Annual conference on IEEE industrial electronics, pp 43–48
18. Ahmad H, Othman NA (2015) HF-Fuzzy logic based mobile robot navigation: a solution to finite escape time. In: International conference on electrical, control and computer engineering

The Identification of Significant Mechanomyography Time-Domain Features for the Classification of Knee Motion

Tarek Mohamed Mahmoud Said Mohamed, Muhammad Amirul Abdullah, Hasan Alqaraghuli, Rabiu Muazu Musa, Ahmad Fakhri Ab. Nasir, Mohd Azraai Mohd Razman, Mohd Yazid Bajuri, and Anwar P. P. Abdul Majeed

Abstract Stroke is the third leading cause of long term disability in the world. More often than not, the patients who suffer from such cerebrovascular disease endure restricted activities of daily living (ADL). Rehabilitation is deemed necessary to improve ones ADL, especially in the early stages of stroke. This study presents the classification of knee motion; particularly extension and flexion, based on muscle signals that could be utilised by an exoskeleton for rehabilitation purpose. A total of 20 subjects participated in the present investigation. The mechanomyography (MMG) signals were collected by accelerometers placed on four of the muscles that control the knee joint, namely, Rectus Femoris, Gracilis, Vastus Medialis, and Biceps Femoris, respectively. Eight statistical features were extracted from the raw data, i.e., root mean square (RMS), variance (VAR), mean, standard deviation (STD), kurtosis, skewness, minimum, and maximum along all x, y and z-axes. The Chi-Square (χ^2) feature selection technique was used to identify significant features, in which 30 was identified amongst the 96 extracted features. A 10-fold cross-validation technique

T. M. M. S. Mohamed · M. A. Abdullah · A. F. Ab. Nasir · M. A. Mohd Razman · A. P. P. Abdul Majeed (✉)
Innovative Manufacturing, Mechatronics and Sports Laboratory (IMAMS), Universiti Malaysia Pahang, Darul Makmur, 26600 Pekan, Pahang, Malaysia
e-mail: amajeed@ump.edu.my

H. Alqaraghuli
School of Electrical Engineering, Faculty of Engineering, Universiti Teknologi Malaysia, 81300 Bahru, Johor, Malaysia

R. M. Musa
Centre for Fundamental and Liberal Education, Universiti Malaysia Terengganu (UMT), Darul Iman, 21030 Kuala Nerus, Terengganu, Malaysia

M. Y. Bajuri
Department of Orthopaedics and Traumatology, Faculty of Medicine, Universiti Kebangsaan Malaysia Medical Centre, Kuala Lumpur, Malaysia

A. P. P. Abdul Majeed
Centre for Software Development & Integrated Computing, Universiti Malaysia Pahang (UMP), 26600 Pekan, Pahang, Malaysia

A. F. Ab. Nasir et al. (eds.), *Recent Trends in Mechatronics Towards Industry 4.0*, Lecture Notes in Electrical Engineering 730, https://doi.org/10.1007/978-981-33-4597-3_29

was employed in training a Support Vector Machine (SVM) model on a dataset that was partitioned with a ration of 80:20 for train and test, respectively. It was demonstrated in the present investigation that through the reduction of features, the test accuracy increased from 83.3 to 90%, suggesting the importance of the selected features. The findings from the study could pave the way for its adoption on a knee-based exoskeleton for rehabilitation.

Keywords Knee motion · Machine learning · Feature selection · Mechanomyography · Rehabilitation

1 Introduction

Stroke is cerebrovascular accidents (CVAs) which is the third common leading cause of long-term adult disability worldwide and mortality after cancer and ischemic heart disease. There are many different causes of stroke; haemorrhage, ischemia (occlusion of blood vessels) by clot or bursts, or any sudden interruption of blood flow to a particular area of the brain [1, 2]. Spasticity, foot drop, and paralysis are the most common disabilities resulting from stroke. It affects the body's ability to coordinate the movement of the muscles, body posture, walking, and balance amongst others. The employment of robotics, in particular, the use of exoskeletons could facilitate rehabilitation efforts significantly as opposed to conventional therapy that is often deemed both labour and cost-intensive [3].

Researchers have actively investigated a myriad of methods to analyse muscles activities noninvasively; namely, surface electromyogram (sEMG), sonomyogram (SMG), tensiomyogram (TMG), and mechanomyography (MMG). These types of signals are normally a function of time and can be analysed in its amplitudes, frequency and phase. The EMG signals are one of the useful biological signals which directly reflect the human motion intentions. It consists of the action potentials from groups of muscles' fibres. It is organised into functional units called motor units (MUs). The sEMG is detected by sensors located directly on the surface of the skin. It can also be digitalised by needles or wire sensors inserted into the muscle tissues invasively [4].

Nonetheless, EMG signals contain fuzziness and it is often difficult to record the same EMG signals for the same motion even of the same person. For a certain motion, each muscle activity is highly non-linear regarding the variant responses that muscle according to the joint angle [5]. On the other hand, MMG has been proposed as another tool to study muscle mechanical activity. MMG is the superficial measurement of the axial vibrations or the sound elicited by contracting muscles [6]. The term mechanomyography represents a technique by which the mechanical activity of the muscle is detected using specific transducers to record muscle surface oscillations due to the mechanical activity of the motor units. MMG signals can be detected using several types of transducers including piezoelectric contact sensors

(PIZ), microphones (MIC), accelerometers (ACC) and also laser distance sensors (LDS) [7].

Hitherto, there are a number of research that has employed the notion of machine learning towards the classification of knee motion. Wu et al. investigated the efficacy of MMG signals detected on clothes in classifying knee motion [8]. The signals were acquired from rectus femoris, biceps femoris, semitendinosus, and gracilis muscles of 8 able-bodied participants. The features extracted from the signals for the machine learning evaluation are mean absolute value (MAV), standard deviation (STD), variance (VAR), root mean square (RMS) as well as a new feature known as the difference of mean absolute value (DMAV). It was shown from the present investigation that the Support Vector Machine (SVM) model that employs a leave-one-out (LOOV) cross-validation technique can yield a classification accuracy of 88% by using the traditional features. However, by including the additional feature, the classification accuracy is improved to 91%.

In a different study, Wu et al. evaluated the efficacy of MMG signals detected with no direct skin contact in classifying knee motion [9]. There were 6 knee motions recognised (static standing, knee flexion, knee extension, knee external rotation, knee internal rotation, and the pause at the end of knee flexion, external rotation or internal rotation). 4-channel mechanomyography signals were detected from the thigh muscles. Using the convolutional neural network (CNN)-SVM combined model, the features could be automatically extracted from the signals. It was shown from the proposed investigation that the CNN model that employs a 5-fold cross-validation technique was able to achieve a classification accuracy of 88.45 $\pm$ 2.79% by using the traditional features. However, using CNN-SVM combined model, the classification accuracy is improved to 94.04 $\pm$ 1.10%.

The present study sought to investigate the efficacy of an SVM classifier in classifying the flexion and extension of the knee by taking into consideration different statistical-based time-domain features extracted from a set of muscles by means of MMG. In addition, a set of significant features are also identified through a feature selection technique to discern its ability to further improve the efficacy of the evaluated classifier.

2 Methods

A total number of 20 subjects participated in the data collection process that was carried out at the Innovative Manufacturing, Mechatronics and Sports Laboratory, Universiti Malaysia Pahang. The age, weight, height as well as the number of hours spent for sports are between 21 and 30 years old, 45–115 kg, 159–183 cm, and 0–10 h/week, respectively [10]. The subjects were made aware of the research purpose and procedures by a simple briefing, and informed consent was signed by each subjects. The ethical approval for this study was obtained via an institutional research ethics committee (FF-2013-327).

A device based on ADXL335 accelerometers were developed for the data collection purpose. The sensors were placed over four different muscles bellies, i.e., Rectus femoris, Gracilis, Vastus Medialis, and Biceps Femoris, respectively as labelled in Fig. 1 and Table 1. By the time the device is turned on and the serial communication initialised, the subjects that were initially rested in the sitting position as shown in Fig. 2 were required to do five repetitions of knee extension and flexion (A total of 200 instances were collected). A developed software was used to display the dataset and store it into.CSV or.mat files. Eight statistical features, viz. root mean square (RMS), variance (VAR), mean, standard deviation (STD), kurtosis (Kurt), skewness

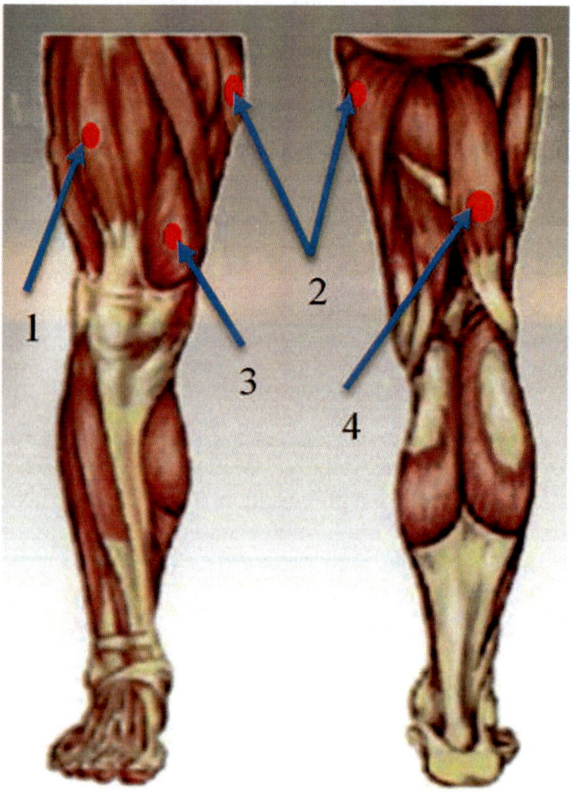

Fig. 1 The location of the MMG sensor placements

Table 1 Muscles and their functions

No.	Muscle	Function
1	Rectus Femoris (RF)	Knee extension
2	Gracilis (G)	Knee flexion and internal rotation
3	Vastus Medialis (VM)	Knee extension
4	Biceps Femoris (BF)	Knee flexion and external rotation

Fig. 2 The data collection procedure

(Skew), minimum (Min), and maximum (Max) along all x, y and z-axes, on the four identified muscles listed in Table 1 were extracted from the signals.

The feature selection is performed through the implementation of Chi-Square (χ^2). This technique evaluates the interdependency between the features towards the response and identifies significant features. Subsequently, Support Vector Machine (SVM) classifier is used to evaluate its performance in classifying the knee motion based on all the extracted features as well as the identified features. An open-source cross-platform integrated development environment (IDE), Spyder 4.1.4 is used to develop the model as well as identifying the significant features. The scikit-learn library 0.23.2 was evoked and the default hyperparameters of the SVM model was used in the present investigation. The dataset was split into 80:20 ratio for test and train data, in which the 10-fold cross-validation technique [11] was used to train the SVM models and the classification accuracy is used to evaluate the performance of the models or rather the significance of the features evaluated.

3 Results and Discussion

Thirty significant features were identified via Chi-Square (χ^2) feature selection technique amongst the 96 extracted features. The features identified on Muscle 1 (BF) along the z-axis are Skew, STD, VAR, Min, Kurt, and for x-axis is Kurt. For Muscle 2 (G) along the x-axis are STD, VAR, Min, Max, whilst for z-axis is Skew, VAR and STD, respectively. Conversely, for Muscle 3 (VM) along the z-axis are Kurt and Skew, whilst for y-axis Kurt, Skew, and Max. In addition along the x-axis for VM are Kurt, Skew and Max, respectively. The features identified on Muscle 4 (BF) along the z-axis are Min, Max, Kurt, STD and VAR, whilst for x-axis the features are Kurt, Max, mean, and Min, respectively. It could be seen from the bar chart depicted in Fig. 3, that the classification accuracy for training by considering all and the selected features are identical, i.e., 84.1%. Nonetheless, it could be seen that the test accuracy

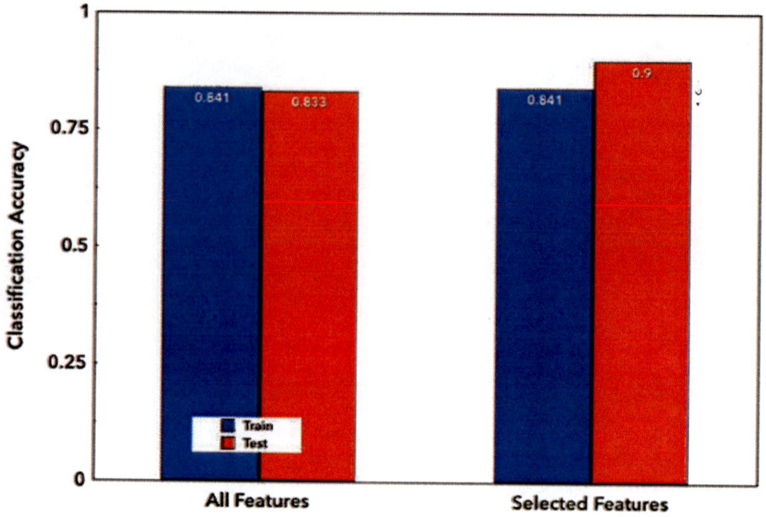

Fig. 3 The performance of the evaluated features on the SVM classifier

improved by 6.7% considering the identified 30 features. In addition, it could be observed that by considering all features, the model appears to overfit, as a lower classification accuracy on the test dataset is noticed, suggesting that some features induce noise to the model [12]. Moreover, it could be seen that even with reduced features, a reasonable classification accuracy could be achieved, implying that in the event a real-time implementation is to be executed, the computational expense could be reduced.

4 Conclusion

The present investigation evaluated a number of features for the classification of knee extension and flexion. Eight main statistical features were extracted namely, RMS, STD, VAR, Mean, Min, Max, Kurtosis, and Skewness from the MMG placed on four muscles. The Chi-Square ($\chi 2$) feature selection technique was used to identify the significant features out of the 96 extracted features, in which 30 was found. It was shown that the selection of the features could provide a relatively good classification accuracy via the SVM classifier. The findings of the present study are non-trivial as it demonstrated that even with a significantly reduced number of features, the classifier is still able to distinguish well the classes. Moreover, it also implies the possible employment of the model in real-time in the control of an assistive exoskeleton for rehabilitation purpose. Future studies will evaluate other forms of feature selection technique, classifiers as well as the identification of optimised hyperparameters of the classifiers.

Acknowledgements The authors would like to acknowledge Universiti Malaysia Pahang for funding this study via RDU180321.

References

1. Paterson J, Paterson J (2009) Common medical conditions and problems. Teach Pilates Postural Faults, Illn Inj 50–77. https://doi.org/10.1016/B978-0-7506-5647-4.50009-3
2. Hoffman R, Benz EJ, Silberstein LE, Heslop H, Weitz JI, Anastasi J, Salama ME, Abutalib SA (2018) Hematology : basic principles and practice
3. Ab Patar MNA, Said AF, Mahmud J, Abdul Majeed APP, Razman MA (2014) System integration and control of Dynamic Ankle Foot Orthosis for lower limb rehabilitation. In: ISTMET 2014—1st International symposium on technology management and emerging technologies, proceedings, pp 82–85
4. De Luca CJ, Adam A, Wotiz R, Gilmore LD, Nawab SH (2006) Decomposition of surface EMG signals. J Neurophysiol 96:1646–1657. https://doi.org/10.1152/jn.00009.2006
5. Kiguchi K, Tanaka T, Fukuda T (2004) Neuro-Fuzzy control of a robotic exoskeleton with EMG signals. IEEE Trans Fuzzy Syst 12:481–490. https://doi.org/10.1109/TFUZZ.2004.832525
6. Silva J, Chau T (2003) Coupled microphone-accelerometer sensor pair for dynamic noise reduction in MMG signal recording. Electron Lett 39:1496. https://doi.org/10.1049/el:20031003
7. Cho YJ, Kim JY (2012) The effects of load, flexion, twisting and window size on the stationarity of trunk muscle EMG signals. Int J Ind Ergon 42:287–292. https://doi.org/10.1016/J.ERGON.2012.02.004
8. Wu H, Wang D, Huang Q, Gao L (2018) Real-time continuous recognition of knee motion using multi-channel mechanomyography signals detected on clothes. J Electromyogr Kinesiol 38:94–102. https://doi.org/10.1016/j.jelekin.2017.10.010
9. Wu H, Huang Q, Wang D, Gao L (2018) A CNN-SVM combined model for pattern recognition of knee motion using mechanomyography signals. J Electromyogr Kinesiol 42:136–142. https://doi.org/10.1016/J.JELEKIN.2018.07.005
10. Križaj D, Šimunič B, Žagar T (2008) Short-term repeatability of parameters extracted from radial displacement of muscle belly. J Electromyogr Kinesiol 18:645–651. https://doi.org/10.1016/J.JELEKIN.2007.01.008
11. Taha Z, Razman MAM, Adnan FA, Abdul Ghani AS, Abdul Majeed APP, Musa RM, Sallehudin MF, Mukai Y (2018) The identification of hunger behaviour of lates calcarifer through the integration of image processing technique and support vector machine. In: IOP conference series: materials science and engineering
12. Razman MAM, Susto GA, Cenedese A, Abdul Majeed APP, Musa RM, Abdul Ghani AS, Adnan FA, Ismail KM, Taha Z, Mukai Y (2019) Hunger classification of lates calcarifer by means of an automated feeder and image processing. Comput Electron Agric 163. https://doi.org/10.1016/j.compag.2019.104883

Parameter Estimation of Lorenz Attractor: A Combined Deep Neural Network and K-Means Clustering Approach

Nurnajmin Qasrina Ann, Dwi Pebrianti, Mohamad Fadhil Abas, and Luhur Bayuaji

Abstract This research is mainly aimed at introducing a deep learning approach to solve chaotic system parameter estimates like the Lorenz system. The reason for the study is that because of its dynamic instability, the parameter of the chaotic system cannot be easily estimated. Moreover, due to the complexity of chaotic systems based on existing approaches, some parameters may be difficult to determine in advance. Therefore, it is crucial to assess the parameter of chaotic systems. To solve the issue of parameter estimation for a chaotic system, deep learning is utilized. After that, it has been suggested to improve the efficiencies in the Deep Neural Network (DNN) model by combining the DNN with an unsupervised machine learning algorithm, the K-Means clustering algorithm. This study constructs the flow of DNN based method with the K-Means algorithm. DNN techniques is suitable in solving nonlinear and complex problem. The most popular method to solve parameter estimation problem is using optimization algorithm that easily trap to local minima and poor in exploitation to find the good solutions. Due to the flow, 80% of training and 20% test sets for each class are divided between the Lorenz datasets. Accuracy by using 80:20 ratio of training and test data gives result 98% of accurate training data, and 73% of test data are predicted with the proposed algorithm while 91 and 40% of the DNN models are predicted in training and test data.

Keywords Machine learning · Chaos system · Deep neural network · K-means clustering · Parameter estimation

N. Q. Ann (✉) · D. Pebrianti
College of Engineering, Universiti Malaysia Pahang, 26300 Gambang, Malaysia
e-mail: qasrinaann@gmail.com

M. F. Abas
Faculty of Electrical and Electronics Engineering Technology, Universiti Malaysia Pahang, 26600 Pekan, Malaysia

L. Bayuaji
Faculty of Computing, Universiti Malaysia Pahang, 26300 Gambang, Malaysia

A. F. Ab. Nasir et al. (eds.), *Recent Trends in Mechatronics Towards Industry 4.0*, Lecture Notes in Electrical Engineering 730, https://doi.org/10.1007/978-981-33-4597-3_30

1 Introduction

The theory of chaos is one of the most revolutionary discoveries of the twentieth century, and so far, it has been a hot field of research. Chaos is a kind of randomness generated by the deterministic system. Chaos Theory refers to the toolbox that provides solutions to nonlinear dynamical systems that are sensitive to initial conditions. Chaos is characterized as the chaotic, unpredictable behavior of deterministic, nonlinear dynamic systems. Simultaneous stretching and folding in the system dynamics are essential to have chaotic behavior.

To control and synchronize the Lorenz chaotic system, it is critical to capture the information of parameters due to the complexity of the system [1]. However, determining them is a challenging task in real applications. Numerous research work on the control and synchronization of chaotic systems has been reported in the literature. Most of them have a disability due to the parameters of the chaotic systems that been assumed to be known [2]. However, in practice, the parameters of the systems cannot be precisely known beforehand, which are generally unknown and unavailable. Parameter estimation is a prerequisite to accomplish the control and synchronization of chaos [3]. At present, it is more significant to study the parameter estimation problem of the complex, chaotic systems, such as the fractional-order chaotic system and the high-dimensional chaotic system [4]. The complex chaotic systems have more estimated parameters and more complex dynamical behaviors. Notably, the fractional-order chaotic system can describe physical phenomena more accurately [5].

Deep Learning is the cutting edge of Artificial Intelligence (AI). Deep Learning uses algorithm layers to process data, understand human speech, and visually recognize objects. Information is passed through each layer, with the output of the previous layer providing input to the next layer. Deep Learning enables the machine to train itself to process and learn from data. Deep learning is applicable to resolution and application in several fields, such as research, business, and government.

Nowadays, the pattern of training deep learning models has moved from a single model to a combination or hybrid model. As far as author knowledge is concerned, researchers are able to combine supervised and unsupervised learning (semi-supervised learning), a deep learning model with optimization algorithms, and a deep learning model with traditional machine learning algorithms such as Support Vector Machine (SVM). This is because the combination and hybridization of the algorithm technique will enhance the performance and functionality of the algorithm to help to solve the complexity of parameter estimation in chaotic systems.

2 Related Works

In the literature, there are two approaches to solve parameter estimation of chaotic systems. The most common is to get the initial values from the original system.

Another one is the parameter estimation of a chaotic system with unknown initial values [4, 6]. The second approach claimed to be more effective than the traditional approach because most of the real applications do not have initial values. It has not been widely explored yet by researchers.

Currently, there are two kinds of parameter estimation approaches, which are synchronization methods and metaheuristic algorithms as stated in [4]. The first one is the synchronization methods [7], which proposes the updating law of parameter estimation according to the stability analysis of a chaotic system. However, its techniques and sensitivities must depend on the considered system, so the design encounters difficulty arising from the complexity of the chaotic system. Simulation is demonstrated

The other one metaheuristic algorithms are prevalent for the parameter estimation of a chaotic system, including hybrid Particle Swarm Optimization (PSO) algorithm [8], improved Differential Evolution algorithm (DE) algorithm [9], Fractional-order Firefly Algorithm (FA) [2], hybrid [3, 10] Cuckoo Search Algorithm, Artificial Bee Colony algorithm [11], improved boundary Chicken Swarm optimization [12], Hybrid Flower Pollination algorithm [13], Hybrid Jaya-Powell algorithm [1], improved Teaching-Learning-based Optimization (TLDO) [14] and improved Bird Swarm Optimization [15]. Compared with the synchronization method, it is simple and easy to be implemented. However, those mentioned classical metaheuristics algorithms are susceptible to being trapped into local optima and suffering from premature convergence. As seen on the trend of the algorithms nowadays, these methods had been improved and hybridized. The developed hybrid algorithms can achieve satisfactory estimation performance, which relies on fine-tuned algorithm-specific parameters.

In [2], the authors discussed about two fundamental manners can be categorized to estimate the system parameters. The first contains the gradient-based methods [16] and the second includes the evolutionary algorithms (EA)-based methods. In EAs method, the unknown parameters are assumed as independent variables and the parameters estimation problem is converted into a combinatorial optimization problem by defining a suitable objective function. EAs methods are not sensitive to initial points, derivative-free and straightforward to implement. EAs are one type of the metaheuristic algorithms.

A hybrid algorithm named Jaya-Powell algorithm, which combines the Jaya algorithm and the Powell algorithm to balance the exploration and exploitation in parameter estimation of the Lorenz chaotic system [1]. The main idea of the Jaya algorithm is offered the probability for candidate solutions to move close to the best solution. In contrast, the Powell algorithm is employed to exploit the local information based on the approximately global optimum obtained by the Jaya algorithm. The proposed algorithm is successfully avoided to fall into the local optimum by balancing exploration and exploitation compared with seven algorithms, including Jaya, Powell, Teaching-learning-based optimization, PSO, Genetic Algorithm (GA), Covariance Matrix Adaptation Evolution Strategy (CMA-ES), and Cluster-Chaotic Optimization algorithms. Lazzus et al. introduce a hybrid swarm intelligence algorithm consist of Particle Swarm Optimization (PSO) and Ant Colony Optimization

(ACO) algorithms. In [8], the proposed hybrid algorithm had been compared with PSO and GA, and the result obtained that PSO-ACO converges much faster than the other two algorithms. In this way, the accuracy shown by PSO-ACO can be used on climatological models based on the Lorenz-type system.

The CNN and K-means clustering combination is implemented in [17] by Dong et al. to solve the Short-Term Load Forecasting (SLF) in the smart grid is one of the example of combination of Deep Learning method. This is because the data set used is substantial and increases the scalability of the device. First, the dataset is clustered into subsets using the K - means algorithm. Then, it will be trained under the CNN model. However, the authors mentioned the effects of different configuration parameters, such as cluster number and iteration, which need further investigation.

The remaining paper is structured as follows: Sect. 3 introduces this study's proposed methodology. Then, Sect. 4 discusses results and analysis, and finally, Sect. 5 concludes this paper.

3 Methodology

In this section discussed the proposed methodology in the study. The combination of a Deep Learning architecture and a Machine Learning algorithm is introduced to enhance the performance of the model. The proposed method is applied to estimate Lorenz system. This is because Lorenz system is a nonlinear system that bounded unstable dynamic behavior that exhibits sensitive to initial conditions.

3.1 Deep Neural Network (DNN) Architecture

Deep learning reflects the Artificial Intelligence (AI) subset. Using deep learning the machine trains itself to process and learn from data. The Artificial Neural Network (ANN) is inspired by the human brains of connected neuron networks and consists of three layers, an input layer, a hidden layer, and a layer of output. That is ANN's purest form. When ANN consists of more than three layers, Deep Neural Network (DNN) is called this. *Keras* Phyton Libraries was implemented in this research to build the network models of multilayer perceptron. The *Keras* Phyton Deep learning library focuses on model creation as a sequence of layers.

Relu and *sigmoid* had been implemented into the algorithm in this experiment. The last layer uses *sigmoid*, while another layer uses *relu*. The Rectified Linear Unit or *Relu* is disabled sparsely. Sparsity results in concise models that are often more predictive and less overfitting. The activation function of sigmoid may disappear from the gradient problem. *Sigmoid* also has a slow rate of convergence, which can avoid trapping in local minima.

The standard neural network has many core layers types for layer types. In this experiment, the fully connected layer was used as ANN needs to be connected all together (input, multiple hidden layers, and output layers).

The model must be compiled with three essential attributes after it has been fully defined. There is an optimizer model, a loss function, and metrics. The optimizer is the search approach that is used to change model weights. Some popular gradient optimizers exist, such as Stochastic Gradient Descent (SGD), RMSprop, and Adam. Adam is an adaptive estimate of the moment that is commonly used for adaptive learning speeds. The second attribute is functioning relating to the model loss. Loss function, also called goal function, is the model equation used by the loss statement. Binary cross-entropy is used in this research and is suited to binary logarithmic loss. The last model attribute that is used is precision. It is to evaluate during training by the model.

Then, the model undergoes the process of preparation. For training, you need to specify the number of epochs and the batch size. Epochs are the number of times the model has been solved to the training dataset. The batch size is also the number of training details shown to the model before carrying out a weight update. Both the epoch and batch size for this research is set at 1000 and 10, respectively. That is because it is suitable for dataset size.

Finally, once the model has been thoroughly trained, the model prediction is established. The test data is used to make predictions to ensure the model is correct or needs modifications.

3.2 K-Means Clustering Algorithm

Clustering is the process of partitioning or grouping a given set of patterns into disjoint "clusters." Clustering is done when the patterns are the same and will be in the same cluster and vice versa. Clustering has been an important issue of research in a variety of applications, including neural networks. In decades ago, the k - means method was proven effective in producing good clustering results for many practical applications. The computational time, however, is directly costly to increase the size of the datasets.

There is a variety of cluster analyses including clustering based on prototypes, clustering based on densities, and hierarchical clustering. It includes the k - means algorithm, LVQ, and mixture-of-Gaussian clustering in prototype-based clustering, and the K-means algorithm is well known for its effectiveness in clustering large scale datasets. Other than that, the authors described in [8] that the K-Mean Clustering algorithm is applied to heterogeneous datasets, and the point variability of each point in the dataset is calculated.

The algorithm is at first assigned randomly to cluster number 1 and denoted c1 based on the datasets, X. Then, the distance to c1 is calculated for all data. After that, X will give the next centroid, c2, at random. The final step is to ensure that all datasets belong to either c1 or c2 by calculating the distance from themselves to the nearest centre. The algorithm is replicated until the entire dataset is complete.

3.3 A Combined Deep Neural Network (DNN) and K-Means Clustering Algorithm

In this study, a novel method of combining DNN and clustering techniques is proposed for estimating the parameter of Lorenz's chaotic system. Lorenz dataset that obtained from Professor Roberto [18] consists of enormous data to be processed. The previous study has about processed the dataset using only Deep Neural Network (DNN) architecture. When applying only Deep Neural Network, the result is only 60%. This result has been submitted to the 4th International Conference on Electric, Electronic, Communication, and Control 2019 (ICEECC'19).

Figure 1 is a flowchart explaining the step-by-step of the proposed method. The first step is data acquisition and algorithm development. Obviously, there are two algorithms involved in this method, which are the K-Means clustering algorithm and DNN. Each of the algorithms has its own hyperparameters that had been tuned accurately. The tuning process is using heuristic tuning or based on the trial-and-error process. Secondly, the data will be processed through the K-Means Algorithm. In this algorithm, the user will give any value of K, represent the cluster number. Then, Silhouette analysis is employed to make sure the K value is suitable for the dataset or not.

Silhouette analysis is to determine the degree of separation between clusters. The computation of the Silhouette coefficient is as below.

$$\frac{b^i - a^i}{\max\left(a^i, b^i\right)} \tag{1}$$

where a^i is the value of the average distance from all data points in the same cluster, and b^i is the value of the average distance from all data points in the closest cluster. The coefficient values in the interval [− 1, 1]. The best sample of K value is when the coefficient value more than 0.8.

Thirdly, after the clustering process is done, the dataset will be trained using the DNN model, and all the hyperparameters will be fine-tuned manually. The tuning is already published in a previous research paper. The summary of the tuning will be provided in the next section. Lastly, the result will be validated with the testing set, and the model performance will be presented.

Subsequently, each testing set of the clusters obtained from the Lorenz dataset will be conducted testing process. Parameter estimation is generated by collecting estimation results conducted by DNN models, as shown in Fig. 2.

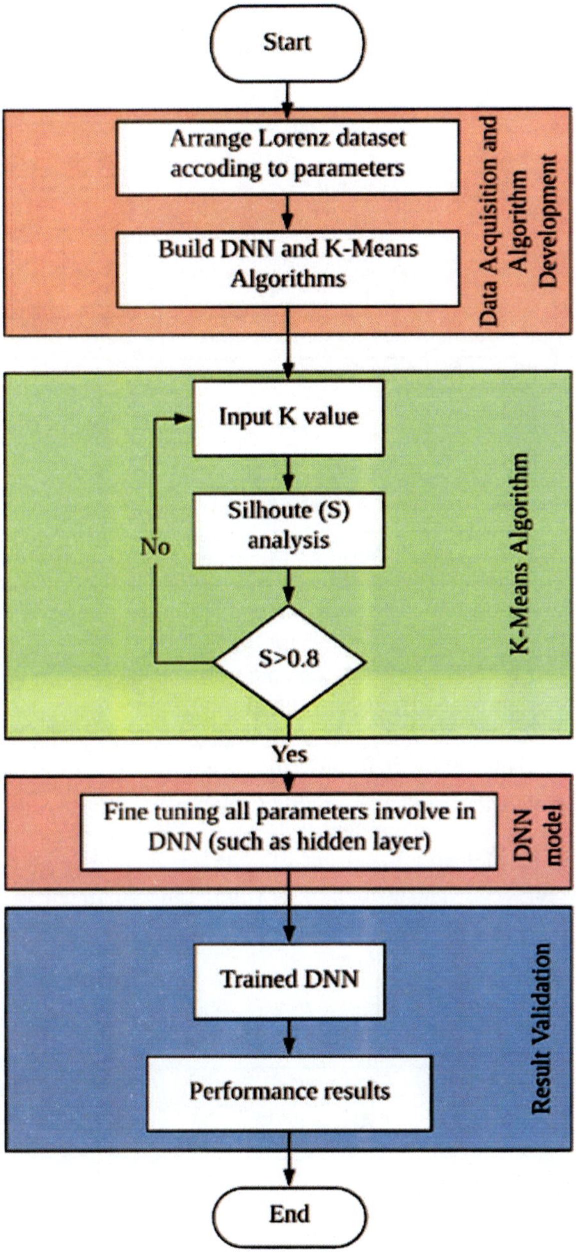

Fig. 1 Flowchart of the combination of deep neural network and K-means clustering algorithm

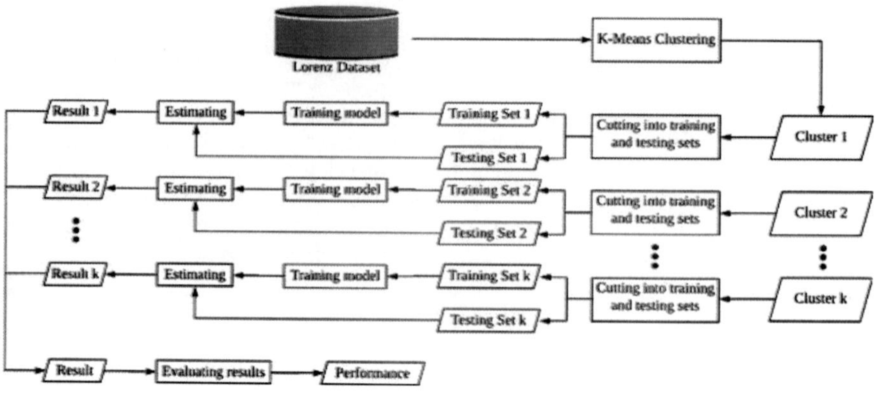

Fig. 2 Flows of constructing the DNN based method with the K-Means algorithm

4 Result and Discussions

For this section, the proposed combined algorithm is used to solve parameter estimation problems. The performance of the proposed combine algorithm will be analyzed in terms of accuracy and loss data. In this analysis, there are three distance metrics had been employed to test the proposed algorithm for the Lorenz database to make sure the effectiveness of the model.

Based on Table 1, it is summarized the mean percentage of accuracy and loss for each clusters with three distance metrics. The result shows that the highest accuracy among all is Cluster 2 using proposed algorithm with cosine function. In a nutshell, the performance index of the proposed algorithm is affected by the number of data in each cluster. From the table, it is proven that the number of data in cluster affect the performance of the algorithm. The smaller the number of data in cluster, the performance index is increased.

Table 1 Summary of the accuracy and loss mean percentage

No. of cluster	Distance metrics	Euclidean squared	Cosine	SAD
Cluster 1	No. of data	2800	3211	1046
	Accuracy (%)	62.03	62.52	60.77
	Loss (%)	65.73	65.25	66.98
Cluster 2	No. of data	799	388	580
	Accuracy (%)	68.89	70.02	68.63
	Loss (%)	58.79	58.25	57.69
Cluster 3	No. of Data			1973
	Accuracy (%)			62.47
	Loss (%)			65.60

Table 2 The comparison between DNN and proposed algorithm according to accuracy value

Model	DNN model	Combination of DNN and K-means clustering algorithm (proposed algorithm)		
		Euclidean squared	Cosine	SAD
Mean (%)	64.49	65.46	66.27*	63.96
Maximum	0.6440	0.7243	0.7287*	0.7142
Minimum	0.6216	0.6104	0.6150	0.6033*

* Best value

Table 3 The comparison between DNN and proposed algorithm according to loss value

Model	DNN model	Combination of DNN and K-Means Clustering Algorithm (Proposed Algorithm)		
		Euclidean Squared	Cosine	SAD
Mean (%)	63.31	62.26	61.75*	63.42
Maximum	0.6182	0.6713	0.6708	0.6771*
Minimum	0.6332	0.5507	0.5537	0.5353*

* Best value

Next is the comparison between DNN and the proposed algorithm is discussed. The result comparison is divided into two parts, which are accuracy and loss value with three different measuring metrics—the value with an asterisk (*) mark is the best value among all. Based on Table 2, the tabulated result is based on accuracy value. There are three metrics that will be compared. Firstly, the best value is the proposed algorithm using the Cosine function with 66.27%. After that, the highest maximum value is also a Cosine function by the proposed algorithm, while the lowest minimum value is the SAD function by the proposed algorithm too. From Table 3, it is visualized the result according to the loss value. The result does not show a significant difference from the previous table. The best value for mean is by the proposed algorithm using Cosine function, while the proposed algorithm using the SAD function gives the best value for maximum and minimum values.

As a summary from the previous analysis (Tables 2 and 3), the K-Means Clustering algorithm did the role very well. With the dataset already been clustered using the well-known clustering algorithm, K-Means Clustering algorithm, the algorithm is helped to improve the accuracy and minimize the loss of the model.

Experimental results for both training and testing datasets by using DNN and proposed algorithms are tabulated in Table 4. From the table, the proposed algorithm using cosine function as distance metric is outperformed other methods with 98% correct prediction for training data and 73% of testing data. On the other hand, the proposed algorithm with the Euclidean squared distance metric gives the lowest accuracy of training data. This is maybe caused by the splitting data process is not optimum by using Euclidean squared compared with Cosine and SAD distance metric function.

Table 4 Experimental result for training and testing dataset by using DNN and proposed algorithms

Model	DNN model	Combination of DNN and K-means clustering algorithm (proposed algorithm)		
		Euclidean squared	Cosine	SAD
Training data (%)	91	89	98*	92
Testing data (%)	40	55	73*	71

5 Conclusion

The new combination algorithm, DNN with K-Means Clustering algorithm, is proposed in this study. As the literature says, an excellent profound model of learning must involve both supervised and unsupervised to produce an impactful algorithm. That is why the combination algorithm had been introduced in this study, and it is clearly improved the percentage of accuracy of training and testing data. Accuracy by using 80:20 ratio of training and test data gives result 98% of accurate training data, and 73% of test data are predicted with the proposed algorithm while 91 and 40% of the DNN only model are predicted in training and test data. The extensive work is to train the proposed algorithm for recommendations in the real fields of engineering and non-engineering. It is to verify that further modifications are required in the algorithm developed. Additionally, an optimization algorithm will be used to optimize all the hyperparameters available in deep learning.

References

1. Zhuang L, Cao L, Wu Y, Zhong Y, Zhangzhong L, Zheng W, Wang L (2020) Parameter estimation of lorenz chaotic system based on a hybrid Jaya-Powell algorithm. IEEE Access 8:20514–20522
2. Mousavi Y, Alfi A (2018) Fractional calculus-based firefly algorithm applied to parameter estimation of chaotic systems. Chaos Solitons Fractals 114:202–215
3. Wang J, Zhou B (2016) A hybrid adaptive cuckoo search optimization algorithm for the problem of chaotic systems parameter estimation. Neural Comput Appl 27:1511–1517
4. Peng Y, Sun K, He S, Yang X (2018) Parameter estimation of a complex chaotic system with unknown initial values. Eur Phys J Plus 133
5. Huang Y, Liu YF, Peng ZM, Ding YJ (2015) Research on particle swarm optimization algorithm with characteristic of quantum parallel and its application in parameter estimation for fractional-order chaotic systems. Wuli Xuebao/Acta Phys Sin 64. https://doi.org/10.7498/aps.64.030505
6. Chen Z, Yuan X, Wang X, Yuan Y (2019) Parameter estimation of chaotic systems based on extreme value points. Pramana J Phys 92:1–19
7. Cui R, Wei Y, Chen Y, Cheng S, Wang Y (2017) An innovative parameter estimation for fractional-order systems in the presence of outliers. Nonlinear Dyn 89:453–463
8. Lazzús JA, Rivera M, López-Caraballo CH (2016) Parameter estimation of Lorenz chaotic system using a hybrid swarm intelligence algorithm. Phys Lett Sect A Gen At Solid State Phys 380:1164–1171
9. Ho WH, Chou JH, Guo CY (2010) Parameter identification of chaotic systems using improved differential evolution algorithm. Nonlinear Dyn 61:29–41

10. Wei J, Yu Y (2017) An effective hybrid cuckoo search algorithm for unknown parameters and time delays estimation of chaotic systems. IEEE Access 6:6560–6571
11. Gu W, Yu Y, Hu W (2017) Artificial bee colony algorithmbased parameter estimation of fractional-order chaotic system with time delay. IEEE/CAA J Autom Sin 4:107–113
12. Chen S, Yan R (2016) Parameter estimation for chaotic systems based on improved boundary chicken swarm optimization. Infrared Technol Appl Robot Sens Adv Control 10157:101571K
13. Xu S, Wang Y, Liu X (2018) Parameter estimation for chaotic systems via a hybrid flower pollination algorithm. Neural Comput Appl 30:2607–2623
14. Zhang H, Li B, Zhang J, Qin Y, Feng X, Liu B (2016) Parameter estimation of nonlinear chaotic system by improved TLBO strategy. Soft Comput 20:4965–4980
15. Xu C, Yang R (2017) Parameter estimation for chaotic systems using improved bird swarm algorithm. Mod Phys Lett B 31:1–15
16. Mariño IP, Míguez J (2006) An approximate gradient-descent method for joint parameter estimation and synchronization of coupled chaotic systems. Phys Lett Sect A Gen At Solid State Phys 351:262–267
17. Xishuang D, Lijun Q, Lei H (2017) Short-term load forecasting in smart grid: a combined CNN and K-means clustering approach. In: 2017 IEEE international conference on big data and smart computing (BigComp), pp 119–125
18. Barrio R, Dena A, Tucker W (2015) A database of rigorous and high-precision periodic orbits of the Lorenz model. Comput Phys Commun 194:76–83

Design and Performance Analysis of Body Worn Textile Antenna Using 100% Polyester at 2.4 GHz for Wireless Applications

Shehab Khan Noor, Nurulazlina Ramli, Najah Najibah Zaini, and N. H. Abd Rahman

Abstract International Mobile Telecommunications-2020 (IMT-2020) is focused on creating a mobile ecosystem with a reasonable price and user friendly. With these reasons, the need for a communications framework that could be deployed at an affordable cost, compact and ease of mobility brought forward the concept of wearable technology. Wearable devices such as textile antennas are being developed with the potential to track, notify and demand attention where hospital emergencies are necessary. However, conventional antenna designs have a rigid structure, limited bandwidth, costly metallization and lack effectiveness. Therefore, in this paper, a simulation of a textile wearable antenna using 100% polyester as a substrate is designed for a wireless application at 2.4 GHz frequency. The antenna performance is observed in terms of reflection coefficient, bandwidth, Voltage Standing Wave Ratio (VSWR), gain and radiation pattern along with a thinner substrate compared to previous works to justify the validity of the current design proposed. The research paper has many possibilities for the future and could assist with when designing and manufacturing flexible and comfortable wearable devices for everyday use.

Keywords On-body communication · 100% polyester antenna · Textile antenna · Wearable devices · Wireless technology

1 Introduction

Antennas are now so significant where human beings can communicate and share information through air and space to each other through wireless communication [1]. Wearable fabric-based antennas have become one of the predominant research topics

S. K. Noor (✉) · N. Ramli · N. N. Zaini
Faculty of Engineering and the Built Environment, Centre for Advanced Electrical and Electronic Systems (CAEES), SEGi University, Kota Damansara, 47810 Petaling Jaya, Selangor, Malaysia
e-mail: shehabrabid97@gmail.com

N. H. Abd Rahman
Faculty of Electrical Engineering, Antenna Research Centre (ARC), Universiti Teknologi MARA, Shah Alam, Malaysia

A. F. Ab. Nasir et al. (eds.), *Recent Trends in Mechatronics Towards Industry 4.0*, Lecture Notes in Electrical Engineering 730, https://doi.org/10.1007/978-981-33-4597-3_31

in a body-centered communication system. Typically, wearable antenna specifications involve lightweight, low cost, and almost maintenance-free with no installation required for all modern applications [2]. There are various specialized branches of the profession that apply body-centric communication systems which include paramedics, firefighters, and armed forces. As per the report published by Analysis Mason [3], sales of wearable devices are expected to hit US$ 22 billion in 2020 while revenues in 2014 were below US$ 3 billion.

Over the last few years, wearable wireless communication systems have become well-known. The previous publications focused on the discussion on the design, fabrication and applications of wearable antennas. In [4] the authors have proposed textile wearable antennas using polyester fabric as the substrate to operate at 2.4 GHz and at 2.45 GHz in [5]. Meanwhile, the investigation of jeans as a substrate for antenna design and its performance is studied in [6, 7]. Besides that, the quantum leaps use of leather as antenna substratum has also been studied [8, 9].

In deciding the substrate of the antenna, three key parameters that needed to be considered are in [4–9]. They are electromagnetic properties of textile materials which include thickness, loss tangent value, and permittivity. Material such as jeans used in [6, 7] has permittivity value of between 1.59 and 1.70. Those materials have a thickness value of 1 mm and 1.6 mm along with a loss tangent value of 0.024 and 0.025. Besides, the proposed antennas in [4, 5] using polyester fabric as a substrate have a permittivity value of 1.4 and 1.748 along with a thickness value of 2.85 mm and 0.28 mm. The loss tangent value is 0.01 and 0.0044 respectively. Moreover, leather fabric in [8, 9] has a permittivity value of 1.8, thickness value 0.5 mm, and 1.39 mm. However, the authors have not provided the data for loss tangent. In all these previous works, the authors have analyzed antenna performance in terms of gain especially along with parameters such as reflection coefficient and Voltage Standing Wave Ratio (VSWR) to make sure the antenna meets the requirements which are shown in Sect. 4 along with the results of this work.

From the previous researchers [4, 5], the antenna substrate used is polyester. However, the gain achieved was considerably low while operating at 2.4 GHz with the gain of 3 dB [4] and operating at 2.45 GHz with the gain achieved was 2.51 dB [5]. In [6, 7], Jeans fabric as the substrate has been used and the gain achieved was below 3.4 dB. Although, authors in [9] achieved 5.94 dB as gain at 2.44 GHz by using leather fabric with the thickness of the fabric was 1.39 mm. Higher the gain, the antenna can transfer more power in the desired direction and reduces receive interference from other (low gain) directions. However, the overall geometric dimension of the antenna increases because of the thicker substrate [10]. This can make the wearer uncomfortable and makes it difficult to integrate into clothing.

In accordance with all these above-mentioned facts, the aim of this research is to design a wearable textile antenna that operates at 2.4 GHz with a better performance in terms of gain using 100% polyester as an antenna substrate. The Computer Simulation Tool (CST Studio 2019) is used for virtual simulation purposes. Section 2 describes the methodology which comprises the textile material that has been used for the proposed design and the numerical dimensions of the antenna. Section 3 focuses on the results and discussions of the proposed design, which includes the reflection coefficient, VSWR, radiation pattern (3D and 2D) and the gains achieved. The conclusion of this research paper is set out in Sect. 4.

2 Methodology

This section discusses the methodological aspect of this paper.

2.1 Wearable Antenna Material

Textile materials are an interesting substrate for flexible antennas as fabric antennas can easily be incorporated into clothing. Textile materials have generally very low dielectric constant which reduces the loss of the surface wave and eventually improving the antenna impedance bandwidth [7]. In this paper, a textile-based wearable antenna was developed and analyzed. Fabrics such as polyester is considered in this paper to analyze its suitability for use as substrate materials. The textile material proposed for the substrate is 100% polyester fabric with good durability is strong yet light and is easily available in the market [11]. However, the conductive parts are made of a 0.035 mm thick copper considering its low cost, wide availability, and durability. Such features together make the antenna flexible in nature. Nevertheless, it is important to determine its characteristics in order to model a fabric that is later used in the making of a wearable antenna. The relative permittivity, thickness, and loss tangent value for the polyester fabric were measured. Three several positions (in the diagonal pattern) have been measured using the probe test to achieve the most optimal and reliable value as shown in Fig. 1. The results obtained at three different textile positions are summarized in Table 1. The average from the measurements is used as the final characteristics of the permittivity, tangent loss, and the thickness for the polyester textile material.

2.2 Design Specification of Proposed Antenna

Prior to pre-designing, the desired parameter is considered significantly because it will considerably influence the antenna's overall performance. The key concept was to build a wearable antenna that can radiate through a fabric where the antenna will be operating at 2.4 GHz. The antenna must have a minimum return loss (S11) value below −10 dB, VSWR below or equals 2, and the impedance of the line to match at 50 Ω as shown in Table 2.

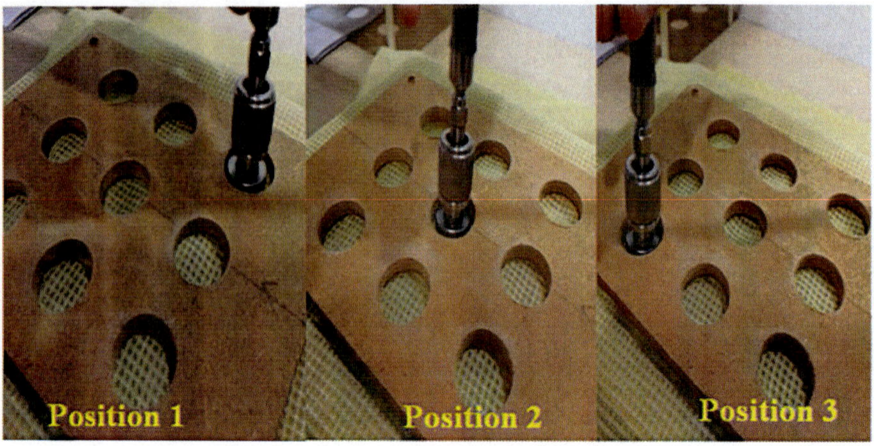

Fig. 1 Textile properties measurement in three different positions

Table 1 Dielectric properties of 100% polyester fabric

Textile properties	Position 1	Position 2	Position 3	Average
Loss tangent	0.143	0.2099	0.1571	0.17
Relative permittivity (εr)	1.437	1.5074	1.5753	1.51
Thickness (mm)	0.27	0.27	0.27	0.27

Table 2 Design specification of the antenna

Specification	Values
Frequency (f)	2.4 GHz
Reflection coefficient (S11)	Less than −10 dB
VSWR	1:2
Input Impedance (Z)	50 Ω
Copper thickness (mm)	0.035 mm

2.3 Antenna Dimension

Figure 2 shows the design structure and parameters of the proposed antenna. Firstly, the operating frequency of the antenna at 2.4 GHz and polyester as a substrate material has been decided to be used in this research paper. Moreover, the copper with 0.035 mm thickness is used as a radiating element and forms the base of the ground structure. From Sect. 2.1, the final specification of the polyester substrate has a permittivity of 1.51, a tangent loss of 0.17, and the 0.27 mm thickness. Next, the dimension of the proposed antenna parameters is calculated by using the formulas (1)–(7) as shown below [12].

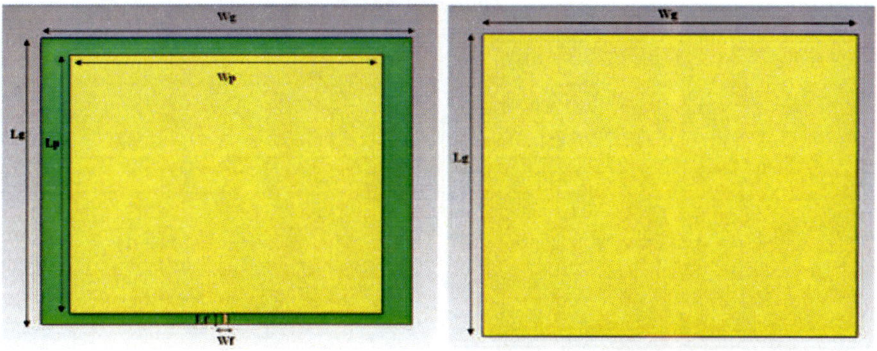

(a) (b)

Fig. 2 CST simulated design structure of the proposed antenna **a** front view; **b** back view

The width, *Wp* and the length *Lp*, of the patch are calculated using the equations as given below:

$$Wp = \frac{c}{2fo\sqrt{\frac{(\varepsilon r+1)}{2}}} \tag{1}$$

Here $C = 3 \times 10^8$ m/s (light speed), $\varepsilon r = 1.51$ and fo = 2.4 GHz

$$L_p = L_{\text{reff}} - 2\Delta L \tag{2}$$

where L_{reff} can be found using:

$$L_{\text{reff}} = \frac{c}{2fo\sqrt{\varepsilon_{\text{reff}}}} \tag{3}$$

$$\varepsilon_{\text{reff}} = \frac{\varepsilon r+1}{2} + \frac{\varepsilon r-1}{2}\left[1+12\frac{h}{w}\right]^{\frac{-1}{2}} \tag{4}$$

Moreover, ΔL can be found using:

$$\Delta L = 0.412h\left(\frac{\varepsilon_{\text{reff}}+3}{\varepsilon_{\text{reff}}-0.258}\right)\left(\frac{\frac{w}{h+0.264}}{\frac{w}{h+0.813}}\right) \tag{5}$$

The width and length of the ground plane can be found using:

$$Wg = 6h + Wp \tag{6}$$

$$Lg = 6h + Lp \tag{7}$$

Table 3 Finalized antenna dimensions

Parameter	Notation	Calculated value (mm)	Optimized value (mm)
Patch width	Wp	55.79	59
Patch length	Lp	50.63	49.75
Ground width	Wg	57.37	70
Ground length	Lg	52.25	55
Feedline width	Wf	1.8	1.01
Feedline length	Lf	2.8	2.00

here h = height of the substrate.

The formulas and the final dimensions of the polyester antenna are tabulated in Table 3. The numerical results were not achieved at 2.4 GHz. Thus, the optimization has been done until the frequency is resonant at 2.4 GHz. Overall, the antenna's dimension (Wg and Lg) has increased by 28.43% to achieve the resonant frequency. In addition, the dimension of the patch (Wp and Lp) has increased slightly by 3.9%. However, after optimizing the antenna to achieve 2.4 GHz as the resonant frequency, dimension of the feedline (Wf and Lf) has decreased from 5.05 mm to 2.02 mm. In other words, the feedline area is reduced by 60%.

3 Results and Discussion

3.1 Reflection Coefficient and Bandwidth

Once the antenna design achieved the desired operating frequency of 2.4 GHz, the simulation results of the proposed antenna were analyzed. The most important specification is the reflection coefficient or known as S11. The reflection coefficient means the amount of power wasted in the load and does not return as a reflection. The acceptance values of S11 must be below than −10 dB [8]. Figure 3 illustrates the simulated reflection coefficient with −10.13 dB at a resonant frequency of 2.4 GHz. Besides that, the bandwidth of 3.7578 MHz has been achieved for the proposed textile patch antenna which has also been demonstrated in Fig. 3. Having narrow bandwidth such as 3.7578 MHz is important for applications which require longer range, high power efficiency, high security and low system complexity.

3.2 Voltage Standing Wave Ratio (VSWR)

The Voltage Standing Wave Ratio (VSWR) is known as the ratio of the maximum voltage in a standing wave pattern to the minimum voltage while reflected from a

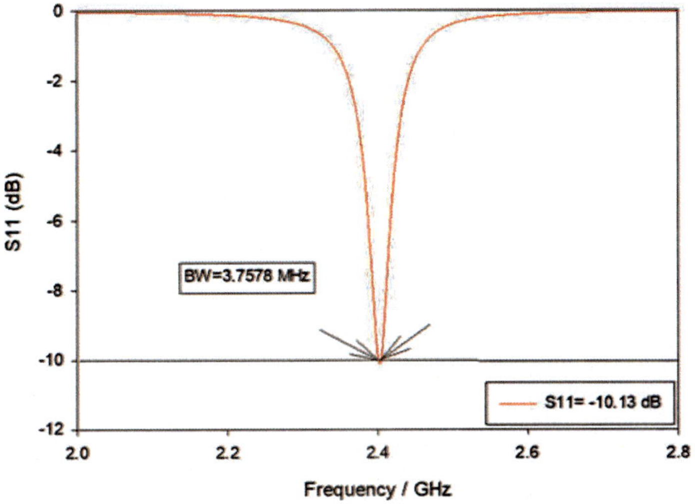

Fig. 3 Simulation results S11 and bandwidth of 100% polyester antenna at 2.4 GHz

load where a standing wave is formed. For the ideal configuration of the feed line to the designed antenna, the VSWR value must be less than 2 [8]. In Fig. 4, it can be seen that the VSWR is acceptable since it is less than 2, where the obtained was valued at 1.91 when at 2.4 GHz.

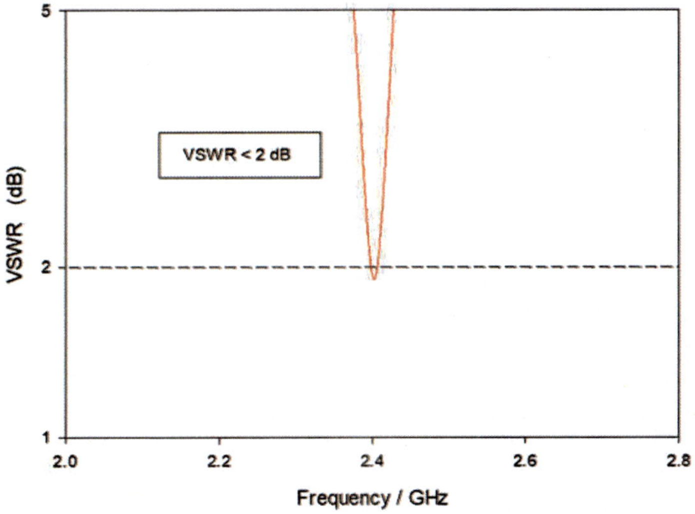

Fig. 4 Simulation results VSWR of 100% polyester antenna at 2.4 GHz

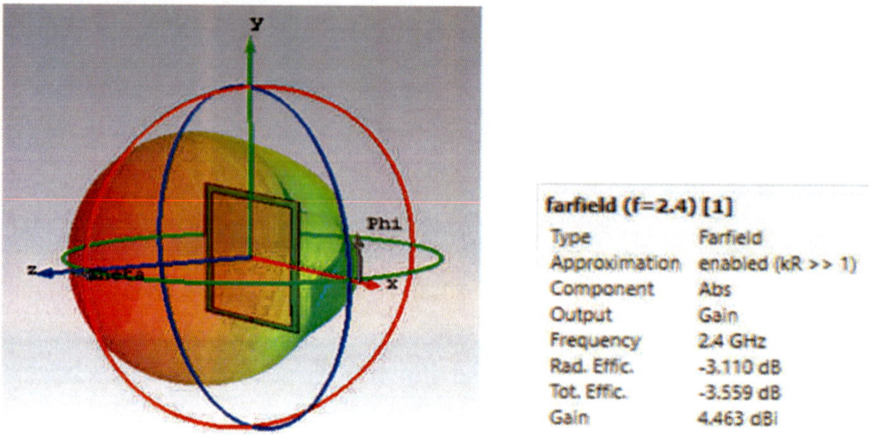

Fig. 5 3D view of radiation pattern for 100% polyester antenna

3.3 Gain

The gain is an indicator of the antennas potential to direct the input power in the single direction into radiation, which is determined at just the peak radiation intensity [8]. Figure 5 demonstrates the 3D view of the designed antenna radiation pattern at its desired frequency. The highest radiation intensity is the red color seen in Fig. 5 which indicated that the gain obtained was 4.463 dB, which is relatively high since the dielectric constant value of the fabric was relatively low. Since the proposed design has relatively higher gain value of 4.463 dB hence the antenna would allow more of the transmitted power to be sent in the direction of the receiver as it is transmitted, improving the signal strength received. In addition, the antenna of this research work has the potential to capture more of the signal while transmitting, again increasing signal strength.

The radiation pattern in the 2D viewpoint is also known as the polar pattern. Figure 6 displays the radiation pattern at a frequency of 2.4 GHz for the 100% polyester fabric. The simulation software provided data on the frequency, magnitude of the main lobe, the direction of the main lobe, angular width (3 dB) and depth of the side lobe. In Fig. 6, the value for the magnitude of the main lobe was 4.47 dB and the main lobe direction of a signal was at 3.0°. In addition, the angular width moved to 81.2°. Lastly, the side lobe level was noted to be 15.0 dB. As illustrated in Fig. 6, the proposed antenna design has directional radiation pattern hence the antenna is capable to cover long range in one particular direction.

Table 4 shows the overall results of the simulated antenna. The proposed antenna achieved a high gain value of 4.463 dB when operated at 2.4 GHz. Furthermore, the results obtained in terms of Reflection Coefficient and VSWR indicated that the antenna is suitable for wireless applications. The research work of this paper has been evaluated by comparing with other works published by [4–9] and can be seen

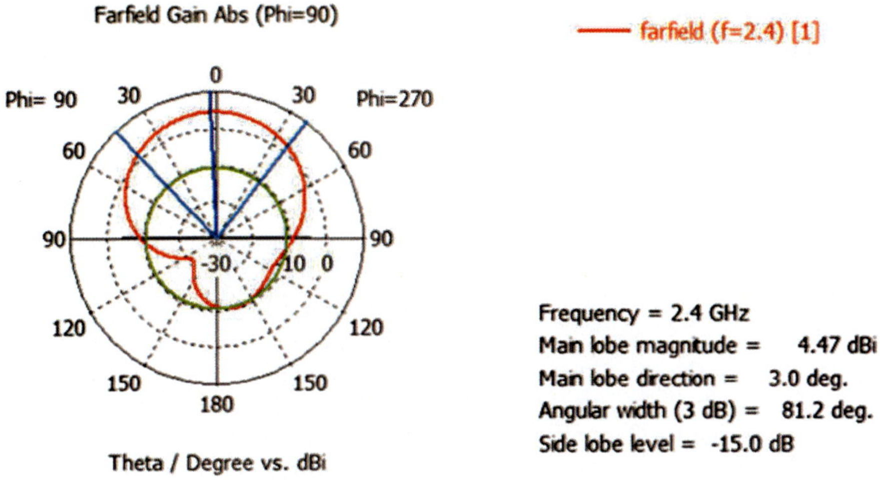

Fig. 6 Gain of 100% polyester antenna in 2D view

Table 4 Overall obtained results for 100% polyester antenna

Specification	Values
Material	100% polyester
Frequency (GHz)	2.4
S_{11} (dB)	−10.13
VSWR	1.911
Gain (dB)	4.463
Bandwidth (MHz)	3.7578

in Table 5. From the table, the fabrics are in the range of 0.27 mm (the proposed antenna) to 2.85 mm of thickness and the gains were achieved about 2.51–5.94 dB. Based on these data, the proposed design of this paper is much more flexible since the thickness is less than 0.27 mm whilst achieving a higher gain of 4.463 dB when

Table 5 Comparison of the proposed antenna with others reference antennas

Reference	Frequency (GHz)	Substrate thickness (mm)	Antenna gain (dB)
[4]	2.4	2.85	3
[5]	2.45	0.28	2.51
[6]	2.44	1.0	2.6
[7]	2.0–3.0	1.0	2.6 – 3.6
[8]	2.45	0.5	4.79
[9]	2.44	1.39	5.94
This paper	2.4	0.27	4.463

compared against other published papers [4–9]. The proposed antenna can be used in health-care, sports and armed forces where wireless body tracking is required.

4 Conclusion

In this paper, the simulation and performance analysis of body worn textile antenna using 100% polyester as substrate was successfully designed to operate at 2.4 GHz. The gain of the antenna is considerably higher at 4.463 dB where the thickness was only 0.27 mm. The 2D and 3D radiation patterns also provided good signal radiation when the main lobe direction was at 3.0° and the main lobe direction at 4.447 dBi with the signal coverage will be about 81.2°. However, the performance of the antenna in terms of the reflection coefficient of −10.13 dB and the VSWR of 1.911 should be improved to maintain a good antennas' performance especially during the measurement of fabricated antenna or in real implementation. The results concluded that for a wearable telecommunications device to be functioning accurately, the device would need to be in close contact with the human body. In the future, the care and attention would have to be paid to the Specific Absorption Rate (SAR) to prevent damage to the human body. Overall, the proposed antenna design has the potential to be used in wireless application technology with further improvements needed.

References

1. Al Kharusi KWS, Ramli N, Khan S, Ali MT, Abdul Halim MH (2020) Gain enhancement of rectangular microstrip patch antenna using air gap at 2.4 GHz. Int J Nanoelectronics Mater 13(Special Issue):211–224
2. Rais NHM, Soh PJ, Malek F, Ahmad S, Hashim NBM, Hall PS (2009) A review of wearable antenna. In: Loughborough antennas and propagation conference, Loughborough, pp 225–228
3. Sabri S, Sam SM, Kamardin K, Daud SM, Salleh NA (2016) Review of the current design on wearable antenna in medical field and its challenges. J Teknologi 111–117
4. Jaiswal P, Sinha P (2018) design of wearable textile based microstrip patch antenna for bandwidth enhancement. Int J Appl Eng Res 13(18):13647–13651
5. Mersani A, Osman L (2016) Design of dual-band textile antenna for 2.45/5.8-GHz wireless applications. In: 5th International conference on multimedia computing and systems (ICMCS), pp 397–399, Marrakech
6. Yang H-C, Azeez HL, Wu CK, Chen WS (2017) Design of a fully textile dual band patch antenna using denim Fabric. In: IEEE international conference on computational electromagnetics (ICCEM), pp 185–187, Kumamoto
7. Osmana MAR, Rahima MKA, Samsuria NA, Elbasheera MK, Alia ME (2012) UWB wearable textile antenna. J Teknologi Sci Eng 58(1):39–44
8. Jeyakumar S, Sakthimurugan K (2017) Wearable textile antenna for ISM band with different dielectric substrate materials. Int J Electron Eng Res 9(8):1259–1266
9. Ahmed I, Rahman M (2017) Design and optimization of a textile antenna for wearable communication. In: IEEE international conference power, control, signals and instrumentation engineering (ICPCSI), pp 0–4

10. Salvado R, Loss C, Gonçalves R, Pinho P (2012) Textile materials for the design of wearable antennas: a survey. Sensors 12:15841–15857
11. Ramachandran T, Sampath MB, Senthilkumar M (2009) Micro polyester fibers for moisture management. Ind Text J 21
12. Balanis CA (1992) Antenna theory: a review. Proc IEEE 80(1):7–23

Simulation on Circularly Polarization Cotton Textile Antenna for Wireless Communication System

Taher Khalifa, Nurulazlina Ramli, Anis Fariza Md. Pazil, N. H. Abd Rahman, and Ahmad Jais Alias

Abstract The fast development of the wearable textile antennas in recent times due to the modern miniaturization of the electronic devices, which included all civil and military fields. It is a part of the clothing used for the communication purposes such as tracking, navigation and fitness, as well as for medical purposes inside hospitals. In this paper, the design of the wearable textile antenna that operates at 2.4 GHz for wireless the communication system is discussed. The antenna is proposed with plus-shaped is employed on the substrate of cotton textile with dielectric constant, $\varepsilon r = 1.6$, the tangent delta loss, $\delta = 0.0400$ dB and the thickness of the substrate is $h = 0.27$ mm with a single microstrip feed line impedance at 50 Ω. The center frequency at 2.4 GHz with the reflection coefficient of −20.01 dB, the gain of 7.44 dB and the efficiency is about 56.79% of the proposed antenna. The left-hand circular polarization (LHCP) has obtained in the cotton textile antenna as AR is successfully achieved below 3 dB. CST used to the design and the simulation results of the antennas performance. The proposed cotton textile antenna is suites to be applied for the future wireless communication system.

Keywords Axial ratio · Circular polarization · Wearable · Wireless

T. Khalifa (✉) · N. Ramli · A. F. Md. Pazil
Centre of Advanced Electrical and Electronic Systems (CAEES), Faculty of Engineering, Built Environment and Information Technology, SEGi University, Kota Damansara, 47810 Petaling Jaya, Selangor, Malaysia
e-mail: tahersunossi@gmail.com

N. H. Abd Rahman
Antenna Research Centre (ARC), Faculty of Electrical Engineering, Universiti Teknologi MARA, 40450 Shah Alam, Selangor, Malaysia

A. J. Alias
Cerna Minda Institute, 43650 Bandar Baru Bangi, Selangor, Malaysia

A. F. Ab. Nasir et al. (eds.), *Recent Trends in Mechatronics Towards Industry 4.0*, Lecture Notes in Electrical Engineering 730, https://doi.org/10.1007/978-981-33-4597-3_32

1 Introduction

Have you ever dreamed of having a personal correspondent in your clothes? This dream may be closer to reality. Recently an antenna has been updated that can be wearable textile materials on your shirt of a rectangular or square or even circular patch that contains personal communication capabilities and GPS tracking of the preferred [1]. The biggest challenge was building an antenna on a material that was not only functional, but also flexible. Because the patch is based on a flexible conductive material such as the polyester material for the rectangular antenna [2]. Wearable textile antennas have long the characterized smart technology with textiles, which are actually the garments. Smart textiles are the natural development of the wearable textile antennas industry [3]. Wearable textiles antennas are trending to the fore in the wearable technology is the future and are likely to be cheaper than you think.

The wearable textile antenna characteristics, which include the dielectric constant and tangent delta loss, Which has an important role in the propagation of the electromagnetic waves through the electrical constant of the substrate and its direct effect on the textile antenna performance [4]. Specific requirements for the wearable textile antennas are a flat structure and flexible construction materials. Some characteristics of the different materials affected the wearable textile antenna [5]. The different materials used to design and develop the wearable textile antenna are not affect by heavy action, bending of the human body or even body humidity because that are designed for that [6]. Then the electrical characteristics of the material had to be measure using a number of techniques. The next goal was to determine the performance of the antenna when the wearer is moving or bending [7]. On the other hand, there are some the disadvantages in the wearable textile antenna design, which is choice of the center frequency of an antenna there is a direct influence on the antenna design dimensions, when selecting the resonant frequency, there is an inverse proportion between the frequency and the dimensions of the antenna design. While, the fast development of this research, have made researchers obtain thin threads of the textile yarn for use in designing the wearable textile antennas for the fitness clothes and the medical purposes [8].

The cotton is one of the most widely produced textiles today. It has an excellent electrical characteristic, which the hygroscopic and sweat releasing characteristic blocked the cotton materials of developing the static electricity [9]. For it's the thermal conductivity characteristics, the cotton materials have the capability to behavior heat and minimize any destructive heat accumulation ratio [10]. This paper study the different types of wearable textiles antennas in order to provide the background information and the application ideas for the designing antennas as shown in Table 1.

The different materials using in the development of wearable textile antennas, their fabrication methods, as well as the antenna types and there application fields were decreased [16]. In this paper, a circular polarization cotton antenna is designed at 2.4 GHz operating frequency for wireless communication systems. An important requirement for different parameters such as bandwidth, reflection coefficient, axial

Table 1 Dielectric textile material characteristics

Textile material	εr	tan δ	References
Cordura	1.90	0.0098	[11, 12]
100% cotton	1.60	0.0400	[13, 14]
100% polyester	1.90	0.0045	[11]
Quartzel fabric	1.95	0.0004	[12]
Cordura/Lycra	1.50	0.0093	[15]

ratio, VSWR, the radiation pattern characteristics, the directivity, and gain had been analyse. The analysis are conducted through the simulation process using CST software. In Sect. 2, methodology of designing a cotton textile antenna has been explain and Sect. 3 focuses on the results and discussions. The conclusion from this research is explained in Sect. 4.

2 Methodology

n this paper, a cotton textile material has been used as a substrate in the antenna design while the radiating element of patch and ground are made from the copper tape with a thickness of 0.00018 mm. The properties of the cotton material have been measured first using the coaxial probe test method at the Antenna Research Centre, Universiti Teknologi MARA. Nine points places have been tested on the cotton textile material and the average has been taken in order to determine its properties. The cotton textile material consists of dielectric constant, εr = 1.60, the tangent delta loss, δ = 0.0400 dB, and the thickness of the substrate is h = 0.27 mm. The antenna is fed with the microstrip feed line of 50 Ω impedance. The proposed antenna must meet the design specification as shown in Table 2.

Once the cotton textile has been decided as a substrate and the properties have been confirmed, the dimensions of the antenna design can be calculated by using the formulas in [15]. The microstrip feedline of Wf * LF or (1.01 * 26.11) mm was used in the design to make sure the RF power is fed directly to the radiating element. The basic of the rectangular antenna is designed in the CST software, but the operating frequency was not dropped at the desired 2.4 GHz. This is because the formulas just considered the properties of the materials without considering the other losses in the design. Therefore, the antenna was optimized again until the frequency dropped

Table 2 Cotton textile antenna design specification

Properties	Value
Frequency	2.4 GHz
Reflection coefficient	<−10 dB
Copper thickness	0.00018 mm
Line impedance	50 Ω

Table 3 Cotton Textile Antenna Dimensions

Parameters	Dimensions (mm)	Description
W	27.51	Patch width
L	24.76	Patch length
Wg	55.02	Substrate width
Lg	49.53	Substrate length
Wf	1.01	Feed line width
Lf	26.11	Feed line length
h	0.27	Substrate thickness

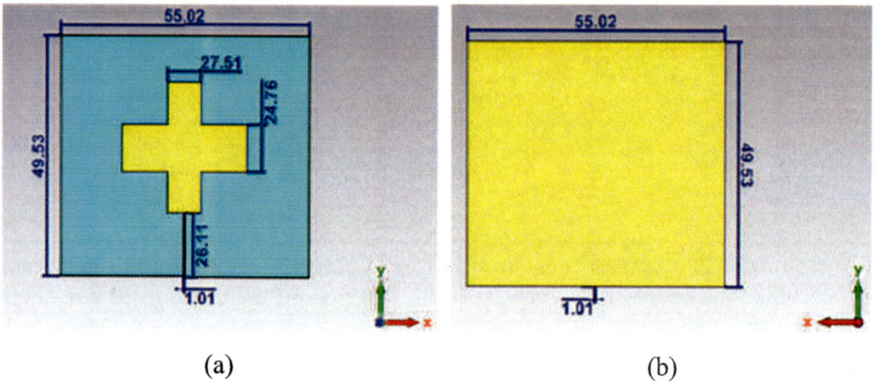

Fig. 1 Structure of cotton textile antenna dimensions operating at 2.4 GHz **a** front view, **b** back view

at 2.4 GHz. The final dimension of the cotton textile antenna is shown in Table 3. Figure 1 shown the structure for the wearable textile antenna that has been proposed in this project.

3 Results and Discussion

Figure 2 shown the S11-parameters, S11 for the proposed cotton antenna where the value is less than −10 dB that indicates the amount of power wasted in load and does not return as a reflection. From the graph, clearly shown that the antenna was resonant at 2.4 GHz with the reflection coefficient of −20.01 dB. The bandwidth of 48 MHz was achieved with the operating frequencies of 2.16–2.64 GHz.

In the antenna design, the value of VSWR must be less than 2 or the ideal matching of the feed line transmission. From Fig. 3, VSWR of the antenna is 1.22 at 2.4 GHz operating frequency. Further investigation of the current distribution on the antenna design has been conducted to ensure that the radiating element is concentrating on

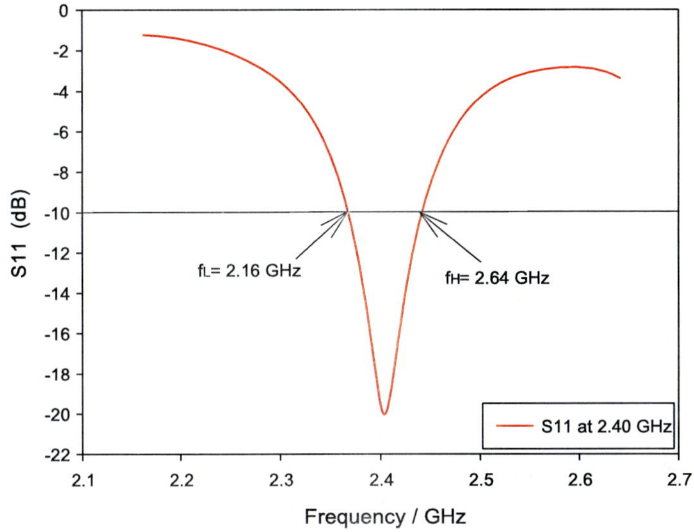

Fig. 2 Simulation results S11-parameters of cotton textile antenna

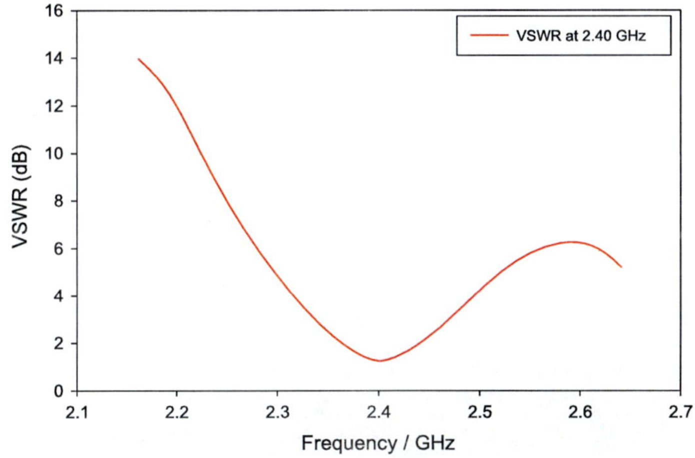

Fig. 3 Simulation results VSWR of cotton textile antenna

the signals. Figure 4 shown that the current distribution is concentrated at the edges of the plus-shaped radiating element of the proposed antenna. This current rotates clockwise with time at the four sides with equal dimensions of the plus-shaped radiating element. As a result, the proposed cotton textile antenna radiates as Left Hand Circular Polarization (LHCP), which uses the polarization on the left side. This shown the proposed cotton antenna has a good impedance matching of 50 Ω in the system.

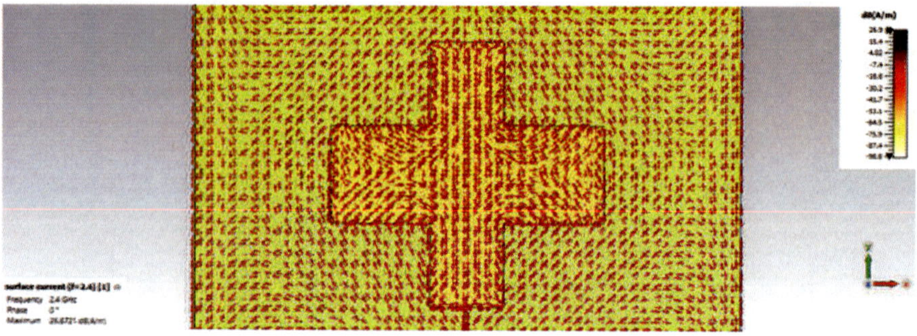

Fig. 4 Surface current distribution of cotton textile antenna

Radiation pattern of the wearable textile antenna consists of various parts, which are knows as lobes. The main lobe or main beam or major lobe is defined as the lobe containing the direction of the maximum radiation pattern. It is known as a quasi-isotropic radiation pattern, in direction of E-plane or H-plane. It is clearly shown that the directivity of the antenna is 8.27 dB and it produced an overall efficiency of about 56.79%. Figure 5 shown 3D of the radiation pattern for the wearable textile antenna that has been designed in this research.

The detailed information about the signal radiation can be observed from 2D view from the H-plane (Phi = 0.0°) and E-plane (Phi = 90.0°) as shown in Fig. 6. Both views provide detailed information about the signal radiation. H-Plane (Phi = 0.0°) shown the main lobe direction is direct at 0° and the angular width of 3 dB which is the coverage signal which covers about 74.1° where the side lobe level is equal to −22.6 dB as shown in Fig. 6a. On the other hand, at E-Plane (phi = 90.0°) the main

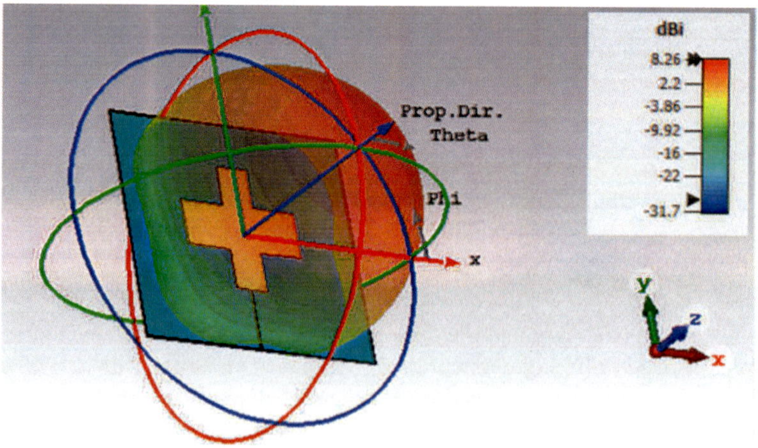

Fig. 5 3D radiation pattern of cotton textile antenna

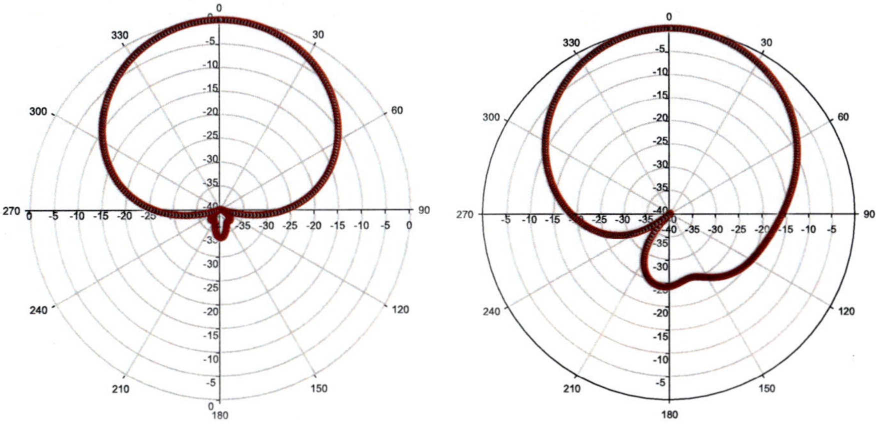

(a) (b)

Fig. 6 Normalized simulation results for 2D radiation pattern of cotton textile antenna at **a** H-plane (Phi = 0°) and **b** E-plane (Phi = 90°)

lobe direction of direct at 0.0° with the angular width 3 dB which is the coverage signal which equals 74.1. While the sidelobe level is equal to −22.4 dB as shown in Fig. 6b.

Besides that, the main lobe magnitude of the gain curve for a cotton antenna design is 7.44 dB as shown in Fig. 7. In order to prove that the proposed cotton antenna can operate at circular polarization, the axial ratio has been confirmed and checked as

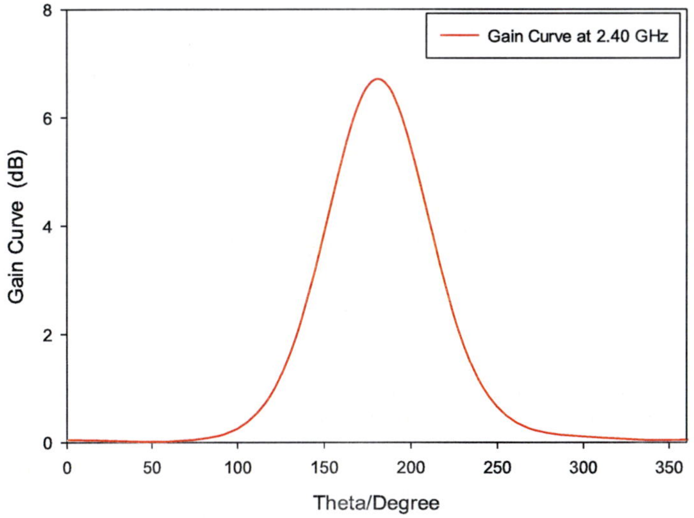

Fig. 7 Simulation results gain curve of cotton textile antenna

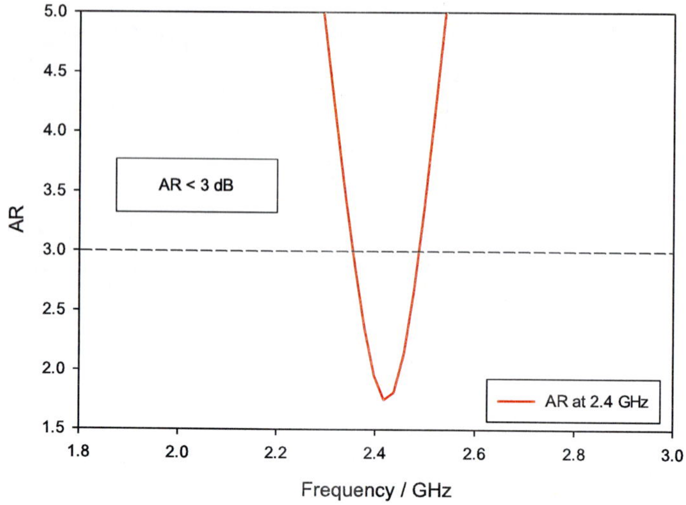

Fig. 8 Simulation results axial ratio of cotton textile antenna

Table 4 Overall performance of simulated cotton textile antenna

Parameter	Simulated results
S11	−20.01 dB
Gain	7.44 dB
VSWR	1.22
Directivity	8.27 dB
Efficiency	56.79%
Bandwidth	48 MHz
AR	1.9
Polarization	LHCP

shown in Fig. 8. The acceptable value of the axial ratio must be lower than 3 dB [15]. According to the graph, AR of 0.012 dB at the center frequency of 2.4 GHz. The overall performance of the simulated cotton textile antenna has been summarised in Table 4.

4 Conclusion

This paper presented a simulation on circular polarization cotton textile antenna at 2.4 GHz for a wireless communication system. Cotton textile material has been used as a substrate in designing the antenna. The antenna uses an LHCP as the signal was radiated equally at the plus-shaped radiating element in a clockwise direction. Besides that, it has been observed that the antenna resonant at 2.4 GHz frequency with the reflection coefficient of −20.01 dB, VSWR of 1.22, the gain for the proposed antenna of 7.44 dB with 56.79% efficiency. In the future, the validation of the proposed antenna between the simulation results and the measurement results must been conducted in order to compare the performance of the antenna. Besides that, the studies would be need to ensure that, the wearable antenna could been worn directly on the human body. In conclusion, the proposed cotton textile antenna has the potential to integrate on the garment and can be used for wireless communication systems.

Acknowledgements The authors would like to thank SEGi University for sponsoring the financial support. The appreciation also had been given to Antenna Research Centre, UiTM and Cerna Minda Institute for support in obtaining the materials and the developments of the antennas designed.

References

1. Atanasova GL, Atanasov NT (2020) Impact of electromagnetic properties of textile materials on performance of a low-profile wearable antenna backed by a reflector. In: 2020 international workshop on antenna technology (iWAT). https://doi.org/10.1109/iwat48004.2020.1570609739
2. Aprilliyani R, Dzagbletey PA, Lee JH, Jang MJ, So JH, Chung JY (2020) Effects of textile weaving and finishing processes on textile-based wearable patch antennas. IEEE Access 8:63295–63301. https://doi.org/10.1109/ACCESS.2020.2984934
3. Xu B (2018) A short review of textile applications in antenna design. Trends Text Eng Fash Technol 1(5):119–121. https://doi.org/10.31031/tteft.2018.01.000522
4. Martinez I, Werner DH (2018) Circular-polarized textile based antenna for wearable body area networks. In: 2018 IEEE international symposium on antennas and propagation and USNC/URSI national radio science meeting, APSURSI 2018—proceedings, pp 1195–1196. https://doi.org/10.1109/apusncursinrsm.2018.8609131
5. Mao C, Werner PL, Werner DH, Vital Di, Bhardwaj S (2019) Dual-polarized armband embroidered textile antenna for on-/off-body wearable applications. In: 2019 IEEE international symposium on antennas and propagation and USNC-URSI radio science meeting, APSURSI 2019—proceedings, pp 1555–1556. https://doi.org/10.1109/apusncursinrsm.2019.8889041
6. Costanzo S, Ammendola G (2020) Compact textile wearable antenna for security applications, pp 1–3. https://doi.org/10.1109/icecom48045.2019.9163616
7. Oshin SP, Amit S (2018) Design and analysis of high gain UWB textile antenna for wearable application. In: RTEICT 2017—2nd IEEE international conference on recent trends in electronics, information and communication technology, vol. 2018(2), pp 200–204. https://doi.org/10.1109/rteict.2017.8256585
8. Gangopadhyay S et al (2017) Design and development of electro-conductive rectangular textile antenna using polypropylene fabric. In: 2017 4th IEEE Uttar Pradesh section international conference on electrical, computer and electronics (UPCON), vol. 2018, pp 518–520. https://doi.org/10.1109/upcon.2017.8251103
9. Agbor I, Biswas DK, Mahbub I (2018) A comprehensive analysis of various electro-textile materials for wearable antenna applications. In: Proceedings of 2018 texas symposium on wireless and microwave circuits and systems (WMCS), pp 1–4. https://doi.org/10.1109/wmcas.2018.8400628

10. Zhong J, Kiourti A, Sebastian T, Bayram Y, Volakis JL (2017) Conformal load-bearing spiral antenna on conductive textile threads. IEEE Antennas Wirel Propag Lett 16:230–233. https://doi.org/10.1109/LAWP.2016.2570807
11. Wang KH, Li JS (2018) Jeans textile antenna for smart wearable antenna. In: 2018 12th international symposium on antennas, propagation and EM theory,, ISAPE 2018—proceedings, pp 1–3. https://doi.org/10.1109/isape.2018.8634337
12. Embong ENFSE, Rani KNA, Rahim HA (2018) The wearable textile-based microstrip patch antenna preliminary design and development. In: 2017 IEEE 3rd international conference on engineering technologies and social sciences (ICETSS), vol 2018, pp 1–5. https://doi.org/10.1109/icetss.2017.8324149
13. Usha P, Nagamani K (2018) Design, simulation and implementation of UWB wearable antenna. In: 3rd IEEE international conference on recent trends in electronics, information and communication technology (RTEICT)—proceedings, (2), pp 1041–1044. https://doi.org/10.1109/rteict42901.2018.9012446
14. Li SH, Li JS (2019) Smart patch wearable antenna on Jeans textile for body wireless communication. In: 2018 12th international symposium on antennas, propagation and EM theory, ISAPE 201—proceedings, pp 1–4. https://doi.org/10.1109/isape.2018.8634084
15. Potey PM, Tuckley K (2018) Design of wearable textile antenna with various substrate and investigation on fabric selection. In: 2018 3rd international conference on microwave and photonics (ICMAP), vol 2018, pp 1–2. https://doi.org/10.1109/icmap.2018.8354539
16. Yan S, Poffelie LAY, Soh PJ, Zheng X, Vandenbosch GAE (2016) On-body performance of wearable UWB textile antenna with full ground plane. In: 2016 10th European conference on antennas and propagation (EuCAP), EuCAP 2016, pp 1–4. https://doi.org/10.1109/eucap.2016.7481477

Development of a 6-DOF 3D Printed Industrial Robot for Teaching and Learning

Adamu Yusuf Abdullahi, Hwa Jen Yap, Mukhtar Fatihu Hamza, and Musa Mohammed Bello

Abstract The trend in modern manufacturing has led to a paradigm shift from the conventional high method to heavily automated processes. This comes with a high cost for those who want to meet up to the increasing pace of this revolution as the machines and the knowledge to control them is becoming expensive. Two major areas that have contributed in shaping this achievement of the industry 4.0 are additive manufacturing and robotics. With the drop in the prices of 3D printers and availability of off-the-shelf electronics components, the industrial experience is brought nearer to those who cannot afford it, especially students who need learn the adequate skill-sets in order to efficiently perform in this fourth industrial revolution. To this end, this paper aims at building a low-cost 3D printed robotic arm for teaching and learning purpose. The robotic arm was printed using three different kinds of 3D printers, with two most common materials (PLA and ABS). The forward and inverse kinematics of the robotic arm were presented. Finally, the developed robotic arm was controlled using the Arduino mega 2560. A simple pick and place experiment were conducted using the robotic arm. Also, the joint accuracies of three major joints were calculated. It is believed that the robotic arm in this paper will serve as an object of study and a means of tutoring students on robotics education.

Keywords 3D printing · Robot education · Robot kinematics · Robot control · STEM

1 Introduction

As the world drive towards a new age of technology particularly the industry 4.0 and IoT, the demand for competent in terms of technical and communication skills

A. Y. Abdullahi · H. J. Yap · M. F. Hamza (✉)
Department of Mechanical Engineering, University of Malaya, 50603 Kuala Lumpur, Malaysia
e-mail: mfhamza.mct@buk.edu.ng

A. Y. Abdullahi · M. F. Hamza · M. M. Bello
Department of Mechatronics Engineering, Bayero University, 3011 Kano, Nigeria

A. F. Ab. Nasir et al. (eds.), *Recent Trends in Mechatronics Towards Industry 4.0*, Lecture Notes in Electrical Engineering 730, https://doi.org/10.1007/978-981-33-4597-3_33

engineers is rapidly on the increase [1]. However, the supply for such qualified engineers is falling short of the demand. New devices, machines, software, and various technologies are being born on a daily basis, which places a huge responsibility on schools and vocational centres to train and produce engineers and technician to use these new technologies and also try to catch up the pace [2].

One of the areas where the above-mentioned training and development need urgent attention is in the field of robotics. Robots have been touching human lives for almost six decades, now since the first robot was put to service in 1961 at General Motors plant [3]. These robots impacted our lives behind the scene as we only enjoyed the already made parts without knowing how they were made. This was due to the fact that the first of the robots to be put to use were the industrial robots [3].

A robot can have degrees of freedom (DOF) ranging from one (1) to six (6), which allows it to perform various forms of movements. There are those with more than six DOF known as the redundant robots as they have more than enough freedom to carry out different movements [4]. Some applications where a 6-DOF robotic arm can be seen being put to use include; pick and place in industries, teaching and learning, writing, surgical operations and applications, hazardous working conditions, spot welding, painting etc [3]. A robot can either be a serial, parallel or hybrid robot. Also, the modelling of robot kinematics comprises direct and inverse kinematics which are used to solve for robot parameters such as position, orientation, and velocity of the end effector with respect to the base coordinate and other parts of the robot [4].

In the area of education, robots have been playing important roles to develop young competent engineers to fit in today's world of rapidly growing technology. With the increased interest and recent widespread of the Science, Technology, Engineering, and Mathematics (STEM) programs, robotics education being an important component of STEM is a promising tool to get students engage in the STEM programmes [5]. The inflow of researchers and students into robotics education over the past few years cannot be overemphasized. Quite a number of researches have shown that educating students of nowadays using robots will not only boost their interests of programming, computational thinking and robotics but also improve their science literacy [5]. Robotics education can serve to enhance students in three major ways which are; being an object of study, being a tool for cognition, and be a means of teaching, development and upbringing of students [2]. In many countries, robot education is in the process of being incorporated into the school curricula so that students in preschool, primary school, high school, and college can get involved in robotics from an early stage in life.

Additive manufacturing (AM) is building process where objects or parts are generated given their computer-aided design (CAD) model or any form of AM file by successively adding material layer by layer [6]. Additive manufacturing has many advantages over the conventional traditional methods especially in the area of producing parts with complex geometry, material management and time-to-market reduction [6]. In recent times, additive manufacturing is becoming widely known as 3D printing. Since emerging technologies are heavily tapping from the advantages of additive manufacturing and 3D printing, the adoption of the aforementioned is swiftly growing.

In this paper, a 6-DOF (degree of freedom) industrial robotic arm is developed using 3D printers. Thus, this marriage between robotics and 3D printing which exploits the benefits of additive manufacturing and combining it with robotics will give birth to a system which will serve to be useful in robotic and manufacturing education. The developed robot is aimed in easing learning for students in the area of robot development, kinematics, controlling of robots, electronics components and programming. The developed robot arm will serve as (1) an object of study; to teach and learn the basic robotics principles such as kinematics and dynamics, and (2) as a means of developing and tutoring. The model is expected to fulfil some industrial tasks alongside being an aid for teaching and learning. It is majorly focused on the pick and place application which is common industrial task used during packaging and palletizing.

2 The Robotic Arm Modelling

The robotic arm model used in this project is obtainable at https://www.thingiverse.com/wondertiger/designs. The robotic arm has 6 DOF including the base and the end effector (5 + 1 DOF) adopting the design of a standard robotic arm known as SCORBOT [4]. The CAD model of the robot manipulator is given in Fig. 1.

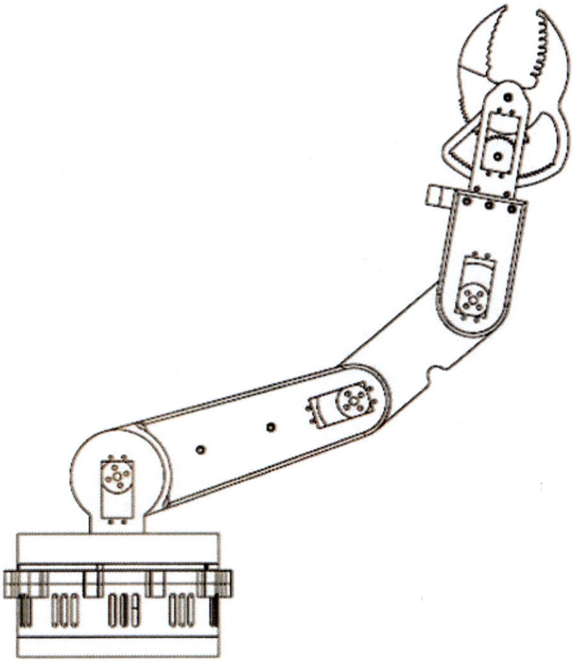

Fig. 1 Proposed robotic arm model

2.1 Modelling of Robotic Arm

To fully grasp the complicated joint analysis, velocities, forces, and torques [7], prior mathematical knowledge is paramount. There are three methods that can be used to model the kinematics of a robotic arm. They are; analytical, numerical, and hybrid methods. The analytical method is used in this paper.

Coordinate Assignment
In this paper the Denavit Hartenberg (DH) method will be adopted due to its simplicity and suitability [4]. The Cartesian coordinate frame was assigned using the DH method. More details on coordinate frame assignment can be found in [4, 8]. The coordinate for this project's robotic arm is shown in Fig. 2.

Homogeneous Transformation Matrix
The DH method employs a 4 × 4 transformations matrix in order to transform from one coordinate frame to another. It is called the homogeneous transformation matrix (HTM) and the general form is given in equation below.

$$H = {}^{i-1}T_i = T(z,\theta)T(z,d)T(x,a)T(x,\alpha)$$

$${}^{i-1}T_i = \begin{pmatrix} c\theta_i & -c\alpha_i s\theta_i & s\alpha_i s\theta_i & a_i c\theta_i \\ s\theta_i & c\alpha_i c\theta_i & -s\alpha_i c\theta_i & a_i s\theta_i \\ 0 & s\alpha_i & c\alpha_i & d_i \\ 0 & 0 & 0 & 1 \end{pmatrix}$$

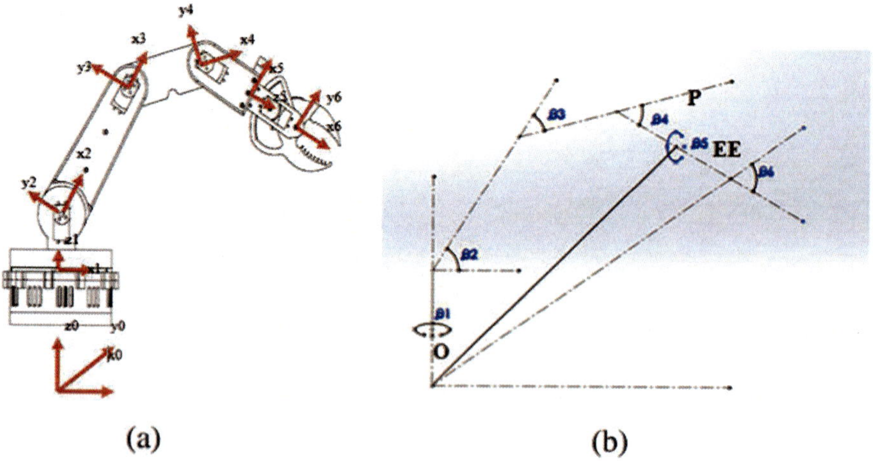

Fig. 2 **a** Coordinate frame assignment, **b** free body diagram of the robot arm

2.2 Forward Kinematics

The DH parameters for the robotic arm is given in Table 1.

Since the forward kinematics involves finding the position of the end-effector depending on the joint angels, a general transformation matrix can be formed by multiplying the corresponding HTMs for all the joints. Knowing a point in space say P_b given in vector form P_b (P_x, P_y, P_z) and a fixed coordinate frame at the origin say P_a (0, 0, 0). Then the position of the end-effector with respect to the fixed frame can be computed using the following equation.

$$AP = {}^{0}T_{n}P \tag{1}$$

where ${}^{0}T_{n}$ is the overall transformation from the base frame to the end-effector frame. It is given by

$$^{0}T_{n} = \sum^{i-1} A_{n} \ and \ 1 = 1, 2, \ldots, n \tag{2}$$

The total number of joint is n. Therefore,

$$^{0}T_{n} = {}^{0}A_{1}\,{}^{1}A_{2}\,{}^{2}A_{3}\,{}^{3}A_{4}\,{}^{4}A_{5}\,{}^{5}A_{6} = {}^{0}A_{6} \tag{3}$$

The convention for the trigonometry adopted from [4] is given below

$$\cos\theta_i = c_i \ and \ \sin\theta_i = s_i$$

Notice that joints 2, 3, and 4 are parallel to each other. This chain can be modelled as a 3-link planar manipulator. The overall transformation is obtained by multiplying equations all the transformation matrices for each joint. Hence, the forward kinematics for the robotic arm is given in Eq. 4.

Table 1 DH parameters for the robotic arm

Joint value (i)	Rotation, z (θ)	Translation, z (d)	Translation, x (a)	Rotation, x (α)
1	θ_1	d_1	0	$-\pi/2$
2	θ_2	0	a_2	0
3	θ_3	0	a_3	0
4	θ_4	0	a_4	$-\pi/2$
5	θ_5	d_5	0	$-\pi/2$
6	θ_6	0	a_6	0

$$^{0}T_n = {}^{0}A_n = \begin{pmatrix} u_x & v_x & w_x & ee_x \\ u_y & v_y & w_y & ee_y \\ u_z & v_z & w_z & ee_z \\ 0 & 0 & 0 & 1 \end{pmatrix} \quad (4)$$

where;

$$u_x = c_1 c_{234} c_5 c_6 + s_1 s_5 s_6 + c_1 s_{234} s_6 \quad (5)$$

$$u_y = c_1 c_{234} c_5 s_6 + s_1 s_5 s_6 + c_1 s_{234} c_6 \quad (6)$$

$$u_z = -c_1 c_{234} s_5 + s_1 c_5 \quad (7)$$

$$v_x = s_1 c_{234} c_5 c_6 - c_1 s_5 c_6 + s_1 s_{234} s_6 \quad (8)$$

$$v_y = -s_1 c_{234} c_5 s_6 + s_1 s_{234} c_6 \quad (9)$$

$$v_z = -s_1 c_{234} s_5 - c_1 c_5 \quad (10)$$

$$w_x = -s_{234} c_5 c_6 + c_{234} s_6 \quad (11)$$

$$w_y = s_{234} c_5 s_6 + c_{234} c_6 \quad (12)$$

$$w_z = s_{234} s_5 \quad (13)$$

$$\begin{aligned} ee_x = {} & a_2 c_1 c_2 + a_3 c_1 c_{23} + a_4 c_1 c_{234} - d_5 c_1 s_{234} \\ & + a_6 s_1 s_5 c_6 + a_6 c_1 c_{234} c_5 c_6 + a_6 c_1 s_{234} s_6 \end{aligned} \quad (14)$$

$$\begin{aligned} ee_y = {} & a_2 s_1 c_2 + a_3 s_1 c_{23} + a_4 s_1 c_{234} - d_5 s_1 s_{234} \\ & + a_6 c_1 s_5 c_6 + a_6 s_1 c_{234} c_5 c_6 + a_6 s_1 s_{234} s_6 \end{aligned} \quad (15)$$

$$ee_z = d_l - a_2 s_2 - a_3 s_{23} - a_4 s_{234} - d_5 c_{234} + a_6 c_{234} s_6 - a_6 s_{234} c_5 c_6 \quad (16)$$

2.3 Inverse Kinematics

The set of angles needed to position the wrist center at any position in the Cartesian plane are; θ_1, θ_2, θ_3, and θ_4. The joint angles in order of which they were derived are given below;

$$\theta_1 = \tan^{-1}\left(P_x/P_y\right) \tag{17}$$

$$\theta_5 = \sin^{-1}\left(-u_x s_1 + u_y c_1\right) \tag{18}$$

$$\theta_{234} = A\ \tan 2\left[(-u_z/c_5),\ \left(u_x c_1 + u_y s_1\right)/c_5\right] \tag{19}$$

$$\theta_3 = \cos^{-1}\left[\left(k_1^2 + k_2^2 - a_2^2 - a_3^2\right)/(2a_2a_3)\right] \tag{20}$$

$$k_1 = (a_2 + a_3c_3)c_2 - (a_2s_3)s_2 \tag{21}$$

$$k_2 = (a_3s_3)c_2 + (a_2 + a_3c_3)s_2 \tag{22}$$

$$\theta_2 = A\ \tan 2(s_2, c_2) \tag{23}$$

$$\theta_4 = \theta_{234} - \theta_2 - \theta_3 \tag{24}$$

$$\theta_6 = A\ \tan 2(s_6, c_6) \tag{25}$$

3 Materials and Method

3.1 Materials

The materials used for the printing are Polylactic Acid (PLA) and Acrylonitrile Butadiene Styrene (ABS) which are otherwise known as consumables. They are the most commonly used plastic materials in 3D printing. Both materials have distinguishing properties giving one an advantage over the other. PLA has the advantages of strength and flexibility over ABS while ABS has the advantages of structural integrity and mechanical usage over PLA. The parts needing high precision were printed using ABS while other parts were printed using PLA.

Table 2 Features of the printers used

Features	3D Printer		
	Prusa mk3 I3	Flashforge Creator Pro	Cr8 3D printer
Bed size	250 m × 250 mm	200 m × 200 mm	200 m × 150 mm
Nozzle size (mm)	0.5	0.4	0.4
Consumables	PLA (orange)	ABS (black)	PLA(yellow)
Print speed max (mm/s)	200	100	100
Print quality	Medium	High	High

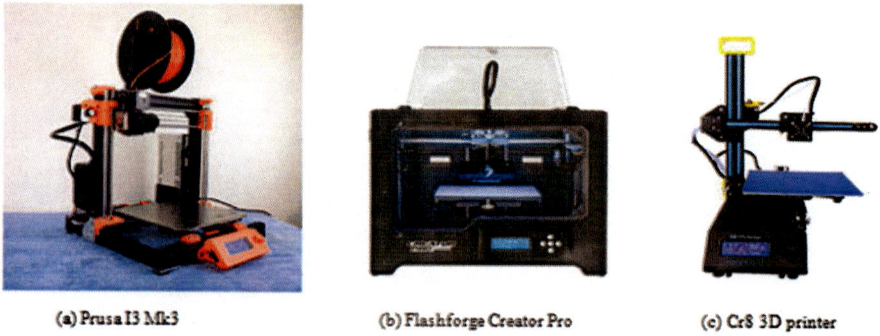

Fig. 3 The pictures of 3D printers used

3.2 *Equipment*

Three different products of 3D printers were used to print the entire parts of the robot arm. The features of the printers are given in Table 2 (Fig. 3).

3.3 *Software*

Three software were utilized in this paper. The first is the SOILDWORKS software which took care of the designs. The files in SOLIDWORKS are saved as STL files before they are further processed for printing. The STL design files needed to be converted to gcodes ready to be fed on the printers. The machines are custom made to read files in .gcode format. For that reason, the two other software were used for converting the STL files to .gcodes which are readable by the printers. CURA 3D

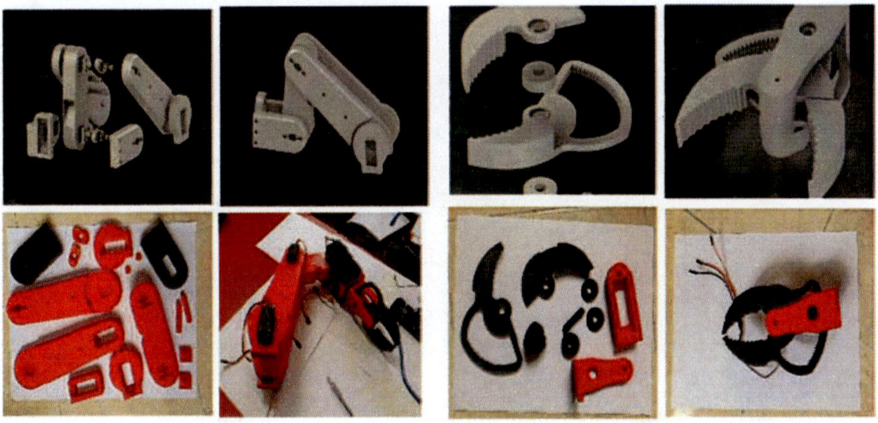

(a) (b)

Fig. 4 **a** The arm assembly, **b** the end-effector

was used for both Prusa MK3 and Cr8 3D, while the flashforge had its custom-made slicing software known as Flashprint.

3.4 3D Printed Parts

The printed robotic arm is divided into three separate components which are the base, the arm, and the end-effector. They were printed individually and then assembled to give the full robotic arm. The orange parts were printed using the Prusa mk3 while the yellow part using Cr8 3D. The black parts were printed using the Flashforge. The printed and assembled parts of the arm and end-effector is shown in Fig. 4. The base and the fully assembled robotic arm are shown in Fig. 5. The figures show inmages from the CAD model on top, while images of the printed parts are below.

3.5 Electronics and Electrical Connection

Designing using off-the-shelf (OTS) electronics components has the advantages of cost reduction, accessibility [9–11]. Table 3 gives a comprehensive list of all the electronics components used for the entire project and areas where they have been put to use.

In controlling this robot arm, an Arduino Mega 2560 microcontroller is used. It is connected serially to the PC, which gives it power supply. From the Arduino IDE, the stepper and servo libraries were used to code the commands sent to the

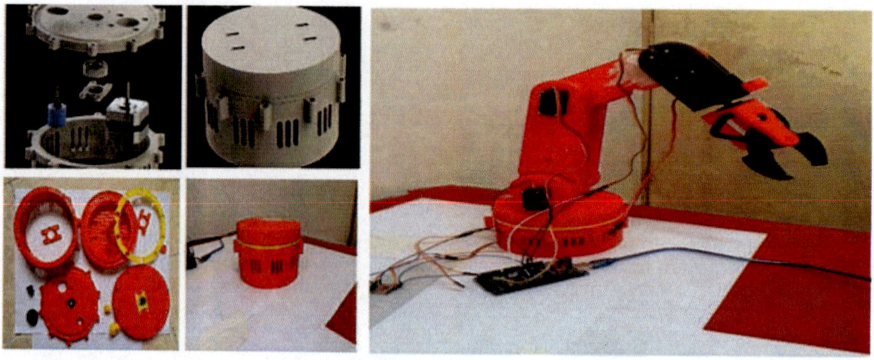

Fig. 5 **a** The base of the robot, **b** the fully assembled robotic arm

Table 3 The list of electronics components used in the project

Components	Description
Arduino 2560	Serves as the microcontroller.
Stepper motor × 1	Controls the base of the robot
Servo motor × 6	Controls the joint angles
Potentiometer × 1	Regulates the speed of the stepper motor
12v ac-dc adaptor	Power supply for the stepper motor
A4988 module	Driver module to control the stepper motor
Connecting wires	Used to establish connections

motors controlling all joints. The control signals are sent via a pulse width modulation (PWM) which is transferred to the board. The stepper motor which requires a higher voltage is controlled using an A4988 stepper motor driver. Figure 6 shows the electrical wiring of the project. The first joint at the base is controlled by the stepper motor. All other joints are being controlled by the servo motors. The servo motors are being controlled by individual 10 KΩ potentiometer.

4 Results and Discussion

The robotic arm is controlled by tuning the motors on all joints until it gets the end-effector to the desired position. The stepper motor is controlled directly from the codes through A4988 while the servo motors are mapped onto potentiometers. The code is tuned until the robot arm gets the end-effector to a desired location.

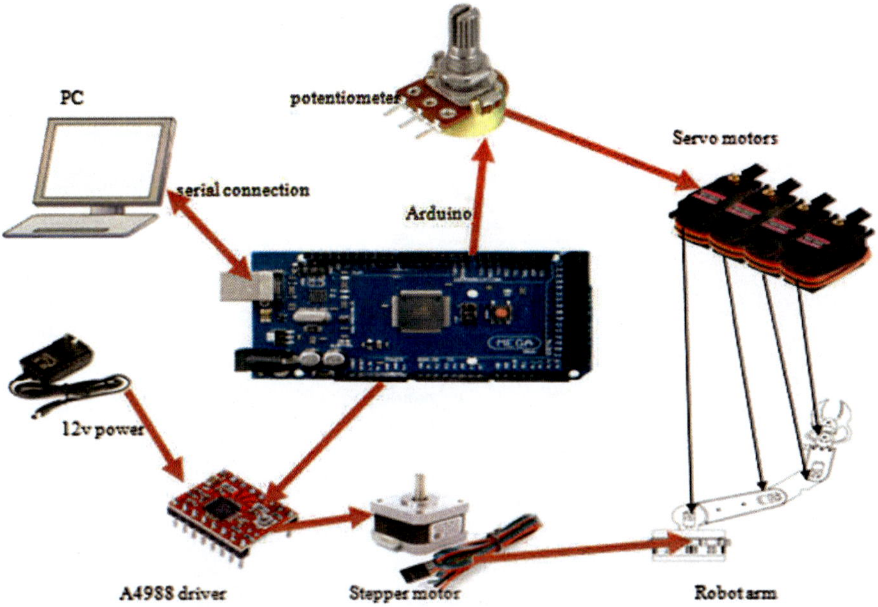

Fig. 6 Electrical wiring of the project

4.1 *Joint Accuracy*

The efficiency of the robot can only be ascertained if the robot follows the motion as specified by the values of the joint angles that are sent to it. In order to determine the accuracy of joint angles, command input angles ranging from 15 to 90° were given with an interval of 15°. The resultant angles were also recorded. Graphs for the joint accuracies is given in Fig. 7.

The percentage accuracy of each of the joint corresponds to the slope of the graphs. The efficiencies of the three arms is seen to be approximately unity. In terms of the angles, the resultant angle for the lower arm is lagging by a 2.06°. The centre and upper arms are having leads of approximately 3.0° both. There was poor stability which was caused by the weight of the robotic arm. The motors were not powerful enough to withstand instability.

4.2 *Pick and Place Experiment*

In order to make school children more acquainted with the industrial related task, robotic arm comes into play [7]. Experiments performed using robotic arm mimics the industrial conditions, this will prepare the students well enough before they graduate. In this paper, a simple pick and place experiment is performed. The set

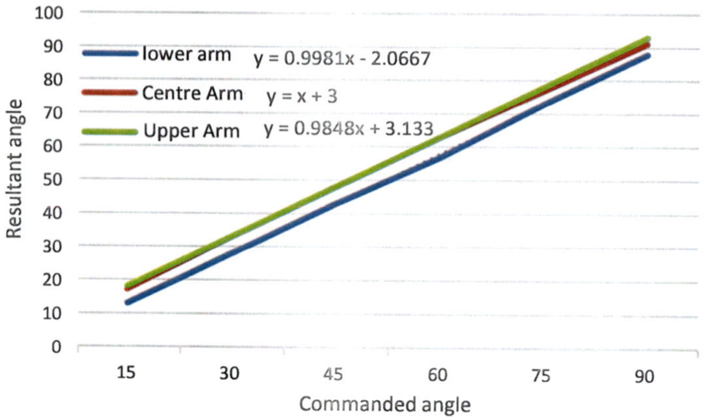

Fig. 7 Joint accuracies for the robotic arm

up for the experiment is the same as shown in Fig. 8. Four edges of a square metre paper were marked as A, B, C, and D. Then a reference point was marked at centre as O. Objects were then placed at these marked points which were used for the pick and plae experiment. An empty water bottle was used as the object in this case. Controlling the robot arm through potentiometers interfaced with each arm,

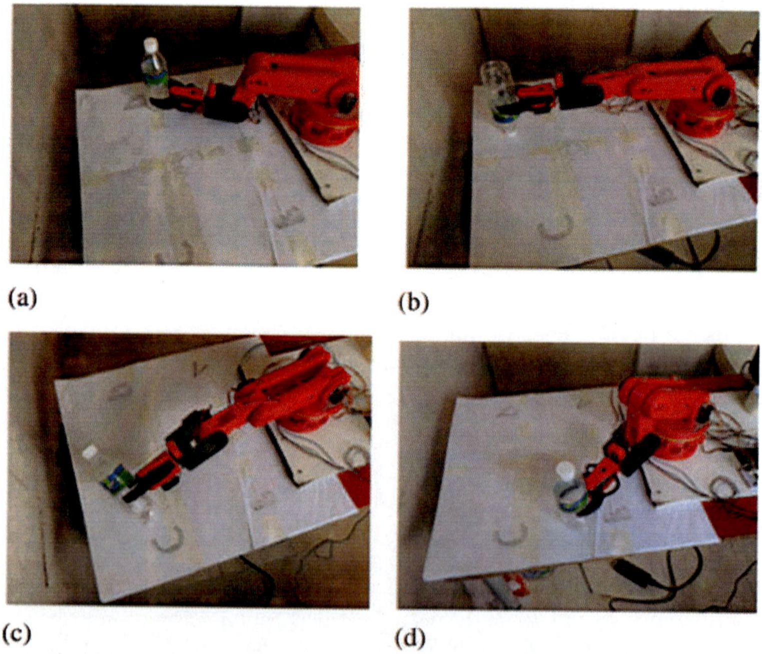

Fig. 8 Pick and place experiment using plastic bottle

the object is picked from one marked point to another. Although there was some level of instability due to the weight of the robot overpowering the servo motors, the water bottle was successfully moved between the various marked points. Attached are photos taken from the experiment.

5 Conclusion and Future Work

In this paper, a robotic arm has been successfully developed and tested. The robot arm was built using three different 3D printers; Prusa I3 Mk3, Flashforge creator Pro, and Cr8 3D. The forward kinematics of the robot arm was derived using the DH convention. The inverse kinematics was derived using the geometrical approach. In controlling the robot arm, Arduino mega 2560 was used. The program is written in the Arduino IDE which is suited for c++ programming, then this program is sent to the board which in turns control the motors. The joint accuracies for three arms were calculated. A simple pick and place experiment is performed to implement the paper.

It is hoped that this robot arm will serve as an educational tool for teaching and learning robotics and additive manufacturing. This combination of 3D printing and robotics makes a good mechatronics system as it comprises the mechanical component (additive manufacturing) and robotics component (electronics) Also it will expose the students to some common industrial based applications such as machine tending, pick and place in packaging. As 3D printers are now affordable and accessible, this paper could also serve as a pivot for building of printed robots especially in learning institutions of developing countries as they have little or no access to real time robots due to their costs.

The robotic arm printed was heavy for the RC servos that were used. The design of the arm can be optimized in order to reduce the weight of the robotic arm. Also the design can be modified to accommodate other better motors in the joints rather than the RC servos used.

The end effector is meant for only pick and place purpose. This can also be upgraded by replacing it with tools such as drill, paint brush, grinding wheel, among others.

References

1. Manzoor S et al (2014) An open-source multi-DOF articulated robotic educational platform for autonomous object manipulation. Rob Comput Integr Manuf 30(3):351–362
2. Ospennikova E, Ershov M, Iljin I (2015) Educational robotics as an inovative educational technology. Procedia Soc Behav Sci 214:18–26
3. Hagele M, Nilsson K, Pires JN (2008) Industrial robotics, in springer handbook of robotics. Springer, Berlin, Heidelberg

4. Tsai L-W (1999) Robot analysis: the mechanics of serial and parallel manipulators. Wiley, Canada
5. Xia L, Zhong B (2018) A systematic review on teaching and learning robotics content knowledge in K-12. Comput Edu 127:267–282
6. Li L, Haghighi A, Yang Y (2018) A novel 6-axis hybrid additive-subtractive manufacturing process: design and case studies. J Manuf Process 33:150–160
7. Sergey F et al (2017) Teaching robotics in secondary school examples and outcomes. Int Fed Autom Control (2017)
8. Mark WS, Seth H, Vidyasagar M (2006) Robot modeling and control. Wiley
9. Barbieri L et al (2018) Design, prototyping and testing of a modular small-sized underwater robotic arm controlled through a Master-Slave approach. Ocean Eng 158:253–262
10. Neto MBdaS, de Mendonça JMC, de Sena APC (2015) Development and control of a prototype manipulator SCARA type as teaching tool. IFAC-PapersOnLine 48(19):209–213
11. Timo O et al (2011) Robot competition as a teaching and learning platform. In: International federation of automatic control. Milano, Italy

Mathematical Model for Planar Reflectarray Antenna Design

M. Inam and M. Y. Ismial

Abstract Reflectarray can be used in many high gain advanced antenna applications such as radar, radio astronomy, satellite tracking, deep space communication and 5G communication systems. One of the most critical factors in designing a reflectarray antenna is the determination of the required reflection phase of the individual elements of the array. The desired reflection phase is then used to form a progressive phase distribution, using the accurate placement of the individual unit cells in the array environment and hence, a planar waveform is achieved. This work presents a robust mathematical model that can be used for determining the required phase of the reflectarray unit cells. The developed model can be applied to any geometrical configuration, any frequency band, any number of array elements, any feeding position (off-set or center feed), and also caters the effect of the dielectric and conductive properties of the material used for designing the reflectarray antennas. To demonstrate the validity, the proposed model is used to obtain the required phase of X-band (8–12 GHz) and Ku-band (12–18 GHz) reflectarrays using various dielectric materials. The progressive phase distribution has been demonstrated for 8 × 8, 12 × 12 and 15 × 15 elements reflectarrays with different locations of the feed horns. It has been demonstrated that the developed model significantly reduces the complexity of the process and the time required to design a reflectarray antenna.

Keywords Reflectarray · Mathematical model · Phase distribution

M. Inam (✉)
Faculty of Electrical and Electronic Engineering Technology (FTKEE), Universiti Teknikal Malaysia Melaka (UTeM), 76100 Durian Tunggal, Melaka, Malaysia
e-mail: inamabbasi@utem.edu.my

M. Y. Ismial
Faculty of Electrical and Electronic Engineering, Universiti Tun Hussein Onn Malaysia (UTHM), 86400 Batu Pahat, Johor, Malaysia
e-mail: yusofi@uthm.edu.my

A. F. Ab. Nasir et al. (eds.), *Recent Trends in Mechatronics Towards Industry 4.0*, Lecture Notes in Electrical Engineering 730, https://doi.org/10.1007/978-981-33-4597-3_34

1 Introduction

The need for a planar reflector or reflectarray antenna is evident from its potential modern-day and traditional high gain antenna applications. A reflectarray antenna consists of two main parts: a planar reflector and a feed antenna. The feeding part is usually covered by a horn antenna or some other high gain antenna. However, the planar reflector part has drawn the attention of many researchers in recent years due to its capabilities and margins of improvements. The planar reflector consists of resonant patch elements with specific design configurations printed on a dielectric substrate. The patch elements design configuration is vital for the required performance as each of the element governs the reflection phase, which is required to achieve the progressive phase distribution. Moreover, the planar reflectors are also responsible for the active and passive nature of the reflectarray antenna.

In passive reflectarrays, each of the reflector resonant elements is designed to provide a constant phase shift which helps to form a planar wave in front of the reflectarray aperture. In the past few years, several different techniques have been used in order to design passive reflectarrays. Some of the techniques that have been used include the elements of same shape and size with variable stub lengths [1], rectangular elements of different lengths [2], elements of same size with different slots on the patch [3], different rotation angle of loaded stubs [4] and ring resonator elements [5].

In active reflectarrays, the reflected phase from each of the resonant element can be controlled either mechanically or electronically. The reflected beam can be directed in the desired direction, which makes a reflectarray capable of achieving a wide-angle electronic beam scanning. Such beamforming approach can have many advantages over traditional tunable antenna array architectures, including a significant reduction in hardware required per element and increased efficiency [6]. Researchers have investigated different techniques for beam steering antennas such as the integration of Radio Frequency Micro Electro Mechanical Systems (RF MEMS) as switches [7], loading varactor diodes with the patch elements and varying the varactor capacitance by using various biasing [8] and using aperture coupled elements where the tuning circuit can be located on the non-resonating surface of the element in order to control the contributed phase from each element [9].

Although the potential of using reflectarrays in advanced applications such as 5G communications is enormous [10, 11] and design perspective of the reflectarray antennas are really attractive, but the design procedure requires precision and the margins of errors are very lean, especially for higher frequencies. Therefore, in this work, a mathematical model has been presented which can be used to design reflectarray with accuracy using limited time and resources. The proposed model can be used generally for a variety of design configurations and frequency ranges regardless of the intended applications.

2 General Progressive Phase Distribution and the Proposed Model

In order to design a reflectarray antenna, the most imperative scrutiny is the analysis of the total electric field on the patch elements printed on the dielectric substrate with a conductive ground plane on the other side. The total electric field on a reflectarray consists of the incident field and the reflected and scattered elements of the field. Therefore, the total electric field can be given by:

$$\bar{E}_t = \left[1 + \overline{\overline{\mathrm{R}}}(\theta_i, \varphi_i) + \overline{\overline{S}}(\theta_i, \varphi_i, L_i, W_i)\right].\bar{E}_0 e^{jk_0(xu_i + yv_i - z\cos\theta_i)} \tag{1}$$

In the case of a waveguide simulator technique, general relation for the total electric field excited in the Y-direction can be written as:

$$\bar{E}_{tw} = \overline{\overline{G}}_{YY}.\overline{J}_Y + \bar{E}_{Yinc}\left(l + \overline{\overline{G}}_{YY}\right) \tag{2}$$

where $\overline{\overline{G}}$ is the Green's function, $\overline{J}$ is the current density and l is the length of the unit cell patch element and $\overline{J}_Y$ can be given by:

$$\overline{J}_Y = \sum_n A_n \varphi_n(x, y) \tag{3}$$

where, A_n is the unknown vector coefficient and φ_n is the required phase from an individual patch element of a reflectarray in order to form a progressive phase distribution.

In order to calculate the phase shift for the elements on X-axis, trigonometric identities can be used as:

$$\varphi = -\frac{2\pi}{3}\cot^{-1}\frac{f}{x_i \pm \Delta X} \tag{4}$$

where, f is the vertical distance of feed from surface of the array and x_i is the distance between the centre of ith element and the point perpendicular to the feed and φ is in degrees. Once φ, is calculated for different values of x = xi, y = 0, the phase shift for all the array elements can be obtained. This method simplifies the calculation of the required phase shift from each of the array elements and reduces the complexity and time required for the periodic reflectarray design. The above technique can also be used for the progressive phase distribution with off-set feed reflectarrays as shown in Fig. 1. In the case of off-set feed reflectarrays, ΔX has to be introduced as the distance between the feed and the line perpendicular to the array centre. Figure 1 shows the geometry of the centre feed and off-set feed reflectarrays for different

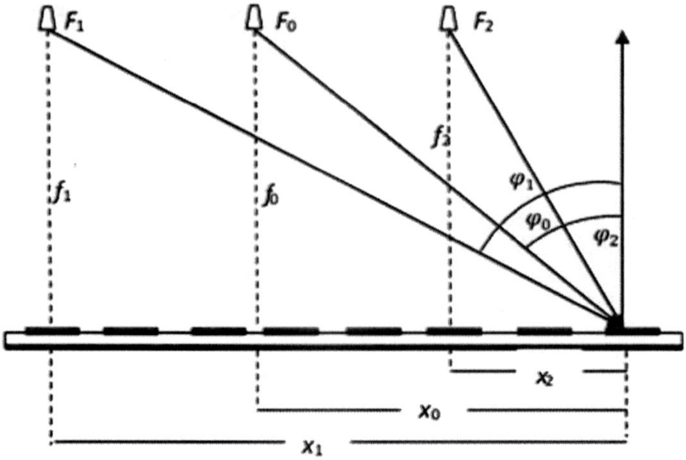

Fig. 1 Reflection phase from different feed points in a planar reflector

planar reflector designs. The feeds F1 and F2 are placed at the off-set distance of $\Delta X = X_0 - X_2$ and $\Delta X = X_0 - X_2$ respectively.

The above analysis provides a general formula for the design of a planar reflector with progressive phase distribution for any dielectric material and either centre or off-set feed configuration. In order to obtain the progressive phase distribution of planar reflector designed with different dielectric substrates, the material properties should be incorporated in Eq. (4). The material properties affect the reflection coefficient (Γ) which affects the reflection phase of the planar reflector. In the case of reflectarray antennas, Γ depends on the attenuation due to dielectric and conductor loss which are given by:

$$\alpha_d = \frac{\omega}{2}\sqrt{(\mu_0 \epsilon_0 \epsilon_r)}\tan\delta \tag{5}$$

$$\alpha_c = \frac{8.68}{W Z_m}\sqrt{\left(\frac{\omega \mu_0}{2\sigma_c}\right)} \tag{6}$$

where, α_d and α_c are attenuation due to dielectric and copper loss respectively.

After incorporation of the effects of dielectric and copper attenuation on Γ and reflection phase of the planar reflector, Eq. (4) can be written as:

$$\varphi = -\frac{2\pi}{3}\cot^{-1}\frac{f}{K\left(x_i \pm \Delta X_f\right)} \tag{7}$$

In Eq. (7), K is a variable which relates φ with Γ and depends on resonant frequency and material properties which affect the radiated and scattered fields. The value of K will be higher for the materials with higher values of dielectric permittivity and loss tangent. Therefore K is directly proportional to attenuation due to dielectric and conductor and K can be given by:

$$K = C.(\alpha_d + \alpha_c) \tag{8}$$

where, C is a compensation variable and varies with different design requirements and materials used. Finally, Eq. (7) can be written as:

$$\varphi = -\frac{2\pi}{3}\cot^{-1}\frac{f}{C.(\alpha_d + \alpha_c)\left(x_i \pm \Delta X_f\right)} \tag{9}$$

Figure 2 shows the reflection phase curves for different materials and different feed positions. It can be observed from Fig. 2 that as the value of K is increased, the slope of the reflection loss curve gets steeper, which shows a lower bandwidth value in planar reflector design. Moreover, for lower values of K, the reflection phase range decreases much lower than 360°, which indicates the rise of phase errors. This trade-off between bandwidth and phase errors for the periodic array follows the same trend as given in [12] for the unit cell reflectarray elements designed in an infinite array environment.

In order to implement the mathematical model for periodic arrays designed using different dielectric materials, the compensation factor C can be expanded by relating C, with different material properties, and finally the reflection phase of the individual elements can be given by:

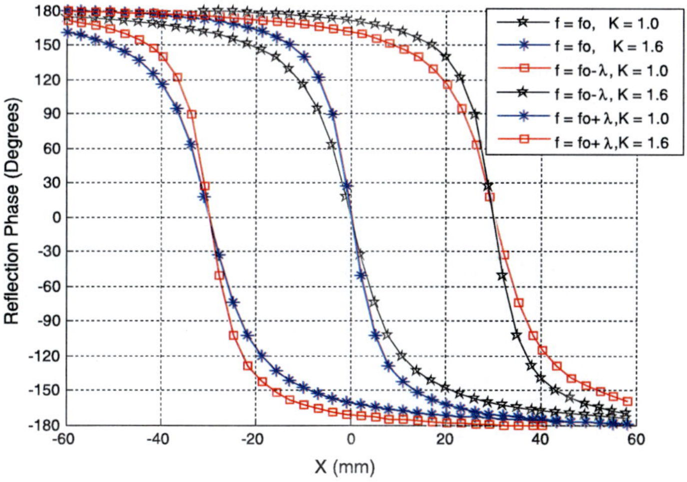

Fig. 2 Effect of different material properties on the reflection phase obtained for different feed positions

$$\varphi = -\frac{2\pi}{3}\cot^{-1}\frac{f}{x\left(\frac{\varepsilon_r}{\tan\delta}\right)(\alpha_d + \alpha_c)(x_i \pm \Delta X_f)} \quad (10)$$

where, x is a constant that has to be derived for different materials. Values of C have been estimated as a function of material properties and the desired frequency range.

3 Implementation of the Proposed Mathematical Model

This section provides the implementation of the proposed algorithm for X-band and Ku-band frequency ranges. Centre frequencies of 10 and 15.2 GHz are chosen for the demonstrative implementation of the proposed algorithm. For the relation in Eq. (10), x is calculated for X-band frequency range as shown in Table 1.

Using the average values of x, the values of C and K can be calculated. Table 1 also shows a summary of the calculated values of C and K for different materials used for the investigations. These values can be used in Eq. (9) to calculate the required reflection phase for the periodic array design. MATLAB $V^{7.5}$ has been used to obtain the required reflection phase curves for different dielectric materials as shown in Fig. 3.

It can be observed from Table 1 that the materials can be divided into two groups according to the values of x. Mean values of x for these two groups have been calculated for operation at 10 GHz as:

$$x_{avg} = \begin{cases} 0.00080, & 2.08 \leq \varepsilon_r \leq 6.5 \\ 0.00040, & 6.50 < \varepsilon_r \leq 13 \end{cases} \quad (11)$$

It can be observed from Fig. 3 that the materials with low dielectric permittivity values covers lesser phase range but shows broader bandwidth because of the gentle

Table 1 Material properties and calculated values of x at 10 GHz

Material	ε_r	$\tan\delta$	$x = c\left(\frac{\tan\delta}{\varepsilon_r}\right)$	$C = x_{avg}(\varepsilon_r/\tan\delta)$	$K = C.(\alpha_d + \alpha_c)$
Teflon	2.08	0.0004	0.00076	4.21	0.25
Taconic TLP5A	2.17	0.0004	0.00078	4.39	0.29
Rogers RT/d 5880	2.20	0.0009	0.00087	1.98	0.27
CEM	4.50	0.0250	0.00088	0.14	0.77
Beryllia	6.50	0.0004	0.00076	13.16	1.39
Alumina	9.75	0.0003	0.00047	10.23	1.33
Silicon	11.9	0.0040	0.00042	1.24	1.78
Gallium Arsenide	13.0	0.0060	0.00037	0.91	2.05

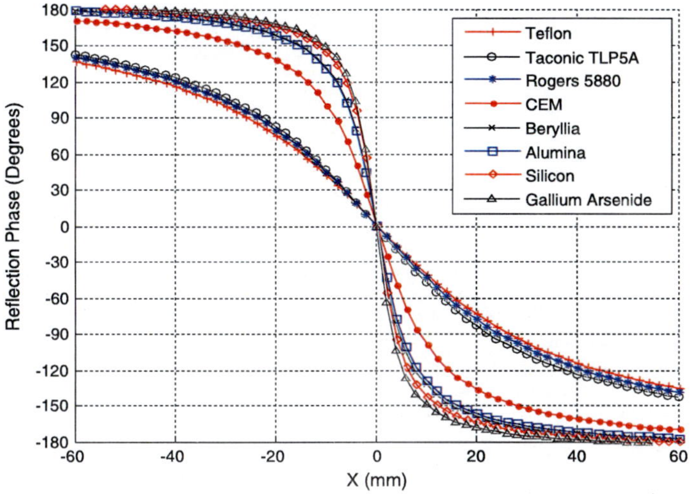

Fig. 3 Reflection phase curves for planar reflectors modelled with different substrates at 10 GHz

slope of reflection phase curve. On the other hand, the materials with a higher dielectric constant are observed to offer lesser bandwidth due to steeper reflection phase curve. These results are in line with the 3D electromagnetic simulations and measured results of reflectarray antennas presented in [13, 14].

A similar procedure was used to obtain the values of x for the design of reflectarray antennas operating in the Ku-band frequency range. The average x values for planar reflector designed with different materials operating at 15.2 GHz are calculated as:

$$x_{avg} = \begin{cases} 0.00055, 2.08 \leq \varepsilon_r \leq 6.5 \\ 0.00028, 6.50 < \varepsilon_r \leq 13 \end{cases} \tag{12}$$

Table 2 FoM for different materials used in X-band and Ku-band frequency ranges

Material	FoM (°/mm)	
	X-Band	Ku-Band
Teflon	3.90	3.98
Taconic TLP5A	4.40	4.47
Rogers RT/d 5880	5.20	5.28
CEM	10.8	10.89
Beryllia	20.0	20.08
Alumina	20.3	21.01
Silicon	22.6	23.30
Gallium Arsenide	24.0	24.99

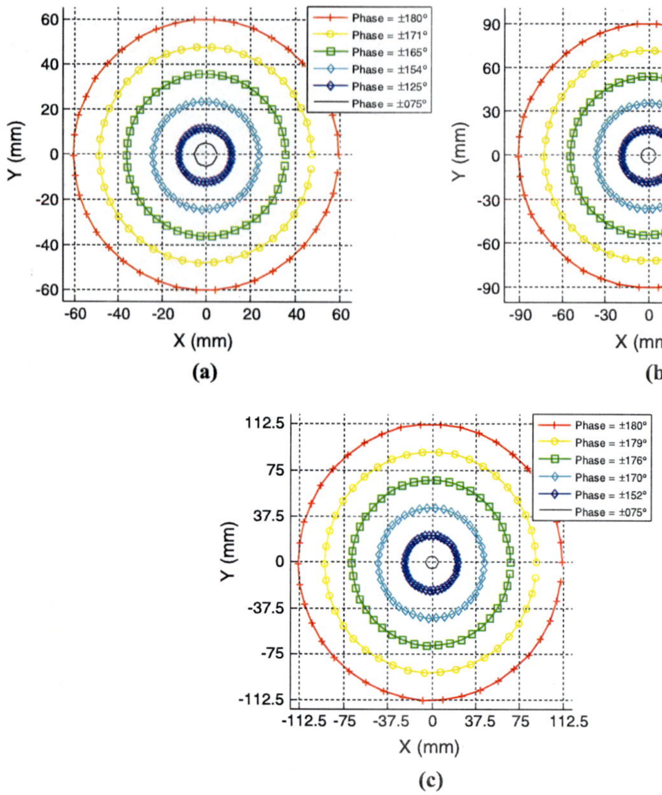

Fig. 4 Required reflection phase for **a** 8×8 element planar reflector array, **b** 12×12 element planar reflector array, **c** 15×15 element planar reflector array

Table 2 shows the comparison of FoM for different materials in X-band and Ku-band frequency ranges. Moreover, it can be validated from Figs. 3 and 4 that materials with lower permittivity and low loss tangent values provide a smoother phase curve showing a broader bandwidth and lower phase range. On the other hand, materials with high dielectric permittivity and loss tangent values offer a steeper phase. Hence phase range is increased, but the bandwidth of the planar reflector is decreased by using higher dielectric permittivity and loss tangent materials, which shows the trade-off between phase range and bandwidth performance.

4 Progressive Phase Distribution in the Array Environment

The proposed mathematical model can be used to obtain the required reflection phase for a planar reflector unit cells placed at particular positions in the array. This section provides the design of planar reflectors with different array sizes using mathematical modeling. Figure 4 shows the required reflection phase for array sizes of 8×8, 12×12 and 15×15 elements designed in X-band frequency range using Rogers RT/d 5880 dielectric substrate.

The feed used in this case is centre feed, and it can be observed from Fig. 4 that in all three cases, the reflection phase is constant on the circle with a fixed radius. The constant required phase, represented by the different colour of circles in Fig. 4, is obtained using the proposed mathematical model. Moreover, in this design, the inter-element spacing is kept at $\lambda/2$, and the array's size is obtained for a different number of elements. The mathematical modelling was also applied to the planar reflector design with off-set feed configurations, as shown in Fig. 5.

An off-set of 0.5λ (15 mm for 10 GHz) has been used for five different feed positions placed at $f = f_o$, $f = f_o - \lambda$, $f = f_o + \lambda$, $f = f_o - 1/2\,\lambda$, $f = f_o + 1/2\,\lambda$. The required reflection phases for progressive phase distribution of planar reflectors with different off-set feed positions are shown in Fig. 5. It can be observed from Fig. 5 that a phase shift is required because of an off-set in the feed position. However, for a particular design (constant number of elements and material properties), the phase values obtained by mathematical modelling for a radius of the circle are independent of the feed positions.

5 Conclusion

Reflection phase is one of the critical factors for obtaining the progressive phase distribution in the design of a reflectarray antenna. A comprehensive and accurate mathematical model has to be developed in order to design the reflectarray antennas for advanced high gain antenna applications. The mathematical modelling can help to replace the complex and time consuming 3D Electromagnetic simulations. However, the robustness and accuracy of the mathematical modelling need much improvement in order to substitute the widely used 3D Electromagnetic simulations. In future, the mathematical modelling for the loss performance of the reflectarrays will also be integrated with the proposed algorithm in order to calculate the beam position and gain of the reflectarray antennas.

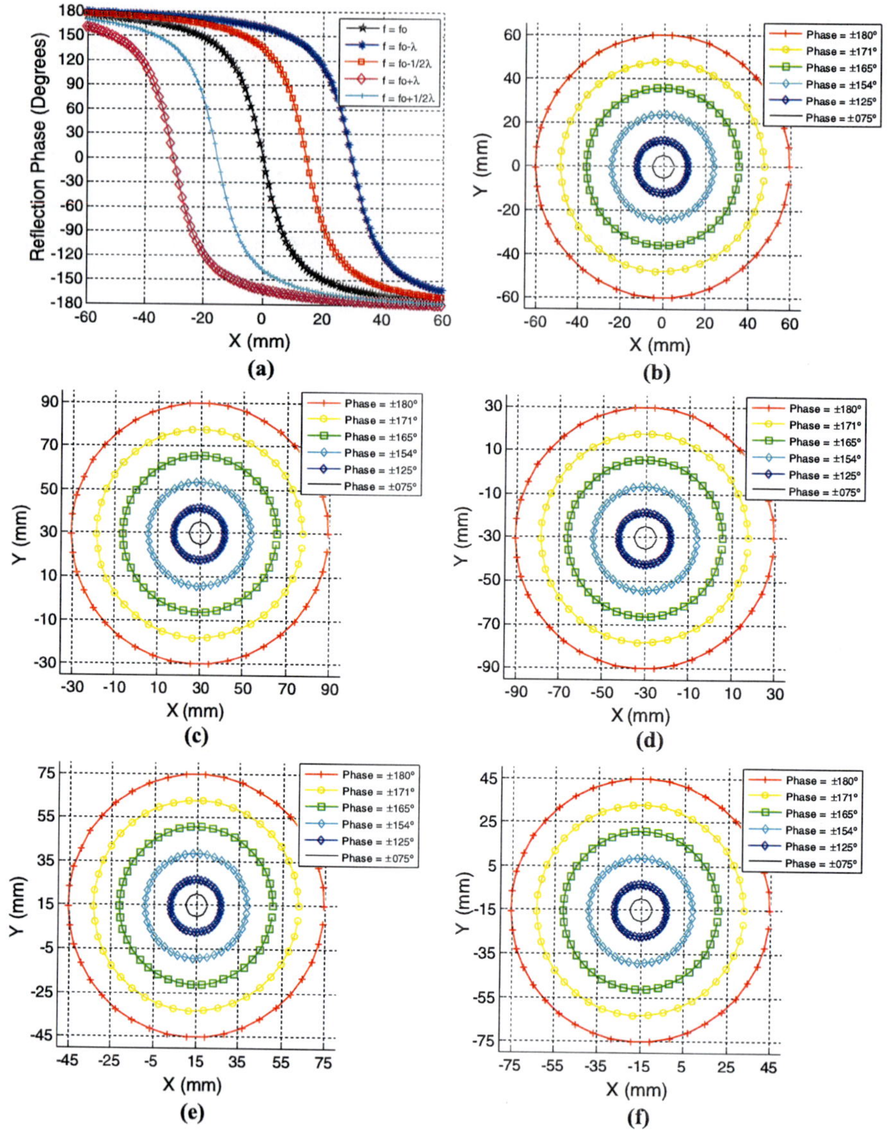

Fig. 5 Required Reflection phase curves for a planar reflector (8 × 8 elements) **a** with different feed positions, **b** f = f0 (Centre feed), **c** f = f0 − λ, **d** f = f0 +λ, **e** f = f0 − 1/2 λ, **f** f = f0 +1/2 λ

References

1. Huang J (1991) Microstrip reflectarray. In: Antennas and propagation society symposium 1991 digest, pp 612–615

2. Pozar DM, Metzler T (1993) Analysis of a reflectarray antenna using microstrip patches of variable size. Electron Lett 29(8):657–658
3. Ismail MY, Inam M (2010) Performance improvement of reflectarrays based on embedded slots configurations. Prog Electromagn Res 14:67–78
4. Huang J, Pogorzelski RJ (1997) Microstrip reflectarray with elements having variable rotation angles. In: IEEE antennas and propagation society international symposium 1997. Digest, vol 2, pp 1280–1283
5. Strassner B, Han C, Chang K (2004) Circularly polarized reflectarray with microstrip ring elements having variable rotation angles. Antennas Propag 52(4):1122–1125
6. Hum SV, Okoniewski M, Davies RJ (2005) Realizing an electronically tunable reflectarray using varactor diode-tuned elements. IEEE Microw Wirel Compon Lett 15(6):422–424
7. Tahir FA, Aubert H, Girard E (2010) Equivalent electrical circuit for designing MEMS-controlled reflectarray phase shifters. Prog Electromagn Res 100:1–12
8. Hum SV, Okoniewski M, Davies RJ (2007) Modeling and design of electronically tunable reflectarrays. IEEE Trans Antennas Propag 55(8):2200–2210
9. Riel M, Laurin JJ (2007) Design of an electronically beam scanning reflectarray using aperture-coupled elements. IEEE Trans Antennas Propag 55(5):1260–1266
10. Dahri MH, Inam M, Jamaluddin MH, Kamarudin MR (2018) A review of high gain and high efficiency reflectarrays for 5G communications. IEEE Access 6(1):5973–5985
11. Inam M, Dahri MH, Jamaluddin MH, Seman N, Kamarudin MR, Sulaiman NH (2019) Design and characterization of millimeter wave reflectarrays for 5G communication systems. Int J RF Microw Comput Aided Eng e21804
12. Abbasi MI, Ismail MY (2011) Reflection loss and bandwidth performance of X-band infinite reflectarrays: simulations and measurements. Microw Opt Technol Lett 53(1):77–80
13. Ismail MY, Inam M, Zain AFM, Mughal MA, Abdullah MFL, Ubin A (2010) Phase and bandwidth enhancement of reconfigurable reflectarray antennas with slots embedded patch. In: 2010 IEEE APS/URSI symposium, July 2010, Toronto, Ontario, Canada
14. Ismail MY, Inam M, Abdullah J (2010) Design optimization of reconfigurable reflectarray antenna based on phase agility technique. In: Third conference on antenna and RCS measurement (ATMS 2010), February 2010, Delhi, India

Broadband Reflectarray Antenna Based on Highly Conductive Graphene

M. Y. Ismial and M. Inam

Abstract This work presents a broadband reflectarray antenna where graphene has been practically used for the first time as the conducting material. The conductivity of graphene has been predicted in the millimeter wave frequency range to be used for the X-band reflectarray antenna design. The graphene produced by the Hummers method has been processed to improve its conductivity by adding Poly Ethylene Glycon (PEG). This improved conductivity of graphene has been used for the design of reflectarray unit cells based on different dielectric substrates. It has been demonstrated that the glass substrate where graphene has been used as conducting patch as well as ground plane provides better reflection loss of 2.46 dB with 100 MHz of 10% bandwidth. On the other hand, Rogers 5880 and FR4 are shown to offer 3.4 dB loss with 40 MHz of bandwidth and 9.8 dB loss with 60 MHz bandwidth, respectively.

Keywords Conductive graphene · Reflectarray antenna · Waveguide simulator · Reflection loss

1 Introduction

Reflectarray has emerged as one of the potential high gain antenna that has the capability to replace the bulky parabolic reflectors and expansive phased array antennas. A microstrip reflectarray antenna was introduced as a high gain antenna [1], which consists of a flat reflector and an illuminating feed horn. The flat reflector contains an array of resonant patch elements designed in such a way that the reflected field forms a planar wavefront in the far-field region [2]. Apart from passive reflectarrays, effective beam steering can also be achieved in reflectarray antennas by various

M. Y. Ismial (✉)
Faculty of Electrical and Electronic Engineering, Universiti Tun Hussein Onn Malaysia (UTHM), 86400 Batu Pahat, Johor, Malaysia
e-mail: yusofi@uthm.edu.my

M. Inam
Faculty of Electrical and Electronic Engineering Technology (FTKEE), Universiti Teknikal Malaysia Melaka (UTeM), 76100 Durian Tunggal, Melaka, Malaysia
e-mail: inamabbasi@utem.edu.my

A. F. Ab. Nasir et al. (eds.), *Recent Trends in Mechatronics Towards Industry 4.0*, Lecture Notes in Electrical Engineering 730, https://doi.org/10.1007/978-981-33-4597-3_35

methods such as the use of non-linear dielectric materials [3–5], the integration of Radio Frequency Micro Electro Mechanical Systems (RF MEMS) as switches [6], [7] and using aperture coupled elements [8].

However, the main concerns of a reflectarray antenna are its limited bandwidth and high loss performance as compared to the parabolic reflector antennas [9–11]. Different configurations have been proposed by researchers in the past few years for the bandwidth and loss performance improvement of reflectarray antennas [12, 13] but considerable efforts are still required to improve the bandwidth performance of reflectarrays.

One of the main factors that limit the bandwidth of reflectarrays is the limited bandwidth of the individual resonant elements. While an improvement to some extent in bandwidth can be achieved by strategic design configuration [14], the investigation to optimize the conducting patches can also significantly improve the performance of the reflectarrays. One of the emerging new materials that can be utilized for this purpose is graphene. It is a 2-D carbon nanomaterial and has attracted much attention in the scientific community due to its amazing electrical, optical, magnetic, and mechanical properties [15]. Graphene is an atom-thick carbon-based material that has many intriguing physical and chemical properties and can be produced by either chemical exfoliation of graphite in surfactant solution [16] or by radiation-induced reduction of graphene oxide by using gamma rays [15].

In this work, the characterization of graphene has been done in order to use it for the design of a reflectarray antenna with improved bandwidth performance in the X-band frequency range. Graphene has been produced using the Hummer method and deposited on different dielectric substrates. Scattering parameter measurements of graphene based reflectarray unit cells have been carried out in order to characterize the reflectarray performance in terms of bandwidth and reflection loss.

2 Conductivity of Graphene

The use of graphene for the design of antennas and other electromagnetic passives could bring significant benefits such as extreme miniaturization, monolithic integration with graphene RF Nano-electronics, efficient dynamic tuning, and even transparency and mechanical flexibility. However, the precise manufacturing/fabrication of graphene based antenna is challenging because small particles of carbonaceous impurities can stick to a single layer of graphene and can change its properties. These small particles can decrease the conductivity of graphene and hence cause an increase in the conductor loss of the antenna. Moreover, the characterization of the conductivity of graphene material, especially at Microwave frequency ranges, still need to be done in order to design graphene based antennas for a number of applications.

This work is carried out to provide a theoretical characterization of the conductivity of graphene using mathematical model development for different frequency bands in GHz frequency range. A number of research works have been reported for the numerical calculation of conductivity of graphene [17–19]. Kubo's formula is most commonly used for this purpose in which conductivity of graphene can be given as a function of frequency (ω), electrostatic bias or chemical potential (μ), phenomenological electron scattering rate (Γ) and temperature (T) as:

$$\sigma(\omega,\mu_c,\Gamma,T)=\frac{je^2(\omega-j2\Gamma)}{\pi\hbar^2}\left[\frac{1}{(\omega-j2\Gamma)^2}\int_0^{\infty}\varepsilon\left(\frac{\partial f_d(\varepsilon)}{\partial\varepsilon}-\frac{\partial f_d(-\varepsilon)}{\partial\varepsilon}\right)d\varepsilon - \int_0^{\infty}\frac{f_d(-\varepsilon)-f_d(\varepsilon)}{(\omega-j2\Gamma)^2-4(\varepsilon/\hbar)^2}d\varepsilon\right] \quad (1)$$

where, e is the charge of an electron, ħ is the reduced Planck's constant, f_d is the Fermi-Dirac distribution, and it has been assumed that no external magnetic field is present. So the local conductivity is isotropic, i.e., there is no Hall conductivity. The first term in Kubo's formula is due to intra-band contributions (σ(ω)), and the second term is due to inter-band contributions (σi(ω)) which can be separated within the random-phase approximation as given in [10]:

$$\sigma(\omega)=\frac{2e^2}{\pi\hbar}\frac{K_BT}{\hbar}\ln\left[2\cosh\left(\frac{\mu_c}{2K_BT}\right)\right]\frac{i}{\omega+i\tau^{-1}} \quad (2)$$

$$\sigma_i(\omega)=\frac{e^2}{4\hbar}\left(H\left(\frac{\omega}{2}\right)+i\frac{4\omega}{\pi}\int_0^{\infty}\frac{H(\varepsilon)-H(\omega/2)}{\omega^2-4\varepsilon^2}d\varepsilon\right) \quad (3)$$

where, τ = 1 ps is the relaxation time [9], T is the temperature, μc is the chemical potential, KB is the Boltzman's constant and H(ε) is defined as:

$$H(\varepsilon)=\frac{\sinh(\hbar\varepsilon/K_BT)}{\cosh(\mu_c/K_BT)+\cosh(\hbar\varepsilon/K_BT)} \quad (4)$$

It has been identified that below 5 THz frequency range, the intra-band contribution is dominant. Therefore, in this region, the inter-band contribution can be neglected, and the conductivity of graphene can be assumed to be purely intra-band. The intra-band equation has been separated for both real and imaginary parts of the conductivity, which can be used to plot the conductivity of graphene material.

$$\sigma_{real}(\omega)=\frac{2e^2}{\pi\hbar}\frac{K_BT}{\hbar}ln\left[2\cosh\left(\frac{\mu_c}{2K_BT}\right)\right]\frac{(1/\tau)}{\omega^2+(1/\tau)^2} \quad (5)$$

$$\sigma_{img}(\omega)=\frac{2e^2}{\pi\hbar}\frac{K_BT}{\hbar}ln\left[2\cosh\left(\frac{\mu_c}{2K_BT}\right)\right]\frac{i\omega}{\omega^2+(1/\tau)^2} \quad (6)$$

The above equations have been used to plot the real and imaginary conductivity of graphene using MATLAB with the following constants:

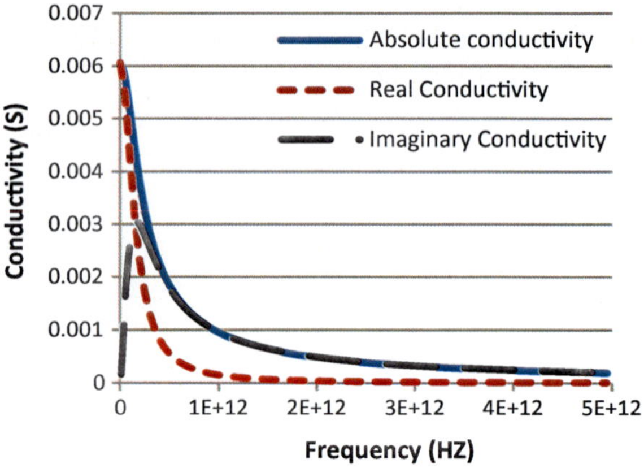

Fig. 1 Real and imaginary part of the intra-band conductivity at room temperature ($T = 300$ K), zero electrostatic bias ($\mu c = 0$ eV) and relaxation time of 1 ps ($\tau = 1 \times 10^{-12}$ s)

Resonant Frequency: $f = 4$ to 5 THz
Charge of Electorn: $e = 4.803e{-}10$ (statCoulombs)
Planks Constant: $\hbar = 6.582e{-}16$ (ev.s)
Boltzman Constant: $KB = 8.617e{-}5$ (ev/K)
Temperature: $T = 300$ (K)
Electrostatic bias or Chemical potential $\mu c = 0$ (ev)
Relaxation time: $\tau = 1 \times 10^{-12}$ (s).

Figure 1 shows the plot of real and imaginary intra-band conductivity of graphene plotted against the frequency using numerical relationships. It has been demonstrated in Fig. 1 that the real conductivity of graphene, under the given conditions, decreases as the frequency is increased while the imaginary part of conductivity increases initially but starts decreasing after a frequency of 0.15 THz. It can also be observed from the above equations that the values of conductivity depend on different parameters.

Therefore, the effect of electrostatic bias and relaxation time on the conductivity of graphene has been investigated. In order to observe the effect of relaxation time (τ) on the conductivity of graphene, (τ) has been varied from 0.3×10^{-13} S to 1.2×10^{-13} S. Figure 2 shows the effect of realization time on the real and imaginary conductivity of graphene. It can be observed from Fig. 2 that the maximum real conductivity increases from 1.8×10^{-4} S to 7.2×10^{-4} S, and maximum imaginary conductivity increases from 8.3×10^{-5} S to 3.6×10^{-4} S as τ is varied from 0.3×10^{-13} S to 1.2×10^{-13} S.

Electrostatic bias (μc) also affects the conductivity of graphene significantly. Therefore μc is varied from 0 to 100 eV, and its effect on the conductivity is observed. The maximum real conductivity increases from 4.5×10^{-4} S to 8.9×10^{-4} S as and maximum imaginary conductivity increases from 2.2×10^{-4} S to 4.4×10^{-4} S as μc is varied from 0 to 100 eV. The results also provide the feasibility of variation of conductivity of graphene, which can be utilized for reconfigurable antenna design.

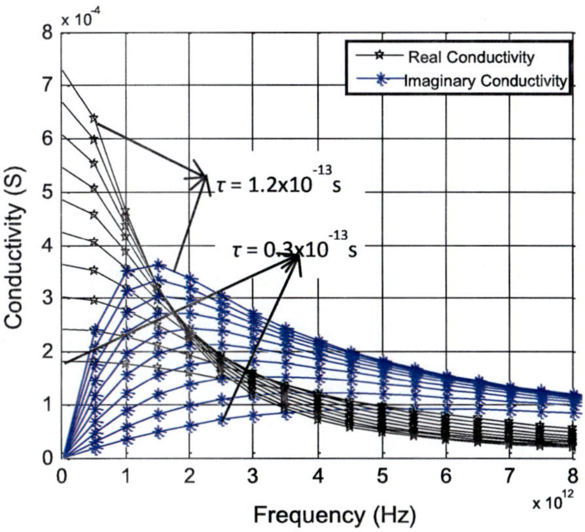

Fig. 2 Real and Imaginary conductivity of graphene in intra-band region with variable τ (0.3×10^{-13} S to 1.2×10^{-13} S) and μc = 0 eV

The effect of Electrostatic bias (μc) on real and imaginary conductivity is shown in Fig. 3.

As this work is based on the design of reflectarray unit cells within the X-band frequency range, the conductivity at standard conditions for 8–12 GHz has also been calculated as shown in Fig. 4. It can be observed from Fig. 4 that the real conductivity varies from 0.00606 to 0.00604 S while the imaginary part of conductivity varies from 0.0030 to 0.0045 S as the frequency is increased from 8 to 12 GHz. These comparatively low values of conductivity of graphene can be improved by adding Poly Ethylene Glycon (PEG) to the graphene. It has been demonstrated in [20] that

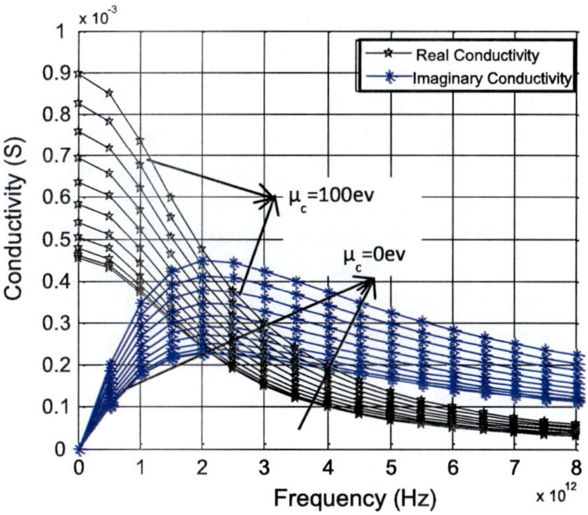

Fig. 3 Real and Imaginary conductivity of graphene in intra-band region with τ = 0.75×10^{-13} S and variable μc (0 to 100 eV)

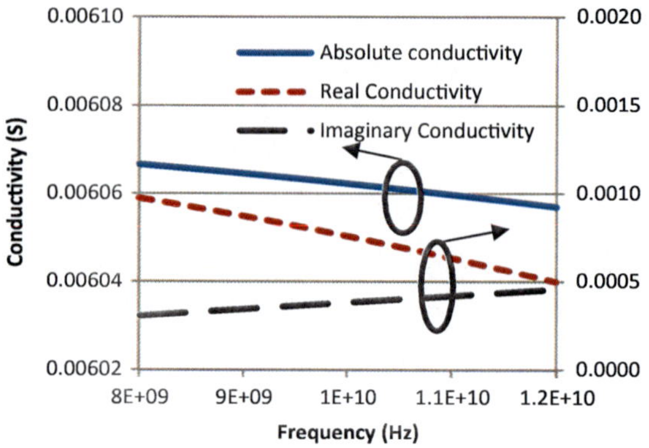

Fig. 4 Calculated conductivity of graphene in X-band frequency range at room temperature (T = 300 K), zero electrostatic bias (μc = 0 eV) and relaxation time of 1 ps ($\tau = 1 \times 10^{-12}$ S)

the electrical and optical properties of graphene can be modified by spin coating PEG solution on to the graphene oxide film followed by the thermal reduction process.

3 Fabrication of Graphene Based Reflectarray Antenna

Graphite Oxide flakes, the initial material for the preparation of multilayer graphene, has been produced by using a modified Hummers method with a relatively small amount of oxidizing agent and short-time processing at ambient temperature. The process of formation of Graphite Oxide flakes from Graphite Powder was completed in 18 h. Graphite Oxide flakes were then mixed with 1 ml/10 mg of de-ionized water and manually steered for 30 min in the presence of ultrasonic waves. The solution was then left for automatic steering for 2 h.

Spin coating of graphene oxide solution was done on the prepared Glass, Rogers Rt/d 5880, and FR4 substrates with rectangular patch configurations. Each sample was coated for 30 s with a spinning speed of 2500 RPM. Approximately 90 μl of graphene oxide solution was used for each sample consisting of two rectangular patch elements. In order to decrease the sheet resistance or to increase the conductivity of graphene, Poly Ethylene Glycon (PEG) was spin-coated on the prepared samples. Figure 5a shows the concept of graphene based reflectarray antenna while Fig. 5b represents a unit cell consisting of two patch elements. This configuration also considers the effect of mutual coupling, which can degrade the performance of the reflectarray antenna.

Figure 5c, d show the graphene based patch elements fabricated on FR4 and Rogers 5880 Rt/Duroid, respectively. The ground plane for these two substrates was fabricated using copper while in the case of a glass substrate, the ground plane, as well as the resonant patches, were fabricated using graphene, as shown in Fig. 5e.

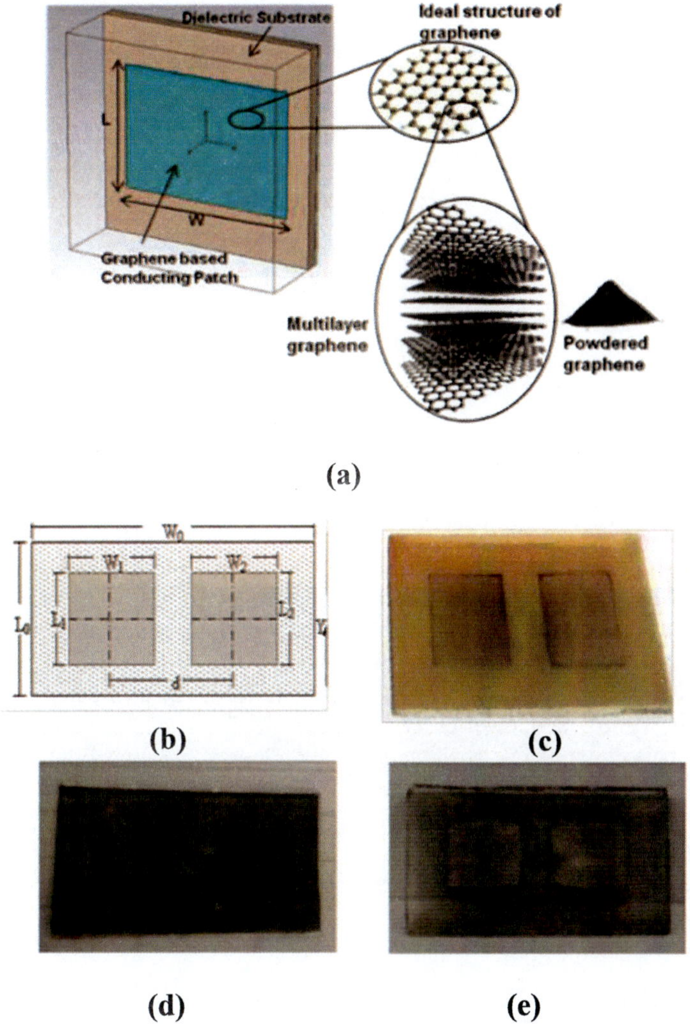

Fig. 5 Graphene based reflectarray unit cells **a** conceptual representation, **b** geometry of two patch elements unit cell, **c** graphene on FR4, **d** graphene on Rogers 5880, **e** graphene on glass dielectric substrate

4 Scattering Parameter Measurements

Scattering parameter measurements of unit cells reflectarray elements were carried out using a waveguide simulator and vector network analyzer as explained in [21]. Figure 6 shows the scattering parameter measurement results obtained for the reflection loss of graphene based reflectarrays using different dielectric substrates. It has been demonstrated in Fig. 6 that a clear resonance occurred at 9.84, 11.28, and

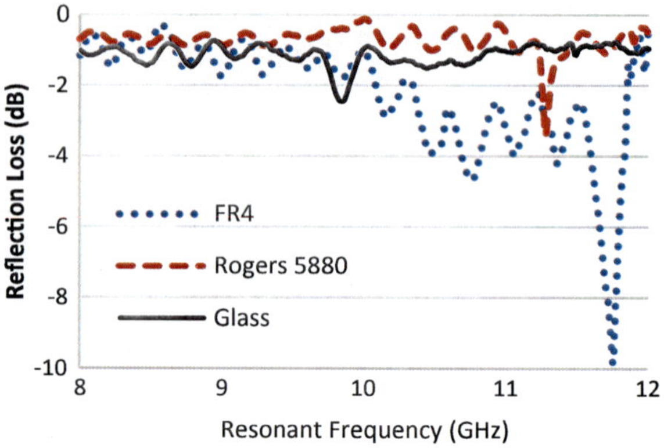

Fig. 6 Scattering parameter results of reflection loss for graphene based reflectarray unit cells fabricated on different substrates

11.76 GHz for glass, Rogers 5880 and FR4 dielectric substrates. The reason for different resonant frequencies is due to the material properties of the dielectrics as the dimensions of the patch were kept the same for all designs.

It can also be observed from Fig. 6, that the reflection loss is highest (9.8 dB) in the case of FR4, which is due to its high dielectric permittivity (ε_r) of 4.3 and high loss tangent (tan δ) of 0.025. On the other hand, although glass has higher dielectric permittivity ($\varepsilon_r = 4.8$) and higher loss tangent (tan $\delta = 0.0054$) as compared to Rogers 5880 ($\varepsilon_r = 2.2$, tan $\delta = 0.0009$), yet reflectarray unit cells fabricated on glass substrate provided lower reflection loss (2.46 dB) as compared to Rogers 5880 (3.4 dB). This effect can be attributed to the amount of highly conductive graphene used for the fabrication, which is much higher in the case of a glass substrate. The graphene also affected the bandwidth performance of the reflectarray, where 10% bandwidth was observed to be highest (100 MHz) in the case of glass substrate as compared to Rogers 5880 and FR4 which provided 10% bandwidths of 40 MHz and 60 MHz respectively.

5 Conclusion

Kubo's formula can be used to predict the conductivity of graphene in the millimeter wave frequency range. This predicted conductivity can effectively be used for the design of antennas based on graphene as conducting material. Graphene produced by the Hummers method can be spin-coated on different dielectric substrates within the required specification of the antenna design configuration. The highly conductive graphene can be used to improve the reflection loss and bandwidth performance of the antennas.

References

1. Munson RE, Haddad H, Hanlen J (1987) Microstrip reflectarray antenna for satellite communication and RCS enhancement or reduction. U.S. Patent 4684952
2. Huang J, Encinar JA (2007) Reflectarray antennas, Wiley, Interscience
3. Ismail MY, Hu W, Cahill R, Fusco VF, Gamble HS, Linton D, Dickie R, Rea SP, Grant N (2007) Phase agile reflectarray cells based on liquid crystals. Proc IET Microw Antennas Propag 1(4):809–814
4. Hu W, Ismail MY, Cahill R, Gamble HS, Dickie R, Fusco VF, Linton D, Rea SP, Grant N (2006) Tunable liquid crystal patch element. IET Electron Lett 42(9)
5. Mossinger, Marin R, Mueller S, Freese J, Jakoby R (2006) Electronically reconfigurable reflectarrays with nematic liquid crystals. IET Electron Lett 42(16)
6. Rajagopalan H, Rahmat Y, Imbriale WA (2008) RF MEMES actuated reconfigurable reflectarray patch-slot element. IEEE Trans Antennas Propag 56(12):369–3699
7. Tahir FA, Aubert H, Girard E (2010) Equivalent electrical circuit for designing MEMS-controlled reflectarray phase shifters. Prog Electromagn Res 100:1–12
8. Riel M, Laurin JJ (2007) Design of an electronically beam scanning reflectarray using aperture-coupled elements. IEEE Trans Antennas Propag 55(5):1260–1266
9. Ismail MY, Inam M (2012) Resonant elements for tunable reflectarray antenna design. Int J Antennas Propag 2012:6, Article ID 9148686
10. Biallowski ME, Encinar J (2007) Reflectarray: potential and challenges. In: International conference on electromagnetics in advanced applications (ICEAA), pp 1050–1053
11. Pozar DM, Targonski SD (1999) A shaped-beam microstrip patch reflectarray. IEEE Transa Antennas Propog 47(7):1167–1173
12. Sze KY, Shafal L Analysis of phase variation due to varying patch length in a microstrip reflectarray. IEEE Trans Antennas Propag 46(7):1134–1137
13. Hum SV, Okoniewski M, Davies RJ (2005) Realizing an electronically tunable reflectarray using varactor diode-tuned elements. IEEE Microw Wirel Compon Lett 15:422–424
14. Ismail MY, Inam M (2010) Performance improvement of reflectarrays based on embedded slots configurations. Prog Electromagn Res C 14:67–78
15. Zhang B, Li L, Wang Z, Xie S, Zhang Y, Shen Y, Yu M, Deng B, Huang Q, Fan C, Li J (2012) Radiation induced reduction: an effective and clean route to synthesize functionalized graphene. J Mater Chem 22:7775–7781
16. Ahmed AF, Humaizah F, Moin F, Mohd H, Ramadan S, Mohamed F, Hua C, Othman N, Ramli S, Rahim I (2013) Graphene colloidal dispersion in various organic solvents. Malaysian J Anal Sci 17(3):475–480
17. Hanson GW (2006) Dyadic green's functions and guided surface waves for a surface conductivity model of graphene. J Appl Phys 103:064302–064302
18. Fallahi A, Perruisseau-Carrier J (2012) Design of tunable biperiodic graphene metasurfaces. Phys Rev B 86:195408
19. Chen PY, Argyropoulos C, Alu A (2013) Terahertz antenna phase shifters using integrally-gated graphene transmission-lines. IEEE Trans Antennas Propag 61(4):1528–1537
20. Umar MIA, Yap CC, Awang R, Salleh MM, Yahya M (2014) The effect of spin-coated polyethylene glycol on the electrical and optical properties of graphene film. Appl Surf Sci 313:883–887
21. Inam M, Ismail MY (2011) Reflection loss and bandwidth performance of X-band infinite reflectarrays: simulations and measurements. Microw Opt Technol Lett 53(1):77–80

Stability Derivative Identification Using Adaptive Robust Extended Kalman Filter for Multirotor Unmanned Aerial Vehicle (M-UAV)

Danial Rosli, Erwin Sulaeman, Ari Legowo, and Alia Farhana Abdul Ghaffar

Abstract The invention of Unmanned Aerial Vehicle (UAV) in the early 1900 for military purposes and UAV applications in commercial purposes afterwards in the early 2000 accelerate its research in many engineering fields. Multirotor UAV such as quadrotor is usually unstable without a flight controller. Consequently, a proper and accurate model of the UAV dynamics is essential for its system stability. System identification allows the researchers to create an accurate parameter to the mathematical model of a dynamic system based on measured data. This work emphasizes in designing a robust and adaptive filter to develop an accurate mathematical model based on Newton-Euler method which includes aerodynamic drag and moment which are necessary in determining the correct model prediction. The focus of the present work is mainly on Kalman filter development for parameter estimation. While Kalman filter is only efficient in linear problem, an extended version of the filter itself deals with the nonlinear problem in which most real problem is actually nonlinear. This work investigates the performance of the extended version of the Kalman filter relative to parameter estimation. The performance of the filters are evaluated based on their estimation with the actual recorded flight data and presented based on data overlapping using Root Mean Square Error (RMSE). Ardupilot APM is used to acquire the actual flight test data and MATLAB is utilized to carry out the state estimation. To evaluate the performances of the filters, Goodness of Fit (GOF) approach was used. It is found that the GOF index of the present approach is 0.853 which is 25% higher than that of the Robust Extended Kalman Filter approach for the present flight test result.

Keywords System identification · Kalman filter · Multirotor UAV

D. Rosli · E. Sulaeman (✉) · A. F. A. Ghaffar
Department of Mechanical Engineering, International Islamic University Malaysia, Gombak, Malaysia
e-mail: esulaeman@iium.edu.my

A. Legowo
Aviation Engineering Division, Higher Colleges of Technology, Abu Dhabi, UAE

A. F. Ab. Nasir et al. (eds.), *Recent Trends in Mechatronics Towards Industry 4.0*, Lecture Notes in Electrical Engineering 730, https://doi.org/10.1007/978-981-33-4597-3_36

1 Introduction

Multi rotor Unmanned Aerial Vehicles (UAV), such as quadrotor UAV, has attracted the interest in current researches. Contributions of the research works in military and commercial applications on UAV technology [1–3] have been accelerated rapidly in concert with the advances in information and communication technology. Significant development of the rotors-based UAVs has grown over years [4–7]. To perform a variety of inspection, surveillance and rescue missions, quadrotor must be capable of maintaining its flight attitude with higher precision. Preserving its stability and maneuverability is one of the most complicated part in dynamic modelling and the most common approach in aerial vehicle system is nonlinear dynamical models.

To assess the stability of the system under extreme condition, a proper dynamic model of the whole system need be to be known a priori. However, a reasonable model is not straightforwardly obtained without the comprehension of the parameters. Traditionally, Newton's law for dynamics of rigid body in three dimensional coordinates [8] can be used to determine the dynamic of the system models. Development studies utilizing test rig methods have been conducted in previous years to find these parameters. Many literatures present the kinematic derivation of dynamic modelling. However, this method is limited to the flight maneuver envelope and is not time efficient. In addition, prior knowledge of the parameters of the dynamic model is required in the classical system identification method. Typically, the parameters are only partially known [9]. The real system may also maneuver hazardously where, to solve this problem, instead of using the real system, the evaluation of the model parameter is performed by simulation [10].

As an alternative way to parameter estimation, system identification technique is one of the answer to obtain the dynamic models. System identification is an approach to predict the dynamic parameters of the flight vehicle system based on the acquired flight data. To show the suitability of the model with the system identification method, a validation to the estimated data must be applied. The structure of system identification is outlined in [11] that covers four important aspects including experimental planning, selection of model structure, parameter estimation and validation.

System identification has been successfully applied to various aerial vehicles. Various system identifications have been tested and experimented by a number of literatures and a critical review on the literatures was done by [12]. However, finding the robustness of the system identification is the limitation to recent researches. Most of the tests were conducted at hovering condition which limits the flight envelope of the multi-rotor UAV for convoluted maneuver. There are some methods inhibiting very sensitive to high noises which can be a big limitation to parameter estimation. A good control should be adaptive, robust and precise so that the dynamic of UAV can vary to wider flight envelope. Surveyed on various application of system identification on fixed-wing, one rotor, and multirotor was recorded and provided in guideline [13].

The aims of this research are to develop an adaptive system identification method that robust enough in uncertainties modelling and to evaluate the performances of such methods to identify the parameters of interest.

2 Methodology

The unknown parameters can be identified from flight test data. The dynamic model parameters are treated as the states and identified through filtering method. Many literatures used the Extended Kalman filter (EKF) to provide the system identification of UAV's parameters. The present work proposes to use an Adaptive Robust Extended Kalman filter (AREKF) for the system identification of the UAV's parameters. In this paper, AREKF estimation performance are compared to Extended Kalman filter (EKF) and Robust Extended Kalman Filter (REKF). To achieve the objectives of the research, a proper guideline of important steps are carefully considered.

2.1 Robust Mathematical Model of M-UAV

Many researchers have developed the model for M-UAV and the derivation is explained in [14]. However, the focus is to develop the robust model. Therefore, the uncertainties model for the dynamics of M-UAV must be developed beforehand.

The dynamics of the model is derived based on two coordinate systems which are reference coordinate frame (XYZ) and body-fixed frame (xyz) as presented in Fig. 1. The subscript n and b on each parameter indicate that the parameter is applied in the reference and body-fixed coordinate systems, respectively. The vehicle gravity load G is assumed to center at O_b, the origin of the body fixed frame. Each individual rotor i of the quadrotor provides thrust T_i and drag force D_{xi}, D_{yi} which are centered at each rotor location. The resultant of the drag forces F_{bx}, F_{by} are acting in Z axis

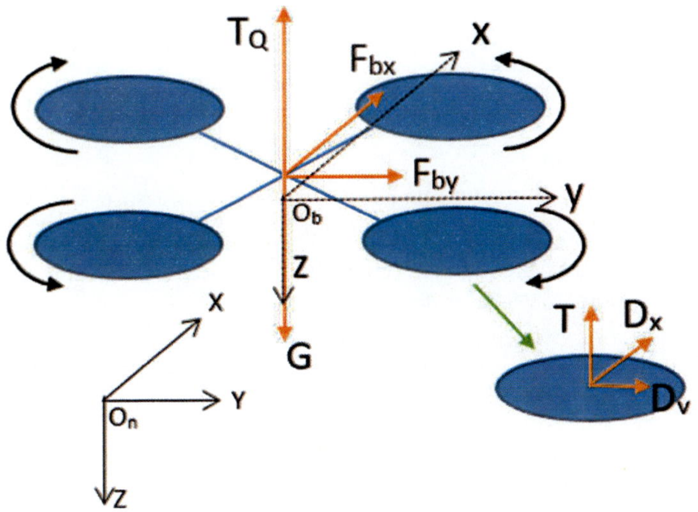

Fig. 1 The coordinate systems and the free body diagram of the quadrotor model

at the distance h from the centre of mass O_b. The resultant of the thrust force T_Q, acting along the Z axis as shown in Fig. 1.

Quadrotor Dynamic Model The derivation of the robust mathematical model can be found in [15]. The complete mathematical model of the quadrotor's dynamic is presented as follow.

$$a_x^b = -k_x \sum_{i=1}^{4} \omega_i V_{nbx}^b - g \sin\theta + (r V_{nby}^b - q V_{nbz}^b) \tag{1}$$

$$a_y^b = -k_y \sum_{i=1}^{4} \omega_i V_{nby}^b + g \cos\theta \sin\varphi + (p V_{nbz}^b - r V_{nbx}^b) \tag{2}$$

$$a_z^b = -k_T \sum_{i=1}^{4} \omega_i + g \cos\varphi \cos\theta + (q V_{nbx}^b - p V_{nby}^b) \tag{3}$$

$$\dot{p} = k_p(\omega_2^2 + \omega_3^2 - \omega_1^2 - \omega_4^2) + k_p D \sum_{i=1}^{4} \omega_i V_{nby}^b \tag{4}$$

$$\dot{q} = k_q(\omega_1^2 + \omega_3^2 - \omega_2^2 - \omega_4^2) + k_q D \sum_{i=1}^{4} \omega_i V_{nbx}^b \tag{5}$$

$$\dot{r} = k_r(\omega_1^2 + \omega_2^2 - \omega_3^2 - \omega_4^2) \tag{6}$$

In the governing equations above, k_x and k_y are designated as drag coefficients in x and y directions, respectively. Other parameters are consolidated into coefficient $k_T, k_p, k_{pD}, k_q, k_{qD}$, and k_r. $\left[V_{nbx}^b \; V_{nby}^b \; V_{nbz}^b\right]$ are the components of V_{nb}^b which stands for the velocity of the b frame with respect to the n frame resolved in the b frame.

2.2 *Robust Adaptive System Identification*

The adaptive system of AREKF is developed in this process. To develop the dynamic model of the system, the state vector is arranged as

$$x = [p \; q \; r \; V_{nbx}^b \; V_{nby}^b \; V_{nbz}^b \; k_x \; k_y \; k_T \; k_p \; k_q \; k_r \; k_{pD} \; k_{qD}] \tag{7}$$

where $[p \; q \; r]^{\mathrm{T}}$ are the angular velocity components in the body-fixed coordinate system. The state equation can be formulated as

$$\dot{x} = \begin{bmatrix} 0_{3\times3} & 0_{3\times3} & 0_{3\times3} & L_1 & L_2 \\ 0_{3\times3} & L_4 & L_5 & 0_{3\times3} & 0_{3\times2} \\ 0_{8\times3} & 0_{8\times3} & 0_{8\times3} & 0_{8\times3} & 0_{8\times2} \end{bmatrix} x + \begin{bmatrix} 0_{3\times1} \\ L_3 \\ 0_{8\times1} \end{bmatrix} g + W_Q \tag{8}$$

where W_Q is the zero mean white process noise vector and

$$L_1 = \begin{bmatrix} (\omega_2^2 + \omega_3^2 - \omega_1^2 - \omega_4^2) & 0 & 0 \\ 0 & (\omega_1^2 + \omega_3^2 - \omega_2^2 - \omega_4^2) & 0 \\ 0 & 0 & (\omega_1^2 + \omega_2^2 - \omega_3^2 - \omega_4^2) \end{bmatrix},$$

$$L_2 = +\begin{bmatrix} e_1 & 0 & 0 \\ 0 & e_2 & 0 \\ 0 & 0 & 0 \end{bmatrix}, L_3 = \begin{bmatrix} -\sin\theta \\ \cos\theta\sin\phi \\ \cos\theta\cos\phi \end{bmatrix}, L_4 = \begin{bmatrix} 0 & r & -q \\ -r & 0 & p \\ q & -p & 0 \end{bmatrix}$$

$$L_5 = -\begin{bmatrix} e_1 & 0 & 0 \\ 0 & e_2 & 0 \\ 0 & 0 & e_3 \end{bmatrix}, e = \left[V_{nbx}^b \sum_{i=1}^{4} \omega_i \; V_{nby}^b \sum_{i=1}^{4} \omega_i \; \sum_{i=1}^{4} \omega_i^2 \right]$$

The measurement vector of the system is chosen as

$$z = \left[p_{gyro} \; q_{gyro} \; r_{gyro} \; V_{nbx\,GPS}^b \; V_{nby\,GPS}^b \; V_{nbz\,GPS}^b \; f_{bx} \; f_{by} \; f_{bz} \right]^T \tag{9}$$

where $\left[p_{gyro} \; q_{gyro} \; r_{gyro} \right]^{\mathrm{T}}$ are the gyroscopes' outputs, $\left[V_{nbx\,GPS}^b \;\; V_{nby\,GPS}^b \;\; V_{nbz\,GPS}^b \right]$ are supplied by the GPS, $\left[f_{bx} \; f_{by} \; f_{bz} \right]$ are the accelerometers outputs. The measurement equation can be expressed as

$$z = \begin{bmatrix} I_{6\times6} & 0_{6\times3} & 0_{6\times5} \\ 0_{3\times6} & L_5 & 0_{3\times5} \end{bmatrix} x + W_R. \tag{10}$$

The development of AREKF is well written in the literatures [16]. In this stage, the parameter of AREKF is tuned based on the preferences that the literatures have outlined to formulate the adaptive model for process and measurement error covariance.

Time Update

$$\hat{x}_{k|k-1} = f_k(\hat{x}_{k-1|k-1}) \tag{11}$$

$$P_{k|k-1} = F_k P_{k-1} F_k^T + Q_k \tag{12}$$

$$\sum_{k|k-1} = \left(P_{k|k-1}^{-1} - \gamma^2 L_k^T L_k \right)^{-1} \tag{13}$$

Measurement Update

$$P_{y,k} = H_k \sum_{k|k-1} H_k^T + R_k \tag{14}$$

$$K_k = \sum_{k|k-1} H_k^T P_{y,k}^{-1} \tag{15}$$

$$\hat{x}_k = \hat{x}_{k|k-1} + K_k\left[y_k - h\left(\hat{x}_{k|k-1}\right)\right] \tag{16}$$

$$P_k = \left(\sum_{k|k-1}^{-1} + H_k^T R_k^{-1} H_k\right)^{-1} \tag{17}$$

where both F_k and H_k are the Jacobian matrix of $f(x_{k-1})$ and $h(x_k)$ respectively. In the regular Kalman filter, L_k is set as the identity matrix I, but in AREKF L_k is formulated as follows:

$$L_k = \gamma\left(P_{k|k-1}^{-1} - \lambda_k^{-1} P_{k|k-1}^{-1}\right)^{1/2} \tag{18}$$

where λ_k is the tuning parameter which is selected in such that it satisfies the following condition:

$$\bar{\sum}_{k|k-1} < \lambda_k P_{k|k-1} \tag{19}$$

By tuning λ_k the filter obtains a better robust behaviour [16]. Nonetheless, too much emphasized is placed on the upper bound $\lambda_t P_{t\,|\,t-1}$ could degrade the optimality of the filter when the worst case scenario happened which is the largest linearization error. Therefore, to enhance the stability of the filter without decreasing the accuracy of the filter, an adaptive scheme to adjust $\sum_{t\,|\,t-1}$ given the unknown response to the changing environment is adopted to the filter.

$$\sum_{t|t-1} = \begin{cases} P_{t|t-1}, & P_{y,t} > \alpha \bar{P}_{y,t} \\ (P_{t|t-1}^{-1} - \gamma^{-2} L_t^T L_t)^{-1}, & otherwise \end{cases} \tag{20}$$

where $\bar{P}_{y,t} = E(\tilde{y}_t \tilde{y}_t^T | \tilde{x}_{t-1})$. The parameter $\alpha > 0$ is introduced to yield an extra degree of freedom to the tune the threshold during implementation to avoid the filter from resetting when the error is rather small.

The test was conducted at an open field with no building approximately 100 m around its radius. Hovering condition with minimal wind was observed during the flight test and the data was acquired using APM Ardupilot flight controller. An X-form of quadrotor is used in this work because it provides a better controllability

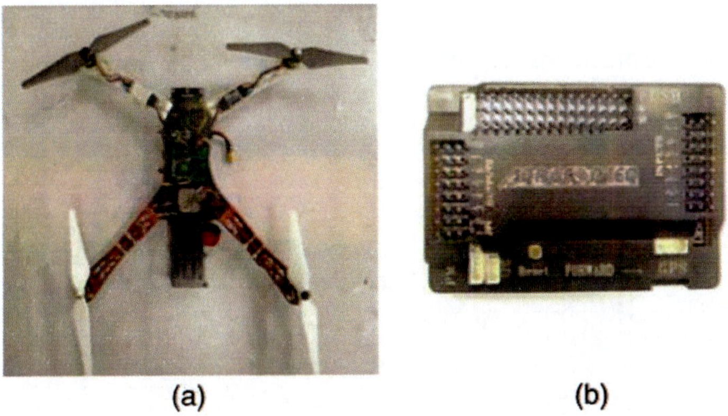

Fig. 2 **a** Quadrotor X-configuration, **b** Ardupilot flight controller (APM)

and stability. The quadrotor and the flight controller are illustrated in Fig. 2. The research work is currently in progress for a long hovering duration as well as other maneuvers.

2.3 *Evaluation and Analysis of Recorded Flight Test Data*

The adaptive robust identification method is implemented in this process for M-UAV modelling. The recorded data are used in AREKF for stability derivative identification. The simulated data is then compared to the actual flight data and its Mean Square Error is calculated. Above all, the performance of the filters are evaluated using goodness of fit (GOF) relative to the identification accuracy. The GOF parameter is defined as

$$GOF = 1 - \frac{\sum_{i-1}^{n}(y_i - f_i)^2}{\sum_{i}^{n}(y_i - y_{mean})^2} \quad (21)$$

where y_i is the actual flight test data, f_i is the estimated flight simulation data, and y_{mean} is the average of y_i. Therefore the GOF of a statistical model measures how well it fits a set of observations. This parameter is used in the present work to compare between the proposed AREKF method with other Kalman Filter methods.

3 Result and Discussion

The performance of the filters was analyzed and compared to the flight data. This section presents the preliminary result for the hovering maneuver which shows promising outcome. From Figs. 3, 4 and 5, the performance of EKF, REKF and AREKF are illustrated, respectively. The EKF shown in Fig. 3 has a big divergence in the estimation at early time step. This is due to the filter instability to high noises.

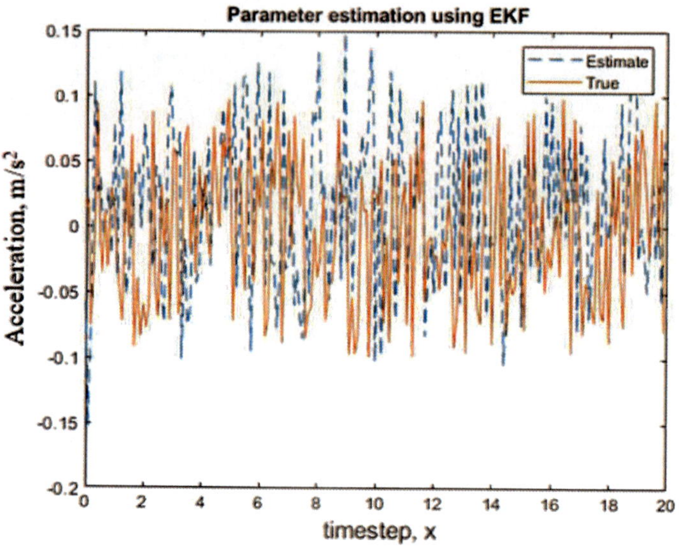

Fig. 3 Comparison of estimation value by EKF with flight data

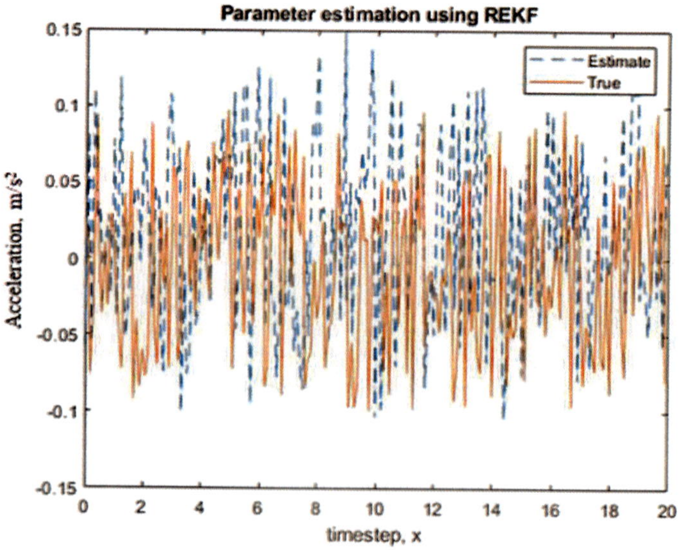

Fig. 4 Comparison of estimation value by REKF with flight data

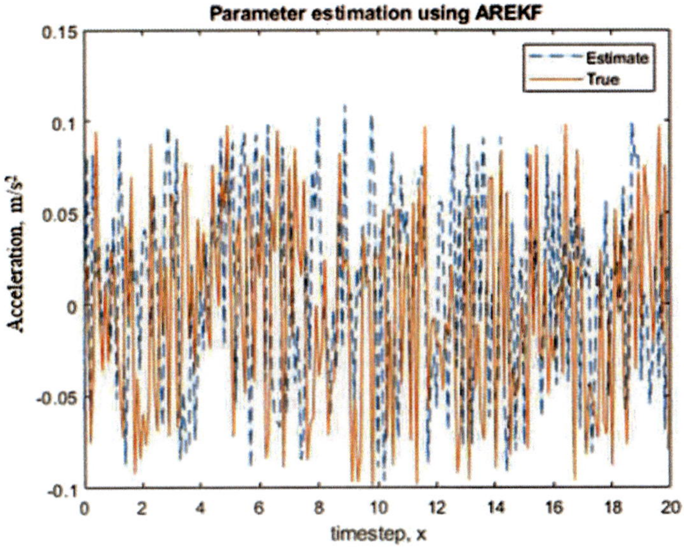

Fig. 5 Comparison of estimation value by AREKF with flight data

The filter cannot compensate to the initial assumption of the error covariance. From Fig. 4, REKF shows a good estimation to the true value where there is no filter divergence occurred. However, at some points, the estimation exceeds the true value from the flight data. This shows that REKF has the incapability to adapt with data divergence and the accuracy of the estimation is reduced at robust behaviour.

Moreover, the performance of AREKF shown in Fig. 5 is considerably accurate where the estimation does not diverge to the robustness of the filter and the value does not exceed the true value from the flight data. This is due to the ability of AREKF to compensate to filter divergence and can adapt to instability at large disturbances. AREKF can achieve the robust behavior of the filter without decreasing the accuracy of the estimation. AREKF combines both EKF and REKF filtering methods to adapt and perform efficiently at certain condition. The filter will check if there is a large innovation by the filter and will revert to EKF. Moreover, AREKF will acts as REKF if there is a need for a big innovation in the filter to compensate to filter divergence.

The GOF of the acceleration model are calculated and shown in Table 1. From the illustration, the GOFs of the model using AREKF identified parameter are higher than EKF and REKF. It can be said that the predicted value agree well with the true value and AREKF perform at best compared to the other two filter identified parameters.

Table 1 The GOF models using the parameters identified by EKF, REKF and REKF

	EKF	REKF	AREKF
GOF	0.664	0.678	0.853

The error mainly come from some error in model derivation and simplification and measurements sensors' errors.

4 Conclusion

The present work demonstrates that by using the data from the flight controller, the parameters estimation can perform well using the AREKF filtering technique. The conducted flight test for hovering condition shows that the dynamic parameters can be well identified using the method. Preliminary result demonstrates promising outcomes of AREKF when comparing with EKF and REKF methods. The goodness of fit index of AREKF is 0.853 which are 25% higher than REKF and EKF in the present case. Further test and improvement in the data noises will be conducted in the next studies.

Acknowledgements The support of Malaysian Ministry of Education under research grant FRGS17-036-0602 is gratefully acknowledge.

References

1. Altug E, Ostrowski JP, Mahony R (2002) Control of a quadrotor helicopter using visual feedback. In: Proceedings 2002 IEEE international conference on robotics and automation (Cat. No.02CH37292), vol 1, pp 72–77
2. Sugiura R, Fukagawa T, Noguchi N, Ishii K, Shibata Y, Toriyama K (2003) Field information system using an agricultural helicopter towards precision farming. In: Proceedings 2003 IEEE/ASME international conference on advanced intelligent mechatronics (AIM 2003), vol 2, pp 1073–1078
3. Bouabdallah S, Murrieri P, Siegwart R (2005) Towards autonomous indoor micro VTOL. Auton Rob 18(2):171–183
4. Corban JE, Calise A, Prasad JVR (1999) Implementation of adaptive nonlinear control for flight test on an unmanned helicopter, vol 4
5. Mettler B et al (1999) System identification of small-size unmanned helicopter dynamics
6. Sanders CP, DeBitetto PA, Feron E, Vuong HF, Leveson N (1998) Hierarchical control of small autonomous helicopters. In: Proceedings of the 37th IEEE conference on decision and control (Cat. No. 98CH36171), vol 4, pp 3629–3634
7. Shim H, Koo TJ, Hoffmann F, Sastry S (1998) A comprehensive study of control design for an autonomous helicopter. In: Proceedings of the 37th IEEE conference on decision and control (Cat. No. 98CH36171), vol 4, pp 3653–3658
8. Phillips WJWS (2004) Mechanics of flight. Aerospace/Engineering, p 1138
9. Hemakumara P, Sukkarieh S (2013) Learning UAV stability and control derivatives using Gaussian processes. IEEE Trans Rob 29(4):813–824
10. Koos S,.Mouret J, Doncieux S (2009) Automatic system identification based on coevolution of models and tests. In: 2009 IEEE congress on evolutionary computation, pp 560–567
11. Dumont G (2012) Basic of system identification. University of British Columbia
12. Abas N, Legowo A, Ibrahim Z, Rahim N, Kassim AM (2013) Modeling and system identification using extended Kalman filter for a quadrotor system. 313–314

13. Hoffer NV, Coopmans C, Jensen AM, Chen Y (2013) Small low-cost unmanned aerial vehicle system identification: a survey and categorization. In: 2013 international conference on unmanned aircraft systems (ICUAS), pp 897–904
14. Abas N, Legowo A, Akmeliawati R (2011) Parameter identification of an autonomous quadrotor. In: 2011 4th international conference on mechatronics (ICOM), pp 1–8
15. Lyu P, Bao S, Lai J, Liu S, Chen Z (2018) A dynamic model parameter identification method for quadrotors using flight data. Proc Inst Mech Eng Part G J Aerosp Eng 233(6):1990–2002
16. Xiong K, Zhang H, Liu L (2009) Adaptive robust extended Kalman filter. In: Kalman Filter Recent Advances and Applications, InTech

A Novel BiGRUBiLSTM Model for Multilevel Sentiment Analysis Using Deep Neural Network with BiGRU-BiLSTM

Md. Shofiqul Islam and Ngahzaifa Ab Ghani

Abstract In multilevel sentiment classification task, there is a challenging task of limited coherence, contextual and semantic information. This paper proposes a new hybrid deep learning architecture for multilevel text sentiment classification with less training and simple network structure for better performance and can handle the implicit semantic information and contextual meaning of text. In this research the proposed hybrid deep neural network architecture made with Bidirectional Gated Recurrent Unit (BiGRU) and Bi-Directional Long Term Short Memory(BiLSTM) of Recurrent Neural Network (RNN) for multilevel text sentiment classification and this performs better with higher accuracy than other methods compared. This proposed method BiGRUBiLSTM model outperformed the traditional machine learning methods and the compared deep learning models with about average of 1% margin accuracy on different datasets.

Keywords GRU · Bi-GRU · BiLSTM · LSTM · Sentiment analysis · Text classification · Machine learning · Natural language processing

1 Introduction

Nowadays with the horizons of social media keep expanding, the impacts they have on people are huge. Many businesses are taking advantages on the input from social media to advertise to specific target market. This is done by detecting and analyzing the sentiment in social media about any particular topic or product related texts. In order to classify the sentiment, sentiment analysis methods are used. Sentiment analysis helps to understand better of user opinions of users in social media or on other reviews. In recent year, the sentiment analysis, opinion mining,

Md. S. Islam · N. A. Ghani (✉)
Faculty of Computing, Universiti Malaysia Pahang, Pekan, Malaysia
e-mail: zaifa@ump.edu.my

N. A. Ghani
Centre for Software Development & Integrated Computing, Universiti Malaysia Pahang, Gambang, Malaysia

A. F. Ab. Nasir et al. (eds.), *Recent Trends in Mechatronics Towards Industry 4.0*, Lecture Notes in Electrical Engineering 730, https://doi.org/10.1007/978-981-33-4597-3_37

emotion recognition are important research area in the field of NLP. The role by Internet users are increasing with the growth of social networks. Users share their views increasingly on the internet with text. Such texts provide a wide range of information that is very helpful when analyzing social media or products for all the consumers. Throughout the area of sentiment analysis, there are still many studies that have obtained successful outcomes. While previous approaches have done well but the opinion mining remains a great challenge for many reasons such as of limited handling of coherence, contextual and semantic information. Many approaches of machine learning excessively depend on training features, adaptation, time-consuming and costly. Some machine learning methods also depends on manual dictionary. Some other lexicographical method analysis the textual sentiment with the position-based technique of grammar words relations. There are also some methods in machine learning named as: K-means, Naive Bayes (NB), Support Vector Machines (SVM), those approaches use frequency and features for classification. In comparison to traditional terminology, short summary texts are rather colloquial, have non-standard syntax constructs and have poor sentence validity. This is also more difficult to correctly determine the polarity of the sentiment. While the offer of deep learning in recent years has improved greatly, such neural network methods throughout the short text do not encode and effectively learn the part-all relationship. In recent years, a lot of neural network-based models have already developed with better performance than all other supervised techniques. Convolutional Neural Network (CNN) [1], Long Short Term Memory (LSTM) [2], BiLSTM [3], Gated Recurrent Unit (GRU) [4] based model are performing better for text classification in recent years However, in past years, the most significant form of sentimental research has been to use deep neuronal network architectures focused on convolution either recursion, like Kim's convolution neural network (CNN) [1] as well as the Xu's neural long short term memory (LSTM) [5].

Such approaches, which mentioned above for sentiment analysis challenges, are commonly employed in several sentiment analysis activities and therefore have shown outstanding performance evaluation tests. Nonetheless, most conventional approaches are restricted to use language abilities so often, or document features may be challenging to interpret in depth. In addition, in conventional models, the learning and training phase after encoding of features is not adequate. The quality of the measurement findings, accuracy is not fully sufficient. The role of sentiment analysis is however indeed facing big challenges.

To solve above limitations, we propose a novel hybrid BiGRU-BiLSTM model for multilevel sentiment analysis using deep neural network with BiGRU-BiLSTM with higher accuracy among compared methods. This model extract grammatical and syntactic features to enhance the generalization ability of the network to enhance the accuracy for multilevel sentiment classification.

The rest of the article is structured according to the following. In Sect. 2 is for related works and Sect. 3 describes the design of the theoretical deep learning platform while sub section of Sect. 3 includes tests and overview of the dataset. The findings with result, discussion and comparison of the studies carried out are presented and explored in Sect. 4. Finally, the conclusion is in Sect. 5.

2 Related Works

In recent years researchers have developed a lot of automated approach for abusive, sarcasm, toxic detection and classification. Some related deep learning approaches are presented below with their findings, research gaps, advantages, and limitations:

Now a day's newer approaches using deep learning architecture performed with higher accuracy. Based on large-scale dataset, many deep learning approaches will systematically obtain the theoretically somaticized and syntactic features of texts with higher accuracy. Like CNN for sentence classification [1], In this method, the textual features can be extracted easily and relational research has made a lot of progress in sentiment analysis. In 2015, Zhang uses character level CNN for text classification [6]. However, CNN is not also fully satisfactory because this method cannot capture long ranged feature, but it does not recognize the important dependent features information of its function and spatial position details. The above limitations of handling long ranged dependent feature of information have been almost solved by the Recurrent Neural Network(RNN) [3, 7, 8]. An approach with BERT [9] for sentiment classification with target dependency.

In recent year, RNN has been widely used in different fields for classification purposes with better result. By studying the previous knowledge or features, RNN may predict results with long-range features. Authors have suggested a variety of RNN variants like LSTM: The LSTM is able to monitor long-term dependencies in sequences with storage units with gate structures which determine how data in storage can be used and modified in order to get more data to improve model calculation advantages [7, 10], Dai introduces sequence learning in recurrent neural network for pretrained LSTMs model [11], Recently, using BiLSTM [12], in target-related task such as classifying relationships based on the various evaluation objects, the deep neural network framework coupled with attention mechanism have provided better performance than conventional approaches [3, 13]. For the tasks of sentiment research, attention processes are primarily used for grouping of aspects based on emotions. Galassi suggested a modern computational architecture that utilizes lexical sentiment tools readily accessible. In order to enhance model efficiency, two forms of attention are provided, namely lexicon dependent contextual attention and contrasting co-attention [14, 15]. GRU [4, 16], Sayeed developed a deep neural networks with GRU and maximum pooling to classify the overlapping sentiments with high accuracy Bi-GRU [17]. A recent method with BERT [9] method uses GloveVector, Word2Vec, WordPiece tokenizer, segment, and position embeddings and encoder layer for classification to find target sentiment. This approach focusses on the target terms instead of the whole sentence but does not perform better with mixed sentiment polarities towards different aspects.

To solve some of limitations mentioned above we propose a novel BiGRU-BiLSTM model for multilevel sentiment analysis using deep neural network with BiGRU-BiLSTM. This model extract grammatical and syntactic features to enhance the generalization ability of the network to enhance the accuracy for multilevel sentiment classification. At first textual words are represented to vectors by embedding approach of Glove vectors that represents semantic information. The resulting vectors are forwarded to the BiGRU. The BiGRU and BiLSTM shorten the long distance of semantic features.

3 Methodology

Multi-class classification is the function of classification for more than two classes that's helps to understand better of text but conventional sentiment classifier classify text in two or three category that is not sufficient enough to understand textual information clearly; each mark is mutually exclusive. The classification does the supposes that every sample is set to only one level. The Multilevel classification assigns a set of aim labels for each sample of textual information. This can be known to forecast a data point which does not mutually exclusive to each other for example Tim Horton are sometimes classified as a bakery or coffee shop. Multilevel text classification does have many uses in practice like to understand better of textual sentiment with different emotions or opinions, in classification of Yelp firms or classification of films into one or more categories of use. We have focused on multilevel text classification. We have done our work on multilevel datasets D. For training and text purposes. We carry out four of data-sets tests on the problem for multilevel sentiment classification. For example, the toxic comments from Jigsaw verified by Kaggle to assess the efficacy of our model. For more description about this dataset see subjection I of this methodology section. The dataset consists documents with comments and each comment set with level of vector.

$$\mathrm{D} = \left\{(\mathrm{C}, \mathrm{T}) | \mathrm{C} \in \text{Documents}, \mathrm{T} \in (0, 1)^{\mathrm{L}}\right\}$$

Here C is the comments in the documents in the dataset T is vector which has six levels and each level L indicates the types of toxicity namely toxicity, severe threat, toxicity, insult, obscenity and identity hate. The goal of our research with BiGRUBiLSTM neural network model to predict the level of sentiment for four dataset name: Kaggle toxic comments dataset, MR, IMDB, SST. Our model consists of following steps namely: Text embedding, BiGRUBiLSTM based learning model and evaluation.

3.1 Text Embedding

The designed model cannot take the inputted data directly, for this reason data need to be embedded. In our method we use GloVe embedding. This embedding layer makes a vector presentation for each word of input text based on semantic relationship with related words. It maps the words on worthwhile space by word similarity of distance and semantic. Each tokenized word is mapped by vector of the corresponding index of word from the embedding matrix. The output of the processed vector by the Glove vector feds to the our learning model named BiGRU-BiLSTM.

3.2 *Learning Model*

In this proposed method, two types of neural networks, first one is BiGRU and second one is BiLSTM of Recurrent Neural Network. Our model has an embedding layer at the beginning with transferring tokenization of word. Then BiGRU takes sequence input with one dimensional spatial dropout to extract dependent features with long distance. Then its output transfers to the BiLSTM layer that understands long-term, bidirectional dependencies among time series as well as sequence data steps. These dependencies could be useful if you want the model to learn from the entire time series every step of the way from the previous output of BiGRU. Finally the drop out layer and dense layer linked with Binary cross entropy, Adam optimizer and ROC-AUC evaluations to predict multilevel class of sentiment.

Long Short Term Memory (LSTM) LSTM [18] is an RNN prolongation that requires inputs to be stored for a long period. LSTM has an advanced memory, as opposed to RNN's basic internal memories it suffered from vanishing point. The memory may be gated in LSTM. It has three gates named as input, forget and the output gates. Basic equation of LSTM is as follows: $i_t = \sigma(w_i[h_t - 1, x_t] + b_i)$, $f_t = \sigma(w_f[h_t - 1, x_t] + b_f)$, $o_t = \sigma(w_t[h_t - 1, x_t] + b_o)$, Here, i_t is used for input gate, o_t for output gate and f_t denotes forget gate, σ represents the activation function, w_x is used to indicate weights of different gates(x), $h_t - 1$ is the output from the previous LSTM with time stamp $t - 1$, x_t is the current time stamp, b used for bias value in different gates.

Gated Recurrent Unit (GRU) GRU [7] performs better than LSTM. GRU does have the capacity to keep memory from prior detections, enabling it to keep track of functions for a long period with back propagation. GRU has two high performance gates those are reset and update gate. Here $r_t j$ is the reset gate and update gate $z_t j$ to the hidden state $h_t j$ for the variable of time t. Here x is used for input and r for reset gate and h for hidden state and z for update and sigma for $r_t = \sigma(W_{xr}x_t + W_{hr}h_{t-1} + br)$, $z_t = \sigma(W_{xz}x_t + W_{hz}h_{t-1} + b_z)$, $\hat{h}_t = \tanh(W_{xh}x_t + W_{hh}(r_t \odot h_{t-1}) + b_h)$, $\hat{h}_t = z_t \odot h_{t-1} + (1 - z_t) \odot \hat{h}_t$, here W represents matrices, b represents model parameters, σ represents element wise sigmoid function and $\odot$ for element wise multiplication.

BiLSTM layer BiLSTMs are proven particularly helpful when the context of its input is required. It's very helpful for works like classifying sentiments. In unidirectional LSTM data flows back to forward from backwards. In Bi-directional LSTM data, on the contrary, flows not just to backwards towards forward and also backwards using two hidden states. Thus the Bi-LSTMs best know the context [12]. BiLSTMs have been used to scale up the chunk of network-usable input information. Figure 1 shows basic structural operation of BiLSTM.

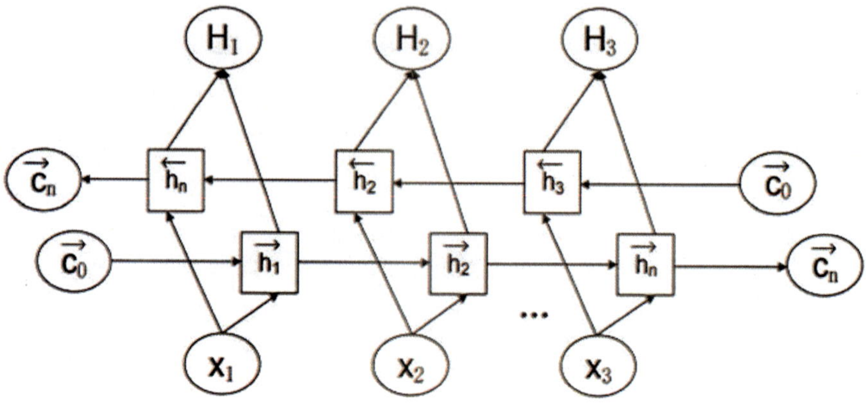

Fig. 1 BiLSTM model. *Source* [19]

BiGRU Layer Bidirectional configurations can manipulate this information with both the 'left' as well as 'right' context on each place for sequential patterns. The bidirectional gated loop unit (BiGRU) consists of two unidirectional, oppositely oriented GRUs, which can effectively improve the context information loss problem [20]. The structural model of BiGRU is shown in Fig. 2.

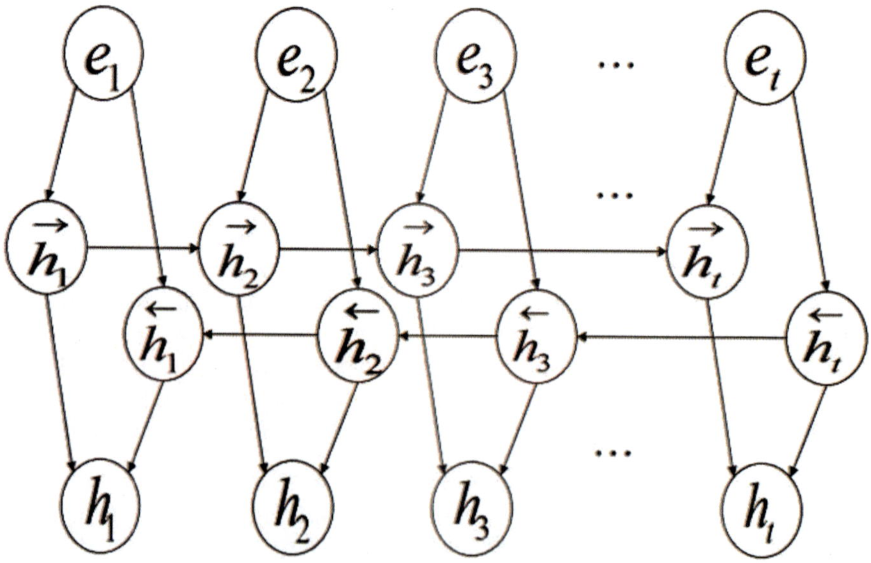

Fig. 2 BiGRU model. *Source* [20]

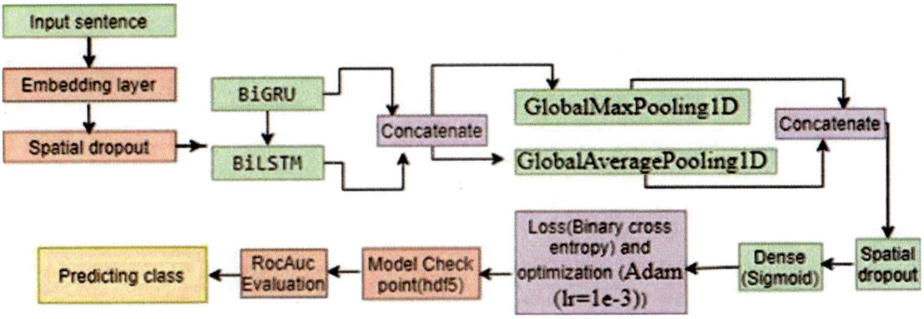

Fig. 3 BiGRU-BiLSTM model

3.3 Overall Structure of BiGRU-BiLSTM Model

Our proposed model as shown in Fig. 3 consists Glove vector at the embedding layer and uses Unicode text and matrix value as float. We set embedding size and max features to and maximum length. In the embedding layer we use one dimensional spatial dropout with 0.2. After that we use BiGRU with 128 units for the first and 64 units for the second. Then BiLSTM with 128 units with max and average pooling with concatenation. In this model we use 4 epochs as neuron size and batch size 128 for the classification purposes.

3.4 Pseudocode for Proposed Algorithm

Multilevel sentiment classification algorithm using BiGRU-BiLSTM model.

Input: Data from dataset
Output: The distribution of probability of different level of sentiments
1.Set some predefined variables: max_features=100000,maxlen=200,embed_size=300
2. Embedding the input data sequences with 300 Dimensional vector of Glove Vector and produce output as tokenized index of embedded words
3. Separate data for training and testing
4. Set epoch_size N=5,SEEDS=10, batch_size(Predicting)=512, LSTM_Units=128, Dense_ Size=4* LSTM_Units, random_state=233,training_size = 0.90
3. For each epoch 1 to 1..... N do
Select M training sample with training size with the random state.
For training sample i=1...M do
Connect dropout layer with 0.25 size with the flatten output with the embedding output
Feed training sample to Bidirectional Gated Unit (BiGRU) and get the output
$BIGRU_i=\{bgru^*_1......, bgru^*_t......, bgru^*_n\}$
Connect Bidirectional LSTM layer with the output from BiiGRU layer
$blstm^*_1= BIGRU_1, blstm^*_2= BIGRU_2........., blstm^*_n= BIGRU_n$
$BiLSTM_j=\{blstm^*_1......, blstm^*_t......, blstm^*_n\}$
Concatenate layer of GlobalMaxPooling1D and GlobalAveragePooling1D with the BiGRU output
Connect dense layer with size 6 and sigmoid function with the dropout output with the BiGRU output
Do the model compilation with following parameter:
loss='binary_crossentropy,optimizer=adam,metrics=['accuracy']
Calculate the probability distribution of sentiment class by the model
End For
End For

3.5 Evaluation

In order to evaluate the architectural outputs, in this evaluation section, the exact accuracy and also the mean are used by us with Receiving Operating Characteristics (ROC) curve value, mean precision, mean recall as well as a mean F1 score.

The accuracy is measured by the following equation.

$$Accuracy = \frac{Number\ of\ exacttly\ matched\ instances}{Total\ instances}$$

3.6 Data Information

To evaluate the performance of our proposed method we carry out a four following dataset tests on the problem for multilevel text-based sentiment classification.

Toxic comments dataset First dataset for toxic comments from Jigsaw verified by Kaggle (https://www.kaggle.com/c/jigsaw-unintended-bias-in-toxicity-classification). The dataset contains feedback via Wikipedia. This dataset consists of 159,571 comments six categories of toxicity.

IMDB IMDB consists of 50,000 movies on the internet and extreme division reviews. This dateset includes 25,000 positive plus 25,000 negative results. The fine period is 294 [21].

SST This is a MR dataset expansion, which contains a single training set, study set and study set with a minimum of 11,855 words. Sata labels are classified into five groups as per the sentinet nature of sentences: extremely positive, positive, negative, and very negative. The median duration of the sentence is 19 [22].

MR MR is a series of English film reviews from Cornell University that were recorded on www.rottentomatoes.com. 5331 film comments were positive and 5331 critical film opinions were negative. The length of the typical word is 20 [23].

4 Experimental Result

We compared our performance with some recent advanced neural network model for multilevel text classification. and eleven features for features interest and the Fig. 4a, b show the accuracy and loss performance of or method with four epochs for toxic comment classification dataset.

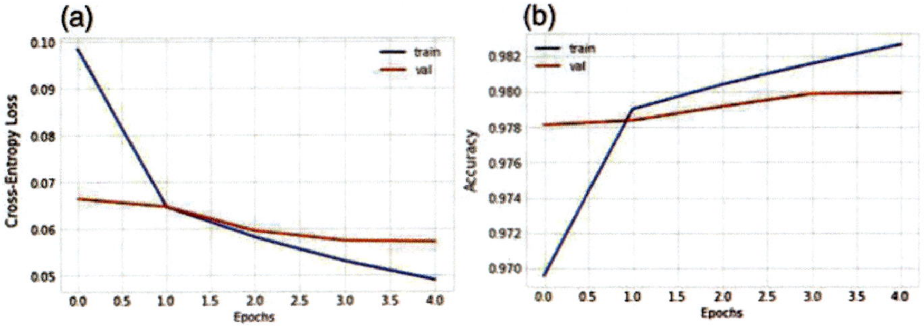

Fig. 4 **a** Shows the accuracy train versus test in four epochs. **b** Shows the loss train versus test in four epochs

Table 1 Comparison of several deep learning approaches with our BiGRU-BiLSTM model

Authors	Model name	Toxic comment classification dataset	MR	IMDB	SST
[24]	SVM	81.11	76.14	86.9	36.03
[1]	CNN-multifilter	95.16	76.10	86.84	46.90
[7]	LSTM	97.71	75.90	87.64	45.60
[8]	CNN-LSTM	97.64	76.60	88.20	47.25
[3]	Bi-LSTM with maxpol	97.35	79.30	88.81	46.50
[26]	Hieachical ConvNets	97.42	76.25	87.90	46.20
[28]	CapsNet dynamic routing	98.48	82.30	89.80	47.10
[27]	ToWE-CBOW	98.46	65.10	90.80	46.80
Proposed	Bi-GRU-LSTM	98.68	82.10	91.98	49.25

Here we present all the compared base line algorithms, SVM for sentiment classification by Pang [24], CNN-multifilter done by kim [1] and this methods uses different filters to extract n gram features, LSTM task done by Cho [7]. Wang developed a CNN-LSTM model using CNNs and RNNs [8]. Bidirectional Long Short Term Memory with maximum pooling by Lai [3]. Feed forward attention based network by Raffel [25]. Hieachical ConvNets done by Yin in 2017, this model uses hierarchical abstraction of input text and performed well on short text by this method [26]. Task-oriented word embedding for text classification by Liu [27]. Bi-LSTM max pooling related work by Lai [3] and Capsule network with dynamic routing based work done by Srivastava. See Table 1 for performance comparison of our model with other models on similar dataset.

5 Conclusion

Our main two contributions on this proposed method are: "1. BiGRU layer in the BiLSTM model used to make short distance of the grammatical and syntactic features. 2. We use a bidirectional LSTM (BiLSTM) layer to understand long-term, bidirectional dependencies among time series as well as sequence data steps. These dependencies could be useful if you want the model to learn from the entire time series every step of the way". In our research we developed a capsule network based BiGRU-BiLSTM model using neural network for multilevel text sentiment classification. This proposed method BiGRU-BiLSTM model outperformed with 1% margin accuracy on different datasets than the traditional machine learning methods and the compared deep learning models. Proposed BiGRU-BiLSTM method achieved accuracy of 98.68% for Kaggle comments classification, 82.10 for IMDB and 91.98 for MR and 49.25 for SST datasets. This article proposed a persistent multi-level textual classification using capsule network dependent architecture with recurrent, convolutional neural network. Multilevel sentiment analysis helps better understanding of text than all other conventional sentiment analysis. We have done simple data pre-processing by our research, but in future we will try to develop additional data pre-processing strategies such as data augmentation by translation strategies and misspelled word methods. We will try to handle more than six classification of sentiment with multiple and multimodal datasets.

Acknowledgement This research is supported in part by grants from the Fundamental Research Grant Scheme (FRGS) by the Government of Malaysia to Universiti Malaysia Pahang. The grant numbers are FRGS/1/2018/ICT02/UMP/02/15 and FRGS/1/2019/ICT02/UMP/02/8.

References

1. Kim Y (2014) Convolutional neural networks for sentence classification. arXiv preprint arXiv:1408.5882
2. Yadav A, Vishwakarma DK (2019) Sentiment analysis using deep learning architectures: a review. Artif Intell Rev 1–51
3. Lai S et al (2015) Recurrent convolutional neural networks for text classification. In: Twenty-ninth AAAI conference on artificial intelligence
4. Du Y et al (2019) A novel capsule based hybrid neural network for sentiment classification. IEEE Access 7:39321–39328
5. Xu J et al (2016) Cached long short-term memory neural networks for document-level sentiment classification. arXiv preprint arXiv:1610.04989
6. Zhang X, Zhao J, LeCun Y (2015) Character-level convolutional networks for text classification. In: Advances in neural information processing systems
7. Cho K et al (2014) Learning phrase representations using RNN encoder-decoder for statistical machine translation. arXiv preprint arXiv:1406.1078
8. Wang X, Jiang W, Luo Z (2016) Combination of convolutional and recurrent neural network for sentiment analysis of short texts. In: Proceedings of COLING 2016, the 26th international conference on computational linguistics: Technical papers
9. Gao Z et al (2019) Target-dependent sentiment classification with BERT. IEEE Access 7:154290–154299
10. Mousa A, Schuller B (2017) Contextual bidirectional long short-term memory recurrent neural network language models: a generative approach to sentiment analysis. In: Proceedings of the 15th conference of the european chapter of the association for computational linguistics, vol 1, Long Papers
11. Dai AM, Le QV (2015) Semi-supervised sequence learning. In: Advances in neural information processing systems
12. Schuster M, Paliwal KK (1997) Bidirectional recurrent neural networks. IEEE Trans Sig Process 45(11):2673–2681
13. Tai KS, Socher R, Manning CD (2015) Improved semantic representations from tree-structured long short-term memory networks. arXiv preprint arXiv:1503.00075
14. Galassi A, Lippi M, Torroni P (2019) Attention, please! a critical review of neural attention models in natural language processing. arXiv preprint arXiv:1902.02181
15. Yanase T et al (2016) bunji at semeval-2016 task 5: Neural and syntactic models of entity-attribute relationship for aspect-based sentiment analysis. In: Proceedings of the 10th international workshop on semantic evaluation (SemEval-2016).
16. Chung J et al (2014) Empirical evaluation of gated recurrent neural networks on sequence modeling. arXiv preprint arXiv:1412.3555

17. Saeed HH, Shahzad K, Kamiran F (2018) Overlapping toxic sentiment classification using deep neural architectures. In: 2018 IEEE international conference on data mining workshops (ICDMW). IEEE
18. Hochreiter S, Schmidhuber J (1997) Long short-term memory. Neural Comput 9(8):1735–1780
19. Li J, Xu Y, Shi H (2019) Bidirectional LSTM with hierarchical attention for text classification. In: 2019 IEEE 4th advanced information technology, electronic and automation control conference (IAEAC). IEEE
20. Zhou L, Bian X (2019) Improved text sentiment classification method based on BiGRU-Attention. J Phys Conf Ser. IOP Publishing
21. Maas AL et al (2011) Learning word vectors for sentiment analysis. In: Proceedings of the 49th annual meeting of the association for computational linguistics: human language technologies, vol 1. Association for Computational Linguistics
22. Socher R et al (2013) Recursive deep models for semantic compositionality over a sentiment treebank. In: Proceedings of the 2013 conference on empirical methods in natural language processing
23. Pang B, Lee L (2005) Seeing stars: exploiting class relationships for sentiment categorization with respect to rating scales. In: Proceedings of the 43rd annual meeting on association for computational linguistics. Association for Computational Linguistics
24. Pang B, Lee L, Vaithyanathan S (2002) Thumbs up? Sentiment classification using machine learning techniques. In: Proceedings of the ACL-02 conference on Empirical methods in natural language processing, vol 10. Association for Computational Linguistics
25. Raffel C, Ellis DP (2015) Feed-forward networks with attention can solve some long-term memory problems. arXiv preprint arXiv:1512.08756
26. Yin W et al (2017) Comparative study of CNN and RNN for natural language processing. arXiv preprint arXiv:1702.01923
27. Liu Q et al (2018) Task-oriented word embedding for text classification. In: Proceedings of the 27th international conference on computational linguistics
28. Srivastava S, Khurana P, Tewari V (2018) Identifying aggression and toxicity in comments using capsule network. In: Proceedings of the first workshop on trolling, aggression and cyberbullying (TRAC-2018)

A Comparative Study on Nonlinear Control of Induced Sit-to-Stand in Paraplegia with Human Mass Variation

Mohammed Ahmed, M. S. Huq, B. S. K. K. Ibrahim, Nura Musa Tahir, Zainab Ahmed, and Garba Elhassan

Abstract Presented is an assessment of the effects of changes in global human for the regulation of Functional Electrical Stimulation (FES) revived sit-to-stand (STS) using three nonlinear control methods. The techniques are the Sliding Mode Control (SMC), Feedback Linearized Control (FLC), and Back Stepping Control Approach. Literature indicates the necessity of improving the system of which applications control has shown such ability. The analytical STS model uses the four segments. Six order polynomial was harnessed for trajectory planning. The work was aimed at relaxing the effect of the variation in human masses such that there will be no need changing parameters when using the device for different subjects. The results indicated the ability of the control schemes to curtail the perturbations and to regulate the system with minimal errors while maintaining the stability as well. Responses closeness to the desired transition trajectories during the movement and without any

M. Ahmed (✉)
Department of Electrical and Electronics Engineering, Abubakar Tafawa Balewa University, P. M. B. 0248 Bauchi, Nigeria
e-mail: inunugoloma@yahoo.com

M. S. Huq
Department of Mechatronics Engineering, Institute of Technology Sligo, Sligo, Ireland
e-mail: saif.huq@gmail.com

B. S. K. K. Ibrahim
School of Mechanical, Aerospace and Automotive Engineering, Coventry University, Coventry, UK
e-mail: ad1465@coventry.ac.uk

N. M. Tahir
Department of Mechatronics and Systems Engineering, Abubakar Tafawa Balewa University, P. M. B. 0248 Bauchi, Nigeria
e-mail: nuratahir85@gmail.com

Z. Ahmed
Yelwan Makaranta Opposite School of Agriculture, Bauchi, Nigeria
e-mail: zainabahmadjak@gmail.com

G. Elhassan
National Space Research and Development Agency, Abuja, Nigeria
e-mail: gzeenol@yahoo.com

A. F. Ab. Nasir et al. (eds.), *Recent Trends in Mechatronics Towards Industry 4.0*, Lecture Notes in Electrical Engineering 730, https://doi.org/10.1007/978-981-33-4597-3_38

disturbance are evident from the results obtained. It was also accomplished within the recommended settings of the stimulation parameters with high levels of robustness, which are highly greater than that of a 2% error margin usually considered. Therefore, indicating the likelihood of the methods for the FES induced STS system for greater performance with the BSC being the best.

Keywords Nonlinear control · Sit-to-Stand movement · Functional electrical stimulation · Robustness evaluation

1 Introduction

Movement restoration by application of suitable forms of electrical signals is known as Functional Electrical Stimulation (FES). It is also applied to other scenarios like in movement therapy and movement rehabilitation. It is implemented for the purposes mentioned earlier in subjects having nervous system ailments resulting from hazards or sickness [1, 2]. Literature shows more improvement is required despite the recorded successes with the application of control systems [3, 4].

The different categories of control systems have been proposed for the FES aided STS movement. In reality, there are two basic control schemes; the linear and nonlinear control methods. Intelligent control schemes are nonlinear control methods using intelligent techniques and mostly a mathematical representation of the control law is not readily available. It may therefore not be suitable for some important analysis such as stability evaluation [5]. Linear methods comprise of the works of; Tsukahera et al. [6] and Fattah et al. [7] utilizing the Proportional-Derivative (PD) controllers, Yu et al. [8], Previdi et al. [9] and Pobonoruic [10] the Proportional Integral Derivative (PID) control. Scholars inferred that the linear control approaches appropriate for the systems under consideration [11].

Works on Intelligent control methods include the research of Davoodi and Andrews [12, 13] where the Adaptive Network-based Fuzzy Inference System (ANFIS) and the Genetic Algorithm, tuned Adaptive Network-based Fuzzy Inference System (GA-ANFIS), Hussain et al. [14] and Massoud [15] proposed combined PID and ANFIS (PID-ANFIS) and applied by Afzal et al. [16] is combined PID and Artificial Neural Networks (PID-ANN). The deficiency of mathematical models makes intelligent based control approaches inferior. In the case of nonlinear control was by Espanjani and Towhidkhah [17] using the Model Predictive Control method. The extremely nonlinear behavior of the system, achievements of nonlinear control techniques, and competency for scrutiny.

The work aims to compare the system response using three nonlinear control methods viz: the Sliding Mode Control (SMC), Feedback Linearized Control (FLC), and Back Stepping Control Approach. Part of the objectives is the effort to measure the level of the robustness as well the phase trajectories addition of which gives the newness of study. Measurements of other response parameters form a portion of the objectives which are clearly defined in the text.

2 Methodology

The model of the paraplegic was approximated using the four segments as presented by Ahmed [18]. Equation (1) is the model in compact form; τ is 1-by-4 matrix which gives the torques of the four joints, $D(q)\ddot{q}$ is a 4-by-4 matrix and it gives the inertia vector, $C(q, \dot{q})\dot{q}$ is ta 4-by-4 matrix which gives the Corriolis vector of the system and $\dot{g}(q)$ is a 1-by-4 matrix which takes of the gravity vector.

$$\tau = D(q)\ddot{q} + C(q, \dot{q})\dot{q} + \dot{g}(q) \tag{1}$$

The muscle model is from the works of Ferrarin and Pedotti [19] and is carefully inserted into the model. The first Eq. (2) gives the damping torque, where β is the coefficient and $\dot{\theta}$ the joint angle velocity. The second Eq. (2) is the joint stiffness torque, where γ and E are the exponential coefficients, θ is the joint angle and ω is the joint resting angle. Equation (3) is the resulting torque produced at the joint due to the electrical stimulation, where P_W is the pulse width, G is the gain, and δ the time constant.

$$\tau_d(t) = \beta\dot{\theta} \tag{2}$$

$$\tau_s(t) = \gamma e^{-E\left(\theta+\frac{\pi}{2}\right)}\left(\theta + \frac{\pi}{2} - \omega\right) \tag{2}$$

$$\tau_a(s) = \frac{P_{w(s)}G}{1 + s\delta} \tag{3}$$

Trajectories were synthesized as proposed by Williams II [20] using the six order polynomial as given by Eq. (4), where θ gives the angle, a_n the coefficients and t^n the times.

$$\theta(t) = a_0 + a_1t + a_2t^2 + a_3t^3 + a_4t^4 + a_5t^5 + a_6t^6 \tag{4}$$

The global human mass variation was obtained from the literature [21–23]. Evaluation of was achieved the proposed method by Alagoz et al. [24]. The approach is similar to the signal to noise ratio measurement employed in communication systems. In this case, it is referred to as a reference to disturbance ratio (RDR). The ratio of the desired signal level to the error signal level is used to evaluate the level of disturbance rejection as given by Eq. (5). The greater the ratio above unity, the better the result obtained.

$$RDR = \frac{y(t)}{e(t)} \tag{5}$$

The controller designs were achieved as obtainable from standard texts. The limitation was that it was for FES supported STS in paraplegic subjects and also is a simulation study using SIMULINK/MATLAB software.

3 Obtained Results

Under this subsection, conducted is the examination of the system response in the absence of disturbance as well as with disturbance effect. The disturbance effect is referred to as the changes in human mass between the considered minimum and maximum.

3.1 Results Without Changes

Figure 1a, b shows the tracking response and errors of the system without any disturbance effect. The tracking response of the system is very close according to the Figure. The tracking errors during the transition resulted in Root Mean Square Errors (RMSEs) of 0.8828, 0.8406, and 0.0039 m° for the SMC, LFC, and BSC respectively. It means the scheme with the least error is the BSC, followed by the FLC and finally the SMC.

Figure 2a indicates the plots of ITEs and ITTEs during the STS movement, the figure portrays the amounts of the ITEs and ITTEs respectively as; 15.4445°, 61.7781° s, 1.7973°, 7.1901° s, 0.0580°, 0.2319° s for the SMC, LFC, BSC. The BSC has the least amount this makes it the best followed by the FLC and SMC

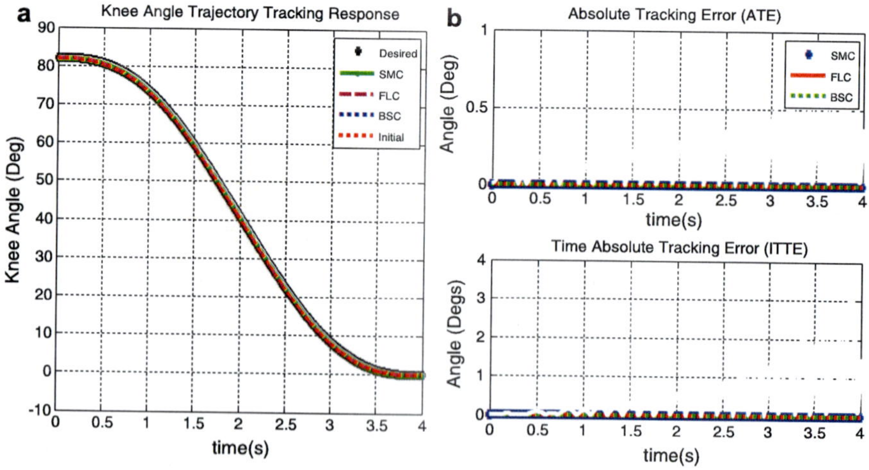

Fig. 1 **a, b** Tracking response and errors without disturbance

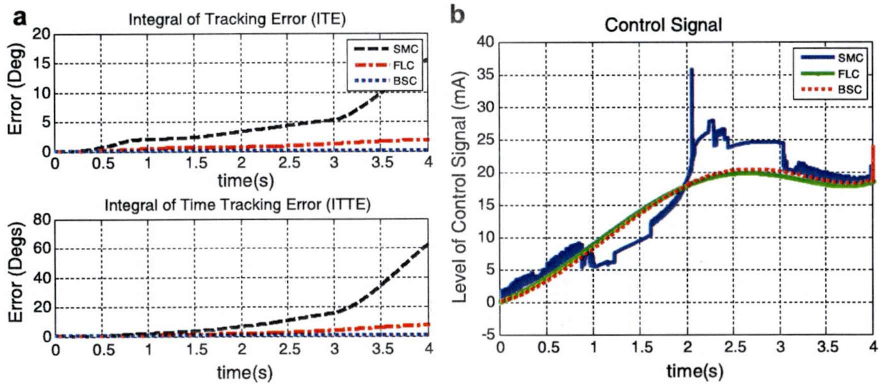

Fig. 2 **a** and **b** are the integral of tracking errors and control signals without disturbance

scheme. Figure 2b is the plots of the control signals for the SMC, FLC, and BSC for the movement duration. It shows that the maximum values are obtained as 35.7763, 19.8116, and 24.0208 mA for the SM, FL, and BS controllers respectively. In this regard, the LFC is leading to having the least amount of current, followed by the BSC and finally the SMC.

Figure 3a shows the IRCSCs/IRCTSCs of the control schemes without disturbances, which indicates that the IRCSCs/IRCTSCs the amount utilized by the SMC controller is the worst, followed by the BSC and the FLC which produces the least and was the best. Figure 3b indicates the resulting ISCs/ITSCs from the FES induced STS system and it shows that SMC utilized the highest amount of stimulation current, the FLC comes next, and BSC utilizes the least. The BSC was the best option regarding the amount of current usage, followed by the FLC and finally the SMC.

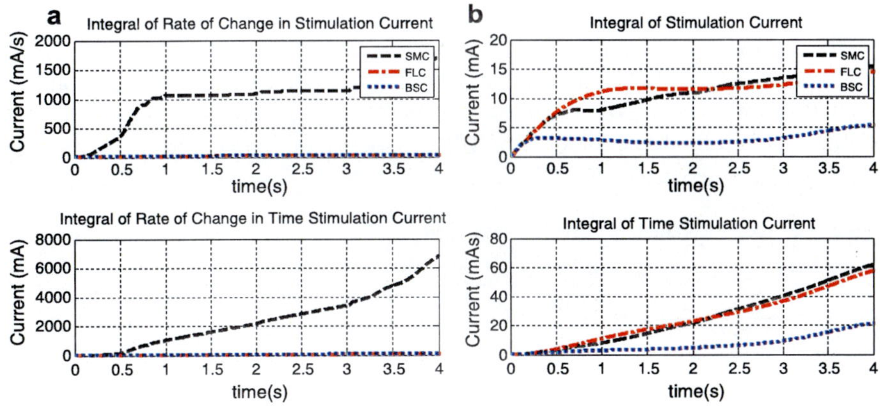

Fig. 3 **a** and **b** are an integral rate of change in currents and currents without disturbance

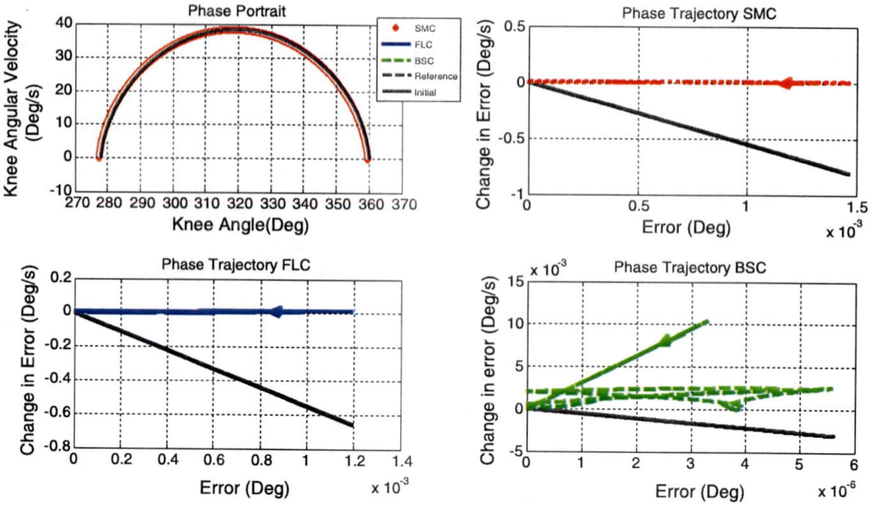

Fig. 4 Phase portrait without disturbance

Figure 4 shows the Phase Plane (PP) plot and the various plane trajectories of the different controllers are given by Fig. 4b–d. The Phase Plane portrait as shown in Fig. 4 indicates that the system plots with the SMC, LFC, and the BSC are very close to the reference or desired trajectory which is regarded as the equilibrium. The various plane trajectories indicate that they all move towards the reference for the different controllers. It portrays the stability of the system with the control schemes considered. The reference states' trajectory is regarded as the equilibrium state for the FES aided STS system.

Table 1 portrays the summary of the response parameters in Figs. 1a, b, 2a, b and 3a, b, while Fig. 4 gives the phase portrait and respective trajectories of the control schemes.

The results from Table 1 show that the level of the ITE was the least with the BSC followed by that of FLC, and then that of the SMC. It shows that the BSC produces the best result and the SMC the worst. The FLC has the lead regarding the IRCSC results as it had the least amount, the BSC follows it, and finally SMC. Regarding

Table 1 Response parameters without disturbance

Parameter	Control Scheme		
	SMC	LFC	BSC
RMSE (°)	0.8828 m	0.8406 m	0.0039 m
ITE/ITTE (°/°s)	15.4445/61.7781	1.7973/7.1901	0.0580/0.2319
MCS (mA)	35.7763	19.8116	24.0208
IRCSC/IRCTSC (mA^{-s}/mA)	1700.2000/6800.0000	22.6749/90.6998	29.8138/119.2554
ISC/ITSC (mA/mA s)	15.4397/61.7586	14.4031/57.6123	5.2633/21.0534

the ISC, the FLC was the best having the least amount, next is the SMC and the last is FLC.

The system phase plane graphs show that the SMC, FLC, and the BSC responses are very close to the reference. Supporting the fact is from their amounts of RMSE Hence, the results of stability assessments all been favorable for the nonlinear control schemes confirm stability.

3.2 *Results with Mass Changes*

Accomplished is an exploration of the effect of changing the mass of the subject between minimum and maximum. The values are 45 kg and 107 kg respectively and are referred to as; 53% mass and 126% mass mostly in the report. Tables 2 and 3 illustrate the summary of the response parameters with 53% and 126% mass respectively and Figs. 5 and 6 are the phase portraits.

Tables 2 and 3 showed a comparison of the parameters. The results portrayed enhancements in the ITE, IRCSC, and ISC parameters of the nonlinear control schemes. To further assess, the level of robustness the responses of the control

Table 2 Response parameters with 53% mass

Parameter	Control Scheme		
	SMC	FLC	BSC
RMSE (°)	0.0598 m	0.0487 m	0.0039 m
ITE/ITTE (°/° s)	1.6820/6.7278	0.2954/1.1815	0.0580/0.2319
MCS (mA)	15.5675	10.5305	10.8019
IRCSC/IRCTSC (mA^{-S}/mA)	1073.3000/4293.3000	12.0013/48.0051	13.6848/54.7392
ISC/ITSC (mA/mA s)	5.7985/23.1939	5.9543/23.8172	2.7887/11.1547
RDR	16,721.0000	20,533.0000	256,409.0000

Table 3 Response parameters with 126% mass

Parameter	Control Scheme		
	SMC	FLC	BSC
RMSE (°)	0.0598 m	0.0487 m	0.0039 m
ITE/ITTE (°/° s)	1.6792/6.7167	0.2954/1.1815	0.0580/0.2319
MCS (mA)	38.6843	25.0346	29.3535
IRCSC/IRCTSC (mA^{-S}/mA)	2586.4000/10346.0000	28.5138/114.0553	37.6315/150.5260
ISC/ITSC (mA/mAs)	13.9390/55.7558	14.1555/56.6218	6.6314/26.5256
RDR	16,721.0000	20,533.0000	256,409.0000

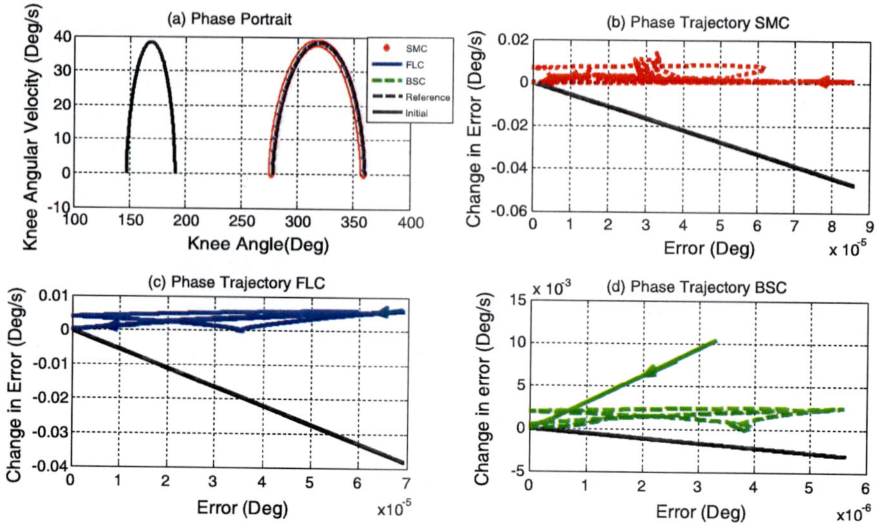

Fig. 5 Phase portraits with 53% mass

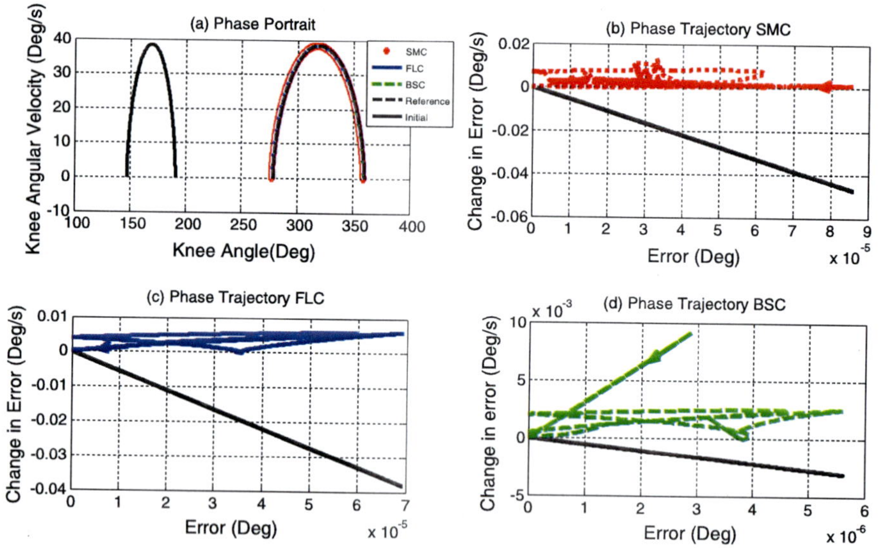

Fig. 6 Phase portraits with 126% mass

schemes, executed were a comparison of the RDRs. The RDRs are 16,721, 20,533, and 256,409 with the SMC, FLC, and BSC respectively. The same results are obtained for both 53 and 126% mass. Results indicated that the BSC is the best having the highest value, followed by FLC and finally the SMC.

Therefore, results show that despite the changes in the subject mass from 53% (minimum) to 126% (maximum). The nonlinear control schemes possess a high level of improvement in the system despite the mass changes. They also maintained the enhancement capabilities when performance is compared with the situation whereby there is no disturbance hitting the system. These measures as well as from the RDR results illustrate high levels of the robustness of the controllers. It shows that the BSC has the best robustness performance which is followed by the FLC and then the SMC.

Figure 5a indicates how the state trajectory due to the 53% mass. It can be seen that despite what happened the control schemes can make necessary compensation. As shown the resulted trajectories show that for all the three control schemes they all move towards the reference. Figure 5b–d show how the trajectories converged to the origin or reference. Therefore, the system is stable using any of the control schemes. Figure 6a shows the phase portrait which indicated that the trajectories using the nonlinear control schemes considered, move towards the reference. Figure 6b–d is the phase trajectories of the SMC, FLC, and BSC respectively. The phase trajectories indicated that they all converge to the reference which indicated that the system is stable despite the change in mass.

4 Discussion of Results

The response parameters of the FES induced STS movement without uncertainties show that the SMC, FLC, and BSC are achieved with a very low amount of RMSE. The values of the RMSE obtained are below 2%, normally targetted as the steady-state error in control systems. The errors are far below that, for instance with 2% error will produce an RDR of 50. The RDR for the respective control schemes is by far greater than that, which implies the resulting errors are by far less than 2%. The Maximum Control Signal (MCS) is equivalent to or gives the Maximum Stimulation Current (MSC). They may be used in the text interchangeably, so also ° and Deg. The ITE, IRCSC, and ISC are also low. Hence, the control objective is achieved. Other results shown are that of stability and is evaluated using the phase plane. The results obtained indicate the stability of the system. The above are the response and stability performance of the system (the FES induced STS movement) without disturbance or nominal performance and nominal stability of the system.

The examination of robustness performance starts with the evaluation of the system performance due to changes in the subject mass. The parameters considered are the most likely expected global human masses. Employed are the least and highest expected human masses of 45 kg and 107 kg respectively. Despite the changes in the system tried to remain close to without disturbances, portraying remarkable improvements in the system. The phase plane portraits also indicate stability almost similar to the situation without changes.

5 Conclusion

The FES induced STS movement does require refinement to attain clinical acceptance. The nonlinear control schemes maintained good results despite disturbances imposed. The nonlinear control schemes showed a high level of robustness and also stable. All these signify the tendency of the scheme to pave the way for clinical passage with appropriate modifications. The study novelty to the best of the authors was in the sense no such comparative study with emphases on evaluating the level of robustness was presented. Furthermore, the nonlinear control techniques have not been earlier explored for the FES assisted STS movement and finally the approach for designing control schemes which considered global changes in subject mass. The robustness analysis was evaluated by considering the minimal and maximum amount of the expected human masses. All the schemes performed very well with the best being the BSC with the best RDR as well as others, followed by the FLC and then the SMC.

References

1. Braz GP, Russold M, Davis GM (2009) Functional electrical stimulation control of standing and stepping after spinal cord injury: a review of technical characteristics. Neuromodulation Technol Neural Interface 12(3):180–190
2. Howlett OA et al (2015) Functional electrical stimulation improves activity after stroke: a systematic review with meta-analysis. Arch Phys Med Rehabil 96(5):934–943
3. Ibitoye MO et al (2014) Mechanomyographic parameter extraction methods: an appraisal for clinical applications. Sensors 14(12):22940–22970
4. Ibitoye MO et al (2014) Mechanomyography and muscle function assessment: a review of current state and prospects. Clin Biomech 29(6):691–704
5. Huq M, Tokhi M (2012) Genetic algorithms based approach for designing spring brake orthosis—Part II: control of FES induced movement. Appl Bion Biomech 9(3):317–331
6. Tsukahara A et al (2010) Sit-to-stand and stand-to-sit transfer support for complete paraplegic patients with robot suit HAL. Adv Robot 24(11):1615–1638
7. Fattah A, Hajiaghamemar M, Mokhtarian A (2008) Design of a semi-active semi-passive assistive device for sit-to-stand tasks. In: 16 th annual international conference on mechanical engineering—ISME2008, 14–16 May 2008. Shahid Bahonar University of Kerman, Iran
8. Yu N-Y, Chen J-JJ, Ju MS (2001) Closed-loop control of quadriceps/hamstring activation for FES-induced standing-up movement of paraplegics. J Musculoskelet Res 5(03):173–184
9. Previdi F et al (2005) Closed-loop control of FES supported standing up and sitting down using virtual reference feedback tuning. Control Eng Prac 13(9):1173–1182
10. Poboroniuc MS (2007) New experimental results on feedback control of FES-based standing in paraplegia. AL.I.CUZA Univ Sci Annals Biophys Med Phys Environ Phys III:83–89
11. Lynch C, Popovic M (2005) Closed-loop control for FES: Past work and future directions. In: 10th annual conference of the international FES society
12. Davoodi R, Andrews BJ (1998) Computer simulation of FES standing up in paraplegia: a self-adaptive fuzzy controller with reinforcement learning. IEEE Trans Rehabil Eng 6(2):151–161
13. Davoodi R, Andrews BJ (1999) Optimal control of FES-assisted standing up in paraplegia using genetic algorithms. Med Eng Phys 21(9):609–617
14. Hussain R, Massoud R, Al-Mawaldi M (2014) ANFIS-PID control FES-supported sit-to-stand in paraplegics: (Simulation study). J Biomed Sci Eng 7(4):208

15. Massoud R (2014) The influence of control design on energetic cost during FES induced sit-to-stand. J Biomed Sci Eng 7(14):1096
16. Afzal T, Khan L, Tokhi M (2010) Simulation of a patient driven strategy for FES supported sit-to-stand movement. In: 2010 international conference on information and emerging technologies (ICIET). IEEE
17. Esfanjani RM, Towhidkhah F (2006) Application of nonlinear model predictive controller for FES-assisted standing up in paraplegia. In 27th annual international conference of the engineering in medicine and biology society, 2005. IEEE-EMBS 2005. IEEE
18. Ahmed M (2019) Nonlinear robust control of functional electrical stimulation system for paraplegia. In: Electrical and Electronic Engineering. University Tun Hussein Onn Malaysia
19. Ferrarin M, Pedotti A (2000) The relationship between electrical stimulus and joint torque: a dynamic model. IEEE Trans Rehabil Eng 8(3):342–352
20. Williams II RL (2013) Simplified robotics joint-space trajectory generation with a via point using a single polynomial. J Rob
21. Fryar CD et al (2016) Anthropometric reference data for children and adults; United States, 2011–2014
22. Walpole SC et al (2012) The weight of nations: an estimation of adult human biomass. BMC Public Health 12(1):439
23. Chang AK, Choi JY (2015) Factors influencing BMI classifications of Korean adults. J Phys Ther Sci 27(5):1565–1570
24. Alagoz BB et al (2015) Disturbance rejection performance analyses of closed loop control systems by reference to disturbance ratio. ISA Trans 55:63–71

The Study of Time Domain Features of EMG Signals for Detecting Driver's Drowsiness

Faradila Naim, Mahfuzah Mustafa, Norizam Sulaiman, and Noor Aisyah Ab Rahman

Abstract Fatigue or drowsiness is one of the major causes of traffic accidents in Malaysia. Physiological signals such as EMG is a useful input to detect drowsiness in drivers. The time domain features are easy to compute and well researched in the field of EMG hand motion detection. The focus of this paper is to find the best set of time domain features to detect drowsiness in drivers' EMG signal from biceps brachii muscle. This study analyzes the time domain features of EMG signals in detecting the drowsiness in drivers during a 2 h simulated driving session. Nine time-domain features are applied to all 15 samples and classified using six classifiers. The best single feature for the long duration signal is the mean absolute value slope (MAVS) with 80% accuracy using Naïve Bayes (NB) classifiers. All features combined gives the highest accuracy of 85% using linear discriminant analysis (LDA) classifier.

Keywords Driver drowsiness · Time-domain features · EMG · Signal processing · Biceps brachii · MAVS · NB · LDA

1 Introduction

A driver's wakeful state during a driving session is very crucial to prevent unwanted road traffic accidents. Road traffic accidents depending on its fatality can cause death or severe injury to the drivers and passengers. Malaysia was ranked third highest death rate from road traffic accidents amongst its neighboring countries ASEAN [1]. One of the causes of road traffic accidents are fatigue as reported by the Inspector General of Police Tan Sri Muhamad Fuzi Harun [2]. Hence, there is a need to detect drowsiness during driving among drivers so that prevention measures can be enforced.

Most of the fatigue or driver's drowsiness studies are done in a driving simulator where inputs such as; physiological sensors, mechanical sensors and behavioral inputs are used to detect the level of drowsiness during a simulated driving. Physiological sensors are the more popular inputs because its highest accuracy compared to

F. Naim (✉) · M. Mustafa · N. Sulaiman · N. A. A. Rahman
Fakulti Teknologi Kejuruteraan Elektrik and Elektronik, Universiti Malaysia Pahang, 26600 Pekan, Pahang, Malaysia
e-mail: faradilan@ump.edu.my

A. F. Ab. Nasir et al. (eds.), *Recent Trends in Mechatronics Towards Industry 4.0*, Lecture Notes in Electrical Engineering 730, https://doi.org/10.1007/978-981-33-4597-3_39

the mechanical factors from the vehicle and environment [3]. Physiological sensors such as electromyography (EMG), electroencephalogram (EEG), electrocardiogram (ECG), electrooculography (EOG) and galvanic skin response (GSR) are among the popular sensors used in the study of driver's drowsiness detection [4].

EMG sensor is one of the non-invasive physiological sensors, less expensive, less noisy and more reliable data compared to other physiological sensors such as ECG and EEG [5]. There are a few studies including Mahmoodi et al. and Zahari et al. that conducted an experiment of drivers 'drowsiness detection using only EMG sensors that acquires high accuracy [6, 7]. Both studies are analysed the arm muscles namely biceps brachii, deltoid, and pectoralis in a driving simulator.

Time domain features can be the simplest form of features that might grant a good EMG pattern recognition result. It is also yield the most number of features for any EMG signals, as Phinyomark listed 26 time-domain features compared to the 11 frequency-domain features [8]. Phinyomark concluded that the time domain features has better classification compared to its frequency domain features since frequency domain features are limited to the same source of its power spectral density values. Phinyomark did an extensive EMG features analysis of hand motion detection. The EMG signal of the hand motion comprise of short time EMG signals [8].

Another EMG sensor study which placed on biceps brachii during a simulated driving were analyzed to classify three classes drowsiness level gives a better separation in time domain features compared to its frequency domain features. The findings state that the average normalized mean of the time domain signal proven to separate three classes; non-fatigue with 0.5004, mild-fatigue with 0.497 and fatigue with 0.494 [6].

Based on these previous researches, there is a limited study on extensive time domain features of EMG signals for driver's drowsiness detection. Hence, this study analyses the EMG signals to find the best time-domain feature or features set of EMG signal to detect the drowsiness level of a driver during driving.

2 Methodology

Figure 1 shows the basic process flow of EMG signal processing for driver's drowsiness detection; starting with EMG data collection following an experimental setup (discussed in Sect. 2.1), then the raw EMG signal will be pre-processed for further processing to extract meaningful features (discussed in Sect. 2.2). These features will be classified to gain the results to see how accurately the machine learning model designed to classify data (discussed in Sect. 2.3).

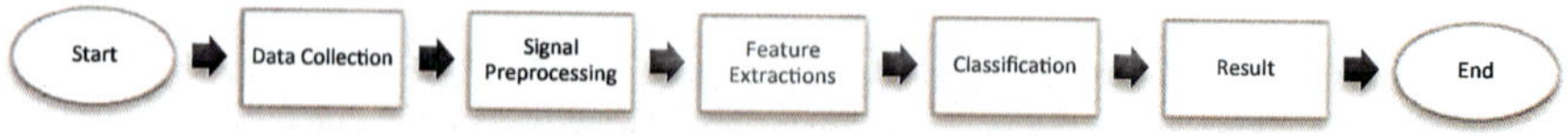

Fig. 1 Process flow of EMG signal processing for driver's drowsiness detection

2.1 Experimental Setup

The data collected among 15 subjects (7 males and 8 females) age between 20 and 26 years old. Each participant is required to sleep for 6–8 hour the night before experiment is conducted and must not consume any caffeinated drinks to induce the natural drowsiness state during experiments. The experiments are conducted between 2 pm and 5 pm since this is the setting where drowsiness will likely to occur [9]. This research is an offline study where the data collected are kept as the research database and later will be analysed in MATLAB.

During the day of experiment, the participants are first required to fill in their basic information and assessment that compliance with the prerequisite setting mentioned. Participant will sit in front of a LCD monitor with a steering wheel that controls the driving simulator game. In this experiment, Need for Speed game is used as the simulator. EMG sensor (Shimmer TM Model) is attached to the right hand (biceps brachii) of the participant. Each participant will drive the simulator for 2 h without interruption as the EMG data is collected.

Each participant must fill in the Fatigue Assessment Scale (FAS) questionnaire before the experiment started to assess the overall drowsiness level during the session. The FAS is a simple questionnaire with 10 level of fatigue level that each participant will declare their overall fatigue level for the duration of the simulated driving. The FAS score will later be used to label the participants' as fatigue or non-fatigue data for classification labels.

2.2 Feature Extractions

The raw EMG data are filtered using Butterworth bandpass filter between 10 and 20 Hz [10] then, sampled at 512 Hz to remove unwanted noise. The total points data is approximately 3 million points are processed further for feature extractions.

These nine features are the most efficient time domain features as previously studied by researchers such as Hudgins and Englehart [11]. These features are also the most common features for EMG signals across all application, and widely used in hand motion detection of EMG signals [12]. The summary of all the time domain features applied in this study are listed in Table 1.

2.3 Classifications

All eleven features extracted are to be classified to two class; fatigue and non-fatigue samples using Matlab. These features are classified using multiple classifiers listed in Table 2; *k*-nearest neighbor (kNN), linear discriminant analysis (LDA), decision tree

Table 1 Summary of each time domain features in the study

Name	Description	Parameters	Ref
Mean absolute value (MAV)	The average of absolute value of each sample divided by the total number of samples N		[15]
Waveform length (WL)	The sum of absolute value of the difference between each two samples		[16]
Integrated EMG (IEMG)	The sum of absolute values of each sample		[17]
Variance (VAR)	The sum of each sample squared divided by total samples less by one		[18]
Zero crossing (ZC)	Calculates the number of sign changes between adjacent samples	$T = 10^{-6}$	[19]
Slope sign changes (SSC)	Calculate the number of sign changes between two adjacent slopes of each sample	$T = 10^{-6}$	[20]
Simple square integral (SSI)	Extension from variance which takes the sum of absolute value of sample squared		[8]
Mean absolute value slope (MAVS)	The differences between adjacent mean absolute values for all samples. (Note: the average of all MAVS is calculated)		[11, 12]
Root mean square (RMS)	The square root value of the mean absolute value		[21]

(DT), Naïve Bayes (NB), random forest (RF) and support vector machine (SVM). The description of each classifiers is listed in Table 2.

K-fold cross validation is applied to each classifier model to ensure that all the features will be thoroughly used as training and tests data since this study has limited number of samples. A tenfold-cross validation ($k = 10$) is selected for 15 samples that attribute to 10% selection of test sample for each iteration of validation. That means, the training samples for each iteration is 90%. Accuracy for each classifier are collected for comparison. With k-fold cross validation, all data are efficiently used for classification.

Table 2 Common classifiers used in EMG signal processing

Classifiers	Description	Parameters	Ref
kNN	Is a lazy classifier where it classifies an input with the majority class of its labelled neighbors point k	$k = 3$	[22]
LDA	Determine a linear separator (2 classes problem) that distributes all labelled data to its classes such that when it is mapped to its labelled axis, the samples are separated following its classes (N=number of classes)	Number of split =50	[23]
NB	Based on the Bayes probability which each sample is independent of other samples' attribute. Probability of each sample to its class is based on the label is given to the sample		[24]
DT	This classifier constructs the tree decision for all features given based on the allocated labels given from the top most significant features down to the micro valued feature		[25]
SVM	Similar to LDA, where in LDA the hyperplane chosen is the one that maximize the classes mean, whereas SVM's hyperplane maximize the margin between the two classes boundaries with minimal error		[23]
RF	Classify a sample based on the results of trained data from all features in decision trees to minimize variance	Number of bags = 50	[26]

3 Results and Discussion

3.1 Data

The collected EMG samples are given labels based on the FAS score which resulted in 9 fatigue samples and 6 non-fatigue samples. The raw data (See Fig. 2a) filtered with bandpass filter removing the artefacts (See Fig. 2b). The noises seen as uneven and random spikes in raw signals has been removed after applying filters. The filtered signal is ready for time domain feature extractions.

3.2 Time Domain Features

Figure 3 shows the distribution of all features for 15 subjects taken. These maps are a good preliminary observation of class separability between these features. Overall, most features scatter all over the space and not one feature shows a clear separable cluster between the two classes.

MAVS and SSI feature maps has one to two non fatigue feature samples that lie away from the other feature samples. Also, the fatigue sample features for MAVS and

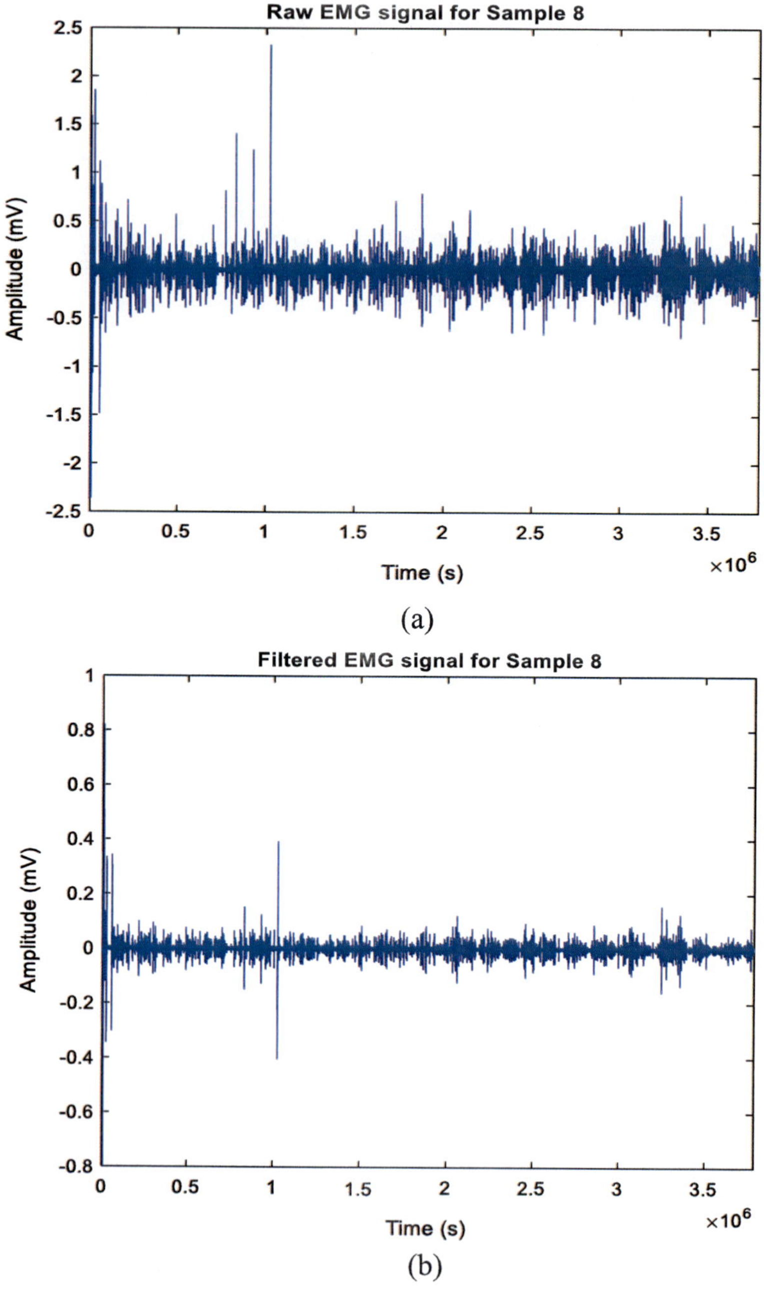

Fig. 2 **a** Raw EMG data for Sample 8 with frequency sampling of 512 Hz, **b** filtered EMG signal

SSI is quite consistent scatter close to each other. These may give an early prediction that these two features might be the best features for the drowsiness classification.

Other than that, the rest of the features maps does not gives any predictor of which feature can classify the two classes efficiently. These features will have to be trained and feed into classifiers for further analysis.

3.3 Classification

As mentioned in Sect. 2.3, all the features extracted are classified using 6 different classifiers; kNN, LDA,NB, SVM, DT, and RF. Along with the classifier, k-fold cross validation is applied to ensure that all feature samples have the opportunity to be trained rather than one iteration of classification which reduced the bias of classification.

Initially, five values of k = 2, 5, 7, 10, 13 for cross validation is applied to choose the best k in selected model of classification (classification of all features,

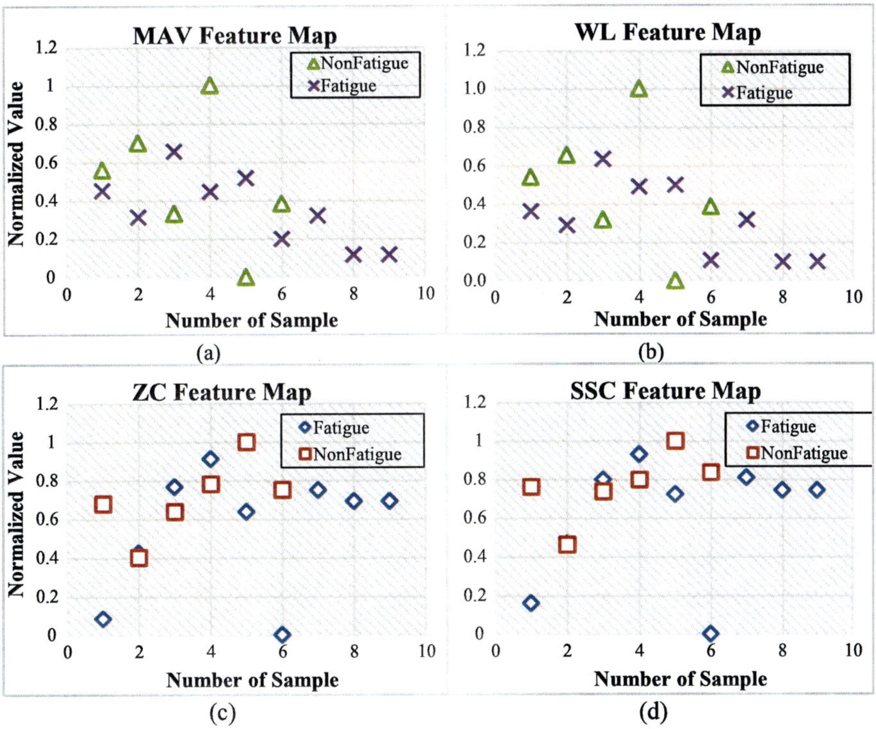

Fig. 3 Feature maps for all samples

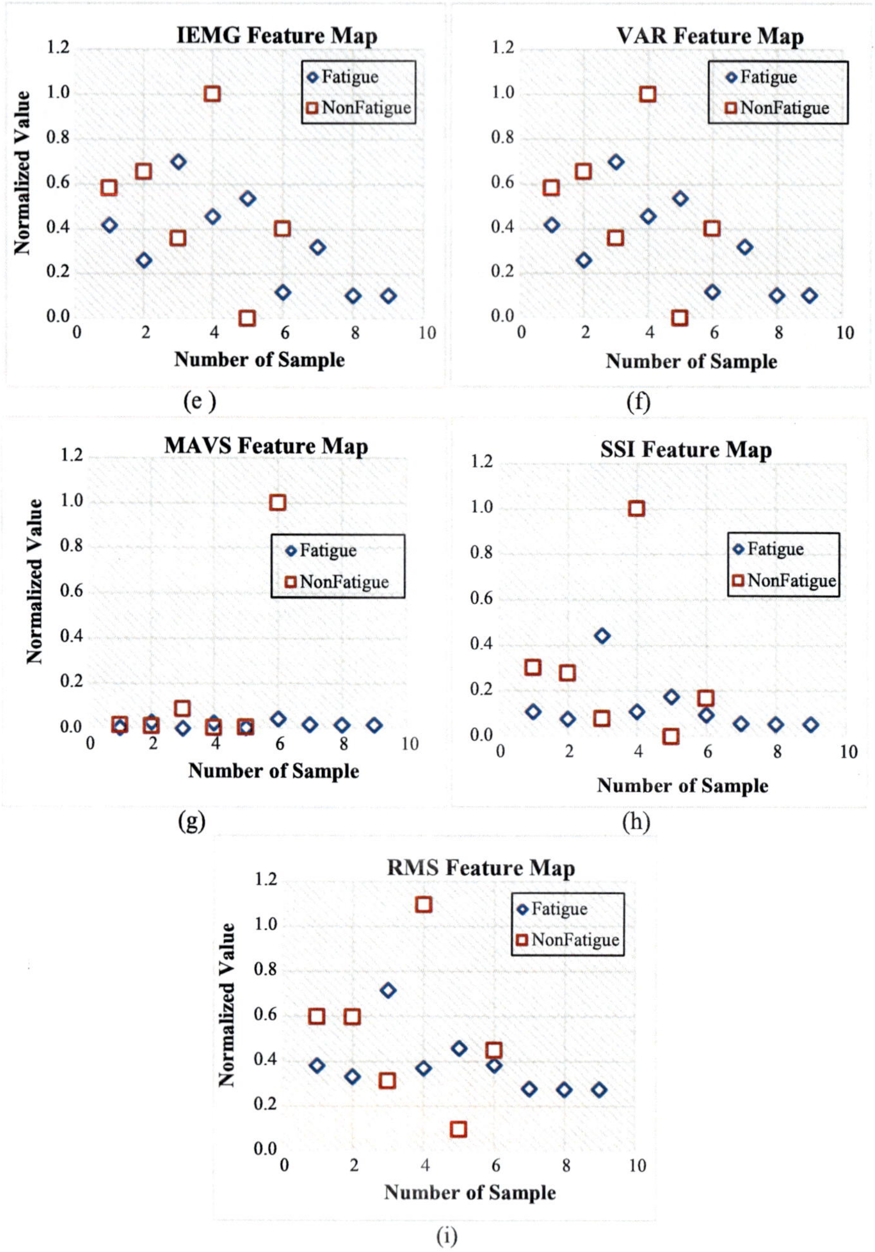

Fig. 3 (continued)

Table 3 The best performance of single feature with its classifier(s)

Features	Classifier (s)	Accuracy (%)
MAVS	NB	80
SSI	SVM/kNN	70
ZC	NB	65
WL	LDA/NB	65
VAR	KNN/SVM	65
RMS	SVM/kNN	65
MAV	LDA/NB	65
IEMG	LDA/kNN/NB	65
SSC	SVM/LDA	60

classification of single feature). In all classification results, $k = 10$ gives the highest accuracy across all 6 classifiers, hence the tenfold-cross validation is selected.

Table 3 lists the accuracy of single feature classification for all features. MAVS gives the highest classification accuracy with 80% using NB classifier. Followed by SSI with 70% using SVM/kNN. This shows that the predicted best feature classified is consistent with their feature maps as previously discussed (Sect. 3.2).

These features then are grouped based on past researches recommendations for [8, 11] determining the best minimal features classification results. Table 4 shows the accuracy of grouped features based on; 1) past researches (noted as reference in the Tables 4 and 3 accuracy of single feature (not referenced in Table 4).

Grouping 3-features; MAVS-SSI-RMS and VAR-MAVS-SSI acquired 75% accuracy. These features are grouped based on the rank of single feature accuracy for all classifier (refer Table 2). The average accuracy for each single feature classification in this group is between 65 and 80%. Both groups gain similar accuracy since their single classification accuracy is the same.

The same method is also applied to 4-features group (VAR-MAVS-SSI-RMS) with 75% accuracy giving a better performance compared to the most commonly used 4-features group (Hudgins) [25]. In a study of hand motion detection of EMG signals, these combinations resulted in 96.78% accuracy using SVM [13].

Table 4 Accuracy of grouped features

Groups of features	Accuracy (%)
All 9 features	85
5-Features (MAV, WL, SSC, ZC, MAVS) [8]	76.2
5-Features (MAV, VAR, MAVS, SSI, RMS)	75
4-Features (VAR, MAVS, SSI, RMS)	75
4-Features (MAV, WL, SSC, ZC) [27]	60
3-Features (MAVS, SSI, RMS)	75
3-Features (VAR, MAVS, SSI)	75

Other benchmark feature group is in Phinyomark's study where it is claimed that the 5-features set is the best combination of EMG time domain features in general Phinyomark's [8] 5-features group (MAV, WL, SSC, ZC, MAVS) give slightly higher accuracy with 76.2% compared to this study 5-features group selection (MAV, VAR, MAVS, SSI, RMS) that gives 75% accuracy. The 5-features group and 4-features group in this study proves that these time domain features are among the efficient features in EMG signal drowsiness detection. The low accuracy in driver's drowsiness detection might be due to the signal length differences, which hand motion signals only comprise of 6 s data compared to this paper having 2 h data. Since this is a preliminary study, more time domain features can be included in this study to find the best features set with minimal number of features for driver's drowsiness detection.

The final feature group classification is to use all 9 features to all classifiers. Accuracy of 85% using NB is acquired and it gives the best performance among all feature groups.

4 Conclusions

This study aims of getting the minimal number of features for good EMG signal's drowsiness detection ended with having to use all 9-time domain features for good classification. Nevertheless, these 9 features are efficient and easy to implement for EMG signals. These nine features have been studied extensively in EMG hand motion detection [8], but not in driving drowsiness detection. The frequency domain features for this study was done by Rahman [14] that acquired the range of energy spectral density's mean for three classes of driver's drowsiness. There is lack of extensive analysis of EMG time domain features for driver's drowsiness detection. Hence, an extensive study of all available time domain features of EMG signals for driving drowsiness detection is needed.

In the future, this study can include more samples to increase the result's reliability. Also, comparison of results to other features domain such as the frequency domains and time-frequency domains can be done for validation purposes.

Acknowledgements This research is supported by the Fundamental Research Grant Scheme (FRGS) research grant number FRGS/1/2019/TK04/UMP/02/7 (RDU1901167). Thank you to the Ministry of Education for the funding granted.

References

1. Milton L (2019, May 14) We have the third highest death rate from road accidents. The Star. Retrieved from https://www.thestar.com.my/

2. Kumar M (2018, June 8) Fatigue, mobile phone use among top causes of road accidents. The Star. Retrieved from https://www.thestar.com.my/
3. Reddy B, Kim Y, Yun S, Seo C, Jang J (2017) Real-time driver drowsiness detection for embedded system using model compression of deep neural networks. In: 2017 IEEE conference on computer vision and pattern recognition workshops (CVPRW). Honolulu, HI, pp. 438-445
4. Takalahti AT (2016) Learning to detect driver's drowsiness. Master's dissertation, University of Helsinki, Findland. Retrieved from https://helda.helsinki.fi/handle/10138/165912
5. Dou dou M, Bouabdallah A, Berge-Cherfaoui V (2020) Driver drowsiness measurement technologies: current research, market solutions, and challenges. Int J ITS Res 18:297–319
6. Mahmoodi M, Nahvi A (2019) Driver drowsiness detection based on classification of surface electromyography features in a driving simulator. Proc Inst Mech Eng 233(4):395–406. https://doi.org/10.1177/0954411919831313
7. Mohd Azli MAS, Mustafa M, Abdubrani R, Abdul Hadi A, Syed Ahmad SNA, Zahari ZL (2019) Electromyograph (EMG) signal analysis to predict muscle fatigue during driving. In: Md Zain Z et al (eds) Proceedings of the 10th national technical seminar on underwater system technology 2018. Lecture notes in electrical engineering, vol 538. Springer, Singapore
8. Taffese TB (2017) A review of using EEG and EMG psychophysiological measurements in user experience research. Master's Thesis, University of Oulu. Available at http://jultika.oulu.fi/files/nbnfioulu-201706022478.pdf
9. Phinyomark A, Phukpattaranont P, Limsakul C (2012) Feature reduction and selection for EMG signal classification. Exp Syst Appl 39(8):7420–7431 Elsevier
10. Muhammad Amzar Syazani MA (2018) Electromyography (EMG) signal analysis to predict muscle fatigue during driving. Unpublished final year project's thesis. Universiti Malaysia Pahang, Malaysia
11. Tkach D, Huang H, Kuiken TA (2010) Study of stability of time-domain features for electromyographic pattern recognition. J Neuro Eng Rehabil 7(21):1–13
12. Phinyomark A, Scheme E (2018) EMG pattern recognition in the era of big data and deep learning. Big Data Cogn Comput 2:21
13. Too J, Abdullah AR, Saad NM (2019) Classification of hand movements based on discrete wavelet transform and enhanced feature extraction. Int J Adv Comput Sci Appl 10(6):83–89
14. Abbaspour S, Lindén M, Gholamhosseini H et al (2020) Evaluation of surface EMG-based recognition algorithms for decoding hand movements. Med Biol Eng Comput 58:83–100
15. Zecca M, Micera S, Carrozza MC, Dario P (2002) Control of multifunctional prosthetic hands by processing the electromyographic signal. Crit Rev Biomed Eng 30:459–485. https://doi.org/10.1615/critrevbiomedeng.v30.i456.80
16. Hudgins BS, Parker P, Scott RN (1993) A new strategy for multifunction myoelectric control. IEEE Trans Biomed Eng 40:82–94
17. Huang HP, Chen CY (1999) Development of a myoelectric discrimination system for a multi-degree prosthetic hand. In: Proceedings of IEEE international conference on robotics and automation, vol 3, pp 2392–2397
18. Tenore FV, Ramos A, Fahmy A, Acharya S, Etienne-Cummings R, Thakor NV (2009) Decoding of individuated finger movements using surface electromyography. IEEE Trans Biomed Eng 56:1427–1434
19. Zardoshti-KermaniM Wheeler BC, Badie K, Hashemi RM (1995) EMG feature evaluation for movement control of upper extremity prostheses. IEEE Trans Rehabil Eng 3:324–333
20. Micera S, Carpaneto J, Raspopovic S (2010) Control of hand prostheses using peripheral information. IEEE Rev Biomed Eng 3:48–68
21. Ahsan R, Ibrahimy MI (2009) EMG signal classification for human computer interaction: a review. Eur J Sci Res 33(3):480–501
22. Sahayadhas A, Sundaraj K, Murugappan M (2014) Electromyogram signal based hypovigilance detection. Biomed Res (India) 25(3):281–288
23. Kim KS, Choi HH, Moon CS, Mun CW (2011) Comparison of k-nearest neighbor, quadratic discriminant and linear discriminant analysis in classification of electromyogram signals based on the wrist-motion directions. Curr Appl Phys 11:740–745

24. Spiewak C (2018) A comprehensive study on EMG feature extraction and classifiers. Open Access J Biomed Eng Biosci
25. Chen J, Wang H, Hua C (2018) Assessment of driver drowsiness using electroencephalogram signals based on multiple functional brain networks. Int J Psychophysiol 0–1. Elsevier
26. Jacobé C et al (2017) Detection and prediction of driver drowsiness using artificial neural network models. Accid Anal Prev 126:95–104. https://doi.org/10.1016/j.aap.2017.11.038 Elsevier
27. Rahman NAA, Mustafa M, Samad R, Abdullah NRH, Sulaiman N (2019) Energy spectral density analysis of muscle fatigue. In: Md Zain Z et al (eds) Proceedings of the 10th national technical seminar on underwater system technology 2018. Lecture notes in electrical engineering, vol 538. Springer, Singapore. https://doi.org/10.1007/978-981-13-3708-6_37

The Classification of Skateboarding Tricks by Means of the Integration of Transfer Learning Models and *K*-Nearest Neighbors

Muhammad Nur Aiman Shapiee, Muhammad Ar Rahim Ibrahim, Mohd Azraai Mohd Razman, Muhammad Amirul Abdullah, Rabiu Muazu Musa, Noor Azuan Abu Osman, and Anwar P. P. Abdul Majeed

Abstract The skateboarding scene has reached new heights, especially with its first appearance at the now postponed Tokyo Summer Olympic Games. Therefore, owing to the scale of the sport in such competitive games, advanced innovative assessment approaches have increasingly gained due attention by relevant stakeholders, especially with the interest of a more objective-based evaluation. We employed pre-trained Transfer Learning coupled with a fine-tuned *k*-Nearest Neighbor (*k*-NN) classifier to form several pipelines to investigate its efficacy in classifying skateboarding tricks, namely Kickflip, Pop Shove-it, Frontside 180, Ollie and Nollie Front Shove-it. From the five skateboarding tricks, a skateboarder would repeatedly perform it for five successful landed tricks captured by YI action camera. From that, the images would be feature engineered and extracted through five Transfer Learning models, namely VGG-16, VGG-19, DenseNet-121, DenseNet-201 and InceptionV3, then classified by employing the *k*-Nearest Neighbor (*k*-NN) classifier. It is demonstrated from the preliminary results, that the VGG-19 and DenseNet-201 pipeline, both attained a classification accuracy (CA) of 97% on the test dataset, followed by the DenseNet-121 and InceptionV3, in which both obtained a test CA of 96%. The least performing pipeline is the VGG-16, where a test CA of 94% is recorded. The result from the current study validated it could providing an objective judgment for judges in classifying skateboard tricks for the competition.

M. N. A. Shapiee · M. A. R. Ibrahim · M. A. Mohd Razman · M. A. Abdullah ·
A. P. P. Abdul Majeed (✉)
Innovative Manufacturing, Mechatronics and Sports Laboratory, Faculty of Manufacturing and Mechatronic Engineering Technology, Universiti Malaysia Pahang, 26600 Pekan, Pahang Darul Makmur, Malaysia
e-mail: amajeed@ump.edu.my

R. M. Musa · N. A. Abu Osman
Universiti Malaysia Terengganu, Terengganu, Malaysia

A. P. P. Abdul Majeed
Centre for Software Development & Integrated Computing, Universiti Malaysia Pahang, 26600 Pekan, Pahang Darul Makmur, Malaysia

A. F. Ab. Nasir et al. (eds.), *Recent Trends in Mechatronics Towards Industry 4.0*, Lecture Notes in Electrical Engineering 730, https://doi.org/10.1007/978-981-33-4597-3_40

Keywords Classification · Skateboarding tricks · Image processing · Transfer learning · Machine learning

1 Introduction

The skateboarding industry total worth has been reported as approximately $USD 4.8 billion in 2010 [1]. In addition, the debut of skateboarding to the now postponed Tokyo Summer Olympic Games (owing to the devastating COVID-19 pandemic) signifies the acceptance of this form of extreme sports into the mainstream arena. Owing to the scale of such competitive games, advanced techniques for the evaluation of the tricks are non-trivial, especially with the aim of assisting judges in providing a more objective-based judgement. Attempts have been made by researchers at large in the classification of different sporting activities ranging from traditional machine learning to advanced deep-learning methods and its variation.

For instance, Convolution Neural Networks (CNN) has been utilised for activity recognition through a wrist-worn accelerometer [2]. In the study, four male subjects been employed, which each of them would perform 20 individual activities basically activity of making a cup of tea. Total of six type of methodology been compared for the study namely CNN, Support Vector Machine (SVM), Linear Discriminant Analysis (LDA) and *k*-means clustering with three evaluation method; cross-validation, inter-subject and hybrid be considered. All the experiment was implemented in Keras with Tensorflow via Python environment. It was reported that the proposed CNN achieved a 99.8% classification accuracy (CA), followed by *K*-means clustering, SVM and LDA with a CA of 87.5, 53.75 and 45%, correspondingly.

Castro et al. [3] have also utilised the use of CNN for daily human activity recognition from egocentric images. Thirty volunteers participated for the data collection process where a smartphone was used to acquire data for 19 activity classes, such as eating, working, cooking, driving, watching TV and spending time with family for over six month period. Total of 40,103 images been collected that retain from 80% of the data as the remaining 20% been removes for privacy and null classes. The authors proposed Late Fusion Ensemble which is a combination of CNN and Random Decision Forests (RDF) that would be compared with the conventional machine learning classifiers, i.e., *k*-NN and RDF alone. As the result from the study, the fusion CNN+RDF pipeline could achieve an 83.07% of CA outclassed the traditional CNN that achieved 78.56% of accuracy. Nevertheless, it is worth to note that CNN requires extensive hyperparameter tuning to attain desirable results and hence, recent research trends have shifted towards the use of pre-trained CNN models via transfer learning.

The employment of Transfer Learning (TL) paradigm, have been employed for the study of X-ray object classification for baggage security purpose [4]. The researches rises two type of situation, first, classify a problem for two class firearm (no gun vs. gun). Second, object classification problem for several classes, namely camera, laptop, ceramic knives, knives, firearm and firearm-component. All images obtained from AlexNet and GoogLeNet database. The classification of the first case study was carried out via two classifier; Random Forest (RF) and SVM, and it was shown from the study that the SVM model could achieve a true positive rate (TPR) of 85.81% whilst, the RF recorded a TPR of 80.74%. As for the second case study, the authors fine-tuned two CNN structure by AlexNet [5] and GoogLeNet [6] to evaluate the problem. The proposed method achieved a mean average precision rate of 95.26% and 98.40%, respectively.

Although, there are studies that have been carried out on the use of machine learning and to a certain extent TL on sports and activity recognition, nonetheless, limited research has been carried out on skateboarding. Image processing techniques as well as Inertial Measurement Unit (IMU) based studies have been investigated in [7–12]. It was reported that commendable classification accuracies of different skateboarding tricks through the use of k-Nearest Neighbor (k-NN) and SVM classifiers, was attainable via the use of Inception V3 for features extraction [9]. It is worth noting at this juncture that the hyperparameter of the classifier was not optimised and the default sklearn library parameters were used in the study.

This paper evaluated one traditional Machine Learning (ML) classifier; k-NN to classifying multiple class of skateboarding tricks namely, Kickflip, Ollie, Frontside 180, Pop Shove-it and Nollie Front-shove that firstly been features extracted through five different pre-trained CNN models, viz. VGG-19, VGG-16, DenseNet-121, DenseNet-201 and InceptionV3 correspondingly. The results from the present study may serve valuable, especially for the judges towards a further progressively objective-based assessment, in addition, to provide a tool for skateboarders to track their performance on the execution of the aforesaid tricks.

2 Methodology

2.1 Experimental Setup

For the study, single YI Action Camera was utilized to capture the execution of the skateboarding tricks. The camera come with specifications of; (Full HD 1920 × 1080 (60 FPS), Lens: F2.8 aperture/155° wide-angle, weight: 72 g, dimension: 60 mm × 42 mm × 21 mm). Furthermore, the arrangement of the experiment, with the camera direction perspective, is depicted in Fig. 1 and the captured images can be seen in Fig. 2.

2.2 Data Collection

A male skateboarder with 5 years' experience (Weight: 53 kg, Height: 170 cm) was recruited to executed the data collection of the study [11]. The skateboarder

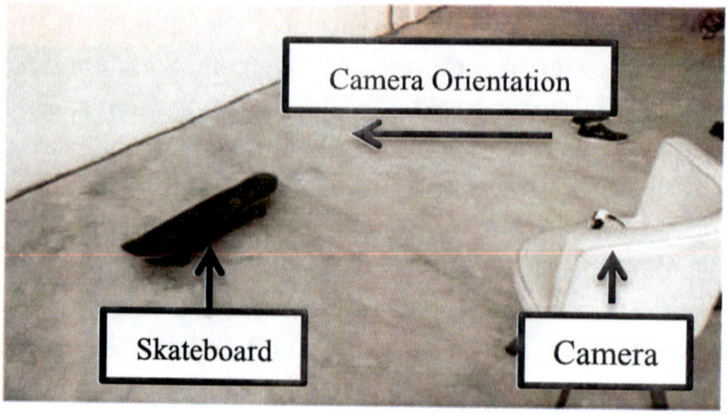

Fig. 1 Experimental setup

Fig. 2 Execution of the tricks

would perform different skateboarding tricks with five (5) times per trick for each successful execution. The tricks were selected based on the skateboarder's experience. The University Malaysia Pahang as the venue for the data collection phase. The description of the tricks could be found in [11].

Table 1 Individual of transfer learning models

Model	Size (MB)	Top-1 accuracy	Top-5 accuracy	Depth	Input size	Flatten size
VGG16	528	0.713	0.901	23	224 × 244	7 × 7 × 512
VGG19	549	0.713	0.900	26	224 × 224	7 × 7 × 512
InceptionV3	92	0.779	0.937	159	299 × 299	8 × 8 × 2048
DenseNet121	33	0.750	0.923	121	224 × 224	7 × 7 × 1024
DenseNet201	80	0.773	0.936	201	224 × 224	7 × 7 × 1920

2.3 Image Processing

The VLC media player 2.2.6. were employed to engineered the captured video as the videos have variety length of time. The video were consequently utilised for the image extraction (for a period between 2 and 3 s) of the execution of the tricks only. Afterward, video to jpg converter V.5.0.101 been utilised to obtain the frame per frame images and then followed by Caesium software to resized the dimension of the images (1080 × 1920 to 300 × 300) pixels [13]. The extraction of the images for each video was set to 30 frames. From the 25 videos captured from the data collection, about 750 images were extracted during this image processing phase.

2.4 Feature Extraction Through Transfer Learning

Transfer Learning (TL) is the state-of-the-art Deep Learning strategy that is often beneficial for limited dataset in training models. Moreover, such a technique could possibly prevent the concept of overfitting, especially from the traditional CNN other than become computationally time consuming [14, 15]. There are several common family models available, namely, VGG [16], DenseNet [17] and Inception [18]. The summary of some essential parameters of different TL models is tabulated in Table 1. It is worth noting that the default input size is the size of the input shape of each model, and each of it has to be (* × * × 3) as RGB ought being precisely three inputs dimension. The flatten size as the last fully connected layers, a combination of these features create a model, a complete working CNN architecture.

2.5 Classifiers

The ML classifier, namely *k*-NN is employed after the TL features extraction phase [19]. The detailed expressions of k-NN model could be followed in [20, 21] literature respectively. Hyperparameter tuning was engaged to improve the CA of the models developed. By the current investigation, two hyperparameters are examined, i.e., the number of neighbours, *k*, which was varied from 20 to 30 as well as distance metrics specifically Minkowski, Euclidean, Cosine and Manhattan, correspondingly. The train, validate to test ratio employed is 60:20:20, respectively. The classification accuracy (CA) and confusion matrix are employed as performance metrics to evaluate the efficacy of the evaluated pipelines [4]. All experiment is conducted on Keras API with TensorFlow by architectures based on Python on a laptop with the following computing ability; Intel Core i5-3217U 1.8 GHz CPU and 4 GB RAM [22–25]

Table 2 The classification accuracy for transfer learning models with *k*-NN classifier

	CA		
	Phase		
Model + kNN	Training (%)	Validation (%)	Testing (%)
VGG16	100	93	94
VGG19	99	97	97
InceptionV3	99	98	96
DenseNet121	98	97	96
DenseNet201	99	97	97

3 Results and Discussion

The successful landings were recorded as overall of 40 trick trials were accomplish by the skateboarder. From that, only 25 tricks counted as successful, and the data (herein the images, i.e., 149 for FS180, 150 for K, 146 for O and 150 for PS and 127 for NFS) retrieve were then fed into the VGG-19, VGG-16, DenseNet-121, DenseNet-201 and InceptionV3 to extract its features prior to its classification via an optimised *k*-NN classifier.

The best TL pipeline was observed to be the VGG19 and DenseNet201 with an optimised *k*-NN model that employs the Manhattan distance metrics with 20 number neighbours. Table 2 lists the CA performance of the TL pipelines whilst in Fig. 3 provides the pictorial evidence of the performance of the evaluated pipelines. It is evident from the results presented that the VGG19 + *k*-NN and DenseNet201 + *k*-NN pipelines could yield a CA for training, validation, and testing of 99.00%, 99.00%, and 97.00% respectively. This is followed by the InceptionV3 + *k*-NN model, in which it achieved a CA of 99.00% on training, 98.00% on validation, and 96.00% on testing, respectively. The DenseNet121 + *k*-NN pipeline attained a CA of 98.00% on training, 97.00% on validation and the same CA for testing as the InceptionV3 model. The least performing pipeline is the VGG16 + *k*-NN model, where the CA for training, validation, and testing attained with 100.00%, 93.00%, and 94.00%, respectively which demonstrates that the model overfits during the training phase.

As we observe confusion matrix for deeper evaluation of the VGG19, DenseNet201, and InceptionV3 models in (Figs. 4, 5 and 6, correspondingly). Firstly, from Fig. 4a, c the PS images trick was misclassified as O and O misclassified as K for c only. Meanwhile, the misclassification for b was arising from K misclassification as O and O misclassified as PS. Furthermore, it is evident from Fig. 5a, b both have misclassification on the K and PS as it misclassified as O. For a added misclassified on K as FS. Other misclassified for b separated across FS, K, and PS

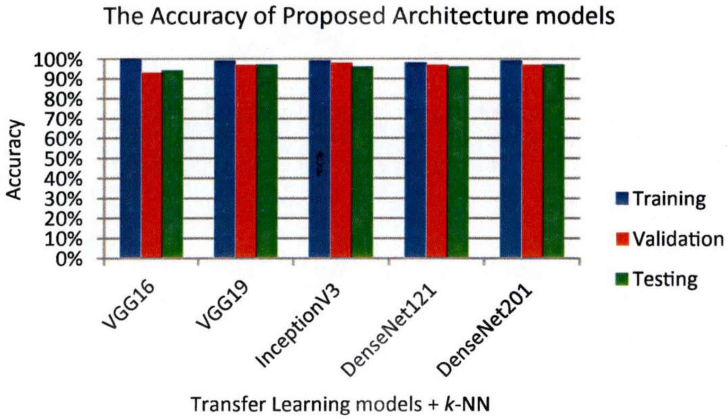

Fig. 3 The CA of proposed architecture via training, validation and testing phase

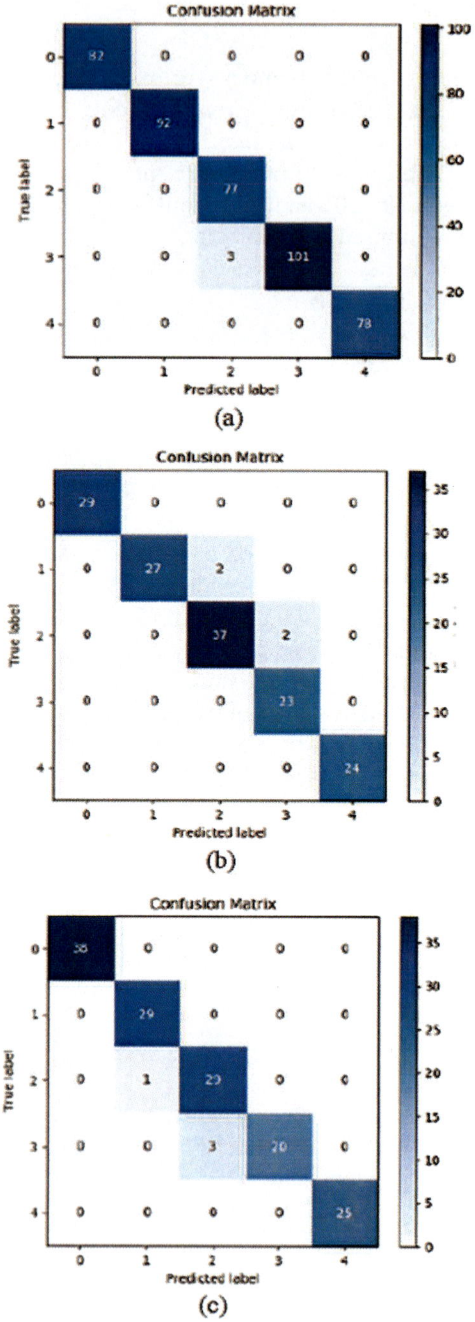

Fig. 4 The confusion matrix for **a** training, **b** validation, **c** testing of VGG19 *k*-NN model

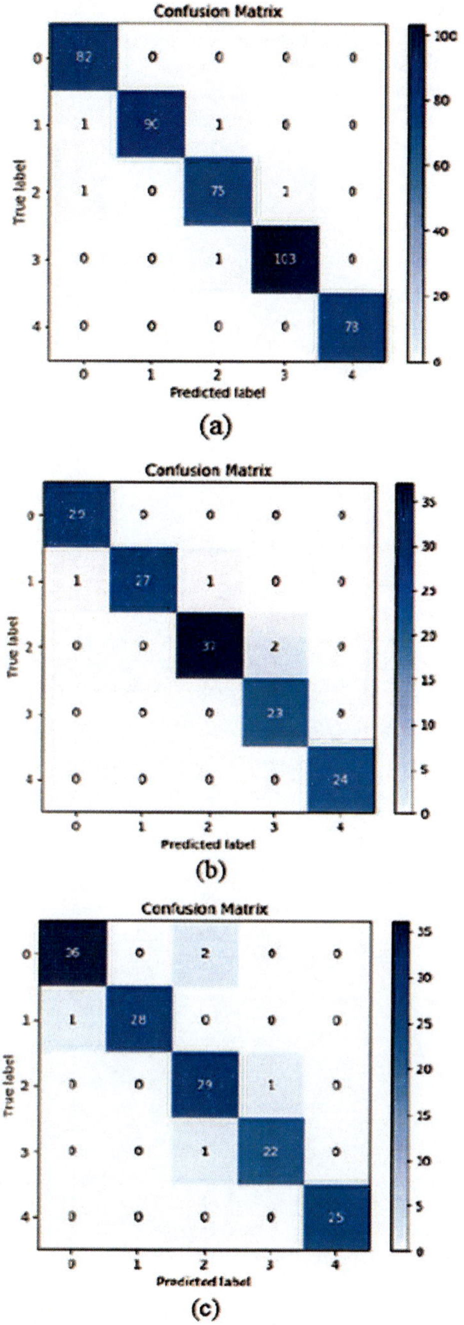

Fig. 5 The confusion matrix for **a** training, **b** validation, **c** testing of DenseNet201 *k*-NN model

Fig. 6 The confusion matrix for **a** training, **b** validation, **c** testing of InceptionV3 *k*-NN model

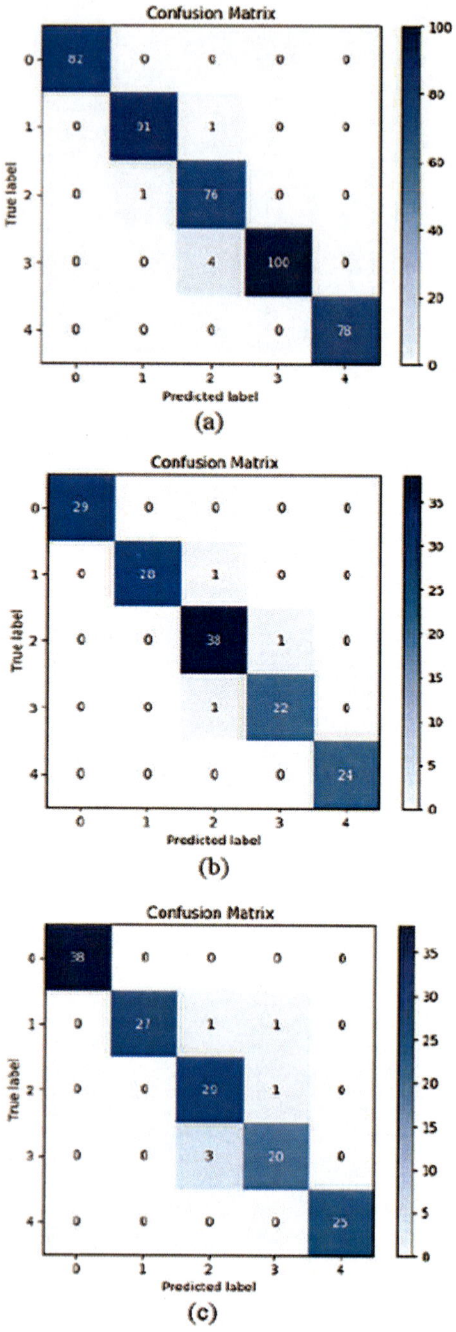

classes. Then, for c the misclassification arises from the O misclassified as FS and K, then PS misclassified as O. Moreover, from the confusion matrix in Fig. 6 for the a, b, and c it could be observed as the misclassification arises for all classes except for the FS and NFS classes. Conclusively, it could be shown by the proposed study, that the VGG19 and DenseNet201 with a *k*-NN pipelines could provide a good classification on the skateboarding tricks.

4 Conclusion

From the preliminary study, we could conclude that the image processing technique with support of ML and TL pipelines display significant positive results in evaluating of the skateboarding tricks. It is evident that the suggested VGG19+*k*-NN and DenseNet201+*k*-NN pipeline could provide reasonably accurate classification of the tricks by achieved 97% of CA, correspondingly. Moreover, the present study would be further investigated by employ particular ML models, namely SVM and RF as well as its associated hyperparameters. The results from the current research demonstrate its suitability for an objective-based decision on skateboarding tricks in competitions.

Acknowledgements The authors would like to acknowledge Universiti Malaysia Pahang and the Ministry of Education Malaysia for supporting and funding this study (FRGS/1/2019/TK03/UMP/02/6 & RDU1901115).

References

1. Corrêa NK, César J, Lima M De, Russomano T, Araujo M (2017) Development of a skateboarding trick classifier using accelerometry and machine learning. 33:362–369. https://doi.org/10.1590/2446-4740.04717
2. Panwar M, Dyuthi SR, Prakash KC, Biswas D, Acharyya A, Maharatna K, Gautam A, Naik GR (2017) CNN based approach for activity recognition using a Wrist—worn accelerometer. In: 2017 39th 2017 39th annual international conference of the ieee engineering in medicine and biology society (EMBC), pp 2438–2441. https://doi.org/10.1109/embc.2017.8037349
3. Castro D, Essa I, Hickson S, Abowd G, Christensen H (2015) Predicting daily activities from egocentric images using deep learning. 75–82
4. Kundegorski ME, Devereux M, Breckon TP (2016) Transfer learning using convolutional neural networks for object classification within X-ray baggage security imagery. https://doi.org/10.1109/ICIP.2016.7532519
5. Krizhevsky A, Hinton GE (2012) ImageNet classification with deep convolutional neural networks. 1–9
6. Szegedy C, Liu W, Jia Y, Sermanet P, Reed S, Anguelov D, Erhan D, Vanhoucke V, Rabinovich A (2015) Going deeper with convolutions. 1–9
7. Groh BH, Kautz T, Schuldhaus D (2015) IMU-based trick classification in skateboarding. In: KDD work large-scale sports analysis

8. Groh BH, Fleckenstein M, Kautz T, Eskofier BM (2017) Classification and visualization of skateboard tricks using wearable sensors. Pervasive Mob Comput 40:42–55. https://doi.org/10.1016/j.pmcj.2017.05.007
9. Shapiee MNA, Ibrahim MAR, Mohd Razman MA, Abdullah MA, Musa RM, Hassan MHA, Abdul Majeed APP (2020) The classification of skateboarding trick manoeuvres through the integration of image processing techniques and machine learning. In: Lecture notes in electrical engineering. Springer, pp 347–356
10. Shapiee MNA, Ibrahim MAR, Razman MAM, Abdullah MA, Musa RM, Abdul Majeed APP (2020) The classification of skateboarding tricks by means of the integration of transfer learning and machine learning models. Lect Notes Electr Eng 678:219–226. https://doi.org/10.1007/978-981-15-6025-5_20
11. Abdullah MA, Ibrahim MAR, Shapiee MNA Bin, Razman MAM, Musa RM, Majeed APPA (2019) The classification of skateboarding trick manoeuvres through the integration of IMU and machine learning. In: Symposium on intelligent manufacturing and mechatronics. Springer, pp 67–74
12. Ibrahim MAR, Abdullah MA, Nur, Shapiee MNA, Mohd Razman MA, Musa RM, Zakaria MA, Abu Osman NnA, Majeed APPA (2020) The classification of skateboarding trick manoeuvres : a K-nearest neighbour approach. In: Proceedings of the 2019 movement, health and exercise (MoHE) and international sports science conference (ISSC), pp 341–347
13. Su Y, Chiu T, Yeh C, Huang H, Hsu WH (2015) Transfer learning for video recognition with scarce training data for deep convolutional neural network. 1–12
14. Jin KH, Mccann MT, Froustey E, Unser M (2017) Deep convolutional neural network for inverse problems in imaging. 26:4509–4522
15. Kruthiventi SSS, Ayush K, Babu RV, Member S (2015) DeepFix : a fully convolutional neural network for predicting human eye fixations. 1–11
16. Simonyan K, Zisserman A (2015) Very deep convolutional networks for large-scale image recognition. In: 3rd international conference on learning representations, ICLR 2015—conference track proceedings, pp 1–14
17. Huang G, Liu Z, Van Der Maaten L, Weinberger KQ (2017) Densely connected convolutional networks. In: Proceedings of 30th IEEE conference on computer vision and pattern recognition, CVPR 2017, pp 2261–2269
18. Szegedy C, Liu W, Jia Y, Sermanet P, Reed S, Anguelov D, Erhan D, Vanhoucke V, Rabinovich A (2014) Going deeper with convolutions
19. Dudani SA (1978) Distance-weighted k-nearest neighbor rules. IEEE Trans Syst Man Cybern SMC 8:311–313. https://doi.org/10.1109/tsmc.1978.4309958
20. Taha Z, Razman A, Majeed APPA (2018) The classification of hunger behaviour of Lates Calcarifer through the integration of image processing technique and k-nearest neighbour learning algorithm. https://doi.org/10.1088/1757-899X/342/1/012017
21. Mohd Razman MA, Antonio G, Cenedese A, Abdul APP, Muazu R, Shahrizan A, Ghani A (2019) Hunger classification of Lates calcarifer by means of an automated feeder and image processing. Comput Electron Agric 163:104883. https://doi.org/10.1016/j.compag.2019.104883
22. Temel D, AlRegib G (2018) Traffic signs in the wild: highlights from the IEEE video and image processing cup 2017 student competition [SP competitions]. IEEE Signal Process Mag 35:154–161
23. Koushik J (2016) Understanding convolutional neural networks. 1–6
24. Rawat W (2017) Deep convolutional neural networks for image classification : a comprehensive review. 2449:2352–2449. https://doi.org/10.1162/NECO
25. Abadi M, Barham P, Chen J, Chen Z, Davis A, Dean J, Devin M, Ghemawat S, Irving G, Isard M, Kudlur M, Levenberg J, Monga R, Moore S, Murray DG, Steiner B, Tucker P, Vasudevan V, Warden P, Wicke M, Yu Y, Zheng X, Brain G, Osdi I, Barham P, Chen J, Chen Z, Davis A, Dean J, Devin M, Ghemawat S, Irving G, Isard M, Kudlur M, Levenberg J, Monga R, Moore S, Murray DG, Steiner B, Tucker P, Vasudevan V, Warden P, Wicke M, Yu Y, Zheng X (2016) TensorFlow: a system for large-scale machine learning

An Improved Grey Wolf Optimizer with Hyperbolic Tangent Updating Mechanism for Solving Optimization Problems

Mohd Zaidi Mohd Tumari, Mohd Ashraf Ahmad, and Mohd Helmi Suid

Abstract The original algorithm of Grey Wolf Optimizer (GWO) has a common problem which is too soon to trap in local optima. This paper presents the Improved Grey Wolf Optimizer (IGWO) by modifying the updating mechanism of the original GWO. The main idea of the new improvement is by introducing a nonlinear updating mechanism based on the hyperbolic tangent function to improve the efficiency of the exploration and the exploitation phase and to decrease the probability of trapping in local optima. The effectiveness of the new approach is evaluated on 30 well-known benchmark functions, and the results are compared with the original GWO. The preliminary findings show that the IGWO algorithm is able to obtain very competitive results in terms of objective functions minimization compared to original GWO algorithms.

Keywords Grey Wolf Optimization · Optimization · Benchmark function · Improved Grey Wolf Optimization

1 Introduction

Fundamentally, an optimization problem is defined as an algorithm with a sequence of the process to find the possible optimal solutions for maximizing or minimizing the objective function in the same time satisfying the problem constraints. There are two main approaches to optimization, which are deterministic and stochastic. In deterministic approaches such as Newton and quasi-Newton methods [1], they are highly dependent on gradient information of the search space which requires the continuity of the objective function and the variables, and they need a good starting point as well. These are the major drawbacks of the deterministic approach since, in

M. Z. M. Tumari
Faculty of Electrical and Electronics Engineering Technology, Universiti Teknikal Malaysia Melaka, Hang Tuah Jaya, 76100 Durian Tunggal, Melaka, Malaysia

M. A. Ahmad (✉) · M. H. Suid
Faculty of Electrical and Electronics Engineering Technology, Universiti Malaysia Pahang, 26600 Pekan, Pahang, Malaysia
e-mail: mashraf@ump.edu.my

A. F. Ab. Nasir et al. (eds.), *Recent Trends in Mechatronics Towards Industry 4.0*, Lecture Notes in Electrical Engineering 730, https://doi.org/10.1007/978-981-33-4597-3_41

most real control design problems, the objective functions are not smooth, and there are many local optimum solutions. Thus, the probability of the solution to stuck in local optimum is very high. That's why many researchers nowadays have shifted from the traditional deterministic approach to stochastic or meta-heuristics approach due to flexibility and effectiveness in solving the complex problems. In the stochastic method, it does not require any gradient information of the search space because the variables are always searching for global optimum stochastically or randomly. Therefore, there is a high chance for the solutions to find the global optimum and avoid suffered from stagnation in local optima. Stochastic or meta-heuristic optimization has two categories which are multi-agent-based and single-agent-based.

There are various single-agent-based algorithms such as Safe Experimentation Dynamics (SED) [2], Simultaneous Perturbation Stochastic Approximation (SPSA) [3] and game-theoretic [4]. For multi-agent-based, there are three classes: evolutionary-based, swarm-based and physics-based algorithm. Evolutionary-based algorithms are based on the evolution of biology/natural, such as Evolution Strategies (ESs) [4] and Genetic algorithms (GAs) [4]. Swarm-based algorithms are motivated by social behaviour of insects, birds, fishes and animal groups such as Particle Swarm Optimization (PSO) [5] and ant colony optimization (ACO) [6]. The last class is a physics-based algorithm which mimics certain physical phenomena or chemical laws such as Gravitational Search algorithm (GSA) [7], simulated annealing (SA) [8] and Henry Gas Solubility Optimization (HGSO) [9].

A Grey Wolf Optimizer (GWO) introduces by [10] is a swarm-based optimization inspired by social behaviour of groups of animals (grey wolves). This algorithm has been widely applied to many optimization problems. For instance, improving wind plant production [11], PID tuning for autonomous underwater vehicle [12], solving optimal reactive power dispatch problem [13], automatic generation control of interconnected power system [14] and liquid slosh system identification [15]. GWO algorithm is based on the hunting behaviour of grey wolves. The hunting behaviour follows the social hierarchy that can be divided into four groups, which are alphas, beta, delta and omega. These grey wolves are hunting the prey by following methods; (1) Locating, hunting and getting prey, (2) Bordering and teasing prey till it stops and (3) Attacking the prey. Like other meta-heuristic algorithms, the GWO has become a popular choice for researchers to solve their optimization problems because of simplicity, flexibility, and few parameters to adjust. Unfortunately, a significant problem with the GWO for some global optimization problems is it suffers from premature convergence which is the solution is not optimum. For that reason, this paper is proposed to improve the classical GWO algorithm to obtain better minimization solutions.

The main objective of this work is to investigate methods for improving the solution for minimization problems of the 30 well-known benchmark functions based on the modification of original GWO, named as Improved Grey Wolf Optimizer (IGWO). The original GWO uses a linear updated mechanism in the algorithm to balance the exploration and exploitation phase. The exploration phase defines as a capability of the algorithm to search for new areas as large as possible in constrains search space. The algorithm also has to prevent the solution to be trapped in a local optimum. Meanwhile, the exploitation phase needs the algorithm to have a local search capability in order to achieve high accuracy solution. In our modification, we purpose a new nonlinear updated mechanism based on the hyperbolic tangent function. This modification can solve the minimization problems by providing an optimal portion of exploration and exploitation. There have been multiple previous attempts to improve original GWO such as modified control parameter for the exploration process [16], nonlinear updating mechanism based on cosine function [17], nonlinear convergence factor with an adaptive location updating strategy [18], fuzzy GWO [19], GWO with modified augmented Lagrangian (MAL) multiplier [20], adaptive GWO [21] and weighted distance GWO [22]. Notably, none of the work discussed considers the hyperbolic tangent function for updating mechanism of GWO algorithm except one in [23] that applied GWO to tune the controller parameter of wind plant. Therefore, the study on the new updating mechanism based on the hyperbolic tangent function method could be interesting to solve global benchmark functions.

The structure of this paper is as follows: Sect. 2 described the original GWO algorithm; Sect. 3 explains the proposed improvise GWO, Sect. 4 presents the experimental results and discussion and finally, Sect. 5 concludes this paper.

2 Overview of Grey Wolf Optimizer (GWO)

The original GWO algorithm is introduced by [10] that inspired by the grey wolves hunting behaviour. This section outlines the brief explanation of the original GWO. Let $g : \mathbf{R}^n \to \mathbf{R}$, N and v_i $(i = 1, 2, \ldots, N)$ are the cost function, the number of agents and design parameter, respectively. Therefore, the equation for the cost function minimization problem can be expressed as:

$$\arg \min_{v_i(1), v_i(2), \ldots} g_i(v_i(t)) \tag{4}$$

for each agent i and the iteration $t = 1, 2, \ldots$ Next, to update the design parameter for each agent, Eq. (5) is implemented for every iteration.

$$v_i(t+1) = \frac{v^1 + v^2 + v^3}{3} \tag{5}$$

where

$$v^1 = v^\alpha - A \cdot |C \cdot v^\alpha - v_i(t)|$$

$$v^2 = v^\beta - A \cdot \left|C \cdot v^\beta - v_i(t)\right|$$

$$v^3 = v^\delta - A \cdot \left|C \cdot v^\delta - v_i(t)\right| \tag{6}$$

for each agent *i*. The vectors $\boldsymbol{A}$ and $\boldsymbol{C}$ can be expressed as Eqs. (7) and (8).

$$\boldsymbol{A} = 2\boldsymbol{a} \cdot \boldsymbol{r}_1 - \boldsymbol{a}, \tag{7}$$

$$\boldsymbol{C} = 2 \cdot \boldsymbol{r}_2, \tag{8}$$

where $\mathbf{r}_1$ and $\mathbf{r}_2$ are pseudorandom numbers in the range [0, 1] and components of $\boldsymbol{a}$ is linearly decreased from 2 to 0 throughout iteration by implementing Eq. (9).

$$\boldsymbol{a} = 2\left(1 - \frac{t}{T}\right), \tag{9}$$

where T is the maximum iterations for the algorithm. The random vectors $\mathbf{r}_1$ and $\mathbf{r}_2$ of $\mathbf{A}$ and $\mathbf{C}$ in (6), are generated independently between v^1, v^2 and v^3. Moreover, the optimal solution is based on the wolves hierarchy, which is a priority is from v^α, followed by v^β and v^δ.

3 Improved Grey Wolf Optimizer (IGWO)

The improvement of the original GWO algorithm is explained in this section. In swarm-based optimization, the outstanding capability of the algorithm in balancing the exploration and the exploitation phase is essential to get global optimum well. According to Eq. (7), the exploration and exploitation process depends on parameters $\boldsymbol{a}$ and $\mathbf{A}$. When the $\mathbf{r}_1$ is in the range $[-1, 1]$, the wolves attack the victim, indicates that the process to search the local optima. When $|\mathbf{A}| > 1$, the wolves are moved to search for the global optimum. Therefore, the exploration and exploitation phase is highly dependent on the parameter $\boldsymbol{a}$. However, as stated in (9), the value of $\boldsymbol{a}$ is linearly decreased from 2 to 0 throughout the iteration. The linearity of the original updating mechanism claims to give a right balance between exploration and exploitation phase. But, the exploration process of GWO is complicated and nonlinear. Thus, the linear updating mechanism of parameter $\boldsymbol{a}$ may not reflect this process in truth. Moreover, in (9), there is no parameter to regulate in order to vary the percentage of exploration and exploitation. Thus, this setting is only limited to specific optimization problems. Therefore, a nonlinear equation for updating mechanism $\boldsymbol{a}$ with parameters that can be adjusted seems to be an excellent choice to cater to this problem. Equation (10) shows a nonlinear updating mechanism based on the hyperbolic tangent function of GWO.

$$\boldsymbol{a} = \sigma(\tanh(\mu(t - T)))^\lambda \tag{10}$$

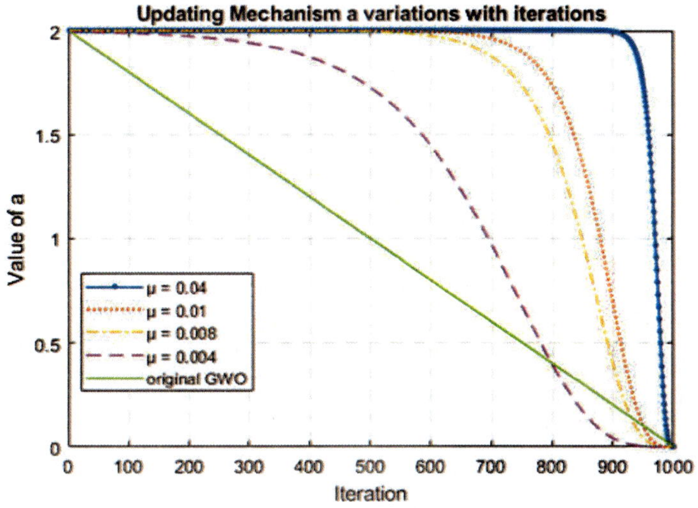

Fig. 1 Value of **a** for different μ ($\sigma = 2$, $\lambda = 4$)

In (10), the parameter λ, μ and σ are the constant introduce to adjust the quantity of exploitation and exploration phase. Figure 1 shows the proposed nonlinear updating mechanism $\boldsymbol{a}$ variations with iterations for different values of μ, σ equal to 2 and λ is set to 4. As expected, our new updating mechanism provides flexibility in adjusting the exploration and exploitation phase, whereas the linear equation is just dividing the exploration and exploitation phase into half. Furthermore, this modification can cover more optimization problems. Finally, the IGWO has the same procedure with original GWO except for the Eq. (9) is replaced with Eq. (10).

4 Experimental Results and Discussion

In this section, we present a detailed evaluation of our results. The first question in this study sought to solve the minimization problems of 30 well-known benchmark functions. These benchmark functions are categories as unimodal benchmark function ($F_1 - F_{10}$), multimodal benchmark function ($F_{11} - F_{20}$) and fixed-dimension multimodal benchmark function ($F_{21} - F_{30}$). The unimodal functions consist of only one global minimum. Thus, it is suitable to test the exploitation phase of our algorithm. The multimodal functions consist of many local minima, so, it is useful to test the exploration phase of the algorithm. For the fixed-dimension multimodal functions, it consists of a vast number of local optima. Thus, these functions use to test the capability of the algorithm to prevent the solution trapping in local optima by balancing the exploration and exploitation phase. Table 1 gives a detailed description of the benchmark functions. Notably, Dim, F_{min}, and Range and represent the

Table 1 Benchmark functions

Function	Dim	Range	F_{min}
Unimodal functions			
F_1 (Chung Reynolds)	30	[−100, 100]	0
F_2 (Sphere)	30	[−5.12, 5.12]	0
F_3 (Powell Singular 1)	30	[−4, 5]	0
F_4 (Powell Sum)	30	[−1, 1]	0
F_5 (Schwefel 2.20)	30	[−100, 100]	0
F_6 (Schwefel 2.21)	30	[−100, 100]	0
F_7 (Schwefel 2.22)	30	[−100, 100]	0
F_8 (Schwefel 2.23)	30	[−10, 10]	0
F_9 (Step 1)	30	[−100, 100]	0
F_{10} (Sum Squares)	30	[−10, 10]	0
Multimodal functions			
F_{11}(Ackley)	30	[−35, 35]	0
F_{12} (Alpine)	30	[−10, 10]	0
F_{13} (Brown)	30	[−1, 4]	0
F_{14} (Cigar)	30	[−100, 100]	0
F_{15} (Exponential)	30	[−1, 1]	−1
F_{16} (Griewank)	30	[−600, 600]	0
F_{17} (Mishra 1)	30	[0, 1]	2
F_{18} (Mishra 2)	30	[0, 1]	2
F_{19} (Mishra 11)	30	[0, 10]	0
F_{20} (Rastrigin)	30	[−5.12, 5.12]	0
Fixed-dimension multimodal functions			
F_{21} (Ackley 2)	2	[−32, 32]	−200
F_{22} (Bartels Conn)	2	[−500, 500]	1
F_{23} (Cross-in-Tray)	2	[−10, 10]	−2.06261218
F_{24} (Hartman)	6	[0, 1]	−3.32236
F_{25} (Matyas)	2	[−10, 10]	0
F_{26} (Rump)	2	[−500, 500]	0
F_{27} (Rotated Ellipse)	2	[−500, 500]	0
F_{28} (Sawtoothxy)	2	[−20, 20]	0
F_{29} (Scahffer 6)	2	[−100, 100]	0
F_{30} (Trecanni)	2	[−5, 5]	0

dimension of the benchmark function, the optimal value and the boundary of the search space, respectively.

The experimental works use the software MATLAB 2020a on the personal computer with Microsoft Window 10, 8 GB RAM and Intel Core i7-6700 Processor (3.41 GHz). The comparative assessment for IGWO with original GWO has been done by setting the maximum number of iterations, T to be 500 and number of agents, N to 50 that's mean the total number of function evaluations is 25,000. The coefficients of IGWO is set as $\mu = 0.01$, $\lambda = 4$ and $\sigma = 2$ based on preliminary investigations. For a fair comparison, the N and T for original GWO are set identical with IGWO. The number of trials is set to 30 trials. Table 2 shows the statistical results of IGWO and GWO. Indicate that, the boldface results are the superior results.

It can be seen that the proposed method drastically improves all unimodal functions (F_1 to F_{10}) for both mean value and standard deviation except for F_9 where the results are identical. From the results, it is proved that IGWO exhibited an excellent exploitation ability compared to original GWO. Meanwhile, we can see that the IGWO algorithm has outmatched the original GWO algorithm on all multimodal functions except F_{15} and F_{18}. For F_{15}, the GWO and IGWO produce the same mean value, but the standard deviation for GWO is better, indicates that GWO is more robust compare to IGWO for this function. While for F_{18}, GWO produces a better result compare to IGWO. This situation proved the No Free Lunch (NFL) theorem that stated no optimization algorithm could solve all optimization problems [24]. Therefore, it is clear that the proposed algorithm also has a better exploration ability and able to overcome premature convergence.

The results for fixed-dimension multimodal benchmark function is obtained in functions $F_{21} - F_{30}$. For function F_{21} and F_{22}, both original GWO and IGWO produce the same results. For F_{23}, the mean value for both algorithms is identical, but, the standard deviation for GWO is better, which indicates the GWO has great robustness over 30 runs. Meanwhile, for function F_{24}, despite the IGWO produces a better mean value, its standard deviation is worse compared to original GWO. For this particular function, IGWO is unable to provide a consistent result. For the rest of the functions (F_{25} to F_{30}), IGWO has surpassed the results of original GWO for mean and standard deviation. Experiments confirmed the superiority of the proposed technique by balancing the exploration and exploitation phase and prevent the trapping in local optima. Figure 2 shows some of the convergence curves obtained from the benchmark functions for original GWO and IGWO. From the results, the IGWO has a faster convergence rate, higher convergence accuracy and less iteration rate for most of the benchmark functions.

Generally, statistical results for 30 runs have a drawback which is it did not compare each of the runs. Thus, there will be a small possibility that superiority happens during 30 runs. Hence, a non-parametric statistical test is performed to compare the results of each run and decide whether the better algorithms is significant or not. In this study, the significance of the results are tested using the Wilcoxon rank-sum. Wilcoxon rank-sum test stated that the better algorithm should be significant if the p-value less than 5%. Table 3 shows the p-values obtained from the test. From the results, it is shown that the significant superiority of IGWO compared

Table 2 Minimization results of benchmark functions

Function	GWO	IGWO
	Mean (Std. Dev.)	Mean (Std. Dev.)
F_1	1.2241e−71 (3.0841e−71)	**2.1616e−82 (7.8165e−82)**
F_2	4.8675e−39 (7.9065e−39)	**1.6327e−44 (3.5187e−44)**
F_3	0.0001779 (0.00027049)	**9.3672e−05 (0.00022384)**
F_4	7.9588e−124 (4.3591e−123)	**1.3206e−138 (5.1455e−138)**
F_5	5.4661e−21 (5.5702e−21)	**1.7251e−23 (1.917e−23)**
F_6	4.0254e−09 (5.9422e−09)	**1.5689e−11 (1.9371e−11)**
F_7	1.3659e−20 (1.0825e−20)	**2.6688e−23 (2.6648e−23)**
F_8	2.4701e−120 (1.1148e−119)	**2.1868e−137 (1.0982e−136)**
F_9	0 (0)	0 (0)
F_{10}	2.2732e−37 (7.1029e−37)	**2.1663e−42 (5.5314e−42)**
F_{11}	0.014856 (0.056536)	**1.581e−14 (2.2899e−14)**
F_{12}	0.00021767 (0.00041381)	**8.1993e−06 (4.491e−05)**
F_{13}	2.9217e−39 (6.8062e−39)	**1.6745e−44 (4.0353e−44)**
F_{14}	1.3026e−30 (2.4937e−30)	**4.2621e−36 (8.4043e−36)**
F_{15}	−1 (**3.5709e−17**)	−1 (1.0713e−16)
F_{16}	0.0025211 (0.0076038)	**0.00060701 (0.0023102)**
F_{17}	6.1769 (11.6037)	**4.2513 (3.2478)**
F_{18}	**6.9039 (10.9747)**	7.6814 (14.2444)
F_{19}	4.6565e−10 (2.8444e−10)	**7.2424e−14 (1.0285e−13)**
F_{20}	2.028 (5.2422)	**0 (0)**
F_{21}	−200 (0)	−200 (0)
F_{22}	1 (0)	1 (0)
F_{23}	−2.0626 (**6.0456e−09**)	−2.0626 (7.1245e−06)
F_{24}	−3.2456 (**0.076372**)	**−3.2568** (0.081684)

(continued)

Table 2 (continued)

Function	GWO	IGWO
	Mean (Std. Dev.)	Mean (Std. Dev.)
F_{25}	2.643e−152 (1.4476e−151)	**9.0212e−190 (0)**
F_{26}	5.9978e−07 (2.3955e−06)	**9.0669e−12 (3.1088e−11)**
F_{27}	3.5745e−239 (0)	**1.6814e−253 (0)**
F_{28}	1.8273e−261 (0)	**3.9408e−264 (0)**
F_{29}	0.0038864 (0.0048412)	**0.00032386 (0.0017739)**
F_{30}	5.8205e−08 (1.3603e−07)	**2.5672e−08 (8.2602e−08)**

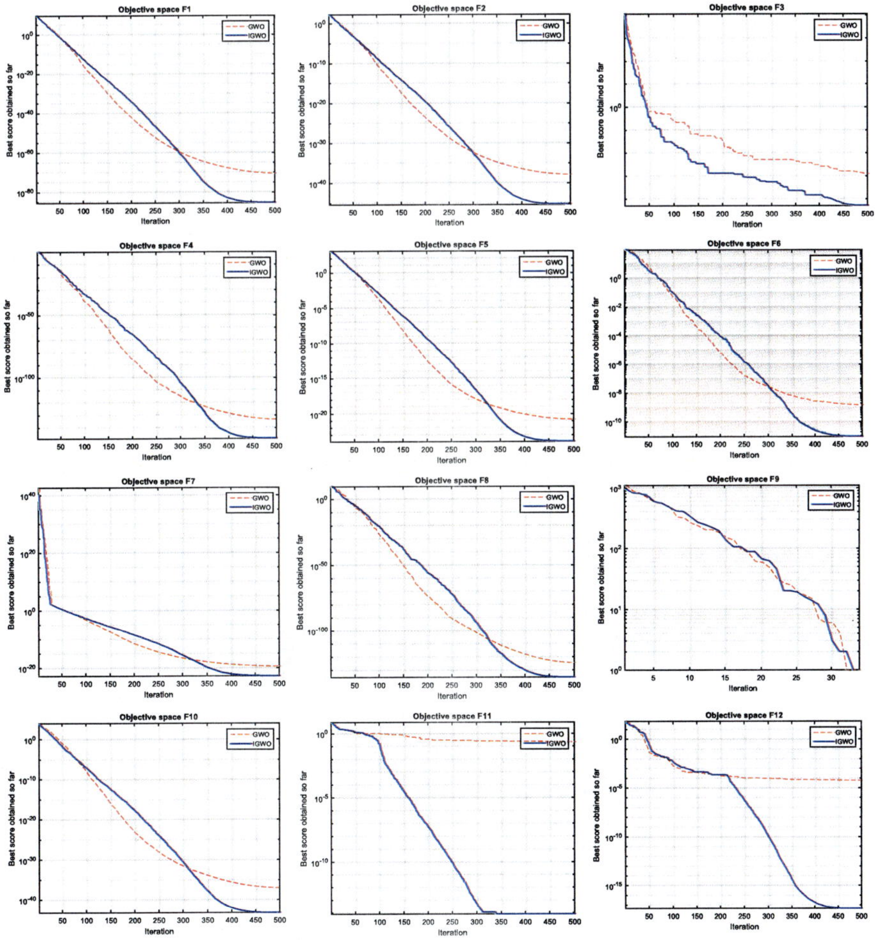

Fig. 2 Convergence curve of GWO and IGWO for selected benchmark functions

Table 3 Results from Wilcoxon rank-sum test

Function	p-value (<0.05)
F_1	3.02e−11
F_2	3.02e−11
F_3	0.0058282
F_4	8.15e−11
F_5	3.02e−11
F_6	3.02e−11
F_7	3.02e−11
F_8	3.02e−11
F_9	**NaN**
F_{10}	3.02e−11
F_{11}	2.74e−10
F_{12}	8.20e−07
F_{13}	3.02e−11
F_{14}	3.02e−11
F_{15}	1.57e−02
F_{16}	**0.22758**
F_{17}	0.0010035
F_{18}	0.032651
F_{19}	3.02e−11
F_{20}	1.11e−07
F_{21}	**NaN**
F_{22}	**NaN**
F_{23}	5.25e−09
F_{24}	0.028129
F_{25}	6.70e−11
F_{26}	1.47e−07
F_{27}	0.02236
F_{28}	2.00e−06
F_{29}	0.00046949
F_{30}	0.0063765

to original GWO for all benchmark functions based on the p-values that less than 0.05 except function F_{16}. For F_{16}, despite IGWO produces a better mean value and standard deviation, from Wilcoxon rank-sum evaluation, this result is not significantly improved the original GWO. For F_9, F_{21} and F_{22}, p-value is Not a Number (NaN) because both GWO and IGWO produce the same mean and standard deviation. As we can see, these statistical results indicate that the IGWO has improved the original GWO for most of 30 well-known benchmark functions and can apply to other optimization problems by adjusting the coefficient of IGWO.

5 Conclusion

This study set out to improve the original GWO by introducing the new nonlinear updated mechanism based on hyperbolic tangent function name as Improved GWO (IGWO) to solve the minimization problems of 30 well-known benchmark functions. This study has shown that the proposed IGWO has produced better statistical results for all benchmark functions compared to the original GWO. From the non-parametric statistical test, the IGWO has a significant superiority compared to original GWO for the majority of the benchmark functions. These initial findings show that IGWO has good potential in solving real application optimization problems.

Acknowledgements Universiti Malaysia Pahang financed this research under the research grant PGRS1903141.

References

1. Shanno DF (1970) Conditioning of quasi-Newton methods for function minimization. Math Comput 24(111):647–656
2. Shukor NSA, Ahmad MA, Tumari MZM (2017) Data-driven PID tuning based on safe experimentation dynamics for control of liquid slosh. In: 2017 IEEE 8th control and system graduate research colloquium, ICSGRC 2017—proceedings, pp 62–66
3. Ahmad MA, Azuma SI, Sugie T (2014) A model-free approach for maximizing power production of wind farm using multi-resolution Simultaneous Perturbation Stochastic Approximation. Energies 7(9):5624–5646
4. Marden JR, Ruben SD, Pao LY (2012) Surveying game theoretic approaches for wind farm optimization. In: 50th AIAA aerospace sciences meeting including the new horizons forum and aerospace exposition
5. Kennedy J (2017) Particle swarm optimization. In: Encyclopedia of machine learning and data mining. Springer US, Boston, MA, pp 967–972
6. Dorigo M, Di Caro G (1999) Ant colony optimization: a new meta-heuristic. In: Proceedings of the 1999 congress on evolutionary computation, CEC 1999, vol 2, pp 1470–1477
7. Rashedi E, Nezamabadi-pour H, Saryazdi S (2009) GSA: a gravitational search algorithm. Inf Sci (Ny) 179(13):2232–2248
8. Kirkpatrick S, Gelatt CD, Vecchi MP (1983) Optimization by simulated annealing
9. Hashim FA, Houssein EH, Mabrouk MS, Al-Atabany W, Mirjalili S (2019) Henry gas solubility optimization: A novel physics-based algorithm. Futur Gener Comput Syst 101:646–667
10. Mirjalili S, Mirjalili SM, Lewis A (2014) Grey Wolf Optimizer. Adv Eng Softw
11. Tumari MZM, Suid MH, Ahmad MA (2020) A modified grey wolf optimizer for improving wind plant energy production. Indones J Electr Eng Comput Sci 18(3)
12. Tumari MZM, Zainal Abidin AF, Yusof AA, Mohd Aras MS, Mustapha NMZA, Ahmad MA (2020) Model-free pid controller based on Grey Wolf Optimizer for hovering autonomous underwater vehicle depth control. In: InECCE2019, Springer, Singapore, pp 25–35
13. Sulaiman MH, Mustaffa Z, Mohamed MR, Aliman O (2015) Using the Gray Wolf Optimizer for solving optimal reactive power dispatch problem. Appl Soft Comput J
14. Gupta E, Saxena A (2016) Grey wolf optimizer based regulator design for automatic generation control of interconnected power system. Cogent Eng
15. Ahmad MA, Musa Z, Suid MHS, Tumari MZM (2020) Grey Wolf Optimizer for identification of liquid slosh behavior using continuous-time hammerstein model. Bull Electr Eng Inform 9(2)

16. Seema, Kumar V (2016) Modified Grey Wolf algorithm for optimization problems. In: Proceedings of 2016 international conference on inventive computation technologies (ICICT), vol 2016, pp 428–433
17. Long W, Cai SH, Jiao JJ, Bin Wu T (2019) An improved Grey Wolf Optimization algorithm. Tien Tzu Hsueh Pao/Acta Electron Sin 47(1):169–175
18. Majeed MM, Patri SR (2018) An enhanced grey Wolf optimization algorithm with improved exploration ability for analog circuit design automation. Turkish J Electr Eng Comput Sci 26(5):2605–2617
19. Rodriguez L, Castillo O, Soria J (2016) Grey Wolf Optimizer with dynamic adaptation of parameters using fuzzy logic. In: 2016 IEEE congress on evolutionary computation (CEC), pp 3116–3123
20. Long W, Liang X, Cai S, Jiao J, Zhang W (2017) A modified augmented Lagrangian with improved grey wolf optimization to constrained optimization problems. Neural Comput Appl 28:421–438
21. Dudani K, Chudasama AR (2016) Partial discharge detection in transformer using adaptive grey wolf optimizer based acoustic emission technique. Cogent Eng
22. Malik MRS, Mohideen ER, Ali L (2016) Weighted distance Grey Wolf Optimizer for global optimization problems. In: 2015 IEEE international conference on computational intelligence and computing research (ICCIC)
23. Mohd Tumari MZ, Zahar MM, Ahmad MA (2020) Optimal tuning of a wind plant energy production based on improved Grey Wolf Optimizer. Bull Electr Eng Inform 9(6):1–8
24. Wolpert DH, Macready WG (1997) No free lunch theorems for optimization. IEEE Trans Evol Comput

Non-dominated Sorting Manta Ray Foraging Algorithm with an Application to Optimize PD Control

Ahmad Azwan Abdul Razak, Ahmad Nor Kasruddin Nasir, Nor Maniha Abdul Ghani, Shuhairie Mohammad, Mohd Falfazli Mat Jusof, and Nurul Amira Mhd Rizal

Abstract This paper presents a multi-objective (MO) version of Manta Ray Foraging Algorithm (MRFO) by using components in Non-Dominated Sorting Genetic Algorithm (NSGAII). MFRO is a recent developed algorithm which was inspired from behavior of a cartilaginous fish called Manta Ray. MRFO search solution by using three strategies of manta ray which are chain foraging, cyclone foraging and somersault foraging. However, this algorithm solves only single-objective problem and can be improved to solve multi-objective problem. Thus, non-dominated sorting (NS) strategies including crowding distance (CD) were adopted into MRFO. NS is a sorting technique based on Pareto's game. It is a fast strategy to develop a good characteristic of Pareto's front (PF). Meanwhile, CD is a strategy to preserve good distribution of solutions along the PF. This proposed algorithm is called NSMRFO. It is tested using several benchmark functions and its performance is compared to its parent by using statically analysis of hypervolume indicator. Then, it is applied to a Proportional-Derivative (PD)-controller for an Inverted Pendulum System (IPS) in order to know its performance on real-world application. Result of the NSMRFO on benchmark functions shows that it outperforms NSGAII and satisfactorily optimizes PD-control for the IPS.

Keywords Multi-objective algorithm · Genetic algorithm · Spiral · NSGAII · Manta ray foraging algorithm · PD-Controller

A. A. Abdul Razak (✉) · A. N. K. Nasir · N. M. Abdul Ghani · S. Mohammad · M. F. M. Jusof · N. A. M. Rizal
Faculty of Electrical and Electronics Engineering Technology, Universiti Malaysia Pahang, 26600 Pekan, Pahang, Malaysia
e-mail: ahmadazwan.ar@gmail.com

A. N. K. Nasir
e-mail: kasruddin@ump.edu.my

N. M. Abdul Ghani
e-mail: normaniha@ump.edu.my

A. F. Ab. Nasir et al. (eds.), *Recent Trends in Mechatronics Towards Industry 4.0*, Lecture Notes in Electrical Engineering 730, https://doi.org/10.1007/978-981-33-4597-3_42

1 Introduction

Optimization algorithm is a tool which frequently applied to solve complex real-world problems [1]. These problems are including problems in science, engineering and social science studies. By applying algorithm, a more reliable and accurate solution can be obtained. Optimization algorithm can be grouped into two types which are single-objective (SO) and multi-objective (MO). SO algorithm is used to solve a single objective problem while MO solve two or more conflicting objective problem. MO is usually derived from SO algorithm and they have more complex structure of programming.

There are several well-known SO algorithms that widely used in various applications such as Particle Swarm Optimization (PSO) [2], Genetic Algorithm (GA) [3], Spiral Dynamics Algorithm (SDA) [4] while Manta Ray Foraging Optimization (MRFO) [5] is a recently new population-based algorithm. MRFO inspired by a cartilaginous fish called manta ray to hunt for foods. It is a large marine creature and most of the time live in group [6]. Their body are large and flat from top to bottom beside a pair of pectoral fins which able them to swims as birds that freely fly. There is also a cephalic lobe that extend in front of their giant mouth. Manta ray hunt and feed on plankton of microscopic organism from the water. By using their cephalic lobe, they funnel the plankton into their mouth. Then, they filter their prey from the water using their gill. Manta ray have two distinct species. First species is reef manta ray which can be found mostly in Indian Ocean and second species is giant manta ray which can be found mostly in tropical, subtropical and warm temperate oceans. A matured manta ray requires up to 5 kg plankton per day. On the other side, the plankton is not found evenly dispersed or live in grouping at a certain location but they follow the flow of the tides of changes in seasons. Excitingly, manta ray always able to find those plankton in large quantity. They succeed because of their special behavior of foraging strategy that includes chain foraging, cyclone foraging and somersault foraging [5].

Chain foraging is an act of a number of manta rays which form a line, one behind another [5]. During this action, they will feed the plankton in their ways and those plankton which were missed will be feed by manta ray behind them. Cyclone foraging is a grouping of manta ray when they detect a large number of plankton [5]. In this action, they link their tails with other member's head in spiral to form a spiraling tides in order to force plankton move to the water surface while feeding the plankton. Then, somersault foraging is the action of manta ray do a sequence of backwards somersault by doing cyclical movement in order to draw plankton towards them. Somersault is a random, repetitive, local and cyclical exercise which help manta ray in food intake [5].

In literature, there exist several MO algorithms based on SO algorithms such as Multi-Objective PSO (MOPSO) [7], Multi-Objective SDA (MOSDA) [8, 9] and Fast-elitist Non-dominated Sorting Genetic Algorithm (NSGAII) [10]. NSGAII is one of the well-known MO algorithms. Those algorithms were programmed using different strategy of MO thus have different performance. MOPSO and MOSDA implement external memory or repository to store global best agent. It also includes grid index strategy to preserve good distribution of solution along Pareto Front (PF). NSGAII implements a fast-elitist non-dominated sorting (NS) and crowding distance (CD) which incorporated to produce good PF diversity. These algorithms are proven to solve numerous real-world applications. However, research on MRFO is still at prior level. Thus, there is no existing MO algorithm based on this MRFO.

This paper proposes a new variant of MRFO to solve MO problem called NSMRFO. It compares the performance of MO version of MRFO and GA in utilizing NS and CD strategies. On top of that, NSMRFO is applied to deal with a PD-controller design for an inverted pendulum system (IPS). This paper is organized as follows. Section 2 explains IPS while Sect. 3 explains NSMRFO in details. Section 4 discusses the performance of NSMRFO on benchmark functions. Section 5 presents the application of NSMRFO to PD-controller of the IPS. Lastly, Sect. 6 presents the conclusion of the paper.

2 Inverted Pendulum System

Figure 1 presents an inverted pendulum (IPS) or a cart pole system that used in this study [9]. IPS is a built of a pole that can rotate freely for 360° and a cart that moves backwards and forwards in translational movement. One end of the pole will be able to move freely while another end is pivoted on the cart. Under stagnant, the pole is pointing downwards. Simultaneously, the pivoted pole moves linearly when the cart moves while the other end rotates freely around the pivoted end. IPS is a common tool to test new controller developed and it is a single-input multi-output (SIMO) system [11]. Regularly, voltage input is treated as its input and its linear position of the cart and the angle of the pole are treated as the outputs of the system. Typically, the control system required to keep the pole pointing upwards while the cart moving of stagnant condition. This is a controlling challenge of this system due to its high instability and complexity. In real-world, IPS simulates two-wheel Segway transporter, two-wheels chair for disable and space rocket.

Table 1 presents the physical parameters for the mechanical and electrical components for IPS.

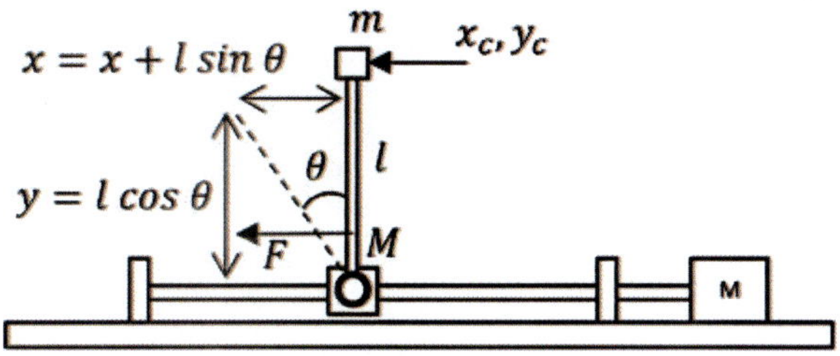

Fig. 1 An inverted pendulum system

Table 1 Physical parameters for IPS

Item		Specification
Mechanical specification	Dimension, $W \times L \times H$	1300 mm × 200 mm × 250 mm
	Length of pendulum, l	400 mm
	Pendulum weight, w	0.2 kg
	Cart weight, M	1 kg
	Translational distance, x	900 mm
	Ball screw pitch	12.7 mm
Electrical specification	Motor input voltage	0–5 V
	Motor output	24 V, 60 W
	Encoder pulse	4000 pulse (×4)
	Motor speed	3800 rpm

Based on the Fig. 1 and Table 1, the IPS can be derived. These derivations were based on the Newton 2nd Law of Motion and can be represented as follow.

$$(M + m)\ddot{x} + ml\theta + F_r\dot{x} = F_v V \tag{1}$$

In Eq. 1, F_r and F_v are as follows.

$$F_r = b + \left(\frac{2\pi}{r}\right)^2\left(\frac{K_m K_b}{R}\right) \tag{2}$$

$$F_v = \left(\frac{2\pi}{r}\right)\left(\frac{K_m}{R}\right) \tag{3}$$

$$m\ddot{x} + ml\ddot{\theta} = mg\theta \tag{4}$$

From Eqs. 2 to 4, b is friction coefficient of the cart, r is the transfer distance per rotation of the ball screw, K_m is motor torque, K_b is the back emf constant and R is motor armature resistance. These derivations furthered derived and converted to state space representation as follow.

$$\begin{bmatrix} \dot{\theta} \\ \ddot{\theta} \\ \dot{x} \\ \ddot{x} \end{bmatrix} = \begin{bmatrix} 0 & 1\,0 & 0 \\ \frac{(M+m)g}{Ml} & 0\,0 & \frac{F_r}{Ml} \\ 0 & 0\,0 & 1 \\ -\frac{mg}{M} & 0\,0 & -\frac{F_r}{M} \end{bmatrix} \begin{bmatrix} \theta \\ \dot{\theta} \\ x \\ \dot{x} \end{bmatrix} + \begin{bmatrix} 0 \\ -\frac{F_v u}{Ml} \\ 0 \\ \frac{F_v u}{M} \end{bmatrix} \tag{5}$$

$$[y] = \begin{bmatrix} 1\,0\,0\,0 \\ 0\,0\,1\,0 \end{bmatrix} \begin{bmatrix} \theta \\ \dot{\theta} \\ x \\ \dot{x} \end{bmatrix} \tag{6}$$

In Eqs. 5 and 6, u is the voltage input, θ is the angle of the pole and x is the cart position.

3 Multi-objective Manta Ray Foraging Optimization Algorithm (NSMRFO)

3.1 *Manta Ray Foraging Optimization (MRFO)*

MRFO was developed inspired from a cartilaginous fish called manta ray. It simulates three hunting behaviors of manta rays which are chain foraging, cyclone foraging and somersault foraging. In this section, the mathematical expression of these behaviors was presented. Please denote that, the location point of plankton at high concentration is assumed as the best solution. This point is the direction of manta ray want to approach to hunt and feed.

Chain Foraging
Manta ray live in groups. When they hunting, they form a chain. They link up in a line and scoop all the plankton on the ways. In this strategy, the leader of manta ray, which is the best solution so far, will guide the link. The mathematical expression is as follows [5].

$$x_i(k+1) = \begin{cases} x_i(k) + r_1(x_{best} - x_i(k) + \alpha(x_{best} - x_i(k)), i = 1 \\ x_i(k) + r_1(x_{i-1}(k) - x_i(k) + \alpha(x_{best} - x_i(k)), i = 2, 3, \ldots, n \end{cases} \quad (7)$$

$$\alpha = 2r\sqrt{|\log(r_1)|} \quad (8)$$

Equations (7) and (8), $x_i(k+1)$ is the position of the searching agent, $x_i(k)$ is the current position of ith individuals at kth iteration, r_1 is a random value between [0, 1], α is weight coefficient, x_{best} is the location where the plankton with high concentration.

Cyclone Foraging
When a location of high concentration plankton was found, manta rays will move towards the food in spiracle and linked-up in chain simultaneously. At the initial state, manta rays put onto random position. The mathematical expression of this behavior is stated in (9) [5].

$$x_i(k+1) = \begin{cases} x_{rand} + r_2(x_{rand} - x_i(k) + \beta(x_{rand} - x_i(k))), i = 1 \\ x_{rand} + r_2(x_{i-1}(k) - x_i(k) + \beta(x_{rand} - x_i(k))), i = 2, 3, \ldots, n \end{cases} \quad (9)$$

Later, manta rays will move towards the best individual using (10) and (11). Denote that, β is the weight coefficient, $k_{\max}$ is the maximum iterations and r_2 is the random number between [0, 1].

$$x_i(k+1) = \begin{cases} x_{best} + r_2(x_{rand} - x_i(k) + \beta(x_{best} - x_i(k))), i = 1 \\ x_{best} + r_2(x_{i-1}(k) - x_i(k) + \beta(x_{best} - x_i(k))), i = 2, 3, \ldots, n \end{cases} \quad (10)$$

$$\beta = 2e^{r_2 \frac{k_{\max}+1}{k_{\max}}} \sin(2\pi r_2) \quad (11)$$

Somersault Foraging

The last strategy of manta rays is somersault foraging. In this plan, location of the best agent is considered as a pivot. Searching agent swim towards and backwards around the pivot then somersault to a new location. Later on, the searching agent update their new location about the x_{best} at the moment. Mathematical presentation of this somersault, S is described in (12) [5]. Denote that, r_3 and r_4 are random number between [0, 1].

$$x_i(k+1) = x_i(k) + S(r_3 x_{best} - r_4 x_i(k)) \quad (12)$$

3.2 *Fast Elitist Non-dominated Sorting Genetic Algorithm (NSGAII)*

In concise, NSGAII is a well-known MO algorithm based on GA [10]. GA consist of three main components including crossover, mutation and tournament selection. In favor turning GA into MO algorithm, NS-II was integrated into GA. It is used to rank an individual respect to two different objectives function using Pareto's optimal. This Pareto's concept makes no other event that makes every individual at least as well off and at least one individual strictly better off. NS-II is efficient in bringing a set of Pareto solution with good accuracy. However, the addition of CD is also required to form a well-distributed solution along the PF.

3.3 *Non-dominated Sorting Manta Ray Foraging Optimization*

NSMRFO is an extended version of MRFO. The usage of solely MRFO, however is not enough to perform well. This is due to deterministic approach in cyclone model of MRFO. In this study, four components of NSGAII were embraced in developing

NSMRFO. From literatures, it is clear that NS-II and CD are good techniques to produce well-PF. Thus, NS-II and CD were used in this study to find the best individual, x_{best} which guides other searching agents towards accuracy. Over and above, mutation and crossover also partially applied to create more randomness. Therefore, the deterministic approach of MRFO, which lead to local optima can be overcame. Figure 2 shows the pseudocode of NSMRFO. In this study, $nPop$ is set to 50 individuals, $k_{\max}$ is set to 667, NFE is equal to 10,000 times. Mutation and crossover rate in this study is set to a 0.5 and 0.2 respectively.

Step 0: Preparation
Select a number of manta rays in a population, $nPop$, maximum number of iterations, k_{max} and function evaluation (NFE).

Step 1: Initialization
Initialize the location points of manta rays in random position $x_i\ (k)$, between feasible region, $x_i(k) = x_l + rand(x_u - x_l)$, where x_l is lower bound and x_u is upper bound.

Step 2: Evaluate Fitness
Evaluate fitness, $f_i = f(x_i(k))$ of manta rays for each location and acquire best solution so far as x_{best}.

Step 3: Apply *Non-Dominated Sorting* and *Crowding Distance.*

Step 4: Apply Foraging Strategies Model of Equations.
Choose a random number, r_1. If r_1 smaller then k/k_{max}, then do ***cyclone foraging***. Otherwise, do ***chain foraging.***

a) ***Cyclone foraging***
Choose a random number, r. If $r < k/k_{max}$, then move individuals by using Equation 9.
Else if $r > k/k_{max}$, then use Equation 10.

b) ***Chain foraging***
Move individuals by using Equation 7.

Evaluate $f_i = f(x_i(k))$. Choose a new best manta ray, x_{best}. If $f\big(x_i(k+1)\big) < f(x_{best})$, then, $x_{best} = x_i(k+1)$.

Step 5: Apply somersault and partial *Mutation* and *Crossover.*
Then apply *somersault foraging* using Equation 12. Choose a random number, r_2. If $r_2 < 0.5$, apply *Mutation* and *Crossover.*

Step 6: Apply *Non-Dominated Sorting* and *Crowding Distance.*

Step 7: Sort population, *pop* and display Pareto Front.

Step 8: Checking termination criterion
Check if termination criterion, k_{max} is met, then stop. Otherwise, return to Step 2.

Fig. 2 Pseudocode of NSMRFO

Table 2 List of benchmark function

Function	Number of Variables	Search Range
Schaffer, f_1	1	$[-10^3, 10^3]$
Fonseca, f_2	3	$[-4, 4]$
Kursawe, f_3	3	$[-5, 5]$
Poloni, f_4	2	$[-\pi, \pi]$

4 Performance Test with Benchmark Functions

Table 2 shows number of variables and search range of the benchmark functions [9]. NSGAII and NSMRFO was performed on four benchmark multi-objective problems. Then, NSGAII and NSMRFO performance to provide good PF with good accuracy and diversity will be evaluated. On top of that, number of function evaluation (*NFE*) was used instead of number of iterations in order to make a fair and square comparison. In this study, every algorithm was evaluated for 10,000 times of *NFE* and run for 25 times [8]. Therefore, the plotted PFs and hypervolume test will be based on average of these 25 times number of runs. Denote that, hypervolume indicator test is a common tool to measure performance of PF [12]. It measures the portion volume of the objective space bounded from below by a reference point. Thus, it covers both accuracy and diversity performance. In this study, an outermost point in PF plot was selected as a reference point. Larger value of hypervolume value indicates better performance which value of 0.00 indicates worst performance while value of 1.00 indicates the best.

5 Application to Optimize PD-Controller

PD-controller is a distinguished linear type controller that has been applied in many applications in industry and academic studies [9]. It has two gains called proportional gain, K_p and a derivative gain, K_d as expressed in (13). As seen in the equation, both gains are multiplied with an error, e and derivative of error, *de/dt*.

$$PD = K_p e + K_d \frac{de}{dt} \tag{13}$$

Figure 3 represents the block diagram for the controller of IPS system. Two PD-controllers will be applied to regulate the IPS. PD_1 will be used to control angle, θ and PD_2 for the cart positioning control, x. NSMRFO was applied to optimized K_ps and K_ds for both controllers simultaneously. Therefore, there will be four dimension of variables in the optimization, which are $K_p(1)$, $K_d(1)$, $K_p(2)$ and $K_d(2)$. From the figure, In is the desired input, x and θ are the output of the system. Difference between x and In is treated as the error and as input for PD_1. Considering θ must be zero, therefore, there is no desired angle should be set in the system. This simulation ran

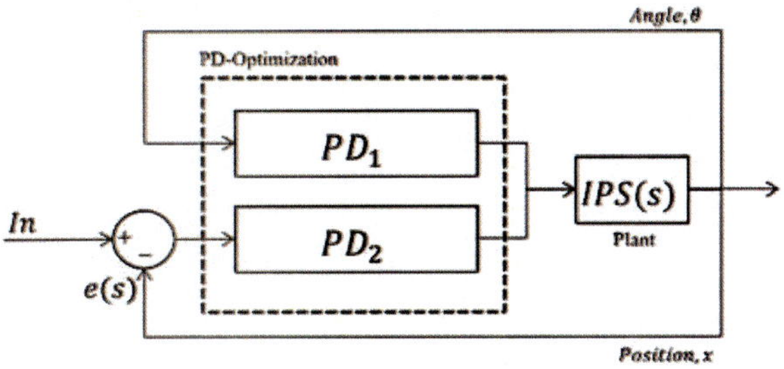

Fig. 3 Application of NSMRFO to optimize PD-controller for IPS

for 10,000 times of *NFE*. Due to long time required to finish the *NFE*, the simulation only ran for three times. Thus, the plotted PF and hypervolume test will be based on the average of this run.

6 Result and Discussion

This section explains the results of the performance test of the proposed NSMRFO on benchmark function and PD-controllers of the IPS. There are three measures that will be analyzed about the performance of the proposed algorithm including the ability of finding the correct PF, the hypervolume indicator to test accuracy and the distribution of solutions along the PF simultaneously.

Figure 4 shows the result of NSMRFO on benchmark functions compared to NSGAII. From these figures, proposed NSMRFO and NSGAII succeed to find optimal solution for the MO problems. For accuracy and diversity, both NSMRFO and NSGAII were seen as even based on the PF produced for all MO problems. The work continued to using numerical analysis. Table 3 presents the result for hypervolume indication test. From the table, Obviously, NSMRFO outperformed NSGAII by producing larger hypervolume for 3 over 4 MO problems. For f_3, both algorithms succeed to produce PF at the same level. These values were tally with the figures in Fig. 4.

Table 4 shows the parameter values that has been optimized for all K_p s and K_d s. Figure on the left in Fig. 5 shows the PF plot for the application of NSMRFO on the PD-controller of the IPS based on 10,000 times of *NFE*. From the figure, both NSMRFO and NSGAII able to find the PF successfully. However, in term of diversity, NSMRFO obviously produced a better distribution of solution along the PF. The solution produced by NSGAII is restrained between [9.213e−05, 1.723] while NSMRFO is restrained between [4.81e−07, 2.301].

For step-response graph, obviously NSMRFO optimized-PD-parameters produced much stable controller for the IPS. Even though rising time for NSMRFO slightly slower than NSGAII, it can provide faster settling time (5 s) compared to NSGAII (13 s). NSMRFO also provide a better overshoot step-response (0.2%) than NSGAII (49%). However, both of them recorded almost 0.0% steady state error

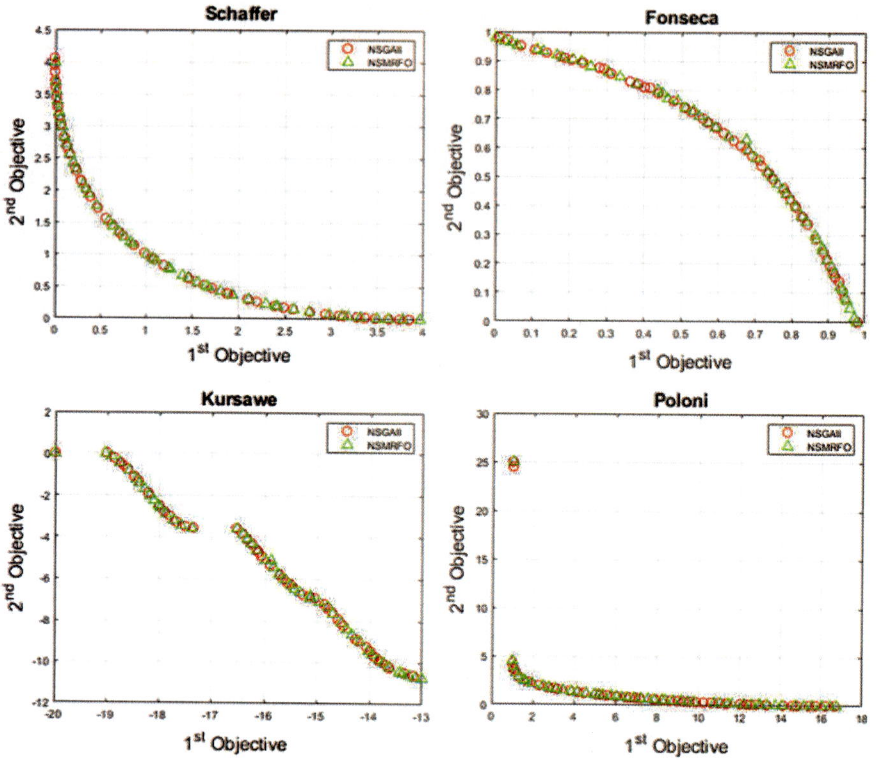

Fig. 4 Pareto fronts produced for f_1, f_2, f_3 and f_4

Table 3 Hypervolume indicator for NSGAII and NSMRFO

	Hypervolume Ind.	
	NSGAII	NSMRFO
Schaffer, f_1	0.40	**0.72**
Fonseca, f_2	0.68	**0.80**
Kursawe, f_3	**1.00**	**1.00**
Poloni, f_4	0.36	**0.40**

Table 4 Values of K_p s and K_d s for both angle and position controller that optimized by the algorithms

	NSGAII	NSMRFO
$K_p(1)$	2.3771e−4	11.1923
$K_d(1)$	8.9556	0.6939
$K_p(2)$	1.6328	8.8898
$K_d(2)$	5.4480	0.4359

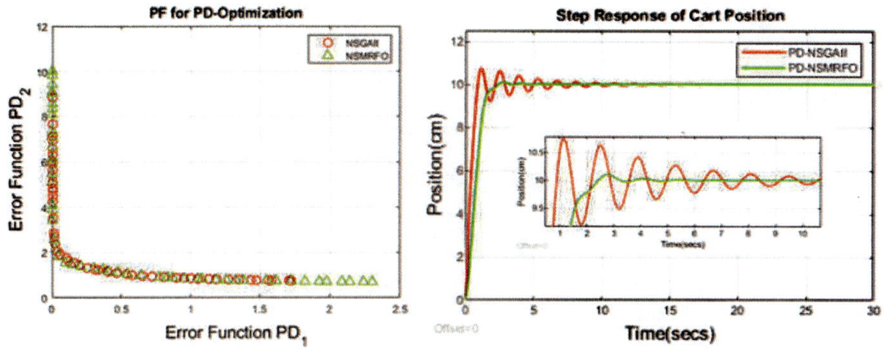

Fig. 5 Figure on the left shows the PF for application of NSGAII and NSMRFO in optimizing PD-Controller for IPS while figure on the right shows the step response of the cart position

Table 5 Step-response analysis of the cart position

	NSGAII	NSMRFO
Settling time, t_s (s)	13	**5**
Rise time, t_r (s)	**0.6**	2.0
Percent Overshoot, % OS	49	**0.2**
Steady state error, e_{ss}	0.03	**0.01**

after reached their settling time. Table 5 presents the summary of the step response analysis.

7 Conclusion

A multi-objective algorithm based on MRFO called NSMRFO has been developed successfully using fast-elitist non-dominated sorting (NS-II) and crowding distance (CD) strategies. These both strategies are originally from NSGAII. NSMRFO has been tested with four MO benchmark problems. Result of the test shows that NSMRFO successfully searches for PFs for all of the MO problems. It has high accuracy and diverse well along the PFs. It has been proved with an hypervolume indicator test. The proposed NSMRFO has been adopted to optimize PD-controller for a position tracking problem of an inverted pendulum system. The result has been presented in time-domain response. Both algorithms have regulated the IPS and resulted a good performance.

Acknowledgements This work has been financially supported by the Universiti Malaysia Pahang (UMP) internal grant with a university reference number RDU1903148. It has been awarded by the Research and Innovation Department, UMP Malaysia.

References

1. April J, Glover F, Kelly J, Laguna M (2001) Simulation/optimization using" real-world" applications." In: Proceeding of the 2001 Winter Simulation Conference (Cat. No. 01CH37304), vol 1. IEEE, pp 134–138
2. Kennedy J, Eberhart R (1995) Particle swarm optimization. In: Proceedings of ICNN'95—international conference on neural networks, vol 4. IEEE, pp 1942–1948
3. Whitley Darrell (1994) A genetic algorithm tutorial. Stat Comput 4(2):65–85
4. Nasir ANK, Osman Tokhi M, Sayidmarie O, Raja Ismail RMT (2013) A novel adaptive spiral dynamic algorithm for global optimization. In: 2013 13th UK workshop on computational intelligence (UKCI). IEEE, pp 334–341
5. Zhao W, Zhang Z, Wang L (2020) Manta ray foraging optimization: An effective bio-inspired optimizer for engineering applications. Eng Appl Artif Intell 87:103300
6. Kitchen-Wheeler A-M (2010) Visual identification of individual manta ray (Manta alfredi) in the Maldives Islands, Western Indian Ocean. Marine Biol Res 6:351–363
7. Coello Coello CA, Salazar Lechuga M (2002) MOPSO: a proposal for multiple objective particle swarm optimization. In: Proceedings of the 2002 congress on evolutionary computation. CEC'02 (Cat. No. 02TH8600), vol 2. IEEE, pp 1051–1056
8. Razak A, Azwan A, Nasir ANK, Sha'akmal S, Sawal MAR (2018) MOSDA: a proposal for multiple objective spiral dynamics algorithm. J Telecommun Electron Comput Eng (JTEC) 10(1–3):15–19
9. Nasir ANK, Razak A, Azwan A, Ashraf AM, Falfazli MJM (2017) A multi-objective spiral dynamic algorithm and its application for PD design. In: 2017 IEEE 8th control and system graduate research colloquium (ICSGRC). IEEE, pp 1–6
10. De, K, Agrawal A, Pratap A, Meyarivan T (2000) A fast elitist non-dominated sorting genetic algorithm for multi-objective optimization: NSGA-II. In: International conference on parallel problem solving from nature. Springer, Berlin, Heidelberg, pp 849–858
11. Nasir ANK, Raja Ismail RMT, Ahmad MA (2010) Performance comparison between sliding mode control (SMC) and PD-PID controllers for a nonlinear inverted pendulum system. World Acad Sci Eng Technol 71:400–405
12. Zitzler E, Brockhoff D, Thiele L (2007) The hypervolume indicator revisited: on the design of pareto-compliant indicators via weighted integration. In: International conference on evolutionary multi-criterion optimization. Springer, Berlin, Heidelberg, pp 862–876

Review on Effects of Adverse Sonic Environment in Hospital and Control Solutions

Khin Fai Chen

Abstract The range of effects from adverse sonic environment widespread from physical, physiological, and psychological in the form of NIHL, hypertension and anxiety and delirium to name a few. It also has immediate effects that could potentially lead to errors and other long-term impacts. Hospital is a sensitive space with patients with various range of illnesses that are directly affected by the hospital's environment; sonic environment being an important and unavoidable part of it. This paper is written after having reviewed different studies done in medical, industrial, and educational field, related to noise and suggests methods to mitigate and control noise in hospital.

Keywords Hospital noise · Noise hazards · Absorption coefficient · Noise control

1 Introduction

Sound plays an important role in our daily life from the aspect of quietness and sonic comfort. Noise is the opposite of comfort and is regarded as unwanted sound [1]. Noise is known to cause distraction, disturbance, and irritation. Exposure to loud explosive or low-level continuous noise could cause permanent hearing damage which is detrimental to physiological and psychological well-being [2]. Prevalence of noise is defined by the sound pressure level (SPL), reverberation time (RT), and background noise.

SPL is a logarithmic measure of the effective pressure of a sound (P), relative to the threshold of human hearing (P_0), 20 μPa as shown in Eq. (1) and measured in dB. The aim would be reduction of SPL, either by the noise source or reflecting boundary wall.

$$\text{SPL(dB)} = 10 \cdot \log_{10}(P/P_0) \tag{1}$$

K. F. Chen (✉)
Guangzhou Soundbox Acoustic Tech. Co. Ltd, Guangzhou 511447, China
e-mail: khinfai@soundbox.hk

A. F. Ab. Nasir et al. (eds.), *Recent Trends in Mechatronics Towards Industry 4.0*, Lecture Notes in Electrical Engineering 730, https://doi.org/10.1007/978-981-33-4597-3_43

RT is the measurement of time needed for the SPL to drop 60 dB after the event of noise excitation. RT can be estimated using Eq. (2).

$$RT_{60}(s) = 0.161 \cdot V/A \tag{2}$$

where, V is the volume of the air enclosed in the space and A is the total absorption within the space. Higher RT leads to poor communication intelligibility.

Background noise is the indirect active noise that is heard as a part of environment, such as clattering of key on computer or HVAC. Background noise is basically sound pressure level measured as an average over a period of time and is presented as equivalent level L_{eq}, as shown in Eq. (3).

$$L_{eq} = 10 \cdot \log_{10}\left(\frac{1}{T}\int_{0}^{T}\frac{P^2(t)}{P_0}dt\right) \tag{3}$$

T is the total duration of measurement, P(t) and P_0 are instantaneous effective pressure and reference pressure, respectively. Background noise should be as low as possible since it is one of the main contributor to the total noise level.

The World Health Organization (WHO) guideline states that hospitals should maintain an average equivalent noise level, L_{eq} of 30–35 dBA. However, a few studies have found that these guidelines are not maintained and adhered to [3–5]. It is reported that the noise level in hospital has been rising steadily from 1960 to 2005 to the magnitude of 10 dBA which resulted in patients' slow progress of recovery and low effectiveness of medical supplements [6]. The adverse effects of noise are evident on healthcare workers as well and it has been related to response related autonomic stress, which contribute to burnout, job dissatisfaction and increased absenteeism [7].

A research conducted in John Hopkins Hospital showed that there is no location within the hospital facilities that comply with WHO guidelines. It is worth noting that, in majority of surgical departments at John Hopkins Hospital, the SPL_{peak}, as high as 110 dB, was reported [8]. The noise spectra were found to be generally flat over the 63 to 1000 Hz frequency range [9]. This is due to constant occurrence of chatter in hospitals and mobile sources whose spectrum coincide with the observed spectral characteristics of the noise. The spectra curving upwards in at frequencies below 63 Hz was attributed to the air flow in the building. The quietest room had 37 dBA of noise floor during the off-peak hours and 50 dBA during peak hours, despite being vacant. A similar study conducted in Dankook University Hospital, South Korea [4] found the L_{eq} to be 63.6 dBA (peaking at 86.6 dBA). Similar trend is observed in Imam Reza Teaching Hospital, Iran [5] where its pediatric intensive care unit (ICU) and cardiovascular emergency recorded noise level of 65 dBA and 70 dBA, respectively. The quietest location was found to be the critical care unit (CCU) which has a noise level of 55 dBA with level peak at 60 dBA between 10:30 am and 12:00 pm. Further data can be reviewed from Table 1.

Table 1 Shows the L_{eq} of various facilities in Imam Reza Teaching Hospital, Iran

Ward (Time)	0800	0900	1000	1100	1200
General emergency	59	61	63	66	60
Surgical emergency	58	62	65	63	63
Cardiovascular emergency	57	66	63	58	63
CCU1	57	56	56	60	55
CCU2	55	53	52	57	56
General ICU	59	62	59	58	64
NICU	58	89	58	64	60
ICU	60	58	62	60	59
Nephrology	66	65	67	63	66
Transplant unit	48	54	59	62	60

The objective of this paper is thus to address the issue of noise in hospital and healthcare settings and to discuss possible noise control measures in such sensitive environment.

2 Effects of Noise

2.1 Noise Induced Hearing Loss (NIHL)

NIHL is a major concern that comes to mind when considering noise. Being exposed to high intensity impulsive noise usually results in a temporary shift of threshold of hearing. The European Parliament (EP) has set the maximum noise exposure level for workers in the EU member state to 87 dBA over a duration of 8 h period, in their Directive 2003/10/EC. However, research has shown that continuous exposure to noise of 70 dBA over an 8 h period caused a shift in the threshold of hearing temporarily, and raised blood pressure [10]. This shows that, aside from loudness and sudden occurrence of noise, duration is a contributing factor for developing NIHL.

The greater noise risk at hospitals is due to the medical equipment noise which is generally dominant at higher frequencies between 1 and 8 kHz. This is the range where human hearing is the most sensitive as shown by the equal loudness contour, shown in Fig. 1. It clearly shows that even low noise levels between 1 and 6 kHz is enough to be heard as loud as those at lower frequencies because human auditory sensitivity, as there are fewer hair cells that correspond higher frequencies, compared to low frequencies [11]. If damaged, lower redundancy at this range could lead to permanent loss of hearing. Therefore, a prolonged exposure could raise the threshold of hearing at these frequencies thus reducing the ability to understand spoken word.

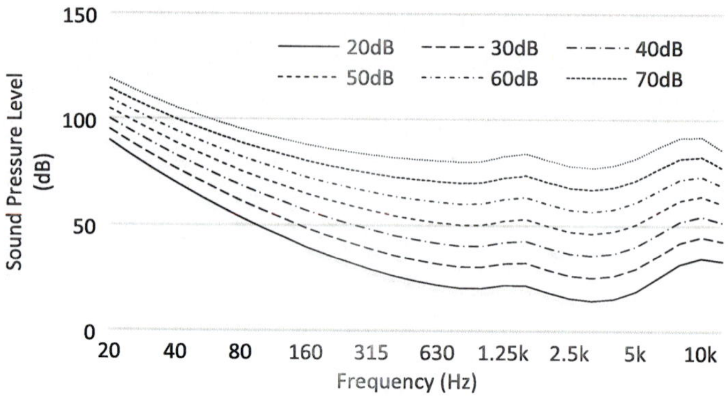

Fig. 1 Equal loudness contour, created using MATLAB code based on formulation given in Sect. 4.1 of ISO226

2.2 *Health Effects and Sleep Disturbance*

Stress reactions to traffic noise were found to be similar to that of stress from occupational environments [12], which lead to increase blood pressure, this is due to development of structural vascular changes in the peripheral resistance vessels which was seen to accelerate due to constant high noise levels exposure of more than 95 dBA [13].

A significant level of risk of coronary heart disease was seen to persist in an 18-year long-term follow-up study, done on industrially employed men who were exposed to industrial noise in Finland [14]. Such noises during sleep, even at low levels, can trigger a stress response which will lead to sleep deprivation, cardiovascular diseases and increases inflammatory response. Patients in post-operative ward experiencing noise tend to require higher dose of sedatives during bed rest [15]. It is also found that there is a causality in sleep deprivation and delirium in ICU patients which exhibit "sleep fragmentation": an increase in light sleep [16]. Noise could trigger anxiety especially in those who have been weakened by a critical illness and under ventilator support [17]. A survey conducted shows that noise is one of the primary disturbances in ICU [18].

2.3 *Speech Interference and Privacy*

Intentional speech, like a conversation or message from public announcement (PA) system are susceptible to interference due to background noise. These disturbances and distortion could degrade the message information as known as signal's intelligibility. When the logarithmic ratio of level of speech signal to noise (SNR) is lower than 15 dB, the intelligibility of the intended speech sound suffers. Therefore, in a noisy environment, with levels above 45–50 dB, people tend to feel the urge to raise their voices, this could result in voice fatigue, mostly in healthcare workers.

In order to openly tell one's problems, patients have to feel comfortable enough which requires a private environment. It is easy to shy away in a noisy place filled with strangers and unable to communicate effectively, out of fear of being overheard when raising their voice. The alternative would be closer communication distance which increases the SPL of speech. However, verbal communication regarding health is more efficient when it is accompanied by body gestures that might provide crucial information of patient's health issues [19]. Therefore, closer discussion proximity increases the risk of physician missing out on these cues.

Overhead paging and PA system announcements suffer in the presence of noise and contribute to noise level in enclosed space. Since hospital design requires having reflective surfaces for their ease of cleaning and sanitation, it allows for sound energy to be entrapped and build up the noise level. Thus, the whole system is cyclic in nature which goes back to affecting audibility of speech in patient physician conversation which could lead to erroneous diagnosis.

When tending to a critical patient, healthcare workers are required to focus on the task. Any distraction, including auditory, could be fatal. There have been a substantial number of health care receivers who have suffered injuries related to treatment of their illnesses [20] indicating errors are not uncommon. Rising stress and occupational burnout in healthcare workers triggered by noise cause subjective disturbance in work, which could potentially lead to errors [21], as measured by Jones's Staff Burnout Scale for Health Professionals ($r = 0.369$, p less than 0.001) and the emotional exhaustion subscale of Maslach's Burnout Inventory ($r = 0.300$, p less than 0.01). These chances of errors could be amplified under the noise level exceeding of 40 dBA. There is already a 10% medication errors rate that arises from confusion of drug names and this rate will only increase when physician intelligibility is interfered [22].

3 Sources of Noise

The most common source of noise disturbance is human: other patients, visitors, and caregivers. Complaints were lodged on snoring, toilet flushing, telephone, television, personal medical devices and disturbances from conversation and procedure of caregivers and visitors. Hospital have a lot of mechanical equipment such as bed carts, operational lifts and wheelchairs which are identified as noise source. Operating this hardware, coupled with human operation introduces human footfall noise. Medical machines such as computerized tomography (CT) scanner, magnetic resonance imaging (MRI) scanner, and high oscillation respirator are vital for their ability to aid in diagnosis and treatment of patients, despite their noise range between 80 and 120 dB [23]. Table 2 show some typical noise level of hospital equipment.

Another source of noise come from building services equipment such as heating, ventilation, and air conditioning (HVAC) system. Development of less noisy HVAC

Table 2 Shows list of different noise sources found in hospitals and their corresponding A-weighted sound level (L_{Aeq})

Source of noise	Level (dB(A))
Items falling onto the floor	Up to 92
Equipment movement (e.g. bed)	90
Connection of gas supply	88
Door closure	85
Pager	84
Talking	75–85
Ventilator alarm	70–85
Nebulizer	80
Telephone	70–80
Television	79
Oximeter	60–80
Monitor alarm	79
Ventilator	60–78
IV infusion alarm	65–77
Endotracheal aspiration unit	50–75

systems has come forth and been used in conjunction with numerous health building guidelines and acoustic standards. However, it is easy to fail to achieve the targeted noise level even with new low noise HVAC systems if the building is overrun which would require the system to perform more than it is designed for. The final type of noise sources would come from external sources as well, such as traffic, construction, and community activities.

4 Noise Controls

The lack of awareness in acoustic engineering and proper coordination during the planning and designing phase of medical facilities and between different functional bodies such as urban planning, architecture, engineering services, construction services and related professional disciplines contribute greatly to haphazard acoustical conditions in hospitals and open spaces [24]. Existing standards and guidelines provided by American National Standard Institution (ANSI), American Society for Testing and Materials (ASTM) and International Organization of Standardization (ISO) regarding materials suitable for such environments, and measurement procedures to be carried out at all stages of constructions, and periodic measurements after completion should be strictly adhered to during the designing and construction phases of medical facilities.

4.1 Human Activities

Human activities are rather preventable with human intervention and proper environmental management. Introduction of quiet time has also shown to improve noise and lessen sleep disturbance to a great extent. Noise levels were generally recorded at 65 dB, attributing to speech sounds, which was also found disruptive by patients in adjoining areas as well [25]. Proper education, planning and implementation of quiet time, where practices like maintaining distance between patient and conversing staff, dimming down lights during night time, and keeping doors closed, have shown to improve noise by up to 50% and patients' hospital stay experience as well.

4.2 Equipment

The medical machines are themselves susceptible to noise. The images from the fMRI scan tend to get distorted when it experiences low frequency (below 100 Hz) vibrations while at work. A proper planning for its location will prove to be a tremendous benefit both ways. Care must be taken to locate such equipment on a separate floor with a good amount of impact insulation. A damping rubber pad for heavier machines to rest on, further reducing the transfer of noise and vibration to and from the machine. For smaller devices, using an enclosure or a noise barrier between the machine and the receiver could reduce the amount of noise at the receiver.

Timely servicing and repairing of patient transportation equipment such as stretchers, wheelchairs, and other transportations equipment like carts, are of vital importance to reduce noise emerging from this hardware which correlates to impact insulation class (IIC). A minimum IIC rating for floor structure is 50 and a recommended rating is 70. An example of the minimally rated floor structure would be an added layer of 6 mm acoustic underlayment, on a concrete deck, which then supports a hard surface on the top. This structure essentially gives an IIC value between 50 and 60 which means that the impact insulation it provides is significant. Adding another floating concrete and using spring isolation hanger to mount the gypsum ceiling of the floor below improves the IIC value by almost additional 20 points.

4.3 Reverberation

Reverberation could mask important speech sound, render it unintelligible and enhance the level of noise. Proper application of sound absorption panel, while considering its cleanability, reduces the amount of lingering energies hence allowing the primary speech sound to be prominent. A reduction in noise level by 5-6 dB and the reverberation time from 0.9 to 0.4 s in the patient room was observed with use of such panels. It shows that for myocardial infarction patients, there is a statistical association between acoustics and patient pulse amplitude, with higher pulse amplitude in worse acoustics. In the literature, pulse amplitude has only been used to a limited extent as a biological stress marker and its differences may influence stroke volume, and if so, this may in part be due to an increased activity of the beta-2 receptors of the sympathetic nervous system during the period of bad acoustics. There was a higher incidence of rehospitalization at both one and three months in the group of bad acoustics compared to that in good acoustics. This difference was statistically significant at 3 months ($p < 0.01$) [26].

Generally, a reverberation time of 0.3-0.5 s reduces the chances of masking, but having shorter reverberation is more ideal. Usage of smooth, hard surfaces as lining at the enclosure boundary contributes to reverberation. For such conditions, acoustical tile and composite panels, for ceiling and wall, made from glass fiber, mineral fiber, plywood, gypsum board etc., are viable solutions. While selecting such product, it is important to consider noise reduction coefficient (NRC) rating or ceiling attenuation coefficient (CAC). A product with NRC rating above 0.9 or CAC rating above 35 is highly recommended. It is beneficial in terms of speech intelligibility as well since it removes any extraneous sound energies that mask important speech sound.

4.4 *Partition and Space Planning*

Providing separate rooms is very effective in giving patients a comfortable hospital stays but such ventures are improbable for masses. Proper planning of placement, orientation of bed and furniture could improve the noise environment, for instance, locating a restroom space near the entrance of the room allows for a noise barrier from the corridor. Installing class A sound absorbers on the ceiling reduced the noise within the room as a result of which EEG arousal of unaware participants sleeping with different sounds being played, from 5.1 to 3.0 [27].

4.5 *Noise Masking*

Utilization of masking noise as background noise at 40 to 60 dBA is a viable option to reduce gap between a base level of noise and intrusive noise so as to reduce disturbance. An ongoing research is the use of static noise and masking noise such as white noise to help with patients' get sleep [28]. Multitudes of devices and mobile apps have been developed based on this, and many sounds occurring in nature to help not just patients but adults having difficulty to fall asleep, as well as that for babies. Usage of masking noise, developed from static noise with spectrum shaped as that of a Long-Term Average Speech Spectrum (LTASS) has been used to block out conversation sounds, both to not disturb patients as well as to mask conversation to be heard unintentionally. For the purpose of the former, this works since it is of continuous nature without many fluctuations so as to cause distraction [29]. It is believed that due to this very property, that raises the base level of noise and causes the fluctuation due to other noise seem less intrusive.

4.6 *Outdoor Noise*

Measures must be taken to prevent outdoor noise from entering the building as well. Collection of noise data during site visits before designing allows for preparation of noise level in future. This should be done in coordination with different functional bodies such as urban planning, architecture, engineering services, construction services and related professional disciplines. Prevention of noise entering from outside could be done in a variety of ways from considering the thickness of the outer facade, usage of noise barriers, acoustic door, and double-glazed windows. The design specifications will not be discussed in this paper.

5 Conclusion

Noise affects the sonic environment in hospital and pose as an adversity for both patients and healthcare workers. There is a wide range of effects due to noise which includes sleep disruption leading to psychological effects, triggering anxiety, healthcare worker burnout leading to absenteeism, increases risk of NIHL leading to hypertension, increases in risk of erroneous diagnosis due to poor sonic environment. Sources like human activities, transportation traffic and HVAC systems, and acoustics of building and room have been known to be the culprits of noise in hospitals. Solutions include planning and designing with close attention to detail and hidden potential causes of noise. Locating noisy medical equipment in isolation from noise sensitive areas with adequate amount of insulation around it, and use of quieter better versions of HVAC system greatly help to reduce noise issues. Strict adherence to standards like ANSI, ASMT, and ISO from design to construction and testing will ensure that the results cohere to guidelines provided by WHO.

References

1. Fink D (2019) A new definition of noise: noise is unwanted and/or harmful sound. Noise is the new 'secondhand smoke'. Proc Mtgs Acoust
2. Kryter KD (1980) Physiological acoustics and health. J Acoust Soc Am 68(10)
3. Busch-Vishniac IJ, West JE, Barnhill C, Hunter T, Orellana D, Chivukula R (2005) Noise levels in John Hopkins hospital. J Acoust Soc Am 48(3629):118
4. Park MJ, Yoo JH, Cho BW, Kim KT, Jeong WC (2014) Noise in hospital rooms and sleep disturbance in hospitalized medical patients. Environ Health Toxicol 29

5. Khademi G, Roudi M, Farhat AS, Shahabian M (2011) Noise pollution in intensive care units and emergency wards. Iran J Otorhinolaryngol 23(4)
6. Konkani A, Oakley B (2012) Noise in hospital intensive care units—a critical review of a critical topic. J Crit Care 27(5):522.e1–522.e9
7. Astrakianakis G, Chow Y, Hodgson M (2014) Noise-indued stress among primary care workers in long term care facilities in British Columbia, Canada. Occupational Envion Med 71(A54)
8. Kracht JM, Busch-Vishniac IJ, West JE (2007) Noise in the operating rooms of John Hopkins Hospital. J Acoust Soc Am 121(5 pt1):2673–2680
9. Kahn D, Cook TE, Carlisle CC, Nelson D, Kramer N, Millman R (1998) Identification and modification of environmental noise in an ICU setting. Chest 114(2):535–540
10. Yamamura K, Itof F (1981) The effecs of variable noise level on man with peak levels below 75 dBA. Eur J Appl Physiol 46(1):69–76
11. Falk SA (1972) Combined effects of noise and ototoxic drugs. Environ Health Persp 5–22
12. Babisch W (2003) Stress hormones in the research on cardiovascular effects of noise. Noise Health 5(18):1–11
13. Fogari R, Zoppi A, Corradi L, Marasi G, Vanasia A, Zanchetti A (2001) Transient but not sustained blood pressure increaments by occupational noise. an ambulatory Blood pressure measurement study. J Hypertension 1020–1027
14. Virkkunen H, Kauppinen T, Tenkanen L (2005) Long-term effect of occupational noise on the risk of coronary heart disease. Scand J Work Environ Health 31(4):291–299
15. Moore M, Nguyen D, Nolan S (1998) Interventions to reduce devibel levels in patient care units. Am Surg 64:894–898
16. Figueroa-Ramos MI, Arroyo-Novoa CM, Lee KA, Padilla G, Puntillo KA (2009) Sleep and delirium in ICU patients: a review of mechanisms and manifestations. Intensive Care Med 35(5):781–795
17. Chlan L, Savik L (2011) Patterns of anxiety in critically ill patients receiving mechanical ventillatory support. Nurs Res 60(3 Suppl):s50–s57
18. Tembo A, Parker V (2009) Factors that impact on sleep in intensive care patients. Intensive Crit Care Nurs 25(6):314–322
19. Woloshynowych M, Davis R, Brown R, Vincent C (2007) Communication patterns in a UK emergency department. Ann Emerg Med 50(4):407–413
20. Leape LL (2015) Patient safety in the Era of healthcare reform. Clin Orthop Relat Res 473(5):1568–1573
21. Topf M, Dillon E (1988) Noise-induced stress as a predictor of burnout in critical care nurses. Heart Lung J Criticl Care 17(5):567–574
22. Aronson JK (2004) Medication errors resulting from the confusion of drug names. Exp Opinion Drug Saf 3(3):167–172
23. Berens R, Weigle C (1995) Clinical investigations in critical care: articles. Chest 108(4):1026–1029
24. Westman J, Walters J (1981) Noise and stress: a comprehensive approach. Environ Health Persp 41:291–309
25. Creating a culture of quiet in hospitals, Simple changes reduce noise levels and improve patient outcomes. Quietyme
26. Hagerman I, Rasmanis G, Blomkvist V, Ulrich R, Eriksen CA, Theorell T (2005) Influence of intensive coronary care acoustics on the quality of care and physiologial state of patients. Int J Cardiol 98(2):267–270
27. Berg S (2001) Impact of reduced reverberation time on sound-induced arousals during sleep. Sleep 24(3):289–292
28. Stanchina ML, Abu-Hijleh M, Chaudhry BK, Carlisle CC, Millman RP (2005) The influence of white noise on sleep in subjects exposed to ICU noise. Sleep Med 6(5):423–428
29. Hughes R (2014) Auditory distraction: a duplex-mechanism account. Psych J 3(1):30–41

Evaluation of the Convolutional Neural Network's Performance in Classifying Steel Strip's Surface Defects

Tan Kai Wen, Nur Safwati Mohd Nor, Teoh Wen Kang, Nor Akmal Fadil, Intan Zaurah Mat Darus, Ahmad Hafizal Mohd Yamin, and Fazila Mohd Zawawi

Abstract Steel strip plays a vital role in many industrial fields. Its defects will impact the manifestation of the product and also reduce the features of the product, resulting in a huge economic loss. Deep learning algorithms, such as Convolutional Neural Network (CNN), have successfully been applied to image classification, while featuring a great level of abstraction and learning capabilities. These features are keys to detect and classify surface defects in a robust and reliable manner. The images used for training and testing the model are obtained from the NEU Surface Defect Database which contains six kinds of typical surface defects of steel strips that are rolled-in-scale, patches, crazing, pitted surface, inclusion and scratches. These images are pre-processing to enhance them and extract some useful information from them. After that, the CNN models are trained and tested with these images to evaluate their performance. The specific hyperparameters for the CNN model which are tuned are number of epochs, batch size, number of convolutional layers, input image size and kernel size. For each hyperparameter, the CNN model is trained and tested several times using different values of that hyperparameter. The training accuracy, testing accuracy and training time are recorded and analyzed. Lastly, the final CNN model with high performance is produced. The final CNN model based on the optimum hyperparameters is produced. It has a very high training accuracy of 95.12% and a fairly high testing accuracy of 85.43%. This paper focused on the application of CNN in the classification of the steel strip's surface defects. The performance of the CNN models with different values of hyperparameters are also evaluated.

Keywords Steel strip · Deep learning algorithms · Convolutional neural network · Surface defects · Performance

T. K. Wen · N. S. M. Nor (✉) · T. W. Kang · N. A. Fadil · I. Z. M. Darus · A. H. M. Yamin · F. M. Zawawi
Faculty of Engineering, School of Mechanical Engineering, Universiti Teknologi Malaysia, Johor Bahru, Johor, Malaysia
e-mail: nursafwati@utm.my

A. F. Ab. Nasir et al. (eds.), *Recent Trends in Mechatronics Towards Industry 4.0*, Lecture Notes in Electrical Engineering 730, https://doi.org/10.1007/978-981-33-4597-3_44

1 Introduction

The hot-rolled steel strip plays a vital role in many industrial fields including aircraft, automobiles, shipbuilding and machinery manufacturing [1]. However, there are some unavoidable defect in hot-rolled steel such as scratch, rolled in scale, crack, edge defect and pits [2]. These defects will affect the appearance of the product and also reduce the features of the product including corrosion resistance, abrasion resistance, and fatigue strength, resulting in a huge economic loss [3]. Therefore, it is vital to detect and classify these surface defects of steel strip to improve its quality. Surface defects are borders that divide a material into several regions that has the same structure but different orientations, resulting in corrosion and coating failure. The corrosion resistance and mechanical properties of a material can be affected by the presence of surface defects. A material can be destroyed by a combination of corrosion and cyclic loading, causing a product to become harmful and decreasing its service life. Early detection of severe surface defects is important in preventing properties from damaging and maintaining the reliability of the structure.

Initially, manual inspection is used to detect surface defect but it is very labor intensive, costly and has less efficiency. Because of the rough industrial settings and human errors, the accuracy of the defect detection is also lower. Hence, big data, deep learning and artificial intelligence play an important part in the automated defect detection system. Deep learning is a type of machine learning techniques that exploit many layers of non-linear information processing for supervised or unsupervised feature extraction and transformation, for pattern analysis and classification [4]. Deep learning algorithms are algorithms that are evolved from artificial intelligence. Deep learning has appeared as a brand-new field of machine learning research since 2006 and it can be used to solve many challenging problems in areas of image processing, speech recognition and signal processing. Data is filtered through multiple layers as shown in Fig. 1, each consecutive layer uses the output from the preceding one to inform its results in deep learning models. Convolutional Neural Networks (CNN) which is a type of deep learning algorithms offer ways to overcome the current constraints in surface defect detection using image processing. CNN have been utilized in image classification successfully, while presenting a great level of abstraction and learning capabilities [5]. These features are keys to detect and classify surface defects in a robust and reliable manner.

In this paper, we focused on the application of CNN in the classification of the steel strip's surface defects. The performance of the CNN models with different values of hyperparameters are also evaluated in the experiment.

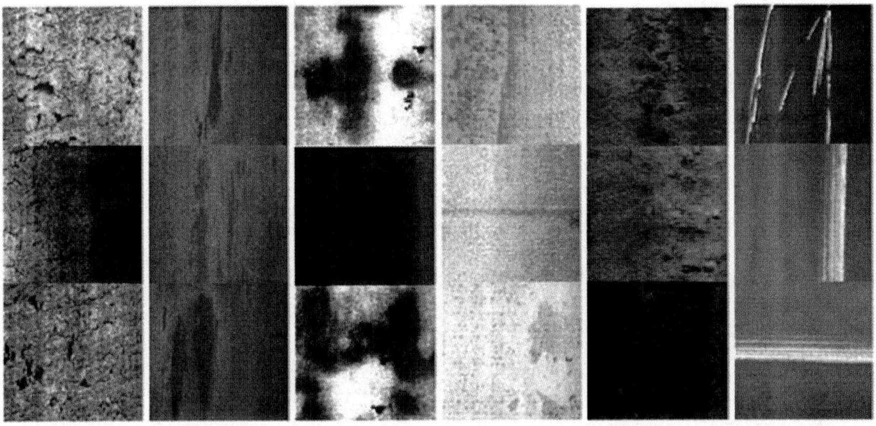

Fig. 1 Images form the NEU surface defect database

2 Related Works

2.1 Artificial Neural Network

Artificial Neural Network (ANN) model solves problems by following the patterns of the human brain. It consists of interconnected units with simple functions but when combining these units, we can build a complex classification function with many layers. Every layer is a set of nodes. The first layer is the input layer and the last layer is the output layer. The layers that exist between the input layer and the output layer are the hidden layers. Layers are a group of nodes and they are connected to the nodes in the following layers. These connections are weighted so they are referred to as weights. Given an input, neural network nodes have outputs, which are real numbers. The output is calculated by applying a function to the outputs of the previous layers. The output of a layer is equal to the input for the next layer. Hence, we can calculate the output of the output layer by calculating the previous layers' outputs consecutively and this process is called inference.

2.2 Convolutional Neural Network

CNN is a deep learning algorithm that can take in an input images and classify them. The pre-processing needed in CNN is much lower compared to other classification algorithms. It is inspired by the human vision and the primary visual cortex in the brain. Y. LeCun introduced the concept of CNN in 1989 [6]. CNN uses the weights sharing method instead of connecting every neuron in each layer to every neuron in the next layer. The weights are connected locally, meaning one node only connects to spatially adjacent nodes in the next layer; the network is considering the spatial structure of the data. CNN consists of various layers including convolutional layer, pooling layer and fully-connected layer.

2.3 *Classification of Surface Defects*

CNN is proved to be efficient in classifying surface defects on steel sheet by learning a discriminative feature vector and a single multi-class classifier of softmax regression simultaneously during the training process [7]. The improved You Only Look Once (YOLO) network can also carry out real time detection of steel strip's surface defects [8]. A method combining the use of discrete shearlet transform (DST) and the gray-level co-occurrence matrix (GLCM) is presented to classify surface defects of hot-rolled steel strips [9]. Few-shot learning with Siamese Neural Network using CNN structure with contrastive loss is also used to classify defects with a small number of steel surface defect images [10]. A high-resolution automated optical inspection (AOI) system based on parallel computing is developed to achieve fast inspection and classification of surface defects [11].

3 Methods

3.1 *Software and Hardware*

Python programming language, version 3.7 is used to write the code and open source library in python such as keras is used during the project. Keras is an open source neural network library that is able to run on top of TensorFlow. The implementation of code for training and testing is done using a laptop equipped with NVIDIA GEFORCE 930M graphic card. The laptop has 12 GB DDR4 2133 RAM and uses Intel Core i5-7200U Processor, 2.5 GHz.

3.2 *Dataset*

The images used for training and testing the model are obtained from the NEU Surface Defect Database. The dataset consists of six types of surface defects of hot-rolled steel strips. 1800 200 $\times$ 200 images are split equally into six categories that are rolled-in-scale, patches, crazing, pitted surface, inclusion and scratches, which means that each category has 300 images. For each of the category, the 300 images are split so that 250 images are used for training while 50 images are used for testing purpose.

3.3 Hyperparameters Optimization

The specific hyperparameters for the CNN model which are studied and tuned in this paper are number of epochs, number of convolutional layers, batch size, input image size and kernel size. For each hyperparameter, the model is trained and tested several times using different values of that particular hyperparameter. The training accuracy, testing accuracy and training time are recorded and analyzed. For the training time, the time for loading and pre-processing the images are included in the training time.

3.4 Structure of the CNN

The CNN model has mainly 5 layers including input layer, convolutional layer, max pooling layer, fully-connected layer and output layer. For the convolutional and the fully-connected layer, the activation function used is the ReLU function. For the output, softmax is used as the activation function. The dimension used for the max pooling is 2×2 while the optimizer of the model is adam. The initial learning rate for the model is 0.01 and the loss function used is categorial cross-entropy.

4 Results and Discussion

4.1 Number of Epochs

From Fig. 2, we can see that the training and testing accuracy of the CNN model increases with the increasing number of epochs. This is because when the number epochs increase, the more times that we pass our entire dataset through our CNN model. The weight of the model is updated more times when the number of epochs is higher. The accuracy gain after the third epochs becomes very small and it can be neglected. there is no significant evident to show the overfitting of model to the dataset because there is no sign of decreasing in the testing accuracy of the model

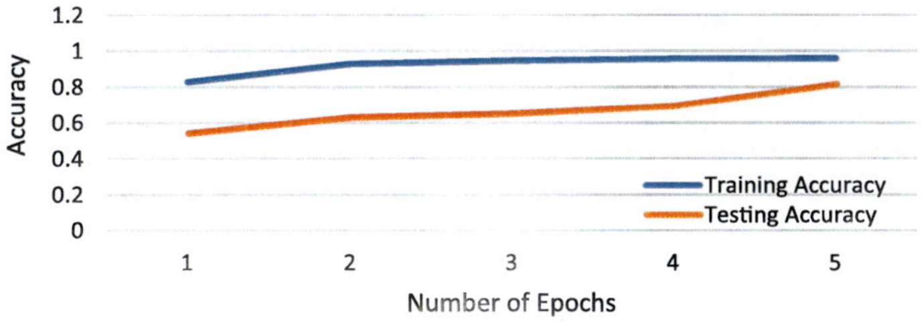

Fig. 2 Graph of accuracy versus number of epochs

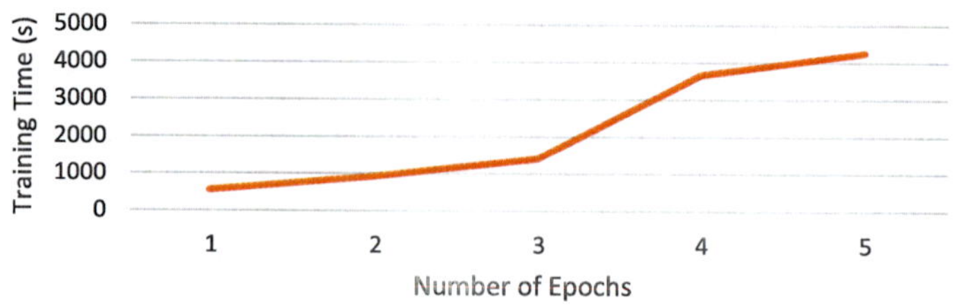

Fig. 3 Graph of training time versus number of epochs

after five epochs. From Fig. 3, the training time increases with the number of epochs but the most significant increase is from the third epoch to the fourth epoch. As the number of epochs increases, it will take a longer time to process as the model go through the whole dataset. Hence, the optimum number of epochs is three as it has a decent training accuracy and testing accuracy. Also, the training time increases rapidly after the third epoch.

4.2 *Number of Convolution Layers*

Based on Fig. 4, the training accuracy is almost constant with the increasing number of convolutional layers. On the other hand, the testing accuracy increases with the increasing number of convolutional layers. The reason for the low accuracy for the model with one convolutional layer is because the one convolutional layer can only extract lower layer features form the images such as gradient orientation and edges. Hence, the model cannot have a wholesome understanding of the images in the dataset. However, with added convolutional layers, it allows the model to capture higher level features which can improve the classification power of the model. As a result, the accuracy of the model will be higher. Based on Fig. 5, the training time for the model with different number of convolutional layers are almost the same. Based on the results, the optimum number of convolutional layers is three because it has the highest training accuracy. Although the model with four convolutional layers has

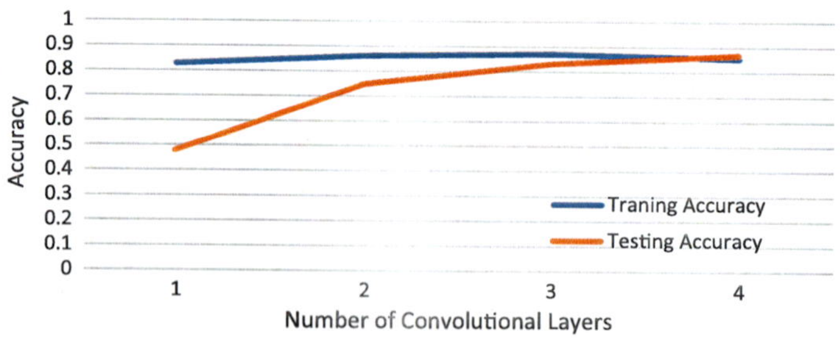

Fig. 4 Graph of accuracy versus number of convolutional layers

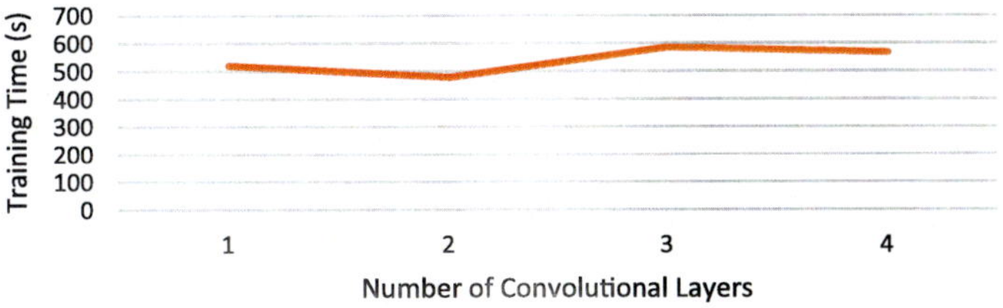

Fig. 5 Graph of training time versus number of convolutional layers

a slightly higher testing accuracy, it is too complex for the classification job. The model that we want is a simple yet effective model.

4.3 Batch Size

From Fig. 6, the training accuracy increases as the batch size increases while the testing accuracy also increases until the batch size of 16 and then drops a bit. A larger batch size can provide a stable estimate of the gradient of the full dataset. Hence, the model can converge to the global optima more easily. From Fig. 7, the training time increases as the batch size increases except a sudden dip at the batch size of 8. While a bigger batch size means fewer total updates in each epoch, it also means that each batch will take more time to process, Also, while making the batch size larger and the total number of batches smaller, it also might result in slower convergence. The optimum value for the batch size is 16 because it has the highest testing accuracy and a decent enough training accuracy. Although the training time is at a fairly high value, but it can be compensated by its high accuracy.

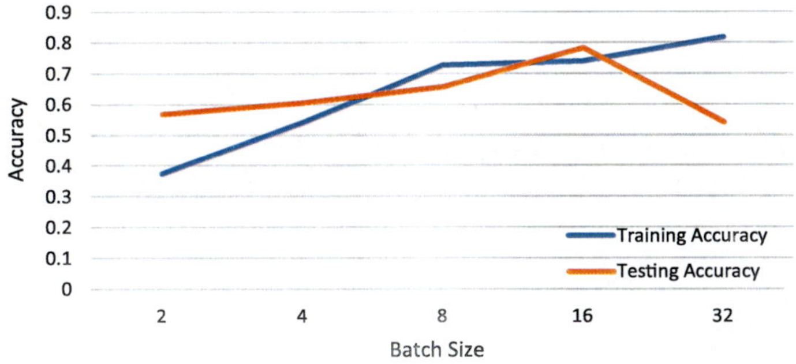

Fig. 6 Graph of accuracy versus batch size

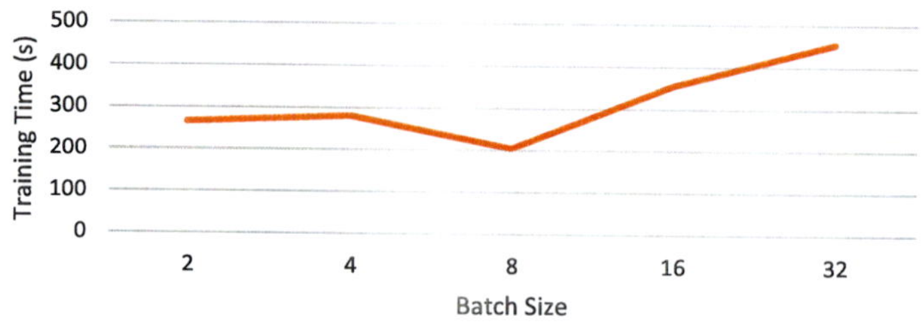

Fig. 7 Graph of training time versus batch size

4.4 Input Image Size

Based on Fig. 8, the training accuracy increases as the size of input image increases while the testing accuracy fluctuates as the size of input image increases. If the input image is too small, the relevant features of the image may be lost and remain undetectable by the model. However, larger input image also has its downside as it only increases the accuracy of the model slightly but increases the training time required greatly. Based on Fig. 9, the training time increases with the size of input image. The amount of training time for the 64 × 64 input image is more than the triple amount of training time for the 32 × 32 input image. This is because the larger

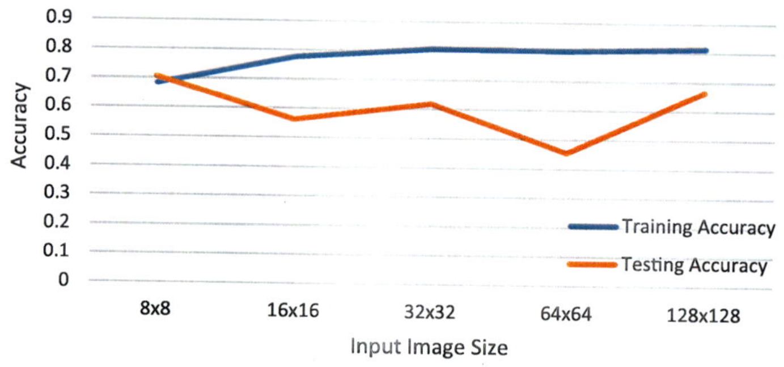

Fig. 8 Graph of accuracy versus input image size

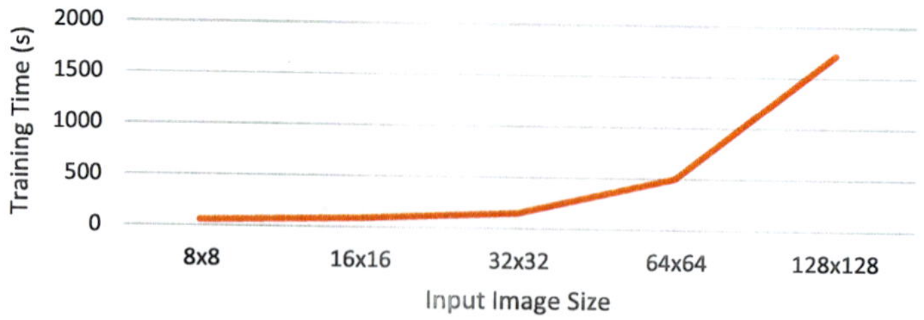

Fig. 9 Graph of training time versus input image size

the input image, the more features the model needs to extract form the image, hence the longer training time. Based on the results, the optimum input image size is 32 × 32. This is mainly because it has a relatively short training time and a fairly high accuracy.

4.5 *Kernel Size*

From Fig. 10, the training accuracy increases first and the drops gradually with increasing kernel size. Meanwhile, the testing accuracy fluctuates with increasing kernel size. The kernel size has a vital role in finding the key features in the image. A larger size kernel can overlook at certain features and skip the essential details in the images. In the other hand, a smaller size kernel can provide too much information for the model, leading to more confusion. Hence, choosing a moderate size kernel is important to achieve a high accuracy in the model. From Fig. 11, the training time also fluctuates with the increasing kernel size. The training times are almost the same for the first four model while the model with 5 × 5 kernel size requires the longest training time. Hence, the optimum kernel size is decided to be 3 × 3 because it has

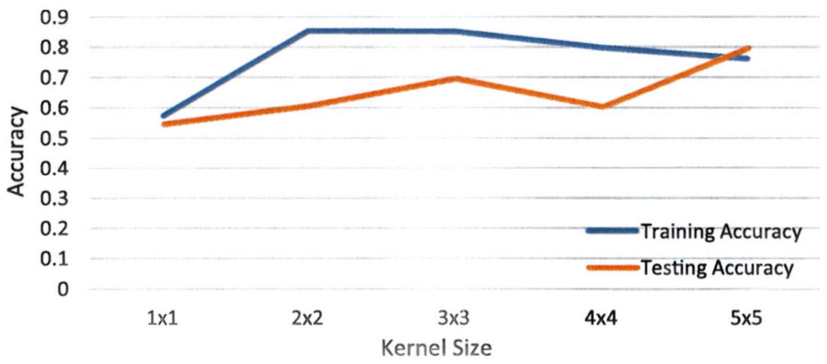

Fig. 10 Graph of accuracy versus kernel size

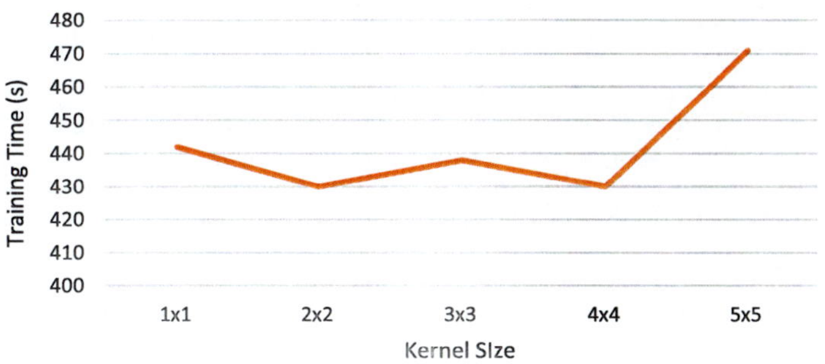

Fig. 11 Graph of training time versus kernel size

Table 1 The optimum value for each hyperparameter

Hyperparameters	Optimum values
Number of epochs	3
Number of convolutional layers	3
Batch size	16
Input image size	32 × 32
Kernel size	3 × 3

relatively high training and testing accuracy. The reason not to choose the 5 × 5 kernel that has the highest testing accuracy is because it has a relatively high training time.

5 Conclusion

In this paper, we use the CNN model to classify different types of steel strip's surface defects. The CNN models are trained and tested with different hyperparameters to obtain the final model with the best performance. The final CNN model based on the optimum hyperparameters has a very high training accuracy of 95.12% and a fairly high testing accuracy of 85.43%. The training time is also at an acceptable value that is 254 s. The values of the hyperparameters for the final CNN model is shown below (Table 1).

References

1. Zhao X, Luo S (2008) Research on system of quality assessing and defects' information managing of strip steel surface. Adv Mater Res 46:173–180
2. Neogi N, Mohanta D, Dutta P (2014) Review of vision-based steel surface inspection systems. Eurasip J Image Video Process 50:1–19
3. Song K, Yan Y (2013) A noise robust method based on completed local binary patterns for hot-rolled steel strip surface defects. Appl Surf Sci 285:858–864
4. Deng L, Yu D (2013) Deep learning: methods and applications. Found Trends Sig Process 7(3–4):197–387
5. da Silva WRL, de Lucena DS (2018) Concrete cracks detection based on deep learning image classification. In: Proceedings vol 2, p 5387. https://doi.org/10.3390/icem18-05387
6. LeCun Y (1989) Generalization and network design strategies. In: Technical Report CRG-TR-89-4, University of Toronto Connectionist Research Group, June 1989. a shorter version was published in Pfeifer, Schreter, Fogelman and Steels (eds) 'Connectionism in perspective', Elsevier 1989
7. Zhou S, Chen Y, Zhang D, Xie J, Zhou Y (2017) Classification of surface defects on steel sheet using convolutional neural networks. Materiali in Tehnologije 51:123–131. https://doi.org/10.17222/mit.2015.335
8. Li J, Geng J, Yin Y (2019) Real-time detection of steel strip surface defects based on improved YOLO detection network. IFAC-PapersOnLine 51:6–81. https://doi.org/10.1016/j.ifacol.2018.09.412

9. Ashour MW, Khalid F, Abdul Halin A et al (2019) Surface defects classification of hot-rolled steel strips using multi-directional shearlet features. Arab J Sci Eng 44:2925–2932
10. Kim MS, Park T, Park P (2019) Classification of steel surface defect using convolutional neural network with few images. In: 2019 12th Asian control conference (Ascc), Kitakyushu-shi, Japan, pp 1398–1401
11. Ye R, Chang M, Pan C-S, Chiang C, Gabayno J (2018) High-resolution optical inspection system for fast detection and classification of surface defects. Int J Optomechatronics 12:1–10

Investigation of Time-Domain and Frequency-Domain Based Features to Classify the EEG Auditory Evoked Potentials (AEPs) Responses

Md. Nahidul Islam, Norizam Sulaiman, Mamunur Rashid, Mahfuzah Mustafa, and MohdShawal Jadin

Abstract The auditory evoked potentials (AEPs) are a kind of electroencephalographic (EEG) signal that is produced by an acoustic stimulus from the region of the brain. The people who are unable to maintain the verbal communication and behavioral response through the sound stimulation, EEG based brain-computer interface (BCI) technology could be an effective alternative to rehabilitate their hearing ability. In this paper, the AEP responses of three distinct English words namely bed, please and sad have been recognized. The EEG features in terms of Fast Fourier Transform (FFT), power spectral density (PSD), spectral centroids, standard deviation, Log energy entropy, mean, skewness, kurtosis has been selected as a feature extraction method. Support Vector Machine (SVM), Linear discriminant analysis (LDA) and K-Nearest Neighbors (K-NN) have been employed to classify the extracted features. Among all these features, power spectral density with SVM classification has achieved the best accuracy. Different performance measures were evaluated to identify the best set of features as well as model. The best classification accuracy was demonstrated by the developed SVM model was observed as 82.86% which clearly indicates that the method provides a very encouraging performance for detecting the AEPs responses.

Keywords EEG · BCI · Auditory evoked potential · SVM · Machine learning

Md. N. Islam (✉) · N. Sulaiman · M. Rashid · M. Mustafa · M. Jadin
Faculty of Electrical and Electronics Engineering Technology, Universiti Malaysia Pahang, Pekan, Pahang 26600, Malaysia
e-mail: nahidul76.edu@gmail.com

N. Sulaiman
e-mail: norizam@ump.edu.my

M. Rashid
e-mail: mamun110218@gmail.com

M. Mustafa
e-mail: mahfuzah@ump.edu.my

M. Jadin
e-mail: mohdshawal@ump.edu.my

A. F. Ab. Nasir et al. (eds.), *Recent Trends in Mechatronics Towards Industry 4.0*, Lecture Notes in Electrical Engineering 730, https://doi.org/10.1007/978-981-33-4597-3_45

1 Introduction

The brain-computer (BCI) interface is a direct communication mechanism that enables the human brain to connect to external devices by using digital computers. To monitor this activity, most BCIs rely on electrical measures of brain activity and sensors located over the head. To record electrical activity from the scalp with electrodes, Electroencephalography (EEG) is widely used. A typical BCI system consists of four phases: signal acquisition, feature extraction, classification, and output translation [1]. Figure 1 shows the complete BCI system with all components.

For better communication and navigation, hearing is very essential for our life. AEP is a sort of EEG signal produced by the brain when auditory stimuli are transmitted. AEP signal consists of positive or negative reproducible peaks, length, amplitude, and behavioral correlation. AEPs produces significantly lower amplitudes as compared to EEG [2].

Early researchers developed various techniques for detecting auditory brain-system response (ABR) peak voltage (V) using spectral energy [3], matched filter [4], spectral analysis [5], and wavelet analysis [6]. In [7], consonant recognition tests have shown that hearers with hearing impairments are much more consonant than listeners with normal hearing. It has also been reported in [8], that listeners with normal hearing usually need less contrast between spectral peaks and valleys than listeners with hearing impairments to reliably distinguish vowels. Lirong et al. [9] proposed a method for the combined analysis of sMRI and fMRI image data and the local features were detected and specified using the Invariant Feature Transform (SIFT) algorithm. The accuracy of sMRI and fMRI with SVM classifier is 72% and 74%, respectively. Mosqueda et al. [10] established an automated system for the detection and classification of hearing loss conditions. The absolute values of brain auditory evoked potential (BAEP) features such as amplitudes and latencies are extracted and used for classifying multilayer perceptron (MLP).

In this paper, the AEP signals have been analyzed using different time and frequency-domain features, and the extracted features have been classified by the SVM, LDA, K-NN classifiers. This paper was structured in the following sections

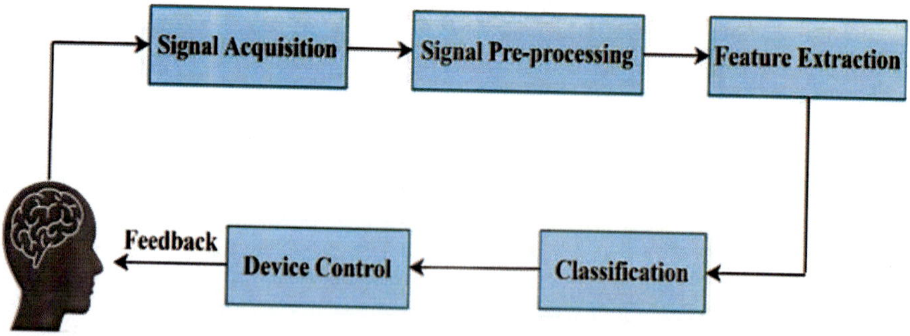

Fig. 1 Complete BCI system

i.e. Sects. 2, 3, and 4 discusses issues related to methodology, classification & results and discussion respectively; finally, Sect. 5 deals with the conclusion.

2 Methodology

In order to make any BCI system, some fundamental components that need to be followed and these steps are data collection, pre-processing, feature extraction, classification and device control. Feature extraction method plays a vital role among these steps. BCI researchers have developed various types of feature extraction methods into the BCI system. Figure 2 Represents a full methodology flow chart.

In this study FFT, PSD, log energy entropy, spectral centroid, standard deviation, mean, kurtosis, and skewness have been investigated and classified with SVM, K-NN, and LDA classifier.

2.1 Details About Experimental Data

The auditory evoked potentiality is a response to auditory stimuli through electrical activity. The auditory stimuli data has been collected from online (Centre de recherche du Centre hospitalier universitaire de Sherbrooke (CRCHUS), Sherbrooke (Quebec), Canada), and preprocessed the data [11]. Experiments were carried out on eight healthy subjects without any declared hearing impairment. The EEG recordings were conducted using the Version I & II actiCAP active electrode device (Brain Products GmbH, Germany), which comprises 64Ag/AgCl electrodes. With the help of Brain-Amp MR amplifiers the signal was amplified and captured by means of Vision Recorder software. The electrodes were placed with a traditional 10–20 layout. The experiments were carried out on eight normal subjects with no reported hearing loss.

Figure 3 shows the individual 64 channels location on the human scalp. The subjects received audio signals via earphones while EEGs were recorded. Each trial was randomly repeated at least 80 times during the experiment. In each trial, a stimulus was provided at random that lasted about 9 s.

2.2 Data Preprocessing

The dataset consists of EEG recordings that were collected while subjects listened to various sound: a, i u, bed, please, sad. Signals were recorded on eight healthy participants. For that reason, we only work with three-classes (bed, please, sad) data and the rest of the data are eliminated. The data were resampled at 500 Hz and band-pass filter (between 0.1 and 45 Hz) have been used to extract the relevant frequency

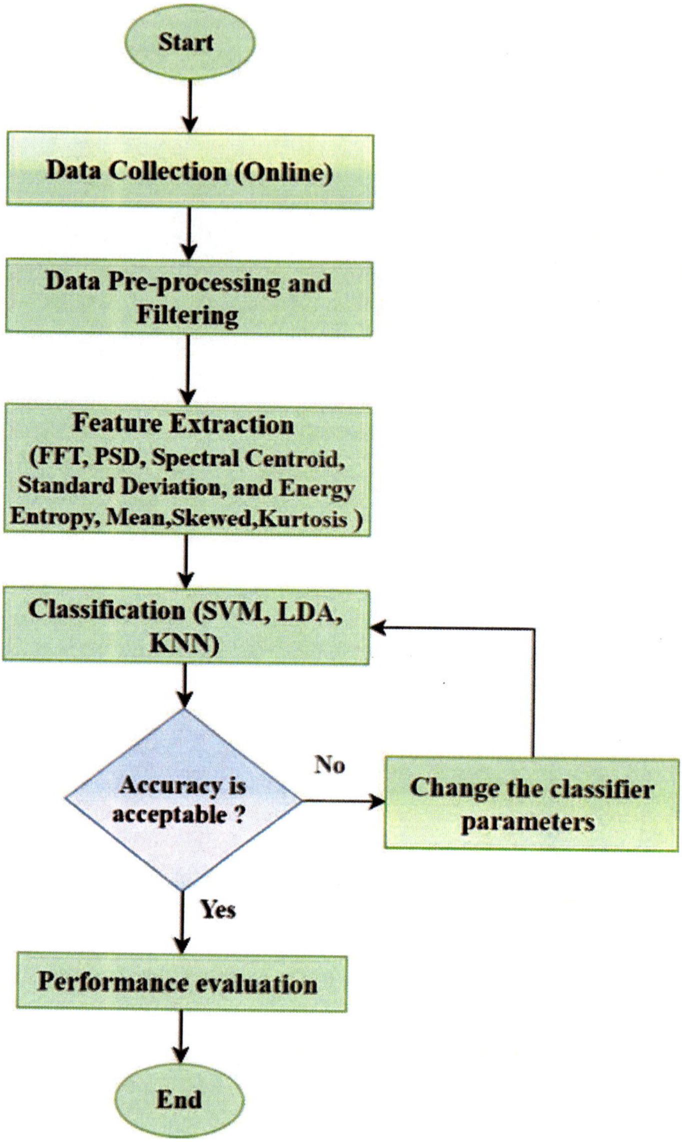

Fig. 2 Methodological flow chart

bands for this analysis. The trail has been removed when signal amplitude exceeded 75 V. After that the baseline has been cleared and replaced bad channels by an average of its neighbors and 525 epochs have been obtained. Finally achieved 64 channels data and several feature extraction methods have been used to extract the features. The AEP signal data of subject-1, channel-1 in time domain have been shown in

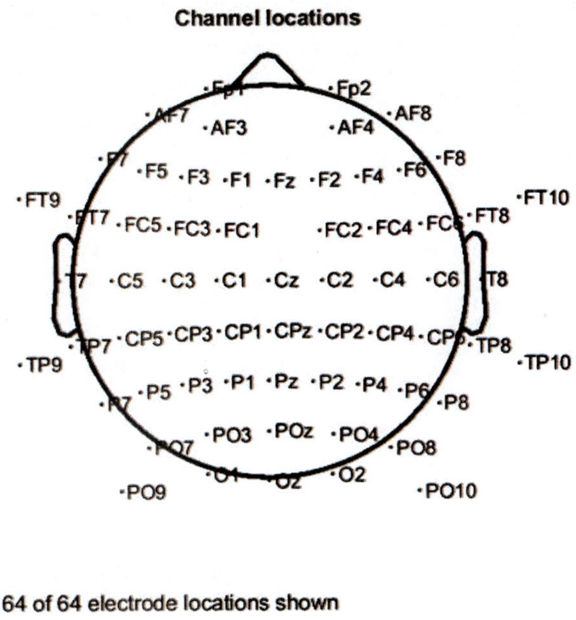

Fig. 3 Location of individual 64 channels

Fig. 4 when the subject hears the 3-types of sound defined as three labels. The other 63 channels data can be plotted in the same way.

2.3 Feature Extraction

Extraction of features plays a critical role in any EEG-based BCI program, because any incorrect collection of EEG features would cause the devices to receive incorrect commands. So, selecting the best feature is very important to obtain the best output from the devices. In this research, different frequency-domain (FT) and time-domain (TD) features including FFT, PSD, Spectral Centroids, Log energy entropy, standard deviation, mean, skewness and kurtosis have been extracted [12–14]. All the feature extraction method has been applied in our study and among the entire feature the most suitable feature has been selected (Table 1).

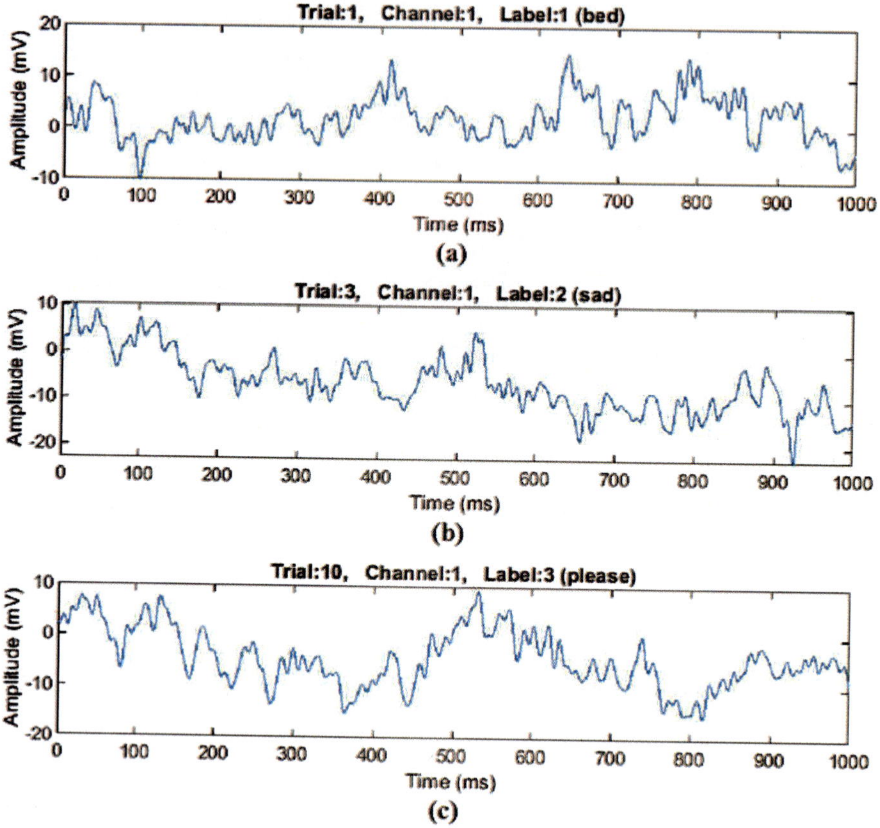

Fig. 4 Plotting of AEP signals in time-domain for subject-1, Channel-1, and All Labels

3 Classification

3.1 *Support Vector Machine (SVM)*

SVM was first introduced in 1992 by Boser, Guyon, and Vapnik in COLT-92. SVMs are a set of associated supervised learning methods used to classify and regress [11]. SVM is a classification and regression prediction tool which utilizes machine learning theory to optimize predictive accuracy, while automatically preventing data overfitting. SVM is a discriminative classifier formally defined by the hyperplane separation. The SVM algorithm uses information as input and outputs a line that, where possible, distinguishes these groups. The marginal vectors called support vectors defined this optimal hyperplane, such as the SVM allows linear decision limits to be labelled as linear SVM. This classifier has always been effectively employed in a fairly disproportionate number of synchronous BCI issues [15]. However nonlinear boundaries for decisions can be set using a kernel function. In BCI research the widely used

Table 1 The equation of all the features

Features	Formula
FFT	$f(x) = \sum_{n=1n=1}^{N-1}\left(\mathrm{X(n)}W_N^{kn}\right)$ where, $W_N = e^{j\frac{2\pi}{N}}$
PSD	$PSD = \lvert\mathrm{X(K)}\rvert^2 = \left\lvert\sum_{n=0}^{N-1} X(nTs)e^{-j\frac{2n\pi k}{N}}\right\rvert^2$
Spectral centroids	$C = \frac{\int xg(x)dx}{\int g(x)dx}$
Log energy entropy	$H_{\log En} = -\sum_{i=0}^{N-1}\left(\log_2(P_i(x))\right)$
Standard deviation	$SD = \sqrt{\frac{1}{N-1}\sum_{i=1i=1}^{N}(x_i - \mu)^2}$ Where, μ = mean
Mean	$Mean = \frac{1}{N}\sum_{i=1}^{N} x(i)$
Skewness	$Skewness = \frac{\frac{1}{N}\sum_{i=1}^{N}(x_i-\mu)^3}{\left(\frac{1}{N}\sum_{i=1}^{N}(x_i-\mu)^2\right)^{\frac{3}{2}}}$
Kurtosis	$Kurtosis = \frac{\frac{1}{N}\sum_{i=1}^{N}(x_i-\mu)^4}{\left(\frac{1}{N}\sum_{i=1}^{N}(x_i-\mu)^2\right)^2}$

Gaussian or Radial Base Function (RBF) kernel is familiarized as Gaussian SVM or RBF SVM. According to [16], The general output of the SVM classification algorithm can be calculated using the following expression:

$$y = \sum_{i=1}^{N}\left(\alpha_i y_i k\left(x_i, x_j\right) + b\right) \tag{1}$$

Here $\alpha_i \geq 0$ are Lagrangian multipliers obtained by solving a quadratic optimization problem, b is the bias, and the kernel function is $k(x_i, x_j)$. In this study, the most popular kernel function namely Gaussian RBF function is used and given by

$$k\left(x_i, x_j\right) = \exp\left(\frac{-\|x_i - x_j\|^2}{2б^2}\right) \tag{2}$$

where б a parameter specified by the user which shows the width of the function of the kernel.

3.2 Linear Discriminant Analysis (LDA)

LDA is deployed to find linear combinations of feature vectors that define the corresponding signal characteristics. LDA attempts to differentiate between two or more sets of different object or event types. To accomplish this task, it uses hyperplanes. A projection is obtained to maximize the distance between class media, whilst the interclass variance is reduced to minimum the isolating hyperplane. This technique involves very little computation and is easy to use. LDA was subsequently widely used in BCI systems such as P300 speller, BCI-based motor imagery and BCI multiclass [17].

3.3 K-Nearest Neighbors (K-NN)

KNN is a case-based learning method and in order to classification, it retains all the training data. In this method it is important to assign the most nearby neighbors in the training to an unknown point in the dominant class [18]. K-NN can approximate any function that empowers it to define nonlinear decision boundaries with a sufficiently high value of k and enough training samples. K in KNN is a parameter referring to the number of nearest neighbors to be used in most voting processes. In this study, the value of K was 10 and distance metric was Euclidean.

3.4 The Performance Measure

The performance of the developed models are Eqs. (3)–(8) [19, 20].

$$Accuracy = \frac{TP + TN}{TP + TN + FP + FN} \times 100\% \quad (3)$$

where TP is true positive and FN is false negative, TN is true negative, and FP is false positive.

$$\Pr ecision = \frac{TP}{TP + FP} \times 100\% \quad (4)$$

$$Sensitivity = \frac{TP}{TP + FN} \times 100\% \quad (5)$$

$$Specificity = \frac{TN}{TN + FP} \times 100\% \quad (6)$$

$$\text{Recall} = \frac{TP}{TP + FN} \quad (7)$$

$$F1 = \frac{1 * Precision * Recall}{Precision * Recall} \quad (8)$$

4 Results and Discussion

This section represented the performance of the proposed method based on EEG data from auditory stimulation. The entire AEP data has been divided into training and test sets using k-fold cross-validation to estimate the efficiency of our approach. The training set has been used to determine the classification model and the test set has been used to evaluation. This study has established 10-fold cross-validation, which is a popular evaluation process. After training and testing the entire dataset with SVM classifier, the performance of the model has been evaluated. With testing dataset, the confusion matrix has been calculated to appraise the performance of the proposed technique where class 1, class 2 and class 3 represent bed, sad and please respectively. Figure 5a shows the confusion matrix whereas Fig. 5b represented the true positive rate (TPR) and false negative rate (FNR).

It could be seen from the confusion matrix that the TPR for class one, two, and three are 91.3%, 87%, and 70.5% respectively whereas the FNR for class one, two, and three are 8.7%, 13%, and 29.5% respectively presented in Fig. 5b. In the proposed approach testing accuracy, sensitivity, specificity, precision, recall, and F1 score has been evaluated. These values have been computed using Eqs. (3)–(8).

True Class	Confusion Matrix			
	1	157	11	4
	2	16	154	7
	3	29	23	124
		1	2	3
		Predicted Class		

(a)

True Class	True Positive and False Negative Rate					
1	91.3%	6.40%	2.30%		91.30%	8.70%
2	9%	87%	4%		87%	13.00%
3	16.50%	13%	70.50%		70.50%	29.50%
	1	2	3		TPR	FNR
	Predicted Class					

(b)

Fig. 5 **a** Confusion matrix. **b** True positive and false negative rate for data

Table 2 Three class AEP signal accuracy with different feature extraction methods

Total class	Feature	Feature type	Classifier	Accuracy (%)
3	Combine (mean, skewness, kurtosis)	TD	SVM	63.25
			K-NN	67.5
			LDA	59.7
3	Standard deviation	TD	LDA	70
			K-NN	68
			SVM	75
3	FFT	FD	K-NN	64
			LDA	60.3
			K-NN	55.6
3	Spectral centroid	FD	K-NN	55.6
			SVM	67.35
			LDA	60
3	Log energy entropy	FD	K-NN	59.5
			SVM	70.5
			LDA	65
3	**PSD**	**FD**	K-NN	75.3
			SVM	**82.86**
			LDA	68.5

The testing accuracy, sensitivity, specificity, precision, recall, and F1 measure of the proposed SVM model are 82.86%, 91.28%, 78.75%, 67.67%, 91.28%, and 77.72% respectively.

We also analyzed the different types of features with multiple classification models to determine the best model for the dataset. Every feature has been classified individually after feature selection. In order to classify these feature, SVM, K-NN and LDA classifier have been used in the proposed method. The several types of EEG feature extraction methods have been analyzed from filtered AEP signal with multiple classification models. It is evident from Table 2 that the performance of the proposed PSD feature with the SVM model is better compared to other classifiers evaluated with the AEP data which is highlighted in bold font.

5 Conclusion

Different types AEP signals together with different feature selection as well as classifiers have been investigated in the present study. The best classification accuracy has been achieved with power spectral density using SVM classifier. Thus, the result of the classification will be applied to the BCI application translation algorithm, such as better communication of the hearing loss patent t. Future work will also include the use of data collected over longer periods to validate the significance of those observations statistically.

Acknowledgements The author would like to acknowledge Universiti Malaysia Pahang for providing FRGS to support this research via RDU190109.

References

1. Bashashati A, Fatourechi M, Ward RK, Birch GE (2007) A survey of signal processing algorithms in brain-computer interfaces based on electrical brain signals. J Neural Eng 4. https://doi.org/10.1088/1741-2560/4/2/R03
2. Plourde G (2006) Auditory evoked potentials. Best Pract Res Clin Anaesthesiol 20:129–139. https://doi.org/10.1016/j.bpa.2005.07.012
3. Grönfors T (1993) Peak identification of auditory brainstem responses with multifilters and attributed automaton. Comput Methods Programs Biomed 40:83–87. https://doi.org/10.1016/0169-2607(93)90002-3
4. Boston JR (1989) Automated interpretation of brainstem auditory evoked potentials: a prototype system. IEEE Trans Biomed Eng 36:528–532. https://doi.org/10.1109/10.24254
5. Vannier E, Adam O, Motsch JF (2002) Objective detection of brainstem auditory evoked potentials with a priori information from higher presentation levels. Artif Intell Med 25:283–301. https://doi.org/10.1016/S0933-3657(02)00029-5
6. Popescu M, Papadimitriou S, Karamitsos D, Bezerianos A (1999) Adaptive denoising and multiscale detection of the V wave in brainstem auditory evoked potentials. Audiol Neuro-Otology 4:38–50. https://doi.org/10.1159/000013818
7. Phatak S, Allen J (2006) Consonant and vowel confusions in speech-weighted noise. In: INTERSPEECH 2006 9th international conference on spoken language processing. INTERSPEECH 2006—ICSLP, vol 3, pp 1515–1518. https://doi.org/10.1121/1.2642397
8. Billings CJ, Bennett KO, Molis MR, Leek MR (2012) Cortical encoding of signals in noise: effects of. 32:53–60. https://doi.org/10.1097/AUD.0b013e3181ec5c46.CORTICAL
9. Tan L, Chen Y, Maloney TC, Caré MM, Holland SK, Lu LJ (2013) Combined analysis of sMRI and fMRI imaging data provides accurate disease markers for hearing impairment. NeuroImage Clin 3:416–428. https://doi.org/10.1016/j.nicl.2013.09.008
10. Mosqueda Cárdenas E, de la Rosa Gutiérrez JP, Aguilar Lobo LM, Ochoa Ruiz G (2019) Automatic detection and classification of hearing loss conditions using an artificial neural network approach. In: Lecture notes in computer science (including subseries lecture notes in artificial intelligence and lecture notes in bioinformatics), Springer Verlag, pp. 227–237. https://doi.org/10.1007/978-3-030-21077-9_21
11. Moinnereau M-A, Whittingstall K, Plourde E (2019) Electroencephalogram (EEG) recordings obtained when simultaneously presenting audio stimulations. IEEE Dataport. http://dx.doi.org/10.21227/e90n-sa08
12. Goyal D, Choudhary A, Pabla BS, Dhami SS (2020) Support vector machines based non-contact fault diagnosis system for bearings. J Intell Manuf 31:1275–1289. https://doi.org/10.1007/s10845-019-01511-x
13. Rashid M, Sulaiman N, Mustafa M, Khatun S, Bari BS, Hasan MJ, Al-Fakih NMMA (2020) Investigating the possibility of brain actuated mobile robot through single-channel EEG headset. Lect Notes Electr Eng 632:579–590. https://doi.org/10.1007/978-981-15-2317-5_49
14. AydIn S, Saraoğlu HM, Kara S (2009) Log energy entropy-based EEG classification with multilayer neural networks in seizure. Ann Biomed Eng 37:2626–2630. https://doi.org/10.1007/s10439-009-9795-x

15. Hortal E, Iáñez E, Úbeda A, Planelles D, Costa Á, Azorín JM (2014) Selection of the best mental tasks for a SVM-based BCI system. In: Proceedings of IEEE International conference on systems, man, and cybernetics. pp 1483–1488 (2014-January). https://doi.org/10.1109/smc.2014.6974125
16. Rashid M, Sulaiman N, Mustafa M, Bari BS, Sadeque MG, Hasan MJ (2020) Wink based facial expression classification using machine learning approach. SN Appl Sci 2. https://doi.org/10.1007/s42452-020-1963-5
17. Cui G, Zhao Q, Cao J, Cichocki A (2014) Hybrid-BCI: classification of auditory and visual related potentials. In: 2014 joint 7th international conference on soft computing and intelligent systems, SCIS 2014 and 15th international symposium on advanced intelligent systems, ISIS 2014, pp 297–300. Institute of Electrical and Electronics Engineers Inc. https://doi.org/10.1109/SCIS-ISIS.2014.7044768
18. Bose R, Khasnobish A, Bhaduri S, Tibarewala DN (2016) Performance analysis of left and right lower limb movement classification from EEG. In: 3rd international conference on signal processing and integrated networks, SPIN 2016, pp 174–179. https://doi.org/10.1109/SPIN.2016.7566683
19. Rashid M, Sulaiman N, Mustafa M, Jadin MS, Najib MS, Bari BS, Khatun S (2020) Analysis of EEG features for brain computer interface application. Lect Notes Electr Eng 632:529–540. https://doi.org/10.1007/978-981-15-2317-5_45
20. Zhu W, Zeng N, Wang N (2010) Sensitivity, specificity, accuracy, associated confidence interval and ROC analysis with practical SAS® implementations. Northeast SAS Users Gr. 2010 health care and life science, pp 1–9

Suppression of Tremors in Parkinson's Patients Using a Dynamic Vibration Absorber

Naheed Rihan, Asan G. A. Muthalif, Thaer M. I. Syam, and N. H. Diyana Nordin

Abstract Parkinson's disease is a central nervous system disorder, whose effects and symptoms are evident in the physical movement and appearance of the patient. The frequency range of 2–10 Hz is mostly where the balanced muscles movement of the patients happens. This paper seeks to create a wearable device that is capable of suppressing these tremors within a specific frequency range by examining the system with and without dynamic vibration absorber (DVA) is done for three main positions, normal, short, and long positions; so that the stiffness and the natural frequency changes. The dynamic vibration absorber, tremor suppression will be achieved. The dynamic vibration absorber is set to be a spring-mass system. From the findings of the experimental studies realizing such a device is possible and that varying the absorber parameters of the device can achieve different levels and degrees of tremor suppression. The proposed DVA showed more than 60% of the vibration suppression. The proposed method is reliable for vibrations suppression for tremors of Parkinson's disease and it is simple without using multiple devices so that it is more comfortable for patients.

Keywords Parkinson's disease · Tremor suppression · Vibration absorber · Acceleration amplitude

1 Introduction

One of the most neurodegenerative disorders that affect human arm movement is Parkinson's disease [1]. In 1817, the first person who argued this disease was Dr James Parkinson in "An Essay on the Shaking Palsy" [2]. Shaking and trembling are the primary initial symptoms of this disease. Involuntary movements of muscles are

N. Rihan · A. G. A. Muthalif (✉)
Smart Structures, Systems and Control Research Lab (S3CRL), Department of Mechatronics Engineering, Kulliyyah of Engineering, International Islamic University, Kuala Lumpur, Malaysia
e-mail: drasan@qu.edu.qa

T. M. I. Syam · N. H. Diyana Nordin
Department of Mechanical and Industrial Engineering, College of Engineering, Qatar University, Doha, Qatar

A. F. Ab. Nasir et al. (eds.), *Recent Trends in Mechatronics Towards Industry 4.0*, Lecture Notes in Electrical Engineering 730, https://doi.org/10.1007/978-981-33-4597-3_46

obstacles in the daily activities for patients suffering from tremors. There are two main types of tremors; action tremor (AT) and resting tremor (RT). Action tremor occurs during voluntary movements, [3] mentioned that the frequency range for Parkinson's disease tremors is in the range of 2–10 Hz.

In recent time, Parkinson's disease treatment is still not found [4]. However, surgeries and medications may slow down the progress rates, which lead to disability. The benefits of the drug called L-Dope used as a remedy for this disease gradually reduce [5]. There are other medical treatments for Parkinson's disease through deep brain stimulation (DBS), drugs, and pallidotomy. However, patients are vulnerable to the side effects of such medications, and the dosage of the cure needs to be increased gradually as the patient' body adapt to ordinary dose [6, 7].

Researchers have developed some devices that may assist Parkinson's disease patients which are; however, costly and usually function-oriented. For instance, [8, 9] invented a spoon to help patients for self-feeding [10, 11]. Modelled and developed a mobility device to enhance patients' walking gait. [12] used a closed-loop FES for out of phase activation of muscles linked to tremorgenic. Therefore, they are neutralising the tremor's effect. Tremors of 2–5 Hz were successfully mitigated. [13] invented a device to combat a disease. [14] suppressed essential tremors when a person holds a tool. [15] have designed an intelligent glove using the gyro effect to suppress vibrations due to resting tremors. This paper aims to develop a wearable device capable of absorbing the vibrations produced by these tremors regardless of the magnitude and frequency of the tremor (within a specific range) in three different orientations of the spring; which depends on the number of coils, to help the patient with their disease.

This article is outlined as follows; Sect. 1 introduces the concept of the Parkinson's disease and its main symptoms as well as reviewing different strategies in how to reduce its harmful effects. Section 2 shows the idea of the DVA which is used in work, and the derivation of the mathematical model of the system is discussed. Section 3 discusses the methodology used and shows the experimental setup. Also, it presents the experimental results obtained. Finally, this paper ends with Sect. 4 which concludes the main findings and mentions the contribution of this work.

2 Vibration Suppression of Tremors Utilizing Self-tuneable DVA

2.1 Dynamic Vibration Absorber (DVA)

DVA is effective when the resonant frequency of absorber and the primary system is harmonized. Unlike tuned mass damper (TMD), DVA must have the ability to change its frequency to match the primary system's natural frequency, either by varying mass or stiffness. Hence, a device that is capable of adjusting its natural frequency (absorber) to match tremor frequency experienced by the patient is needed for an effective DVA.

2.2 *Mathematical Modelling of the DVA*

This study has acquired the values for the mass, spring stiffness and damping factor of the human arm. The DVA and human arm were modelled as a 2-DOF mass-spring-damper system, as shown in Fig. 1. Without DVA, the patient's arm modelled as:

$$F(t) = m_1 x''(t) + b_1 x'(t) + k_1 x(t) \tag{1}$$

With DVA, the Eq. (1) become (without external force):

$$m_2 y''(t) + b_2\left[y'(t) - x'(t)\right] + k_2[y(t) - x(t)] = 0 \tag{1}$$

where m_1 is the mass of the arm, b_1 is the damping coefficient of the arm, k_1 is the arm's stiffness, m_2 is the DVA's mass, b_2 is the damping coefficient of m_2 due to m_1, and k_2 is the stiffness of m_2 due to m_1.

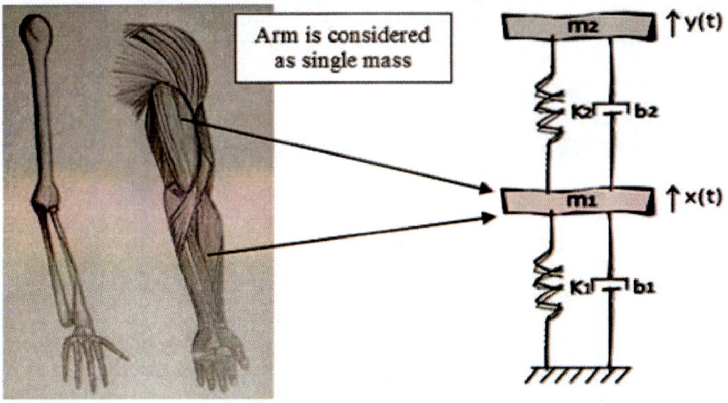

Fig. 1 Simplified 2-DOF spring-mass-damper system with a comparison to human arm [16]

3 Self-tunable DVA-Experimental Investigation

3.1 Experimental Setup

The setup used for the study is shown in Fig. 2 which comprises of the dynamic signal analyzer (DSA), an accelerometer, an exciter amplifier, an inertial shaker and a mannequin hand wearing the DVA system which includes of a semi-circular part with a spring. The accelerometer captured vibration signals which were analyzed by the DSA. DSA also provided an excitation signal for the shaker. The excitation frequency for the model is set within 2–12 Hz. DSA measured accelerometer reading to the input excitation signal (g/V). Transfer function plot with and without DVA showed the effect of the absorber at sweeping frequencies. The parameters used in designing the DVA are listed in Table 1.

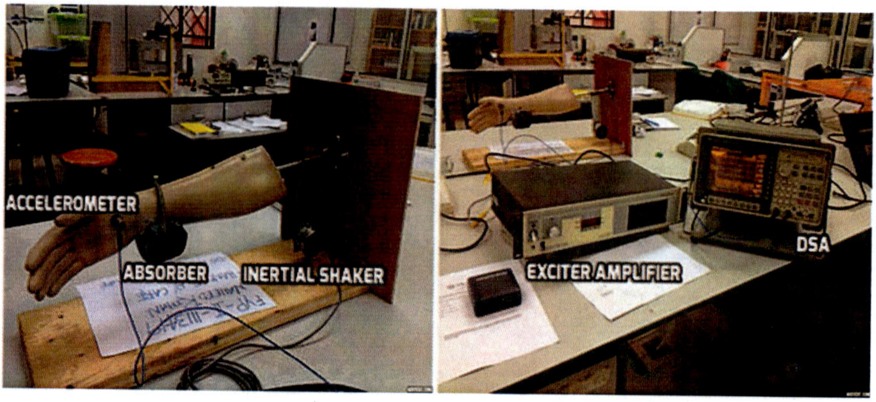

Fig. 2 Experimental setup

Table 1 Parameters of the DVA and mannequin hand

Parameters	Value
Mass of mannequin hand	0.53 kg
Length of the mannequin hand	0.12 m
Mass of absorber	0.07 kg
Spring inner diameter	1 mm
Spring outer diameter	7 mm
Spring stiffness	22.15 N/m

3.2 Experimental Results and Discussion

Stiffness of the DVA is varied by adjusting the length of spring, i.e. effective number of spring coils. An experiment of the system with and without DVA is done for three main positions, normal, short, and long positions. The position is identified by stretching or compression the spring so that stiffness is varying and the natural frequency so that it is matched for vibration suppression for Parkinson's disease which is 2–10 Hz [3]. Since the exact spring was not obtained, the spring length used for the tests was varied in its position such that different numbers of turns of coils were involved in the test. By showing it could be shown that for different turns of coils, the absorber will absorb different tremors corresponding to those turns of coils. For the normal position of the spring, it is defined when there is neither stretching nor compression on the spring. The orientation of these positions is shown in Fig. 3. The experiment was conducted for Parkinson's disease tremors range 2–10 Hz.

Figure 4 shows FRF with and without DVA at spring normal position. The transfer function is the ratio of the accelerometer reading to the input sine sweep (g/V) the excitation frequency. Amplitude reduction is seen from 6.5 to 8.6 Hz, which is within the tremor range. A drop of more than 60% of the amplitude can be seen.

Figure 5 shows the FRF of the system with and without DVA at spring shorter position. The tremor absorption has taken place between the ranges of 6.2 to 10 Hz, and from 10.4 to 12 Hz. Increase in the natural frequency due to increase if stiffness in shorter. Approximately 20% of the suppression of vibration can be observed.

Figure 6 shows the FRF of the system with and without DVA at spring longer position. Tremor absorption between 6.2 and 10.4 Hz is seen in Fig. 6. More than 50% of the maximum amplitude is suppressed in this position.

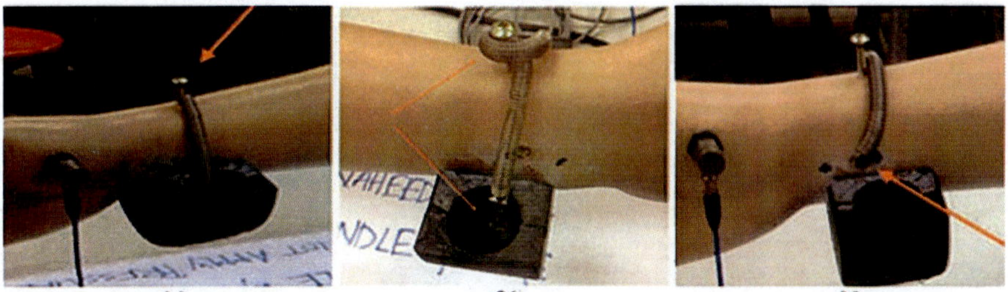

(a) (b) (c)

Fig. 3 The orientation of the spring for difference cases; **a** normal position, **b** shorter position and **c** longer position

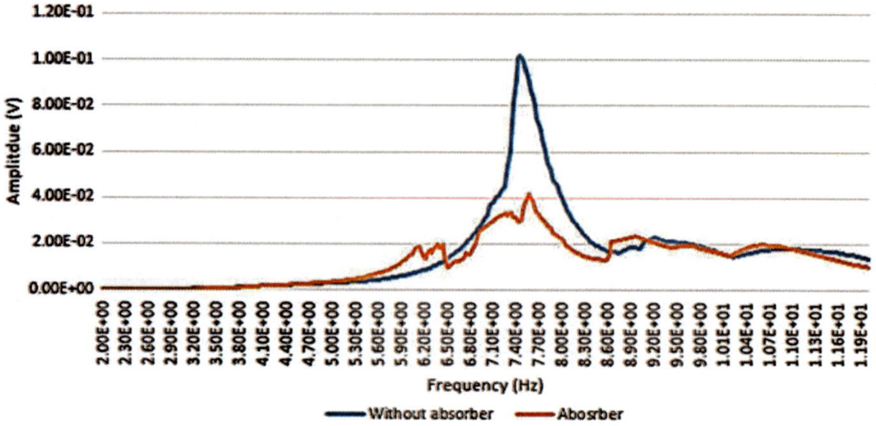

Fig. 4 Vibration amplitude with and without absorber for frequency sweep of 2–10 Hz at normal position of the spring

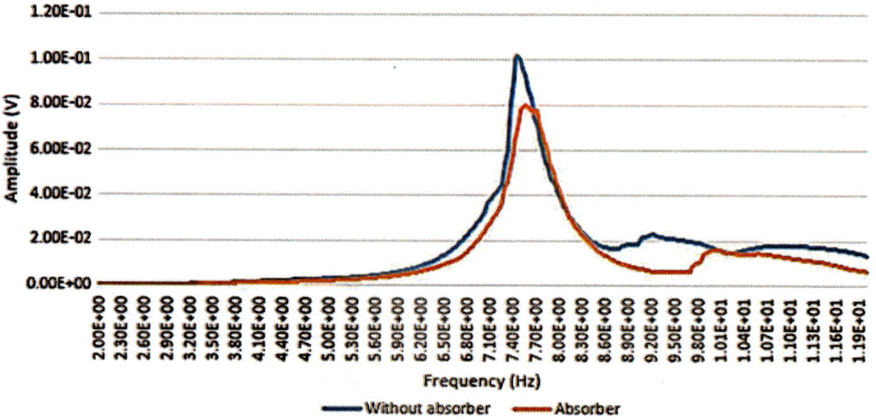

Fig. 5 Vibration amplitude with and without absorber for frequency sweep of 2–10 Hz at a shorter position of the spring

4 Conclusion

The tunable DVA described in this paper is suitable to reduce tremors for Parkinson's disease patients that occurs at a frequency of 2–10 Hz. Three positions of the spring are considered, normal, short and long positions were the number of coils of the spring is adjusted. The FRF is obtained and plotted for the excitation frequency which is a swept sine at 2–10 Hz. Reduction of more than 50% was achieved with DVA for normal and long positions where 20% reduction for the short position. The

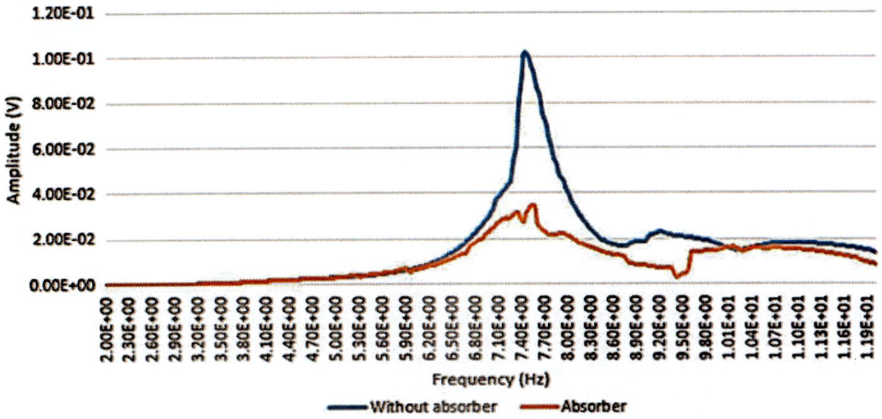

Fig. 6 Vibration amplitude with and without absorber for frequency sweep of 2–10 Hz at longer position of the spring

proposed method is reliable for vibrations suppression for tremors of Parkinson's disease and it is simple without using multiple devices so that it is more comfortable for patients.

References

1. Anon. n.d. Hall, W. D.: 'Hand-Held Gyroscopic Device', US Patent 5562707 Google Scholar (1996)
2. Bächlin M, Plotnik M, Roggen D, Maidan I, Hausdorff JM, Giladi N, Tröster G (2010) Wearable assistant for Parkinsons disease patients with the freezing of gait symptom. IEEE Trans Inf Technol Biomed 14(2):436–446
3. Björklund LM, Rosario S-P, Chung S, Andersson T, Chen IYC, McNaught KSP, Brownell AL, Jenkins BG, Wahlestedt C, Kim KS, Isacson O (2002) Embryonic stem cells develop into functional dopaminergic neurons after transplantation in a Parkinson rat model. Proc Natl Acad Sci U S A 99(4):2344–49 (2002)
4. Camara C, Isasi P, Warwick K, Ruiz V, Aziz T, Stein J, Bakštein E (2015) Resting tremor classification and detection in Parkinson's disease patients. Biomed Sig Process Control 16:88–97
5. Hashemi SM, Golnaraghi MF, Patla AE (2004) Tuned vibration absorber for suppression of rest tremor in Parkinson's disease. Med Biol Eng Compu 42(1):61–70
6. Hermann RP, Phalangas AC, Mahoney RM, Alexander MA (1999) Powered feeding devices: an evaluation of three models. Arch Phys Med Rehabil 80(10):1237–1242
7. Kiguchi K, Hayashi Y (2013) Upper-Limb tremor suppression with a 7DOF exoskeleton power-assist robot. In: Proceedings of the annual international conference of the IEEE engineering in medicine and biology society, EMBS, pp 6679–82
8. Ladipo I, Muthalif AGA, Rozaime A, Asyraf A (2011) Design and development of an active mass damper for broadband vibration control. In: 2011 4th international conference on mechatronics: integrated engineering for industrial and societal development, ICOM'11—conference proceedings (2011)

9. Levine CB, Fahrbach KR, Siderowf AD, Estok RP, Ludensky VM, Ross SD (2003) Diagnosis and treatment of Parkinson's disease: a systematic review of the literature. Evid Rep/Technol Assess (Summary) 57:1–4
10. Olgac N, Huang C (2000) New method for multiple frequency vibration absorption. In Proceedings of the American control conference, vol 3, pp 2097–2101. IEEE
11. Palacios-Sánchez L, Torres Nupan M, Botero-Meneses JS (2017) James Parkinson and his essay on 'Shaking Palsy', two hundred years later. Arquivos de Neuro-Psiquiatria 75(9):671–72
12. Prochazka A, Elek J, Javidan M (1992) Attenuation of pathological tremors by functional electrical stimulation I: method. Ann Biomed Eng 20(2):205–224
13. Smith SL, Shannon K (1997) Vector-based analysis of motor activities in patients with Parkinson's disease. In: Conference proceedings of the EUROMICRO, pp 50–55
14. Teixeira CJ, Bicho E, Rocha LA, Gago MF (2013) A self-tunable dynamic vibration absorber: Parkinson's disease's tremor suppression. In: 3rd Portuguese bioengineering meeting, ENBENG 2013—book of proceedings
15. Zulkefli AM, Muthalif AGA, Nordin DNH, Syam TMI (2019) Intelligent glove for suppression of resting tremor in Parkinson's disease. Vibroengineering Procedia 29

Design of Condition Monitoring System on Stairlift Based on Internet of Things (IoT) for Physical Data Acquisition Using Multi Sensors

Rohmat Setiawan, Danardono A. Sumarsono, and Wahyu Sulistiyo

Abstract This paper discusses the monitoring system on the stairlift using the internet of things, where the system is embedded in the stairlift physical using sensors installed in the stairlift component and then integrated into the computer through the internet of things. Through the acquisition of multi-sensor information, a lot of information can be accessed such as vibration, speed, load passenger, motor temperature, and power consumption. With a lot of information that can be monitored in real-time, it can create centralized monitoring, and failure of stairlift operations can be prevented as early as possible through the early warning system (EWS). Also, this system can provide technical support in maintenance systems and stairlift safety regulations. Based on the analysis of monitoring results obtained, the prototype stairlift is feasible to be developed for industrial scale, operationally it meets ASME A18.1 and ISO 2372 standards. This is indicated in the variation of passenger loads for stairlifts up to a maximum of 150 kg obtained an average maximum speed 0.2 m/s, maximum vibration velocity (RMS) of 4.5 mm/s, the maximum motor temperature of 75 °C and maximum power consumption of 600 W.

Keywords Stairlift · Monitoring system · Internet of things · Early warning system (EWS)

R. Setiawan (✉) · D. A. Sumarsono · W. Sulistiyo
Department of Mechanical Engineering, Universitas Indonesia, 16424 Depok, Jawa Barat, Indonesia
e-mail: rohmat.setiawan77@gmail.com

D. A. Sumarsono
e-mail: dasumarsono@gmail.com

R. Setiawan
UPT of Manufacturing, Politeknik Manufaktur Astra, 14330 Jakarta Utara, DKI Jakarta, Indonesia

W. Sulistiyo
B2TKS, BPPT 15314 Tangerang Selatan, Banten, Indonesia

A. F. Ab. Nasir et al. (eds.), *Recent Trends in Mechatronics Towards Industry 4.0*, Lecture Notes in Electrical Engineering 730, https://doi.org/10.1007/978-981-33-4597-3_47

1 Introduction

The development of facilities in buildings to complement human needs in moving places is growing, including elevators, escalators, and travelators [1]. In addition to the stairlift which is specifically intended for people with disabilities, this has been explained in the ASME-A18.1 standard [2]. Technological advances make people with disabilities play roles independently and more productively in society. The stairs in the building are a barrier for those who have limited mobility. A stairlift is one of the means of transportation or facilities to help facilitate the mobility of people with disabilities, this facility is installed on the stairs in the building. Because the stairlift is specifically for people with disabilities, it is very important to be equipped with safety, smoothness, and comfort features in the stairlift operation [3].

One way to ensure the safety of transportation equipment is with good maintenance management [4]. Likewise, on the stairlift, there is a need for a comprehensive maintenance system in order to be able to detect early in the event of an operational failure. To support the running of a good maintenance system, this study proposed a monitoring system of parameters that supports technical instruments to facilitate decision making in maintenance and action in the event of a failure in operations [5].

With the rapid development in modern sensors and network technology, the internet of things (IoT) technology has been widely used in monitoring systems for building transportation, such as elevators [6]. IoT is one of the most sophisticated signal sensing technologies and can gather very good detailed information data and network platforms, and can ensure real-time and reliable data transmission. This monitoring system will acquire technical information data in real-time, including vibration [7], speed [8], passenger load [9], motor temperature [10] and power consumption [11]. Data information will be integrated with computers through the internet of things [12].

As the development of system technology, the development of IoT in terms of data acquisition, is increasingly advanced, including using microcontroller nodeMCU ESP32 [13]. This ESP32 is used as a technical instrument for data acquisition from sensors installed [14] on the stairlift device, then the data is communicated to the ThingSpeak cloud platform IoT so that data can be retrieved for further analysis of the condition of the stairlift [15].

2 Methods

2.1 Internet of Things on the Stairlift

The Internet of Things on stairlift is used for real-time monitoring systems and as a safety system for users. If this IoT has been applied in the stairlift then it can realize intelligent stairlift management [16]. The IoT architecture diagram on the stairlift

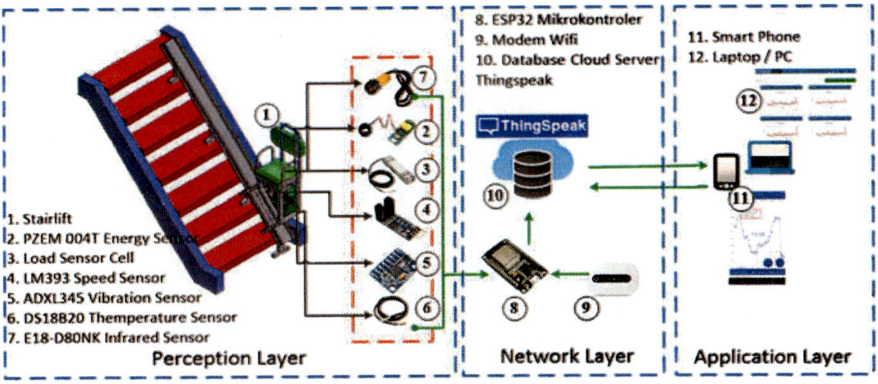

Fig. 1 The architecture diagram of stairlift Internet of Things [17]

(see Fig. 1). The following will explain the process flow monitoring system on the stairlift based on the internet of things:

1. Multi-sensors mounted on the stairlift to retrieve physical data, the sensors installed are as follows:
   (a) Power Meter: to measure the amount of electricity consumed by the stairlift during operation.
   (b) Load Cell Sensor: to determine the passenger load, so that it can be detected if it exceeds the specified passenger load standard.
   (c) Speed Sensor: to find out the speed of the stairlift during operation, so it is easier to know when the speed decreases.
   (d) Vibration Sensor: to find out the mechanical vibrations in the stairlift, to make it easier to detect abnormal mechanical vibrations.
   (e) Temperature Sensor: to determine the temperature of the motor when operating, so it is easier to detect excessive temperatures in the motor.
2. The multi-sensor is then connected to the ESP32 microcontroller, this ESP32 interacts with the environment controlled via I/O. The ESP32 input receives signals from the installed multi-sensor, then is communicated to the thingspeak cloud database via a wifi module installed on the ESP32 microcontroller [14].
3. Thingspeak cloud database is a data storage located in the cloud/internet. This database can be accessed by clients from cloud services that are distributed to users via the internet by cloud providers. The data stored in the cloud will be retrieved and processed through a platform that is available at ThingsPeak which is accessed from a mobile device or Personal Computer (PC) [15].
4. Web Service Monitoring (thingspeak) [15] is a software system designed to support being able to interact between systems on a network. Web service stores information data in XML/CSV/JSON format, so that this data can be accessed by other systems even though different platforms and operating systems. For communication using HTTP. In this Web Service Monitoring will later process

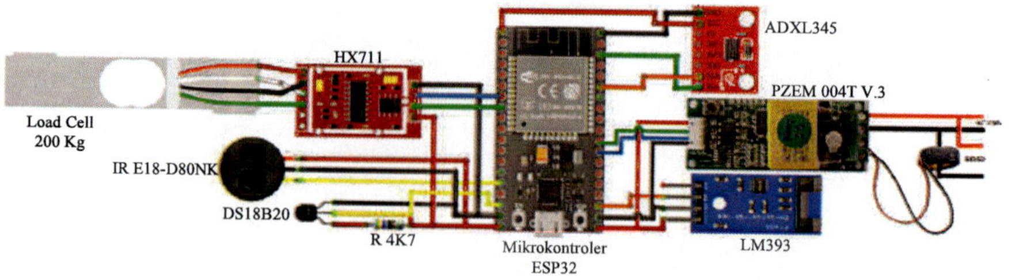

Fig. 2 The circuit schematic ESP32 microcontroller with multi sensors

existing physical data to be displayed in graphical form of motor speed, temperature, passenger load, power consumption and mechanical vibration over time, so that the data can be presented in real-time.

3 Result and Discussion

3.1 Implementation of Prototype Multi Sensor with ESP32 Microcontroller (Perception Layer)

In this study, as many as six sensors are connected to the nodeMCU ESP32 microcontroller with power supply 5 V DC used is utilizing a 5000 mAh power bank. The design of six sensors and nodeMCU ESP32 microcontroller is designed based on the movement of the staircase lift up and down. As mentioned earlier, in this study the ADXL345 vibration sensor, LM393 speed, DS18B20 temperature, PZEM004T V.3 energy consumption, load cell, and E18-D80NK usage cycle where the outputs of these sensors are digital data. Figure 2 shows the circuit schematics of multiple sensors connected to the ESP32 microcontroller module board.

After testing the prototype multi-sensor shows that the multi-sensors can work well by displaying data readings from the six sensors that are desired. This test has also been performed on each sensor node router. Figure 3 shows the installation process of installing a multi-sensor prototype on a stairlift.

3.2 Implementation of Design Physical Data Transmission Multi Sensors Using Wifi Modem (Network Layer)

After ensuring the multi-sensor prototype can work properly and can be displayed on the serial monitor, the next step is to conduct data acquisition wirelessly, in this case, using a WiFi modem, the detected value data on each sensor is sent to the thingspeak cloud server by node-MCU ESP32 when getting a wifi network. Next,

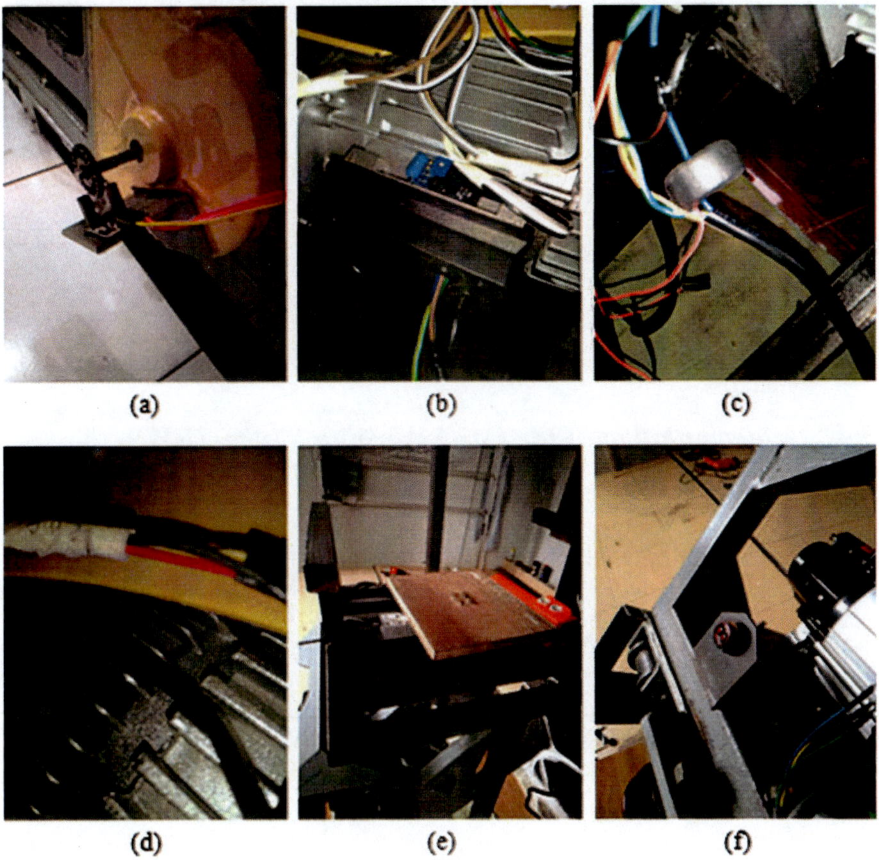

Fig. 3 Installation of multi-sensor installation on stairlift: **a** LM393 speed sensor **b** vibration sensor ADXL345 **c** current sensor PZEM004T **d** temperature sensor DS18B20 **e** weighing sensor load cell 200 kg **f** duty cycle sensor IR E18-D80NK

it is processed by a thingspeak cloud server and displayed in the form of a web application. Figure 4 is the module of installing ESP32 node-MCU and wifi modem which is then communicated with the thingspeak platform.

3.3 Implementation of Interface Based on Platform Thingspeak (Application Layer)

At this stage the implementation of software interface based on a web application that is implemented to display data in. Figure 5 is a web service display of the ThingsSpeak platform which presents physical data acquisition on the stairlift with real-time.

Fig. 4 Installing a wifi modem with an ESP32 microcontroller

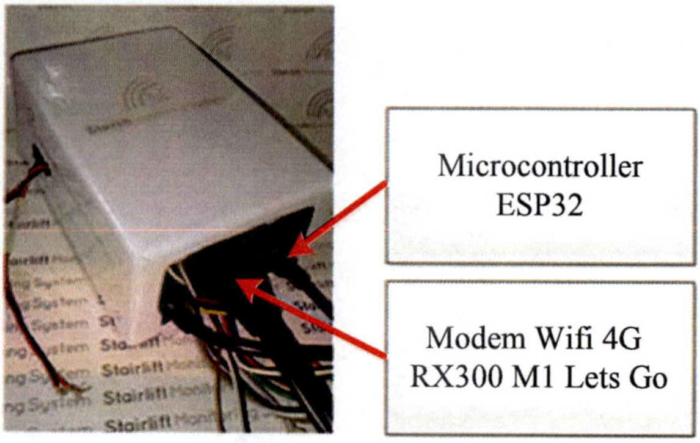

3.4 Implementation of the Early Warning System (EWS) via Telegram

At this stage an early warning system (EWS) is implemented in the form of notifications if conditions occur outside the predetermined standard, the notification is sent using the Telegram software/application to the PC/smartphone. This is done so that the user or technician can get data early if there is an operation failure on the stairlift. The following is a detailed table of predetermined size standards (Table 1).

Based on the above table, then if the operational stairlift exceeds the limits specified, then thingspeak will send a notification to the telegram as early as possible. Figure 6 is a notification implementation via telegram.

3.5 Discussion of Physical Data Acquisition Results on Stairlift

In this discussion to see the effect of passenger load (0, 50, 100, 150 kg) and the operating time of the stairlift to the motor temperature, vibration, speed, and power consumption on the stairlift. This was done in order to evaluate the prototype stairlift that was being made. The data displayed is the average of physical data acquisition from operational stairlift, which is then displayed in graphical form. The following are detailed chart data based on multiple sensors mounted on the stairlift.

1. Motor temperature

Based on the Fig. 7, it appears that the higher the weight of the passenger, the faster the motor temperature will rise. This happens because the motor will work harder so a high electric current is needed to increase the power consumption of the motor. In the Fig. 7, it can be seen that the highest average motor temperature is at a passenger weight of 150 kg which is 52.4 °C, and the lowest is without passengers which is 33.2 °C.

Fig. 5 Display realtime data on the thingspeak platform

Table 1 Standard operating safety measures

Properties	Standard size	References
Speed	0.2 m/s	[2, 8]
Motor temperature	75 °C	[18]
Vibration (RMS)	4.5 mm/s	[7, 19]
Power consumption	600 W	[7]
Load passenger	115 kg	[2]

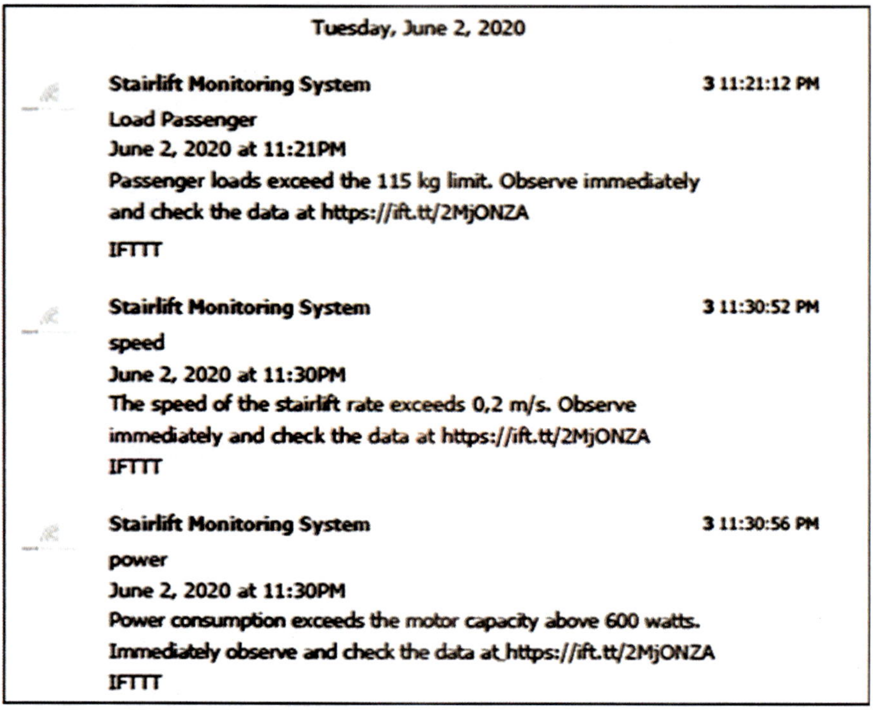

Fig. 6 Early warning system (EWS) implementation via telegram

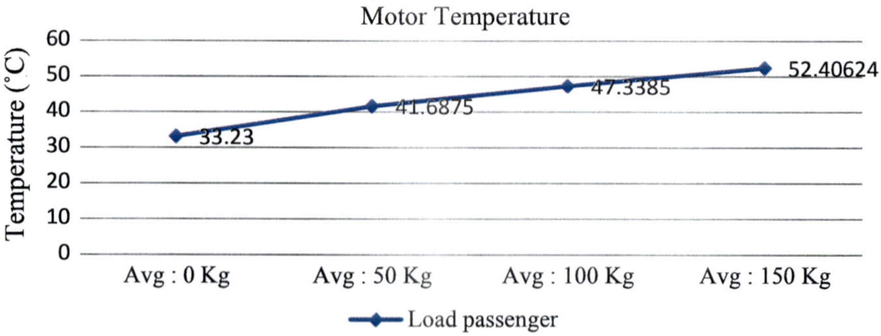

Fig. 7 Results of the physical data acquisition of the motor temperature

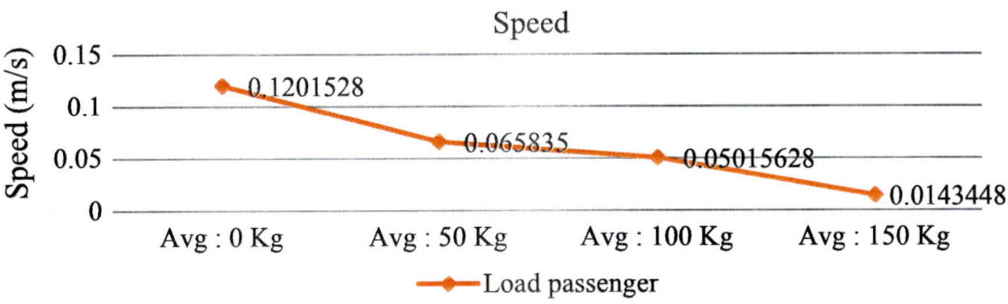

Fig. 8 Results of the physical data acquisition of the speed

2. Speed of the Stairlift

Based on the Fig. 8, it appears that the higher the weight of the passenger, the slower the speed of the stairlift. This happens because of the frictional force on the stairlift wheel is getting bigger and the wire rope is also getting heavier in rubbing against the roller. In the Fig. 8, it can be seen that the highest speed of the stairlift is without passengers which is 0.12 m/s, and the lowest is at the weight of 150 kg passengers which is 0.014 m/s.

3. Power consumption

Based on the Fig. 9, it appears that the higher the weight of the passenger, the higher the power consumption will be. This happens because the motor will work harder so a high electric current is needed to increase the power consumption of the motor. In the Fig. 9, it can be seen that the highest average power consumption is at 150 kg passenger weight is 551.38 W, and the lowest is at no passenger is 307.6 W.

4. Vibration

Based on the Fig. 10, it appears that the higher the weight of the passenger, the lower the vibration produced. This happens because the movement speed of the stairlift is getting slower and tends to be stable because of the weight of the passenger. In the Fig. 10 it can be seen that the highest average vibration is without passengers which is 0.96 mm/s, and the lowest is at passenger weight 150 kg which is 0.04 mm/s.

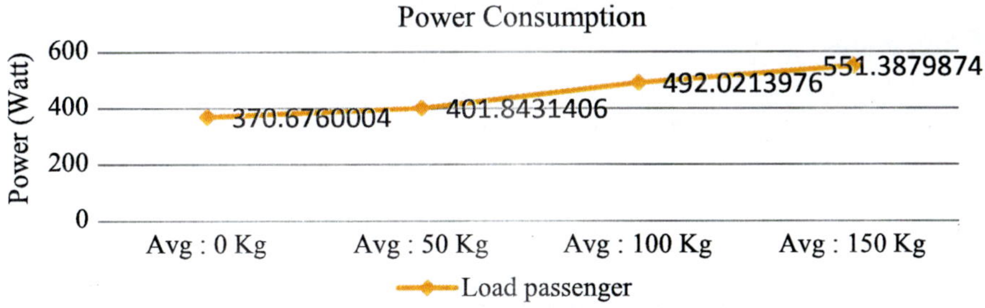

Fig. 9 Results of the physical data acquisition of the power consumption

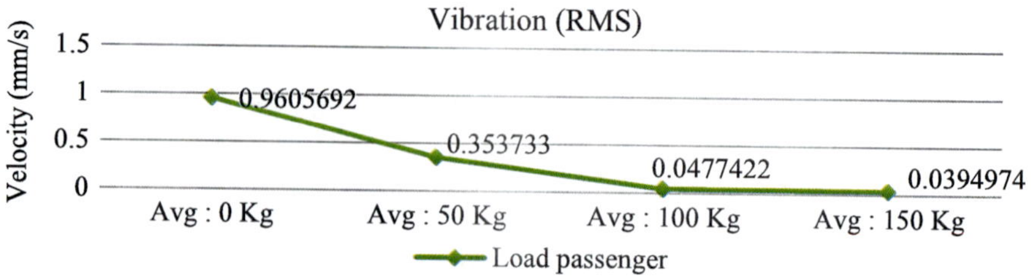

Fig. 10 Results of the physical data acquisition of the vibration

4 Conclusions

In conclusion, the results of this study can be concluded that a prototype multi-sensor using nodeMCU ESP32 was successfully achieved, and also a wifi network connection to transmit physical data to the thingspeak cloud server can run. The main objective of this research is to study the operational conditions of IoT technology-based stairlift by comparing the standard measurements on ASME A18.1 and other standards if conditions are based on data above the predetermined limits, the early warning system (EWS) works successfully by sending notifications to the telegram so that it can be immediately observed, analyzed and handled as early as possible. This research is also an improvement of existing tools to be more advanced and efficient in their functions. Finally, the purpose of this study is to analyze stairlift performance. Stairlift performance based on the analysis of existing data can be categorized in accordance with ASME A18.1 standards and other standards, but must still be improved in the future for better stairlift performance. This research can be enhanced by more efficient methods by combining the application of failure classification studies to detect and find the location of failures in a more precise and efficient way.

Acknowledgements The authors would like to thanks Universitas Indonesia, for funding this research. In addition to that, Politeknik Manufaktur Astra has provided scholarships to the first authors to continue their education at the Strata 2 level.

References

1. Zhou Y, Wang K, Liu H (2018) An elevator monitoring system based on the internet of things. Procedia Comput Sci 131:541–544
2. Balmer DCJAT (2010) Impact of the A18. 1 ASME Standard on platform lifts and stairway chairlifts on accessibility and usability. Assistive Technology 22(1):46–50
3. Navya K et al (2018) Smart stair lift for disabled and elderly. Int J Pure Appl Math 120(6):4647–4660
4. Du C, Zhu S (2012) Research on urban public safety emergency management early warning system based on technologies for the internet of things. Procedia Eng 45:748–754

5. Olalere IO, Dewa M, Nleya B (2018, May) Remote condition monitoring of elevator's vibration and acoustics parameters for optimised maintenance using IoT technology. In: 2018 IEEE Canadian Conference on Electrical & Computer Engineering (CCECE), IEEE, pp 1–4
6. Chooruang K, Meekul K (2018) Design of an IoT energy monitoring system. In: 2018 16th international conference on ICT and knowledge engineering (ICT&KE). IEEE
7. Riley CM, Lin BK, Habetler TG, Kliman GB (1997, February) Stator current-based sensorless vibration monitoring of induction motors. In: Proceedings of APEC 97-Applied Power Electronics Conference, vol. 1. IEEE, pp 142–147
8. Padte AK, Bajaj R (2017) A proposed framework for realtime monitoring of speed and location of a vehicle. International Journal of Advance Research, Ideas and Innovations in Technology (IJARIIT) 3(6)
9. HanumYahaya F, Shauri RLA, Bakar SA (2019) Dorper BSI monitoring with load cells and raspberry PI. IEEE
10. Ping L et al (2007) A design of the temperature test system based on grouping DS18B20. IEEE
11. Tantidham T et al (2018) Energy consumption collection design for smart building. IEEE
12. Hafizi MSM et al (2018) Developing a monitoring system for tripping fault detection via IoT. IEEE
13. Babiuch M, Foltýnek P, Smutný P (2019) Using the ESP32 microcontroller for data processing. IEEE
14. Carducci CGC et al (2019) Enabling ESP32-based IoT applications in building automation systems. IEEE
15. Parida D (2019) Real-time environment monitoring system using ESP8266 and Thingspeak on Internet of Things Platform. IEEE
16. Stannah ANR, Titchener N (2013) Stairlift obstruction monitoring. Google Patents
17. Zhou Y, Wang K, Liu H (2018) An elevator monitoring system based on The Internet Of Things. Procedia Comput Sci 131:541–544
18. Sung G-M (2019) Internet-of-Things based controller of a three-phase induction motor using a variable-frequency driver. IEEE
19. Tavner PJ, Penman J (1987) Condition monitoring of electrical machines, vol 1. Research Studies Press Letchworth

Non-vibrate Palm Oil Tree Harvesting Cutter Using DC Motor

Zamzuri Hamedon, Ammar Zakwan Abdullah, Ismayuzri Ishak, and Hasnulhadi Jaafar

Abstract Long exposure use of palm oil tree harvesting cutter powered by fuel engine can give risk of Hand Arm Vibration Syndrome (HAVS) disease, due to the transmitted vibration produced from the motorized cutter. In this paper, a new design of harvesting cutter which was powered by DC motor was proposed. Results shows that the vibration value were significantly low compare to the conventional motorized cutter; ranging between 0.04 and 0.29 m/s^2.

Keywords Harvesting cutter · Palm oil · Vibration · Hand arm vibration syndrome

1 Introduction

Palm oil is one of production that have very high demand in global as vegetable oils where it is one-quarter in global consumption and in international trade its lead by 60% and for usage in food productions lead by 74% and for industrial usage, it lead 24% [1]. In economic, production of palm oil increases strongly where it reached 128%, which is 58 million tons per year for last decades, Malaysia is one of main country produced palm oil [2]. As understood, the demand of palm oil is very high in world market and the production must be fast especially in harvesting thus, labor workforce very important in this industry. In harvesting oil palm, to take the bunch of fruit of oil palm, there are many ways such as manual harvesting, lifting platform, motorized cutting tool, vacuum operated cutters and advanced robotic.

Common problem for cutting mechanisms devices for palm oil harvesting is it have large vibration. The source of the vibration is coming from the transmitted of potential energy to kinetic energy and vice versa. This vibration can cause health problem to the worker that using this cutting mechanism devices. The health problem

Z. Hamedon (✉) · A. Z. Abdullah · I. Ishak
Faculty of Manufacturing and Mechatronic Engineering Technology, University Malaysia Pahang, Pekan, Pahang, Malaysia
e-mail: Zamzuri@ump.edu.my

H. Jaafar
School of Manufacturing Engineering, University Malaysia Perlis, Kangar, Perlis, Malaysia

A. F. Ab. Nasir et al. (eds.), *Recent Trends in Mechatronics Towards Industry 4.0*, Lecture Notes in Electrical Engineering 730, https://doi.org/10.1007/978-981-33-4597-3_48

that related to this vibration is Hand Arm Vibration Syndrome (HAVS) when the worker using the cutting mechanisms devices with frequently which mean every day uses [3].

Usually for this HAVS symptoms, the worker feel numbness around their finger and they cannot feel object properly. If continued, their hand will loss strength and if this is happened, the productivity of the workers is drop and this affected the rate of production for palm oil [4]. Although electrical motor either AC or DC are popular in the application of the hand tools, however it is never been used as harvesting cutter in the palm industry. This paper presents the newly invented of palm oil plantation harvesting cutter of using DC motor in order to overcome the vibration issue.

2 Preliminary Design

The initial ideas are generated from the application of cordless hand tool that are widely been used in various sector. The hypothesis of using the electrical motor cutter is, it may give to less vibration to the tool as compare to the motor powered by fuel engine. Initial design is sketched as shown in Fig. 1. The flow of the process is shown in Fig. 2. It begins with defining the problem, designing until actual prototype testing.

The first design is utilize the 24 V lithium battery with build-in wire inside the pole. The height or length of pole can be adjusted by assembly and disassembly by mean of telescopic approach. While the cutting head is using sickle placed at the end of pole. The jigging movement of sickle generate by DC motor that located at cutting head and convert from rotational movement by using bevel gear. However, this design produces high vibration when the sickle jigging to cut the bunches of fruit or the fronds. This action will create high vibration as motorized cutter product [5] and also need high torque motor and high RPM motor to support the sickle since the weight of sickles approximately around 6–8 kg. Another issue was the non-rigid telescopic pole may add the extra vibration to the devices.

The second design is also generated by 24 V lithium battery power source with build in wire inside the pole. However, the telescopic pole is replaced by single non-adjustable pole for this design while the cutting action is carried out by the circular blade saw and directly mount at the motor install to the attachment at the cutting head. The cover plate is design and placed around the circular blade to ensure the safety of the user.

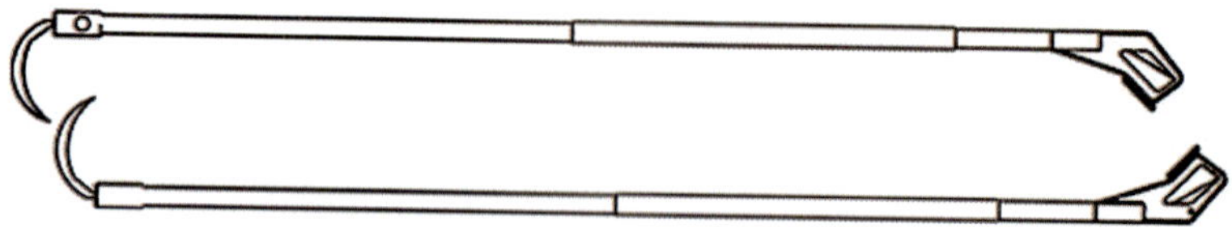

Fig. 1 Initial design of harvesting cutter

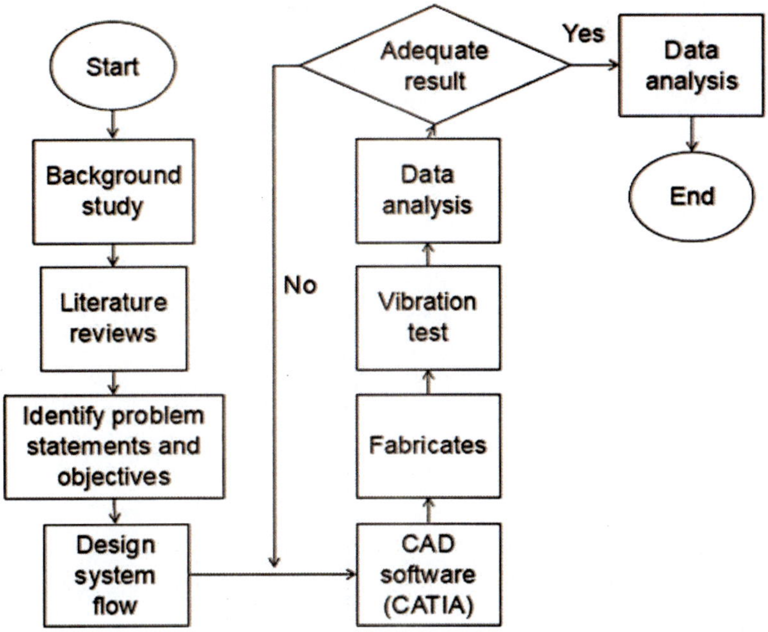

Fig. 2 Flow chart

2.1 *Prototype Fabrication*

Prototype was built base on the design 2 is shown in Fig. 3. Initially, this prototype has problems with the bevel gear, where it is not easy to align the bevel gear perfectly contact with motor shaft. The uneven contact between the gear produce unsmooth rotation that caused high vibration to occur. After fixing is done properly and a preliminary vibration test is carried out. The vibration is very high. The gear is then removed and the vibration test is repeated. Vibration is reduced tremendously when the cutter is mounted directly the motor's shaft. The cutter is then fixed to the motor by using special connector and rubber padding is added to reduce vibration as shown in Fig. 5. Rubber padding will act as anti-vibration or damper to reduce the transmitting of vibration from the pole to the hand of user. Other part and components are listed in the Table 1.

2.2 *Experimental Set-up*

To detect the small vibration or shock the Model KS903 triaxial accelerometer is used. Setup of the experiment is shown in Fig. 6. For this experiment, only single point is measured. The triaxial accelerometer is located at the place where the user will grip

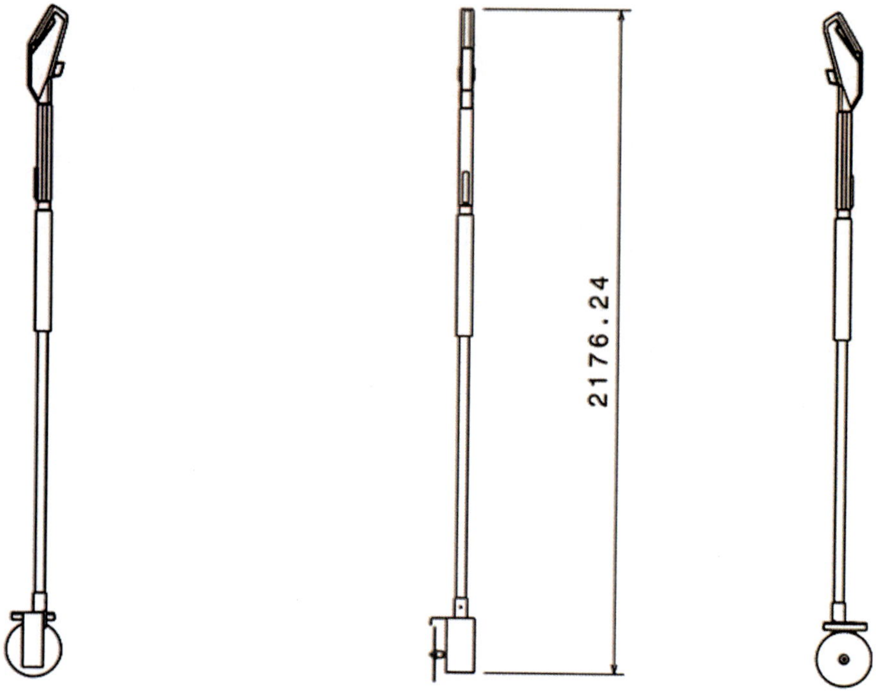

Fig. 3 Design 2

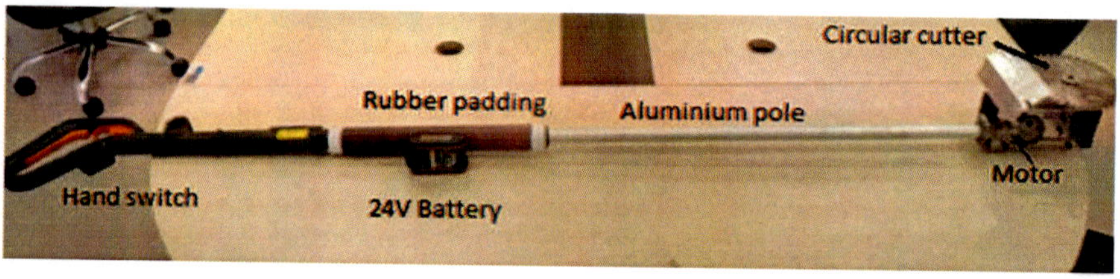

Fig. 4 Actual prototype

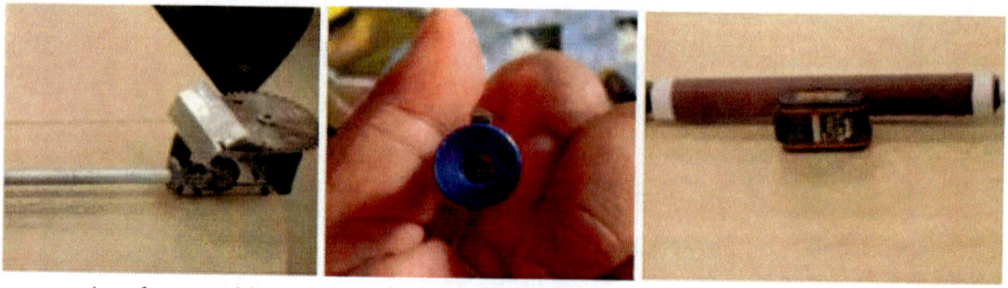

a. Attachment with motor

b. Connector for cutter

c. 24V battery and hand padding

Fig. 5 Closed up view of the components

Table 1 List of components

No.	Components	Descriptions
1.	DC motor	• Speed: 195 rpm • Torque: 55 kg cm • Voltage: 24 V • Quantity: 1 unit
2.	Pole	• Aluminum hollow t = 2 mm • Length 1.8 m • Diameter 50 mm
3.	Hand padding	• Synthetic rubber
4.	Handle	• ABS with build-in 24 V battery

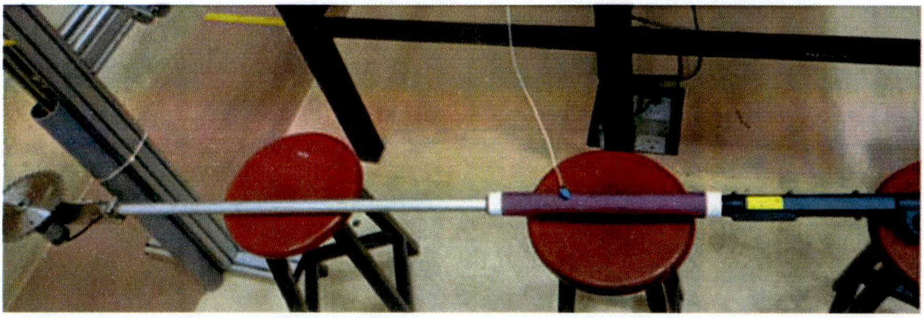

Fig. 6 Location triaxial accelerometer node

the pole. The triaxial accelerometer stick on the rubber padding using dry formable adhesive and connected to the DASY-lab for reading and sync the MyAcceleration software for data logging. Both software is link to each other to generate the result. The experiment set-up is same as the experiment that made to know the vibration of motorized cutter [6]. The result obtained is the compared with others.

3 Results and Discussion

3.1 Vibration Test

Few tests are performed with different conditions; with/without cutter, with/without hand padding, with/without cutter connecter. The whole ideas of having different condition of the testing are to understand the source of the vibration.

First test is carried with the condition of full assembly without hand padding. The result is shown in Fig. 7 where the maximum acceleration detected on the pole are measured. The maximum vibration without hand padding is 0.29 m/s^2. To reduce the vibration, the hand padding was added on the pole. The vibration for the device with hand padding is reduce to 0.18 m/s^2. It showed that the rubber hand padding

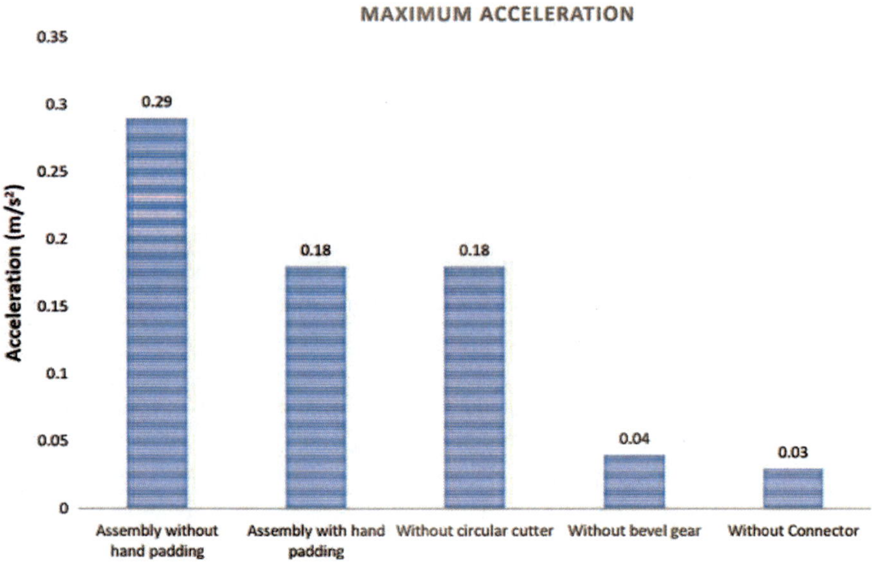

Fig. 7 Maximum acceleration

mat reduce the vibration by acting as a damping. The prototype is then been tested to detect the cause of vibration. Started with removing the circular cutter. The triaxial accelerometer is place at the same position. The vibration of the device without circular cutter is 0.18 m/s^2. It showed that the movement from the cutter gave no effect to the vibration of the device.

Next test is carried out by removing the bevel gear. The maximum vibration for the device without bevel gear is only 0.04 m/s^2. It showed that the bevel gear is the source for the vibration. The test is continued by removing the connector. Only slight drop of vibration value to 0.03 m/s^2 is detected by triaxial accelerometer. This to confirm that the bevel gear gave significant effect to the vibration of the devices [7, 8].

4 Conclusion

The design of the harvesting cutter of palm oil using DC motor is able to reduce the vibration. This may reduce stiffness and fatigue of the workers thus may increase the productivity. The used of the 24 V battery also make this device more portable and lighter to use. The battery is rechargeable and duplicable. However, the cost of producing the actual device might be higher due to the cost of the battery.

Acknowledgements This work was supported by the Universiti Malaysia Pahang Research Grant Scheme *RDU1703214*. The authors would also like to acknowledge the assistance provided by Faculty of Manufacturing and Mechatronic Engineering Technology of University Malaysia Pahang.

References

1. Ferdous Alam ASA, Er AC, Begum H (2015) Malaysian oil palm industry: Prospect and problem. J Food Agric Environ 13(2):143–148
2. World Bank (2010) World Development Report 2010: Development and Climate Change. Washington, DC. © World Bank. https://openknowledge.worldbank.org/handle/10986/4387 License: CC BY 3.0 IGO
3. Kattel BP, Fernandez JE (1999) The effects of rivet guns on hand-arm vibration. Int J Ind Ergon 23:595–608
4. Su TA, Hoe VCW, Masilamani R, Awang Mahmud AB (2010) Hand-arm vibration syndrome among a group of construction workers in Malaysia. Occup Environ Med 68(1):58–63
5. Salleh SM, Rahim EA, Ghazali IH, Azmi K, Jelani AR, Ismail MF, Ahmad MR (2013) Hand-arm vibration analysis of palm oil fruit harvester machine. Appl Mech Mater 315:621–625
6. Jelani AR, Ahmad MR, Azaman MIH, Gono Y, Mohamed Z, Sukawa S, Aduka A, Aziz A, Bakri A, Mohamed A, Selamat MB, Ismail A, Deraman MS, Shuib AR, Kushairi NKA (2018) Development and evaluation of a new generation oil palm motorised cutter (Cantas Evo). J Oil Palm Res 30:276–288
7. Hamedon Z, Zahid MNO, Tomadi SH, Noor MM, MJ MR (2008) Design and fabricate the prototype of a motorized cutter for harvesting palm fruit, Malaysian Science and Technology Congress, MSTC08, 16-17 Dec, KLCC Malaysia 2008, pp 599–603
8. Azaman MIH, Jelani AR, Ahmad MR, Shuib AR, Ismail A (2018) Vibration isolator for the oil palm motorised cutter, MPOB information series, ISSN 1511-7871

Feasibility Study of CO, CO_2, NO_2, and O_2 Sensors for Hazardous Gas Detection System in Vehicle Cabin

Cheow Shek Choon and Ismayuzri Bin Ishak

Abstract CO, CO_2, and NO_2 are gases that are commonly found from the emission of a vehicle exhaust pipe. They normally exist in the form of odorless, colorless, and tasteless. Large amount inhalation of these gases will cause serious health effects such as fatigue, respiratory problems, drowsiness, and death. These toxic gases will flow into the vehicle cabin once leakage happens and the O_2 gas inside the vehicle cabin will decreases and oxygen deficient occurs. The objectives of this research is to study the feasibility of CO, CO_2, NO_2, and O_2 gas sensors for hazardous gas detection system in vehicle cabin and the suitable location for the hazardous gas detection device to place inside the vehicle cabin. Two experiments were carried out to determine the amount of CO, CO_2, NO_2, and O_2 gas released from the exhaust pipe and inside the vehicle cabin by using two different vehicles. The gas concentration collected from these two experiments were compared to the safe limit. The CO concentration released from the exhaust pipe was over the safe limit and the O_2 percentage collected was too low compared to the safe limit. On the other hand, the concentration of gases collected inside the vehicle cabin were at a safe level. The concentration of CO_2 and NO_2 collected from the exhaust pipe and inside vehicle cabin were below the safe limit where these two gases are at safe level. Therefore, it is proven that the CO and O_2 sensors are important for hazardous gas detection system. This can be concluded that the CO and O_2 sensors are suitable for the hazardous gas detection system in vehicle cabin.

Keywords Carbon monoxide (CO) · Oxygen (O_2) · Hazardous gas detection device · Vehicle cabin

C. S. Choon · I. B. Ishak (✉)
Faculty of Manufacturing & Mechatronic Engineering Technology, Universiti Malaysia Pahang (UMP), 26600 Pekan, Pahang Darul Makmur, Malaysia
e-mail: ismayuzri@ump.edu.my

C. S. Choon
e-mail: cheowshekchoon@gmail.com

A. F. Ab. Nasir et al. (eds.), *Recent Trends in Mechatronics Towards Industry 4.0*, Lecture Notes in Electrical Engineering 730, https://doi.org/10.1007/978-981-33-4597-3_49

1 Introduction

According to the Malaysia Automotive Association, registration of new vehicles records a total of 598,714 units in the year 2018 [1]. Other than that, the total of newly registered vehicles around the world hit an amount of 78,959,555 in the year 2018 [2]. The increase in vehicles around the world will bring further noxious impact such as pollutants into the atmosphere. Such contaminants will have a serious impact on individuals' health. More than 30% of the dangerous gases released to the atmosphere are contributed by the motor vehicles [3].

There are two common types of engine that is used by the vehicles, which is the diesel engine and the gasoline engine. The diesel engine uses diesel fuel to operate and the gasoline engine uses petroleum fuel to work. Different fuel combustion will release different types of toxic gases. These toxic gases emission is emitted through the exhaust pipe once the internal combustion of the engine takes place and it is the main reason for the presence of high concentration toxic substances in the atmosphere [4].

There are a few types of gases that emitted from the exhaust pipe when the combustion of petroleum takes place, such as Carbon Monoxide (CO), Carbon Dioxides (CO_2), Oxygen (O_2), and Nitrogen Dioxides (NO_2) [5]. CO, CO_2, and NO_2 gases are likely to cause discomfort, irritation of eyes, fatigue, and drowsiness. It will reduce the awareness of the drivers and increases the possibility of an accident. Exposure to a high concentration of these hazardous gases will cause chronic effects and death.

The quality of air inside the vehicle cabin also plays an important role in ensuring the safety of the driver or passenger. Vehicle cabin is a small and enclosed space where the air is not easy to flow out of the vehicle. Once the particulates entered the cabin through ventilation, it will cause health effects to the driver. The oxygen inside the cabin will also be displaced when leakage happens, leaving the environment oxygen deficient. Besides, chances of leakage of exhaust from the vehicle into the cabin while falling asleep could increase the risk of death. Therefore, these four types of gases were chosen to be studied in this research.

Carbon monoxide (CO) exist in the form of odorless, colorless, and tasteless that is very harmful to the human when a large amount is inhaled. Carbon monoxide is released when incomplete combustion of all carbon type fuel is undergoing [6]. It is normally found from the emission of an exhaust which contributes nearly 60 percent [7]. Inhaling carbon monoxide approximate higher than the safety level, it will cause irritation of eyes, headaches, and also vomit. Other than that, it will also cause drivers to have confusion and visual disturbance that will lead to a serious accident when driving. Breathing in a high concentration of CO will reduce the amount of oxygen that needs to be transported to other critical organs such as the heart and brain [8]. The safe level of CO according to the ASHRAE standard is 9 ppm [9].

Carbon Dioxide (CO_2) is a colorless gas that has an odorless smell and a sour taste. It is one of the minor components, around 0.03–0.04% that can be found in Earth's atmosphere according to The International Council on Combustion Engines CIMAC [10]. CO_2 is the main component produced from the exhaust of the engine.

According to the current OSHA standard, the safe limit for CO_2 is 5000 ppm for an 8-h of daily workplace exposure and 1000 ppm for regular level in occupied space [11].

Nitrogen dioxide (NO_2) found in the atmosphere is mainly produced by the combustion of fossil fuels. NO_2 gas is a reddish-brown or dark brown to yellowish gas that comes with a pungent smell. Inhalation of NO_2 will cause minor and major health effects on human beings. Short-term exposure in NO_2 concentration will increase the risk of respiratory mortality [12, 13]. The chronic health effects that cause by NO_2 is that mutations will occur and reproductive hazard appear [9]. The safe limit set by the NIOSH is 1 ppm [4].

In the year 2000, a paper focus on monitoring air quality inside the cabin using CO and O_2 sensors by Galatis et al. [14]. The research is focusing on on-road testing using an Australian manufactured Toyota Camry (1997, 2.2 litre) and a simulation suicide experiment using Ford Falcon (1996, 4.0 litre). A hose of 30 mm diameter was connected to the exhaust outlet of the vehicle and then directed into the vehicle cabin for simulation suicide experiment. Previous research only used two types of gas sensors which are CO and O_2 gas sensors. Other than that, previous research is focusing on CO and O_2 concentration dependence on vehicle speed, driving scenario, and suicide simulation.

This research is mainly focusing on the study of the suitability of CO, CO_2, NO_2, and O_2 sensors as hazardous gas detection device and the appropriate locations to place the sensors inside the vehicle cabin. Besides, the previous work is done in dynamic condition where this research is done on static condition. It can be seen that it is a different kind of research as compared with the previous research.

2 Materials and Methods

To study the CO, CO_2, NO_2, and O_2 sensors feasibility to be used as the hazardous gas detection system in vehicle cabin. The selection of CO, CO_2, NO_2, and O_2 sensors are based on the sensitivity and the detectable range as compared with the safe limit. There are two types of experiments are being performed with two vehicles using CO, CO_2, NO_2, and O_2 sensors. The two vehicles are Grand Livina 1.6L A/T built in the year 2008 and Perodua Kancil 660EX which was built in the year 1998. Both vehicle experiments are performed under stationary mode.

2.1 Sensing

The sensing stage consists of four sensors that are directly exposed to air released from the exhaust pipe and air inside the vehicle cabin. The four sensors involved in this research are carbon dioxide sensor (MG-811), carbon monoxide sensor (MQ-7), air quality sensor (MICS-6814), and grove oxygen sensor (ME-O2). These four

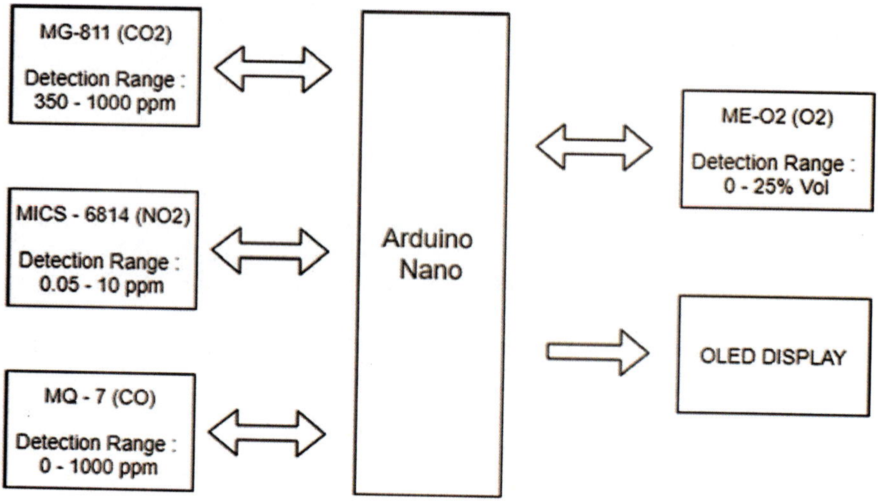

Fig. 1 Block diagram for testing device

Table 1 Exposure limits for CO, CO_2, NO_2, and O_2 gases

Types of gas	Exposure limits	Administration
CO	9 ppm	ASHRAE standard
CO_2	1000 ppm	OSHA
NO_2	1 ppm	NIOSH
O_2	19.5%	OSHA

sensors are used to detect the presence of carbon dioxide, carbon monoxide, nitrogen dioxide, and oxygen gas.

From Fig. 1, these four sensors are chosen based on the detection range and the sensors' sensitivity. The detection range of these sensors are suitable to be compared with the safety limits. The safety limits of each gases is shown in Table 1. Other than that, the sensitivity of the sensors are also take into account. MQ-7 is highly sensitive to CO gas, MG-811 is highly sensitive to CO_2 gas, and ME-O2 is highly sensitive to O_2 gas. While the MICS-6814 is an air quality sensor that can detect NO_2 gas. In short, these four sensors are suitable for this research due to their sensitivity and the detection range.

2.2 *Sensor Calibration*

These four sensors are preheated for 48 h before using for calibration. All of these sensors are receiving analog data. In order to convert the analog reading into their units such as ppm, a few samples of known concentration in ppm was used. The analog readings received from the sensors are converted into their units based on the datasheet and compared to the samples of known concentration gas.

2.3 *Experimental*

In this research, an integrated hardware is used for data collection. There are two types of data collection are carried out. The first experiment is performed at the exhaust pipe and the second is done inside the vehicle cabin. The vehicle cabin experiment is based on the five different locations as shown in Fig. 2. The CO, CO_2, NO_2, and O_2 sensors are connected to the Arduino Nano and the readings are displayed using an OLED display. The data was collected based on the Nissan Grand Livina 1.6L A/T and Perodua Kancil 660EX. Both types of vehicles are working based on gasoline engines. The experiments take place when both vehicles are in stationary mode. Both vehicle exhaust pipe were checked and inspected before undergo the experiment. The final selection of gas sensor used for designing the hazardous gas detection device is based on the results obtained.

3 Results

The data collected from the exhaust pipe are tabulated into a graphical form for a better and clearer view. On the other hand, the data collected from the vehicle cabin are summarized into 6 samples and 120 time steps. This is to calculate the average of gas concentration collected and tabulated into table form for clearer view. All the data collected from these two experiments are being compared with the safe limit.

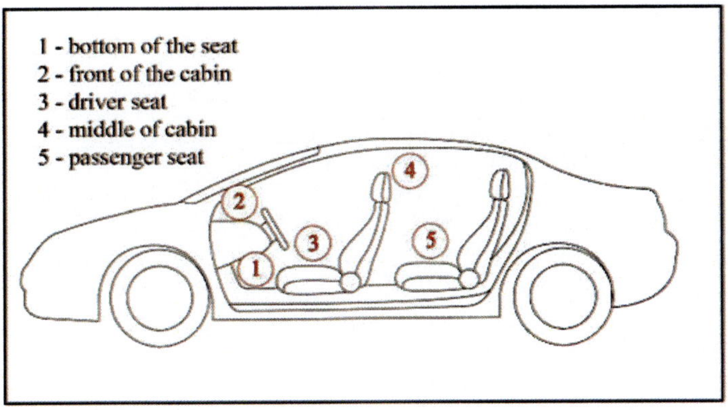

Fig. 2 Positions that are tested inside the cabin

3.1 Exhaust Pipe

It can be seen that the CO concentration obtained in Fig. 3 for both cars tested are above the safe limit (9 ppm). The highest CO concentration recorded for Grand Livina is 32.63 ppm and the highest CO concentration recorded for Perodua Kancil is 23.85 ppm. Other than that, the oxygen percentage obtained in Fig. 6 for both cars tested are also lower than 19.5%. The percentage of oxygen recorded for both vehicles slumped to almost 0% of oxygen. On the other hand, the NO_2 concentration and the CO_2 concentration obtained in Figs. 4 and 5 are at a safe level. The CO_2 concentration recorded for both vehicles are in the range between 340 and 410 ppm. The highest NO_2 concentration recorded for Grand Livina and Perodua Kancil is 0.0792 and 0.371 ppm.

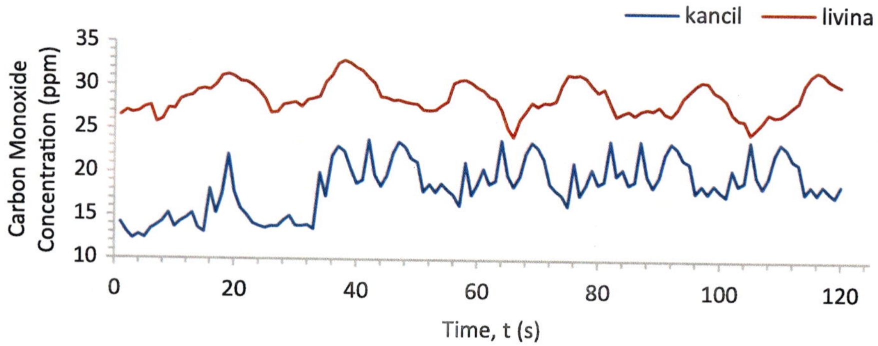

Fig. 3 Carbon monoxide concentration

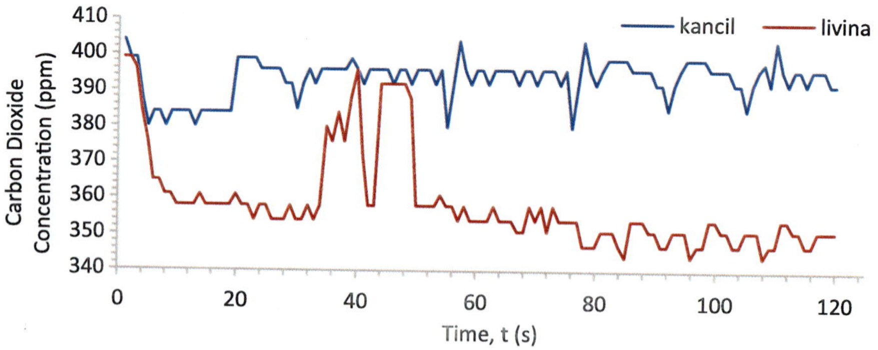

Fig. 4 Carbon dioxide concentration

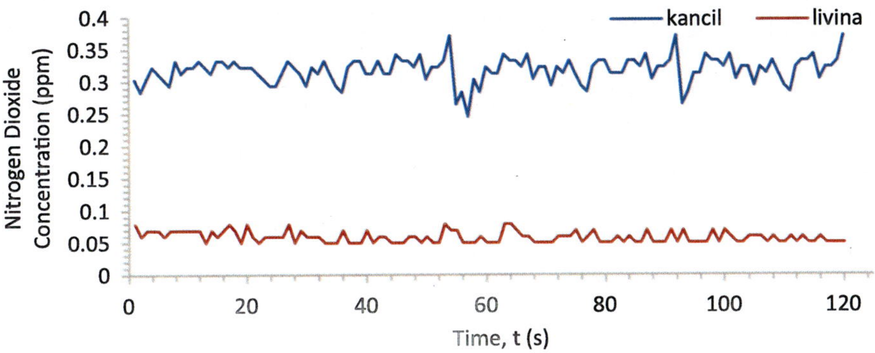

Fig. 5 Nitrogen dioxide concentration

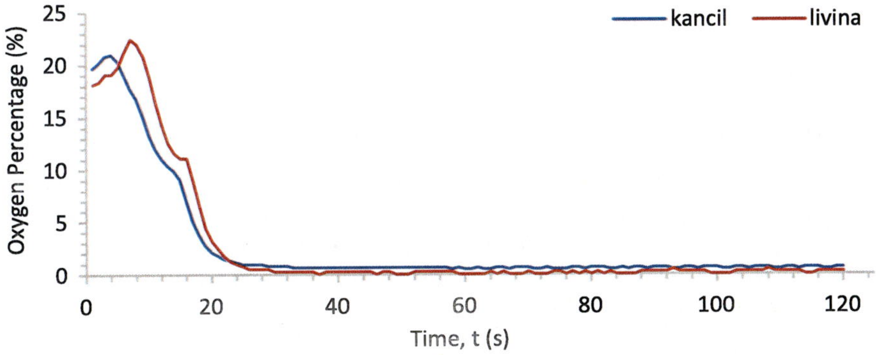

Fig. 6 Oxygen percentage

3.2 Vehicle Cabin

Tables 2 and 3 shows the average of gas concentration collected at different cabin locations for two vehicles. All types of gas obtained from the Grand Livina shown in Table 2 are at the safe level, where the concentration of CO, CO_2, NO_2 are below

Table 2 Average of gas concentration collected at different location for Grand Livina

Location of cabin	CO (ppm)	CO_2 (ppm)	NO_2 (ppm)	O_2 (%)
Bottom	5.390 ± 0.061	390.833 ± 32.233	0.056 ± 0.012	20.532 ± 0.317
Middle	5.218 ± 0.040	518.167 ± 20.392	0.056 ± 0.008	20.582 ± 0.048
Driver	5.320 ± 0.041	386.000 ± 4.195	0.060 ± 0.012	20.978 ± 0.148
Passenger	5.593 ± 0.065	375.000 ± 9.839	0.063 ± 0.140	20.765 ± 0.038
Front	5.755 ± 0.172	504.500 ± 143.987	0.063 ± 0.012	20.812 ± 0.069

Table 3 Average of gas concentration collected at different location for Perodua Kancil

Location of cabin	CO (ppm)	CO_2 (ppm)	NO_2 (ppm)	O_2 (%)
Bottom	8.257 ± 0.346	369.000 ± 14.085	0.512 ± 0.008	20.662 ± 0.041
Middle	6.535 ± 0.237	520.667 ± 5.538	0.309 ± 0.012	20.765 ± 0.096
Driver	7.762 ± 0.215	523.500 ± 12.341	0.321 ± 0.018	20.507 ± 0.129
Passenger	6.828 ± 0.371	429.167 ± 56.908	0.267 ± 0.015	20.777 ± 0.105
Front	8.880 ± 0.577	373.833 ± 6.047	0.542 ± 0.034	20.685 ± 0.070

the safety limits and the percentage of O_2 is above 19.5%. On the other hand, the concentration of CO for Perodua Kancil at front of the cabin shown in Table 3 is the highest and almost reach the safe limit. Other gases collected shown in Table 3 for Perodua Kancil are at the safe level where CO_2 below 1000 ppm, NO_2 below 1 ppm, and O_2 is above 19.5%.

4 Discussions

It can be seen that the CO concentration obtained from exhaust pipe in Fig. 3 for both cars tested are above the safe limit (9 ppm). The peak reading of CO concentration observed for Grand Livina between 30 to 50 s is due to cold engine. The engine tends to consume more fuel to ram up the rpm in order to become stable. The O_2 percentage obtained in Fig. 6 for both cars tested are also lower than 19.5%. Besides, the CO concentration and O_2 percentage obtained the largest concentration value compared to CO_2 and NO_2. Nevertheless, the CO concentration and O_2 percentage shows the biggest difference when comparing the exhaust pipe and vehicle cabin results. On the other hand, the NO_2 concentration and the CO_2 concentration obtained in Figs. 4 and 5 are at a safe level. The higher value of CO_2 and NO_2 for Perodua Kancil compare to Grand Livina is due to the older car engine which is manufactured in year 1998. Grand Livina implemented with a modern engine which is manufactured in year 2008. Modern engine implemented a lot of technologies which helps to reduce the consumption of fuels while increasing the power. Older engine are designed to work continuously and harder either the car is in ideal or accelerating. The fuel consumption for older car engine will be higher and more emission will be released. This can be concluded that the CO and O_2 sensors are suitable to be used for hazardous gas detection device.

The average of gas concentration obtained from the Grand Livina shown in Table 2 are at the safe level, where the concentration of CO, CO_2, NO_2 are below the safety limits and the percentage of O_2 is above 19.5%. The average CO and NO_2 gas concentration recorded from the front of the cabin is the highest among all location. Meanwhile, the average concentration of CO_2 recorded from the middle of the cabin is the highest and the lowest percentage of O_2 is the bottom of the seat.

On the other hand, the average of CO_2, NO_2, and O_2 gas concentration detected from five different location for Perodua Kancil that shown in Table 3 are on the safe level. The CO_2 below 1000 ppm, NO_2 below 1 ppm, and O_2 is above 19.5%. Only the average of CO gas concentration for Perodua Kancil at front of the cabin recorded the highest value (8.880 ppm) which is almost reach the safe limit (9 ppm). Other than that, the highest average NO_2 gas concentration recorded are also from the front of the cabin. Simultaneously, the highest average CO_2 gas concentration obtained is from the driver seat. Whereby the location of vehicle cabin that recorded the lowest average percentage of O_2 is at the driver seat. It can be seen that the CO level collected from front of the cabin, driver seat, and bottom of the seat are slightly high. This is due to emission of CO is coming from the air ventilation. The ventilation intake air from the outside of vehicle. Therefore higher CO gas concentration are being intake and released at the front of the vehicle through the ventilation. Next, the CO_2 levels are lowest for bottom of seat and front of cabin. This can be said that the CO_2 are coming from the back of the car and results in lower CO_2 concentration at front of the vehicle.

The location of the cabin that suitable for the detection device to be place based on results in Tables 2 and 3 are front of the cabin. This is because there is two highest reading obtained from Grand Livina is located at front of the cabin. Other than that, the concentration of CO recorded from Perodua Kancil shown in Table 3 is almost reach the safe limit. Furthermore, the exhaust pipe emission is being collected from the windshield's base outside of vehicle. The air will be blown by the blower into the cabin from the ventilation which is located at front of the vehicle. The bottom of the seat is not considered due to inappropriate locations for a driver to be in that position. At the same time, the location such as the middle of the cabin, driver seat, and passenger seat are rejected due to low concentration of gas obtained according to the results. Consequently, the gases are likely to gather at the front of the cabin. Figure 7 shows the location of the hazardous gas detection device placed inside the vehicle cabin.

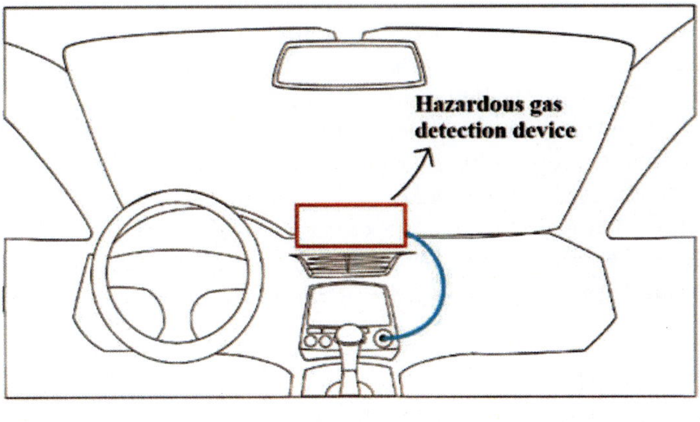

Fig. 7 Location of hazardous gas detection device placed inside the vehicle cabin

The final hazardous gas detection device proposed is based on an Arduino Nano microcontroller board and combined with the MQ-7 (CO) and ME-O2 (O_2) sensors. The device also comes with a 0.96 inch OLED display which is used to display the real-time readings of the sensors and a buzzer to alert the users once over the safe limit. At the same time, this device also implemented with a Bluetooth device that enables the user to connect to the android phone to observe the real-time sensors readings.

5 Conclusion

In this study, CO, CO_2, NO_2, and O_2 were used to identify the total amount of gases released from the exhaust pipe and inside the vehicle cabin. From the results shown, it is proved that CO and O_2 sensors are suitable to be used as hazardous gas detection device for vehicle cabin. Furthermore, the suitable location for the hazardous gas detection device place inside the cabin based on results is front of the cabin. A final device is proposed to detect the hazardous gas inside vehicle cabin and it can alert the users once the gases are over the safe limit. It comes with a small size and portable weight which allows the users to carry. The proposed device should implement with the MQ-7 (CO) and ME-O2 (O_2) sensors. It can comes with an OLED display that displays the sensors' real-time readings. Other than that, a buzzer need to be embedded in the proposed device so that it can alert the users once the gas inside the vehicle cabin is over the safe limit. The proposed device would be better if it can enables the user to connect to the android phone via the Bluetooth module. This allows the users to check the sensors' readings through a smartphone without checking the device monitor. Lastly, the proposed device need to be power on by using direct power from the car supply adapter.

In summary, CO and O_2 sensors are suitable to be used as hazardous gas detection device for vehicle cabin. This hazardous gas detection device is an important device that can ensure the safety of the driver or passenger inside the vehicle when hazardous gas inside increases and oxygen deficient occurs. The sensors should be mounted at central of the front cabin to avoid exhaled gas and humidity from the users that may not reflect the average air quality of the cabin. The CO and O_2 sensors should be highly durable and can withstand high temperature. This is due to the location of the device is front of the cabin which is highly exposed by the sunlight.

The methodology used in this study can be extended with vehicles based on diesel engines. This is because the composites released from diesel combustion might be different from the gasoline combustion. This research can be improved by having an experiment during the vehicles is in motion mode. It can be compared with the results obtained in the above section and get a better view.

Acknowledgements The authors would like to acknowledge Universiti Malaysia Pahang for funding this study.

References

1. Malaysia Automotive Association (2019) Market_Review_2018.pdf, p 19
2. European Automobile Manufacturers Association (2019, February) Economic and market report: EU automotive industry full-year 2018, pp 1–29
3. Ramya V, Palaniappan B (2014) Embedded technology for vehicle cabin safety monitoring and alerting system. Middle East J Sci Res 20(12):2210–2212
4. Sassykova LR, Aubakirov YA, Sendilvelan S, Tashmukhabetova ZhKh, Faizullaeva MF, Bhaskar K, Batyrbayeva AA, Ryskaliyeva RG, Tyussyupova BB, Zhakupova AA, Sarybayev MA (2019) The main components of vehicle exhaust gases and their effective catalytic neutralization. Orient J Chem 35(1):110–127
5. Ajith S, Harivishnu B, Vinesh TK, Sooraj S, Prasad G (2017, January) Automated gas pollution detection system. In: 2017 2nd international conference for convergence in technology, I2CT 2017, vol 2017, pp 483–486
6. Vakkilainen EK (2017) Solid biofuels and combustion. In: Steam generation from biomass. Elsevier, pp 18–56
7. Miller BG (2011) Coal-fired emissions and legislative action. In: Clean coal engineering technology. Elsevier, pp 301–374
8. Ibrahim AA (2019) Carbon dioxide and carbon monoxide level detector. In: 2018 21st International Conference of Computer and Information Technology (ICCIT) 2018, pp 1–5
9. New Jersey Department of Health and Senior Sevices (1989) Hazardous substance fact sheet, nitrogen dioxide
10. The International Council on Combustion Engines CIMAC (2008) Guide to diesel exhaust emissions control of NOx, SOx, particulates, smoke and CO_2. no 28
11. Kane International Limited (2019) What are safe levels of CO and CO_2 in rooms? https://www.kane.co.uk/knowledge-centre/what-are-safe-levels-of-co-and-co2-in-rooms. Last accessed 20 Nov 2019
12. Renjie C, Peng Y, Meng X, Wang L, Liu C, Yue N, Zhijing L, Yunning L, Jiangmei L, Jinlei Q, Jinling Y, Haidong K, Maigeng Z (2018) Associations between ambient nitrogen dioxide and daily cause-specific mortality. Epidemiol **29**(4) 482–489
13. Mills IC, Atkinson RW, Kang S, Walton H, Anderson HR (2015) Quantitative systematic review of the associations between short-term exposure to nitrogen dioxide and mortality and hospital admissions. In: BMJ Open, vol 5, no 5
14. Galatsis K, Wlodarski W, Li YX, Kalantar-Zadeh K (2000, January) Vehicle cabin air quality monitor using gas sensors for improved safety. In: COMMAD conference on optoelectronic and microelectronic materials and devices, vol 2000, pp 65–68

Impact Analysis of Harassment Against Women in Bangladesh Using Machine Learning Approaches

Busrat Jahan, Bahar Uddin Mahmud, Abdullah Al Mamun, Md. Mujibur Rahman Majumder, and Mahbubul Alam

Abstract Violence against women is a major threat in Bangladesh, where reports of harassment of women and girls have spread an alarming rate. Unfortunately, despite significant achievements in women's development and bearing a magnanimous history of women's movement, incidences of harassment against women is still a burning issue. The majority of women are harassed by their relatives, friends and other people. From the survey we have tried to give an idea about the types, causes and impacts of harassment against women in Bangladesh. This is a survey-based paper and using Apriori algorithm to analyze the impacts of harassment against women and girls of Bangladesh. For these reasons, we have selected 2300 respondents to identify the impact of harassment. This study aims to find out the impact of violence in our society and cohere it with our social norms and values. The impact of sexual harassment on these outcomes (Anxiety, Intense fear, Ongoing fears, Depressions, Disrupted work life, Degradation of performances in study or work, Face difficulties with communication, Intimacy and enjoyment of social activities, Sleep disturbances or Nightmares) among different age's women/school and college going girls was compared to the outcomes among each other. In this study, according to comparison we find out that teenager, age below 18, is most vulnerable to harassment.

Keywords Sexual harassment · Survey · Sampling · Apriori algorithm · Impact · Of harassment

B. Jahan (✉) · B. U. Mahmud · Md. Mujibur Rahman Majumder · M. Alam
Department of CSE, Feni University, Trunk Road, Feni, Bangladesh
e-mail: hossenbipasa980@gmail.com

A. A. Mamun
Faculty of Engineering and Technology, Multimedia University, Melaka, Malaysia

A. F. Ab. Nasir et al. (eds.), *Recent Trends in Mechatronics Towards Industry 4.0*, Lecture Notes in Electrical Engineering 730, https://doi.org/10.1007/978-981-33-4597-3_50

1 Introduction

Sexual harassment is a global issue that is experienced by society in developed and developing countries alike. Sexual harassment is one of the most common forms of violence against women until women worked outside the home; the problem became a problem [1]. It has recently debuted as a critical social issue important involvement for individuals, organizations and society as a whole [2]. With the growing participation of women in the academic and the labor force in the official and unofficial sections in some of developing countries, a huge portion of teenage girls and young women come to school and work regularly. In Bangladesh, where violence against women is widespread [3, 4] day by day. Verbal abuse of sexual harassment against adolescents and women, from threat to rape [5, 6]. Harassment is often occurred not only school going girls but also working women, housewives and even older women. The basis on the Bangladesh National Women Lawyers Association (BNWLA), the percent of women in Bangladesh have become the sufferers of eve-teasing at some point in their life is 91% [7]. Incidents of sexual harassment are ubiquitous and can occur in urban areas as well as in rural areas in Bangladesh and anywhere else in the world. The combination of violence and social crime is a form of sexual harassment that makes various problems for a woman or a girl and basically acts as an obstacle between them and their full potential. A small survey conducted in Bangladesh reported that 45% of female students are between the ages of aged 12 to 16 years thought at times about stopping their education because of the indecent behavior of male perpetrators on the way to school and 73% of parents decided to stop their daughters' education half way [5]. There is no estimate of the national prevalence of sexual harassment in public spaces in Bangladesh. However, Action Aid found that in 2015, they surveyed (among 800 women and girls) surveyed 84% of women reported sexual harassment in public [8]. BRAC University conducted a study where they found that 94% of women face sexual harassment while using public transport, such as buses and three-wheelers [9]. Considering the above scenarios, we have been analyzing the impact of sexual harassment against women in Bangladesh by using association rules.

1.1 *Objective of the Study*

In this study, the sample data was collected for measuring the harassment impact over the women and girls in Bangladesh. The exact objectives were given here:

- Conduct to analyze the impact of sexual harassment that a women and girls face in many cases.
- Find the impact of harassment against all age's women via association rule mining.
- Impact of harassment among several age groups.
- Persons involved in harassment.
- Places where frequently faces harassment.

- Find association rules between several impacts and get frequent item sets from our collected data.

2 Data Collection

We have been collected both qualitative and quantitative data to measure the impact analysis of harassment against women in Bangladesh. We collected data from 2300 participants of school and college going students and women who are housewife and service holder by creating different types of questionnaire. The first step of data was collected randomly from online by using survey monkey. Another data was preferred to make a comparison group from the nearby districts.

2.1 Sampling Data Set

After collection of data we will sort the data to make fit for machine.

Following steps have been followed to get the expected output:

(i) collecting raw data; (ii) preprocessing the raw data; (iii) converting the "raw file: into "csv format"; (iv) sorting the data as our system requirement; (v) appling Apriori algorithm-based.

3 Proposed Method

The proposed method is involved in these basic steps: (i) read each item in a transaction, (ii) support of every item is calculated, (iii) compare to Support item and min_support, if no then remove the item otherwise insert items to frequent item-set, (iv) find confidence, for each non-empty sub-set, (v) again compare to confidence and min_ confidence, if yes then insert to strong rules otherwise remove sub-set. The workflow of the proposed method is shown in Fig. 1.

3.1 A Priori Algorithm-Based Classification

A priori algorithm depth preferences selection process is a step-by-step solution to handle frequent item sets that are combined with Association Rules (AR). Association rule mining is a powerful technique to find out relationships among several datasets in large databases and extracts more frequent data set based on given conditions. It generally correlates one set of item with other sets of item in the database. A minimum support threshold and a minimum confidence threshold, find all the rules

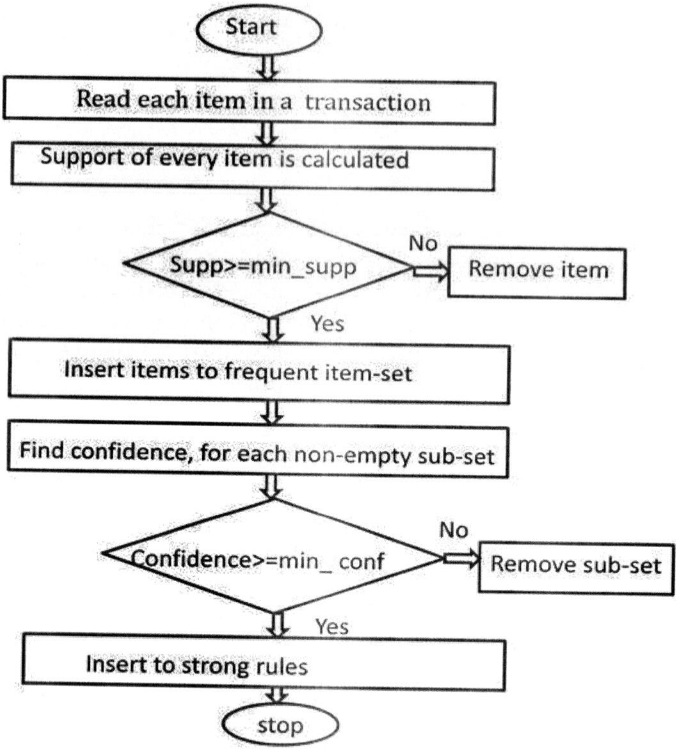

Fig. 1 Workflow of the proposed method

in the dataset that satisfy the support and confidence thresholds. Association rules may be represented by this common notation:

Let,

$$\mathrm{X} \rightarrow \mathrm{Y}[S, C] \tag{1}$$

Where S is support and C is confidence for two item sets X, Y. Support refers to the frequency of rules in transactions. Support consists of two values high and low.

$$\text{Support } (X \rightarrow Y[S, C]) = \mathrm{P}(X \cup Y) \tag{2}$$

The confidence is, out of all the transactions that contain X, the percentage that contain Y as well. Confidence of a rule can be computed as support X Y Support X. The confidence of a rule measures the strength of the rule (correlation between the antecedent and the consequent) while the support measures the frequency of the antecedent and the consequent together. Then we can be defined confidence as:

$$\text{Confidence } (X \rightarrow Y) = P\left(\frac{Y}{X}\right) = \frac{Support(X, Y)}{Support(X)} \tag{3}$$

Confidence can also be expressed in percentage (%). An iterative approach known as level-wise search used by apriori and prior data is analyzed to get later frequent item sets of data such as K-item sets are used to find out (K + 1)-item sets. It is obvious that all nonempty subsets of a frequent item set must also be frequent. There are two-step processes in Apriori algorithm first one is the join step and second one is the prune step. Join step denotes that to find further frequent item set suppose L_K, a set of candidate K-item sets is generated by joining L_{k-1} with it and it is indicated as C_k. Suppose to find the item sets L_3 we have to consider all L_{3-1} item sets and join with L_3. Prune step denotes to reduce the size of set of candidates C_k. C_k is superset of L_{k-1}. That means item sets in C_k may or may not be frequent but all frequent sets are in there. Now after getting frequent item sets we have generated the strong association rules. Those the rules are satisfied both minimum support and minimum confidence. From minimum support we will get frequent item sets. Here is the process how to deal with constraint minimum confidence. For getting association rules we follow following steps: All nonempty subsets "s" of frequent item sets l must be generated. For every nonempty subset s of l, output "s => (l – s)" If support_count (l)/support_count(s) > = minconf (1), where minimum confidence is given threshold value.

4 Experimental Works

Several python libraries were used such as Pandas, Numpy, CSV etc. for the experiment. Below the explanation of our algorithmic approach has given:

(i) List generation of each individual item C_1 (): Read the data from CSV file and made a list of it then counts of all individual data have been stored in a dictionary.
(ii) Generation of one frequent item sets—L_1 (): After listed out all items now all counts are compared with minimum support and the values had a support greater than support threshold stored in a new dictionary.
(iii) Generation of two frequent item sets-C_2 (): Now the system automatically traversed through all item sets of C1 to find two item sets that are identical.
(iv) Generation of L_2 (): All item sets found in C_2 further check if they are existed in an individual item set then they are added to list L_2 and then threshold by minimum support. And then a function was needed to find out the length of all two frequent item sets.
(v) Generation of L_2 (C_2, data): C_{2_i}—L_{2_i}—L (length of individual two frequent item) The process repeated for three and more frequent item sets and every time generate a new dictionary by appending the old dictionary values and comparing with threshold value.

Table 1 Participant distribution with age groups of facing harassment

Total participants 2300	
Age of facing harassment	Participant distribution
>18	1169
18–24	577
25–34	180
Above 35	154
Don't remember	220

(vi) Association rules generation (): Now for all frequent items in final list generated their total support and the all combination of frequent items have been considered by splitting in left and right manner and generated support of these combinations. If total support/combination support is greater than minimum confidence value, these are added to list of rules and finally an output file of rules was generated.

Using the minimum support new dictionary of items have been created from old dictionary. Assume that the minimum support is 200 and minimum confidence is 45%.

Here Table 1 represent the participant distribution for facing harassment as their age.

Tables 2 and 3 represent the data with total support and data with minimum support respectively. Now go through all item sets of L_k to find two item sets that are identical. Then the data has been stored in a list C_k in a sorted manner and make

Table 2 Impact of harassment with total support_count

Total participants 2300			
Attributes	Support_count	Attributes	Support_count
Anxiety	1060	Lack of intimacy and enjoyment of social activities	287
Intense fear	618	Degradation of performances in study or work	508
Ongoing fears	860	Face difficulties with communication	309
Ongoing guilt feeling	168	Degradation of performances in study or work	508
Depressions	837	Under 18	1169
Sleep disturbances or nightmares	420	18–24	577
Avoidance behaviors	84	25–34	180
Headaches	168	Above 35	154
Disrupted work life	419	Don't remember	220

Table 3 Generate group of one frequent item with minimum support_count

Total participants 2300	
Attributes	Support_count
Anxiety	1060
Intense fear	618
Ongoing fears	860
Depressions	837
Sleep disturbances or nightmare	420
Disrupted work life	419
Face difficulties with communication	309
Lack of intimacy and enjoyment of social activities	287
Degradation of performances in study or work	508
Under 18	1169
18–24	577
Don't remember	220

a set of C_k to avoid repetition. If item set in C_k belongs to an individual item list, it has been added to list Ct, and its support is updated by 1.

4.1 Association Rule

Total support has been calculated for each item sets in the frequent item list then all possible combination of item sets has been made by splitting them and support has been generated for this combination from dictionary then it has been added as a rule, then when the calculated support_count(s) is greater than minimum confidence, it is written in a list. Considering minimum confidence threshold is 45%. The resulting association rules are shown in Table 5.

Table 4 represents two frequent items with support and two frequent data with minimum support_count (Table 5).

5 Experimental Results and Discussions

At first data was represented as Pandas data frame format and then it is converted in list format with the help of Numpy library for making the data compatible with python then apriori and FP-growth algorithm have been applied to the data set.

In Fig. 2, shows to organize all impacts based on frequency of occurrences by using minimum support value.

Table 4 Two frequent Attribute sets with support_count and the compare with minimum support_count

Total participants 2300				
Attributes	**Support_count**		**Attributes**	**Support_count**
('Degradation of performances in study or work', 'Disrupted work life')	220		('Degradation of performances in study or work', 'Disrupted work life')	220
(' intimacy and enjoyment of social activities', 'Ongoing fears')	63		('Intense fear', 'Under 18')	396
('Face difficulties with communication', 'Intense fear')	105	**Compare with minimum support_count**	('Anxiety', 'Degradation of performances in study or work')	353
('Intense fear', 'Under 18')	396		....	....
('Ongoing fears ', 'Under 18')	189		....	....
('Anxiety', 'Degradation of performances in study or work')	35		('Anxiety', 'Under 18')	639
....	....		('Depressions', 'Intense fear')	264
....	....		('Ongoing fears', 'Under 18')	595
('Anxiety', 'Under 18')	639		('18-24', 'Anxiety')	289
('Depressions', 'Intense fear')	264		(' Intimacy and enjoyment of social activities', 'Face difficulties with communication')	287
('Ongoing fears', 'Under 18')	595	→	....	....
('18-24', 'Anxiety')	289		....	....
(' intimacy and enjoyment of social activities', 'Face difficulties with communication')	287		('Anxiety', 'Ongoing fears')	506
....	....		('Anxiety', 'Intense fear')	418
....	....		....	....
('Intense fear', 'Sleep disturbances or Nightmares')	42		....	....
(Anxiety', 'Ongoing fears')	506		('Depressions', 'Under 18')	595
('Anxiety', 'Intense fear')	418		('Anxiety', 'Depressions')	529
('Anxiety ', 'I dont remember')	21		('Depressions', 'Sleep disturbances or Nightmares')	221
....	....			
....	....			
('Anxiety', 'Face difficulties with communication')	168			
('Degradation of performances in study or work', 'I dont remember')	42			
('Depressions', 'Under 18')	595			
('Anxiety', 'Depressions')	529			
('Anxiety ', 'Ongoing fears')	126			
('Depressions', 'Sleep disturbances or Nightmares')	221			

In Fig. 3 shows the age frequency of respondent and represents the period when they mostly face harassment. Based on our research it is clear that teenager, age below 18, is most vulnerable to harassment. Based on our experiment, we have shown the relationship between impacts and respondent age. From Fig. 4, it is clear that impacts mostly dominate over teenager and respectively to other age groups.

Figure 5 shows the group of association rule (yellow circle) between several harassment impacts and age group (green circle). It is clear that several impacts made strong association rules among them and with age groups generated by association rule mining.

Table 5 Shows the association rules between deferent impacts and age group

Rules	Confidence	Results Status
['Ongoing fears'] => ['Under 18']	(0.69)	Accepted
Lack of intimacy and enjoyment of social activities'] => ['Face difficulties with communication']	(1.0)	Accepted
['Under 18'] => ['Depressions']	(0.38)	Rejected
....	....	....
....	....	....
['Intense fear'] => ['Anxiety']	(0.67)	Accepted
['Anxiety'] => ['Under 18']	(0.60)	Accepted
['Ongoing fears'] => ['Anxiety']	(0.58)	Accepted
['Anxiety', 'Depressions'] => ['Under 18']	(0.49)	Accepted
['Anxiety', 'Intense fear'] => ['18-24']	(0.36)	Rejected
['Anxiety','ongoinging fears'] => ['18-24']	(0.29)	Rejected
['Intense fear', Under 18] => ['Anxiety']	(0.58)	Accepted
['Under 18'] => ['Depressions']	(0.38)	Rejected
....	....	....
....	....	....
['Depressions', 'Under 18'] => ['Anxiety']	(0.61)	Accepted
['Anxiety', 'Ongoing fears'] => ['Depressions']	(0.56)	Accepted
....	....	....
....	....	....

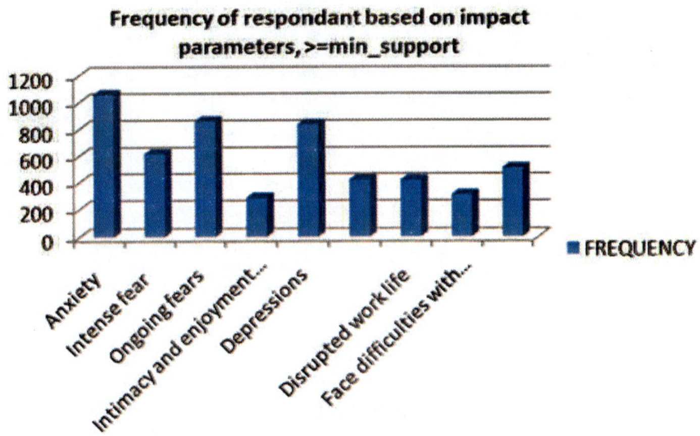

Fig. 2 Age frequency of respondent and represent the period when they mostly face harassment

6 Conclusion

Over the past few decades, research, activities, and funding have been dedicated to recruiting, retaining, and improving women in a variety of workplaces, especially science, engineering and medicine. However, as women increasingly enter these fields they face bias and barriers and it is not surprising that sexual harassment is one of these barriers. We are used association rules in this research. From both association rules generated by Apriori technique and FP-Growth technique, it is clear

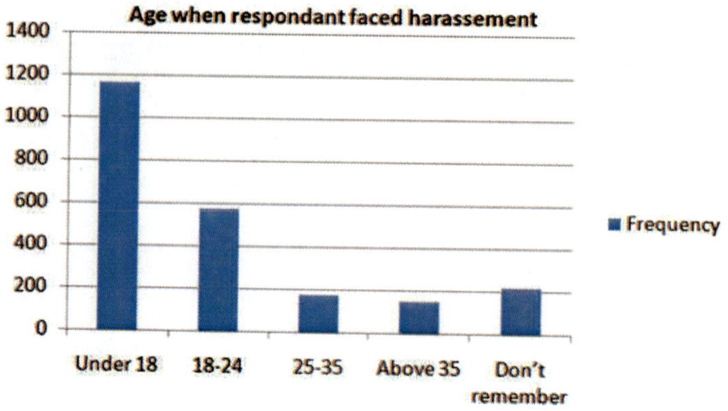

Fig. 3 All impacts based on frequency of occurrences by using minimum support value

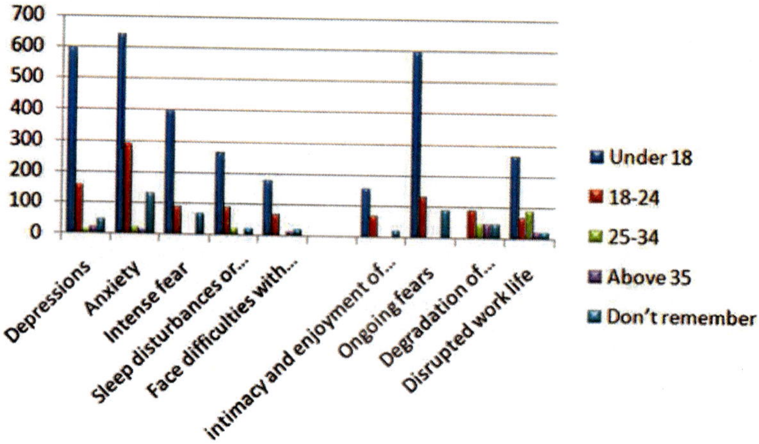

Fig. 4 Impacts mostly dominates over teenager

that most vulnerable age group is teenager that faces harassment most. It is also shown that anxiety, depressions, intense fear, face diffculties on communication are most frequent impacts that happened and generate strong association rules. According to the results generated by algorithm it is shown that anxiety, depressions, ongoing fear have strong association with age group under 18 individually and again these impacts made association with other frequent impacts and with age group. However, large datasets have ignored during the analysis and development. Data domain area is only eight divisions City Corporations. In addition, Apriori algorithm has some drawbacks such as excessive iterations, uniform minimum support count, lack of partitioning and sampling and inability to find rare material. Future research will attempt to address the limitations of the Apriori algorithm, the larger data from greater region.

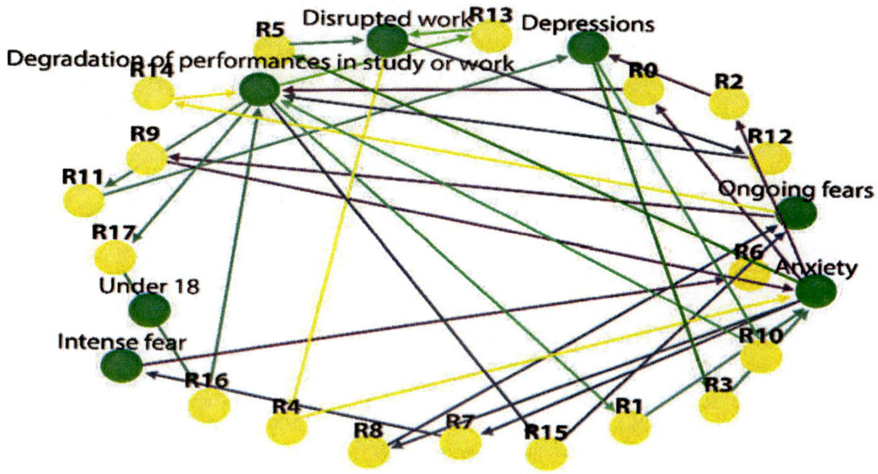

Fig. 5 Group of association between several harassment impacts and age groups

References

1. Fitzgerald LF, Shullman SL, Bailey N, Richards M, Swecker J, Gold Y, Weitzman L (1988) The incidence and dimensions of sexual harassment in academia and the workplace. J Vocat Behav 32(2):152–175
2. Gelfand MJ, Fitzgerald LF, Drasgow F (1995) The structure of sexual harassment: a confirmatory analysis across cultures and settings. J Vocat Behav 47(2):164–177
3. Bates LM, Schuler SR, Islam F and Islam MK (2004) Socioeconomic factors and processes. Associated with domestic violence in rural Bangladesh. Int Fam Plann Perspect 30(4):190–199
4. Koenig MA, Ahmed S, Hossain MB, Mozumder AKA (2003) Women's status and domestic violence in rural Bangladesh: individual-and community-level effects. Demography 40(2):269–288
5. Karim S (2005) Gendered violence in education: realities for adolescent girls in Bangladesh: Dhaka: Action Aid Bangladesh
6. Timmerman G (2003) Sexual harassment of adolescents perpetrated by teachers and by peers: an exploration of the dynamics of power, culture, and gender in secondary schools. Sex Roles 48(5–6):231–244
7. BNWLA (2013) Ending impunity: monitoring report for the implementation of the domestic violence (Prevention and Protection) Act 2010. Bangladesh National Woman Lawyers' Association, Dhaka
8. Senthilingam M (2017) Sexual harassment: how it stands around the globe—CNN. Retrieved 15 Oct 2018 from https://edition.cnn.com/2017/11/25/health/sexual-harassment-violence-abuse-global-levels/index.html
9. BRAC (2011) Eve teasing is undeniably a form of sexual harassment-unite against it! Retrieved 16 Jan 2019 from http://www.brac.net/latest-news/item/469-eve-teasing-is-undeniably-a-form-of-sexual-harassment-unite-against-it

Fuzzy Logic Controller Optimized by MABSA for DC Servo Motor on Physical Experiment

Nurainaa Elias and Nafrizuan Mat Yahya

Abstract This paper represents the control system of the DC servo motor using fuzzy logic controller optimized by MABSA. The fuzzy logic controller that can use in a wide range is well known to the industry application and control system. However, there are still problem with the speed and position control of DC servo motor. Both speed and position cannot be balanced when loading and unloading materials. Therefore, the fuzzy logic controller will be designed using the Matlab toolbox and then will be optimized by the modified adaptive bats sonar algorithm (MABSA) to solve this problem. The best position of the range of the membership functions will be generated through the algorithm and then will be inserted in the membership function of the designed fuzzy logic controller. After the proposed design is fully developed, an experiment will be carried out to test the performance. The performance will be in terms of rising time, settling time and percentage of overshoot. The experiment will be using Arduino as the microcontroller and encoder as the feedback. The experiment will be compared with the cases that use the fuzzy logic controller only without the optimization of MABSA. The result shows that the proposed design gives a 19% improvement in rising time and 8% in settling time. In conclusion, the proposed design of FLC optimized by MABSA is better compared to FLC without optimization.

Keywords Fuzzy logic controller · Bat algorithm · DC servo motor · Optimization · Matlab and simulink

1 Introduction

DC motor is the most widely used as an actuator for continuous motion production and whose speed of rotation can be easily controlled [1]. The function of the DC motor that has the stationary part (stator) and the rotating part (rotor) making them ideal for use in applications where speed and position control are required [2]. There are

N. Elias (✉) · N. Mat Yahya
Universiti Malaysia Pahang, Gambang, Malaysia
e-mail: nurainaa_elias@yahoo.com

A. F. Ab. Nasir et al. (eds.), *Recent Trends in Mechatronics Towards Industry 4.0*, Lecture Notes in Electrical Engineering 730, https://doi.org/10.1007/978-981-33-4597-3_51

three types of DC motors which are the brushed, brushless and servo motor [2]. The brushed motor generates a magnetic field in the wound rotor by passing an electrical current through a commutator and the carbon brush [3]. Commonly brushed DC motors are inexpensive, small in size and simple to control. The brushless DC motor generates a magnetic field in the rotor by utilizing a permanent magnet connected to it. The brushless DC motor is typically smaller but more costly than conventional brushed DC motor since they use "Hall effect" switches in the stator [4]. DC servo motor is sort of the brushed DC motor type with some kind of feedback control connected to the rotor shaft [5]. DC servo motor is chosen to be used in this research since it provided feedback function to the system.

Adding a controller to the DC servo motor system will make the response time to become faster. The most commonly be used in the industry nowadays is the proportional-integral-derivative (PID) controller [5]. The advantages of PID are that it has a simple structure and can operate in a wide range of conditions [5]. However, PID cannot get better optimum value during optimization when a swarm intelligence algorithm is applied [6]. Therefore, the fuzzy logic controller is often mentioned as an alternative to PID [6]. The fuzzy logic controller implemented by Mamdani works based on if-then rules. The fuzzy logic works by human thinking making it easier to be used for optimization [7]. The limits of fuzzy sets may also be undefined and uncertain, rendering them valuable for model approximations. The first step in the process of the fuzzy controller synthesis is to identify the fuzzy controller input and output variables [8]. This is achieved along with the controller's intended feature. There are no universal guidelines for choosing such variables, but usually, the variables selected are the regulated machine states, their errors, variance of errors and/or aggregation of errors [8]. There are important structures in the fuzzy controller which are fuzzification, membership functions, rule-based and also defuzzification. The optimization by swarm intelligence algorithm will be done in one of the structures of the fuzzy logic controller.

A swarm intelligence algorithm is the collective behavior of autonomous, self-organized, natural or artificial processes [9]. Examples of swarm intelligence involve social insect foraging parties, mutual travel, social insect spawning, and collaborative sorting and clustering [9]. The well-known algorithm such as Genetic Algorithms (GA), Ant Colony Optimization (ACO), Particle Swarm Optimization (PSO), Differential Evolution (DE), Artificial Bee Colony (ABC), Glowworm Swarm Optimization (GSO), and Bat Algorithm (BA). However, BA is commonly used for solving the optimization problem involving DC servo motor nowadays [10]. Since the modified adaptive bats sonar algorithm (MABSA) is the new developed from BA, MABSA is used for the optimization of the fuzzy logic controller [11]. MABSA is formulated after modifying the search procedures of the original BA and adding a new component to it [11]. As shown in Fig. 1, MABSA works by the concept of echolocation of a colony of bats to find prey.

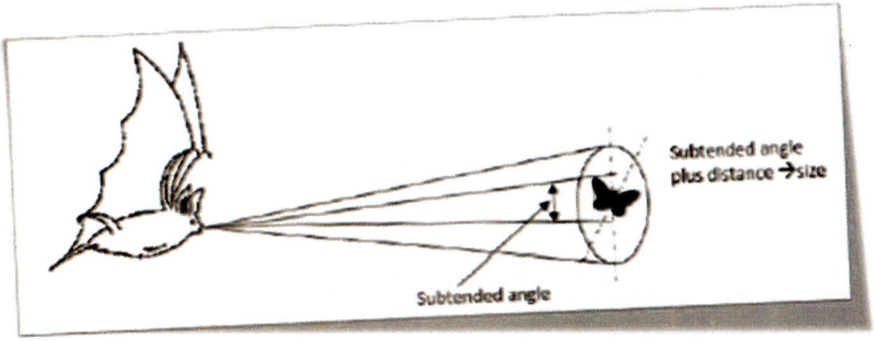

Fig. 1 Bat echolocation for finding prey

2 Methodology

For the experiment, a prototype of DC servo motor control is used for analyzing the result of different parameters. The first parameter is by running the prototype without using the fuzzy logic controller (FLC). The second one is by using FLC and lastly is by using FLC that has been optimized by the modified adaptive bats sonar algorithm (MABSA). Therefore the first step is by generating Arduino coding for running the prototype of the position control. The coding for the position control of the DC servo motor is shown in Fig. 2.

Since the output will be the position, the direction of the motion should be in the forward direction and also backward direction. Hence, the angle used in this experiment will be 0° until 90° (forward) and −90° and 0° (backward).

For the second parameter, the FIS file for designing the controller as shown in Fig. 3 is converted into the Arduino coding. The FIS file is from the fuzzy logic controller (FLC) that has been developed. The FIS file is developed by designing the fuzzy logic controller first. The inputs of the fuzzy are error and change of error while the output will be the position. After the FLC is fully developed in the Matlab toolbox, the FIS file will be built in the Matlab m-file. The input will be the error and change of error. Since there is no turning method used in this parameter, the fuzzy system will employ normalized fuzzy sets, which require that the membership values of all fuzzy sets sum up to unity. Therefore, the value for input 1 and input 2 will be the same for all seven membership functions.

For the third parameter, the FIS file for designing a controller that has been optimized by the modified adaptive bat sonar algorithm as shown in Fig. 4 is converted into the Arduino coding. The input will be the error and change of error. The iteration runs for 30 times and the best fitness choice is where the value is nearer to zero. So 4 values for the best position are chosen to identify the center points of normalized, triangular membership functions for classify a variable's entire fuzzy partition. This property is preserved by a description which codes instead of its absolute position the distances between the adjacent fuzzy sets. After getting the new location of center

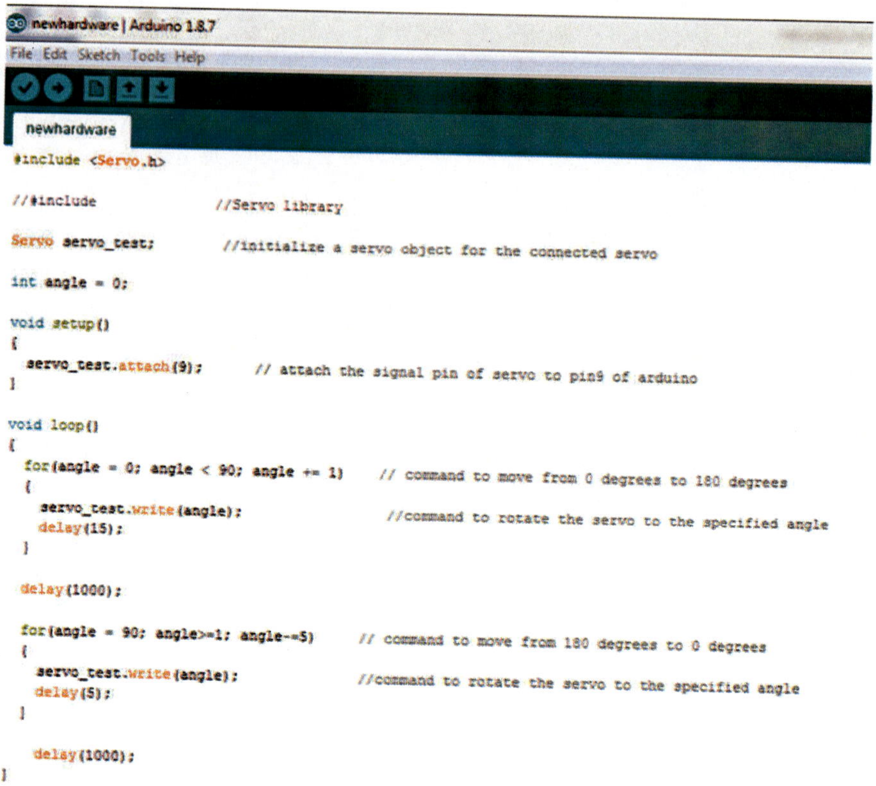

Fig. 2 Arduino coding for DC servo motor

points, a new designing fuzzy logic controller will be built. The steps after getting a new design of FLC is the same as the second parameters where the m-file need to be converted into the Arduino file.

The Arduino coding then will be tested for the functionality and also the real experiment will take part. For the setup of the experiment, Arduino Mega 2560 is used as the microcontroller of the DC servo motor position control. Potentiometer then will be used to control the rotation of the servo motor starting from 0° to 90° positive and also negative. Since servo motors use input for deciding the shaft's location, the location is quite accurately controlled. As a consequence, servo motors are required for monitoring the target orientation, rotate objects, shift legs, arms or robots hands, and move highly accurate sensors. Servo motors are small in size and they could be securely attached to an Arduino, since they have built-in circuitry for regulate their movement. Many servo motors have three black/brown ground wire connections, red power wire (around 5 V) and yellow or white PWM wire as well. For this experiment, the power and ground pins are directly connected to the Arduino 5 V and GND pins respectively. The PWM input will be connected to one of the Arduino's digital output pins which are PWM pin 9.

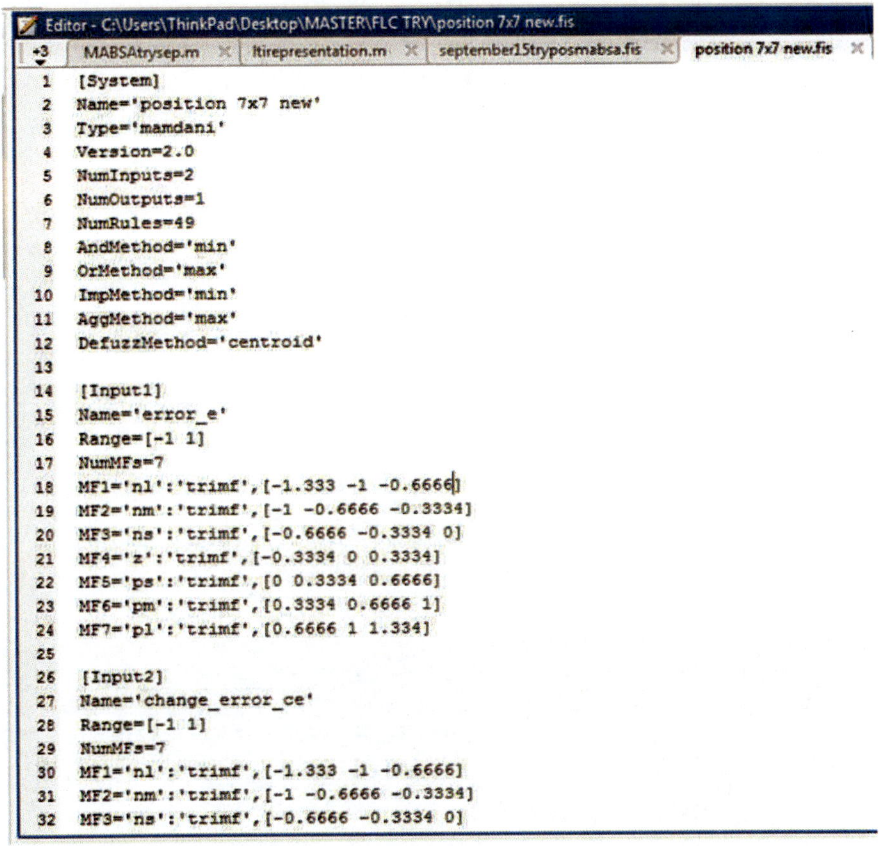


```
[System]
Name='position 7x7 new'
Type='mamdani'
Version=2.0
NumInputs=2
NumOutputs=1
NumRules=49
AndMethod='min'
OrMethod='max'
ImpMethod='min'
AggMethod='max'
DefuzzMethod='centroid'

[Input1]
Name='error_e'
Range=[-1 1]
NumMFs=7
MF1='nl':'trimf',[-1.333 -1 -0.6666]
MF2='nm':'trimf',[-1 -0.6666 -0.3334]
MF3='ns':'trimf',[-0.6666 -0.3334 0]
MF4='z':'trimf',[-0.3334 0 0.3334]
MF5='ps':'trimf',[0 0.3334 0.6666]
MF6='pm':'trimf',[0.3334 0.6666 1]
MF7='pl':'trimf',[0.6666 1 1.334]

[Input2]
Name='change_error_ce'
Range=[-1 1]
NumMFs=7
MF1='nl':'trimf',[-1.333 -1 -0.6666]
MF2='nm':'trimf',[-1 -0.6666 -0.3334]
MF3='ns':'trimf',[-0.6666 -0.3334 0]
```

Fig. 3 The FIS file of fuzzy logic controller

When the angle is 0°, the motor does not rotate. When the angle is positive 90°, the motor rotates in the forward direction else in the negative direction. From the investigation of simulation, the newly developed fuzzy logic that optimized by MABSA is more reliable and accurate when comparing to conventional fuzzy. And to prove the validity of the simulation, the hardware implementation is made up. Figure 5 shows the photo view of hardware implementation.

3 Results and Discussion

For validating the simulation results, an experiment is done. In this experiment, FLC was developed by using Arduino Mega 2560, DC servo motor, Potentiometer and also the prototype of the solar panel.

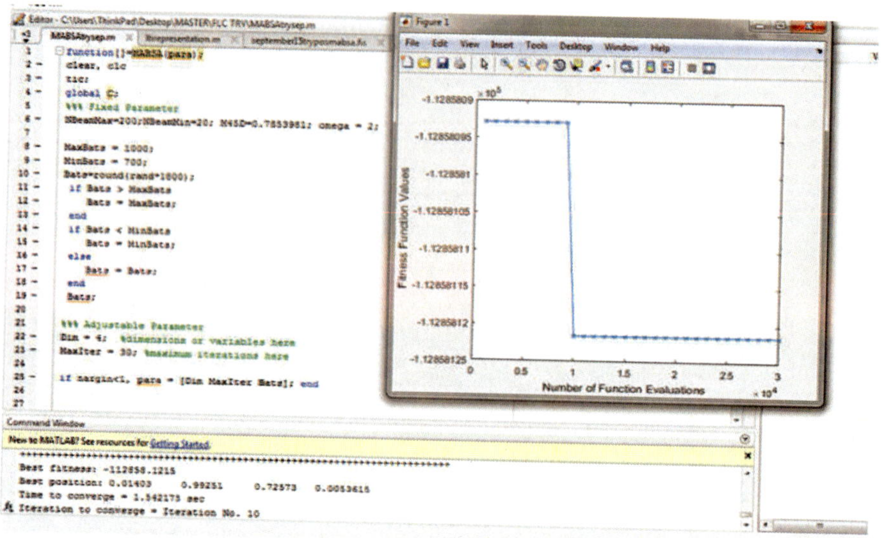

Fig. 4 Arduino coding for DC servo motor position control

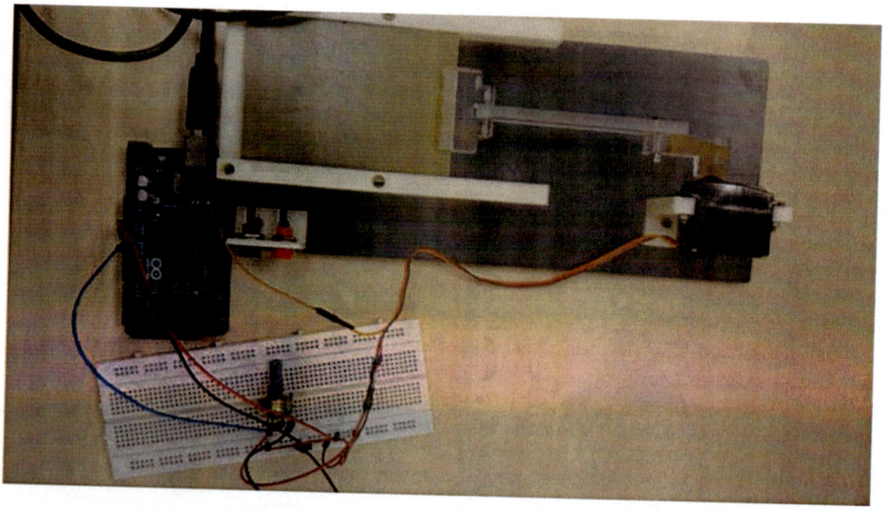

Fig. 5 Experiment setup photo view

The results for the step response for the hardware implementation is shown in Fig. 6. The blue color from the graph represents the reference point while the orange straight line is the DC motor system that did not use the controller for the opening and closing of the solar panel. The grey line has been implemented with the fuzzy logic controller and the yellow line is the line that represents the system that used the FLC optimized by MABSA. In the experiment, the step response graph shows

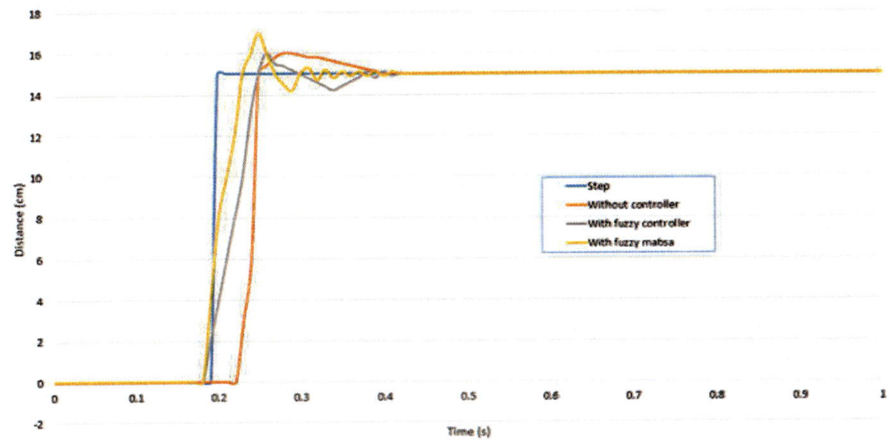

Fig. 6 Experimental result for 3 cases

Table 1 Hardware results comparison

Metrics	Values		
Parameter	Without fuzzy controller	With fuzzy controller	With fuzzy controller optimized by MABSA
Rise time (s)	0.31	0.27	0.25
Settling time (s)	0.41	0.40	0.37
Overshoot (%)	0.67	0.67	13.3

almost the same pattern with the simulation graph. For instance, the FLC has a better performance compared to the system that did not use the FLC but still, the system that used FLC optimized by MABSA has the best step response performance. The experimental results were displayed as shown in Table 1. The values of the response parameters similar to the simulation result also being analyzed. Through the implementation of hardware, the subsystems that used fuzzy logic controller optimized by MABSA show the fastest response than the system without a controller.

3.1 *Cases When Without FLC Is Used as the Reference Point*

For better comparing the result, a case in introduced. The case is included so that the percentage of every parameter can be analyzed more specifically and the result obtained will be more obvious in terms of step response performance. The experimental results that used without FLC as the reference point are displayed as shown in Table 2. The values of the response parameters are converted into a percentage.

Table 2 The experimental results when without FLC is used as the reference point

Metrics	Values	
Parameter	With FLC	With FLC optimized by MABSA
Rise time (s)	12.9	19.4
Settling time (s)	2.4	9.8
Overshoot (%)	0	18.9

As displayed in the table above, almost 19% of the rising time by the system that used FLC optimized by MABSA comparing without using FLC for the experimental result when without FLC is used as the starting point. The most difference in percentage is when the parameter is settling time. The difference is 7.4 for the experiment when comparing the system with FLC and system with FLC optimized by MABSA. And of course, the pattern is still the same where the system that has been employed by FLC optimized by MABSA give better result compared to the system with FLC only.

3.2 *Cases When with FLC Only Is Used as the Reference Point*

The comparison for these cases is to shows that the system that has the controller has a faster step response compared system that did not use the controller. The experimental results that used only FLC as the reference point are displayed as shown in Table 3. The values of the response parameters are converted into a percentage.

The negative sign means that the percentage (time) of the system without FLC is less compared to the system with FLC. The difference between the system without the controller and with the controller can be seen clearly as the highest difference is when the rise time between both of them is 22.2 for the experiment. Summarily, the system with FLC optimized by MABSA has the best step response performance for all cases.

From the overall graph and tabulated results, both show that the system with optimized MABSA gives better performance in terms of the rising time and settling time. Comparing with the system that did not use fuzzy logic controller, MABSA is 0.6 s faster in the rise time and also settling 0.3 s getting to achieve settling time

Table 3 The experimental results when with FLC only is used as the reference point

Metrics	Values	
Parameter	Without FLC	With FLC optimized by MABSA
Rise time (s)	−14.8	7.4
Settling time (s)	−2.5	7.5
Overshoot (%)	0	18.9

compared to the system that has fuzzy logic but did not optimize by MABSA. For the results for cases when the system without FLC is used as the reference point, FLC optimized by MABSA shows 19.4% more in rising time and also 9.8% in settling time. While the system with FLC only gives 12.9 and 2.4% for rising time and settling time respectively. The performance results also the same when the system with FLC only is used as the reference point. 7.4% and 7.5% in rising time and settling time respectively for the system with FLC optimized by MABSA. However, the system without FLC provides a result that worse than with FLC only because the value shows a negative sign which is −14.8 and −2.5 for rising time and settling time respectively. Therefore, based on the results, the proposed FLC with MABSA gives faster response time and settling time without exceeding the maximum percentage of overshoot.

4 Conclusion

The experimental setup for verifying the performance of the DC servo motor with the fuzzy logic controller optimized by MABSA has been presented. The goal of this project was accomplished because the MABSA optimized Fuzzy Logic Controller (FLC) is operational efficiency and provided the result based on a user's requirement (fuzzy rules). The results also show that the proposed FLC optimized by MABSA gives an improvement in rising time and settling time compared to the system without FLC and the system without MABSA. For future works, the proposed FLC optimized by MABSA can be compared with other types of swarm intelligence algorithms like PSO or GA for better validation and comparison.

Acknowledgements This research is funded by the Fundamental Research Grant Scheme (FGRS) from Ministry of Higher Education Malaysia under research grant RDU190178. The authors would like to thank the funder for encouraging this research.

References

1. Krishnan R (2017) Permanent magnet synchronous and brushless DC motor drives. CRC Press, Natick, MA
2. Xia CL (2012) Permanent magnet brushless DC motor drives and controls. Wiley, Singapore
3. Xue XD, Cheng KW, Cheung, NC (2008) Selection of electric motor drives for electric vehicles. In: 2008 Australasian Universities power engineering conference, pp 1–6. IEEE
4. Song JH, Choy I (2004) Commutation torque ripple reduction in brushless DC motor Drives Using A Single DC current sensor. Int J Power Electron 19(2):312–319
5. Bindu R, Namboothiripad MK (2012) Turning of PID controller for DC servo motor using genetic algorithm. Int J Emerg Technol Adv Eng 2(3):310–314
6. Liu J, Zhang P, Wang F (2009) Real-time DC servo motor position control by PID controller using Labview. In: International conference on intelligent human-machine system and cybernetics, pp 206–209

7. Elias N, Yahya NM (2019) Comparison of DC motor position control simulation using MABSA-FLC and PSO-FLC. In: 15th international colloquium on signal processing & its applications (CSPA). Penang, pp 39–42
8. Elias N, Yahya NM (2018) Simulation study for controlling direct current motor position utilizing fuzzy logic controller. Int J Autom Mech Eng 15(4):5989–6000
9. Parpinelli R, Lopes HS (2011) New inspirations in swarm intelligence: a survey. Int J Bio-Inspired Comput 3(1):1–16
10. Xue F, Cai Y, Cao Y, Cui Z, Li F (2015) Optimal parameter settings for bat algorithm. Int J Bio-Inspired Comput 7(2):125–128
11. Azlan NA, Yahya NM (2019) Modified adaptive bats sonar algorithm with doppler effect mechanism for solving single objective unconstrained optimization problems. In: 15th international colloquium on signal processing & its applications (CSPA), pp 27–30. Penang

Effect of Different Signal Weighting Function of Magnetic Field Using KNN for Indoor Localization

Caceja Elyca Anak Bundak, Mohd Amiruddin Abd Rahman, Muhammad Khalis Abd Karim, and Nurul Huda Osman

Abstract The present work aimed to investigate the signal weighting function based on magnetic field (MF) models to obtain accurate location estimates for indoor positioning system. We compare the state-of-the-art Euclidean distance and three proposed different signal weighting function namely actual weight, square weight and square root weight which used to estimate location using MF. Additionally, the effect of signal weighting function is investigated further using multiple k value of K nearest neighbor (KNN) algorithm. According to the results, the square root weighting function have low position error of 8.156 m than Euclidean distance with improvement of 5.5%. We also found that the use of ($k = 5$) of KNN for square weight of my distance measure give the lowest mean estimation error of 7.188 m.

Keyword Indoor positioning · magnetic field · KNN · euclidean distance

1 Introduction

Recently, indoor localization is receiving increased attention due to widespread use of the smartphone and location-based services application. In order to implement indoor localization services, researchers have studied various indoor positioning techniques using different technologies, including Wi-Fi [1], Bluetooth [2] and Zigbee [3]. System based on Wi-Fi and Bluetooth are energy-consuming which mobile user location tracking requires continuous access point scanning. In addition, WiFi signal is weak or not available in some scenarios, such as underground parking areas.

C. E. A. Bundak · M. A. Abd Rahman (✉) · M. K. Abd Karim · N. H. Osman
Faculty of Science, Universiti Putra Malaysia, 43400 UPM Serdang, Malaysia
e-mail: mohdamir@upm.edu.my

C. E. A. Bundak
e-mail: gs52905@student.upm.edu.my

M. K. Abd Karim
e-mail: mkhalis@upm.edu.my

N. H. Osman
e-mail: nurulhuda@upm.edu.my

A. F. Ab. Nasir et al. (eds.), *Recent Trends in Mechatronics Towards Industry 4.0*, Lecture Notes in Electrical Engineering 730, https://doi.org/10.1007/978-981-33-4597-3_52

In contrast, signals obtained from the geomagnetic field are distorted by large iron structures and present stable characters. Indoor magnetic field (MF) is a pervasive field of geomagnetic-induced anomalies. Study in [4] shows that the magnetic fields produced in the indoor environment were relatively stable and reproducible. Moreover, the values of magnetic field signal (MFS) at different positions are different and hence it is possible to position a smartphone by fingerprint mapping of MFS measurements. To incorporate, fingerprinting technique, the measurements must be stored in a database. The database contains coordinates of reference points (RPs) and the magnetic field measure along a smartphone's three axes vector (x, y, z). The observed magnetic field fingerprint is compared to the one stored in the database during the online phase, and then the coordinate with the closest match is determined as the user's estimate location. Due to the stability and uniqueness of the magnetic field, many researchers have developed a number of magnetic field-based positioning systems. MaLoc [5], an indoor localization system based on magnetic fingerprints proposes reliability enhanced particle filter. They include an estimate of the complex step-length and heuristic sampling. Consequently, the tests of their systems in a large building obtained an average precision of 12 m. Research by [6] use fusion algorithm utilizes the extended Kalman filter (EKF) before the particle filter (PF) scheme to combine pedestrian dead reckoning (PDR) with the magnetic fingerprint, using a small covariance ellipse rather than the whole localization area. Their algorithm improves the inherent blindness and solve particles degradation issues in the traditional PF scheme. The localization accuracy for the fusion algorithm are 1-2 m in a large building when the user walks slowly. However, all of these methods estimate track based on the path of the corridor and not exact discrete location position in the building. Additionally, existing PF algorithms have several disadvantages which are re-sampling of the classical PF scheme may cause the problem of particle degradation and high computational cost.

Since the KNN algorithm is relatively simple, consumes nearly zero training effort, and still provides comparable accuracy to the more complex techniques, therefore it is still widely being used as a benchmark positioning algorithm. The difference of signal level is usually calculated using Euclidean distance. There are many alternative functions to be used with KNN which in [7] use the information of each AP signal at each fingerprint to weight the signal distance calculation in KNN algorithm. In [8] KNN is applied to accelerometer and gyroscope data to estimate k paths by adding points that meet the distance and orientation constraints between the new and previously predicted points. On the other hand, [9] use KNN algorithm to test their UJIIndoorLoc-Mag database by using discrete location method and continuous location method using KNN algorithm. However, they did not compare distance/similarity measurements for the KNN algorithm and magnetic fingerprinting. In [10] create an algorithm of conducts suitable selection and partitioning on the target positioning area before calculating the KNN. Their methods focus on the training data where they divide the area according the MF signals uniqueness at that area.

Although there is various KNN based on magnetic field algorithm utilized by different indoor localization systems, the effect of a signal weighting function in magnetic field localization have yet to be studied. Therefore, this paper using signal difference in indoor location estimation study on the effect of weighting function on the magnetic field.

This paper makes the following contributions. (i) To study the performance of three different weight factor for different signal weighting function with multiple K value for KNN algorithm, (ii) To analyses the changes in the positioning accuracy by using different similarity measure on specific MF vector, (iii) To investigate the magnetic field vectors difference effect on the signal weighting function.

2 Magnetic Fingerprint-Based Map

Each fingerprint in the magnetic map consists of MF readings along a smartphone's three axes (x, y, z), reference point RP, and coordinate of the point (x, y). The magnetic fingerprint at the ith RP in the magnetic map is represented as

$$MF_i = \left[RP_i, Coor_x, Coor_y, m_x, m_y, m_z\right] \tag{1}$$

where MF_i is the magnetic fingerprint data at RP_i, $Coor_x$ and $Coor_y$ are the x and y coordinate of RP_i, and m_x, m_y, m_z are the of MF strength in three vector at RP_i.

The fingerprinting localization system is generally divided into two phases: a training phase (offline phase) and a testing phase (online phase). In the training phase, features of the MF reading at each RP location, are collected, and stored to a database. In the testing phase, localisation algorithm is used to estimate the position by comparing the testing data with the recorded fingerprints from the training data. The testing data can be used to either positioning or evaluating the wellness of the algorithm.

2.1 *Nearest Neighbor Technique*

There are various localisation algorithms could be used to estimate position. The start of the art Nearest Neighbor (NN) localisation algorithm is a matching algorithm use to measure distance $d(m_1, m_2)$ between two signals, $m_1 = \left[m_{x1}, m_{y1}, , m_{z1}\right]$ and $m_2 = \left[m_{x2}, m_{y2}, m_{z2}\right]$ using Euclidean distance formula shown in Eq. 2. The estimation location chooses by the minimum distance between the signals.

$$d(m_1, m_2) = \sqrt{(d_{mx})^2 + \left(d_{my}\right)^2 + (d_{mz})^2} \tag{2}$$

where $d(m_1, m_2)$ is Euclidean distance and d_{mx}, d_{my}, and d_{mz} represents the distance at X, Y and Z vector directions.

K Nearest Neighbor (KNN). Unlike the NN algorithm, KNN is to find k locations with k minimum signal distances to estimate the location. The estimation position signal is determined by taking the average of all those K neighbors locations based on k minimum distance.

$$(x, y) = \frac{1}{k}\sum_{i=1}^{k}(x_i, y_i) \tag{3}$$

where (x_i, y_i) represents the coordinate of the ith RP for estimation position selected by k value and (x, y) represents average estimation position.

2.2 *Weighted MF Signal Distance Similarity Measure Techniques*

The magnetometer in most smartphones can measure the magnetic field, in the form of a vector with three individual elements (m_x, m_y, m_z). as shown in Fig. 1 for which describes the magnetic field component for north, east, and vertical directions, respectively. At different locations the intensity of the Earth's magnetic field is not the same. In addition, the magnetic intensity in vectors of X, Y and Z and the overall magnitude are mostly unique in the various environments.

In this paper, we apply different signal weighting function when two elements are fix based on the Euclidean equations while one vector intensity is change with three different weight. The distance measure between test points (TPs) and RPs of magnetic vector signal at the vector axis X, Y and Z which can be express as m_{f1}, m_{f2} and m_{f3} respectively. The Eq. (2) is converted as follows:

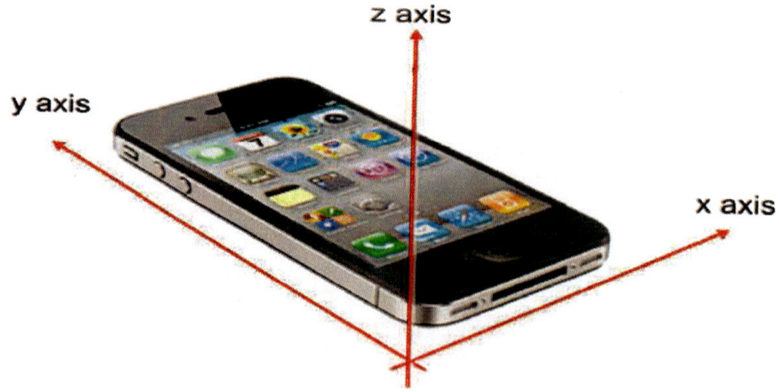

Fig. 1 MF vector axis on smartphone

$$d = \sqrt{\left(d_{mfs1}\right)^2 + \left(d_{mfs2}\right)^2 + (w_i)} \tag{4}$$

where d represents the distance measures, d_{mfs1} and d_{mfs2} represents the fix magnetic vector different of either m_x and m_y, my and m_z or m_x and m_z using the Euclidean distance and w_i represents the distance magnetic field component according to the weight factor at i th magnetic vector distance.

Three different weight factors used are described as follows:

$$\text{w}(i) = \begin{cases} d_{mfi} & ActualWeight \\ (d_{mfi})^2, & SquareWeight \\ \sqrt{d_{mfi}}, & SquareRootWeight \end{cases} \tag{5}$$

where $\text{w}(i)$ represents the weighting factor and d_{mfi} represents the distance magnetic field component. Finally, after introduction of $\text{w}(i)$, the three weight factor become as follows:

$$d_{actual} = \sqrt{\left(d_{mfs1}\right)^2 + \left(d_{mfs2}\right)^2 + \left(d_{mfi}\right)}, \tag{6}$$

$$d_{square} = \sqrt{\left(d_{mfs1}\right)^2 + \left(d_{mfs2}\right)^2 + \left(d_{mfi}\right)^2}, \tag{7}$$

$$d_{SquareRoot} = \sqrt{\left(d_{mfs1}\right)^2 + \left(d_{mfs2}\right)^2 + \sqrt{\left(d_{mfi}\right)}} \tag{8}$$

3 Experimental Data and Mapping

We evaluate the proposed weighting function using real experiments in an indoor environment. The data are collected at the third floor of faculty of science building 1, Universiti Putra Malaysia UPM, Malaysia. The plan is illustrated in Fig. 2. Each map is split equally into train data with 72 RPs (green dot) and testing data with 72 TPs (blue dot) in the experimental area in the space of 21 m × 33 m. The data collection application, installed on an LG Nexus 5 smartphone use to collect the MF fingerprints in four orientations, namely, North, South, East, and West for each point, 1 m apart at a 0.5 m height. The data provided by the magnetometer and the orientation of the device is stored 14 times per second. So, the device frequency band is 0.07 Hz. Table 1 shows the MF database. For 1 RPs is stored 40 magnetic field measurements and for each direction is stored 10 data in 1 location. The total data sample are 2880 data.

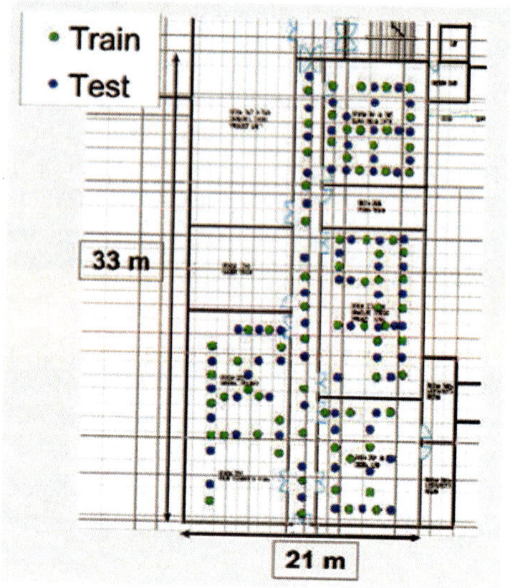

Fig. 2 Floor map of the MF database

Table 1 MF database description

Total reference point	Total direction at 1 location	Total data at 1 direction	Total data at 1 location	Total overall data
72	4	10	40	2880

4 Experimental Results and Analysis

The evaluation metrics used are mean error, minimum and maximum error and cumulative distribution function (CDF) of the position error which is distance between true and estimated position.

4.1 Comparison with Multiple Distance

Distance measures are calculated using different methods, Euclidean, Manhattan, Square and Square Roots and the equation define as follows:

$$d_{Euclidean} = \sqrt{(m_x)^2 + (m_y)^2 + (m_z)^2} \tag{9}$$

$$d_{Manhattan} = m_x + m_y + m_z \tag{10}$$

$$d_{Square} = (m_x)^2 + (m_y)^2 + (m_z)^2 \tag{11}$$

Table 2 Mean error for different distance measure

Euclidean distance	Manhattan distance	Square root distance	Square distance
8.637 m	8.860 m	8.860 m	8.637 m

$$d_{SquareRoot} = \sqrt{m_x + m_y + m_z} \quad (12)$$

where d is the distance measure and m_x, m_y and m_z represents the distance components of magnetic field in each direction.

The comparison results are shows in Table 2 using mean errors as the evaluation. From the table, it can be seen that Euclidean distance and Square Distance have same mean errors with 8.637 m, whereas for Manhattan and Square root are 8.860 m. Euclidean and Square Distance have low errors compare to Manhattan and Square Root distance. Manhattan and Euclidean distance could be used to compare for three elements magnetic vector of online and offline MF signal [11]. Their results show there are not much difference using different intensity distance (Manhattan or Euclidean). Therefore, we use Euclidean distance for our analysis effect different of weight function distance measures.

4.2 *Results of Different Signal Weighting Function for Each MF Signal*

Table 3 shows the mean error for 1NN different signal weighting function for each MF Signal. As can be seen from Table 3, square root weight from Eq. (8) for m_z vector result has lowest mean error with 8.156 m compared to the others, for m_z using square weight has the greatest error with 9.079 m. It can be noted that different signal weighting function that use square root weight has lower location estimation error for m_x and m_z compare to other signal weight. The maximum error is 28.96 m shown in Table 4 while the minimum error is 0.805 m shown in Table 5.

Figure 3 shows the comparison between the weight factor of different signal vector in terms of distance error compare with Euclidean distance lines. The accuracy of actual weight almost the same for all the magnetic vector distance. So, actual weight does not affect much on the changes of positioning accuracy and from Table 3 result

Table 3 Mean Error 1NN difference distance similarity measure for each MF signal

MF Vector distance	Equations			
	(6)	(7)	(8)	(9)
m_z	8.585	9.079	8.156	8.637
m_x	8.685	8.493	8.191	
m_y	8.517	8.357	8.705	

Table 4 Maximum error 1NN difference distance similarity measure for each MF signal

MF vector distance	Equations			
	(6)	(7)	(8)	(9)
m_z	28.960	28.960	28.960	28.960
m_x	28.213	28.960	28.960	
m_y	28.960	28.960	28.960	

Table 5 Minimum error 1NN difference distance similarity measure for each MF signal

MF vector distance	Equations			
	(6)	(7)	(8)	(9)
m_z	0.887	0.887	0.845	0.887
m_x	0.845	0.887	0.887	
m_y	0.805	0.805	0.887	

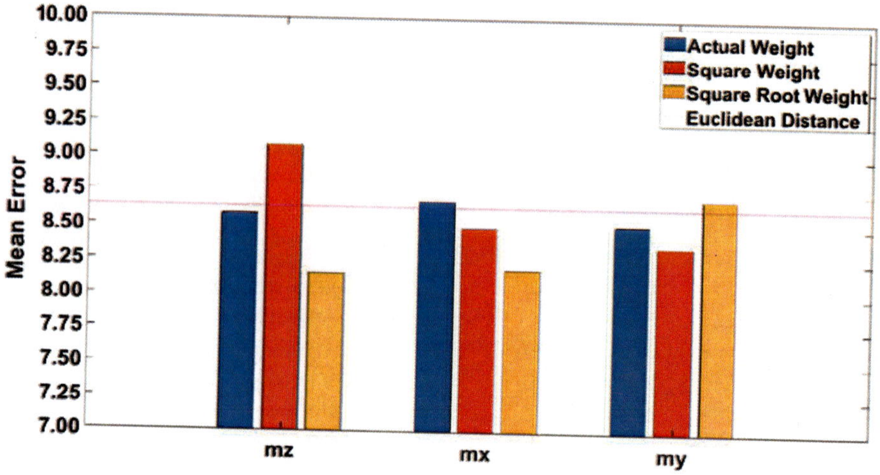

Fig. 3 Performance comparison between the weight factor of different signal vector in terms of distance error

of the actual weight is also slightly similar with Euclidean distance. The accuracy of m_z distance when applied with each weight factors shows changes more than m_x and m_y.

We also vary the k value to see how the different signal weighting function performs as tabulate in Table 6 . It is noted that when each time we increase nearest neighbor for k value, the signal weighting function and Euclidean performs more better. The lowest error achieved is 7.188 m when square weight applies on my distance for (k = 5), whereas the highest error is 9.079 m for square weight of m_z distance at (k = 1). Overall, all the results show that Eq. (6) is best for weight m_x,

Table 6 Mean Error multiple KNN difference distance similarity measure for each MF signal

MF signal vector	K value	Weight function equations			
		(6)	(7)	(8)	(2)
mz	1	8.585	9.079	8.156	8.637
	2	8.185	8.754	8.284	8.074
	3	7.757	8.721	8.067	7.951
	4	7.553	8.581	7.857	7.737
	5	7.416	8.460	7.697	7.430
mx	1	8.685	8.493	8.191	8.637
	2	8.290	8.282	8.422	8.074
	3	7.880	7.900	7.957	7.951
	4	7.835	7.804	7.786	7.737
	5	7.567	7.644	7.548	7.430
my	1	8.517	8.357	8.705	8.637
	2	7.668	7.937	8.254	8.074
	3	7.492	7.406	8.073	7.951
	4	7.355	7.374	8.073	7.737
	5	7.351	7.188	7.832	7.430

Eq. (8) for weight m_z and Eq. (7) for weight m_y. We observe that the error could be reduce when the weights are changes. The researcher could try different weight configuration according to their environment to obtain best result because there is no equation better than other. However, if we change only one weight, it will give better results compared to using distance metrics such as Euclidean, Manhattan, Square and Square root.

We present the accuracy results of the signal weighting function m_z vector for the three weight factor compare with Euclidean distance. Figure 4 shows the results for CDF Actual MF different signal weighting function. As can be seen where within 8 m, the location error is 60% for Square Root weight function, while less than 60% for both Actual and Square Weight. Square root signal weighting function outperforms the others after when the error is larger than 5 m.

5 Conclusion

We presented an analysis on the effect of different signal weighting function. We studied that for 1NN different signal weighting function that use square root weight has lower location estimation error for m_x and m_z compare to other signal weight. We found that lowest error achieved is 7.188 m when square weight applies on m_y distance for (k = 5). We showed that square root signal weighting function outperforms the others after when the error is larger than 5 m by presenting the graph

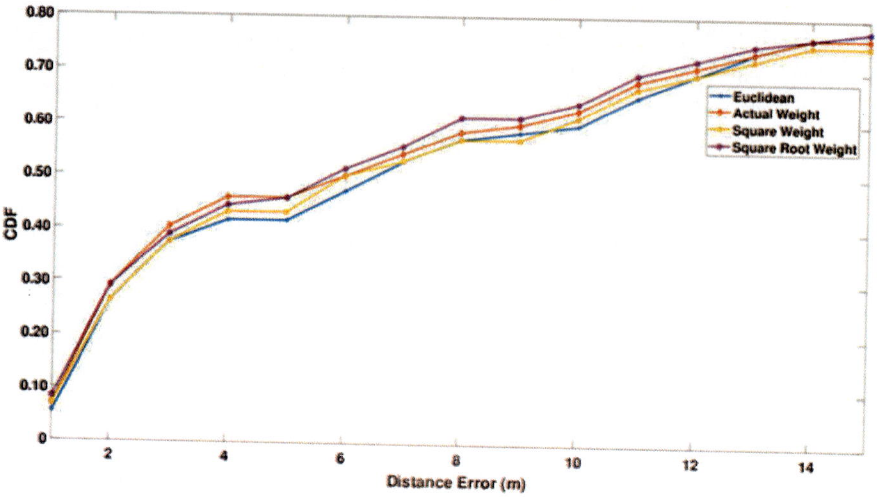

Fig. 4 CDF different signal weighting function of MF signal for 15 m distance error

CDF with different similarity measures for Z magnetic vector. Our analysis under three weighting function demonstrated the effectiveness of the different similarity measures.

Acknowledgements This work is financially supported by Universiti Putra Malaysia and the Ministry of Higher Education, Malaysia under Fundamental Research Grant Scheme (FRGS)—FRGS/1/2017/TK04/UPM/02/5.

References

1. Garcia-Valverde T et al (2013) A fuzzy logic-based system for indoor localization using WiFi in ambient intelligent environments. IEEE Trans Fuzzy Syst. https://doi.org/10.1109/TFUZZ.2012.2227975
2. Wang Q, Feng Y, Zhang X, Sun Y, Lu X (2016) IWKNN: an effective bluetooth positioning method based on Isomap and WKNN. Mob Inf Syst 8765874. https://doi.org/10.1155/2016/8765874
3. Dong ZY, Xu WM, Zhuang H (2018) Research on ZigBee indoor technology positioning based on RSSI. Procedia Comput Sci 154:424–429
4. Gozick B, Subbu KP, Dantu R, Maeshiro T (2011) Magnetic maps for indoor navigation. IEEE Trans Instrum Meas 60:3883–3891
5. Xie H, Gu T, Tao X, Ye H, Lu J (2016) A reliability-augmented particle filter for magnetic fingerprinting based indoor localization on smartphone. IEEE Trans Mob Comput 15:1877–1892
6. Wang G, Wang X, Nie J, Lin L (2019) Magnetic-based indoor localization using smartphone via a fusion algorithm. IEEE Sens J 19:6477–6485
7. Abd Rahman MA, Abdul Karim MK, Anak Bundak CE (2019) Weighted local access point based on fine matching k-nearest neighbor algorithm for indoor positioning system. In 2019

AEIT international annual conference, AEIT 2019. https://doi.org/10.23919/AEIT.2019.8893365
8. Ma Z, Poslad S, Hu S, Zhang X (2018, October) A fast path matching algorithm for indoor positioning systems using magnetic field measurements. In: IEEE international symposium on personal, indoor, and mobile radio communications PIMRC 2017, pp 1–5 (2018)
9. Torres-Sospedra J, Rambla D, Montoliu R, Belmonte O, Huerta J (2015) UJIIndoorLoc-Mag: a new database for magnetic field-based localization problems. In: 2015 international conference on indoor positioning and indoor navigation (IPIN 2015), pp 13–16. https://doi.org/10.1109/ipin.2015.7346763
10. Du Y, Arslan T (2018) A segmentation-based matching algorithm for magnetic field indoor positioning. In: 2017 International Conference on Localization and GNSS (ICL-GNSS), pp 1–5. https://doi.org/10.1109/ICL-GNSS.2017.8376237
11. Li B, Gallagher T, Dempster AG, Rizos C (2012) How feasible is the use of magnetic field alone for indoor positioning? In: 2012 international conference on indoor positioning and indoor navigation (IPIN 2012). https://doi.org/10.1109/IPIN.2012.6418880

The Classification of Electrooculogram (EOG) Through the Application of Linear Discriminant Analysis (LDA) of Selected Time-Domain Signals

Farhan Anis Azhar, Mahfuzah Mustafa, Norizam Sulaiman, Mamunur Rashid, Bifta Sama Bari, Md Nahidul Islam, Md Jahid Hasan, and Nur Fahriza Mohd Ali

Abstract Recently, Human Computer Interface (HCI) has been studied extensively to handle electromechanical rehabilitation aids using different bio-signals. Among various bio-signals, electrooculogram (EOG) signal have been studied in depth due to its significant signal pattern stability. The primary goal of EOG based HCI is to control assistive devices using eye movement which can be utilized to rehabilitate the disabled people. In this paper, a novel approach of four classes EOG has been proposed to investigate the possibility of real-life HCI application. A variety of time-domain based EOG features including mean, root mean square (RMS), maximum, variance, minimum, medium, skewness and standard deviation have been explored. The extracted features have been classified by the linear discriminant analysis (LDA) with the classification accuracy of training accuracy (90.43%) and testing accuracy (88.89%). The obtained accuracy is very encouraging to be utilized in HCI technology in the purpose of assisting physically disabled patients. Total 10 participants have been contributed to record EOG data and the range between 21 and 29 years old.

Keywords Human computer interface · HCI · Electrooculogram · EOG · Machine learning

F. A. Azhar (✉) · M. Mustafa · N. Sulaiman · M. Rashid · B. S. Bari · M. N. Islam
Faculty of Electrical and Electronics Engineering Technology, Universiti Malaysia Pahang, 26600 Pekan, Pahang, Malaysia
e-mail: farhan.anis97@gmail.com

M. J. Hasan
Faculty of Manufacturing and Mechatronics Engineering Technology, Universiti Malaysia Pahang, 26600 Pekan, Pahang, Malaysia

N. F. M. Ali
School of Civil Engineering, Universiti Sains Malaysia, Engineering Campus, 14300 Nibong Tebal, Pulau Pinang, Malaysia

A. F. Ab. Nasir et al. (eds.), *Recent Trends in Mechatronics Towards Industry 4.0*, Lecture Notes in Electrical Engineering 730, https://doi.org/10.1007/978-981-33-4597-3_53

1 Introduction

Some motor related disease, such as injured-vertebra, apoplexy and motor neuron disease (MND) extremely limit the patients' peripheral activities or speech [1]. These physically impaired patients significantly increase the cost of disease burden in the economy of any developed or developing country. To limit the cost due to disease burden, the physically impaired patients should be rehabilitated. Assistive devices could be an excellent initiative in this purpose as these rehabilitation devices are used to improve the quality of the life of disabled people. Recently, speech recognition based HCIs are widely investigated by the researchers, but the research outcome did not meet the expected level due to the surrounding noise of targeted human voice [2]. On the other hand, vision-based head/hand gesture HCI is very promising technology for indoor environment as its performance significantly fall down at outdoor environment [3]. The existing challenge of HCI could be overcome through the EOG signal based HCI technology as the EOG signal precisely detect the eye movement activities [4]. The identification of eye movements for a wide range of uses, such as wheelchair control [5], mobile robot control [6], cursor mouse control [7], eye writing recognition [8] and eye exercise recognition [9].

EOG is the most convenient eye movement detection technique used to analyze the electrical activity around the upward, downward, right and left side of the eyes. These eye movement activities are captured through the appropriate electrode's placements on the head. Together with EOG signal, sometimes the captured signal conveys noise due to the physical movement or external stimulus [10]. Generally, five different phases in HCI need to be follows [11]. First phase is acquisition EOG signal from human's forehead, second is pre-processing of EOG signals to remove noise, third is to extract the most effective features, fourth is to classify the EOG features utilizing machine learning algorithms and the final phase is to translate the classified result into device commands [12, 13]. Among all phases, feature extraction is the most important because inappropriate EOG features will increase the misclassification rate that may create wrong device commands [11]. This could result in a malfunction of the HCI system, which could damage disabled people. HCI researchers have explored many EOG features in the last one decade. Combination of seven feature extraction techniques including maximum peak amplitude value (PAV), the maximum valley amplitude value (VAV), the maximum peak amplitude position value (PAP), the maximum valley amplitude position value (VAP), the area under curve value (AUC), the number threshold crossing value (TCV) and variance of EOG signal (VAR) has been proposed in [14] where the EOG data has been captured through the two channels. These features were utilized to classify the eight classes eye movement activities. Another study in [15] the combination of different time domain features from EOG signal has been proposed.

Here, fourteen features have been extracted and the extracted features have been classified by the K-Nearest Neighbour (K-NN) and Artificial Neuron Network (ANN). Wavelet transform based feature extraction technique has been proposed in [16] to classify the EOG signal where, the wavelet transform analyses the localized variation of power within a time series. By breaking down a time domain arrangement into time–frequency domain, the amplitudes of a wavelet include enormous changes inside an assigned timeframe and incredibly little qualities outside of that time while being band-constrained regarding their recurrence content. In [17], the features in terms of power spectral density (PSD) and autoregressive parameter (AR) have been computed from wavelet coefficients to investigate the EOG data. The extracted features have been classified using K-NN and feed forward Neural Network (NN) with the accuracy of 69.17 and 80% respectively. Here, different EOG activities including straight, up, down, right, left and eye blinking has been recognized. In this paper, five class EOG data have been analyzed. Eight distinct features in time domain have been investigated and these features have been classified by the LDA. The performance of classifier has been evaluated by a wide range of evaluation metrics. This paper has been organized in the following sections i.e. Sects. 2 and 3 discusses issues related to methodology, results and discussion respectively; finally, Sect. 4 deals with the conclusion.

2 Methodology

There are several important components in every HCI system including signal acquisition, pre-processing of signals, feature extraction and classification [11]. Figure 1

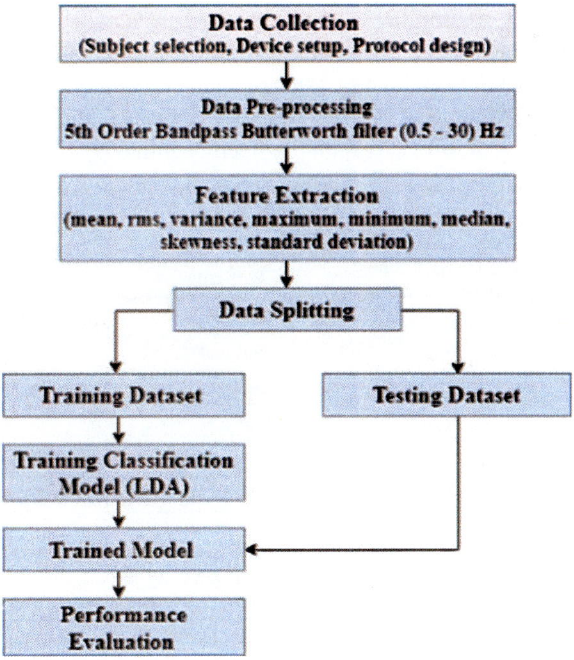

Fig. 1 Complete methodology of the proposed study

shows the complete flow diagram of the conducted experiment where the first stage is the EOG data collection. The captured EOG data was filtered in the pre-processing stage by the 5th-order Butterworth filter with a frequency range of 0.5–30 Hz. The features in terms of different statistical features have been extracted from the filtered EOG data. Finally, the extracted features have been classified using LDA based machine learning algorithm.

2.1 EOG Data Collection

The entire data collection session has been carried out in the AppECE laboratory, Faculty Technology of Electrical & Electronics Engineering, Universiti Malaysia Pahang, Malaysia. Total 10 healthy participants (5 males and 5 females) have been contributed to record EOG data. Table 1 shows details description about participants. Before data collection, all the participants have been asked to answer a list of questionnaires. All the EOG contributing participants were healthy and they did not have any history of physical or mental illness. They are not currently taking any medication that will affect their EOG signal. They were clearly explained about their contribution in the EOG data collection and trained earlier the data collection day. From 10 subjects, a total of 1080 observations have been recorded. Total 108 observation have been recorded for 1 subject with 27 observation each movement eyes upward, right left and downward.

The EOG data has been recorded utilizing the Mindata chip (Multifunction Biosensor Module). The sampling rate of this chip is 256 Hz. In this experiment, five electrodes have been utilized to capture EOG data. Figure 2 shows the electrode name and their position. To place the electrode, the internationally recognized method namely 10/20 system of electrode position has been followed [18].

The entire EOG data has been captured by following a data acquisition protocol shown in Fig. 3. To stimulate the EOG data, a power point slide has been prepared

Table 1 Description of participant for data collection

Participant	Sex	Age
Subject-1	Male	22
Subject-2	Male	22
Subject-3	Male	24
Subject-4	Male	25
Subject-5	Male	21
Subject-6	Female	22
Subject-7	Female	27
Subject-8	Female	26
Subject-9	Female	29
Subject-10	Female	22

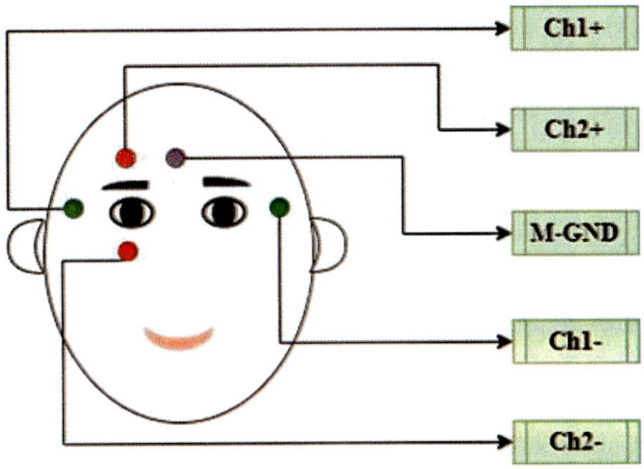

Fig. 2 Electrode positioning system for the proposed EOG data collection

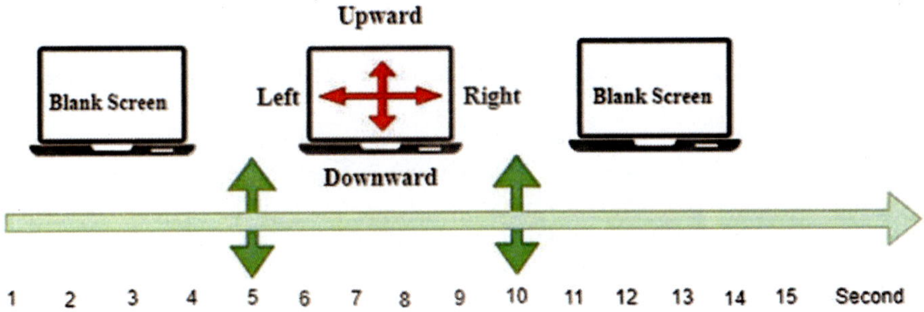

Fig. 3 EOG data acquisition protocol

and the duration of this slide was 15 s. The first and last five seconds was blank screen whereas the middle five seconds displayed an arrow symbol (upward, right, left and downward). The participant has been asked to follow the slide of middle five seconds by moving the eyeball. The stimulating slide has been placed about 0.45 m in front of the participant. During this procedure, the participant was instructed to restrain voluntary blinking as much as possible specially during middle five seconds.

2.2 *Feature Extraction*

After data collection, the EOG has been preprocessed through time window selection and filtering. During time window selection, the samples from first five seconds and last five seconds have been eliminated. The samples from middle five seconds have been selected for further analysis. Then the selected EOG has been filtered through

Table 2 Formulas for statistical features

Features	Formula
Root mean square (RMS)	$\mathrm{RMS_K} = \sqrt{\frac{1}{N}\sum_{n=1}^{N} x_n^2}\mathrm{RMS_K} = \sqrt{\frac{1}{N}\sum_{n=1}^{N} x_n^2}$
Mean	$Mean = \frac{1}{N}\sum_{n=1}^{N} x_n$
Variance	$\mathrm{RMS_K} = \sqrt{\frac{1}{N}\sum_{n=1}^{N} x_n^2}\mathrm{RMS_K} = \sqrt{\frac{1}{N}\sum_{n=1}^{N} x_n^2}$
Skewness	$Skew = \frac{\frac{1}{N}\sum_{n=1}^{N}(x_n-\mu)^3}{\left(\frac{1}{N}\sum_{n=1}^{N}(x_n-\mu)^2\right)^{\frac{3}{2}}}$
Standard deviation	$SD = \sqrt{\frac{1}{N-1}\sum_{n=1}^{N}(x_n-\mu)^2}$ Where, $\mu = \boldsymbol{mean}$

the 5th order Butterworth bandpass filter. The frequency of bandpass filter range was 0.5–30 Hz.

In this study, different time-domain based statistical features including maximum, root mean square (RMS), mean, variance, minimum, medium, skewness and standard deviation have been extracted. Table 2 shows the equation for RMS, mean, variance, skewness and standard deviation [19].

2.3 *Classification: Linear Discriminant Analysis (LDA)*

LDA is used to find the linear feature vector combinations that define the features of the corresponding signal. LDA attempts to divide two or more sets with various types of objects or events. To accomplish this task, it uses hyperplanes. Isolation is acquired by searching for the projection that maximizes the gap between the means of classes while the interclass variation is minimized. This technique has a very limited requirement for computing and is easy to use. Subsequently, In BCI systems such as P300 speller, motor imagery based BCI, LDA has been used successfully [20].

2.4 *Performance Evaluation*

In the proposed system, initially, the performance has been evaluated by the classification accuracy which is calculated by Eq. (12) [11].

$$Accuracy = \frac{TP + TN}{TP + TN + FP + FN} \times 100\% \quad (1)$$

where TP is true positive and FN is false negative, TN is true negative, and FP is false positive. The sensitivity, specificity, precision, error rate, recall, F1-score, false positive rate and kappa formulas are indicated in the Eq. (2–6) [21] respectively.

$$Sensitivity = \frac{TP}{TP + FN} \times 100\% \quad (2)$$

$$Precision = \frac{TP}{TP + FP} \times 100\% \quad (3)$$

$$Specificity = \frac{TN}{TN + FP} \times 100\% \quad (4)$$

$$\text{Error Rate} = (100 - \text{accuracy})\text{Error Rate} = (100 - \text{accuracy}) \quad (5)$$

$$F1 = \frac{1 * Precision * Recall}{Precision * Recall} \quad (6)$$

3 Result and Discussion

In the proposed method, four classes EOG signal has been classified using the LDA classifier. During classification, the four classes EOG data including upward, right, left, downward have been denoted by 1, 2, 3 and 4 respectively. In order to assess the performance, the confusion matrix has been computed. Figure 4. Represent the confusion matrix for training and testing. From the confusion matrix of testing dataset, total 23 observations have been predicted correctly out of 26 observations in class 1. Similarly, in class-2, class-3 and class-4, the correctly predicted observations were 25, 22 and 26 out of 29, 26 and 27 respectively in testing observations.

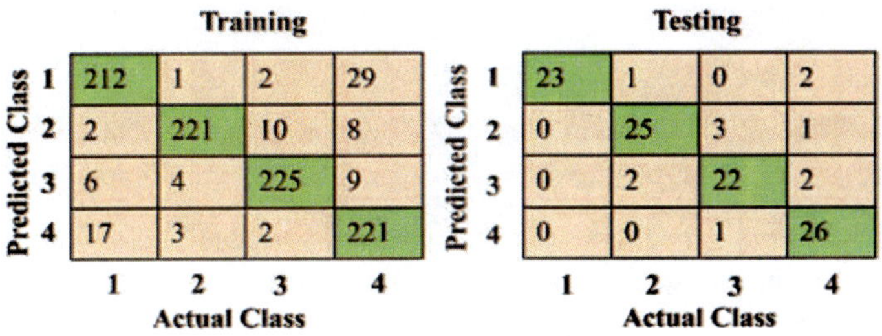

Fig. 4 Confusion matrix of training and testing performance

Table 3 Training and testing performance of classifier

Performance evaluation metrics	Training performance (%)	Testing performance (%)
Accuracy	90.43	88.89
Error	9.57	11.11
Sensitivity	90.44	88.90
Specificity	96.81	96.29
Precision	90.72	89.44
F1_score	90.51	88.97

Besides, confusion matrix, the performance of the LDA classifier has been evaluated by the wide range of metrics including accuracy, error rate, sensitivity, specificity, precision and F1-score shown in Table 3. From the table, it is observed that the training accuracy (90.43%) is comparatively higher than the testing accuracy (88.89%).

4 Conclusion

In the proposed study, four types of EOG signals including upward, right, left, downward has been classified. Initially, the combination of time domain based statistical features have been extracted and the extracted features have been classified by the LDA. Although the achieved classification accuracy is encouraging, some issues need to overcome to make the system suitable for real-life HCIs. Since the main aim of HCI technology is to support the physically disabled people, the data should be collected from the target users. Moreover, Data from more subjects should be collected and there should be an increased number of trials per subject. The result of the classification should be translated into system commands and in the meantime the entire experiment in real time should be developed.

Acknowledgements We thank FTKEE, UMP for resources support in terms of material and laboratory. We thank the Ministry of Education for supporting this study via FRGS/1/2019/TK04/UMP/02/7 (RDU1901167).

References

1. Lv Z, Wang Y, Zhang C, Gao X, Wu X (2018) An ICA-based spatial filtering approach to saccadic EOG signal recognition. Biomed Signal Process Control 43:9–17. https://doi.org/10.1016/j.bspc.2018.01.003
2. Gupta R, Laghari K, Banville H, Falk TH (2016) Using affective brain-computer interfaces to characterize human influential factors for speech quality-of-experience perception modelling. Human-Centric Comput Inf Sci 6. https://doi.org/10.1186/s13673-016-0062-5

3. Hasan H, Abdul-Kareem S (2014) Human–computer interaction using vision-based hand gesture recognition systems: a survey. https://doi.org/10.1007/s00521-013-1481-0
4. Kim MR, Yoon G (2013) Control signal from EOG analysis and its application. Int J Electr Comput Energ Electron Commun Eng 7(10)
5. Barea R, Boquete L, Mazo M, López E (2002) System for assisted mobility using eye movements based on electrooculography. IEEE Trans Neural Syst Rehabil Eng 10:209–218. https://doi.org/10.1109/TNSRE.2002.806829
6. Kim Y, Doh NL, Youm Y, Chung WK (2007) Robust discrimination method of the electrooculogram signals for human-computer interaction controlling mobile robot. Intell Autom Soft Comput 13:319–336. https://doi.org/10.1080/10798587.2007.10642967
7. Norris G, Wilson E (2002, November) The eye mouse, an eye communication device. https://doi.org/10.1109/nebc.1997.594960
8. Tsai J-Z, Lee C-K, Wu C-M, Wu J-J, Kao K-P (2008) A feasibility study of an eye-writing system based on electro-oculography. J Med Biol Eng 28:39–46
9. Bulling A, Ward JA, Gellersen H, Tröster G (2011) Eye movement analysis for activity recognition using electrooculography. IEEE Trans Pattern Anal Mach Intell 33:741–753. https://doi.org/10.1109/TPAMI.2010.86
10. Kher RK, Shah S (2016) Design of electrooculogram based wheelchair control design of electroOculoGram based wheelchair control. Conf Natl Conf Recent Trends Comput Commun Technol (RTCCT 2016), no. May, 2016
11. Rashid M et al (2020) Analysis of EEG features for brain computer interface application. In: Kasruddin Nasir A et al (eds) InECCE2019. Lecture Notes in Electrical Engineering, vol 632. Springer, Singapore. https://doi.org/10.1007/978-981-15-2317-5_45
12. Sanei S, Chambers JA (2007) EEG signal processing. Wiley, West Sussex, England. https://doi.org/10.1002/9780470511923
13. Nicolas-Alonso LF, Gomez-Gil J (2012) Brain computer interfaces, a review. https://doi.org/10.3390/s120201211
14. Aungsakul S, Phinyomark A, Phukpattaranont P, Limsakul C (2012) Evaluating feature extraction methods of electrooculography (EOG) signal for human-computer interface. Procedia Eng 32:246–252. https://doi.org/10.1016/j.proeng.2012.01.1264
15. Noor NMM, Ahmad S (2013) Analysis of different EOG-based eye movement strength levels for wheelchair control. Int J Biomed Eng Technol 11:175–196. https://doi.org/10.1504/IJBET.2013.055043
16. Reddy MS, Narasimha B, Suresh E, Rao KS (2010) Analysis of EOG signals using wavelet transform for detecting eye blinks. In: 2010 international conference on wireless communications & signal processing, pp 1–4. https://doi.org/10.1109/WCSP.2010.5633797
17. Banerjee A, Datta S, Pal M, Konar A, Tibarewala DN (2013) Classifying electrooculogram to detect directional eye movements. Procedia Technol 10:67–75. https://doi.org/10.1016/j.protcy.2013.12.338
18. Homan RW, Herman J, Purdy P (1987) Cerebral location of international 10-20 system electrode placement. Electroencephalogr Clin Neurophysiol 66:376–382. https://doi.org/10.1016/0013-4694(87)90206-9
19. Altın C, Er O (2016) Comparison of different time and frequency domain feature extraction methods on elbow gesture's EMG. Eur J Interdiscip Stud 5:35–44. https://doi.org/10.26417/ejis.v5i1
20. Cui G, Zhao Q, Cao J, Cichocki A (2014) Hybrid-BCI: classification of auditory and visual related potentials. In: 2014 joint 7th international conference on soft computing and intelligent systems, SCIS 2014 and 15th international symposium on advanced intelligent systems, ISIS 2014, pp 297–300. Institute of Electrical and Electronics Engineers Inc. (2014). https://doi.org/10.1109/SCIS-ISIS.2014.7044768
21. Liang L, Chen J, Shewokis P, Getchell N (2016) Developmental and condition-related changes in the prefrontal cortex activity during rest. J Behav Brain Sci 6:485–497. https://doi.org/10.4236/jbbs.2016.612044

Experimental Study of the Effect of Vehicle Velocity on the Ride Comfort of a Car on a Road with Different Types of Roughness

Kazem Reza Kashyzadeh and Nima Amiri

Abstract Vehicle speed control is one of the most important factors at increasing safety, reducing the number and severity of road accidents. Also, other factors such as inappropriate designing, the overuse of speed bumps, nonconformity to the prevailing standards of the world in construction, maintenance, and repair of roads can cause damage to the car suspension and consequently brings the discomfort of car occupants when crossing the actual road under different driving conditions (various speeds and maneuvers). In the present paper, the effect of different vehicle velocities on the comfort of occupants was investigated empirically. To this end, a passenger vehicle was driven at different constant speeds of 30, 40, and 50 km/h on the actual road which has various types of speed bumps and uneven such as sewer door. Also, the accelerometer sensor of mobile was used to extract the acceleration time histories entered on occupants in three different directions X, Y, and Z. Finally, the BS 6841 and ISO 2631-1 standards were used to compare the laboratory-measured vertical acceleration at different speeds with the vertical acceleration that occupants feel comfortable. The results indicated that the maximum speed for not feeling the discomfort of occupants is 67 km/h. Moreover, the maximum human tolerance was obtained approximately 178 min of continuous travel.

Keywords Ride comfort · Vehicle velocity · Road roughness · Speed bump · Road condition · Discomfort feeling of occupants

K. Reza Kashyzadeh (✉)
Department of Mechanical and Instrumental Engineering, Academy of Engineering, Peoples' Friendship University of Russia (RUDN University), 6 Miklukho-Maklaya Street, Moscow 117198, Russian Federation
e-mail: reza-kashi-zade-ka@rudn.ru

N. Amiri
School of Mechanical Engineering, Sharif University of Technology, Tehran, Iran

A. F. Ab. Nasir et al. (eds.), *Recent Trends in Mechatronics Towards Industry 4.0*, Lecture Notes in Electrical Engineering 730, https://doi.org/10.1007/978-981-33-4597-3_54

1 Introduction

The comfort of occupants is one of the important factors in the design of various vehicles (including road, rail, or air). Many parameters affect the comfort of car occupants such as condition and pressure tire, characteristics and performance of the suspension, road roughness, vehicle maneuvers (acceleration, braking, etc.), and the driving behavior. However, previous studies most have focused on the effects of road roughness and vertical vibrations on the vehicle occupants. In some published papers, researchers have pursued to optimize the characteristics of the suspension components (spring stiffness and damping coefficient) to increase the occupant comfort.

Uys et al. have optimized the spring stiffness and damping coefficient to increase the ride comfort of an off-road vehicle under different road profiles and various speeds [1]. Marzbanrad et al. have optimized the characteristics of a passive suspension system for ride comfort improvement using the Design of Experiment (DOE) method [2]. The most important achievement of this research was that the ride comfort is most sensitive to a change in rear spring stiffness.

The smartphone has been used to estimate road roughness conditions using the relationship between road surface features and vibration acceleration [3]. Also, it has been found that acceleration data from smartphones has a linear relationship with road roughness conditions. However, this relationship is directly dependent on the speed (for speeds <60 km/h).

Gonzalez et al. have presented a convenient, fast, and economical method for road condition estimation using vehicle accelerometers data [4]. In this regard, the Power Spectral Density (PSD) of a road profile can be obtained from the PSD of the axle or body accelerations measured over the road profile. Moreover, a combinatorial optimization technique has been presented to determine the pavement profile heights based on measured accelerations at the vehicle axle. [5].

The acceleration data have been double integrated numerically to obtain a pavement profile [6]. The obtained results have been compared with the data collected during different tests (an inertial profiler and the newly developed smartphone application). The findings of this study showed that the International Roughness Index (IRI) values measured with smartphone applications were similar to those collected with the inertial profiler (only for the range of low to medium roughness). But, for the range of high roughness, the IRI values obtained utilizing smartphone applications are lower than the IRI values collected with the inertial profiler.

Chugh et al. have published a comprehensive review of the various systems of road condition detection [7]. They found that smartphone sensors are gaining importance in this field as they are economically cost-effective and also increase scalability. Perttunen et al. have developed a pattern recognition system for detecting road conditions from accelerometer and GPS readings [8]. In this research, spectral analysis of multi-channel acceleration signals has been utilized to get valid road surface anomaly labels. Also, a new speed dependence removal approach has been proposed to obtain more accurate features. Aleadelat and Ksaibati have estimated the Pavement Serviceability Index (PSI) directly by using the variance value of time series of acceleration data at two different vehicle velocities (40 and 50 mi/h) [9]. A new algorithm has been developed for road bumps and potholes detection by using the analysis results of the vertical acceleration impulse [10].

In summary, the published results demonstrated that the condition of the road surface can be effectively identified and the transient events can be properly detected using sensed data from mobile sensors and GPS [11–13]. Furthermore, many studies have been conducted on the effect of vehicle velocity on the evaluation of road surface quality and road roughness simulation by Finite Element Analysis (FEA) [14, 15]. Also, some scholars examined the simultaneous effects of vehicle velocity and road roughness on fatigue phenomenon in the automotive industry [16, 17].

Karen et al. have presented a new ride comfort simulation based on the full-vehicle suspension which the seat-back, seat-surface, and feet acceleration values collected from different tests considered as inputs [18]. Also, the optimization process has been added to MATLAB/Simulink simulation to find the optimum values of each parameter.

The stationary wavelet transforms analysis has been used to remove the effects of speed, slopes, and drifts from mobile phone sensor signals using the envelope technique [19]. Jin et al. have studied the effect of unsprung mass to sprung mass on the ride comfort of vehicles driven by in-wheel motors [20]. It is found that the vehicle ride comfort decreases by raising the unsprung mass. To improve the ride comfort, they proposed to use a semi-active air-suspension proportional–integral–derivative control system. Ahlin and Granlund have presented a new relationship between road roughness (IRI), speeds, and vertical human Whole-Body Vibration (WBV) using a quarter car model [21].

In this study, the effect of vehicle speed on the comfort of car occupants was investigated on a real road including various bumps and sewer doors. The innovations of this research can be mentioned as follows:

1. Consideration of real suspension characteristics and its performance in the car (there are some differences in the characteristics of the front and rear suspension systems such as spring stiffness);
2. Since the tests were performed on a real passenger vehicle, the mass effects of different components such as motors, doors, etc. are considered in this study. Besides, the inertia effects (e.g., gearboxes) are considered that distinguishes it from simplified models in Adams and Matlab software;
3. Experimental study of the actual road including speed bumps, sewer door, and 90-degree turn in the path, measuring the acceleration time histories in different directions, and repetition of the process at different vehicle speeds;
4. Using different techniques (BS and ISO standards) to assess the comfort of car occupants;
5. Providing a new classification of occupant comfort in terms of vehicle velocity using experimental data and applying Linear Regression (LR) technique.

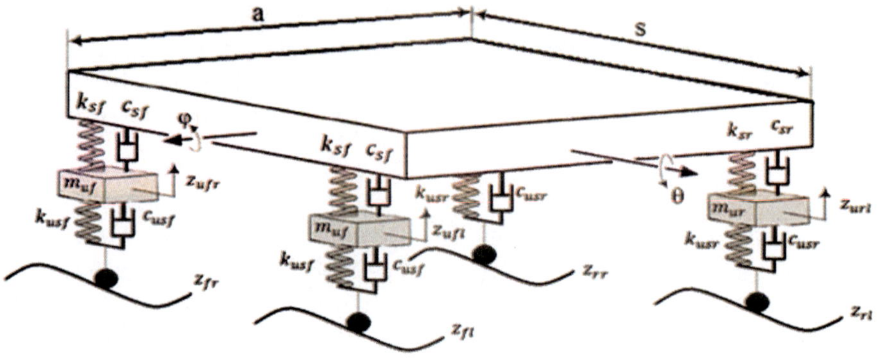

Fig. 1 Schematic of the characteristics of the passenger car suspension system used in this study

2 Experimental Procedure

The mobile accelerometer sensor of the iPhone 7 plus with the data rate of 30 Hz was used to measure the applied acceleration on vehicle occupants in different directions (X, Y, and Z represent the longitudinal, transverse, and vertical acceleration, respectively). To this end, a rigid weight (75 kg) was used as the front passenger of the car, which is firmly attached to the seat. Also, the cell phone is fastened to the rigid weight using a strap so that it does not move. The vertical acceleration measurement was performed without considering gravitational acceleration.

Experiments and data recording were carried out at different speeds (30, 40, and 50 km/h) on the actual road. In this study, a passenger vehicle was used with a 1205 kg dry weight and the Gross vehicle weight of 1494 kg (Curb weight plus CNG gas capsule with full capacity and considering two occupants with a standard mass). The specifications and values of suspension system parameters are presented in Fig. 1 and Table 1, respectively.

The Google Road Map (GRM) database is a common technique that was used to find the path. Moreover, the Geographic Information System (GIS) was utilized to record geographic information and analyze mapped information. The details of the actual road path and the geographic coordinates of the start and endpoint of the travel are demonstrated in Figs. 2 and 3, and Table 2, respectively.

3 Results and Discussion

The acceleration entered on occupants was measured in the different directions for moving the car at various constant speeds. The statistical characteristics of each recorded data including mean value, standard deviation, and Root Mean Square (RMS) were calculated using MATLAB software and are presented in Table 3.

Table 1 Values of suspension system parameters

Parameters	Unit	Value
Vehicle mass + fuel + double passenger	kg	1494
Weight of each occupant	kg	75
Fuel weight (gasoline and compressed natural gas: CNG)	kg	139
Roll moment of inertia	kg m^2	460
Pitch moment of inertia	kg m^2	2160
Tires weight	kg	59
Front suspension stiffness	N/m	35,000
Rear suspension stiffness	N/m	38,000
Damping coefficient of suspension in traction mode (both front and rear)	N s/m	1000
Damping coefficient of suspension in compression mode (both front and rear)	N s/m	720
Tires spring stiffness	N s/m	190,000
Damping coefficient of tires	N s/m	10
Wheelbase	m	2.671
Car length	m	4.502
Car width	m	1.72
Car height	m	1.46

The results indicated that as the vehicle speed increases, the RMS parameter increases in different directions. For example, as the vehicle speed increased from 30 to 40 and 50 km, the vertical RMS parameter increased to 41.5% and 75.5%, respectively. Also, the increase rate of the RMS in the vertical direction is much more than the increase rate in transverse and longitudinal directions. In other words, the increasing speed of the car had the greatest effect on the vertical acceleration compared to other acceleration components. Therefore, to investigate the comfort of the vehicle car, the greatest attention was paid to the vertical acceleration component. The time history of the recorded vertical acceleration according to the road unevenness is shown in Fig. 4. Next, to examine the ride comfort, the definitions of both standards BS-6841 and ISO 2631-1 were used in this research. The British standard has presented the classification of acceptable ride quality based on the RMS value of vertical acceleration (Table 4).

Comparison of the results for the vertical acceleration in actual tests and various speeds with BS 6841 showed that in all cases, the occupants completely comfort. In other words, the suspension characteristics of car, car speed, and various unevenness and bumps were under standard conditions. Then, the critical speeds at each of the occupant comfort intervals were calculated using the linear regression method (Eq. 1) and a new classification of occupant comfort in terms of speed is presented in Table 5.

$$\mathrm{V} = 27.9 + 116 \times A + 24.1 \times A^2 - 3.4 \times A^3 \quad (1)$$

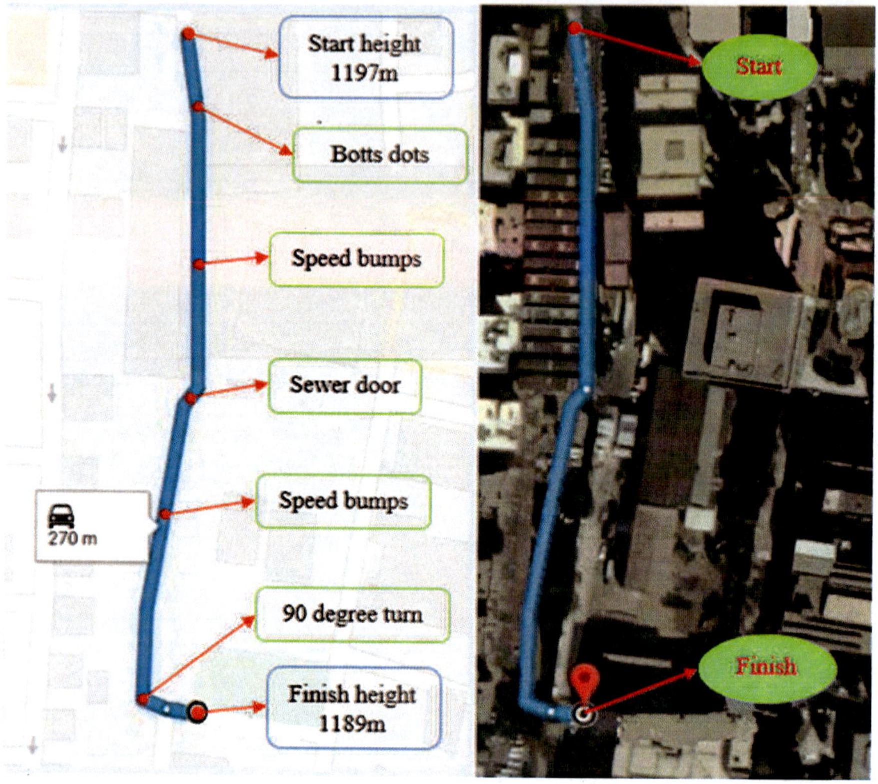

Fig. 2 Details of a real road

where V is vehicle velocity (km/h) and A represents RMS of vertical acceleration (m/s^2). The results showed that if the passenger car moves at its maximum speed on the test road with the specifications outlined above, the occupant may sometimes feel uncomfortable but it is tolerable. Also, the critical speed was obtained as 67 km/h for the discomfort starts.

Besides, the ISO standard is not limited to the vertical acceleration and takes into account all acceleration components. Also, it has reported occupant discomfort in terms of travel time. In other words, using this standard can be declared that after what time on the specified road, the occupants feel uncomfortable (Fig. 5).

To compare the laboratory results with the ISO standard, the data were converted from the time domain to the frequency domain using Fourier analysis. The results showed that there are no discomfort conditions for the occupants. The standard has specifically stated that the value of vertical acceleration RMS must be less than 0.33 for an operating frequency of 20–60 Hz. The results of the statistical analysis of the car in this frequency range reveal that the driver will feel uncomfortable after 8 h driving at the speed of 71 km/h. Eventually, the diagram of critical velocities on the

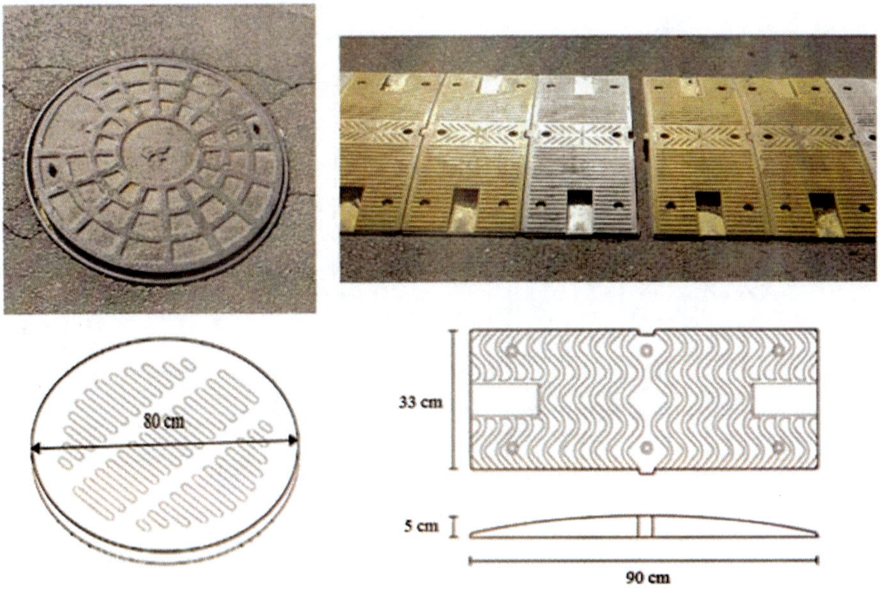

Fig. 3 Dimensions of speed bump and sewer door

Table 2 Geographic coordinates of the used road

Location	Longitude	Latitude
Start point	51°21′2.371″	35°42′12.189″
Endpoint	51°21′12.189″	35°42′4.179″

Table 3 Statistical characteristics of recorded acceleration time histories

Acceleration components	Vehicle speed (km/h)	Mean	Standard deviation	Root mean square (RMS)
Longitudinal acceleration (X)	30	−0.0298	0.1610	0.1637
	40	−0.0171	0.2216	0.2223
	50	−0.0005	0.2988	0.2988
Transverse acceleration (Y)	30	−1.0054	0.1455	1.0159
	40	−1.0029	0.2088	1.0244
	50	−1.0021	0.2272	1.0275
Vertical acceleration (Z)	30	−0.0211	0.1269	0.1286
	40	−0.0205	0.1808	0.1819
	50	−0.0087	0.2256	0.2257

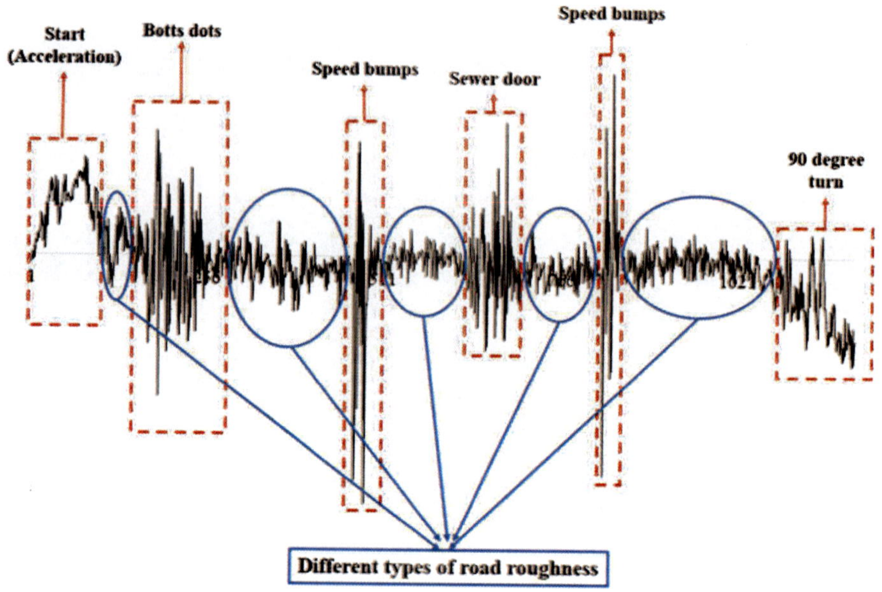

Fig. 4 Details of vertical acceleration history related to each uneven of test road

Table 4 Levels of acceptability of ride quality proposed by the British Standard Institution [22]

Range of RMS vertical acceleration (m/s^2)	Acceptability
<0.315	Not uncomfortable
0.315–0.63	A little uncomfortable
0.5–1	Fairly uncomfortable
0.8–1.6	Uncomfortable
1.25–2.5	Very uncomfortable
>2	Extremely uncomfortable

Table 5 Levels of acceptability of ride quality based on the vehicle velocity

Speed (km/h)	Acceptability
<67	Not uncomfortable
68–109	A little uncomfortable
109–165	Fairly uncomfortable
165–261	Uncomfortable
261–415	Very uncomfortable
>415	Extremely uncomfortable

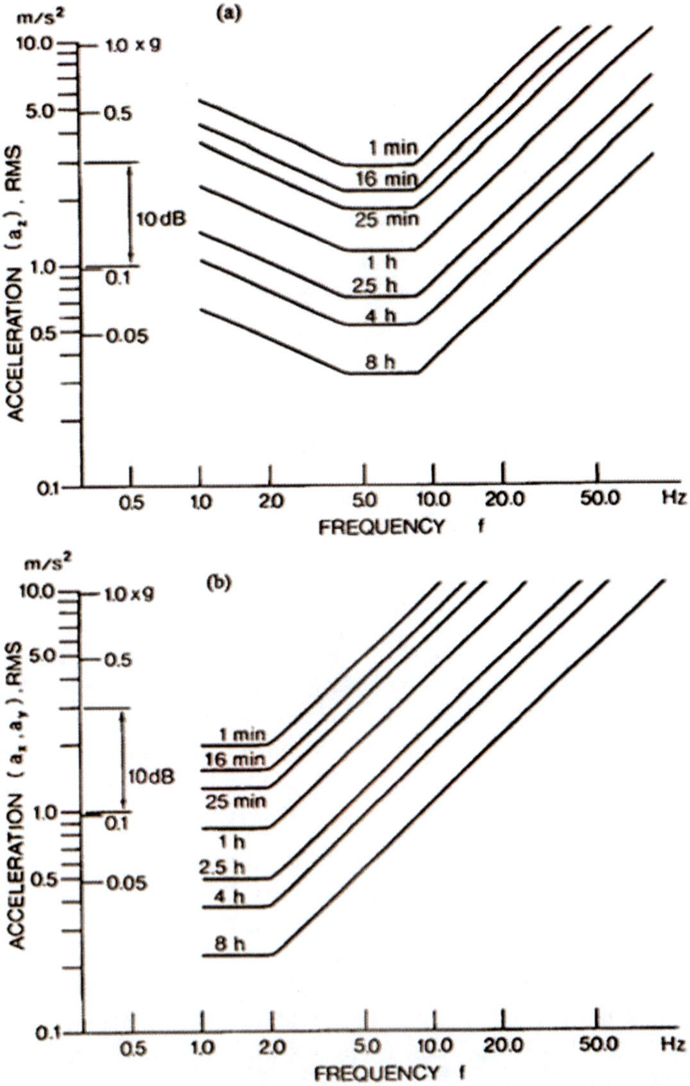

Fig. 5 Limits of whole-body vibration for decrease proficiency recommended by ISO: **a** vertical direction and **b** transverse direction [23]

test road in terms of the human endurance time for feeling discomfort is depicted in Fig. 6.

The results showed that due to the allowed speed on different roads of the country, the maximum occupant comfort time is 178 min (for the used car). Therefore, it is recommended to take short breaks for long trips before reaching this time.

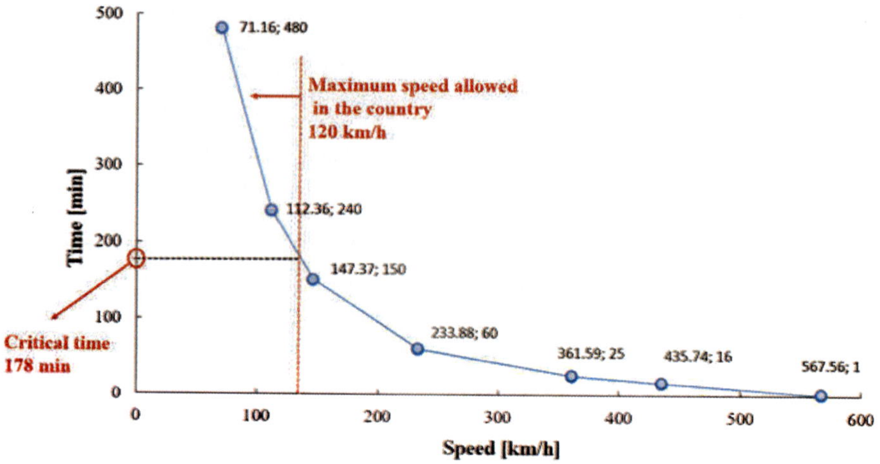

Fig. 6 Diagram of human endurance time for feeling discomfort via car speed

4 Conclusion

In the present paper, the main aim is to investigate the effect of vehicle speed on occupant comfort. To this end, the acceleration time histories entered on the occupant were measured in a real test road by utilizing mobile sensors and applications. Then, the statistical characteristics of acceleration components were calculated. Finally, the obtained data were analyzed using BS 6841 and ISO 2631-1 standards. The achievements of this research can be mentioned as follows:

1. Analysis of acceleration time histories collected at different directions during the actual road test showed that no discomfort would be experienced by occupants until the car travels at speeds of less than 67 km/h.
2. The results indicated that the car never causes extreme discomfort to its occupants, even if it is driven at maximum speed (according to the BS 6841).
3. Comparison of the laboratory results with ISO 2631-1 reveal that if the car was driven at a speed of 71 km/h, the occupants will feel uncomfortable after 8 h of continuous motion.
4. The maximum occupant comfort time with the maximum speed on the roads of the country is 178 min. Hence, strongly is advised that take a short break at least every 150 min to avoid the discomfort of occupants including headaches, dizziness, and nausea on the long trips.

Acknowledgements The publication has been prepared with the support of the "RUDN University Program 5-100".

References

1. Uys PE, Els PS, Thoresson M (2007) Suspension settings for optimal ride comfort of off-road vehicles travelling on roads with different roughness and speeds. J Terramech 44:163–175
2. Marzbanrad J, Mohammadi M, Mostaani S (2013) Optimization of a passive vehicle suspension system for ride comfort enhancement with different speeds based on design of experiment method (DOE) method. J Mech Eng Res 5(3):50–59
3. Douangphachanh V, Oneyama H (2013) A study on the use of smartphones for road roughness condition estimation. J Eastern Asia Soc Transp Stud 10:1551–1564
4. González A, O'brien EJ, Li YY, Cashell K (2008) The use of vehicle acceleration measurements to estimate road roughness. Veh Syst Dyn 46(6):483–499
5. Harris NK, González A, Obrien EJ, McGetrick P (2010) Characterisation of pavement profile heights using accelerometer readings and a combinatorial optimisation technique. J Sound Vib 329:497–508
6. Islam S, Buttlar WG, Aldunate RG, Vavrik WR (2014) Measurement of pavement roughness using android-based smartphone application. Transp Res Rec 2457:30–38
7. Chugh G, Bansal D, Sofat S (2014) Road condition detection using smartphone sensors: a survey. Int J Electr Electron Eng 7:595–602
8. Perttunen M, Mazhelis O, Cong F, Kauppila M, Leppänen T, Kantola J, Collin J, Pirttikangas S, Haverinen J, Ristaniemi T, Riekki J (2011) Distributed road surface condition monitoring using mobile phones. In: International conference on ubiquitous intelligence and computing. Springer, Berlin, Heidelberg, pp 64–78
9. Aleadelat W, Ksaibati K (2017) Estimation of pavement serviceability index through android-based smartphone application for local roads. Transp Res Rec J Transp Res Board 129–135
10. Vittorio A, Rosolino V, Teresa I, Vittoria CM, Vincenzo PG (2014) Automated sensing system for monitoring of road surface quality by mobile devices. Procedia-Soc Behav Sci 111:242–251
11. Alessandroni G, Klopfenstein L, Delpriori S, Dromedari M, Luchetti G, Paolini BD, Seraghiti A, Lattanzi E, Freschi V, Carini A, Bogliolo A (2014) Smartroadsense: collaborative road surface condition monitoring. In: The eighth international conference on mobile ubiquitous computing, systems, services and technologies, UBICOMM
12. Allouch A, Koubâa A, Abbes T, Ammar A (2017) Roadsense: smartphone application to estimate road conditions using accelerometer and gyroscope. IEEE Sens J 17:4231–4238
13. Li X, Goldberg DW (2018) Toward a mobile crowdsensing system for road surface assessment. Comput Environ Urban Syst 69:51–62
14. Loprencipe G, Zoccali P, Cantisani G (2019) Effects of vehicular speed on the assessment of pavement road roughness. Appl Sci 9(9):1783–1801
15. Reza Kashyzadeh K, Ostad-Ahmad-Ghorabi MJ, Arghavan A (2013) Study effects of vehicle velocity on a road surface roughness simulation. Appl Mech Mater 372:650–656
16. Reza Kashyzadeh K (2020) A new algorithm for fatigue life assessment of automotive safety components based on the probabilistic approach: the case of the steering knuckle. Int J Eng Sci Technol 23(2):392–404
17. Reza Kashyzadeh K, Farrahi GH, Shariyat M, Ahmadian MT (2018) Experimental accuracy assessment of various high-cycle fatigue criteria for a critical component with a complicated geometry and multi-input random non-proportional 3D stress field. Eng Fail Anal 90:534–553
18. Karen I, Kaya N, Ozturk F, Korkmaz I, Yildizhan M, Yurttas A (2012) A design tool to evaluate the vehicle ride comfort characteristics: modelling, physical testing, and analysis. Int J Adv Manuf Technol 60:755–763
19. Seraj F, Vander Zwaag BJ, Dilo A, Luarasi T, Havinga P (2014) Roads: a road pavement monitoring system for anomaly detection using smart phones. In: Big data analytics in the social and ubiquitous context. Springer, Berlin
20. Jin L, Yu Y, Fu Y (2016) Study on the ride comfort of vehicles driven by in-wheel motors. Adv Mech Eng 8(3):1–9
21. Ahlin K, Johan Granlund NO (2002) Relating road roughness and vehicle speeds to human whole body vibration and exposure limits. Int J Pavement Eng 3(4):207–216

22. British Standard Institution (1987) British standard guide to measurement and evaluation of human exposure to whole body mechanical vibration and repeated shock, BS 6841
23. International Standards Organisation (1997) International standard mechanical vibration and shock—evaluation of human exposure to whole body vibration, 2nd edn. ISO 2631: Part 1

Performance Evaluation of BPSO & PCA as Feature Reduction Techniques for Bearing Fault Diagnosis

Atik Faysal, Ngui Wai Keng, and M. H. Lim

Abstract Vibration-based signal processing is the most popular and effective approach for fault diagnosis of bearing. In this paper, time-frequency domain analysis, i.e. empirical mode decomposition (EMD) was applied to the raw vibration signal. Intrinsic mode function (IMF) containing the characteristics of vibration data was analysed to obtain 90 statistical features. Two feature reduction algorithms, namely principal components analysis (PCA) and binary particle swarm optimiser (BPSO) were applied individually for feature reduction. The reduced feature subsets were 12 and 35 for PCA and BPSO, respectively. K-Nearest Neighbours (K-NN) was used as an intelligent method for fault diagnosis. K-NN was applied to the entire feature set and individually on the selected feature subset of PCA and BPSO. The reduced feature subset with PCA performed the finest in all the measurements taken. For BPSO, although it effectively reduced the feature dimension and classification time, the testing accuracy was slightly lower. Comparing the output accuracy of the K-NN classifier for the selected methods demonstrated the effectiveness of PCA and BPSO as efficacious feature reduction techniques.

Keywords Feature reduction · Principal components analysis · Binary particle swarm optimiser · Empirical mode decomposition

1 Introduction

Bearing is one of the most used machinery parts of industries that undergoes a huge load of friction, erosion, and different types of failure [1]. It is important to develop a fast and reliable fault diagnosis method to monitor the machine health condition. The incorporation of artificial intelligence with signal processing has been proven to

A. Faysal · N. W. Keng (✉)
Faculty of Mechanical and Automotive Engineering Technology, Universiti Malaysia Pahang, 26600 Pekan, Pahang, Malaysia
e-mail: wkngui@ump.edu.my

M. H. Lim
Institute of Noise and Vibration, Universiti Teknologi Malaysia, Jalan Sultan Yahya Petra, 54100 Kuala Lumpur, Malaysia

A. F. Ab. Nasir et al. (eds.), *Recent Trends in Mechatronics Towards Industry 4.0*, Lecture Notes in Electrical Engineering 730, https://doi.org/10.1007/978-981-33-4597-3_55

be efficient in condition monitoring. Among the different signals, vibration signals have been widely used along with various machine learning classifiers as intelligent diagnosis methods.

Among many signal processing methods, time domain, frequency domain, and time-frequency domain analysis are the most popular ones [2]. The advantage of time-frequency domain analysis over other methods is it provides both time and frequency information. Among the many popular time-frequency domain analysis, short time Fourier transform, wavelet transform, and empirical mode decomposition (EMD) have been widely used. In recent years, a lot of research has been conducted by incorporating EMD with artificial intelligence techniques. Lin et al. [3] combined EMD with FFT to analyse the oil-leakage fault signal of the gearbox of wind turbines. K-NN was applied for fault diagnosis. Pandya et al. [4] processed acoustic emission using EMD and Hilbert–Huang Transform. For classification, a modified K-NN algorithm, where $k = \sqrt{n}$ was proposed based on the asymmetric proximity function. This present study used EMD [5] for feature extraction from vibration signal. By obtaining different statistical features from IMFs, a large number of data can be gathered. In that sense, EMD appears to be suitable for machine learning algorithms. However, a large number of features raise the condition for feature reduction. Two feature reduction approaches, namely PCA and BPSO were applied separately in the current work.

Principle component analysis (PCA) can be a handy tool to perform feature reduction, where the first few columns contain most of the variance of the feature table. PCA is combined with machine learning algorithm to classify different fault classes. Sharma et al. [6] used PCA on the feature of rolling element bearings and the performance was evaluated using weighted K-NN classifier. For three bearing conditions, an accuracy of 100% was achieved using the Mahalanobis distance metric. Among the population-based approaches, the application of particle swarm optimiser (PSO) to features selection has attracted a lot of attention lately. In [7], Binary PSO (BPSO) was implemented with regularised Fisher's criterion, which was used as a fitness function. Later, PCA and BPSO were individually applied to focus on the performance of BPSO. Fei [8] proposed an improved BPSO to select the parameter K of K-NN. The traditional K-NN was trained by varying the parameter K to obtain better accuracy. However, such literature on the comparison between PCA and BPSO using the same classifier is rare. This paper tries to evaluate the performance of PCA and BPSO as effective feature reduction methods. The performance of the reduced feature subset is measured using K-NN classifier. In the end, the performance of the classifiers using the reduced feature subset obtained from the two feature reduction techniques is discussed.

2 Theoretical Background

2.1 PCA

PCA [9] transfers the possible correlated variables into a lower number of uncorrelated variables known as principal components. The first principal component contains the highest possible variability and the following components contain the same and so on. PCA can be an effective tool for dimensionality reduction. It can also be used to reduce the complexity of signal processing. The steps of PCA are as follows:

Step 1: Take the feature set $X_{n \times m}$ where n is the number of data samples and m is the dimension of the feature space.
Step 2: Normalise the data of the entire feature set, $norm(X)$.
Step 3: Take the covariance matrix of the feature set, $cov(X)$.
Step 4: Calculate the eigenvectors and eigenvalues of the covariance matrix.
Step 5: Choose components and form a feature vector.
Step 6: Derive the new data of m-dimension.

2.2 PSO

PSO was proposed by Kennedy and Ebenhart in 1995 [10]. A group of particles (like bird or fish swarm) is assumed and their position and velocity are constantly updated. The velocity and position of the particles are updated according to the following equation:

$$v_{id}^{t+1} = \omega v_{id}^{t} + c_1 r_1 \left(Pbest_i^t - x_{id}^t\right) + c_2 r_2 \left(Gbest_g^t - x_{id}^t\right) \tag{1}$$

$$x_{id}^{t+1} = x_{id}^{t} + v_{id}^{t+1} \tag{2}$$

where v_{id}^t denotes the velocity and x_{id}^t is the position of particle after t-th iteration. $Pbest_i^t$ is the particles current best position and $Gbest_g^t$ is the global best position. ω is the inertia weight and d is the dimension of population. c_1, c_2 are the acceleration coefficients. r_1, r_2 are the random numbers uniformly distributed between [0, 1].

2.3 K-Nearest Neighbours

K-NN classifier is a widely used pattern recognition tool. The basic idea of the algorithm is that features belonging to different classes will form clusters in their individual feature space. Neighbours are the features that are close to each other [11].

In the case of a new vector, the K-NN algorithm calculates the distance with the training data using distance matrices, such as Minkowsky, Correlations, Manhattan, Chi-square, and Euclidean distance. From the result, the nearest neighbours can be calculated and the output class can be obtained. A class is assigned to the new vector from the class pool of its k nearest neighbours, here, k represents the number of the nearest points considered.

3 Data Collection

In this study, for fault diagnosis, the vibration data were obtained from the online dataset of the Case Western Reserve University Bearing Data Center [12]. Acceleration data were measured from a 2 hp Reliance Electric motor bearings. The collected vibration signals included the following operating conditions: (1) normal condition/baseline condition, (2) inner race fault, (3) outer race fault, and (4) ball fault. Each fault condition included three different fault sizes, i.e. 0.007, 0014, and 0.021 in.

4 Feature Extraction

For feature extraction, 1202 data points were taken per sample and the number of total samples was 900. The number of samples was 90 for the baseline condition and each fault condition had 270 samples individually. The time-frequency domain analysis, i.e. EMD simply decomposes a vibration signal into several IMFs. The IMF function must meet two basic conditions:

(a) The number of extrema and zero crossings in the whole vibration series must be the same or differ at most by one.
(b) The mean of upper and lower envelops must be zero at any given point.

It is observed that for the later values of IMFs, the strength of the signal decreases. Therefore, in this experiment, only the first five IMFs were taken for feature extraction from every observation. Different statistical and time-domain features were extracted from the obtained IMFs of drive end and fan end vibration data. From each IMF, nine different features were extracted, which were mean, root mean square, variance, standard deviation, kurtosis, skewness, peak, crest factor, and energy. From the five IMFs, 45 features were extracted in total from the drive end and fan end individually. Juxtaposing all the features, the size of the feature vector became 900×90 for all number of observations. All features were normalised between (0, 1) to ensure the elimination of overfitting while classification.

5 Feature Reduction

The output accuracy of the classifier algorithm depends largely on the features of the dataset. However, all of the features do not have an equal contribution to the performance of the classifier. There might be some features that are not as important as others. Moreover, for a large number of features, the feature set may become redundant and overlapping may occur between different classes. This may result in a reduction of the performance of the classifier with longer computational time. For this reason, the elimination of insignificant features is important. Two different feature reduction methods were used in this study, namely PCA and BPSO. These two methods will be discussed elaborately in the following subsection.

5.1 Feature Reduction Using PCA

PCA was applied to the raw feature set and the values of the score and explained variance were obtained. The score represents an equal number of new variables in a compact feature space and the explained variance is the weight of importance [13]. From the percentage of the explained variance, the optimal number of principal components was selected. Later, by increasing the number of the principal components, K-NN was applied to the selected feature subset and each output was recorded. The threshold level of accuracy for the minimal number of principal components was obtained thereby.

5.2 Feature Reduction Using BPSO

The proposed BPSO model, which combines K-NN for its fitness function evaluation was used to select the best features from the feature set to improve the classification ability. In BPSO, the number of initial population is taken the same as the number of features, where the positions of features are initialised with random 0's and 1's. For example, X = [0 1 1 0 1 0 0 1 … 1] is a position vector, where bit 1 denotes that the feature is selected and bit 0 denotes that the feature is not selected. Therefore, new feature subsets are generated and for a population of N_p features, N_p feature subsets is generated. The ultimate goal is to find the best feature subset that maximises the class distinguishability.

The training error of K-NN classifier was used as the fitness function of the BPSO algorithm and the model was trained to minimise this fitness function. The parameter setting for K-NN was such that the number of neighbours was two and 2-fold cross-validation (CV) was used for data partitioning. The 2-fold CV divided the training sample into two equal subsets, among which one was used for training the model.

The training accuracy of K-NN, A_i, was obtained and by using that the fitness of ith particle was defined as follows:

$$fit_i = 1 - A_i \tag{3}$$

The training error was evaluated in each iteration and the personal best, $Pbest$ and the global best, $Gbest$ were updated.

In the present work, to select the most desired features from the features pool (90), BPSO was implemented. To measure the accuracy of the selected feature subset, K-NN was applied. The BPSO was initialised with the following parameters:

- The initial population is the same as the particle size, i.e. 90 particles, where particle size is the number of all features (as recommended by [14])
- $\omega_{\min} = 0.4$, $\omega_{\max} = 0.9$, $v_{min} = -2$, $v_{\max} = 2$, $c_1 = 2$, $c_2 = 2$, r_1 and r_2 are randomly generated between 0 and 1.
- Number of iterations, $N_i = 200$.

5.3 *K-NN Modelling*

In this study, the Euclidean distance metric was considered for the K-NN model and the number of neighbours was two [15]. To avoid overfitting due to the large number of samples (900), 50% holdout cross-validation (CV) [16] was used, meaning half of the dataset was used for training the classifier and the other half was used for testing. The computational platform used was MATLAB 2017b and the computation unit used was Intel core i5, CPU 2.3 GHz, 4 GB RAM.

6 Results and Discussion

6.1 *Implementation of K-NN on the Raw Feature Set Without Feature Reduction*

Firstly, K-NN was applied to the entire feature set without passing through any feature reduction methods. The objective was to observe the performance of the classifier without having the feature dimension reduced and later, this result was used to compare the performance of the selected feature reduction methods. From the output of K-NN, the inner race fault and outer race fault conditions were completely successful to be classified with around 5% error for both the baseline and ball fault conditions. The overall testing accuracy of the classifier was 97.78%.

6.2 Implementation of PCA

PCA was applied to the entire feature set for feature reduction. Figure 3 was obtained from the cumulative summation of the explained variance for the increasing number of principal components. This helped to find the minimum number of the principal components that should be considered for classification.

Based on Fig. 1, it can be observed that at 20 principal components, more than 95% of the feature set is covered. At 40, the variance is over 99% and at around 70, the variance level reaches 100%.

Next, the K-NN was applied to the first 40 principal components because it covered almost all the variance. The effectiveness of these components on the classifier performance was determined. The accuracy of every single component was stored and the results in Fig. 2 were obtained. It was evident that there was a sharp increase in accuracy until the first four principal components. The accuracy was 98% at only four principal components and it reached 99.56% at 12 principal components. After this point, there was not much change in the accuracy except for a slight variation. Therefore, it is safe to assume that at 12 principal components, the classifier provides the best accuracy.

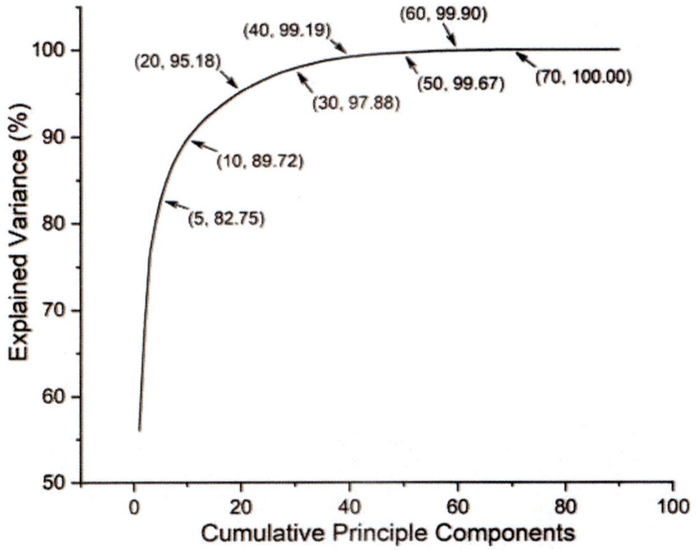

Fig. 1 The rate of explained variance of individual principal components

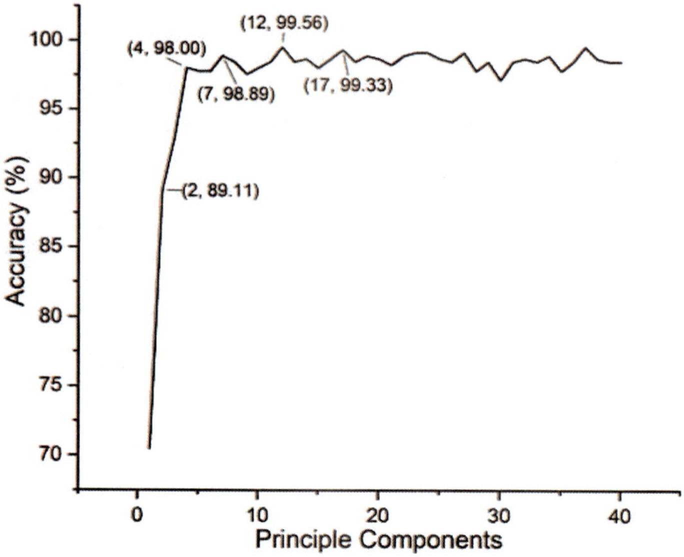

Fig. 2 Accuracy of K-NN using the different numbers of principal components

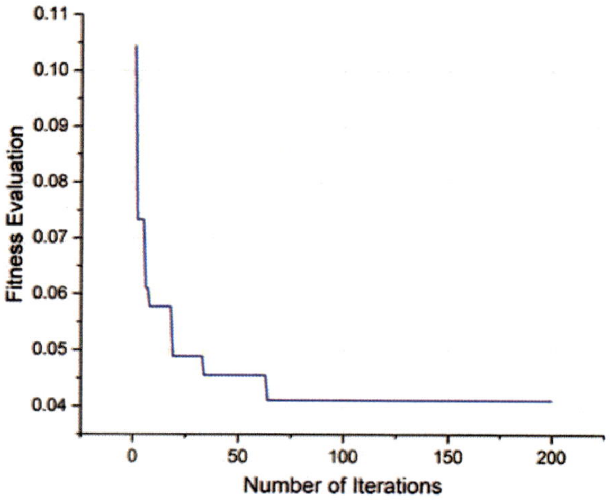

Fig. 3 Convergence curve for the fitness function of BPSO

6.3 *Implementation of BPSO*

By looking at the convergence curve of the proposed BPSO + K-NN based features selection algorithm, the result can be analysed. In Fig. 3, the Y-axis and the X-axis represents the number of iterations. It shows that around 55 generations, the algorithm reaches the global best solution. This proves that an adequate number of iterations is provided.

Table 1 Comparison of test accuracy and classification time of different methods

Method	No. of features	Accuracy (%)		Classification time (s)
		Training	Testing	
K-NN	90	99.33	97.78	0.111
PCA & K-NN	12	100	99.56	0.076
BPSO & K-NN	35	99.78	97.11	0.086

After applying BPSO, the number of selected features was 35, which implied that among the 90 initial features, the selected ones were the most efficient. These selected features were used for classification using K-NN, where the parameters of K-NN were the same for the entire feature set. The results will be discussed in the next section.

In this part, the output accuracy of the K-NN classifier was listed with the reduced feature subset of PCA and BPSO for comparison. In summary, three different methods were applied in the present work, i.e. (1) classification with no feature reduction, (2) classification with PCA, and (3) classification with BPSO. The following table depicts the accuracy and computation time of the methods.

Based on Table 1, it can be observed that the K-NN method, while taking all the features, requires the longest computation time. For PCA, with only 12 principal components, the highest accuracy of training and testing is obtained. The training accuracy is 100% for PCA, whereas the testing accuracy is 1.78% higher than the one with no feature reduction. Therefore, the number of dimensions is reduced from 90 to only 12 with less computation time using PCA. This result suggests that the feature subset obtained by PCA is more effective for classification than using the entire feature set.

For BPSO, the 35 features subset provided accuracy close to the accuracy of that of the entire features with a significant reduction of computation time. The testing accuracy of the BPSO was only 0.67% less than the performance of the K-NN with no feature reduction. Not only the features were significantly reduced but also it provided better training accuracy. The possible explanation for the reduction of testing accuracy is, unlike PCA, BPSO does not transfer the features by projecting them onto a set of orthogonal axes. Instead, it chooses the most dominant features using a metaheuristic algorithm, which includes the stochastic elements. Although being the least significant, the unselected features carry some information that contributes to the classification. However, for its significant reduction in dimension and computational time, a slight reduction of accuracy might be considered.

7 Conclusion

In this paper, IMFs from EMD were used to obtain different statistical features. The IMFs provide many details of vibration signals, thus producing sufficient features.

The proposed two feature reduction methods, i.e. PCA and BPSO proved to be effective in terms of reducing features, improving accuracy, and deducting computational time. The threshold accuracy of the K-NN classifier was obtained using only 12 principal components from 90 features. PCA seemed to do well comparatively in terms of the number of reduced features and classification accuracy. The number of selected features using BPSO was 35, which proved to be adequate for classification. Although BPSO successfully reduced the number of features and classification time, its testing accuracy was slightly less. This leaves the scope for improving BPSO by modifying its parameters. In the future, the authors wish to use different fitness functions with more suitable initialisation parameters. The results in this study justified the importance of PCA and BPSO as effective feature reduction methods.

Acknowledgements The authors are grateful to the Government of Malaysia for the financial support provided under the FRGS/1/2018/TK03/UMP/02/24 grant. Additional funding for this research also came from Universiti Malaysia Pahang (PGRS200303) and the Institute of Noise and Vibration, Universiti Teknologi Malaysia (RDU192303).

References

1. Liu Z, Zhang L (2020) A review of failure modes, condition monitoring and fault diagnosis methods for large-scale wind turbine bearings. Meas J Int Meas Confed 149:107002
2. Wei Y, Li Y, Xu M, Huang W (2019) A review of early fault diagnosis approaches and their applications in rotating machinery. Entropy 21(4). https://doi.org/10.3390/e21040409
3. Lin DF, Chen PH (2012) Fault recognition of wind turbine using EMD analysis and FFT classification. In: Proceedings of 2012 3rd international conference on digital manufacturing & automation (ICDMA 2012), pp 414–417
4. Pandya DH, Upadhyay SH, Harsha SP (2013) Fault diagnosis of rolling element bearing with intrinsic mode function of acoustic emission data using APF-KNN. Expert Syst Appl 40(10):4137–4145
5. Cheng Y, Wang Z, Chen B, Zhang W, Huang G (2019) An improved complementary ensemble empirical mode decomposition with adaptive noise and its application to rolling element bearing fault diagnosis. ISA Trans 91:218–234
6. Vernekar K, Kumar H, Gangadharan KV (2017) Engine gearbox fault diagnosis using empirical mode decomposition method and Naïve Bayes algorithm. Sadhana Acad Proc Eng Sci 42(7):1143–1153
7. Ziani R, Felkaoui A, Zegadi R (2017) Bearing fault diagnosis using multiclass support vector machines with binary particle swarm optimization and regularized Fisher's criterion. J Intell Manuf 28(2):405–417
8. Fei SW (2019) The HYBRID METHOD of VMD-PSR-SVD and improved binary PSO-KNN for fault diagnosis of bearing. Shock Vib 2019:1–8
9. Nimityongskul S, Kammer DC (2010) Frequency domain model reduction based on principal component analysis. Mech Syst Signal Process 24(1):41–51
10. Kennedy J, Eberhart R (1995) Particle swarm optimization—IEEE conference publication. In Proceedings of ICNN'95—international conference on neural networks, vol 4. IEEE, pp 1942–1948
11. Rashid M, Sulaiman N, Mustafa M, Bari BS, Sadeque MG, Hasan MJ (2020) Wink based facial expression classification using machine learning approach. SN Appl Sci 2(2):1–9

12. Loparo KA (2013) Bearing data center. Case Western Reserve University. https://csegroups.case.edu/bearingdatacenter/pages/download-data-file
13. Shin EC, Craft BD, Pegg RB, Phillips RD, Eitenmiller RR (2010) Chemometric approach to fatty acid profiles in Runner-type peanut cultivars by principal component analysis (PCA). Food Chem 119(3):1262–1270
14. Samanta B, Nataraj C (2009) Use of particle swarm optimization for machinery fault detection. Eng Appl Artif Intell 22(2):308–316
15. Sharma A, Jigyasu R, Mathew L, Chatterji S (2018) Bearing fault diagnosis using weighted K-nearest neighbor. In 2018 2nd International Conference on Trends in Electronics and Informatics (ICOEI), Tirunelveli, pp 1132–1137
16. Křížek P, Kittler J, Hlaváč V (2007) Improving stability of feature selection methods. In Lecture notes in computer science (including subseries lecture notes in artificial intelligence and lecture notes in bioinformatics), vol 4673. LNCS, pp 929–936

A Study on Different Techniques in ALPR System: The Systems Performance Analysis

Gan Vi Vi and Ahmad Afif bin Mohd Faudzi

Abstract License plate recognition and localization are crucial steps in most transportation applications. For instance, trace for stolen vehicles, speed or airport gate monitoring, road traffic monitoring and car parking control. This is required a system to extract automatically and recognize the character of the license plate from the image captured. The automatic license plate recognition (ALPR) system has aroused great interest in the research community, because in certain regions, cities or countries have certain limitations and lack of similarity between different license plates. The ALPR system includes three important components: first step is license plate localization, second step is character segmentation and third step is character recognition. This paper is to give comprehensive reviews of the localization and recognition techniques involved in the license plate recognition system and compares existing effective solutions.

Keywords License plate localization · Edge detection · OCR · Template matching · Deep learning · OpenALPR

1 Introduction

The city's traffic condition and the management of parking lots attract people's attention inevitably with the increasing number of vehicles. As one of the important features of vehicle, the vehicle license plate plays an important role and become the core of the intelligent transportation system.

In 1976, the Automated License Plate Recognition (ANPR) was invented at the Police Scientific Development Branch in the U.K. However, in the past ten years,

G. V. Vi (✉) · A. A. b. M. Faudzi
Faculty of Electrical and Electronics Engineering, University Malaysia Pahang, 26000 Pekan, Pahang, Malaysia
e-mail: vivi.gan35@gmail.com

A. A. b. M. Faudzi
e-mail: afif@ump.edu.my

A. F. Ab. Nasir et al. (eds.), *Recent Trends in Mechatronics Towards Industry 4.0*, Lecture Notes in Electrical Engineering 730, https://doi.org/10.1007/978-981-33-4597-3_56

with the improvement of digital cameras and the increase of computing power, it has aroused great interest [1].

The idea behind this system is the ability to detect and recognize the number plate of a vehicle from an image automatically. This system consists of a camera which is one of the important elements that will take the image of the vehicle and extract the place of the number in the image. Extracting the character from the image taken depends on how clear of the image. The image processing will translate the extracted number to digital numbers through the pixels of the image that has been taken [2].

The ALPR used for security purpose especially in the building of the government and military field. It also can be used in public area such as parking lots system. Other than that, it can be implemented in public street to detect the vehicles that exceed speed limit and may detect stolen car if found [3].

In Numbeo's 2020 Crime Index, the data shows that Malaysia ranks fifth in the world. Crime rates such as intrusion and theft have risen to the point where street people are concerned about their safety and well-being. As a result, the market is now highly demanding on property within the a-gated and a-guarded community. Therefore, license plate detection for ALPR system is the most important step to be concerned [4].

This paper emphasizes on license plate localization and recognition based on various existing techniques likes OCR-based method, template matching, deep learning, Harris Corner algorithm and then compared with OpenALPR HAAR and LBP algorithms.

1.1 License Plate Localization

License plate localization is an important step of the system, because it can locate the license plate containing alphabet and number. The concept of plate localization is to define the edge of license plate because the image contains a lot of data [5].

The way to extract more than one license plate from complex background image is significant. There is various process was taken to extract the edge of license plate [6]. License plate localization techniques is reviewed based on template matching and Harris Corner method.

Template Matching Template matching is an image processing method which find the location of the specific image (template) or prototype image. The image in the scene image (large image). Finds a location of the template in the scene, views the template as a sliding window, and then move it to all points in the image and find the similarity value of template for each point in the scene image [7]. The location of the template is the point with the largest similarity value in the image. A new method of matching through the normalized cross correlation method in the shortest time is introduced. Template matching by correlation is performed between the template w and the image f where position of template is to be determined in the image [8, 9].

Harris Corner Harris corner algorithm used in license plate localization mainly is to extract the license plate from the vehicle image. Harris Corner Detector is a corner detection operator to detect the license plate corner. Chris Harris and Mike Stephens improved Moravec's corner detector in 1988 and launched the product for the first time. Compared with the previous one, Harris' corner detector directly considers the direction, take it into account with the difference in corner scores, instead of using shift patches for every 45-degree angle, and has proven to be more effective in distinguishing edges from edges accurate corner. Since then, it has been improved and adopted by many algorithms to pre-process images for ALPR system [10, 11].

OpenALPR (HAAR and LBP) The captured vehicle image consists of HAAR and LBP features that can be extracted and train them with cascade classifiers to localize the license plate. In order to select the number of cascade classifier optimally, verification process is performed. Then use the extracted license plate for character recognition [12]. Haar wavelet consists of low and high interval, so called single-wavelength square wave. A square wave is a pair of adjacent rectangles in two dimensions, which is dark and bright. This method is to separate the obvious differences which similar to binary method [13].

The actual rectangular combination is not a real Haar wavelet which used for visual. Instead, they are modified to adapt visual recognition tasks. Because of this difference, these functions are called Haar functions or Haar-like functions [15]. The existence of Haar-like features is determined by subtracting the average dark area pixel value from the average bright area pixel value. If the difference is higher than the threshold (set during learning), the function is considered to exist. An example of Haar-like features is shown in Fig. 1.

Local Binary Pattern (LBP) is very effective but a simple texture operator which marks the pixels which is thresholding the other pixel with centre pixels value and change result as binary form. Due to its computational simplicity and discriminative ability, LBP texture operators have become a popular method in extracting application [13].

The LBP texture provides good results in many applications. In practical applications, its robustness to monotonic grayscale changes is the most important characteristic of the LBP operator which is illumination changes. Another that it can analyze image in real-time environment with its computational simplicity [16].

1.2 Character Recognition

Final step in ALPR system is license plate recognition. Recognition process is important since it determine the accuracy of recognition rate. The process is to recognize character and number of license plate [17].

Normalization is to standardize the character into a block, which does not contain any extra pixels on all four edges of the character. At this stage, it is necessary to

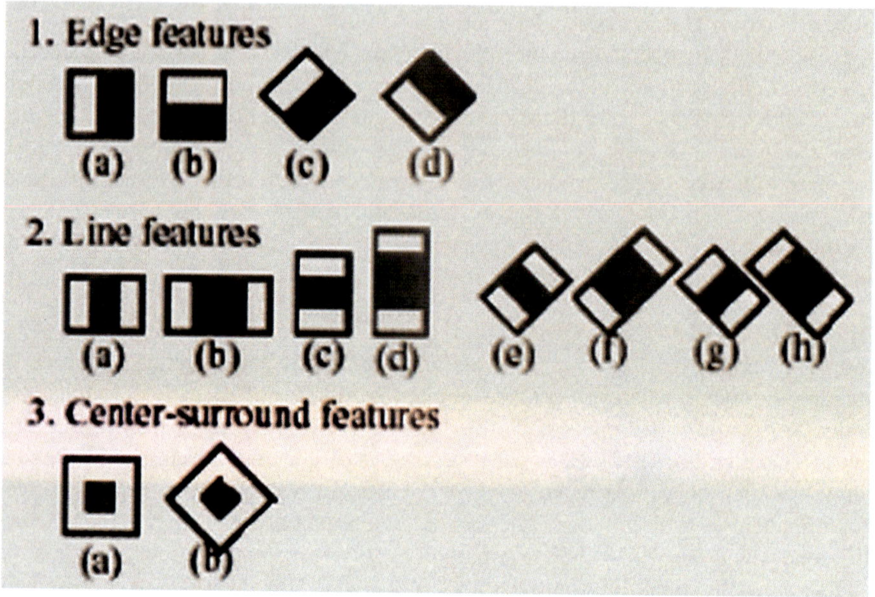

Fig. 1 HAAR-like features [14]

recognize the character image of the license plate taken from the license plate image. Character recognition of license plates can be performed through OCR-based method and deep learning.

OCR-based Method Optical Character Recognition used in license plate detection by generating the license plate image through simulation process. Optical character recognition (OCR) compacts with the problem of recognizing optically processed characters. OCR is performed offline. It can recognize manually printed and printed characters, but performance directly depends on the quality of the input document.

Generally, the OCR system has five components, which are optical scanning, position segmentation, feature extraction, post-recognition processing and pre-processing [18]. Especially in feature extraction, the robustness criteria are as follows: noise (sensitivity to broken line segments, bumps, gaps, filled rings, etc.), deformation (to local changes (such as rounded corners, inappropriate protrusions, expansion And contraction), style changes (sensitivity to style changes, such as using different shapes to represent the same character or using serifs, slashes, etc.), translation (sensitivity to movement of the entire character or its constituent parts), rotation (sensitivity to changing the direction of those characters) [19].

Deep Learning Deep learning mimics the work of the human brain such as decision making by creating pattern and processing data. Deep learning develops synchronously with the digital era. The digital era has brought explosive growth of various forms of data. This data (referred to as big data for short) comes from resources such as social media, internet search engines. Therefore, in license recognition topic, deep learning is implemented in develop automatic license plate reader system [2].

The final step in ALPR system development is recognition phase. Therefore, it closes down all the processes passing through image acquisition, followed by the position of the license plate until segmentation. Segmentation phase provide input (segmented character) to the recognition phase. The learning model which can read the image corresponding to characters is used for this recognition phase [21].

To make most of the data available for learning, used the same image processing steps before applying license plate segmentation and cut each character separately by scaling each character to a square [22].

2 Related Work

Various existing techniques and system performances will be discussed and reviewed in this section. The localization techniques consist of template matching, Harris Corner and OpenALPR (HAAR and LBP). The recognition techniques consist of template matching, OCR-based method and deep learning.

2.1 Review for License Plate Localization Using Template Matching, Harris Corner, Haar and LBP Algorithms

Automatic License Plate Recognition System using Web Technologies and Image Processing for Real Time Retrieval of Information The proposed technique demonstrates the necessity of using automated systems to keep vehicle information. In the proposed algorithm, an effective method for localizing Indian license plates has been designed [23].

Technique used: A vehicle number plate detection (VNPD) system based on template matching and normalized cross correlation is proposed. It also uses enhanced Otsu's method for threshold partitioning.

System performance: The VNDP system able to hit accuracy of 98.07% when testing on Indian license plate.

Performances Analysis of Vehicle Number Plate Recognition System Using Template Matching The proposed technique had been experimented to measure the system performance by comparing the system accuracy using template matching algorithm [24].

Technique used: To recognize license plate by template matching technique by using normalized cross correlations algorithm and phase correlation algorithm.

System performance: Accuracy of normalized cross correlation hits 67.98% and the accuracy of phase correlation hits 63.46%.

New Morphology-Based Method for Robust Iranian Car Plate Detection Image isolation is the main stage of the system. In different situations, the camera will be separated from the digital image. First, after applying morphological operators to image positioning, the image is pre-processed for further recognition. Morphological operators are based on shapes. Character segmentation is done by morphological process. It will remove all small noise elements. Then separate the characters by dilation process from each other. Then perform a partition scan to perform character segmentation. The proposed method had been done experiment to measure the system performance by comparing the system accuracy using template matching algorithm [25].

Technique used: A new real-time robust license plate detection method based on morphology and template matching.

System performance: It provides better performance for license plate localization. It can correctly find the position of the board, up to 97.3%.

License Plate Detection using Harris Corner and Character Segmentation by Integrated Approach from An Image In this paper, the problem of license plate localization is solved through the integrated segmentation method. License plates provide a variety of systems to recognize that each has its own special interests and obstacles. The important step of the ALPR system is to limit the license plate number accurately, segmentation, and recognition [26].

Technique used: This paper proposes the Harris Corner algorithm, which is ultimately robust under changing motion and lighting conditions. The accuracy of the license plate limit is forwarded to the segmentation stage. Segmentation is done by a connected component analysis method that combines the number of pixels, aspect ratio, and character height.

System performance: The total amount of correctly recognized license plates is 96.92%. The skipped plates eventually reflected the sunlight, making it difficult to locate. The wrongly segmented region usually intersected with other elements and the success rate of the segmentation stage is 93.84%.

Real-Time Implementing of Indian License Plate Recognition System This article introduces a simple and fast technique for Indian license plate recognition system. Implementing Raspberry Pi and OpenALPR software framework to develop a real-time Indian license plate recognition system. This system can be used for some application purposes [27, 28].

Technique used: HAAR and LBP algorithms features in OpenALPR are extracted from acquired vehicle image and subjected to training of cascade classifier in order to localize license number plate.

System performance: For HAAR feature, processing time required 9 s and give accuracy of 95%. For LBP feature, processing time required 1.5632 s and give accuracy of 87.5%.

2.2 *Review for License Plate Recognition Using Template Matching, OCR-Based Method and Deep Learning*

Automatic Vehicle Identification by Plate Recognition The author proposes an ALPR with simple algorithm. ALPR works in different parts: license plate area extraction, character segmentation and recognition. Template matching plays a role in recognition process. First, the plate is converted into a binarized image, and then the simple algorithm is applied to search the plate area. After that, the morphology operator is applied to obtain the position. Then use morphological functions to do segmentation. Then separate characters with dilation operator. The characters are normalizing before recognition [29].

Technique used: Normalization to refine characters into blocks that do not contain spaces, and then adjust characters to equal size for template matching. The cross-correlation method is used for character recognition.

System performance: The plate detection is obtained with higher accuracy. It shows that the accuracy of the plate extraction is 97.6%, the segmentation is 96%, and the character recognition rate is 98.8%.

Impact of Character Spacing on the Performance of Automatic Number Plate Recognition (ANPR) Systems through Simulation In this article, the author determined the effect of illegal spacing between license plate characters on ANPR performance. The reasons for the inaccurate reading of ANPR data were examined in detail and suggestions were made on how to make improvements to minimize the risk of misreading [30].

Technique used: In this paper, the key data set is generated through a simulation process, which will generate a license plate number image. In this paper, optical character recognition (OCR) is used.

System performance: The variable spacing do not have adverse effect on the ALPR engine. As a result, there is only 45.07% accuracy rate. It cannot predict the syntax rules or illegal spacing that lead to misread or capture inability. Therefore, more work must be done.

Thai License Plate Recognition Based on Deep Learning The paper emphasizes on license plate recognition by doing research into Thai motorcycle license plate. Thai license plate is a triple-line type which contains three lines of character sets, representing categories, registered location and registered numbers respectively from top to bottom [31].

Technique used: There are training step and recognizing step. First, using MobileNet (deep learning) to segment and train 3 lines of each license plate. Second, segmenting and training Thai letters and numbers.

System performance: Accuracy of line recognition hits 96.94% and accuracy of character recognition hits 91.76%.

Linux Embedded System for Vehicle License Plate Recognition This paper is seeking to develop a low-cost embedded system to monitor the flux of vehicles by recognizing license plate. It uses the Linux embedded system Raspberry Pi 2 B+ to segment the pictures taken by the camera, send commands through the Arduino Uno board, and then follow the signs to control the operation of the car [28, 32].

Technique used: OpenALPR library and the Python programming language is implemented to recognize license plate. A Linux embedded system based on Raspbian and Raspberry Pi platform.

System performance: The application is satisfactory because the software developed achieved good results with a hit rate of approximately 92%, making the project highly feasible for applications with Linux embedded systems on the Raspberry Pi platform.

2.3 *Review Summary*

In this section, localization and recognition accuracy rate are summarized based on various techniques reviewed in Table 1. Detection rate and recognition rate is formulated as Eq. 1:

Table 1 Performance analysis based on different detection and recognition techniques

Research topic	Localization techniques	Detection rate (%)
1	Normalized cross correlation [23]	98.07
2	Normalized cross correlation [24]	67.98
	Phase correlation [24]	63.46
3	Template matching [25]	97.30
4	Harris corner [26]	96.92
5	HAAR [27]	95.00
	LBP [27]	87.50
Research topic	**Recognition techniques**	**Recognition rate (%)**
1	Template matching [29]	98.80
2	OCR-based method [30]	45.07
3	Deep learning [31]	91.76
4	OpenALPR [32]	87.50

$$\text{Recognition Rate} = \frac{\text{Sample successfully recognized}}{\text{Total sample}} \times 100\% \quad (1)$$

3 Conclusion

License plate detection in different areas such as automatic parking, automatic toll gates, border control, law enforcement and so on, the license plate recognition process requires very high accuracy. This article gives an overview of license plate recognition technology and describe the way of these techniques help to recognize license plates and enhance the detection rate and recognition rate. This article summarizes the performance and concepts of license plate localization and recognition techniques.

Acknowledgements The authors would like to thank Universiti Malaysia Pahang (UMP) for providing the facilities and fund to complete this project. This work is supported by UMP Internal Grant of RDU1703140.

References

1. Xing J, Li J, Xie Z, Liao X, Zeng W (2016) Research and implementation of an improved radon transform for license plate recognition. In: 2016 8th international conference on intelligent human-machine systems and cybernetics (IHMSC), Aug 2016, vol 01, pp 42–45
2. Ahn C-S, Lee B-G, Yang S-S, Park S-C (2017) Design of car license plate area detection algorithm for enhanced recognition plate. In: 2017 4th international conference on computer applications and information processing technology (CAIPT), Aug 2017, pp 1–4
3. Feng Y, Li S, Pang T (2018) Research and system design of intelligent license plate recognition algorithm. In: 2018 37th Chinese control conference (CCC), Jul 2018, pp 9209–9213
4. Crime index by city 2019 mid-year. https://www.numbeo.com/crime/rankings.jsp. Accessed 12 Dec 2019
5. Pingping L, Wenjing X (2019) Research on recognition technology of license plate image. In: 2019 international conference on computer network, electronic and automation (ICCNEA), Sep 2019, pp 52–56
6. Chao H, haiying l, huiyuan Z, Valentin (2018) License plate recognition and location algorithm based on compressive sensing. In: 2018 IEEE international conference on information and automation (ICIA), Aug 2018, pp 1593–1598
7. Lin G, Xue B, Xu B, Chen C (2019) License plate recognition based on mathematical morphology and template matching. In: 2019 Chinese automation congress (CAC), Nov 2019, pp 405–410
8. Template matching—an overview | ScienceDirect topics. https://www.sciencedirect.com/topics/engineering/template-matching. Accessed 24 May 2020
9. Pinthong T, Yimyam W, Chumuang N, Ketcham M (2018) License plate tracking based on template matching technique. In: 2018 18th international symposium on communications and information technologies (ISCIT), Sept 2018, pp 299–303
10. Sánchez J, Monzón N, Salgado A (2018) An analysis and implementation of the Harris corner detector. Image Process On Line 8:305–328

11. Falih E (2017) Improvement of Corner detection algorithms (Harris, FAST and SUSAN) based on reduction of features space and complexity time, vol 35, p 112
12. Symmetry features for license plate classification—IET Journals & Magazine. https://ieeexplore-ieee-org.ezproxy.ump.edu.my/document/8548625. Accessed 12 Jun 2020
13. Kadir K, Kamaruddin M, Nasir H, Safie S, Bakti Z (2014) A comparative study between LBP and Haar-like features for Face Detection using OpenCV, pp 335–339
14. Fig. 3. Examples of Haar Features used in OpenCV. ResearchGate. https://www.researchgate.net/figure/Examples-of-Haar-Features-used-in-OpenCV_fig3_221317469. Accessed 24 May 2020
15. Comparison of vehicle detection using Haar-like feature, LBP and HOG technique for feature extraction in cascade classifier. Int J Adv Sci Technol. http://sersc.org/journals/index.php/IJAST/article/view/953. Accessed 12 Jun 2020
16. Local binary patterns—Scholarpedia. http://www.scholarpedia.org/article/Local_Binary_Patterns. Accessed 24 May 2020
17. Ingole SK, Gundre SB (2017) Characters feature based Indian vehicle license plate detection and recognition. In: 2017 international conference on intelligent computing and control (I2C2), Jun 2017, pp 1–5
18. Sabu AM, Das AS (2018) A survey on various optical character recognition techniques. In: 2018 conference on emerging devices and smart systems (ICEDSS), Mar 2018, pp 152–155
19. Omran SS, Jarallah JA (2017) Iraqi car license plate recognition using OCR. In: 2017 annual conference on new trends in information communications technology applications (NTICT), Mar 2017, pp 298–303
20. Hargrave M (2020) How deep learning can help prevent financial fraud. Investopedia. https://www.investopedia.com/terms/d/deep-learning.asp. Accessed 24 May 2020
21. Abedin MdZ, Nath AC, Dhar P, Deb K, Hossain MS (2017) License plate recognition system based on contour properties and deep learning model. In: 2017 IEEE region 10 humanitarian technology conference (R10-HTC), Dec 2017, pp 590–593
22. Khazri A (2019) Automatic License Plate Detection & Recognition using deep learning. Medium, 10 Oct 2019. https://towardsdatascience.com/automatic-license-plate-detection-recognition-using-deep-learning-624def07eaaf. Accessed 24 May 2020
23. Shevale K, Gite S (2016) Automatic license plate recognition system using web technologies and image processing for real time retrieval of information. /paper/Automatic-License-Plate-Recognition-System-using-of-Shevale-Gite/62136e9ba5c10161c55a55ec5d3c24dd2d098f5e. Accessed 24 May 2020
24. (PDF) Performance analysis of vehicle number plate recognition system using template matching techniques. https://www.researchgate.net/publication/325362213_Performance_Analysis_of_Vehicle_Number_Plate_Recognition_System_Using_Template_Matching_Techniques. Accessed 12 Dec 2019
25. Iranian license plate detection using cascade classifier—IEEE Conference Publication. https://ieeexplore-ieee-org.ezproxy.ump.edu.my/document/8786468. Accessed 12 Jun 2020
26. Panchal T, Patel H, Panchal A (2016) license plate detection using Harris corner and character segmentation by integrated approach from an image. Procedia Comput Sci 79:419–425
27. Desai GG, Bartakke PP (2018) Real-time implementation of Indian license plate recognition system. In: 2018 IEEE Punecon, 2018, pp 1–5
28. Luai Taha Ahmed A-M, Nurhafizah AT, Syamimi S, Mohamad Shaiful AK, Ahmad Afif MF (2018) Development of automated gate using automatic license plate recognition system. Presented at the 10th National Seminar on Underwater System Technology 2018 (NUSYS'18), UMP Pekan, Aug 2018, pp 1–8 [online]. Available at: http://umpir.ump.edu.my/id/eprint/22871/. Accessed 13 Jun 2020
29. Zhang Z, Wan Y (2019) Improving the accuracy of license plate detection and recognition in general unconstrained scenarios. In: 2019 IEEE symposium series on computational intelligence (SSCI), 2019, pp 1194–1199
30. Tejas B, Omkar D, Rutuja D, Prajakta K, Bhakti P (2017) Number plate recognition and document verification using feature extraction OCR algorithm. In: 2017 international conference on intelligent computing and control systems (ICICCS), 2017, pp 1317–1320

31. Puarungroj W, Boonsirisumpun N (2018) Thai license plate recognition based on deep learning. Procedia Comput Sci 135:214–221
32. Batista J, Imaculada R, César S (2018) Linux embedded system for vehicle license plates recognition. Int J Comput Appl 182:43–46

Disturbance Rejection Performance Evaluation of GA Optimized PI Controller for Brushed DC Motor for Cart Follower Application

A. A. M. Zahir, S. S. N. Alhady, W. A. F. W. Othman, A. A. A. Wahab, and M. F. Ahmad

Abstract PI controller, a form of PID controller without derivative term, is one of the most widely used controller in industrial application. PI controller was being selected due to the presence of derivative component that could yield instability in the presence of noise. PI controller is tuned by using GA-Original ITSE, GA-Modified ITSE, CHR, and AMIGO. The modification is crucial as original error criteria produce insignificant improvement in performance parameters such as settling time and overshoot percentage. A tuning method with balance set-point tracking performance and disturbance rejection properties is needed as normally controller with good set-point tracking is associated with poor disturbance rejection properties. The plant is tuned by stated tuning methods with and without disturbance. The controller is then being implemented and the step responses are being evaluated. Two type of tests, load test and surface inclination angle test are conducted. Disturbance rejection performance is evaluated for all tuning methods. By Simulation GA-Modified ITSE produce the fastest settling time, improving 36.47–44.47% compare with CHR and AMIGO. GA-Modified ITSE also performed the best in reducing overshoot percentage to 0.0361%, with percentage difference of 198.681% compared with second best performer, AMIGO. GA-Modified ITSE shows the best performance in disturbance rejection properties with settling time of 5.5 s. Experimentally, GA-Modified performed the best in most of step response parameters, overshoot percentage and settling time with 3.7833% and 8.1301 s respectively. During disturbance rejection test, AMIGO and GA-Modified ITSE shows satisfactory performance with better stability compared with CHR and GA-Original ITSE that yield oscillatory response.

Keywords Brushed DC motor · PI controller · Genetic algorithm · Disturbance rejection evaluation

A. A. M. Zahir · S. S. N. Alhady (✉) · W. A. F. W. Othman · A. A. A. Wahab · M. F. Ahmad
School of Electric and Electronic, Engineering Campus Universiti Sains Malaysia, 14300 Nibong Tebal, Penang, Malaysia
e-mail: sahal@usm.my

A. F. Ab. Nasir et al. (eds.), *Recent Trends in Mechatronics Towards Industry 4.0*, Lecture Notes in Electrical Engineering 730, https://doi.org/10.1007/978-981-33-4597-3_57

1 Introduction

PID controller is widely being used in various application due to its simplicity and good performance [1]. PI controller is one type of PID controller without derivative terms. PID and PI controller are so versatile that only 5–10% control loops could not be controlled by SISO PI or PID controllers [2]. In industry, 90% of PID controllers have their derivative component eliminated or turned off [3].

The elimination of derivative component made the implementation of PI controller simple and less susceptible to noise [4]. Derivatives component are often eliminated as sometimes it produced random (high frequent) noise measurement, resulting in huge variation in control signal [5]. The implementation of derivative component is not recommended in the presence of noise in the feedback readings [6]. PI controller is workable with 2 type of system, first order and second order system [6].

There are various tuning method available to tune the controller ranging from classical design formulae to meta-heuristic tuning methods. Selected tuning methods that were being selected are Chien, Hrones and Reswick (CHR), Approximate M-Constrained Integral Gain Optimization (AMIGO), and Genetic Algorithm (GA).

For meta-heuristic tuning method, error criteria such as ITSE, ITAE, IAE, and ISE are widely being used as objective functions [7, 8]. Authors in [8] evaluated the performance of original error criteria using GA optimization to tune PID controller. The results show that by using original error criteria, there is still some overshoot present, with insignificant in improvement in terms of settling time compare with Ziegler-Nichols and Iterative method.

Research paper in [9, 10] show that the performance of step response performance parameters in simulation were improved by modifying the original error criteria, adding the terms of settling time, rise time, overshoot percentage and steady state error as GA by using original error criteria focused solely on minimizing the error function, e(t) without considering other parameters.

Disturbance rejection properties is one of the important parameters in evaluating the performance of the controller. Research in [11] shows that by using Virtual Disturbance Feedback Tuning (VRFT), disturbance rejection properties are improved, but yields poor tracking performance. IMC-PID controller produced good set point tracking, but sluggish disturbance rejection properties [12]. A system that have balanced performance between set-point tracking and disturbance rejection properties is desirable, as usually when a system have a good set-point tracking ability, the disturbance rejection is traded off.

The objectives of this paper is to evaluate the performance of CHR, AMIGO, GA by using original ITSE, and GA by using modified ITSE in terms of output step response parameter such as settling time, rise time, overshoot percentage, and steady state error, together with disturbance rejection properties. The system is simulated with and without disturbance, and then implemented into the prototype. Two type of experiment, load test and surface inclination angle test are being done in order to evaluate the performance of all tuning methods under the presence of disturbance experimentally.

2 Methodology

PI controller was optimized by using Genetic Algorithm optimization and the results are being compared with AMIGO and CHR. PI controller regulate the velocity of the cart follower by acting on the feedback fed by the ultrasonic sensors. Targeted velocity decided was 0.5 m/s, average velocity for wheelchair user as stated in the previous work in [13].

Controller performance evaluation could be divided into two categories, tracking performance and disturbance rejection performance. Tracking performance parameters include settling time, rise time, steady state error and overshoot percentage. Disturbance rejection evaluation involved general trend of the step response, and settling time of controller effort when being subjected to disturbance.

The subject of tuning is the brushed DC motor equipped to propel the cart follower system. Details of conceptual design process and system identification process was published in [14]. The plant exhibit second order system as in Eq. 1:

$$\frac{10.01}{s^2 + 2.553s + 10.91} \tag{1}$$

2.1 PI Controller Architecture

Basic form of PI controller is stated in Eq. 2. The original form of the controller includes derivative component but in case of PI controller, the derivative component was being eliminated or equated to zero. Derivative component could improve the stability of the system and reduced overshoot percentage caused by integral component, the drawback that it could made the system prone to instability in case of the presence of noise from the feedback [6].

The validation on controller selection was done in authors work in [13] resulting in the selection of PI controller, due to the volatile reading of velocity at the first half of transient response when there is the presence of derivative component.

$$C(s) = K_P + \frac{K_i}{s} \tag{2}$$

where K_p, K_i proportional and integral gain respectively. The gains value were calculated by using AMIGO, CHR, and Genetic Algorithm. The performance of classical formulae and genetic algorithm were being evaluated and analyzed.

2.2 Genetic Algorithm Optimization

The basic concept of genetic algorithm is based on the survival of the fittest, where the best fit solutions was being selected based on the score value of the objective function.

The first step before initiating the optimization was to formulize the objective function. Integral Time Squared Error was the best fit error criterion for this research [9]. The basic form of the ITSE is shown in Eq. 3.

ITSE function was then modified by adding performance parameters such as rise time, overshoot percentage, settling time, and steady state error. The modification improved settling time and reduce overshoot percentage [9]. Modified ITSE is described as in Eq. 4:

$$\mathrm{ITSE} = \int_0^{\tau} \mathrm{te(t)^2dt} \tag{3}$$

$$ITSE_{Modified} = \int_0^{\tau} \mathrm{te(t)^2dt} + \alpha O_v + \beta SS_e + \delta t_s + \Upsilon t_r \mathrm{t} \tag{4}$$

where e(t) is error signal, $\mathrm{O_v}$ is percentage of overshoot, $\mathrm{t_s}$ is settling time, $\mathrm{t_r}$ is rise time, and $\mathrm{SS_e}$ represent steady state error. α, β, δ, Υ are weightage defined by users, in his case the values are 30, 30, 25, and 15 respectively based on the performance observed during simulation as in [9].

Genetic Algorithm optimized the gains, identified as two unknown parameters, and the solutions were evaluated by using objective function as in Eqs. 3 and 4.

The process started with the creation of 50 initial population where the upper and lower boundary are [2 5] and [0 0] respectively. The population was then evaluated by using specified objective function.

The stopping criteria decided was the average change in value of objective function should be less than 1×10^{-6} for minimum 50 generations, or the simulation reach maximum 300 generations, whichever comes first. Crossover fraction and elite count used are 0.8 and 0.1 respectively.

After the solutions were evaluated and did not satisfied stopping criteria, the fittest chromosomes were being selected and undergone crossover and mutation process which produced new population. The process repeated until the optimization met stopping criteria.

2.3 *Disturbance Rejection Performance Evaluation—Simulation*

Disturbance rejection properties indicates how well the system reacts in case of the presence of disturbance. A good controller should be able to make the system turn back to steady state condition when there is disturbance in the system. A block diagram model of the system in the presence of input disturbance is shown in Fig. 1.

In order to further evaluate the system, the block diagram in Fig. 1 was being reduced to a condition where the disturbance was made to become the input, and the error was made to become the output. The block diagram was simplified to a condition where the PI controller acted as the feedback to the plant so that the controller effort graph could be created in order to evaluate the disturbance properties of the system [15]. The simplified version of the block diagram is depicted in Fig. 2.

A SIMULINK model based on the block diagram was created in order to simulate a situation where there is input disturbance. The system was being subjected to a square wave input disturbance with the amplitude of 15 and time delay of 0.4 s. The response of the system under input disturbance for every tuning method were plotted.

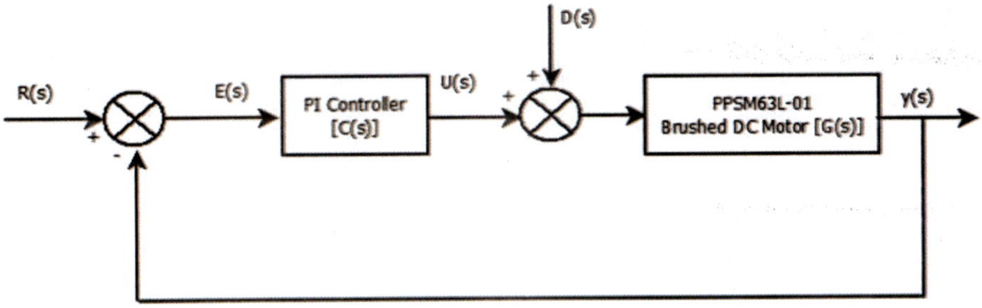

Fig. 1 System with input disturbance

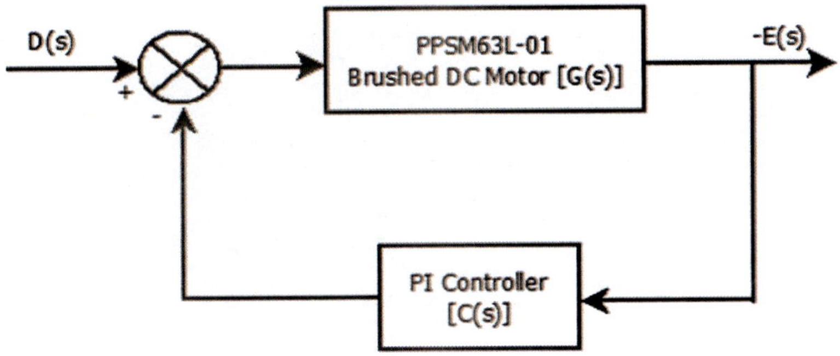

Fig. 2 Simplified block diagram

2.4 Disturbance Rejection Performance Evaluation—Experiment

PI controller was implemented into the cart follower system as in previous work in [13]. Crucial hardware in implementing the PI controller include a rotary encoder which acted as a feedback to the controller, MD30C motor driver that functions to regulate the motor by using Pulse Width Modulation (PWM), and Arduino Mega 2560 that contained PI controller coding, calculating needed signal to be produced via the motor controller.

After implementation of PI controller into the cart follower system, the system was then subjected into a condition where there is the presence of disturbance. Two type of experiment were conducted, which is load test and inclination angle test.

Load test was being done to evaluate the ability of the system to retain the targeted velocity under the presence of load. Load with weight of 20 and 40 kg was applied on the cart follower at $t = 10$ s.

The step response of load disturbance test was plotted for all tuning methods. Figure 3 depicted the cart follower with the presence of load.

Surface inclination angle test was done in order to test the system under climbing phase and descending phase without additional load. American with Disabilities Act (ADA) recommended ramp angle is at 4.8° at public facilities [16]. Cart follower was subjected to ramp angle of 6°, 8°, and 10°, a lot steeper than ADA recommendation in order to test the system under extreme ramp condition.

Fig. 3 The prototype under the presence of additional load

Fig. 4 Phase of surface inclination angle test

Fig. 5 Fabricated ramp

The ramp designed was divided into 3 phase which is inclination, horizontal, and declination phase. The illustration of ramp phases is depicted in Fig. 4 and fabricated ramp is shown in Fig. 5. The ramp angle is adjustable by using hinge so that the same test rig could fulfill the angle condition of 6°, 8° and 10°. The system reaction towards abrupt change in surface angle was recorded and evaluated.

3 Results and Discussion

Table 1 described all of the gains obtained by different tuning method. CHR produced the biggest value of proportional gain while AMIGO produced the lowest proportional gain. In terms of integral gain CHR produced the biggest gain and GA-modified ITSE yield the lowest value of integral gain.

Both CHR and AMIGO produced bigger integral gain due to the fact that CHR and MIGO were a classical design formulae that were being derived from ZN method.

Table 1 Gains tuned by various tuning method

Methods	K_p	K_i
Chiens-Hrones-Reswick	1.733	3.494
AMIGO	0.5994	2.0727
GA-Original ITSE	1.8436	2.0217
GA-Modified ITSE	0.8435	1.5535

Those formulae intended to produced quick response, thus resulting in a big value of integral gain, thus tend to produce some overshoot and oscillation in certain application.

3.1 Step Response Performance

Figure 6 depicts step response of different tuning methods obtained from the simulation and Table 2 describe results of performance parameters. CHR and MIGO shows some oscillatory response at simulated step response. CHR produced the biggest value of overshoot, with 30.06% and the fastest rise time with 0.2895 s. This is due

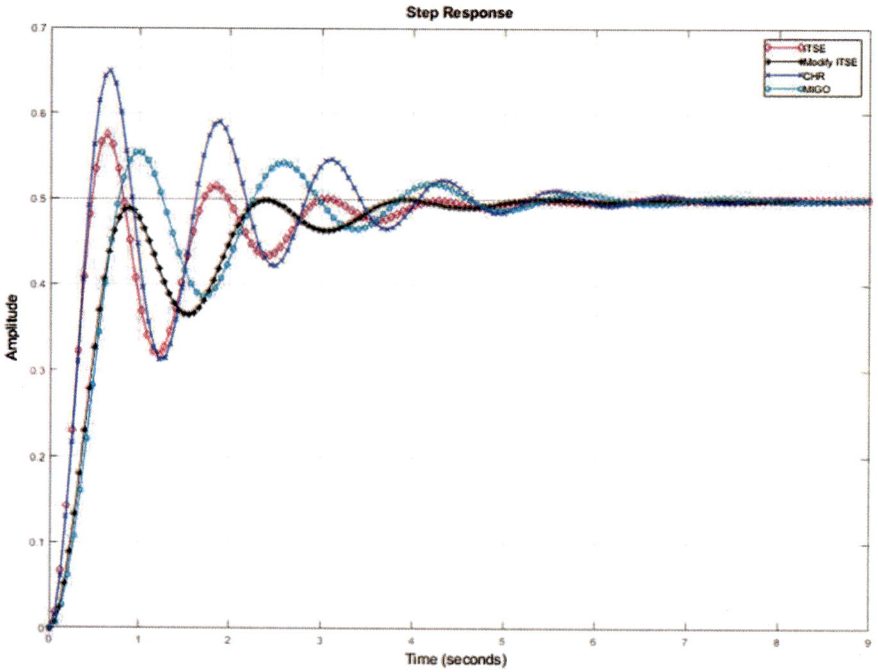

Fig. 6 Step response of various tuning methods—simulation

Table 2 Performance of PI tuning methods—simulation

Tuning method	Overshoot (%)	Rise time (s)	Settling time (s)	Steady-state error
CHR	30.06	0.2895	5.610	0.001400
AMIGO	10.91	0.4967	5.161	0.001500
GA-Original ITSE	14.97	0.3145	3.964	0.000078
GA-Modified ITSE	0.0361	0.5265	3.569	0.004800

to CHR induced big values of proportional and integral gain at the same time, thus increasing the overshoot and yields oscillatory response. Even though, GA-Original ITSE was the one with the biggest proportional gain, it overshoot was smaller than CHR due to fact that it produced lower integral gain than CHR.

Both tuning by using Genetic Algorithm produced better settling time than classical design formulae, with the GA-Original ITSE yields improvement of 26.24% and 34.38% than AMIGO and CHR respectively.

GA-Modified ITSE was the best performer in 2 out of 4 performance parameter, overshoot percentage and settling time. GA-Modified ITSE nearly totally eliminated the overshoot percentage, producing 0.0361%. The percentage difference of overshoot percentage between GA-Modified ITSE with second best performer in this sector, AMIGO is 198.681%. It produce the best in terms of settling time with 3.569 s, improvement of 10.49% from GA-Original ITSE. Good performance of overshoot percentage, rise time, and steady response, jeopardized the performance of this method in terms of rise time as GA-Modified ITSE produced the slowest rise time among all of the tuning method due to the fact that a small proportional gain yielded from the weightage assigned at the objective function during optimization.

GA-Original ITSE have the smallest steady state error with 0.004800 and GA-Modified ITSE seems struggle in reducing the steady state error, producing the biggest steady state error. The modification of objective function failed to improve the steady state error at the same time with improving the overshoot percentage and settling time, although the similar value of weightage assigned for both 3 parameters. The function tried to reduce the overshoot percentage and settling time, resulting to the settlement of the final value of the response earlier. Even though GA-Modified ITSE performed the worst in terms of steady state error, the performance of all tuning methods are satisfactory, due to all of the steady state error values are near to 0, making the difference between each performance are not substantial.

Figure 7 shows step response of the system with no load obtained experimentally while Table 3 described the numerical results of performance parameters. PI controller tuned by GA-Modified ITSE outperformed all of the tuning methods in 2 out of 3 performance parameter. CHR produced the lowest rise time among all of tuning methods, while GA-Modified ITSE was the best on settling time and overshoot percentage.

From the step response in Fig. 7, CHR and GA-Original ITSE produce oscillatory response nearly at the first quarter of the response time, due to big values of proportional and integral gain, whereas AMIGO did not show oscillatory performance although its integral gain value is nearly the same as GA-Original ITSE, due to the fact that it has a low value of proportional gain.

By using 5% tolerance band in defining the final steady state value of the settling time, all of the experimental readings of tuning methods shows deviation between 77.97 and 98.1227% from the simulation. Graphically, based on Fig. 7 all of the tuning methods started to converge to steady state condition at t = 3 s to t = 5 s. The stringent definition of settling time of 5% tolerance band indicated that if any of the response exceeded 5% value of predetermined targeted steady state value (0.5 m/s), the response was not considered as achieving steady state condition. In

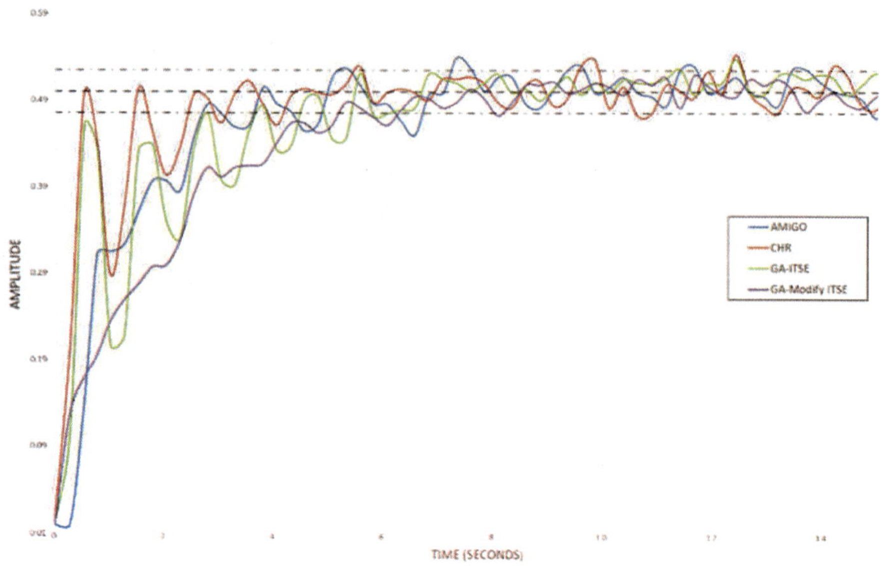

Fig. 7 Step response of various tuning methods—experiment

Table 3 Performance of PI tuning methods—experiment

Tuning method	Overshoot (%)	Rise time (s)	Settling time (s) (5% tolerance band)
CHR	8.662	0.4181	14.3110
AMIGO	7.8952	2.1812	15.1026
GA-Original ITSE	6.8880	0.3839	12.491
GA-Modified ITSE	3.7833	4.0416	8.1301

case of hardware implementation, there is the presence of noise that could make the reading exceeded 5% tolerance band at some point. Although 5% tolerance band was quite stringent considering the presence of noise, the deviation values in this experiment are acceptable based on related researches and literatures.

3.2 *Disturbance Rejection Performance*

Disturbance rejection properties is the abilities of the system to track back to targeted velocity in case of the presence of disturbance. Figure 8a depicts the conditions of the system when input step disturbance applied to the system in terms of square wave with amplitude of 15 as in Fig. 8b.

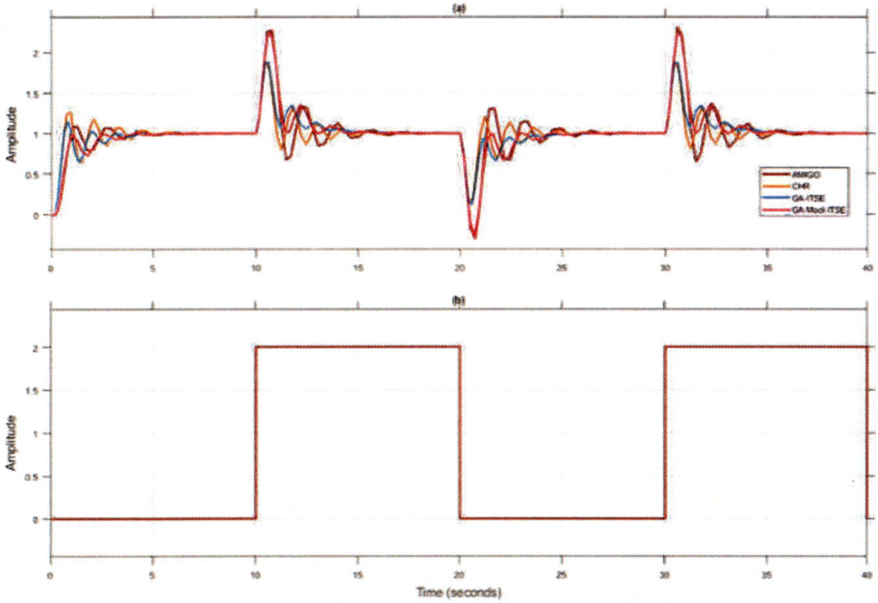

Fig. 8 **a** Step response of various tuning methods subjected to step input disturbance; **b** square wave of step input disturbance

The response indicates that CHR and GA-ITSE response faster compare with AMIGO and GA-Modi ITSE. This is due to the fact K_P for both CHR and GA-Original ITSE are bigger, as K_P is directly proportional to the error signal e(t) as bigger error made the system with bigger proportional gain react faster.

From the response, CHR and AMIGO seems to oscillate before settled down to desired velocity. Thus is due to the bigger value of integral gain were bigger for both tuning method compared with both GA tuning, resulting the systems tuned by CHR and AMIGO tend to react more when there is error over time that caused offset to the steady state value, resulting in oscillation phenomenon.

Figure 9 indicates the ability of the tuned system to settle down when there is input disturbance. The error response in Fig. 9 was generated based on the block diagram in Fig. 2.

Based on the response, GA-Modify ITSE settle down the fastest with settling time 5.26 s followed by GA-Original ITSE, CHR, and AMIGO with 5.5 s, 5.63 s, and 6.48 s respectively.

Figures 10 and 11 show the response of the systems when the load were applied at approximately t = 10 s. Load applied were 20 kg for results in Fig. 10 and 40 kg for results in Fig. 11. The track back response for all of the systems seems comparable, where the response of GA-Original ITSE contains oscillation when 20 kg load was applied.

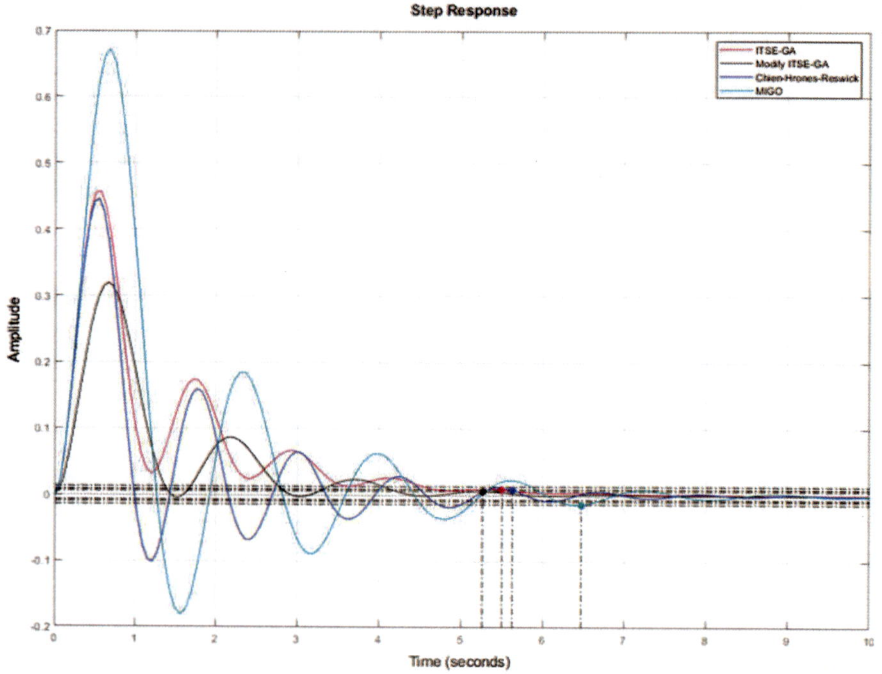

Fig. 9 Disturbance rejection properties of different tuning methods

For 40 kg load test response as in Fig. 11, CHR shows extreme oscillatory response for almost 3 s, due to big value of proportional gain and integral gain that caused instability and oscillation. GA-Original ITSE, GA-Modified ITSE and AMIGO shows similar performance in tracking response. When tracking back to desired velocity, percentage of overshoot produced by AMIGO, GA-Original ITSE, and GA-Modified ITSE are 4.13%, 7.08%, and 11.11% respectively.

Based on the load test of 20 and 40 kg, it could be concluded that GA-Modified ITSE and AMIGO produced satisfactory performance in terms of disturbance rejection without the presence of oscillation whereas CHR are the most prone to oscillatory response when there is the presence of load disturbance.

Figures 12, 13, and 14 show the response of the system when being subjected to disturbance of surface inclination angle of 6°, 8°, and 10° respectively, without the presence of additional load. This test, was divided into 3 phase, ascend, horizontal, and descend. Response of the system especially at crucial phase such as ascend phase and descend phase were observed. The best performer in terms of response during the test, GA-Modified and AMIGO were highlighted with markers and continuous line as in Figs. 12, 13 and 14.

The cart follower entered the climbing phase at approximately t = 5 s. The time for descending phase was not uniform due to every tuning methods' responses behave differently when being subjected to disturbance at climbing phase, thus affecting the

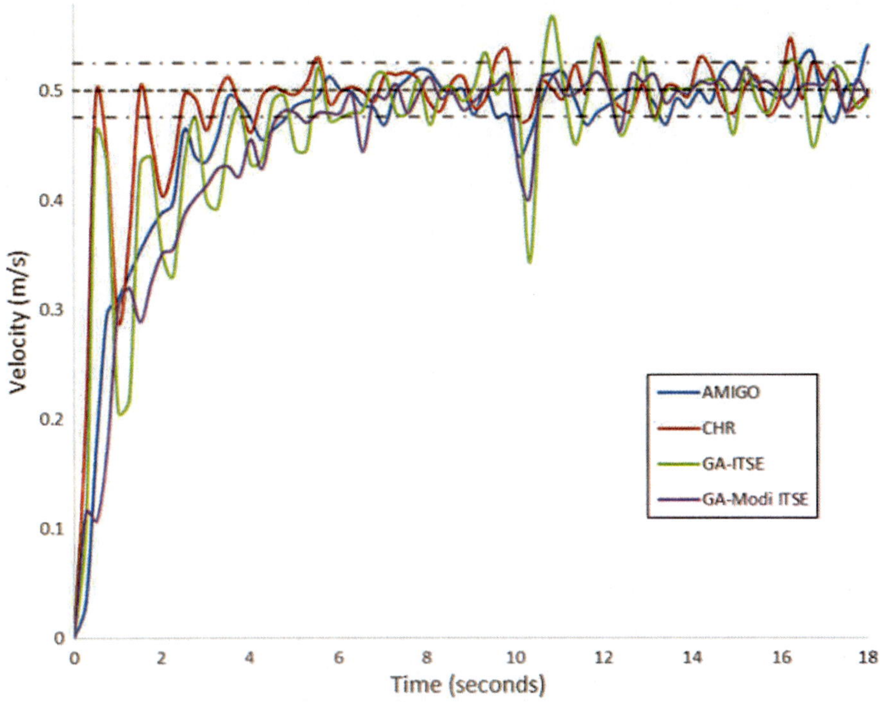

Fig. 10 Response of the system with additional load of 20 kg

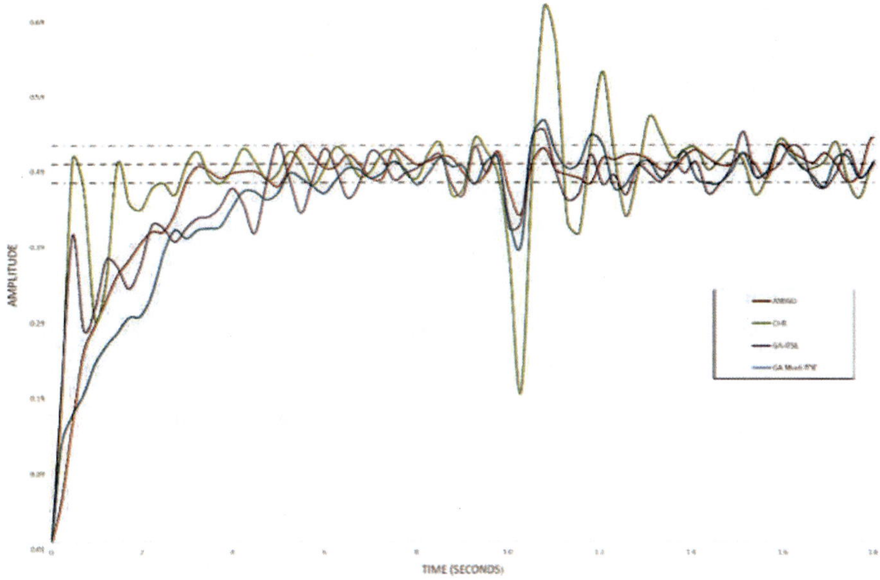

Fig. 11 Response of the system with additional load of 40 kg

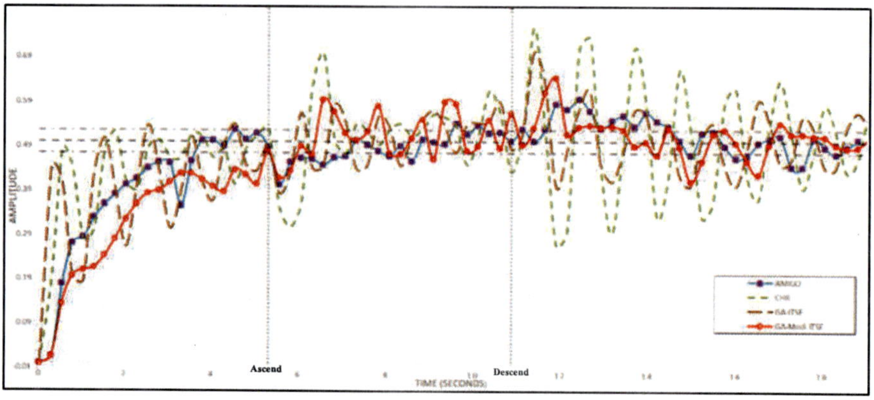

Fig. 12 Response of the system at inclined surface angle of 6°

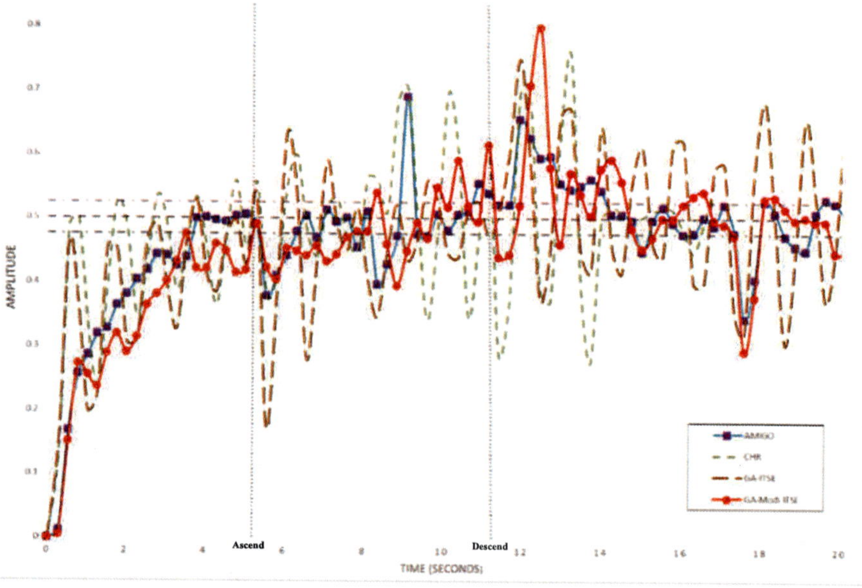

Fig. 13 Response of the system at inclined surface angle of 8°

velocity of the cart entering the horizontal and descend phase. From the response it could be conclude that the cart was entering the descend phase at approximately t = 11 s for all tuning methods.

As in Fig. 12 for 6° inclination angle, CHR produced oscillatory response at climbing phase. There is no significant abnormality of the response at horizontal phase. At descending phase, CHR produce extreme oscillation, due to the system tried to slow down the velocity affected by declination angle and momentum caused

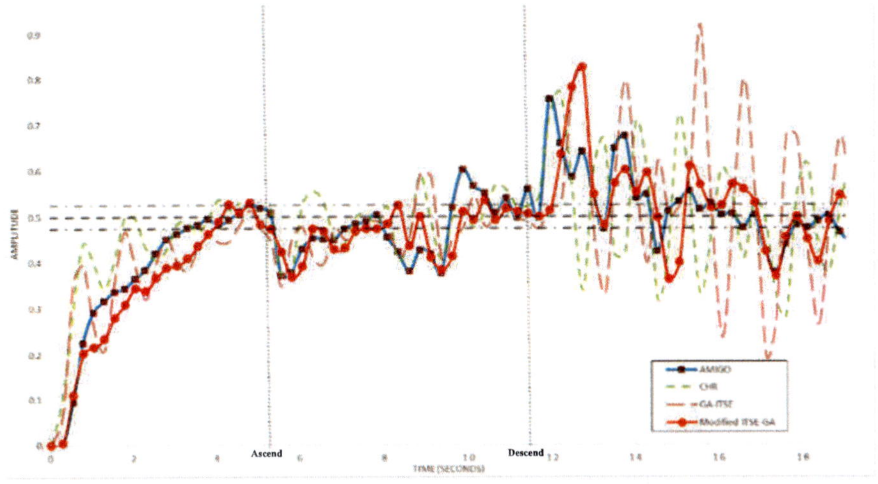

Fig. 14 Response of the system at inclined Surface angle of 10°

by the cart weight. It could be observed that the response of CHR are very slow in converging the velocity to the targeted velocity compare with other tuning methods. AMIGO and GA-Modified ITSE show favorable performance at descending phase as it does not produce oscillation.

Same pattern of response are observed when the inclination angle were increased to 8° and 10° as in Figs. 13 and 14 respectively, where CHR and GA-Original ITSE show significant oscillatory response compared with GA-Modified ITSE and AMIGO.

As a conclusion based on the experiment of disturbance rejection test, CHR and GA-ITSE have tendency and possibility to produced oscillatory response due to high number of K_P as the proportional gain react to the error extremely that caused instability to the system. Both tuning method that induced high value of K_P and K_I at the same time, yields oscillation, as the integral gain could cause the system to oscillate.

4 Conclusion

This paper intend to compare the performance of various tuning method of PI controller by simulation and experiment. PI controller was tuned by using CHR, AMIGO, and Genetic Algorithm. For Genetic Algorithm, 2 objective function, ITSE and Modified ITSE were being used. Modified ITSE includes parameters such as settling time, rise time, overshoot percentage, and steady state error, together with error function as objective functions. The system was then simulated in the presence of disturbance to evaluate disturbance rejection properties. The performance of all tuning methods were being evaluated. The PI controller was then being implemented into the prototype. Step responses were recorded and plotted without disturbance and with disturbance. Load test and surface inclination test were conducted to evaluate the performance of the controller for all tuning methods subjected to disturbance during real implementation. During simulation, GA with modified ITSE produced the best results in 2 out of 4 parameters, which is settling time and overshoot percentage, while GA with original ITSE yield the least steady state error. At simulated disturbance condition, GA-modified produce the fastest settling time in input disturbance condition. In real application, GA-Modified ITSE obtained the best results in 2 out of 3 parameters, settling time and overshoot percentage. During load test and surface inclination test, AMIGO and GA-Modified ITSE produce better response compare with CHR and GA-Original ITSE, which have possibility to oscillate when there is disturbance. As conclusion GA by using Modified ITSE have the best response among all the tuning methods in step response performance and disturbance rejection response.

Acknowledgements This work was supported by Ministry of Science, Technology and Innovation (MOSTI) Malaysia; under grant number USM/305/PELECT/6013112.

References

1. Abdulameer A, Aras M, Saleem D (2016) GUI based control system using PID controller for education. Indones J Electr Eng Comput Sci 3(1):91–101
2. O'Dwyer A (2000) A summary of PI and PID controller tuning rules for processes with time delay. Part 1; PI controller tuning rules. IFAC Workshop on Digital Control: Past, Present and Future of PID Control 33(4):159–164
3. Bissel C (1994) Control engineering. Chapman and Hall, New York
4. Shinskey (1998) Process control systems—application, design, and tuning. McGraw-Hill, New York
5. Haugen F (2012) The good gain method for simple experiment tuning of PI controllers. Model Identif Control 33(4):141–152
6. Paz MA, Ramirez-delReal TA, Garibo SC, Araiza-Illan D, DeLuna-Ortega CA, Aguilar-Justo MO (2017) Adaptive Proportional-Integral controller using OLE for process control for industrial applications. Int J Adv Rob Syst 14(5):1–11
7. Sungthong A, Assawinchaichote W (2016) Particle Swam Optimization Based Optimal PID parameters for air heater temperature control system. Procedia Comput Sci 86:108–111
8. Mirzal A, Yoshii S, Furukawa M (2012) PID parameters optimization by using genetic algorithm. arXiv preprint arXiv:1204.0885
9. Zahir AAM, Alhady SSN, Othman WAFW, Zhiling L, Wahab AAA (2020) Objective functions modification of GA optimized PID controller for brushed DC motor. Int J Electr Comput Eng 10(3):2426–2433
10. Aly AA (2011) PID parameters optimization using genetic algorithm technique for electrohydraulic servo control system. Intell Control Autom 2(02):69
11. Eckhard D, Campestrini L, Boeira EC (2018) Virtual disturbance feedback tuning. IFAC J Syst Control 3:23–29
12. Gopi Krishna Rao PV, Subramanyam MV, Satyaprasad K (2014) Design of internal model control-proportionalintegral derivative controller with improved filter for disturbance rejection. Syst Sci Control Eng: Open Access J 2:583–592

13. Zahir AAM, Alhady SSN, Othman WAFW, Wahab AAA, Ahmad MF (2019) Genetic algorithm optimization and implementation of velocity control PI controller for cart follower application. Int J Adv Trends Comput Sci Eng 8(5):1886–1892
14. Zahir AAM, Alhady SSN, Othman WAFW, Wahab AAA, Ahmad MF (2020) Conceptual design of autonomous cart follower for wheelchair user. Platform J Eng 4(1):2–11
15. Nise NS (2007) Control systems engineering, 6th edn. Wiley, Hoboken, NJ
16. Architectural US, and Transportation Barriers Compliance Board. Americans with Disabilities Act (ADA), Accessibility Guidelines for Buildings and Facilities (1992)

Localised Muscle Contraction Predictor for Steering Wheel Operation in Simulated Condition

Nor Kamaliana Khamis, Dieter Schramm, Mohd Anas Mohd Sabri, and Muhamad Syukri Abdul Khalid

Abstract Evaluation of the steering wheel control is important to optimise the posture of the driver. The purpose of this study was to determine the relationship between muscle contraction at the shoulder and anthropometric variables when performing steering wheel task. Participants were recruited to perform multiple steering wheel actions. The surface electromyogram (SEMG) evaluation and anthropometric parameter measurement of individuals were recorded simultaneously during the experiment. For the statistical analysis, the anthropometric parameter was selected as an independent variable, while muscle activity based on SEMG measurement was chosen as the dependent variable. The results reveal that the left deltoid muscle showed the highest contraction at the right turn with high degree of turning. The SEMG and anthropometric data were positively correlated, and the predictive model shows the validity of the proposed model with the R^2 value nearly 0.50. This finding recommends that driver's anthropometric parameter may provide a good reference in a real driving task for controlling the steering wheel. Thus, some of potential utilization from this research is the optimizing in changing the vehicle design for allowing an independent adjustment to the relative distance between the driver seat and the steering wheel.

Keywords Health risk · Safety · Driver · Electromyography · Deltoid · Anthropometric

N. K. Khamis (✉) · M. A. M. Sabri
Department of Mechanical and Manufacturing Engineering, Faculty of Engineering and Built Environment, Universiti Kebangsaan Malaysia, Bandar Baru Bangi, Selangor, Malaysia
e-mail: kamaliana@ukm.edu.my

D. Schramm
Faculty of Engineering, University of Duisburg-Essen, Duisburg, Germany

M. S. A. Khalid
Malaysian Institute of Road Safety Research (MIROS), Taman Kajang Sentral, Kajang, Selangor, Malaysia

A. F. Ab. Nasir et al. (eds.), *Recent Trends in Mechatronics Towards Industry 4.0*, Lecture Notes in Electrical Engineering 730, https://doi.org/10.1007/978-981-33-4597-3_58

1 Introduction

Driving activity exerts physical demands to the driver. These demands, in turn, may affect the driver's discomfort level, alertness level and fatigue level [1]. Sammonds et al. [2] noted that long driving and discomfort have an impact on driving performance and health issues such as pain in buttock, neck and back. There are various factors that cause discomfort. Among them are physiological factors and ergonomic environment while driving a car. Physiological factors that can be experienced by the driver are due to muscle activity and sensory stimuli. Meanwhile, ergonomic environmental factors refers to external surroundings experienced by the driver, such as vibration, and noise due to driving condition [3, 4].

1.1 Driving Task and Its Effect

The activity associated with driving a car is different to that of sitting. It requires the driver to concentrate while driving and interacting with the steering wheel in a constrained space and a restricted position. The driver's hands and arms when maneuvering the steering wheel are noticeably higher than if positioned while working at a desk.

While driving, the body of the driver will experience movement due to uneven or irregular road surfaces and while turning to the right or to the left. This may induce physical stress on the body (and driver), leading to discomfort and muscular injury, commonly known as repetitive stress injury [5]. Hence, the main aim of this study was to identify factors that influenced driving task particularly when focusing in steering wheel operation. The next sub section elaborate the assessment on driver's condition while interacting with steering wheel. Extensive reviews has been carried out to compile findings related to this issue.

1.2 Assessment on Driver's Condition in Steering Wheel Operation

Substantive research have been performed to evaluate the driver's condition when operating the steering wheel [6–8]. Previous studies have mainly focused on the various aspects and issues related to the driver, while several studies have investigated the steering wheel design and steering wheel preferences [8–10]. Morioka and Griffin [11] and Li et al. [12] for instance, focused on issue on vibration when controlling the steering wheel. Recent study conducted by Eriksson et al. [13] evaluated the driving issues on self-driving vehicles. In this study, comparison between simulator and actual road condition was performed.

Several studies have shown that there is a good relationship between body posture, SEMG measurement and body measurement [14–16]. The change in the joint angle results in the change of muscle length. Joint angle normally change due to movement such as turning steering wheel with different angle. Based on the literature and limitations in prior studies, this article presents the localised muscle contraction prediction model based on anthropometric body measurement. The next sub section will be explained regarding this topic.

1.3 Localised Muscle in Steering Wheel Operation

Regarding the steering wheel operation, this involves the upper part of the body to move while rotating the steering wheel control in varying degrees. This operation requires specific muscles in the shoulder and arm to engage. Table 1 shows the common muscle parts investigated in past studies. It shows muscles for the upper body part only consisting of shoulder, neck, head, arm and hand. Based on Table 1, past studies implemented different design of experiment in term of posture setting and driving or sitting action setting.

Table 1 Muscle parts selection in past studies

References	Shoulder/neck/head				Arm/hand		
	T	D	ST	SC	BB	TB	ER
Balasubramanian and Adalarasu [17]	✔	✔		✔			
Grujicic et al. [18]		✔					
Franz et al. [15]	✔						
Vilimek et al. [19]		✔			✔	✔	
Saginus et al. [20]	✔	✔					
Gao et al. [21]		✔				✔	
Liu et al. [22]		✔				✔	
Khamis et al. [23]	✔	✔			✔		
Lin et al. [24]	✔	✔					
Ismail et al. [25]	✔						

T trapezius, *D* deltoids, *ST* sternocleidomastoid, *SC* splenius capitis, *BB* biceps brachii, *TB* triceps brachii, *ER* extensor carpi radialis

1.4 Relationship Between SEMG and Anthropometric Body Measurement in Steering Wheel Operation

Up to this date, a few studies have combined the subjective or objective assessment method with body measurement when engaged with steering wheel. For instance, Mossey et al. [10] determined relationship between steering wheel and body part parameters.

Pertaining to association between objective and subjective assessment via SEMG measurement, a significant relationship was found between muscle contraction value and sitting comfort [15]. Based on this study, there is an inverse relationship between comfort and muscle contraction. In this case, when the participants feel more comfortable, it will lead to reduction of muscle activation. Furthermore, reach parameters also one of factors that may influence the driver's comfort [14]. All in all, this article focus on muscle sequence pattern at the left deltoid muscle in determining the driver's condition while controlling the steering wheel. Comparison of the muscle reaction towards steering wheel control based on different actions were conducted in this study. Statistical analysis was done to determine relationship between anthropometric and muscle contraction value. There were two hypotheses in the study:

Hypothesis 1—Comparison between steering wheel actions according to direction and degree of turning:

H_0: There is no significant difference between actions for steering wheel task.

H_1: There is significant difference between actions for steering wheel task.

Hypothesis 2—Relationship between anthropometric and muscle contraction measurement:

H_0: There is no correlation between both variables.

H_1: There is strong correlation between both variables.

2 Materials and Methods

2.1 Sampling

Eleven female participants with age in between 23 and 33 years old were recruited to participate in this study. The inclusion criteria required that all participants must have a full Malaysia driving license with at least three years of driving experience. An approval letter of Ethical Practice was obtained from the Ethical Committee at Universiti Kebangsaan Malaysia with reference number UKM PPI/111/8/JEP-2016-200.

All participants were instructed to familiarise with the car simulator setup and the driving instruction prior to the experiment. Information related to the simulator driving procedures and the assessment methods to be used were given to all participants. Consent form was signed by the participant after receiving all relevant information on the study. The fore arm length and shoulder grip length were measured using an anthropometer.

2.2 *Equipment Setup*

Figure 1 shows the car simulator that employed in this study. This simulator was place at the Ergonomics Laboratory, Faculty of Engineering and Building Environment, UKM. A seat back can be inclined upwards and backwards as shown in Fig. 1. However, the simulator's seat height cannot be adjusted. It can be adjusted to the front, up to 35 cm or to the back, up to 45 cm.

SEMG assessment was utilised in this study to examine the muscle reaction when controlling the steering wheel. A Trigno™ Personal Monitor with Parallel-Bar Sensors from Delsys Incorporation was used to record the muscle signal. The SEMG electrodes were attached directly on the left deltoid muscle after careful preparation of skin surface and parallel to its muscle fibers [26]. To place the SEMG electrode, the participant required to sit on the stable chair (provide a good foot and back support). After placement of the electrodes on the specific muscle, a trial test could be conducted to ensure the accurate signal is gathered.

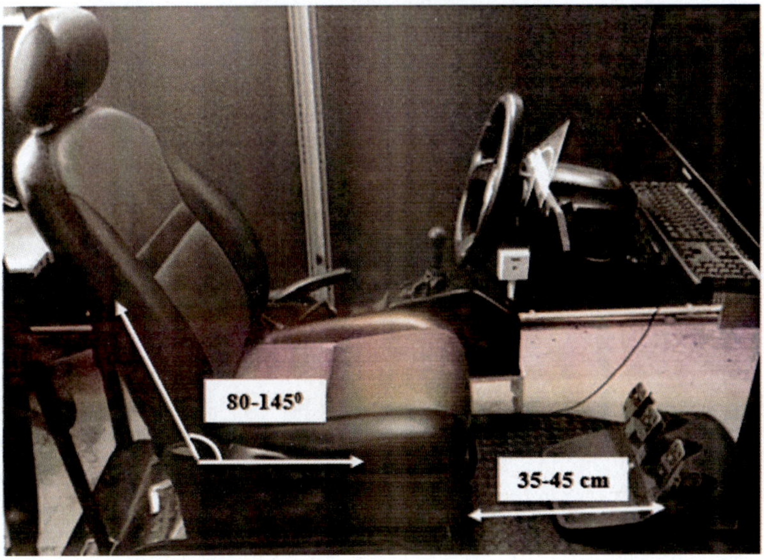

Fig. 1 Simulator setup

Fig. 2 Sensor assembly

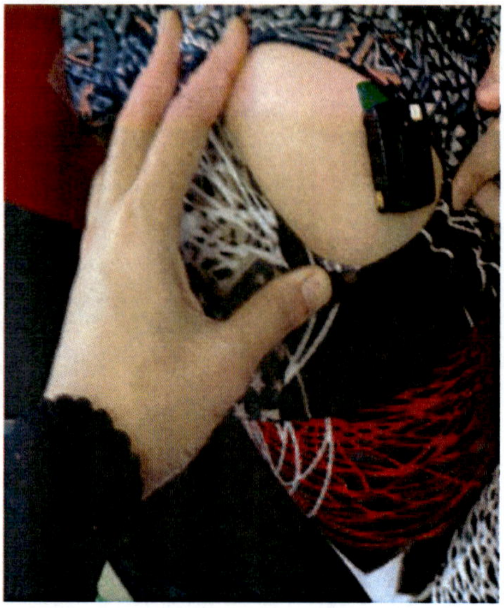

2.3 *Experiment Design and Procedure*

The participant is required to be seated in the simulator at the furthest distance from the steering wheel (mean knee angle = $135 \pm 40°$) with the seat back inclination at 100°. During the recording procedure to observe muscle contraction, the participant was required to hold the steering wheel at 10 and 2 o'clock position. This position was suggested by the past studies as one of good position when gripping the steering wheel [27, 28]. During experiment, the participant was required to perform a series of steering wheel actions, with different degree of turning (i.e.: 10° and 45°) and different direction of turning (i.e.: left and right). Figure 2 illustrates the position of the SEMG sensor on the selected muscle.

2.4 *Muscle Contraction Analysis*

Root mean square (RMS) refers to amplitude signal. RMS is a common value used to analyse muscle contraction value from SEMG. In this study, the SEMG signal was filtered by using high pass, low pass and Notch filter. These filters were used to reduce artifact and noise in the signal.

2.5 *Statistical Analysis*

Correlation analysis was carried out before developing prediction model in this study. Correlation analysis can determined a strength of relationship between two variables. Pearson rank analysis was used in this study to determine correlation between variables. In order to provide prediction model, regression analysis was used by considering the effect of DV towards IV. In this study, the DV is based on the localised muscle contraction, while the IV refers to the body measurement, which is referred to the body measurement.

3 Results

3.1 *Pattern of Muscle Contraction*

According to direction of turning, there is an inverse contraction value between left turning and right turning at the left deltoid. The left deltoid muscle contraction shows the lowest contraction when performing the left turning. In contrast, when executing right turning, the muscle illustrates the greatest contraction. With respect to degree of turning, 45° turning depicts the highest contraction, compared to 10°. Referring to the muscle condition, all steering wheel action including right turn at 45°, right turn 10°, left turn 45° and right turn 10° demonstrates high activation, the RMS value is more than 5 μV. In this case, the left deltoid not in a rest condition for these actions.

3.2 *Relationship Between Muscle Contraction Measurement and Anthropometric Body Measurement*

Correlation analysis was conducted to determine relationship between muscle contraction at the left deltoid and anthropometric variable. Based on the muscle contraction pattern, it shows that 45° of the right turn produce the highest contraction compared to other actions. Hence, this action is used for the next statistical analysis. This is due to the highest muscle contraction shows the severity of discomfort among the driver. Referring to correlation analysis, depicts the forearm length and high degree of turning action are strongly correlated, $r(11) = 0.068$, $p < 0.05$. However, a weak correlation was found between shoulder grip length with 45° of the right turn action ($r = 0.309$).

In order to foresee the driver's condition level based on localised muscle contraction value at the left deltoid and fore arm length, regression method was used. Based on regression analysis, R Square shows 0.463, which indicate 46.3% of the localised muscle contraction can explained by the fore arm length. Table 2 presents the coefficient table for this analysis. This table shows the regression model predicts the

Table 2 Coefficient table for the estimation model for steering wheel

Variable	Unstandardized coefficients		Standardized coefficients	t	Sig.
	B	Std. error	Beta		
Constant, m	−190.234	80.301		−2.369	0.042
Fore arm length	5.174	1.859	0.680	2.784	0.021

Table 3 Linear regression validation

Assumption	Muscle activity	Fore arm length
Normality	df(11) = 0.898, p > 0.05	df(11) = 0.947, p > 0.05
Linearity	r(11) = 0.680, positive (linear)	
Auto-correlation	DW test: 1.5 < 2.330 < 2.5	
Heteroscedasticity	Beta = 0.090, p > 0.05	
Multicollinearity	Tolerance (1.0 > 0.1) and VIF (1.0 < 10.00)	

localised muscle contraction significantly well with $p < 0.05$. The significant level for fore arm length, which indicated as IV was less than 0.05, indicating the possibility to obtain the t value for the constant which was −2.369. The slope for the fore arm length was 2.784.

Hence, at the confidence level 95%, the second hypothesis in this study would be rejected. There is significant relationship between anthropometric measurement and localised muscle contraction for steering wheel task at the highest degree of action. Based on Table 2, the equation model to estimate the driver's condition level based on localised muscle contraction value and fore arm length can be used as given in Eq. (1):

$$y = 5.174x - 190.234 \quad (1)$$

Model validation was tabulated in Table 3. Table 3 shows that this model complied with Best Linear Unbiased Estimator criteria where all assumptions related to linear regression are fulfilled.

4 Discussion

This study demonstrates the principles of shoulder muscle function and reaction when turning the steering wheel. Different turning direction and degree of the steering wheel shows different activation value. Right turning produces the highest contraction at the left shoulder. It is due to hand grip activity on the steering wheel may activate this body part particularly the deltoid muscle to experience high contraction

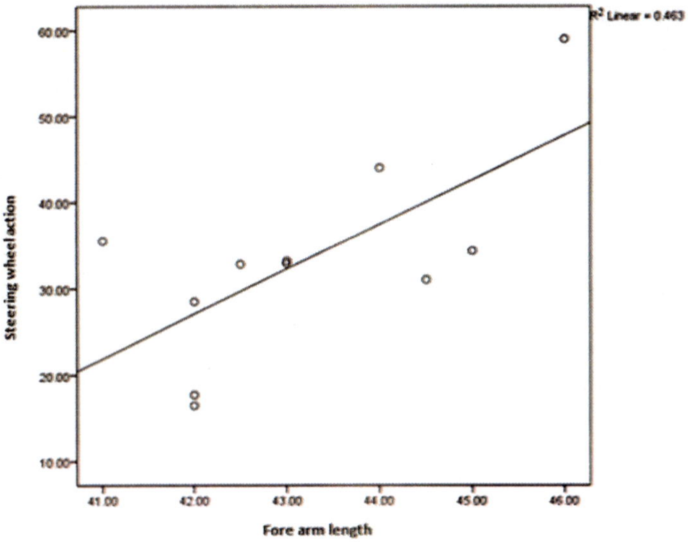

Fig. 3 Prediction model

when this gripping and turning the steering wheel require the shoulder to abduct and flex.. This principle is supported by other studies conducted by Balasubramanian and Adalarasu [17] and Pandis et al. [7] which highlighted that the shoulder muscle produce the active contraction while placing the hand on the steering wheel. Based on this analysis, the deltoid is the most dynamic muscle in a raised position for the arm. This muscle demonstrates double activation compared to the other muscle for the upper body parts [7]. This activity if occurs for long duration, may lead to muscle fatigue [29, 30].

Figure 3 shows the linear line of the prediction model for this study. According to Fig. 3, there is positive linear relationship between the highest degree of action and fore arm length with R Square = 0.463. It shows that fore arm length measurement is a moderate predictor for muscle contraction while controlling the steering wheel. This parameter presents extension and retraction motion of the shoulder and hand when the driver reach the steering wheel to maneuver the car. The fore arm length is among the frequent variables to identify the working distance and reach [31]. Driving posture will be change according to driver's activity. This change can be measured by analyzing the muscle activation by using SEMG. It may provide a good indicator on comfort level among drivers based on different driving position.

5 Conclusion

Steering wheel operation activates shoulder body part to control and maneuver the vehicle. The deltoid muscle depicts significant role in controlling the car. In this study, comparison of different steering wheel actions has been conducted. The muscle demonstrates the highest contraction pattern when the steering wheel is turning to the opposed direction with the muscle. In this case, the left deltoid shows the highest value of contraction when performing the right turn. Regarding comparison between degrees of turning, the largest contraction was depicted with the highest degree. Furthermore, this study also highlighted a strong correlation between the muscle contractions with fore arm hand length. It is in agreement with the fact that reach parameter to the car controls may influence value of muscle contraction. However, the fore arm hand length is not the best predictor to estimate the muscle contraction due to moderate relationship by referring to R Square which is nearly to 0.5 only.

Acknowledgements The authors would like to acknowledge all participants involved in this research.

Funding This work was supported by the Universiti Kebangsaan Malaysia and Ministry of Higher Education, Malaysia under Fundamental Research Grant Scheme (FRGS/1/2018/TK03/UKM/03/2).

Conflicts of interest The authors declare no conflict of interest.

References

1. Zhang T, Chan AH, Xue H, Zhang X, Tao D (2019) Driving anger, aberrant driving behaviors, and road crash risk: testing of a mediated model. Int J Environ Res Public Health 16(3):297–300
2. Sammonds GM, Fray M, Mansfield NJ (2017) Effect of long term driving on discomfort and its relationship with seat fidgets and movements. Appl Ergon 58:119–127
3. Griffin MJ (2012) Frequency-dependence of psychophysical and physiological responses to hand-transmitted vibration. Ind Health 50(5):354–369
4. Parsons KC (2000) Environmental ergonomics: a review of principles, methods and models. Appl Ergon 31(6):581–594
5. NEA (National Education Association) (2004) Repetitive stress injuries handbook. American Labor Education Center
6. Auberlet J-M, Rosey F, Anceaux F, Aubin S, Briand P, Pacaux M-P, Plainchault P (2012) The impact of perceptual treatments on driver's behavior: from driving simulator studies to field tests-first results. Accid Anal Prev 45:91–98
7. Pandis P, Prinold JAI, Bull AMJ (2015) Shoulder muscle forces during driving: sudden steering can load the rotator cuff beyond its repair limit. Clin Biomech 30(8):839–846
8. Li Z, Li SE, Li R, Cheng B, Shi J (2017) Online detection of driver fatigue using steering wheel angles for real driving conditions. Sensors 17(3):495
9. Döring T, Kern D, Marshall P, Pfeiffer M, Schöning J, Gruhn V, Schmidt A (2011) Gestural interaction on the steering wheel. In: Proceedings of the 2011 annual conference on human factors in computing systems, p 483

10. Mossey ME, Xi Y, McConomy SK, Brooks JO, Rosopa PJ, Venhovens PJ (2014) Evaluation of four steering wheels to determine driver hand placement in a static environment. Appl Ergon 45(4):1187–1195
11. Morioka M, Griffin MJ (2009) Equivalent comfort contours for vertical vibration of steering wheels: effect of vibration magnitude, grip force, and hand position. Appl Ergon 40(5):817–825
12. Li W, Zhang M, Lv G, Han Q, Gao Y, Wang Y, Tan Q, Zhang M, Zhang Y, Li Z (2015) Biomechanical response of the musculoskeletal system to whole body vibration using a seated driver model. Int J Ind Ergon 45:91–97
13. Eriksson A, Banks VA, Stanton NA (2017) Transition to manual: comparing simulator with on-road control transitions. Accid Anal Prev 102:227–234
14. Adler S (2007) The relation between long-term seating comfort and driver movement. Friedrich-Schiller-Universitat Jena, Dipl.-Sportwiss
15. Franz M, Zenk R, Vink P, Hallbeck S (2011) The effect of a lightweight massage system in a car seat on comfort and electromyogram. J Manip Physiol Ther 34(2):107–113
16. Fazlollahtabar H (2010) A subjective framework for seat comfort based on a heuristic multi criteria decision making technique and anthropometry. Appl Ergon 42(1):16–28
17. Balasubramanian V, Adalarasu K (2007) EMG-based analysis of change in muscle activity during simulated driving. J Bodyw Mov Ther 11(2):151–158
18. Grujicic M, Pandurangan B, Xie X, Gramopadhye AK, Wagner D, Ozen M (2010) Musculoskeletal computational analysis of the influence of car-seat design/adjustments on long-distance driving fatigue. Int J Ind Ergon 40(3):345–355
19. Vilimek M, Horak Z, Petr K (2011) Optimization of shift lever position. J Chem Inf Model 53:1689–1699
20. Saginus K, Marklin R (2013) The effects of mobile computer location in a vehicle cab on muscle activity and body posture of large and small drivers. Proceedings of the human factors and ergonomics society annual meeting 57(1):902–906
21. Gao ZH, Fan D, Wang D, Zhao H, Zhao K, Chen C (2014) Muscle activity and co-contraction of musculoskeletal model during steering maneuver. Bio-Med Mater Eng 24(6):2697–2706
22. Liu Y, Ji X, Hayama R, Mizuno T, Nakano S (2014) Method for measuring a driver's steering efficiency using electromyography. Proc Inst Mech Eng Part D: J Autom Eng 228:1170–1184
23. Khamis NK, Deros BM, Nawawi R, Jaya P, Saleh C (2017) Muscle activity of the upper and lower body part in car gearing action: a preliminary study. J Mech Eng SI 3(2):157–165
24. Lin MY, Barbir A, Dennerlein JT (2017) Evaluating biomechanics of user-selected sitting and standing computer workstation. Appl Ergon 65:382–388
25. Ismail AH, Khamis NK, Md Deros B (2018) The effect of grasping the steering wheel while positioning the shoulder closer to the body. Malays J Public Health Med 18(Special issue 2):123–127
26. Soderberg GL, Knutson LM (2000) A guide for use and interpretation of kinesiologic electromyographic data. Phys Ther 80:485–498
27. Sanders MS, McCormick EJ (1993) Human factors in engineering and design, 7th edn. McGraw Hill, New York
28. Walton D, Thomas JA (2005) Naturalistic observations of driver hand positions. Transp Res Part F: Traffic Psychol Behav 8(3):229–238
29. Lieber RL, Friden J (1993) Muscle damage is not a function of muscle force but active muscle strain. J Appl Physiol 74(2):520–526
30. Proske U, Morgan DL (2001) Muscle damage from eccentric exercise: mechanism, mechanical signs, adaptation and clinical applications. J Physiol 537(2):333–345
31. Kee D, Lee I (2012) Relationships between subjective and objective measures in assessing postural stresses. Appl Ergon 43(2):277–282

Experimental of CVT Ratio Control Using Single Actuator Double Acting Electro-mechanical Continuously Variable Transmission

Nur Rashid Mat Nuri, Khisbullah Hudha, and Muhammad Luqman Hakim Abd Rahman

Abstract The continuously variable transmission (CVT) provides smooth acceleration and good vehicle fuel consumption due to its superior advantage that offers an infinite number of transmission ratios within its range. In order to overcome the drawbacks in the current CVT systems, a new electro-mechanical CVT is established. Single actuator double acting electro-mechanical (SADAEM) CVT system consists of two actuators to vary the variable pulleys via top linkages and power screw mechanisms. Two closed loop of Proportional-Derivative (PD) controller are set to regulate the actuators in varying the radius of drive and driven pulleys to control the SADAEM CVT system ratio. The SADAEM CVT system is analyzed based on the CVT ratio performance in term of transient response in up-shift and down-shift patterns. The SADAEM CVT system has been evaluated through simulation and experimental works, where both simulation and experimental works shows acceptable results in tracking the desired trajectory step inputs with a slight peak-to-peak error less than 5% and time delay below 0.3 s. Furthermore, based on the maximum settling time, one complete cycle to return back to original pulley position is approximately 15 s.

Keywords Electro-mechanical CVT · PID controller · CVT ratio controller

1 Introduction

The improvement of vehicle performance in terms of low fuel consumption and gas emission has become essential for engineers and researchers in the automotive industries. The latest trend in automotive research is focusing on the engine and transmission efficiencies to increase vehicle performance. Since improving the

N. R. M. Nuri (✉)
Department of Mechanical Engineering Technology, Faculty of Mechanical and Manufacturing Engineering Technology, Universiti Teknikal Malaysia Melaka, Melaka, Malaysia
e-mail: nrashid@utem.edu.my

K. Hudha · M. L. H. A. Rahman
Department of Mechanical Engineering, Faculty of Engineering, National Defence University of Malaysia, Kuala Lumpur, Malaysia

A. F. Ab. Nasir et al. (eds.), *Recent Trends in Mechatronics Towards Industry 4.0*, Lecture Notes in Electrical Engineering 730, https://doi.org/10.1007/978-981-33-4597-3_59

engine efficiency is difficult, researchers are more interested in developing the efficient transmission that allows the engine speed to be maintained under various vehicle load conditions, leading to better fuel consumption [1–3]. Typically, the type of transmission available either in manual transmission (MT), automatic transmission (AT) or continuously variable transmission (CVT).

The CVT has a significant advantage over the MT and the AT in reducing the fuel consumption because this type of transmission offers infinite number of CVT ratio coverage within its limit. This unique characteristic allows the vehicle engine to run within its optimum operating range for various vehicle load conditions. In addition, CVT also provides a smooth vehicle acceleration performance without gear shift shock as experienced by MT or AT [4–7]. The existing CVT system available in the market uses an internal hydraulic system, namely electro-hydraulic-mechanical (EHM) actuation system to actuate the axial movement of the movable pulleys on the drive and driven sides. Moreover, the belt-type CVT is the most widely implemented in an automotive application.

The existing vehicle with CVT has several disadvantages such as heavy electro-hydraulic components, high energy consumption, possibility of leakage in the hydraulic line and difficulty to control due to the non-linearity of the compressible hydraulic oil [8–10]. In order to overcome these drawbacks, an alternative way of changing and maintaining the CVT ratio using an electro-mechanical (EM) actuation system was initiated. Based on the previous studies, there are developed EM CVT with different mechanisms such as Electro-mechanical pulley actuation (EMPAcT) CVT [11–13], hollow ball screw unit for CVT pulley developed by NTN [14], Electro-mechanical dual acting pulley (EMDAP) CVT [15–17] and Slider crank electro-mechanical (SCEM) CVT [18–20].

Overall, the EHM actuation CVT system has a CVT ratio range from 0.39 to 2.52 and is widely used in the market [21, 22]. A single actuator double acting electro-mechanical (SADAEM) CVT system was designed to overcome the high value of the minimum CVT ratio in the EM actuation system compared to EHM actuation system [23–25]. The SADAEM CVT system consists of the primary DC motor to simultaneously vary the CVT ratio and the secondary DC motor acting as the input torque and the speed to the system. With the assistance of the power screw and the top linkage mechanisms, the axial movement of the dual acting pulleys is expanded to achieve a broader CVT ratio of between 0.3 and 2.9.

Since the SADAEM CVT system introduces a new mechanism, the dynamic performance of the CVT ratio in several desired input responses is still unknown. Based on this reason, the development of a CVT ratio control algorithm is crucial, such that the SADAEM CVT system can accurately and smoothly achieve the target CVT ratio. This paper focuses on the performance of CVT ratio control for SADAEM CVT system. First section describes the importance of EM CVT and brief background on CVT technology. The second section briefly presents the design and experimental setup on SADAEM CVT test rig by detailing the hardware configurations between mechanical and electronics parts. Then, the overall results and conclusions are presented in the third and the last section, respectively.

2 Description of SADAEM CVT System

The SADAEM CVT system comprises of mechanical parts at the drive and driven sides, namely the left-threaded, the right-threaded power screws and a few parts at the top linkage and pulley mechanisms that act as the mechanical actuators and perform linkage connection from the primary DC motor to the pulleys, as shown in Fig. 1. The power screw converts each rotation to approximately two millimetres of power screw nut movement. To ensure a different movement of the power screw nut for each rotation of the primary DC motor shaft, the left-threaded power screw is placed on the drive side, while the right-threaded power screw is fixed on the driven side. The axial motion of the movable pulleys is related to the axial movement of the right and left hinges linked to the displacement of the power screw nut.

The advantage of using the power screw mechanism is that the axial pulley movement can be controlled by the primary DC motor to move, stop or stay at the desired position. When the pulley is stopped or remains in a particular position, the primary DC motor is in standby mode, thereby reducing the electrical power and keeping the SADAEM CVT system in energy saving mode. Meanwhile, a commonly used variable speed rubber belt is placed on both sides between the pulley sheaves. Furthermore, a spring-loaded belt tensioner is placed on the outer surface of the variable speed rubber belt to increase the belt tension.

In order to prevent the belt slip, the pulley sheaves need a clamping force of about 1 kN to clamp the rubber V-belt [26, 27]. Hence, the power screw mechanism is required to generate a torque of approximately 1.74 Nm. The use of pre-assembled gear head with gear ratio of 10 in the primary DC motor will multiply the torque of the primary DC motor by ten times to fulfil the torque demand by the power screw.

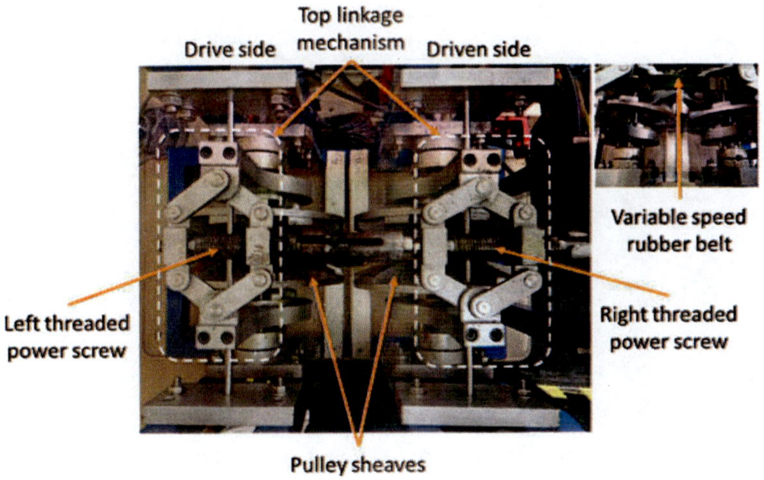

Fig. 1 The SADAEM CVT system that consists of the power screw, top linkage, pulley sheaves and variable rubber belt mechanisms

The primary DC motor generates a torque of about 0.174 Nm. Therefore, the amount of torque supplied by the DC motor is sufficient enough when operating at a rated torque of 0.2 Nm.

The setup process needs to be carried out before any experimental works is performed. The experimental setup for the SADAEM CVT consists of hardware and software configuration to examine the performance of the system. Figure 2 illustrates the hardware set up in the construction of the experimental test rig. Two units of DC power source are needed to power the primary and secondary DC motor system. The motor driver for each DC motor determines the direction and speed of the DC motor shaft controlled by the NI card using the MATLAB/Simulink software. Next,

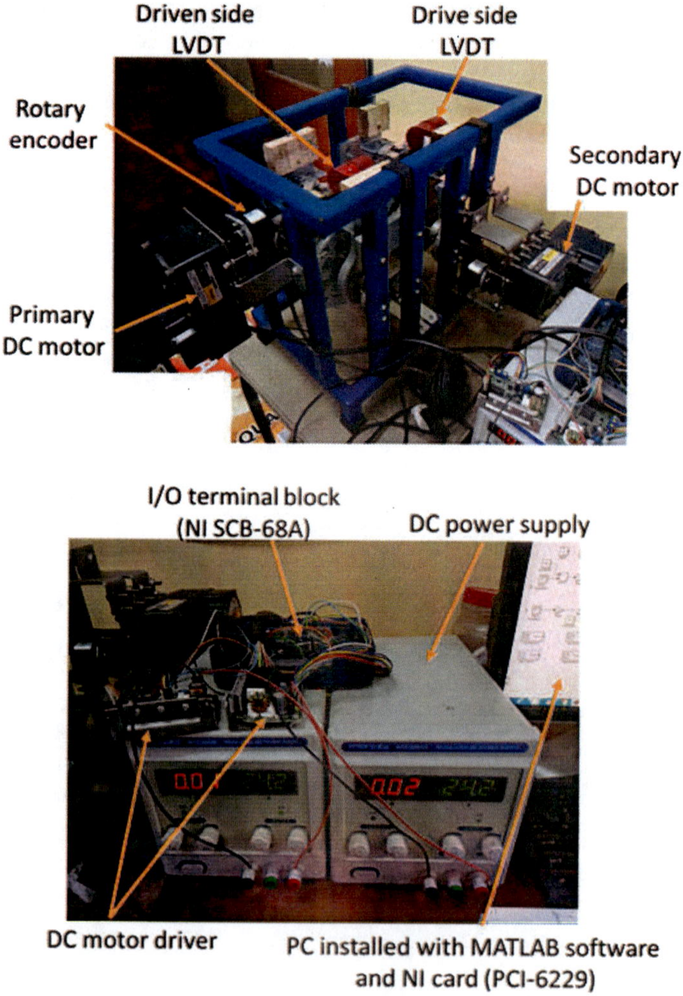

Fig. 2 Experimental test rig of the SADAEM CVT system

the drive side input shaft and the driven side output shaft are attached to the rotary encoders to record the rotation of the pulley shaft.

Additionally, the rotary encoder is installed at the primary DC motor shaft to monitor the rotation of the power screw. The input shaft on the drive side is connected to the secondary DC motor via the gearhead with a ratio of 5. With a constant speed of 500 rpm on the secondary DC motor, an input shaft on the drive side rotates at a constant speed of 100 rpm. Meanwhile, the axial position of the pulleys is measured using the LVDT sensors attached on both the drive and driven sides.

The MATLAB/Simulink software is configured to the real-time windows target to embed the hardware system into the software system. The software system is set to run for 40 s using the *Bogacki-Shampine* solver with a fixed step size of 0.01 s.

A PD-based control algorithm is applied to control the SADAEM CVT ratio positions in both Software-in-the-Loop (SIL) and Hardware-in-the-Loop (HIL) configurations.

3 Results and Discussion

One type of preferred trajectory, known as step is applied to verify the performance of an actual and validated SADAEM CVT system model in tracking the CVT ratio positions. The step input leads to transient response performance. The step input represents two trajectory patterns of the CVT ratio, namely up-shift and down-shift. The CVT ratio will start at 2.9 for the up-shift pattern and move to the desired CVT ratio of 0.3. Unlike the up-shift pattern, the down-shift pattern will begin from CVT ratio of 0.3 and shift to the desired CVT ratio of 2.9.

Additionally, a sensitivity analysis is performed for the step input by varying the controller parameters K_p and K_d to determine the optimum values of the PD controller based on the transient response performance of the SADAEM CVT system. Figures 3 and 4 show the step response for the SIL and HIL with a variation of K_d values in the up-shift and down-shift conditions for fixed K_p value of 300, respectively. This fixed K_p value is obtained by trial and error method whereby the acceptable transient response performance is considered. In these figures, the dashed line represents the desired CVT ratio positions, the dotted line indicates the results using SIL method with K_d value of 2.5, the green solid line shows the CVT ratio with K_d value of 5 and the blue solid line represents the CVT ratio with K_d value of 9. The performance of the CVT ratio response is evaluated with regards to the rise time, maximum overshoot percentage, settling time and percentage of steady-state error as listed in Table 1.

In detail, the rise time, maximum overshoot, settling time and steady-state error of 3.65 s, 7.2%, 4.5 s and 2.3%, respectively for the up-shift condition using the HIL method and K_d value of 5. Similarly, the rise time, maximum overshoot, settling time and steady-state error is improved to 3.49 s, 5.5%, 4.4 s and 1.3%, respectively for K_d with the value of 9. Meanwhile, for down-shift condition using HIL method, the transient response with the K_d value of 9 shows better performance compared to the

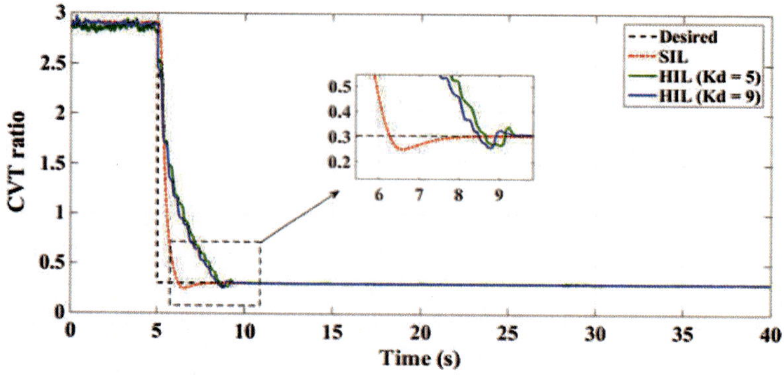

Fig. 3 CVT ratio response in up-shift step condition ($K_p = 300$)

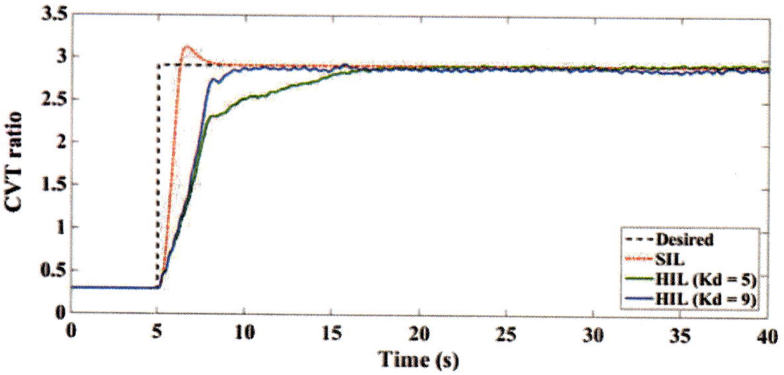

Fig. 4 CVT ratio response in down-shift step condition ($K_p = 300$)

Table 1 Transient response of CVT ratio tracking control in different K_d values for up-shift and down-shift conditions ($K_p = 300$)

Condition	Method/K_d value	Rise time (s)	Maximum overshoot (%)	Settling time (s)	Steady-state error (%)
Up-shift	SIL/2.5	1.27	12.8	4.0	0.0
	HIL/5	3.65	7.2	4.5	2.3
	HIL/9	3.49	5.5	4.4	1.3
Down-shift	SIL/2.5	1.34	7.3	5.0	0.0
	HIL/5	12.00	NA	14.0	0.7
	HIL/9	5.00	NA	10.0	2.0

transient response with the K_d value of 5 except for the steady-state error percentage. Moreover, the findings for the SIL method indicate better transient response for both K_d values in all parameters except for the maximum overshoot compared to the HIL method.

The fixed K_p value is increased to 400 for further analysis of the controller performance. Similarly, several K_d values with fixed K_p value are set. The controller performance in step input with another fixed proportional gain value is shown in Figs. 5 and 6. Likewise, the CVT ratio transient responses are tabulated in Table 2. For the up-shift condition and HIL method, the transient response for a derivative gain of 9 is better than the derivative gain of 5 in terms of the rise time, maximum overshoot, settling time and steady-state error. However, for down-shift condition, the K_d value of 5 performs much better or equal to the K_d value of 9 for all parameters of the transient response. Overall, these results indicate that both SIL and HIL methods are able to follow the desired step input with acceptable transient response performance.

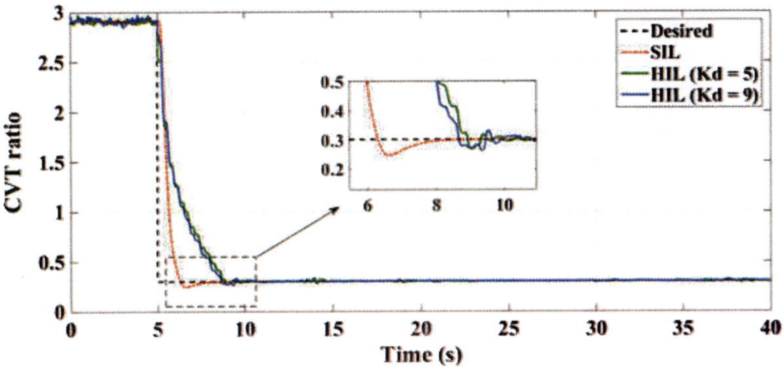

Fig. 5 CVT ratio response in up-shift step condition ($K_p = 400$)

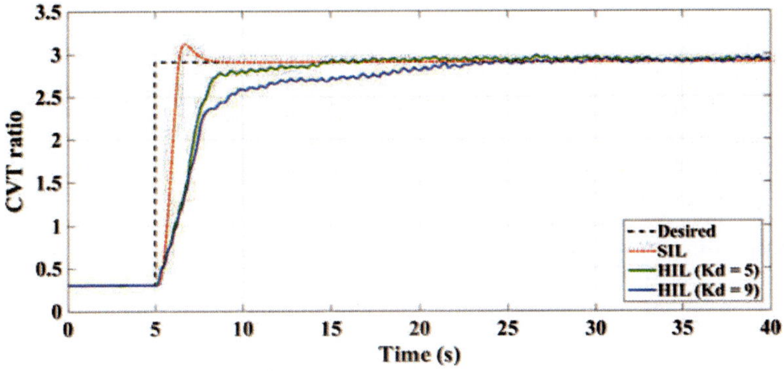

Fig. 6 CVT ratio response in down-shift step condition ($K_p = 400$)

Table 2 Transient response of CVT ratio tracking control in different K_d values for up-shift and down-shift conditions ($K_p = 400$)

Condition	Method/K_d value	Rise time (s)	Maximum overshoot (%)	Settling time (s)	Steady-state error (%)
Up-shift	SIL/2.5	1.27	12.5	4.0	0.0
	HIL/5	3.91	6.6	4.9	2.3
	HIL/9	3.65	5.6	4.9	1.3
Down-shift	SIL/2.5	1.34	7.3	5.0	0.0
	HIL/5	9.00	NA	10.0	0.9
	HIL/9	20.00	NA	21.0	0.9

Next, the maximum settling time for the CVT ratio to shift from 2.9 to 0.3 and return to 2.9 in one complete cycle is approximately 15 s by referring to the transient response of step input and the best results from each fixed proportional gain value of 300 and 400.

4 Conclusion

By using the PD-based control algorithm technique, the performance evaluation of CVT ratio tracking control has been carried out with step position trajectories. The finding of the step input has indicated the response of the transient and steady-state error. Lastly, both the SIL and HIL methods have shown acceptable results that can match the desired trajectory inputs with a slight PTP error of below 5% and time delay below 0.3 s. In future works, the new configuration of SADAEM CVT will be implemented in the automotive powertrain system for application of the vehicle speed control system.

Acknowledgements The authors would like to thank the Universiti Teknikal Malaysia Melaka (UTeM), Universiti Pertahanan Nasional Malaysia (UPNM) and Malaysia-Japan International Institute of Technology (MJIIT) Universiti Teknologi Malaysia (UTM) for their financial support, technical advises and allowing the use of their research facilities.

References

1. Pfiffner R, Guzzella L, Onder CH (2003) Fuel-optimal control of CVT powertrains. Control Eng Pract 11(3):329–336
2. Takiyama T (2002) Engine-CVT consolidated control using LQI control theory. JSAE Rev 20(2):251–258
3. Damiani L, Repetto M, Prato AP (2014) Improvement of powertrain efficiency through energy breakdown analysis. Appl Energy 121:252–263

4. Ide T (2000) Effect of belt loss and oil pump loss on the fuel economy of a vehicle with a metal V-belt CVT. SAE Technical Paper No. 2000-05-0130
5. Lee H, Kim H (2003) CVT ratio control for improvement of fuel economy by considering powertrain response lag. KSME Int J 17(11):1725–1731
6. Srivastava N, Haque I (2009) A review on belt and chain continuously variable transmissions (CVT): dynamics and control. Mech Mach Theory 44(1):19–41
7. Julio G, Plante JS (2011) An experimentally-validated model of rubber-belt CVT mechanics. Mech Mach Theory 46(8):1037–1053
8. Shibayama T, Yada H, Morita Y, Fujikawa M (2008) Introduction of the latest hydraulic control system for automatic transmission. In: Proceedings of the JFPS international symposium on fluid power. 2008. The Japan Fluid Power System Society, vol 2008, No. 7-1, pp 137–142
9. Aparow VR, Hudha K, Ahmad F, Jamaluddin H (2014) Model-in-the-loop simulation of gap and torque tracking control using electronic wedge brake actuator. Int J Veh Saf 7(3–4):390–408
10. Bradley TH, Frank AA (2002) Servo-pump hydraulic control system performance and evaluation for CVT pressure and ratio control. VDI BERICHTE 1709:35–42
11. Van de Meerakker KG, Rosielle PC, Bonsen B, Klaassen TW (2004) Design of an electromechanical ratio and clamping force actuator for a metal V-belt type CVT. SAE Technical Paper, No. 2004-40-0008
12. Klaassen TW (2007) The EMPACT CVT: dynamics and control of an electromechanically actuated CVT. Ph.D. thesis, Technische Universiteit Eindhoven
13. Klaassen TW, Bonsen B, Van De Meerakker KG, Vroemen BG, Veenhuizen PA, Veldpaus F, Steinbuch M (2008) The Empact CVT: modelling, simulation and experiments. Int J Model Identif Control 3(3):286–296
14. Umemoto T (2013) NTN module technology contributes to energy efficiency and CO_2 reduction in automobiles. NTN Tech Rev 81:12–21
15. Tawi KB, Mazali II, Supriyo B, Husain NA, Kob C, Salman M, Abidin YZ (2014) Pulleys' axial movement mechanism for electro-mechanical continuously variable transmission. Appl Mech Mater 663:185–192
16. Supriyo B, Tawi KB, Jamaluddin H (2013) Experimental study of an electro-mechanical CVT ratio controller. Int J Autom Technol 14(2):313–323
17. Supriyo B, Tawi KB, Jamaluddin H, Hussein M (2014) Experimental study of electro-mechanical dual acting pulley continuously variable transmission ratio calibration. J Teknol 71(2):121–127
18. Rahman ML, Hudha K, Kadir ZA, Amer NH, Aparow VR (2018) Modelling and validation of a novel continuously variable transmission system using slider crank mechanism. Int J Eng Syst Model Simul 10(1):49–61
19. Rahman ML, Hudha K, Kadir ZA, Amer NH, Murrad M (2018) Simulation study on the vehicle speed control in longitudinal direction using a new continuously variable transmission (CVT) system. J Mech Eng 7(1):127–147
20. Hudha K, Rahman ML, Amer NH, Kadir ZA (2018) Ratio tracking control of slider crank based electromechanical CVT system. In: Proceeding in the 57th annual conference of the society of instrument and control engineers of Japan (SICE). IEEE, Nara, Japan, pp 1530–1537
21. Kwak Y, Cleveland C (2017) Continuously variable transmission (CVT) fuel economy. SAE Technical Paper, No. 2017-01-2355
22. Aoyama T, Takahara H, Kuwabara S, Miyata H, Nakayashiki M, Kasuga S (2014) Development of new generation continuously variable transmission. SAE Technical Paper. No. 2014-01-1728
23. Nuri NR, Hudha K, Mazlan SA (2019) Design and simulation of a new single actuator double acting electro-mechanical continuously variable transmission. Int J Mech Eng Robot Res 8(1):114–120
24. Nuri NR, Hudha K, Mazlan SA, Harun MH (2018) Vehicle speed control strategy using Fuzzy-PID controller for continuously variable transmission system. Proc Mech Eng Res Day 2018:32–33
25. Hudha K, Nuri NR, Mazlan SA (2019) Multi-objective optimization of vehicle speed control using gravitational search algorithm for electro-mechanical continuously variable transmission.

In: IOP conference series: materials science and engineering, vol 530, no 1. IOP Publishing, p 012031
26. Messick MJ (2018) An experimentally-validated V-belt model for axial force and efficiency in a continuously variable transmission. MSc thesis, Virginia Polytechnic Institute and State University
27. Bertini L, Carmignani L, Frendo F (2014) Analytical model for the power losses in rubber V-belt continuously variable transmission (CVT). Mech Mach Theory 78:289–306

A Summarization of Image and Video Databases for Emotion Recognition

Arselan Ashraf, Teddy Surya Gunawan, Farah Diyana Abdul Rahman, and Mira Kartiwi

Abstract In the past decades, human-computer interaction has become increasingly significant in our day to day lives. Many kinds of research have been conducted in the area of memory research, depression detection, and behavioural insufficiency identification, lie detection (covered up), and emotion recognition. Thus, the quantity of nonexclusive feeling and image/video databases or those custom-made to precise requirements have become enormous. In this way, an extensive yet minimal guide is expected to assist analysts with finding the most appropriate image/video database and comprehend what sorts of image/video databases as of now exist. In this paper, different elicitation strategies are discussed, and the databases are basically categorized into a slick and enlightening comparison table in order to be used for emotion recognition model.

Keywords Emotion recognition · Image emotion databases · Video emotion databases

1 Introduction

In past years, emotion recognition has gotten one of the main points in the field of Machine Learning and Artificial Intelligence. The gigantic increment in the improvement of sophisticated human-computer cooperation advances has additionally supported the pace of progress in this field. Facial activities pass on the feelings which thusly pass on an individual's character, state of mind and intentions.

A. Ashraf · T. S. Gunawan (✉) · F. D. A. Rahman
Electrical and Computer Engineering Department, International Islamic University Malaysia, Kuala Lumpur, Malaysia
e-mail: tsgunawan@iium.edu.my

T. S. Gunawan
Fakultas Teknik dan Ilmu Komputer, Universitas Potensi Utama, Medan, Indonesia

M. Kartiwi
Information Systems Department, International Islamic University Malaysia, Kuala Lumpur, Malaysia

A. F. Ab. Nasir et al. (eds.), *Recent Trends in Mechatronics Towards Industry 4.0*, Lecture Notes in Electrical Engineering 730, https://doi.org/10.1007/978-981-33-4597-3_60

With facial acknowledgement and human-computer interaction getting progressively conspicuous as time passes, the measure of databases related to both face discovery and outward appearances has been developed enormously [1]. A crucial part in making, training and in any event, assessing supervised emotion recognition models is an all-around labelled database of visual data fit for the ideal application. For instance, emotion recognition has a wide range of utilizations running from straightforward human-robot PC communication [2, 3] to mechanized depression identification [4].

Many scholarly works are dedicated to describing different face and emotion detection databases. However, there is a lack of providing a thorough and broad overview of the prominent databases pertaining to emotion recognition using images/videos. Although there as of now are a lot of gathered databases out there that fit numerous particular models [5], it is essential to perceive that there are a few distinct angles that influence the substance of the database. The choice of the members, the technique used to gather the information and what was gathered all significantly affect the exhibition of the last model [6]. The social foundation of members just as their state of mind during recordings can influence the consequences of the database to be explicit to a specific gathering of individuals. Since most calculations take an adjusted and trimmed face as an input, the most fundamental type of datasets is an assortment of portrait pictures or expertly edited faces, with uniform lighting and foundations. In any case, real-time situations are increasingly convoluted, requiring the creators to try different things with various lighting, head posture and impediments. When evaluating an emotion recognition system, one of the essential factors to be considered is the level of naturalness of the database used to gauge the recognition performance. If a poorly made database is used, fallacious conclusions may be established [7]. Hence, it is imperative to have a well-fitted dataset. With the growing phase in the emotion recognition scholarly works there is a vital need to have a well-directed and proper database-oriented work which in turn will help the researchers to overview various image/video-based emotion datasets without any hustle. This paper is an effort to contribute towards providing an analysis of the image/video-based emotion datasets.

This paper presents a detailed summarization of different image/video-based emotion recognition databases. It is divided into seven sections. Section 2 outlines various elicitation methods for data acquiring. Section 3 presents the various categories of emotions taken into consideration while formulating the dataset. Section 4 presents various image and video databases available for emotion recognition. Section 5 summits the paper with a brief conclusion.

2 Elicitation Methods of Image/Video Data

A significant decision to make in acquiring data for emotion acknowledgement databases is the way to bring out various feelings in the participants. This is the motivation behind why facial feeling databases are isolated into three fundamental

Fig. 1 Presented expressions from WSEFEP database

classes, Presented, Actuated, and Unconstrained [8]. Inspiring articulations should be possible in a few distinct manners, and shockingly, they yield fiercely various outcomes.

2.1 Presented

Emotions carried on dependent on conjecture or with the direction from on-screen characters, or experts are called presented expressions [9]. Most facial feeling databases, particularly the early ones comprised simply of posed facial appearances, as it is the most effortless to accumulate. Notwithstanding, they likewise are the minimal agent of genuine bona fide feelings as constrained feelings are frequently over-misrepresented or missing unobtrusive subtleties. Because of this, human articulation examination models made using presented databases frequently have inferior outcomes with certifiable data [10], as shown in Fig. 1.

2.2 Actuated

This strategy for elicitation shows increasingly certifiable feelings as the members ordinarily cooperate with others or are dependent upon varying media to conjure genuine feelings. Actuated feeling databases have gotten increasingly necessary as of late because of the constraints of presented expressions. The exhibition of the models, all things considered, is extraordinarily improved, since overemphasized and fake expressions do not block them, making them progressively natural, as found in Fig. 2. Even though actuated databases are superior to the presented ones, they despite have a few issues with honesty. Since the feelings are regularly conjured in a lab setting with the management of definitive figures, the subjects may subliminally hold their expressions under control [9, 10].

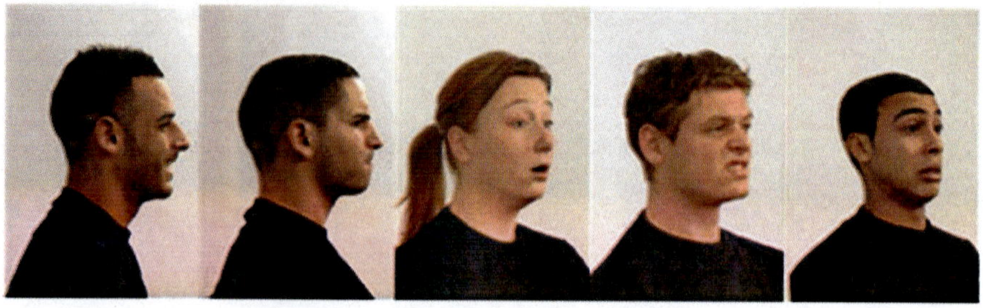

Fig. 2 Actuated expressions from Radboud Faces Database (RaFD)

2.3 *Unconstrained*

Unconstrained feeling datasets, as shown in Fig. 3, are viewed as the nearest to real-life situations. However, since genuine feeling can only be found, when the individual does not know about being recorded, they are hard to gather and label. The procurement of data is as a rule in strife with protection or morals, while the naming must be done physically, and the analyzer must speculate the genuine feeling. This challenging errand is both tedious and erroneous [13], having a sharp diverge from presented and actuated datasets, where names are either predefined or can be obtained from the elicitation content. There still exist a few databases out there that comprise of data obtained from films, YouTube recordings, or even TV arrangement. Nevertheless, these databases have innately fewer examples in them than their presented and actuated partners.

3 Categories of Emotion

The feelings characterize the reason for a database spoke to in it. The more significant part of the databases selects to catch the six essential feeling types: anger, disgust, fear, happiness, sadness and surprise. Frequently creators will, in general, add contempt to these, shaping seven essential feelings and often neutral is incorporated. Be that

Fig. 3 Unconstrained expressions from SFEW_2 dataset [14]

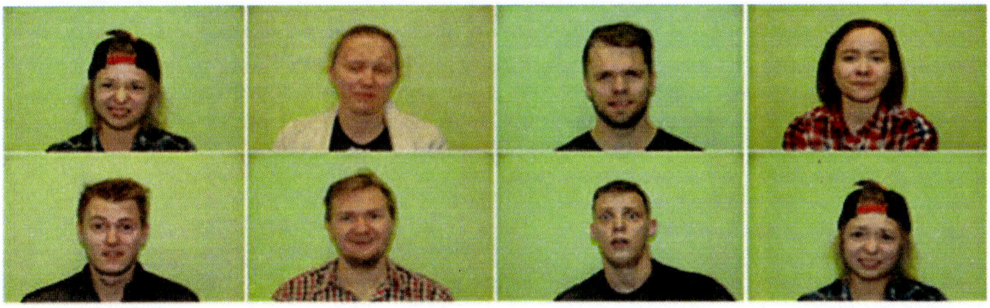

Fig. 4 iCV-MEFED database sample pictures

as it may, they cover an exceptionally little subcategory of every single imaginable feeling, so there have been endeavours to join more [15]. Aside from anger and disgust inside the six essential feelings, researchers have attempted to catch other negative articulations, for example, boredom, disinterest, pain, embarrassment and depression. Sadly, these classes are more earnestly to evoke than different kinds of feelings.

The primary motivation behind why most databases have few classifications is that the more feelings included, the more troublesome they are to mark, and more information is required to train a model appropriately. Moderately more current databases have started recording progressively unobtrusive feelings holed up behind other constrained or predominant feelings, which centres around passionate giggling and various kinds of chuckling, and others that attempt to record feelings taken cover behind a neutral or straight face. One of the later databases, the iCV-MEFED [16] database, as shown in Fig. 4, adopts on an alternate strategy by presenting shifting blends of feelings at the same time, where one feels plays the regular job, and the other is complimentary.

4 Image and Video Databases for Emotion Recognition

To get any significant elevated level semantic data, for example, feelings, essential data about the position and direction of body parts must be acquired first. Face appearances can be gained from picture and video datasets. All emotion recognition strategies play out the best on very much adjusted facial pictures. Facial landmarks (eyes, nose, mouth, and certain places of their parts) are significant prompts for comprehension of appearances. They can be additionally utilized for head pose estimation, frontalization, identification and numerous different assignments. Emotion recognition from images/videos have a full application, for instance, in human-PC association [17], and suggestion frameworks [18]. Notwithstanding, the examination in emotion recognition is hampered by the accessible datasets which are typically centred around a situation or circumstance. In this manner, existing strategies prepared on such datasets are unable, to sum up to different applications. Further, the undertaking of emotion recognition is inalienably hard since dependable ground truth is complicated to make because of the concealed nature and intricacy of feelings.

4.1 Image Databases for Emotion Recognition

Most new facial appearance databases, like the CK [19], just comprise of frontal portrait pictures taken with straightforward RGB cameras. More current databases attempt to structure assortment strategies that fuse information, which is nearer to real-life situations by utilizing various points and impediment. To build the precise and accurate human emotion recognition models, databases have extended the edge from a portrait to the whole upper body area. Picture databases are the most established and most typical sort. This way, it is justifiable that they were made with the most assorted of objectives, fluctuating from expression observation to neuropsychological research [20], and have a broad scope of data gathering styles, including self-photography. Picture databases, as a rule, have the most significant number of members and a greater example size. While it is moderately simple to discover a database appropriate for the job needing to be done, classifications of feelings are very constrained, as image databases just spotlight on six essential feelings or smile/neutral identification.

4.2 Video Databases for Emotion Recognition

The most advantageous organization for catching instigated and unconstrained feelings is video. This is because of the absence of a clear beginning and end focuses on non-presented feelings. Posed video databases will, in general, be tiny in the number of members, for the most part around 10, and frequently proficient entertainers have been utilized. Not at all like with still pictures, researchers have attempted to profit by voice, discourse or some other sort of articulations for emotion recognition. Numerous databases have additionally attempted to assemble smaller scale articulations, as they do not appear on still pictures or are harder to get. The presented video databases have substantially centred around six essential feelings and a neutral expression. Media actuated databases have a more significant number of members, and varying media typically incite the emotions. Since the feelings in these databases are actuated utilizing outside methods, this configuration is extraordinary for social affair counterfeit [21] or covered up feelings. Interaction induced video databases are increasingly one of a kind method of gathering data, like kid robot interaction or exploring past recollections. This kind of databases sets aside fundamentally longer effort to make [22], yet this does not appear to influence the sample size. Practically the entirety of the unconstrained databases is in video group from other media sources, simply since they are so hard to gather. Table 1 shows a comparison of around 20 image and video-based emotion databases.

5 Conclusions

With the fast increment of computing power and size of data, it has become increasingly more achievable to recognize emotions, distinguish individuals, and check trustworthiness dependent on video, audio or image input, stepping forward in human-PC collaboration, yet also in psychological maladjustment discovery, clinical research, and security. In this paper, an outline of the existing image and video databases in fluctuating classifications has been given. Different data gathering techniques have been discussed for emotion database creation, along with the categories of emotions.

Table 1 Comparison of image/video emotion database

No.	Database name	Emotions	Number of subjects	Number of images or videos	Image types	Type
1	Warsaw Set of Emotional Facial Expression Pictures (WSEFEP) [11]	Joy, Anger, Disgust, Fear, Surprise and Sadness	30	210 images	Colour	Posed
2	Radboud Faces Database (RaFD) [12]	Anger, disgust, fear, happiness, sadness, surprise, contempt, and neutral	67	8040 images	Colour	Posed
3	Static Facial Expressions in the Wild (SFEW) [14]	Angry, Disgust, Fear, Happy, Neutral, Sad and Surprise	The database contains an extensive age range of subjects from 1 to 70 years	700 images	Colour	Spontaneous
4	MMI Database [23]	79 facial displays including single AUs or combinations of AUs	43	250 images	Colour	Posed
5	Japanese Female Facial Expression (JAFFE) Database [24]	Neutral, sadness, surprise, happiness, fear, anger, and disgust	10	213 images	Eight-bit gray	Posed

(continued)

Table 1 (continued)

No.	Database name	Emotions	Number of subjects	Number of images or videos	Image types	Type
6	Extended Cohn-Kanade Dataset (CK+) [25]	Neutral, sadness, surprise, happiness, fear, anger, contempt and disgust	123	593 images	Mostly gray	Posed and spontaneous
7	IMPA-FACE3D [26]	Neutral frontal, joy, sadness, surprise, anger, disgust, fear, opened, closed, kiss, left side, right side, neutral sagittal left, neutral sagittal right, nape and forehead	38	534 images	Colour	Posed
8	FEI Face Database [27]	Neutral, smile	200	2800 images	Colour	Posed
9	The Bosporus Database [28]	Anger, Disgust, Fear, Happiness, Sadness and Surprise	105	4666 images	Colour	3D and 2D human face
10	Indian Movie Face database (IMFDB) [29]	Anger, Happiness, Sadness, Surprise, Fear, Disgust	100	34512 images	Colour	Spontaneous
11	Amsterdam Dynamic Facial Expression Set—Bath Intensity Variations (ADFES-BIV) [30]	Anger, disgust, fear, sadness, surprise, happiness, pride, contempt, embarrassment, and neutral	7 males and 5 females	370 videos	Coloured	Posed
12	Ryerson Audio-Visual Database of Emotional Speech and Song (RAVDESS) [31]	Calm, happy, sad, angry, fearful, surprise, disgust, and neutral	24	7356 videos	Colour	Posed

(continued)

Table 1 (continued)

No.	Database name	Emotions	Number of subjects	Number of images or videos	Image types	Type
13	Belfast Database [32]	Set 1 (disgust, fear, amusement, frustration, surprise)	114	570 videos	Colour	Spontaneous
		Set 2 (disgust, fear, amusement, frustration, surprise, anger, sadness)	82	650 videos	Colour	Spontaneous
		Set 3 (disgust, fear, amusement)	60	180 videos	Colour	Spontaneous
14	iSAFE (Indian Semi-Acted Facial Expression Database) [33]	Happy, Sad, Fear, Surprise, Angry, Neutral, Disgust	44	395 videos	Colour	Spontaneous
15	Indian Spontaneous Expression Database (ISED) [34]	Sadness, surprise, happiness, and disgust	50	428 videos	Colour	Spontaneous
16	The Second Emotion Recognition in The Wild Challenge and Workshop (EmotiW 2014) dataset [35]	Anger, Disgust, Fear, Happiness, Neutral, Sadness and Surprise	–	578 training videos and 383 validation videos	Colour	Spontaneous
17	Cohn-Kanade AU-Coded Facial Expression Database [36]	23 facial displays including single AUs or combinations of AUs	100	486 videos	Eight-bit gray	Posed
18	Denver Intensity of Spontaneous Facial Action (DISFA) Database [37]	The database includes 66 facial landmark points	27	4845 videos	–	Spontaneous

(continued)

Table 1 (continued)

No.	Database name	Emotions	Number of subjects	Number of images or videos	Image types	Type
19	eNTERFACE '05 [38]	Anger, disgust, fear, happiness, sadness, surprise	42	1166 videos	Colour	Induced
20	Surrey Audio-Visual Expressed Emotion (SAVEE) database [39]	Anger, disgust, fear, happiness, sadness, surprise, neutral	4	480 utterances	Colour	Acted

Image and Video databases have been sorted out into the table to give the reader a simple method to discover valuable information. This paper ought to be a decent beginning stage for any individual who thinks about preparing a model for emotion recognition. To summit up all it is better to state that the possibility of advancements in image and video databases are highly promising, and so is the scope for research in the field of image/video emotion recognition.

Acknowledgements The authors would like to express their gratitude to the Malaysian Ministry of Education (MOE), which has provided research funding through the Fundamental Research Grant, FRGS19-076-0684 (FRGS/1/2018/ICT02/UIAM/02/4).

References

1. Noroozi F, Marjanovic M Njegus A, Escalera S, Anbarjafari G (2017) Audio-visual emotion recognition in video clips. IEEE Trans Affect Comput
2. Daneshmand M, Abels A, Anbarjafari G (2017) Real-time, automatic digi-tailor mannequin robot adjustment based on human body classification through supervised learning. Int J Adv Robot Syst 14(3)
3. Bolotnikova A, Demirel H, Anbarjafari G (2017) Real-time ensemble based face recognition system for NAO humanoids using local binary pattern. Analog Integr Circuits Signal Process 92:467–475
4. Valstar MF, Schuller BW, Smith K, Eyben F, Jiang B, Bilakhia S, Schnieder S, Cowie R, Pantic M (2013) The continuous audio/visual emotion and depression recognition challenge. In: Proceedings of the 3rd ACM international workshop on audio/visual emotion challenge—AVEC 13
5. Athanaselis T, Bakamidis S, Dologlou I, Cowie R, Douglas-Cowie E, Cox C (2005) ASR for emotional speech: clarifying the issues and enhancing performance. Neural Netw 437–444
6. Jaimes A, Sebe N (2005) Multimodal human computer interaction: a survey. Springer, Beijing, China, pp 1–15
7. Qadri SAA, Gunawan TS, Alghifari MF, Mansor H, Kartiwi M, Janin Z (2019) A critical insight into multi-languages speech emotion databases. Bull Electr Eng Inform

8. Wu C-H, Lin J-C, Wei W-L (2014) Survey on audiovisual emotion recognition: databases, features, and data fusion strategies. APSIPA Trans Signal Inf Process
9. Sebe N, Cohen I, Gevers T, Huang TS (2005) Multimodal approaches for emotion recognition: a survey. In: Electronic imaging 2005. International Society for Optics and Photonics, pp 56–67
10. Zeng Z, Pantic M, Roisman GI, Huang TS (2009) A survey of affect recognition methods: audio, visual, and spontaneous expressions. IEEE Trans Pattern Anal Mach Intell 31:39–58
11. Olszanowski M, Pochwatko G, Kuklinski K, Scibor-Rylski M, Lewinski P, Ohme RK (2015) Warsaw set of emotional facial expression pictures: a validation study of facial display photographs
12. Langner O, Dotsch R, Bijlstra G, Wigboldus DHJ, Hawk ST, van Knippenberg A (2010) Presentation and validation of the Radboud Faces Database. Cogn Emot 24:1377–1388
13. Kirouac G, Dore FY (1985) Accuracy of the judgment of facial expression of emotions as a function of sex and level of education. J Nonverbal Behav 9:3–7
14. Dhall A, Goecke R, Lucey S, Gedeon T (2011) Static facial expression analysis in tough conditions: data, evaluation protocol and benchmark. In: IEEE ICCV 2011 workshop BEFIT
15. Petridis S, Martinez B, Pantic M (2013) The mahnob laughter database. Image Vis Comput 31:186–202
16. Loob C, Rasti P, Lüsi I, Jacques JCS, Baró X, Escalera S, Sapinski T, Kaminska D, Anbarjafari G (2017) Dominant and complementary multi-emotional facial expression recognition using c-support vector classification. In: 12th IEEE international conference on automatic face & gesture recognition (FG 2017), pp. 833–838
17. Kanade T, Cohn JF, Tian Y (2000) Comprehensive database for facial expression analysis. In: Proceedings of fourth IEEE international conference on automatic face and gesture recognition, pp 46–53
18. Ekman P, Freisen W (1976) Pictures of facial affect. Consulting Psychologists, Palo Alto
19. Pantic M, Valstar M, Rademaker R, Maat L (2005) Web-based database for facial expression analysis. In: Proceedings of IEEE international conference on multimedia and expo (ICME05)
20. Lyons MJ, Akamatsu S, Kamachi M, Gyoba J (1998) Coding facial expressions with Gabor wavelets. In: Proceedings of FGR98, pp 200–205
21. Lucey P, Cohn JF, Kanade T, Saragih J, Ambadar Z, Matthews I (2010) The Extended Cohn-Kanade Dataset (CK+): a complete facial expression dataset for action unit and emotion-specified expression. In: 3rd IEEE workshop on CVPR for human communicative behavior analysis
22. Mena-Chalco J, Marcondes R, Velho L (2012) FacesDB. VISGRAF 2012 [online]. Available at: http://app.visgraf.impa.br/database/faces/documents/. Accessed 02 June 2020
23. Thomaz CE, Giraldi GA (2010) A new ranking method for Principal Components Analysis and its application to face image analysis. Image Vis Comput 28:902–913
24. Savran A, Sankur B (2017) Non-rigid registration based model-free 3D facial expression recognition. Comput Vis Image Underst 162:146–165
25. Setty S, Husain M, Beham P, Gudavalli J, Kandasamy M, Vaddi R, Hemadri V, Karure JC, Raju R, Rajan B, Kumar V, Jawahar CV (2013) Indian Movie Face Database: a benchmark for face recognition under wide variations. In: National conference on computer vision, pattern recognition, image processing and graphics (NCVPRIPG)
26. Corneanu CA, Escalera S, Baro X, Hyniewska S, Allik J, Anbarjafari G, Ofodile I, Kulkarni K (2017) Automatic recognition of deceptive facial expressions of emotion
27. Henry SG, Fetters MD (2012) Video elicitation interviews: a qualitative research method for investigating physician-patient interactions. Ann Fam Med 10(2):118–125
28. Wingenbach TSH, Ashwin C, Brosnan M (2016) Validation of the Amsterdam Dynamic Facial Expression Set—Bath Intensity Variations (ADFES-BIV): a set of videos expressing low, intermediate, and high intensity emotions
29. Russo L (2018) The Ryerson Audio-Visual Database of Emotional Speech and Song (RAVDESS): a dynamic, multimodal set of facial and vocal expressions in North American English

30. Sneddon I, McRorie M, McKeown G, Hanratty J (2012) The Belfast induced natural emotion database. IEEE Trans Affect Comput 3(1):32–41
31. Singh S, Benedict S (2020) Thampi SM, Hegde RM, Krishnan S, Mukhopadhyay J, Chaudhary V, Marques O, Piramuthu S, Corchado JM (eds) Indian Semi-Acted Facial Expression (iSAFE) dataset for human emotions recognition. Springer, Singapore, pp 150–162
32. Happy SL, Patnaik P, Routray A, Guha R (2016) The Indian spontaneous expression database for emotion recognition. IEEE Trans Affect Comput
33. Dhall A, Goecke R, Lucey S, Gedeon (2011) Acted facial expressions in the wild database. Research School of Computer Science, College of Engineering and Computer Science, The Australian National University
34. Kanade T, Cohn JF, Tian Y (2000) Comprehensive database for facial expression analysis. In: Proceedings of FGR00, pp 46–53
35. Mavadati SM, Mahoor MH, Bartlett K, Trinh P, Cohn J (2013) DISFA: a spontaneous facial action intensity database. IEEE Trans Affect Comput 4(2):151–160
36. Martin O, Kotsia I, Macq B, Pitas I (2006) The eNTERFACE'05 audiovisual emotion database. In: International conference on data engineering workshops
37. Haq S, Jackson PJB (2009) Speaker-dependent audio-visual emotion. In: International conference on auditory-visual speech processing, pp 53–58
38. Brown S, Pepper M (2014) The emotion reading robot, TECHNOLOGY
39. Berkovsky S (2015) Emotion-based movie recommendations: how far can we take this? In: Proceedings of the 3rd workshop on emotions and personality in personalized systems

Speech Emotion Recognition Using Feature Fusion of TEO and MFCC on Multilingual Databases

Syed Asif Ahmad Qadri, Teddy Surya Gunawan, Mira Kartiwi, Hasmah Mansor, and Taiba Majid Wani

Abstract In the speech signal, emotion is considered one of the most critical elements. For the recognition of emotions, the field of speech emotion recognition came into existence. Speech Emotion Recognition (SER) is becoming an area of research interest in the last few years. A typical SER system focuses on extracting features such as pitch frequency, formant features, energy-related features, and spectral features from speech, tailing it with a classification quest to foresee different classes of emotion. The critical issue to be addressed for a successful SER system is the emotional feature extraction, which can be solved by using different feature extraction techniques. In this paper, along with Teager Energy Operator (TEO) and Mel Frequency Cepstral Coefficients (MFCC) a trailblazing feature extraction method, a fusion of MFCC and TEO as Teager-MFCC (T-MFCC) is used for the recognition of energy-based emotions. We have used three corpora of emotions in German, English, and Hindi to develop the multilingual SER system. The classification of these energy-based emotions is done by Deep Neural Network (DNN). It is found that TEO achieves a better recognition rate compared to MFCC and T-MFCC.

Keywords Speech emotion recognition · Deep neural network · Multilingual database · TEO · MFCC

S. A. A. Qadri · T. S. Gunawan (✉) · H. Mansor · T. M. Wani
Electrical and Computer Engineering Department, International Islamic University Malaysia, Gombak, Malaysia
e-mail: tsgunawan@iium.edu.my

T. S. Gunawan
Fakultas Teknik Dan Ilmu Komputer, Universitas Potensi Utama, Medan, Indonesia

M. Kartiwi
Information Systems Department, International Islamic University Malaysia, Gombak, Malaysia

A. F. Ab. Nasir et al. (eds.), *Recent Trends in Mechatronics Towards Industry 4.0*, Lecture Notes in Electrical Engineering 730, https://doi.org/10.1007/978-981-33-4597-3_61

1 Introduction

Speech is a form of articulated communication for people to interact with each other. It is the most natural and fastest way of interchanging information among humans. This assurance has incited scientists to consider speech a productive and viable strategy for informativeness between the humans and machine and prompted the inclination of developing a system which can recognize speech without entailing any technical proficiency from the users. Speech recognition imparts the interface to recognize speech. The conversion of speech signals to word sequences by implementing an algorithm is known as speech recognition. Despite the vast advancement carried out in speech recognition, the natural interaction between machines and humans is still very far. The purpose of this is a machine that cannot comprehend the emotional state of the speaker. This has prompted the presentation of a generally new research field to be specific speech emotion recognition (SER), characterized as the emotional state extraction of a speaker from his/her speech.

The detection of human emotion from the speech is a challenging matter, thus makes it a vast area of interest for many researchers in recent years. The predominant subject is the speech feature extraction in speech emotion recognition which adequately distinguish the content of speech emotion and simultaneously are independent of the verbal content and the speaker. Since, various speech features are scrutinized in SER, yet the best features for the task are not discovered by the researchers. Speech features may be classified among four classes: qualitative features, spectral features, continuous features, and TEO (Teager energy operator)-based features. Human emotions can be classified as happy, joy, sad, neutral, and angry. In the past century, some of the speech features constituted the fundamental frequencies, linear prediction cepstrum coefficient (LPCC), Mel frequency cepstrum coefficient (MFCC), and Teager Energy Operator (TEO), which are the building blocks of speech processing [1]. MFCC is tremendously utilized in speech recognition and speech emotion recognition systems as the recognition rate is very high. LPCC describes the aspects of a channel of a person. Moreover, the features of the channel get changed as per different emotions with less computation. The nonlinear airflow in the vocal system is the basis for the generation of speech, as per the experimental studies carried out by TEO.

Extracting various features from the speech leads to an accurate recognition of emotions. Emotion can be recognized in two measurements: activation and valence [2]. Activation has alluded to the measure of energy wanted to pass on a particular emotion. Though, emotions like happiness and anger commensurate to high activation yet co different affect. The valence dimension specifies the dissimilarity. There happen changes in the speaking rate, pitch, energy, and spectrum as per diverse emotional states. Every emotion relates to different bits of the expressed utterances. Consequently, it is tough to recognize these parts of utterance [3].

Although many kinds of research have been conducted on SER, most of the researchers are only on recognizing one or two speech databases. Therefore, in this paper, multilingual databases, including English, German, and Hindi, spoken by both females and males, will be examined [4]. For the SER system, Mel-Frequency Cepstral Coefficients (MFCC) and Teager Energy Operator (TEO) are the most used features. The fusion of both TEO and MFCC features, as well as various structures of deep neural networks, will be evaluated. The recognition rate is used for indicating the performance criteria. The rest of the paper is organized as follows. Section 2 discusses the speech emotion recognition and feature extraction techniques followed by the classifier. Results and discussion are shown in Sect. 3. Lastly, Sect. 4 concludes the paper.

2 Speech Emotion Recognition

The SER system is kindred as a pattern recognition system. The speech emotion recognition system involves extraction, feature selection, classification, and recognized emotional output [5]. The examination of the SER is grounded on the dimension of straightforwardness of the database, is utilized as a contribution to the SER framework. The significant impediment in SER system is the requirement to recognize a lot of essential emotions to be arranged by an automatic emotion recognizer.

A fundamental arrangement of emotions comprises 300 passionate states, and hence it is hard to recognize a vast number of emotions. According to the "Palette hypothesis," any emotion can be disintegrated into essential emotion like how any shading is a blend of some fundamental hues. Essential emotions are fear, joy, sadness, anger, and disgust [2]. Feature processing for speech recognition is an efficient tool to be used for model building and recognition, which can be done by extracting the information which is speaker-dependent primarily. Many parameters are used in the determination of any specific emotion present in the speech. As such, the change in the parameters would represent the changes in the emotions. The challenge in the process stands in the fact that suitable features need to be extracted very carefully in order to characterize the emotions correctly. Speech signals are not steady in border sense either; hence we need to take up piecewise smaller speech signal fragments. These small speech signals are known as frames. Hence with each frame, we consider the speech signal to be almost stationary. Some of the speech features which can be called as prosodic features and contain pitch and energy like characteristics can be extracted from every frame, thus known as local features.

Feature extraction is the most noteworthy phase of speech emotion recognition and contains numerous strategies. Feature extraction has a critical role to play in the process of recognition. Here the annihilation of any irrelevant data present in the given output is performed, and required information is retained back at the same time. Thus, we can say this process is one of those that helps in feature reduction or dimension reduction. The emotions present in the speaker's speech can be spoken to by the vast number of attributes that are available inside the speech, and a change in these qualities will bring about comparing the change in emotions. In this manner, the extraction of these speech features winds up a standout amongst the most fundamental parameter in SER and accordingly portrays emotions in speech. A portion of

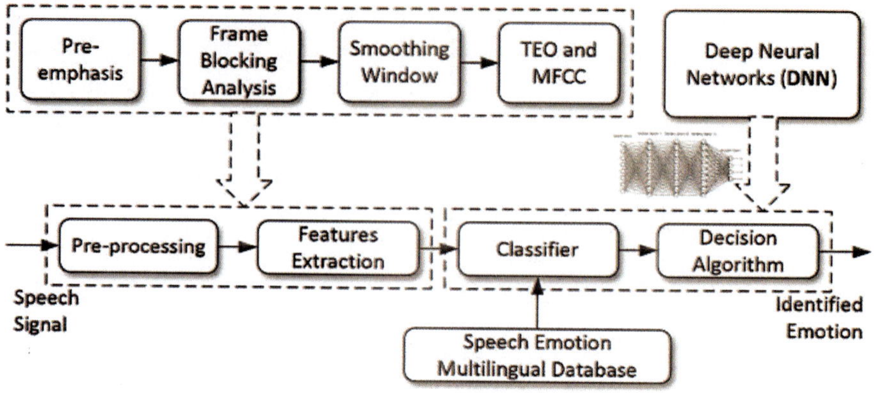

Fig. 1 Proposed speech emotion recognition system

the parametric portrayals is Mel-frequency cepstrum coefficients (MFCC), Teager Energy Operator (TEO), and Linear Prediction Coefficients (LPC).

Several classifiers can be used for the classification of emotions, including Artificial Neural Network (ANN), Hidden Markov Model (HMM), Gaussian Mixture Model (GMM), Deep Neural Network (DNN) and Support Vector Machine (SVM) [6]. In this paper, DNN is used for the process of emotion classification. Also, the two most popular audio features will be extracted, including TEO and MFCC, as shown in Fig. 1.

2.1 *Mel-Frequency Cepstral Coefficients*

Mel frequency cepstral coefficients were introduced in 1980 [7]. It is hinged on the human peripheral auditory system. MFCC provides better performance of recognition as it is less susceptible to noise [8]. MFCC is broadly used in audio classification experiments because of its excellent performance. It is used for the extraction of features from the speech signal. The perception of speech or frequency of a tone is measured as per the unit "Mel" [9]. Mel scale, a nonlinear frequency scale which is formulated on the auditory perception is utilized by MFCCs. The measure of perceived frequency or pitch of a tone is known as a Mel. The conversion of frequency scale to Mel scale can be done using eq. (1).

$$f_{mel} = 177 \ln\left(1 + \frac{f_{Hz}}{700}\right) \tag{1}$$

where f_{mel} is the frequency in Mels and f_{Hz} is the regular frequency in Hz, M filters are the filter banks which contain the triangular shaped filters and spaced uniformly on the Mel scale are generally used to calculate MFCCs, as depicted in Eq. (2).

$$H_m[k] = \begin{cases} 0 & k < f[m-1] \\ \frac{k-f[m-1]}{f[m]-f[m-1]} & f[m-1] < k \le f[m] \\ \frac{f[m+1]-k}{f[m+1]-f[m]} & f[m] < k \le f[m+1] \\ 0 & k > f[m+1] \end{cases} \tag{2}$$

where $m = 0, 1, \ldots, M-1$. The log-energy Mel spectrum is then calculated as follows:

$$S[m] = \ln\left[\sum_{k=0}^{N-1} |X[k]|^2 H_m[k]\right] \quad m = 0, 1, \ldots, M-1 \tag{3}$$

where $X[k]$ is the discrete Fourier transform (DFT) of a speech input $x[n]$.

Since, inverse discrete Fourier transform (IDFT) are utilized by traditional cepstrum, Mel frequency cepstrum is typically implemented using discrete cosine transform (DCT) since $S[m]$ is even as shown in Eq. (4).

$$\hat{x}[n] = \sum_{k=0}^{N-1} S[m] \cos\left[\left(m + \frac{1}{2}\right)\frac{\pi n}{M}\right], m = 0, \ldots, M-1 \tag{4}$$

Normally, 20–40 M filters, and 13 kept coefficients are utilized. One of the researches divulged that with 32–35 filters the performance of speech recognition and speaker identification systems reached a peak [1].

2.2 *Teager Energy Operator*

The Teager Energy Operator (TEO) gives a proportion of the energy of a speech signal. It is a simple algorithm which is derived, for enabling on-the-fly calculation of the energy required. TEO is also used to calculate the signal energy [10]. The TEO for a continuous-time signal is defined as:

$$\psi(x(t)) = \frac{d}{dt}x^2(t) - x(t)\left(\frac{d^2}{dt^2}x(t)\right) \tag{5}$$

whereas the TEO for a discrete signal is defined as:

$$\psi(x(n)) = x^2(n) - x(n+1)x(n-1) \tag{6}$$

where $x(n)$ is the sampled speech signal, ψ is the Teager energy operator.

The Teager energy operator is a nonlinear operator which was introduced to calculate instantaneous energy with improved signal to noise ratio (SNR). There is always some noise associated with the recording and processing equipment. With TEO, the

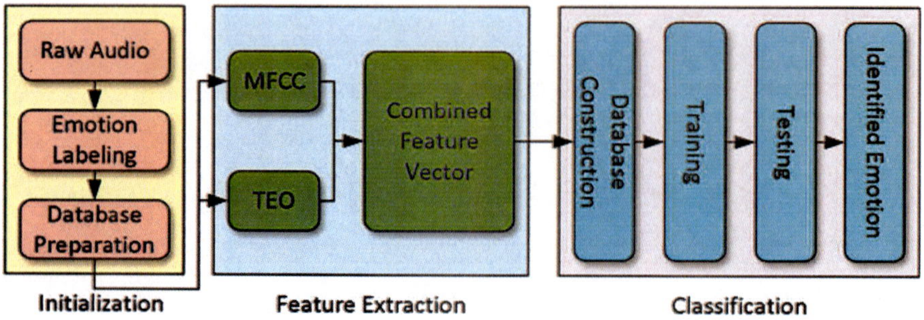

Fig. 2 Implementation flow for fusion features (MFCC-TEO)

noisy part present in the signal is suppressed while calculating its corresponding energy. Contrary to this, the typical squared energy operator takes the input signal along with the noise present and finds the energy.

2.3 *The Fusion of TEO and MFCC (TMFCC)*

Combined MFCC and TEO are used for feature extraction, a combination of features from MFCC and TEO are used for training the neural network. Also, in [11], a fusion of TEO and MFCC was used for the recognition of stressed emotions from the speech signal, and it was found that TEO performed much better for stressed emotions as compared to MFCC.

MFCC features set extracted from the speech signal, and then using a multi-rate filter bank, the speech signal was first divided into nonuniform sub-bands in Mel-scale. The Teager energies of the sub-signals are then estimated. Finally, the feature vector is constructed by a combination of both. Figure 2 shows the implementation flow for fusion (MFCC-TEO).

2.4 *Deep Neural Network Classifier*

The classifier that is employed in this paper is the Deep Neural Network. The previous conventional neural network consist of the input layer, a hidden layer, and an output layer. A deep neural network grows the potential outcomes by including more layers in the hidden layer segment. Deep Neural Network has been used with great success in recognizing emotions. In recent times, deep networks have illustrated representation and discriminative learning abilities over an extensive varieties of significance. ML researchers are growing the extents of deep learning by looking for them as future applications in other various fields like in speech emotion recognition [1, 12].

DNN takes as input the traditional acoustic features inside a speech portion and creates fragment level emotion state probability distributions, where from utterance-level features are developed and used to decide the utterance-level emotion state. Since the fragment level output already gives extensive emotional information, and the classification of utterance-level does not include much training, it is not worthy of utilizing DNNs for the utterance-level classification. DNN is trained to predict the probabilities of each emotional state with the segment-level features. The DNN can be treated as a segment-level emotion recognizer.

An exciting characteristic of DNNs is that they have the capability of learning high-level invariant features from raw data. One of the researches that have implemented DNN is conducted by [13] that extracted the features autonomously.

3 Results and Discussion

In this section, the experimental set up of speech emotion recognition will be discussed, followed by the preparation of the speech emotion database and some optimum parameters used. Also, the experiments for the performance analysis done over the deep neural network structures are evaluated for the proposed SER system.

3.1 Experimental Setup and Database

For processing, a simple performance system was utilized i.e., a multicore system with Intel Core i5 5200U 2.20 GHz, 8 GBytes RAM, and 1 TBytes hard disk, installed with Windows 8.1 operating system and MATLAB 2018b with signal processing and neural network toolbox. Other running applications were minimized during the simulation as much as possible.

For this paper, we have used three corpora of emotions in German, English, and Hindi to develop the proposed multilingual SER system. The English corpus is Toronto's emotional speech set (TESS), the German corpus is the well-known Berlin Emo-DB, and The Hindi corpus is Indian Institute of Technology Kharagpur Simulated Emotion Hindi Speech Corpus (IITKGP-SEHSC). There are two emotions considered for this paper, which are anger and neutral. The total utterances used in this paper are shown in Table 1.

Table 1 Total utterances for multilingual databases

Emotion	German	English	Hindi	Total
Anger	100	100	150	350
Neutral	63	100	150	313

3.2 Experiments on DNN Structures

The feature extraction, training, and testing were run over a loop for Multiple Configurations, such as Number of Hidden Layers [14], Neurons in each Hidden layer, its configuration, and Test/Train Ratio. The ratio was as per the standard procedure 70:15:15, in which 70 samples are for training, 15 for testing, and 15 for validation. The activation function used for the hidden layers and the output layer was tan sigmoid. Neural Network Pattern recognition toolbox from Matlab was used with scaled conjugate gradient backpropagation as the learning algorithm. Table 2 shows the performance results for MFCC, TEO, and Fusion (MFCC-TEO) feature extraction techniques. The loop was run for various neural network configurations.

As can be seen from Table 2, with two hidden layers, TEO shows the highest recognition rate of 90.5% with neural network configuration [1515 87]. The configurations [20 10] and [500 200 100 50] are slightly behind the best performance of TEO with a recognition rate of 90% and 90.1%. For the fusion of features, the best performance is obtained with the configuration [4750 856 154 28] with the recognition rate of 86.7%, at four hidden layers. The highest recognition rate of MFCC is obtained with the configuration [7751 2278 670 197 58 17], which is 83.7% with six hidden layers (Fig. 3).

The recognition rate (%) is formulated in the form of a confusion matrix for MFCC, TEO, and Fusion, as shown in Fig. 4. It is evident by the classification

Table 2 Performance results for MFCC, TEO, and fusion (TEO + MFCC)

Neural network configuration	Hidden layer(s)	Recognition rate %		
		MFCC	TEO	FUSION
[10]	1	83	89	81
[20]	1	82	86	78
[30]	1	81	87	82
[20 10]	2	79	90	78
[10 20 15]	3	81.1	87	80
[10 20 30 10]	4	66	75	76
[40 35 20 15 10]	5	77	88	75
[200 100 50 10]	4	83	89	80.3
[500 200 100 10]	4	80	90.1	78
[1000 300 100 50 10]	5	82	88	82
[363]	1	78	88	79.2
[1515 87]	2	79.2	**90.5**	79
[3094 363 43]	3	82	87	75
[4750 856 154 28]	4	81	89	**86.7**
[6320 1515 363 87 21]	5	79	86	84
[7751 2278 670 197 58 17]	6	**83.7**	85	83

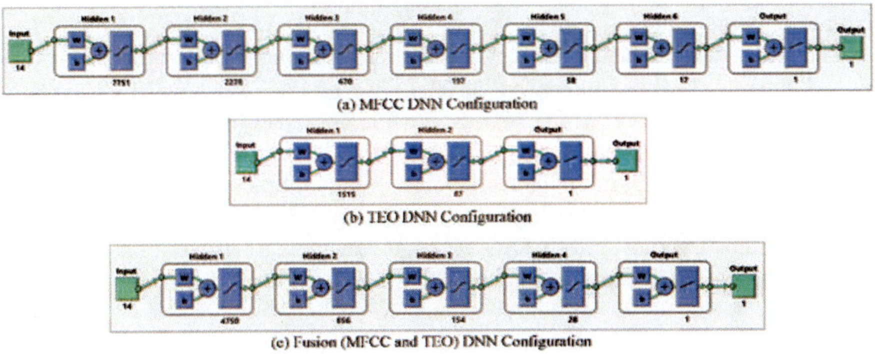

Fig. 3 Optimum configuration for MFCC, TEO, and fusion

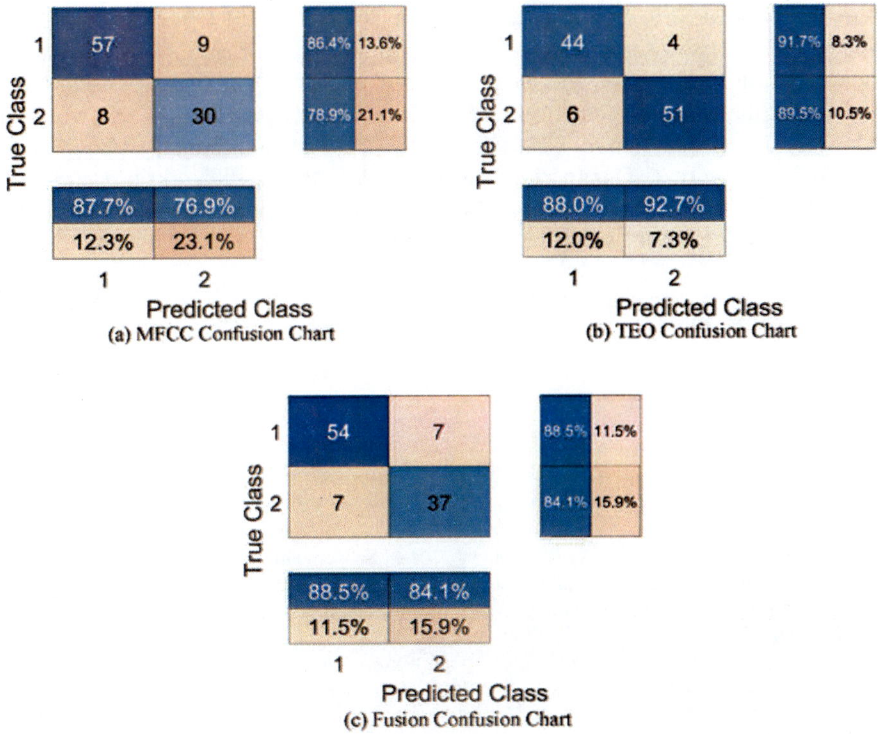

Fig. 4 Confusion matrix for MFCC, TEO, and fusion

accuracy, and energy emotions are detected best by TEO, followed by Fusion and MFCC. The main reason that TEO provides the best results is that it works on the energy present in the speech signal. It shows the instantaneous and frequency changes of the signal amplitude, which is very sensitive towards the minute changes. Also, if there is a change in the hidden layer neurons, the overall classification accuracy does not change; instead, TEO increases the performance of classification. In terms of the recognition rate, TEO is followed by Fusion (TEO + MFCC) and MFCC. The fusion has higher accuracy than MFCC because Fusion consists of TEO, as TEO being a better feature extraction technique for energy-based emotions than that of MFCC. In speech waveforms, there might be some variations from time to time, which depends on the physical condition of the speakers' vocal cord, MFCC is less susceptible to these variations as compared to TEO. Thus, it has the least recognition rate (%) in terms of energy-based emotions.

4 Conclusions

The multilingual SER system uses TEO, MFCC, and TMFCC for the classification of energy-based emotion, angry and neutral. The multilingual dataset (German database, English database, and Hindi database) and DNN classifier are used in the analysis of the work. A recognition rate of 90.5% is obtained using TEO, which is relatively higher than that of T-MFCC and MFCC, with an accuracy of 86.7% and 83.7%, respectively. TEO outperforms MFCC and TMFCC as it works on the energy present in the signal, shows the instantaneous and frequency changes of the signal amplitude, which is very sensitive towards minute changes.

Acknowledgements The authors would like to express their gratitude to the Malaysian Ministry of Education (MOE), which has provided research funding through the Fundamental Research Grant, FRGS19-076-0684 (FRGS/1/2018/ICT02/UIAM/02/4).

References

1. Alghifari MF, Gunawan TS, Kartiwi M (2018) Speech emotion recognition using deep feedforward neural network. Indones J Electr Eng Comput Sci 10(2):554–561
2. Fernandez R (2004) A computational model for the automatic recognition of affect in speech. Massachusetts Institute of Technology.
3. El Ayadi M, Kamel MS, Karray FJPR (2011) Survey on speech emotion recognition: features, classification schemes, and databases. Pattern Recogn 44(3):572–587
4. Qadri SAA, Gunawan TS, Alghifari MF, Mansor H, Kartiwi M, Janin Z (2019) A critical insight into multi-languages speech emotion databases. Bull Electr Eng Inform 8(4):1312–1323
5. Ingale AB, Chaudhari D (2012) Speech emotion recognition. Int J Soft Comput Eng 2(1):235–238
6. Swain M, Routray A, Kabisatpathy P (2018) Databases, features and classifiers for speech emotion recognition: a review. Int J Speech Technol 21(1):93–120
7. Chandrasekar P, Chapaneri S, Jayaswal D (2014) Emotion recognition from speech using discriminative features. Int J Comput Appl 101:16
8. Gilke M, Kachare P, Kothalikar R, Rodrigues VP, Pednekar M (2012) MFCC-based vocal emotion recognition using ANN. In: International conference on electronics engineering and informatics (ICEEI 2012). IPCSIT

9. Lanjewar RB, Mathurkar S, Patel N (2015) Implementation and comparison of speech emotion recognition system using Gaussian Mixture Model (GMM) and k-Nearest Neighbor (k-NN) techniques. Procedia Comput Sci 49:50–57
10. Jena B, Singh SS (2018) Analysis of stressed speech on Teager energy operator (TEO). Int J Pure Appl Math 118(16):667–680
11. Bandela SR, Kumar TK (2017) Stressed speech emotion recognition using feature fusion of Teager energy operator and MFCC. In: 2017 8th international conference on computing, communication and networking technologies (ICCCNT). IEEE, pp 1–5
12. Han K, Yu D, Tashev I (2014) Speech emotion recognition using deep neural network and extreme learning machine. In: Fifteenth annual conference of the international speech communication association
13. Zhou X, Guo J, Bie R (2016) Deep learning based affective model for speech emotion recognition. In: 2016 international IEEE conferences on ubiquitous intelligence & computing, advanced and trusted computing, scalable computing and communications, cloud and big data computing, internet of people, and smart world congress. IEEE, pp 841–846
14. Gunawan TS, Kartiwi M (2018) On the use of edge features and exponential decaying number of nodes in the hidden layers for handwritten signature recognition. Indones J Electr Eng Comput Sci 12(2):722–728

Normal Forces Effects of a Two In-Wheel Electric Vehicle Towards the Human Body

Nurul Afiqah Zainal, Muhammad Aizzat Zakaria, K. Baarath, Anwar P. P. Abdul Majeed, Ahmad Fakhri Ab. Nasir, and Georgios Papaioannou

Abstract Traditionally, in order to comprehend the impact of vibration on human and vehicle ride comfort, past research often models the human biodynamic and vehicle models individually. Recent trends suggest that a better understanding of the behaviour could be achieved by fusing the models instead of analysing it separately. The present study evaluates the impact of the normal forces on specific parts of the human body. A human biodynamic model with five degrees of freedom is modelled together with a two in-wheel electric car model travelling at a speed of 10 km/h to investigate the effect of the normal forces. From the present investigation, it could be observed that the proposed model could highlight the impact of the normal forces on the body parts when the car is travelling either on a straight path or in taking corners.

Keywords Vehicle vibration · Human biodynamic model · Ride comfort

1 Introduction

Car driver and passengers are susceptible to vibrations that originate largely from the vehicle owing to the interaction between the road and the vehicle itself. Although it is not apparent, it has been reported that prolonged exposure does to a certain degree bring about harmful effects towards human health, for instance, back problem, effects on the digestive system and decrease in the alertness level which in turn, could lead to road accidents amongst others [1]. It is also worth mentioning that, abrupt manoeuvre or braking may deteriorate the human comfort level and even injury during travelling

N. A. Zainal · M. A. Zakaria (✉) · K. Baarath
Autonomous Vehicle Laboratory, Automotive Engineering Centre (AEC), Universiti Malaysia Pahang, 26600 Pekan, Pahang, Malaysia
e-mail: maizzat@ump.edu.my

N. A. Zainal · M. A. Zakaria · K. Baarath · A. P. P. Abdul Majeed · A. F. Ab. Nasir
iMAM's Laboratory, Faculty of Manufacturing and Mechatronic Engineering Technology, 26600 Pekan, Pahang, Malaysia

G. Papaioannou
School of Aerospace, Transport and Manufacturing, Cranfield University, Cranfield, UK

A. F. Ab. Nasir et al. (eds.), *Recent Trends in Mechatronics Towards Industry 4.0*, Lecture Notes in Electrical Engineering 730, https://doi.org/10.1007/978-981-33-4597-3_62

[2, 3]. It is understandable that the vibration transferred through via the vibrated surface affects the human comfort level of the human body.

More often than not, the human biodynamic model focuses more on the human model without coupling to the vehicle model. In a recent study, by exciting specific vibration values on the lumped masses of the human model, the effect of vibration towards human comfort can be investigated. The lumped masses of the human model are connected by springs and dampers which characterise the human body muscles and joints [4]. The full vehicle model takes into account the effect of both the lateral and longitudinal forces as the weight shift is considered [5]. The Dugoff's model is used to calculate the lateral and longitudinal forces as it considers the tire stiffness independently for both directions.

It has been remarked in a previous study that in the quarter car and half car model, road conditions at rear tires cannot be incorporated [6]. This is due to the road conditions are not included in all tires of the vehicle. However, a study done by [7, 8] shows that the half car model is able to show the comfort ride level by manipulating the tire spring and damper parameters. The study demonstrated that nonlinear parameter of tire spring and dampers should be considered instead of linear parameters for the ride performance and handling of a passive suspension vehicle. Mitra et al. had used full car model to analyse the performance of vehicle dynamics for different road profiles which focused more on road with bump, where it was demonstrated through the study that the synthetic tire bump could affect human health as the vehicle body acceleration is very high even at speed of 10 km/h [9].

Furthermore, as for the full car model, research in [10] had indicated that this model is suitable to be used when the longitudinal acceleration as well as the braking and cornering influence on the vehicle pitch are taken into consideration. Consequently, this paper attempts to examine the effect of vibration of a five degrees of freedom (5 DOF) human biodynamic model [11, 12] coupled with a two in-wheel electric car model. The results demonstrated from the present investigation is non-trivial towards understanding the effect of normal forces owing from different drive maneuvers towards the human body.

2 Human-Vehicle Model

A schematic of a seated 5 DOF human biodynamic model is depicted in Fig. 1. This model consists of masses, m_i with dampers, c_i and springs, k_i which represents the muscles and joints of the human body which are derived by using Equation of Motion (EOM). Table 1 lists the biodynamic parameters of the 5 DOF model.

EOM of the human biodynamic model are derived based on Fig. 1 which depends on each of the human body parts. This vertical model consists of five degrees of freedom which are lower torso, m_1, viscera, m_2, upper torso, m_3, skull, m_4 and brain, m_5. This 5 DOF human model can be derived as below;

$$[m]\{\ddot{z}\} + [c]\{\dot{z}\} + [k]\{z\} = \{F\} \tag{1}$$

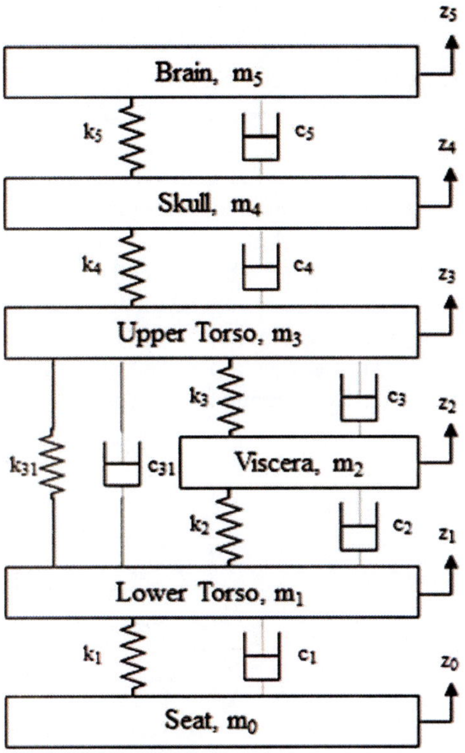

Fig. 1 Schematic of 5 DOF model [11, 12]

Table 1 5 DOF human model parameters [11, 12]

Mass (kg)	Damping (Ns/m)	Stiffness (N/m)
$m_1 = 36.0, m_2 = 5.5, m_3 = 15.0, m_4 = 3.5, m_5 = 1.5$	$c_1 = 2475.0, c_2 = 330.0, c_{31} = 909.1, c_3 = 200.0, c_4 = 450.0, c_5 = 340.0$	$k_1 = 49340.0, k_2 = 20000.0, k_{31} = 192000.0, k_3 = 10000.0, k_4 = 1800000.0, k_5 = 156000.0$

where,

$$m_1\ddot{z}_1 + c_1(\dot{z}_1 - \dot{z}_0) + k_1(z_1 - z_0) - c_{31}(\dot{z}_3 - \dot{z}_1) - k_{31}(z_3 - z_1) - c_2(\dot{z}_2 - \dot{z}_1) - k_2(z_2 - z_1) = 0 \quad (2)$$

$$m_2\ddot{z}_2 + c_2(\dot{z}_2 - \dot{z}_1) + k_2(z_2 - z_1) - c_3(\dot{z}_3 - \dot{z}_2) - k_3(z_3 - z_2) = 0 \quad (3)$$

$$m_3\ddot{z}_3 + c_{31}(\dot{z}_3 - \dot{z}_1) + k_{31}(z_{31} - z_1) + c_3(\dot{z}_3 - \dot{z}_2) + k_3(z_3 - z_2) - c_4(\dot{z}_4 - \dot{z}_3) - k_4(z_4 - z_3) = 0 \quad (4)$$

$$m_4\ddot{z}_4 + c_4(\dot{z}_4 - \dot{z}_3) + k_4(z_4 - z_3) - c_5(\dot{z}_5 - \dot{z}_4) - k_5(z_5 - z_4) = 0 \quad (5)$$

$$m_5\ddot{z}_5 + c_5(\dot{z}_5 - \dot{z}_4) + k_5(z_5 - z_4) = 0 \quad (6)$$

The consequential forces acting on the vehicle are produced by the forces and moments from the road via the tires of the vehicle. Dugoff's model is used due to the deformable properties of the tires. This model considers the tire stiffness independently of both the lateral and longitudinal directions.

Lateral and longitudinal model normally are used independently without considering the wheel orientation. As for this study, these models are combined which resulted in the direction of the tires change causing the normal force, F_z to be shifted following the steering angle. This also applied with the lateral force that acting on the tire.

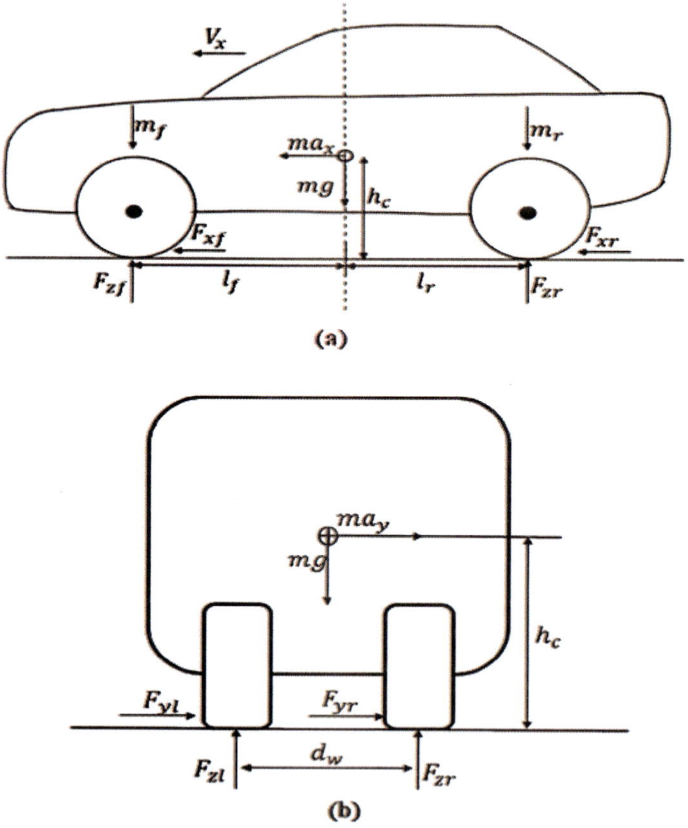

Fig. 2 **a** Forces and mass distribution acting on the vehicle, **b** vehicle's load transfer [5]

Figure 2 shows the normal force, F_z that acting on the vehicle model. This normal force is acting on each of the tire which been represented by F_{zfl}, F_{zfr}, F_{zrl} and F_{zrr} which indicates the normal forces on front left, front right, rear left and rear right, respectively. The total mass of the vehicle is divided into two parts, i.e., m_r and m_f for the rear and the front, respectively as shown in Fig. 2a. Conversely, Fig. 2b shows the load transfer in the lateral motion. Based on the figure, Eqs. (7)–(10) are used to obtain the vehicle mass shift from the of vehicle lateral and longitudinal acceleration [5].

$$F_{zfr} = \left(m_{Sij} + load_{ij}\right) g \frac{l_r}{l} + \frac{m_f a_y}{d_w} - m a_x \frac{h_c}{l} \tag{7}$$

$$F_{zfl} = \left(m_{Sij} + load_{ij}\right) g \frac{l_r}{l} - \frac{m_f a_y}{d_w} - m a_x \frac{h_c}{l} \tag{8}$$

$$F_{zrr} = \left(m_{ij} + load_{ij}\right) g \frac{l_r}{l} + \frac{m_f a_y}{d_w} + m a_x \frac{h_c}{l} \tag{9}$$

$$F_{zrl} = \left(m_{ij} + load_{ij}\right) g \frac{l_r}{l} - \frac{m_f a_y}{d_w} + m a_x \frac{h_c}{l} \tag{10}$$

where,

$l = l_f + l_r$, $l_f = l_r = 0.575$ m, g $= 9.81$ m/s^2, $m_r = 163.1$ kg, $m_f = 159$ kg, $d_w = 1.43$ m and $h_c = 0.105$ m.

The simulation is carried out by using MATLAB Simulink commercial software package. Figure 3 shows the schematic diagram of the human-vehicle model employed in the present investigation. Tire deflection, k_t is added into the model to normalize the error from the vehicle model [10]. Tire deflection is due to the behavior of the tire itself that acting like a spring in response to vertical forces which leads

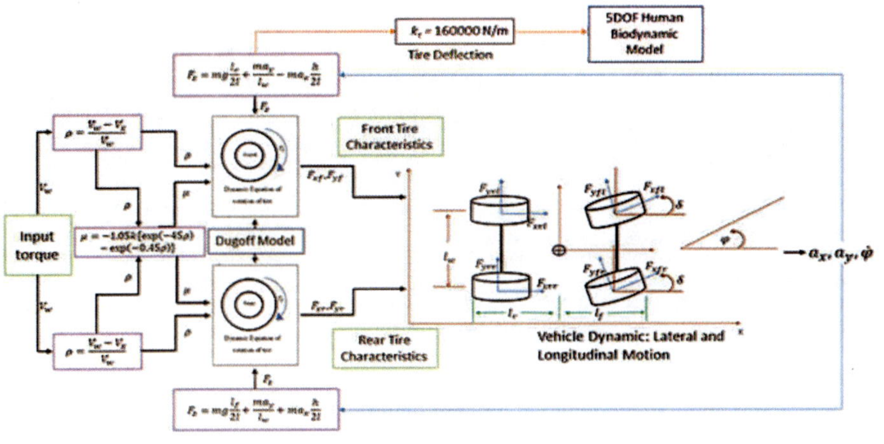

Fig. 3 Schematic diagram of the human-vehicle system

towards the variations in normal tire load. The car speed is set to 10 km/h in the simulation to align with the usage of a scaled down in-wheel vehicle used in the experimental set-up.

Figure 4 illustrate the steering input begins to make a turn at 1 s. It indicates the vehicle is in the cornering mode before it moves forward. The displacement output from F_{zfr} of the vehicle model was used as biodynamic model input.

Based on the steering input defined, Fig. 5 depicts the normal forces, F_z that acting on each of the vehicle tire at rear and front position. When the vehicle is on cornering mode, rear and front tire both shows opposite results acting on which depends on the

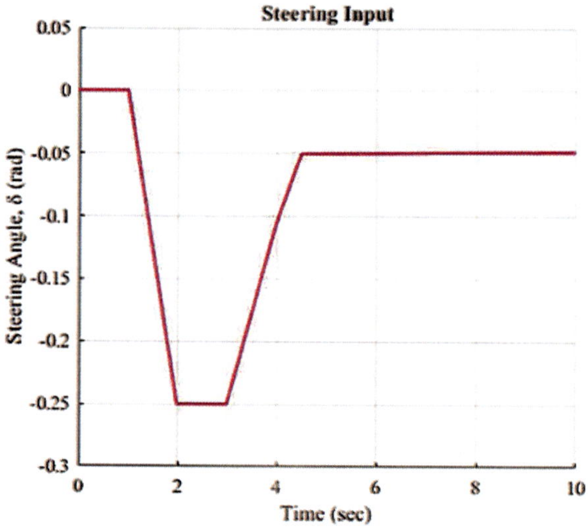

Fig. 4 Steering input

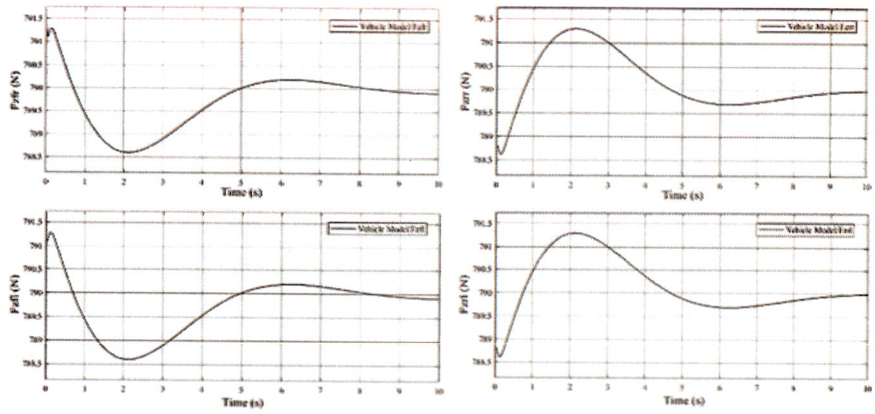

Fig. 5 Normal forces, F_z acting on each tire

vehicle direction. This is due to high forces acting in the opposing direction on either side of the vehicle which affect the stability of the vehicle. As for this study, when the vehicle turns to the right side, the normal force acting on the right is increased while the normal force acting on the left is decreased.

Figure 6 shows the relationship between the human-vehicle model on the front right of the vehicle tire. The normal force output at the front right of the vehicle, F_{zfr} was taken as an input to the human biodynamic model as to mimics the usual location of the driver that seat at the front right of the vehicle which is right-hand driver. Besides that, it is worth to note that this study is more focused on the human comfort level of the driver as compared to the passengers.

As the vehicle turns slightly following the steering input, a significant displacement on the body parts are recorded as shown in Fig. 7. The figure shows the close-up displacement that occur on the human body parts between 2 and 2.5 s. The displacement occurs rapidly as the vehicle is moving and in cornering mode. A clear displacement between the brain and the skull could be seen, suggesting a repeated collision between the two vital components of the head transpires [13].

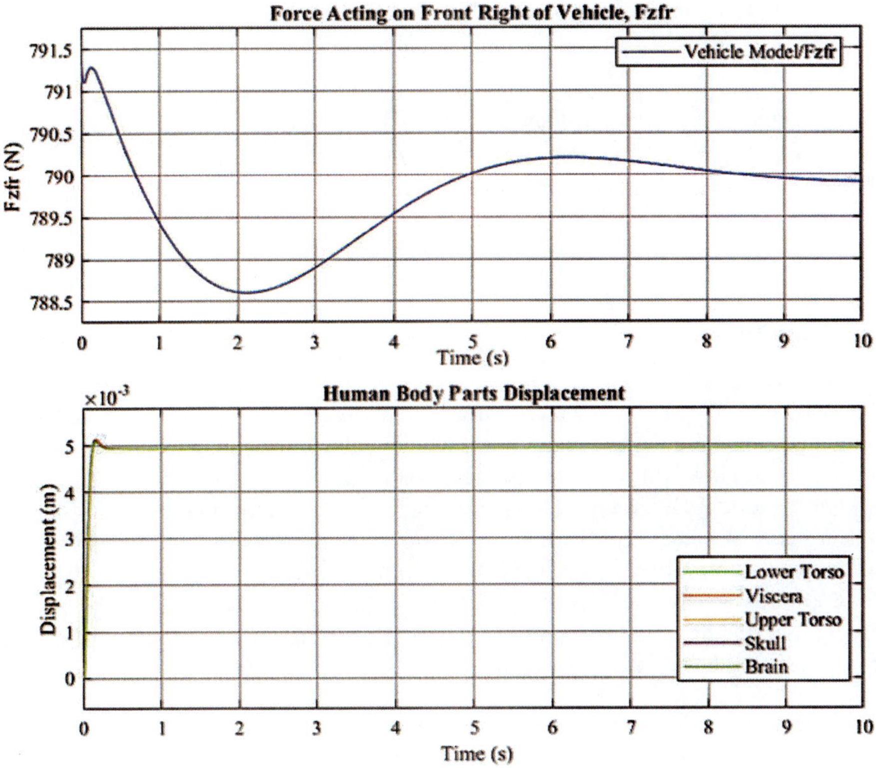

Fig. 6 Relationship between human-vehicle model on the front right of the vehicle

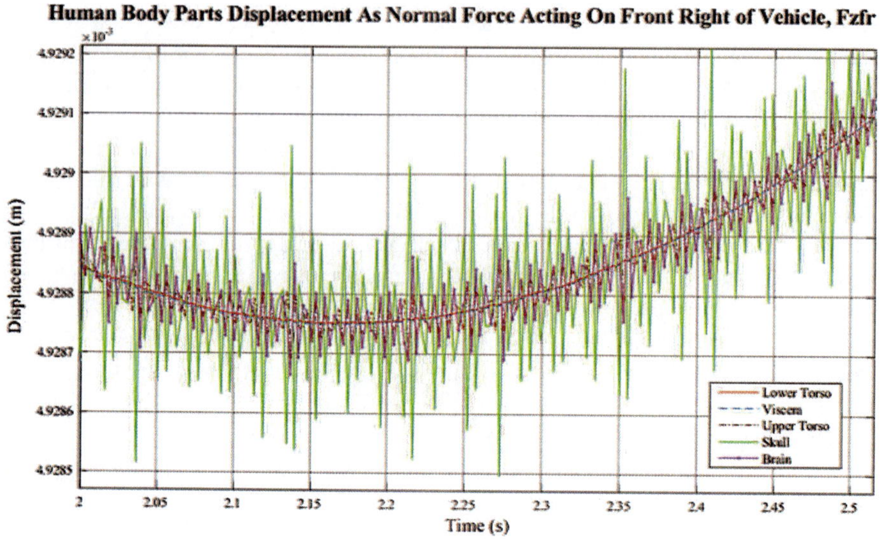

Fig. 7 Human body parts displacement

It is apparent, that in the event that this observation repeatedly occurs, in which it does, the human comfort level whilst travelling is somewhat affected. Consequently, the human health is affected in the long term owing to the prolonged undesirable human comfort level [14, 15] although the displacement value recorded is relatively small.

3 Conclusion

It could be concluded from the present investigation that the effect of normal forces acting on vehicle does affect the human body parts in terms of comfort when the vehicle is moving. Although the displacement recorded are relatively small, this will affect the comfort level especially while travelling on a long journey, uneven road surface terrain and at high speeds. As the present investigation only considers the effect of a single tire, further study on the impact of normal forces with different load towards the human body parts acting on different tire location of the vehicle is required. In addition, the vibrational effect, i.e., the harmonics from the aforesaid forces towards health shall be examined.

Acknowledgements The author would like to thank Ministry of Higher Education Malaysia (KPT) and Universiti Malaysia Pahang (www.ump.edu.my) for financial supports given under FRGS/1/2018/TK08/UMP/02/1, RDU190328 and PGRS190371.

References

1. Katu US, Desavale RG, Kanai R (2003) Effect of vehicle vibration on human body—RIT experience. In: NaCoMM, pp 1–9
2. Zakir U et al (2017) Autonomous emergency braking system with potential field risk assessment for frontal collision mitigation, pp 15–17
3. Hamid UZA et al (2017) Autonomous emergency braking system with potential field risk assessment for frontal collision mitigation. In: Proceedings—2017 IEEE IEEE conference on systems, process and control, ICSPC 2017, Dec 2017, pp 71–76
4. Zainal NA, Zakaria MA, Baarath K (2018) A study on the exposure of vertical vibration towards the brain on seated human driver model. ISBN 9789811087875
5. Baarath K, Zakaria MA, Zainal NA (2018) An investigation on the effect of lateral motion on normal forces acting on each tires for nonholonomic vehicle. ISBN 9789811087875
6. Mahala MK, Gadkari P, Deb A (2004) Mathematical models for designing vehicles for ride comfort, vol i, pp 168–175
7. Patole SS, Sawant PSH (2015) Ride performance analysis of half car vehicle dynamic system subjected to different road profiles with wheel base delay and nonlinear parameters. *International* J Adv Res Sci Eng 3(4):552–561. ISSN 2321-3124
8. Ferdek U, Łuczko J (2016) Vibration analysis of a half-car model with semi-active damping. J Theor Appl Mech 54:321
9. Mitra A, Benerjee N, Khalane H (1997) Simulation and analysis of full car model for various road profile on a analytically validated MATLAB/SIMULINK model. IOSR J Mech Civ Eng 22–33
10. Rajamani R (2006) Vehicle dynamics and control
11. Liang CC, Chiang CF (2006) A study on biodynamic models of seated human subjects exposed to vertical vibration. Int J Ind Ergon 36(10):869–890
12. Taha Z, Arif Hassan MH, Hasanuddin I (2015) Analytical modelling of soccer heading. Sadhana Acad Proc Eng Sci 40(5):1567–1578
13. Afiqah N, Muhammad Z, Zakaria A, Anwar KB, Majeed PPA (2020) The normal vehicle forces effects of a two in-wheel electric vehicle towards the human brain on different road profile maneuver. SN Appl Sci
14. Azizan MA, Fard M (2014) The influence of vibrations on vehicle occupant fatigue. Internoise Conf 62(2631):1–15
15. Sezgin A, Arslan YZ (2012) Analysis of the vertical vibration effects on ride comfort of vehicle driver. J Vibroeng 14(2):559–571

On the Effect of Feature Compression on Speech Emotion Recognition Across Multiple Languages

Muhammad Fahreza Alghifari, Teddy Surya Gunawan, Nik Nur Wahidah Nik Hashim, Mimi Aminah binti Wan Nordin, and Mira Kartiwi

Abstract The ability of computers to recognize emotions from the speech is commonly termed as speech emotion recognition (SER). While in recent years, many studies have been performed, the golden standard has yet to be achieved due to many parameters to consider. In this study, we investigate the effect of speech feature compression of Mel-frequency Cepstral Coefficient (MFCC) across four languages—English, French, German, and Italian. The classification was performed using a deep feedforward network. The proposed methodology has shown to have significant results when tested using a network which was trained in the same language, and up to an accuracy rate of 80.8% when trained using all four languages.

Keywords Speech emotion recognition · Average MFCC · Neural network

1 Introduction

In human-computer interaction (HCI), recognizing the user's emotion is one prospective challenge to achieve. An accurate emotion recognition system can be applied in many existing industries to improve their service, such as detecting emotional stability in driving, checking the mental health of a patient, customer service, as well as the entertainment industry.

M. F. Alghifari · T. S. Gunawan (✉) · M. A. W. Nordin
Electrical and Computer Engineering Department, International Islamic University Malaysia, Kuala Lumpur, Malaysia
e-mail: tsgunawan@iium.edu.my

T. S. Gunawan
Fakultas Teknik dan Ilmu Komputer, Universitas Potensi Utama, Medan, Indonesia

N. N. W. N. Hashim
Mechatronics Engineering Department, International Islamic University Malaysia, Kuala Lumpur, Malaysia

M. Kartiwi
Information Systems Department, International Islamic University Malaysia, Kuala Lumpur, Malaysia

A. F. Ab. Nasir et al. (eds.), *Recent Trends in Mechatronics Towards Industry 4.0*, Lecture Notes in Electrical Engineering 730, https://doi.org/10.1007/978-981-33-4597-3_63

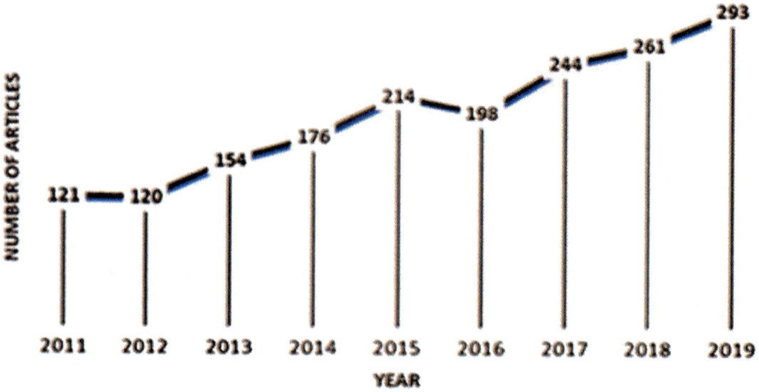

Fig. 1 Amount of search results per year from IEEE Explore related to the keyword "Speech Emotion Recognition."

There are various sources mediums for emotion recognition to be performed on, from discerning it from human facial expressions, analyzing text data and performing natural language processing, EEG, heart rate, and even body movements [1–5]. Although these methods have been extensively studied and yielded encouraging results, there are two underlying problems. Certain methods, such as EEG involve the need to use sophisticated equipment, which is not practical to implement in a daily scenario. The other problem is that some methods are intrusive, which may affect the ability of the subject to display emotions. Due to these reasons, emotion recognition using speech signals is one of the most popular methods of emotion recognition. Figure 1 displays the growth of speech emotion recognition (SER) papers.

Despite the growth and a large number of publications, there are still many areas that can be explored. Recent new researches leverage brand new machine learning models and algorithms, which can greatly improve the accuracy, or highlight a newly created dataset. However, in all cases, choosing the right speech feature which a model learns on is something that SER needs. Hence in this paper, our main contribution is to showcase the effect of speech feature compression on SER across multiple languages. Unlike other studies that combine many different types of features, we only use a single compressed feature, the Averaged Mel-frequency Cepstral Coefficients (AMFCC). Our side contribution is to provide accuracy benchmarks for other papers, particularly in language datasets that have not been visited enough. Last but not least, we attempt to answer the question of whether an SER model trained in one language can be used for another language.

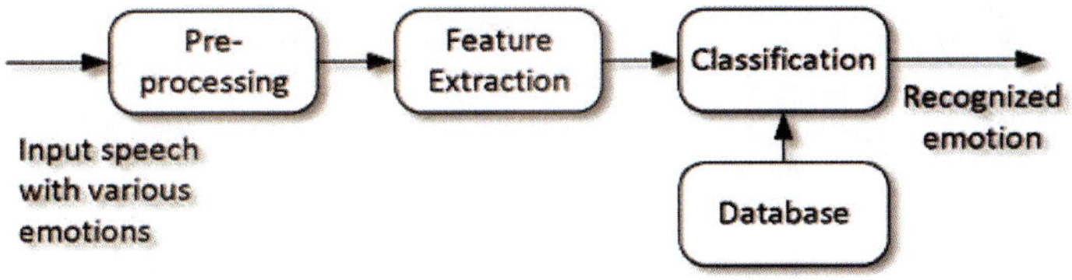

Fig. 2 General SER flow

2 Related Works

As previously mentioned, the SER has vastly been explored, and therefore the following section will highlight mostly in recent years. Figure 2 showcases the general flow of an SER system.

There are several speech features that can be extracted from a speech signal. According to [6], they can be divided into prosodic, glottal, spectral, cepstral, and TEO-based features. Although prosodic features such as pitch, intensity, and speech rate are commonly used [7], the spectral feature MFCC is one of the most popular speech features extracted, used in [8–10]. The MFCC can then be further processed to include the delta and delta of acceleration [10].

These speech features are then fed into a classifier. Although some old algorithms such as SVM are still used until today [7], advancements in computational technology and resources enabled the machine learning field to improve further and be applied in SER. With the rise of deep learning, various models have been utilized, such as 1D-CNN [8] and RNN, Gated Recurrent Units (GRU) [10]. Furthermore, even in similar model architecture, an augmented layer may improve the accuracy, such as the study by Xu et al. [9], implementing a self-attention layer and head fusion.

To train a classifier, a database is needed. The specifications of the said database can vary, from the number of emotions recorded to the language used. Until today, newly developed datasets are released, such as the Arabic dataset by Abdel-Hamid et al. [7], and the Thai dataset by Mekruksavanich et al. [8]. Finally, to classify the emotions, typically, the output will be the number of emotions that the dataset has. However, some studies opt to focus on certain emotions that have more practical applications [11]. Other researches, such as [8], grouped the emotions into positive and negative emotions and shown to have very high accuracy results.

3 Proposed Speech Emotion Recognition System

The SER is performed using Matlab R2018b on an Intel (R) Core i5-7200U CPU @ 2.50 GHz. VOICEBOX [12] is used for MFCC extraction as well as VAD. The overall methodology can be viewed in Fig. 3. Before extracting any speech features,

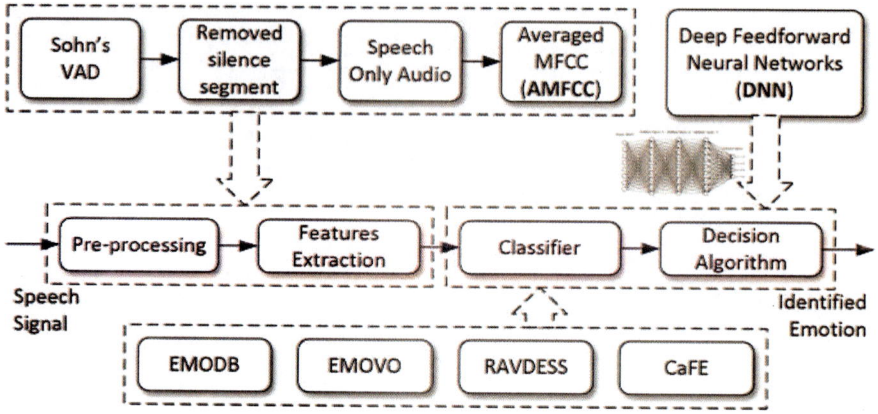

Fig. 3 Proposed SER system

we first pass the speech signals through a Voice Activity Detection (VAD) filter. VAD has shown to improve SER accuracy in our previous study [13].

The speech feature used in our system is Average MFCC (AMFCC), which can be obtained by averaging MFCC across the time domain. The feature has shown a degree of success in depression prediction [14]. Utilizing AMFCC has two benefits—it greatly reduces the amount of computation needed for the training and testing in terms of thousands for a large sampling rate voice signal. The other is the advantage to forgo padding or length discrepancies between voice samples, which is particularly important for simple classifiers that have a fixed input. In this study, we have tested the robustness of AMFCC across four languages using four datasets, as shown in Table 1.

Since we would like to showcase the effect of AMFCC, we have opted to use one classifier configuration throughout the experiments. As MFCC is suitable for

Table 1 Database specification

Dataset	Language	Emotions	Speakers	Audio samples
EMODB [16]	German	Anger, boredom, disgust, fear, happiness, sadness, neutral	10 (5 male 5 female)	535
RAVDESS [17]	English	Calm, happy, sad, angry, fearful, surprise, and disgust	24 (12 male 12 female)	7356
EMOVO [18]	Italian	Disgust, fear, anger, joy, surprise, sadness, neutral	6 (3 male 3 female)	588
CaFE [19]	French	Sadness, happiness, anger, fear, disgust, surprise, neutral	12 (6 male 6 female)	936

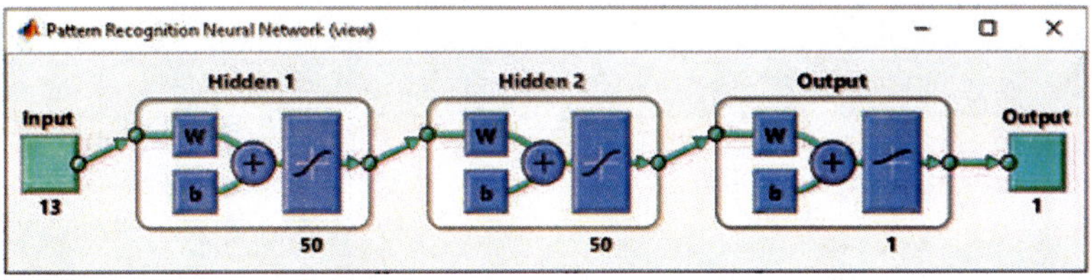

Fig. 4 Feedforward configuration

multiclass networks [15], a deep feedforward network with the configuration of [50 50] is used, as shown in Fig. 4.

The hyperbolic tangent sigmoid transfer function is used in the hidden layers, while Softmax is used in the output layer. The scaled conjugate gradient back-propagation function algorithm is used for network training. To evaluate the model performance, we adopt a simple accuracy evaluation as the distribution of emotions are balanced across the datasets. It is given by

$$Accuracy = \frac{TruePositives + TrueNegatives}{AllSamples} \tag{1}$$

For emotion recognition, we have opted to choose joy, anger, sad, fear, and neutral (JASFN). JASN was chosen for practical application in mind, such as the customer service sector, while the fear and sadness emotion is useful for depression prediction and suicide prevention.

4 Results and Discussion

The following section showcases the study results in 4 sections. The ratio used for training-validation-testing is 8:1:1. To ensure that the results are accurate, we adopt to use a balanced randomizer. The training and testing are then repeated ten times, and the average accuracy is recorded. The average standard deviation is also recorded to showcase the model accuracy stability.

4.1 Speech Emotion Recognition Using AMFCC Across Languages

In the first step of the study, we firstly test the accuracy results by passing the raw AMFCC to the classifier. The results can be shown in Table 2.

From the results, we can observe a very good recognition result for the German EMODB dataset, followed by the Italian EMOVO dataset. Figure 5 displays the

Table 2 Classification of 4 language datasets (J = joy, A = anger, S = sad, F = fear, N = neutral)

Dataset	Emotions	Train set	Validation set	Test set	Std. Dev
EMODB	JAS	0.923	0.873	0.900	0.051
	JASF	0.878	0.818	0.797	0.072
	JASFN	0.875	0.773	0.798	0.057
RAVDESS	JAS	0.785	0.728	0.728	0.085
	JASF	0.673	0.624	0.573	0.080
	JASFN	0.676	0.600	0.576	0.058
EMOVO	JAS	0.934	0.88	0.824	0.0574
	JASF	0.886	0.785	0.723	0.0848
	JASFN	0.843	0.764	0.652	0.115
CaFE	JAS	0.842	0.682	0.682	0.158
	JASF	0.635	0.576	0.438	0.095
	JASFN	0.649	0.528	0.472	0.130

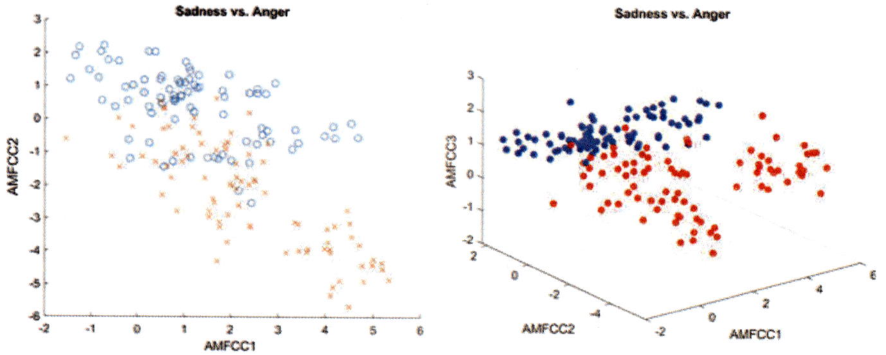

Fig. 5 2D and 3D representation of AMFCC coefficient Distribution between the sad (blue) and angry (red) emotion in the EMOVO dataset

distribution of AMFCC 1–2. We can easily observe the distinction between sad and angry coefficients.

Although generally, when the number of emotions increases, generally the recognition rate decreases, one interesting observation is how the neutral emotion is consistently more correctly classified by the classifier to the point where it boosts the overall accuracy boosts when the JASF configuration is compared to the JASFN. This demonstrates the sensitivity of AMFCC to distinguish the neutral emotion region or the overlapping of the fear coefficients with the other emotions.

4.2 *Speech Emotion Recognition Using Voice Activity Detection*

The following results are the accuracy obtained when passing the voice sample through a VAD filter (Table 3).

From the above results, we can observe a 1–5% accuracy improvement across languages, except in the CaFE French dataset. Although this could imply that using a VAD filter might not be suitable for all languages, we consider the number of samples used is not enough for a proper conclusion. The AMFCC distribution of the CaFE dataset is showcased in Fig. 6. As observed, the coefficient regions overlap more compared to the EMOVO dataset in Fig. 5.

Table 3 Classification of 4 language datasets using VAD (J = joy, A = anger, S = sad, F = fear, N = neutral)

Dataset	Emotions	Train set	Validation set	Test set	Std. Dev
EMODB	JAS	0.938	0.885	0.835	0.063
	JASF	0.893	0.812	0.803	0.081
	JASFN	0.892	0.812	0.832	0.066
RAVDESS	JAS	0.791	0.766	0.717	0.052
	JASF	0.724	0.656	0.651	0.052
	JASFN	0.666	0.616	0.577	0.050
EMOVO	JAS	0.929	0.836	0.828	0.058
	JASF	0.862	0.741	0.741	0.076
	JASFN	0.889	0.769	0.774	0.067
CaFE	JAS	0.710	0.705	0.641	0.113
	JASF	0.634	0.590	0.483	0.082
	JASFN	0.700	0.572	0.536	0.067

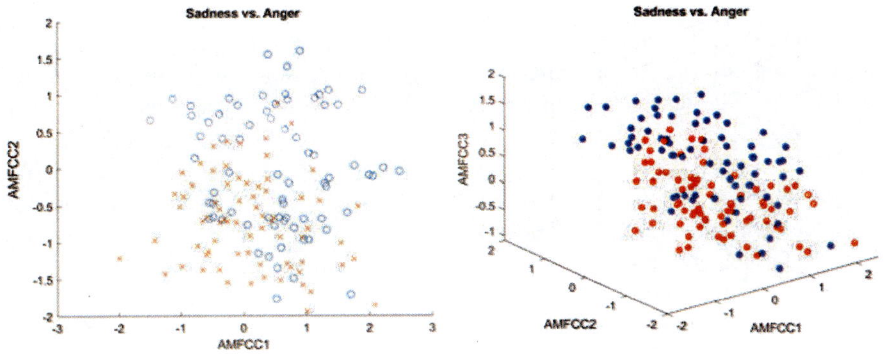

Fig. 6 2D and 3D representation of AMFCC coefficient distribution between the sad (blue) and angry (red) emotion in the CaFE dataset

Table 4 Cross-language SER recognition results

		Network			
		Ger	Eng	Ita	Fre
Input	Ger	0.981	0.345	0.409	0.403
	Eng	0.446	0.899	0.421	0.477
	Ita	0.562	0.476	0.944	0.454
	Fre	0.196	0.446	0.444	0.926

4.3 *Cross-Language Speech Emotion Recognition*

From the models created in Sect. 4.2, we selected the best performing JAS networks in each language and tested the accuracy results when fed with another language. The results are recorded in Table 4.

From the results obtained, it can be concluded that although it can perform very well when trained and tested using its own language, the current methodology performs poorly when tested with another set. The German language, in particular, although it boosts the best accuracy with self-trained, it performed the worst when fed in other languages, none above the 45% accuracy in a 3-class prediction system. The German network, however, has shown to have a recognition rate of 56% when tested with Italian, a significant correlation that may be explored further in future research.

4.4 *Multilingual Speech Emotion Recognition Model*

The final experiment conducted is building a multilingual network. All 4 datasets are combined and used as training and testing data using the same configuration as the previous steps, with a train—val—test ratio of 8-1-1. The results are displayed in Fig. 7.

From the results obtained, we achieved multilingual SER with a combined accuracy rate of 80.8%. The training and testing result itself was 82.8% and 73.1%, respectively. The distribution of Joy-Anger-Sad AMFCC coefficients is plotted in Fig. 7. We can observe that although the Joy (green) coefficients are spread across the 3D-plane, the Anger (red) and Sad (blue) regions are considered distinguishable.

Although our results are not as good as [20], which is 93%, their system only considers two languages, German and Polish, while our system considers German, English, French and Italian. One thing to consider is that the datasets were recorded with different environments, which may affect the levels of noise introduced to the sample. In our system, achieving 80% is acceptable if it can accommodate this variance. Another thing to note is the usage of specific datasets. As far as we know, the only paper which cites the French dataset was the study by Jia et al. [21]. As the

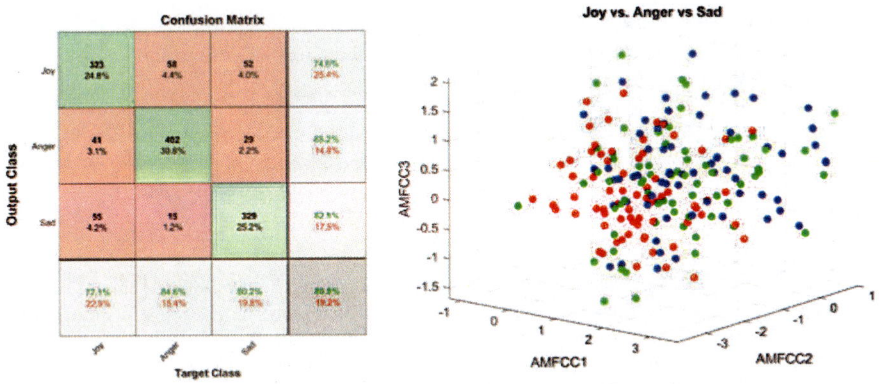

Fig. 7 Confusion matrix and 3D representation of AMFCC coefficient distribution between joy (green), anger (red), sad (blue) and emotions in the multilingual network

evaluation parameter differs, we cannot directly compare the system performance. However, we hope that our study can be used for the benchmark for future papers.

5 Conclusions and Future Works

This study showcases the robustness of AMFCC in speech emotion recognition across languages. Despite the very high compression in information, AMFCC has shown to be able to recognize emotions when trained and tested using their own dataset. Unfortunately, the feature shows to have performed rather poorly when tested across language datasets, which encourages the idea that multiple datasets are needed in order to build a multilingual SER system using AMFCC. The final network that built has an accuracy rate of 80.8%.

The authors would like to acknowledge the following limitations. The model trained and tested in the study were from acted emotion datasets, which may not reflect the emotions in the life which may not be as expression or clean as datasets collected in a quiet environment. The next is the choice of [50 50] for training the number was heuristically chosen so that it can be considered a deep neural network (more than a single layer) but not extensive to the point of overrepresenting features, balancing between underfitting and overfitting. It is possible for the study to investigate the optimum number of neurons for each language; however, it will generate too many variations to consider and will pull the attention of the paper away from the original speech compression effect.

For future researches, more speech features can be considered alongside AMFCC to improve the SER accuracy rate. More The CaFE French dataset also has the potential to be explored further, to which we hope that our paper can be a benchmark for future studies.

Acknowledgements The authors would like to express their gratitude to the Malaysian Ministry of Education (MOE), which has provided research funding through the Fundamental Research Grant, FRGS19-076-0684 (FRGS/1/2018/ICT02/UIAM/02/4).

References

1. Pranav E et al (2020) Facial emotion recognition using deep convolutional neural network. In: 2020 6th international conference on advanced computing and communication systems (ICACCS)
2. Park S, Bae B, Cheong Y (2020) Emotion recognition from text stories using an emotion embedding model. In: 2020 IEEE international conference on big data and smart computing (BigComp)
3. Hwang S et al (2020) Subject-independent EEG-based emotion recognition using adversarial learning. In: 2020 8th international winter conference on brain-computer interface (BCI)
4. Du G, Long S, Yuan H (2020) Non-contact emotion recognition combining heart rate and facial expression for interactive gaming environments. IEEE Access 8:11896–11906
5. Ahmed F, Bari ASMH, Gavrilova ML (2020) Emotion recognition from body movement. IEEE Access 8:11761–11781
6. El Ayadi M, Kamel MS, Karray F (2011) Survey on speech emotion recognition: features, classification schemes, and databases. Pattern Recogn 44(3):572–587
7. Abdel-Hamid L, Shaker NH, Emara I (2020) Analysis of linguistic and prosodic features of bilingual Arabic–English speakers for speech emotion recognition. IEEE Access 8:72957–72970
8. Mekruksavanich S, Jitpattanakul A, Hnoohom N (2020) Negative emotion recognition using deep learning for Thai language. In: 2020 joint international conference on digital arts, media and technology with ECTI northern section conference on electrical, electronics, computer and telecommunications engineering (ECTI DAMT & NCON)
9. Xu M, Zhang F, Khan SU (2020) Improve accuracy of speech emotion recognition with attention head fusion. In: 2020 10th annual computing and communication workshop and conference (CCWC)
10. Koo H et al (2020) Development of speech emotion recognition algorithm using MFCC and prosody. In: 2020 international conference on electronics, information, and communication (ICEIC)
11. Alghifari MF, Gunawan TS, Kartiwi M (2018) Speech emotion recognition using deep feedforward neural network. Indones J Electr Eng Comput Sci 10(2)
12. Brookes D (2010) VOICEBOX: a speech processing toolbox for MATLAB. Available from: http://www.ee.ic.ac.uk/hp/staff/dmb/voicebox/voicebox.html. Cited 14 Feb 2019
13. Alghifari MF et al (2019) On the use of voice activity detection in speech emotion recognition. Bull Electr Eng Inform 8(4)
14. Alghifari MF et al (2019) On the optimum speech segment length for depression detection. In: 2019 IEEE international conference on smart instrumentation, measurement and application (ICSIMA)
15. Gunawan TS et al (2018) A review on emotion recognition algorithms using speech analysis. Indones J Electr Eng Inform (IJEEI) 6(1):12–20
16. Burkhardt F et al (2005) A database of German emotional speech. In: Ninth European conference on speech communication and technology
17. Livingstone SR, Russo FA (2018) The Ryerson Audio-Visual Database of Emotional Speech and Song (RAVDESS): a dynamic, multimodal set of facial and vocal expressions in North American English. PLoS ONE 13(5)

18. Costantini G et al (2014) EMOVO corpus: an Italian emotional speech database. In: International conference on language resources and evaluation (LREC 2014). European Language Resources Association (ELRA)
19. Gournay P, Lahaie O, Lefebvre R (2018) A Canadian French emotional speech dataset. In: Proceedings of the 9th ACM multimedia systems conference
20. Absa AHA, Deriche M, Mohandes M (2018) A bilingual emotion recognition system using deep learning neural networks. In: 2018 15th international multi-conference on systems, signals & devices (SSD)
21. Jia X et al (2019) ET-GAN: cross-language emotion transfer based on cycle-consistent generative adversarial networks. arXiv preprint arXiv:1905.11173

Real-Time Power Quality Disturbance Classification Using Convolutional Neural Networks

Budi Yanto Husodo, Kalamullah Ramli, Eko Ihsanto, and Teddy Surya Gunawan

Abstract There is a growing interest in disturbance monitoring to maintain power quality. This paper developed a real-time power quality disturbance (PQD) detection system using convolutional neural networks (CNN) due to its fast and accurate feature extraction and classification. First, 29 classes of power quality disturbance were synthetically generated around 5000 samples for each type. Second, an efficient CNN structure was developed to extract unique features. Next, the output of CNNs was then inputted into a fully connected layer with a softmax and classification layer to act as the classifier for 29 classes of PQD signals. Our proposed algorithm was then trained using 80% of the synthetic signals, while 20% of the synthetic signals were used for testing. Experimental results showed that the proposed algorithm produced a good result with the classification accuracy of 97.52% trained using 100 epochs. Furthermore, it requires only 80.96 μs to detect each 16 ms segment of PQD signals.

Keywords Power quality disturbance · Recurrent neural network · Bidirectional long short-term memory · Time-frequency based feature extraction · Classification

1 Introduction

Nowadays, the development of electric power systems dominates by smart grids design, where distributed power systems such as photovoltaics and wind power

B. Y. Husodo · K. Ramli
Electrical Engineering Department, Universitas Indonesia, Depok, Indonesia
e-mail: budi_yanto@mercubuana.ac.id

B. Y. Husodo · E. Ihsanto
Electrical Engineering Department, Universitas Mercu Buana, West Jakarta, Indonesia

T. S. Gunawan (✉)
Electrical and Computer Engineering Department, International Islamic University Malaysia, Kuala Lumpur, Malaysia
e-mail: tsgunawan@iium.edu.my

School of Electrical Engineering & Telecommunications, UNSW, Sydney, Australia

A. F. Ab. Nasir et al. (eds.), *Recent Trends in Mechatronics Towards Industry 4.0*, Lecture Notes in Electrical Engineering 730, https://doi.org/10.1007/978-981-33-4597-3_64

plants, which generate electric power intermittently, are joined into the grid. Moreover, various non-linear loads due to the rapid development of power electronics were causing power quality issues, such as distortion of voltage, current, and frequency. In the end, it will severely affect the quality of the existing power system [1, 2]. Accurate positioning and identification of power quality disturbances is the premise of power quality analysis and governance.

Power quality disturbance (PQD) raises severe problems in various parts of the electric power system. Voltage sag, which is usually due to electric power system faults leading to small variations of induction motor speed and reduction of capacitor bank output. In contrast, voltage swell may cause severe damage to voltage-sensitive appliances such as computers and electronic gadgets or failures of adjustable speed drives and automatic controllers [3]. Variation of system frequency as the consequence of intensive integration of intermittent renewable energy-based power generations may cause problems such as time deviation of clocks, a variation of motor speed, and risk of under frequency tripping of protective devices [2]. Thus, PQD recognition and evaluation becomes very urgent to ensure safe operation as well as standard power quality.

The classification study of power quality disturbance (PQD) is divided into three stages, pre-processing, feature extraction, and classifier design [4, 5], as shown in Fig. 1. In general, the PQD signals are nonstationary. Most of the PQDs are a combination of two or more disturbances, which result in unique features. Common feature extraction methods including Kalman filter (KF) [6], short-time Fourier transform (STFT) [7], discrete wavelet transform (DWT) [5], Hilbert-Huang transform (HHT) [8], and Stockwell transform (ST) [4]. The next step after feature extraction is a classifier. Common disturbance classification methods including Support Vector Machine [4], extreme learning machine (ELM) [5, 8], decision tree (DT) [7, 9], Deep Convolutional Neural Network [10], and Long Short-Term Memory (LSTM) [11]. However, as evident in ECG signal processing [12, 13], CNN could be used for automatic feature extraction as well as for classifiers. By applying the same design principles, we will apply the CNN architecture for PQD signal classification.

Many types of research have been conducted using various features and classifiers for PQD classification. However, the classification accuracy could still be improved for effective and efficient electrical power disturbance mitigation [14]. Therefore, the objective of this paper is to develop a fast and accurate PQD classification using CNN architecture. First, 29 PQD synthetic signals will be generated based on IEEE-1159 standard. Next, CNN architecture, which combines feature extraction and classifier,

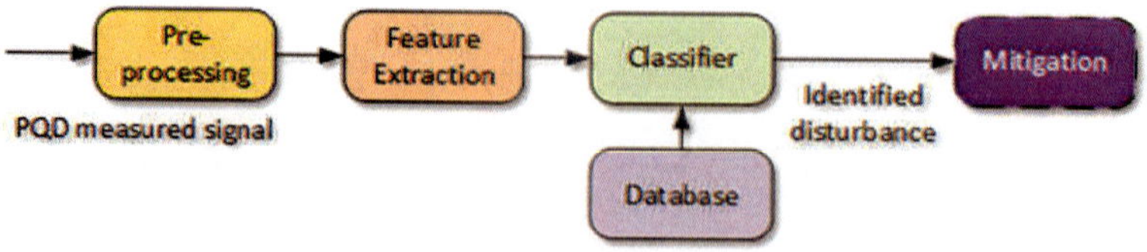

Fig. 1 Basic power quality disturbance classification and mitigation

will be developed. The proposed deep learning architectures will be trained and tested for its training time and accuracy.

2 Power Quality Disturbance

Power quality disturbance is known as any sudden change in the voltage, as well as current and frequency operation, which may cause malfunction or failure of electrical appliances [14]. Twenty-nine PQD signals, as shown in Table 1, will be generated based on the parametric equations [15]. The parameter variation of the PQD equation conforms to the parametric equation of IEEE Std 1159-2019 [16].

We generated the PQD synthetic signal using the following parameters, including amplitude 1 pu, duration $t = 0.16$ s, total period $T = 8$, total sampling point 1024, and sampling frequency 6.4 kHz. Each type of PQD simulation generates 5000 signals to give a total of 145,000 signals. A simulation sample of power quality disturbance signals is shown in Fig. 2 for class 1, 7, 22, 23, and 28.

3 CNN Architecture and Experimental Setting

3.1 CNN Architecture for PQD Classification

The core of machine learning is establishing a network with specific weights and deviations trained using input data and labeled output to classify a new input. Figure 3 shows our proposed algorithm adopted from [12, 13], which combine feature extraction and classifier. Besides, Table 2 shows the CNN model in more detail.

From Table 2, it can be derived that the total parameters were 43,965 with 43,709 and 256 for the trainable and non-trainable parameters, respectively. It can be concluded that as described in [12, 13], the CNN model is efficient, fast, and accurate as it combines the feature extraction and classification. Classification accuracy, which reflects the performance of our proposed algorithms, is summarized in Eq. 1 [12].

$$Acc = \frac{TP + TN}{TP + TN + FP + FN} \quad (1)$$

where (true positive) is the estimated PQD type that is correctly classified, (true negative) is another PQD type that is correctly identified as the corresponding PQD, (false positive) is another PQD that is classified incorrectly, and (false negative) is the estimated PQD type that is classified incorrectly.

Table 1 Power quality disturbance classes adopted from [15]

Class	Disturbances
1	Pure sinusoidal
2	Sag
3	Swell
4	Interruption
5	Transient/impulse/spike
6	Oscillatory transient
7	Harmonics
8	Harmonics with sag
9	Harmonics with swell
10	Flicker
11	Flicker with sag
12	Flicker with swell
13	Sag with oscillatory transient
14	Swell with oscillatory transient
15	Sag with harmonics
16	Swell with harmonics
17	Notch
18	Harmonics with sag with flicker
19	Harmonics with swell with flicker
20	Sag with harmonics with flicker
21	Swell with harmonics with flicker
22	Sag with harmonics with oscillatory transient
23	Swell with harmonics with oscillatory transient
24	Harmonics with sag with oscillatory transient
25	Harmonics with swell with oscillatory transient
26	Harmonics with sag with flicker with oscillatory transient
27	Harmonics with swell with flicker with oscillatory transient
28	Sag with harmonics with flicker with oscillatory transient
29	Swell with harmonics with flicker with oscillatory transient

3.2 *Experimental Setup*

The proposed algorithm has been implemented in Python with Tensorflow [17] and Keras [18] libraries. The experiments are performed on a computer with Intel Core i7-7700 CPU with eight logical processors, memory of 8 GBytes, graphic card Nvidia

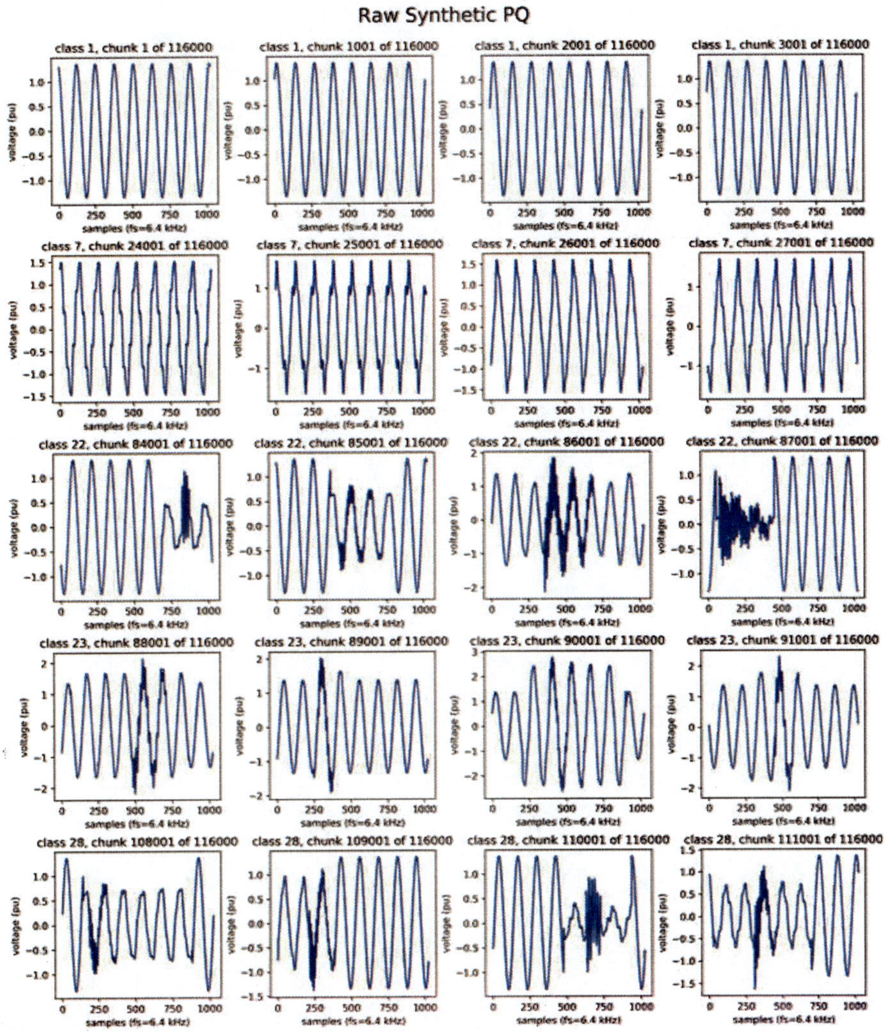

Fig. 2 Samples of power quality disturbance signals for class 1, 7, 22, 23, and 28 (each row represent four samples taken from 116,000 samples)

GeForce GTX 1060 6 GB DDR5 (1280 CUDA cores), using Microsoft Windows 10 64 bits operating system.

Adopted from [5], the PQD signal specifications are amplitude 1pu, duration $t = 0.16$ s, total period $T = 8$, total sampling point 1024 and sampling frequency 6.4 kHz. The PQD signal was synthetically generated as described in [15]. Each type of PQD simulation generates 5000 signals to give a total of 145,000 signals. Out of 5000 signals, 80% will be used for training, and 20% will be used for testing.

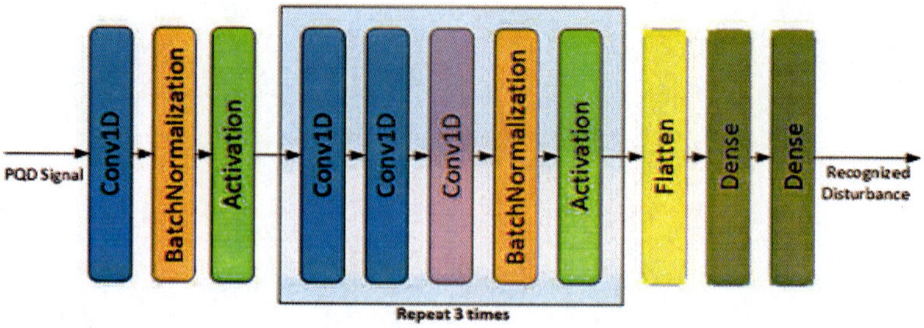

Fig. 3 Proposed CNN Architectures for PQD classification

Table 2 Proposed CNN models in details

Layer	Features shape	Param #
input_1	(1024, 1)	0
conv1d_1	(512, 32)	192
batch_normalization_1	(512, 32)	128
activation_1	(512, 32)	0
conv1d_2	(512, 32)	5152
conv1d_3	(512, 32)	1056
conv1d_4	(128, 32)	5152
batch_normalization_2	(128, 32)	128
activation_2	(128, 32)	0
conv1d_5	(128, 32)	5152
conv1d_6	(128, 32)	1056
conv1d_7	(32, 32)	5152
batch_normalization_3	(32, 32)	128
activation_3	(32, 32)	0
conv1d_8	(32, 32)	5152
conv1d_9	(32, 32)	1056
conv1d_10	(8, 32)	5152
batch_normalization_4	(8, 32)	128
activation_4	(8, 32)	0
flatten_1	(256)	0
dense_1	(32)	8224
dense_2	(29)	957

4 Experimental Results and Analysis

This section described the results obtained from various experiments on Deep CNN architectures with training time and accuracy measurements.

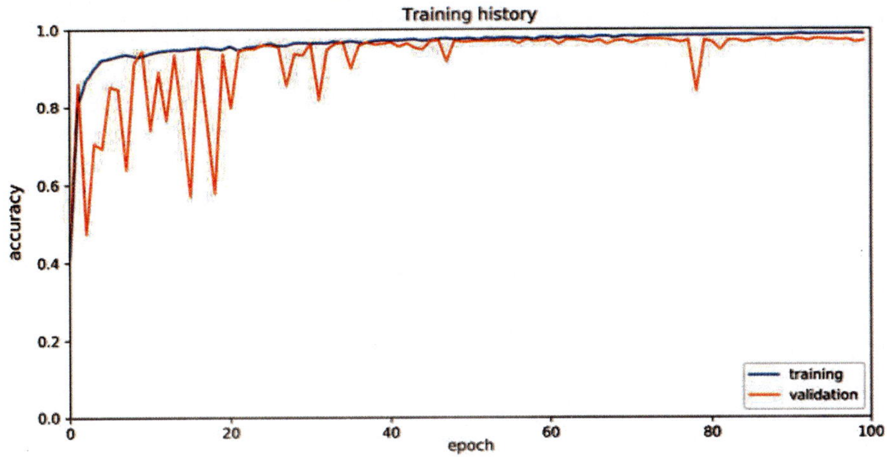

Fig. 4 Training history of 100 epoch iteration of the CNN model described in Table 2

4.1 Training Process

Figure 4 shows the training history for 100 epoch using the CNN model described in Table 2 and Fig. 3. The recorded training time for 100 epoch was 2582 s. Figure 4 shows the difference between the training and the testing accuracy, which is only about 1%. The orange curve in Fig. 4 also shows that there are overfitting happens, which could be due to the difficulty of recognizing complex PQD signals. However, this occurred only at the beginning of the testing process. After 50 epochs, the accuracy was relatively stable at around 97.52%. Meanwhile, the training accuracy was recorded to be 98.74%.

4.2 Performance Evaluation

The recorded testing time (executing only the trained parameters) was 2322 ms for testing 29,000 segments (1000 for every 29 classes). Therefore, our proposed system could be implemented in a real-time application. The time to detect the disturbance is only 80.96 μs. The testing accuracy for each class of the PQD signals is given in Table 3.

As is shown in Table 3, the average accuracy for all 29 classes was 97.52%. Only two classes had accuracy lower than 90%, i.e., class 22 and 23 (highlighted in bold in Table 3). Without the two classes, the average accuracy was 98.36%. Further improvement to the CNN architecture could be performed by adding more layers or different deep learning types, like Bidirectional Long Short-Term Memory (BiLSTM) or autoencoder.

Table 3 Testing the accuracy of various disturbance signals

Disturbance classes	Accuracy (%)
1—Pure sinusoidal	100.00
2—Sag	100.30
3—Swell	100.00
4—Interruption	99.80
5—Transient/impulse/spike	100.00
6—Oscillatory transient	99.97
7—Harmonics	100.00
8—Harmonics with sag	100.00
9—Harmonics with swell	100.00
10—Flicker	100.00
11—Flicker with sag	99.90
12—Flicker with Swell	100.00
13—Sag with oscillatory transient	96.40
14—Swell with oscillatory transient	97.70
15—Sag with harmonics	96.30
16—Swell with harmonics	95.60
17—Notch	100.00
18—Harmonics with sag with flicker	100.00
19—Harmonics with swell with flicker	100.00
20—Sag with harmonics with flicker	95.20
21—Swell with harmonics with flicker	93.80
22—Sag with harmonics with oscillatory transient	85.50
23—Swell with Harmonics With Oscillatory transient	86.90
24—Harmonics with sag with oscillatory transient	98.90
25—Harmonics with swell with oscillatory transient	99.70
26—Harmonics with sag with flicker with oscillatory transient	100.00
27—Harmonics with swell with flicker with oscillatory transient	99.50
28—Sag with harmonics with flicker with oscillatory transient	90.70
29—Swell with harmonics with flicker with oscillatory transient	91.40
Average	97.52%

5 Conclusions

We have presented our proposed algorithm to identify complex power quality disturbances using deep CNN. First, 29 classes of power quality disturbance were synthetically generated around 5000 samples for each type. Second, an efficient CNN structure was developed to extract the unique features and to classify 29 classes of PQD signals. Our proposed algorithm was then trained using 80% of the synthetic signals, while 20% of the synthetic signals were used for testing. Experimental results showed that the proposed algorithm produced a good result with the classification accuracy of 97.52%, trained using 100 epochs. Furthermore, to detect the electrical disturbance, our algorithm requires only 80.96 μs providing a way for real-time implementation. Further research is including adding a different type of deep neural networks, signal denoising using an autoencoder, evaluation using real PQD signals, and mitigation of electrical disturbance.

Acknowledgements The authors would like to express their gratitude to the Malaysian Ministry of Education (MOE), which has provided research funding through the Fundamental Research Grant, FRGS19-076-0684 (FRGS/1/2018/ICT02/UIAM/02/4). The authors would like to acknowledge as well, support from International Islamic University, University of New South Wales, Universitas Indonesia, and Universitas Mercu Buana.

References

1. Luo L et al (2017) Design and application of power quality monitoring system for the smart substation based on IEC 61850. CIRED-Open Access Proc J 2017(1):577–580
2. Bollen MH, Gu IY (2006) Signal processing of power quality disturbances, vol 30. Wiley, New Jersey
3. Mishra M (2019) Power quality disturbance detection and classification using signal processing and soft computing techniques: a comprehensive review. Int Trans Electr Energy Syst 29(8):e12008
4. Li J et al (2016) Detection and classification of power quality disturbances using double resolution S-transform and DAG-SVMs. IEEE Trans Instrum Meas 65(10):2302–2312
5. Wang J, Xu Z, Che Y (2019) Power quality disturbance classification based on DWT and multilayer perceptron extreme learning machine. Appl Sci 9(11):2315
6. Xi Y et al (2018) Detection of power quality disturbances using an adaptive process noise covariance Kalman filter. Digit Signal Process 76:34–49
7. Singh U, Singh SN (2017) Application of fractional Fourier transform for classification of power quality disturbances. IET Sci Meas Technol 11(1):67–76
8. Sahani M, Dash PK (2018) Automatic power quality events recognition based on Hilbert Huang transform and weighted bidirectional extreme learning machine. IEEE Trans Ind Inform 14(9):3849–3858
9. Zhong T et al (2019) Power quality disturbance recognition based on multiresolution S-transform and decision tree. IEEE Access 7:88380–88392
10. Wang S, Chen H (2019) A novel deep learning method for the classification of power quality disturbances using deep convolutional neural network. Appl Energy 235:1126–1140
11. Junior WLR et al (2019) Classification of power quality disturbances using convolutional network and long short-term memory network. In: 2019 international joint conference on neural networks (IJCNN). IEEE

12. Ihsanto E et al (2020) An efficient algorithm for cardiac arrhythmia classification using ensemble of depthwise separable convolutional neural networks. Appl Sci 10(2):483
13. Ihsanto E et al (2020) Fast and accurate algorithm for ECG authentication using residual depthwise separable convolutional neural networks. Appl Sci 10(9):3304
14. Moreno-Muñoz A (2007) Power quality: mitigation technologies in a distributed environment. Springer, London
15. Igual R et al (2018) Integral mathematical model of power quality disturbances. In: 2018 18th international conference on harmonics and quality of power (ICHQP). IEEE
16. IEEE (2019) IEEE recommended practice for monitoring electric power quality. IEEE Std 1159-2019. IEEE Power and Energy Society
17. Abadi M et al (2016) Tensorflow: a system for large-scale machine learning. In: 12th USENIX symposium on operating systems design and implementation (OSDI 16)
18. Chollet F (2015) Keras: deep learning library for Theano and TensorFlow, vol 7(8), p T1. https://keras.io/k

Power Quality Disturbance Classification Using Deep BiLSTM Architectures with Exponentially Decayed Number of Nodes in the Hidden Layers

Teddy Surya Gunawan, Budi Yanto Husodo, Eko Ihsanto, and Kalamullah Ramli

Abstract In recent years, there is growing interest in automatic power quality disturbance (PQD) classification using deep learning algorithms. In this paper, the average of instantaneous frequency and the average of spectrum entropy were used as time-frequency based feature extraction due to its discriminatory nature. Bidirectional Long Short-Term Memory (BiLSTM) architectures with exponentially decayed number of nodes in deep multilayers were utilized as Deep Recurrent Neural Network (DRNN) classifier. We experimentally generated fifteen classes of synthetic PQD signals. Each class contains 1000 samples divided randomly into training, validation, and testing. Results showed that four hidden layers of BiLSTM with exponentially decayed nodes interleaved with dropout layers provided the best classification accuracy of 99.23%.

Keywords Power quality disturbance · Recurrent neural network · Bidirectional long short-term memory · Time-frequency based feature extraction · Classification

1 Introduction

In the past few decades, renewable energy gaining much interest in sustainable energy generation and distribution. Many power electronic converters have been utilized in residential, commercial, and industrial sectors. It will create various unwanted power quality disturbances (PQD) [1] when various power electronic converters

T. S. Gunawan (✉)
Electrical and Computer Engineering Department, International Islamic University Malaysia, Kuala Lumpur, Malaysia
e-mail: tsgunawan@iium.edu.my

School of Electrical Engineering & Telecommunications, UNSW, Sydney, Australia

B. Y. Husodo · K. Ramli
Electrical Engineering Department, Universitas Indonesia, Depok, Indonesia
e-mail: budi_yanto@mercubuana.ac.id

B. Y. Husodo · E. Ihsanto
Electrical Engineering Department, Universitas Mercu Buana, West Jakarta, Indonesia

A. F. Ab. Nasir et al. (eds.), *Recent Trends in Mechatronics Towards Industry 4.0*, Lecture Notes in Electrical Engineering 730, https://doi.org/10.1007/978-981-33-4597-3_65

are connected to the primary power grid. The low quality of power created various problems, such as increased power loss, less efficient distribution system, and equipment damage. The primary task to mitigate PQD is identification and disturbance parameter estimation. PQD classification can be divided into at least three stages, pre-processing, feature extraction, and classifier [2, 3], as shown in Fig. 1.

In general, the PQD signals are nonstationary, and most of the disturbances in the power systems are a combination of several disturbances, which require unique features. Common feature extraction methods including Kalman filter (KF) [4], short-time Fourier transform (STFT) [5], discrete wavelet transform (DWT) [3], Hilbert-Huang transform (HHT) [6], and Stockwell transform (ST) [2]. Instantaneous frequency is one of the powerful signal analysis to understand the detailed mechanisms for nonlinear and nonstationary processes [7], like PQD signals. On the other hand, spectral entropy has been successfully utilized to analyze biomedical signals [8] as it can quantify the spectral complexity of an uncertain system.

As shown in Fig. 1, the next step after feature extraction is a classifier. Common disturbance classification methods including Support Vector Machine [2], extreme learning machine (ELM) [3, 6], decision tree (DT) [5, 9], Deep Convolutional Neural Network [10], and Long Short-Term Memory (LSTM) [11]. To the best of our knowledge, bidirectional LSTM (BiLSTM) has not been utilized for PQD classification. A BiLSTM learns long-term bidirectional dependencies between time steps of time series.

Many types of research have been conducted using various features and classifiers for PQD classification. However, the classification accuracy could still be improved for effective and efficient electrical power disturbance mitigation [12]. First, fifteen PQD synthetic signals will be generated based on IEEE-1159 standard. Next, time-frequency based features extraction will be used, including the average of instantaneous frequency and the average of spectral entropy. Next, multilayers BiLSTM will be used as the classifier. The proposed deep learning architectures will be trained, validated, and tested for its training time and accuracies.

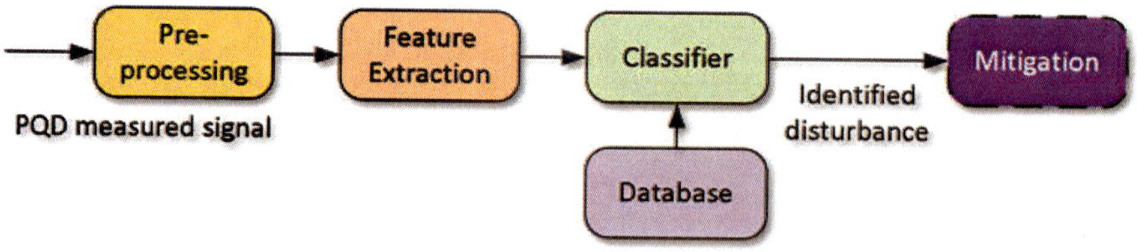

Fig. 1 Basic power quality disturbance classification and mitigation

2 Electric Power Disturbance

Power quality disturbance may be defined as any sudden change in the voltage, current, frequency operation that causes a malfunctioning or electrical appliance failure [12]. Fifteen PQD signals will be generated based on the parametric equations [13], as shown in Table 1. The PQD synthetic equations confirm to the parametric equation of IEEE Std 1159-2019 [14].

Table 1 Power quality disturbance parametric equations adopted from [13]

No.	Disturbances
1	**Pure sinusoid**: $v(t) = A\sin(\omega t - \varphi)$
2	**Sag**: $v(t) = A(1 - \alpha(u(t - t_1) - u(t - t_2)))\sin(\omega t - \varphi)$
3	**Swell**: $v(t) = A(1 + \beta(u(t - t_1) - u(t - t_2)))\sin(\omega t - \varphi)$
4	**Interrupt**: $v(t) = A(1 - \rho(u(t - t_1) - u(t - t_2)))\sin(\omega t - \varphi)$
5	**Harmonics**: $v(t) = A\left(\sin(\omega t - \varphi) + \sum_{n=3}^{7} \alpha_n \sin(n\omega t - \vartheta)\right)$
6	**Oscillatory transient**: $v(t) = A\left(\sin(\omega t - \varphi) + \beta e^{-\frac{t-t_1}{\tau}} \sin(\omega_n(t - t_1) - \varphi)(u(t - t_1) - u(t - t_2)))\right)$
7	**Flicker**: $v(t) = A\left(1 + \lambda\sin\left(\omega_f t\right)\right)\sin(\omega_t - \varphi)$
8	**Notch**: $v(t) = A\left(\sin(\omega t - \varphi) - \sin\left(\sin(\omega t - \varphi)\sum_{n=0}^{Nc-1} k(u(t - (t_c + sn)) - u(t - (t_d + sn)))\right)\right)$
9	**Spikes**: $v(t) = A\left(\sin(\omega t - \varphi) - \psi\left(e^{-750(t-t_a)} - e^{-344(t-t_a)}\right)(u(t - t_a) - u(t - t_b))\right)$
10	**Sag with harmonics**: $v(t) = A\left(\sin(\omega t - \vartheta_1) + (-\alpha(u(t - t_1) - u(t - t_2)))\sum_{n=1}^{5} \alpha_n \sin(n\omega t - \vartheta_n)\right)$
11	**Swell with harmonics**: $v(t) = A\left(sin(\omega t - \vartheta_1) + (\beta(u(t - t_1) - u(t - t_2)))\sum_{n=1}^{5} \alpha_n \sin(n\omega t - \vartheta_n)\right)$
12	**Interruption with harmonics**: $v(t) = A(1 - \rho(u(t - t_1) - u(t - t_2)))\sin(\omega t - \varphi) + \sum_{n=1}^{5} \alpha_n \sin(n\omega t - \vartheta_n)$
13	**Harmonics with flicker**: $v(t) = A\left(\sum_{n=3}^{7} \alpha_n \sin(n\omega t - \vartheta)\right)\left(1 + \lambda\sin\left(\omega_f t\right)\right)$
14	**Flicker with sag**: $v(t) = A\left(1 + \lambda sin\left(\omega_f t\right) - \alpha(u(t - t_1) - u(t - t_2))\right)\sin(\omega t - \varphi)$
15	**Flicker with swell**: $v(t) = A\left(1 + \lambda sin\left(\omega_f t\right) + \beta(u(t - t_1) - u(t - t_2))\right)\sin(\omega t - \varphi)$

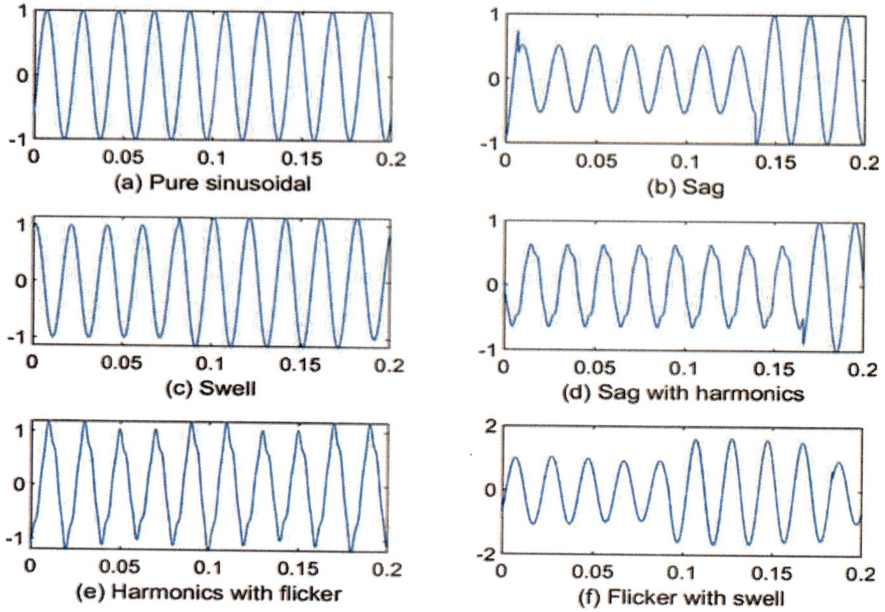

Fig. 2 Samples of power quality disturbance signals

To simplify PQD synthetic signal generation, we fixed the amplitude to be 1pu, duration $t = 0.2$ s, total period $T = 10$, total sampling point 1280, and sampling frequency 6.4 kHz. For each type of PQD signal, we generate a randomized 1000 signals producing a total of 15,000 synthetic signals. A simulation sample of power quality disturbance signals is shown in Fig. 2.

3 Deep BiLSTM Architectures for PQD Classification

The primary function of machine learning is to train and optimize a proposed network architecture with specific weights and biases. The training process was using input data and labeled output to classify a new input. Figure 3 shows our proposed algorithm, in which subsequent sections will elaborate each part in more detail.

3.1 Pre-processing and Time-Frequency Based Feature Extraction

As shown in Fig. 3, the first step is pre-processing, which includes segmentation and noise reduction (if required). In our case, we will segment the signal every 0.2 s (1280

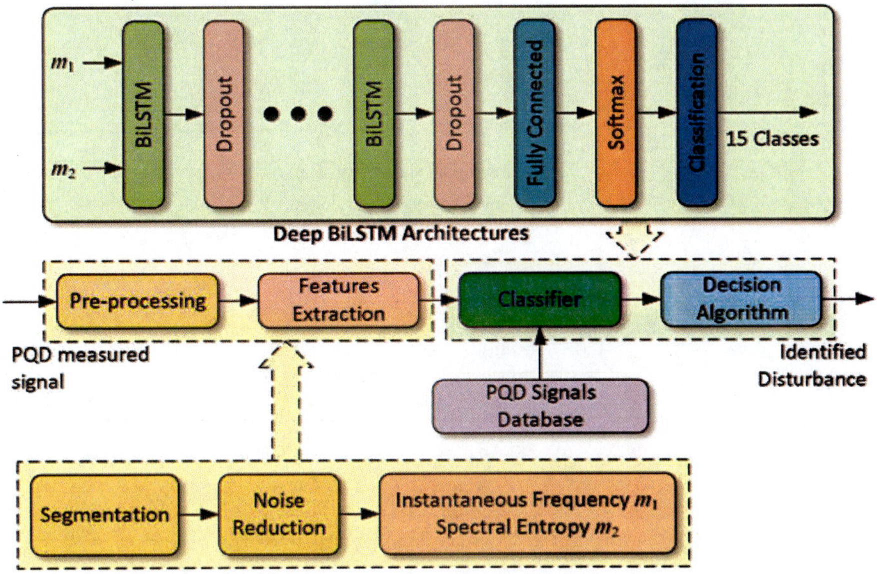

Fig. 3 Proposed deep BiLSTM architectures for PQD classification

samples for a sampling frequency of 6400 Hz). For noise reduction, Wiener filter, wavelet denoising, or other noise reduction techniques could be used to clean a noisy signal. Next, two time-frequency based features will be extracted from the time-series signal, i.e., the average of instantaneous frequency and the average of spectral entropy. Both instantaneous frequency and spectral entropy were selected due to their ability to extract unique information [7, 8]. If $S(t, f)$ is the power spectrum of input signal $x(t)$, then the instantaneous frequency, $f_{inst}(t)$, can be calculated as follows:

$$f_{inst}(t) = \frac{\int_0^\infty f P(t, f)}{\int_0^\infty P(t, f)} \tag{1}$$

The spectral entropy of a signal is a measure of its spectral power distribution. For a signal $x(n)$, the power spectrum is $S(m) = |X(m)|^2$, where $X(m)$ is the discrete Fourier transform of $x(n)$. To compute the instantaneous spectral entropy given a time-frequency power spectrogram $S(t, f)$, the probability distribution at time t is:

$$P(t, m) = \frac{S(t, m)}{\sum_f S(t, f)}. \tag{2}$$

Then, the spectral entropy at time t is:

$$H(t) = -\sum_{m=1}^{N} P(t.m) \log_2 P(t, m). \tag{3}$$

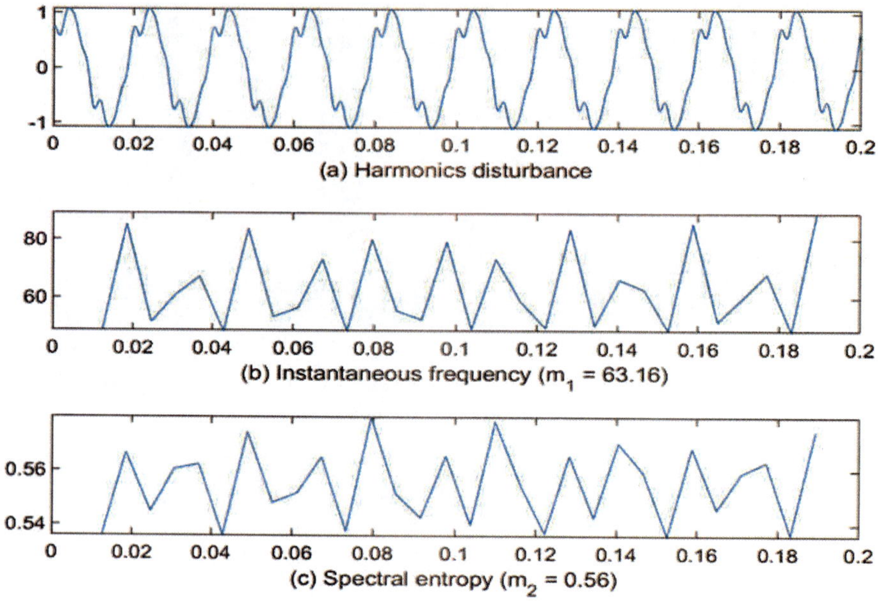

Fig. 4 Samples of instantaneous frequency and spectral entropy for harmonics disturbance

Figure 4 shows the sample of harmonics disturbance signal, instantaneous frequency, and spectral entropy. The instantaneous frequency and spectral entropy produced 30 bins across the time when the original input signal was 1280 samples with 6400 Hz sampling frequency. To reduce the computational load, two average feature vectors were derived for each signal as follows:

$$\mathbf{m} = \begin{bmatrix} m_1 \\ m_2 \end{bmatrix} = \begin{bmatrix} \bar{f}_{inst} \\ \bar{H} \end{bmatrix} \tag{4}$$

3.2 *Deep Bidirectional Long-Short Term Memory Architectures*

The Recurrent Neural Network (RNN) is suited for modelling sequential data, like PQD signals. BiLSTM architecture is the extension to the single LSTM network, in which they are capable of learning long-term dependencies and maintain contextual features from the past and future states. As shown in Fig. 5, BiLSTM comprises two separate hidden layers that feed-forward to the same output layer. It calculates the hidden forward sequence, $\vec{h}$, the hidden backward sequence $\overleftarrow{h}$. Moreover, the output sequence y by iterating over the following equations:

Fig. 5 A bidirectional LSTM architecture

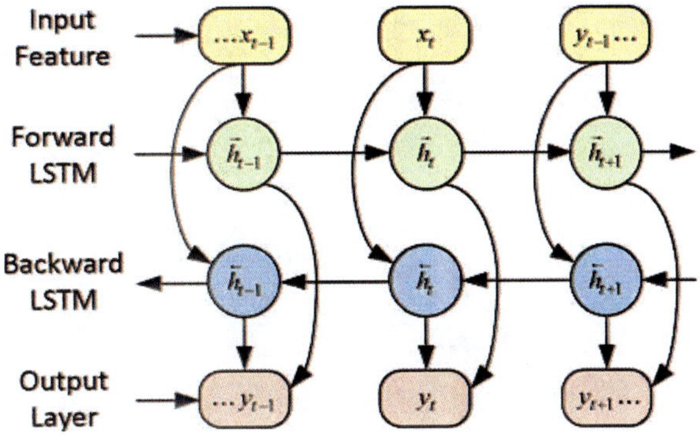

$$\vec{h}_t = \sigma\left(W_{x\vec{h}}x_t + W_{\vec{h}\vec{h}}\vec{h}_{t-1} + b_{\vec{h}}\right) \quad (5)$$

$$\overleftarrow{h}_t = \sigma\left(W_{x\overleftarrow{h}}x_t + W_{\overleftarrow{h}\overleftarrow{h}}\overleftarrow{h}_{t-1} + b_{\overleftarrow{h}}\right) \quad (6)$$

$$y_t = W_{\vec{h}y}\vec{h}_t + W_{\overleftarrow{h}y}\overleftarrow{h}_t + b_y \quad (7)$$

As shown in Fig. 3, we will experiment with the various number of BiLSTM layers and the various number of nodes in each layer. Of the particular interest, we will use an exponentially decayed number of nodes as proposed in [15] for efficient training with good accuracy. In deep BiLSTM architectures, a BiLSTM network is followed by the dropout layer. Figure 6 summarize the performance metrics utilized to evaluate our proposed algorithm. Of the particular interest is classification accuracy.

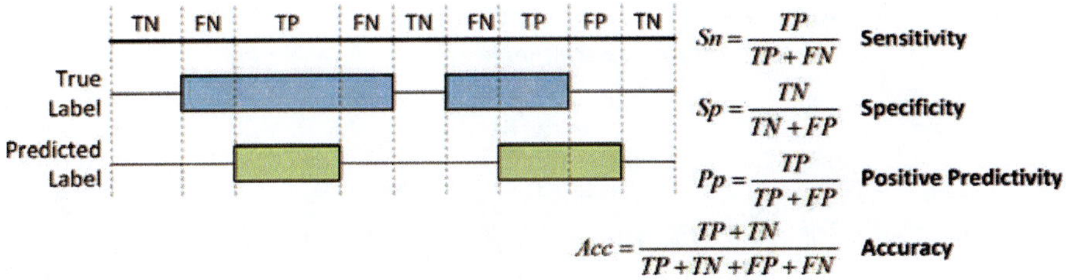

Fig. 6 Performance metrics

Table 2 Training parameters

Parameters	Value
solverName	adam
MaxEpochs	100
L2Regularization	0.0005
Shuffle	Every-epoch
MiniBatchSize	32
ExecutionEnvironment	gpu

4 Experimental Results and Analysis

This section described the experimental setup, various experiments on Deep BiLSTM architectures, and the accuracy of optimum configuration with more extended training.

4.1 Experimental Setup

A high-performance system was used for processing, i.e., a multicore system with AMD Ryzen 7 3700X (8 cores with 16 threads), 32 GB RAM, 256 GB SSD, 512 GB hard disk, and NVidia GeForce GTX 960 with 4 GB GDDR5 (1024 shading units, 64 texture mapping units, and 32 render output units). The system is equipped with the latest Windows 10 operating system and Matlab 2020a with Signal Processing, Deep Learning, and Parallel Computing Toolboxes. The synthetic PQD signals were generated for 1000 samples. It is randomly selected 600 samples for training, 100 samples for validation, and 300 samples for testing. If μ_{train} and σ_{train} is the average and standard deviation of all train features. We performed data normalization for all train, validation, and test features as follows:

$$\mathbf{m}_{normalize} = \frac{(\mathbf{m} - \mu_{train})}{\sigma_{train}} \tag{8}$$

The number of nodes in the BiLSTM layer, as well as the number of BiLSTM layers, will be varied, while the output layer is set to fifteen. Moreover, the deep neural networks were trained using the training options, as shown in Table 2.

4.2 Experiments of Various Deep BiLSTM Architectures

In Table 3, the best recognition rates were highlighted in bold. Using the training parameters as shown in Table 2, the maximum recognition rate is achieved for deep

Table 3 Recognition rate (%) of various BiLSTM architectures

No.	Configuration	Training time (s)	Acc_{train} (%)	Acc_{val} (%)	Acc_{test} (%)
1	[200]	112.18	51.13	42.53	49.10
2	[100]	97.73	63.55	49.40	35.50
3	[50]	88.71	48.51	37.20	43.87
4	[25]	87.25	46.50	34.20	35.97
5	[200 100]	198.80	60.69	63.93	62.13
6	[200 100 50]	277.82	64.49	64.47	63.03
7	[200 100 50 25]	367.07	54.15	48.33	51.67
8	[200 55]	196.36	54.71	59.07	57.43
9	[200 84 36]	284.25	54.98	52.87	50.77
10	[200 105 55 29]	372.65	62.07	58.27	62.87
11	[200 119 71 42 25]	461.66	59.95	62.20	61.50
12	[200 130 84 55 36 23]	542.01	47.47	55.60	52.27
13	[100 50]	181.74	68.99	59.47	66.53
14	[50 25]	171.78	65.14	56.20	60.60
15	[200 50 25]	269.57	60.86	59.20	52.03
16	[300 142 67 32]	425.17	54.23	62.47	58.70
17	[100 62 39 24]	353.62	**74.00**	63.40	**69.00**
18	[100 68 47 32 22]	442.50	71.74	**64.93**	67.93

BiLSTM architectures number 16, i.e. [100 62 39 24]. It required 353.62 s of training time, 74.00% training accuracy, 63.40% validation accuracy, and 69.00% testing accuracy. Note that the maximum epoch for this experiment was set to 100. Other configurations could have experimented, for example, more BiLSTM layers or a higher number of hidden nodes. The only limitation is the capacity of the GPU, as the training using CPU will be extremely slow. Table 3 also showed that the increasing number of BiLSTM layers and the number of hidden nodes were not correlated with higher accuracy.

4.3 Experiment on the Optimum Configuration

The optimum BiLSTM configuration found from the previous experiment is [100 62 39 24] and detailed in Fig. 7. This architecture is then further trained using similar training parameters (see Table 2), while parameter MaxEpochs is changed to 1000 epochs. Figure 8 shows the training process for 1000 epochs in 3440.34 s. We achieved a recognition rate of 99.98%, 98.33%, and 99.23% for training, validation, and testing, respectively. Figure 8 also showed that the training process was stabilized

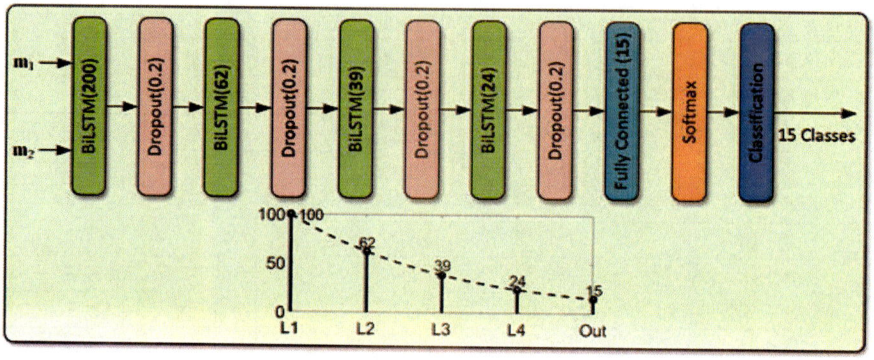

Fig. 7 Optimum BiLSTM architecture

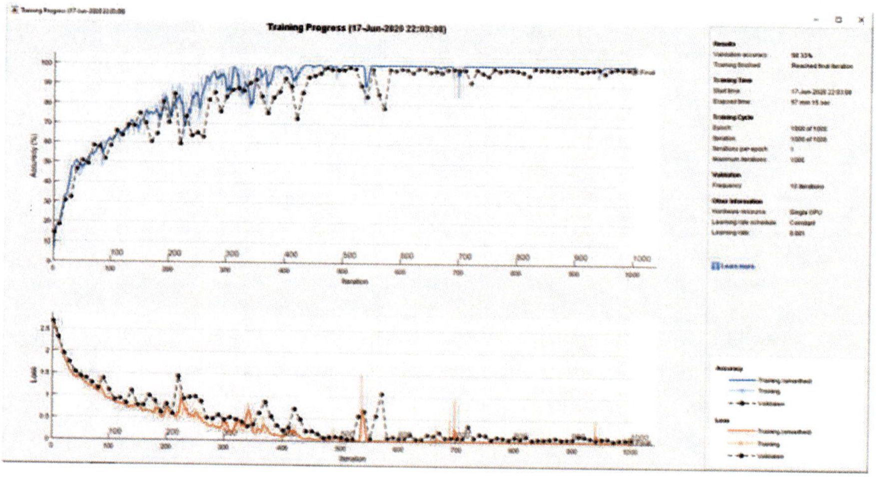

Fig. 8 The training process of optimum BiLSTM architecture

at around 500 epochs. Therefore, it can be concluded that our proposed algorithm can efficiently classify all 15 PQD signals.

Our proposed algorithm will be benchmarked with the recent research [3] due to the PQD synthetic signals similarity. Table 4 shows the comparison in more detail. In [3], the synthetic signals were added Gaussian noise, and then it retrained the system, while in our case, we did not retrain the system. Without retraining, the accuracy of our proposed system will be lower for a lower signal to noise ratio (SNR). To improve the accuracy, pre-processing by denoising the input signal could be performed, such as wavelet denoising, wiener filter, inverse filter, deep autoencoder, etc.

Table 4 Algorithm benchmarking

	DWT with extreme learning [3]	Proposed deep BiLSTM algorithm
Dataset	• 15 PQD synthetic signals • Amplitude 1 pu • Duration $t = 0.2$ s • Total period $T = 10$ • Total sampling point 1280 300 signals for each class (total 4500 signals)	• 15 PQD synthetic signals • Amplitude 1 pu • Duration $t = 0.2$ s • Total period $T = 10$ • Total sampling point 1280 • 1000 signals for each class (total 15,000 signals)
Features extraction	Discrete Wavelet Transform with seven statistical features (7 features)	Average of instantaneous frequency and average of spectral entropy (2 features)
Classifier	Multilayer perceptron extreme learning machine	Deep BiLSTM architectures with exponential decayed number of hidden nodes
Accuracy for 50 dB noisy signals	98.60% (with retraining)	98.77% (without retraining)
Accuracy for 40 dB noisy signals	98.10% (with retraining)	85.90% (without retraining)

5 Conclusions

We have presented our proposed algorithm to identify complex power quality disturbances using deep BiLSTM architecture. First, the average instantaneous frequency and spectral entropy were computed as time-frequency based features. Second, various configuration of deep BiLSTM architectures has experimented. Results show that our proposed algorithm is able to identify the 15 PQD signals by the accuracy of 99.23%. Further research is including replacing feature extraction with convolutional neural network layers, signal denoising using an autoencoder, evaluation using real PQD signals, and mitigation of electrical disturbance.

Acknowledgements The authors would like to express their gratitude to the Malaysian Ministry of Education (MOE), which has provided research funding through the Fundamental Research Grant, FRGS19-076-0684 (FRGS/1/2018/ICT02/UIAM/02/4). The authors would like to acknowledge as well as support from International Islamic University, University of New South Wales, Universitas Indonesia, and Universitas Mercu Buana.

References

1. Bollen MH, Gu IY (2006) Signal processing of power quality disturbances, vol 30. Wiley, Hoboken, NJ
2. Li J et al (2016) Detection and classification of power quality disturbances using double resolution S-transform and DAG-SVMs. IEEE Trans Instr Meas 65(10):2302–2312

3. Wang J, Xu Z, Che Y (2019) Power quality disturbance classification based on DWT and multilayer perceptron extreme learning machine. Appl Sci 9(11):2315
4. Xi Y et al (2018) Detection of power quality disturbances using an adaptive process noise covariance Kalman filter. Digit Signal Process 76:34–49
5. Singh U, Singh SN (2017) Application of fractional Fourier transform for classification of power quality disturbances. IET Sci Meas Technol 11(1):67–76
6. Sahani M, Dash PK (2018) Automatic power quality events recognition based on Hilbert Huang transform and weighted bidirectional extreme learning machine. IEEE Trans Ind Inform 14(9):3849–3858
7. Huang NE et al (2009) On instantaneous frequency. Adv Adapt Data Anal 1(02):177–229
8. Zhang A, Yang B, Huang L (2008) Feature extraction of EEG signals using power spectral entropy. In: 2008 international conference on BioMedical engineering and informatics. IEEE
9. Zhong T et al (2019) Power quality disturbance recognition based on multiresolution S-transform and decision tree. IEEE Access 7:88380–88392
10. Wang S, Chen H (2019) A novel deep learning method for the classification of power quality disturbances using deep convolutional neural network. Appl Energy 235:1126–1140
11. Junior WLR et al (2019) Classification of power quality disturbances using convolutional network and long short-term memory network. In: 2019 international joint conference on neural networks (IJCNN). IEEE
12. Moreno-Muñoz A (2007) Power quality: mitigation technologies in a distributed environment. Springer, London
13. Igual R et al Integral mathematical model of power quality disturbances. In: 2018 18th international conference on harmonics and quality of power (ICHQP). IEEE
14. IEEE (2019) IEEE recommended practice for monitoring electric power quality. IEEE Std 1159–2019. IEEE Power and Energy Society
15. Gunawan TS, Kartiwi M (2018) On the use of edge features and exponential decaying number of nodes in the hidden layers for handwritten signature recognition. Indones J Electr Eng Comput Sci 12(2):722–728

Machine Vision and Convolutional Neural Networks for Tool Wear Identification and Classification

Tiyamike Banda, Bryan Yeoh Wei Jie, Ali Akhavan Farid, and Chin Seong Lim

Abstract Machining is becoming increasingly advanced to satisfy industrial precision standards of the components. With accuracy being analogous to quality, the study of tool wear is growing in importance as wear can notably affect the tool life. In smart manufacturing, machining processes incorporate artificial intelligence in machine vision to improve key process functions through metrology, improving dimensional accuracy and surface integrity of products. By enhancing machine vision or other suitable forms of optical metrology, it is possible to identify different wear types and their corresponding causative mechanisms. In this study, the flank wear region was identified and classified using deep learning features and transfer learning by pre-trained CNN models. Tool wear mechanism's features extracted by Alexnet and VggNet16, have been used to train a Support Vector Machine (SVM) for classification. A comparison of fine-tuned CNN and CNN-SVM models has been made. The fine-tuned models employed backpropagation (BP) and stochastic gradient descent (SGD) to optimise the weights for performance improvement. The fine-tuned CNN models accomplished a higher accuracy compared to a CNN-SVM with an average validation accuracy of at least 95%. Fine-tuned Alexnet had a better performance than VggNet16 with an average validation accuracy of 96.43% and classification time of 0.244 s. With high accuracy and minimum time complexity, fine-tuned Alexnet can be used for online tool wear mechanisms classification. This implies that fine-tuned CNNs are effective and reliable in performing classification of tool wear.

Keywords Machine vision · Tool wear · Convolution neural network · Classification

T. Banda · B. Y. W. Jie · A. A. Farid · C. S. Lim (✉)
Department of Mechanical, Materials and Manufacturing Engineering, University of Nottingham Malaysia, Jalan Broga, Semenyih, Selangor, Malaysia
e-mail: chinseong.lim@nottingham.edu.my

A. F. Ab. Nasir et al. (eds.), *Recent Trends in Mechatronics Towards Industry 4.0*, Lecture Notes in Electrical Engineering 730, https://doi.org/10.1007/978-981-33-4597-3_66

1 Introduction

In machining operations, attaining the highest quality of surface finish is the aim and that is hugely reliant on the condition of the cutting tool insert [1]. Wear, thermal fatigue and corrosion are factors that adversely affect the quality of the machined product. Detecting and identifying faults in high-speed CNC milling processes is becoming more complex, time-consuming and error-prone, thus impeding productivity. By being proactive and employing technology to monitor the efficiency of the machining process, acquired data can provide information to avoid unplanned downtime. This also prevents significant damages and ensuring the efficiency and safety of machines with optimal operations. One machining phenomena which need proper inspection during cutting is tool wear. Tool Condition Monitoring (TCM) is divided into two methods: direct and indirect method.

Indirect TCM measures parameters or signals (vibrations, force, acoustic emission, power and current) of the cutting performance which allows for deductions related to the severity of tool wear [2]. A new method of measuring vibration levels in the form of sound signals was introduced using microphones which largely minimized vibration signals by epoxy coated tool holders [3]. A system was devised with a self-developed smart sensor comprised of an infrared camera for image acquisition and an external Resistance Temperature Detector (RTD) temperature sensor for camera calibration for tool-breakage detection [4]. The analysis of the scatter spectrum of laser light reflection using machine vision was adopted in the classification for surface roughness of machined workpiece [5]. Acoustic Emission (AE) signals could deliver instantaneous feedback to dynamic conditions during contact between the tool and workpiece, making AE sensors suitable for TCM in micro-milling processes [6]. However, the extraction of valid signal features and faults diagnosis in machinery applications is challenging with high noise to signal situations in the industrial environment, impeding progressive wear monitoring and quantification of prediction errors [2, 7].

Developments in present-day digital image processing have shifted direct methods toward machine vision due to its reliability and accuracy [8]. Previous limitations of direct TCM and its incapability to classify tool wear mechanisms has framed a need to develop a technique for online TCM using machine vision. With the implementation of AI models, machining conditions and performance constraints can be predicted, enabling tool wear monitoring and machining parameters optimization. Computer vision and machine learning can be used to develop flank wear and damage detection systems for end milling tools [9].

Similarly, with the development of neural networks now becoming a key focus in artificial intelligence algorithms, the systems can perform tasks beyond wear detection. CNN can be applied in the manufacturing sector by monitoring the operational state of gearboxes and bearings, intelligently identifying and classifying the defects [10]. In the aspect of TCM, the application of Artificial Neural Network (ANN) and image analysis were used in quantifying the wear severity of the tool's cutting edge [11]. Feature extraction was alternatively used as training and testing methods for the

Bayesian network, SVM, k-nearest neighbour regression algorithm and other CNN models for improved classification results [12]. Cutting parameters and extracted features of the wear region are used as inputs for a Wavelet Neural Network (WNN) model to deduce the degree of tool wear [8]. The shape of the tool wear area was able to be determined, extracted and automatically classified into respective categories by a CNN [13, 14].

CNN models are the preferred choice as it constantly achieves high accuracies for numerous pattern recognition tasks of image identification and classification [15]. While CNN models for image classification contain hundreds of layers and achieve better performance than shallow Neural Network (NN) architectures, it is prone to big data sets and computational power for training to build a robust model [16].

Identifying tool wear helps to acknowledge the prevalent mechanisms behind its occurrence. This helps to select the best remedy to minimise the rate of wear progression thus prolonging tool life. However, manual identification and classification are more complex and time-consuming. Therefore, there is a need to establish an automatic tool wear identification and classification method using machine vision and machine learning approach. This study used Alexnet and VGGNet-16 pre-trained neural networks to identify and classify mechanism types of wear. Tool wear images were characterised and categorised by colour and shape for easy identification and classification. Without access to big data sets to build a CNN from scratch, deep feature extraction and transfer learning were adopted to train a custom CNN-based model for wear mechanisms classification.

2 Methodology

2.1 Data Acquisition

Data collection was performed using a microscope camera (OrzBuy 5× Digital Zoom USB Microscope) to manually capture the wear images of different used inserts. Image acquisition for each tool insert was carried out using the setup as shown in Fig. 1.

The captured images were manually categorized according to their corresponding colour, shape boundaries and edges. 4 categories of wear were analysed in this research as shown in Fig. 2.

2.2 Data Pre-processing and Preparation

All images were sharpened by Contrast Limited Adapted Histogram Equalisation (CLAHE) algorithm as a form of data pre-processing. The goal of the image enhancement algorithm is to reduce image noise by equalising the light contrast on each tile

Fig. 1 Data acquisition setup: (1) computer; (2) microscope camera; (3) tool insert

Fig. 2 Sample images of wear types: (1) thermal; (2) adhesion; (3) chip; (4) notch

of tool wear images. An increase in contrast between regions of interest and their edges also allows for improved visibility of image features for the CNN extraction during training. Images were saved into a single folder containing all 4 categories.

2.3 *Transfer Learning Using CNN and SVM*

The training process for CNNs is tolerable for users with a Graphics Processing Unit (GPU) but time-consuming for Central Processing Unit (CPU) operators due to the complicated framework of CNN algoritms. For a CPU method, the combination of CNN and SVM would suffice, where the CNN performs high-quality feature extraction using its convolutional layers while the SVM classifies the test images based on the features extracted by convolutional layers and systematically grouped by fully connected neurones in Fully Connected (FC) layers. The selected layer for feature extraction output for training and testing images were the activations on 'FC 7' which consists of 4096 feature dimensions as it gives a better generalization of the input images. Therefore, a multi-class SVM uses the features from FC7 to classify wear images. The SVM has a linear kernel with scale 1, cost of [0.1; 1.0], prior of [0.5, 0.5], beta of 4096×1, classification method of CECOC and binary learner bias ranging from -2.172 to 1.083. There is no updating of weights and optimisation of hyper-parameters in SVM. Table 1 shows Alexnet CNN, VggNet-16 CNN and SVM general parameters used in this analysis.

Table 1 General parameters of Alexnet, VGG 16 and SVM

Model	Input	Output	Kernels	Max. pooling	Parameters
Alexnet	227 × 227	1000	11 × 11; 5 × 5; 3 × 3	3 × 3	62 million
VGG 16	224 × 224	1000	3 × 3	2 × 2	138 million
SVM	4096	4	Linear		

2.4 *Transfer Learning Using Fine-Tuned CNN Models*

A typical CNN architecture can be summarized in few stages as shown in Fig. 3. The feature extraction phase of CNN contains two groups of layers, the convolutional layers and pooling layers, which are joined to the framework consisting of a Fully Connected (FC) layer and a classification layer. The convolution layers have several weights (filters), where the preceding layer convolutes the input through a batch of weights and is responsible for comparing and detecting features. A pooling layer is followed by a convolution layer to minimize the number of features through sub-sampling, scaling down, computational complexity and producing invariance to minor deviations and misrepresentations. Tool wear distinct features are typically colour, boundaries, edges and surface texture which are extracted in the first convolution layers of CNN. By modifying fully connected layers, tool wear images can be classified through a process called transfer learning.

Transfer learning is a deep learning technique whereby the usage of a pre-trained model as a foundation for another task as it is more beneficial and effective to fine-tune the network. The initial random weights of the CNN are repeatedly trained to extract and learn new features to identify the most functional weights for the specified task, hence minimizing classification error on the set of training images. With this method, the model can achieve higher accuracy with relatively small amounts of data

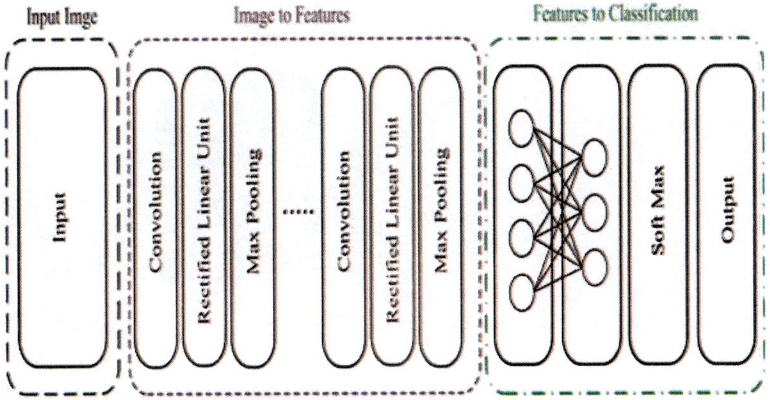

Fig. 3 Architecture of a CNN

and reduce training time as it will not be necessary to undergo as many epochs (a full training cycle on the entire dataset) compared to a newly trained model.

To enrich the capacity for the CNNs to learn, another programmable data augmentation code was done by translating, scaling and rotating images on the training set, whereas only the original images are used for testing. A programmable data augmentation code is a blind technique which is done by a computer program in Matlab.

The models were fine-tuned by using the backpropagation algorithm (BP) and stochastic gradient descent (SGD) to optimize the weights and biases in the CNN architecture. The BP algorithm continually propagates the error term back to the previous layer and update the weights between the two layers for gradual refinement until improvement stops.

Fully connected layers (FC7) were modified from a default value of 1000 down to 4 and the final classification layer was replaced to automatically synchronise the output value of 'FC 8'.

In this paper, Alexnet and VGGNet-16 were employed as both are incredibly powerful models capable of achieving high accuracy on challenging datasets. A standard Alexnet has a total of eight layers, comprised of five convolutional layers, three fully connected layers and a softmax layer which can be used to classify 1000 different classes. The VGGNet-16 has a similar architecture but made up of thirteen convolutional layers and three fully connected layers. The networks feature the usage of Rectified Linear Units (ReLu) which achieves a training error six times faster compared to the usual tanh or sigmoid functions in regular CNNs and helps the optimization converge faster. During the validation stage, misclassified images are flagged by the network during the validation process and saved into a different folder. The summary of transfer learning classification process is illustrated as a flow diagram in Fig. 4.

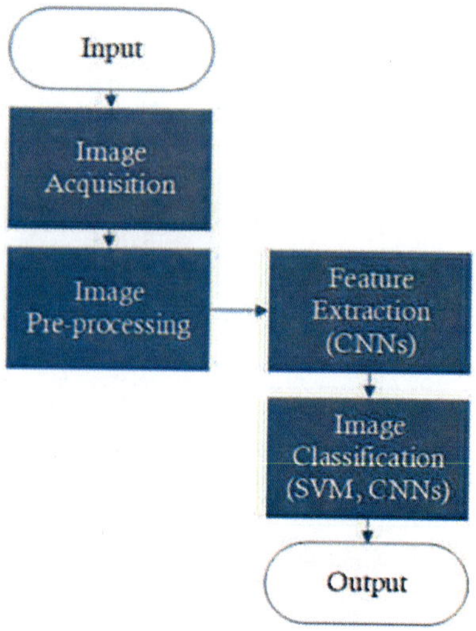

Fig. 4 Flowchart of the proposed method

3 Results and Discussion

The network was then loaded with the pre-processed images. The function was synchronised by equalising the number of images for all 4 categories based on the class with minimum number of images. The data is split into 80% training and 20% testing. Each category had approximately 400 images; 320 for training and 80 for validation. The pre-trained CNNs were modified and ran using the deep network designer provided by MATLAB. The training sets were randomized each time the training was repeated. The training was repeated 20 times for each CNN model to obtain the average performance. There are several performance indicators which were used in this study. These are average training time, training accuracy, validation accuracy, precision, recall, specificity and average recognition time. Average training time is the average time required to complete training the model in the specified number of epochs. Accuracy is the ratio of the correctly classified images to the whole dataset. The training accuracy describes how effective the model learns the new image features while the validation accuracy describes how the validation images are predicted during the training process. The validation accuracy also explains how accurately the actual training features from training images are represented by the features from the validation images. If the validation features do not accurately fit the training features, a problem of overfitting or underfitting occurs. Precision indicates how accurate the prediction is, recall measures how acceptable the correctly labelled images are, and specificity measures the proportion of wrongly labelled images that are correctly identified. Average recognition time is the average time taken for a trained network to individually identify and classify an image when it is given a new image.

3.1 Comparison of CNN-SVM and Fine-Tuned CNN Models

The number of false predictions for chipping class can be linked to the low number of features defining chipping. The results of the Alexnet-SVM model against the fusion of VGGNet-16-SVM are shown in Fig. 5a and b respectively. Alexnet-SVM yielded an accuracy of 90.4% whereas the VGGNet-16-SVM achieved a score of 87.9%. Both models displayed difficulty distinguishing notching from chipping. The number of notch wear images may need to be increased to lessen the likelihood of such mistakes as the net can learn more features and the representation of a class is better generalized.

The models were further fine-tuned with optimum hyper-parameters to optimise the weights using SGD and back propagation methods. The optimum learning rate, momentum, batch size and number of epochs used are 0.0001, 0.9, 32 and 10 after several tests were conducted. The comparison of classification performance for different models is shown in Table 2. Despite fine tuned-Alexnet having a simpler structure, it attained the highest mean validation accuracy of 96.43% across four types

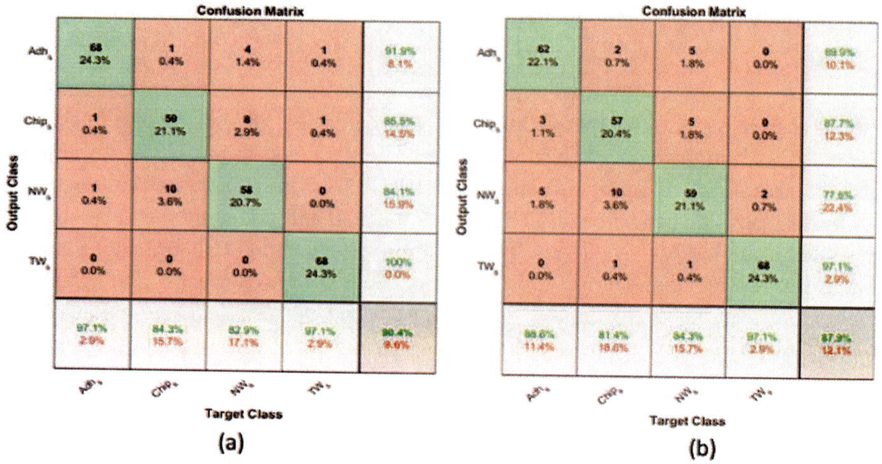

Fig. 5 Confusion matrix of the **a** Alexnet-SVM model and **b** VGGNet-16-SVM

Table 2 Comparison of Performance indicators, average training and classification time for both CNN-SVM and fine tuned CNN models

Network model	Validation accuracy (%)	Precision (%)	Recall (%)	Specificity (%)	Training time (m)	Average time (s)
Alexnet-SVM	90.40	97.14	100	100	0.026	0.264
VGG16-SVM	87.90	97.14	97.14	99.05	0.072	1.972
Tuned-Alexnet	96.43	98.57	100	100	10.38	0.244
Tuned-VGG16	95.36	98.57	97.18	99.05	286.18	1.971

of tool wear, while the fine tuned-VGGNet-16 (VGG16) achieved 95.36% accuracy. With an increased number of layers in VGG16, more features are extracted with no distinct characteristics for a specific class thus creating more false positives for classification.

The average training time was 10.4 min and 286.2 min, and the average recognition time for one sample was 0.244 s and 1.971 s for tuned-Alexnet and tuned-VGG16 respectively. Based on Table 2, it can be deduced that the time difference in training is credited to the architecture of the CNN models. Also, fine-tuning and training CNN improve the model's accuracy substantially. Despite the training procedure being computationally demanding, the model does not require re-training during execution, hence these methods are suited for on-line monitoring applications with a minimum recognition time. Although both tuned-Alexnet and Alexnet-SVM showcased shorter recognition times compared to the VGG models, tuned-Alexnet is more suitable option for its overall higher performance as shown in Table 2.

All four models achieved a minimum of 87.9% accuracy, 97.14% precision, 99.05% specificity and 97% recall, respectively. This reflects the custom

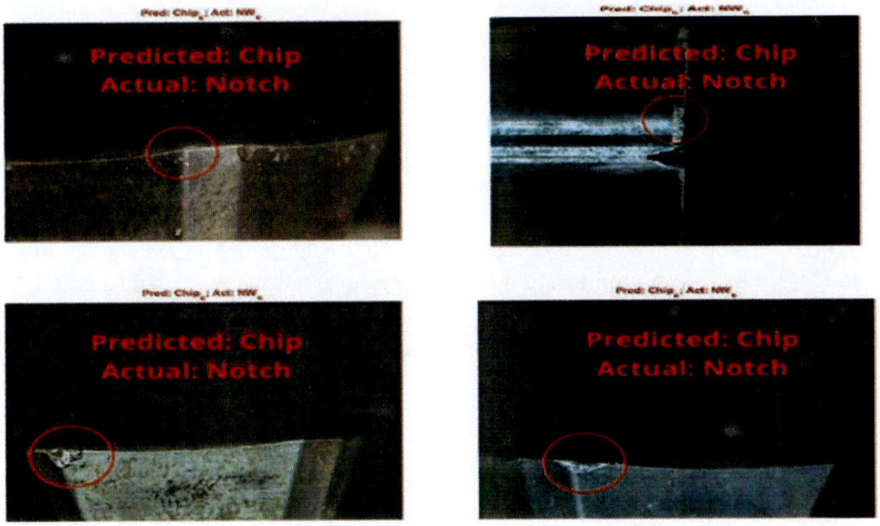

Fig. 6 Misclassified images of notch wear as chip wear

CNN model's ability to accurately predict and classify the relevant images, justifying its capacity for tool wear classification.

3.2 Misclassification

Tool wear classification is a difficult task because of multiple wears existing in a single image. However, there is an assumption that one type of wear dominates in a region due to a specific cause. There is a high possibility that CNN can mistakenly classify one wear for another. This is called tool wear misclassification as shown in Fig. 6. Several causes of tool wear misclassification had been identified in this research. One of the causes of classification errors was the fragmentation of chips which may resemble certain features of wear. However, it was independent of the type of cutting tool and background. Quality of the images do play a pivotal part in deep learning features extraction. The accuracy of the model highly depends on the accurate information of features provided to the classification models.

4 Conclusion

Automatic identification and classification of tool wear types built on CNN, namely Alexnet and VGGNet-16 deep learning frameworks provided promising results. This research has shown that a fine-tuned CNN model attains a higher accuracy compared to a custom CNN-SVM model. Subsequently, between all four models, the Alexnet model is the preferred option, scoring a 96.43% validation accuracy and with an average recognition time of 0.244 s, thus proving the reliability of this model for on-line tool wear classification. Practicality wise, this model is suited for applications with low or minimal hardware scenarios and periodic detection of the tool wear mechanisms across most machining processes such as milling, turning and drilling. The future work will focus on application of custom CNN models for on-machine vision TCM in CNC milling processes. The CNC Machining Parameters Optimisation (MPO) based on minimisation of the prevalent tool wear mechanisms will be the area of focus on achieving a robust on-line machine vision TCM-MPO custom model for advanced smart machining technology.

References

1. Vetrichelvan G, Sundaran S, Kumaran SS, Velmurugan P (2014) An investigation of tool wear using acoustic emission and genetic algorithm. J Vib Control 21(15):3061–3066
2. Al-Obaidi S, Leong M, Hamzah R, Abdelrhman A (2012) A review of acoustic emission technique for machinery condition monitoring: defects detection & diagnostic. Appl Mech Mater 229–231:1476–1480
3. Bejaxhin A, Paulraj G (2019) Experimental investigation of vibration intensities of CNC machining centre by microphone signals with the effect of TiN/epoxy coated tool holder. J Mech Sci Technol 33(3):1321–1331
4. Ramirez-Nunez JA, Trejo-Hernandez M, Romero-Troncoso RJ, Herrera-Ruiz G, Osornio-Rios RA (2018) Smart-sensor for tool-breakage detection in milling process under dry and wet conditions based on infrared thermography. Int J Adv Manuf Technol 97(5–8):1753–1765
5. Gupta M, Raman S (2001) Machine vision assisted characterization of machined surfaces. Int J Prod Res 39(4):759–784
6. Ren Q, Balazinski M, Baron L, Jemielniak K, Botez R, Achiche S (2014) Type-2 fuzzy tool condition monitoring system based on acoustic emission in micromilling. Inf Sci 255:121–134
7. Dutta S, Pal S, Mukhopadhyay S, Sen R (2013) Application of digital image processing in tool condition monitoring: a review. CIRP J Manuf Sci Technol 6(3):212–232
8. Garcia-Ordas MT, Alegre-Gutierrez E, Alaiz-Rodriguez R, Gonzalez-Castro V (2018) Tool wear monitoring using an online, automatic and low cost system based on local texture. Mech Syst Signal Process 112:98–112
9. Klancnik S, Ficko M, Balic J, Pahole I (2015) Computer vision-based approach to end mill tool monitoring. Int J Simul Model 14(4):571–583
10. Chen Z, Li C, Sanchez RV (2015) Gearbox fault identification and classification with convolutional neural networks. Shock Vib 2015:1–10
11. Mikolajczyk T, Nowicki K, Klodowski A, Pimenov DY (2017) Neural network approach for automatic image analysis of cutting edge wear. Mech Syst Signal Process 88:100–110
12. Fatemeh A, Antoine T, Marc T (2018) Tool condition monitoring using spectral subtraction and convolutional neural networks in milling process. Int J Adv Manuf Technol 98:3217–3227
13. Garcia-Ordas MT, Alegre E, Gonzalez-Castro V, Alaiz-Rodriguez R (2017) A computer vision approach to analyze and classify tool wear level in milling processes using shape descriptors and machine learning techniques. Int J Adv Manuf Technol 90:1947–1961
14. Wu X, Liu Y, Zhou X, Mou A (2019) Automatic identification of tool wear based on convolutional neural network in face milling process. Sensors 19(18):3817
15. He K, Zhang X, Ren S, Sun J (2016) Deep residual learning for image recognition. In: 2016 IEEE conference on computer vision and pattern recognition (CVPR), pp 770–778

16. Wei D, Wang K, Heyns S, Zuo MJ (2019) Convolutional neural networks for fault diagnosis using rotating speed normalized vibration. Fields Inst Commun 67–76

TOM: The Assistant Robotic Tutor of Musicianship with Sound Peak Beat Detection

Gareth Hawkins and Esyin Chew

Abstract Most past literatures surrounding musical robots has focused on either the engineering design of physical robots that could produce musical sounds, or on the mixed mechatronics aspects of algorithms enhancement to design robots gaining musicianship skills. The recent combined both these research fields into one is called Robotic Musicianship, which is innovative but with limited literatures. Robot musician involved sound peak beat detection techniques to enable musical perception. Responding to the novelty demand, the study adapted design approach of onset beat detection, blended with positivists' philosophical approaches, and deductive principles associated with these philosophies were adhered to. The study aim to conduct a conceptual design and implement the drum rudiment for a robotic assistant tutor to co-demo electronic snare drum. The design approach of onset beat detection is adapted. The fundamentals of this robot prototype and rudiment work well, although many improvements can be done on the functionality and fluidity, doing so should further enhance engagement and musical development, and achieve the long-term goal to branch out to higher levels of education for music.

Keywords Robotics · Robot application · Robotic assistant tutor · Robotic musician · Sound peak beat detection

1 Motivation and Bibliometric Investigation of Robotic Musicianship

Due to the new norm of the COVID-19 social distancing and the lacking of resources of accessing drum tutor, this research investigates the past literatures of robotic musician, followed by the preliminary design of an assistant robotic musician as a tutor for

G. Hawkins · E. Chew (✉)
EUREKA Robotics Lab, Cardiff School of Technologies, Cardiff Metropolitan University, Western Avenue, Cardiff CF5 2YB, UK
e-mail: echew@cardiffmet.ac.uk

G. Hawkins
e-mail: G.Hawkins@outlook.cardiffmet.ac.uk

A. F. Ab. Nasir et al. (eds.), *Recent Trends in Mechatronics Towards Industry 4.0*, Lecture Notes in Electrical Engineering 730, https://doi.org/10.1007/978-981-33-4597-3_67

beginner drummers. There are two primary research areas within Robotic Musicianship: Musical Mechatronics and Machine Musicianship. Musical mechatronics is focused on constructing physical robots that are able to produce musical sounds with motors, solenoids and gears, come in many shapes and sizes and have been developed to learn many different musical instruments over the years [1]. For examples: (1) StrumBot a six-stringed robotic electric guitar that uses a dual-pick strumming arm in order to pluck guitar strings [2]; (2) Cog is another upper-torso humanoid robot with 21 degrees of freedom, and has various sensory systems, including visual, auditory and tactile senses [3]. Cog was built with compliant arms (each with 6 DoF), instead of using stiff arms, as it was more suitable for a robot that would interact with an unknown environment. These compliant arms were controlled by non-linear oscillators, which provided energy to generate motion [4]. Auditory feedback was used to modify the oscillator behaviour, which moved the arms. After receiving raw audio, it was processed and thresholded to find the sound of the drum. An auditory-motor loop was developed that combined auditory feedback and auditory processing, which made the robot listen to its own drum hits, thus making the robot continuously hit the drums.

Machine Musicianship is focused on developing algorithmic expression, music cognition [5] and its foundation is derived from the fields of music theory, which involves analysing, performing and composing music with computers. Implementing such theoretical models usually ranges from simple musical problems to live interactive musical compositions. Some systems have been developed that employ machine musicianship. Systems such as EMMY and Emily Howell [6, 7] was programmed to find a set of instructions or creation of a musical piece with note-to-note analysis, then creates scale degrees that are similar to the original style it learnt from, in order to produce new styles instead of replicating music style, and could further communicate with users in different languages, to expand collaboration with different cultures [6]. HAILE is the robot percussionist capable of perceptual, physical and socials aspects of musicianship, to analyse musical input from a human, in real-time [7]. HAILE was programmed with improvisation algorithms, which enabled the robot to detect multiple musical qualities, such as pitch, volume and rhythmic elements. To encourage human-robot interaction, the algorithm transforms and modifies human input to create a similar improvised output which a human player would react to, during performances. MahaDeviBot was the combination of ideas taken from music robotics, hyperinstruments and machine musicianship that uses a symbolic Music Information Retrieval (MIR) based approach [1]. The MIR approach enabled the robot to automatically supply its database with drum beats, learnt from pre-programed human input. How the robot performed the drum beats autonomously was from accessing its database of drum beats, using sensor-based retrieval that included a set of queries received from a human sitar player, in real-time. Shimon is an autonomous, improvising Marimba player [8], a change from the usual percussive instruments and is the successor to HAILE. The approach used for Shimon, is different compared to previous works, it uses a gesture-based behaviour system that focuses on physical movement and anticipates said movements for input, instead of the conventional auditory approach [8]. Shimon uses a non-humanoid head for social communication that provides organic movement, and is equipped with a camera-like shutter in order to convey emotion.

The two common approaches used for beat detection are autocorrelation analysis and onset component detection, two robots previously mentioned, HAILE [7] and MahaDeviBot [1] both utilise onset beat detection. Nevertheless, there has been newer approaches within beat detection for robotics, using a NAO robot by combining the two common methods of beat detection which involves estimating a global tempo by analysing autocorrelation of the onset features from musical pieces [9, 10]. A newly proposed method for beat detection was developed for a dancing robot, they proposed a multi-model method that combines both auditory and visual tracking [11]. Their beat tracking method is based upon the work from Murata et al. [12], that involves calculating the autocorrelation of onset features, similar to Xia et al.'s approach [9]. What makes this approach distinctive, is that the inclusion of a visual tracking system which uses a Kinect camera that tracks the skeleton of a dancer and their movement of joints. Both audio and visual tracking systems work together to estimate the tempo and beat time from the music and human dancing. From their experimental results, it showed that their proposed method always outperformed beat-tracking methods, and on average it over performed other related methods too. Although, their results did show that the visual tempo estimation failed on occasions that included small movement with the hands and feet. This could suggest that this method might not be suitable for visual tracking for body movements from musicians that perform small movements, for example movements such as quick alternate guitar strumming, or soft drum hits that require small movements.

2 Research Method and Design

We stand a positivists' philosophical approaches with deductive principles, to conduct a conceptual design and implement the drum rudiment for a robotic assistant tutor to co-demo electronic snare drum. In this study, the conceptual design is grounded on the design approach of onset beat detection that has been observed in Weinberg's, Hoffman's and Kapur's research [1, 7, 8], which shows heavy usage of auditory features within robotics utilising beat detection methods. Figure 1 shows the initial

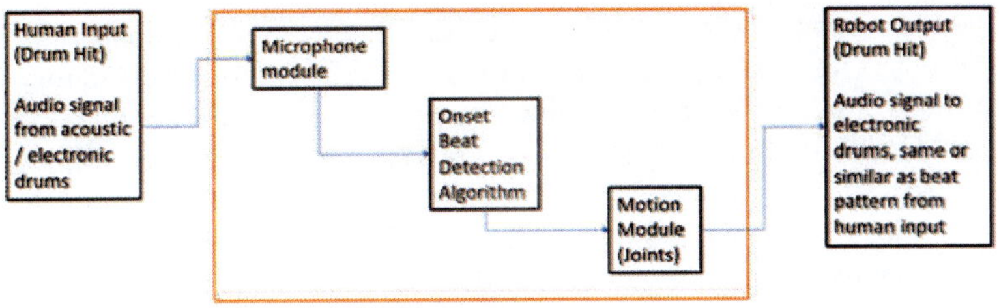

Fig. 1 Conceptual beat detection design for drum tutor assistive robot

conceptual design for the beat detection approach used in this study, and onset beat detection serves as a good departing point for Robotic Musicianship research and development. Other NAO robots have used this approach or a higher level of this approach, as seen from Xia et al. [9], however, this design would be for a different scenario, that being a drum tutor assistant. Other methods were considered such as, the multi-modal method from Ohkita et al. [11] to use for research.

The overall research approach is focused more on machine musicianship design rather than music mechatronics, but understanding the methods used within musical mechatronics has provided useful insights, such as the visual features used for tracking of humans [11], instead that could potentially be used to track the drums in the exploration. Computation and semi-autonomous responses to beat and music will be the main focus, while some psychology could be involved. Since the NAO robot is already built, although construction of the extended 3D printed accessories could be present, if the robot needs it to support the musical process. While robots have been introduced as teaching assistants, and the same with robots being involved with musical tasks, none have been combined both together and seen how introducing a robot into one on one music teaching would affect the learning experience, in a drum lesson scenario.

3 Robot Tom Design and Implementation

3.1 *The Design of an Assistant Robotic Tutor of Musicianship*

A drum lesson scenario is going to be used to help design the robot drum teacher assistant. In the drum lesson scenario, there will be 3 actors, a human teacher, a human student and the robot teacher assistant, TOM. The teacher will begin the drum lesson by powering on TOM. When TOM is powered on, the teacher can control TOM's hands by talking to it. Speech commands such as "open" and "close" will tighten or loosen the grip of the hands (see Fig. 2), this enables the teacher to place drumsticks into TOM's hands. Once the drumsticks are in position, the teacher can proceed onto the next step, which is activating the three different mode that enables TOM to play drums.

The three modes are activated by touching the tactile head sensors on TOM, which will start the sound peak detection module, ready to receive auditory input. The student will provide an auditory input by playing one crochet note on the drum snare, TOM will listen for the peak sound from the student's crochet note, and will then proceed to play a drum rudiment. The sound peak detection will continue to run as this provides the teacher and student to continuously play and practice the drum rudiment. Although, it is important to note, that before moving onto a different mode, the current mode must first be reset, otherwise the sound peak detection module from the previous mode will still be active and will pick up auditory input, thus confusing

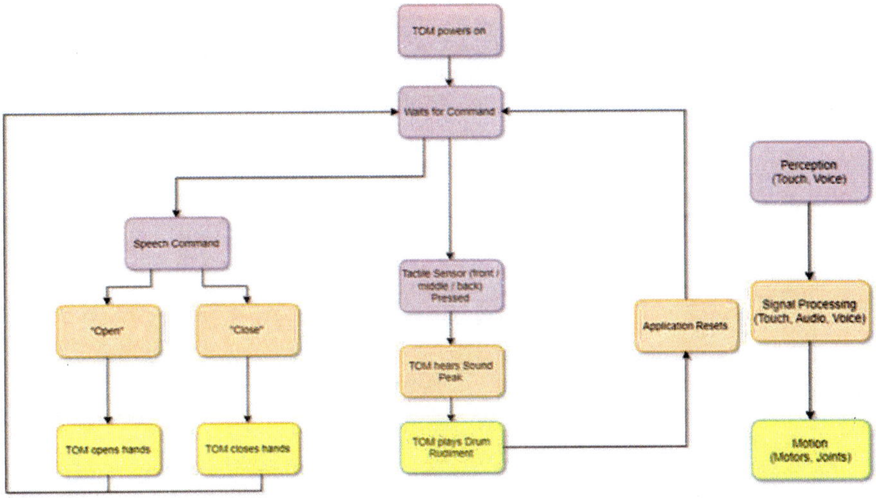

Fig. 2 Drum lesson scenario, robot application overview

TOM on which drum rudiment to play. After TOM has finished playing the drum rudiment, the student is expected to replicate the drum rudiment. When the student is playing the drum rudiment, the teacher can then observe and evaluate the students playing (see Fig. 2).

At current development, the robot only plays simple drum rudiments, this is because drum rudiments are one of the fundamentals of drumming, and is essential to playing drums. This replicates the process of how traditional drum teaching is currently taught and relates to the way a human would learn drums, by starting with the basics. This also provides a small and suitable scope for a robot drum teacher prototype, that can be tested to students who are beginners or completely new to playing drums. The design of this program is not intended to replace a music teacher but rather aid them in their work. The design is meant to encourage human-robot collaboration and interaction for both human actors (teacher and student) and provide an interesting experience for students to help them learn and give them motivation to continue learning. Figure 3 presents the different types of interaction that is used within this application:

3.2 The Implementation of Assistant Robotic Tutor of Musicianship

IDE, Speech Recognition and Tactile Head Sensors—Three Drum Rudiments
In order to implement the above designs onto a NAO robot, the Integrated Development Environment (IDE), Choregraphe 2.8 was used, which is built for developing applications for NAO. To help program NAO, Choregraphe's built-in framework,

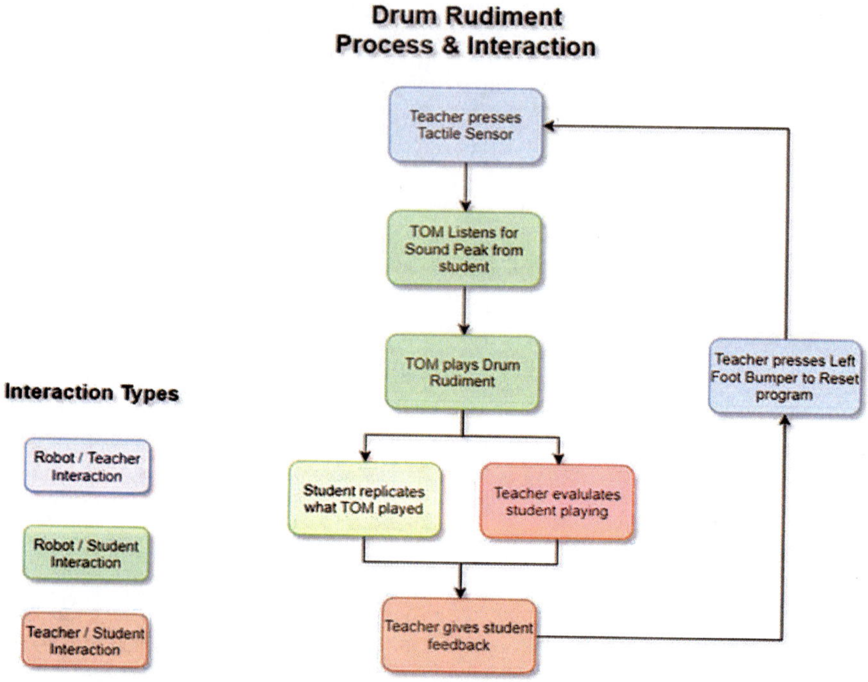

Fig. 3 Drum lesson scenario, robot application overview

NAOqi framework, was used to call many core modules (touch, audio, motion), that can communicate with each other. Alongside using the NAOqi framework to help program the application, the Python Software Development Kit (SDK) was used. To implement Speech Recognition into TOM, the Speech Recognition box was used, which enables TOM to recognise a word from a list of words, e.g. 'open' and 'close'. The outputs on the Switch case box are then connected to custom Python script boxes to control both the hands on TOM, correspondingly labelled 'Open Hands' and 'Close Hands', by using the 'angleInterpolation' method. The parameters for angleInterpolation are as followed: (jointName, targetAngles, targetTimes, isAbsolute). Therefore, the first motionProxy sets the initial position of the hands to 0%, which makes both hands close. The second motionProxy changes the position of the hands to 30% within 0.5 s (see Fig. 4), which opens up or close the hands in order to loosely place a drumstick in the hands. Speech recognition could also be used to help implement new features such as changing the robot's position for different drum grip techniques, or for hitting different drums on a drum kit.

The Tactile Head box, which detects touch for all tactile head sensors, was used to implement the three tactile head sensors functions and make them perform three different drum rudiments. The tactile head box features If statements which states that, if the signal value is greater than 0, transmit a new signal to outputs of the tactile head box. The front tactile head sensor is used to play the first drum rudiment

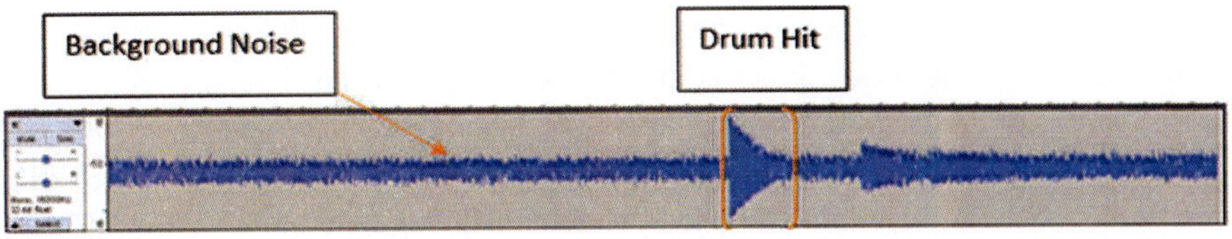

Fig. 4 Drum hit audio waveform recorded from NAO's microphone, visualised in audacity

which enables the robot to play 1 crotchet note on the drums, using the right arm. The middle tactile head sensor is used to play the second drum rudiment, which again uses the right arm, but instead plays 4 crotchet notes on the drums. Finally, the rear tactile head sensor is used for the third rudiment, which enables the robot to play 4 crotchet notes on the drums, with both arms alternating with each crotchet note played. To activate each tactile head sensor, a user must make a tactile gesture. The signals generated from the tactile gestures are sent to the ALTouch module, by using the ALMemory module which triggers the memory events, FrontTactilTouched, MiddleTactilTouched and RearTactilTouched, which are present within the Tactile head box.

Sound Peak Detection
Being able to process human touch to signal the tactile head box, TOM requires a mean to play these rudiments through sound detection methods. This was achieved by using the Sound Peak box, which outputs a signal whenever a sound peak is detected. The Sound Peak box utilises ALSoundDetection, a NAOqi audio module. ALSoundDetection detects and processes significant sounds, such as a drum hit, from the front facing microphone at 16,000 Hz, in OGG format. In conjunction with using this audio module, the core module ALMemory is used, and the memory event 'SoundDetected', is called upon when a significant sound have been detected. Both these modules communicate with each other to process audio signals. The way the sound processing works, is very similar to the Root-Mean-Squared (RMS) method of calculating an average audio signal overtime. When a signal is detected and passed through to the ALSoundDetection module, the raw signal is levelled by calculating the average signal, then the peak sound from this averaged signal is captured (see Fig. 4). An array of sound elements is then generated, which is stored in an 'index', that contains all the indexes of the starting and ending points of the peak sound detected. To use this index, it must be sent through by calling upon the SoundDetected memory event, in the ALMemory module. The SoundDetected event organises sound elements as followed:

[[index_1, type_1, confidence_1, time_1], ..., [index_n, type_n, confidence_n, time_n]] (Doc.aldebaran.com, 2020c).

Where "n" represents the number of sound elements found in the peak sound detected. Each index will have a "type", where type_1 represents the start of the sound, and type_0 represents the end of the sound captured. So, to retrieve the peak sound from this SoundDetected event, each element must be analysed and when type_1 is found, a signal is stimulated. This is achieved from the Array Streaming script that is present in the Sound Peak box. This script selects each element one at a time, and if the sound index equals to 1, then a signal is stimulated to the next box function. The reason for using this RMS method to create an average signal, is that it is easier for TOM to perceive, than a raw audio signal. When analysing raw audio waveforms, a random peak sound could occur when recording and TOM could miss that peak sound and not stimulate any signals. But, by calculating the RMS power level of the continuous waveform, that was created by the drum hit, it makes it easier for TOM to perceive and analyse the audio signal, which makes this function more reliable with sending power signals to other functions.

The output signals that the Sound Peak boxes produce are sent to Timeline boxes. These timeline boxes are used to create custom animations that TOM performs, in order to hit the drums. There are three timeline boxes used in this application, each of which represent the three previously talked about drum rudiments that will be played in the drum lesson (Fig. 5).

Animation and Reset—Arms and Bumpers

The custom animation process focuses on the movement of the arm joints. This involved using the Inner Motion Timeline, Robot View, the Inspector and a Physical Robot. The physical robot was essential to this process, as custom animations cannot be made with a virtual robot in Choregraphe 2.8. When a keyframe is selected in the Inner Motion Timeline, it saves the current position of the robot into that keyframe. The robot view was used to choose an arm joint that wanted to be moved, once chosen the inspector would appear with multiple actuators associated with the arm joint. In the inspector, the actuators were adjusted by using both the sliders and entering degree values to move the actuators into the desired position. Through vigorous testing of multiple actuator positions and keyframe placements, suitable animations and motions were designed and developed for each drum rudiment.

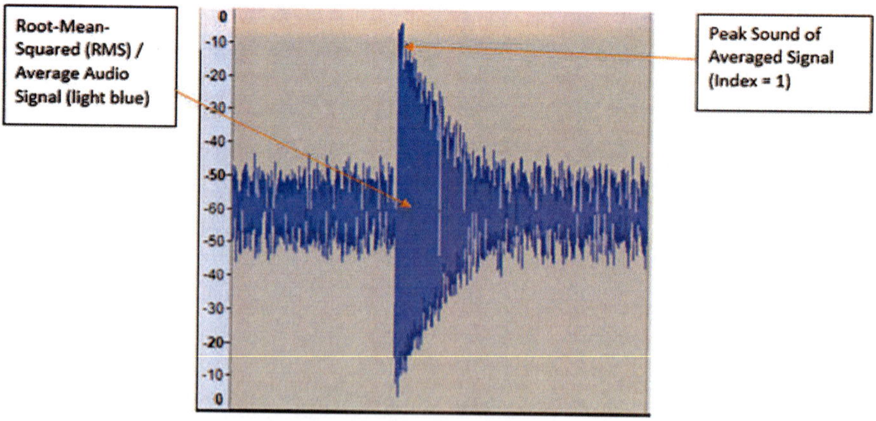

Fig. 5 Close up of drum hit only waveform

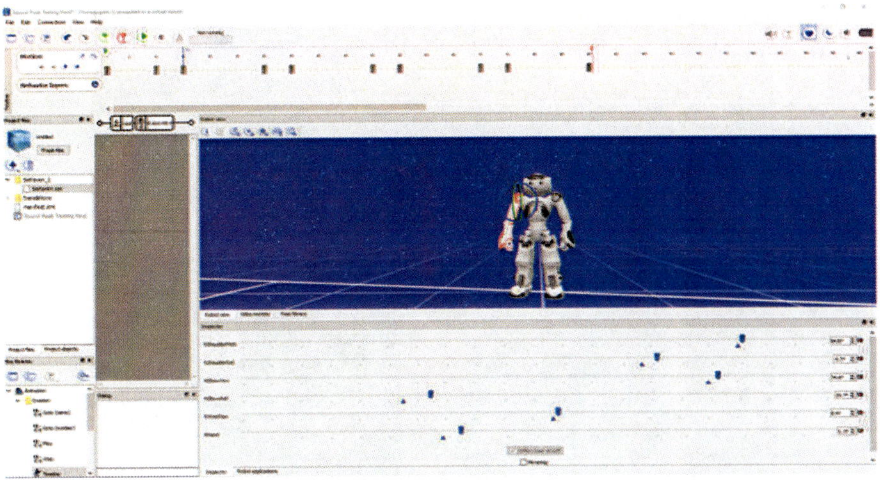

Fig. 6 Inner motion timeline, robot view, joint inspector

By discovering the complexity of creating custom animations with python, comparing it with the previous custom python scripts, and how the Timeline box provide quick and easy to create animations, it made sense to continue using the Timeline boxes to develop the rest of the animations for the other drum rudiments, to help alleviate the time constraints for this project. The bumper sensors located on both feet of TOM were used to implement the Reset function that was needed for this exploration. Once the sound peak boxes are activated, they will continue to run even when moving onto the next drum rudiment. This makes multiple sound peak boxes listen for sound peaks at the same time, which will confuse TOM on which drum rudiment to play, and result with potentially playing the wrong drum rudiment.

The positive and negative feedback was taken into consideration on what features should be focused on, and how the strengths and weaknesses can be addressed in future works (Figs. 6 and 7).

4 Reflection and Conclusion

An investigation has been conducted on how a prototype Robotic Musicianship can be used as a learning tool, which enhances engagement and musical development, in a drum lesson scenario. TOM is capable of beat detection by using basic sound peak detection, on drum hits that a student or teacher performs. The tactile sensors provides different modes that accommodate the multiple drum rudiments that TOM can play [13]. The appearance and basic movements makes TOM suitable and engaging for teaching younger audiences and would be appropriate for lower levels of drum teaching. Additional drum rudiments could also be implemented in the

Fig. 7 Robot Tom played electronic snare drum testing & exported animations to python code, from first drum rudiment testing

future by using new and different tactile gestures. However, while TOM can perform basic drumming techniques, it is however limited or not suitable for advanced drum teaching that requires better functionality and fluidity for advanced drumming techniques. The move and force from TOM, is also limited. This presents less variation in drum hit velocity, which limits the ability of performing different drum styles, and dynamics. The sound detection functionality can be enhanced further, with increasing the accuracy in detection for background noise removal, e.g. only capturing instrumental audio and not conversations from humans. At a concert setting, the issue of unwanted mechanical sounds would not matter much, but within a classroom setting, the noise of the robot could cause distractions. While replacing components worked for Strumbot [2], it could cause more problems than solutions when using a NAO robot, but considering this issue and solution would be useful for the future when constructing an original robot for a long term research and design in the next phase.

References

1. Kapur A (2005) A history of robotic musical instruments. ICMC. 10.1.1.88.4599
2. Vindriis R, Carnegie D (2016) StrumBot—an overview of a strumming guitar robot. In: Proceedings of the international conference on new interfaces for musical expression, Queensland Conservatorium, Griffith University. NIME, pp 146–151
3. Brooks R, Breazeal C, Marjanovic M, Scassellati B, Williamson M (2002) The Cog Project: building a humanoid robot. Lecture notes in artificial intelligence, vol 1562. https://doi.org/10.1007/3-540-48834-0_5
4. Brooks RA, Breazeal C, Marjanović M, Scassellati B, Williamson MM (1999) The Cog Project: building a humanoid robot. In: Nehaniv CL (ed) Computation for metaphors, analogy, and

agents. CMAA 1998. Lecture notes in computer science, vol 1562. Springer, Berlin, Heidelberg
5. Rowe R (2004) Machine musicianship. MIT Press, Cambridge, MA, USA
6. Cope D (2013) The well-programmed clavier. XRDS: Crossroads, The ACM Magazine for Students. Association for Computing Machinery (ACM) 19(4):16
7. Weinberg G, Driscoll S, Parry RM (2007) HAILE—an interactive robotic percussionist, In: HRI'07, 8–11 Mar, Arlington, Virginia, USA
8. Hoffman G, Weinberg G (2010) Shimon: an interactive improvisational robotic marimba player. In: Conference on human factors in computing systems—proceedings, pp 3097–3102. https://doi.org/10.1145/1753846.1753925
9. Xia G, Tay Dannenberg R, Veloso M (2012) Autonomous robot dancing driven by beats and emotions of music. In: AAMAS '12: proceedings of the 11th international conference on autonomous agents and multiagent systems, vol 1, pp 205–212
10. Nao. Softbank. www.softbankrobotics.com/emea/en/nao. Accessed 18 June 2020
11. Ohkita M, Bando Y, Ikemiya Y, Itoyama K, Yoshii K (2015) Audio-visual beat tracking based on a state-space model for a music robot dancing with humans. In: IEEE/RSJ international conference on intelligent robots and systems (IROS), 28 Sept–2 Oct 2015. Hamburg, Germany, pp 5555–5560
12. Murata K, Nakadai K, Takeda R, Okuno HG, Torii T, Hasegawa Y, Tsujino H (2008) A beat-tracking robot for human-robot interaction and its evaluation. In: Humanoids 2008—8th IEEE-RAS international conference on humanoid robots, Daejeon, pp 79–84
13. Tom R (2020) YouTube demo. www.youtube.com/watch?v=56Nl4tGbjAA

Investigation on Accuracy of Sensors in Sensor Fusion for Object Detection of Autonomous Vehicle Based on 2D Lidar and Ultrasonic Sensor

Mohammad Hazrul Ashraf Bin Rosdi and Ahmad Shahrizan Abdul Ghani

Abstract This paper presents the investigation of sensor fusion between two types of sensors for object detection which is used to be equipped in the application of autonomous vehicles (AV) prototype. This research involves usage of low-cost sensors including 2D Lidar (Lidar) and ultrasonic sensor where the investigation is conducted by integrating the collected data from both sensors and then implementing the obstacle detection algorithm method by means of sensor fusion. The main objective of this project is to investigate the sensors capability and accuracy for detection purpose. In addition, the investigation involves the area of detection covered by both sensors based on sensor fusion. Lidar is used for long-range object detection and the ultrasonic sensor is used to detect the object in the short-range distance. The data collected from both sensors are processed and fused to determine the location of the detected object in the front field of view (FOV) of AV prototype. The performance of the implemented method shows good results where both sensors can detect objects in the front area of the prototype at close range as well as far-range.

Keywords Autonomous vehicle · Sensor fusion · Object detection · 2D lidar · Ultrasonic sensor

1 Introduction

Improvements in the automotive field in line with the rapid advances technology has led to the evolution of autonomous vehicles (AV). Over the past few decades, automotive companies and research organizations such as Mercedes-Benz, Audi, Nissan, and Google have developed AV prototypes aimed at creating a future car that can be entirely controlled by systems and sensors without relying on drivers. AV is

M. H. A. B. Rosdi · A. S. Abdul Ghani (✉)
Faculty of Manufacturing and Mechatronic Engineering Technology, Universiti Malaysia Pahang, 26600 Pekan, Malaysia
e-mail: shahrizan@ump.edu.my

M. H. A. B. Rosdi
e-mail: hazrulrosdi95@gmail.com

A. F. Ab. Nasir et al. (eds.), *Recent Trends in Mechatronics Towards Industry 4.0*, Lecture Notes in Electrical Engineering 730, https://doi.org/10.1007/978-981-33-4597-3_68

a car that fully equipped with sensors and the combination of those sensors called "sensor fusion".

One of the tasks an AV can do is it can navigate the road while avoiding any obstacles the vehicle may encounter. The AV can accomplish this task by using sensors to detect where the obstacle located on the route and these sensors vary from close-range infrared sensors to long-ranged high-frequency radar and global positioning satellites (GPS) [1]. From all of the previous researches, most of the methodology that was used by researchers are by combined at least two types of sensors to get an accurate result for object detection. Transportation, especially cars, has become an essential and necessary component in our daily life to perform any tasks. The varieties of models and design that currently existed in the market had been improved over time and the evolution of vehicle had led to the creation of a revolutionary product known as "autonomous vehicle". The main factors that differentiate the AV with other regular vehicles are the implementation of the sensors in each aspect of the vehicle. Furthermore, an AV must be able to detect and accurately track nearby moving objects in real-world driving conditions [2].

In an AV, to process all of the information and data in expeditious time, it required the sensor with high performance to process and give responses towards the obstacles that were detected through the sensor so the vehicle can avoid any collision. In the high-level research, the prototype of AV that was developed has the capability to detect the movement of humans and able to recognize the traffic sign on the road. Therefore, to create the AV prototypes, the main objectives of the prototype must be considered depending on the research purposes. This paper is distributed as follows: Sect. 2 discussed the previous related works. Section 3 discussed the implementation and experiment setup. Section 4 shows the experimental results, and finally, Sect. 5 is where the conclusion and recommendation are discussed.

1.1 Research Background

Recent advances in the development of AV have led to the diversity of integration especially when it involves the use of sensors to enhance the function and effectiveness of AV. Depends on the goal, the selection of the sensor must be suitable based on what needs to be accomplished by the AV because normally it developed with different functions and features. As an example, [3] was implemented the fusion of LiDAR and camera sensors because their work is only focused on pedestrian detection to ensure high detection accuracy.

The combination of several sensors such as camera, Lidar, and radar in AV known as "sensor fusion". Sensor fusion is a technique to integrate data from disparate sources to generate precise information [4]. The resulting data is more accurate compared to the single-use of any sensors. This particularly important when combining different types of information that will provide the most accurate results. The use of various kinds of sensors in a system of an AV is not a problem because the implementation of sensor fusion has made it easy.

However, in implementing the sensor fusion mechanism, the selection of sensors with the highest level performance and function is essential as the characteristics of the sensors would be the top priority to produce the AV that has excellent quality. The parameters that need to consider when selecting sensors are the area of detection, response time, accuracy, and processing speed. These parameters are crucial to consider in choosing the appropriate sensors for sensor fusion implementation. In previous researches, several algorithms and architectures were created by researchers as an alternative to implement the sensor fusion method into their study based on the purpose of the research itself. The previous research study is discussed in Sect. 2.

2 Related Work

The development of an AV requires the use of different type of sensors depends on the main goal of the research itself. In [5], the main research contributes to the detection of obstacles in the front area of the vehicle by using an ultrasonic sensor, Logitech USB web camera, and Raspberry Pi 2B as the processor. The distance between the vehicle and obstacle is measured by the ultrasonic sensor and the camera is used for detecting the moving or stationary objects through the image. By fusing both of these sensors, the obstacles with distances are accurately measured and the system also able to distinguish between moving and static obstacles. For the fusion method, the ultrasonic sensor and the camera are connected to the raspberry pi board and processed through the Open-CV software and the color detection technique was used to process the detected image. The ultrasonic sensor measures the distance from the object and its reading is combined with the image from the camera. The outcome of the research is proven by analyzing the shape and color of the object that was detected and the distance is also measured. Unfortunately, the ultrasonic sensor only able to detect the obstacle in a short-range only and if the distance of the obstacle more than 70 cm, the measured distance will not be accurate.

On top of that, research on the real AV had successfully developed by [6] that capable to operate in narrow two-way roads. The author had applied the low-cost sensor consists of Global Navigation Satellite System (GNSS) BU-353-24, Logitech USB webcam C270, gyroscope, odometer, and encoder. The GNSS receiver is a standard low-cost model with a USB interface. The main contribution of the author is the implementation of two-level fusion. The first level fusion is known as the back lane marking registry (BLMR) that merges data from the lane detector method that was generated by using the camera with a computer vision algorithm and the data from dead reckoning method that strongly depends on the gyroscope and odometer. Dead reckoning is a technique that allows the estimation of the current position by knowledge of the previous position and integration of speed [7]. The second level of fusion is localization performed by a scheme of filter that uses information from the GNSS and BLMR to update the estimation of vehicle position on the map.

In another previous study carried out by [8], it has developed an intelligent monocular vision autonomous car prototype that capable of reaching the destination safely

Fig. 1 The detected road indicated as the blue lines

while avoiding the obstacles. The ultrasonic sensor and Pi camera are the main sensor used for the project along with raspberry pi B as the processor. The lane and obstacle detection algorithm are fused together by the author as the fusion mechanism to provide the necessary control over the AV prototype. A combination of feature and model base is used in the lane detection algorithm consists of seven major methods where the author confirms that all of the methods are valid for any kind of roads whether they are marked with white lanes or not. The detected lane and road are indicated as the blue line as shown in Fig. 1.

For the obstacle detection algorithm, the ultrasonic sensor analyzes the surrounding to calculate the distance of the obstacles from the AV after a fixed interval of time. The author declares that the method and algorithms used have been successfully implemented and the AV prototype able to determine the lane on the road and avoid the obstacles. One other thing, the ultrasonic that was used in the research has the potential to hit the obstacles if the prototype moves too fast so the precaution must be taken to avoid the collision from occurring.

A research done by [9] focused on the sensor fusion technique for Autonomous Mobile Robot (AMR) that required less cost both in terms of economically and computationally. It allows AMR to detect an obstacle, find the distance, and measure the size of the obstacle by using the Canon 550D camera and ultrasonic sensor. The authors have proposed the combination of appearance-based and range-based obstacle detection techniques to measure the obstacle size as well as the distance from AMR. For the range-based using an ultrasonic sensor, the distance of the obstacle can be measured by using the formula in Eq. (1) once the speed of the sound wave is known.

$$Distance = speed \times time \tag{1}$$

Another essential point, the obstacle size measurement has been done by taking visual information of the object from the camera. The image processing technique was processed by the MATLAB software where the author has converted the color image to a grayscale image for computational simplicity. A book entitled "Stephen Johnson on Digital Photography" [10] states that grayscale is an image in which the value of each pixel contains only intensity information and this type of images consist exclusively of gray shades that vary from black at the lowest intensity whereas white to strong intensity. After conversion to gray scale, thresholding has been performed on the image to separate the object from the background. Thresholding is a simple method of segmentation that can be used to convert grayscale images to binary images [11]. Then, the opening and closing process was performed after the thresholding to eliminate all noises from the image. Finally, the dimensions of the object in the image can be measure after isolating the object in the image. The horizontal length and vertical height of the object is found by using the Eqs. (2) and (3) respectively.

$$p = \frac{hq}{m} \tag{2}$$

p object length in real life
h horizontal distance
q object length in image
m horizontal s pixels

$$t = \frac{vs}{n} \tag{3}$$

$$v = 2x \tan\left(\frac{\alpha}{2}\right) \tag{4}$$

t vertical length of the object in real life
s vertical of the object image in terms of pixel
n vertical pixels

For instance, the 3D LiDAR has been used by [12] in their research entitled "Fusion of 3D-LIDAR and Camera Data for Scene Parsing" by using the monocular video camera and Velodyne HDL-64E 3D LiDAR. Velodyne HDL-64E is a great LiDAR scanner on the market where its effective coverage limits within 70 m from the center of the sensor. Nevertheless, due to the fact that 3D LiDAR is more expensive than 2D LiDAR, most previous studies have applied 2D LiDAR into their study to save costs. However, one disadvantage of 2D LiDAR is its senses using only a single horizontal scanning line and one of the efficient solutions to overcome this problem by using sensor fusion [13].

In conclusion, the main thing that should be the focus on the development of AV is the number of sensors that need to be applied to it. As mentioned earlier, the AV supposedly has the capability to detect any object and obstacle in its surrounding area. One of the significant reviews that have been highlighted from the previous research is the sensors that was used is too expensive. Due to the critical problems identified, this research will focus on the main objective of using sensors that are reasonably priced as well as providing good quality for obstacle detection at the desired area accurately. The purpose of this research is also to ensure that the sensor used can detect objects from close range to long-distance.

3 Methodology

Hardware and software used in this method are shown in the Table 1.

Ultrasonic sensor will determines the distance to a target by measuring time lapses between the sending and receiving of the ultrasonic pulse for the short distance object where this data will be used to implement the obstacle detection algorithms. The ultrasonic sensor and Lidar was chosen because the ultrasonic can detect any obstacle between the range 1 cm to 1.5 m practically, and the measuring angle is 15° only. Meanwhile, the Lidar can detect the obstacle from the range 12 cm to 10 m where the measuring angle is 360°. If the obstacle suddenly exists in the range 1 cm to 11 cm, Lidar unable to detect the obstacle and the collision will occur between the AV prototype and the obstacle. So, the ultrasonic will be used to detect short-range obstacles, while Lidar will be used for long-range detection.

3.1 *Block Diagram*

The project system consists of an Arduino Uno board, 2D YD-Lidar X4, and Ultrasonic sensor. The detection of the object through the laser scanning will be done by the LiDAR and the Ultrasonic sensor will detect the object using ultrasonic wave's detection. All of these input are interfaced with the Arduino Uno board. The data from LiDAR and ultrasonic sensor will be process and display in the serial monitor in Arduino software as shown in Fig. 2.

The LiDAR placed on the prototype will be connected directly to the Arduino Uno in order to obtain the live data scanned by LiDAR. The data acquired from LiDAR contain information about the distance of the obstacle and the angle of LiDAR rotation to scan the points. Meanwhile, the Ultrasonic sensor will determines the distance to a target by measuring time lapses between the sending and receiving of the ultrasonic pulse for the short distance object where this data will be used to implement the obstacle detection algorithms.

Table 1 List of hardware and software uses in this investigation

No.	Hardware	Description	Software	Description
1	Arduino UNO Rev3	Microcontroller	Arduino IDE	Programming platform
2	YD Lidar X4	Object detection sensor (long-range)	Arduino IDE	Programming platform
3	HC-SR04 ultrasonic sensor	Object detection sensor (short range)		

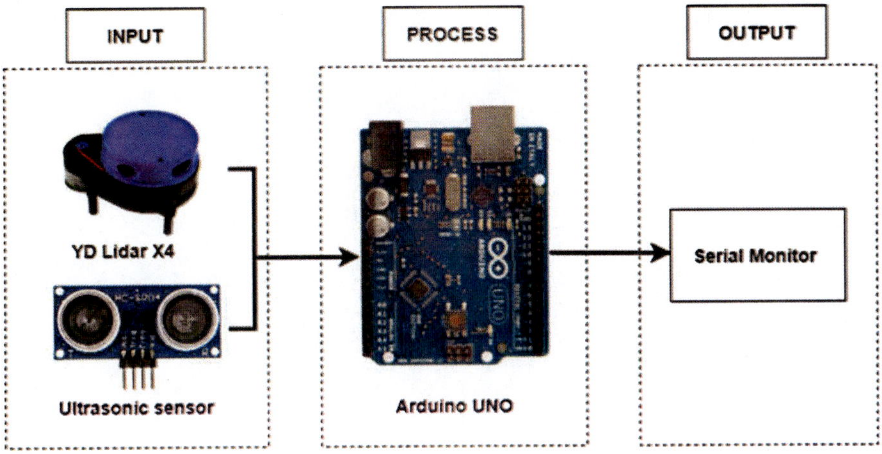

Fig. 2 The block diagram of the system

3.2 *Experiment Setup and Process*

The experiment was conducted on a laboratory-scale AV prototype where the distance of the obstacle was set and manually measured. Then, the sensor used to detect the obstacle distance so that the accuracy of each sensor can be analysed.

Ultrasonic Sensor Testing

The ultrasonic signal experiment is conducted to measure the distance value and compared it with the manual measurements. The distance of the obstacle can be measured by using the formula in Eq. (4) once the speed of the sound wave is known [12].

$$Distance = speed \times time \quad (5)$$

Lidar Testing

Lidar X4 can detect the obstacles from 12 cm to 10 m, and it also can identify the angle location of the obstacles since the lidar can rotate 360°. So, it depends on the user whether the data need to be collected on a particular angle only, and this parameter can be changed in the coding.

Sensor Fusion

The integration of the Lidar and ultrasonic sensor is the important part where both of the data will be compare to make sure both of the sensor detect the obstacle within their range. If the obstacle in the range detection of both sensor, the data for each sensor will be display in the serial monitor and their coding are combined into a single system. Part of the coding shown in the Table 2.

Table 2 Coding for the detection algorithm

```
Algorithm: Obstacle detection algorithm
if(Lidar.Connect()):
    print(Lidar.GetDeviceInfo())
    gen = Lidar.StartScanning()
    t = time.time() # start time
    while (time.time() - t) < 120:
        for angle in range(0,360):
            if(data[angle]>0 and data[an-
gle]<181): #collect data between angle 1 to
180 degree only
                x[angle] = data[angle] *
math.cos(math.radians(angle))
                y[angle] = data[angle] *
math.sin(math.radians(angle))
```

4 Experimental Results

From Table 3, it can be observed that obstacle has been located on four different distances, which are 0.5 m, 1 m, 1.5 m, and 2 m. The three total of trials for each of the distance was set up to see the stability of the detection at different trials. From the observation, it can be perceived that YD-Lidar X4 can detect the obstacles with 96% accuracy in the range of 0.5 m while 4% error was recorded and the ultrasonic sensor

Table 3 Data obtained from YD-Lidar X4 and ultrasonic sensor

Obstacle distance (m)	Number of trials	Measured distance	
		YD-Lidar X4 (m)	HC-SR04 ultrasonic sensor (m)
	1	0.48	0.5
0.5	2	0.48	0.5
	3	0.48	0.5
	1	0.96	0.62
1	2	0.97	0.63
	3	0.96	0.63
	1	1.43	0.89
1.5	2	1.45	0.86
	3	1.44	0.88
	1	1.94	1.07
2	2	1.95	1.08
	3	1.89	0.99

Table 4 Percentage of accuracy and error of YD-Lidar X4 and ultrasonic sensor

Obstacle distance (m)	Percentage of accuracy		Percentage of error	
	YD-Lidar X4 (%)	HC-SR04 ultrasonic sensor (%)	YD-Lidar X4 (%)	HC-SR04 ultrasonic sensor (%)
0.5	96	100	4	0
1	96.33	62.67	3.67	37.33
1.5	96	58.44	4	41.56
2	96.33	52.33	3.67	47.67

was able to identify the obstacles 100% accurate within the same range without any error. The ultrasonic can detect the obstacle 100% without any error because the surrounding environment where no other object detected by the pulses travel from the ultrasonic. If the obstacle located far from the ultrasonic location, the pulses travel through the air away from the transmitter will tend to detect the other object in the surrounding, and it will reduce the accuracy of ultrasonic. Next, the accuracy of the YD-Lidar X4 for 1 m distance is 96.33% and 62.67% for the ultrasonic sensor. The percentage of error recorded by the YD-Lidar X4 and ultrasonic sensor is 3.67% and 37.33%. Then, for the obstacle distance in 1.5 m, the accuracy recorded by the YD-Lidar X4 is 96%, followed by 58.44% for ultrasonic sensor while 4% and 41.56% of error for YD-Lidar X4 and ultrasonic sensor. Finally, the accuracy recorded at 2 m distance was still in the same percentage for the YD-Lidar X4 where 96.33% was recorded similar at the 1 m distance. Meanwhile, the accuracy for the ultrasonic sensor is hugely lower compared to the YD-Lidar X4, where the accuracy only 52.33%. The percentage of errors for both sensors is 3.67% and 47.67% for YD-Lidar X4 and ultrasonic sensors. Table 4 show the comparison percentage of accuracy and error of YD-Lidar X4 and ultrasonic sensor for each tests.

5 Conclusion and Recommendation

In this paper, we presented the research regarding the use of two types of low-cost sensors with the implementation of sensor fusion to evaluate the quality and performance of these sensors to see if it is relevant to be equipped in autonomous vehicles. Based on the result obtained from the experiment, it can be concluded that Lidar and ultrasonic sensors are relevant to be equipped in the prototype of the AV because both of the sensors able to detect the obstacles in the short-range and long-range. The weakness of each sensor manages to be overcome.

The future improvement of the project, the better microprocessor like Raspberry Pi should be considered to be used instead of Arduino because the processing time will affect the quality of detection. Finally, the output such as DC motor should be used as the output where the user may control the direction of the motor to avoid detected obstacles so the scope of research will be more extensive and challenging.

Acknowledgements The authors would like to thanks all reviewers. This work is supported by the internal Universiti Malaysia Pahang Research Grant RDU1903140 entitled 'Analyzing the Effects of Dark Channel Fusion and Pixels Distribution Shifting on Image Histogram for Development of Remotely Surface Vehicle.'

References

1. Wall RW, Bennett J, Eis G, Lichy K, Owings E (2002) Creating a low-cost autonomous vehicle. In: IECON proceedings (industrial electronics conference), vol 4, pp 3112–3116
2. Cho H, Seo YW, Kumar BVKV, Rajkumar RR (2014) A multi-sensor fusion system for moving object detection and tracking in urban driving environments. In: Proceedings—IEEE international conference on robotics and automation, pp 1836–1843
3. Wu TE, Tsai CC, Guo JI (2018) LiDAR/camera sensor fusion technology for pedestrian detection. In: Proceedings—9th Asia-Pacific signal and information processing association annual summit and conference, APSIPA ASC 2017, 2018-Feb (December), pp 1675–1678
4. Kocic J, Jovicic N, Drndarevic V (2018) Sensors and sensor fusion in autonomous vehicles. In: 2018 26th telecommunications forum, TELFOR 2018—proceedings, pp 1–4
5. Kumar RS, Stanley PK, Gandhi AS(2017) Raspberry Pi based vehicle collision avoidance system. In: Proceedings of IEEE international conference on innovations in electrical, electronics, instrumentation and media technology, ICIEEIMT, 2017-Jan (978), pp 211–215
6. Vivacqua R, Vassallo R, Martins F (2017) A low cost sensors approach for accurate vehicle localization and autonomous driving application. In: Sensors (Switzerland), vol 17
7. Denny M (2012) The science of navigation: from dead reckoning to GPS. In: Choice reviews online, vol 50
8. SinghPannu G, Dawud Ansari M, Gupta P (2015) Design and Implementation of autonomous car using raspberry Pi. Int J Comput Appl 113(9):22–29
9. Shahdib F, Mahmud H (2013) Obstacle detection and object size measurement for autonomous mobile robot using sensor. Int J Comput Appl 66:28–33
10. Johnson S (2006) Stephen Johnson on digital photography. Retrieved from http://books.google.com/books?id=0UVRXzF91gcC&pgis=1
11. Gonzalez RC, Woods RE (2007) Digital image processing (3rd edn). In: Prentice-Hall, Inc. Upper Saddle River, NJ, USA ©2006. Retrieved from http://dl.acm.org/citation.cfm?id=1076432
12. Zhao G, Xiao X, Yuan J, Ng GW (2014) Fusion of 3D-LIDAR and camera data for scene parsing. J Visual Commun Image Representation 25(1):165–183
13. Gatesichapakorn S, Takamatsu J, Ruchanurucks M (2019) ROS based autonomous mobile robot navigation using 2D LiDAR and RGB-D camera. In: 2019 1st International symposium on instrumentation, control, artificial intelligence, and robotics, ICA-SYMP 2019, pp 151–154

Decision Support System on Determination of Contraception Tools as an Effort to Suppress the Number of Growth Ratios in Indonesia

Agus Perdana Windarto and Tutut Herawan

Abstract This study aims to recommend the selection of contraception as an effort to reduce the number of growth ratios in Indonesia with the Multi-Objective Optimization method on the basis of Ratio Analysis (MOORA). The data comes from the National Population and Family Planning Agency and the Central Statistics Agency using 5 assessment criteria i.e. security (C1), time period (C2), age (C3), cost (C4), size (C5) and 4 alternatives family planning used i.e. birth control pills (A1), KB injections (A2), implants (A3) and spiral (A4). The sample of this study were female aged 15–49 years and married who are used to family planning program. The results by using the MOORA method showed that from the 4 alternatives used, it was obtained the recommendation of implant contraceptive selection (A3) as the highest rank and birth control pills (A1) as the second rank 2. This study will help the family planner in choosing appropriate and safe family planning program and it further can assist the government in suppressing the number of growth ratios in Indonesia.

Keywords Decision support system · MOORA · Contraception tools · Population number

1 Introduction

In Indonesia, population growth is considered as a wake-up alarm every year. This is marked by the rate of population growth in Indonesia which reaches 1.49% per year from the total population of Indonesia starting in 2015 (URL: https://bps.go.id/). Therefore, the government is trying to suppress the ratio of population growth in Indonesia to a minimum of 1.1%. One of the ways of the government is by educating family planning programs. The Family Planning Program in Indonesia began around 1957. In 1970 the Family Planning program was inaugurated as a government

A. P. Windarto (✉)
STIKOM Tunas Bangsa, Pematangsiantar, North Sumatera, Indonesia
e-mail: agus.perdana@amiktunasbangsa.ac.id

T. Herawan
Sekolah Tinggi Pariwisata Ambarrukmo, Yogyakarta, Indonesia

A. F. Ab. Nasir et al. (eds.), *Recent Trends in Mechatronics Towards Industry 4.0*, Lecture Notes in Electrical Engineering 730, https://doi.org/10.1007/978-981-33-4597-3_69

program. Family Planning is a movement to form healthy and prosperous families by limiting number of births. Based on data from the Central Bureau of Statistics (2014–2017), the female aged 15–49 years and married who are using contraceptives has been decreased. In year 2017, the number of women using contraception is 58.70%. This is characterized by many women having difficulty in choosing the type of contraception. This is not only due to limited knowledge, but also their ignorance of the requirements and safety of using these contraceptives.

Data mining and knowledge discovery (DMKD) model exist to solve many real-life problems, including healthcare and medicine [1–5]. The evidence from previous researches showed that DMKD has been used in solving problems in the field of healthcare [6–9] and medicine [10–16]. Still however, studies of DMKD model in supporting family planning still received less attention.

Motivated from above explanation, we present a DMKD model based on decision support system to solve the problem of family planning. We use the Multi-Objective Optimization method on the basis of Ratio Analysis (MOORA). The MOORA is a multi-objective optimization technique that can be successfully applied to solve various types of complex decision making problems [17–20]. This method has a good level of selectivity in determining an alternative. The MOORA's approach is defined as a process simultaneously to optimize two or more conflicting ones on several controls. This research is expected to help the family community in choosing appropriate and safe contraceptives. Furthermore, the results are also expected to help the government in suppressing the number of growth ratios in Indonesia so that the purpose of the formation of family planning programs in improving family welfare can be realized.

2 Proposed Method

2.1 Decision Support System

Decision Support System (DSS) is a series of procedures that are interconnected in carrying out a particular job which consists of complementary elements because of one purpose, goals and objectives [21]. This is done to improve the process and quality of decision making by integrating data and knowledge to improve effectiveness and efficiency in decision making [22].

2.2 Multi-objective Optimization Method on the Basis of Ratio Analysis (MOORA)

The MOORA method is a system that optimizes two or more conflicting attributes simultaneously. This method has a good level of selectivity because it can determine goals from conflicting criteria. Where criteria can be of benefit value or cost value [17, 18]. The steps to complete the MOORA method are as follows [23, 24]:

a. Identify the attributes used.
b. All attributes that have been identified are formed in the decision matrix. Data is described as x_{ij}. Where x_{ij} is an alternative to i in the attribute to j. Then the system ratio is developed into an alternative comparison, where attributes are compared as denominators. The attribute itself is a representative for all these alternatives.

$$X = \begin{bmatrix} x_{11} & x_{12} & \dots & x_{1n} \\ x_{21} & x_{22} & \dots & x_{2n} \\ \dots & \dots & \dots & \dots \\ x_{m1} & x_{m2} & \dots & x_{mn} \end{bmatrix} \quad (1)$$

c. As the denominator, the square root of the sum of squares of each alternative is chosen. This ratio can be formulized as follows:

$$x_{ij}^{*} = \frac{x_{ij}}{\sqrt{\left[\sum_{i=1}^{m} x_{ij}^{*}\right]}}, \text{ for } j = 1, 2, \dots, n \quad (2)$$

where x_{ij} is the dimension number as interval [0, 1] which is then normalized the calculation from alternative to i in attribute to j.
d. In optimizing objects, they are normalized with the maximum value minus the minimum value. The optimization formula is given as follow:

$$y_i = \sum_{j=1}^{m} x_{ij}^{*} - \sum_{j=g+1}^{n} x_{ij}^{*} \quad (3)$$

where g is the maximum attribute, $(n–g)$ is the number of attributes that will be the minimum value, and y_i is the alternative normaliation value for all attributes. To make attribute values more accurate, it can be done by multiplying the appropriate weight. When calculating the attributes multiplied by the weight, the results are as follows:

$$y_i = \sum_{j=1}^{g} w_j x_{ij}^{*} - \sum_{j=g+1}^{n} w_j x_{ij}^{*} \quad (4)$$

e. Values can be positive or negative depending on the maximum and minimum amounts in the decision matrix. And the final result of the calculation is showing the results of ranking y_i. Thus, the best alternative value has the highest y_i value. While the lowest alternative value has the lowest y_i value.

3 Result and Discussion

In this study five criteria are used: security (C1), time period (C2), age (C3), cost (C4), size (C5) and 4 alternatives such as birth control pills (A1), KB injections (A2), implants (A3) and spiral (A4). While the weight and value of each criterion can be seen in Table 1.

Based on Table 2, the assessment of weight is done by using percentage, where the criteria C1 and C4 are cost, meanwhile C2, C3, and C5 are benefits. In determining benefit or cost criteria, we use the concept of reasoning based on existing facts.

Based on Table 2, the weighting of each criterion is carried out according to the needs of the study. The following is the data used and obtained based on the information from each contraceptive product according to the criteria used as shown in the following Table 3.

Based on Table 3, the data is converted based on the weighting assessment in Table 2 so that the following results are obtained.

The data that has been converted in Table 4 is then processed by the MOORA method. Previously the data was changed into a decision matrix using formula in Eq. (1).

$$X = \begin{bmatrix} 5\ 5\ 5\ 5\ 5 \\ 3\ 1\ 5\ 3\ 3 \\ 3\ 3\ 5\ 3\ 3 \\ 5\ 3\ 5\ 1\ 3 \end{bmatrix}$$

Data that has been changed into the decision matrix will be normalized using formula in Eq. (2).

$$X = \begin{bmatrix} 0.606339\ 0.753778\ 0.5\ 0.753778\ 0.693375 \\ 0.363803\ 0.150755\ 0.5\ 0.452267\ 0.416025 \\ 0.363803\ 0.452267\ 0.5\ 0.452267\ 0.416025 \\ 0.606339\ 0.452267\ 0.5\ 0.150756\ 0.416025 \end{bmatrix}$$

Then optimizing the attribute value by multiplying the value of each alternative for each criterion with the weighting criteria that results in the optimization of attribute values.

Table 1 Weight and Criteria

Criteria	Weight (%)	Description
Security (C1)	30	Cost
Time period (C2)	25	Benefit
Age (C3)	20	Benefit
Cost (C4)	15	Cost
Size (C5)	10	Benefit

Table 2 Weighting assessment

Security (C1)	Value
High	5
Medium	3
Low	1
Age (C3)	Value
>35 years	1
20–35 years	5
<19 years	3
Size (C5)	Value
Big	1
Medium	3
Smal	5
Time Period (C2)	Value
>10 years	5
1–10 years	3
<1 year	1
Cost (C4)	Value
<IDR 100.000	5
IDR 100.000–IDR 500.000	3
>IDR 500.000	1

Table 3 Research Data

Alternatives	Criteria				
	C1	C2	C3	C4	C5
Birth control pills (A1)	High	50 years	20–35 years	IDR 50.000	Small
KB injections (A2)	Medium	3 months	20–35 years	IDR 100.000	Medium
Implants (A3)	Medium	3 years	20–35 years	IDR 250.000	Medium
Spiral (A4)	High	10 years	20–35 years	IDR 500.000	Medium

Table 4 Conversion data

Alternatives	Criteria				
	C1	C2	C3	C4	C5
Birth control pills (A1)	5	5	5	5	5
KB injections (A2)	3	1	5	3	3
Implants (A3)	3	3	5	3	3
Spiral (A4)	5	3	5	1	3

Table 5 Value of Yi and ranking

Alternatives	Maximum	Minimum	Yi	Rank
	C2 + C3 + C5	C1 + C4	Max-Min	
Birth control pills (A1)	0.357782	0.29496847	0.062813642	2
KB injections (A2)	0.179291	0.17698108	0.002310349	4
Implants (A3)	0.254669	0.17698108	0.077688185	1
Spiral (A4)	0.254669	0.20451507	0.050154199	3

$$X = \begin{bmatrix} 0.606339\ 0.753778\ 0.5\ 0.753778\ 0.693375 \\ 0.363803\ 0.150755\ 0.5\ 0.452267\ 0.416025 \\ 0.363803\ 0.452267\ 0.5\ 0.452267\ 0.416025 \\ 0.606339\ 0.452267\ 0.5\ 0.150756\ 0.416025 \end{bmatrix} * \mathrm{W_j} =$$

$$X = \begin{bmatrix} 0.181902\ 0.188444\ 0.1\ 0.113067\ 0.069338 \\ 0.109141\ 0.037688\ 0.1\ \ 0.06784\ \ 0.041603 \\ 0.109141\ 0.113066\ 0.1\ \ 0.06784\ \ 0.041603 \\ 0.181902\ 0.113066\ 0.1\ 0.022613\ 0.041603 \end{bmatrix}$$

Optimization of attribute values is then used to find the value of y to obtain ranking from each alternative. Results of ranking and ranking can be seen in the following Table 5.

Based on Table 5 above, the results showed that from 4 alternatives used, it was found that implants (A3) were the best recommendations for the selection of contraceptives ($Yi = 0.077688185$) and birth control pills (A_1) as the second recommendation on the selection of contraceptives ($Yi = 0.062813642$).

4 Conclusion

This study has presented a DMKD model based on decision support system to solve the problem of family planning. The results stated that the MOORA method can be used to determine contraceptive choices, especially for the family planning community in Indonesia. This as an effort to reduce the number of growth ratios in Indonesia and help the government in creating a prosperous society. Through family planning programs, the 5 assessment criteria used such as security (C1), time period (C2), age (C3), cost (C4), size (C5) and 4 alternatives such as birth control pills (A1), KB injections (A2), implants (A3) and spiral (A4) achieved final result that implants (A3) is the best recommendation for contraceptive selection with the value of $Y_i = 0.077688185$.

References

1. Martinez-Martin N, Insel TR, Dagum P, Greely HT, Cho MK (2018) Data mining for health: staking out the ethical territory of digital phenotyping. NPJ Digit Med 1(1):1–5
2. Yadav P, Steinbach M, Kumar V, Simon G (2018) Mining electronic health records (EHRs) a survey. ACM Comput Surv (CSUR) 50(6):1–40
3. Altaf W, Shahbaz M, Guergachi A (2017) Applications of association rule mining in health informatics: a survey. Artif Intell Rev 47(3):313–340
4. Pawar MP (2016) Review on data mining techniques for fraud detection in health insurance. IJETT 3(2)
5. Kim JC, Chung K (2019) Associative feature information extraction using text mining from health big data. Wireless Pers Commun 105(2):691–707
6. Jalalian A, Tay FE, Liu G (2016) Data mining in medicine: relationship of scoliotic spine curvature to the movement sequence of lateral bending positions. In: Industrial conference on data mining, pp 29–40. Springer, Cham
7. Eickhoff C, Kim Y, White RW (2020) Overview of the health search and data mining (hsdm 2020) workshop. In Proceedings of the 13th international conference on web search and data mining, pp 901–902
8. Abdelghafar S, Darwish A, Hassanien AE (2020) Intelligent health monitoring systems for space missions based on data mining techniques. In: Machine learning and data mining in aerospace technology, pp 65–78. Springer, Cham
9. Suresh A, Kumar R, Varatharajan R (2020) Health care data analysis using evolutionary algorithm. J Supercomputing 76(6):4262–4271
10. Zhang Y, Guo SL, Han LN, Li TL (2016) Application and exploration of big data mining in clinical medicine. Chin Med J 129(6):731
11. Estape ES, Mays MH, Sternke EA (2016) Translation in data mining to advance personalized medicine for health equity. Intell Inf Manag 8(1):9
12. Gatta R, Lenkowicz J, Vallati M, Rojas E, Damiani A, Sacchi L, De Bari B, Dagliati A, Fernandez-Llatas C, Montesi M, Marchetti A (2017) pMineR: an innovative R library for performing process mining in medicine. In: Conference on artificial intelligence in medicine in europe, pp 351–355. Springer, Cham
13. Luque C, Luna JM, Luque M, Ventura S (2019) An advanced review on text mining in medicine. Wiley Interdisc Rev Data Mining Knowl Discovery 9(3):e1302
14. Yang LN, Wu ZL, Yang ZJ Li SG, Ouyang CS (2020) Exploring mechanism of key chinese herbal medicine on breast cancer by data mining and network pharmacology methods. Chin J Integr Med pp 1–8
15. Wang X, Huang Y, Li M, Lin H, Lin C, Yang W, Ye X (2020) Acupoint catgut embedding for insomnia: Protocol for a systematic review and data mining. Medicine 99(9):e19333
16. Hu Y, Yu Z, Chen X, Luo Y, Wen C (2020) A bibliometric analysis and visualization of medical data mining research. Medicine 99(22):e20338
17. Brauers WK, Zavadskas EK (2006) The MOORA method and its application to privatization in a transition economy. Control Cybern 35:445–469
18. Chakraborty S (2011) Applications of the MOORA method for decision making in manufacturing environment. Int J Adv Manuf Technol 54(9–12):1155–1166
19. Ghoushchi SJ, Yousefi S, Khazaeili M (2019) An extended FMEA approach based on the Z-MOORA and fuzzy BWM for prioritization of failures. Appl Soft Comput 81:105505
20. İç YT (2020) A multi-objective credit evaluation model using MOORA method and goal programming. Arab J Sci Eng 45(3):2035–2048
21. Sprague RH, Watson HJ (eds) (1986) Decision support systems: putting theory into practice. Prentice-Hall, Englewood Cliffs, NJ
22. Kukar M, Vračar P, Košir D, Pevec D, Bosnić Z (2019) AgroDSS: a decision support system for agriculture and farming. Comput Electron Agric 161:260–271

23. Arabsheybani A, Paydar MM, Safaei AS (2018) An integrated fuzzy MOORA method and FMEA technique for sustainable supplier selection considering quantity discounts and supplier's risk. J Clean Prod 190:577–591
24. Brauers WK, Zavadskas EK (2009) Robustness of the multi-objective MOORA method with a test for the facilities sector. Technol Economic Dev Economy 15(2):352–375

K-Means Algorithm with Rapidminer in Clustering School Participation Rate in Indonesia

Agus Perdana Windarto and Tutut Herawan

Abstract School participation rate is a measurement of the absorptive capacity of educational institutions towards school-age population. School participation rate is a basic indicator used to assess the access of the population to educational facilities, especially for the school-aged population. The higher the School participation rate, the greater the number of people who have the opportunity to education. However, the increase in School participation rate cannot always be interpreted as increasing the equalization of community opportunities for education. Therefore, this paper employs *K*-Means algorithm which is implemented in data clustering. We cluster the school participation rate of provinces in Indonesia, where the data are obtained from Central Bureau of Statistics. The data used is from 2003–2016 which consists of 34 provinces. The criteria used are School Participation Numbers, which are grouped into four categories. Data will be processed in three clusters i.e. high, medium and low clusters. Centroid data for high cluster are 99.49, 95.74, 77.33 and medium cluster are 44.81, 95.25, 85.88, 59.62 and 16.18, meanwhile for low cluster are 14.06, 13.38, 10.65 and 2.64. We found that the assessment of School participation rate as follow: The highest cluster is achieved by DI Yogyakarta Province; 32 provinces are in the medium cluster and one province in the low cluster i.e. North Kalimantan. These results can be an input to the government in increasing the number of School participation rate as an effort to educate more the school-age population.

Keywords Data mining · *K*-Means · School participation rate · Cluster

1 Introduction

Some indicators can be used as measurement of the level of progress and success of compulsory education programs implemented by local governments. One of them

---

A. P. Windarto (✉)
STIKOM Tunas Bangsa, Pematangsiantar, Sumatera Utara, Indonesia
e-mail: agus.perdana@amiktunasbangsa.ac.id

T. Herawan
Sekolah Tinggi Pariwisata Ambarrukmo, Yogyakarta, Indonesia

A. F. Ab. Nasir et al. (eds.), *Recent Trends in Mechatronics Towards Industry 4.0*, Lecture Notes in Electrical Engineering 730, https://doi.org/10.1007/978-981-33-4597-3_70

is School participation rate [1]. School participation rate is a basic indicator used to assess access of the population to educational facilities, especially for school-aged residents in a region [1, 2]. The higher the School participation rate, the greater the number of people who have the opportunity to receive education. In other words, measuring the amount of society that can be absorbed in education, as well as dropout rates that indicate the number of students who are not in school anymore with various causes.

Data mining and knowledge discovery models has been widely used in education field [3–5]. One of the most popular data mining and knowledge discovery models in solving educational field is Data Clustering [6–9]. Clustering is a method to search and classify data that have similarity characteristics between one data with other data. Clustering is one method of data mining that is unsupervised. The unsupervised method means that this method is applied without training and does not require output target [10–15]. In performing the clustering process, there is an algorithm that is often used because of its relatively fast and easy to adapt i.e. the *K*-Means algorithm [16, 17]. The *K*-Means is one of the simplest unsupervised learning algorithms that solve the well-known clustering problem. The auction follows a certain number of clusters [18–23].

This study aims as a first step in applying data mining and knowledge discovery model in gaining nuggets of School participation rate. The data used is obtained from the Central Bureau of Statistics (URL: https://bps.go.id/) through the National Socio-Economic Survey where the task is to collect data related to socio-economic conditions of the community including: health condition, education, fertility, family planning, employment, housing, and other socio-economic conditions. In summary, this study embarks the following objectives:

a. We employ *K*-Means algorithm which are implemented in data clustering.
b. We classify the school participation rate of provinces in Indonesia.
c. We expect that the obtained results from this study can provide the description of School participation rate and furthermore the government can continue to make equity, especially to educate more the school-age population.

The rest of this paper is organized as follow: Sect. 2 describes presents the proposed method, data collection and experiment design. Section 3 presents the obtained results and following by discussion. Finally, Sect. 4 concludes this work.

2 Proposed Method

This section presents the proposed method, data collection and experiment design.

2.1 K-*Means*

The *K*-means is a part of clustering which includes an unsupervised training algorithm. This algorithm is quite easy to implement and run, easy to adjust, relatively fast and widely used [13, 24, 25]. The *K*-means algorithm starts with the formation of a cluster partition at the beginning then iteratively clustered the partition is repaired until there is no significant change in the cluster partition. Besides the *K*-means algorithm is also called the evolution algorithm where the method of operation has a meaning similar to the name of the algorithm [26]. The algorithm completion steps are as follows:

(a) Determine the number of clusters;
(b) Allocate data into clusters randomly;
(c) Calculate the centroids of the data in each cluster;
(d) Allocate each data to the nearest centroid;
(e) Go back to Step c, if the data is still moving cluster (centroid value changes), do it until there is no more data change (no centroid and cluster value changes) [26, 27].

2.2 Data Collection

Implementation of data mining clustering school participation rate in Indonesia, required relevant data. Sources of research data obtained from the Central Bureau of Statistics of the Republic of Indonesia through the National Socio-Economic Survey. Data used in 2003–2016 which consists of 4 criteria, namely: a. Age 7–12 years, b. Age 13–15 years, c. Age 16–18 years, and d. Age 19–24 years. Data will be processed in 3 clusters of high cluster (C1), medium cluster (C2) and low cluster (C3).

2.3 Data Processing

Before the data enters the process, it is first grouped according to predetermined criteria by taking the average school participation rate in Indonesia from 2003–2016. Grouped data is then processed using *K*-Means using 3 cluster labels. in determining the amount ($k = 3$), the process is carried out using the Davies-Bouldin Index parameter (abbreviated with the term DBI) contained in the Rapidminer software. DBI is a matrix used to evaluate the results of clustering algorithm to the maximum $k = 3$.

2.4 Stage Analysis

This stage, the analysis of school participation figures in Indonesia will be processed with the help of RapidMinner 5.3 software. In the previous stage, cluster labels were determined ($k = 3$) by evaluating using the Davies-Bouldin Index. The three clusters are high cluster (C1), medium cluster (C2) and low cluster (C3).

3 Results and Discussion

The data obtained are calculated in advance based on the number of school participation rates in Indonesia in years 2003–2016 based on 4 assessment criteria i.e. Age 7–12 years, Age 13–15 years, Age 16–18 years, and Age 19–24 years. The process of data calculation is done by finding the average value of school participation rates in each province. The calculation results can be seen in the following Table 1.

The data in Table 1 will be processed into clusters by applying the *K*-Means algorithm using Rapidminner where 3 clusters has been determined, namely: high cluster (C1), medium cluster (C2) and low cluster (C3).

3.1 Centroid Data

In the application of *K*-means algorithm, the midpoint or centroid value is derived from the grouping of data performed. In this case, we form 3 clusters, namely: high cluster (C1), medium cluster (C2) and low cluster (C3). The cluster point determination is done by taking the highest value for the high cluster (C1), the average value for the medium cluster (C2) and the smallest value for the low cluster (C3). The calculation process can be seen below.

a. High cluster (C1):	MAX (average value on criterion 1 (Age 7–12 years)
	MAX (average value on criterion 2 (Age 13–15 years)
	MAX (average value on criterion 3 (Age 16–18 years)
	MAX (average value on criterion 4 (Age 19–24 years)
b. Medium cluster (C2):	AVERAGE (average value on criterion 1 (Age 7–12 years)
	AVERAGE (average value on criterion 2 (Age 13–15 years)
	AVERAGE (average value on criterion 3 (Age 16–18 years)
	AVERAGE (average value on criterion 4 (Age 19–24 years)
c. Low cluster (C3):	MIN (average value on criterion 1 (Age 7–12 years)
	MIN (average value on criterion 2 (Age 13–15 years)
	MIN (average value on criterion 3 (Age 16–18 years)
	MIN (average value on criterion 4 (Age 19–24 years)

Table 1 The average of school participation rate

Province	Average of school participation rate			
	Age 7–12 years	Age 13–15 years	Age 16–18 years	Age 19–24 years
Aceh	92.10	88.30	69.67	23.98
North Sumatra	98.63	91.88	68.72	17.30
West Sumatra	98.16	90.68	70.31	24.97
Riau	98.18	91.10	66.46	16.28
Jambi	98.21	88.08	59.57	14.94
South Sumatra	98.06	86.86	57.11	13.14
Bengkulu	98.46	89.61	64.50	18.97
Lampung	98.24	87.92	56.27	11.84
Bangka Islands	97.30	82.70	52.38	8.86
Riau Islands	84.44	80.62	59.97	9.83
DKI Jakarta	98.91	93.08	65.31	19.09
West Java	98.06	84.26	52.39	12.71
Central Java	98.82	87.88	56.64	13.31
DI Yogyakarta	99.49	95.74	77.33	44.81
East Java	98.16	86.89	57.31	14.45
Banten	98.40	88.53	58.67	14.17
Bali	98.57	90.88	68.95	16.89
West Nusa Tenggara	97.51	87.54	60.11	17.06
East Nusa Tenggara	95.44	83.27	55.67	16.79
West Kalimantan	96.57	85.44	53.74	14.05
Central Kalimantan	98.66	88.14	56.47	13.85
South Kalimantan	97.85	82.61	54.20	13.81
East Kalimantan	98.50	93.07	69.07	18.08
North Kalimantan	14.06	13.38	10.65	2.64
North Sulawesi	98.13	89.26	61.78	15.14
Central Sulawesi	97.18	83.88	56.67	16.45
South Sulawesi	96.46	82.83	56.65	20.53
Southeast Sulawesi	97.52	87.41	61.58	19.63
Gorontalo	95.58	80.18	53.40	16.67
West Sulawesi	75.40	64.23	42.56	11.35
Maluku	98.07	92.71	70.62	23.68
North Maluku	97.69	91.01	65.16	19.39
West Papua	74.37	71.89	52.32	15.60
Papua	80.31	75.53	53.65	15.04

Centroid data for each cluster can be seen in Table 2:

Table 2 Centroid data

	Age 7–12 years	Age 13–15 years	Age 16–18 years	Age 19–24 years
(C1) High cluster	99.49	95.74	77.33	44.81
(C2) Medium cluster	92.99	84.04	58.70	16.63
(C3) Low cluster	14.06	13.38	10.65	2.64

3.2 *Clustering Data*

Employing centroid in Table 2 for iteration 1, the clustering process is done by calculating the distance of the cluster center using Euclidean distance. From the calculation of cluster central distance at iteration 1 on the number of school participation rate, 1 province in high cluster (C1) i.e. DI Yogyakarta, 32 provinces in medium cluster (C2) i.e. Aceh, North Sumatra, West Sumatera, Riau, Jambi, South Sumatra, Bengkulu, Lampung, Bangka Islands, Riau Islands, DKI Jakarta, West Java, East Java, East Kalimantan, West Kalimantan, Central Kalimantan, South Kalimantan, East Kalimantan, North Sulawesi, Central Sulawesi, South Sulawesi, Southeast Sulawesi, Gorontalo, West Sulawesi, Maluku, North Maluku, West Papua and Papua and 1 province low cluster (C3) i.e. North Kalimantan. The calculation process of cluster central spacing and grouping in iteration 1 can be described in Table 3 as following.

To determine the distance of the cluster center on iteration 1 as shown in Table 3 is done by taking the minimum value of 3 clusters. The result of grouping iteration 1 can be seen in the following Table 4.

In Table 4, the *K*-Means process will continue to iterate until the data grouping equals the previous iteration data grouping. In other words, the process will continue iterating until the data in the last iteration is the same as the previous iteration. The iterative grouping 1 can be depicted in the following Figs. 1 and 2.

The iteration process continues to reset the value of the centroid data on iteration 2 of the results of the data grouping in the iteration 1. The value of the centroid on iteration 2, then used to calculate the distance of the cluster center using Euclidian Distance. This iteration process will continue and stop if the last iteration grouping result equals the previous iteration grouping hail. Determination of centroid value, the grouping of data on iteration 2 can be seen in the following Table 5.

After obtaining the centroid value on iteration 2, do the calculation of cluster center distance process using Euclidean distance, as shown in the following Table 6.

To determine the distance of the cluster center on iteration 2 as shown in Table 6, is done by taking the minimum value of 3 clusters. The result of grouping iteration 2 can be seen in the following Table 7.

From Table 7, the result of grouping on iteration 2 obtained the same result with iteration 1. This indicates that the process of grouping the participation of schools in Indonesia ended in iteration 2. The iteration 2 grouping can be depicted in the following Figs. 3 and 4.

Table 3 Calculation of cluster central distances

Province	Average of school participation rate				Cluster			
	Age 7–12 years	Age 13–15 years	Age 16–18 years	Age 19–24 years	C1	C2	C3	Distance
Aceh	92.10	88.30	69.67	23.98	24.55	13.90	125.06	13.90
North Sumatra	98.63	91.88	68.72	17.30	29.09	13.93	130.00	13.93
West Sumatra	98.16	90.68	70.31	24.97	21.68	16.59	130.79	16.59
Riau	98.18	91.10	66.46	16.28	30.91	11.72	128.13	11.72
Jambi	98.21	88.08	59.57	14.94	35.61	6.88	123.31	6.88
South Sumatra	98.06	86.86	57.11	13.14	38.64	6.96	121.34	6.96
Bengkulu	98.46	89.61	64.50	18.97	29.51	10.01	126.89	10.01
Lampung	98.24	87.92	56.27	11.84	39.92	8.45	121.68	8.45
Bangka Islands	97.30	82.70	52.38	8.86	45.71	10.99	116.25	10.99
Riau Islands	84.44	80.62	59.97	9.83	44.50	11.51	109.36	11.51
DKI Jakarta	98.91	93.08	65.31	19.09	28.52	12.91	129.65	12.91
West Java	98.06	84.26	52.39	12.71	42.27	9.00	117.99	9.00
Central Java	98.82	87.88	56.64	13.31	38.50	8.00	122.32	8.00
DI Yogyakarta	99.49	95.74	77.33	44.81	0.00	36.34	142.50	0.00
East Java	98.16	86.89	57.31	14.45	37.45	6.45	121.63	6.45
Banten	98.40	88.53	58.67	14.17	36.61	7.45	123.29	7.45
Bali	98.57	90.88	68.95	16.89	29.57	13.53	129.42	13.53

(continued)

Table 3 (continued)

Province	Average of school participation rate				Cluster			
	Age 7–12 years	Age 13–15 years	Age 16–18 years	Age 19–24 years	C1	C2	C3	Distance
West Nusa Tenggara	97.51	87.54	60.11	17.06	33.73	5.90	122.95	5.90
East Nusa Tenggara	95.44	83.27	55.67	16.79	37.77	3.98	117.19	3.98
West Kalimantan	96.57	85.44	53.74	14.05	40.22	6.79	118.27	6.79
Central Kalimantan	98.66	88.14	56.47	13.85	38.11	7.85	122.35	7.85
South Kalimantan	97.85	82.61	54.20	13.81	40.88	7.35	117.62	7.35
East Kalimantan	98.50	93.07	69.07	18.08	28.13	14.88	130.88	14.88
North Kalimantan	14.06	13.38	10.65	2.64	142.50	117.16	0.00	0.00
North Sulawesi	98.13	89.26	61.78	15.14	34.15	8.08	124.88	8.08
Central Sulawesi	97.18	83.88	56.67	16.45	37.11	4.67	119.11	4.67
South Sulawesi	96.46	82.83	56.65	20.53	34.54	5.74	118.53	5.74
Southeast Sulawesi	97.52	87.41	61.58	19.63	30.91	7.02	123.81	7.02
Gorontalo	95.58	80.18	53.40	16.67	40.27	7.05	114.59	7.05
West Sulawesi	75.40	64.23	42.56	11.35	62.47	31.47	86.27	31.47
Maluku	98.07	92.71	70.62	23.68	22.42	17.11	131.87	17.11
North Maluku	97.69	91.01	65.16	19.39	28.64	10.96	127.56	10.96
West Papua	74.37	71.89	52.32	15.60	51.76	23.15	94.68	23.15
Papua	80.31	75.53	53.65	15.04	47.15	16.16	101.26	16.16

Table 4 Iteration group 1

Province	Cluster		
	C1: High	C2: Medium	C3: Low
Aceh		1	
North Sumatra		1	
West Sumatra		1	
Riau		1	
Jambi		1	
South Sumatra		1	
Bengkulu		1	
Lampung		1	
Bangka Islands		1	
Riau Islands		1	
DKI Jakarta		1	
West Java		1	
Central Java		1	
DI Yogyakarta	1		
East Java		1	
Banten		1	
Bali		1	
West Nusa Tenggara		1	
East Nusa Tenggara		1	
West Kalimantan		1	
Central Kalimantan		1	
South Kalimantan		1	
East Kalimantan		1	
North Kalimantan			1
North Sulawesi		1	
Central Sulawesi		1	
South Sulawesi		1	
Southeast Sulawesi		1	
Gorontalo		1	
West Sulawesi		1	
Maluku		1	
North Maluku		1	
West Papua		1	
Papua		1	

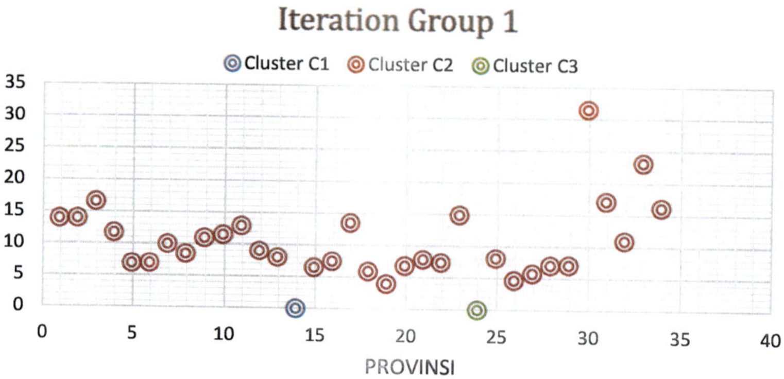

Fig. 1 Group of iteration 1

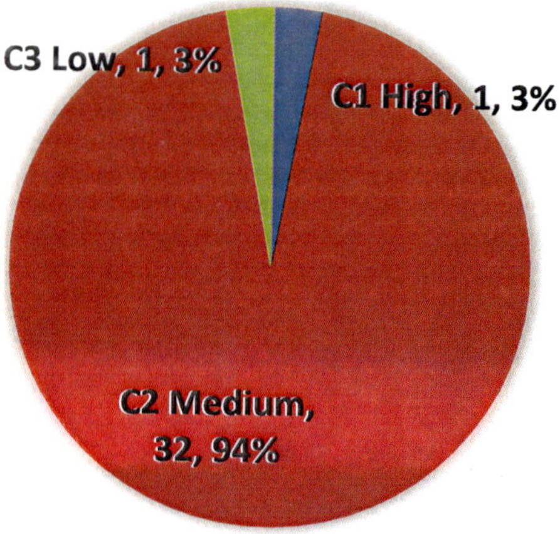

Fig. 2 Clustering results of iteration 1

Table 5 Centroid data iteration 2

	Age 7–12 years	Age 13–15 years	Age 16–18 years	Age 19–24 years
(C1) High cluster	99.49	95.74	77.33	44.81
(C2) Medium	95.25	85.88	59.62	16.18
(C3) Low cluster	14.06	13.38	10.65	2.64

3.3 *Clustering Results with Rapidminer*

In the testing phase of *K*-means algorithm, Rapidminer tools proves the truth in the previous analysis phase and manual testing on grouping the number of school participation rates in Indonesia using 4 assessment criteria. The results obtained by using 3 clusters are 1 province for high cluster (C1), 32 provinces for medium cluster (C2) and 1 province for low cluster (C3) can be seen in the following Fig. 5.

Table 6 Calculation of cluster central distances of iteration 2

Province	Accumulated school participation rate				Cluster			
	Age 7–12 years	Age 13–15 years	Age 16–18 years	Age 19–24 years	C1	C2	C3	Distance
Aceh	92.10	88.30	69.67	23.98	24.55	13.32	125.07	13.32
North Sumatra	98.63	91.88	68.72	17.30	29.10	11.46	130.01	11.46
West Sumatra	98.16	90.68	70.31	24.97	21.69	14.93	130.79	14.93
Riau	98.18	91.10	66.46	16.28	30.91	9.09	128.13	9.09
Jambi	98.21	88.08	59.57	14.94	35.61	3.89	123.31	3.89
South Sumatra	98.06	86.86	57.11	13.14	38.64	4.94	121.34	4.94
Bengkulu	98.46	89.61	64.50	18.97	29.51	7.47	126.89	7.47
Lampung	98.24	87.92	56.27	11.84	39.92	6.57	121.69	6.57
Bangka Islands	97.30	82.70	52.38	8.86	45.71	10.97	116.25	10.97
Riau Islands	84.44	80.62	59.97	9.83	44.50	13.60	109.36	13.60
DKI Jakarta	98.91	93.08	65.31	19.09	28.52	10.30	129.65	10.30
West Java	98.06	84.26	52.39	12.71	42.26	8.65	118.00	8.65
Central Java	98.82	87.88	56.64	13.31	38.50	5.82	122.33	5.82
DI Yogyakarta	99.49	95.74	77.33	44.81	0.00	35.33	142.50	0.00
East Java	98.16	86.89	57.31	14.45	37.45	4.22	121.63	4.22
Banten	98.40	88.53	58.67	14.17	36.61	4.68	123.29	4.68
Bali	98.57	90.88	68.95	16.89	29.57	11.11	129.42	11.11

(continued)

Table 6 (continued)

Province	Accumulated school participation rate				Cluster			
	Age 7–12 years	Age 13–15 years	Age 16–18 years	Age 19–24 years	C1	C2	C3	Distance
West Nusa Tenggara	97.51	87.54	60.11	17.06	33.73	2.98	122.95	2.98
East Nusa Tenggara	95.44	83.27	55.67	16.79	37.76	4.78	117.19	4.78
West Kalimantan	96.57	85.44	53.74	14.05	40.22	6.41	118.27	6.41
Central Kalimantan	98.66	88.14	56.47	13.85	38.11	5.67	122.36	5.67
South Kalimantan	97.85	82.61	54.20	13.81	40.88	7.25	117.62	7.25
East Kalimantan	98.50	93.07	69.07	18.08	28.12	12.45	130.89	12.45
North Kalimantan	14.06	13.38	10.65	2.64	142.50	120.12	0.00	0.00
North Sulawesi	98.13	89.26	61.78	15.14	34.15	5.05	124.88	5.05
Central Sulawesi	97.18	83.88	56.67	16.45	37.11	4.07	119.11	4.07
South Sulawesi	96.46	82.83	56.65	20.53	34.54	6.21	118.53	6.21
Southeast Sulawesi	97.52	87.41	61.58	19.63	30.91	4.82	123.81	4.82
Gorontalo	95.58	80.18	53.40	16.67	40.27	8.46	114.60	8.46
West Sulawesi	75.40	64.23	42.56	11.35	62.46	34.31	86.27	34.31
Maluku	98.07	92.71	70.62	23.68	22.42	15.22	131.87	15.22
North Maluku	97.69	91.01	65.16	19.39	28.63	8.56	127.56	8.56
West Papua	74.37	71.89	52.32	15.60	51.75	26.18	94.68	26.18
Papua	80.31	75.53	53.65	15.04	47.15	19.17	101.26	19.17

Table 7 Iteration Group 2

Province	Cluster		
	C1: High	C2: Medium	C3: Low
Aceh		1	
North Sumatra		1	
West Sumatra		1	
Riau		1	
Jambi		1	
South Sumatra		1	
Bengkulu		1	
Lampung		1	
Bangka Islands		1	
Riau Islands		1	
DKI Jakarta		1	
West Java		1	
Central Java		1	
DI Yogyakarta	1		
East Java		1	
Banten		1	
Bali		1	
West Nusa Tenggara		1	
East Nusa Tenggara		1	
West Kalimantan		1	
Central Kalimantan		1	
South Kalimantan		1	
East Kalimantan		1	
North Kalimantan			1
North Sulawesi		1	
Central Sulawesi		1	
South Sulawesi		1	
Southeast Sulawesi		1	
Gorontalo		1	
West Sulawesi		1	
Maluku		1	
North Maluku		1	
West Papua		1	
Papua		1	

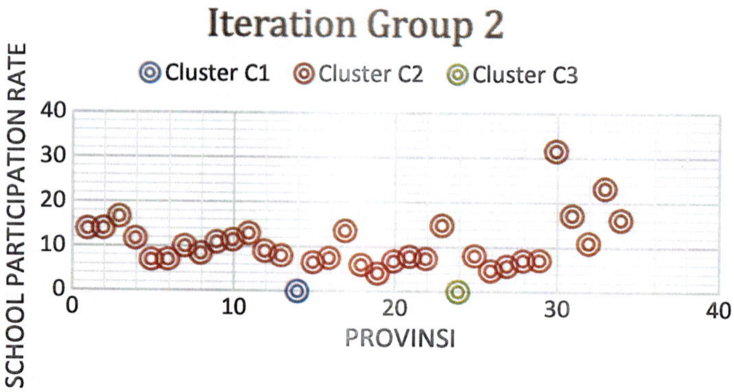

Fig. 3 Group of iteration 2

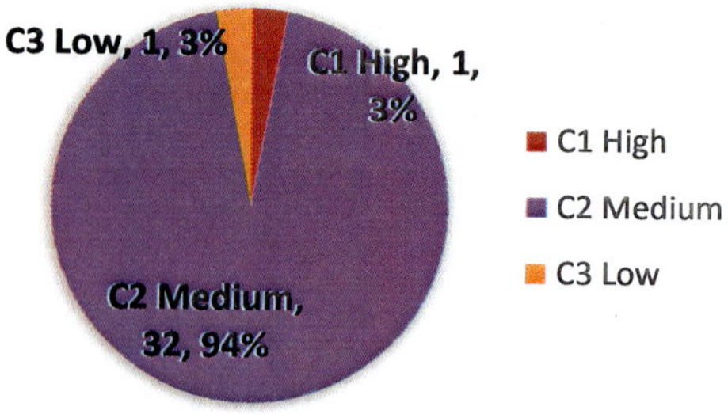

Fig. 4 Clustering results of iteration 2

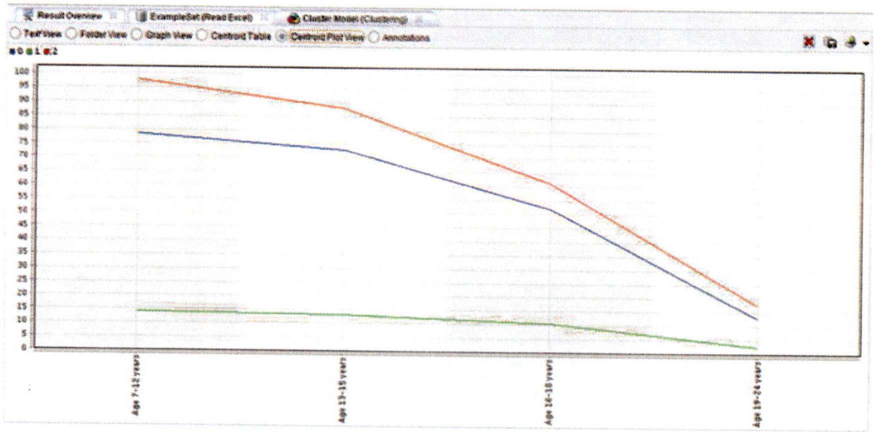

Fig. 5 Clustering results with rapidminer

4 Conclusion and Recommendation

This paper has presented *K*-Means algorithm with Rapidminer in clustering school participation rate in Indonesia. Data is processed to achieve the result of school participation rate clustering in Indonesia which is categorized into 4 criteria, namely: Age 7–12 years, Age 13–15 years, Age 16–18 years, and Age 19–24 years. Data were processed by comparing the results of manual calculations and by using Rapidminner tool using cluster that have been determined, namely: (C1) high cluster, (C2) medium cluster and (C3) low cluster. The results show that 1 province in high cluster(C1), 32 provinces in medium cluster (C2) and 1 province in low cluster (C3). The results of this study can provide input to the government that the number of participation rates in Indonesia is still very low. Further study is needed to explore factors that affect the achievement of education in a region, both in terms of availability of educational facilities and socioeconomic conditions. Some factors such as population density in each province should be also investigated.

References

1. Drèze J, Kingdon GG (2001) School participation in rural India. Rev Dev Econ. https://doi.org/10.1111/1467-9361.00103
2. Afridi F (2011) The impact of school meals on school participation: evidence from rural India. J Dev Stud. https://doi.org/10.1080/00220388.2010.514330
3. Dutt A, Ismail MA, Herawan T (2017) A systematic review on educational data mining. IEEE Access 5:15991–16005. https://doi.org/10.1109/ACCESS.2017.2654247
4. Aghabozorgi S, Mahroeian H, Dutt A, Wah TY, Herawan T (2014) An approachable analytical study on big educational data mining. Lecture Notes Computer Science (including Subser. Lecture Notes Artif Intell Lect Notes Bioinformatics), 858(PART 5):721–737. https://doi.org/10.1007/978-3-319-09156-3_50
5. Zain SJM, Herawan T (2014) Data mining for education decision support: a review. Int J Emerg Technol Learn. https://doi.org/10.3991/ijet.v9i6.3950
6. Herawan T, Deris MM, Abawajy JH (2010) A rough set approach for selecting clustering attribute. Knowl-Based Syst. https://doi.org/10.1016/j.knosys.2009.12.003
7. Qin H, Ma X, Zain JM, Herawan T (2012) A novel soft set approach in selecting clustering attribute. Knowl-Based Syst. https://doi.org/10.1016/j.knosys.2012.06.001
8. Mamat R, Herawan T, Deris MM (2013) MAR: maximum attribute relative of soft set for clustering attribute selection. Know-Based Syst. https://doi.org/10.1016/j.knosys.2013.05.009
9. Yanto ITR, Ismail MA, Herawan T (2016) A modified Fuzzy k-Partition based on indiscernibility relation for categorical data clustering. Eng Appl Artif Intell. https://doi.org/10.1016/j.engappai.2016.01.026
10. Aggarwal N, Aggarwal K, Gupta K (2012) Comparative analysis of k-means and enhanced k-means clustering algorithm for data mining. Int J Sci Eng Res 3(3)
11. Oyelade OJ, Oladipupo OO, Obagbuwa IC (2010) Application of k means clustering algorithm for prediction of students academic performance. Int J Comput Sci Inf Secur 7(1):292–295
12. Arzoo MK, Prof A, Rathod K (2017) K-Means algorithm with different distance metrics in spatial data mining with uses of NetBeans IDE 8. 2. Int Res J Eng Technol 4(4):2363–2368
13. Windarto AP (2017) Implementation of data mining on rice imports by major country of origin using algorithm using k-means clustering method. Int J Artif Intell Res 1(2):26–33

14. Kaur N, Sahiwal JK, Kaur N, Punjab P (2012) Efficient k-means clustering algorithm using ranking method. Int J Adv Res Comput Eng Technol 1(3):85–91
15. Yadav A, Dhingra S (2016) An enhanced K-Means clustering algorithm to remove empty clusters. IJEDR 4(4):901–907
16. Suryawanshi R (2016) A novel approach for data clustering using improved k- means algorithm. 142(12):13–18
17. Singh K, Malik D, Sharma N (2011) Evolving limitations in K-means algorithm in data mining and their removal. IJCEM Int J Comput Eng Manag ISSN 12(April):2230–7893
18. Kodinariya TM, Makwana PR (2013) Review on determining number of cluster in K-Means clustering. Int J Adv Res Comput Sci Manag Stud
19. Prahara A, Yanto ITR, Herawan T (2017) Histogram thresholding for automatic color segmentation based on k-means clustering. Adv Intell Syst Comput 549(AISC):344–354. https://doi.org/10.1007/978-3-319-51281-5_35
20. Hadinata E, Sembiring RW, Kusumasari TF, Herawan T (2017) The algorithm expansion for starting point determination using clustering algorithm method with fuzzy C-Means. In: Advances in intelligent systems and computing. https://doi.org/10.1007/978-3-319-51281-5_51
21. Mohd WMW, Beg AH, Herawan T, Noraziah A, Chiroma H (2019) Multi-dimensional K-Means algorithm for student clustering. In: Proceedings of the international conference on data engineering 2015 (DaEng-2015), vol 520, pp 343–351. https://doi.org/10.1007/978-981-13-1799-6
22. Mohd WMW, Beg AH, Herawan T, Rabbi KF (2012) MaxD K-Means: a clustering algorithm for auto-generation of centroids and distance of data points in clusters. In: Communications in computer and information science. https://doi.org/10.1007/978-3-642-34289-9_22
23. Maseri Binti Wan Mohd W, Beg AH, Herawan T, Noraziah A, Rabbi KF (2013) Improved parameterless K-means: auto-generation centroids and distance data point clusters. In: Information retrieval methods for multidisciplinary applications
24. Supriyadi B, Windarto AP, Soemartono T, Mungad (2018) Classification of natural disaster prone areas in Indonesia using K-means. Int J Grid Distrib Comput 11(8):87–98
25. Windarto AP (2017) Penerapan data mining pada ekspor buah-buahan menurut negara tujuan menggunakan K-Means clustering. Techno.COM 16(4):348–357
26. Sudirman, Windarto AP, Wanto A (2018) Data mining tools | rapidminer: K-means method on clustering of rice crops by province as efforts to stabilize food crops in Indonesia. In: IOP Conference series: materials science and engineering, vol 420, no 1. https://doi.org/10.1088/1757-899x/420/1/012089
27. Widyastuti M, Fepdiani Simanjuntak AG, Hartama D, Windarto AP, Wanto A (2019) Classification model C.45 on determining the quality of custumer service in bank BTN pematangsiantar branch. J Phys Conf Ser 1255(012002):1–6. https://doi.org/10.1088/1742-6596/1255/1/012002

Investigation on Integration of Sensors and Vision-Based Vehicle Detection System for Autonomous Vehicle

Muhammad Aizzat Iqbal Bin Abd Rashid and Ahmad Shahrizan Abdul Ghani

Abstract The vehicle detection system is an essential part of autonomous navigation systems. In this paper, the investigation on the vehicle detection system is conducted by integrating the processed data from the ultrasonic sensor, together with the collected data from the Pixy sensor. The ultrasonic sensor is used as obstacle detection sensors with the optimal performances, while the Pixy camera is used as the vision-based sensor device. The data obtained from both sensors are processed and compared to each other. Results received from both data comparisons conclude the current condition as the safety status of the car was being determined and delivered to the user. In other words, the sensors' data were integrated to validate the accuracy of the information received by the controller to navigate the vehicle. The Pixy camera detects the vehicle in the range of camera view. On the other hand, for the ultrasonic sensor, three stages of distance detection level are being measured. The performance of the proposed method is demonstrated by experimenting statically several times on a prototype scale.

Keywords Autonomous car · Vehicle detection · Data integration · Pixy camera · Ultrasonic sensor

1 Introduction

In today's age of digital technology and intelligent systems, the automotive industries backed by several internet technology companies are facing towards full self-driving automation. In 2014, SAE International and National Highway Traffic Safety Administration (NHTSA) have remarks in detail the levels of driving automation from 'Level 0—No Driving Automation' to 'Level 6—Full Driving Automation'. In 2016, NHTSA followed these levels of automation, and thus now the SAE J3016 Recommended Practice contains the standard meaning related to driving automation systems [1]. Automatic or autonomous vehicle (AV) is embedded with various type

M. A. I. Bin Abd Rashid · A. S. Abdul Ghani (✉)
Faculty of Manufacturing and Mechatronic Engineering Technology, Universiti Malaysia Pahang, 26600 Pekan, Malaysia
e-mail: shahrizan@ump.edu.my

A. F. Ab. Nasir et al. (eds.), *Recent Trends in Mechatronics Towards Industry 4.0*, Lecture Notes in Electrical Engineering 730, https://doi.org/10.1007/978-981-33-4597-3_71

of systems such as orientation, detection, and localization system. An AV contains five essential functions which are localization, vision, planning, vehicle control, and system management [2]. AV is also able to perform multiple functions such as detecting obstacles, taking sharp bends and turns, overtaking, following traffic signals, and turning its lights on under low light conditions [3]. The accuracy and robustness of the road-boundary detection for an intelligent vehicle is a formidable task under a variety of road pavement types, lane structures, weather conditions, and noises [4].

Currently, for the human-crewed vehicle, accidents and collisions usually occur because of the distraction that happens to the driver, especially when the driver is tired or in a rush. In the case of AV, the vehicle detection system continuously monitors the condition of the car to avoid collision and accident. However, selecting an adequate sensor is one of the most crucial parts where the specification and parameters needed to be satisfied. In addition to that, relying on a single sensor is not an option anymore since a single sensor cannot overcome all the surrounding factors by itself. Recently, many researchers have extensively studied sensor integrations. Various types of sensors have been used in making a detection system, such as an infrared sensor and ultrasonic sensor. The behavior of the sensors used is different as the linear characteristic is provided by the ultrasonic sensor, and the nonlinear from the infrared sensor [5]. Nevertheless, influenced by the reflectance surface of the object detected, the feedback gained from the infrared sensor will vary. An angle also needs to be taken care of carefully as slightly different in it will produce different measuring results.

Two types of infrared integration are one of the popular approaches due to its similarity in terms of the working principle of the infrared. The output from both infrared sensors concerning distance measurement is compared by using a long-range and short-range infrared sensor. The distance reading accuracy of the long-range infrared will be confirmed by the short-range infrared's distance reading when the detected object appears in the range of detection [6]. Although the reading accuracy can be improved by using this approach, infrared's principle of reflected light waves will limit the functionality of the detection system as the system will be less efficient under the sunlight besides of the interference from the car's tail light in the dark surrounding.

The most critical part of detecting the vehicle lies in the integration of the sensors. The integration of the data received from the sensors had to be carefully analyzed since it would decide the final output to be delivered to the user. Instead of confusing the user with informing the measuring output from the sensor separately, one sensor's data needed to be compared to the other in the first hand before the final decision will be made to inform the user regarding the current car distance with the vehicle in range. Therefore, this paper has taken the initiative to propose an alternative method in the vehicle detection system. The proposed method is conducted using a vision-based sensor that is a Pixy camera, together with an ultrasonic sensor to create one vehicle detection system. This paper is distributed as follows: Sect. 2 discussed the previous related works. Section 3 discussed the proposed method. Section 4 is where the results are being shown, and lastly, Sect. 5 is where the conclusion will be drawn.

2 Related Work

Studies on self-driving cars have been performed since the 1920 s where promising trials occurred in the 1950 s. Research has continued since then. Various type of method and experiment has been conducted throughout this past several decades. One of the most popular methods nowadays is by using simultaneous localization and mapping (SLAM) method. In a paper prepared by Chul-hong [7], object detection and SLAM were obtained by experimenting on a stereo camera with two different fields of view (FoVs). The standard type of camera used is a wide stereo camera that has more advantage on the SLAM algorithm and narrow FOVs camera for obstacle detection. SLAM consists of three sections (tracking, optimization, and loop closure) where the vanishing point used to measure robot alignment and point features applied rather than line features for tracking, optimization, and loop closure (Fig. 1).

Moreover, a study [8] has been conducted where radar sensor and the camera image is used in a fusion to develop a driver assistance system by implementing Generative Adversarial Networks (GAN) concept, specifically Multi-Generator Generative Adversarial Networks (CMGGANs). Radar data by the CMGGANs can be fused at a pixel level with original camera images of the same scene to increase the robustness of a camera. Besides, their semantic labels can be used to enable an efficient end to end semantic segmentation of the radar data, altogether avoiding costly and challenging labeling of the radar sensor data. In addition to that, there are other studies [9], that experimenting on object detection also by using radar and camera data. Tracks denote targets detected from radar input data with information like range and azimuth associated with them. This information is extracted from the radar. For the camera, the characteristics of the camera are determined by the

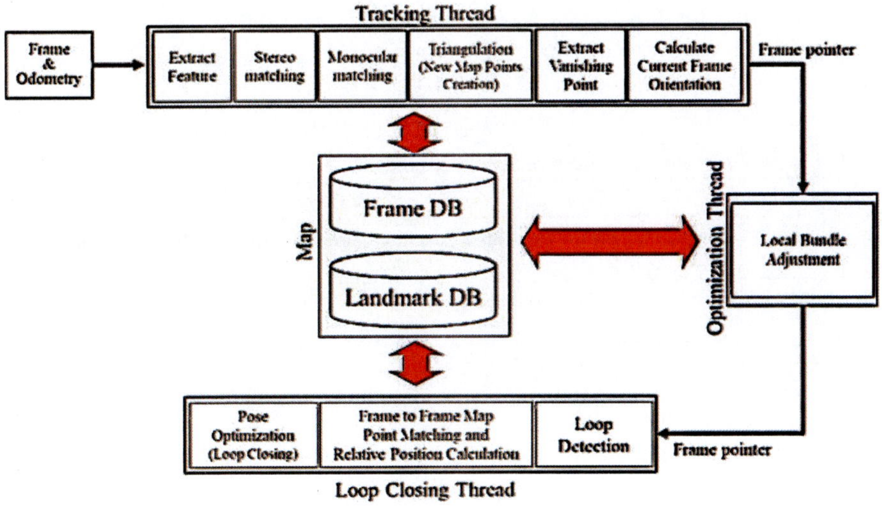

Fig. 1 The proposed Stereo SLAM structure [7]

calibration process. The data coming from the laser scanner are easier to cluster than radar, and they are more accurate.

Moreover, it is easier to quantify the reliability and to model the uncertainties of such data. However, laser sensors fail to overcome some situations such as non-planar road configuration, or a varying pitch angle due to the ego vehicle maneuver depending on an acceleration or road shape variations [10]. From both studies with the application of radar and camera fusion, it can be concluded that both studies can be used to perform an object detection system. However, radar is not a suitable sensor to perform this task as it has a low range of detection, and its accuracy affected profoundly on the presence of light.

3 Methodology

Hardware and software are two parts classified in the proposed method. The software part plays the central role in this project as most of it is used to collect, analyze, and make a decision about the data. In addition to that, choosing compatible hardware is also vital to complete this system. System efficiency depends on factors like a dusty environment and changes in ambient light, which needed to be considered in choosing the hardware and devices. Table 1 shows the list of hardware and software used in this method.

3.1 Block Diagram

Figure 2 shows the flow of the system where the input data is sent to the micro-processor to be processed. Next, the output signal is sent to be delivered to the buzzer.

Table 1 List of hardware and software uses in this method

No.	Hardware	Description	Software	Description
1	Arduino mega 2560 Rev3	Microcontroller	Arduino IDE	Programming platform
2	Pixy 2 CMUcam5	Vision-based sensor	PixyMon v2	Setup tool software for Pixy2
3	HC-SR04 ultrasonic sensor	Object detection sensor		
4	Piezo buzzer	Output indicator		

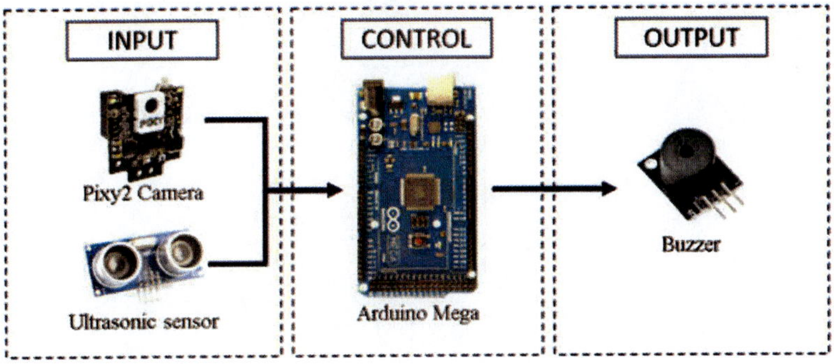

Fig. 2 Block diagram of the components

3.2 Flow Chart

The proposed vehicle detection algorithm can be presented as a flow chart, as shown in Fig. 3. After the system starts, both input will begin collecting data. Pixy starts to detect whether there is a vehicle detected in the range of view, and the ultrasonic sensor starts to detect the vehicle in front and measure the distance if there is any. Both data will then be compared. If the Pixy detects the vehicle insight of view and the ultrasonic sensor also detects the vehicle, it indicates that the data is real. The signal will be sent to the buzzer according to the distance data by the ultrasonic sensor, which is beeping in pulse mode if the distance is between 30 cm and 50 cm. If the vehicle is too close (less than 30 cm), the buzzer will continuously beep without a pulse. However, if only one of the input detects the vehicle, data will be considered as false, and the system will start detecting again.

3.3 Experiment Setup and Process

The experiment is conducted on a laboratory scale. All the components and hardware were assembled on the first hand, as shown in Fig. 4.

Ultrasonic testing

First of all, the ultrasonic signal processing experiment is conducted to measure the distance value and compared it with the manual measurements. Sounds able to travel at 340 meters per second (m/s), which is approximately 29.412 microseconds per centimeter (μs/cm). In sequence to that, the following formula is used to calculate the distance from the ultrasonic sensor [11]:

$$Distance(cm) = \frac{(Time \times Speed\ of\ Sound)}{2} \tag{1}$$

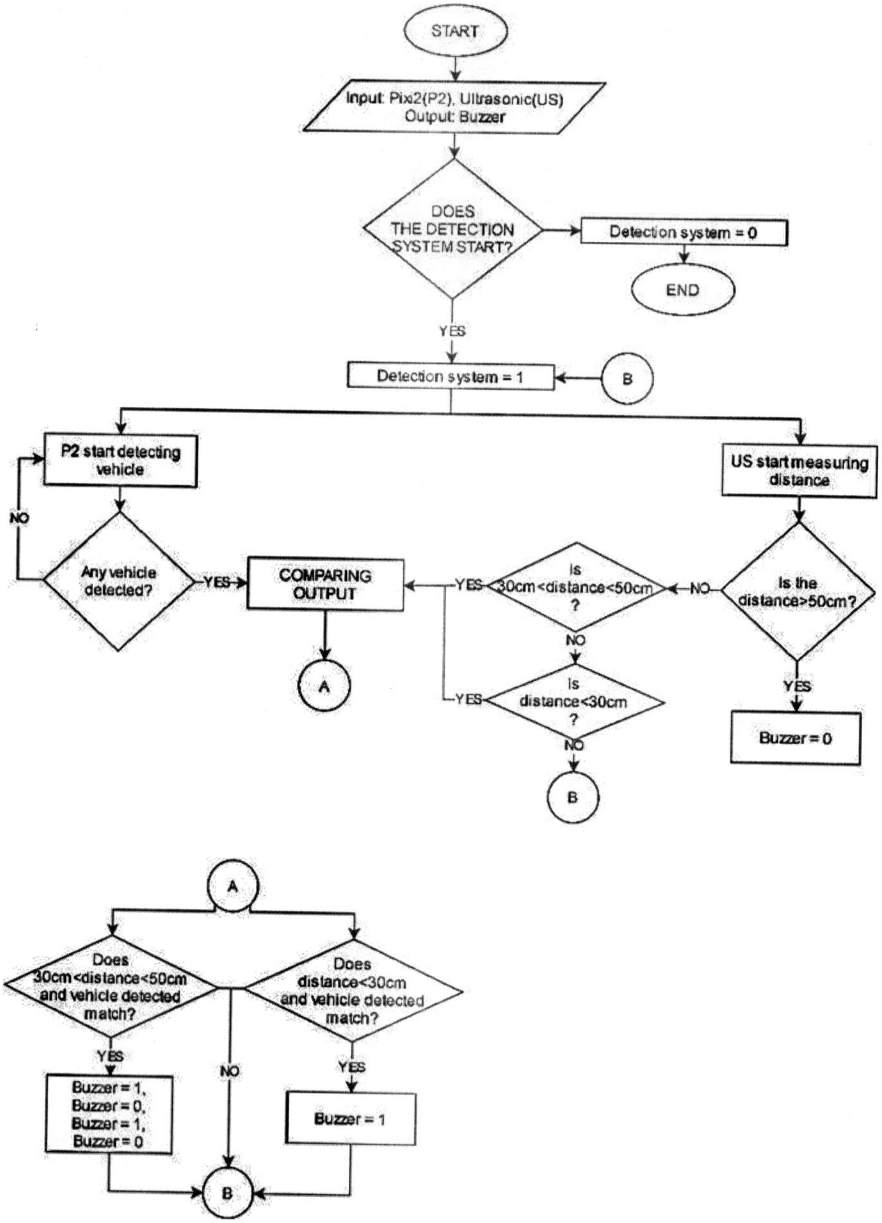

Fig. 3 The flow chart of the proposed algorithm

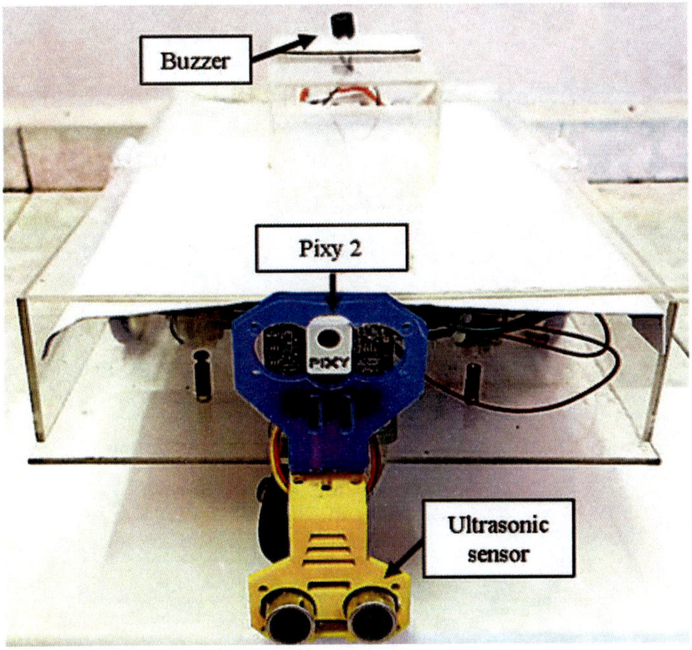

Fig. 4 Picture of the prototype setup

Pixy camera testing

Pixy camera comes with PixyMon v2 that is a setup tool software for Pixy2. Pixy already provides the algorithm library to be used in Arduino IDE. The only setup needed for Pixy is to train the image by using PixyMon v2 so that it will recognize the vehicle. After that, testing is conducted to determine the ability of the Pixy to detect multiple and overlapping images.

Integration of the sensors.

The integration of the ultrasonic sensor and Pixy is the last and most crucial part of this method. After the previous testing shows satisfying results, both sub-system is combined into one single system that compared the data before producing an output that will be the buzzer's signal.

4 Results

Following the ultrasonic sensor and Pixy testing, the results show that the proposed algorithm is effective and achieves its purpose for the vehicle detection system. Figure 5 shows the capability of Pixy image processing in multiple and overlapping visual images of the car. In addition to that, it also shows the ultrasonic sensor data

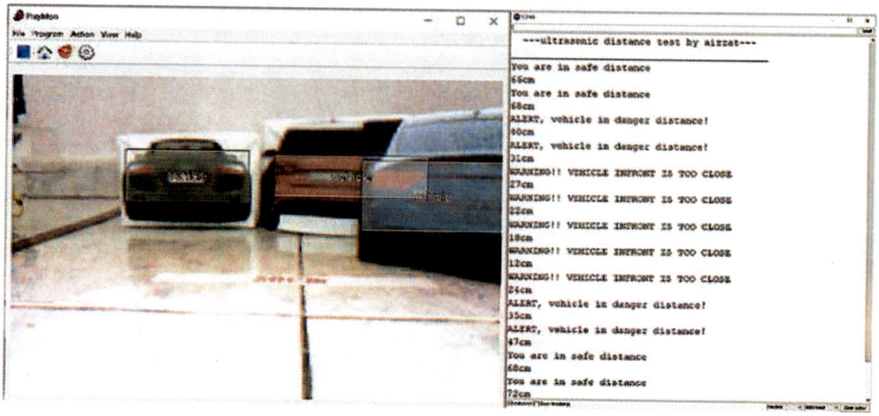

Fig. 5 Pixy 2 trained image and ultrasonic sensor distance data

Table 2 Integration results

Ultrasonic sensor	Pixy	Buzzer	Serial display
<30 cm	Detected	ON	Warning distance
>0 cm	Undetected	OFF	Pixy rescan the vehicle
30–50 cm	Detected	Beeping	Alert distance
Undetected	Detected	Off	Ultrasonic rescan the vehicle
>50 cm	Detected	Off	Safe distance

recorded and monitored from the Arduino serial monitor. Three types of distance data measured by the ultrasonic sensor to determine the condition of the detected vehicle. If the vehicle detected in the distance more than 50 cm, it is classified as 'safe distance.' Alert notification will show if the vehicle is in danger distance between 30 cm to 50 cm and warning notification displayed if the vehicle detected is too close, which has a distance that is less than 30 cm, as shown in Table 2. However, if the ultrasonic sensors do not detect the vehicle while Pixy detected, the ultrasonic sensors will rescan the vehicle. Ultrasonic sensors take approximately 0.6 s to complete one scanning. The same case goes to Pixy. If the Pixy does not detect the vehicle while the ultrasonic sensor detects it, Pixy will rescan at a rate of 60 scannings per second.

5 Conclusion and Recommendation

In this paper, we proposed an algorithm for vehicle detection based on the camera and sensors data. Based on the experimental result, it can be concluded that our method is effective. Some problems in a single input detection system manage to be overcome. An error from a single sensor or camera data that might be affected by other factors like low ambient light can be surpassed.

The future work and features should be on two-folded. From the input aspect, LiDAR sensors are recommended to be used as the primary detection sensors. While LiDAR has a higher detection accuracy, its wider scanning angle, together with a higher sampling rate, will improve the system significantly. It is also able to cover the blind spot that cannot be detected by any single point detection sensor. Besides that, Raspberry Pi is a more powerful microprocessor compared to Arduino and is more suitable to process enormous LiDAR data in sequence with image processing for the camera. Monocular field of view camera with autofocus and higher resolution is also recommended for its accuracy, low-price with the ability to obtain rich color information. In addition, image processing for the catured image or video should be processes in order to improve its color and contrast especially when the vehicles are not clearly visible due to low environtment light [12, 13].

As for the output, instead of using a buzzer, a monitor display will help to deliver the information to the user in graphical ways form. Since there is also a system that uses buzzer or alarm like seatbelt and reverses system, adding another buzzer might distract and confuse the user. So having a monitor display together with a buzzer will improve the user experience.

Acknowledgements The authors would like to thanks all reviewers. This work is supported by the internal Universiti Malaysia Pahang Research Grant RDU1903140 entitled ‘Analyzing the Effects of Dark Channel Fusion and Pixels Distribution Shifting on Image Histogram for Development of Remotely Surface Vehicle.’

References

1. Bécsi T, Aradi S, Fehér Á, Gáldi G (2017) Autonomous vehicle function experiments with low-cost environment sensors. In: 20th EURO working group on transportation meeting, EWGT 2017, 4–6 Septe 2017, Budapest, Hungary, pp 333–340
2. Jo K, Kim J, Kim D, Jang C, Sunwoo M (2014) Development of autonomous car—part I: distributed system architecture and development process. IEEE Trans Ind Electron 61(12):7131–7141
3. Iqbal A, Ahmed SS, Tauqeer MD, Sultan A, Abbas SY (2017) Design of multifunctional autonomous car using ultrasonic and infrared sensors. In: 2017 International symposium on wireless systems and networks (ISWSN), pp 1–5
4. Li Q, Zheng N, Cheng H (2004) Springrobot: a prototype autonomous vehicle and its algorithms for lane detection. IEEE Trans Intell Transp Syst 5(4):300–309
5. Mustapha B, Zayegh A, Begg RK (2013) Ultrasonic and infrared sensors performance in a wireless obstacle detection system. In: 2013 First international conference on artificial intelligence, modelling and simulation, pp 1–6
6. Rathod SM, Apte S (2019) Obstacle detection using sensor based system for an four wheeled autonomous electric robot. In: Proceedings of the fourth international conference on communication and electronics systems (ICCES 2019), pp 493–497
7. Kim C-H, Lee T-J, Cho D-I (2018) An application of stereo camera with two different FoVs for SLAM and obstacle detection. IFAC PapersOnLine 51–22:148–153

8. Lekic V, Babic Z (2019) Automotive radar and camera fusion using generative adversarial networks. Comput Vis Image Underst 1–8
9. Jha H, Lodhi V, Chakravarty D (2019) Object detection and identification using vision and radar data fusion system for ground-based navigation. In: 2019 6th International conference on signal processing and integrated network (SPIN), pp 590–594
10. Gruyer D, Cord A, Belaroussi R (2013) Vehicle detection and tracking by collaborative fusion between laser scanner and camera. In: 2013 IEEE/RSJ international conference on intelligent robots and systems (IROS), pp 5207–5215
11. Saddam (2020) Arduino & ultrasonic sensor based distance measurement, 27 June 2015. (Online). Available: https://circuitdigest.com/microcontroller-projects/arduino-ultrasonic-sensor-based-distance-measurement. Accessed 15 July 2020
12. Sundarajoo S, Abdul Ghani AS (2018) Improvement of auto-tracking mobile robot based on HSI color model. Indonesian J Electr Eng Comput Sci 12(3):1349–1357. ISSN 2502-4752. http://dx.doi.org/10.11591/ijeecs.v12.i3.pp1349-1357
13. Abdul Ghani AS, Mat Isa NA (2015) Homomorphic filtering with image fusion for enhancement of details and homogeneous contrast of underwater image. Ind J Geo-Mar Sci (IJMS) 44(12):1904–1919. ISSN: 0975-1033

Study of Linear-Correlation Based Solar Irradiance Measurement Device Photovoltaic Application

Amirah Nurhafizah Abu Bakar, Nuratiqah Mohd Isa, Mohamad Shaiful Abdul Karim, Ruhaizad Ishak, Suazlan Mt Aznam, and Ahmad Syahiman Mohd Shah

Abstract Solar irradiance is the most fundamental element for energy generation from the photovoltaic system. Huge number of measurement instruments are produced commercially for better reading of this parameter. Nevertheless, there are still no detail studies that have ever been reported regarding the critical relationship between solar irradiance and other electrical parameters such as voltage or current in the literature. This study is essential to determine the correct parameters to be considered in developing a precise measurement of solar irradiance. In this study, a linear-correlation based solar irradiance measurement device has been designed to investigate this relationship. The obtained solar irradiance values from the proposed device exhibit an excellent agreement with those measured by the commercial solar meters/sensors which highly suggest that this study should become a trigger for better improvement of the solar measurement system in the future.

Keywords Solar irradiance · Photovoltaic system · Development of solar irradiance measurement device · Study of Linear-Correlation

1 Introduction

Photovoltaic power generation is the way to utilize the energy from the sunlight [1]. Solar energy is converted into electrical energy through photocell. Energy monitoring stations have been developed in order to maintain the reliability of the photovoltaic system so that any malfunctions arise can be solve immediately. Typically, the solar

A. N. A. Bakar · N. M. Isa · M. S. A. Karim · A. S. M. Shah (✉)
Department of Electrical Engineering, College of Engineering, University Malaysia Pahang, 26600 Pekan, Pahang, Malaysia
e-mail: asyahiman@ump.edu.my

R. Ishak
Faculty of Electrical and Electronics Engineering Technology, University Malaysia Pahang, 26600 Pekan, Pahang, Malaysia

S. M. Aznam
Department of Science in Engineering, Kulliyah of Engineering, International Islamic University Malaysia, 53100 Gombak, Selangor, Malaysia

A. F. Ab. Nasir et al. (eds.), *Recent Trends in Mechatronics Towards Industry 4.0*, Lecture Notes in Electrical Engineering 730, https://doi.org/10.1007/978-981-33-4597-3_72

power intensity that reaches the ground of the earth is varied over time. Although the total amount of power intensity at the top of the atmosphere has a high value of 1367 W/m^2, this radiation is mostly reflected into outer space as it strikes the Earth's atmosphere [2]. There are various types of solar sensor or solar meter that used to measure solar radiation such as pyranometer and solarimeter which have their own construction to produce electrical signals [3]. The measured radiations are then integrated over time to obtain the irradiation values. The sum of diffuse and reflected solar radiations is called global solar radiation. The solar radiation sensor is used to measure this solar radiation. This sensor converts the incident solar radiation into electric current that can be measured by various means which is also used to test the performance of the PV modules [3]. Conversely, the operating temperature of the PV module is highly influenced by the radiation level.

The desired power amount which is harnessed directly from the sunlight might not be generated all the time continuously due to the dependency on various meteorological elements such as relative humidity, cloudiness level, precipitation, aerosol, atmospheric pressure and air temperature [4–6]. These factors lead to degradation of the solar radiation that is transmitting across the atmosphere to the land surface. Up to recent, numerous works have been studied related to the prediction of solar irradiance based on several simulation data such as European Centre for Medium-Range Weather Forecasts (ECMWF) [7], North American Mesoscale (NAM) [8], Meteosat Second Generation/Spinning Enhanced Visible and InfraRed Imager (MSG/SEVIRI) [9, 10], and grid point value (GPV) [4, 11, 12].

Previous studies have carried out proficiency tests to confirm the competency required to conduct primary calibration of reference cell [13]. There are many methods have been introduced to calibrate all 7 PV reference cells. Liu et al. [13] is the only laboratory that uses direct sunlight method to calibrate the reference cell, and this direct sunlight calibration testbed of IEECAS will be used for PV scale traceability and secondary calibration in China. The other researchers have recalibrated the reference cell from previous primary calibration using a round robin method [14]. Majority laboratories have developed programs that will calibrate and design new reference cells, according to WPVS. There are 12 tests have been carried out to prevent or minimize the further problem that might arise later while maintaining the results. There are a few studies that use reference cells to detect partial shading and manage cell interconnection in PV modules to prevent cell damage due to hotspots [15]. The studies have limitations to output estimation and shading detection only. However, there is no further discussion suggested establishing good accuracy from the reference cell. Therefore, to establish the usefulness and accuracy of reference cells in both performance monitoring and optimization, more experimental studies are needed.

Some studies adopt common methods to measure different types of solar irradiation and irradiance in modelling the solar radiation. The use of tree-based ensemble methods in modelling which are gradient boosting, bagging and random forest (RF) algorithms shows that bagging and RF algorithm has better estimation compared to the gradient boosting [16]. These developed ensemble model have also been compared multi-layer perceptron (MLP), support vector regression (SVR), and decision tree (DT). Next, FoBa, leapForward, spikeslab, Cubist, and bagEarthGCV models are recent and productive study which used to solve the problem of solar irradiance forecasting and have been proved its accuracy compared to the conventional methods [17]. In order to improve the process of component separation, gradient boosting machine learning (ML) technique is used which direct and diffuse solar radiation components are estimated from 1-min global horizontal irradiance data that leads to improvement of the performance of single conventional separation models based on ensembles and widen the original geographical or climatological area of application of any single separation model [18].

Despite numerous published studies regarding solar irradiance prediction and solar radiation modelling, it is very rare to find the paper that studies further on the development of solar radiation measurement system since it may be due to the availability of the commercial solar sensors is very common in the market [16–18]. It can be said that those commercial solar sensors differ to each other in term of their features and performances. The most reliable solar meter which can measure solar irradiance in all directions is pyranometer that is usually installed close to PV panel and has wired system to log the data from the PV panel. There are two types of pyranometer which are thermophile and semiconductor [19, 20]. Though many conventional solar meters are commercialized in the market, deep studies based on academic or statistical analysis are rarely discussed. Thus, this matter will be addressed in this paper. In this study, a linear-correlation based solar irradiance measurement device has been developed in term of academical evaluation purposes to measure the solar irradiance and analyze the correlation of involved parameters.

2 Overall Scheme Design

In this section, all the procedure and project description will be discussed. The whole system needs to be properly handled so that it can function appropriately.

2.1 System Design and Configuration

Figure 1 presents the whole system design of this project using Proteus software. In this figure, a DC power supply of V1 acts as a solar cell. A solar cell is used to capture the energy generated from the sunlight. A WSL solar cell with the rated maximum voltage and current of 6 V and 120 mA, respectively, is used in this project as the main component to capture the sunlight and brings the reading of the voltage so that the solar irradiance can be further determined in the preceding section.

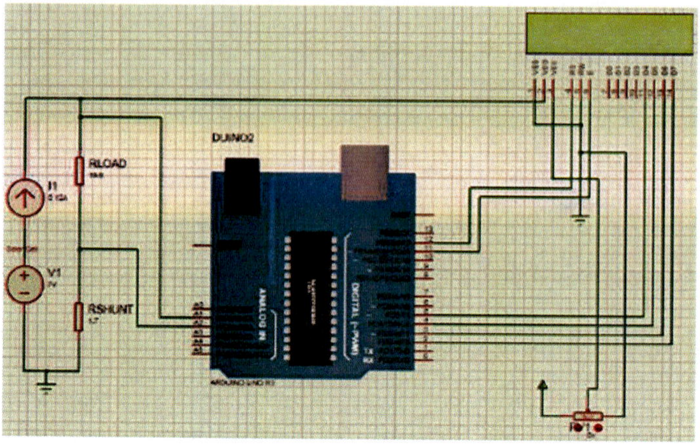

Fig. 1 Design of solar irradiance measurement device using Proteus (This Figure is blurred)

2.2 Software

The main software that is commonly used in this project is the Arduino IDE, which exports any related coding to the Arduino board. All commands of controlling variables and parameters such as current, voltage, solar irradiance, and others are included. Proteus is used to design the circuit of the developed system. All components needed in this project are inserted one by one to simulate whether it can run properly according to what has been planned without any problem that may affect the system. The example of the circuit designed by using Proteus software is demonstrated in Fig. 1.

2.3 Hardware Development

The hardware and coding development are implemented the same as designed in Proteus and Arduino IDE, respectively. The fully developed system of the solar radiation measurement device is represented in Fig. 2. The system consists of a solar cell, LCD, power bank as the external power, Arduino Uno, and also resistors. The connection of the solar cell will go to the voltage divider, and the output of the voltage divider will be connected to the A2 pin of the Arduino, which will produce the shunt voltage. Meanwhile, the voltage from the solar cell will be produced by connecting it to the A1 pin with the input of the voltage divider. From here, LCD will display the values that need to be measured from the system. The brightness of the LCD will be controlled by 5 kΩ-rated potentiometer. This solar radiation sensor demonstrates a concept of voltage divider as a fundamental structure. Based on Fig. 3, V_{Panel}, I_{Panel}, R_{Shunt} and R_{Load} are defined as panel/cell voltage, panel current, shunt

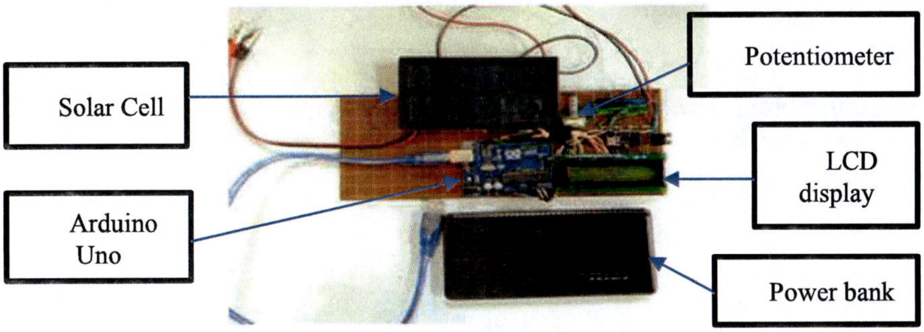

Fig. 2 The full system of the developed solar radiation measurement device with an external power source

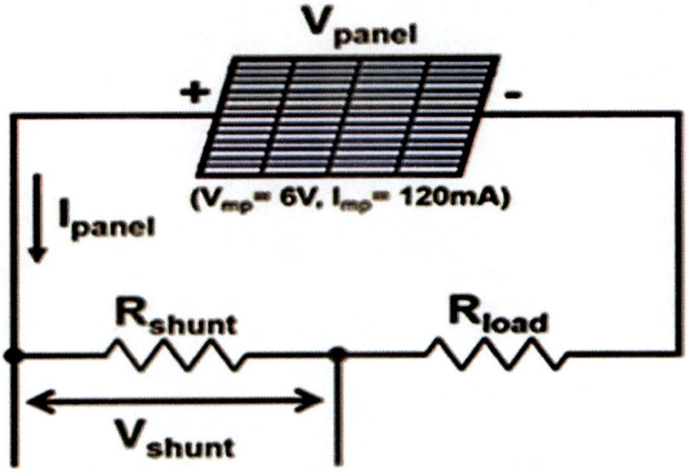

Fig. 3 Circuit of solar irradiance meter based on voltage divider concept

resistance, and load resistance, respectively. The shunt current (I_{Shunt}) is calculated using Eq. (4) where the voltage across the shunt resistor, denoted by shunt voltage (V_{Shunt}) is measured. The output from the solar cell and voltage divider are connected to A1 and A2 pins of the Arduino, respectively. To determine the value of the shunt current (I_{Shunt}) several calculations are considered as follows:

$$I_{Panel} = I_{Load} = I_{Shunt} \tag{1}$$

$$V_{Cell} = V_{Load} + V_{Shunt} \tag{2}$$

$$V_{Panel} = I_{Total} = R_{Total} \tag{3}$$

$$V_{Shunt} = I_{Shunt} \times R_{Shunt} \tag{4}$$

Furthermore, total resistance determined in this project is simply derived using the calculation below:

$$\begin{aligned}
&V_{Cell} = I(R_{Load} + R_{Shunt}) \\
&\text{If } V_{Mp} = 6V;\ I_{Mp} = 120mA \\
&6 = 120m(R_{Load} + R_{Shunt}) \\
&I = 1/20(R_{Load} + R_{Shunt})
\end{aligned} \tag{5}$$

From here, it is suggested that the value of total resistance to be used for R_{Load} and R_{Shunt} should be limited to 20 Ω only. However, to get the accurate reading of the resistance for each resistor, a multimeter is used to measure the actual values. Thus, the actual measured values of R_{Load} and R_{Shunt} are 18.8 Ω and 1.7 Ω, respectively, which totals the resistance value of 20.5 Ω. This total resistance and R_{Shunt} values are then substituted in Eq. (3) and (4) to produce Eq. (6) and Eq. (7), respectively.

$$I_{Total} = V_{Cell}/20.5\Omega \tag{6}$$

$$I_{Shunt} = V_{Shunt}/1.7\Omega \tag{7}$$

3 Experimental Setup

There are two experiments have been carried out in this study to determine the value of solar irradiance and to test the reliability of the system. First, Experiment 1 is carried out whereby only the solar cell and voltage divider are connected in the same circuit. The experiment is carried out in Alternative Energy Laboratory, Faculty of Electrical and Electronics Engineering, University Malaysia Pahang, Malaysia. The circuit is placed under a halogen-based spotlight, as shown in Fig. 4.

The room temperature usually 25 °C but during the experiment there is some heat produced by the halogen spotlight that make the ambient temperature between the cell and spotlight slightly increases. With the help of air ventilation from the fan, we can control the ambient temperature which varies in range of 31 until 34 °C. Two types of commercially-available solar irradiance meters are used to make the comparison of measured solar irradiance values. Both devices, SEAWARD Solar Survey 200R, and Solar Power Meter TES1333 are labelled as Solar Meter 1 and Solar Meter 2, respectively. Both solar meters are placed at the same level of height as the solar cell of the developed device. The value of the shunt voltage is measured directly using a multimeter, and the values of instantaneous solar irradiance from both solar meters are also recorded, simultaneously. The height of the solar cell and

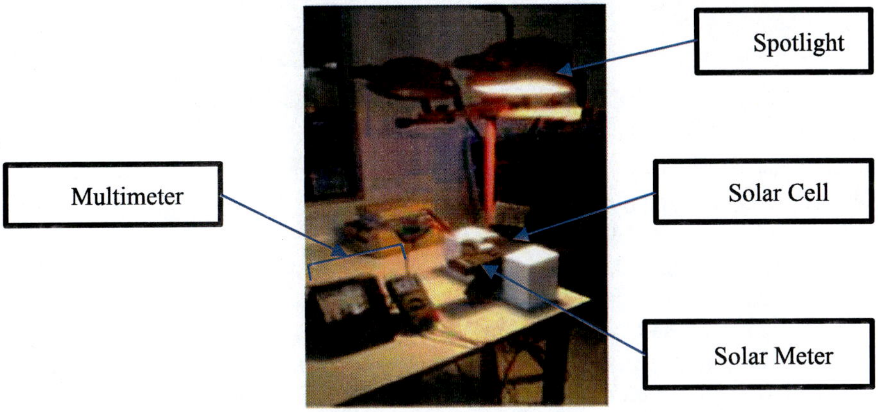

Fig. 4 Experimental setup

solar meters from the spotlight are varied to obtain several sampling series of solar irradiance. The light intensity of the spotlight is also varied by turning on one bulb and both bulbs in sequence to differ the reading of solar irradiance at one particular height.

Based on recorded data, a graph will be plotted for each solar meter to estimate the I_{Shunt} based on Eq. (7) and the trend of this I_{Shunt} will be compared to the solar irradiance measured by both conventional solar meters. From these graphs, a nearly-linear line is expected to be produced and form new equations for both conventional solar meters. These equations are rearranged and included in the Arduino coding so that the value of the solar irradiance of the developed solar irradiance measurement device will be calculated automatically and instantly produced through the LCD. Next, the full system circuit has been set up for Experiment 2. In this experiment, the values of the solar irradiance obtained based on Experiment 1 through our proposed measurement device will be measured and evaluated. The reading of the solar irradiance based on both equations derived from Experiment 1 will be displayed through the LCD and logged to the computer. The process of Experiment 2 is repeated the same as in Experiment 1 for both cases of the light intensity and height setting. The readings of solar irradiance displayed on the LCD and both conventional solar meters are recorded in parallel and tabulated in the next section. On the other hand, statistical elements of correlation coefficient (r), coefficient of determination (R^2) and root mean square error (RMSE) have been taken into account to evaluate the performance of our proposed solar irradiance measurement device based on this experimental result using the equations below [4]:

$$\mathrm{R} = \frac{\sum (Sc - Saveg, c)(Sd - Saveg, d)}{\sqrt{\sum (Sc - Saveg, c)^2} \times \sqrt{\sum (Sd - Saveg, d)^2}} \tag{8}$$

$$R^2 = 1 - \frac{\sum(Sd - Sc)^2}{\sum(Sc - Saveg, c)^2} \quad (9)$$

$$\text{RMSE} = \sqrt{\frac{1}{N}\sum(Sd - Sc)^2} \quad (10)$$

where S_c, S_d, $S_{aveg,c}$, Saveg,d and N are solar irradiance of conventional solar meter, solar irradiance of developed solar meter, average solar irradiance of conventional solar meter, average solar irradiance of developed solar meter and the number of data collected, respectively. Before that, repetitive modification of Arduino coding has been made based on the calculated parameters to get more reliable data by reducing the calculation error. Then the reliability of the data is compared with both conventional solar meters. The value of shunt voltage produced directly from the solar cell is recorded and collected for further analysis. Then, the data is compared to the reading of solar irradiance shown on both conventional solar meters.

3.1 Experiment 1

From Table 1, it can be seen clearly that the reading of solar irradiance for both solar meter 1 and solar meter 2 are quite different. Figure 5 and Fig. 6 demonstrate the correlation graphs between the shunt current and the solar irradiance measured by Solar Meter 1 and Solar Meter 2, respectively. From Fig. 5, it can be suggested that the data is expected to have a correlation near to 1, which means that these two variables, solar irradiance and shunt current, have a strong relationship. Majority dots of sampling are concentrated with the line of the best fit. On the other hand, although the majority dots of sampling in Fig. 6 are seemed to have marginally deviated from the ideal line, further analysis is still necessary to understand the correlation trend between these two variables.

The changing of the output over changing of the quantity being measured is called sensitivity [21]. The sensitivity of this system is calculated by the change of voltage shunt divided by the measured solar irradiance from the developed solar meter. The equation for the sensitivity as below:

$$\text{Sensitivity} = \Delta V_{\text{Shunt}}/\Delta \text{Irradiance} \quad (11)$$

Based on the above Fig. 5 and 6, the correlation between the solar irradiance and shunt current for both Solar Meter 1 and Solar Meter 2 can be interpreted as follows:

Solar Meter 1:

$$\begin{aligned} y &= \left(0.0628 \times 10^{-3}\right)x \\ x &= y/0.0000628 \end{aligned} \quad (12)$$

Solar Meter 2:

Table 1 The measurement result of the shunt resistor, shunt current, solar irradiance of solar meter 1 and solar irradiance of solar meter 2 obtained from Experiment 1

No	Shunt resistor (Ω)	Shunt voltage (mV)	Shunt current (mA)	Solar irradiance of solar meter 1 (W/m²)	Solar irradiance of solar meter 2 (W/m²)
1	1.7000	13.2000	7.7647	135.0000	–
2	1.7000	17.7000	10.4100	162.9500	113.5000
3	1.7000	22.5000	13.2353	206.7000	148.5000
4	1.7000	26.5000	15.5882	244.6000	168.5000
5	1.7000	26.8000	15.8000	273.0000	188.0000
6	1.7000	33.1000	19.4706	272.0000	178.5000
7	1.7000	33.9000	19.9412	322.3000	224.5000
8	1.7000	35.5000	20.8824	309.1500	212.0000
9	1.7000	43.0000	25.2941	408.6000	287.5000
10	1.7000	50.3000	29.5882	476.7500	330.0000
11	1.7000	55.0000	32.3529	501.5000	367.5000
12	1.7000	60.6000	35.6471	568.8000	391.5000
13	1.7000	71.7000	42.1765	671.3000	454.0000
14	1.7000	103.4000	60.8235	913.0000	658.5000
15	1.7000	105.3000	61.9412	1006.5000	908.0000
16	1.7000	182.6000	107.4120	1734.5000	1448.0000

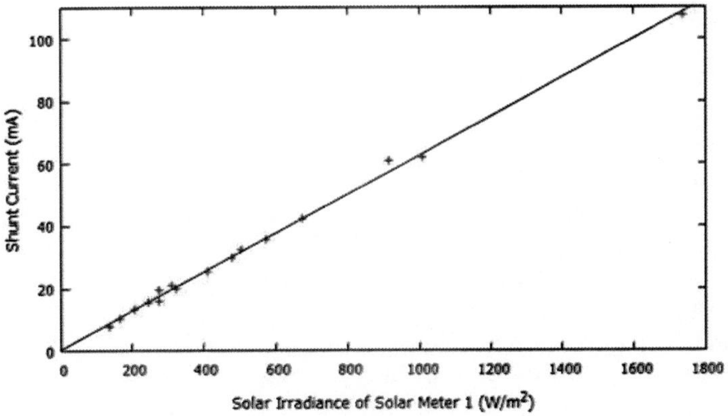

Fig. 5 Correlation between the solar irradiance and shunt current obtained through the measurement using Solar Meter 1 (SEAWARD Solar Survey 200R)

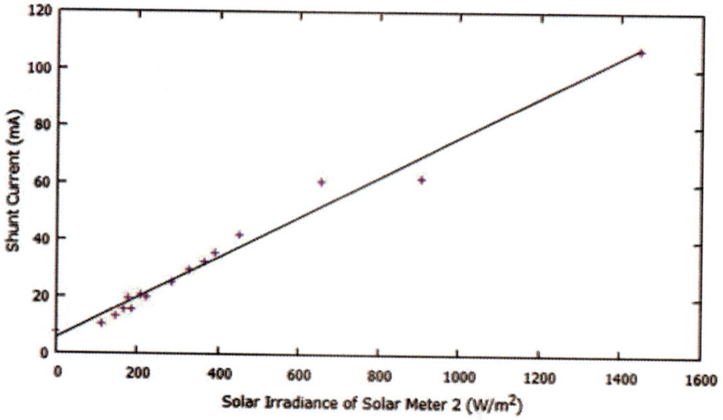

Fig. 6 Correlation between the solar irradiance and shunt current obtained through the measurement using Solar Meter 2 (Solar Power Meter TES1333)

$$\begin{aligned} y &= \left(0.0785 \times 10^{-3}\right)x \\ x &= y/0.0000785 \end{aligned} \tag{13}$$

where x and y are defined as solar irradiance and shunt current, respectively.

3.2 *Experiment 2*

Experiment 2 has been performed by executing Eq. (12) and (13) determined by Experiment 1 into the Arduino coding, parallelly. The readings of solar irradiances are compared between our developed solar meter and both conventional solar meters. Table 2 indicates the comparison result of solar irradiance measured by developed solar irradiance measurement device and conventional solar meters obtained from Experiment 2. Moreover, the correlation between the measured values of solar irradiance for the developed meter and both solar meters (1 and 2) are represented in Figs. 7 and 8, respectively.

Equations (8)–(10) are taken into account to evaluate the performance of the developed system. The correlation coefficient, r shows how strong the data is related. There is either positive value, a negative value, or zero. If the correlation coefficient gives +1 value means it has a strong positive relationship while if the value is -1, it means the data has a strong negative relationship. There is no relationship between the data if the zero value appears. The positive value means both variables get larger, but negative value means that one variable gets larger, and the others get smaller [22]. Therefore, for the case of Solar Meter 1, the evaluation result indicates that the r for Fig. 7 achieves a value of 0.9901. Meanwhile, the r for the case of Solar Meter

Table 2 The comparison results of solar irradiance measured by developed solar irradiance measurement device and conventional solar meters obtained from Experiment 2

No	Solar irradiance of developed meter 1 (W/m^2)	Solar irradiance of developed meter 2 (W/m2)	Solar irradiance of solar meter 1 (W/m2)	Solar irradiance of solar meter 2 (W/m2)
1	178.4345	157.9335	129.0000	105.0000
2	197.4168	337.1272	152.0500	207.5000
3	235.3816	443.4286	188.2000	243.0000
4	296.1253	206.5284	253.8000	114.5000
5	356.8689	209.5656	263.1500	124.0000
6	406.2231	470.7632	306.9500	262.5000
7	459.3738	252.0861	331.0500	153.0000
8	531.5070	552.7672	358.1000	318.5000
9	539.0999	288.5323	416.0000	189.5000
10	634.0118	662.1058	457.3000	380.0000
11	664.3837	440.3914	461.0500	298.5000
12	933.9337	953.6752	619.2500	562.5000
13	1298.3953	583.1390	765.7500	389.0000
14	The value cannot be read due to out of range of the solar meter	874.7085	The value cannot be read due to out of range of the solar meter	580.0000
15	The value cannot be read due to out of range of the solar meter	1801.0490	The value cannot be read due to out of range of the solar meter	1109.5000

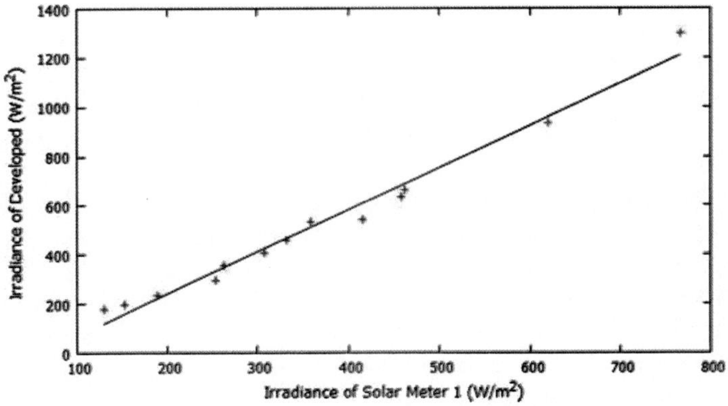

Fig. 7 Correlation between the measured values of solar irradiance for the developed meter with solar meter 1

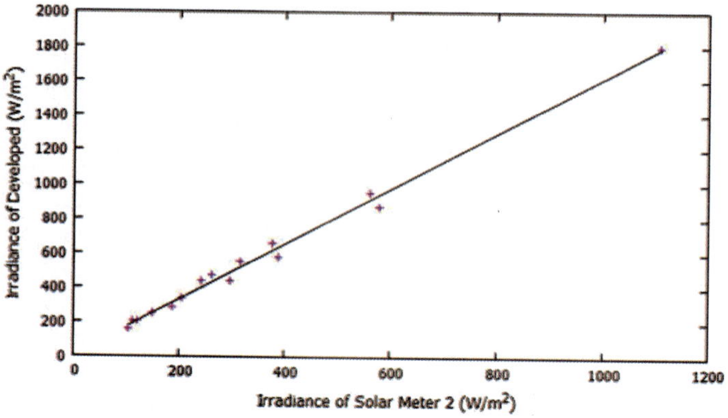

Fig. 8 Correlation between the measured values of solar irradiance for the developed meter with solar meter 2

2 from Fig. 8 has reached the value of 0.9962. From here, it can be suggested that the Solar Meter 2-based model produces the r value slightly higher than that of Solar Meter 1.

The coefficient of determination, R^2 is the output of regression analysis. An R^2 of 0 means that the dependent variable cannot be predicted from the independent variable while an R^2 of 1 means the dependent variable can be predicted without error from the independent variable. Whereas an R^2 between 0 and 1 indicates the extent to which the dependent variable is predictable [23]. From Experiment 2, the value of R^2 is 0.7460 for Fig. 7 and 0.6089 for Fig. 8. An R^2 of 0.7460 means that 74.60% of the variance in Y is predictable from X, while an R^2 of 0.6089 means that 60.89% of the variance in Y is predictable. If the coefficient, R^2 is larger, it shows that the model fits the data collected. Thus, the coefficient of determination of developed Solar Meter 1 is the best model that fits the data collected.

In order to determine the sensitivity of the solar meter or the device, Eq. (11) have been used and the results for Solar Meter 1 and Solar Meter 2 are 0.0002 Vm^2/W and 0.0001 Vm^2/W respectively. These shows that Solar Meter 1 has high sensitivity compared to Solar Meter 2.

Root mean square error, RMSE is used to measure the differences between values predicted by a model or an estimator and the values observed. When compared to two models of developed solar meters, the smaller the value of RMSE, the better the model but that small differences between those RMSE may not be relevant or event significant. RMSE is scale-dependent. It depends on the dependent variable. It does not have any values that can indicate a good RMSE. The resulting model is the best if it gives a good in-sample fit, associated with low error measures and WN residuals and avoids over-fitting by giving the best out-of-sample forecast accuracy [24]. From Experiment 2, the RMSE value of Solar Meter 2 is bigger than that of Solar Meter 1, where their values are 264.4448 Wm^{-2} and 204.5928 Wm^{-2}, respectively. It shows that there is a lot of data concentrated around the line of best fit for the case of Solar Meter 1. Since Solar Meter 1 show the best result compared to Solar Meter 2 which has the reading of R^2 of 0.7460, RMSE of 204.5928 and the sensitivity of 0.0002 Vm^2/W, so Solar Meter 1 can be used as a device in order to measure the solar irradiance.

In order to determine the sensitivity of the solar meter or the device, Eq. (11) have been used and the results for Solar Meter 1 and Solar Meter 2 are 0.0002 Vm^2/W and 0.0001 Vm^2/W respectively. These shows that Solar Meter 1 has high sensitivity compared to Solar Meter 2.

Root mean square error, RMSE is used to measure the differences between values predicted by a model or an estimator and the values observed. When compared to two models of developed solar meters, the smaller the value of RMSE, the better the model but that small differences between those RMSE may not be relevant or event significant. RMSE is scale-dependent. It depends on the dependent variable. It does not have any values that can indicate a good RMSE. The resulting model is the best if it gives a good in-sample fit, associated with low error measures and WN residuals and avoids over-fitting by giving the best out-of-sample forecast accuracy [24]. From Experiment 2, the RMSE value of Solar Meter 2 is bigger than that of Solar Meter 1, where their values are 264.4448 Wm^{-2} and 204.5928 Wm^{-2}, respectively. It shows that there is a lot of data concentrated around the line of best fit for the case of Solar Meter 1. Since Solar Meter 1 show the best result compared to Solar Meter 2 which has the reading of R^2 of 0.7460, RMSE of 204.5928 and the sensitivity of 0.0002 Vm^2/W, so Solar Meter 1 can be used as a device in order to measure the solar irradiance.

4 Conclusion

In a nutshell, a prototype of a solar irradiance measurement device has been successfully developed. The obtained solar irradiance values from the proposed device exhibit very good agreement with those measured by the commercial solar meters/sensors. Impressively, the value of correlation coefficient, r reaches almost ideal value with 0.99. This study should be a kick-start idea for a better improvement system in the future. Further study can be done based on this solar irradiance measurement device by exploring more about the critical relationship between solar irradiance and other electrical parameters such as internal resistance, cell voltage, etc.

Acknowledgements Authors would like to extend their gratitude to Universiti Malaysia Pahang under Fundamental Research Grant (RDU) with reference no. RDU200348. This work also was partly supported by Postgraduate Research Grants Scheme (PGRS) with reference no. PGRS200317.

References

1. Ballestrín J, Roldán MI (2017) Contact sensors for measuring high surface temperature in concentrated solar radiation environments. Meas. J. Int. Meas. Confed. 109:65–73
2. Marzo A et al (2017) Daily global solar radiation estimation in desert areas using daily extreme temperatures and extraterrestrial radiation. Renew. Energy 113:303–311
3. Madeti SR, Singh SN (2017) Monitoring system for photovoltaic plants: A review. Renew Sustain Energy Rev 67:1180–1207
4. Shah ASBM, Yokoyama H, Kakimoto N (2015) High-Precision Forecasting Model of Solar Irradiance Based on Grid Point Value Data Analysis for an Efficient Photovoltaic System. IEEE Trans. Sustain. Energy 6(2):474–481
5. H. Ohtake, K. ichi Shimose, J. G. da S. Fonseca, T. Takashima, T. Oozeki, and Y. Yamada, "Accuracy of the solar irradiance forecasts of the japan meteorological agency mesoscale model for the kanto region, Japan," *Sol. Energy*, vol. 98, pp. 138–152, 2013
6. Almorox J (2011) Estimating global solar radiation from common meteorological data in Aranjuez, Spain. Turkish J. Phys. 35(1):53–64
7. E. Lorenz, J. Hurka, D. Heinemann, and H. G. Beyer, "Irradiance Forecasting for the Power Prediction of Grid-Connected Photovoltaic Systems," *IEEE J. Sel. Top. Appl. Earth Obs. Remote Sens.*, vol. 2, no. 1, pp. 2–10, 2009
8. Mathiesen P, Brown JM, Kleissl J (2013) Geostrophic wind dependent probabilistic irradiance forecasts for coastal california. IEEE Trans. Sustain. Energy 4(2):510–518
9. Geraldi E, Romano F, Ricciardelli E (2012) An advanced model for the estimation of the surface solar irradiance under all atmospheric conditions using MSG/SEVIRI data. IEEE Trans Geosci Remote Sens 50(8):2934–2953
10. Eissa Y, Marpu PR, Gherboudj I, Ghedira H, Ouarda TBMJ, Chiesa M (2013) Artificial neural network based model for retrieval of the direct normal, diffuse horizontal and global horizontal irradiances using SEVIRI images. Sol Energy 89:1–16
11. Saito K et al (2006) The operational JMA nonhydrostatic mesoscale model. Mon Weather Rev 134(4):1266–1298
12. K. Saito *et al.*, "Nonhydrostatic atmospheric models and operational development at JMA," *J. Meteorol. Soc. Japan*, vol. 85 B, pp. 271–304, 2007
13. H. Liu, S. Igari, S. Sang, L. Zhou, and Y. Zhai, "The first proficiency testing for primary calibration of terrestrial photovoltaic reference cells," *2015 IEEE 42nd Photovolt. Spec. Conf. PVSC 2015*, pp. 0–4, 2015
14. Osterwald CR et al (1999) World Photovoltaic Scale: An international reference cell calibration program. Prog Photovoltaics Res Appl 7(4):287–297
15. C. B. Yahya, "Performance monitoring of solar photovoltaic systems using reference cells," *Proc. Int. Conf. Microelectron. ICM*, pp. 445–449, 2008
16. Hassan MA, Khalil A, Kaseb S, Kassem MA (2017) Exploring the potential of tree-based ensemble methods in solar radiation modeling. Appl Energy 203:897–916
17. Sharma A, Kakkar A (2018) Forecasting daily global solar irradiance generation using machine learning. Renew Sustain Energy Rev 82(May):2254–2269
18. Aler R, Galván IM, Ruiz-Arias JA, Gueymard CA (2017) Improving the separation of direct and diffuse solar radiation components using machine learning by gradient boosting. Sol Energy 150:558–569
19. R. Poling, "What is a pyranometer?," *Solar Power World*, 2015. [Online]. Available: https://www.solarpowerworldonline.com/2015/03/what-is-a-solar-pyranometer/
20. "Pyranometer," *wikipedia*, 2017. [Online]. Available: https://en.wikipedia.org/wiki/Pyranometer. [Accessed: 09-Jan-2018]
21. "What is a measuring instrument's sensitivity?," *Quora*. [Online]. Available: https://www.quora.com/What-is-a-measuring-instruments-sensitivity
22. "Correlation," *Creative Research Systems*, 2016. [Online]. Available: www.surveysystem.com/correlation.htm

23. "Coefficient of determination," *Stat Trek.* [Online]. Available: https://stattrek.com/statistics/dictionary.aspx?definition=coefficient_of_determination
24. "What are good RMSE values?," *Stack Exchange.* [Online]. Available: https://stats.stackexchange.com/questions/56302/what-are-good-rmse-values

3D Traffic Sign Detection Using Camera-LiDAR Projection

Wonho Song and Hyun Myung

Abstract Localization using High definition (HD) maps is one of the key solutions for high precision autonomous driving on the urban environment. For accurate HD map-based localization, more features included in HD maps are required to be detected while driving in real-time. Although there were a few studies about detecting road features to match with HD map, mostly used only a simple intensity filter or dimensionality filter, which is not robust in varying environments. This paper proposes a novel traffic sign detection algorithm in the 3D point cloud using a 2D image for robust 3D object detection. Deep learning 2D object detection is the first step, and then by using LiDAR-camera calibration data and outlier removal algorithm, 3D point cloud data of the traffic signs are collected. The performance evaluation was done using Waymo Open Dataset and we compared the proposed method with other 3D object detection results in the challenges board. The proposed method is the robust solution for any feature recognition and 3D detection included in the HD map for precise localization.

Keywords 3D detection · Traffic sign detection · HD map-based localization

1 Introduction

Localization in an autonomous vehicle is one of the most critical and challenging components of an autonomous driving system [1]. HD maps are used for high precision driving on the urban environment [2]. Most of the HD map includes lanes, road markings and also traffic signs. Figure 1 shows one example of features included in HD maps. For lateral correction of the position, lanes and other features are used, and

W. Song (✉) · H. Myung
Korea Advanced Institute of Science and Technology (KAIST), E3-2, 291 Daehak-Ro, Yuseong-Gu, Daejeon 34141, South Korea
e-mail: swh4613@kaist.ac.kr

H. Myung
e-mail: hmyung@kaist.ac.kr

A. F. Ab. Nasir et al. (eds.), *Recent Trends in Mechatronics Towards Industry 4.0*, Lecture Notes in Electrical Engineering 730, https://doi.org/10.1007/978-981-33-4597-3_73

Fig. 1 Example of traffic signs and road marking on vehicle view (left) and HD map (right)

for longitudinal correction cases, sometimes use a road marking and other features on autonomous driving.

The precise 3D location of traffic signs on vehicle view may provide accurate localization results by matching with the location on the HD map. HD maps can provide very precise location information of features on the road. Therefore, the key information needed for HD map localization is a precise object, included in HD maps, detection results on real-world 3D space.

2 Traffic Signs Detection

2.1 Related Work

HD map matching based localization approach has long studied for the solution of calculating precise global localization of autonomous vehicles. In [3], the authors evaluated the absolute pose using road lanes and markings detection algorithm to associate with elements in the high-resolution road maps. They used stereo camera, vehicle odometry, and a low-cost Global Navigation Satellite System (GNSS) module to cover 5 km driving in narrow urban roads under varying weather conditions. The accuracy was 0.08 m in typical road scenarios, and the system was available 98% of times.

One study [4] was about vehicle localization in an HD map using road signs using LiDAR. They were using the front polar grid and de-skewing of a region of interest in 3D points to extract road signs among point cloud data. Particle filtering was used to match with HD map road sign locations, and along-track mean error was 0.93, and cross-track mean error was 0.09.

There was also a study just focused on traffic sign detection in point cloud data. As shown in [5], the traffic signs were detected directly from LiDAR point cloud data. They were just using intensity and dimensionality filter for 3D traffic sign detection in point cloud data, which were not in real-time. The results for traffic sign detection were close to 98% using their own dataset collected in urban areas in Galicia, Spain.

2.2 *Problem Statement*

Localization in an urban area is considered challenging because the data from GNSS results are not good enough due to high buildings. In HD map-based localization, to get better lateral error results as well as longitudinal error in localization, detecting as many road features included in HD map as possible is crucial. Although there were a few studies about detecting road features to match with HD map, when detecting 3D pose in point cloud data, papers mostly used only a simple intensity filter or dimensionality filter, which is not robust in varying environments. Also, most of the detection result was not reliable because they were not evaluated in open dataset, instead used their own data.

2.3 *Solution Approach*

In this paper, we aim to present robust traffic sign detection on 3D point clouds collected by any laser scanner on an autonomous vehicle. It is designed first to detect the signs on given camera image data and find the location of the detected objects on point clouds using outlier removal. This method will be a robust solution of detecting a 3D relative position of any object that can be trained with any certain image dataset.

3 Our Approach

Figure 2 shows the overall framework of our approach. First, detect the traffic sign on image and project lidar point cloud on the image. Second, make the frustum in

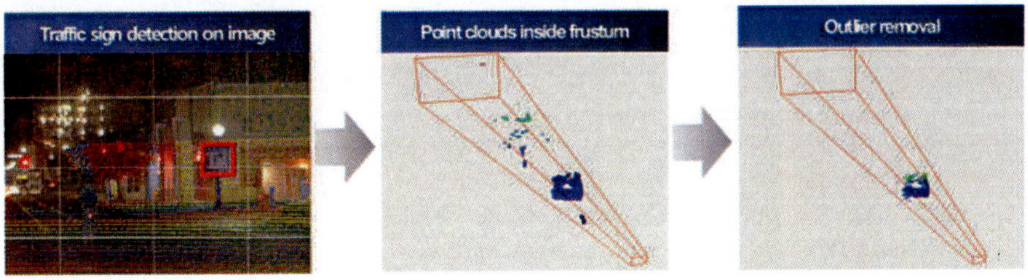

Fig. 2 The framework of our approach

3D space and collect the points included in the frustum. Finally, remove the outlier included in the frustum. The detailed explanation is mentioned in Sects. 3.1 and 3.2.

3.1 2D Image Traffic Sign Detection

To train the traffic signs, German Traffic Sign Recognition Benchmark (GTSRB) dataset [6] are used. GTSRB consists of single-image traffic signs of more than 50,000 images in total. All images are used to train in a single class, and You Only Look Once (YOLO) v3 [7] is used as a training model. About 2000–20,000 iteration weight value is used, and the batch value is set to 64, and the subdivision value is set to 8. AMD Ryzen 7 and GTX 2080 environment is used for training.

3.2 3D Point Cloud Extraction

Waymo Open Dataset [8] will be used for testing the accuracy of 3D traffic sign detection. Other famous datasets such as KITTI dataset [9] do not usually contain 3D bounding boxes of traffic signs in point cloud data. Additionally, Waymo Open Dataset has the most accurate ground truth and calibration value when KITTI dataset sometimes has a problem with ground truth and calibration results. Therefore, Waymo Open Dataset is used for testing the results.

When we get weight values of 2D image object training and when we can detect small traffic signs in the camera image in real-time, the next step is to project LiDAR point clouds on the image on the corresponding location. Waymo Open Dataset calibration values are used for projection. As a projection result in Fig. 3., we got an accurate projection result on the image.

With 2D detection results from 3.1, we made frustums of the detected bounding box of traffic signs in 3D point cloud data. The second step of Fig. 2 shows the result of the example of frustum result of the detected object. Outliers which is not actually point clouds of traffic signs are included in the frustums. The next step is to remove point clouds except for the point clouds of the objects.

An intensity filter was first used to remove outliers. Then Random sample consensus (RANSAC) [10] is used to fit the point clouds on a fitted plane. The result of outlier removal is shown in the third step of Fig. 2. The blue area is showing a flatted result of the detected traffic signs.

4 Evaluation

As mentioned in 3.2, Waymo Open Dataset is used to evaluate 3D detection results. Waymo Open Dataset provides an accurate 3D bounding box of the traffic signs. The

Fig. 3 Point cloud projected on an image plane (waymo open dataset example)

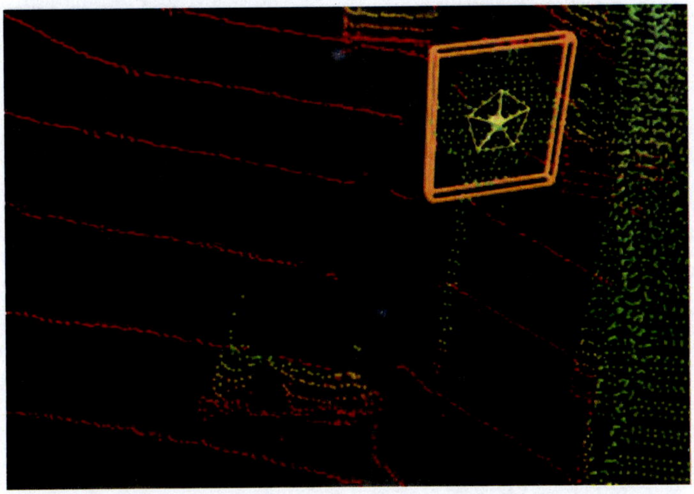

Fig. 4 3D traffic sign detection evaluation with waymo open dataset

results are evaluated using Waymo Open Dataset Challenge. The accuracy result is shown in Table 1 and Fig. 5. Figure 5 shows precision-recall curve for traffic sign as the y-axis shows precision, and the x-axis shows recall (Fig. 4).

Table 1 Accuracy results in waymo open dataset

	AP (average precision)	APH (average precision heading)
TYPE_SIGN	0.6955	0.5943

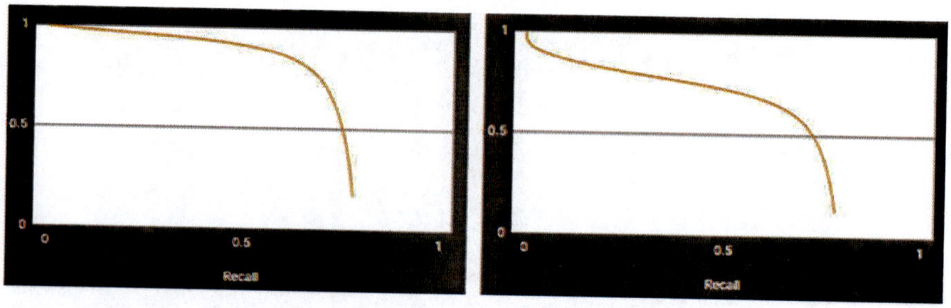

Fig. 5 3D traffic sign detection AP and APH result with waymo open dataset

There are no results for recognition of traffic signs in Waymo Open Dataset Challenge leaderboards. Therefore, by comparing with other 3D pedestrian detection AP, APH results in Waymo Open Dataset Challenge boards, the result from this algorithm is about the 9th most accurate algorithm. When considering most of the top AP results are between 0.67 to 0.74, and traffic signs are more challenging due to the size compare to pedestrian detection, the result of our method is showing considerably excellent performance.

5 Conclusion

In this paper, the robust 3D traffic sign detection algorithm is introduced using frustum from 2D bounding box results and using outlier removal to receive a precise result in 3D space. The algorithm can be applied in any object when it is collected with a single camera and a single LiDAR. It can be applied using any camera and LiDAR. The accuracy result has to be improved up to 0.8 to get reliable accuracy localization result in HD map-based matching. By comparing different networks to improve performance in 2D image detection or using different outlier removal algorithm can enhance the performance of the 3D detection results.

The future work is to add a tracking algorithm to get a continuous 3D position of the object even when the object passes the field of view of the camera. The HD map-based localization should also be added to see if it can be the localization solution in the GNSS denied environments.

Acknowledgements This article was supported by the research project "Development of A.I. based recognition, judgement and control solution for autonomous vehicle corresponding to atypical driving environment," which is financed from the budget of the Ministry of Science and ICT (MSIT) and the Institute for Information and Communication Technology Promotion (IITP), South Korea, Contract No. 2019-0-00399.

References

1. Jo K, Jo Y, Suhr JK, Jung HG, Sunwoo M (2015) Precise localization of an autonomous car based on probabilistic noise models of road surface marker features using multiple cameras. IEEE Trans Intell Transp Syst 3377–3392
2. Bauer S, Alkhorshid Y, Wanielik G (2016) Using high-definition maps for precise Urban vehicle localization. In: IEEE 19th international conference on intelligent transportation systems (ITSC), pp 492–497
3. Poggenhans F, Salscheider NO, Stiller C (2018) Precise localization in high-definition road maps for Urban regions. IEEE/RSJ international conference on intelligent robots and systems (IROS)
4. Ghallabi F, El-Haj-Shhade G, Mittet M, Nashashibi F (2019) LIDAR-based road signs detection for vehicle localization in an HD Map. IEEE Intell Vehs Symp (IV)
5. Soilán M, Riveiro B, Martínez-Sánchez J, Arias P (2016) Traffic sign detection in MLS acquired point clouds for geometric and image-based semantic inventory. ISPRS J Photogrammetry and Remote Sens
6. Stallkamp J, Schlipsing M, Salmen J, Igel C (2011) The German traffic sign recognition benchmark: a multi-class classification competition. In: International joint conference on neural networks
7. Redmon J, Farhadi A (2018) YOLOv3: An incremental improvement. In: Proceedings of the IEEE conference on computer vision and pattern recognition (CVPR), pp 1854–1862, Salt Lake City, UT, USA, 18–22 June 2018
8. Sun P, Kretzschmar H, Dotiwalla X, Chouard A, Patnaik V, Tsui P, Guo J, Zhou Y, Chai Y, Caine B, et al (2019) Scalability in perception for autonomous driving: Waymo open dataset. *arXiv*, arXiv–1912
9. Geiger A, Lenz P, Urtasun R (2012) Are we ready for autonomous driving? The kitti vision benchmark suite. In: IEEE conference on computer vision and pattern recognition CVPR
10. Martin A, Fischler, Robert C (1981) Bolles: random sample consensus: a paradigm for model fitting with applications to image analysis and automated cartography. Commun ACM 24 (6):381–395

Development of Physiotherapy-Treadmill (*PhyMill*) as Rehabilitation Technology Tools for Kid with Cerebral Palsy

Mohd Azrul Hisham Mohd Adib, Rabiatul Aisyah Arifin, Mohd Hanafi Abdul Rahim, Muhammad Rais Rahim, Muhammad Shazzuan Sharudin, Afif Awaluddin Othman, Ahmad Hijran Nasaruddin, Afiq Ikmal Zahir, Idris Mat Sahat, Nurul Shahida Mohd Shalahim, Narimah Daud, and Nur Hazreen Mohd Hasni

Abstract Main problems with the motor function for Cerebral Palsy (CP) kids are delayed or arrested on motor development. Therapeutics typically emphasize sound-inhibitory exercises, balance training, and preparatory tasks while walking, sitting, and standing to enhance the functioning of children with CP. The treadmill training is used for repeated tasks-specific walking. The focus is to increase the strength of lower extremity, walking speed, or endurance. In this study, we developed the physio-treadmill device for CP kids called *PhyMill*. This *PhyMill* is mainly constructed from the aluminum profile connected to connector made by polylactic acid (PLA) material equipped with an automatic control system. The height of the device can be adjusted according to the user's height. *PhyMill* offers three operating modes, the first one allowing you to control the movement of the patient forward and backward. The height of the device can be adjusted automatically according to the user's height is the second mode. The third mode is a special display screen to attracts the attention of

M. A. H. M. Adib (✉) · R. A. Arifin · M. H. A. Rahim · M. R. Rahim · M. S. Sharudin · A. A. Othman · A. H. Nasaruddin · A. I. Zahir
Medical Engineering and Health Intervention Team (MedEHiT), Department of Mechanical Engineering, College of Engineering, Universiti Malaysia Pahang, Lebuhraya Tun Razak, 26300 Kuantan, Pahang, Malaysia
e-mail: azrul@ump.edu.my

I. M. Sahat
Human Engineering Group (HEG), Faculty of Mechanical and Automotive Engineering Technology, Universiti Malaysia Pahang, 26600 Pekan, Pahang, Malaysia

N. S. M. Shalahim
Department of Industrial Engineering, College of Engineering, Universiti Malaysia Pahang, Lebuhraya Tun Razak, 26300 Kuantan, Pahang, Malaysia

N. Daud
Kuantan Physical Therapy, Physiotherapy Center, Ground Floor Lot B1.10, Block B, Bangunan al-Tabari, IM 7/3, Bandar Indera Mahkota 7, 25582 Kuantan, Pahang, Malaysia

N. H. M. Hasni
Family Health Unit, Pahang State Health Department, Jalan IM 4, 25582 Bandar Indera Mahkota, Kuantan, Pahang, Malaysia

A. F. Ab. Nasir et al. (eds.), *Recent Trends in Mechatronics Towards Industry 4.0*, Lecture Notes in Electrical Engineering 730, https://doi.org/10.1007/978-981-33-4597-3_74

the patient. The protection when using this device is also highlighted by supporting the user with an adjustable harness. Treadmill exercise for non-ambulatory children with CP as rehabilitation technology tools is a promising technique for the treatment of partial body weight support.

Keywords Children · Cerebral palsy · Gait · Treadmill training · Rehabilitation

1 Introduction

Cerebral palsy (CP) is a clinical diagnosis of a static and non-progressive disorder caused by brain damage or brain malformation that occurs during brain development—before, during, or after birth affecting one to two children per 2000 live births in Malaysia [1, 2]. Injuries, violence, medical neglect, negligence, infections, and injury are recognized as other risk factors leading to CP [3]. Besides, a genetic contribution to CP is even more commonly recognized than previously believed [4–7].

The Gross Motor Function Classification System (GMFCS) is likely to remain the same for one person's life, but medication and rehabilitation may enhance the performance of patients. However, people with CP may experience a decline in physical capacity without exercise and rehabilitation. [8, 9]. Recently different methods and strategies have been used to rehabilitate children with CP, ranging from very conservative to traditional techniques. However, in Malaysia, there is no technology-based rehabilitation center. This is because of the concerns of the therapeutic doubt about design during treatment that are less safe and cannot meet the CP requirements.

Recently, partial body weight supported by treadmill training (PBWSTT) is an alternative effectively used in children with CP who have not yet independently walked and may enhance gait speed. It consists of an overhead pulley system that has some of its weight supported. [10, 11]. Depending on the extent of disability, one or more physicians may be required to help maintain the correct posture and to physiologically and temporarily shift the legs through a pattern of gait [11]. Besides, researchers reported that repetitive gait training was more effective in improving ambulation velocity, walking endurance, gross motor function, step length, stride length, and cadence can be improved by PBWSTT than traditional physical therapy in children with CP [10–18].

Throughout the development, ergonomics was also taken into consideration. Ergonomics is described as the scientific study of the human-machine relationship which related to the comfort between the man and machine while operating the device. The goal of ergonomics is to make the device suitable for the user instead of making the user adapt to the device.

In this paper, the physiotherapy treadmill called *(PhyMill)* device is well developed which is intended for repetitive walking training for CP's kids [19]. *PhyMill* provides backward gait training on the treadmill which could be viewed as a way to enhance forward gaiting capacity for therapeutic intervention.

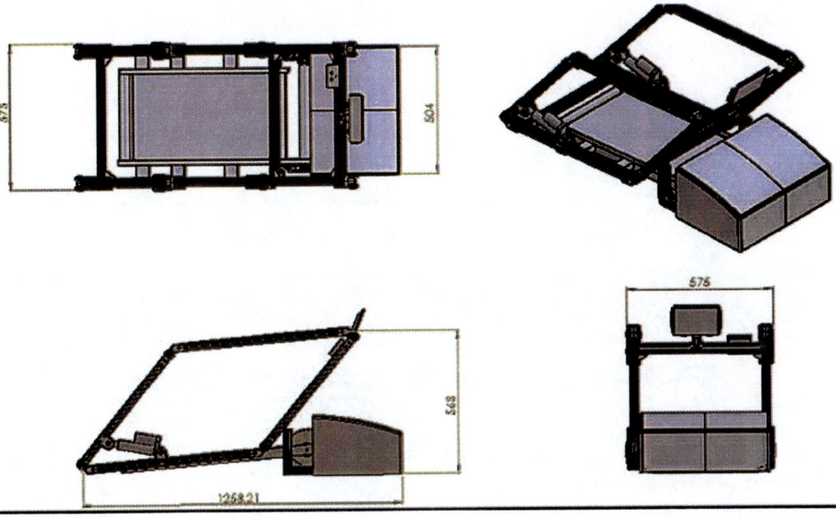

Fig. 1 Design and development of *PhyMill* device with full dimension. [unit: mm]

2 Methodology

2.1 Design and Development of PhyMill

Sympathy for the plight of children with impaired ability to control position and body activation due to damage to the developing brain or better known as Cerebral Palsy has prompted a group of Universiti Malaysia Pahang (UMP) researchers to produce Physio-Treadmill *(PhyMill)* products to help patients undergo walking movement exercises [19]. The *PhyMill* is designed using SolidWorks 2019. The device allows therapy exercises at home or in some other place rather than in a rehabilitation center. Figure 1 shows the isometric view of the final design and the full dimension of the *PhyMill.*

This frame *PhyMill* is mainly built from Aluminium Profile 30 × 30 mm connected with connector made from polylactic acid (PLA) material. Figure 2 shows the workflow of the *PhyMill* device design and development process.

2.2 Automatic Control System

Arduino Setup

Arduino as a microcontroller for navigating the 12 and 24 v systems actuator with this automatic control system. Two automatic control systems are used in this treadmill. Microcontrollers are powered by supply 24–12 v. Arduino Uno board only received

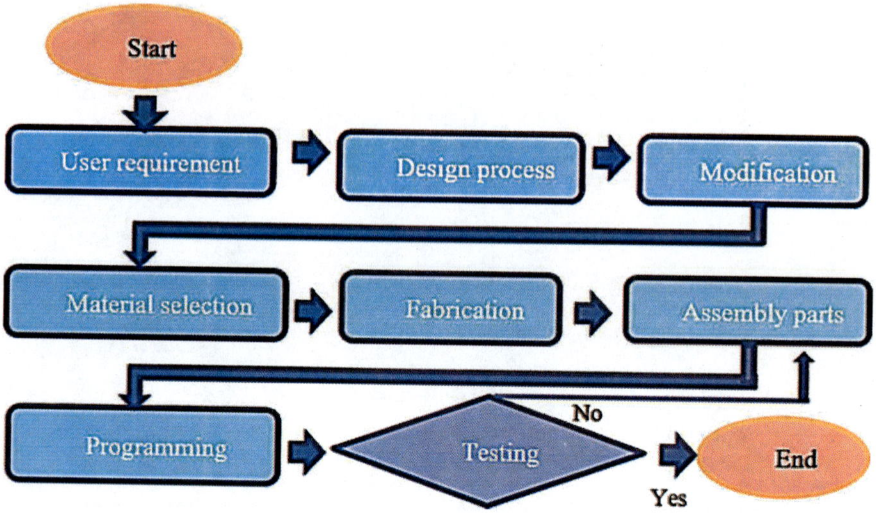

Fig. 2 The workflow of the *PhyMill* device development process

5 v. To get 12–5 v, voltage regulators are needed for this system. The device circuit flows as shown in Fig. 3.

Motor 24 v control system

The potentiometer used by the motor control system as a rotation shaft speed control. Motor drivers are used to connecting the microcontroller Arduino with the motor. The item that used to construct this system is power supply 24v, motor dc 24v, potentiometer, and motor driver. This system is based on how much voltage is supplied to the motor by the potentiometer. The potentiometer is configured by coding Arduino,

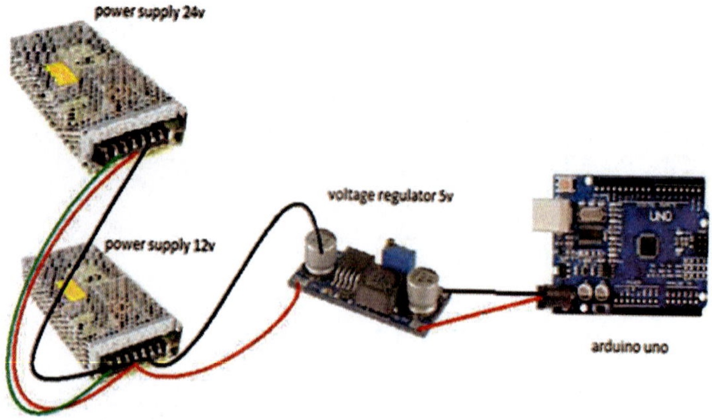

Fig. 3 Circuit flow of system power up arduino 5v

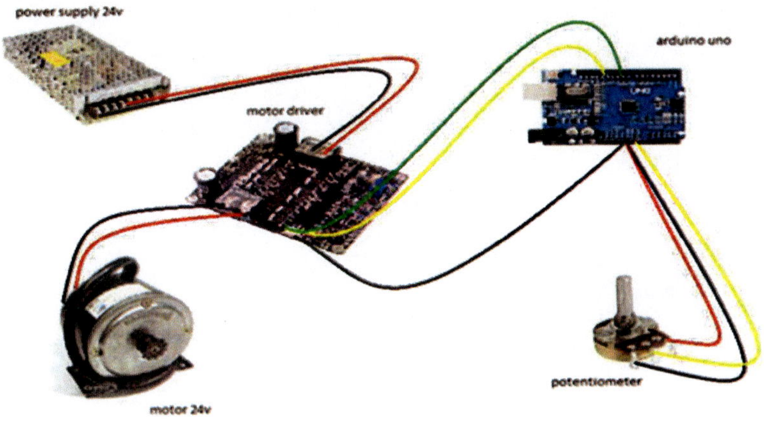

Fig. 4 Motor controller circuit

with the help coding of Arduino the potentiometer can reduce and increase the voltage that supplies the motor driver. Figure 4 shows the flow of the motor system.

Actuator 12 v control system

In this treadmill, the actuator is used for the height adjusted hand holder for children. This system used a simple controller which consists of two switch pushbuttons, Arduino Uno, power supply 12 v, relay 5 v 2 channel and a pair of actuator 12 v. Switch buttons in this system are designed up and down which is one button for up and another one is down. In this process, when the button is pushed the actuator will go up and another one will go down. This is a simple circuit that relies on the relay to switch over the voltage from positive to negative and negative to positive with the help of switch push button. The following Fig. 5 displays the actuator 12 v system (Table 1).

3 Result and Discussion

In this study, the *PhyMill* device is successfully developed as shown in Fig. 6. The device has been designed to allow partial body weight support using a safety harness and to provide the conditions of (1) walking speed between almost zero and fast walking and (2) forward and backward walking. The actuator allows for height adjustment of the treadmill so that it can be used by users of various heights.

Physiotherapy is a form of muscle stress relief in various parts of the body. This method requires the guidance of the physiotherapist, and to resolve this method periodically, a new approach is explored using a partial body weight supported treadmill training such as *PhyMill*. The *PhyMill* will improve the monitoring and quantitative measurements of patient activity in the recovery treatments. Therapy is performed

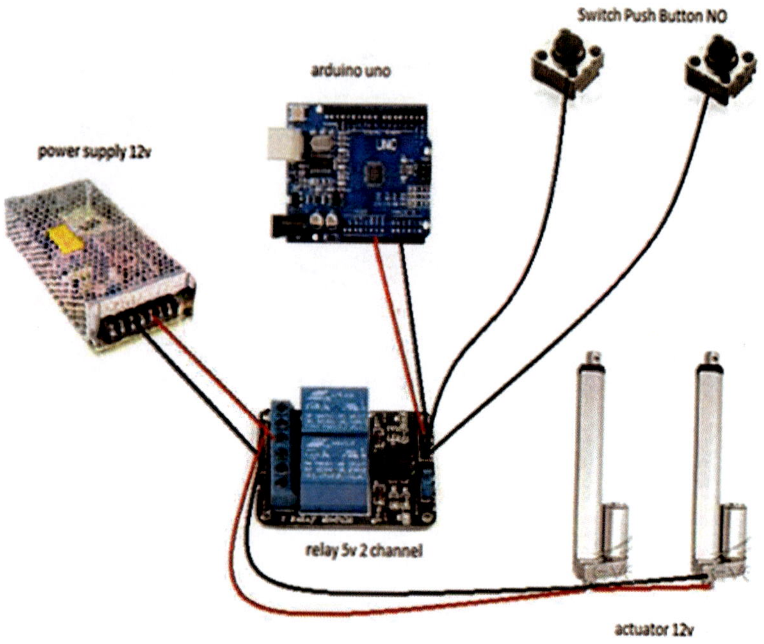

Fig. 5 Actuator circuit system

Table 1 Specification of actuator

model	XDHA12-100
Voltage (volt)	DC 12
Stroke (mm)	100
Speed (mm/s)	20
Push load (N)	500
Duty cycle (%)	20

automatically and can be done at home, based on certain technical concepts. Besides, home traction devices also meant that traction can be applied by patients who had been advised by the therapist about weights and duration of traction treatment.

This *PhyMill* offers three operating modes, the first one allowing you to control the movement of the patient forward and backward. By simply pressing the special button given, it is automatically operated entirely. Speed control is also created to inspire patients' adaptation of their walking pace according to the level of rehabilitation training determined by the therapist. Automatic height adjustment is the second mode. *PhyMill* also offers patients the facilities for determining the holder's level to their body height. In the third mode, a special display screen attracts the attention of

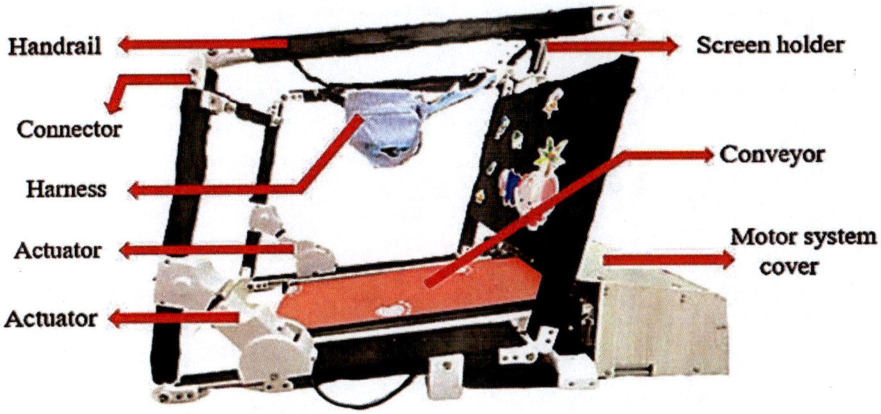

Fig. 6 The final product of *PhyMill*

the patient and avoids the boredom of the patient during the rehabilitation training session.

PhyMill can be used in CP patients with spastic diplegia type which is suitable for GMFCS II to III patients aged as early as 4–7 years old. The treadmill can be used for children between 90 and 100 cm in height and can accommodate loads of up to 30 km. While using this tool the patient gives a positive response and feels happy. No negative feedback or complaints about this device during training. Figure 7 shows clearly that one of the patients with CP is comfortable holding the handrails during exercise.

Fig. 7 Demonstrated the functional *PhyMill* device using real patients with CP

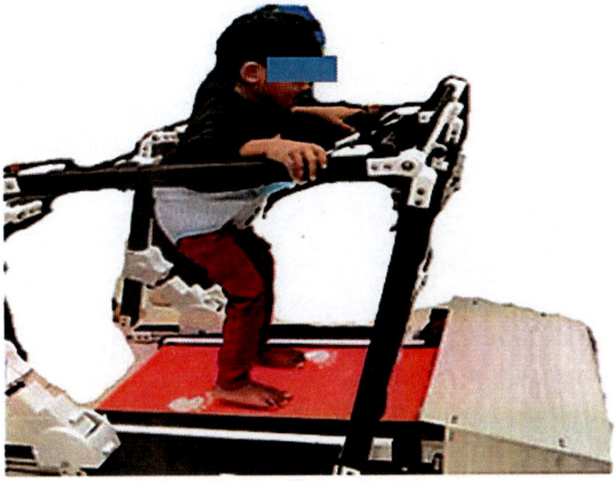

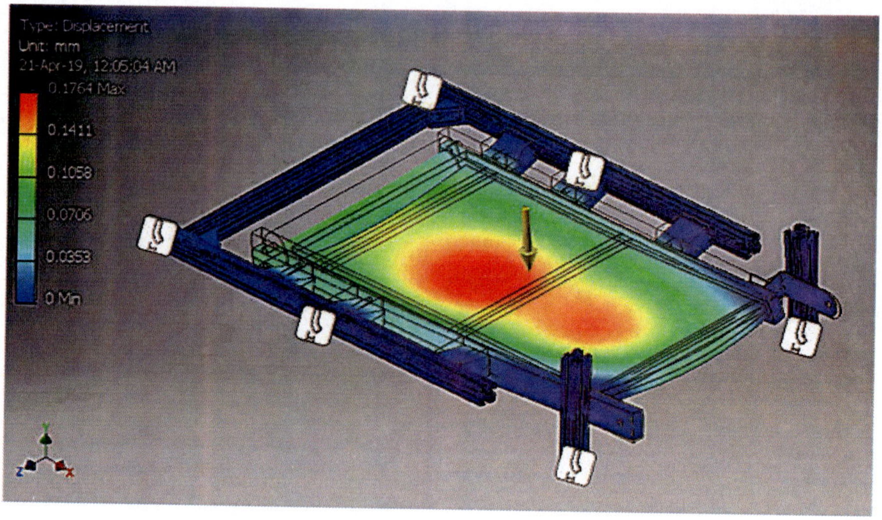

Fig. 8 Displacement result for treadmill base

Critical Structure Analysis

The 3D drawing of the *PhyMill* is developed in SolidWorks software. The critical parts are palm throttle and the *PhyMill* base. The Autodesk Inventor simulation was used, where the material of aluminum 6061-T6, ABS plastic, steel, and wood was chosen. The designed palm throttle and base of the *PhyMill* were first saved as a STEP format and then transferred to the Autodesk Inventor simulation software. The force that is applied is 686 N (70 kg) for the *PhyMill* base and 490 N (50 kg) for palm throttle to the press region.

Figures 8 and 9 shows the maximum displacement for treadmill base and palm throttles are 0.1764 and 0.5089 mm respectively. Based on the results, both critical parts can withstand the force when pressed. The red color refers to the maximum displacement and the blue color refers to the minimum displacement.

4 Conclusion

As a summary, the smart physiotherapy treadmill *(PhyMill)* is well developed. We expect this *PhyMill* device will be widely used at an affordable price in Malaysia. This *PhyMill* exercise for non-ambulatory children with CP as rehabilitation technology tools is a promising good technique of treatment for partial body weight support treadmill. In particular, we have aim to enhance the esthetic value and its activity of the current prototype. By adding a few more specialized functions such as remote control, using light on the patient's foot movements can make this product achieve specifications towards the treatment of the patient's active movement.

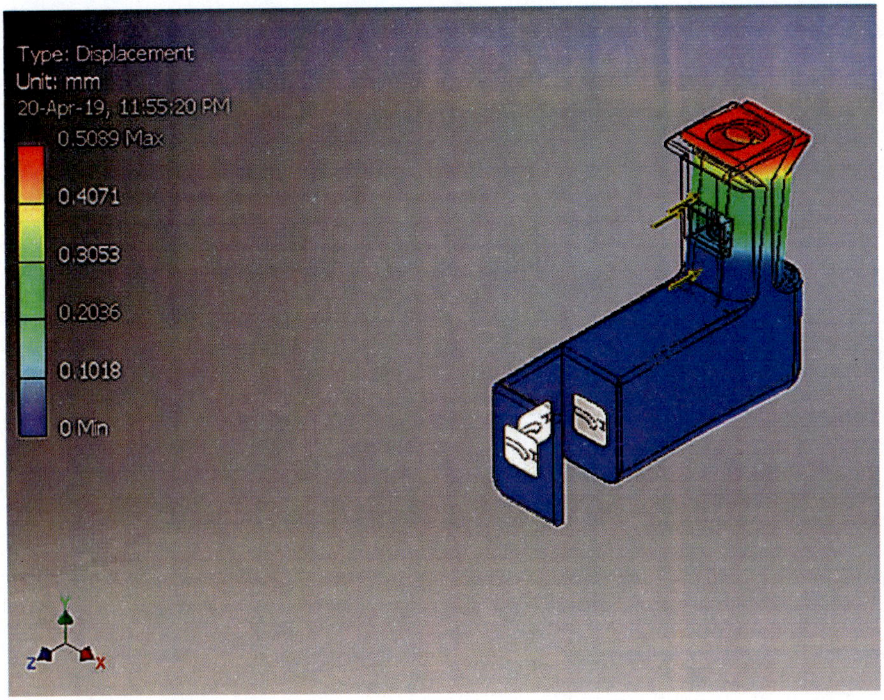

Fig. 9 Displacement result for palm throttle

Acknowlegdement A big thank you dedicated to University Malaysia Pahang (UMP) under grant PGRS2003199 and MedEHiT is gratefully acknowledged for providing us with a good environment and facilities in order to complete these research activities. By this opportunity, we would like to thank Dr. Zakri Ghazali form College of Engineering and Dr. Muhamad Hilmi Jalil from Faculty of Mechanical and Automotive Engineering Technology, Universiti Malaysia Pahang for sharing valuable information in accordance with our research interest. We would face many difficulties without their assistance.

References

1. What is the cause of cerebral Palsy? |CerebralPalsy.org | Cerebral palsy information. (Online). Available: https://www.cerebralpalsy.org/about-cerebral-palsy/cause. Accessed 24 Apr 2020
2. Khoo A (2010) Cerebral palsy—portal myhealth. (Online). Available: http://www.myhealth.gov.my/en/cerebral-palsy/. Accessed 24 Apr 2020
3. Ismail I, Physiotherapy management for cerebral palsy—portal myhealth. (Online). Available: http://www.myhealth.gov.my/en/physiotherapy-management-cerebral-palsy/
4. Segel R et al (2015) Copy number variations in cryptogenic cerebral palsy. Neurology 84(16):1660–1668
5. Moreno-de-Luca A, Ledbetter DH, Martin CL (2012) Genomic insights into the causes and classification of the cerebral palsies. Lancet Neurol 11(3):283–292. (Elsevier)

6. Zarrei M et al (2018) De novo and rare inherited copy-number variations in the hemiplegic form of cerebral palsy. Genet Med 20(2):172–180
7. McMichael G et al (2015) Whole-exome sequencing points to considerable genetic heterogeneity of cerebral palsy. Mol Psychiatry 20(2):176–182
8. Gross motor function classification system (GMFCS) | cerebral palsy. (Online) Available: https://www.abclawcenters.com/cerebral-palsy/gross-motor-function-classification-system/
9. Andreani IM, Kuswanto D (2019) Pengembangan desain treadmill sebagai alat latihan berjalan pada cerebral palsy dengan memanfaatkan realitas virtual. J Sains dan Seni ITS 8(1)
10. Wier LM, Hatcher MS, Triche EW, Lo AC (2011) Effect of robot-assisted versus conventional body-weight-supported treadmill training on quality of life for people with multiple sclerosis. J Rehabil Res Dev 48(4):483–492
11. Willoughby K, Dodd K, Shields N (2009) A systematic review of the effectiveness of treadmill training for children with cerebral palsy. Disabil Rehabil July 2018:1–9
12. Rouse L, Stigler L, Is body weight supported treadmill training more effective than treadmill training at improving gait efficiency and endurance in children with cerebral palsy ? p 88
13. van der Krogt MM, Sloot LH, Buizer AI, Harlaar J (2015) Kinetic comparison of walking on a treadmill versus over ground in children with cerebral palsy. J Biomech 48(13):3577–3583
14. Ameer MA, Fayez ES, Elkholy HH (2019) Improving spatiotemporal gait parameters in spastic diplegic children using treadmill gait training. J Bodyw Mov Ther 23(4):937–942
15. Jung T, Kim Y, Kelly LE, Abel MF (2016) Biomechanical and perceived differences between overground and treadmill walking in children with cerebral palsy. Gait Posture 45:1–6
16. Mall V (2019) Treadmill therapy in cerebral palsy. Eur J Paediatr Neurol 23(4):543
17. Shabkhiz F, Lohrasbi S (2014) The study of the effect of treadmill exercise on spasticity of lower extremity of cerebral palsy children of 3 to 10 years. October
18. Body-Weight-Supported treadmill training: using evidence to guide physical therapy intervention definition
19. Ariffin RA, Adib MAHM, Shalahim NSM, Daud N, Hasni NHM, (2020) An ergonomic perspective of user need on physio-treadmill (PhyMill) criteria: knowledge and awareness of cerebral palsy among future parents. J Phys: Conf Ser 1529:52–71

Pediatrics Technology Applications: Enhance the Bilirubin Jaundice (*BiliDice*) Device for Neonates Using Color Sensor

Mohd Azrul Hisham Mohd Adib, Mohd Hanafi Abdul Rahim, Idris Mat Sahat, and Nur Hazreen Mohd Hasni

Abstract In the few days after birth, many newborn children develop jaundice, a color that turns yellowish on the skin and whites of the eyes. Indeed, in the first few days around half of all newborns have mild jaundices. Jaundice may begin early and last longer in premature babies than in full-term babies. This study focuses on enhancing a portable and economical smart bilirubin jaundice (*BiliDice*) device for neonates. By using the RGB color sensor and Arduino-Uno controller, the system effectively detects three conditions which are normal, mild and critical jaundice. The proposed device uses only one parameter which is reading of bilirubin in mg/dL. The features of the *BiliDice* device output will appear on the LCD based on the level of bilirubin. This present device is well developing so the clinical checking process can be done easily in a short time. The advantage is also lightweight and portable for this prototype device. This device is easy and simple to use. Suitable to improve the physician's ability in Malaysia to treat neonates jaundice.

Keywords Jaundice · Bilirubin · RGB color sensor · Neonates · Pediatrics · Technology · Medical device

M. A. H. M. Adib (✉) · M. H. A. Rahim
Medical Engineering and Health Intervention Team (MedEHiT), Department of Mechanical Engineering, College of Engineering, Universiti Malaysia Pahang, Lebuhraya Tun Razak, 26300 Kuantan, Pahang, Malaysia
e-mail: azrul@ump.edu.my

I. M. Sahat
Human Engineering Group (HEG), Faculty of Mechanical and Automotive Engineering Technology, Universiti Malaysia Pahang, 26600 Pekan, Pahang, Malaysia

N. H. M. Hasni
Family Health Unit, Pahang State Health Department, Jalan IM 4, 25582 Bandar Indera Mahkota, Kuantan, Pahang, Malaysia

A. F. Ab. Nasir et al. (eds.), *Recent Trends in Mechatronics Towards Industry 4.0*, Lecture Notes in Electrical Engineering 730, https://doi.org/10.1007/978-981-33-4597-3_75

1 Introduction

Nowadays in Malaysia, about 60–70% of the healthy newborn baby proficient hyperbilirubinemia were ascending from an augmented production of bilirubin and inadequate capability of the immature liver to collect and expel it [1, 2]. Generally, it occurs during the first week after birth. Hyperbilirubinemia is nearly happening in all newborn babies when total serum bilirubin (TSB) is greater than 15.0 mg/dL nevertheless can be measured as severe jaundice when the total serum bilirubin concentration is greater than 12.9 mg/mL and 10% of the newborn baby population is informed misery this condition [3]. Jaundice in the neonate's population is common [4]. The level of bilirubin will gradually increase if the rigidity is not perceived within the appropriate separation time and if the jaundice is left crude. When a certain level is exceeded there is a likelihood of survival or certain forms of brain injury. [5]. Commonly, the sample of the blood are reserved and many research laboratories are implemented to access the precise bilirubin level [6]. Since the method is repeated, it will cause a bit traumatic to the children and also need help from the expertise. This burdens the detection and monitoring of bilirubin at regular intervals with a noninvasive and easy technique. The non-invasive technology is less painful and traumatic for children [7–9]. Compared to an invasive technique, it also delivers quick results. There are two methods currently in place for non-invasive bilirubin level detection; color comparison and optical method. [10, 11]. However, both methods still under on-going research due to the accuracy and stability of the parameters setup and sensor used. In this paper, the portable and smart device called bilirubin jaundice (*BiliDice*) device is well developed using RGB color sensor automatic detection [12]. The color sensor is used to classified the RGB color component on the neonate's skin. A relationship has been observed between the blue component at the skin level and the bilirubin level. Normally the proportion of the blue color between the RGB color components was premeditated. Based on the previous investigation, the bilirubin level will be calculated consistently and shows the bilirubin level.

2 Methodology

2.1 *Design and Specification*

The device is designed with the help of SolidWorks 2019. Each component and component of product design and drawing is the real product with exact dimensions. There is a clear detail of the components and parts required for production. Figure 1 shows the final design isometric perspective and the whole size of the *BiliDice* device. In SolidWorks, every component is closely linked to create a product design.

The main feature of this device is the PLA material and the Arduino-UNO is the microcontroller. Figure 2 shows the design and development of the *BiliDice* device workflow. Arduino is used as a controller to code and execute the requested command

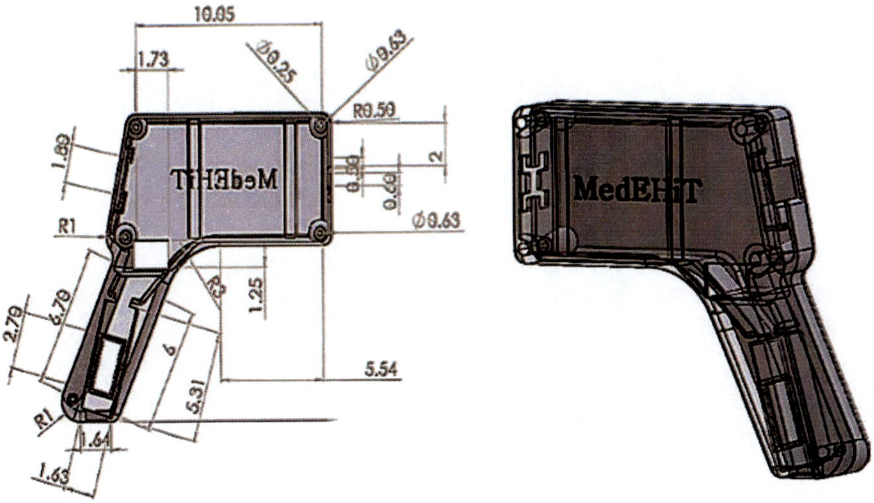

Fig. 1 *BiliDice* device design and development with full size (unit: cm)

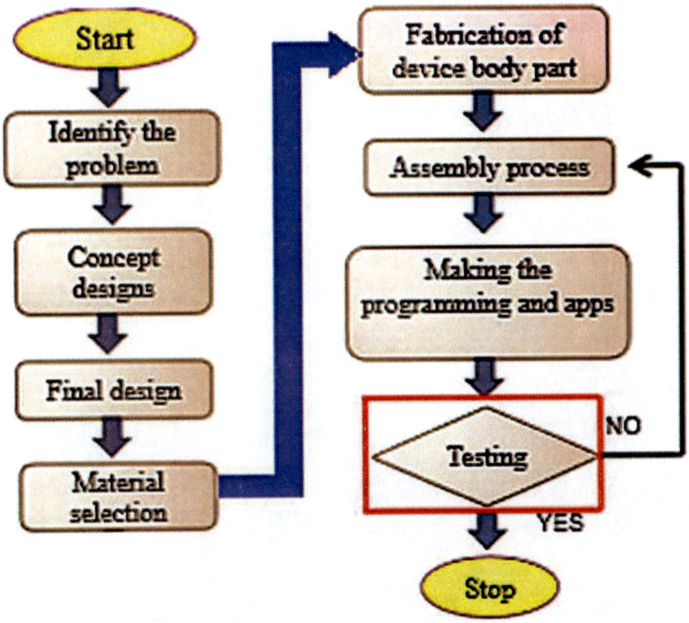

Fig. 2 The workflow of production process of the *BiliDice* device

Table 1 Material specifications

Material	SUNLU PLA
Print temperature (Celcius)	190–220
Length (mm)	330
Diameter (mm)	1.75 ± 0.05
Recommended printing speed (mm/s)	20/60

on the device. The encoding is made with a color sensor RGB to imitate the human skin's modification color. The RGB color sensor is the control mechanism for analog input in the Arduino controller.

2.2 *Materials and Specifications*

The polylactic acid (SUNLU PLA) filament is used in the material selection process in this prototype development to create a complete *BiliDice* device structure. Table 1 shows the particulars of the materials used.

2.3 *Principle Operation and Working*

In Fig. 3, the value of bilirubin based on a blue light array is derived from the integrated color sensor between Arduino-UNO and TCS34725, RGB [12]. The power supply unit supplies 9 V DC to the processing unit in this present device. The unit records the RGB value in a blue light color to exact values. The correct level of

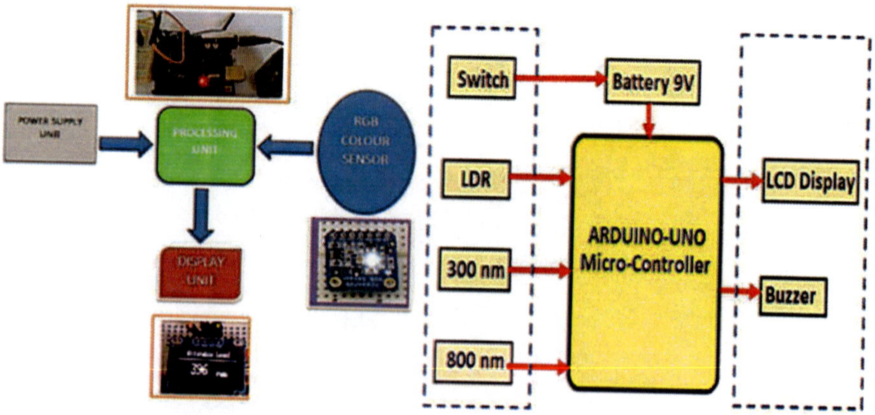

Fig. 3 Represented the operation and working principle of the *BiliDice* device

bilirubin is obtained from this process unit. The formal jaundice is therefore detected and the result is forwarded to the display unit of the LCD.

3 Result and Discussion

3.1 Operation and Working Principle

In this study, the smart bilirubin jaundice (*BiliDice*) device is well developed. The device mainly focuses on the identification of jaundice for neonates on the RGB color sensor models TCS34725, as seen in the Fig. 4. The *BiliDice* device for the correct Red, Green and Blue (RGB) values has been created. The color mark Red, Blue, Green was printed. The Red value labeled "255" and the Red value labeled as "0" were charged to calibrate the Red color component. For the calibration of the Blue and Green color component, the same procedure is implemented. In this study, the blue light color is furthermost effective for phototherapy. The reason is the regular absorber of bilirubin in the range. Thus, the reflection of the skin is minimal in comparison to non-jaundice neonates from jaundiced neonates. The percentage of blue is then configured with the reached RGB value and used to determine the bilirubin level based on the blue color percentage. This bilirubin level and the jaundice status were shown on the LCD (Table 2).

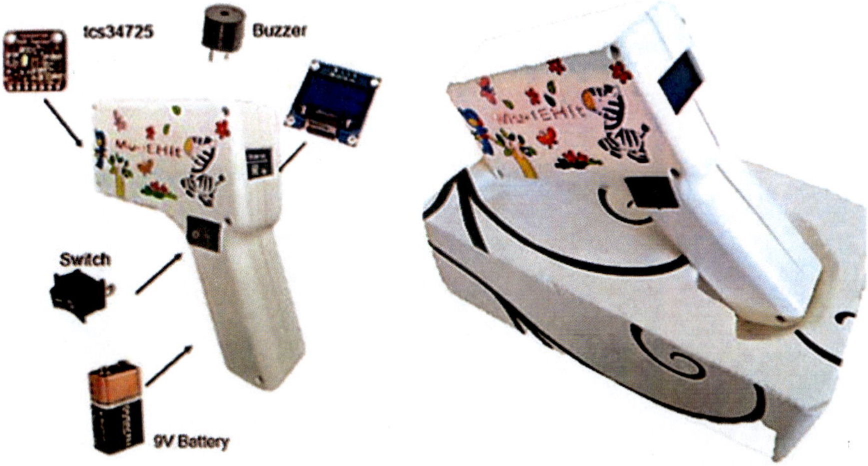

Fig. 4 Left: Components of *BiliDice* device, right: represented the full-prototype of the *BiliDice* device

Table 2 Specification of color sensor

Product	TCS34725
Dynamic range	3,800,000:1
Voltage	3.3 V
Dimension	2.1 cm × 2.1 cm × 0.1 cm
Weight	1.6 g

3.2 *Response for Jaundice and Non-Jaundice Data Analysis*

The preliminary study was based on corresponding labels of skin color as shown in Table 3 for jaundice and non-jaundice. The level of bilirubin is supposed to adhere to a certain standard. The processing unit is controlled based on the state of jaundice and non-jaundice. The final decision on jaundice and non-jaundice is referred to the Table 4.

The results for five patients was shown in Fig. 5. Out of three patients identified as mild jaundice level and one with severe jaundice level. Another patient is classification as non-jaundice. Approximately 6 mg/dl, 9 mg/dl and 10 mg/dl were present in three patients with moderate jaundice. One is, however, serious 18 mg/dl jaundice. The main indication of neonatal jaundice is the coloration of the yellow skin from the appearance of the baby. Normally, jaundice may cause the baby cannot sleep properly and lead to poor feeding. This could mark jaundice worse because of inadequate eating because the baby can be dry. When a baby has jaundice, it will eat white calcareous stool (poo) and make the urine darker than usual. Jaundice is not painful, but severe impediments can occur if elevated bilirubin levels are not managed in a good time.

Table 3 Implementation of color level and bilirubin level

Color Series (CS)	Label of Color	Bilirubin level (mg/dl)
CS-1		8
CS-2		11
CS-3		18
CS-4		22

Table 4 Guide on jaundice and non-jaundice decision-making

Bilirubin level (mg/dl)	Jaundice level
BL < 5	Normal
5 < BL < 11	Mild
11 < BL < 19	Severe
19 < BL	Critical

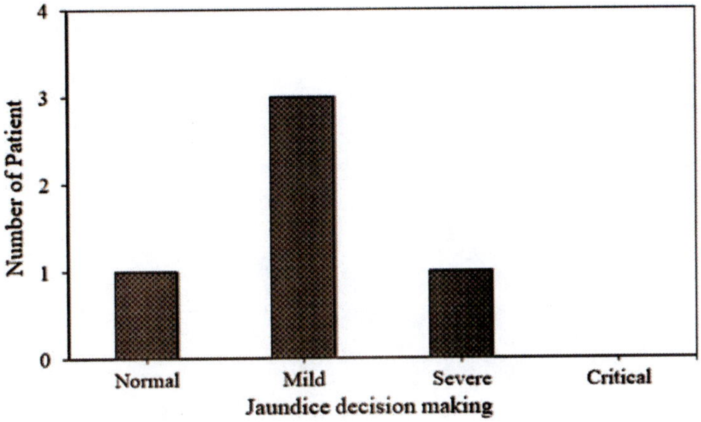

Fig. 5 Preliminary test on five patients with jaundice and non-jaundice

3.3 *Validation of Features*

The prototype was developed as shown in Fig. 6. Preliminary investigation to evaluate the appropriate efficient use of the *BiliDice* device was completed. The device well repeated the patient with jaundice and non-jaundice. Figure 7 clearly shows one of the patients with mild jaundice successfully detected by using the *BiliDice* device. The patient gives a good response and feels comfortable when we used the device. No negative feedback or complained during handling the device. In four different jaundice and non-jaundice patients, the same preliminary tests have been carried out. Referred to the parents, they felt no discomfort or the interaction forces applied by the device and could handle the device freely.

Fig. 6 Illustrated the prototype of the *BiliDice* device

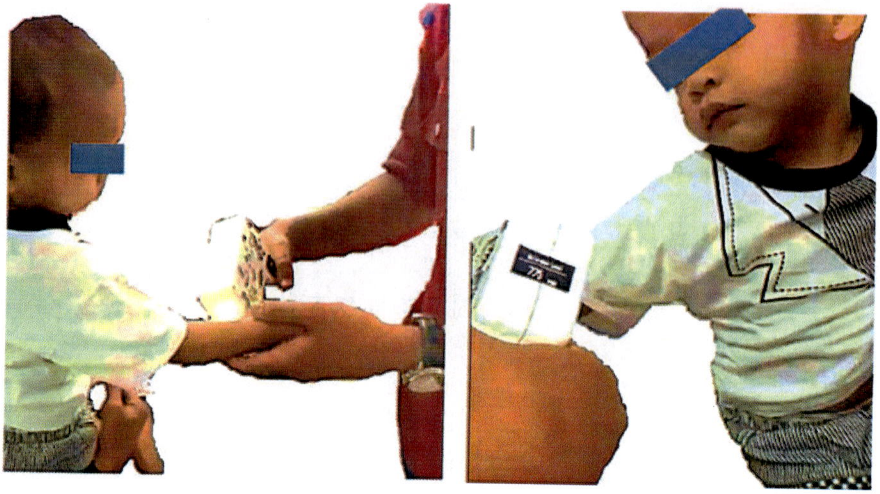

Fig. 7 The functional test on the *BiliDice* device was demonstrated with the help of real patient jaundice

4 Conclusions

Every day a new era of technology is being generated to improve the living quality of human life, making this project an appropriate, crucial innovation for humankind. The main objectives of this paper are achieved by developing and functioning the prototype of a smart bilirubin jaundice (*BiliDice*) device. The low production cost in comparison with similar devices available on the market is one of the main features. However, further improvements can be carried out in the future especially in terms of operation and safety, performance and functionality, and also the design of this device.

Acknowledgements The support of the University Malaysia Pahang under grant PDU203205, PGRS2003200 and MedEHiT are gratefully acknowledged. The authors of this paper also would like to express their gratitude to Dr. Suhaila from Klinik Kesihatan Kurnia for supporting these research activities.

References

1. Mansor M (2012) Jaundice in newborn monitoring using color detection method Procedia Engineering 29 1631–1635; Author F, Author S (2016) Title of a proceedings paper. In: Editor F, Editor S (eds) Conference 2016, LNCS, vol 9999, pp 1–13. Springer, Heidelberg
2. Osman Z, Ahmad A, Muharam A (2014) Rapid prototyping of neonatal jaundice detector using skin optics theory. In: IEEE conference on biomedical engineering and sciences, pp 328–331

3. Chowdhary AK, Dutta S, Ghosh R (2017) Neonatal jaundice detection using color detection method. Int Adv Res J Sci Eng Technol 46:197–203
4. Adebami OJ (2015) Assessment of knowledge on causes and care of neonatal jaundice at the Nigerian primary and secondary health institutions International. J Res Med Sci 10:2605–2612
5. Saleem A, Junaid M, Mohammadi SH, Jebran M, Iram S, Indikar L (2013) Embedded based preemies monitoring system with jaundice detection and therapy. Int J Sci Technol Res 6:153–162
6. Ramy N (2016) Jaundice, phototherapy and DNA damage in full-term neonate's. J Perinatol 2:132–136
7. Greco C, Arnolda G, Boo NY, Iskander IF, Angela A, Rinawati Rohsiswatmo O, Shapiro SM, Richard JW, Wennberg P, Tiribelli C, Zabetta CD (2015) Neonatal jaundice in low and middle-income countries: lessons and future directions don ostrow trieste yellow retreat. Neonatology 110:172–180
8. Aydın M, Hardalaç F, Ural B, Karap S (2016) Neonatal jaundice detection system. Trans Proc Syst 40:166–177
9. Castro-Ramos J, Toxqui-Quit C, Manriquez FV, Orozco-Guillen E, Padilla-Vivanco F, Sánchez-Escobar JJ (2014) Detecting jaundice by using digital image processing. Int Soc Opt Eng 8948:1605–1712
10. Ali N, Muji SZM, Joret A, Amirulah R, Podari N, Dol Risep NF (2015) Optical technique for jaundice detection ARPN. J Eng Appl Sci 20:9929–9933
11. Denney PA, Seidman DS, Stevenson DK (2001) Neonatal hyperbilirubinemia New England. J Med 8:581–590
12. Adib MAHM, Rahim MHA, Hasni NHM (2019) Development of bilirubin jaundice (BiliDice) device for neonates. Proc Mech Eng Res Day 188–189

A Supervised Learning Neural Network Approach for the Prediction of Supercapacitive Energy Storage Materials

Varun Geetha Mohan, Mohamed Ariff Ameedeen, and Saiful Azad

Abstract Material researchers are progressively embracing the utilization of machine learning techniques to find hidden patterns in data and make predictions without explicit human development. Thousands of papers have been published in the use of carbon for supercapacitor applications. The manufacturing conditions for getting highly super-capacitive carbons from bio-wastes could be analyzed from the existing data using proper machine learning techniques. This work aims to provide a solution called feed forward back propagation neural networks, a supervised learning approach for the prediction of super-capacitive energy storage materials. The proposed method is to apply on the prediction of key parameters with the actual data of the two processes. The configuration of Levenberg-Marquardt back-propagation neural network has been given the smallest mean square error (0.002892, 0.006884) with correlation coefficient (0.992, 0.9789) respectively was three-layer artificial neural network with hidden layer with 9 neurons. The ANN results showed that neural network model can be satisfactorily simulate and predict the behavior of the process.

Keywords Energy storage materials · Machine learning · Backpropagation neural network

1 Introduction

The advancement of energy storage and conversion devices is an indispensable for highly dynamic storage devices. These devices are sustainable in nature and it is necessary for reducing the discontinuity and uncertainty of sustainable energy generation. Also, energy storage devices are connected to the daily energy demands and it depends on cost-efficiency as well as science and technology development which can enable effective energy storage [1]. For a broad range of applications, such as new energy vehicles, consumer electronics, and aerospace, rechargeable batteries

V. G. Mohan (✉) · M. A. Ameedeen · S. Azad
Faculty of Computing, Universiti Malaysia Pahang, 26300 Kuantan, Malaysia
e-mail: munnavarun1988@gmail.com

A. F. Ab. Nasir et al. (eds.), *Recent Trends in Mechatronics Towards Industry 4.0*, Lecture Notes in Electrical Engineering 730, https://doi.org/10.1007/978-981-33-4597-3_76

are extremely significant, as a major component of electrochemical energy storage. The rechargeable batteries with better energy density as well as power density having higher demands move forward and with improved life cycle, more secure and reasonable cost to meet the developing needs of these applications. Therefore, for improving the performance of rechargeable batteries, it is mandatory to develop vital rechargeable battery materials encompassing with electrodes and electrolytes [2]. Nowadays, carbon for supercapacitor application becomes one of the most popular research topics as a result, thousands of papers have been published every year. Researchers have been gone through new breakthroughs such as to analyze the existing data using appropriate machine learning and data mining techniques for getting the manufacturing conditions of suitable super-capacitive waste biomass derived carbons [3]. Finally, material science is on the edge of embracing fourth paradigm called Materials 4.0 or data-driven exploration on large-scale data. However, data can generate from calculations or manipulations, literature study, hypothetical study, experimental data, and even from experiments which have been failed and finally, by interpolation of existing data leads to new data [3, 4].

Machine learning (ML) is the core of artificial intelligence (AI) and data science, is a versatile topic lying in computer science as well as in statistics [4, 5]. Although, ML is a rapidly innovative technology and due to that there should be some techniques outdated in ML. Even though, there is still importance for using scientific reasoning to find out reliable structure, property, and processing and it will never be expired [4]. There are some major steps in ML—data preparation, descriptor selection, model selection and prediction, and model application [6] and in order to apply these processes to material discovery and design, a complete cycle is required. It is fulfilled through experimental process data collection, then by performance prediction and end up with experimental validation. ML techniques provide efficient toolsets for extracting important correlations among material phenomena [7]. Moreover, researchers are searching for connections in materials datasets to discover new compounds [8], make performance predictions [9], to quicken computational model advancement [10] and gain new experiences from characterization techniques [11].

With the development in computational technology and computer science, the application of machine learning techniques for the material synthesis has become the common trend [12, 13]. The quality of data analysis and data mining has been affected by the data quality of the materials database as well as directly affects the application of data that are mined [4]. However, as per best knowledge, no more machine learning techniques have been developed for identifying the super-capacitive energy storage materials. Since, the materials science data are heterogenous and complex and there is a big task to organize data from many sources and diverse datasets, then make into readable and searchable as well as controlling and analyzing. The aim of this study is to obtain the best parameters by increase the diversity of the search. Therefore, to achieve this aim, the paper deals with the implementation of backpropagation neural network (BPNN) to prediction the best super-capacitive energy storage materials. Backpropagation (BP) [14] technique is a kind of supervised learning technique that the network is trained by providing it with input and matching output patterns.

BP is a popular technique used in training neural networks due to its simplicity in implementation and efficiency as well as its tremendous power to manipulate big amount of data.

The rest of the paper has been arranged as methodology and results & discussions section. Methodology starts with the introduction of parameters and followed by the concepts as well as implementation of techniques. Finally, the experimental procedures are elaborated as well as the obtained results are compared and included as tables and graphs in result and discussion.

2 Proposed Solution and Methodology

The carbon is the ultimate element for manufacturing supercapacitors. There are some activation processes to improve the efficiency of carbon by increasing its surface area as well as its specific capacitance as shown in Fig. 1. Chemical activation and electrochemical characterization are the two major process for getting high

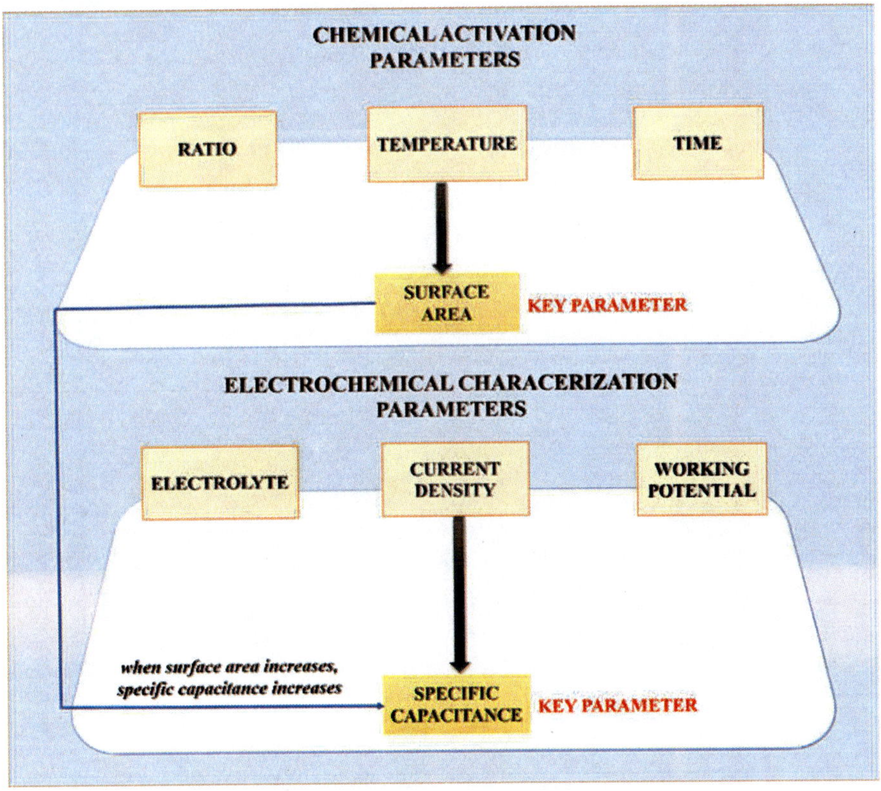

Fig. 1 Parameters for identifying the manufacturing conditions of super-capacitor

surface area and high specific capacitance respectively. Therefore, surface area and specific capacitance are considered as the key parameters. Carbon/Chemical Mass Ratio, Temperature and Time are the other input parameters in the chemical activation process. Similarly, in the case of electrochemical characterization—Electrolyte, Current density and Working potential are the other parameters. Finally, there is a relation between two key parameters is that surface area is directly proportional to the specific capacitances.

The proposed method is to apply on the prediction of key parameters (surface area & specific capacitance) with the actual data of the two processes (chemical activation and electrochemical characterization). The twenty experimental data sets were used and has been separated into input matrix and target matrix. The input variables were carbon(C)/chemical(K) mass ratio (C: K), temperature in '°C' and time in hour (h) for the output surface area with a unit of m^2g^{-1} as shown in Table 1. Similarly, the next input variables such as electrolyte with molar quantity (e.g. 6 M KOH, 1 M

Table 1 Summary of input and output dataset retrieved for the chemical activation process of selected articles

Input parameters				Output parameter	Reference
Article No.	C/K mass ratio	Temperature (°C)	Time (h)	Surface area (m^2g^{-1})	
1	1: 4	200	1	835.4	[15]
2	1: 5	750	2	2484	[16]
3	1: 4	750	1	482	[17]
4	1: 3	800	5	1297.6	[18]
5	1: 1	800	1	2082	[19]
6	1: 4	900	1	2960	[20]
7	1: 3	800	1	2488	[21]
8	1: 3	700	2	514.7	[22]
9	1: 1	700	2	1166	[23]
10	1: 4	800	1	2700	[24]
11	1: 3	800	3	1229	[25]
12	1: 3	700	2	1372.87	[26]
13	1: 5	800	1	2671	[27]
14	1: 3	700	1	2349.37	[28]
15	1: 3	600	1	800	[29]
16	1: 4	500	4	462.1	[30]
17	1: 5	850	1	2696	[31]
18	1: 4	700	1	2509	[32]
19	1: 4	800	1	2718	[33]
20	1: 4	700	4	506	[34]

Na_2SO_4, etc.), current density in A g^{-1} and working potential or potential window will be mentioned in unit volt (V) for the output parameter specific capacitance (C_s) with a unit F g^{-1} as shown in Table 2.

The data sets were divided into training, validation and test subsets, each of which contained 10 (one half), 5 (one fourth), 5 (one fourth) sets, respectively. A three-layer backpropagation neural network (BPNN) was optimized to predict and simulate the output from the actual data. Figure 2 are showing the optimized network of BPNN with a three-layer neural network with tangent sigmoid transfer function (tansig) at hidden layer, neurons and linear transfer function (purelin) at output layer for the chemical activation process. Conversely, the similar topology is taken into consideration in the development of the electrochemical process.

Table 2 Summary of input and output dataset retrieved for the electrochemical process of selected articles

Input parameters				Output parameters	Reference
Article No.	Electrolyte	Current density (A g^{-1})	Potential window (V)	Specific capacitance (F g^{-1})	
1	1 M H_2SO_4	0.5	1	374	[15]
2	2 M KOH	6	0.9	200	[16]
3	1 M KOH	2	1	130	[17]
4	6 M KOH	0.5	0.8	148	[18]
5	6 M KOH	2	1	334	[19]
6	6 M KOH	0.1	0.8	258	[20]
6	6 M KOH	0.1	1	182	[20]
7	6 M KOH	1	1	379	[21]
8	6 M KOH	5	1	289	[22]
9	6 M KOH	0.5	1	367	[23]
10	6 M KOH	0.05	1	250	[24]
11	2 M H_2SO_4	0.5	1	315	[25]
12	2 M KOH	0.5	1	340	[26]
13	1 MLiPF6	10	1.5	158	[27]
14	3 M KOH	1	1	140	[28]
15	1 M H_2SO_4	0.5	1.2	390	[29]
16	1 M KOH	0.5	1	210	[30]
17	6 M KOH	0.1	1	17	[31]
18	1 M H_2SO_4	0.25	0.9	311	[32]
19	1 M Na_2SO_4	1	1	309.6	[33]
20	1 M Na_2SO_4	1	1	150	[34]

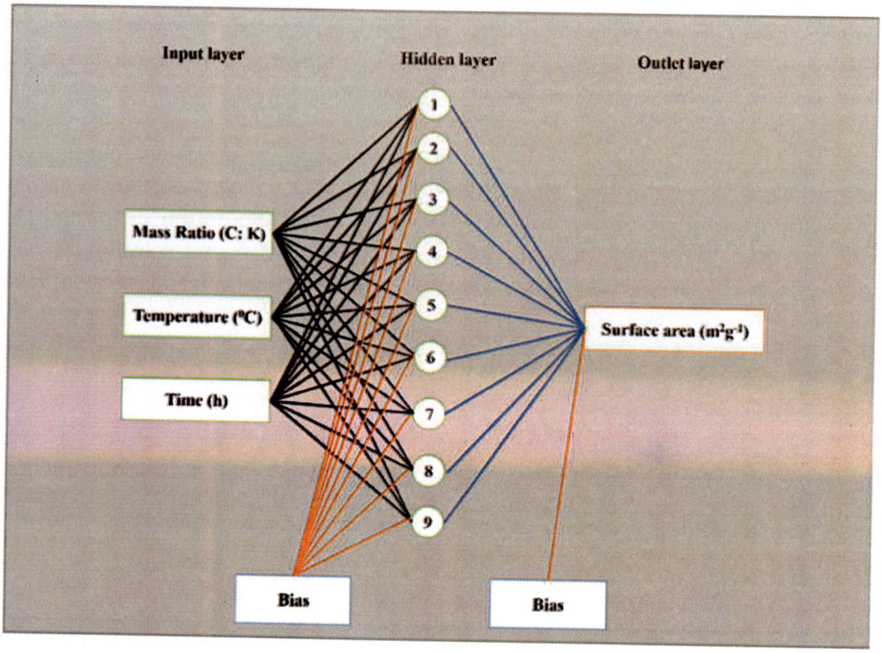

Fig. 2 Backpropagation neural network optimized structure for chemical activation process

3 Results and Discussion

For determining the best BPNN, comparison of different BPNN techniques were studied has been shown in Table 3. Tansig were used at hidden layer as well as a purelin were used at output layer. Further, five neurons were used in the hidden layer as initial value for all BPNN techniques. Levenberg-Marquardt (LM) backpropagation technique was able to have smaller mean square error (MSE) compared to other BPNN techniques. Therefore, LM backpropagation technique was recognized the

Table 3 Comparison of BPNN techniques with 5 neurons in the hidden layer

Backpropagation neural network (BPNN)	Mean square error (MSE)
Levenberg-marquardt backpropagation	0.0091487
Scaled conjugate gradient backpropagation	0.026799
Batch gradient descent	0.586191
Variable learning rate back propagation	0.439428
Batch gradient descent with momentum	0.517261

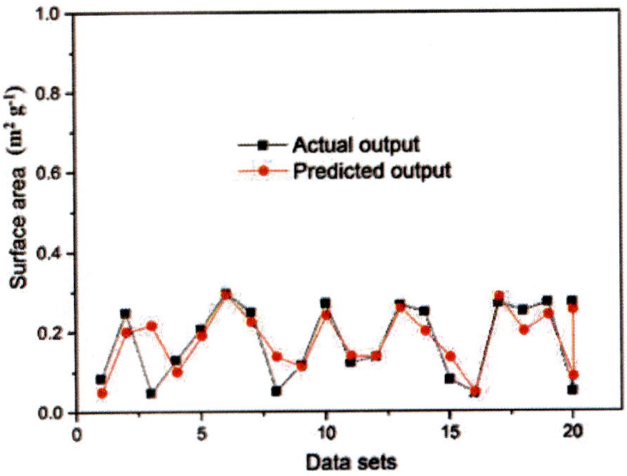

Fig. 3 Comparison of actual and predicted output for the chemical activation process

training technique in this study. Also, the optimum number of neurons was determined based on the minimum value of MSE of the training and prediction set [35, 36]. From this, optimization was done by using LM technique and varying neuron number in the range 1–20 and 9 neurons were selected as the best number of neurons. More than 9 neurons increasing did not significantly decrease the value of MSE.

As per the procedures mentioned above in the methodology section, the model has been trained by boosting the input pattern move forward and then calculates the increasing variation between the target value and the model output. The training should be undertaken by batch mode. Also, the error is back propagated to the model in order to adjust each connection weight within the model. Then, the model has been trained again with the updated weight values and this process is tackle until a combination of optimum weight is gained. Further, weight is showing the error of the model lesser than a specific error point. Finally getting normalized values of actual output (key parameter) as well as predicted output and then mean square error to analyses how close is actual and predicted output.

A graph is drawn to show relation between actual output and predicted output are shown in the Fig. 3. In the end, MSE for chemical activation process from the 20 datasets are 0.002892 with the correlation coefficient (R^2) 0.992. The summary of actual output and predicted output for the data set of electrochemical process has been shown as graph analysis in the Fig. 4. Then, the MSE for the 20 data sets of electrochemical process is 0.006884 having R^2 0.9789.

4 Conclusions

The configuration of the BPNN giving the smallest MSE was three-layer ANN with tangent sigmoid transfer function (tansig) at hidden layer with 9 neurons, linear transfer function (purelin) at output layer and Levenberg-Marquardt BPNN. The

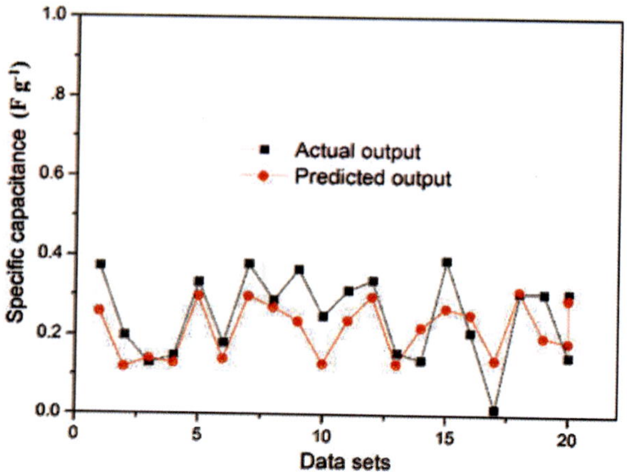

Fig. 4 Comparison of actual and predicted output for the electrochemical process

ANN results showed that neural network modeling could adequately simulate and predict the behavior of the process. The proposed solution has been implemented in sample datasets not in big data. In future, more accuracy can be obtained by using the big data, with more datasets. However, the materials data are not homogenous and complicated moreover, there is a big challenge to retrieve data from numerous sources and different datasets into simplified and accessible as well as controlling and analyzing unstructured to enable an informatic system for materials data. Therefore, the application of the proposed solution to big data is suggested in future work to evaluate its performance.

Acknowledgements This work has been carried out by the Universiti Malaysia Pahang (UMP) from the Grant (TRGS/1/2018/UMP/02/2/2) under Grant No. RDU191802-2. The authors would like to thank the Ministry of Education Malaysia for providing us with the Trans-Disciplinary Research Grant (TRGS) for us to perform this research.

References

1. Shetti NP, Dias S, Reddy KR (2019) Nanostructured organic and inorganic materials for Li-ion batteries: a review. Mater Sci Semicond Process 104:104684
2. Liu Y et al (2020) Machine learning assisted materials design and discovery for rechargeable batteries. Energy Storage Mater
3. Jose R, Ramakrishna S (2018) Materials 4.0: materials big data enabled materials discovery. Appl Mater Today 10:127–132
4. Liu Y et al (2020) Machine learning in materials genome initiative: a review. J Mater Sci Technol 57:113–122
5. Ramprasad R et al (2017) Machine learning and materials informatics: recent applications and prospects. NPJ Comput Mater 3
6. Lu W et al (2017) Data mining-aided materials discovery and optimization. J Materiomics 3(3):191–201

7. Yang Z et al (2018) Deep learning approaches for mining structure-property linkages in high contrast composites from simulation datasets. Comput Mater Sci 151:278–287
8. Hautier G et al (2010) Finding nature's missing ternary oxide compounds using machine learning and density functional theory. Chem Mater 22(12):3762–3767
9. Klanner C et al (2004) The development of descriptors for solids: teaching "catalytic intuition" to a computer. Angew Chem Int Ed 43(40):5347–5349
10. Nelson LJ et al (2013) Compressive sensing as a paradigm for building physics models. Phys Rev B 87(3):035125
11. Belianinov A et al (2015) Identification of phases, symmetries and defects through local crystallography. Nat Commun 6:7801
12. Wang J et al (2020) Recent progress of biomass-derived carbon materials for supercapacitors. J Power Sources 451:227794
13. Jiménez J et al (2018) K <inf> DEEP </inf> : protein-ligand absolute binding affinity prediction via 3D-convolutional neural networks. J Chem Inf Model 58(2):287–296
14. Kannaiyan M, Karthikeyan G, Thankachi Raghuvaran JG (2020) Prediction of specific wear rate for $LM25/ZrO_2$ composites using levenberg–marquardt backpropagation algorithm. J Mater Res Technol 9(1):530–538
15. Zhu X et al (2018) Sustainable activated carbons from dead ginkgo leaves for supercapacitor electrode active materials. Chem Eng Sci 181:36–45
16. Shree Kesavan K, Surya K, Michael MS (2018) High powered hybrid supercapacitor with microporous activated carbon. Solid State Ionics 321:15–22
17. Shen L et al (2017) Phytoplankton derived and KOH activated mesoporous carbon materials for supercapacitors. Mater Lett 205:98–101
18. Chen H et al (2017) An activated carbon derived from tobacco waste for use as a supercapacitor electrode material. New Carbon Mater 32(6):592–599
19. Han Y et al (2017) Fish gill-derived activated carbon for supercapacitor application. J Alloy Compd 694:636–642
20. Yang C-S, Jang YS, Jeong HK (2014) Bamboo-based activated carbon for supercapacitor applications. Curr Appl Phys 14(12):1616–1620
21. Qu S et al (2018) Promising as high-performance supercapacitor electrode materials porous carbons derived from biological lotus leaf. J Alloy Compd 751:107–116
22. Enock TK et al (2017) Biogas-slurry derived mesoporous carbon for supercapacitor applications. Mater Today Energy 5:126–137
23. Hu X et al (2018) Facile synthesis of microporous carbons with three-dimensional honeycomb-like porous structure for high performance supercapacitors. J Electroanal Chem 823:54–60
24. de Paula FGF et al (2018) High value activated carbons from waste polystyrene foams. Microporous Mesoporous Mater 267:181–184
25. Liang J et al (2018) Microwave assisted synthesis of camellia oleifera shell-derived porous carbon with rich oxygen functionalities and superior supercapacitor performance. Appl Surf Sci 436:934–940
26. Chen Y et al (2018) Synthesis of porous carbon spheres derived from lignin through a facile method for high performance supercapacitors. J Mater Sci Technol
27. Sun W et al (2016) Hemp-derived activated carbons for supercapacitors. Carbon 103:181–192
28. Yu K et al (2018) High surface area carbon materials derived from corn stalk core as electrode for supercapacitor. Diam Relat Mater 88:18–22
29. Karnan M et al (2017) Electrochemical studies on corncob derived activated porous carbon for supercapacitors application in aqueous and non-aqueous electrolytes. Electrochim Acta 228:586–596
30. Misnon II et al (2015) Electrochemical properties of carbon from oil palm kernel shell for high performance supercapacitors. Electrochim Acta 174:78–86
31. Teo EYL et al (2016) High surface area activated carbon from rice husk as a high performance supercapacitor electrode. Electrochim Acta 192:110–119
32. Li X et al (2011) Preparation of capacitor's electrode from sunflower seed shell. Biores Technol 102(2):1118–1123

33. Su X-L et al (2018) Three-dimensional porous activated carbon derived from loofah sponge biomass for supercapacitor applications. Appl Surf Sci 436:327–336
34. Vijayan BL et al (2019) Facile fabrication of thin metal oxide films on porous carbon for high density charge storage. J Colloid Interface Sci
35. Elmolla ES, Chaudhuri M, Eltoukhy MM (2010) The use of artificial neural network (ANN) for modeling of COD removal from antibiotic aqueous solution by the Fenton process. J Hazard Mater 179(1):127–134
36. Yetilmezsoy K, Demirel S (2008) Artificial neural network (ANN) approach for modeling of Pb(II) adsorption from aqueous solution by Antep pistachio (Pistacia Vera L.) shells. J Hazardous Mater 153(3):1288–1300

Two-Steps Approach of Localization in Humanoid Robot Soccer Competition

Anhar Risnumawan, Miftahul Anwar, Rokhmat Febrianto, Cipta Priambodo, Mochamad Ayuf Basthomi, Puguh Budi Wasono, Hendhi Hermawan, and Tutut Herawan

Abstract Humanoid robot has gained much interest in a soccer game. The teams have to prepare the best strategy for winning the match. The strategy, such as forming a formation during attacking and defending, could potentially achieve a significant result, yet the strategy genuinely demands a robust localization algorithm. Existing works of localization on humanoid robot soccer favor to focus on precisely locating the robot position while this is opposed to how humans play soccer. Human play soccer needs to know roughly the position while attacking or defending. In this paper, we present a practical localization approach of humanoid soccer robot. The approach consists of two steps, namely local movement and correction. The local movement is calculated using a well-proven Kalman filter algorithm with the input joint odometry of robot kinematics. The correction step is only performed when either kick-off or the two goalposts are observed. The steps are developed mainly based on soccer game rules, computation, and accuracy in minds. Experiments show encouraging results and best suited for the real humanoid soccer robot competition,

A. Risnumawan (✉) · M. Anwar · R. Febrianto · C. Priambodo · M. A. Basthomi · P. B. Wasono · H. Hermawan
Politeknik Elektronika Negeri Surabaya, Surabaya, Indonesia
e-mail: anhar@pens.ac.id

M. Anwar
e-mail: Miftahulanwar@ce.student.pens.ac.id

R. Febrianto
e-mail: rokhmat@me.student.pens.ac.id

C. Priambodo
e-mail: ciptapriambodo07@ce.student.pens.ac.id

M. A. Basthomi
e-mail: ayuf@ee.student.pens.ac.id

P. B. Wasono
e-mail: gioalusio4720@me.student.pens.ac.id

H. Hermawan
e-mail: hendhi@pens.ac.id

T. Herawan
Sekolah Tinggi Pariwisata Ambarrukmo, Yogyakarta, Indonesia

A. F. Ab. Nasir et al. (eds.), *Recent Trends in Mechatronics Towards Industry 4.0*, Lecture Notes in Electrical Engineering 730, https://doi.org/10.1007/978-981-33-4597-3_77

with the error rate in the x-axis is 3.3 cm, while on the y-axis is 9.2 cm. We believe they could bring many benefits for the research in the related fields.

Keywords Two-steps approach · Localization · Humanoid robot soccer · Kalman filter · Local movement · Correction

1 Introduction

Humanoid robot soccer has been active research in the past decade. The research is highly supported by the RoboCup soccer competition, which is held annually around the world, with the primary goal of the robots able to win against the winning world cup championship in 2050. The game rules have been created to imitate the real soccer game.

During the robot soccer competition, teams have to prepare their best strategy for winning the match. Forming a particular formation while attacking and defending, covering players, counter-attacking, pressing, zone marking, and passing requires the robots to know their respective position, which can be achieved by the localization algorithm, as shown in Fig. 1. Without this, all the robots simply pursue the ball and highly possibly collide with other robots. The robot moves individually without collaboration with other robots, which is opposed to real soccer and can result in unpredictability.

Existing works of localization of humanoid robot soccer mainly employ particle filter and Monte Carlo methods. Particle-based methods [1–3] have shown the performance by modifying the algorithm to boost accuracy and solve a common kidnapping problem. Monte Carlo based methods [4–6] have been proposed and can handle the kidnapping problem. However, the Monte Carlo method requires enough samples to

Fig. 1 Our platform of humanoid soccer robot forming a particular formation (a goalkeeper, two robots as side-midfielder, and a robot as striker) while attacking. Thus, a practical localization algorithm plays a great role to support this strategy

get satisfactory performance, while in the real robot, it is hard always to satisfy the required samples. Robot kinematics odometry and visual landmarks are often used as input to the algorithms. It is interesting that, in robot soccer competition, solving a complete kidnapping problem is highly likely not needed, and pursuing only the localization accuracy tends to increase the computational cost, which is not suitable for the low computing power of humanoid robot embedded platform.

In humanoid robot soccer, the kidnapping problem occurs when a robot is picked up by the team, and the robot is started in a new unknown position that can massively disrupt the convergence of the particle-filter-like algorithms, especially when the new position is quite a far away from the current position. The robot is picked up, such as due to the disabled robot from error, collision with others that makes the robot unable to move, rules violation, inactive for a relatively long time, the robot moves to the outside of the soccer field. Therefore, the robot must quickly estimate the position when such a condition occurs so that the soccer strategy can still be maintained. The strategy, such as positioning, attacking, and defending, can be performed without disrupting other teammate robots. However, in the real robot soccer competition, solving a complete kidnapping problem is highly likely not needed, as the robot is often just moved away a few centimeters from the current position. Thus, an algorithm that can benefit from such robot soccer rules and computation is preferable, with the trade-of of the position accuracy.

In this paper, we present an efficient localization system that is better suited for a humanoid robot soccer competition. We consider three criteria for developing our method, that are soccer game rules, computation, and accuracy. The proposed method does not rely upon much to the precise localization as in analogy with how humans playing soccer. Human play soccer only needs to know roughly the position while attacking or defending. Soccer game rules such as when the robots collide with one another and unable to move, the referee will move the robot away few centimeters, the robots also enter the field from the middle-side, no throw-in, no corner kick, all of these rules also do not rely much on precise localization, and the kidnapping problem rarely happens.

Therefore, the proposed method consists of two steps, that are local movement and correction, while considering those criteria. The local movement is obtained using a Kalman filter with the input joint odometry of robot kinematics. We found that the localization error will be generated gradually while using odometry alone. Thus, the system has to be equipped with global position correction using visual landmarks detector. Ball position during kick-off and goalposts are used for the landmarks. The correction is only performed when kick-off or the goalposts is observed. In this way, the computation can be reduced while maintaining the correct position. The correction starts by measuring the distance to the observed landmarks and the orientation of the robot, then the global robot position is determined by the triangle equation. Experiments have been performed using our humanoid robot platform, namely EROS and tested in the real humanoid robot soccer. Experiments also show encouraging results, and we believe this could bring many benefits for the research in the related fields.

2 Related Works

Existing works on humanoid robot soccer localization can be mainly categorized into two methods: particle filter and Monte Carlo. The works by [3] proposed a particle filter based localization on humanoid robot soccer. The work puts much effort into solving the kidnapping problem by modifying the filter during updating particles. Several conditions have been added, such as when the field borders are observed. The conditions such as two field boundaries, goalposts, a large variance for far away distance, are used. These conditions are used during updating particles to solve a common kidnapping problem.

The work by [2] proposed a particle filter for localization and determining ball position to the robot. The work obtained the accuracy by integrating robot kinematics odometry with the goalposts and field borders. Landmarks detector for locating the goalposts and field borders have been developed. Similarly, the work by [1], a ROS based system, has been developed and using particle filter localization method. Robot kinematics odometry, angle estimation using gyro, and accelerometer are used as input to the particle filter to determine precisely the robot's global position.

The work by [4] has proposed Monte Carlo localization of humanoid soccer robot using vision such as a camera. The work has shown quite well performance tested in simulation. However, experiments using a real humanoid robot is essential to be tested. The work by [5] modified the Monte Carlo method for localization in order to solve the kidnapping problem. Odometry input is integrated with visual landmarks to boost the localization accuracy. A hybrid method of Monte Carlo and UKF has been proposed by [6] with the intention to solve both the kidnapping problem and unimodal of Kalman filter, which only works for local.

The works mentioned above established much focus on precise localization and solving the kidnapping problem. Using Monte Carlo for localization potentially solve the kidnapping problem, but it requires sufficient samples for better performance. It is also interesting that Kalman filter still mainly been used for humanoid robot localization due to lower computation though it has kidnapping problems.

In practice, in order to develop better localization for real humanoid robot soccer, there are three things to consider, that are soccer game rules, computation, and accuracy. Solving a complete kidnapping problem is highly likely not needed based on soccer rules; e.g., robots enter the field from the middle-side, no throw-in, no corner kick, the referee moves away from the robots few centimeters when the collision occurs. Computation is limited to the embedded platform being used and CPU slots for other processes such as servo controller, vision processing. Pursuing only the localization accuracy tends to increase computation while slowing other processes, which is not practical.

3 Proposed Method

Our approach consists of two steps, that are local movement and correction. We consider three criteria for developing our approach, that are soccer game rules, computation, and accuracy. Differing from the previous works which rely on precise localization with a quite high computation, our approach does not rely upon much to the precise localization as in analogy with how human playing soccer. Human play soccer only needs to know roughly the position while attacking or defending. Most of the computation of our approach is gained from a Kalman filter, which is known to be computationally fast. The overall block diagram of the proposed system is shown in Figs. 2 and 3.

Let the robot global pose $g \in \mathcal{G}$ lies in the space of the field coordinate $\mathcal{G} \subseteq \mathcal{N} \times \mathbb{R}$ containing coordinate x, y, and orientation θ. Zero pose (0, 0, 0) is located at the center field with a positive x orients to the right attacking direction while positive y orients to the forward attacking direction. Note that the robot is assumed to be at zero pose (0, 0, 0) at the center-field facing enemy goal for calculating the global reference. In order to save some computation, we make a grid of the soccer field $\mathcal{N} \subseteq \mathbb{N}^2$, where the grid $\mathcal{N}$ has a range $\{[-6, 6], [-4, 4]\}$, as shown in Fig. 4.

3.1 Local Movement

Kalman filter is the core of this step. Kalman filter is a well-proven algorithm that can combine multiple measurements that contain noise to estimate system state (see Fig. 5). The noise can be from the slight dimension mismatch of the designed robot model with the real robot assembly or the servo measurement feedback. Measurements are taken from robot kinematics odometry, while the system state is a global robot position. In humanoid robot soccer practice, using Kalman filter is computationally fast and less memory usage since it does not store all the preceding state

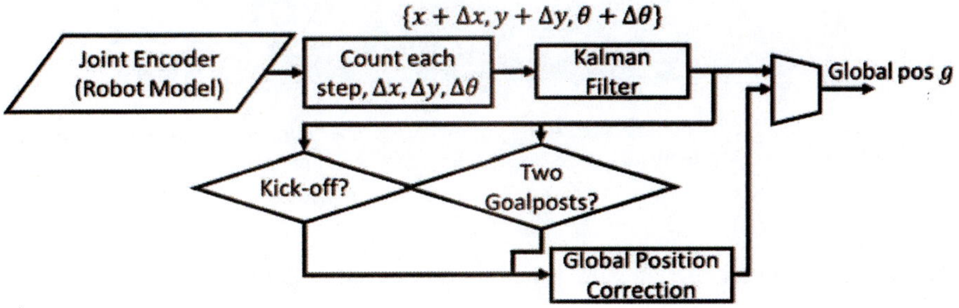

Fig. 2 Block diagram of the localization system. It consists of two steps; local movement using Kalman filter and correction using visual landmark detectors (ball detector during kick-off and two goalposts)

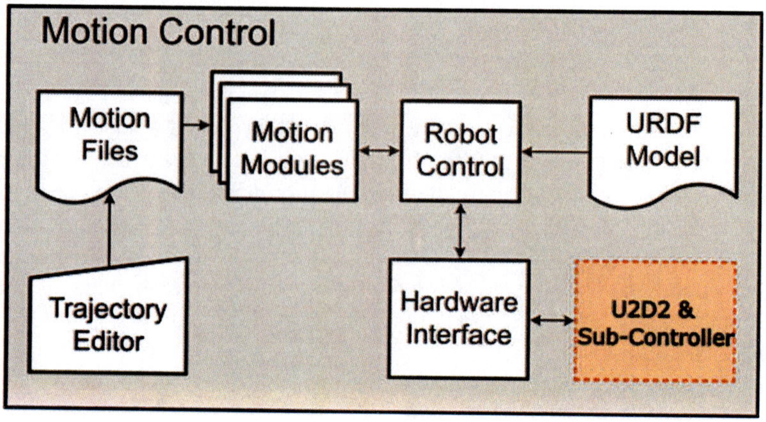

Fig. 3 The overall architecture of ROS in our robot. Each block is a package in ROS terminology. Robot kinematic and dynamic properties are supplied to the Robot Control package using the URDF model. Most of the proposed algorithm is implemented in the Robot Control package. Motion modules contain joints configuration for each robot motion, such as kicking and get-up. The robot servos are connected using USB-to-Dynamixel (U2D2) serial communication in a multipoint fashion. The rest of the sensors are connected with the sub-controller board

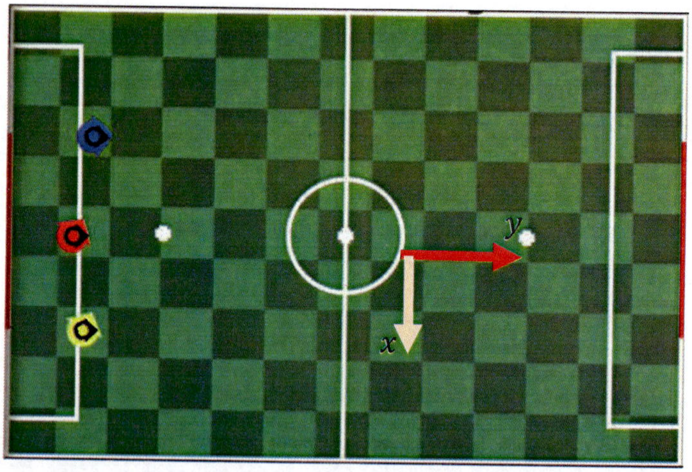

Fig. 4 Grid of the soccer field to save some computation. The field has 6 × 9 m. The grid size is 13 horizontals and 9 verticals. Blue, red, and yellow markers indicate the robots position. (Best viewed in color)

information, it only calculates the previous state. These are crucial for the real-time performance of humanoid robot embedded platforms.

Let the robot joints servo feedbacks are denoted as δ_i with the joint number i. The function $\mathscr{k}$ is used to calculate robot pose using Kalman filter and is defined as,

```
Input : Q, R, s_t, u_t, g_t-1, P_t-1
Output: ġ_t, P_t

    1: ġ_t ← f(g_t-1, u_t)
    2: P_t ← F_t P_t-1 F_t^T + Q
    3: k_t ← P_t H_t^T (H_t P_t H_t^T + R)^-1
    4: ġ_t ← ġ_t + k_t * z_t - h(ġ_t)
    5: P_t ← (I - k_t * H_t) * P_t
```

Fig. 5 Pseudo-code for calculating Kalman filter

$$\mathcal{k}: (\delta_i \in \mathbb{R}, \text{robot\_model}) \mapsto \mathcal{G}. \quad (1)$$

We assume that the humanoid robot is a single entity moving in a 2D space x and y-direction, $x, y \in \mathcal{N}$. From Newton's law of force, mass, and acceleration, $F = m.a$, and the differentiation of the velocity and acceleration, $v = dx/dt$ and $a = d^2x/dt^2$, the state-space model of the robot is,

$$\dot{g} = \begin{bmatrix} 0\,1\,0 \\ 0\,0\,1 \\ 0\,0\,0 \end{bmatrix} \boldsymbol{g} + \begin{bmatrix} 0 \\ 0 \\ 1 \end{bmatrix} \boldsymbol{u} + \begin{bmatrix} 1\,0\,0 \\ 0\,1\,0 \\ 0\,0\,1 \end{bmatrix} \boldsymbol{w} \quad (2)$$

where the state vector $\boldsymbol{g} = \begin{bmatrix} g \\ \dot{g} \\ \ddot{g} \end{bmatrix}$ contains global pose, velocity, and acceleration, an input vector $\boldsymbol{u}$, and process noise vector $\boldsymbol{w}$. The sensor is modeled using a velocity sensor from the servo feedback.

$$\mathcal{s} = [0 \quad 1 \quad 0]\boldsymbol{g} + [0]\boldsymbol{u} + [0 \quad 0 \quad 0]\boldsymbol{w} + \boldsymbol{v} \quad (3)$$

where $\mathcal{s}$ is the sensor output, and $\boldsymbol{v}$ is measurement noise. We assume that the model has no process noise, and the sensor slip error is 0.01, thus the covariance matrix of the process noise vector, $Q = \begin{bmatrix} 0\,0 \\ 0\,0 \end{bmatrix}$, the covariance matrix of the measurement noise vector, $R = [0.01]$, and it does not have a correlation between the process and measurement noises, $N = \begin{bmatrix} 0 \\ 0 \end{bmatrix}$.

Where:

- Q and R are process and measurement covariance matrix
- $\mathcal{s}_t$ is measured data
- u_t is a control plane
- f and h are a non-linear function to compute predicted state and measurement
- F_t and H_t defined by following jacobians

$$F_t = \left.\frac{\partial f}{\partial x}\right|_{x_{t-1|t-1}, u_t} \tag{4}$$

$$H_t = \left.\frac{\partial h}{\partial x}\right|_{x_{t|t-1}} \tag{5}$$

In order to calculate forward and inverse kinematics and dynamics from the robot model, Rigid Body Dynamics Library (RBDL) [7] is employed. This library requires the robot model is loaded in a Unified Robot Description Format (URDF) format. The URDF robot model contains the kinematic properties of each joint and link, together with dynamics properties such as the link weight and inertia matrix. The robot kinematics is used to send absolute position commands to the respective joint servos, while the robot dynamic is used to send commands to the respective servo torques. Newton-Euler inverse dynamics algorithm of the RBDL library is used to estimate the joint torques. We found that using only the robot kinematics to drive the servos could potentially reduce the servo lifetime as the servos are forced to move to the desired position with the fixed torque values.

3.2 *Correction*

The global position error from the local movement step might gradually increase as the robot moves around the soccer field. This error, for example, such as when the referee moves the corresponding robot away for a few centimeters as the robot collides with other robots. Therefore, a correction step is crucial to reduce global position error. In this step, there is no need for precise localization, which is in analogy with how humans play soccer. Human estimates the position roughly during soccer game both for attacking and defending.

The correction step is activated either during kick-off, or the two goalposts are observed. The soccer game rules state that kick-off is observed by the robots when the kick-off button of the game controller is clicked. All the robots must return to their respective home field. The robots then perform positioning using localization to return to home. The referee then will put the ball to the center-field, and the robot's global position is corrected using the ball diameter size. In practice, we found that using the detected ball diameter is the best practice to perform correction instead of detecting the circle center field lines. The circle-center field lines detector is less robust as compared to the ball detector. This is because the field grass is now fully synthetics, and the white lines are less contrast with the green grass due to blurry white color painted on synthetic grass. More specifically, ball detector using LBP features, [8], and cascaded classifiers are employed to locate the ball, as in [9]. Then the ball diameter size is calculated (see Fig. 6).

We create a map $\mathcal{m}_{ball}: (b \in \mathbb{R}) \longmapsto (d_{ball} \in \mathbb{R})$ for each ball diameter size b against the actual distance of the robot to the ball d_{ball}, and perform interpolation $\mathcal{i}_{ball}: (\mathcal{m}_{ball}(b) \in \mathbb{R}) \longmapsto (d_{ball} \in \mathbb{R})$ when the ball size is not on the map. With

```
Input : ballDiameter
        robotTheta
Output: robotPos ← x,y

    BallDistance ← regression(ballDiameter)
    localBall.x ← ballDistance * sin(robotTheta)
    localBall.y ← ballDistance * cos(robotTheta)
    robotPos.x ← -localBall.x
    robotPos.y ← -localBall.y
```

Fig. 6 Pseudo-code for determining robot position using ball diameter during the kick-off

the information of both the ball distance d_{ball} and robot orientation θ, the global robot position is calculated using a function g: $(d_{ball}, d_{goalposts}, \theta) \in \mathbb{R}^3 \mapsto \mathcal{G}$. The function g is calculated using a simple triangle equation. The final robot global position g is then defined as in Eq.,

$$g := \begin{cases} g(d_{ball}, 0, \theta), \text{ if kick-off, or both are observed} \\ g(0, d_{goalposts}, \theta), \text{ if the goalposts are observed} \\ \text{Eq. 1, otherwise} \end{cases} \quad (6)$$

The two goalposts detector is calculated using Hough transform to detect white vertical lines from the edge of the green field. Nearly closed distances of the detected vertical lines are merged to find a single goalpost. If the two goalposts are detected, the length between goalposts l is calculated. We create a map $m_{goalposts}: (l \in \mathbb{R}) \mapsto (d_{goalposts} \in \mathbb{R})$ for each length l against the actual distance to the robot $d_{goalposts}$, and perform interpolation $i_{goalposts}: (m_{goalposts}(l) \in \mathbb{R}) \mapsto (d_{goalposts} \in \mathbb{R})$ when the length is not on the map. With the information of both the goalposts distance $d_{goalposts}$ and robot orientation θ, the global robot position is calculated using the above Eq.

We found that during the soccer game competition, the error is significantly accumulated during the local movement step. This error is due to the frequent falling of the robot during movement. Due to imperfect gait calibration, the robot falls frequently. In order to deal with this, a robust balancing algorithm is employed using accelerometer and gyro sensors by the slight movement to the force direction, as in [10].

4 Experiments

We tested the proposed system using our humanoid robot soccer platform, EROS. The platform uses an embedded platform of Intel NUC Core i3 as the primary computing and sub-controller for the hardware interface. The specification of the EROS platform is shown in Fig. 7. We implement our C ++ code using ROS based framework [11] in Ubuntu Kinetic operating system.

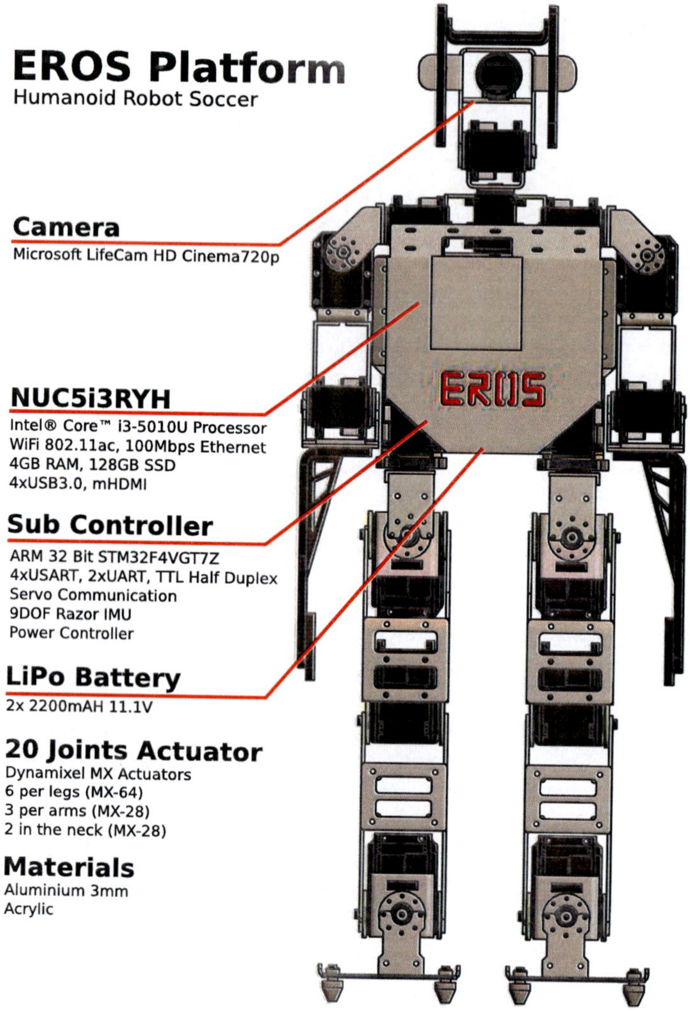

Fig. 7 EROS platform specification

In order to perform monitoring from the robot state, related data are sent wirelessly to a standard desktop PC. A game controller is a dedicated software for referees to control all the robots in the field. The game controller contains several buttons, such as the start-stop of each robot, kick-off, and free-kick buttons. A GUI interface for monitoring is depicted as in Fig. 8. The experiment using the only joint encoder of robot kinematic as odometry is shown in Fig. 9. The error could gradually generate using only odometry information. Thus, the global robot position correction step is essential to be added.

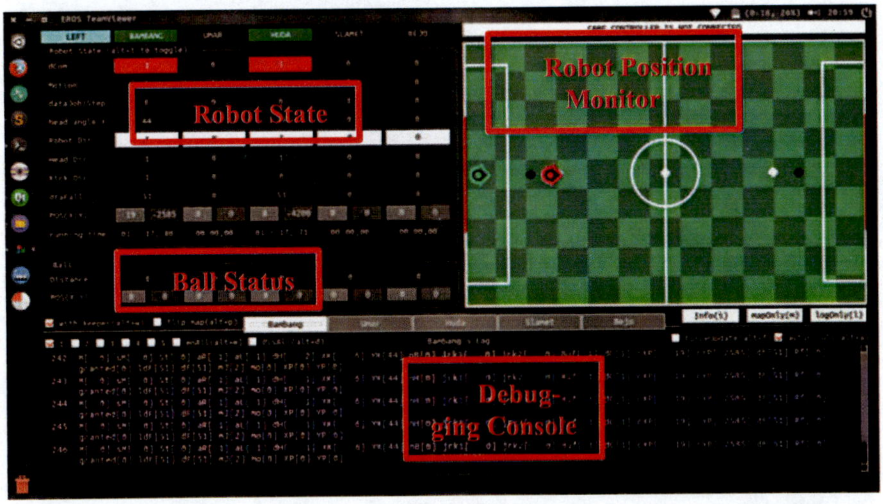

Fig. 8 Monitoring GUI interface of our platform. It consists of four screens; Robot state of each robot, robot position monitor that is useful for localization, ball status, and debugging console. (Best viewed in color)

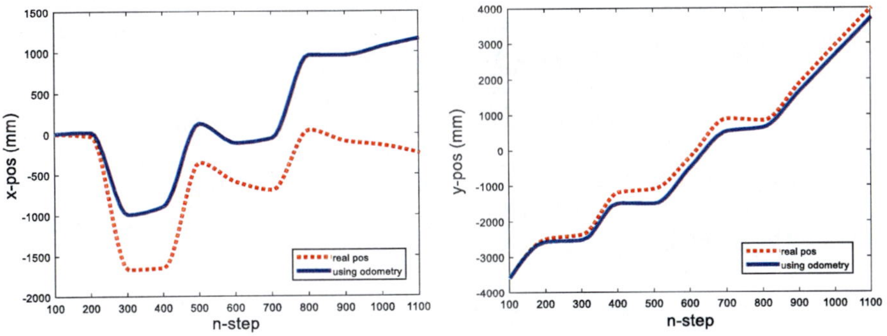

Fig. 9 Plot of the robot trajectory between the ground truth and using only odometry. Using only joint encoder of robot kinematic as odometry can gradually accumulate error as mainly indicated in the left figure. Thus, the global robot correction step is highly important to be added

The performance of the proposed system is shown in Fig. 10. The robot moves in a predefined trajectory (ground truth), then the proposed system is tested, with and without correction. Our system shows fairly well for following the ground truth trajectory, which is indicated by a nearly fitted curve around the ground truth. The error rate in the x-axis is 33 mm, while on the y-axis is 92 mm. The reason why the y-axis error is higher than the x-axis is that most of the robot movements are going in the forward direction (such as while attacking and defending). Thus, the error is accumulated more on the y-axis. It is interesting to note that using only joint odometry of robot kinematic, the graph shows at some points it encounters

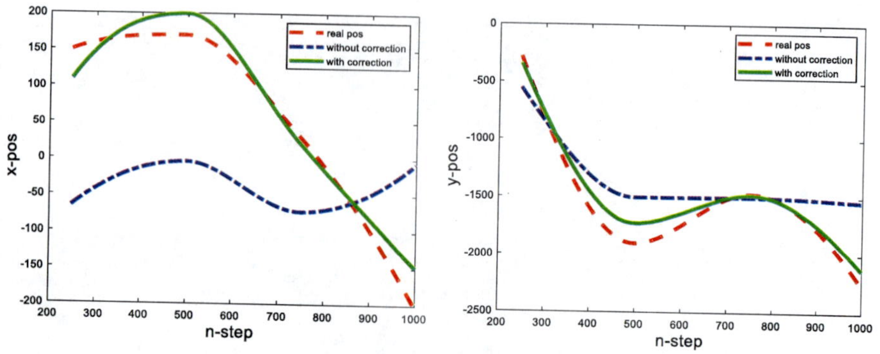

Fig. 10 Trajectory plot comparison between ground truth (real pos), without correction, and with correction trajectories. The proposed system with correction shows fairly well to follow on the correct path. Y-axis is in millimeters

oscillation due to the error that is gradually accumulated then eventually deteriorates the global robot position. We also found that the error is quickly accumulated due to robot falling frequently. Unfortunately, our strategy of using robot balancing strategy may seem not enough to counter unexpected disturbance such as collision with other robots. Our landmark detector for locating the goalposts shows quite well and thus making the correction quite robust to lighting, contrast, and similar color with the background. Unfortunately, unexpected problems might occur in the real robot soccer game. For example, the light suddenly off due to electricity trip, some persons might stand behind the goalposts that visually have similar color, or the goalposts dimension might not satisfy the expected rule due to assembly error from technicians. It is interesting to perform goalposts detector with a more sophisticated machine learning algorithm, such as a deeper network using a pre-trained model followed by fine-tuning [9, 12–14].

5 Conclusion

During a robot soccer game, forming a particular formation while attacking or defending plays a significant role in winning the match. To obtain the formation, the localization algorithm that satisfies soccer game rules, computation, and accuracy has been developed and tested in real humanoid robot soccer. More specifically, two steps of approach have been developed that are local movement and correction. The local movement primarily employs a joint encoder of robot kinematics odometry and the correction step to improve locating the robot position. Experiments show the system runs quite well in the real humanoid soccer robot with the error rate in the x-axis is 33 mm, while on the y-axis is 92 mm. Our team has successfully achieved the second winner in Indonesian humanoid robot soccer 2019. We believe that this work could bring benefits to the research in the related fields. For future work, an in-depth investigation of errors while the robot frequently falls during movement is refreshing to be solved using an approximate algorithm of robot joint encoder.

References

1. Allgeuer P, Schwarz M, Pastrana J, Schueller S, Missura M, Behnke S (2018) A ROS-based software framework for the NimbRo-OP humanoid open platform. arXiv preprint arXiv:1809.11051
2. N'Guyen S, Passault G, Pirrone A, Rouxel Q (2018) Rhoban football club: robocup humanoid kid-size 2017 champion team paper. In: RoboCup 2017: robot world cup XXI, vol 11175, p 423
3. Qian Y, Lee DD (2016) Adaptive field detection and localization in robot soccer. In: Robot world cup, pp 218–229
4. Almeida AC, Costa AH, Bianchi RA (2017) Vision-based monte-carlo localization for humanoid soccer robots. In: 2017 Latin American robotics symposium (LARS) and 2017 Brazilian symposium on robotics (SBR), pp 1–6
5. Muzio A, Aguiar L, Máximo MR, Pinto SC (2016) Monte carlo localization with field lines observations for simulated humanoid robotic soccer. In: 2016 XIII Latin American Robotics Symposium and IV Brazilian Robotics Symposium (LARS/SBR), pp 334–339
6. Teimouri M, Salehi ME, Meybodi MR (2016) A hybrid localization method for a soccer playing robot. In: 2016 Artificial intelligence and robotics (IRANOPEN), pp 127–132
7. Felis ML (2017) RBDL: an efficient rigid-body dynamics library using recursive algorithms. Auton Robots 41(2):495–511
8. Viola P, Jones M (2001) Rapid object detection using a boosted cascade of simple features. CVPR 1(1):511–518
9. Rizgi AK et al (2018) Improving field and ball detector for humanoid robot soccer EROS platform. In: 2018 International electronics symposium on engineering technology and applications (IES-ETA), pp 266–269
10. Rizgi AK et al (2018) Implementation of balance recovery by slight movement in humanoid robot soccer. In: 2018 International electronics symposium on engineering technology and applications (IES-ETA), pp 95–99
11. Quigley M et al (2009) ROS: an open-source robot operating system. In: ICRA workshop on open source software, vol 3, p 5
12. Risnumawan A, Sulistijono IA, Abawajy J (2016) Text detection in low resolution scene images using convolutional neural network. In: International conference on soft computing and data mining, pp 366–375
13. Sulistijono IA, Risnumawan A (2016) From concrete to abstract: multilayer neural networks for disaster victims detection. In: 2016 International electronics symposium (IES), pp 93–98
14. Rizgi AK et al (2020) Visual perception system of EROS humanoid robot soccer. Int J Int Inf Technol (IJIIT) 16(4):1548–3657

Evaluation of the Transfer Learning Models in Wafer Defects Classification

Jessnor Arif Mat Jizat, Anwar P. P. Abdul Majeed, Ahmad Fakhri Ab. Nasir, Zahari Taha, Edmund Yuen, and Shi Xuen Lim

Abstract In a semiconductor industry, wafer defect detection has becoming ubiquitous. Various machine learning algorithms had been adopted to be the "brain" behind the machine for reliable, fast defect detection. Transfer Learning is one of the common methods. Various algorithms under Transfer Learning had been developed for different applications. In this paper, an evaluation for these transfer learning to be applied in wafer defect detection. The objective is to establish the best transfer learning algorithms with a known baseline parameter for Wafer Defect Detection. Five algorithms were evaluated namely VGG16, VGG19, InceptionV3, DeepLoc and Squeezenet. All the algorithms were pretrained from ImageNet data-base before training with the wafer defect images. Three defects categories and one non-defect were chosen for this evaluation. The key metrics for the evaluation are classification accuracy, classification precision and classification recall. 855 images were used to train and test the algorithms. Each image went through the embedding process by the evaluated algorithms. This enhanced image data numbers then went through Logistic Regression as a classifier. A 20-fold cross-validation was used to validate the score metrics. Almost all the algorithms score 85% and above in terms of accuracy, precision and recall

Keywords Transfer learning · VGG16 · Wafer defect detection

J. A. Mat Jizat (✉) · A. P. P. Abdul Majeed · A. F. Ab. Nasir
Faculty of Manufacturing and Mechatronic Engineering Technology (FTKPM), Universiti Malaysia Pahang, 26600 Pekan, Pahang, Malaysia
e-mail: jessnor@ump.edu.my

Z. Taha
Fakultas Teknologi Industri, Universitas Islam Indonesia, Jl. Kaliurang KM. 14,5 Sleman, Yogyakarta 55584, Indonesia

E. Yuen · S. X. Lim
Ideal Vision Integration Sdn Bhd, 02-25 Level 2, Setia Spice Canopy, Jln Tun Dr Awang, 11900 Bayan Lepas, Penang, Malaysia

A. F. Ab. Nasir et al. (eds.), *Recent Trends in Mechatronics Towards Industry 4.0*, Lecture Notes in Electrical Engineering 730, https://doi.org/10.1007/978-981-33-4597-3_78

1 Introduction

In a semiconductor industry, wafer defect detection is key to achieve a better yield for manufacturing process. However, training people to manually detect the defect can prove to be time consuming for the industry as it normally take about 6–9 months of training to achieve 90% accuracy. However, within 15 months after the training had been completed, manual defect detection can dropped to between 70 and 85% accuracy due to several factors such as demotivation due to repetitive work, process advancement, or increased difficulties [1]. Thus many industry players adopting an automated defect detection system using machine vision and machine learning capabilities. One of the technique in machine learning that driving the auto-mated defect detection system is transfer learning. Transfer learning is a technique in machine learning where the model/algorithm was pre-trained on other image data-base and storing the "knowledge" gain to be used in other application area such as wafer defect detection.

1.1 Related Work

Automated optical inspection machine had evolved from labour-intensive manual inspection towards fully automated machine with minimum human intervention. This fully automated machines have the ability to make decision on the quality of the product based on image captured. Naturally, image processing algorithm can be employed for such machine [2–4]. However, the industry especially semiconductors industry requires faster decision machine. With the advancement of computing technologies, Convolution Neural Network (CNN) becomes available. CNN offer an improved performance in terms of training time and features extraction from the images [5]. Thus, myriad of CNN algorithms had been employed in the automated optical inspection machine. CNN is a deep learning algorithm that extract specifics features from input images and classify them based on the extracted features.

CNN rely heavily on extracted features from the images for making decision and requiring hundreds of images to be analysed in order to code a reliable algorithms. Thus, a lot of image databases exist on the internet. The biggest database currently is ImageNet where it host millions of images in under various classifying terms [6]. From here, transfer learning becomes a part of machine learning algorithm where the algorithms were pre-trained with the images in a database, and using the knowledge gained try to predict the image classification.

VGG16, VGG19 and InceptionV3 are among the most popular transfer learning algorithms used to predict an image classification especially in wafer defect detection [7–9]. In most cases, it can reliably classify images with high accuracy over 95% accuracy [8]. Due to the fact that these algorithm can learned from existing images in database, an update with small amount of images for training is sufficient to make image classification in a short time. However, these algorithms are open to be tuned by programmer to better suit the application requirements [9, 10]

Thus the objectives of this paper is establish the best transfer learning algorithms with known baseline parameter for wafer defect detection. This is important so that in researcher can focus on improving the evaluated metrics of a single algorithm by tuning its hyperparameter. This algorithm then will be deployed into a machine vision system that can reliably detect wafer defect.

2 Methodology

2.1 *Transfer Learning Model*

Five transfer learning models were chosen to be evaluated. Each models was pre-trained using ImageNet dataset. ImageNet is an open source image database organized by Stanford University and Princeton University. The description of the model are as follows:

- **VGG16**—is a convolutional neural network architecture developed by Visual Geometry Group from University of Oxford. It consists of 16-layers image recognition model with fixed size 224 × 224 RGB as an input [11].
- **Inception V3**—is Google's deep neural network for image recognition. It consists of 48-layers deep convolution neural network. Default size for input image is 299 × 299 RGB images [12].
- **Deeploc**—is a deep neural networks algorithm developed mostly to analyse yeast cell images. It was trained on 21882 images of single cell [13].
- **VGG19**—is similar to VGG16 but with a 19-layers deep convolution neural network [11].
- **Squeezenet**—is a deep model for image recognition that achieves high accuracy with fewer parameters. It is based on Convolution Neural Network with smaller architecture than AlexNet [14].

The model used were standard and untuned as tuning the hyperparameter is not of the scope of this paper. The scores were evaluated based on classification accuracy, precision and recall of their classification result. Classification accuracy mainly discuss on the capability of the model to correctly predict the images in all of its classes. Meanwhile, classification precision deals with the images that was predicted as positives, how many had been correctly predicted. On the other hands, classification recall deals with the proportion of the actual positive images and the misclassification of the actual positive images. All these metrics will give a thorough analysis of the model predictive capability.

2.2 Hardware and Software

The training and testing were conducted using a laptop with Intel(R) Core(TM) i5-6200U CPU 2.30 GHz processor with 8 GB DDR3 RAM and NVIDIA GeForce 930MX graphic card. It was run using a rapid machine learning software Orange [15] with Image Analytics add-on which based from Python programming language.

2.3 Dataset

The dataset is acquired from a semiconductor industry using industrial machine vision platform jaeger developed by Idealvision Sdn Bhd. The data set for training consists of images from 3 categories of most common defect namely Bump, Burnt Mark, and Foreign Object with 154, 146, 154 images respectively. Another category of non-defects with 253 images were also added into the training. On the other hands another dataset, which is totally separated from the training dataset, was used for testing. This dataset consist of 168 images mix from all four categories. The 168 images were chosen at random from total available images. In Fig. 1, shows sample of images used in the paper.

2.4 Training and Testing

After the dataset was loaded to the software, each images went through embedding process. This embedding process was using transfer learning model to calculate feature vector for each image thus representing the images in terms of data numbers. This enhanced image data numbers then went through Logistic Regression, a machine learning classifier, for further classification process. Logistic Regression was chosen because it is the best performance classifier from previous study to classify this particular image dataset. The data then went through 20-fold cross validation to avoid overfitting of the data. After this the software generates a confusion matrix to indicate the performance of the training.

In terms of testing, the testing dataset was loaded into the software, and went through embedding process using transfer learning model. Using the same Logistic Regression Classifier, each images class/category was predicted and the result was tabulated into confusion matrix and the performance was compared to other transfer learning model.

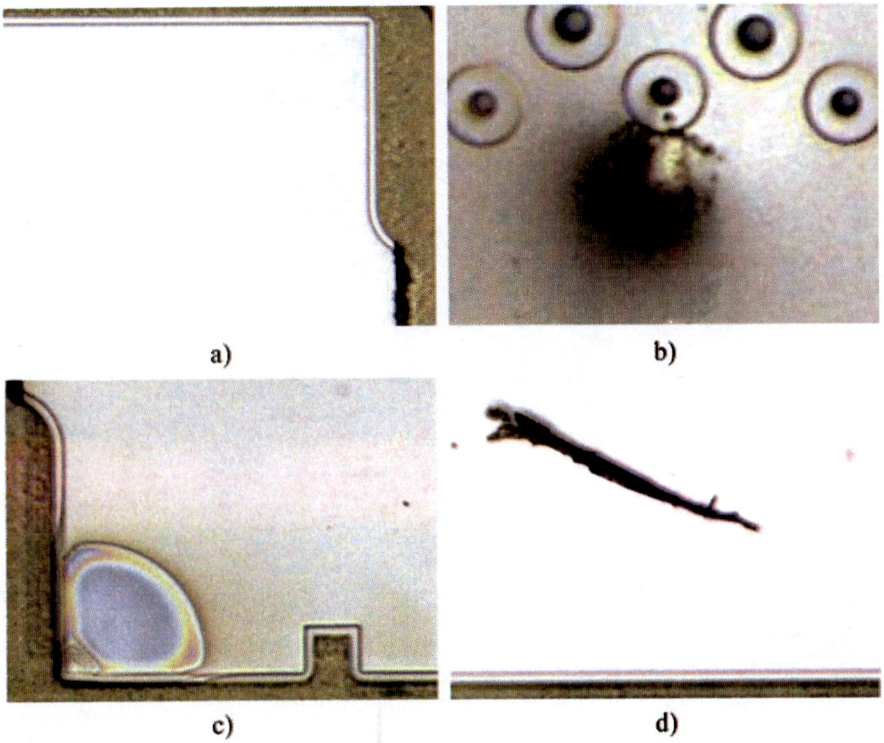

Fig. 1 Sample images from each categories **a** Good **b** Bump **c** Burnt mark **d** Foreign object

3 Result and Discussion

Figure 2 shows the comparison of the average accuracy performance of 5 different transfer learning models in terms of wafer defect classification. The accuracy performance give an indication of reliability of the models in terms of predicting the classification of the wafer defects. Out of the five transfer learning model tested, VGG16 model give the best classification accuracy with 91.7% during training and 92.3% during testing. All other models achieved more than 85% classification accuracy for training and testing except for DeepLoc model with about 75% classification accuracy for testing and training which indicate the worst performance of the models. Even though in training VGG16 and VGG19 achieved a relatively close classification accuracy, the dropped in performance during testing by VGG19 made VGG16 a clear winner in terms of wafer defect detection performance.

As wafer defects are a relatively seldom occurrence due to fact each wafer manufac-turing process strive for zero defect, a large number of images will be classified as good compared to images that will be classified as defects. This mean in wafer defect detection system, a good classification will be on bias. Thus, an accuracy prediction is not enough to evaluate the system. The system must be also

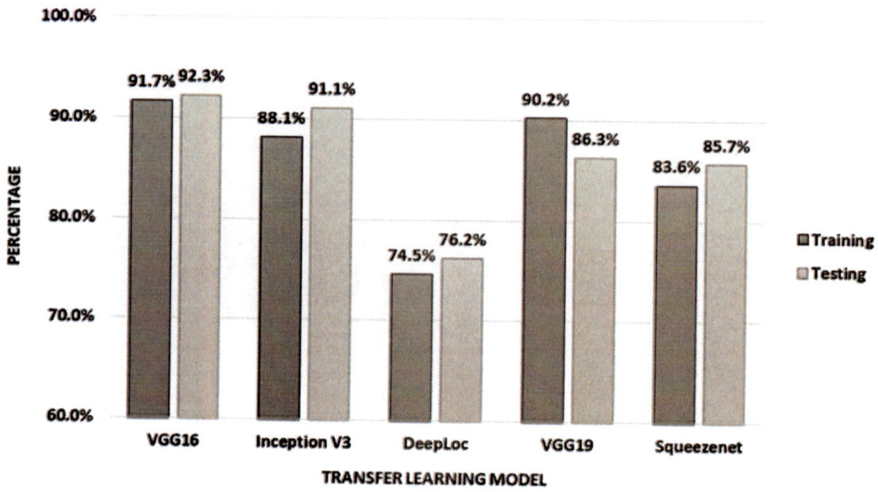

Fig. 2 Comparison of average classification accuracy performance

be evaluated in terms of Precision-Recall. Figures 3 and 4 show the training mean precision-recall per-formance and testing mean precision-recall performance across the categories of the transfer learning models respectively.

From Figs. 3 and 4 we can learn that VGG16 model give a balance precision recall both in training and testing. This shows that for VGG16 the misclassification between false negative and false positive is quite balance. For the training, other models give a lower recall than precision within 0.1% to 0.4%. This result shows that all other models are bias toward false negative. False negative means some of

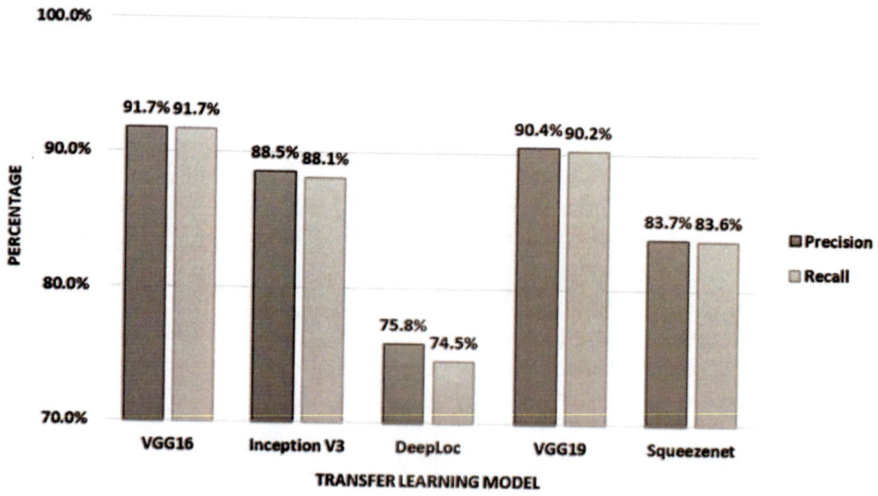

Fig. 3 Training mean precision-recall performance for transfer learning model

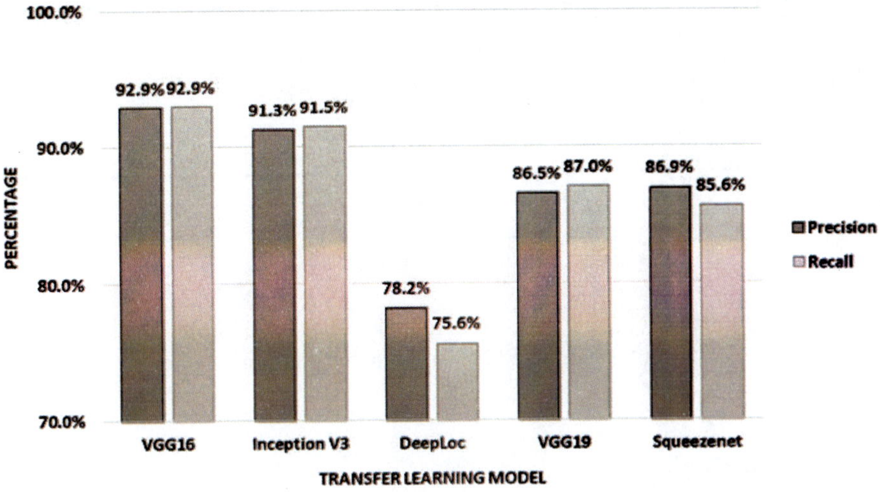

Fig. 4 Testing mean precision-recall performance for transfer learning model

the good wafer being wasted and may effected the wafer yield. However, during testing, only Deeploc and Squeezenet exhibits lower recall than precision within 0.3–0.4% while InceptionV3 and VGG19 exhibit higher recall than precision within 0.3–0.5%. A higher recall indicates that the model now bias towards false positive. False positive may indicates that more defect will escape inspection thus creating a defect product from manufacturing process.

In order to understand the performance of VGG16 better, further analysis on the confusion matrix can be conducted. Hence, Figs. 5 and 6 shows the confusion matrices for VGG16 in training and testing respectively.

From Figs. 5 and 6 we can see that VGG16 had a tendency to misclassify each defects as a good classification. From training 29 defects images had been misclassify as good images while 6 defect images had been misclassify as good images in testing. On the other hand, good images were tend to be misclassify as Burnt Mark, which score the highest instances compared to Bump and Foreign Object both in training and testing. The good images is less misclassify as Foreign Object both in

		Predicted				
		Bump	Burnt Mark	Foreign Object	Good	Σ
Actual	Bump	149	0	0	5	154
	Burnt Mark	0	129	1	16	146
	Foreign Object	3	4	139	8	154
	Good	5	13	4	231	253
	Σ	157	146	144	260	707

Fig. 5 Training confusion matrix for VGG16

		Predicted				
		Bump	Burnt Mark	Foreign Object	Good	Σ
Actual	Bump	34	0	0	0	34
	Burnt Mark	0	32	0	2	34
	Foreign Object	0	0	30	4	34
	Good	3	4	0	59	66
	Σ	37	36	30	65	168

Fig. 6 Testing confusion matrix for VGG16

training and testing. Thus, the researchers believed that Foreign Object images are very distinct, and easily recognized by the Transfer Learning models.

4 Conclusion

From the experiment above, it can be established that the transfer learning model VGG16 is the best model to run a wafer defect detection at 92.3% accuracy with Logistics Regression classifier and 20-fold cross validation procedure. However, it is best to remember that the specific defects are known prior to the classification proce-dure is conducted. A new class of defect may not be detected other than the previ-ously known defect class.

Acknowledgements The authors would like to thanks IdealVision Sdn Bhd for providing the image dataset to make this evaluation possible.

References

1. Tuv E, Guven M, Ennis P, Lee DHL (2018) IT@Intel white paper: faster, more accurate defect classification using machine learning
2. Huang SH, Pan YC (2015) Automated visual inspection in the semiconductor industry: a survey. https://doi.org/10.1016/j.compind.2014.10.006
3. White KP, Kundu B, Mastrangelo CM (2008) Classification of defect clusters on semiconductor wafers via the hough transformation. IEEE Trans Semicond Manuf 21:272–277. https://doi.org/10.1109/TSM.2008.2000269
4. Lin HD, Chiu SW (2011) Flaw detection of domed surfaces in LED packages by machine vision system. Expert Syst Appl. https://doi.org/10.1016/j.eswa.2011.05.080
5. LeCun Y (1989) Others: generalization and network design strategies. In: Connectionism in perspective
6. Russakovsky O, Deng J, Su H, Krause J, Satheesh S, Ma S, Huang Z, Karpathy A, Khosla A, Bernstein M, Berg AC, Fei-Fei L (2015) ImageNet large scale visual recognition challenge. Int J Comput Vis. https://doi.org/10.1007/s11263-015-0816-y
7. Shen Z, Yu J (2019) Wafer map defect recognition based on deep transfer learning. In: IEEE international conference on industrial engineering and engineering management. https://doi.org/10.1109/IEEM44572.2019.8978568

8. Lee JH, Lee JH (2019) A reliable defect detection method for patterned wafer image using convolutional neural networks with the transfer learning. In: IOP conference series: materials science and engineering. https://doi.org/10.1088/1757-899X/647/1/012010
9. Hou D, Liu T, Pan YT, Hou J (2019) AI on edge device for laser chip defect detection. In: 2019 IEEE 9th annual computing and communication workshop and conference, CCWC 2019. https://doi.org/10.1109/CCWC.2019.8666503
10. Kang S (2018) On effectiveness of transfer learning approach for neural network-based virtual metrology modeling. IEEE Trans Semicond Manuf. https://doi.org/10.1109/TSM.2017.2787550
11. Simonyan K, Zisserman A (2015) Very deep convolutional networks for large-scale image recognition. In: 3rd International conference on learning representations, ICLR 2015—conference track proceedings
12. Szegedy C, Vanhoucke V, Ioffe S, Shlens J, Wojna Z (2016) Rethinking the inception architecture for computer vision. In: Proceedings of the IEEE computer society conference on computer vision and pattern recognition. https://doi.org/10.1109/CVPR.2016.308
13. Almagro Armenteros JJ, Sønderby CK, Sønderby SK, Nielsen H, Winther O (2017) DeepLoc: prediction of protein subcellular localization using deep learning. Bioinformatics. https://doi.org/10.1093/bioinformatics/btx431
14. Iandola FN, Moskewicz MW, Ashraf K, Han S, Dally WJ, Keutzer K (2016) SqueezeNet. arXiv
15. Demšar J, Curk T, Erjavec A, Gorup Č, Hočevar T, Milutinovič M, Možina M, Polajnar M, Toplak M, Starič A, Štajdohar M, Umek L, Žagar L, Žbontar J, Žitnik M, Zupan B (2013) Orange: data mining toolbox in python. J Mach Learn Res 14:2349–2353

Transitioning into a Deregulated Energy Market for Sabah: Strategies and Challenges for Generators

Tze Wei Lim, Andrew Huey Ping Tan, Eng Hwa Yap, Kim-Yeow Tshai, and Wei Kong

Abstract With an increasing energy demand, energy security is progressively becoming crucial for Malaysia's socio-economic growth. In addition, with the ratification of the Paris Climate Change Agreement, policies for the energy sector, which are the prime factors of global greenhouse gas emissions should be given thorough attention, especially for a developing economy like Malaysia. This paper is concerned with the design and development of a systems model to predict the behaviours of relevant energy market key indicators in transitioning Sabah's energy sector towards a deregulated model. The dynamic and systemic nature of the electricity market necessitates a systems approach to model the scenarios projecting across envisaged possibilities in the transition process. Results show that 15% renewable energy penetration in Sabah will provide benefits of improving system performance, reduce fossil fuel dependence and harmful emissions, and increase system sustainability.

Keywords Energy market · Deregulation · Sabah · System dynamics

1 Introduction

During the early days, the electricity markets (EM) in various parts of Malaysia were mostly vertically integrated with a monopolistic supply and demand core. Liberalization of energy supply started in the early 1990s when Independent Power Producers (IPP) could generate and sell their power to the utility company. However, a major shift happened in 2010 when the government approved the National Renewable

T. W. Lim · A. H. P. Tan · K.-Y. Tshai
Faculty of Science and Engineering, University of Nottingham Malaysia, 43500 Semenyih, Selangor, Malaysia

E. H. Yap (✉)
School of Intelligent Manufacturing Ecosystem, Xi'an Jiaotong-Liverpool University, Suzhou, Jiangsu 215123, People's Republic of China
e-mail: EngHwa.Yap@xjtlu.edu.cn

W. Kong
Centre for Foundation and General Studies, Infrastructure University Kuala Lumpur, 43000 Kajang, Selangor, Malaysia

A. F. Ab. Nasir et al. (eds.), *Recent Trends in Mechatronics Towards Industry 4.0*, Lecture Notes in Electrical Engineering 730, https://doi.org/10.1007/978-981-33-4597-3_79

Energy Policy and Action Plan giving more focus to the generation of renewable energy (RE). Feed-in-Tariffs (FiT) scheme was implemented by the Sustainable Energy Development Authority in peninsular Malaysia and extended to Sabah in 2014. Energy markets across Malaysia were opened for competition and economic modelling methods became applicable in determining the relationships between each factor to better understand the necessary frameworks required for transitioning the electricity generation market. Robust analyses of the implementation and effectiveness of the new markets' rules have become crucial for both the federal and state governments, policy-makers and regulators. The focus of the review lies on aspects of transitioning of deregulation on the generation side of the electricity market.

The power industry has been a closed non-linear feedback system [1]. The electricity market exhibits dynamic behaviour in areas of management, technology, progress, industry configuration, consumer policy, supply-demand management and governmental policies [2]. Most of the time, the electricity market (EM) offers incomplete information, where uncertain decision-making could mean irreversible implications inflicted upon the deregulation plans of the EM. This study is hence concerned with understanding the dynamic complexities of transitioning into a deregulated energy market for Sabah, by using system dynamics, which could provide insights into the behaviour of identified key indicators during the transition process. Section 2 provides a brief review into Sabah's energy situation and deregulation. Section 3 details the methodology used for this research. Section 4 provides the model constructed and verification and validation for them. Section 5 provides the results and discussion stemming from the simulation of the models. Section 6 concludes the study and proposes future works.

2 Sabah's Energy Outlook

2.1 Background

Energy security has been a key indicator for Malaysia's socio-economic growth towards becoming a high-income nation by 2030. To ensure the stable energy security with affordable prices from the depleting indigenous fossil fuel resources, Five-Fuel Diversification Policy has been implemented. Under the Tenth Malaysia Plan, FiT is introduced as part of initiative to increase RE sources [3].

Sabah has an estimated population of over 3.5 million, and a growing economy based on the exports of natural resources and fossil fuels as well as tourism. In 2014, Sabah has approximately 900 MW of total capacity, from which peak demand utilized in the region of 700 MW. With an annual electricity growth rate of 7%, Sabah's electricity sector also faces problems of aging, expensive, and unreliable diesel generators [4]. Due to geographical constraints, 25% of Sabah east coast's electricity demand is reliant on the west coast grid, as the east coast still relies heavily on the aging and expensive diesel generation plants [5]. Among the 17 major plants,

8 are diesel plants, 3 are combined cycle gas plants and only one major hydropower plant in Tenom Pangi [5], which relies on inconsistent natural stream flow to generate power. Four biomass power plants were built under the Small Renewable Energy Production (SREP) programme as alternatives to the generation mix. However, the disconnect of transmission grid between Sarawak and Kalimantan hinders the import and export of energy generation between those states. These indicate the needs of ensuring continuous secure electricity supply. A liberalised market will encourage the involvement of IPPs, which could secure the electricity supply.

2.2 *Generation Mix*

Gas power is set to be dominant for the near future due to a lack of viable alternatives. Gas-fired capacity will be 1,486 MW by the end of 2023 with an approximate increase of 132% while diesel will be projected to deplete by the end of 2023. The implementation of FiT will increase the share of RE power plants and provides impetus for the development of new plants. The implementation of Five Fuel Policy and New Energy Policy encouraged the inclusion of RE and several programs were launched such as Malaysia Building Integrated Photovoltaic Technology Application (MBIPV), SREP, and Biomass Power Generations Project. Under the 11th Malaysia Plan, the significance of sustainable energy sources alongside relieving dependence on fossil fuels whilst maintaining stability power supply was highlighted. As one of the signatories of Kyoto Protocol, Malaysia has been working towards a low carbon emission economy and community with the aim of greenhouse gas reduction, pledging to decrease 40% of CO_2 emissions intensity [3].

2.3 *Deregulation and the Policy*

Previously, the electricity supply industry was a vertically integrated utility, where mechanisms of generation, transmission, and distribution were controlled by a single entity. The reformation of the system strongly relied on governmental policies and participation of IPPs to open up the electricity market, which increased competition and private sector share in the electricity market. To stimulate a competitive electricity energy market, key-enablers include deregulation of electricity prices, creation of electricity trading arrangements, separation of generation, transmission, and distribution, privatisation of electricity market, as well as attracting foreign direct investments [6].

In developed countries, electricity market evolved from the single buyer model to the electrical wholesale market and further developed to retail market model. This allows the transition from vertically integrated electricity market to fully unbundled and competitive markets. The single buyer entity acts as a middleman between generators and distributors. It reduces the cost and institutionally demanding processes with the avoidance of third-party access to transmission. For deregulation, it happens at distinct levels with various degree of deregulation. In Sabah, the liberalization of the electricity market is transiting from a fully regulated model to single buyer generation where IPPs involve as owner-operator to generate and bid the power into distribution company.

In terms of policy, Energy Commission has created Malaysia Grid Code (MGC) and the Malaysian Distribution Code (MDC) as benchmarks for ensuring the standard service delivery. With the overhaul of the MGC, standards of the electricity supply industry (ESI) has been heightened by having separate the roles of the various players. The New Enhanced Dispatch Arrangement (NEDA) provided opportunity for non-PPA generators such as RE generators and cogenerates to operate as merchant generators to sell energy to the single-buyer. The NEDA implemented two phases, namely PPA and non-PPA generators. Its principle is based on the competitive bid, complemented the current cost-based bidding system with optional price-based bidding.

3 Methodology and Materials

3.1 System Dynamics

Figure 1 shows the research process flow used as adapted from Sterman's [2] modelling process. The research starts with identifying the problem and forming a hypothesis. Causal loop diagrams (CLD) are constructed in step 2, where interrelationships between variables are formed. Next, step 3 includes interview with key stakeholders and gathering of data. The data consist of two portions—namely inputs by stakeholders and data obtained as secondary from literature reviews from various sources [4, 7–11]. Once the information is sufficient, stock flow diagrams (SFD) are constructed where verification and validation take place at the same time. Verification and validation methods were adopted from Sterman [2], where necessary and sufficient procedures were carried out to ensure the model is working properly as intended. The verification and validation steps include boundary adequacy, dimensional consistency, parameter assessment, extreme condition, and integration error tests [2]. Once the model is finalised, simulation and interpretation of results happen in step 5. The data from which the models used to simulate are obtained from various relevant sources [12–15].

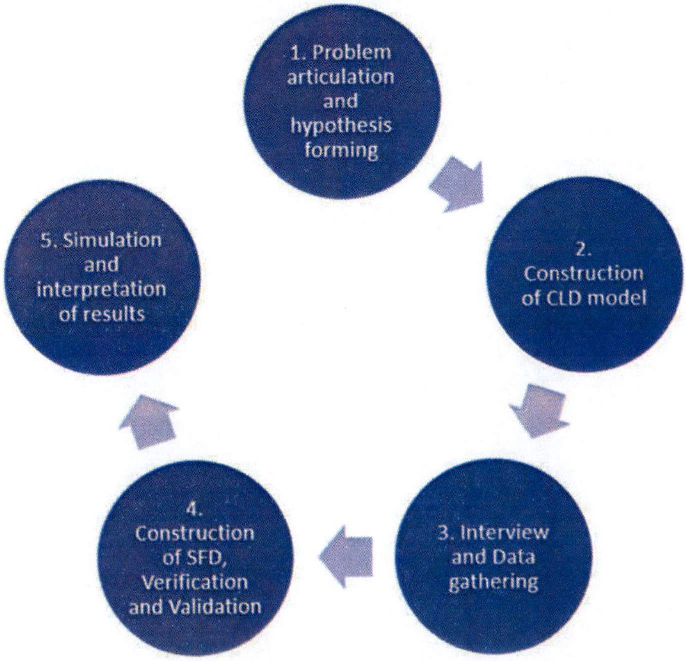

Fig. 1 Research process flow

4 Modelling and Validation

4.1 Backbone of Model

As system dynamic emphasizes on correlating factors, the aim and boundary of the model are first defined. The electricity market for generation is strongly influenced by electricity demand, reserve margin, energy gap, total GHG emission, and total energy needed. Figure 2 shows the backbone of the causal loop diagram (CLD), which includes three balancing loops, B1, B2, and B3. Figure 3 shows the complete CLD model, which acts as the base for conversion into Stock Flow Diagram (SFD). Figure 4 shows the subsystem for condition statement used to ensure that RE share is at least 20% and GHG emission to not exceed 3 million tCO_2 [16].

Generation subsystems in Fig. 5 calculates the reserve margin and energy gap using forecast demand as input. Emission subsystems in Fig. 6 calculates the GHG emission for both NON-RE and RE energy generation capacity.

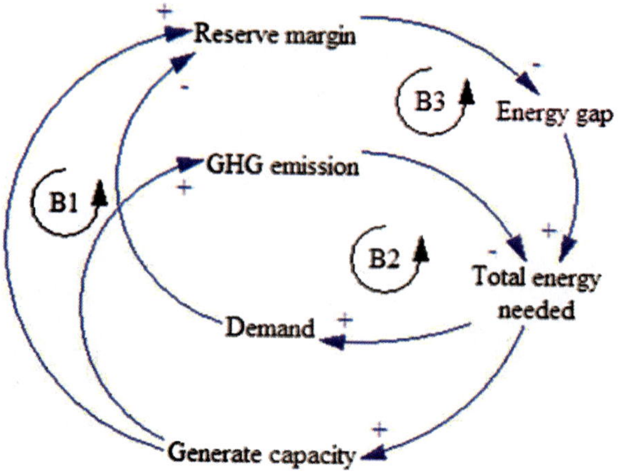

Fig. 2 Backbone of CLD model

5 Results and Discussion

5.1 Scenarios and Results

For this study, 4 scenarios were simulated. For each scenario, the duration of the simulation projects from 2010 to 2030. The scenarios used are business as usual (BASE), RE penetration of 5% and 15%, and both RE and Non-RE penetration of 5%. From Fig. 7, it shows that for the Non-RE & RE scenario, the reserve margin is the highest, followed by 5% RE penetration and the lowest is 15% RE penetration. Ideally, the reserve margin of electricity generation should be in the region of 20% of total generation. A higher percentage would result in over-production and causes wastage of resources and lower will result in low energy security (AV). Despite the over-production peak at 2016, the reserve margin gradually stabilizes and decreases after 2020. This can be attributed to increased new generation plants. For the total GHG emission rate, shown in Fig. 8, the 15% RE penetration scenario achieve the lowest emission rate, followed by 5% RE penetration, and RE & NON-RE scenario. Naturally, a high penetration of RE will have a relatively lower GHG emission.

Figure 9 shows that 5% RE penetration gives rise to the best efficiency followed by 15% RE penetration and Non-RE & RE scenario has the lowest efficiency due to the high reserve margin, where there is an over-production of electricity. The 3A framework [17] is implemented in the electricity sector in Malaysia for evaluating the energy security in term of availability(AV), applicability(AP) and acceptability(AC), and Table 1 shows that scenario 2 is the most favourable when these three factors are taken into account.

From the results obtained above, several key challenges are anticipated for the future of Sabah's energy market:

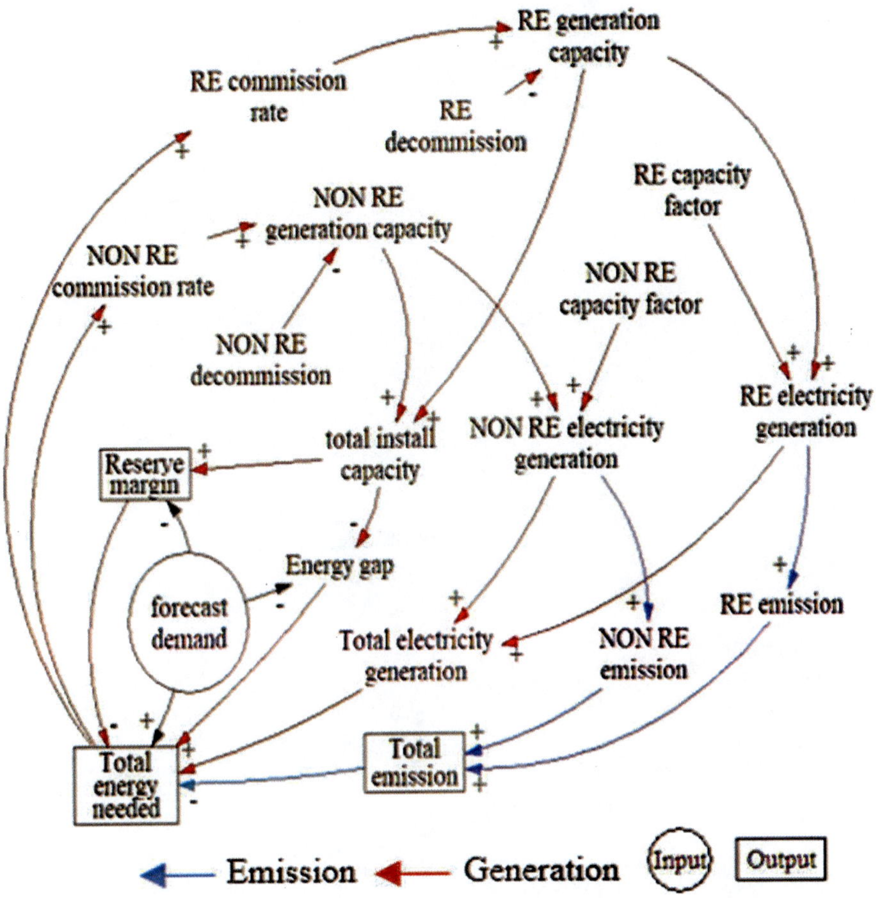

Fig. 3 Full CLD model

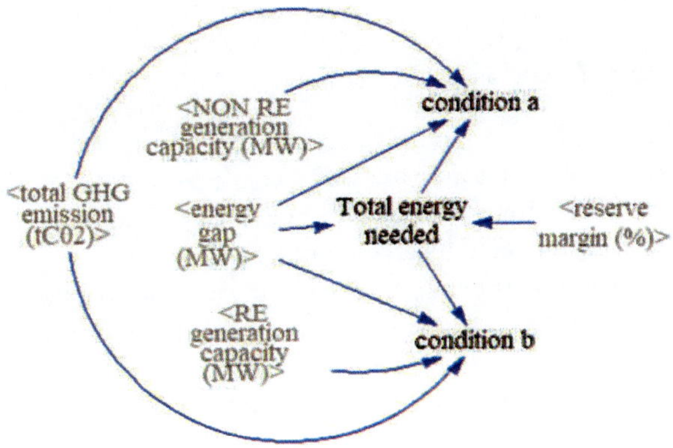

Fig. 4 Condition statement subsystem

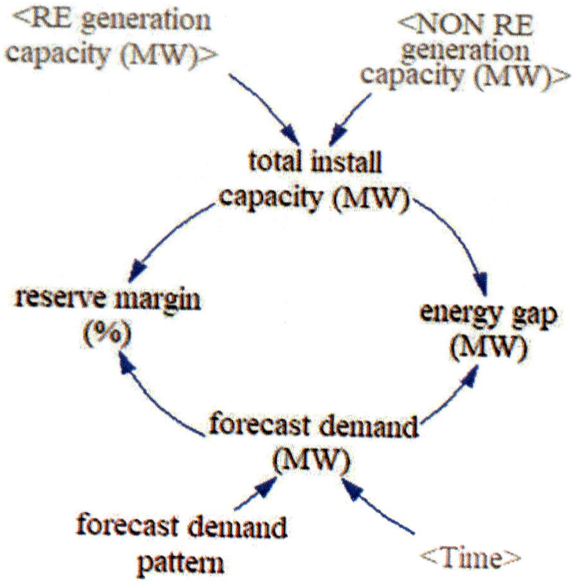

Fig. 5 Generation subsystems

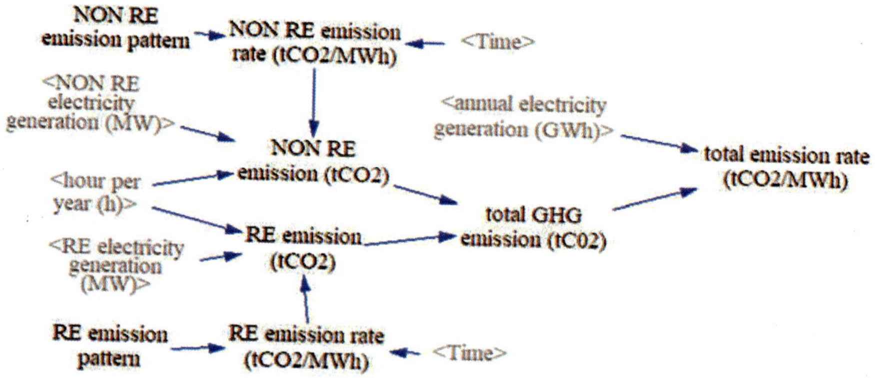

Fig. 6 GHG emission subsystems

1. Reserve margin stresses, where prediction of peak demand reaches 1,500 MW, translating into a growth of 3.7% per annum on the electricity generation but is accompanied by an electricity demand increase of 8–10% [18].
2. The investment choices mainly depend on the overall energy security policy, imported resources from Sarawak interconnection, environmental policy, and market power in EM.
3. The RE generation is still ineffective in terms of economic viability and technological efficiency, as the investments in RE do not match the rewards reaped.

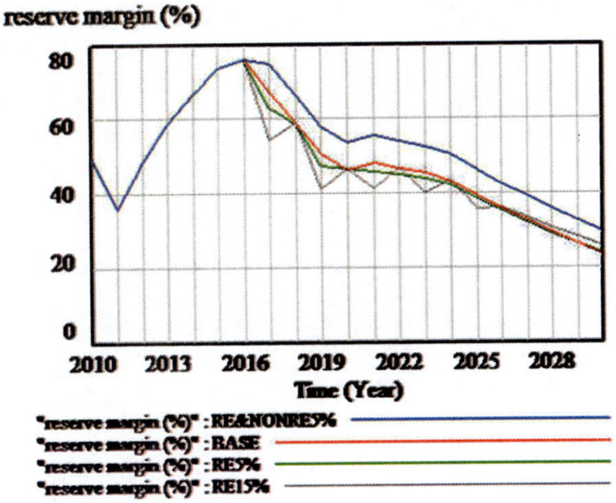

Fig. 7 Reserve margin (%)

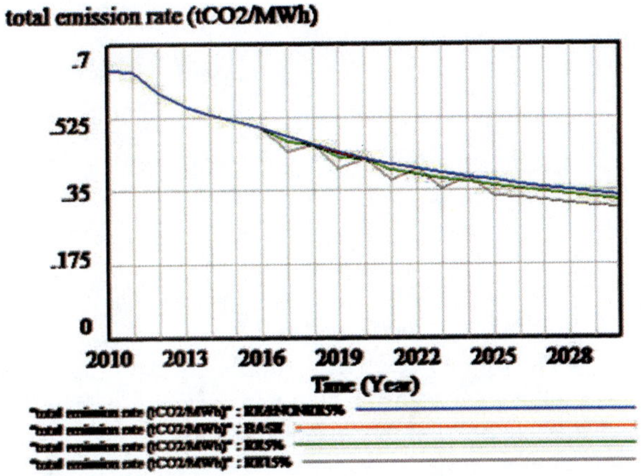

Fig. 8 Total GHG emission rate

To achieve a better energy security in transitioning to a deregulated electricity market with higher RE penetration, robust strategies are required to:

1. Introduce an appropriate regulatory framework. For example, feed-in-tariff (FiT) can act as a catalyst for the entry of RE generation and attract investment for RE powerplants. Apart from that, advocacy programmes can help to increase the awareness of all stakeholders on the benefits and advantages of utilising RE.
2. Creating a single buyer role in SESB organisational structure with continued monitoring of all generator performances to ensure sustained efficiency from aging powerplants, as well as to avoid over-capacity of electricity generation, which causes inefficiency.
3. Preparation of PPA for SESB power stations and IPPs to have agreement to facilitate the monitoring of operation efficiency.

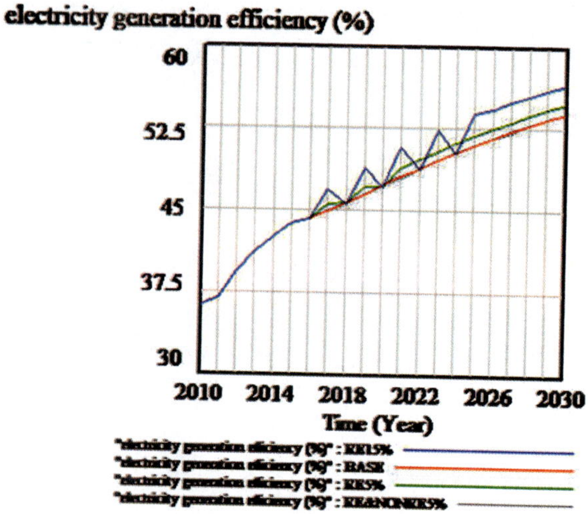

Fig. 9 Graph of electricity generation efficiency (%)

Table 1 Table of simulation results to 2030

2030	BASE	S1	S2	S3
Reserve margin (%)	23.17	23.20	23.51	29.30
Emission rate (tCO_2/MWh)	0.337	0.333	0.326	0.338
Efficiency	53.84	54.20	54.90	53.82

4. Determine and set efficiency targets for financial and technical performance of utilities concerned so as to facilitate the regulatory processes of the Commission in creating an electricity supply industry that is competitive and viable.
5. Enhance the transmission lines between east and west Sabah, as well as Sarawak interconnection to improve the energy security.
6. Capping of funding on the RE generation below 10 MW to ensure full utilisation of the limited fund for RE generation.

6 Conclusions and Future Work

In conclusion, a successful transition into a deregulated energy market for Sabah is unlikely without a suitable extent of RE penetration. This is necessary to complement existing fossil fuel generators, reduce GHG emissions, and increase the reliability of energy generation resources, as seen from the results. From the simulation, it was shown that the inclusion of 15% RE resources in the generation mix improves system performance, minizes dependence on fossil fuel, and reduces harmful emissions. This also results in increased system sustainability. This research recommends a focus on refining a reliable and suitable energy generation mix for Sabah's electricity market with the consideration of both GHG emissions and energy security.

References

1. Ford A (1995) System dynamics and the sustainable development of the electric power industry. Syst Dyn 95 I
2. Sterman JD (2000) Business dynamics: systems thinking and modeling for a complex World. McGraw-Hill
3. Tan CS, Maragatham K, Leong YP (2013) Electricity energy outlook in Malaysia. In: IOP Conference series: earth and environmental science, vol 16, pp 012126
4. McNish T, Kammen DM, Gutierrez B, Biographies Daniel Kammen AM (2010) Clean energy options for Sabah an analysis of resource availability and unit cost
5. Contacts A, Van Nee C, Ratings U, Poh A, Analyst S, Ratings U (2015) Malaysian power sector—energising a steady growth path
6. Database RP et al (2015) Malaysia (2012). No. 2012
7. Suruhanjaya Tenaga Energy Commission (2017) Peninsula Malaysia electricity supply outlook 2017
8. Energy Commision (2016) Peninsular Malaysia electricity supply industry outlook 2016, p 73
9. Suruhanjaya Tenaga Malaysia and Energy Commision (2014) Peninsular Malaysia electricity supply industry outlook 2014
10. Halabi LM, Mekhilef S, Olatomiwa L, Hazelton J (2017) Performance analysis of hybrid PV/diesel/battery system using homer: a case study Sabah, Malaysia. Energy Convers Manag 144:322–339
11. Suruhanjaya Tenaga Malaysia (2015) Maklumat Prestasi Dan Statistik Industri Pembekalan Elecktrik Di Malaysia
12. Ahmad S, Mat R, Muhammad-sukki F, Bakar A (2016) Application of system dynamics approach in electricity sector modeling.pdf. 56:29–37
13. Petitet M, Finon D, Janssen T (2017) Capacity adequacy in power markets facing energy transition: a comparison of scarcity pricing and capacity mechanism. Energy Policy 103:30–46
14. Momodu A, Oyebisi TO, Obilade TO (2012) Modelling the Nigeria's electric power system to evaluate its long-term performance. In: Proceedings of the 30th international conference of the system dynamics society, pp 1–31
15. Davies EG, Simonovic S (2009) Energy sector for the integrated system dynamics model for analyzing behaviour of the social-economic-climatic model
16. Kettha (2008) National Renewable energy policy and action plan—Malaysia. Nat Renew Energy Policy 90
17. Commission E (2017) Electricity market deregulation and energy security
18. Oh TH, Hasanuzzaman M, Selvaraj J, Teo SC, Chua SC (2017) Energy policy and alternative energy in Malaysia: issues and challenges for sustainable growth—an update. Renew Sustain Energy Rev 1–11

Rain Classification for Autonomous Vehicle Navigation Using Machine Learning

Abdul Haleem Habeeb Mohamed, Muhammad Aizzat Zakaria, Mohd Azraai Mohd Razman, Anwar P. P. Abdul Majeed, Mohamed Heerwan Bin Peeie, Choong Chun Sern, and Baarath Kunjunni

Abstract Autonomous vehicles (AV) has gained popularity in research and development in many countries due to the advancement of sensor technology that is used in the AV system. Despite that, sensing and perceiving in harsh weather conditions has been an issue in this modern sensor technology as it needs the ability to adapt to human behaviour in various situations. This paper aims to classify clear and rainy weather using a physical-based simulator to imitate the real-world environment which consists of roads, vehicles, and buildings. The real-world environment was constructed in a physical-based simulator to publish the data logging and testing using the ROS network. Point cloud data generated from LiDAR with a different frame of different weather are to be coupled with three machine learning models namely Naïve Bayes (NB), Random Forest (RF), and k-Nearest Neighbour (kNN) as classifiers. The preliminary analysis demonstrated that with the proposed methodology, the RF machine learning model attained a test classification accuracy (CA) of 99.9% on the test dataset, followed by kNN with a test CA of 99.4% and NB at 92.4%. Therefore, the proposed strategy has the potential to classify clear and rainy weather that provides objective-based judgement.

Keywords Rain modelling · Autonomous vehicle · Random forest · K-nearest neighbour · Naïve Bayes · Machine learning

---

A. H. Habeeb Mohamed · M. A. Zakaria (✉) · M. A. M. Razman · A. P. P. Abdul Majeed · C. C. Sern · B. Kunjunni
Innovative Manufacturing, Mechatronic and Sports Laboratory, Faculty of Manufacturing and Mechatronic Engineering Technology, Universiti Malaysia Pahang, Pekan, Malaysia
e-mail: maizzat@ump.edu.my

A. H. Habeeb Mohamed · M. A. Zakaria · M. H. B. Peeie · B. Kunjunni
Autonomous Vehicle Laboratory, Automotive Engineering Centre, Universiti Malaysia Pahang, Pekan, Malaysia

A. F. Ab. Nasir et al. (eds.), *Recent Trends in Mechatronics Towards Industry 4.0*, Lecture Notes in Electrical Engineering 730, https://doi.org/10.1007/978-981-33-4597-3_80

1 Introduction

The development of autonomous vehicles (AV) is gaining popularity nowadays due to the rapid advancement of sensor technology. Although current automobiles include various driver assistance systems, electronics, and communications, requirements of automation are creating a whole new level of demands [6]. The AV has acquired a recent interest among researchers worldwide. The primary goal is to reduce the number of human errors and traffic levels by giving the intelligent machine more freedom to function autonomously [1, 5].

AV development in Malaysia is not at par with other developed countries mainly because of the local climate that includes heavy rainfall thus making it challenging to establish the technology [7]. Malaysia is divided into Peninsular Malaysia and Borneo which experience mean annual one minute rainfall rates that are higher in the latter, with values between 84.7 and 153.9 mmh-1 for 0.01% exceedance, and between 81.8 and 143.8 mmh-1 in the former [8].

Currently, there is a lack of studies that test and validate sensors in the Malaysian climate. One of the many difficulties involved in developing an autonomous vehicle system is the sensing and perceiving in adverse weather conditions, where the rain becomes the ultimate obstacle to the progress of driverless technology [4]. This becomes a more serious problem when the software created to drive an automated car requires the ability to adapt to various outdoor conditions like that of humans [6].

Light detection and ranging (LiDAR) system is one of the important systems for autonomous vehicles [3]. It is a sensor that utilises reflected light to measure the geometry of the environment near the sensor and uses a system based on time of flight (TOF) principle [4]. It is essential in scanning the outdoor environment which generates large scale point clouds at one per second in high resolution and 3D maps [10].

However, it is challenging to accurately operate and test the performance of a sensor in rainy conditions because of the unpredictable weather and its randomness. Therefore, this study aims to simulate an accurate real-world environment and evaluate the classifying ability of various machine learning models based on LiDAR data.

2 Methodology

Figure 1 illustrates the overall flow of the investigation in a Linux-based operating system, where the open-source physically based simulator Carla created scenarios based on the real-world environment [2]. LiDAR data was gathered using ros-bridge to communicate with ROS as Carla controls the dynamic changes of the simulator.

As shown in Fig. 2a, b, the environment simulated by Carla imitates the real-world environment accompanied by LiDAR that was mounted on top of the vehicle in a static position. The data was then published using a ros-bridge to communicate with

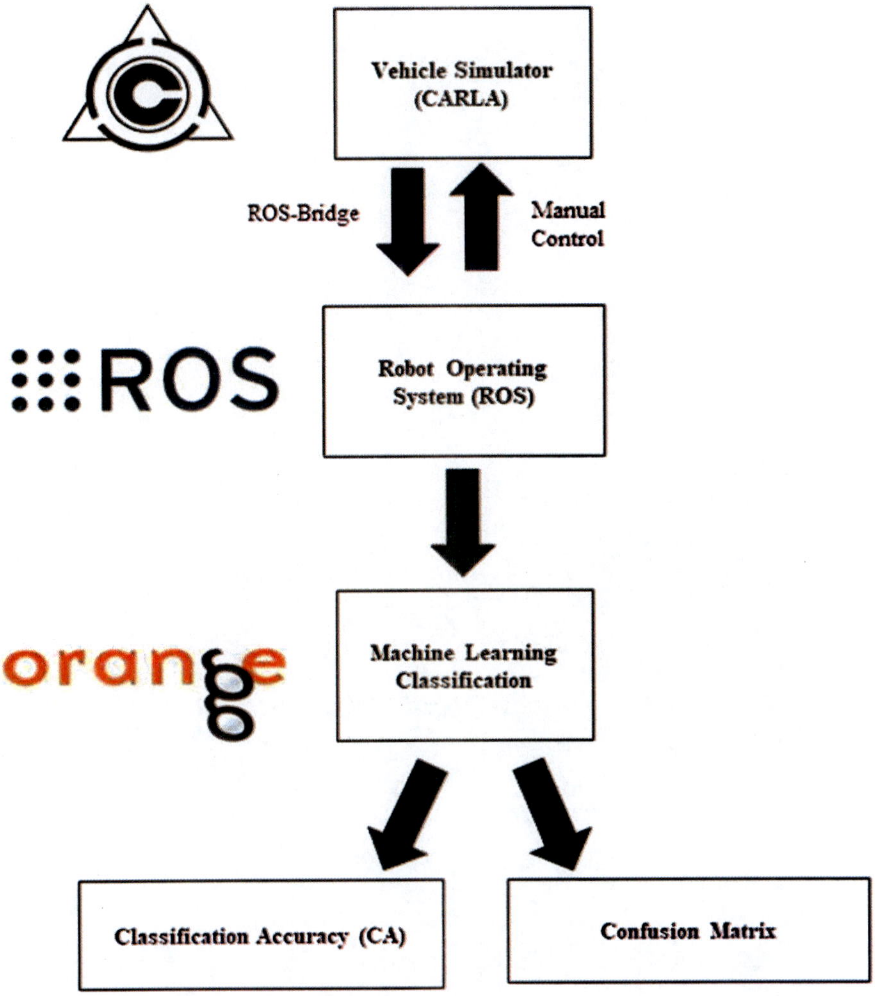

Fig. 1 Block diagram for overall methodology

the ROS environment so that the data can be subscribed by it in a point cloud as exhibited in Fig. 2. The parameters that are to be set in Carla are detailed in Table 1 for daytime sets, Table 2 for weather changes and Table 3 for LiDAR parameters.

ROS was used as a data logger, where it gathered the LiDAR data from Carla using the ros-bridge pipeline to publish the data. The data is then recorded in *bag format. The recorded raw data of LiDAR was used as the features in Orange for machine learning model classification which utilised the point cloud data location and the intensity (X, Y, Z, I). Using the features selected, it is then to be run with a machine learning models including Naïve Bayes, Random Forest and k-Nearest Neighbor as classifiers which possess the best algorithm for reduced dimensionality

a

Fig. 2 **a** The Carla test scenario at daytime in clear weather and point cloud data. LiDAR displayed using ROS, respectively. **b** The Carla test scenario at daytime in rainy weather and point cloud data. LiDAR displayed using ROS, respectively

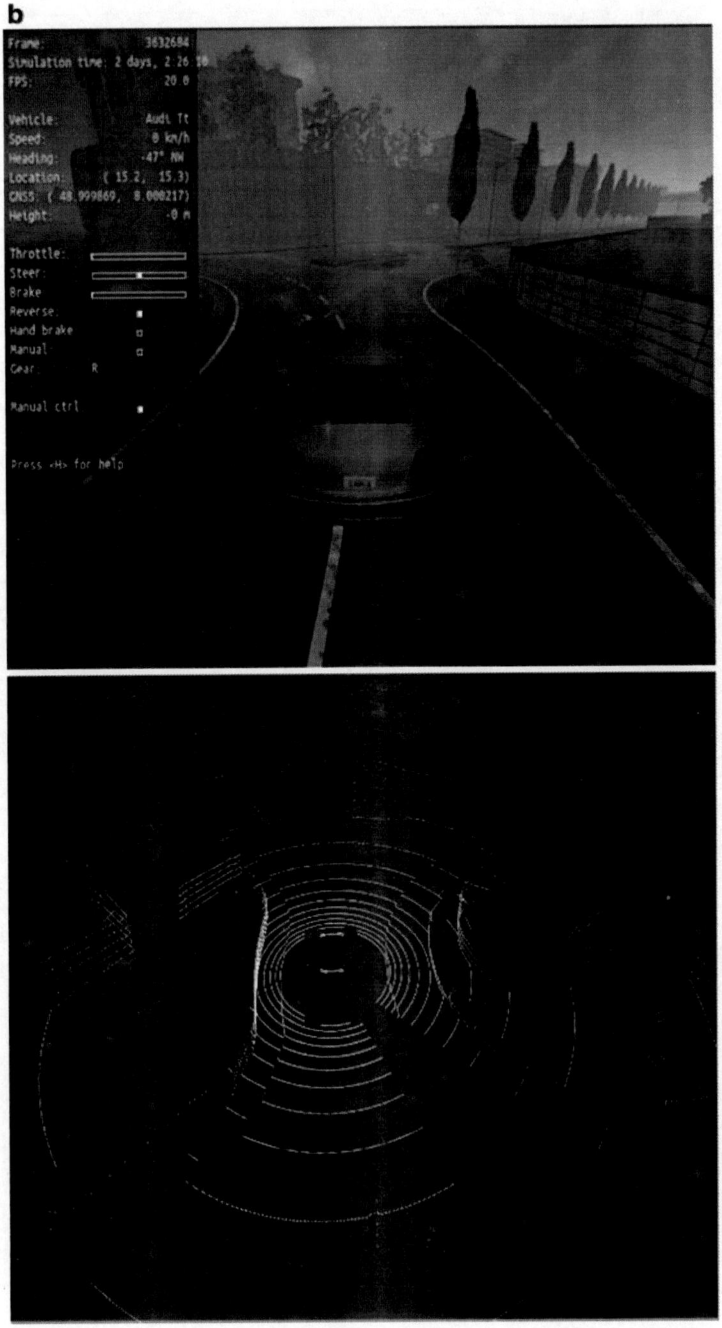

Fig. 2 (continued)

Table 1 Daytime parameters

Sun presets	Day
Altitude angle	60.0°
Azimuth angle	0.0°

Table 2 Clear and raining weather parameters

Weather preset	Clear	Rain
Cloudiness	10.0	100.0
Rain precipitation	0.0	80.0
Puddles precipitation	0.0	90.0
Wind intensity	5.0	100.0
Fog density	0.0	20.0
Fog distance	0.0	0.0
Wetness	0.0	100.0

Table 3 LiDAR parameters

Parameter	Value
Channels	16
Opening angle	15°
Range	150 m
Upper field of view	2.0°
Lower field of view	30.0°
Rotation Frequency	20 Hz

and multisource of datasets [9, 10]. Table 4 shows the generated LiDAR data sample. For this investigation, three frames from clear and rainy weather with a total of 40,960 point cloud were grouped to the location of point cloud with the intensity of 10,420 datasets. A total of 70% of the dataset data was used to train the machine learning models while the remaining 30% was used to test its accuracy.

Table 4 Data samples generated from LiDAR

Frame per second	20
Tick time (s)	0.05
Point cloud per frame	7500
Point cloud per second (s)	150,000

3 Results and Discussions

Two types of weather were set to be simulated by Carla which are clear and rainy weather in the daytime and 3 frames of data were taken for each weather. In this study, all the features listed in the previous section were used to develop the model, where 70% of the dataset's data was utilised to train the machine learning models. From Fig. 3, the NB model achieved the lowest classification accuracy of 92.5% while RF charted the highest classification accuracy at 99.7% and kNN recorded 99.5% of classification accuracy with a slight decimal difference from RF. From the results gathered, it was evident that the ML models performed well due to their inherent nature in understanding the data provided. The trained models were then tested with the remaining 30% of datasets data to further observe their accuracy. For the classification accuracy, NB achieved the lowest classification accuracy of 92.4%, RF had the best classification accuracy of 99.9% while kNN recorded 99.4% of classification accuracy. RF model demonstrated high potential as a good Machine learning model since its test accuracy was higher than the train data accuracy.

Further inspection was carried out on the confusion matrix of the machine learning models. Based on Fig. 4, misclassification recorded by the NB confusion matrix in clear weather was 53 and rainy weather was 180. In Fig. 5, the misclassification of RF confusion matrix was only detected from twice rainy weather thus making it the best classifier. Meanwhile, the misclassification recorded by kNN confusion matrix in clear weather was 8 and 9 in rainy weather as detailed in Fig. 6. The phenomenon of misclassification may be improved by increasing the data collection, evaluating the features selected for the classification accuracy, and optimising the hyperparameter of the developed model.

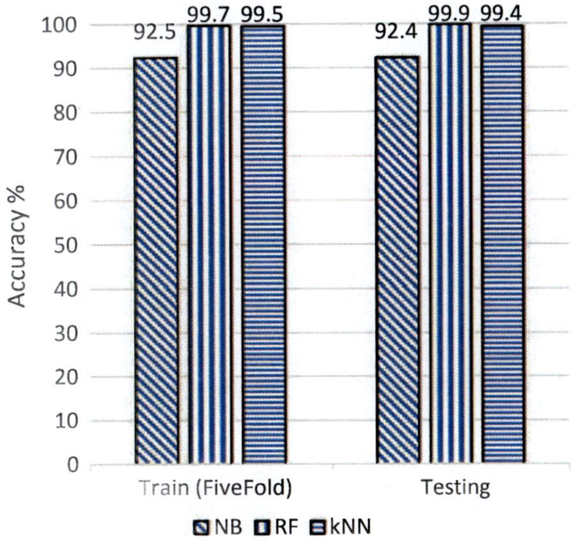

Fig. 3 Efficacy of developed ML models

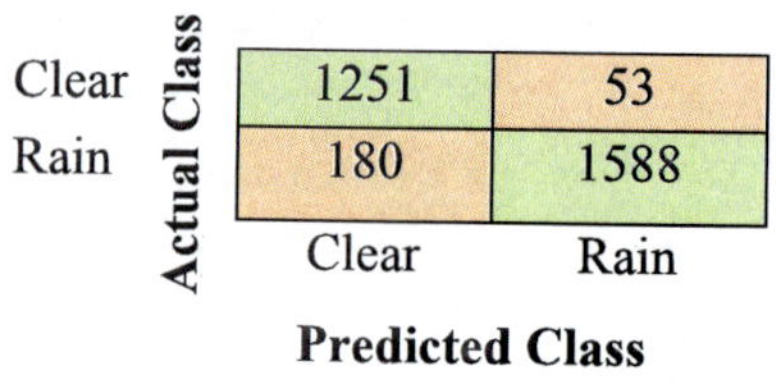

Fig. 4 NB confusion matrix

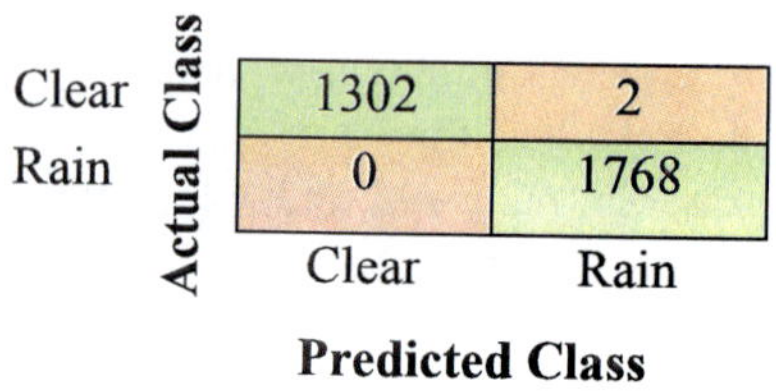

Fig. 5 RF confusion matrix

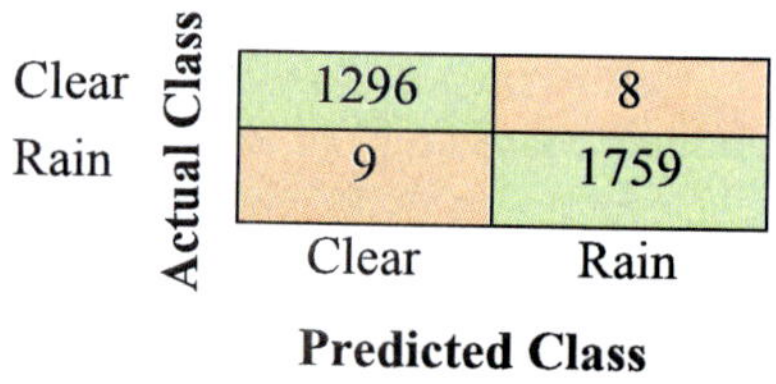

Fig. 6 kNN confusion matrix

4 Conclusion

In this preliminary investigation, it was found that the rain classification system that was developed can provide a reasonable classification accuracy of the evaluated clear and rainy weather by utilising the selected features and machine learning models. Further studies can be carried out by including the sun reflection angle and the reflectivity of different materials to imitate the real-world environment noises on the best machine learning model which in this case was the Random Forest (RF) that achieved 99.9% of classification accuracy. The preliminary results obtained suggest encouraging objective-based judgement when using LIDAR that measures the reliability of the sensor in the presence of rain.

Acknowledgements The author would like to thank Ministry of Higher Education Malaysia (KPT) and Universiti Malaysia Pahang (www.ump.edu.my) for financial supports given under **FRGS/1/2018/TK08/UMP/02/1** and **RDU1903139**. The authors also thank the research team from Autonomous Vehicle Laboratory AEC, Innovative Manufacturing, Mechatronics and Sport Laboratory (iMAMS); who provided insight and expertise that greatly assisted in the present research work.

References

1. Bimbraw K (2015) Autonomous cars: past, present and future: a review of the developments in the last century, the present scenario and the expected future of autonomous vehicle technology. In: ICINCO 2015—12th international conference on informatics in control, automation and robotics, proceedings, 1(August), pp 191–198. https://doi.org/10.5220/0005540501910198
2. Dosovitskiy A, Ros G, Codevilla F, Lopez A, Koltun V (2017) CARLA: an open urban driving simulator (CoRL), pp 1–16. Retrieved from http://arxiv.org/abs/1711.03938
3. Filgueira A, González-Jorge H, Lagüela S, Díaz-Vilariño L, Arias P (2017) Quantifying the influence of rain in LiDAR performance. Measur J Int Measur Confederation 95:143–148. https://doi.org/10.1016/j.measurement.2016.10.009
4. Goodin C, Carruth D, Doude M, Hudson C (2019) Predicting the influence of rain on LIDAR in ADAS. Electronics 8(1):89. https://doi.org/10.3390/electronics8010089
5. Hamid UZA, Zakuan FRA, Zulkepli KA, Azmi MZ, Zamzuri H, Rahman MAA, Zakaria MA (2017) Autonomous emergency braking system with potential field risk assessment for frontal collision mitigation. In: 2017 IEEE conference on systems, process and control (ICSPC) (December), pp 71–76. https://doi.org/10.1109/SPC.2017.8313024
6. Kutila M, Pyykönen P, Ritter W, Sawade O, Schäufele B (2016) Automotive LIDAR sensor development scenarios for harsh weather conditions, pp 265–270. https://doi.org/10.1109/ITSC.2016.7795565
7. Kwan MS, Tanggang FT, Juneng L (2011) Projected changes in future climate extremes in Malaysia. Nat Symp Clim Change Adapt 42(8):1051–1058
8. Omotosho TV, Mandeep JS, Abdullah M, Adediji AT (2013) Distribution of one-minute rain rate in Malaysia derived from TRMM satellite data, pp 2013–2022. https://doi.org/10.5194/angeo-31-2013-2013
9. Sai Tarun GB, Sriram JV, Sairam K, Sreenivas KT, Santhi MVBT (2019) Rainfall prediction using machine learning techniques. Int J Innovative Technol Exploring Eng 8(7):957–963. https://doi.org/10.13140/RG.2.2.26691.04648
10. Teri SS, Musliman IA (2019) Machine learning in big lidar data: a review. Int Arch Photogrammetry Remote Sens Spat Inf Sci—ISPRS Arch 42(4/W16):641–644. https://doi.org/10.5194/isprs-archives-XLII-4-W16-641-2019

Development of an Innovative Inferior Alveolar Nerve Block (IANB) Simulator Kit with Data Visualization and Internet of Things (IoT)

S. Z. Zainudin, M. H. M. Ramli, T. I. B. T. Jamaluddin, and S. A. Abdullah

Abstract This paper presents the continuous work of an improved version of the Local Anaesthesia Simulator Kit (LASK), an inferior alveolar nerve block (IANB) clinical simulator kit for simulation-based training. The main objective of this project is to facilitate anaesthesia administration via digital sensing mechanism and training session in a virtual environment. The implementation of internet of things (IoT) provides wireless machine to machine communication to transmit real-time sensor data into web application (web-app) and data analysis purpose. To ensure seamless data transmission, a dedicated wireless protocol namely web-socket protocol is used for two-way communication between the simulator kit and the developed web-app. The developed web-app provides access of real-time training interface and a dashboard for displaying the current training score including overall performance trend. This real-time training interface also displays graphical statistical data analysis as well as 3D data visualization. In this regard, an open-sourced 3D rendering module is integrated into the web-app to enhance the clinical training sensation.

Keywords IANB simulator kit · Local anaesthesia simulator kit (LASK) · Data visualization · Simulation-based learning · Clinical simulation

S. Z. Zainudin · M. H. M. Ramli (✉) · S. A. Abdullah
Faculty of Mechanical Engineering, Universiti Teknologi MARA Selangor, Shah Alam, Malaysia
e-mail: haniframli@uitm.edu.my

T. I. B. T. Jamaluddin
Faculty of Dentistry, Universiti Teknologi MARA Selangor, Shah Alam, Malaysia

A. F. Ab. Nasir et al. (eds.), *Recent Trends in Mechatronics Towards Industry 4.0*, Lecture Notes in Electrical Engineering 730, https://doi.org/10.1007/978-981-33-4597-3_82

1 Introduction

In the dental profession, local anaesthesia is an indispensable part of dental treatment to reduce or eliminate pain for patients. One of the most frequent techniques used in achieving this is via performing an inferior alveolar nerve block (IANB) which targets the inferior alveolar nerve as shown in Fig. 1 [1].

Techniques in performing local anaesthesia (injection) is incorporated in worldwide dental curriculum and compulsory for undergraduate students to train and master so that they will become competent prior to treating patients in their clinical years. Clinical skills and competency are crucial for these undergraduates as they are required to perform clinical procedures in their clinical years in order to be certified as competent clinicians upon graduation. These undergraduates are required to learn the theoretical part in their pre-clinical years with pre-clinical simulation training. In practice, these undergraduates will conduct clinical training which requires them to perform their first injection by performing local anaesthesia to a fellow student, a conventional method used in worldwide dental curriculum [2, 3].

Nowadays, many dental institutions adopt and use enhanced simulation-based learning to further provide undergraduates with some set of skills before engaging in clinical procedures that involve human interaction. There is a sense of maturity in usage of simulation for dental education as technology and IoT is quite reliable and is received well among undergraduates [4]. Simulation based learning such as virtual patients, interactive mannequin simulators and computer simulation offer users with knowledge and experience by replicating actual procedures [5]. In Malaysia, amendments to the New Dental Officer Programme (NDOP), Ministry of Health Malaysia (2016) requires a more effective and innovative clinical simulation training at undergraduate level, especially for specific clinical skills such as performing a mandibular

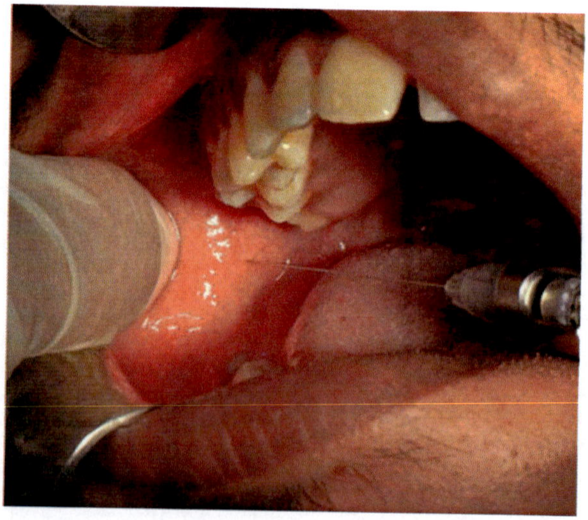

Fig. 1 Illustration of dentist performing an IANB (injection) on a patient

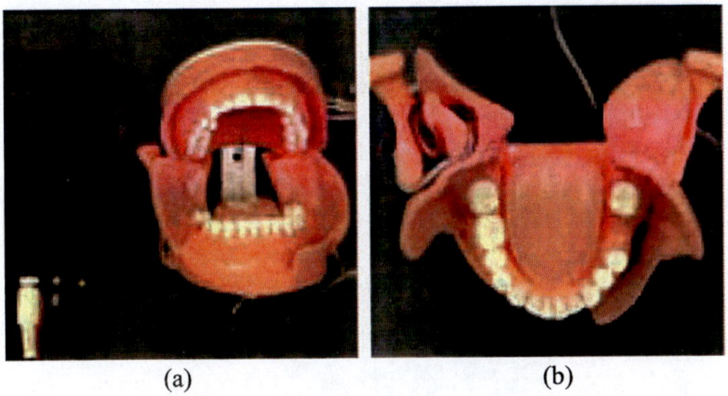

Fig. 2 Illustration of physical upper and lower jaw model with feedback system via buzzer and lighting device. Lower jaw seen in (**b**). *Source* Adapted from [6]. **a** Front view: buzzer and lighting device on the left side. **b** The electrode for sensing mechanism of injection

regional anesthesia (IANB). This has resulted in the introduction of further innovative methods for simulation-based learning to provide undergraduates with sufficient experience prior to their first injection. Here, we examine some of the related research on this subject. Most basic training kit for simulation training is using physical lower jaw (mandibular) model. Improvements have been made since 25 years back in upgrading existing simulator kits by providing feedback such as buzzing sound and lighting as shown in Fig. 2 [6].

Simulation-based training has been improved by implementation of newer techniques and technologies such as augmented reality (AR) and virtual reality (VR) as shown in Figs. 3 and 4 respectively [7, 8].

Advancement of technology should be utilized in order to improve simulation-based training. This motivates the need for an alternative approach for simulation-based training to provide real-time data feedback and an interesting environment to end-users. We also explore available technologies to be applied for the simulation-based learning to maximize users' experience (UX) gained by both end-users; dental undergraduates and dental educators.

Based on the above findings and research works, a new IANB simulator kit – LASK (Local Anaesthesia Simulator Kit) was developed to cater and expedite skill mastering among dental undergraduates. First, a lower jaw (mandible) block is materialized by using Digital Imaging and Communication in Medicine (DICOM) data rendered and converted into 3D model as shown in Fig. 5. Next, this lower jaw 3D model is printed and fused with electronic parts to provide real-time data feedback as shown in Fig. 6 [9].

In this present work, we extend our previous work by further enhancement of the LASK with integration of IoT as well as establishment of 3D and statistical data visualization environment. The work can be broadly separated into two parts, the first is improvement of the physical simulator kit, the second is the development of

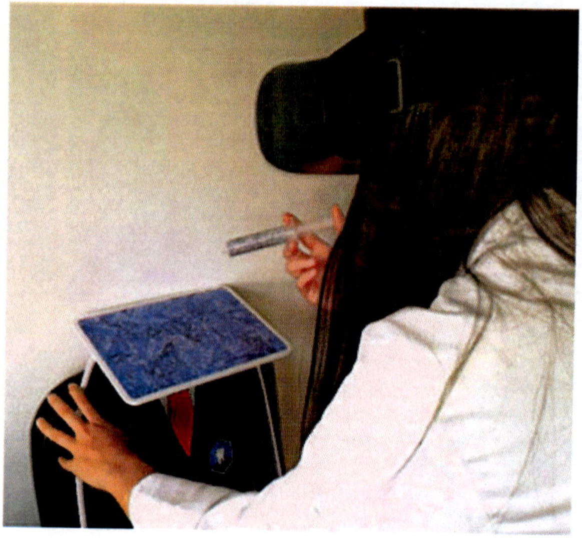

Fig. 3 Illustration of a person conducting local anaesthesia training using AR mode with VR goggles on a plastic model. *Source* Adapted from [7]

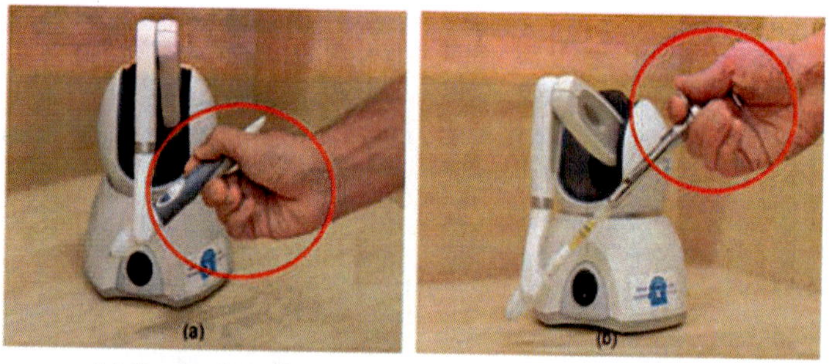

(a) Device handler (b) Carpule syringe

Fig. 4 Illustration of device handler for the VR system. **a** Original stylus. **b** Carpule syringe. *Source* Adapted from [8]

the software for the simulator kit. In this regard, the software part consists of web application deployed on a web server.

2 Improvement of the Simulator Kit

This section discusses the improvement in development of the LASK. In this regard, the aim is to develop a modular simulator kit, so later on, it can be mounted to a

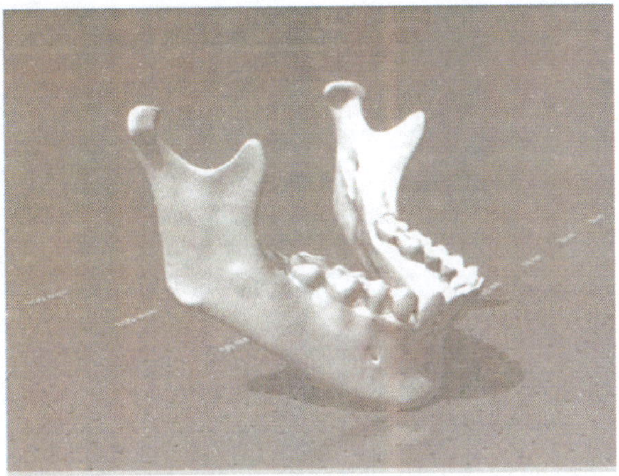

Fig. 5 3D model of lower jaw rendered using DICOM data

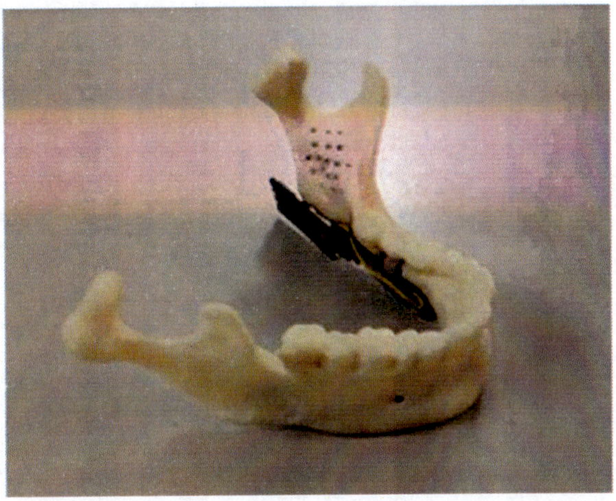

Fig. 6 Illustration of 3D printed lower jaw model fused with electronic parts

readily available mannequin in dental schools to improve the learning experience by replicating the real physical procedure of performing an IANB. Apart from that, improvement on the sensing mechanism is required to solve the arising issues from the previous prototype, where the sensor could not cover all the targeted areas for injection due to limited sensing nodes in the previous sensing module. As a result, some areas are left undetected eventhough injections are applied to that areas. To overcome this problem, a circular shape capacitive sensing module is used to cover the targeted areas of IANB. This sensing module also uses less input ports (3 input ports) compared to the previous module (10 input ports) and ultimately enhances the processing time. Thus, greatly improves the responsiveness of the simulator kit. As can be seen in Fig. 7, the area covered by both sensing modules are quite similar but

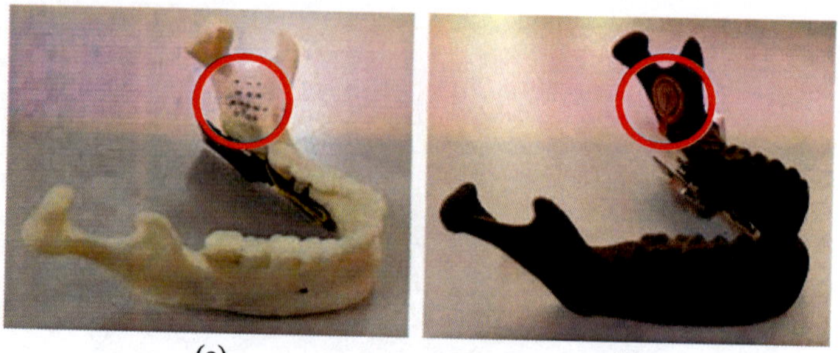

Fig. 7 Comparison between the previous sensing and the current sensing modules. a Previous simulator kit with 10 points of sensing nodes. b Latest simulator kit with circular sensing nodes

the latter sensor is more effective as it only maps 3 points rather than 10 points in the previous sensor.

The implementation of IoT enables machine to machine communication wirelessly and eliminates the need of data cable to transfer the sensor data. However, it is quite challenging to handle wireless data transmissions, after a series of testing and fine-tuning, a radio transceiver: nRF24L01 is found to be optimal to kit and is adopted to transmit the sensor data via a dongle receiver with Wi-Fi module. And then data is transferred continuously using socket.io library and web-socket protocol to a dedicated server. This protocol provides a continuous two-way communication between the LASK and the web server, and therefore facilitates real-time data transmission between a server, a web application and a database concurrently. This process enables the web application to seamlessly display real-time analyzed data from the database. The overall process flow of this system is shown in Fig. 8.

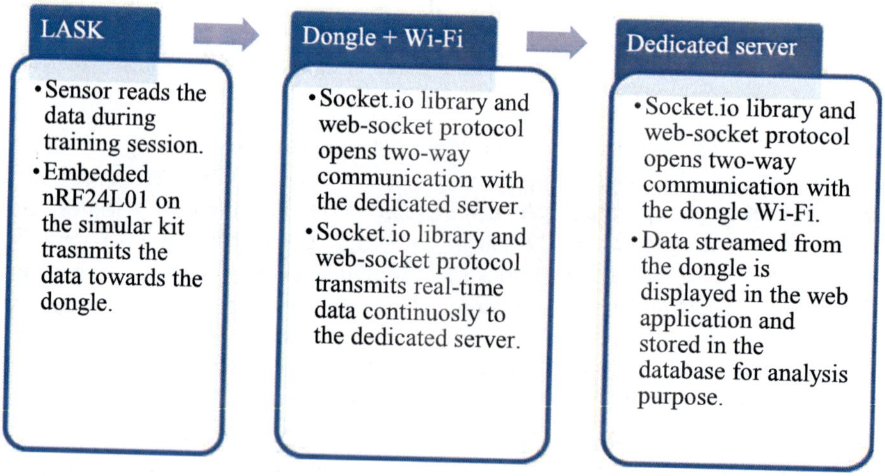

Fig. 8 Flow process of the data transmission

In addition, a standalone power supply is embedded to the new LASK, that uses lithium polymer battery to run the kit which promotes portability and eliminates the use of additional cables. The new kit also received new circuit layouts to enhance the processing time, responsiveness, and better stability when in operation in longer periods of time. The key requirements for software development is taken into consideration at this stage. The software should be able to be accessed by end-users via personal devices such as laptop or smartphone to ease the usage of the LASK. In this regard, the development of web application is the most feasible choice because of abundant resources including open-sourced JavaScript library and many other web development tools such as Vue.js and Nuxt.js.

3 Development of Web Application

This section discusses the development of web application. The web application is designed for the undergraduates to login to their account and view the overall performance of their training sessions as well as real-time training dashboard to monitor their current scores and performance during a particular real-time training session, while dental educators are able to login to observe concurrently or at a later time of their students' performances in pre-clinical simulation training. Figure 9 shows the illustration of the login page of the web-app.

The real-time training dashboard is designed with typical data visualization environment of performing an IANB including pressure data, location of injection data, and success rates in graphical forms. To virtually simulate what is going on while performing an IANB, a 3D model of IANB is embedded to the web-app with an animation-based on orientation of the needle and the injected locations performed

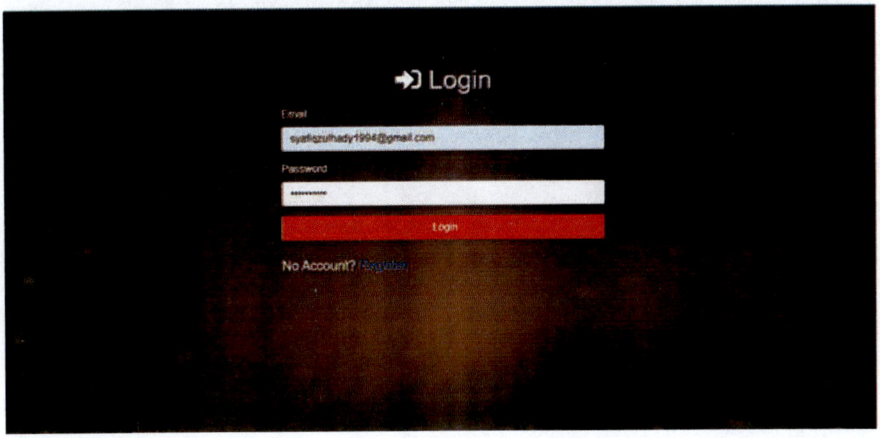

Fig. 9 Illustration of login page of the web application for the LASK

by end-user to promote an enhanced clinical simulation as if they are performing IANB on real patients.

3.1 3D IANB Visualization

In this regard, JavaScript library is used in the web browser to display the 3D graphics. Firstly, the lower jaw model in STL format is converted to gltf format in which it is supported by JavaScript 3D rendering module Three.js. This conversion process requires a conversion software package namely Blender software. The orientation of the model is fixed, and loaded into web-application (using gltf format) using Three.js rendering module as shown in Fig. 10.

Visualization environment displays the sensor data that is mapped onto the lower jaw model in 3D graphical representation with animation. By using this concept, proper design is made to display the visualization environment with graphical data analysis using another JavaScript library such as Chart.js. Figure 11 shows the illustration of the completed interface for training sessions with data visualization.

This will enhance user experience from both end-users as students can easily monitor their performance for each training sessions by referring to the interface, and dental educators can assess their students' pre-clinical skills. Responsive data is shown when injection is made during the training sessions. Figure 12 shows the zoom in of the virtual space on the interface indicating the position of injection during a training session. Apart from real-time data, all data obtained from each training session is stored in a database for further analysis.

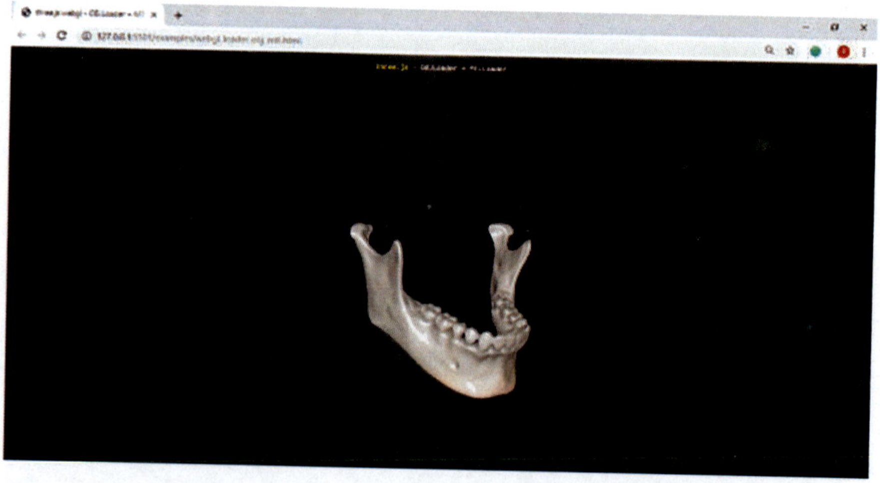

Fig. 10 Illustration of lower jaw model in visualization environment through a web browser

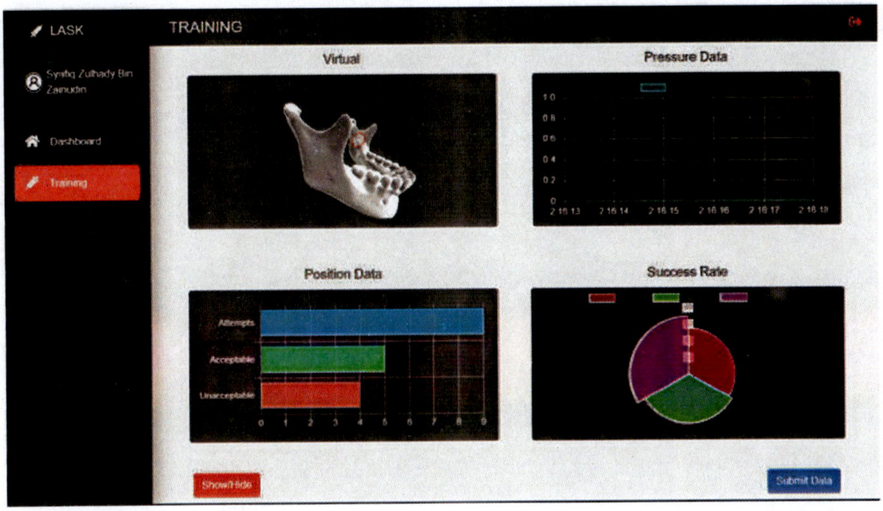

Fig. 11 Illustration of real-time training dashboard interface

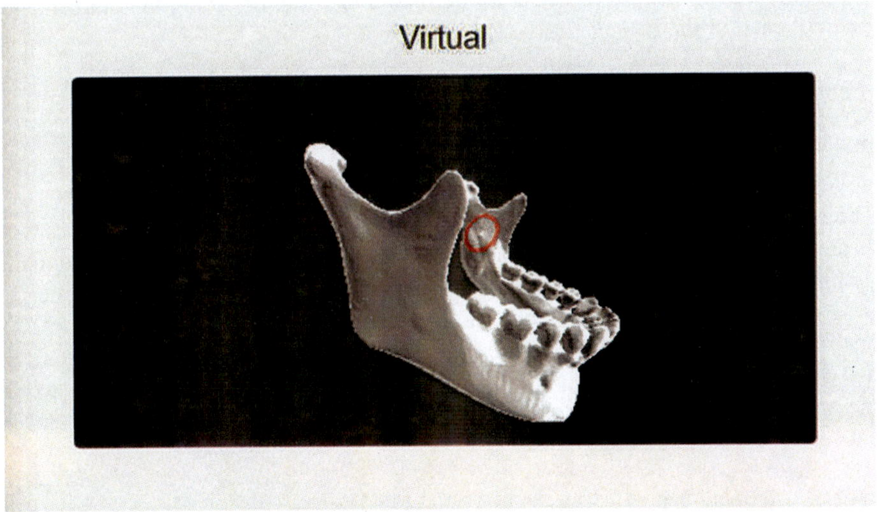

Fig. 12 Illustration of zoom in at the virtual space in the real-time training dashboard

3.2 Database Management System

In this regard, a dedicated server of MySQL database, which is a relational type database is used. This configuration contributes to easier data management and analytic purpose. The data stored in the database is queried by the web-app to display the performance trends based on analysis made for each student when login to his

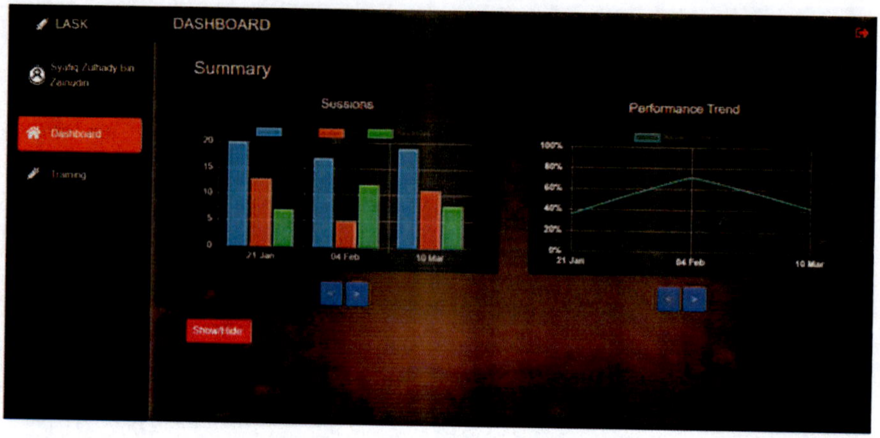

Fig. 13 Illustration of dashboard tab interface showing summary of each training session of the students

account. The data analysis is depicted as a summary in the dashboard tab as shown in Fig. 13. Assessment for the students could be done by monitoring the performance trend of each of them.

4 Conclusion

The innovative LASK for IANB training provides physical training method as well as an interface showing real-time data including data visualization environment and graphical data analysis which offers better UX to the undergraduates. Utilization of technologies for data management such as internet of thing and web-socket protocol enables continuous data transfer. Hence, providing real-time data transfer of analytic and displaying purpose. Web application provides flexibility for end-users to access the software using laptop or smartphone, therefore offering better UX for dental educators and dental undergraduates.

References

1. Kim C, Hwang K, Park C (2018) Local anesthesia for mandibular third molar extraction. J Dent Anesth Pain Med 18(5):287–294
2. Hossaini M (2011) Teaching local anesthesia in dental schools: opinions about the student-to-student administration model. J Dent Educ 75(9):1263–1269
3. Balamanikandasrinivasan C (2014) "Dental students" perception and anxiety levels during their first local anesthetic injection. 21(6):45–51
4. Pawar AM, Anushka T (2017) Simulation in dentistry. Ec Dent Sci 12(3):115–121

5. Burden A, Pukenas EW (2018) Use of simulation in performance improvement. Anesthesiol Clin 36(1):63–74
6. Masura K (2017) The development of a simulator from 1992 to 2016 for inferior alveolar nerve block injection and skill education for delivering local injections of dental anesthetics. J Dent Soc Hokkaido Med Univ 36(1):1–22
7. Won Y-J, Kang S-H (2017) Application of augmented reality for inferior alveolar nerve block anesthesia: a technical note. J Dent Anesth Pain Med 17(2):129–134
8. Tori R (2017) Virtual reality simulator for dental anesthesia training in the inferior alveolar nerve block. 25(4):357–366
9. Zainudin SZ, Mohd Ramli MH, Tengku Jamaluddin TIB, Abdul Wahab H, Husin A, Abdullah SA (2019) The development of sensing architecture for inferior alveolar nerve block clinical simulator kit the development of sensing architecture for Inferior Alveolar Nerve Block clinical simulator kit. In: IOP conference series: materials science and engineering, vol 637

Optimization of CNG Direct Injector Parameters Using Model-Based Calibration Framework

Mohamad Hafidzul Rahman Alias, Mohd Fadzil Abdul Rahim, and Rosli Abu Bakar

Abstract This paper presents an optimization study conducted based on a direct injector model running on compressed natural gas. The purpose of the study is to identify the optimal setup for the selected input parameters which deliver maximum injection quantity at the lowest solenoid current. The optimized injector input parameters were the injection pressure, injection duration, and input voltage. The optimization study was conducted using MATLAB's Simulink, Model-Based Calibration (MBC) Toolbox and injector test rig. The optimization data is generated by a validated, zero-dimensional, first principle injector model. The optimize calibration results were implemented in the injector experiment for verification. It was found that the simulation result of the mass flow rate for baseline versus optimization shows an increment of 15.64%. In comparison, the experimental result for baseline versus optimization shows an increase of 35.79%. Additionally, a comparison between the baseline work for simulation versus experiment produced RMSE of 0.2467 while the optimization work for simulation versus the experiment provides an RMSE value of 0.1860.

Keywords Compressed natural gas · Direct injection · Model-based calibration · Optimization

1 Introduction

Compressed natural gas (CNG) has already been regarded as the most suitable fuel to succeed gasoline and diesel. CNG fuel price is 20–40% lower than conventional fuel [1] with a massive reserve and produces cleaner combustion [2]. It also has higher

M. H. R. Alias · M. F. A. Rahim (✉) · R. A. Bakar
Faculty of Mechanical Engineering, Universiti Malaysia Pahang, Pekan, Pahang, Malaysia
e-mail: mfadzil@ump.edu.my

M. F. A. Rahim
Innovative Manufacturing and Mechatronics and Sports (IMAMS) Laboratory, Faculty of Manufacturing Engineering, Universiti Malaysia Pahang, Pekan, Pahang, Malaysia

A. F. Ab. Nasir et al. (eds.), *Recent Trends in Mechatronics Towards Industry 4.0*, Lecture Notes in Electrical Engineering 730, https://doi.org/10.1007/978-981-33-4597-3_83

thermal efficiency, higher knock resistance [3, 4], and higher octane number [5, 6]. It allows the fuel to be operated at lean and stoichiometric conditions [7].

The technology of direct injection (DI) can resolve the engine's power, emission and fuel consumption problem when using CNG as fuel. DI system boosts the engine's volumetric efficiency [2, 8], reduce the need for throttling control [6], and lower the pumping and heat transfer losses, which leads to low fuel consumption [7]. It enhances the injection strategy, which accurately manages the control of fuel injection, such as injection timing and the injection duration [9]. The latest gasoline direct injection (GDI) system can be operated at 80–200 bar of pressure based upon its operating setting [8]. GDI engines can run on both homogeneous stoichiometric charge and stratified charge [8]. During homogeneous charge mode, the spray-induced flow might increase combustion instability. The uniformities of fuel blend can be deteriorated because of the reduction in the period for the air-fuel mixture. In the stratified charge mode, to have better controllability over fuel injection, the DI system required a very high fuel pressure for injector operation which is a complex task to solve because of the leaking issue of CNG happen around the nozzle area [2]. It is found that CNG-DI engine has the best capability to improve fuel flow and ignition process that stimulates a higher engine performance and lessen the fuel usage [10] as well as emissions [5, 8].

Practically all conventional fuel injectors are solenoid drive unit. The electrical energy supplied to the injector will magnetize and demagnetize the injector solenoid. It will then translated into mechanical energy. The mechanical energy is defined by the movement of the needle in the injector as it open and close. The dynamic motion controls the fuel flow through the nozzle. The needle movement is called as the injector temporal characteristic. The temporal aspect derives injection parameters such as the injection duration, injection frequency, rate shape, opening delay time and closing delay. All are controlled by the engine control unit (ECU). In feedforward control, predefined optimal setting is stored in the base maps of the ECU and will send the control signal to the power driver of the fuel injector.

The challenge for CNG is as there is neither dedicated CNG DI injector nor commercialize DI injector for gaseous fuel. Hence, we resorted to the conversion of the conventional DI injector. The changes in fuel properties from liquid to gases affect the injector characteristic benchmarked by the manufacturer. For this reason, the injector characterization for the natural gas engine is desired by researchers. DI injector researches were also analyzed utilizing modelling approaches expanded from the injector physical modelling reaching to a complete injection system modelling. DI injector primarily modelled using an analytical approach. A coupled multi-physics problem was either fully or just partially included in the numerical models [11].

The Model-Based Calibration (MBC) Toolbox was used comprehensively to develop a test plan, fit statistical models, study input-output relationships and generate optimal calibrations of the engine's electronic control unit (ECU) [12]. Instead of experimenting with a real engine, the MBC actualize a calibration procedure by using a mathematical-physical model which were widely applied for diesel and gasoline engines [13]. The model response can be easily simulated together with control system parameters in a desktop computer [14]. The coupling of different parameters can be solved when combined with the MBC experiment resulting in high efficiency and strong repeatability [15]. Based on Environmental Protection Agency (EPA) federal test, an MBC method has been applied to minimize the engine emissions and fuel consumption [16]. The economic efficiency and fuel consumption of MBC were improved and more productive than traditional calibration [17]. MBC is more appealing due to the cost and time merits [18, 14].

The Calibration Generation (CAGE) Toolbox was used as an optimization tool to construct the response between decided variables like injection timing, ignition timing, valve timing and injection strategy and the desired objectives namely maximizing torque and power, minimizing emissions and fuel economy of an engine [19]. The MBC Toolbox and CAGE Toolbox were adapted into feedforward control design for the engine control strategy in detail [12]. A zero-dimensional, first principle injector model, consisting of an electromagnetic, mechanical and flow sub-models have been built explaining how the model simulated DI injector operate using CNG as fuel [20].

This study was conducted to determine the optimal calibration setup for influential input parameters on the output mass flow rate of injector utilizing CNG fuel. For this optimization study, it will be conducted using MATLAB Simulink software and MATLAB MBC Toolbox software. A zero-dimensional, first principle injector model was used to generate data for optimization. The selected input parameters that have been analyzed to be optimized were the injection pressure, injection duration, input voltage. Each input parameters are varied in the prescribed range based on the literature. Next, these optimize calibration results will be implemented in injector experiment for validation purpose.

2 Methodology

The optimization of injector was divided into four different sections; design of experiment, data modelling, calibration generation and verification. Figure 1 shows the flowchart of this study.

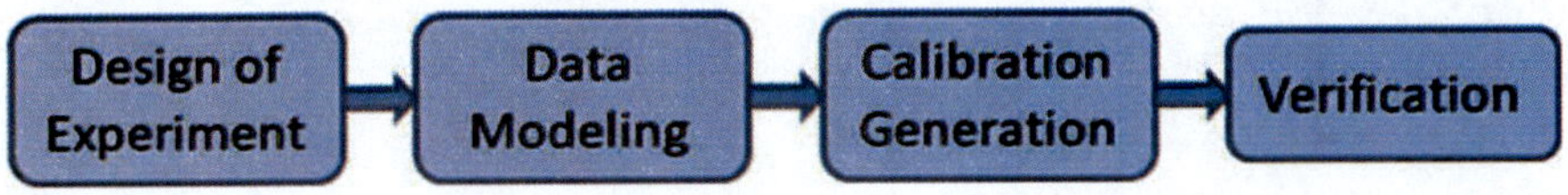

Fig. 1 Optimization flowchart

2.1 Design of Experiment

In this design of the experiment, the Simulink model [20] was used to provide sampling data for optimization calibration. The mechanical inputs and electromagnetic inputs of the injector parameter are based on Zhang et al. [21]. The Simulink model was run with 100 different excitations of input parameter for representing a comprehensive data collection. The selected input parameters that have been excited were the injection pressure, injection duration, input voltage, nozzle diameter, armature mass, and the spring constant whereas the mass flow rate was set as the output parameter. The acquired data are exported to the MBC toolbox.

2.2 Data-Driven Modeling

The core activity of data-driven modelling was on the selection of the most suitable modelling technique. The Model Browser of MBC toolbox offered numerous modelling technique. In the current study, a one-stage test plan is chosen as the model's data structure. The data collected was divided into two parts: training data and validation data. 70% of data was allocated as training data, and the rest 30% was set as validation data. The radial basis function (RBF) model was found to produce the least RMSE and validation RMSE over other modelling technique. The typical RBF equation [19] is shown below:

$$z(x) = \varnothing(\|x - \mu\|) \tag{1}$$

where

x n-dimensional vector
μ n-dimensional vector called the centre of the RBF
‖ Denote Euclidean distance
Ø Characteristic of RBF.

The best model was chosen based on the RMSE, predicted error sum of squares (PRESS) RMSE, maximum error, residual and normal plots lastly cross-validation statistics [22, 12]. The equation of RMSE is shown below:

$$RMSE = \sqrt{\frac{1}{n}(x_i - y_i)^2} \tag{2}$$

where

n number of data points
x predicted values
y actual values.

RMSE calculates the actual correlation value and predicted value to find the error magnitude [12]. PRESS RMSE decides the predictive nature of the model as the data run for validation against the model prediction [23]. The objective is to find model fitting with the lowest RMSE value [24, 25]. A significant amount of data inputs are required to ensure the model can predict as low as possible the RMSE value.

2.3 Calibration Generation

In MBC toolbox framework, CAGE was utilized to generate an optimized solution based on the selected model. The model developed in the MBC Toolbox was imported into CAGE environment [12]. CAGE has proven able to generate an optimal solution for injection parameters, rail pressure and the Exhaust Gas Recirculation (EGR) [26]. In the CAGE toolbox, numbers of optimization algorithm are offered. Generally, they can be divided as heuristic and non-heuristic type. It also categorized as a single objective and multiple objectives algorithm [27]. Multiple objective optimization solver was chosen for this study for maximization of injector mass flow rate (MFR) and minimization of the solenoid operating current. The objective functions of the study can be written as:

Maximization

$$MFR_{cng} = f\left(D_{noz}, M_{arm}, t_{dur}, V_{sol}, p_{cng}, k_{spr}\right) \tag{3}$$

Minimization

$$I_{sol} = V_{sol}/I_{sol} \tag{4}$$

where

MFR_{cng}	the mass flow rate of CNG across the injector nozzle
D_{noz}	the nozzle diameter
M_{arm}	the mass of the injector armature and needle
t_{dur}	injection duration
V_{sol}	the voltage across solenoid
p_{cng}	the CNG pressure from the tank to the injector
k_{spr}	The injector spring constant.

The free and fixed variables and the corresponding range constraint and mid-values are presented by Tables 1 and 2, respectively.

The injector nozzle, D_{noz}, the armature and needle mass, M_{arm}, and the spring contant, k_{spr} parameters can be easily adjusted and optimized during the simulation and optimization stage. However, in practical, these properties are hardly modified and considered as infeasible for further optimization with regarding to the current

Table 1 The free variables and the corresponding range constraint

Variables	Minimum	Minimum
V_{sol} (V)	1	60
p_{cng} (s)	1	30
t_{dur} (bar)	45	60
I_{sol}	1	10A

Table 2 The fixed variables and corresponding mid-values

Variables	Mid-value
D_{noz}	0.00068 m
M_{arm}	0.003 kg
k_{spr}	12,140 N/m

testing facilities [28]. The algorithm chosen for this study is Normal Boundary Intersection (NBI) method. Based on NBI algorithm, the intersection of the feasible objective region with a normal to the convex unifications of the columns of the pay-off matrix generating Pareto optimal solutions [29].

2.4 *Verification*

The verification of optimal setup was done by using simulation and experimental testing. Figure 2 shows the schematic layout of the experiment test rig that was used for the verification stage. In the experiment routine, each set of data were repeated three times to get the average measured data.

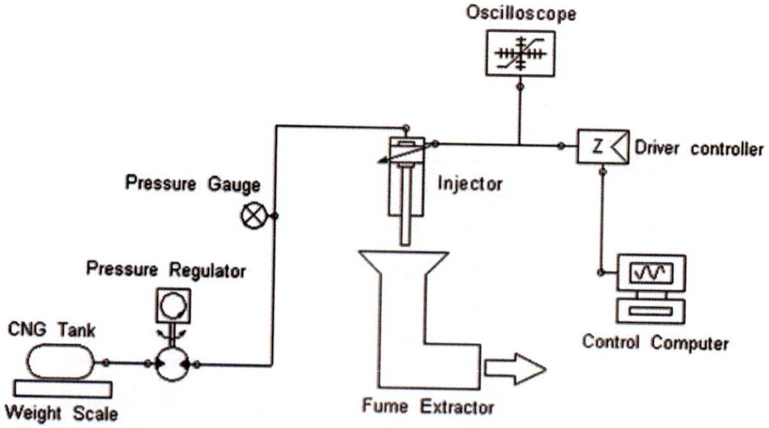

Fig. 2 Schematic layout of the experimental setup [28]

3 Result

3.1 MBC Model Fitting and CAGE Optimal Result

Data-driven modelling using a one-stage test plan and Interpolating RBF produced an RMSE of 0 and validation RMSE 0.0249. A model gives excellent execution if the validation RMSE value is less than 2.5% [24]. RMSE gives the insight on how well the model fits the data, while validation RMSE predicts the model nature indicating overfitting if ever happened [12]. Figure 3 presents the predicted and calculated mass flow rate produced by the model. All predicted data accurately matched the calculated data.

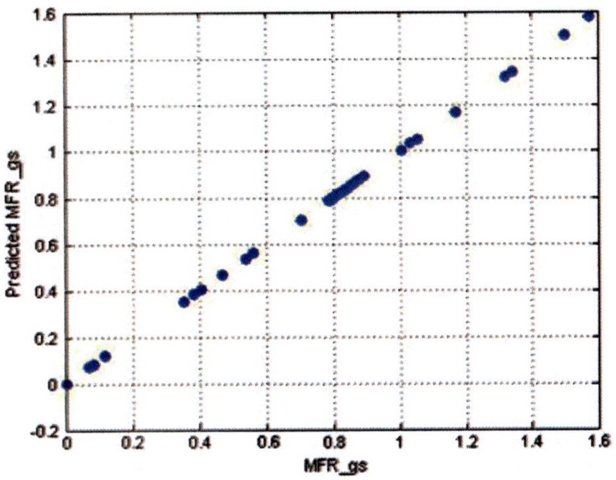

Fig. 3 Predicted and measured mass flow rate produced by data-driven modelling

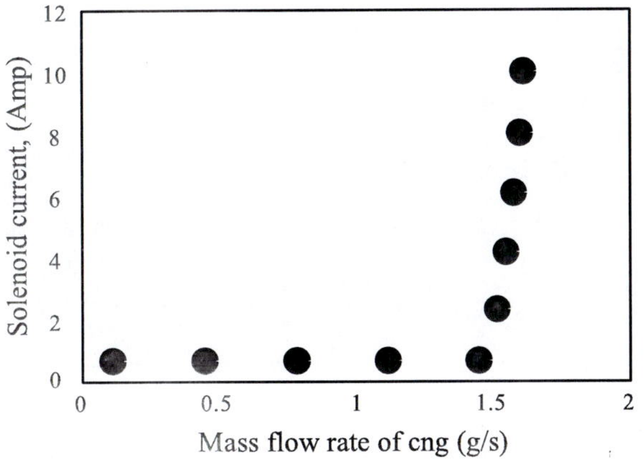

Fig. 4 The pareto rank for the two competitive objectives

CAGE optimization produced an overall of ten solutions. All the proposed solutions are presented in Fig. 4. The highest-ranked MFR recorded by the optimal calibration 1.6067 g/s with the setup of injection duration at 30 ms, injection pressure at 60 bar, input voltage 15 V, and operating current at 10 A. Based on the best solution, the solver proposed a voltage increment of 12–15 V, use of the highest pressure of 60 bar and the use of maximum current of 10 A. We also notified that the range of possible maximum mass flow rate is in the range of 1.5–2 g/s.

3.2 *MBC Optimization Verification Result*

Baseline Simulation Versus Optimize Simulation

Figure 5 presents the graph of baseline simulation versus optimize simulation result. Both the baseline and optimize simulation result show a steady linear increasing of mass flow rate. The calculated value of RMSE is 0.1615, MAE is 0.1564, MAPE is 37.07%, MPE is −37.07%, and the average modelling error for this graph is 15.64%. Based on this comparison, it is expected that the optimization setup is able to produce the increment of mass flow rate at about 15.64% by using an experimental method.

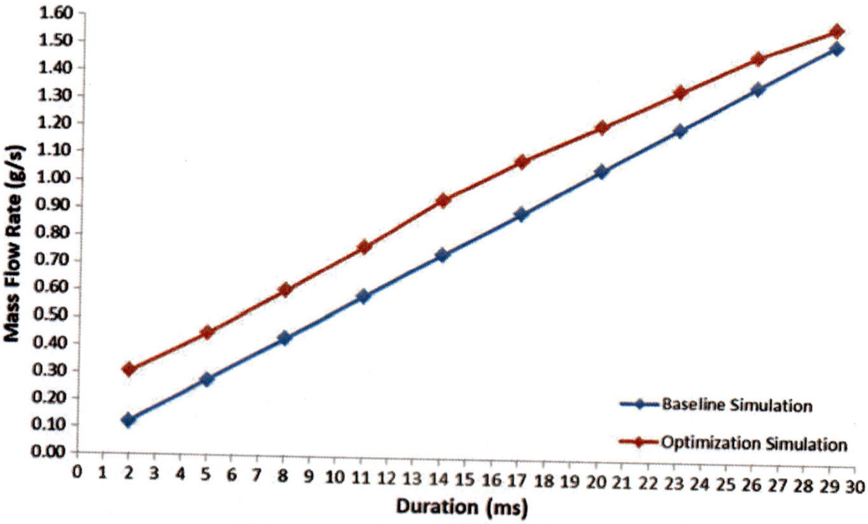

RSME	MAE	MAPE	MPE	Model
0.1615	0.1564	37.07%	-37.07%	15.64%

Fig. 5 Baseline simulation versus optimize simulation graph

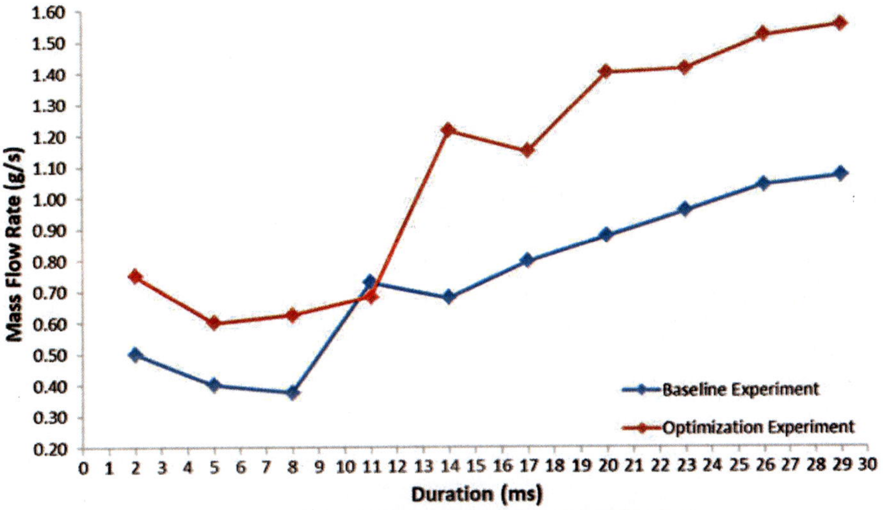

RMSE	MAE	MAPE	MPE	Model
0.3907	0.3579	49.55%	-48.30%	35.79%

Fig. 6 Baseline experiment versus optimize experiment graph

Baseline Experiment Versus Optimize Experiment

Figure 6 presents the graph of baseline experiment versus optimize experiment. Both results show an increasing pattern and a similar, slightly fluctuated trend during the increment. The baseline experiment produces the highest fluctuation at about 11 ms injection duration. In contrast, the optimized experiment provides the highest variation at 14 ms. The results show that the optimum setup is multiplying the fluctuation to become more visible. The calculated RMSE is 0.3907, MAE is 0.3579, MAPE is 49.55%, MPE is −48.30%, and the average modelling error for this graph is 35.79%. It can be concluded that the optimum setup can increase the amount of mass flow rate by 35.79%. It is about two times greater than the increment predicted by the simulation.

Baseline Experiment Versus Baseline Simulation

Figure 7 presents the graph of experimental baseline versus simulated baseline cases. Both graphs demonstrate increasing patterns at increased injection duration. It is visible that the simulation result shows a steady, linear growing trend for mass flow rate. However, the experimental result indicates a slightly fluctuated, and its rate of increment is lower than the simulated graph. It suggests that the model is unable to capture the injector dynamics accurately. The value of RMSE is 0.2467, MAE is 0.2089, MAPE is 28.55%, MPE is −3.41%, and the average modelling error for this graph is 20.89%.

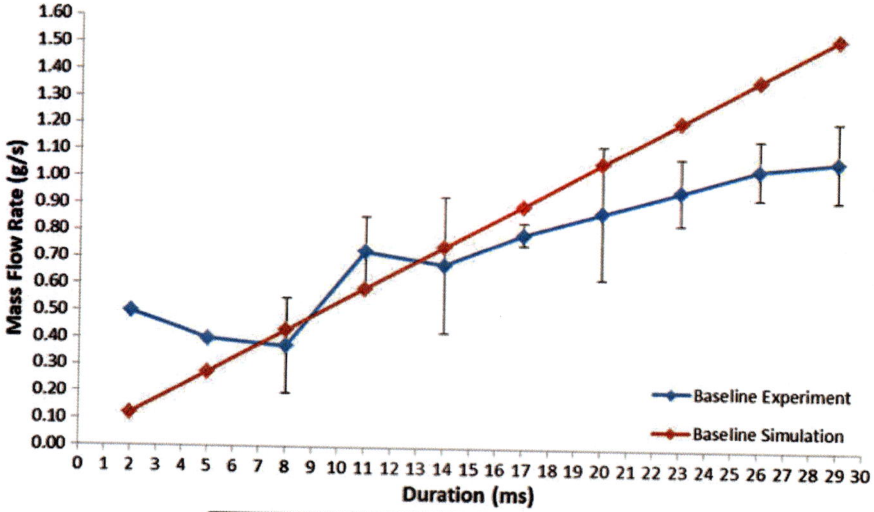

RMSE	MAE	MAPE	MPE	Model
0.2467	0.2089	28.55%	-3.41%	20.89%

Fig. 7 Baseline experiment versus baseline simulation graph

Optimize Experiment Versus Optimize Simulation

Figure 8 presents the graph of optimize experiment versus optimize simulation result. Both plots show an increasing pattern as the duration increased. However, the trends of both graphs differ. The simulation produced a steady linear increasing graph for the mass flow rate. However, the experimental result posses a slight fluctuation in the mass flow rate at increased injection duration. It shows that the simulation can predict the trend of the mass flow rate but cannot detect the fluctuation and its location. The value of RMSE is 0.1860, MAE is 0.1353, MAPE is 15.06%, MPE is 12.12%, and the average modelling error for this graph is 13.53%. The low error produced suggests that the simulation for this optimize setup is considered as acceptable. Additionally the model can be used to perform an early check on the optimal setup.

4 Conclusion

An optimization study in an MBC framework has been executed to predict the optimal set up for the selected injector parameters. The optimization results are verified using simulation and laboratory testing. The optimization suggests that the maximum achievable mass flow rate is 1.607 g/s. It is obtained at an injection duration of 30 ms, the injection pressure of 60 bar, the input voltage of 15 V, and operating current of 10 Amp. At the verification stage, the optimization results show a significant increment of mass flow rate. From the simulation work of baseline versus optimized

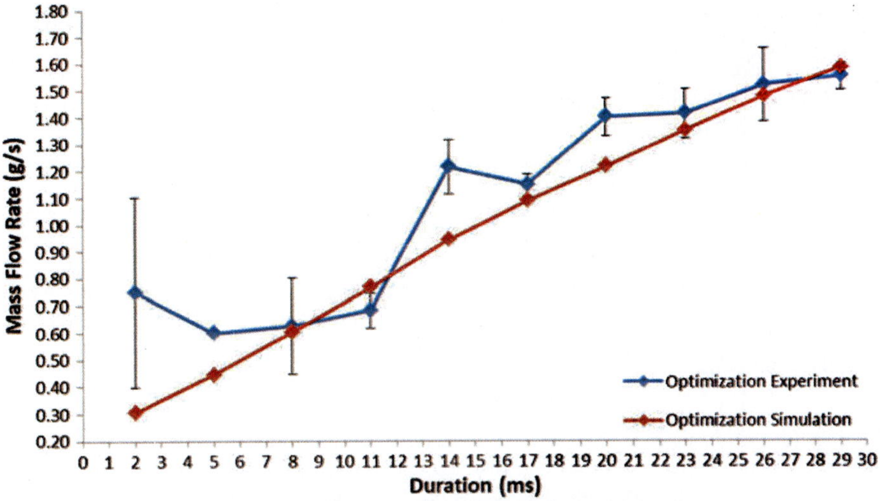

RMSE	MAE	MAPE	MPE	Model
0.1860	0.1353	15.06%	12.12%	13.53%

Fig. 8 Optimize experiment versus optimize simulation graph

case, it is expected to have an increase in the mass flow rate of about 15.64%. Whereas from the experimental work of baseline versus optimize, it is found that the increment of the mass flow rate is approximately 35.79%. The simulation versus experimental work for the baseline case produced a calculated RMSE of 0.2467. Whereas, the simulation versus experimental work for the optimization case provides an RMSE of 0.1860. The reduce of RMSE error shown that the use of optimal set up has improved the controllability of the injector; hence the prediction accuracy has increased. The MBC optimization framework is able to provide optimal solutions hence increased the mass flow rate for CNGDI injector. A detailed study is required to find the source of mass flow rate fluctuation and suitable methods to dampen the fluctuation effect.

Acknowledgements The authors would like to thank to the Ministry of Higher Education (MOHE) and Universiti Malaysia Pahang for providing the funding for the project under grant scheme RDU1703145 and FRGS 2017-1 reference code FRGS/1/2017/TK03/UMP/03/2 number RDU170127.

References

1. Hao H, Liu Z, Zhao F, Li W (2016) Natural gas as vehicle fuel in China: a review. Renew Sustain Energy Rev 62:521–533
2. Moon S (2018) Potential of direct-injection for the improvement of homogeneous-charge combustion in spark-ignition natural gas engines. Appl Therm Eng 136(January):41–48

3. Wang T, Zhang X, Zhang J, Hou X (2017) Numerical analysis of the influence of the fuel injection timing and ignition position in a direct-injection natural gas engine. Energy Convers Manag 149:748–759
4. Martins F, Souza F (2010) Mixture formation analysis in a direct injection spark ignition (DISI) engine. SAE Technical Papers
5. Choi M, Song J, Park S (2016) Modeling of the fuel injection and combustion process in a CNG direct injection engine. Fuel 179:168–178
6. Liu Y, Yeom J, Chung S (2013) A study of spray development and combustion propagation processes of spark-ignited direct injection (SIDI) compressed natural gas (CNG). Math Comput Model 57(1):228–244
7. Cho HM, He BQ (2007) Spark ignition natural gas engines—a review. Energy Convers Manag 48(2):608–618
8. Xu H, Wang C, Ma X, Sarangi AK, Weall A, Krueger-venus J (2015) Fuel injector deposits in direct-injection spark-ignition engines. Prog Energy Combust Sci 50:63–80
9. Chen W, Pan J, Fan B, Liu Y, Peter O (2017) Effect of injection strategy on fuel-air mixing and combustion process in a direct injection diesel rotary engine (DI-DRE). Energy Convers Manag 154(October):68–80
10. Kalam MA, Masjuki HH (2011) An experimental investigation of high performance natural gas engine with direct injection. Energy 36(5):3563–3571
11. Taha Z, Rahim MFA, Mamat R (2017) Injection characteristics study of high-pressure direct injector for Compressed Natural Gas (CNG) using experimental and analytical method. IOP conference series: materials science and engineering, vol 257, no 1, p 12057
12. Everett RV (2011) An improved model-based methodology for calibration of an alternative fueled engine. The Ohio State University. Thesis
13. Roepke K (2014) Design of experiments for engine calibration. J Soc Instrum Control Eng 53(4):322–327
14. Wang Y, Dqg F, Lwv G, Wr S, Txlfno P (2015) Model based calibration: a case study for calibrating control systems for downsized boosted engines. Focus Dyn Syst Control no December:19–21
15. Prucka RG (2008) An experimental characterization of a high degree of freedom spark-ignition engine to achieve optimized ignition timing control. The University of Michigan. Thesis
16. Farrell P, Foster D, Ghandhi J, Moskwa J, Reitz R (2001) Diesel engine injection rate-shape optimization using genetic algorithms and multidimensional modeling. Project Report
17. Zeng Q, Liu B, Shi X, Zhang C, Hu J (2018) Model based calibration for improving fuel economy. Therm Sci 22(3):1259–1270
18. Nikzadfar K, Shamekhi AH (2019) Investigating a new model-based calibration procedure for optimizing the emissions and performance of a turbocharged diesel engine. Fuel 242(August):455–469
19. Ho T, Karri V (2011) Hydrogen powered car: two-stage modelling system. Int J Hydrogen Energy 36(16):10065–10079
20. Alias MHR, Rahim MFA, Bakar RA (2019) Single hole direct injector simulation validation and parametric sensitivity study. In: 5TH International Conference On Mechanical Engineering Research (ICMER) 2019
21. Zhang X, Palazzolo A, Kweon C-B, Thomas E, Tucker R, Kascak A (2014) Direct fuel injector power drive system optimization. SAE Int J Engines 7(3):2014-01-1442
22. Morton T, Connors R, Maloney P, Sampson D (2003) Model-based optimal calibration of a dual independent variable valve-timing engine. Des Exp der Mot 77–85
23. Guerrier M, Cawsey P (2004) The development of model based methodologies for gasoline ic engine. SAE Tech Pap Ser 01(1466)
24. Jiang S, Nutter D, Gullitti A (2012) Implementation of model-based calibration for a gasoline engine. SAE Int 01(0722)
25. Beham M, Yu DL (2004) Modelling a variable valve timing spark ignition engine using different neural networks. IMechE 218(April):1159–1171

26. Cameretti MC, Landolfi E, Tesone T, Caraceni A (2019) Virtual calibration method for diesel engine by software in the loop techniques. Int J Automot Mech Eng 16(3):6940–6957
27. Jingqi X (2013) The research and summary of evolutionary multi-objective optimization algorithm. Intell Comput Evol Comput 505–512
28. Alias MHR, Rahim MFA, Rodzi MHMI, Bakar RA (2018) Effect of injection pressure, injection duration, and injection frequency on direct injector's mass flow rate for compressed natural gas fuel. In: MATEC web of conferences, vol 225, no 02008
29. López Jaimes A, Zapotecas-Martínez S, Coello C (2011) An introduction to multiobjective optimization techniques. In: António Gaspar-Cunha JAC (ed) Optimization in polymer processing. Nova Science Publishers, pp 29–57

The Application of Modified Equipment in Retention of Motor Task Performance Amongst Children of Low and High Working Memory Capacity

Rabiu Muazu Musa, Mohsen Afrouzeh, Pathmanathan K. Suppiah, Anwar P. P. Abdul Majeed, Mohammad Sadegh Afroozeh, and Mohamad Razali Abdullah

Abstract The ability of children to learn and retain motor-related tasks could ease the pathway of mastering sport-specific skills that are non-trivial in spurring children's athletic development. Modification of equipment may facilitate the acquisition of complex motor tasks with respect to children's specific characteristics. The influence of modified equipment in retention ability of motor task amongst children with low and high working memory capacities is investigated in this study. Forty children aged 9–10 years were recruited and the Wechsler Intelligence Scale for Children was used to determine the working memory capacity of the children. High and low working memory (HWM), (LWM) were identified and allotted into 4 different groups of 10 children each Viz. (A) HWM with standard mini basketball equipment, (B) LWM with standard mini basketball equipment, (C) HWM with modified mini basketball equipment and (D) LWM with modified mini basketball equipment. Basketball throw from the free-throw line in pre and post-tests were used as the acquisition and retention tasks ability respectively. There was a significant

R. M. Musa (✉)
Centre for Fundamental and Continuing Education, Universiti Malaysia Terengganu, 21030 Kuala Nerus, Terengganu, Malaysia
e-mail: rabiumuazu86@gmail.com

M. Afrouzeh
Informetrics Research Group, Ton Duc Thang University, Ho Chi Minh City, Vietnam

Faculty of Sports Science, Ton Duc Thang University, Ho Chi Minh City, Vietnam

P. K. Suppiah
Faculty of Psychology and Education, Universiti Malaysia Sabah, Kota Kinabalu, Malaysia

A. P. P. A. Majeed
Innovative Manufacturing, Mechatronics and Sports Laboratory, Universiti Malaysia Pahang, 26600 Pekan, Pahang Darul Makmur, Malaysia

M. S. Afroozeh
Faculty of Humanities, Department of Sport Sciences, Jahrom University, Jahrom, Iran

M. R. Abdullah
East Coast Environmental Research Institute, Universiti Sultan Zainal Abidin, 21300 Kuala Nerus, Terengganu, Malaysia

A. F. Ab. Nasir et al. (eds.), *Recent Trends in Mechatronics Towards Industry 4.0*, Lecture Notes in Electrical Engineering 730, https://doi.org/10.1007/978-981-33-4597-3_84

main effect of memory and equipment towards children's retention ability. A significant difference was also observed across both HMC and LWM when subjected to the use of modified equipment (F, 1, 24.025) = 23.958, $p < 0.001$. No statistically significant difference was detected between the HWM and LWM in their ability to retention tasks $p > 0.05$. The usage of modified basketball equipment could enhance mastery of motor tasks in children irrespective of their memory capacity.

Keywords Retention skill · Motor acquisition · Modified equipment · Working memory capacity · Motor learning

1 Introduction

Basketball is regarded as one of the popular and most viewed sport in the world with a wider range of amateur and professional players partaking in the sport. Children could begin to play basketball as early as 5 years old. There are exist several benefits for playing basketball, especially amongst children. Participation in a sport of basketball could foster children's development through active movement as well as promote the acquisition of basic gross and motor skills such as coordination, balance, and agility [1]. There are many skills that are required to play the game of basketball, nonetheless, for a beginner and children alike, learning to score points (i.e. shooting baskets) could be fun that many would prefer to begin with. This skill would be preferred by both beginners and children since it is a skill that is directly leading to a score. Moreover, it was reported that the skill is the most favourite amongst many participants [2]; and has also been connected to the attainment of fun and satisfaction in a group of children [3].

The utilization of standard equipment in the sport of basketball might provide hindrance towards the acquisition of basic motor skills in children and younger athletes. Similarly, the size of the ball could also affect the quality of the shot in children. Previous researchers revealed that the use of smaller balls could assist the children in the execution of effective shots [4]. The authors further inferred that that changing the ball mass could also lead to a better shot. Hitherto, the results of a similar study showed that a decrease in the height of the ring was effective in throwing performances amongst children, but no positive effect was found when the ball demission was reduced [5]. Moreover, children were reported to be more motivated when the balls, as well as the sizes of the rings, are lowered. Consequently, the mastery of the skill in shooting is observed through the execution of quality shots which in turn brings about high motivation in the participants [5–8].

Working memory capacity is one of the major challenges in the children learning process. This process lasts more than adults. Therefore, the involvement of working memory capacity should be used less than it is used to [6–11]. Considering the fact that working memory plays an important role in explicit motor learning in adults, it is likely that the development of working memory during childhood will increase explicit information processing for obvious information processing. Going by this hypothesis, then the children might learn more in a situation that does not employ more memorization and use of working memory capacity [12, 13].

A number of researchers investigating modelling and children's motor skill performance were majorly centred upon the theory of preceding researchers as a framework for examining cognitive-developmental characteristics and the modelling process [14]. Although, the work of the earlier researcher who examined the effects of modelling and verbal self-instruction on children's capability to recall a progressive motor task demonstrated that older children could also gain when projected with a verbal or silent model [15]. Nonetheless, it was observed that younger children's performance was only improved when presented with only a verbal model. It is worth noting that, through the breaking down of skills into smaller components, children might likely learn better by means of unconscious process. Moreover, the usage of equipment (e.g. a heavy basketball) may increase the difficulty level of performing the skills which in turn reduces the willingness to use conscious processes [16]. It also appears that the effect of working memory capacity on children's retention of motor learning is high, although its exact role has not been proven. Hence, the present study was intended to test the effectiveness of modified equipment on the retention ability of 9–10 years old children with high or low working memory capacity in the performance of basketball shots.

2 Methodology

2.1 Participants

The participants were 40 children without any previous basketball experience. All selected participants were right-handed, between the ages of 9–10 and reported no previous injury. Participants were briefed on the physical risks and informed consent was received from parents and participants. The study was approved by the Ethics Committee of the Ferdowsi University of Mashhad (IR.FUM.ICBS.97.1435). Based on their scores in the Wechsler Intelligence Scale for Children—Memory for Digit Span [17], the participants were divided into two categories of High Working Memory Capacity (HWMC) and Low Working Memory Capacity (LWMC). The participants were tested for their performance in the basketball free-throw test before being randomly distributed using the matched-pair design. All groups attempted the free throw 4.25 m from the backboard.

2.2 Experimental Procedure

Each practice session in the acquisition phase (8 days) consisted of 4 blocks of 15 trials per block. Participants were allowed to rest for 10 min before beginning with the

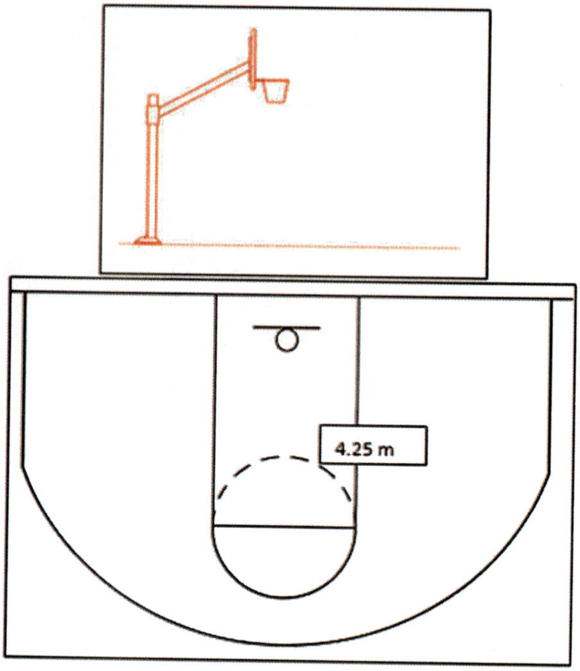

Fig. 1 Layout of the experimental task

next block. A day before the first practice session, participants underwent a pretest for the experimental task; attempting to score from the free-throw line. A day after the final practice session, participants underwent the acquisition test and 7 days after the acquisition test a retention test was carried out.

The experimental task for the groups HWMC-Standard and LWMC—Standard performed the basketball free throw under standard mini basketball regulations; from a line 4.25 m behind a backboard with a rim circumference of 45 cm at a height of 2.60 m. The ball used weighed 485 g with a circumference between 69 and 71 cm. The HWMC—Modified and LWMC—Modified groups used modified equipment; rim circumference 45 cm at a height of 2.00 m for the experimental task. The ball used had a smaller mass 440 g and a circumference between 69 and 71 cm. The experimental layout is projected in Fig. 1.

2.3 *Evaluation of the Skills*

For the purpose of assessing the learning of free-throw basketball skill, the performance in each attempt was determined accordance with the American Alliance for Health, Physical Education, Recreation and Dance (AAPEHRD) basketball test based on which each throw was scored as follows: a 3 point score if the ball hit the basket without hitting the hoop or the board, a 2 point score if the ball hit the board or the hoop of the basket, and a 1 point score if neither the board nor the hoop was hit by the ball.

2.4 Data Analysis

Before undertaking the full data analysis in the present study, normality distribution assessment was carried out using Kolmogorov–Smirnov tests in order to ensure that the study variables are normally distributed in line with the requirement of the statistical test needed in the study. After confirming the normality distribution of the data acquired, split-plot also known as mixed-design ANOVA, with equipment (standard and modified) and retention ability as the between-subject factors was conducted. A value of $P < 0.05$ was considered as significant level. Statistical software SPSS version 20.00 was used as a platform for conducting the data analysis.

3 Results and Discussion

Table 1 shows the descriptive statistics of the variables experimented in the current study. The two types of working memory and the equipment, number of observations, mean as well as standard deviations are displayed. It could be observed from the table that the mean retention ability of both high and low working memory has increased when the children utilized modified equipment.

Figure 2 displays the box plots of comparative analysis of the variables experimented in the current investigation. It could be seen from Figure (a) that the mean performances of HWM on retention ability have greatly increased when the modified equipment is used. Interestingly in Figure (b), the mean performances of the LWMC group were observed to equally increased when the modified equipment is used as compared to standard equipment. It is worth to mention that no statistically significant difference was detected between the HWM and LWM in their ability to retention tasks $p > 0.05$.

Table 1 Descriptive statistics of the study variables

Equipment	Memory type	N	Mean	Std. deviation
Standard	High	10	9.200	1.033
	Low	10	7.000	1.054
Modified	High	10	14.600	0.699
	Low	10	13.700	1.160

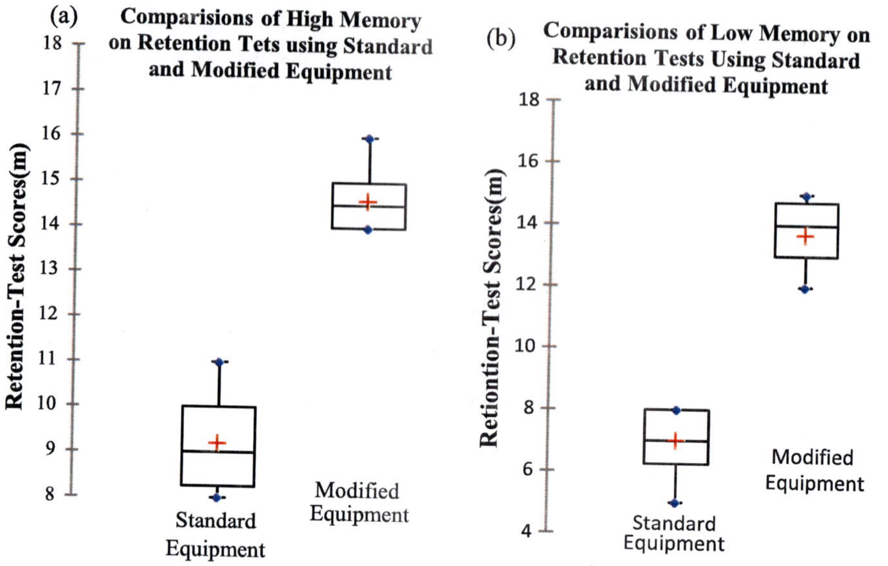

Fig. 2 Comparative efficacy of modified and standard equipment on retention ability

Figure 3 projected the interactions of working memory types and equipment on children's retention capacity. It could be observed from the figure that a statistically significant difference was found across both HWM and LWM when subjected to the use of modified equipment as opposed to the standard equipment $p < 0.001$. These findings provide evidence that the use of modified equipment could enhance the retention of motor tasks in children irrespective of their working memory capacity types.

Fig. 3 Interactions of working memory types and equipment on children's retention capacity

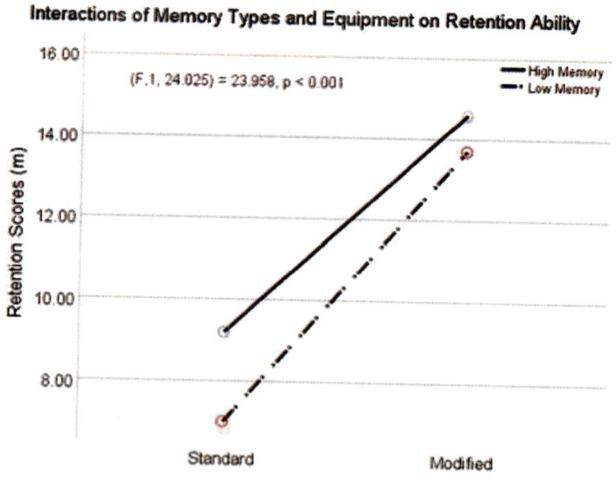

The findings of the experiment carried out demonstrated that the adjustment of the equipment was effective in enhancing the children's retention capacity. Similarly, the findings revealed that the retention ability of the children whilst using modified equipment was better as compared to using standard equipment (Table 1 and Fig. 2). This finding has substantiated the theoretical believe that modification of equipment might help children to perform better. The finding is therefore incongruent with the proponents of the constraints-led approach who infer that modifying the task allows children to search for new solutions by exploring the practice environment, which ultimately facilitates unconscious processes of learning [18]. In a nutshell, modification of equipment could lead to a positive change in the performance of the children by minimizing working memory capacity in motor learning. Reducing the working memory capacity is an important factor that pushes the children to focus more on the execution of the motor tasks rather than memorizing the techniques needed to execute the tasks which could take longer to master and consequently inhibit faster learning.

Although, no statistically significant difference was detected between the two groups with respect to working memory and retention ability, the performances of the HWMC group was found to be slightly better than the LWMC group (Fig. 2a, b). Evidence suggests that children with HWMC are more likely to develop skills faster or better than children with low memory capacity [19]. It is believed that children with higher working memory capacity can acquire and use new skills better than those with low working memory capacity whereas children with LWMC are likely to forget skill training and thus are unable to perform the skill correctly [20, 21]. Nonetheless, it is worth to stress that the lack of differences observed between the two groups of working memory capacities have accentuated the benefit of using customized equipment when training children to acquire new motor tasks irrespective of their working memory capacities.

It was also found from the findings of the study that both HWMC and LWMC have significantly improved in their retention of motor task performances when subjected to the use of modified equipment as opposed to the standard equipment (Fig. 3). The more a situation coincides with a test environment, the better the performance will become. These two factors are interrelated [22, 23]. For the acquisition of new skills, the environment and skill interact with each other to help the learner acquire new skills and execute them effectively [24, 25].

4 Conclusion

It has been demonstrated from the present study that the employment of modified equipment could be a great technique in accelerating the learning as well as retention motor of tasks. The importance of adjusting sports equipment, especially for children could have a positive impact on learning new motor skills regardless of their working memory capacity level. It is inferred from the findings of the present study that customizing learning experiences that do not rely on working memory capacity, could have a positive long-term effect for retention of motor skills in children. It is therefore, recommended that physical education teachers, trainers, as well as coaches, should consider modifying equipment and environment with respect to the children specific characteristics during practice in order to accelerate learning of motor tasks which would go a long way in enhancing retention of skills that could foster better execution of performance.

Acknowledgements The authors would like to acknowledge the participants for their cooperation towards the accomplishment of this study.

References

1. Gaggioli A, Morganti L, Mondoni M, Antonietti A (2013) Benefits of combined mental and physical training in learning a complex motor skill in basketball. Psychology 4:1–6. https://doi.org/10.4236/psych.2013.49A2001
2. Palao J, Ortega E, Olmedilla A (2004) Technical and tactical preferences among basketball players in formative years. Iber Congr Basketb Res 4:38–41. https://doi.org/10.2466/ICBR.4.38-41
3. Piñar MI, Cárdenas D, Conde J, Alarcón F, Torre E (2007) Satisfaction in mini-basketball players. Iber Congr Basketb Res 4:122–125. https://doi.org/10.2466/ICBR.4.122-125
4. Arias JL, Argudo FM, Alonso JI (2011) Effect of two different forms of three-point line on game actions in girls' mini-basketball. S Afr J Res Sport Phys Educ Recreat 33:9–22
5. Chase MA, Ewing ME, Lirgg CD, George TR (1994) The effects of equipment modification on children's self-efficacy and basketball shooting performance. Res Q Exerc Sport 65:159–168. https://doi.org/10.1080/02701367.1994.10607611
6. Grawer R, Grawer R, Rains ST (2003) Youth basketball skills and drills. In: Coaches Choice. Coaches Choice
7. Hanlon TW (2005) Absolute beginner's guide to coaching youth baseball. Que Pub
8. Piñar López MI (2005) Incidence of the change of a set of game rules on some of the variables that determine the training process of minibasket players (9–11 years)
9. Alloway TP, Gathercole SE, Pickering SJ (2006) Verbal and visuospatial short-term and working memory in children: are they separable? Child Dev 77:1698–1716. https://doi.org/10.1111/j.1467-8624.2006.00968.x
10. Luciana M, Conklin HM, Hooper CJ, Yarger RS (2005) The development of nonverbal working memory and executive control processes in adolescents. https://doi.org/10.1111/j.1467-8624.2005.00872.x
11. Thomason ME, Race E, Burrows B, Whitfield-Gabrieli S, Glover GH, Gabrieli JDE (2009) Development of spatial and verbal working memory capacity in the human brain. J Cogn Neurosci 21:316–332. https://doi.org/10.1162/jocn.2008.21028
12. Berry DC, Broadbent DE (1988) Interactive tasks and the implicit-explicit distinction. Br J Psychol 79:251–272. https://doi.org/10.1111/j.2044-8295.1988.tb02286.x
13. Hammond J, Smith C (2006) Low compression tennis balls and skill development. J Sports Sci Med 5:575
14. Yando R, Seitz V, Zigler E (1989) Imitation, recall, and imitativeness in children with low intelligence of organic and familial etiology. Res Dev Disabil 10:383–397. https://doi.org/10.1016/0891-4222(89)90039-5
15. Weiss MR (1983) Modeling and motor performance: a developmental perspective. Res Q Exerc Sport 54:190–197. https://doi.org/10.1080/02701367.1983.10605293

16. Capio CM, Poolton JM, Sit CHP, Holmstrom M, Masters RSW (2013) Reducing errors benefits the field-based learning of a fundamental movement skill in children. Scand J Med Sci Sport 23:181–188. https://doi.org/10.1111/j.1600-0838.2011.01368.x
17. Wechsler D (1974) Manual for the Wechsler intelligence scale for children, revised. Psychol Corporation
18. Renshaw I, Davids K, Savelsbergh G (2010) Motor learning in practice. https://doi.org/10.4324/9780203888100
19. Buszard T, Farrow D, Reid M, Masters RSW (2014) Scaling sporting equipment for children promotes implicit processes during performance. Conscious Cogn 30:247–255
20. Engle RW, Carullo JJ, Collins KW (1991) Individual differences in working memory for comprehension and following directions. J Educ Res 84:253–262
21. Gathercole SE, Durling E, Evans M, Jeffcock S, Stone S (2008) Working memory abilities and children's performance in laboratory analogues of classroom activities. Appl Cogn Psychol Off J Soc Appl Res Mem Cogn 22:1019–1037
22. Magill RA (1998) Knowledge is more than we can talk about: implicit learning in motor skill acquisition. Res Q Exerc Sport 69:104–110
23. Charles MAG, Abdullah MR, Musa RM, Kosni NA, Maliki ABHM (2017) The effectiveness of traditional games intervention program in the improvement of form one school-age children's motor skills related performance components. J Phys Educ Sport 17:925–930
24. Musa RM, Abdul Majeed APP, Taha Z, Chang SW, Nasir AFA, Abdullah MR (2019) A machine learning approach of predicting high potential archers by means of physical fitness indicators. PLoS One 14. https://doi.org/10.1371/journal.pone.0209638
25. Maliki ABHM, Abdullah MR, Juahir H, Muhamad WSAW, Nasir NAM, Musa RM, Mat-Rasid SM, Adnan A, Kosni NA, Abdullah F, Abdullah NAS (2018) The role of anthropometric, growth and maturity index (AGaMI) influencing youth soccer relative performance. In: IOP conference series: materials science and engineering. Institute of Physics Publishing. https://doi.org/10.1088/1757-899X/342/1/012056

Firefly Algorithm for Functional Link Neural Network Learning

Yana Mazwin Mohmad Hassim, Rozaida Ghazali, Norlida Hassan, Nureize Arbaiy, and Aida Mustapha

Abstract Functional Link Neural Network (FLNN) is a type of single layer feedforward neural network that is less complex than other multilayer feedforward networks. The network uses less weight parameters which make the network training using Backpropagation algorithm less complicated. However, due to the less numbers of weight parameters, FLNN network is prone to yield inconsistent results. This work proposed Firefly algorithm (FA) as learning algorithm for FLNN. The results from the present investigation suggest that the FA algorithm is able to train the FLNN network and enhance its performance with better accuracy rate.

Keywords Firefly algorithm · Functional link neural network · Network training

1 Background

Functional Link Neural Networks (FLNN) is a type of single layer feedforward neural network introduced by Klassen and Pao [1, 2]. FLNN network is less complex as compared to other multilayer feedforward models such as the Multilayer Perceptron (MLP) as it has no hidden layers [2]. FLNN has been demonstrated to be able to handle the nonlinear mapping tasks [3, 4].

FLNN has been able to perform well in many applications involving with complex prediction and classification tasks [5–11]. The nonlinear input-output mapping tasks are enabled by its nonlinear activation functions which are controlled by the weight parameters of the network. These weight parameters can be adjusted until the activation functions progressively approximate the network output to the desired output signal based on its input patterns [12]. All weight parameters and bias parameters are tuned with an effort to reduce the network error. In neural network, this procedure is known as network training.

Y. M. M. Hassim (✉) · R. Ghazali · N. Hassan · N. Arbaiy · A. Mustapha
Faculty of Computer Science and Information Technology, Universiti Tun Hussein Onn Malaysia (UTHM), 86000 Parit Raja, Batu Pahat, Johor, Malaysia
e-mail: yana@uthm.edu.my

A. F. Ab. Nasir et al. (eds.), *Recent Trends in Mechatronics Towards Industry 4.0*, Lecture Notes in Electrical Engineering 730, https://doi.org/10.1007/978-981-33-4597-3_85

Standard FLNN models mostly employ the Backpropagation (BP) learning algorithm for it network training [13–18]. The BP-learning algorithm is a type of first-order iterative optimization algorithm also known as gradient descent. However, such a technique has a number of drawbacks which may affect the performance of FLNN [14–16, 18]. Therefore, this work attempts to employ Firefly algorithm to train the FLNN network to overcome such drawbacks.

2 Standard Learning Algorithm of FLNN

FLNN has been able to perform well in many applications involving with complex prediction and classification tasks [5–11]. The nonlinear input-output mapping tasks are enabled by its nonlinear activation functions which are controlled by the weight parameters of the network. These weight parameters can be adjusted until the activation functions progressively estimate the network to the desired output signal based on its input patterns [12]. This method is also known as network training, where all weight parameters and bias parameters are tuned with an effort to reduce the network error. Standard FLNN model mostly used Backpropagation (BP) algorithm as it training algorithm [13–18]. In BP-learning, the FLNN weight parameters are adjusted based on the Delta rule to minimize the network error measured based on Mean Squared Error, *MSE*:

$$MSE = \frac{1}{n}\sum_{i=1}^{n}\left(\hat{Y}_i - Y_i\right)^2 \tag{1}$$

where Y_i is the network output, $\hat{Y}_i$ is the desired output of the ith input pattern, and n is the total number of input patterns. The network output Y_i is calculated based on Eq. (1). The method calculates the gradient of network error and then fed them back to the algorithm to update the weight parameters. Throughout the training phase, the algorithm will continuously update weight parameters and bias parameters to minimize the network and actual output until the termination requirement is met. The FLNN network used for this work is the tensor model as illustrated in Fig. 1.

Nonetheless, it is worth noting that there are some limitations of using the BP-learning algorithm in FLNN training, i.e., it is highly dependent to the error surface [12] which tends to get trapped in local minima. This may cause FLNN to yield inconsistent results especially when they are used in solving non-linear separable classification problems where weight space is multimodal and have many local minima.

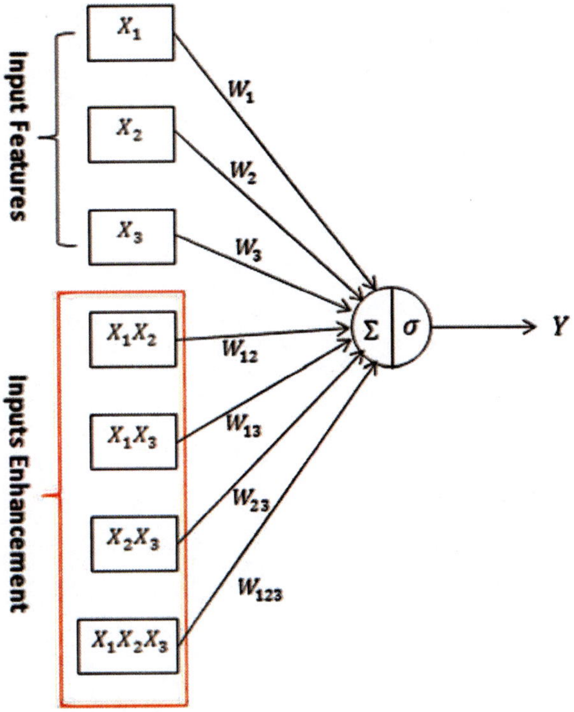

Fig. 1 FLNN with tensor model

3 FLNN Training with Firefly Algorithm

Firefly Algorithm (FA) is a metaheuristic algorithm proposed by Yang [19] inspired by the flashing patterns and the behavior of fireflies. The flashing light of fireflies was designed associated with the objective function in optimization problems [20]. Until recent, FA has successfully solve many optimization problems [21–25] which motivates this work to apply it as learning algorithm to ascertain the optimal weight parameters for FLNN. The two key parameters in FA which are: (1) the light intensity I_i defined as:

$$I(r) = I_0 e^{-\gamma r^2} \tag{2}$$

and (2) the attractiveness of fireflies β defined as:

$$\beta(r) = \beta_0 e^{-\gamma r^2} \tag{3}$$

in FA algorithm, r is the distance between each firefly. The distance, r_{ij} between two fireflies i and j at x_i and x_j is determined based on:

```
Begin
1) Initialize Maximum Cycle Number, MCN
2) Set FLNN weights as objective function: f(X), X = (x1,x2,...xd);
3) Generate an initial population of fireflies xi (i = 1,2,...d);
4) Evaluate light intensity Ii for firefly at xi by using equation (2)
5) Cycle = 1;
   while (Cycle < MCN)
         for i = 1 : n (all n fireflies)
            for j = 1 : n (all n fireflies)
               if (Ij > Ii)
                  Update distance using equation (5)
                  Move firefly i toward j in d-dimension
                  Evaluate new solution and update Ii
               end if
         end for j
       end for i
   rank fireflies and find current best
   end while
End
```

Fig. 2 Firefly algorithm for FLNN training

$$r_{ij} = \|x_i - x_j\| = \sqrt{\sum_{k=1}^{D} (x_{i,k} - x_{j,k})^2} \tag{4}$$

where $x_{i,k}$ is the kth component of the spatial coordinate x_i of ith firefly. The movement of a firefly i attracted to brighter firefly j at x_i and x_j is determined by Eq. (5).

$$x_i = x_i + \beta_0 e^{-\gamma r^2}(x_j - x_i) + \alpha \varepsilon_i \tag{5}$$

In this work, we employ FA for FLNN training. In this training process, the FLNN weights and biases needed to be transformed into objective function with its training dataset before it can be trained by FA algorithm. The Firefly algorithm for FLNN training is shown as in Fig. 2.

4 Experimentation and Results

In this simulation experiments, FLNN model trained with Backpropagation (FLNN-BP), FLNN model trained with Firefly algorithm (FLNN-FA) and MLP model trained with Backpropagation (MLP-BP) were considered for comparison. The parameter setting for each model is summarized in Table 1. The result of the FLNN-BP and FLNN-FA were compared with MLP-BP. Table 2 lists, five public benchmark datasets retrieved from UCI Machine Learning Repository [26] were used for the evaluation

Table 1 Parameters setting

Parameter	MLP-BP	FLNN-BP	FLNN-FA
Learning rate	0.1–0.5	0.1–0.5	–
Momentum	0.1–0.9	0.1–0.9	–
Epoch	1000	1000	1000
Minimum error	0.001	0.001	0.001

Table 2 Public benchmark datasets [26]

Datasets	Number of instances	Number of attributes	Number of classes
Breast cancer wisconsin	699	9	2
Indian liver patient dataset	583	10	2
PIMA Indians diabetes	768	8	2
Thyroid disease	215	5	3
Mammographic mass	961	5	2
Lymphography	148	18	4

of all network models investigated. Conversely, Tables 3, 4, 5, 6, 7 and 8 illustrates the efficacy of the proposed model against conventional means.

It could be seen from the results tabulated in Table 3 through Table 8, the FLNN model trained with FA algorithm (FLNN-FA) model gives better results than the conventional FLNN-BP model and MLP-BP model in Breast Cancer Wisconsin,

Table 3 Breast cancer wisconsin

Breast cancer wisconsin (original) dataset			
Learning algorithm	Best network structure	Number of tunable weights	Accuracy (%)
MLP-BP	9-6-1	67	96.86
FLNN-BP	45-1	46	96.25
FLNN-FA	45-1	46	96.96

Table 4 ILPD (Indian liver patient dataset)

ILPD (Indian liver patient dataset) dataset			
Learning algorithm	Best network structure	Number of tunable weights	Accuracy (%)
MLP-BP	10-9-1	109	70.61
FLNN-BP	55-1	56	69.63
FLNN-FA	55-1	56	70.73

Table 5 Pima Indians diabetes

Pima Indians diabetes dataset			
Learning algorithm	Best network structure	Number of tunable weights	Accuracy (%)
MLP-BP	8-9-1	91	75.77
FLNN-BP	36-1	37	73.93
FLNN-FA	36-1	37	75.97

Table 6 Thyroid disease

Thyroid disease dataset			
Learning algorithm	Best network structure	Number of tunable weights	Accuracy (%)
MLP-BP	5-5-3	48	95.27
FLNN-BP	15-3	45	93.46
FLNN-FA	15-3	45	94.78

Table 7 Mammographic mass

Mammographic mass dataset			
Learning algorithm	Best network structure	Number of tunable weights	Accuracy (%)
MLP-BP	5-5-1	36	50.84
FLNN-BP	15-1	16	58.50
FLNN-FA	15-1	16	82.25

Table 8 Lymphographic

Lymphographic dataset			
Learning algorithm	Best network structure	Number of tunable weights	Accuracy (%)
MLP-BP	18-9-4	211	74.24
FLNN-BP	171-4	468	60.57
FLNN-FA	171-4	468	84.96

ILPD, PIMA, Mammographic and Lymphographic dataset. The result also demonstrates that FLNN-FA model perform better than MLP trained with BP algorithm model with better classification accuracy with a reduced number of weight parameters. Although for the Thyroid dataset, the conventional MLP-BP could provide a better classification accuracy, nonetheless, it is worth noting that the difference with the proposed FLNN-FA is rather negligible with a notable reduced computational expense owing the reduced number of tunable weights.

5 Conclusion

The experiment has showed that the proposed FA Algorithm can effectively train the FLNN network with better accuracy value. Therefore, from the present investigation, it could be deduced that FA could be employed to train FLNN models. In future, further investigation on prediction and multiclass classification dataset will be conducted to verify the practicability of the proposed training algorithm.

Acknowledgements The authors would like to express their gratitude to Universiti Tun Hussein Onn Malaysia (UTHM) for supporting the present study.

References

1. Klassen M, Pao YH, Chen V (1988) Characteristics of the functional link net: a higher order delta rule net. In: IEEE international conference on neural networks 1988, vol 501, pp 507–513
2. Pao YH (1989) Adaptive pattern recognition and neural networks. Addison-Wesley Longman Publishing Co., Inc.
3. Ismail A (2001) Training and optimization of product unit neural networks. Master thesis University of Pretoria
4. Pao YH, Takefuji Y (1992) Functional-link net computing: theory, system architecture, and functionalities. Computer 25:76–79
5. Emrani S, Salehizadeh SMA, Dirafzoon A, Menhaj MB (2010) Individual particle optimized functional link neural network for real time identification of nonlinear dynamic systems. In: The 5th IEEE conference on Industrial Electronics and Applications (ICIEA), 2010, pp 35–40
6. Teeter J, Mo-Yuen C (1998) Application of functional link neural network to HVAC thermal dynamic system identification. Ind Electron IEEE Trans 45:170–176
7. Purwar S, Kar IN, Jha AN (2007) On-line system identification of complex systems using Chebyshev neural networks. Appl Soft Comput 7:364–372
8. Patra JC, Bornand C (2010) Nonlinear dynamic system identification using Legendre neural network. In: The 2010 International Joint Conference on Neural Networks (IJCNN), pp 1–7
9. Patra JC, Kot AC (2002) Nonlinear dynamic system identification using Chebyshev functional link artificial neural networks. IEEE Trans Syst Man Cybern B: Cybern 32:505–511
10. Marcu T, Köppen-Seliger B (2004) Dynamic functional-link neural networks genetically evolved applied to system identification. In: ESANN, pp 115–120
11. Nanda SJ, Panda G, Majhi B, Tah P (2009) Improved identification of nonlinear MIMO plants using new hybrid FLANN-AIS model. In: Advance computing conference, 2009. IACC 2009 IEEE international, pp 141–146
12. Samarasinghe S (2006) Neural networks for applied sciences and engineering. Auerbach Publications
13. Haring S, Kok J (1995) Finding functional links for neural networks by evolutionary computation. In: Van de Merckt T et al (eds) BENELEARN 1995, proceedings of the fifth Belgian–Dutch conference on machine learning, pp 71–78
14. Haring S, Kok J, Van Wesel M (1997) Feature selection for neural networks through functional links found by evolutionary computation. In: ILiu X et al (eds) Advances in intelligent data analysis (IDA-97). LNCS 1280, pp 199–210
15. Sierra A, Macias JA, Corbacho F (2001) Evolution of functional link networks. IEEE Trans Evol Comput 5:54–65
16. Abu-Mahfouz I-A (2005) A comparative study of three artificial neural networks for the detection and classification of gear faults. Int J Gen Syst 34:261–277

17. Misra BB, Dehuri S (2007) Functional link artificial neural network for classification task in data mining. J Comput Sci 3:948–955
18. Dehuri S, Mishra BB, Cho S-B (2008) Genetic feature selection for optimal functional link artificial neural network in classification. In: Proceedings of the 9th international conference on intelligent data engineering and automated learning. Springer-Verlag, Daejeon, South Korea, pp 156–163
19. Yang X-S (2008) Nature-inspired metaheuristic algorithms. Luniver Press
20. Alweshah M (2014) Firefly algorithm with artificial neural network for time series problems. Res J Appl Sci Eng Technol 7:3978–3982
21. Hassanzadeh T, Vojodi H, Moghadam AME (2011) An image segmentation approach based on maximum variance Intra-cluster method and Firefly algorithm. In: 2011 Seventh International Conference on Natural Computation (ICNC), pp 1817–1821
22. Mohd Noor MH, Ahmad AR, Hussain Z, Ahmad KA, Ainihayati AR (2011) Multilevel thresholding of gel electrophoresis images using firefly algorithm. In: 2011 IEEE International Conference on Control System, Computing and Engineering (ICCSCE), pp 18–21
23. Horng M-H (2012) Vector quantization using the firefly algorithm for image compression. Expert Syst Appl 39:1078–1091
24. Abedinia O, Amjady N, Naderi MS (2012) Multi-objective environmental/economic dispatch using firefly technique. In: 2012 11th International Conference on Environment and Electrical Engineering (EEEIC), pp 461–466
25. Gandomi A, Yang X-S, Talatahari S, Alavi A (2013) Firefly algorithm with chaos. Commun Nonlinear Sci Numer Simul 18:89–98
26. Lichman M (2013) UCI machine learning repository http://archive.ics.uci.edu/ml. University of California, School of Information and Computer Science, Irvine, CA

Kinematic Variables Defining Performance of Basketball Free-Throw in Novice Children: An Information Gain and Logistic Regression Analysis

Mohsen Afrouzeh, Ferman Konukman, Rabiu Muazu Musa, Pathmanathan K. Suppiah, Anwar P. P. Abdul Majeed, and Mohd Azraai Mohd Razman

Abstract The current investigation is designed to determine the relevant kinematic variables (KV) that could define the performance of basketball free-throw in novice children via the application of machine learning analysis. A number of seven different KV were examined from 15 children (mean age 9.93 ± 0.55 years) that constituted actions from the shoulder, elbow, wrist, knee, hand velocity, flexion as well as extension stages. The children completed 4 blocks of 15 trials of basketball free-throw tasks from a standing position 3 meters away from the front of the board using modified equipment. The data of the kinematics variables were collected in a controlled laboratory environment with 2D dimensional video data acquisition process. An information gain (IG) analysis is applied to extract the KV that could best describe successful and fail throw performance whilst Logistic Regression model (LR) was used to ascertain the predictability of the extracted KV in defining the performance of the throws. The IG extracted a set of 4 kV that could best describe the successful and fail throw performances namely, shoulder movement, knee, elbow as well as wrist kinematics. The LR model was able to provide a reasonably good prediction rate of 88% with respect to the extracted KV. The approach utilised in the present study provides useful information in identifying kinematics patterns that could best

---

M. Afrouzeh
Informetrics Research Group, Ton Duc Thang University, Ho Chi Minh City, Vietnam

Faculty of Sports Science, Ton Duc Thang University, Ho Chi Minh City, Vietnam

F. Konukman
College of Arts and Sciences, Sport Science Program, Qatar University, Doha, Qatar

R. M. Musa (✉)
Centre for Fundamental and Continuing Education, Universiti Malaysia Terengganu, 21030 Kuala Nerus, Terengganu, Malaysia
e-mail: rabiumuazu86@gmail.com

P. K. Suppiah
Faculty of Psychology and Education, Universiti Malaysia Sabah, Sabah, Malaysia

A. P. P. Abdul Majeed · M. A. M. Razman
Innovative Manufacturing, Mechatronics and Sports Laboratory, Universiti Malaysia Pahang, 26600 Pekan, Pahang Darul Makmur, Malaysia

A. F. Ab. Nasir et al. (eds.), *Recent Trends in Mechatronics Towards Industry 4.0*, Lecture Notes in Electrical Engineering 730, https://doi.org/10.1007/978-981-33-4597-3_86

define successful and fail basketball free throws performances in novice children. This finding may assist the coaches in modifying training strategies to ensure easier as well as the successful execution of basketball free-throws.

Keywords Basketball · Free-throw performance · Information gain · Logistic regression · Kinematic variables

1 Introduction

In a game of basketball, a free-throw can be seen as an unopposed action put forward by a basketball player to score points through shooting from behind a designated point (free-throw line) marked at the end of the restricted area. A successful free-throw shooting demands good accuracy, precision, suitable biomechanical integration as well as concentration [1]. An understanding and application of motion mechanics is required to use the "good technique" as stated by the previous researchers and to allow the ability of athletes to be fully developed [2]. Several studies propose that the shooting performance of a player can be improved by proper training by means of a systematic approach, such as the use of implicit learning methods, especially analogy learning [3, 4]. Moreover, the importance of developing good shooting technique by applying kinematic movement has been highlighted in the previous studies [4, 5].

It has been reported that one of the causes of low rate of free-throw success is the failure of most players to learn the correct technique at an early stage [6]. As such, the detection of the essential variables pertaining to the success in the execution of free-throw shooting is vital for the provision of proper feedback as well as the improvement of training of technique in the learning of novice basketball players. As pointed out by the previous investigators, that the present challenge in learning of motor skill relatively lies in making some recommendations for modifying of equipment and games to help cultivate the passion of the sport as well as developing skills acquisition particularly in children and beginners [7]. Therefore, experimental evidence is needed for developing task and equipment modification to aid the learning process of beginners in this sport. The purpose of the current investigation is therefore to identify the related critical kinematic variables that could characterize the successful and unsuccessful outcome of basketball free-throw through the application of a non-conventional method, i.e. machine learning analysis.

2 Methodology

2.1 Participants

The participants of the current investigation involved a total number of 15 right-handed male healthy children (mean age 9.93 ± 0.55 years; height 1.39 ± 4.16 m;

body mass 31.65 ± 3.23 kg). The participants are a novice to the sport of basketball. It is worth mentioning that before the commencement of the study, the participants were informed about the purpose of the research and all the risks or otherwise emanating from the study has been communicated to the participants and their guardians. All the procedures were approved by the University Ethics Committee for the ethical use of human subjects, in accordance with the Helsinki Declaration.

2.2 Kinematic Variables Measurement and Tasks Evaluation

Each participant performed basketball free-throw task in standing position, 3 m away from the front of the basket board using a small ball (i.e., 440 g, 69–71 cm) and a rim of 45 cm circumference with an adapted height of 2 m. This changed task was undertaken to minimize the duration of the learning process (due to the shorter shooting distance and lower rim height compared to the normal free-throw position) and to enable the collection of kinematic data in a controlled laboratory setting. A total number of 10 free throws were performed in standing position using the right hand for each child, and their kinematics variables i.e. shoulder, elbow, wrist, knee, hand velocity, flexion as well as extension stages were recorded in a biomechanical laboratory with two-dimensional (2D) video data collection (i.e., using 240 Hz camera resolution). Moreover, the final results of the throws either successful or fail were also recorded for further analysis.

2.3 Features Identification via Information Gain Approach

Information Gain (IG) is known to be an entropy-based variable assessment approach that is frequently used in the field of machine learning [8]. The IG approach is primarily designed to extract information provided by one or more variables with regard to a given categorical dependent variable [9]. It is essential to note that the IG is key in evaluating the variables that may provide information in order to assess the value of such a variable for the tasks of classification or discrimination. In the present investigation, the IG is used to retrieve information that shows the value of the KV, i.e. the activity of the shoulder, elbow, wrist, knee, hand velocity, flexion, as well as extension stages in defining the successful and unsuccessful free-throw performances.

2.4 Application of the Logistic Regression Classifier

Logistic regression (LR) is considered as one of the widely used learning algorithms often applied for solving categorization problems. The LR is derived from the function utilised at the foundation of the technique also known as the logistic function. The Logistic function which otherwise referred to as the sigmoid function was initially formed by a statistician for the purpose of determining the elements of population growth in an ecology setting that were perceived as rapidly expanding to the threshold of the actual carrying limit of the environment. The logistic function is normally an S-pattern curve that could consider any real-valued number and map it into value within the range of 0 and 1 [10]. To get more insights on the LR algorithm, the readers could refer to the subsequent literature [11–13].

In the present study, the LR algorithm is applied to classify the successful as well as unsuccessful basketball free throw with regards to the kinematic variables of the children previously described. The data were separated into a ratio of 70:30 for both training and testing. A 10-fold cross-validation method was used whilst the Ridge (L2) algorithm was employed with the c set at 1. The kinematic variables were treated as the independent variables whilst the free-throw performances i.e. successful and unsuccessful were used as the dependent variables. The efficacy of the model in predicting the study variables was assessed via the Classification Accuracy (CA), the Area under the curve (AUC), Precision, Recall as well as F1. All the data evaluation is carried out through Orange statistical software version 3.26 for Windows.

3 Results and Discussion

Table 1 displays the characteristics of the information collected with respect to the kinematic variables analyzed in the experiment. The table shows the relevant details on the ranking of variables, precisely the gain details, the gain ratio, the total variance, and the chi-square values. It could be detected that there is a sum 4 kinematic variables namely, shoulder, elbow, wrist as well as knee actions are rated as the most contributed kinematic variables that could define the successful and fail free-throw performance of the children under investigation.

Table 1 Features gain information values on the kinematic variables assessed

Kinematic variables	Infor. gain	Gain ratio	Variance	χ^2
Shoulder	**0.053**	0.026	9.756	9.775
Elbow	**0.044**	0.022	13.506	8.82
Wrist	**0.041**	0.02	11.414	5.773
Knee	**0.034**	0.017	5.101	4.311
Flexion stage	0.026	0.013	0.526	0.578
Extension stage	0.021	0.01	3.072	2.409
Hand velocity	0.011	0.006	1.705	0.638

The values in bold denote the important kinematic variables

Ranking techniques via the employment of information gain analysis has been portrayed as successful in practical application due to its effectiveness in the extraction of valuable information from a set of variables [14]. In this approach, a suitable ranking measure is used to rank the variables while the threshold is used to delete the variables that fall below the preset limit [15]. It is important indicating that the main objective of the variable ranking is to assess the importance of the variables with regard to the class attributes, i.e. the success and failure of the free throws. It basically stresses that if a variable is to be important, It can be distinct of the data set but could not be exclusive of the class labels; that is to say, it is possible to remove a variable that may not influence the class labels [9]. The aim of using a feature ranking by means of an analysis of the information gain in the current investigation is to classify one or more variables that could best be used to ascertain the influence of the class features i.e. successful and fail free throws shots.

Figure 1 exhibits the grouping of the KV in the study based on 4 identified variables. It could be observed from the figure that the grouping of the children performances in the free-throws is well segregated with respect to the outcomes of the free-throws i.e. fail or success.

Figure 2 highlights the confusion matrix generated from the LR algorithm. It could be detected that 8 misclassifications are observed from the successful free-throws whereas, 3 misclassifications were found in the fail free-throw group. It is worth highlighting that the confusion matrix is generated from the performance of the model developed in the current investigation. It is important to note that the model was able to demonstrate a rather reasonable prediction ability based on the

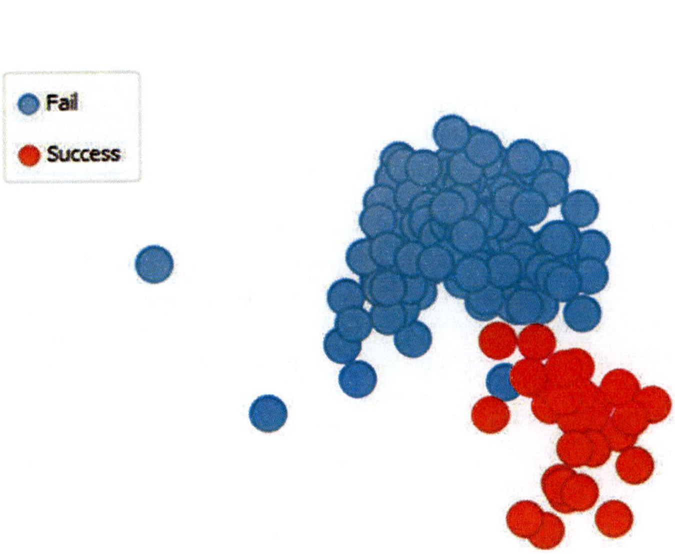

Fig. 1 Categorization of the children performances based on the 4 sets of kinematic variables after feature extractions

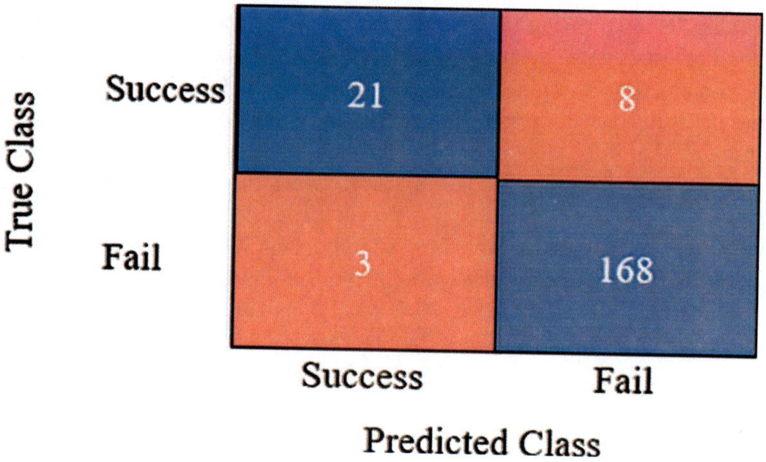

Fig. 2 Confusion matrix of the logistic regression model developed

parameters evaluated. A total of 88% CA, 80% AUC, 87% Precision, 88% Recall, as well as 86% F1, was demonstrated by the model.

The CA is the overall classification accuracy retrieved from the model whilst the AUC (area under the curve), is the measure of the separation ability of the model which normally fells within the category of 0 (no separation) to 1 (high separation). The precision stands for the ratio of the actual predicted positive observations to the overall predicted observations. The Recall (sensitivity) evaluates the ratio of correctly predicted observations to all the observations in the actual class. The F1 is the weighted average of both Precision and Recall which considers the false positive as well as false negatives into account [16–18]. In a nutshell, it can be inferred that the performance of the model developed in the present investigation is rather non-trivial when considering the outputs of the aforesaid model parameters assessed. Therefore, it is tempting to reveal that the extracted kinematic variables measured in the current investigation could potentially predict the outcome of the basketball free-throw performances of the children.

The kinematic performance variables identified in the present study specifically, the shoulder, elbow, wrist as well as knee actions are demonstrated to be non-trivial in the categorization of the consequence of the free-throws in novice children. It has been reported previously that to guarantee good performance in the precision and accuracy motor task, the athletes are needed to prepare themselves with appropriate kinematic techniques [19–22]. Moreover, the identified variables have been reported to play a significant role in determining the success of basketball free-throws [5, 23, 24]. Basketball players are required to be precise when aiming at the basket since the baskets are relatively small, as such the players need to exert appropriate kinematics actions in order to consistently make successful shoots.

4 Conclusion

The present investigation has successfully ascertained the essential kinematic variables that could distinguish between successful and fail basketball free-throw performances through the application of a machine learning technique. The research has shown that the kinematic variables constituting the hip, elbow, wrist and knee movements are influential in affecting the quality of a free throw in this game. The Knowledge Gain Approach, in conjunction with the Machine Learning Model Analysis, is advantageous in providing information on the associated kinematic variables that may be used to improve children's success in free-throw basketball. The results of the present investigation may be of benefit to administrators, team managers and sports commentators in mapping out training strategy that might go a long way in aiding children to improve their performances in this game. It is recommended that further research should be carried out to ascertain the effect of different learning instructions of motor skills and its associations with the kinematic actions of children through machine learning models.

Acknowledgements The authors wish to appreciate the participants for their help in carrying out this analysis.

References

1. Lam W-K, Lee WC-C, Lee WM, Ma CZ-H, Kong PW (2018) Segmented forefoot plate in basketball footwear: does it influence performance and foot joint kinematics and kinetics? J Appl Biomech 34:31–38
2. Arias JL (2012) Influence of ball weight on shot accuracy and efficacy among 9-11-year-old male basketball players. Kinesiology 44:52–59
3. Brancazio PJ (1981) Physics of basketball. Am J Phys 49:356–365. https://doi.org/10.1119/1.12511
4. Burns FT (1990) Teaching components for shooting improvement in wheelchair basketball—tidbits of information about shooting a basketball. In: Proceeding of the national wheelchair basketball symposium for coaches, athletes and officials, pp 79–83
5. Hudson JL (1985) Prediction of basketball skill using biomechanical variables. Res Q Exerc Sport 56:115–121. https://doi.org/10.1080/02701367.1985.10608445
6. Owen E (1982) Playing and coaching wheelchair basketball. University of Illinois Press, Urbana
7. Farrow D, Reid M (2010) The effect of equipment scaling on the skill acquisition of beginning tennis players. J Sports Sci 28:723–732. https://doi.org/10.1080/02640411003770238
8. Musa RM, Abdul Majeed APP, Musa A, Abdullah MR, Kosni NA, Razman MAM (2021) An Information Gain and Hierarchical Agglomerative Clustering Analysis in Identifying Key Performance Parameters in Elite Beach Soccer. Presented at the (2021). https://doi.org/10.1007/978-981-15-7309-5_26
9. Alhaj TA, Siraj MM, Zainal A, Elshoush HT, Elhaj F (2016) Feature selection using information gain for improved structural-based alert correlation. PLoS One 11
10. Musa RM, Suhaimi MZ, Mohamad Razali Abdullah APPAM, Maliki ABHM (2020) Predicting students academic performance from wellness status markers using machine learning techniques. Indian J Sci Technol 13:2047–2055. https://doi.org/10.17485/IJST/v13i29.999

11. Musa RM, Majeed APPA, Kosni NA, Abdullah MR (2020) Machine learning in team sports: performance analysis and talent identification in Beach Soccer & Sepak-takraw. Springer, Berlin
12. Muazu Musa R, Majeed PPA, Abdullah MR, Ab. Nasir AF, Arif Hassan MH, Mohd Razman MA (2019) Technical and tactical performance indicators discriminating winning and losing team in elite Asian beach soccer tournament. PLoS One 14:e0219138
13. Musa RM, Abdul Majeed APP, Taha Z, Chang SW, Nasir AFA, Abdullah MR (2019) A machine learning approach of predicting high potential archers by means of physical fitness indicators. PLoS One 14. https://doi.org/10.1371/journal.pone.0209638
14. Chandrashekar G, Sahin F (2014) A survey on feature selection methods. Comput Electr Eng 40:16–28
15. Lei S (2012) A feature selection method based on information gain and genetic algorithm. In: 2012 International Conference on Computer Science and Electronics Engineering, pp. 355–358. IEEE
16. Musa RM, Taha Z, Maje Anwar PPA, Abdullah MR (2019) Machine learning in sports : identifying potential archers. Springer Singapore, Singapore. https://doi.org/10.1007/978-981-13-2592-2
17. Musa RM, Abdullah MR, Maliki ABHM, Kosni NA, Mat-Rasid SM, Adnan A, Juahir H (2018) Supervised pattern recognition of archers' relative psychological coping skills as a component for a better archery performance. J Fundam Appl Sci 10:467–484
18. Taha Z, Musa RM, Abdul Majeed APP, Abdullah MR, Abdullah MA, Hassan MHA, Khalil Z (2018) The employment of support vector machine to classify high and low performance archers based on bio-physiological variables. IOP Conf Ser Mater Sci Eng 342:012020. https://doi.org/10.1088/1757-899X/342/1/012020
19. Musa RM, Abdullah MR, Juahir H, Maliki A, Mat-Rasid SM, Kosni NA, Adnan A, Alias N, Eswaramoorthi V (2018) A multidimensional analysis of physiological and mechanical variables among archers of different levels of expertise. J Fundam Appl Sci 10:18–32
20. Eswaramoorthi V, Abdullah MR, Musa RM, Maliki ABHM, Kosni NA, Raj NB, Alias N, Azahari H, Mat-Rashid SM, Juahir H (2018) A multivariate analysis of cardiopulmonary parameters in archery performance. Hum Mov 19:35–41. https://doi.org/10.5114/hm.2018.77322
21. Taha Z, Musa RM, Abdullah MR, Razman MAM, Lee CM, Adnan FA, Abdullah MA, Haque M (2017) The application of inertial measurement units and wearable sensors to measure selected physiological indicators in archery. Asian J Pharm Res Heal Care 9:85–92
22. Suppiah PK, Kiet TWK, Musa RM, Abdullah MR, Lee JLF, Maliki ABHM (2017) The effectiveness of a core muscles stability program in reducing the postural sway of adolescent archers: a panacea for a better archery performance. Int J Physiother 4:296–301
23. Ogawa M, Hoshino S, Fujiwara M, Nakata H (2019) Relationship between basketball free-throw accuracy and other performance variables among collegiate female players. J Phys Fit Sport Med 8:127–136
24. Wong DW-C, Lam W-K, Chen TL-W, Tan Q, Wang Y, Zhang M (2020) Effects of upper-limb, lower-limb, and full-body compression garments on full body kinematics and free-throw accuracy in basketball players. Appl Sci 10:3504

The Identification of Significant Time-Domain Features for Wink-Based EEG Signals

Tang Jin Cheng, Jothi Letchumy Mahendra Kumar, Mamunur Rashid, Rabiu Muazu Musa, Mohd Azraai Mohd Razman, Norizam Sulaiman, Rozita Jailani, and Anwar P. P. Abdul Majeed

Abstract Brain-Computer Interface (BCI) is said to be a system that can measure and convert the brain activity into readable outputs. These outputs are said to be beneficial to the people who face physical challenges in carrying out their daily life as the outputs can be employed to control the BCI-based assistive device. Electroencephalography (EEG) is one of the electrophysiological monitoring techniques that record the brain's electrical activity. Informative attributes can be extracted from the massive outputs of EEG signal and help in increasing the effectiveness of the BCI-based device. This study aims to discover the significant statistical time-domain features that can be used in the classification of the left wink, right wink and no wink utilising EEG signals. EMOTIV Insight was used as the EEG recording device to obtain the EEG signals triggered from the winking motion of the left and right wink. Six healthy subjects that ranged between 23 years old to 27 years old were involved in the wink-based EEG recordings. Nine statistical time-domain features were extracted, namely mean, median, standard deviation, variance, root-mean-square (RMS), minimum (Min), maximum (Max), skewness and kurtosis. The identification of the significant features is attained via a filter method known

T. J. Cheng · J. L. Mahendra Kumar · M. A. Mohd Razman · A. P. P. Abdul Majeed
Innovative Manufacturing, Mechatronics and Sports Laboratory, Faculty of Manufacturing and Mechatronics Engineering Technology, Universiti Malaysia Pahang (UMP), 26600 Pekan, Pahang Darul Makmur, Malaysia

M. Rashid · N. Sulaiman
Faculty of Electrical and Electronics Engineering Technology, Universiti Malaysia Pahang (UMP), 26600 Pekan, Pahang Darul Makmur, Malaysia

R. M. Musa
Centre for Fundamental and Liberal Education, Universiti Malaysia Terengganu (UMT), 21030 Kuala Nerus, Terengganu Darul Iman, Malaysia

R. Jailani
Faculty of Electrical Engineering, Universiti Teknologi MARA (UiTM), 40450 Shah Alam, Selangor Darul Ehsan, Malaysia

A. P. P. Abdul Majeed (✉)
Centre for Software Development & Integrated Computing, Universiti Malaysia Pahang (UMP), 26600 Pekan, Pahang Darul Makmur, Malaysia
e-mail: amajeed@ump.edu.my

A. F. Ab. Nasir et al. (eds.), *Recent Trends in Mechatronics Towards Industry 4.0*, Lecture Notes in Electrical Engineering 730, https://doi.org/10.1007/978-981-33-4597-3_87

as information gain ratio. The ratio of training data to testing data was set to 70:30. The selected features for classification of winking is fed into various types of classifiers to observe the effect of this feature selection method on the performance of the classification, i.e. k-Nearest Neighbour (k-NN), Support Vector Machine (SVM), and Decision Tree. It was established from the present investigation that Standard Deviation, Variance and Min from channel AF4 were found to be significant. The classification accuracy (CA) for both train and test data with the filter feature selection method is observed to be comparably equal to the CA obtained from utilising all features. The findings from the study are non-trivial towards the realisation of a real-time BCI-based system.

Keywords Electroencephalogram (EEG) · Winking · Brain-computer interface (BCI) · Time-domain features · Machine learning · Feature selection · Classification

1 Introduction

Stroke is a severe medical condition that causes the interruption of blood flow to and within the brain. The lack of oxygen due to the artery blockage or rupture subjected to the sudden death of brain cell which said to be fatal to human [1]. It is considered as the major causes of long-term disability worldwide, and this incidence tends to happen as the age increases [2]. It has been revealed that from the population of new stroke patients, about three-quarters of the new stroke patients confronted with disabilities and to a certain extent, it became an obstacle for these patients in executing their activities of daily living (ADL) [3]. Recent technological development and innovation, especially in the area of BCI and supportive devices have become a piece of great news for the patients, whereby it assists the patients in improving their quality of life with the aids of robotics [4–7]. From this aspect, it is crucial to investigate the control of BCI, particularly in the EEG signals identification and classification.

Facial expressions are a form of nonverbal communication and somehow, it can convey different meanings in various contexts. There are various types of facial expressions, including look towards a direction, smiling, blinking, and winking [8]. The eye gestures-based EEG signals classification is worth to investigate, especially the wink-based EEG signals. Although there is much literature regards the EEG signals classification, nonetheless limited studies have been reported on wink related BCI. Kowalcyzk and Sawicki [9] reviewed on the detection of blink and wink as a tool to control. The authors reported that winking is said to be the intentional single eye movement, in other words, it is unnatural since it involves only the movement of one eye. The significance of winking can be greatly captured in the EEG signals and used as the BCI-control method [9].

Rashid et al. [10] researched on the classification of winking by using machine learning approach. Three male subjects and two female subjects, ranging from

23 years old to 27 years old have participated in the EEG signal recording with utilising the EMOTIV Insight device. Two different sessions of EEG signal recording are held on two different days. The authors employed the Fast Fourier Transform (FFT) feature selection method to transform the data into the frequency domain. The feature of the sample range is also extracted from the data. The ratio of training data to the testing data is set to 75:25. In the case of classifiers, Linear Discriminant Analysis (LDA), Support Vector Machine (SVM), and *k*-nearest Neighbor (*k*-NN) were used. The results showed that LDA performs better if the feature extraction technique of FFT is used, while SVM and *k*-NN had the same results if the sample range feature is used.

It is been observed other studies that the wink-based EEG signals are comparatively limited, and the feature extraction method is not exploited much in wink-based EEG signals. Hence, this study aims at identifying the significant statistical time-domain based features of wink-based EEG signals by using a filter type method and evaluate the efficacy of different machine learning models in the classification of winking.

2 Methodology

2.1 Data Acquisition and EEG Recording Device

For the EEG signal recording, a Bluetooth non-invasive device named Emotiv Insight (as shown in Fig. 1) was used, particularly in the wink-based EEG signal acquisition. This device comprised of five main channels and two reference channels. The material of the sensors are Hydrophilic semi-dry polymer and the main sensors are attached to channel named AF3, AF4, T7, T8 and Pz according to the International 10-20 system. The sampling rate of the device is 128 Hz. The bandwidth ranged from 0.5 to 43 Hz and the resolution for each channel is 0.51 μV.

Fig. 1 EMOTIV insight

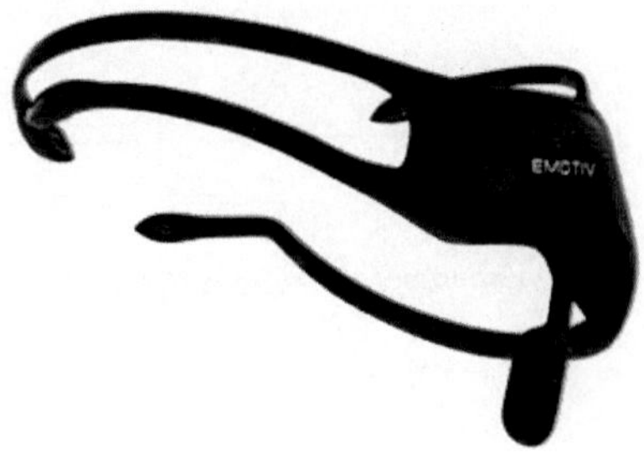

2.2 Subjects

A total of six healthy subjects (four male and two female) volunteered to take part in the EEG recording sessions. The age of the subjects is between 23 years to 27 years old. The participants have normal visions and none of the participants is subjected to any neurological disorder. The ethical approval for this study was obtained via an institutional research ethics committee (FF-2013-327).

2.3 Experimental Setup

The participants were asked to sit on an ergonomic chair in a comfortable position and relax their body. The environment is ensured silence throughout the recording sessions to prevent any external disturbances from the environment to be recorded. Before the experiment, the participants were asked to focus throughout the recording sessions. The subjects were also told to perform only the winking action according to the cue displayed on the monitor.

In each cycle, a black blank screen is appeared from t = 0 to t = 3 to minimize the occurrence of the artifacts. At t = 3, an acoustic stimulus occurred on the monitor to inform the subject ready for the winking. At the fifth seconds, an arrow cue that is pointed to the left or to the right is displayed on the screen and the subject was instructed to wink with the corresponding eye. This trial is repeated for five times in each trial, which contributed to a length of 60 s for one cycle. A total of 6 trials, which comprised of 90 samples, are acquired with 30 samples are left wink, right wink, and no wink, respectively. All these recordings are obtained on the same day (Fig. 2).

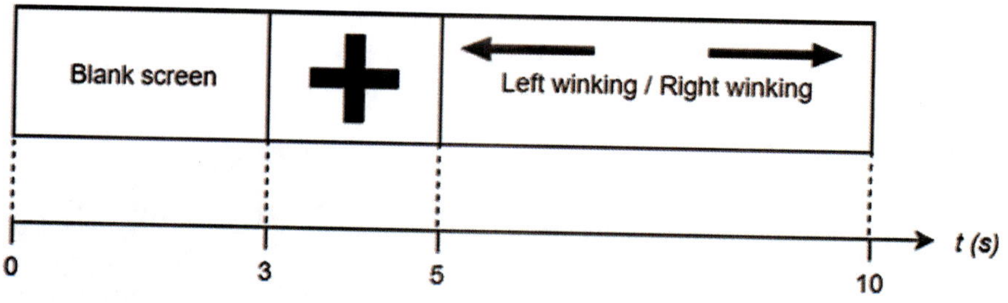

Fig. 2 Experiment paradigm of the EEG recording

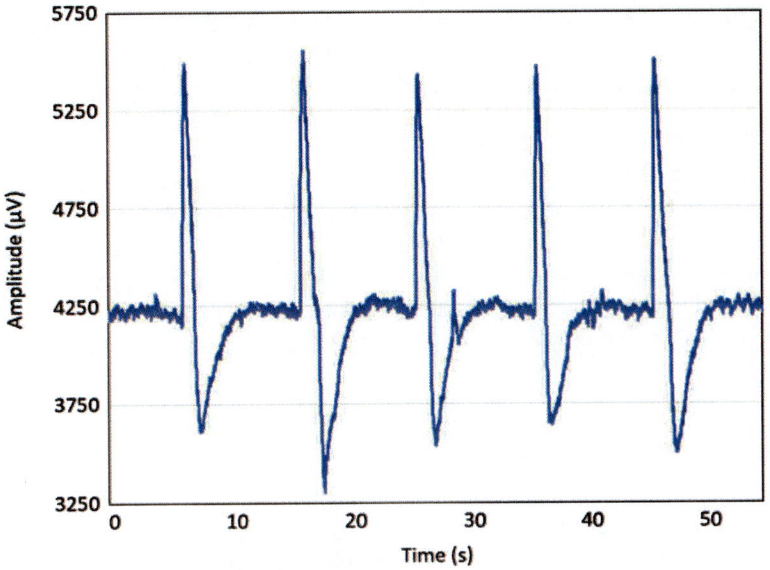

Fig. 3 Pre-processed left wink EEG signal

2.4 Signal Preprocessing, Feature Extraction and Selection

A fifth-order digital Sinc and notch filter of 50 Hz is used to filter the raw EEG signals for the preprocessing phase. Figure 3 represents one of the samples of the pre-processed EEG signal for five trial of the right wink. The pre-processing phase is followed by the epoch splitting section. The filtered signals were separated into two categories, which are winking signals and no winking signals respectively. For each trial, 3200 samples can be collected for which the subjects are under winking motion. After the pre-processing phase, nine statistical time-domain based features were extracted from the filtered samples [11–13]. The extracted features are mean, median, standard deviation, variance, root-mean-square (RMS), minimum (Min), maximum (Max), skewness and kurtosis from five channels, respectively. In this study, the filter method is applied in the feature selection stage. In this case, the information gain ratio is selected as the scoring method.

2.5 Classification

Various classifiers were applied to observe the effect of the features selected by the information gain ratio on the results. k-Nearest Neighbor (*k*-NN), Support Vector Machine (SVM), and Decision Tree were employed for the classification of winking. The *k*-NN is a supervised ML algorithm which relies on the distance and the number of neighbours for classification [14–16]. In this study, the Euclidean distance and neighbour's number has been set to five which were used to build the k-NN model. SVM is a supervised ML algorithm that uses kernel tricks and regularisation method to classify the datasets accordingly [11, 17]. In this study, the Radial Basis Function (RBF) kernel was employed. Whereas, Decision Tree is a non-parametric supervised learning method that employs a tree-like model of decisions. The parameters that were used to build Decision Tress model are minimum sample leaf which has been set to one and minimum sample split which has been set to two. Orange Data Mining version 3.23 was utilised to develop the classification model. The training to testing ratio was adjusted to 70:30 ratio from 90 samples. The evaluation of models is carried out by observing the Classification Accuracy (CA).

3 Result and Discussion

The weightage of the features obtained from the information gain ratio is illustrated in Fig. 4. It is observed that Min_AF4, Standard Deviation_AF4 and Variance_AF4 are the most significant features with a weightage of 0.563. This finding is in agreement to a certain extend with the findings reported by Rashid et al. [10]. The CA of

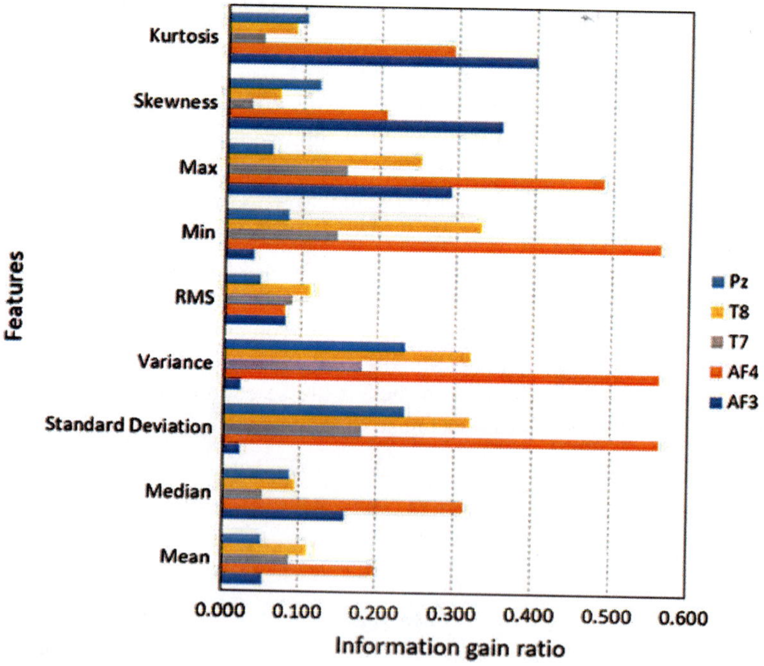

Fig. 4 The identified significant features via information gain ratio

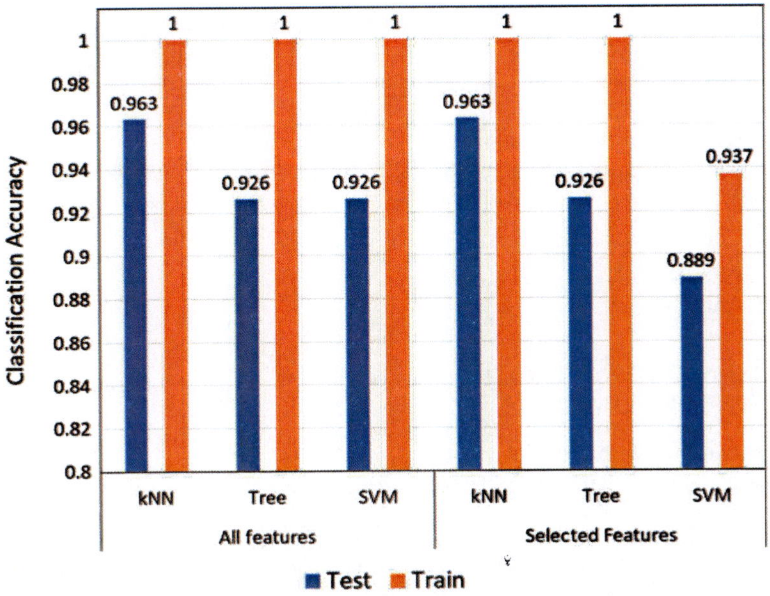

Fig. 5 Classification accuracy of the evaluated models based on different feature set

both training data and test data for various types of classifiers evaluated are depicted in Fig. 5. It can observe that for both selected features or all features, the test CA and train CA of k-NN and Tree are comparable. However, for SVM, both train CA and test CA are decreased with selected features, this could probably be caused by the unoptimised selection of the hyperparameters of the model. It is evident from the present study that the k-NN classifier is able to provide a reasonably accurate classification of the different types of winks evaluated. In addition, it is also apparent that the selection of the features is non-trivial towards achieving the reported accuracies. The reduction of the features from 45 to 3 further implies the possible reduced computational expense in the event that real-time execution is implemented on a BCI system.

4 Conclusion

The present study evaluated the identification of significant statistical time-domain features towards the classification of wink-based EEG signals. Through the implementation of a filter technique, namely information gain ratio, three features were found to be significant amongst the 45 features extracted. A number of classifiers were used to evaluate the importance of the features in classifying the winks. It was shown that comparable classification accuracy is attainable even with the use of only three features. With the reduction of features it is postulated that the computational expense could be reduced for its application on real-time BCI systems. Future study will investigate different feature selection techniques as well as the optimisation of the hyperparameters of different models.

Acknowledgements The authors would like to acknowledge Universiti Malaysia Pahang for funding this study via RDU180321.

References

1. Johnson W, Onuma O, Owolabi M, Sachdev S (2016) Stroke: a global response is needed. Bull World Health Organ 94:634
2. Kooi Cheah W, Peng Hor C, Abdul Aziz Z, Looi I (2016) A review of stroke research in Malaysia from 2000–2014. Med J Malaysia 71
3. Katan M, Luft A (2018) Global burden of stroke. Semin Neurol 38:208–211. https://doi.org/10.1055/s-0038-1649503
4. Millán JDR, Rupp R, Müller-Putz GR, Murray-Smith R, Giugliemma C, Tangermann M, Vidaurre C, Cincotti F, Kübler A, Leeb R, Neuper C, Müller KR, Mattia D (2010) Combining brain-computer interfaces and assistive technologies: state-of-the-art and challenges. Front Neurosci 4:1–15. https://doi.org/10.3389/fnins.2010.00161
5. Gao Q, Zhao X, Yu X, Song Y, Wang Z (2018) Controlling of smart home system based on brain-computer interface. Technol Heal Care 26:769–783. https://doi.org/10.3233/THC-181292
6. Tang J, Liu Y, Hu D, Zhou ZT (2018) Towards BCI-actuated smart wheelchair system. Biomed Eng Online 17:1–22. https://doi.org/10.1186/s12938-018-0545-x
7. Tariq M, Trivailo PM, Simic M (2018) EEG-based BCI control schemes for lower-limb assistive-robots. Front Hum Neurosci 12. https://doi.org/10.3389/fnhum.2018.00312
8. Raheel A, Majid M, Anwar SM (2019) Facial expression recognition based on electroencephalography. In: 2nd international conference on computing, mathematics and engineering technologies iCoMET 2019, pp 1–5. https://doi.org/10.1109/ICOMET.2019.8673408
9. Kowalczyk P, Sawicki D (2018) Blink and wink detection as a control tool in multimodal interaction
10. Rashid M, Sulaiman N, Mustafa M, Bari BS, Sadeque MG, Hasan MJ (2020) Wink based facial expression classification using machine learning approach. SN Appl Sci 2:183
11. Chatterjee R, Bandyopadhyay T (2016) EEG based motor imagery classification using SVM and MLP. In: Proceedings of international conference computational intelligence and networks, pp 84–89. https://doi.org/10.1109/CINE.2016.22
12. Altın C, Er O (2016) Comparison of different time and frequency domain feature extraction methods on elbow gesture's EMG. Eur J Interdiscip Stud 5:35–44. https://doi.org/10.26417/ejis.v5i1
13. Nazmi N, Rahman MAA, Yamamoto SI, Ahmad SA, Zamzuri H, Mazlan SA (2016) A review of classification techniques of EMG signals during isotonic and isometric contractions. Sensors (Switzerland). 16:1–28. https://doi.org/10.3390/s16081304
14. Khairuddin IM, Na'im Sidek S, Majeed APPA, Puzi AA (2019) Classifying motion intention from EMG signal: a k-NN approach. In: 2019 7th International Conference on Mechatronics Engineering (ICOM), pp 1–4. IEEE
15. Razman MAM, Majeed APPA, Musa RM, Taha Z, Susto GA, Mukai Y (2020) Time-series identification on fish feeding behaviour. In: Machine learning in aquaculture, pp 37–47. Springer, Berlin

16. Letchumy J, Rashid M, Musa RM (2020) The classification of wink-based EEG signals : the identification of significant time-domain. Springer Singapore. https://doi.org/10.1007/978-981-15-7309-5
17. Chengaiyan S, Retnapandian AS, Anandan K (2020) Identification of vowels in consonant–vowel–consonant words from speech imagery based EEG signals. Cogn Neurodyn 14:1–19. https://doi.org/10.1007/s11571-019-09558-5

Hyper-Heuristic Strategy for Input-Output-Based Interaction Testing

Fakhrud Din and Kamal Z. Zamli

Abstract Software testing aims at exploring faults within software in order to ensure it meets all necessary specifications. Test case design strategies play key role in software testing. Classical test case design strategies, however, do not sufficiently include support for exploration of faults due to interaction between parameter values. New strategies known as t-way strategies (where t expresses interaction strength) have been developed for finding interaction faults. However, existing t-way strategies for input-output-based relationship (IOR) interaction testing mostly adopt greedy algorithms which often generate poor quality test data. Therefore, this paper presents the design of a new IOR test suite generation strategy called IOR_HH based on the exponential Monte Carlo with counter (EMCQ) hyper-heuristic. EMCQ is a parameter free hyper-heuristic which works as controller of the three implemented low-level meta-heuristic operators, namely crossover, peer learning and global pollination in the proposed IOR_HH strategy. Experimental results demonstrate the impact of the proposed strategy against existing computational strategies for IOR interaction testing.

Keywords Input-output interaction · Hyper-heuristic · Exponential monte carlo

1 Introduction

Dependency on software systems in every walk of human life makes software testing one of the crucial stages within software development life cycle [1]. Software testing

F. Din (✉) · K. Z. Zamli
Faculty of Computer Systems and Software Engineering, Universiti Malaysia Pahang, 26300 Kuantan, Pahang, Malaysia
e-mail: kamalz@ump.edu.my

K. Z. Zamli
e-mail: fakhruddin@uom.edu.pk

F. Din
Department of Computer Science & IT, University of Malakand, Lower Dir District, KPK, Pakistan

A. F. Ab. Nasir et al. (eds.), *Recent Trends in Mechatronics Towards Industry 4.0*, Lecture Notes in Electrical Engineering 730, https://doi.org/10.1007/978-981-33-4597-3_88

has become more challenging owing to interaction faults which occur when two or more input parameters of a software-under-test (SUT) are used in combination [2]. To detect interaction faults, t-way test data generation technique is found very effective. This technique generates minimum possible test cases for covering enormous number of interactions (i.e., the combinatorial explosion problem) among input parameters of a SUT [3]. Although useful, existing t-way testing strategies offer little support for complex input output relationships between a SUT's parameter values [4].

Input-output-based relationship (IOR) interaction testing as part of t-way testing, generates test data which consist of interactions among several input parameters with common output [5, 6]. Software tester utilizes specification of SUT to identify which input parameter values interact and effect which outputs. IOR interaction testing has shown capacity to generate few test cases by avoiding unnecessary interactions and eliminating redundant test data. Moreover, IOR interaction testing not only reduces testing efforts it shows success in detecting faults.

Mostly, existing t-way strategies for IOR interaction testing are based on greedy algorithms. Union [5], Greedy [7], ReqOrder [8], ParaOrder [8], Density [9], Integrated t-way Test Data Generator (ITTG) [10] are some early well-known strategies for IOR interaction test suite generation. Following these strategies, AURA [11] and General Variable Strength (GVS) [12] have included support for IOR interaction testing. Although useful as far as test suite generation time is concerned, these strategies often generate test suites with large sizes. Based on the effectiveness of meta-heuristic algorithms [13–19] and their hybrid forms [20–23] for other types of t-way testing, the adoption of a hyper-heuristic for IOR interaction testing can further increase the optimality of IOR test suites.

Hyper-heuristics are promising methodologies for solving complex optimization problems [24]. These new optimization methods support solution of problems with different nature for promoting generality of problem solvers as meta-heuristics are problem dependent. Moreover, meta-heuristics require extensive parameter-tuning prior to generate good solutions. Hyper-heuristics address these issues while increase the potential of existing meta-heuristics by exploiting their strengths whereas compensating their weaknesses at the same time. Motivated by these alluring features of hyper-heuristics, this paper presents the design of a new strategy called IOR_HH for addressing the IOR test suite generation problem. The proposed strategy adopts exponential Monte Carlo with Counter (EMCQ) [25] as its high-level heuristic whereas it implements genetic algorithm (GA) [26] crossover operator, teaching learning-based optimization (TLBO) [27] peer learning operator and flower pollination algorithm (FPA) [28] global pollination operator as its low-level heuristics. EMCQ hyper-heuristic selects one of these operators based on performance, whereas the operators actually generate IOR test suites.

The rest of the paper is organized as follows. Section 2 overviews IOR interaction testing. Section 3 outlines the related work. Section 4 highlights the proposed IOR_HH for IOR test suite generation. Section 5 synthesis results by discussion. Finally, Sect. 6 concludes this paper.

2 Overview of Input-Output-Based (IOR) Interaction Testing

Input-output-based (IOR) interaction testing generates interaction test suite in order to cover which combination of inputs influence which outputs of a software under test (SUT). To elaborate IOR interaction testing, a hypothetical example is presented in Fig. 1. The SUT consists of four two-valued inputs whereas it has three outputs influenced by two inputs (please refer to Fig. 1a). Output f_1 is influenced by two inputs *A* and *B*. Output f_2 is influenced by inputs *A* and *C*. Output f_3 is influenced by inputs *C* and *D*. The notation *IOR(N, C, Rel)* can be used to express a SUT. Here, *N* represents the number of test cases, *C* represents the configuration in the form V^P where V is for values of inputs and P is for number of inputs (or *C* will be in the form: $(V_1^{P_1}, V_2^{P_2}, \ldots, V_n^{P_n})$ in case SUT's inputs have different number of values) and Rel denotes all the interactions as a multi set input relationships that effects the outputs in the form $\{\{R_1\}, \{R_2\}, \ldots, \{R_k\}\}$. For the SUT shown in Fig. 1a, *IOR* $(N, 2^4, \{\{A, B\}, \{A, C\}, \{C, D\}\}$ notation can be used.

There are total 12 interactions between the values of the four given inputs that effect their respective outputs as depicted in Fig. 1b. IOR interaction testing attempts to cover these interactions with minimum possible test cases to avoid exhaustive testing of the SUT with 12 test cases for the running example. Table 1 shows only four test cases obtained by the proposed strategy in order to cover all the required IOR interactions. This substantially reduces testing efforts for SUTs with hundreds of complex input-output relationships.

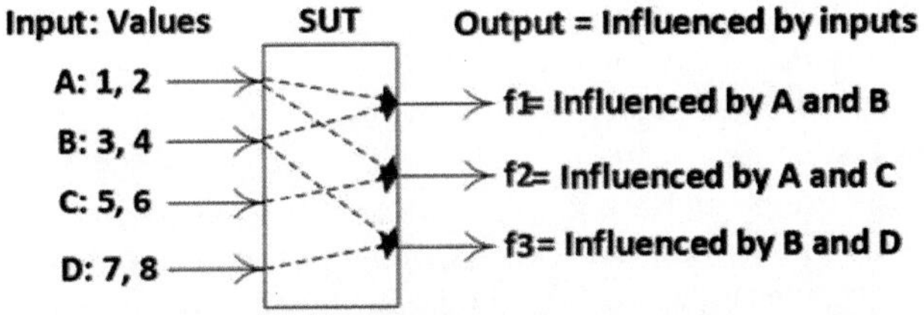

(a) Software-under-test (SUT) with inputs and outputs

A	B	C	D
1	3	--	--
1	4	--	--
2	3	--	--
2	4	--	--

$f_1 = (1100)$

A	B	C	D
1	--	5	--
1	--	6	--
2	--	5	--
2	--	6	--

$f_2 = (1010)$

A	B	C	D
--	--	5	7
--	--	5	8
--	--	6	7
--	--	6	8

$f_3 = (0011)$

(b) All combinations of inputs' values influence respective outputs

Fig. 1 Software-under-test (SUT) with four two-valued inputs and three outputs influenced by two inputs

Table 1 IOR test suite for the SUT in Fig. 1a

Test Case	A	B	C	D	IOR interactions covered
ID = TC_1	1	3	5	7	{{1,3}, {1, 5}, {5, 7}}
ID = TC_2	2	3	6	7	{{2,3}, {2, 6}, {6, 7}}
ID = TC_3	1	4	6	8	{{1,4}, {1, 6}, {6, 8}}
ID = TC_4	2	4	5	8	{{2,4}, {2, 5}, {5, 8}}

3 Related Works

As part of the uniform and variable strength t-way testing, many existing strategies support IOR features. Most of these strategies employ greedy algorithms and iterative processes to generate IOR test suites [6]. Two early strategies for IOR interaction testing are Union [5] and Greedy [7]. Union first starts by generating partial test cases based on the given relationships between inputs and outputs. For completing these test cases, Union then assigns random values to missing inputs not present in the given input–output relationships. Greedy strategy is based on the principles of the Union strategy. However, test cases completion in Greedy is performed greedily with the aim of achieving maximum coverage of interactions.

ReqOrder [8] is an extension of the Union strategy. In ReqOrder, 'don't care' positions are filled only when coverage requirement corresponding to interactions for these positions is encountered. This helps ReqOrder to generate test suites with smaller sizes than other strategies. ParaOrder [8] proposed is based on in-parameter-order method that generates an initial test suite for a sub-system considering small number of factors. The strategy then extends the system by accumulating new interactions to obtain a test suite for new sub-system. The extending procedure is performed repeatedly till the inclusion of all factors into that system. For IOR interaction testing, ParaOrder offers competitive test suite sizes against all the previous algorithms. Density [9] strategy adopts heuristics for IOR test suite generation. This strategy uses global and local density values associated with input-output relationships. While generating a test case, the strategy selects a relationship having greater local density value. Then from the exhaustive set of interactions of the selected IOR, the interaction having greater global density value is added to the test case. The Density strategy repeats this process till the generation of final test suite.

AURA [11] and General Variable Strength Strategy (GVS) [12] are two relatively new strategies based on greedy algorithms that support IOR test suite generation. At the beginning, AURA generates an empty test case. It then randomly generates test cases that cover maximum number of IOR interactions. This process is repeated till the enumerated list of IOR interactions is not empty. Unlike AURA, there is no enumeration of interaction elements in GVS which makes it a memory efficient strategy for IOR interaction testing.

Although helpful in terms of IOR test suite generation time, all the aforementioned strategies generate poor quality test data (i.e., larger size test suite). To generate test data with maximum optimality, strategies based on new search engine methods such hyper-heuristics need to be introduced in the literature.

4 Design of IOR_HH Strategy

The IOR_HH strategy adopts the Exponential Monte Carlo with Counter (EMCQ) [25] hyper-heuristic for selecting best performer from three implemented low-level meta-heuristic operators as shown in Fig. 2. These operators include crossover, peer learning, and global pollination from genetic algorithm (GA), teaching learning-based optimization (TLBO) algorithm and flower pollination algorithm (FPA), respectively (refer to Fig. 2). As low-level heuristics of the IOR_HH, these operators are responsible for the generation of the optimal IOR test suites. To avoid bulky implementation, only operators from the well-known meta-heuristic algorithms work as low-level heuristics in the proposed IOR_HH strategy.

EMCQ selects a low-level heuristic based on the fulfillment of Monte Carlo criteria. The criteria always favor a low-level heuristic which generates good solution. However, EMCQ introduces probability density to allow the currently running low-level heuristic even if it obtains a poor solution. This feature helps EMCQ to avoid trapping in local optima. The probability density is given in Eq. (1).

$$\psi = e^{-\delta \cdot \frac{T}{q}} \tag{1}$$

where δ represents the difference between fitness value of the current operator and last operator, T denotes iteration counter, and q represents a control parameter for consecutive non-improving iterations.

Like simulated annealing (SA), EMCQ's probability density Ψ decreases as iteration counter T increases. However, unlike SA, there is no use of cooling schedule in case of EMCQ which makes it a parameter free hyper-heuristic. Moreover, the dynamic manipulation of q is another useful feature of EMCQ as this feature enables

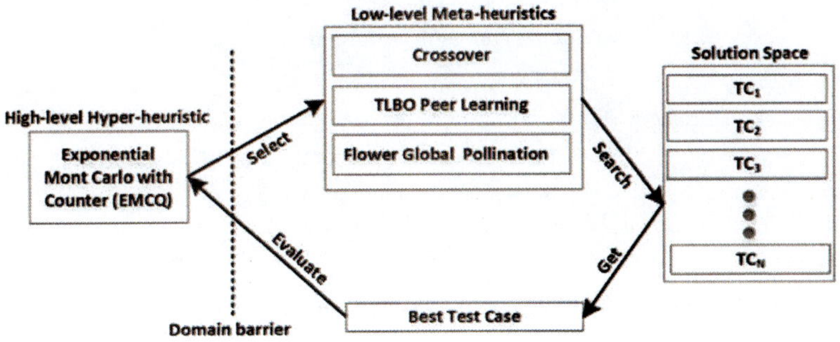

Fig. 2 Framework of IOR_HH

EMCQ to either increase or decrease acceptance probability of poor moves. More specifically, EMCQ always increments q when it replaces the current operator whereas it resets q to 1 when there is no replacement. With the manipulation of q, EMCQ improves the solution diversification. The pseudo code for IOR_HH based on the EMCQ hyper-heuristic is given in Fig. 3.

Referring to the IOR_HH strategy code in Fig. 3, line 1 enumerates the given IOR interactions based on binary representation such as 1100 to specify interaction between first input and second input. These interactions are stored in the hash map (H). Line 2 initializes Θ_{max} and S that define maximum iterations and population size, respectively for the strategy. Line 3 randomly initializes the population of solutions (i.e., test cases). An example a of test case is $Z_1 = \{1, 3, 5, 7\}$ for the SUT given in Sect. 2. Such test cases constitute the population for the low-level meta-heuristic operators. Line 4 sets solution S_0 to null so that at the beginning its fitness function evaluates to zero. The main loop of the strategy starts in line 6 with the main criteria

Input: Parameters (k) and their corresponding values *(v)*
Output: Final IOR Test Suite IOR_{ts}

1. Enumerate all the specified IOR interactions in a Hash Map (H)
2. Initialize maximum number of iterations (Θ_{max}) and population size S
3. Initialize random population of test cases: $Z = \{Z_0, Z_1, ..., Z_{S-1}\}$ for the three low-level heuristics: $LH=\{LH_1, LH_2, LH_3\}$
4. Set the solution S_0 to null: $S_0 \leftarrow null;$
5. While(coverage criteria not met i.e., H is not empty)
6.   Initialize iteration counter: $T \leftarrow 1;$
7.   While($T < \Theta_{max}$)
8.   Apply current LH_i to generate best IOR test case and assign to S_i
9.   Compute $\delta \leftarrow f(S_i) - f(S_0);$
10.   Assign S_i to S_0: $S_0 \leftarrow S_i$
11.   If ($\delta > 0$) // Check whether fitness is improving
12.   Then do not replace current LH_i;
13.   Reset control parameter to 1: $q \leftarrow 1;$
14.   Else
15.   Calculate probability density: $\psi \leftarrow e^{-\delta * T/q};$
16.   If ($random\ (0,1) < \psi$)
17.   Then do not replace current LH_i;
18.   Reset control paramaeter to 1: $q \leftarrow 1;$
19.   Else
20.   Replace current LH_i randomly with LH_{i^*} such that $i^* \neq i;$
21.   Increment control parameter: $q{+}{+};$
22.   Assign S_i to best test case S_{best}: $S_{best} \leftarrow S_i;$
23. $T{+}{+};$
24.   Add S_{best} to IOR_{ts} and remove covered interactions from H
25. Display/save IOR_{ts}

Fig. 3 Pseudo code of IOR_HH based on EMCQ hyper-heuristic for repetition. The criteria state that all the IOR interactions need to be removed from the hash map (H). The iteration counter T is initialized in line 6. The sub while loop begins in 7. Here, EMCQ logic is used for the selection and acceptance of low-level operators. A low-level meta-heuristic operator from the three implemented operators is randomly selected to generate a solution S_i (line 8). Line 9 in the code differentiates between current and previous fitness functions and assigns the results to δ. After assigning current solution (S_i) to previous solution (S_0) as given in line 10, the value of δ is evaluated to check whether the solution generated by the current low-level heuristic is better. possible fitness improvement. If true, EMCQ does not replace current LHi (line 11). In case of no improvement in the fitness evaluation. Probability density Ψ is computed in line 15. A random number in the interval [0 1] is thrown which if less than Ψ, EMCQ still does not replace the current low-level meta-heuristic operator. In both of these scenarios (i.e., no replacement of current LH_i, the control parameter q is set to 1 (line 13 and line 18). EMCQ replaces current LH_{i*} with a different LH_i such that subscript $i*$ is different from subscript i. Control parameter q is increment by 1 whenever a replacement of current LH_i occurs as given in line 21 in the code. Line 22 assigns the current S_i to S_{best}. Iteration counter T of the inner loop is incremented in line 23. At the end of the EMCQ logic for maximum number of iterations, the strategy adds the S_{best} to the IOR_{ts} and removes the covered interactions from H (see line 24). Finally, IOR_{ts} is displayed or saved to file in last line (25) of the pseudo code.

4.1 Low-Level Meta-Heuristic Operators

IOR_HH implements three low-level meta-heuristic operators including crossover from genetic algorithm (GA), peer learning from teaching learning-based optimization (TLBO) algorithm and global pollination from flower pollination algorithm (FPA) under the control of EMCQ hyper-heuristic. The crossover operator swaps two parts of the current test case based on the length specified by α which is set randomly. The newly obtained test case is evaluated against the current test case for replacement if better using fitness function values.

Peer learning operator is taken from learning phase of TLBO. This operator selects two test cases randomly in an attempt to search for better test case in each iteration. The test case with high fitness function evaluation improves the quality of the test case with low fitness function evaluation. FPA's global pollination operator obtains a new test case based on the Lévy Flight motion which updates all the values of the current test case column-wise. Instead of using these operators, other meta-heuristics or their operators can be chosen as low-level heuristics.

5 Results and Evaluation

IOR_HH is benchmarked against existing strategies for IOR interaction testing by adopting the experiments presented in [12]. Windows-based PC having 3.60 GHz Intel Core i7 CPU and 16 GB DDR3 RAM runs the proposed strategy for experimentations. The Java language is used for implementation. Each experiment is independently run for 20 times for statistical significance. Result are reported in tabular form where bold entries denote best test suite sizes obtained by a strategy.

Two experiments considering two systems are adopted for evaluating the performance of the proposed strategy. In both experiments, 60 different input-output relationships are taken as shown below.

Rel = {{1, 2, 7, 8}, {0, 1, 2, 9}, {4, 5, 7, 8}, {0, 1, 3, 9}, {0, 3, 8}, {6, 7, 8}, {4, 9}, {1, 3, 4}, {0, 2, 6, 7}, {4, 6}, {2, 3, 4, 8}, {2, 3, 5}, {5, 6}, {0, 6, 8}, {8, 9}, {0, 5}, {1, 3, 5, 9}, {1, 6, 7, 9}, {0, 4}, {0, 2, 3}, {1, 3, 6, 9}, {2, 4, 7, 8}, {0, 2, 6, 9}, {0, 1, 7, 8}, {0, 3, 7, 9}, {3, 4, 7, 8}, {1, 5, 7, 9}, {1, 3, 6, 8}, {1, 2, 5}, {3, 4, 5, 7}, {0, 2, 7, 9}, {1, 2, 3}, {1, 2, 6}, {2, 5, 9}, {3, 6, 7}, {1, 2, 4, 7}, {2, 5, 8}, {0, 1, 6, 7}, {3, 5, 8}, {0, 1, 2, 8}, {2, 3, 9}, {1, 5, 8}, {1, 3, 5, 7}, {0, 1, 2, 7}, {2, 4, 5, 7}, {1, 4, 5}, {0, 1, 7, 9}, {0, 1, 3, 6}, {1, 4, 8}, {3, 5, 7, 9}, {0, 6, 7, 9}, {2, 6, 7, 9}, {2, 6, 8}, {2, 3, 6}, {1, 3, 7, 9}, {2, 3, 7}, {0, 2, 7, 8}, {0, 1, 6, 9}, {1, 3, 7, 8}, {0, 1, 3, 7}}

In case of first experiment, the system has ten parameters each carrying three values (i.e., $C = 3^{10}$). Table 2 shows the results obtained by IOR_HH and its counterparts for |Rel| = 10, the first 10 IOR interactions from Rel, until all 60 input-output relationships i.e., |Rel| = 60.

Same relationships are used for the second experiment but now the system's configuration is $C = 2^3\ 3^3\ 4^3\ 5$. It means that the first three parameters carry two values, the next three parameters carry three values, the next three parameters carry four values and the last parameter carries five values. Table 3 presents the obtained results.

Table 2 Comparison of generated IOR test suites for IOR(N, 3^3, |Rel|) by the proposed strategy against existing strategies

|Rel|	ParaOrder	Union	Greedy	AURA	GVS	ReqOrder	Proposed (IOR_HH)	
							Best	Mean
10	105	503	104	89	104	153	**84**	85.70
20	103	858	110	99	98	148	**88**	90.60
30	117	1599	122	132	116	151	**107**	107.80
40	120	2057	134	139	117	160	**116**	116.60
50	148	2635	138	147	127	169	**124**	125.60
60	142	3257	143	158	140	176	**132**	136.90

Table 3 Comparison of generated IOR test suites for IOR(N, 2^3 3^3 4^3 5, |Rel|) by the proposed strategy against existing strategies

|Rel|	ParaOrder	Union	Greedy	AURA	GVS	ReqOrder	Proposed (IOR_HH)	
							Best	Mean
10	**144**	144	154	**144**	**144**	154	**144**	144.0
20	**160**	161	187	182	162	187	**160**	160.0
30	165	179	207	200	169	207	**164**	164.6
40	165	181	203	207	170	203	**163**	165.6
50	182	194	251	222	200	251	**184**	184.0
60	197	209	250	230	200	250	**188**	191.8

The results in Table 2 clearly show the dominance of the proposed strategy over other competing strategies. For each experiment out of total six experiments, IOR_HH outperformed all other strategies in terms of generating optimal IOR test suites. Apart from the proposed strategy, GVS offered competitive results as compared to other strategies shown under the GVS column in Table 2.

The strategy also remained on top in case of generating test suite sizes for the second experiment as shown in Table 3. IOR_HH outperformed all its competitors in case of four IOR interactions. Like Table 2, again IOR_HH generated best results for all six problem instances. ParaOrder generated two best sizes, whereas AURA and GVS generated one best test suite size. ParaOrder results are comparable to that of the proposed strategy as far as second experiment is concerned.

6 Conclusion

In this paper, an input-output-based relationship (IOR) interaction testing strategy based on the exponential Monte Carlo with counter (EMCQ) hyper-heuristic called IOR_HH is presented. The EMCQ based IOR_HH controls three low-level meta-heuristic operators that solve the IOR test suite generation problem. The proposed strategy is very useful for the benchmark problem as all the obtained experimental results are very promising. In future, the proposed strategy will be adopted for other optimization problems in software engineering such as software module clustering, constrained test suite generation and parameter optimization of COCOMO II model for software cost prediction.

Acknowledgements The work reported in this paper is funded by Fundamental Research Grant from Ministry of Higher Education Malaysia titled: A Reinforcement Learning Sine Cosine based Strategy for Combinatorial Test Suite Generation. We thank MOHE for the contribution and supports, Grant Number: RDU170103.

References

1. Ahmed BS, Zamli KZ, Afzal W, Bures M (2017) Constrained interaction testing: a systematic literature study. IEEE Access 5
2. Younis MI, Zamli KZ, Isa NAM (2008) MIPOG-Modification of the IPOG strategy for t-way software testing. In: Distributed frameworks and applications, pp 1–6. IEEE, Penang, Malaysia
3. Othman RR, Zamli KZ, Mohamad SMS (2013) T-way testing strategies: a critical survey and analysis. Int J Dig Content Technol Appl 7(9):222
4. Zamli KZ, Alsewari ARA, Hassin MHM (2013) On test case generation satisfying the MC/DC criterion. Int J Adv Soft Comput Appl 5(3)
5. Schroeder PJ, Korel B (2000) Black-box test reduction using input-output analysis. In: International symposium on software testing and analysis, ACM, Portland, USA (2000)
6. Alsewari ARA, Tairan NM, Zamli KZ (2015) Survey on input output relation based combination test data generation strategies. ARPN J Eng Appl Sci 10(18):8427–8430
7. Schroeder PJ, Faherty P, Korel B (2002) Generating expected results for automated black-box testing. In: 17th IEEE international conference on automated software engineering, pp 139–148. IEEE
8. Ziyuan W, Changhai N, Baowen X (2007) Generating combinatorial test suite for interaction relationship. In: 4th international workshop on software quality assurance, pp 55–61. ACM, Dubrovnik, Croatia
9. Wang ZY, Xu BW, Nie CH (2008) Greedy heuristic algorithms to generate variable strength combinatorial test suite. In: Proceedings of the 8th international conference on quality software, pp 155–160. IEEE Computer Society
10. Othman RR, Zamli KZ (2011) ITTDG: integrated t-way test data generation strategy for interaction testing. Sci Res Essays 6(17):3638–3648
11. Ong H, Zamli KZ (2011) Development of interaction test suite generation strategy with input-output mapping supports. Sci Res Essays 6(16):3418–3430
12. Othman RR, Zamli KZ, Nugroho LE (2012) General variable strength t-way strategy supporting flexible interactions. Maejo Int J Sci Technol 6(3):415
13. Ahmed BS, Gambardella LM, Afzal W, Zamli KZ (2017) Handling constraints in combinatorial interaction testing in the presence of multi objective particle swarm and multithreading. Inf Softw Technol 86:20–36
14. Alsewari ARA, Zamli KZ (2011) Interaction test data generation using harmony search algorithm. In: IEEE symposium on industrial electronics and applications, pp 559–564. IEEE, Langkawi, Malaysia
15. Din F, Alsewari ARA, Zamli KZ (2017) A parameter free choice function based hyper-heuristic strategy for pairwise test generation. In: IEEE international conference on software quality, reliability and security companion, pp 85–91. IEEE, Prague, Czech Republic
16. Din F, Zamli KZ (2018) Fuzzy adaptive teaching learning-based optimization strategy for GUI functional test cases generation. In: 7th international conference on software and computer applications, pp 92–96. ACM, Kuantan, Malaysia
17. Nasser AB, Alsewari AA, Tairan NM, Zamli KZ (2017) Pairwise test data generation based on flower pollination algorithm. Malaysian J Comp Sci 30(3):242–257
18. Nasser AB, Zamli KZ, Alsewari AA, Ahmed BS (2018) An Elitist-flower pollination-based strategy for constructing sequence and sequence-less t-way test suite. Int J Bio-Inspired Comput 12(2):115–127
19. Zamli KZ, Din F, Ramli N, Ahmed BS (2019) Software module clustering based on the fuzzy adaptive teaching learning based optimization algorithm. arXiv preprint arXiv:1902.11159
20. Nasser AB, Zamli KZ, Alsewari ARA, Ahmed BS (2018) Hybrid flower pollination algorithm strategies for t-way test suite generation. PLoS ONE 13(5):e0195187
21. Zamli KZ, Alkazemi BY, Kendall G (2016) A Tabu search hyper-heuristic strategy for t-way test suite generation. Appl Soft Comp 44:57–74

22. Zamli KZ, Din F, Kendall G, Ahmed BS (2017) An experimental study of hyper-heuristic selection and acceptance mechanism for combinatorial t-way test suite generation. Inf Sci 399:121–153
23. Din F, Zamli KZ (2018) Hyper-Heuristic based strategy for pairwise test case generation. Adv Sci Lett 24(10):7333–7338
24. Zamli KZ (2018) Enhancing generality of meta-heuristic algorithms through adaptive selection and hybridization. In: International conference on information and communications technology, pp 67–71. IEEE, Yogyakarta, Indonesia
25. Ayob M, Kendall G (2003) A Monte Carlo hyper-heuristic to optimise component placement sequencing for multi head placement machine. In: International conference on intelligent technologies, pp 132–141, Thailand
26. Holland JH (1975) Adaptation in natural and artificial systems. University of Michigan Press
27. Rao RV, Savsani VJ, Vakharia DP (2011) Teaching learning-based optimization: a novel method for constrained mechanical design optimization problems. Comput Aided Des 43(3):303–315
28. Yang XS (2012) Flower pollination algorithm for global optimization. In: International conference on unconventional computing and natural computation, pp 240–249. Springer

Forecasting Daily Travel Mode Choice of Kuantan Travellers by Means of Machine Learning Models

Nur Fahriza Mohd Ali, Ahmad Farhan Mohd Sadullah, Anwar P. P. Abdul Majeed, Mohd Azraai Mohd Razman, Chun Sern Choong, and Rabiu Muazu Musa

Abstract In transportation studies, forecasting users' mode choice in daily commute is crucial in order to manage traffic problems due to high number of private vehicles on the road. Conventional statistical techniques have been widely used in order to study users' mode choice; however, the choice of the most appropriate forecasting method still remains a significant concern. In this paper, we investigate the application of a number of machine learning models, namely Random Forest (RF), Tree, Naïve Bayes (NB), Logistic Regression (LR), k-Nearest Neighbour (k-NN), Support Vector Machine (SVM), as well as Artificial Neural Networks (ANN) in predicting the daily travel mode choice in Kuantan. The data was collected from a survey of Revealed/Stated Preferences (RPSP) Survey among Kuantan travellers in which eight features were taken into consideration in the present study. The classifiers were trained on the collected dataset by using five-folds cross-validation method to predict the daily mode choice. It was shown from this preliminary study that the RF, as well as ANN classifiers, could provide satisfactory classification accuracies to up to 70% in comparison to the other models evaluated. Therefore, it could be concluded that the evaluated features are rather important in deciding the travel model choice of Kuantan travellers.

Keywords Mode choice · Public transport · Private vehicles · Machine learning

---

N. F. Mohd Ali (✉) · A. F. Mohd Sadullah
School of Civil Engineering, Universiti Sains Malaysia, Engineering Campus, 14300 Nibong Tebal, Pulau Pinang, Malaysia
e-mail: fahriza90@gmail.com

A. P. P. A. Majeed · M. A. Mohd Razman · C. S. Choong
Innovative Manufacturing, Mechatronics and Sports Laboratory, Faculty of Manufacturing and Mechatronics Engineering Technology, Universiti Malaysia Pahang, 26600 Pekan, Pahang, Malaysia

R. Muazu Musa
Center for Fundamental and Continuing Education, Universiti Malaysia Terengganu, Kuala Nerus, Terengganu, Malaysia

A. F. Ab. Nasir et al. (eds.), *Recent Trends in Mechatronics Towards Industry 4.0*, Lecture Notes in Electrical Engineering 730, https://doi.org/10.1007/978-981-33-4597-3_89

1 Introduction

Forecasting daily travel demand on users' mode choice is a significant component to managing traffic problems occurred especially during peak hours. It plays an important role in investigating travel demands among users in the matter of time starting from origin to destination and planning the transportation services that meet users' needs, improving the efficiency and attract users to commute by public transport for their daily travel. Forecasting travel demand of users makes it a complex process of the transportation services. It is used to figure out users' preferences on time of all components involved in a public transportation services including walking time, waiting time, and in-vehicle time to make it accessible, reliable, and effective in providing services. Subsequently, policymakers were in charged to provide good transportation services that will satisfy users' needs. Therefore, it is very important to assist policymakers in providing efficient services by forecasting and predicting the daily travel demand of users so that users will choose public transport as their daily mode.

In order to improve current transportation services, a robust and accurate model for predicting daily travel demand of users is strongly required. Prediction of the travel mode is a pattern recognition problem (supervised learning), in which several variables including human characteristics and geographical patterns explain the choices among the travel modes [1]. In this article, we applied a few classifiers in modelling and estimation procedures by studying the quality of their prediction. These models have been introduced as alternatives to presenting the complex behavior modeling and pattern recognition.

In a research conducted by [2] to study work travel mode choice modelling with data mining by using Decision Trees (DT) and Neural Networks (NN), as well as Multinomial Logit (MNL) model, a total of 15,064 residences with 34,680 respondents from nine counties in the San Francisco Bay Area were randomly sampled. This study emphasizes the work trip or trip-to-work mode choice, that is, the primary trip in a home-to-work activity chain before a work activity. The result shows that Neural Network classifier achieved better performance (78.2%) compared to Decision Trees (76.8%) and Multinomal Logit (72.9%) model.

Hagenauer and Helbich [3] in earlier research, managed to conduct a mode choice prediction by using Dutch travel diary data from the years 2010 to 2012, enriched with variables on the built and natural environment as well as on weather conditions. This article compares the predictive performance of seven selected machine learning classifiers for travel mode choice analysis. It was shown form the study that Random Forest model evaluates is able to provide the highest classification accuracy of 91.4%, whilst the other classifiers such as SVM, ANN, NB, and MNL managed to provide classification accuracy of 82.5%, 60.6%, 60.2%, and 56.1%, respectively.

Zhou et al. [4] modelled travel mode in Chicago using machine learning techniques by considering environmental and temporal factors. The authors investigated the spatiotemporal patterns of the bike-sharing system (BSS) and taxi trips in Chicago from 2014 to 2016. The performance of multiple machine learning models was compared namely, the Logit, k-NN, Naïve Bayes, and Random Forest able to predict 80.1%, 78.4%, 72.2%, and 81.7%, respectively. It was shown that the ensemble model (Random Forest) could provide a reasonably well classification accuracy.

In this paper, different machine learning models, i.e., Random Forest (RF), Tree, Naïve Bayes (NB), Logistic Regression (LR), k-Nearest Neighbour (k-NN), Support Vector Machine (SVM), as well as Artificial Neural Networks (ANN) are evaluated in order to identify the daily-travel mode choice of Kuantan City travellers. It is expected that through the present investigation a predictive model could be proposed to policymakers for sustainable transportation planning.

2 Methodology

2.1 Data Collection

The methodology aimed to build a robust prediction model that attains high accuracy for forecasting daily mode choice of users in Kuantan city. It involved three main steps as presented in Fig. 1. The mode choices were categorized as N and P. N choices represented users' mode choice as private vehicles (including soft mode), meanwhile, P choices represented public transport as users' mode choice.

2.2 Data Collection and Pre-processing Phase

In this phase of our proposed methodology, we collect the Revealed/Stated Preference (RP/SP) survey data through the questionnaire form. The data was collected in Kuantan city centre (as shown in Fig. 2) during weekdays from 8 AM until 5 PM. The number of respondents collected was 386 respondents including workers, students, and unemployed users.

The main features used in this dataset are;

- Walking distance from home to the nearest bus stop (WD1),
- Waiting time at the bus stop (WT),
- Sitting time in the public transport (IVT),
- Walking distance from last stop to the destination (WD2), and
- Total travel time of all travel components from origin to destination (TT).

The other features involved in making prediction of users' mode choice in this study are;

- Users' opinion on the dominant factor that influenced their choices (DOM),
- Users' preferences on ticket prices (SP Ticket), and
- Users' origin (Region)

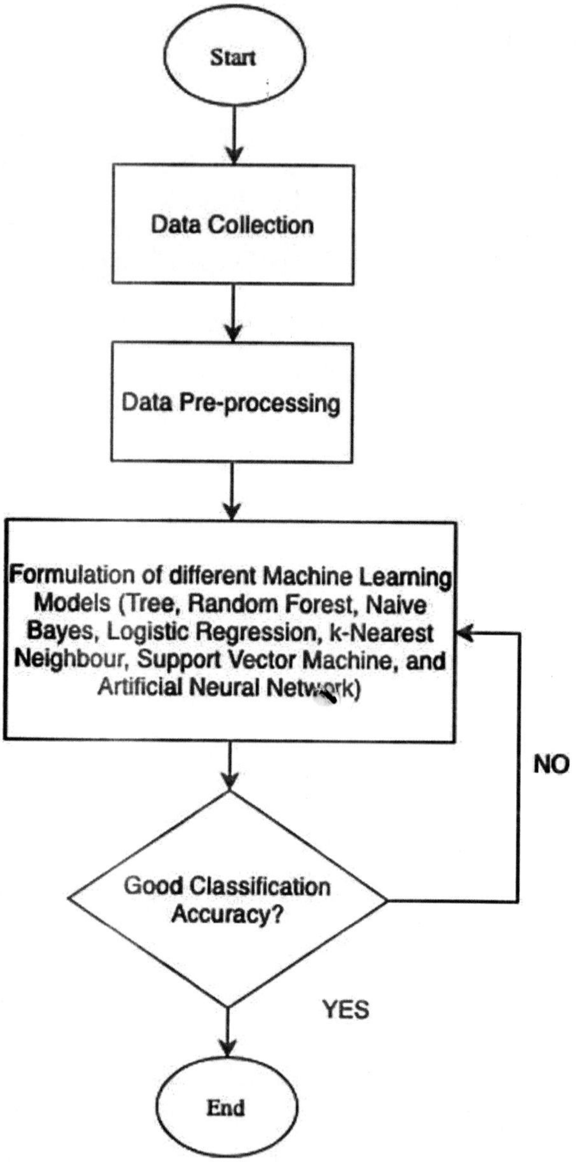

Fig. 1 Research methodology for mode choice forecasting

The distribution of the collected dataset was divided into four regions;

- Region 1: Users who lived ≤5 km,
- Region 2: Users who lived between 6 and 20 km,
- Region 3: Users who lived between 21 and 40 km, and
- Region 4: Users who lived >40 km.

Fig. 2 Data collection area

2.3 Classifier

Several machine learning models were implemented to examine its performance in classifying the daily travel mode choice, i.e., Random Forest (RF), Tree, Naïve Bayes (NB), Logistic Regression (LR), k-Nearest Neighbour (k-NN), Support Vector Machine (SVM), as well as Artificial Neural Networks (ANN). The default settings of the Orange platform are used for the classifiers. Here, we train the classifiers on the collected dataset using five-folds cross-validation technique as it has been shown to mitigate overfitting phenomenon [5]. Three common evaluation measurements are used to evaluate the results. These measurements are the precision (PR), recall (RE), and the classification accuracy (CA).

$$PR = TP/(TP + FP) \tag{1}$$

$$RE = TP/(TP + FN) \tag{2}$$

$$CA = (TP + TN)/(TP + FP + TN + FN) \tag{3}$$

where FP and FN are the false positive and negative rates, whereas TP and TN are the true positive and negative rates.

3 Result and Discussion

3.1 *Results and Comparisons*

The results computed for evaluation are based on the precision, recall, and the accuracy of the testing phase. Six baseline classifiers were adopted to discern the effectiveness of the classifiers. Table 1 and Figs. 3, 4, and 5 illustrates the results of PR, RE, and CA of the classifiers. It could be observed from Table 1 that the Random

Table 1 Results of classification accuracy representation with the seven classifiers

Classifier model	Classification accuracy (CA) %		Weighted average of precision (PR)	Weighted average of recall (RE)
	Training	Testing		
Tree	66.1	65.8	0.658	0.661
Random forest	**69.3**	69.9	0.689	0.693
Naïve Bayes	65.2	64.6	0.656	0.652
Logistic regression	64.6	65.0	0.638	0.646
SVM	46.0	46.8	0.487	0.460
kNN	68.0	70.2	0.678	0.680
Neural network	68.9	68.5	0.685	0.689

Bold indicates the highest training value of the Random Forest model compared to the other classifiers

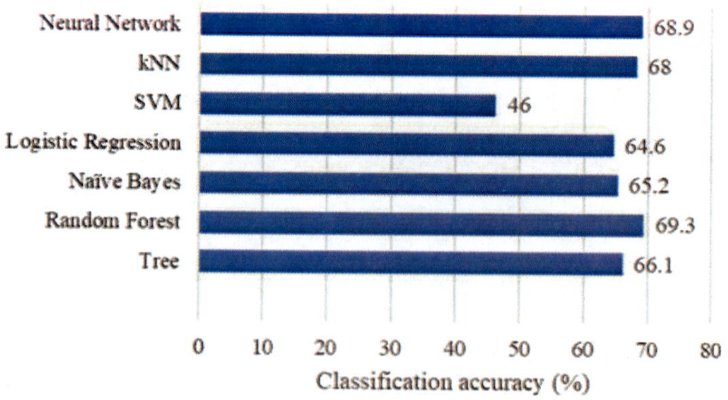

Fig. 3 The classification accuracy of all classifiers

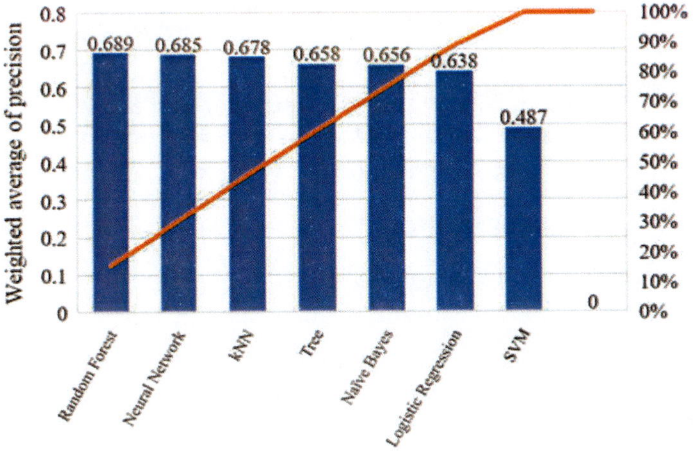

Fig. 4 The weighted average precision of all classifiers

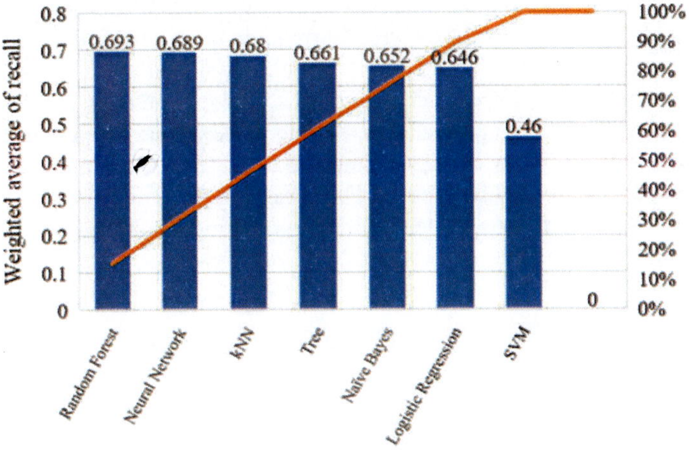

Fig. 5 The weighted average recall of all classifiers

Forest model yields the highest classification accuracy on both train and test data. The worst performing classifier is the SVM model with a CA of 46% and 46.8%, for both train and test data, respectively. Other models demonstrated comparable CA to the Random Forest model. It could be seen that the closest competitor is the kNN model in terms of average CA, nonetheless, the Random Forest triumphs by 0.7%.

The effectiveness of the Random Forest classifier in making predictions is in agreement with previous researchers. In a study conducted by [6], the accuracy result of Random Forest achieved 71.89% while predicting the household travel mode. The study investigated whether large-scale data does contribute towards the improvement of CA in estimating household travel modes. In another study conducted by [7]

		Predicted N	Predicted P	Σ
Actual	N	738	216	954
	P	282	385	667
	Σ	1020	601	1621

(a) Training result

		Predicted N	Predicted P	Σ
Actual	N	329	89	418
	P	120	157	277
	Σ	449	246	695

(b) Testing result

Fig. 6 The confusion matrix of random forest

demonstrated that the Random Forest classifier is able to distinguish well the mode selection of people with disabilities.

Figure 6 reveals that the number of misclassifications observed through the Random Forest model's confusion matrix. It could be seen that 216 "N" choices was misclassified as "P". Meanwhile, 282 "P" choices were misclassified as mode choice "N". This misclassification may be caused by the tendency of users to choose private vehicles although total travel time offered by public transports is reduced, regardless their origin. In normal cases, users will be triggered to switch mode towards public transports with lower travel time offered. Therefore, since human decision making is complex, the mode choices recorded in this study sometimes overlap between those who agreed to choose public transport and those who rejected public transport even though improvements on travel time were offered to this mode of transport. The misclassifications may be reduced by increasing the data collected, evaluating the sensitivity of the features selected, and optimising the hyperparameter of the developed models.

4 Conclusion

In this paper, we evaluated different machine learning models for predicting the daily travel mode choices of travellers in Kuantan city centre. It was shown from the study that the Random Forest classifier performs significantly better than other classifiers investigated. Future research will evaluate other machine learning models as well as investigating the selection of the features that could further improve the CA in predicting daily travel mode choice of travellers in Kuantan city centre. In addition, a sensitivity test by optimising the hyperparameters of the models shall also be investigated.

References

1. Mitchell TM (2017) Tom Mitchell : Naive Bayes and logistic regression. Mach Learn, 1–17. https://doi.org/10.1093/bioinformatics/btq112
2. Xie C, Lu J, Parkany E (2003) Work travel mode choice modeling with data mining: decision trees and neural networks. Transp Res Rec, 50–61. https://doi.org/10.3141/1854-06
3. Hagenauer J, Helbich M (2017) A comparative study of machine learning classifiers for modeling travel mode choice. Expert Syst Appl 78:273–282. https://doi.org/10.1016/j.eswa.2017.01.057
4. Zhou X, Wang M, Li D (2019) Bike-sharing or taxi? Modeling the choices of travel mode in Chicago using machine learning. J Transp Geogr 79:102479. https://doi.org/10.1016/j.jtrangeo.2019.102479
5. Taha Z, Musa RM, Abdul Majeed PPA, Alim MM, Abdullah MR (2018) The identification of high potential archers based on fitness and motor ability variables: A Support Vector Machine approach. Hum Mov Sci 57:184–193. https://doi.org/10.1016/j.humov.2017.12.008
6. Liang L, Xu M, Grant-Muller S, Mussone L (2019) Household travel mode choice estimation with large-scale data—an empirical analysis based on mobility data in Milan. Int J Sustain Transp 0:1–16. https://doi.org/10.1080/15568318.2019.1686782
7. Bantis T, Haworth J (2017) Who you are is how you travel: a framework for transportation mode detection using individual and environmental characteristics. Transp Res Part C Emerg Technol 80:286–309. https://doi.org/10.1016/j.trc.2017.05.003

The Classification of Hallucination: The Identification of Significant Time-Domain EEG Signals

Chin Hau Lim, Jothi Letchumy Mahendra Kumar, Mamunur Rashid, Rabiu Muazu Musa, Mohd Azraai Mohd Razman, Norizam Sulaiman, Rozita Jailani, and Anwar P. P. Abdul Majeed

Abstract Electroencephalogram (EEG) has now become one of the means in the medical sector to detect hallucination. The main objective of this study is to classify the onset of hallucination via time-domain based EEG signals. In this study, significant time-domain features were identified to determine the best features that could yield high classification accuracy (CA) on different classifiers. Emotiv Insight, a 5 channels headset, was used to record the EEG signal of 5 subjects aged between 23 and 27 years old when they are in a hallucination state. Eight statistical-based features, i.e., mean, standard deviation, variance, median, minimum, maximum, kurtosis, skewness and standard error mean from each channel. The identification of the significant features is obtained via Extremely Randomised Trees. The classification performance of all features, as well as selected features, are evaluated through, i.e. Random Forest (RF), *k*-Nearest Neighbours (*k*-NN), Naïve Bayes (NB), Support Vector Machine (SVM), Artificial Neural Network (ANN) and Logistic Regression (LR). The dataset was separated into the ratio of 70:30 for training and testing data. It was shown from the study, that the LR classifier is able to provide excellent CA on both the train and test dataset by considering the identified significant

C. H. Lim · J. L. Mahendra Kumar · M. A. Mohd Razman · A. P. P. Abdul Majeed (✉)
Innovative Manufacturing, Mechatronics and Sports Laboratory, Faculty of Manufacturing and Mechatronics Engineering Technology, Universiti Malaysia Pahang (UMP), 26600 Pekan, Pahang Darul Makmur, Malaysia
e-mail: amajeed@ump.edu.my

M. Rashid · N. Sulaiman
Faculty of Electrical and Electronics Engineering Technology, Universiti Malaysia Pahang (UMP), 26600 Pekan, Pahang Darul Makmur, Malaysia

R. M. Musa
Centre for Fundamental and Liberal Education, Universiti Malaysia Terengganu (UMT), 21030 Kuala Nerus, Terengganu Darul Iman, Malaysia

R. Jailani
Faculty of Electrical Engineering, Universiti Teknologi MARA (UiTM), 40450 Shah Alam, Selangor Darul Ehsan, Malaysia

A. P. P. Abdul Majeed
Centre for Software Development & Integrated Computing, Universiti Malaysia Pahang (UMP), 26600 Pekan, Pahang Darul Makmur, Malaysia

A. F. Ab. Nasir et al. (eds.), *Recent Trends in Mechatronics Towards Industry 4.0*, Lecture Notes in Electrical Engineering 730, https://doi.org/10.1007/978-981-33-4597-3_90

features. The identification of such features is non-trivial towards classifying the onset of hallucination in real-time as the computational expense could be significantly reduced.

Keywords EEG · Hallucination · Classification · Machine learning

1 Introduction

It has been reported that one over 20 people in the general population has experienced hallucination at least once in a lifetime without associating it with drug or alcohol consumption [1]. Hallucination is an experience in which one senses or feels something that does not exist externally, for instance, an illusion due to some syndrome or drug effect. The cause of hallucination can be generally separated into four major groups, namely, psychiatric disorder, neurological condition, drug or alcohol and non-clinical group [2]. The International Pilot Study of Schizophrenia (IPSS) estimated that around 70% of Schizophrenia patients experience hallucination, and most of them experience auditory hallucination, followed by visual hallucination [3, 4]. Hallucination is much harder to be diagnosed compared to physical diseases as it is a mental disease. The diagnosis of hallucination requires a complete medical, neurological and psychiatric evaluation which is a rather complicated process. Blood and urine test, Magnetic Resonance Imaging (MRI), Single Photon Emission Computed Tomography (SPECT), functional MRI (fMRI) are some of the means to diagnose hallucination [5–8]. Electroencephalogram (EEG) has also been reported to be able to diagnose hallucination through brain signals [9].

Although there is an abundance of literature that have reported on the classification of EEG signals, nonetheless, studies on the classification of the onset of hallucination are rather limited. Dwi Saputro et al. [10] investigated the classification of different type of seizure from EEG recording with a sampling frequency between 250 and 2556 Hz. A total of 17 different feature were extracted through different feature extraction methods, namely, Hjorth Descriptor, Mel Frequency Cepstral Coefficients (MFCC) and Independent Component Analysis. It was shown from the study that, a Non-linear SVM model is able to achieve a classification accuracy of 90.5% against a Linear-based SVM model.

Bird et al. [11] aimed at determining discriminative EEG-based features and suitable classification methods for recognising mental state. Statistical features like mean, standard deviation, min, max, derivative, log-covariance and log-energy entropy were extracted in the study. Different feature selection techniques were investigated, i.e., OneR, Information Gain, Correlation, Symmetrical Uncertainty and Evolutionary Algorithm. A number of classifiers were evaluated, i.e., Naïve Bayes, Bayes Net, J48, Random Forest, Random Tree, Multilayer Perception (MLP) and Support Vector Machine (SVM). The tenfold cross validation-technique were utilised in the study. A10-fold cross-validation had been chosen as the cross-validation technique. Random Forest, along with features extracted by OneR was found to be the best model with a CA of 87.16%.

Time-domain features of wink-based EEG signals was investigated by Letchumy et al. [12]. Emotiv Insight mobile EEG system was utilised to collect winking EEG-signals of the left and right eye. An Extremely Randomised Trees method was utilised to identify the significant features from the eleven statistical time-domain features. It has been shown that the three features that were found to be significant, which are maximum, range, and standard error obtained similar classification accuracy compared with the model with all the eleven statistical features. The classification accuracy attained by both the models were 87%. The proposed method established that a comparable classification accuracy is obtainable through the selection of significant features.

The main aim of this study is to identify the significant time-domain features that could yield the highest classification accuracy and determine the best classifiers which are compatible with the classification of hallucination. Several classifiers such as Random Forest (RF), *k*-Nearest Neighbors (*k*-NN), Naïve Bayes (NB), Support Vector Machine (SVM), Artificial Neural Network (ANN) and Logistic Regression (LR) will be used to determine the pre-symptoms of hallucination.

2 Methodology

2.1 DAQ

Emotiv Insight, a wireless headset with five channels which are located at AF3, AF4, T7, T8 and Pz to record brainwaves and transfer it into signal data to detect the hallucination was used in the present study (as depicted in Fig. 1). The resolution capability for each channel is 0.51 μV. The EEG signals will be recorded in the sampling rate of 128 samples per seconds for each channel with a digital notch filter at 50 Hz. The electrical potential differences will detect and read by the electrode from the five channels to collect the hallucination signal. The reference electrode,

Fig. 1 Emotiv insight

Pz, is placed at the left mastoid. The setup was in accordance to the International 10–20 EEG system.

2.2 *Participants*

Five undergraduate student aged between 23 and 27 years old with no epilepsy history condition was recruited for the EEG signal collection sessions. The subjects do not have any medical record in psychiatric, neurological or physical disease and had a good or corrected vision. Informed consent was obtained from the participants before they enrolled in the experiment. The ethical approval for this study was obtained via an institutional research ethics committee (FF-2013-327).

2.3 *Experimental Protocol*

The experiment was conducted in a room at the Faculty of Electrical and Electronics Engineering Technology, Universiti Malaysia Pahang which allows the subjects to carry out the experiment free from any kind of disturbance and keep concentrated during the experiment sessions. The environment is free from noise and in a room temperature to ensure that subjects are relaxed throughout the experiment. The subject was asked to sit in a comfortable position and looking at the computer screen without any physical movement and eye blinking to prevent any noise or unwanted signal generated throughout the experiment [11].

The subjects were given the instructions to look at the computer screen, which located one meter away from them. A presentation slide which consists of five hallucination images was used throughout the experiment. The subjects were instructed to stay focus and look at the computer screen without any blinking when the slide is playing. The first slide (a blank slide, referred as 'Blank' in Fig. 2) indicates that subjects to relax and in the comfortable position for 5 s followed by 5 s hallucination image (refer as 'Hallu' in Fig. 2). The overall duration of the experiment is 60 s for each subject, and the slide will be repeated from blank to hallucination action for every 5 s in 60 s.

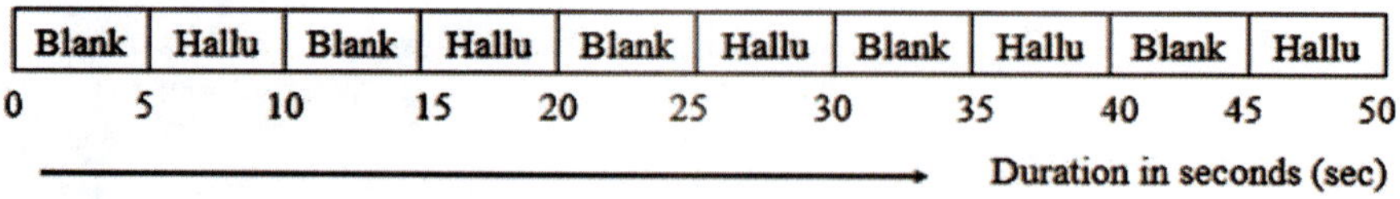

Fig. 2 Experimental paradigm for signal acquisition

2.4 Feature Extraction and Selection

A digital notch filter at 50 Hz was utilised to filter the raw EEG signals prior to feeding it to a fifth-order digital Sinc filter during the preprocessing stage (Fig. 3). The EEG signals are then split into hallucination and blank data for 5 s according to the slide, and thus 5 hallucination and 6 blank data were separated via Spyder IDE. Each hallucination and blank data will have 640 samples each (5 s × 128 samples per second) were then extracted into statistical features for classification [13–15]. There were eight (8) statistical features for each channel being extracted, which are mean, standard deviation, variance, median, minimum, maximum, kurtosis, skewness and standard error mean. Three channels (T7, T8 and Pz) were selected as hallucination usually occurs at the temporal part of the brain; thus 24 features were selected as the original feature set [16, 17].

A Tree-based ensemble method, Extremely Randomised Tree (known as Extra Tree) is an algorithm that selects splits and cut-point to identify the significant features. The best split is calculated by looking for randomly selected features in a subset. The normalised cumulative reduction of the mathematical parameters used in the decision on the split function is determined (Gini Index if the Gini Index is used in the construction of the forest) [18]. The Extremely Randomised Tree is used to distinguish the important features in the present investigation.

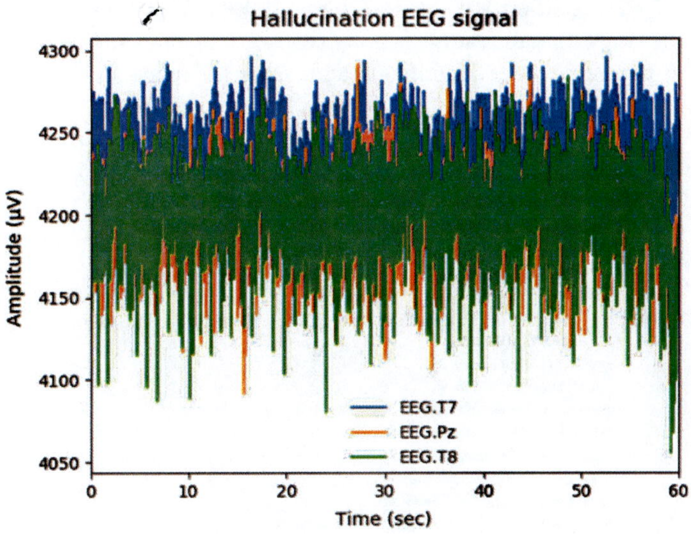

Fig. 3 Preprocess EEG signal for hallucination

2.5 Classification

A number of classifiers are used to assess the performance of the appraised features in identifying the onset on hallucination, namely Random Forest (RF), *k*-Nearest Neighbors (*k*-NN), Naïve Bayes (NB), Support Vector Machine (SVM), Artificial Neural Network (ANN) and Logistic Regression (LR), respectively. The hyperparameters of the models evaluated are the default parameters available in the scikit-learn library. The dataset is split to a ratio of 70:30 for training and testing, correspondingly. The performance of the classifiers is assessed through the classification accuracy (CA). The evaluation is carried out using Python Spyder IDE v4.1.4.

3 Result and Discussion

It could be seen from Fig. 4 that the significant features identified via Extremely Randomised Tree is Mean T7, Median T7 and Min T7, respectively. The findings are not surprising as the T7 electrode is located on the left temporal, which is the brain region where hallucination often occurs [16, 17]. Tables 1 and 2 tabulates the train and test accuracy of the classifiers evaluated, namely RF, *k*-NN, NB, SVM, ANN and LR on both feature sets (all and significant). It is apparent from Table 1 that the SVM and RF classifiers are able to achieve 100% CA on both the train and test dataset in the event that all features are considered. Nonetheless, it is worth noting that by considering the three identified features, the LR model is also able to yield an excellent classification accuracy on both the train and test dataset, suggesting that no apparent misclassification transpired. The present study demonstrates that with the selection of a number of significant features, the onset of hallucination could be

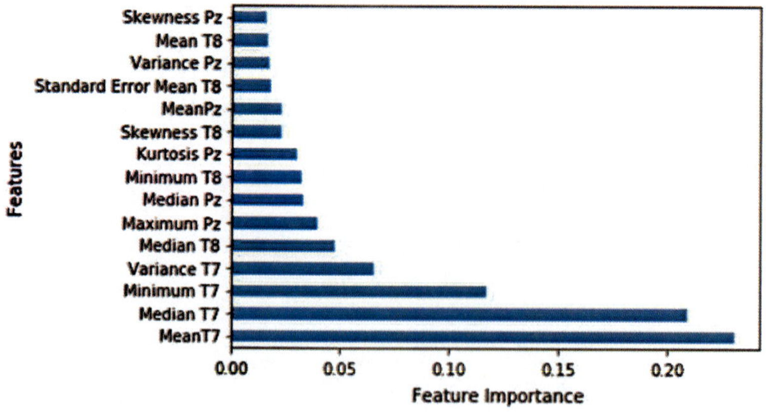

Fig. 4 The identification of significant features

Table 1 The CA of the models based on all features

Classifier	Training accuracy (%)	Test accuracy (%)
k-NN	85.7	72.2
SVM	100	100
RF	100	100
ANN	100	83.3
NB	97.6	88.9
LR	100	94.4

Table 2 The classification accuracy of the models based on the selected features

Classifier	Training accuracy (%)	Test accuracy (%)
k-NN	85.7	72.2
SVM	97.6	100
RF	100	88.9
ANN	100	94.4
NB	95.2	100
LR	100	100

identified and implies that the computational expense could be reduced in the event it is to be implemented in real-time.

4 Conclusion

In this present study, statistical-based time-domain features of EEG signals are used for the classification of the onset of hallucination. It was shown from the study that via the employment of the Extremely Randomised Tress feature selection technique, three features were deemed to be significant, namely, Mean T7, Median T7 and Min T7 amongst 24 features extracted. Subsequently, it was shown that through the selection of the features, an excellent classification accuracy could also be achieved through the LR classifier. Future studies will seek to investigate other types of feature selection technique as well as other forms of classifiers apart from evaluating the effect of hyperparameter optimisation towards the classification of hallucination.

Acknowledgements We would like to gratefully acknowledge Universiti Malaysia Pahang for supporting the present study [RDU180321].

References

1. McGrath JJ, Saha S, Al-Hamzawi A, Alonso J, Bromet EJ, Bruffaerts R, Caldas-De-Almeida JM, Chiu WT, De Jonge P, Fayyad J, Florescu S, Gureje O, Haro JM, Hu C, Kovess-Masfety V, Lepine JP, Lim CCW, Mora MEM, Navarro-Mateu F, Ochoa S, Sampson N, Scott K, Viana MC, Kessler RC (2015) Psychotic experiences in the general population: a cross-national analysis based on 31 261 respondents from 18 countries. JAMA Psychiatry 72:697–705. https://doi.org/10.1001/jamapsychiatry.2015.0575
2. Waters F, Fernyhough C (2017) Hallucinations: a systematic review of points of similarity and difference across diagnostic classes. Schizophr Bull 43:32–43. https://doi.org/10.1093/schbul/sbw132
3. Chaudhury S (2010) Hallucinations: clinical aspects and management. Ind Psychiatry J 19:5. https://doi.org/10.4103/0972-6748.77625
4. Sartorius N, Jablensky A, Korten A, Ernberg G, Anker M, Cooper JE, Day R (1986) Early manifestations and first-contact incidence of schizophrenia in different cultures: a preliminary report on the initial evaluation phase of the WHO collaborative study on determinants of outcome of severe mental disorders. Psychol Med 16:909–928. https://doi.org/10.1017/S0033291700011910
5. Mbbs NK, Mortimer A (2013) Causes, diagnosis and treatment of. 6–9
6. Klotz DM, Penfold RS (2018) Low mood, visual hallucinations, and falls—heralding the onset of rapidly progressive probable sporadic Creutzfeldt-Jakob disease in a 73-year old: a case report. J Med Case Rep 12:1–5. https://doi.org/10.1186/s13256-018-1649-4
7. Paradowski B, Kowalczyk E, Chojdak-ŁUkasiewicz J, Loster-Niewińska A, Słuzewska-Niedźwiedź M (2013) Three cases with visual hallucinations following combined ocular and occipital damage. Case Rep Med 2013:1–6. https://doi.org/10.1155/2013/450725
8. van Lutterveld R, Sommer IEC, Ford JM (2011) The neurophysiology of auditory hallucinations—a historical and contemporary review. Front Psychiatry 2:1–7. https://doi.org/10.3389/fpsyt.2011.00028
9. Jakab A, Kulkas A, Salpavaara T, Kauppinen P, Verho J, Heikkilä H, Jäntti V (2014) Novel wireless electroencephalography system with a minimal preparation time for use in emergencies and prehospital care. Biomed Eng Online 13. https://doi.org/10.1186/1475-925X-13-60
10. Dwi Saputro IR, Maryati ND, Solihati SR, Wijayanto I, Hadiyoso S, Patmasari R (2019) Seizure type classification on EEG signal using support vector machine. J Phys Conf Ser 1201. https://doi.org/10.1088/1742-6596/1201/1/012065
11. Bird JJ, Manso LJ, Ribeiro EP, Ekart A, Faria DR (2018) A study on mental state classification using EEG-based brain-machine interface. In: 9th International conference on intelligent systems 2018 theory, research innovations application IS 2018—procedure, pp 795–800. https://doi.org/10.1109/IS.2018.8710576
12. Letchumy J, Rashid M, Musa RM The classification of wink-based EEG signals : the identification of significant time-domain. Springer, Singapore. https://doi.org/10.1007/978-981-15-7309-5
13. Idayu N, Ali MA, Kasim MS (2020) Intelligent manufacturing and mechatronics. Springer, Singapore. https://doi.org/10.1007/978-981-13-9539-0
14. Ghayab HR Al, Li Y, Abdulla S, Diykh M, Wan X (2016) Classification of epileptic EEG signals based on simple random sampling and sequential feature selection. Brain Inform 3:85–91. https://doi.org/10.1007/s40708-016-0039-1
15. Zhang Y, Ji X, Liu B, Huang D, Xie F, Zhang Y (2017) Combined feature extraction method for classification of EEG signals. Neural Comput Appl 28:3153–3161. https://doi.org/10.1007/s00521-016-2230-y
16. Starr DA (2011) 乳鼠心肌提取 HHS public access. Physiol Behav 176:139–148. https://doi.org/10.1016/j.physbeh.2017.03.040

17. Hugdahl K, Løberg EM, Nygård M (2009) Left temporal lobe structural and functional abnormality underlying auditory hallucinations in schizophrenia. Front Neurosci 3:34–45. https://doi.org/10.3389/neuro.01.001.2009
18. Geurts P, Ernst D, Wehenkel L (2006) Extremely randomised trees. Mach Learn 63:3–42

The Classification of Blinking: An Evaluation of Significant Time-Domain Features

Gavin Lim Jiann Kai, Jothi Letchumy Mahendra Kumar, Mamunur Rashid, Rabiu Muazu Musa, Mohd Azraai Mohd Razman, Norizam Sulaiman, Rozita Jailani, and Anwar P. P. Abdul Majeed

Abstract Stroke is one of the most widespread causes of disability-adjusted life-years (DALYs). EEG-based Brain-Computer Interface (BCI) system is a potential solution for the patients to help them regain their mobility. The study aims to classify eye blinks through features extracted from time-domain EEG signals. Six features (mean, standard deviation, root mean square, skewness, kurtosis and peak-to-peak) from five channels (AF3, AF4, T7, T8 and Pz) were collected from five healthy subjects (three male and two female) aged between 22 and 24. The Chi-square (χ^2) method was used to identify significant features. Six machine learning models, i.e. Support Vector Machine (SVM)), Logistic Regression (LR), Random Forest (RF), Naïve Bayes (NB) and Artificial Neural Networks (ANN), were developed based on all the extracted features as well as the identified significant features. The training and test datasets were divided into a ratio of 70:30. It is shown that the classification accuracy of the evaluated classifiers by considering the fifteen features selected through the Chi-square is comparable to that of the selection of all features. The highest classification accuracy was demonstrated via the RF classifier for both cases. The findings suggest that even that with a reduced feature set, a reasonably high

---

G. L. J. Kai · J. L. Mahendra Kumar · M. A. Mohd Razman · A. P. P. Abdul Majeed (✉)
Innovative Manufacturing, Mechatronics and Sports Laboratory, Faculty of Manufacturing and Mechatronics Engineering Technology, Universiti Malaysia Pahang (UMP), 26600 Pekan, Pahang Darul Makmur, Malaysia
e-mail: amajeed@ump.edu.my

M. Rashid · N. Sulaiman
Faculty of Electrical and Electronics Engineering Technology, Universiti Malaysia Pahang (UMP), 26600 Pekan, Pahang Darul Makmur, Malaysia

R. M. Musa
Centre for Fundamental and Liberal Education, Universiti Malaysia Terengganu (UMT), 21030 Kuala Nerus, Terengganu Darul Iman, Malaysia

R. Jailani
Faculty of Electrical Engineering, Universiti Teknologi MARA (UiTM), 40450 Shah Alam, Selangor Darul Ehsan, Malaysia

A. P. P. Abdul Majeed
Centre for Software Development & Integrated Computing, Universiti Malaysia Pahang (UMP), 26600 Pekan, Pahang Darul Makmur, Malaysia

A. F. Ab. Nasir et al. (eds.), *Recent Trends in Mechatronics Towards Industry 4.0*, Lecture Notes in Electrical Engineering 730, https://doi.org/10.1007/978-981-33-4597-3_91

classification accuracy could be achieved, i.e., 91% on the test set. This observation further implies the viable implementation of BCI applications with a reduced computational expense.

Keywords EEG · Blink · Feature selection · Machine learning

1 Introduction

In Malaysia, stroke is the second leading cause of death, according to the Malaysian National Burden of Disease Study and vital registry system [1]. The prevalence of stroke among Malaysians was estimated to be 0.3% in 2006 and 0.7% in 2011 with 1.7% among those aged 55–59 years, 2% in 60–64 years, 3% in 65–69 years, 3.5% in 70–74 years and 7.8% in 75 years and beyond [2]. Two-thirds of the reported stroke cases were of ischaemic origin, that is a restriction of blood flow to the brain, with the remaining one-third being haemorrhagic origin which is the rupture of blood vessels [3]. More often than not, people who are affected by stroke are left with limited or reduced mobility. Hitherto, researchers are actively investigating the employment of Brain-Computer Interface (BCI) that exploits electroencephalogram (EEG) signals as a means to facilitate the use of robotics rehabilitation. The use of such technology could consequently reduce the burden of physiotherapists [4, 5].

Different machine learning models have been investigated for classifying EEG-based signals in the literature. For instance, Chai et al. [6] present an improvement of classification performance for EEG-based driver tiredness classification between sleepy and wakeful states with the data collected from 43 participants who had obtained a driver's license aged between 18 and 55 years old. The features were extracted using power spectral density (PSD) and fifth-order autoregressive (AR) model. The Sparse-Deep Belief Network (DBN) was used to classify the EEG-based driver fatigue classification amongst other conventional machine learning models. The three-fold cross-validation technique was used to develop the models. A receiver operating characteristic (ROC) graph is used to evaluate the performance of the models. It was shown from the study that the proposed classifier outperformed the rest with an ROC of 84.8% for PSD, whilst 93.1% for AR features, respectively.

Jin et al. [7] introduce a Sparse Bayesian Extreme Learning Machine (SBELM)-based algorithm to improve the classification accuracy of motor imagery. SBELM combines both ELM and sparse Bayesian learning to reduce redundancy of hidden neuron and able to control the model complexity. The BCI Competition IV dataset IIb is used to evaluate the performance of the proposed architecture. The performance of different classifiers was evaluated, i.e., Support Vector Machine (SVM), ELM, BELM, and SBELM on different sample sizes, viz. 30, 50, and 80%. It was shown from the study that the classification accuracy for SBELM is better for the 30 and 50% training sample size, whilst comparable to BELM for 80%.

The present study aims at evaluating the efficacy of different machine learning models in classifying blinks based on a selection of significant statistical-based time-domain features. It is envisaged that through the findings of the study, a possible adaptation for neurorehabilitation with a reduced computational expense could be realised.

2 Methods

This research focuses on classifying three (3) classes of blinking (once, twice and thrice) from EEG data collected from 5-channel Emotiv Insight. Five subjects (three male and two female) volunteered to participate in this study. The subjects are aged between 22 and 24. The participants are seated in a closed room in a comfortable position and were told to relax to reduce artefacts in the signals. They are seated one meter away from a screen. Ethical approval was obtained for the present investigation through an institutional ethics committee (FF-2013-327).

The instructions were presented in the form of visual cues displayed on the screen for each trial with a rest period between each trial as per experiment paradigm shown in Fig. 1. Before the trials, a demonstration is shown to the subjects to make sure they understand the process of data collection. The experiment consists of three parts whereby the subjects are instructed to blink once, twice or thrice on cue. Each session contains five trials (five seconds per trial) which total up to 640 samples (128 samples per second).

The raw data have to be prepared and processed to extract blink data only. A total of 150 instances were finally extracted with 50 for each class. The EEG signal is processed to improve the signal-to-noise ratio and improve the accuracy of the classifiers. In this study, six statistical features are extracted, namely, mean, standard deviation (SD), root mean square (RMS), skewness (Skew), kurtosis (Kur) and peak-to-peak (P2P), respectively. Subsequently, two feature sets are prepared, i.e., all features and features selected via the χ^2 feature selection approach. Six classifiers were used to evaluate the significance of the features towards the classification of the blinks, *k*-Nearest Neighbor (*k*-NN), Support Vector Machine (SVM), Logistic Regression (LR), Random Forest (RF), Naïve Bayes (NB) and Artificial Neural Networks (NN). The default hyperparameters of the models that are available via

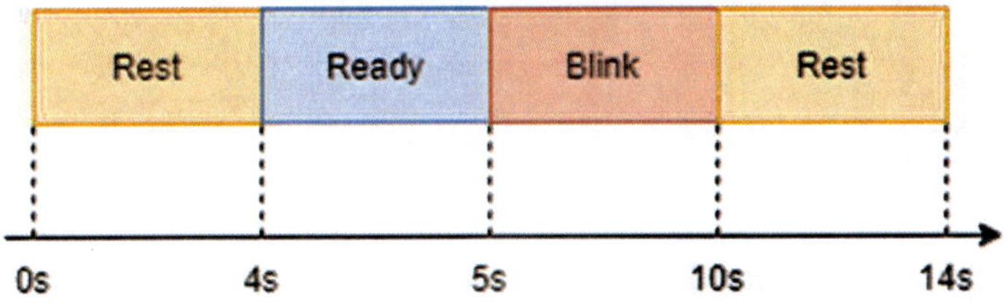

Fig. 1 Experiment paradigm of EEG Blinking signal collection

Table 1 Classifier performance on the different feature sets

	All feature	All feature	15 χ^2	15 χ^2
Model	Train	Test	Train	Test
k-NN	0.72	0.78	0.75	0.78
LR	0.74	0.67	0.73	0.64
NB	0.68	0.42	0.68	0.49
ANN	0.80	0.76	0.79	0.78
RF	0.90	0.84	0.88	0.91
SVM	0.71	0.71	0.70	0.69

the scikit-learn library were used in the present study. The hold out cross-validation technique is used in the present study, in which 70% is used for training whilst the remaining 30% is used for testing. The performance of the features is evaluated via Python Spyder IDE v4.1.4 by observing the classification accuracy (CA) as well as the confusion matrix of the developed models.

3 Result and Discussion

The significant features identified via the χ^2 feature selection technique from the extracted 30 features are P2P, Kur, SD and Skew of AF3, Kur, SD, Skew and P2P of AF4, P2P, SD, Kur and Skew of Pz, Skew and SD of T8 as well as Kur of T7. Table 1 tabulates the performance of the evaluated classifiers on the set of features. A comparable CA performance is observed by considering all features and the identified 15 features across the evaluated classifiers. The RF classifier is able to yield good classification for both all and selected features. This observation is in agreement with previous studies [8]. A noteworthy improvement of the test set is noticed by considering the identified. Moreover, an overfitting phenomenon could also be observed if all features are used for the development of the RF model. However, it is also worth mentioning that the hyperparameters of the models are taken as default, and possible improvements could be made in the event that the hyperparameters are optimised. It could be seen from Fig. 2 and Fig. 3 that the twice blink misclassification is reduced from three to one by selecting the significant features that inadvertently improve the CA.

4 Conclusion

The present study investigated the significance of identifying important time-domain EEG-based features for blinking. It was shown that the χ^2 feature selection technique could identify 15 significant features amongst the 30 extracted features. It was further

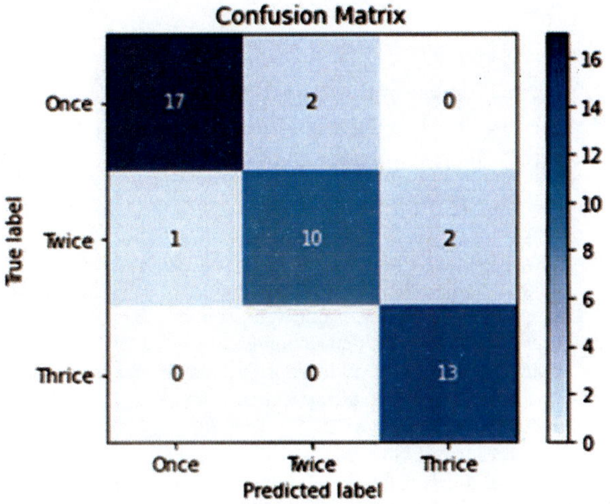

Fig. 2 Confusion matrix of the RF classifier considering all features

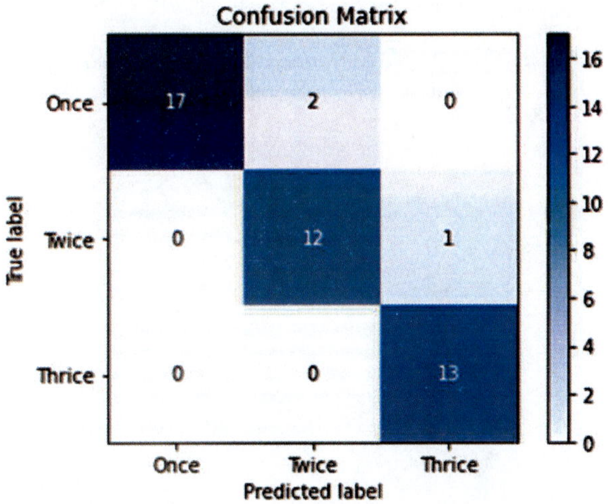

Fig. 3 Confusion matrix of the RF classifier considering the identified features

shown that a comparable classification accuracy could be attained by only utilising the selected features. Future studies shall incorporate the use of other feature selection techniques as well as the inclusion of the frequency-domain spectrum. Moreover, the effect of optimising the hyperparameters in improving the CA should also be investigated. The present study further implies the possible reduction of the computational expense for the possible adaptation for neurorehabilitation.

Acknowledgements The authors would like to acknowledge University Malaysia Pahang for funding this study via RDU180321.

References

1. National Institute of Health (2017) Malaysian burden of disease and injury study 2009–2014
2. Suzana S, Kee C, Jamaludin A, Noor Safiza M, Khor G, Jamaiyah H, Geeta A, Ahmad Ali Z, Rahmah R, Ruzita A, Ahmad Fauzi Y (2012) The third national health and morbidity survey. Asia Pacific J Public Heal 24:318–329. https://doi.org/10.1177/1010539510380736
3. Kooi Cheah W, Peng Hor C, Abdul Aziz Z, Looi I (2016) A review of stroke research in Malaysia from 2000–2014. Med J Malaysia 71
4. Rashid M, Sulaiman N, Majeed APP, Musa RM, Nasir AFA, Bari BS, Khatun S (2020) Current status, challenges and possible solutions of EEG based brain-computer interface: a comprehensive review. Front Neurorobot 14:25
5. Ab Patar MNA, Said AF, Mahmud J, Majeed APPA, Razman MA (2014) System integration and control of Dynamic Ankle Foot Orthosis for lower limb rehabilitation. In: 2014 International Symposium on Technology Management and Emerging Technologies (ISTMET), IEEE, pp 82–85
6. Chai R, Ling SH, San PP, Naik GR, Nguyen TN, Tran Y, Craig A, Nguyen HT (2017) Improving EEG-based driver fatigue classification using sparse-deep belief networks. Front Neurosci 11. https://doi.org/10.3389/fnins.2017.00103
7. Jin Z, Zhou G, Gao D, Zhang Y (2018) EEG classification using sparse Bayesian extreme learning machine for brain-computer interface. Neural Comput Appl. https://doi.org/10.1007/s00521-018-3735-3
8. Letchumy J, Rashid M, Musa RM The classification of wink-based EEG signals : the identification of significant time-domain. Springer, Singapore. https://doi.org/10.1007/978-981-15-7309-5

The Classification of Electrooculography Signals: A Significant Feature Identification via Mutual Information

Phua Jia Hwa, Jothi Letchumy Mahendra Kumar, Mamunur Rashid, Rabiu Muazu Musa, Mohd Azraai Mohd Razman, Norizam Sulaiman, Rozita Jailani, and Anwar P. P. Abdul Majeed

Abstract Stroke is currently known as the third most frequent reason for disability worldwide where the quality of life of its survivors in terms of their daily functioning is seriously affected. Brain-Computer Interface (BCI) is a system that can acquire and transform brain activity into readable outputs. This system is particularly beneficial to the people who encounter physical challenges in carrying out their daily life as the BCI outputs can be applied to BCI-based assistive devices. One of the BCI inputs that are frequently used is the Electrooculography (EOG) signal. EOG signal is the electrical voltage emitted from the movement of our eyeballs. This study aims to extract and identify significant statistical-based time-domain features based on the EOG signals acquired that would facilitate the classification of EOG movements via Support Vector Machine (SVM). The EOG signals were obtained via BioRadio. Five healthy subjects that ranged between 22 and 30 years old were involved in the EOG data acquisition. A total of 7 statistical time-domain features, namely, mean, standard deviation, variance, median, minimum, maximum, and standard error mean were extracted from all four BioRadio channels. The Mutual Information (MI) feature selection technique was employed to identify significant features. The 70:30

P. J. Hwa · J. L. Mahendra Kumar · M. A. Mohd Razman · A. P. P. Abdul Majeed (✉)
Innovative Manufacturing, Mechatronics and Sports Laboratory, Faculty of Manufacturing and Mechatronics Engineering Technology, Universiti Malaysia Pahang (UMP), 26600 Pekan, Pahang Darul Makmur, Malaysia
e-mail: amajeed@ump.edu.my

M. Rashid · N. Sulaiman
Faculty of Electrical and Electronics Engineering Technology, Universiti Malaysia Pahang (UMP), 26600 Pekan, Pahang Darul Makmur, Malaysia

R. M. Musa
Centre for Fundamental and Liberal Education, Universiti Malaysia Terengganu (UMT), 21030, Kuala Nerus, Terengganu Darul Iman, Malaysia

R. Jailani
Faculty of Electrical Engineering, Universiti Teknologi MARA (UiTM), 40450 Shah Alam, Selangor Darul Ehsan, Malaysia

A. P. P. Abdul Majeed
Centre for Software Development & Integrated Computing, Universiti Malaysia Pahang (UMP), 26600 Pekan, Pahang Darul Makmur, Malaysia

A. F. Ab. Nasir et al. (eds.), *Recent Trends in Mechatronics Towards Industry 4.0*, Lecture Notes in Electrical Engineering 730, https://doi.org/10.1007/978-981-33-4597-3_92

hold-out cross-validation technique was used in the study. It was demonstrated from the present investigation that an excellent comparable classification on both train and test dataset is attainable even by utilising the identified features. The findings further suggest the possible application of neurorehabilitation owing to the reduced computational expense resulting from the reduced feature set.

Keywords Electrooculography (EOG) · Eyeball movement · Brain-computer interface (BCI) · Time-domain features · Machine learning · Feature selection · Classification

1 Introduction

Stroke is known as the third most frequent cause for disability worldwide where the quality of life of its survivors in terms of activities of daily living is seriously affected [1]. Hung et al. [2] reported that approximately 80% of stroke survivors are suffering after-effect of damaged motor function of the upper limb, especially on the arm and hand. It has also been well-documented that the most effective period for rehabilitation to be instituted within the first six months post-stroke [3]. Conventional rehabilitative efforts are deemed labour intensive for physiotherapists; hence robotics rehabilitation is seen to be a possible solution to mitigate the issue [4–6].

Brain-Computer Interface (BCI) is seen as a potential bridge between assistive technology (robotics rehabilitation) and the patient as it that converts brain signals into commands communicating with the robot-assisted system that promote the neural plasticity required for recovery of function after stroke [7]. The method to combine hybrid BCI and robot-assisted therapy appears to be a more effective means to assist the patients to recover from stroke compared to the traditional rehabilitation therapy [8].

Different studies have been carried out on investigating biosignals ranging from brain to muscle signals. Ismail et al. [6] investigated the utilisation of electromyogram (EMG) signals from bicep muscles in identifying pre-intention and intention of the flexion and extension of the elbow joint. A number of time-domain features were extracted namely waveform length, mean absolute value, root mean square and standard deviation. It was shown from the study that the *k*-Nearest Neighbour (*k*-NN) model is able to classify better the intention classes against the Support Vector Machine (SVM) model based on the extracted features with a classification accuracy of 96.4%.

Conversely, Vahdani-Manaf, and Pournamdar [9] evaluated the efficacy of different *k*-NN classifiers by varying the number of *k* between 3, 5 and 7 in classifying eye movement signals from Electrooculography (EOG). The data were collected from 28 healthy male and female subjects between the age of 18–26 years old.The feature extraction method used in the study was the extreme point strategy (signal peaks). It was shown from the study that the *k*-NN with 3 neighbours provide a classification accuracy of 98% based on the selected feature.

It is apparent from the literature that the investigation on the identification of significant features, particularly with regards to EOG signals is rather limited. Therefore, the present study sought to investigate the employment of Mutual Information (MI) feature selection technique in order to identify significant features extracted from EOG signals. The findings from the current study may facilitate the realisation of BCI-EOG based neurorehabilitation efforts.

2 Methodology

2.1 Data Acquisition and EOG Recording Device

In the present study, BioRadio is used to collect EOG signal from subjects. The BioRadio is a wireless data acquisition device that is able to record, display, and analyse physiological signals in real-time. The physiological signals are amplified, sampled, and digitised, which then transmitted to a computer via Bluetooth for post-analysis. The BioRadio device is depicted in Fig. 1. Figure 2 illustrates the location of the placement of the electrodes on the subject's face for EOG signal collection.

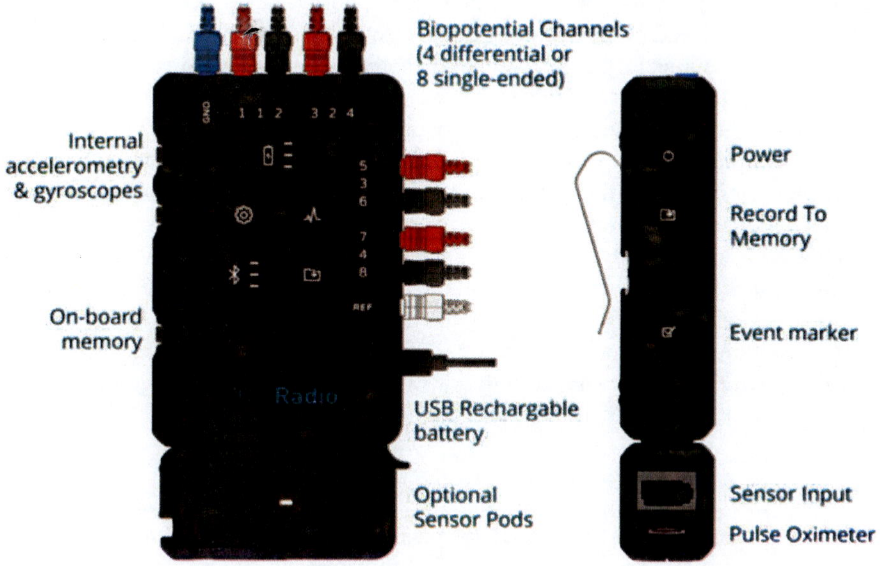

Fig. 1 BioRadio device

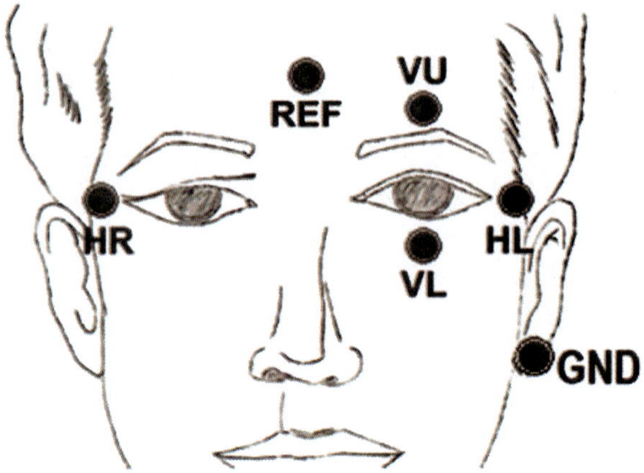

Fig. 2 Cable to connect electrodes with BioRadio

2.2 *Experimental Protocol*

A total of 5 healthy subjects which consist of 3 males and 2 females were volunteered to take part in the EOG signal acquisition. These subjects were in the age group between 22 and 30 years old. All of the subjects were healthy that do not suffer any disease, including genetic diseases. The ethical approval for the present study was obtained (FF-2013-327). Before placing the electrodes on the facial spots (as shown in Fig. 2) of the subjects, the spots were cleaned using alcohol whips. The sampling rate of BioRadio was set to 250 Hz.

The subjects were seated ergonomically on a chair and were given 5 min to relax. After 5 min, the subjects were told to look on a laptop that has a dot in the middle of the monitor. After 4 s, the background will change from white to a blue colour to indicate the ready signal for the subject for 1 s, and the dot will change to an arrow which is pointing upwards, where the subject will have to move their eyeball in the direction similar to the arrow for 5 s, and then back to the middle of the monitor. The process (as depicted in Fig. 3) was repeated up to 60 s, where the subjects will move their eyeball upwards for a total of 5 times. The same process was repeated 3 times, while replacing the upward arrow with downward, left, and right arrow, respectively. Therefore, a total of 4 sets of data were collected.

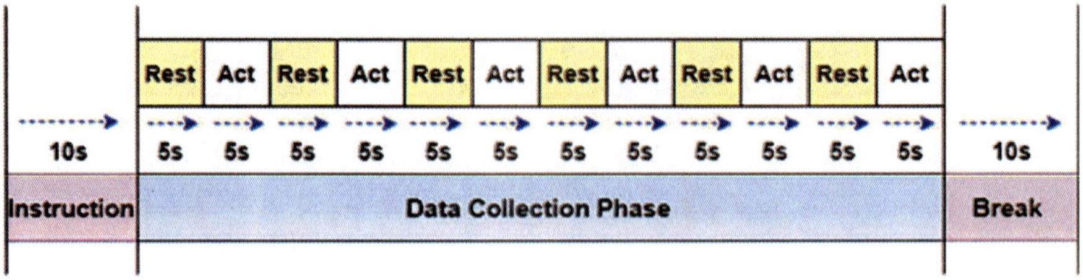

Fig. 3 Experiment paradigm of the EOG recording

2.3 *Signal Preprocessing and Feature Extraction*

The BioRadio has a built-in notch filter and 4th order Butterworth filter, where the filter selections were through its integrated software, BioCapture. As for this research work, the notch filter was set to have a cut off frequency of 50 Hz, and the 4th order Butterworth filter was set to have a band-pass frequency from 0.5 to 40 Hz. The pre-processing phase was done using epoch splitting section. The filtered signal was separated into two sections, which are resting and eyeball movement action. Therefore, there were 2500 samples per trial. After the pre-processing phase, seven statistical time-domain based features were extracted from the filtered samples. The extracted features were mean, standard deviation, variance, median, minimum, maximum, and standard error mean from all four channels, respectively.

2.4 *Features Selection and Classification*

The Mutual Information (MI) features selection technique was employed to select the significant features in the present study. MI selects the best features through the feature gap by comparing the ranking of the MI with the target value of each feature that was extracted. The classifier used to evaluate the significance of the features in classifying the EOG signals is the Support Vector Machine (SVM). It is worth noting that the hyperparameters of the SVM model are not altered and taken as default as provided by the scikit-learn library. The hold-out method, with a ratio of 70:30, is used in the study. The evaluation of the significance of the features is carried out by observing the classification accuracy and the confusion matrix of both feature sets. The analysis was carried out using Python Spyder IDE v4.1.4.

3 Result and Discussion

It could be observed from Fig. 4 that the MI technique identified mean, median, minimum and maximum to be significant features across all channels. The findings are in agreement with studies reported in [12, 13]. The identified features were then employed on the SVM classifier, and it was demonstrated that the classification accuracy by considering all and selected are comparable. Both the train and test accuracy in classifying four eyeball movements, i.e., left, right, up and down were found to be 100% for both feature sets as shown in Figs. 5, 6 and 7 depict the confusion matrix of the evaluated features on the test dataset. The findings further suggest that by considering four features could somewhat reduce the computational expense in the event that a real-time classification is required implying the promising potential it has for BCI-based neurorehabilitation assistive devices.

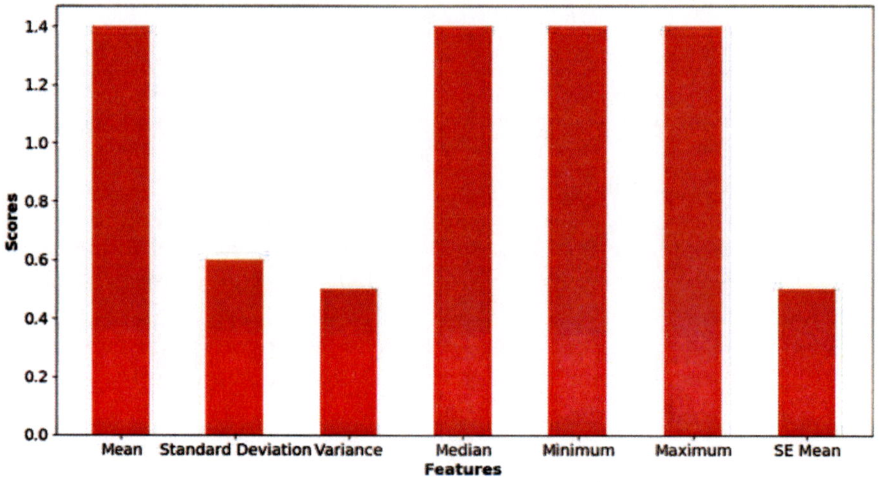

Fig. 4 Identified features via mutual information

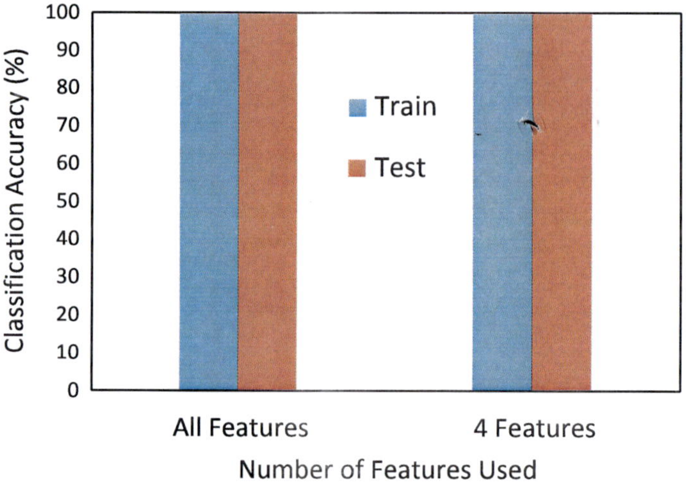

Fig. 5 Classification accuracy based on different feature sets

4 Conclusion

The present study initially evaluated the efficacy of the SVM model in classifying eye movements based on seven statistical-based time-domain features extracted from EOG signals. It was shown from the study that the four was identified to be significant via the Mutual Information (MI) feature selection technique. It was further

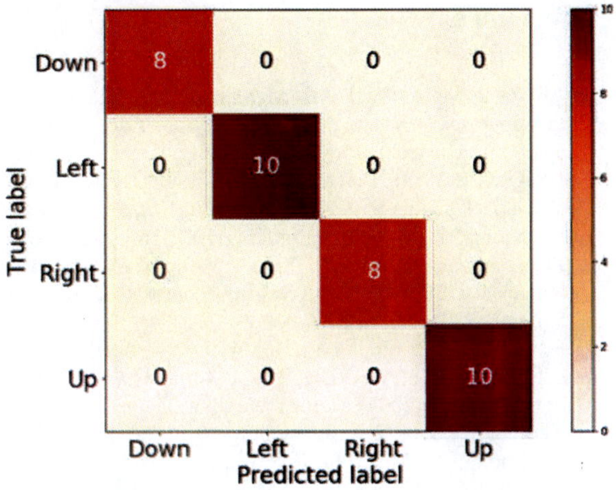

Fig. 6 Confusion matrix of the test dataset by using all features

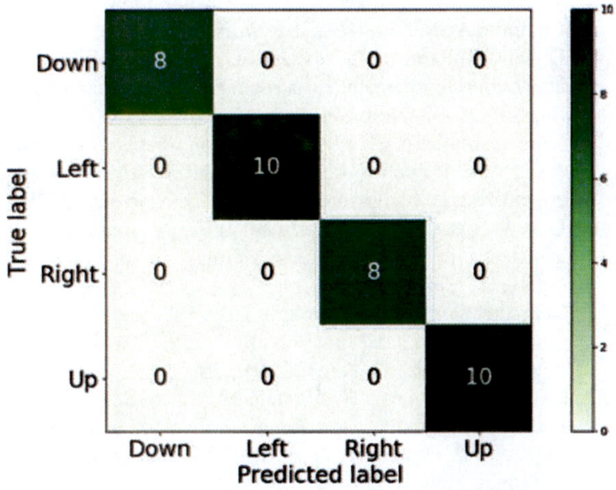

Fig. 7 Confusion matrix of the test dataset by using selected features

demonstrated that by considering the significant features, a comparable classification accuracy is attainable. It is worth to note that the implementation of such a feature selection method would reduce the computational expense in BCI real-time implementation. Future studies shall investigate the efficacy of different types of classification algorithms, feature selection techniques as well as hyperparameter tuning.

Acknowledgements The authors would like to acknowledge University Malaysia Pahang for funding this study via RDU180321.

References

1. Ab Patar MNA, Said AF, Mahmud J, Majeed APPA, Razman MA System integration and control of dynamic ankle foot orthosis for lower limb rehabilitation. In: International symposium on technology management and emerging technologies (ISTMET). IEEE, pp 82–85
2. Hung LC, Sung SF, Hsieh CY, Hu YH, Lin HJ, Chen YW, Yang YHK, Lin SJ (2017) Validation of a novel claims-based stroke severity index in patients with intracerebral hemorrhage. J Epidemiol 27:24–29. https://doi.org/10.1016/j.je.2016.08.003
3. Zhang J, Wang B, Zhang C, Xiao Y, Wang MY (2019) An EEG/EMG/EOG-based multimodal human-machine interface to real-time control of a soft robot hand. Front Neurorobot 13. https://doi.org/10.3389/fnbot.2019.00007
4. Fisher BE, Sullivan KJ (2001) Activity-dependent factors affecting poststroke functional outcomes. Top Stroke Rehabil 8:31–44. https://doi.org/10.1310/B3JD-NML4-V1FB-5YHG
5. Schaechter JD (2004) Motor rehabilitation and brain plasticity after hemiparetic stroke. Prog Neurobiol 73:61–72. https://doi.org/10.1016/j.pneurobio.2004.04.001
6. Khairuddin IM, Na'im Sidek S, Majeed APPA, Puzi AA (2019) Classifying motion intention from EMG signal: a k-NN approach. In: 2019 7th international conference on mechatronics engineering (ICOM). IEEE, pp 1–4
7. McFarland DJ, Wolpaw JR (2011) Brain-computer interfaces for communication and control. Commun ACM 54:60–66. https://doi.org/10.1145/1941487.1941506
8. Dipietro L, Ferraro M, Palazzolo JJ, Krebs HI, Volpe BT, Hogan N (2005) Customized interactive robotic treatment for stroke: EMG-triggered therapy. IEEE Trans Neural Syst Rehabil Eng 13:325–334. https://doi.org/10.1109/TNSRE.2005.850423
9. Vahdani-Manaf N (2017) Classification of eye movement signals using electrooculography in order to device controlling. In: 2017 IEEE 4th international conference on knowledge-based engineering and innovation (KBEI). IEEE, pp 339–342
10. Gray V, Rice CL, Garland SJ (2012) Factors that influence muscle weakness following stroke and their clinical implications: a critical review. Physiother Canada 64:415–426. https://doi.org/10.3138/ptc.2011-03
11. Lum PS, Godfrey SB, Brokaw EB, Holley RJ, Nichols D (2012) Robotic approaches for rehabilitation of hand function after stroke. Am J Phys Med Rehabil 91:242–254. https://doi.org/10.1097/PHM.0b013e31826bcedb
12. Rashid M, Sulaiman N, Mustafa M, Bari BS, Sadeque MG, Hasan MJ (2020) Wink based facial expression classification using machine learning approach. SN Appl Sci 2:183
13. Letchumy J, Rashid M, Musa RM The classification of wink-based EEG signals : the identification of significant time-domain. Springer, Singapore. https://doi.org/10.1007/978-981-15-7309-5

The Classification of Skateboarding Tricks: A Support Vector Machine Hyperparameter Evaluation Optimisation

Muhammad Ar Rahim Ibrahim, Muhammad Nur Aiman Shapiee, Muhammad Amirul Abdullah, Mohd Azraai Mohd Razman, Rabiu Muazu Musa, and Anwar P. P. Abdul Majeed

Abstract The growing interest in skateboarding as a competitive sport requires new motion analysis approaches and innovative ways to portray athletes' results as the conventional technique of the classification of the tricks is often inadequate in providing accurate and often biased evaluation during competition. This paper aims to identify the suitable hyperparameters of a Support Vector Machine (SVM) classifier in classifying five different skateboarding tricks (Ollie, Kickflip, Frontside 180, Pop Shove-it, and Nollie Frontside Shove-it) based on frequency-domain features extracted from Inertial Measurement Unit (IMU). An amateur skateboarder with the age of 23 years old performed five different skateboard tricks and repeated for five times. The signals obtained then were converted from time-domain to frequency-domain through Fast Fourier Transform (FFT), and a number of features (mean, kurtosis, skewness, standard deviation, root mean square and peak-to-peak corresponding to *x*–*y*–*z* axis of IMU reading) were extracted from the frequency dataset. Different hyperparameters of the SVM model were optimised via grid search sweep. It was found that a sigmoid kernel with 0.01 of gamma and regularisation, *C* value of 10 were found to be the optimum hyperparameters as it could attain a classification accuracy of 100%. The present findings imply that the proposed approach can well identify the tricks to assist the judges in providing a more objective-based evaluation.

Keywords IMU sensor · Machine learning · Skateboard · Classification · Trick

M. A. R. Ibrahim · M. N. A. Shapiee · M. A. Abdullah · M. A. M. Razman (✉) ·
A. P. P. A. Majeed
Innovative Manufacturing, Mechatronics and Sports Laboratory, Faculty of Manufacturing and Mechatronic Engineering Technology, Universiti Malaysia Pahang, 26600 Pekan, Pahang Darul Makmur, Malaysia
e-mail: mohdazraai@ump.edu.my

R. M. Musa
Centre for Fundamental and Continuing Education, Department of Credited Co-Curriculum, Universiti Malaysia Terengganu, Terengganu, Malaysia

A. P. P. A. Majeed
IBM Centre of Excellence, Universiti Malaysia Pahang (UMP), 26600 Pekan, Pahang Darul Makmur, Malaysia

A. F. Ab. Nasir et al. (eds.), *Recent Trends in Mechatronics Towards Industry 4.0*, Lecture Notes in Electrical Engineering 730, https://doi.org/10.1007/978-981-33-4597-3_93

1 Introduction

Skateboarding is a sport that falls under the category of extreme game, where someone rides a wooden deck with four wheels. Through the tracking foot momentum and weight in short intervals, the skateboarder could execute different tricks. Therefore, foot rotation is an integral part of skateboarding. Skateboarding is worth approximately USD 4.8 billion in the sports industry. The sport was confirmed in 2016 for its first appearance at the Tokyo Summer Olympic Games 2020 (now shifted to 2021 due to COVID-19 pandemic). Due to the increasing popularity of the sport, the early development of talent is no longer an option.

The recent development of activities detection technologies has resulted in an increased interest in Machine Learning (ML) [1]. This expected to offer specifically detailed information on physical moves in a dynamic situation, such as skating, to help an individual understand and analyse the physical movements [2]. In several studies, factors that have influenced skateboarding tricks have been studied. Nevertheless, it is important to bear in mind, that judges have often subjectively judged the skateboarding tricks based on their past experiences which could lead to biases if incorrect judgments are made.

Hitherto, there is limited literature available in the classification of skateboarding tricks. Basic tricks were carried out for some investigation in order to ease the collection of data and understanding the basis of signal processing on fundamental tricks. Groh et al. [3] performed a study to predict six different tricks namely, Ollie (O), Nollie (N), Kickflip (K), Heelflip (H), Pop Shove-it (PS), and 360-Flip (360) with the involvement of seven male skateboarders (age: 25–29 years old, stance: 3 regular and 4 goofy). Each correctly performed trick was repeated five times. In a similar study on recognition of skateboarding tricks, an earlier investigation was carried out to observe performance while skateboarding through Graphical User Interface (GUI) and to differentiate two skateboarding tricks specifically, Ollie and Frontside 180. Only two tricks were performed in order to ensure the skater can reproduce the tricks consistently. The trick was repeated 20 times each [4].

Groh et al. [3] used a miPod sensor system with built-in of IMU with a 16-bit axis resolution and 200 Hz sampling rate. The sensor has a synchronous timestamp of 150 ms. The IMU comprises ±16 g and ±2000∘/s of the 3D accelerometer and 3D gyroscope to classify manoeuvres skateboarding tricks. In extensive research, Groh et al. [5] integrated a miPod sensor (IMMU) with number of bit axis, sampling rate, and measurement ranges are similar to previous study. The range of 3D Magnetometer is ±1200 μT. On the contrary, Park et al. [2] used a Arduino kit fromKytronix namely, snowboard for sensing pressure matrix and 160 data points were collected from pressure data.

Apart from time-domain features, researchers have also employ frequency-domain features. For instance, Ashqar et al. [6] conducted a research with different classifiers of machine learning using smart phone application to detect transportation modes of walking, running, bus, cycling and as a passenger in a car. The extraction of useful features in order to provide details for classifying the training model of the machine learning algorithm with an understanding of the skateboard's trick is essential. Groh et al. [3] invert the x-axes and z-axes for all goofy rider stance data in the data pre-processing method in order to distinguish tricks for both stance types of skaters. From the time-domain features, relationship between *x*–*y*-axis, *x*–*z*-axis and *y*–*z*-axis were obtained such as the kurtosis, variance, skewness, bandwidth and dominant frequency, resulting in a total of 54 new features were obtained. Additionally, a total of 345 features were extracted from time series and frequency series data based on measure of 10 variability (similar with the 8 derivative variance with additional energy and spectralEntropy) and 8 derivative variability which is the range, inter-quartile range and standard deviation as well as value distribution of max, min, mean, variance and the difference between positive and negative [6].

Machine learning algorithms have shown as a powerful method for classifying not only skateboarding tricks but also be able to distinguish between two stances, goofy and regular. As an illustration, Anlauff et al. [4] utilised a Linear Discriminant Analysis (LDA) and shrinkage method was applied to improve the accuracy of classification by regularise the covariance matrices using lemma. A 10-folds CV was employed to the classifier resulting in correct rate and sensitivity with 96.0% and 97.0% respectively for Ollie trick and 86.0% of correct rate and 90.0% of sensitivity for Frontside 180. Four classifiers: NB, PART, SVM and kNN were compared. The evaluation of all four classifier were based on a leave-one-subject-out cross-validation. NB and SVM were the best classifier with accuracy of 97.8% [3]. Five classifier of supervised classification: NB, RF, LSVM, RB-SVM and kNN were used to divide all the event detection into 11 tricks classes, 1 bail class and 1 rest class. The evaluation of all five classifier were based on a leave-one-subject-out cross validation. RB-SVM was the best classification accuracy with 89.1% for only correctly landing tricks. Classification accuracy of all events, RF was the best classifier with 79.8% [5]. Furthermore, the study on identification of five different transportation modes has shown that the RF-SVM have a maximum classification accuracy of 97.02% [6].

It is interesting to note that although minimal studies have been done with respect to skateboarding, many sports activities and simple daily activities utilising IMU sensors (time-domain and frequency-domain datasets) as well as machine learning have also been well acknowledged and documented [7–13]. This paper intends to evaluate the improvement that could be made on the classification accuracy of SVM model by performing hyperparameter optimisation on a number of extracted frequency-domain features. The present study is an extension from a previous work that equipped with the sk8pro device [14]. This result of this study can be beneficial for a more accurate evaluation by judges and to enhance the skateboard athletes' performance further.

2 Methodology

2.1 Instrumented IMU Device

CATIA software was used to design the architecture of the computer modelling of the IMU device. A Zortrax M200 Plus 3D printer was utilised to print out the device's casing with Acrylonitrile Butadiene Styrene (ABS) as its material. The ABS material was chosen because of its advantageous as in, elevated force resistance and excellent absorbing nature as the device is prone to shock from the tricks presented. The Arduino Pro Mini act as the microcontroller with equipped of signal detection sensor (MPU6050) and a Bluetooth Module (HC-06) which powered by 3.7 V Lithium Polymer battery. The signals obtained from the tricks are derived from the acceleration (m/s2), and the angular velocity (°/s) haul out from the 6D-IMU sensor with a sampling time of 50 ms (Fig. 1).

Figure 2 illustrates the location on the skateboard of the instrumented IMU unit.

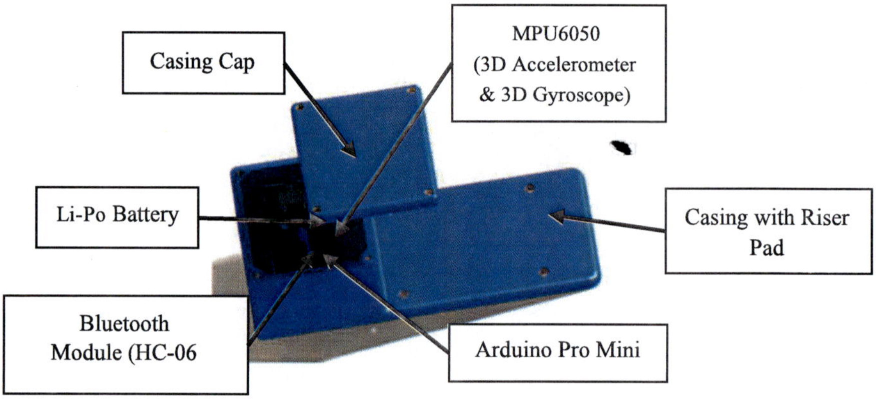

Fig. 1 Sk8pro device

Fig. 2 The sk8pro device attached at bottom front of the board **a** the 3D printed sk8pro device **b** attachment of the sk8pro to the skateboard

Table 1 Executed tricks

Name	Rotation (angle and axis)
Ollie (O)	Nose liftoff (about $45^\circ + x$)
Nollie FS Shuvit (NFS)	Incline spin on the vertical (about $180^\circ - z$)
Frontside 180° (FS180)	Vertical spin about ($180^\circ - z$)
Pop Shove-it (PS)	Clockwise turn on vertical axis ($180^\circ + z$)
Kickflip (K)	Turn whole the board clockwise about longitudinal axis ($360^\circ + y$)

It is positioned at the bottom front of the skateboard (Nose), and behind the front truck is the unit actually set. As the device is built alongside with riser pad, it easy to mount the device on the deck using the existing fastener to ensure the device's stability. The choice of the device's location is non-trivial as it does not impede the skateboarders' movement when executing a particular trick. In fact, the location of the device reduces the risk of damage to the system throughout the process of data collection.

2.2 *Data Collection*

A 23 years old amateur skateboarder with 170 cm tall and a weight of 54 kg from University Malaysia of Pahang was requested to execute five sundry tricks (as shown in Table 1) and to be repeated for five times of each trick as previously investigated [15]. The tricks of the skater carried out were chosen based on his competence and comprehensiveness. All of the tricks performed were in goofy stance direction. In the preprocessing stage, the identification event was done to filter noisy and unnecessary data points. Figure 3 shows an executed trick diagram plus its accelerometer and gyroscope signals.

2.3 *Data Processing*

The time-domain data acquired from the IMU sensor from sk8pro device. Transforming the time domain to frequency domain was applied through Fast Fourier Transform (FFT). We are interested in the magnitude of amplitudes in the frequency responses. Just half the sampling rate (N/2) can be used for frequency signals based upon a Nyquist-Shannon sampling theorem, in order to scrutinise the frequency signals. Each signal has decreased to 10 Hz in frequency. Figure 4 displays the frequency signals of the trick.

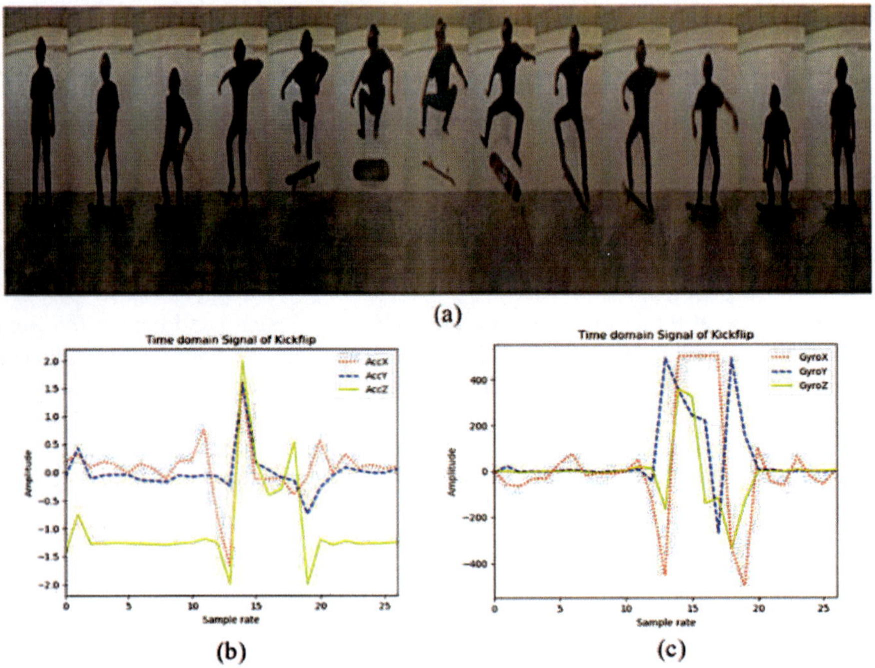

Fig. 3 **a** A Kickflip (K) trick was executed and b, **c** its corresponding signals

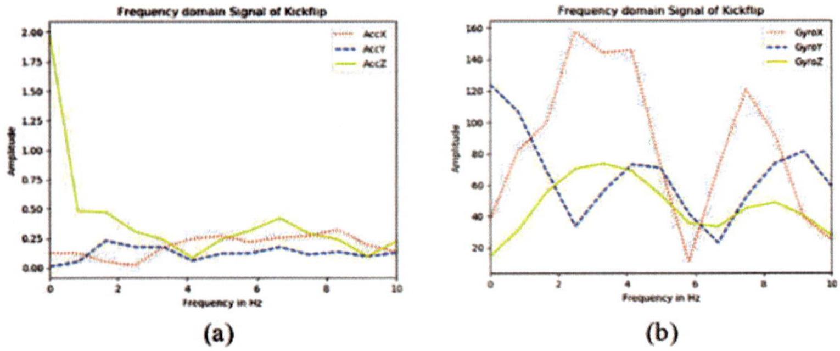

Fig. 4 Converted acceleration and gyro signal for Kickflip

2.4 Machine Learning

The absolute FFT data (frequency domain) are then analysed using MATLAB 2016b for the resulting features in all six degrees of freedoms, of standard deviation, kurtosis, mean, peak to peak, skewness, root mean square and all readings. This is resulting in 36 new features were generated. An SVM with Radial Basis Function (RBF) kernel was initially used in this investigation towards its efficacy in classifying the tricks.

Table 2 Hyperparameter tuning for SVM model

Parameter			
Kernel	RBF	Sigmoid	Linear
Gamma, γ	$1e^{-5}, 1e^{-4}, 1e^{-3}, 1e^{-2}$		No
C (regularisation)	0.001, 0.10, 0.1, 10, 25, 50, 100, 1000		

It should be noted that the classifier default settings are taken from the scikit-learn library [16]. The machine learning models were assessed to determine the accuracy, precision, recall, and the F1 score calculated according to the confusion matrix. In the current study, the leave-one-out (LOO) cross-validation technique was implemented to train the model from the total successful dataset recorded.

A total of 40 successful tricks were recorded, nonetheless, and only 25 tricks were found to be effective as it landed to the standard action of each foot stays on the board. Hence, the collected data were used to train the classifiers' algorithm. A stratified split with 68% for training and 32% for test, respectively was carried out prior invoking the LOO on the training dataset. In order to improve the classification accuracy, the model's hyperparameters were then optimised using grid search sweep in which the twofold technique was used to evaluate the hyperparameters. Three parameters have been taking into account to evaluate its efficacy. The parameter used in grid search is shown in Table 2.

3 Results and Discussion

It is evident that an overfitting phenomenon could be observed on the default SVM model. A drop of 25% could be seen in the testing accuracy, as depicted in Fig. 5. Nonetheless, upon evaluating the features based on the optimised hyperparameters, the 'Optimised' model could attain a classification accuracy (CA) of 100% on both the train and test dataset. The optimised hyperparameters identified is the sigmoid kernel with 0.01 of gamma and regularisation, C value of 10. Table 3 tabulates the comparison between the default SVM and SVM with optimised hyperparameter in terms of CA, F1 score, precision and recall.

Further analysis can be demonstrated on the confusion matrix of default SVM and optimised SVM trained models in Fig. 6a, b respectively and showed that there is no misclassification of tricks (highlighted in blue). Furthermore, the confusion matrix of test results as in Fig. 7a, b illustrate that the misclassification (highlighted in red) reported by the tested default SVM model resulted from the NFS and O tricks which were misclassified as FS180 and K tricks respectively, while there is no misclassification recorded by the tested optimised SVM model. From this investigation, a fair classification performance to predict skateboard tricks can be achieved by tuning the hyperparameter of the model.

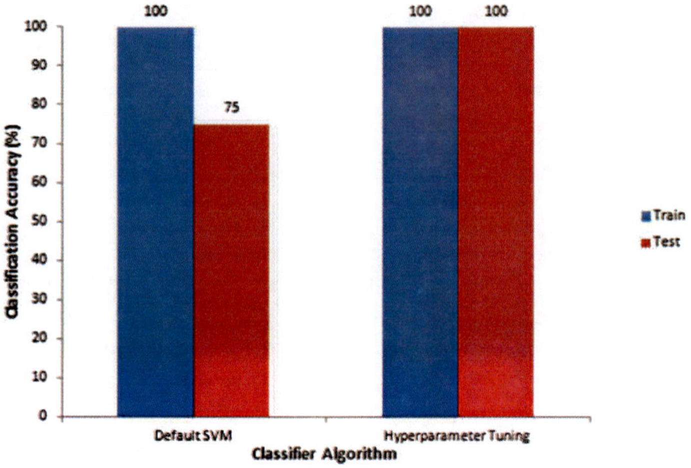

Fig. 5 Graph of comparison of classifiers performance

Table 3 Evaluation of the developed classifiers

Classifier	Evaluation	CA	F1-score	Precision	Recall
RBF-SVM	Train	1.000	1.000	1.000	1.000
	Test	0.750	0.740	0.850	0.750
Hyperparameter Tuning SVM	Train	1.000	1.000	1.000	1.000
	Test	1.000	1.000	1.000	1.000

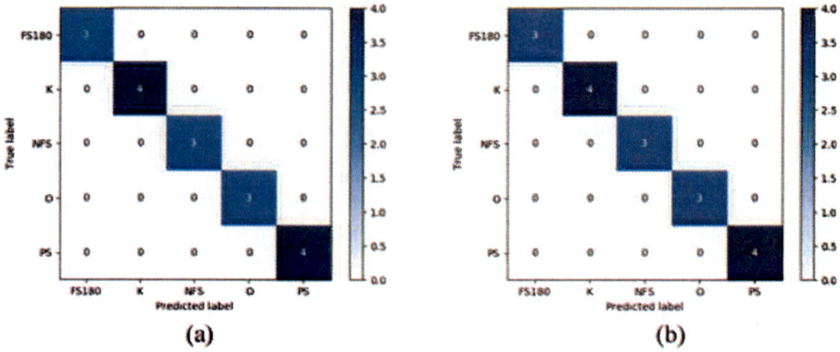

Fig. 6 Confusion matrix of the training of **a** default SVM and **b** optimised hyperparameter SVM model

4 Conclusion

This study investigates the influence of hyperparameter optimisation towards the classification accuracy of skateboarding tricks. It was demonstrated that from the

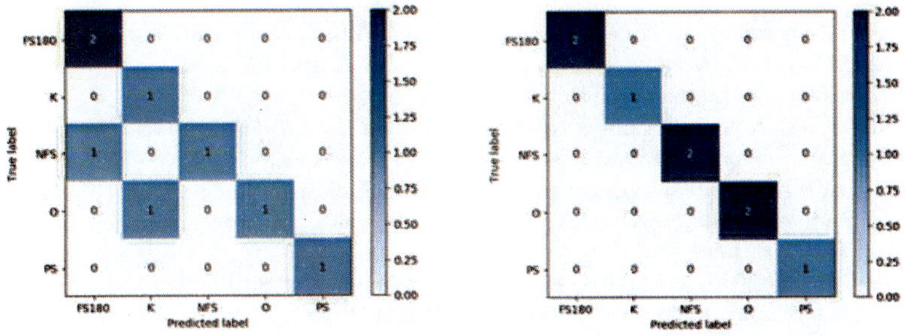

(a) (b)

Fig. 7 Confusion matrix of the testing of **a** default SVM and **b** optimised hyperparameter SVM model

frequency domain features extracted from the instrumented IMU device mounted on the deck, that the optimised SVM model is able to predict accurately both on the train and test dataset, unlike the default SVM model which could not predict well on the test dataset. Future study will be conducted to include more subjects, considering other features, as well as exploiting other feature selection techniques. The findings of the present study could provide a more comprehensive and accurate based appraisal of the tricks performed.

Acknowledgements The authors would like to acknowledge the Ministry of Education, Malaysia and Universiti Malaysia Pahang for supporting and funding this research via FRGS/1/2019/TK03/UMP/02/6 (RDU1901115).

References

1. Camomilla V, Bergamini E, Fantozzi S, Vannozzi G (2018) Trends supporting the in-field use of wearable inertial sensors for sport performance evaluation: a systematic review. Sensors (Switzerland) 18(3)
2. Park HK, Yi H, Lee W (2017) Recording and sharing non-visible information on body movement while skateboarding. In: Proceedings of the 2017 CHI conference on human factors in computing systems—CHI'17, pp 2488–2492
3. Groh BH, Kautz T, Schuldhaus D (2015) IMU-based trick classification in skateboarding. KDD Work Large-Scale Sport Anal
4. Anlauff J, Weitnauer E, Lehnhardt A, Schirmer S, Zehe S, Tonekaboni K (2010) A method for outdoor skateboarding video games. In: Proceedings of the 7th international conference on advances in computer entertainment technology—ACE'10, p 40
5. Groh BH, Fleckenstein M, Kautz T, Eskofier BM (2017) Classification and visualisation of skateboard tricks using wearable sensors. Pervas Mob Comput 40:42–55
6. Ashqar HI, Almannaa MH, Elhenawy M, Rakha HA, House L (2019) Smartphone transportation mode recognition using a hierarchical machine learning classifier and pooled features from time and frequency domains. IEEE Trans Intell Transp Syst 20(1):244–252
7. Stewart TOM, Narayanan A, Hedayatrad L, Neville J, Mackay L, Duncan S (2018) A dual-accelerometer system for classifying physical activity in children and adults, pp 2595–2602

8. Anand A, Sharma M, Srivastava R, Kaligounder L, Prakash D (2018) Wearable motion sensor based analysis of swing sports. In: Proceedings of 16th IEEE international conference on machine learning application ICMLA 2017, pp 261–267, Jan 2018
9. Connaghan D, Kelly P, O'Connor NE, Gaffney M, Walsh M, O'Mathuna C (2011) Multi-sensor classification of tennis strokes. Proceedings of IEEE sensors, pp 1437–1440
10. Friday H, Ying T, Mujtaba G, Al-garadi MA (2019) Data fusion and multiple classifier systems for human activity detection and health monitoring: review and open research directions, vol 46, pp 147–170, June 2018
11. Gani O et al (2019) Journal of network and computer applications a light weight smartphone based human activity recognition system with high accuracy. J Netw Comput Appl 141(May):59–72
12. Groh BH, Fleckenstein M, Eskofier BM (2016) Wearable trick classification in freestyle snowboarding. In: BSN 2016—13th annual body sensing networks conference, pp 89–93
13. Corrêa NK et al (2017) Development of a skateboarding trick classifier using accelerometry and machine learning. Res Biomed Eng 33(4):362–369
14. Ar M et al The Classification of skateboarding trick manoeuvres: a K-nearest neighbour approach, pp 341–347
15. Abdullah MA, Ibrahim MAR, Bin Shapiee MNA, Mohd Razman MA, Musa RM, Majeed APPA (2020) The classification of skateboarding trick manoeuvres through the integration of IMU and machine learning. In: Lecture notes in mechanical engineering, pp 67–74
16. Nur M et al The classification of skateboarding trick manoeuvres through the integration of image processing techniques and machine learning